Mathematics of Quantum Computing

Wolfgang Scherer

Mathematics of Quantum Computing

An Introduction

 Springer

Wolfgang Scherer
Kingston, UK

ISBN 978-3-030-12360-4   ISBN 978-3-030-12358-1   (eBook)
https://doi.org/10.1007/978-3-030-12358-1

Translation from the German edition language edition: *Mathematik der Quanteninformatik* by
Dr. Wolfgang Scherer, © Springer-Verlag GmbH Germany 2016. Published by Springer-Verlag GmbH
Germany is part of Fachverlagsgruppe Springer Science+Business Media. All Rights Reserved.
© Springer Nature Switzerland AG 2019

This Springer imprint is published by the registered company Springer Nature Switzerland AG
The registered company address is: Gewerbestrasse 11, 6330 Cham, Switzerland

Para Negri

und für Matthias und Sebastian

Preface

In the last two decades the digitization of our lives has accelerated at a sometimes breathtaking rate. It is appearing in ever more aspects of our existence to the point of becoming all-encompassing. Ever larger amounts of data are generated, stored, processed and transmitted. This is driven by increases in processing speed and computational performance. The latter is achieved by ever greater miniaturization of circuits and physical memory requirements. As this trend continues rather sooner than later atomic or even sub-atomic scales will be reached. At the latest at that point the laws of quantum mechanics will be required to describe the computational process along with the handling of memory.

Originating from this anticipation as well as from scientific curiosity, many researchers over the last quarter century have thus investigated how information can be stored and processed in systems described by quantum mechanical laws. In doing so they created the science of quantum computing.

Quantum computing is unique in the sense that nowhere else are fundamental questions in physics so closely connected to potentially huge practical implications and benefits. Our very basic understanding of what constitutes reality is challenged by the effects which at the same time seem to enable enormous efficiency gains and to revolutionize computational power and cryptographic protocols.

What is also quite enticing is that quantum computing draws from many 'distinct' branches of mathematics such as, of course, analysis and linear algebra, but to an even greater extent functional analysis, group theory, number theory, probability theory and not least computer science.

This book aims to give an introduction to the mathematics of this wide-ranging and promising field. The reason that despite being an introduction it is so voluminous is that the reader is taken by the hand and led through all arguments step by step. All results are proven in the text in—for the cognoscenti perhaps excruciating —detail. Numerous exercises with their solutions provided allow the reader to test and develop their understanding. Any requisites from branches, such as number theory or group theory, are provided with all stated results proven in the book as well.

For the above reasons this book is eminently suitable for self-study of the subject. The attentive and diligent reader does not have to consult other resources to follow the arguments. The level of mathematical know-how required approximately corresponds to second year undergraduate knowledge in mathematics or physics.

It is very much a text in mathematical style in that we follow the pattern of motivating text—definition—lemma/theorem/corollary—proof—explanatory text and all over again. In doing so all relevant assumptions are clearly stated. At the same time, it provides ample opportunities for the reader to become familiar with standard techniques in quantum computing as well as in the related mathematical sub-fields. Having mastered this book the reader will be equipped to digest scientific papers on quantum computing.

I enjoyed writing this book. I very much hope it is equally enjoyable to read it.

Acknowledgements Throughout my academic life many people have taught, motivated, enlightened and inspired me. I am truly grateful to every one of them and hope they will find this book to their liking.

A very special thanks goes to the organizers and participants of the 2017 Summer Academy of the Studienstiftung des Deutschen Volkes. Their review and critical feedback for the German version has helped to erase several errors and to improve the presentation. It also was a pleasure to spend some time with them in lovely South Tirol. In particular, I am wholeheartedly grateful to Joachim Hilgert who, in addition to giving feedback on the German version, also very swiftly and thoroughly proofread a part of this manuscript.

The largest debt of gratitude is owed to Maria-Eugenia, Matthias and Sebastian, who during many years of this project have taken the back seat and took my retreat from family life in their stride but never wavered in their support and shared my enthusiasm. I am very grateful in particular to Sebastian, who once again proofread large portions of the manuscript. His detailed review detected many errors and his constructive criticism during numerous enjoyable and lengthy sessions helped improve precision and clarity of the exposition.

Needless to say that even his diligence will not have detected every error or shortcoming. Those were still caused by the author.

Kingston Upon Thames, UK   Wolfgang Scherer
March 2019

Contents

1 Introduction 1
  1.1   Some History 1
  1.2   Reader's Manual 4
  1.3   What is not in this Book 7
  1.4   Notation and References 8

2 Basic Notions of Quantum Mechanics 11
  2.1   Generalities 11
  2.2   Mathematical Notions: HILBERT Spaces and Operators 12
  2.3   Physical Notions: States and Observables 29
       2.3.1 Pure States 29
       2.3.2 Mixed States 43
  2.4   Qubits 56
  2.5   Operators on Qubits 63
  2.6   Further Reading 74

3 Tensor Products and Composite Systems 77
  3.1   Towards Qbytes 77
  3.2   Tensor Products of HILBERT Spaces 78
       3.2.1 Definition 78
       3.2.2 Computational Basis 85
  3.3   States and Observables for Composite Systems 90
  3.4   SCHMIDT Decomposition 106
  3.5   Quantum Operations 109
  3.6   Further Reading 125

4 Entanglement 127
  4.1   Generalities 127
  4.2   Definition and Characterization 128
  4.3   Entanglement Swapping 133
  4.4   EINSTEIN–PODOLSKY–ROSEN-Paradox 135

4.5   Bell Inequality.................................... 141
      4.5.1 Original Bell Inequality......................... 141
      4.5.2 CHSH Generalization of the Bell Inequality 146
4.6   Two Impossible Devices 154
      4.6.1 Bell Telephone 154
      4.6.2 Perfect Quantum Copier....................... 158
4.7   Further Reading 160

5   Quantum Gates and Circuits for Elementary Calculations 161
5.1   Classical Gates 161
5.2   Quantum Gates.................................... 168
      5.2.1 Unary Quantum Gates 169
      5.2.2 Binary Quantum Gates......................... 171
      5.2.3 General Quantum Gates 172
5.3   Quantum Circuits 201
5.4   On the Process of Quantum Algorithms 206
      5.4.1 Preparation of Input and Use of Auxiliary Registers 207
      5.4.2 Implementation of Functions and Quantum Parallelism... 208
      5.4.3 Reading the Output Register...................... 212
5.5   Circuits for Elementary Arithmetic Operations 213
      5.5.1 Quantum Adder.............................. 213
      5.5.2 Quantum Adder Modulo N...................... 226
      5.5.3 Quantum Multiplier Modulo N 229
      5.5.4 Quantum Circuit for Exponentiation Modulo N 233
      5.5.5 Quantum Fourier Transform 237
5.6   Further Reading 245

6   On the Use of Entanglement 247
6.1   Early Promise: Deutsch–Jozsa Algorithm 247
6.2   Dense Quantum Coding 251
6.3   Teleportation 253
6.4   Quantum Cryptography.............................. 255
      6.4.1 Ciphers in Cryptography 255
      6.4.2 Quantum Key Distribution without Entanglement 258
      6.4.3 Quantum Key Distribution with Entanglement......... 262
      6.4.4 RSA Public Key Distribution 266
6.5   Shor Factorization Algorithm 271
      6.5.1 Generalities 271
      6.5.2 The Algorithm............................... 273
      6.5.3 Step 1: Selection of b and Calculation of $\gcd(b,N)$ 275
      6.5.4 Step 2: Determining the Period with a Quantum
             Computer 276
      6.5.5 Step 3: Probability of Selecting a Suitable b 290
      6.5.6 Balance Sheet of Steps.......................... 296

6.6   Generalizing: The Abelian Hidden Subgroup Problem 301
6.7   Finding the Discrete Logarithm as a Hidden Subgroup
       Problem 310
6.8   Breaking Bitcoin Signatures 317
6.9   GROVER Search Algorithm 324
       6.9.1   Search Algorithm for Known Number of Objects 324
       6.9.2   Search Algorithm for Unknown Number of Objects..... 337
6.10   Further Reading 341

7   Error Correction 343
7.1   What Can Go Wrong? 343
7.2   Classical Error Correction 345
7.3   Quantum Error Correction............................. 355
       7.3.1   Correctable Errors 355
       7.3.2   Detection and Correction 384
       7.3.3   Stabilizer Formalism 390
7.4   Further Reading 402

8   Adiabatic Quantum Computing 403
8.1   Introduction 403
8.2   Starting Point and Assumptions........................ 404
8.3   Generic Adiabatic Algorithm.......................... 413
8.4   Adiabatic Quantum Search 419
8.5   Replicating a Circuit Based by an Adiabatic Computation..... 440
8.6   Replicating an Adiabatic by a Circuit Based Computation..... 484
8.7   Further Reading 497

9   Epilogue 499

Appendix A: Elementary Probability Theory 501

Appendix B: Elementary Arithmetic Operations 505

Appendix C: LANDAU Symbols 513

Appendix D: Modular Arithmetic 515

Appendix E: Continued Fractions 545

Appendix F: Some Group Theory 559

Appendix G: Proof of a Quantum Adiabatic Theorem 621

Solutions to Exercises 641

References 755

Index 759

Glossary

AHSP   stands for the Abelian Hidden Subgroup Problem, which is the problem to identify a subgroup from a function on the group that leaves the subgroup invariant

AQC   stands for adiabatic quantum computing

BB84   stands for a quantum mechanical cryptographic key distribution method proposed in 1984 by BENNETT and BRASSARD in [1]

CECC   stands for classical error correcting code

CHSH   stands for CLAUSER-HORNE-SHIMONY-HOLT, four authors of a generalization of the BELL inequality given in their joint article [2]

DLP   stands for Discrete Logarithm Problem, which is the problem to find the discrete logarithm d in the context of a group when only given the group elements g and $h = g^d$

DSA   stands for Digital Signature Algorithm, which is any cryptographic protocol that allows to add a digital signature to a digital document so that the signature can be easily verified by anyone but is (almost) impossible to forge

ECDSA   stands for Elliptic Curve Digital Signature Algorithm, which is a DSA based on elliptic curves. It is used by *bitcoins*

EK91   stands for a cryptographic key distribution protocol proposed in 1991 by EKERT [3], which uses the CHSH version of the BELL inequality to detect eavesdropping

EPR   stands for EINSTEIN-PODOLSKY-ROSEN, three authors of a paper [4] in 1935 in which the counter-intuitive effects of quantum mechanics are used to argue for its incompleteness

HSP   stands for Hidden Subgroup Problem

ONB   is an abbreviation for orthonormal basis, the maximal set of linearly independent and pairwise orthogonal unit vectors in a linear space with a scalar product

QECC   stands for quantum error correcting code
QUBO   stands for Quadratic Unconstrained Binary Optimization
RSA   denotes the 'classical' *public key* cryptographic method developed by
              RIVEST, SHAMIR, and ADLEMAN in 1978

Symbols

$:=$	Defining equality, that is, the expression $a := b$ defines a by b
$\mathbb{N}$	Set of natural numbers, that is, $\mathbb{N} := \{1, 2, 3, \ldots\}$
$\mathbb{N}_0$	Set of natural numbers including zero
Pri	Set of prime numbers, that is, Pri $:= \{2, 3, 5, 7, 11, \ldots\} \subset \mathbb{N}$
$\mathbb{Z}$	Set of integers, that is, $\mathbb{Z} := \{0, \pm 1, \pm 2, \pm 3, \ldots\}$
$\mathbb{F}_2$	Field of binary numbers $\{0, 1\}$ with binary addition and standard multiplication
$\mathbb{Q}$	Field of rational numbers $\frac{q}{p}$ with $q \in \mathbb{Z}$ and $p \in \mathbb{N}$
$\mathbb{R}$	Field of real numbers; $\mathbb{R}_+$ denotes the positive real numbers
i	Imaginary unit, that is, $\mathrm{i}^2 = -1$
$\mathbb{C}$	Field of complex numbers $a + \mathrm{i}b$ with $a, b \in \mathbb{R}$ and $\mathrm{i}^2 = -1$
$\bar{z}$	Complex conjugate of $z = a + \mathrm{i}b$ with $a, b \in \mathbb{R}$, that is, $\bar{z} = a - \mathrm{i}b$
$|z|$	Absolute value of the complex number $z = a + \mathrm{i}b$ with $a, b \in \mathbb{R}$, that is, $|z| = \sqrt{z\bar{z}} = \sqrt{a^2 + b^2}$
$f\{S\}$	Image $f\{S\} \subset Y$ of the subset $S \subset X$ of a mapping $f : X \to Y$
$\mathbf{a}$	Vector in $\mathbb{R}^n$, $\mathbb{C}^n$ or $\mathbb{F}_2^n$
$\bar{\mathbf{a}} \cdot \mathbf{b}$	Scalar product of the vectors $\mathbf{a}$ and $\mathbf{b}$ in $\mathbb{C}^n$; for $\mathbf{a}, \mathbf{b} \in \mathbb{R}^n$ this equals $\mathbf{a} \cdot \mathbf{b}$
$|\mathbf{a}|$	Norm of the vector $\mathbf{a} \in \mathbb{C}^n$, that is, $|\mathbf{a}| := \sqrt{\bar{\mathbf{a}} \cdot \mathbf{a}}$
$\mathbf{a} \cdot \sigma$	Linear combination of PAULI matrices $\sigma_1 = \sigma_x$, $\sigma_2 = \sigma_y$, $\sigma_3 = \sigma_z$ with coefficients $\mathbf{a} \in \mathbb{R}^3$, that is, $\mathbf{a} \cdot \sigma := \sum_{j=1}^3 a_j \sigma_j$
X, Y, Z	Alternative notation for the PAULI matrices, that is, $\mathbf{1} = \sigma_0$, $X = \sigma_x = \sigma_1$, $Y = \sigma_y = \sigma_2$, $Z = \sigma_z = \sigma_3$ often used in the context of quantum gates
$\mathbb{H}$	HILBERT space, that is, a complex vector space with a scalar product which induces a norm

$\mathrm{Span}\{v_1, \ldots, v_n\}$ Linear 'span' of a set of vectors $\{v_1, \ldots, v_n\}$ of a linear
space $\mathbb{V}$ over a field $\mathbb{F}$, that is, the subspace of $\mathbb{V}$ created
by all vectors of the form $v = \sum_{j=1}^{n} a_j v_j$, where

$$\left\{ a_j \mid j \in \{1, \ldots, n\} \right\} \subset \mathbb{F}$$

$\mathbb{H}$ Qubit HILBERT space, that is, the HILBERT space $\mathbb{H} \cong \mathbb{C}^2$

$|\psi\rangle$ 'Ket' notation for a vector in a HILBERT space

$\langle\psi|$ 'Bra' notation for a vector in the dual space to a HILBERT space

$\langle\varphi|\psi\rangle$ Scalar product of the vectors $|\varphi\rangle, |\psi\rangle \in \mathbb{H}$

$||\psi||$ Norm of the vector $|\psi\rangle \in \mathbb{H}$, that is, $||\psi|| := \sqrt{\langle\psi|\psi\rangle}$

δ_{xy} KRONECKER delta

$$\delta_{xy} := \begin{cases} 1, & \text{if } x = y \\ 0, & \text{else.} \end{cases}$$

To avoid confusion we sometimes insert a comma
and write, for example, $\delta_{np,mq}$ instead of δ_{npmq}

$\mathbf{1}$ Identity operator in $\mathbb{H}$, or identity matrix in $\mathbb{R}^n, \mathbb{C}^n, \mathbb{F}_2^n$, that is,
$\mathbf{1}|\psi\rangle = |\psi\rangle$ for all $|\psi\rangle \in \mathbb{H}$. To make things more explicit we
may also write $\mathbf{1}^A$ for the identity operator in $\mathbb{H}^A$ to distinguish
it from an identity operator $\mathbf{1}^B$ in a different HILBERT space $\mathbb{H}^B$.
Likewise, we also write $\mathbf{1}^{\otimes n}$ for the identity operator in
$\mathbb{H} = \mathbb{H}^{\otimes n}$

$\mathrm{L}(\mathbb{H})$ Set of linear operators on $\mathbb{H}$, that is,

$$\mathrm{L}(\mathbb{H}) := \{A : \mathbb{H} \to \mathbb{H} \mid A \text{ linear}\}$$

$\mathrm{B}(\mathbb{H})$ Set of bounded linear operators on $\mathbb{H}$, that is,

$$\mathrm{B}(\mathbb{H}) := \{A \in \mathrm{L}(\mathbb{H}) \mid ||A|| < \infty\}$$

A^T Transpose of matrix A, that is, $A_{ij}^T = A_{ji}$

A^* Adjoint operator (or matrix) of A, that is, $\langle\varphi|A\psi\rangle = \langle A^*\varphi|\psi\rangle$
for all $|\varphi\rangle, |\psi\rangle \in \mathbb{H}$

$\mathrm{tr}(A)$ Trace of operator $A \in \mathrm{L}(\mathbb{H})$ or matrix $A \in \mathrm{L}(\mathbb{C}^n)$, that is,
$\mathrm{tr}(A) := \sum_j A_{jj}$, where A_{ij} are the matrix elements in some ONB

$\mathrm{tr}^B(M)$ The partial trace over $\mathbb{H}^B$ for an operator $M \in \mathrm{L}(\mathbb{H}^A \otimes \mathbb{H}^B)$

$\sigma(A)$ Spectrum of $A \in \mathrm{L}(\mathbb{H})$, that is, the set

$$\sigma(A) := \{\lambda \in \mathbb{C} \mid (A - \lambda\mathbf{1})^{-1} \text{ does not exist}\}$$

$\mathrm{Eig}(A, \lambda)$ The linear subspace spanned by all eigenvectors of the operator
$A : \mathbb{H} \to \mathbb{H}$ with eigenvalue λ, that is,

$$\mathrm{Eig}(A, \lambda) := \mathrm{Span}\{|\psi\rangle \in \mathbb{H} \mid A|\psi\rangle = \lambda|\psi\rangle\}$$

$B_{sa}(\mathbb{H})$	Set of bounded self-adjoint operators on a HILBERT space $\mathbb{H}$, that is,

$$B_{sa}(\mathbb{H}) := \{A \in B(\mathbb{H}) | A^* = A\}$$

$\mathcal{U}(\mathbb{H})$	Group of unitary operators on a HILBERT space $\mathbb{H}$, that is,

$$\mathcal{U}(\mathbb{H}) := \{A \in B(\mathbb{H}) | A^* A = \mathbf{1}\}$$

$D(\mathbb{H})$	Convex set of density operators on a HILBERT space $\mathbb{H}$, that is,

$$D(\mathbb{H}) := \{\rho \in L(\mathbb{H}) | \rho^* = \rho, \rho \geq 0, \mathrm{tr}(\rho) = 1\}$$

$D_{\leq}(\mathbb{H})$	Set of self-adjoint positive operators on a HILBERT space $\mathbb{H}$ with trace less than 1, that is,

$$D_{\leq}(\mathbb{H}) := \{\rho \in L(\mathbb{H}) | \rho^* = \rho, \rho \geq 0, \mathrm{tr}(\rho) \leq 1\}$$

$[A, B]$	Commutator of the operators A and B, that is, $[A, B] := AB - BA$
$\mathbf{P}\{\text{Event}\}$	Probability of 'Event' happening
$|\varphi\rangle \otimes |\psi\rangle$	Tensor product of two vectors $|\varphi\rangle, |\psi\rangle \in \mathbb{H}$
$\mathbb{H}^{\otimes n}$	n-fold tensor product of the qubit HILBERT space $\mathbb{H}$
$S_{\mathbb{V}}^r$	The sphere of radius r in the normed vector space $\mathbb{V}$, that is, the set of vectors $\mathbf{v} \in \mathbb{V}$ with $\|\mathbf{v}\| = r$
$B_{\mathbb{V}}^r$	The full ball of radius r in the normed vector space $\mathbb{V}$, that is, the set of vectors $\mathbf{v} \in \mathbb{V}$ with $\|\mathbf{v}\| \leq r$
$|x\rangle$	Vector of the computational basis in $\mathbb{H}^{\otimes n}$ given for each $x \in \mathbb{N}_0$ with $x = \sum_{j=0}^{n-1} x_j 2^j < 2^n$ and $x_j \in \{0, 1\}$ as

$$|x\rangle := |x\rangle^n := \overset{0}{\underset{j=n-1}{\bigotimes}} |x_j\rangle = |x_{n-1}\rangle \otimes \ldots \otimes |x_0\rangle = |x_{n-1}\ldots x_0\rangle$$

$\neg A$	Negation of proposition A
$\lfloor a \rfloor$	Integer part of a real number $a \in \mathbb{R}$, that is,

$$\lfloor a \rfloor := \max\{z \in \mathbb{Z} | z \leq a\}$$

$\lceil a \rceil$	Nearest integer from above a real number $a \in \mathbb{R}$, that is,

$$\lceil a \rceil := \min\{z \in \mathbb{Z} | z \geq a\}$$

$a \bmod n$	Remainder of a after division by n, that is,

$$a \bmod n := a - \left\lfloor \frac{a}{n} \right\rfloor n$$

$x_{n-1}\ldots x_0{}_2$	Binary representation of a number $x \in \mathbb{N}_0$ satisfying $x < 2^n$, that is,

$$x = x_{n-1}\ldots x_0{}_2 := \sum_{j=0}^{n-1} x_j 2^j \qquad \text{with } x_j \in \{0, 1\}$$

$\overset{2}{\oplus}$ Binary addition $a \overset{2}{\oplus} b := (a+b) \bmod 2$

$\overset{2}{\odot}$ Binary scalar product defined as

$$x \overset{2}{\odot} y := x_{n-1} y_{n-1} \overset{2}{\oplus} \ldots \overset{2}{\oplus} x_0 y_0 = \left(\sum_{j=0}^{n-1} x_j y_j \right) \bmod 2$$

for vectors $|x\rangle, |y\rangle \in \mathbb{H}^{\otimes n}$ in the computational basis or vectors $\mathbf{x}, \mathbf{y} \in \mathbb{F}_2^n$

$\boxplus$ Factor-wise binary addition for vectors $|x\rangle, |y\rangle \in \mathbb{H}^{\otimes n}$ in the computational basis; this addition is defined by

$$|x \boxplus y\rangle := \overset{0}{\underset{j=n-1}{\bigotimes}} |x_j \overset{2}{\oplus} y_j\rangle$$

$a|b$ a divides b, that is, there exists a $z \in \mathbb{Z}$ such that $b = az$

$a \nmid b$ a does not divide b, that is, all $z \in \mathbb{Z}$ satisfy $b \neq za$

$\gcd(a_1, \ldots, a_n)$ Greatest common divisor of $a_i \in \mathbb{Z}, i \in \{1, \ldots, n\}$ with $\sum_{i=1}^n |a_i| \neq 0$, that is,

$$\gcd(a_1, \ldots, a_n) := \max\{k \in \mathbb{N} | \forall a_i : k|a_i\}.$$

Note that for $a \neq 0$ one has $\gcd(0, a) = |a|$

$\mathrm{scm}(a_1, \ldots, a_n)$ Smallest common multiple of $a_1, \ldots, a_n \in \mathbb{Z}$ with $\prod_{i=1}^n a_i \neq 0$, that is,

$$\mathrm{scm}(a_1, \ldots, a_n) := \min\{k \in \mathbb{N} | \forall a_i : a_i|k\}$$

$\mathrm{Pri}(n)$ Set of prime factors in the prime factorization of n

$n = \prod\limits_{p \in \mathrm{Pri}} p^{v_p}$ Prime factorization of $n \in \mathbb{N}$, where the exponents of the primes $p \in \mathrm{Pri}$, which do not occur as a prime factor of n, are zero, that is, $v_p = 0$ if $p \notin \mathrm{Pri}(n)$. For $a \in \mathbb{Z} \setminus \{0\}$ we define the prime factorization of $|a| \in \mathbb{N}$ and set $a = \mathrm{sign}(a) \prod_{p \in \mathrm{Pri}(|a|)} p^{|a|_p}$

$\phi(n)$ EULER function

$$\phi : \mathbb{N} \to \mathbb{N}$$
$$n \mapsto \phi(n) := |\{r \in \{1, \ldots, n-1\} | \gcd(n, r) = 1\}|$$

$\mathrm{ord}_N(b)$ The order of b modulo N defined for natural numbers b and N with the property $\gcd(b, N) = 1$ as

$$\mathrm{ord}_N(b) := \min\{n \in \mathbb{N} | b^n \bmod N = 1\}$$

id_A Identity map on the set A, that is, $\mathrm{id}_A : A \to A$ with $\mathrm{id}_A(a) = a$

$o(\cdot)$ Small LANDAU symbol, defined here for functions on $\mathbb{N}$ in the limit $n \to \infty$ as

$$f(n) \in o(g(n)) \text{ for } (n \to \infty)$$
$$:\Leftrightarrow \forall \varepsilon \in \mathbb{R}_+, \exists M \in \mathbb{N} : \forall n > M : |f(n)| \leq \varepsilon |g(n)|$$

$O(\cdot)$	Big LANDAU symbol, defined here for functions on $\mathbb{N}$ in the limit $n \to \infty$ as

$$f(n) \in O(g(n)) \text{ for } (n \to \infty)$$
$$:\Leftrightarrow \exists C \in \mathbb{R}, \ M \in \mathbb{N} \forall n > M : |f(n)| \le C \, |g(n)|$$

$e^A = \exp(A)$	Exponential function applied to A, that is, $e^A = \exp(A) := \sum_{n=0}^{\infty} \frac{A^n}{n!}$, where A can be complex number, a matrix or an operator
$|S|$	Denotes the number of the elements of the set S
$|\mathcal{G}|$	Denotes the order of a finite group $\mathcal{G}$, that is, the number of its elements
$\mathcal{H} \le \mathcal{G}$	For groups $\mathcal{H}$ and $\mathcal{G}$ the expression $\mathcal{H} \le \mathcal{G}$ states that $\mathcal{H}$ is a subgroup of the group $\mathcal{G}$
$\mathcal{H} < \mathcal{G}$	For groups $\mathcal{H}$ and $\mathcal{G}$ the expression $\mathcal{H} < \mathcal{G}$ states that $\mathcal{H}$ is a proper subgroup of the group $\mathcal{G}$
$\mathcal{H} \trianglelefteq \mathcal{G}$	For groups $\mathcal{H}$ and $\mathcal{G}$ the expression $\mathcal{H} \trianglelefteq \mathcal{G}$ states that $\mathcal{H}$ is a normal subgroup of the group $\mathcal{G}$
$\mathcal{H} \cong \mathcal{G}$	States that group $\mathcal{H}$ is isomorphic to group $\mathcal{G}$, that is, there exists a bijective map between them which respects the group operations; for linear spaces the group operation is addition
$\mathcal{P}$	The PAULI group given by

$$\mathcal{P} := \{i^a \, \sigma_\alpha \,|\, a, \alpha \in \{0, \ldots, 3\}\} < \mathcal{U}(\mathbb{H})$$

$\mathcal{P}_n$	The n-fold PAULI group given by

$$\mathcal{P}_n := \{i^a \sigma_{\alpha_{n-1}} \otimes \ldots \otimes \sigma_{\alpha_0} \in L(\mathbb{H}^{\otimes n}) | a, \alpha_j \in \{0, \ldots, 3\}\}$$
$$< \mathcal{U}(\mathbb{H}^{\otimes n})$$

Chapter 1
Introduction

1.1 Some History

The origin of the prefix 'quantum' goes back to the start of the 20th century, when PLANCK in the derivation of the black-body radiation law postulated the existence of a minimal 'quantum' of energy [5]. A few years later EINSTEIN also used this assumption in the theory of the photo-electric effect [6]. Despite these early origins the story of quantum mechanics only came to the fore almost twenty years later in the 'golden twenties' with the works of BOHR, SCHRÖDINGER, HEISENBERG, PAULI, BORN and many others.

Quantum mechanics describes so-called microscopic systems by means of a mathematical formalism that in general allows only statements about *probabilities*. What can be known about a system, that is, its *state*, is described mathematically by a vector in a linear space. This makes it possible that a system can be in a state which is a *linear combination* of other states. Moreover, the mathematical theory of quantum mechanics also provides a statement about which physical quantities of a system—its so-called *observables*—can in principle be determined at the same time and with which maximally possible precision. The HEISENBERG uncertainty principle is perhaps the most well known notion in this context.

Starting from a few basic assumptions, the so-called *Postulates* of quantum mechanics, the theory allows the derivation of many results and statements. Some of these can be difficult to reconcile with our intuition and have become known as *paradoxes*. Prominent examples hereof are the EINSTEIN–PODOLSKY–ROSEN*(EPR)- Paradox* [4] and the often quoted SCHRÖDINGER*'s cat*, which SCHRÖDINGER presented to the world in a review paper [7]. In this article SCHRÖDINGER also coins the term *entanglement*, which describes a quantum mechanical phenomenon that challenges our intuition about what constitutes reality yet at the same time plays an essential role in quantum computing.

Entanglement is also crucial in the context of an inequality for correlations derived by BELL in the sixties under the assumption of the existence of so-called *hidden variables* [8]. Roughly speaking the BELL *inequality* makes assumptions

© Springer Nature Switzerland AG 2019
W. Scherer, *Mathematics of Quantum Computing*,
https://doi.org/10.1007/978-3-030-12358-1_1

that are very much in line with our intuition of what one might call the physical reality of a system. At the same time, however, quantum mechanics predicts that this inequality is violated in certain entangled states. In 1969 a generalization of the BELL *inequality* was derived by CLAUSER et al. [2]. This version was then indeed shown to be violated in an experiment conducted by ASPECT et al. [9]. In other words, rather than following our intuition about reality, nature follows the 'counter-intuitive' predictions of quantum mechanics.

As with the predictions for the BELL inequality quantum mechanics has so far passed all other all other experimental tests. Beyond that it has lead to numerous applications like lasers, transistors, nuclear energy, nuclear resonance tomography and many more, which have fundamentally changed the world and will continue to do so. Insofar it is probably no exaggeration to name quantum mechanics as the most successful scientific theory ever.

The history of *information theory* has its beginning around the forties of the past century with WIENER [10] and SHANNON [11]. Classical information is stored and processed in the form of clearly distinguishable *binary states*. The success of conventional computer science and digitization relies on the fact that any superposition of these binary states is excluded and the system is always in a clearly definable state. As we shall see in due course, this is in stark contrast to what can be said about entangled states utilized in quantum computation.

A first mentioning of the possibly enhanced capabilities of a *quantum computer* compared to the classical computational process is attributed to FEYNMAN [12]. Motivated by the difficulty of classical computers to simulate quantum systems efficiently, he wondered in the early eighties if a computer whose computational operations utilized quantum effects would outperform a classical processor. In doing so he noted that a 'quantum mechanical computer' can indeed simulate quantum systems more efficiently than a classical computer.

An analysis looking at the combination of quantum mechanics with the theory of computational processes à la TURING [13] by BENIOFF [14] appeared simultaneously with FEYNMAN's paper. Also in 1982 appeared the often quoted *Quantum No-Cloning Theorem* by WOOTTERS and ZUREK, stating the impossibility to copy an unknown quantum system [15]. The combination of quantum mechanics with information theory then picked up speed in the eighties with contributions by DEUTSCH, who formalized quantum mechanical computational processes and circuits [16, 17].

This was motivated by the hitherto often ignored fact that information has a physical origin. Consequently, (at first hypothetical) questions were asked, such as: which physical options exist for the storage and processing of information by quantum mechanical systems? Will this lead to knew information theoretic phenomena? Are there efficiency gains?

The rapidly progressing miniaturization of computational storage devices increasingly brought these questions out of the hypothetical realm into the focus of a practical interest. This ever more so because the deliberate control of microscopic systems had advanced considerably in the last decades. By now the class of such microscopic systems encompasses atoms, electrons and photons, which nowadays can be manipulated quite extensively in the laboratory.

The basic tenet of the combination of information and quantum theory is the usage of quantum mechanical states for the storage and processing of information obeying the laws of quantum mechanics. Contrary to the classical binary representation of information, where this is to be avoided, the possibility of a superposition of several states and the probabilistic nature of quantum mechanics allows for the emergence of new and interesting possibilities.

One such possibility was added in the nineties with the curious *teleportation* of quantum states by BENNETT et al. [18]. This was followed by SHOR with his *factorization algorithm* [19, 20] and GROVER with his *search algorithm* [21, 22], which both demonstrated for the first time the potential supremacy of quantum computers in solving real world problems. These three methods lead to a substantial increase in the interest in quantum computers. Soon they were experimentally realized, albeit more as proofs of concept. First in 1997 the teleportation by BOUWMESTER et al. [23]; thereafter in 1998 the search algorithm by CHUANG et al. [24] and in 2001 the factorization of 15 by L. M. VANDERSYPEN et al. [25].

The factorization and search algorithm require functioning quantum computers, if they were to work on large inputs, and so far have only been demonstrated in minimal circumstances in the laboratory. In contrast, teleportation has been achieved over ever larger distances such as 143 km [26] or even up to a satellite orbiting the earth [27].

The nineties also saw important contributions to the computational process such as methods for *error correction*. Given the difficulty to shield quantum systems from unwanted interactions the ability to correct errors becomes crucial. This task is made more difficult in a quantum computer due to the fact that a measurement can impact the state of the system. Nonetheless methods to deal with these problems were proposed by CALDERBANK and SHOR [28] as well as by STEANE [29]. Moreover, the properties of quantum gates, in particular *universality*, were proven by BARENCO et al. [30] and DiVINCENZO [31].

Simultaneous to the progress around the computational process, gates and algorithms new protocols were developed for the public *exchange of cryptographic keys* based on quantum mechanical properties that allowed eavesdropping detection. These protocols do not require the existence of a quantum computer and can thus already be realized with existing hardware. The first of these protocols utilizing quantum properties was presented in 1984 by BENNETT and BRASSARD [1]. Whereas their protocol does use quantum mechanical properties to detect eavesdropping, it does so without recourse to entanglement. A different protocol making explicit use of entanglement properties was proposed at the beginning of the nineties by EKERT [3].

Another way to harness the computational power of quantum mechanics emerged with what has become known as *adiabatic quantum computing*. Rather than using quantum analogs of classical gates, this method exploits the results around the Quantum Adiabatic Theorem. Here the solution to a problem is encoded in an eigenstate (typically the ground state) of a Hamiltonian that is reached by a suitably slow adiabatic evolution from a known initial state. This idea had its precursor in what was called quantum stochastic optimization and quantum annealing introduced in 1989 by APOLLONI et al. [32]. The groundwork for what we now call adiabatic quantum computing was laid by VAN DAM et al. [33] in 2001 and in 2008 AHARONOV et al. [34] showed the efficiency equivalence between the adiabatic and gate based quantum computation.

In the first decade of the present millennium the theory of a *topological quantum computer* began to be developed. Influenced by the works of FREEDMAN et al. [35] and KITAEV [36] this paradigm of quantum computing makes use of the topological properties of quantum systems in two spatial dimensions. This potential path towards a quantum computer would have the benefit of offering greater stability and protection against undue interactions with the environment and thus requiring less error correction.

It seems that the last chapter of quantum computing has not been written yet and this book is also meant as an invitation to perhaps contribute a few lines to that story.

1.2 Reader's Manual

In this book we present the elementary mathematical aspects of quantum computing. It is self-contained in the sense that with only two minor exceptions all stated results are also proven in the book. To master its content it is not necessary to consult other references, making it suitable for autonomous self-study.

The level of presentation is such that students of physics, mathematics or computer science with knowledge of advanced undergraduate mathematics or anyone with comparable mathematical knowledge will be able to digest the material after which they should be able to read scientific papers on the subject.

The form of presentation is generally a repetition of sequences of motivating or explanatory text followed by definition(s), then results in the form of lemmas, propositions or theorems and then their proofs. Often a main result is prepared in smaller packages in the form of several preparatory lemmas leading up to a theorem. Likewise, several consequences of a main result may be packaged in corollaries. Numerous exercises form an essential part of the logic flow. Their solutions are provided in appendices, but the reader is well advised to attempt to solve them, as it will greatly enhance their understanding of and familiarity with the material.

The mathematical objects necessary for quantum mechanics such as HILBERT spaces, operators, and their properties are presented in Chap. 2 together with the basic principles (aka 'Postulates') of quantum mechanics. The background material for some other areas such as results from number theory or group theory has been outsourced to several appendices. Otherwise, we adhere to what we call the proof-as-you-go approach, in other words, when we need a result in a chain of arguments, we will prove it then and there.

Our approach enables us to state the results and their pre-conditions in a mathematically rigorous form. However, as to the level of generality and technical detail applied—in particular in the context of infinite-dimensional HILBERT spaces—some compromises had to be made. The mathematics of quantum computing hardly ever needs more than finite-dimensional spaces. Hence, most results are restricted to this case, but almost all hold in the infinite-dimensional case, albeit with much more technical proofs.

In Chap. 2 we begin our exposition with a brief introduction into the mathematical formalism of quantum mechanics that equips the reader with the mathematical know-how necessary for the understanding[1] of quantum mechanics.

Using the previously made available mathematics we then present in Sect. 2.3 the basic postulates of quantum mechanics and derive some results, such as the uncertainty relation. In Sect. 2.3.1 we consider pure states and in Sect. 2.3.2 mixed states as well before we introduce qubits in Sect. 2.4. At the end of this chapter we present in Sect. 2.5 operators on qubits, which are important in the context of quantum gates.

In Chap. 3 we give a description of two or more 'particles'[2] with the help of tensor products. For this we define tensor products and introduce the useful computational basis in Sect. 3.2.1, before we review states and observables for composite systems in Sect. 3.3. In doing so we also exhibit a number of identities for partial traces that are particularly useful in the context of composite systems. Section 3.4 contains a presentation of the SCHMIDT decomposition. The chapter concludes with a detailed exposition of quantum operations in Sect. 3.5.

In Chap. 4 we discuss entanglement in some detail. We begin this with a general definition in Sect. 4.2 that also holds for mixed states before we present special criteria for entangled pure states. In the subsequent Sect. 4.3 we show the curious possibility of entangling two systems even though they have not interacted. In Sect. 4.4 the EINSTEIN–PODOLSKI–ROSEN-paradox, (aka EPR-paradox) is discussed extensively. After that we turn to the BELL inequality in Sect. 4.5, which we first present in its original form given by BELL in Sect. 4.5.1. The version derived by CLAUSER, HORNE, SHIMONY and HOLT is then presented in Sect. 4.5.2. The sections about EPR and the BELL inequality do not contain results that are essential for quantum computing. Nevertheless, we have included them here, since they illustrate the counter-intuitive aspects of entanglement and since the BELL inequality is used to detect eavesdropping in a cryptographic protocol, which we present later. At the end of this chapter we illustrate two devices that cannot work as intended due to the laws of quantum mechanics. One of them is the proposal to use entangled states to transmit signals instantaneously, which has become known as BELL's telephone. In Sect. 4.6.1 we show that this does not work. Likewise, we cannot build a device that copies arbitrary unknown qubits. This statement, which has become known as the Quantum No-Cloning Theorem, will be proven in Sect. 4.6.2.

In Chap. 5 we turn to quantum gates and circuits. After a short recall of the usual classical gates in Sect. 5.1 we consider quantum gates in Sect. 5.2. In Sect. 5.2.3 we show how arbitrary unitary transformations can be generated with a suitable number of elementary gates. Next, we exhibit in Sect. 5.5 how elementary computational processes, such as addition, modular exponentiation or the quantum FOURIER

---

[1] Although many, including FEYNMAN, say that no one can 'understand' quantum mechanics.

[2] In this book the 'particle' stands synonymous for any object which is described by quantum mechanics, such as electrons or photons.

transform, can be implemented with elementary gates. This chapter stands somewhat apart in that it is not really necessary for the understanding of the following
material. In this sense it can be left out without endangering the reader's grasp of
the other parts of the book.

In Chap. 6 we return to entanglement and look at a few prominent examples
of how useful it can be. As a uniquely quantum mechanical ingredient it allows
effects that are impossible to generate with classical bits. We begin this chapter
in Sect. 6.1 with the DEUTSCH–JOZSA algorithm as an historically early indicator
of the promise of quantum computing. Next in our list is dense quantum coding,
which we treat in Sect. 6.2. This is followed by teleportation, which we present in
Sect. 6.3. After this we turn in Sect. 6.4 to quantum cryptography. Following a brief
introduction to ciphers in cryptography in Sect. 6.4.1 we introduce two protocols in
which the laws of quantum mechanics allow the detection of eavesdropping. The
protocol presented in Sect. 6.4.2, however, does not make use of entanglement and
requires the transport of particles from sender to receiver. In contrast, the protocol
presented in Sect. 6.4.3 does utilize entanglement and avoids the exchange of particles, if both parties already have a supply of qubits entangled with those of the other
side. In Sect. 6.4.4 we exhibit the basics of the RSA public key distribution protocol, which is then shown to be vulnerable in a detailed look at SHOR's factorization
algorithm in Sect. 6.5. A generalization of this algorithm consisting of finding hidden subgroups of abelian groups is presented in Sect. 6.6. In Sect. 6.7 we show how
the general hidden subgroup algorithm may be used to find the discrete logarithm
in abelian groups. This is of particular relevance for cryptographic protocols such
as the Elliptic Curve Digital Signature Algorithm that rely heavily on finding the
logarithm being difficult. In Sect. 6.8 we have closer look at this protocol which is
used for signing *bitcoin* transactions. The chapter concludes with a detailed look at
GROVER's search algorithm in Sect. 6.9. This is not only a prime example of the so
called amplitude amplification quantum algorithms but also one of the few where
the underlying problem is easily understood by the uninitiated.

In Chap. 7 we introduce the basic notions of error correction. We begin this
by means of an overview of possible error sources for quantum computation in
Sect. 7.1. In Sect. 7.2 we exhibit the essentials of classical linear error correcting
codes. This is done with a view towards similar structures in quantum error correcting codes to which we turn in Sect. 7.3. Quantum codes, error and recovery operators are introduced in Sect. 7.3.1 along with theorems providing conditions under
which a quantum code can detect and correct a given set of errors. In Sect. 7.3.2
we define error detection by means of syndrome extraction and give an error detection and correction protocol. A compact and elegant formulation of quantum error
correcting codes is presented in Sect. 7.3.3 with the stabilizer formalism.

In Chap. 8 a detailed exposition of adiabatic quantum computing is given starting
with a brief introduction in Sect. 8.1. In Sect. 8.2 we state the assumptions underlying the adiabatic method and derive important results about the quality of the adiabatic approximation, which is at the heart of this method. In doing so, we make
use of the quantum Adiabatic Theorem for which we give a thorough proof in
Appendix G. A generic version of the adiabatic method is presented in Sect. 8.3.
As an application of this we then look in Sect. 8.4 at a search algorithm using the

adiabatic method. There we also show that with a suitably adapted algorithm the efficiency of the GROVER search algorithm of Sect. 6.9 can be obtained. Since both the adiabatic and circuit based method produce the final state, the question arises, if one method might be more efficient than the other. In Sects. 8.5 and 8.6 we shall show that the two approaches are indeed equivalent in their efficiency.

At times the auxiliary results we use to prove a statement themselves require rather lengthy proofs. Whenever this becomes to voluminous or it seems best to gather several of such related results, we have opted to outsource them to appendices in order not to interrupt the flow of arguments too much.

In Appendix A we thus collect a few definitions from probability theory.

The algorithms presented in Appendix B are formalized binary versions of normal addition and subtraction. With their help we can then verify that the quantum circuits defined in Sect. 5.5 indeed implement these two elementary operations.

In Appendix C we briefly give our definitions of the LANDAU symbols.

All modular arithmetic necessary for our exposition of cryptography and the factorization algorithm is defined and proven in Appendix D.

The same holds for the results from continued fractions, which are presented in Appendix E.

In Appendix F we present those elements of group theory that we need for our exposition of some quantum algorithms such as the hidden subgroup problem in Sect. 6.6 as well as the stabilizer formalism for quantum error correction in Sect. 7.3.3.

Appendix G contains a rigorous proof of the Quantum Adiabatic Theorem making use of resolvent operators and their properties. This result is then used in analyzing the adiabatic method in Chap. 8.

Finally, even though *solutions to all exercises* can be found in Appendix G.3 the reader is encouraged to try to solve these problems. Attempting to do so will facilitate the learning process even if such attempts are unsuccessful.

1.3 What is not in this Book

Since this book is meant to be an introduction, not all aspects of the large and still growing realm of quantum computing can be presented here. The following list gives some of those topics that are *not covered* in this book.

Methods of quantum mechanics Those who would like to learn something about the methods and results of quantum mechanics in analyzing physical systems, such as atomic spectra, symmetry groups and representations, perturbation theory, scattering theory or relativistic wave equations, are better advised to consult one of the numerous textbooks on quantum mechanics [37–40].

Interpretations of quantum mechanics Even though we consider the EPR paradox and the BELL inequality and try to convey how these bring to the fore some counter-intuitive phenomena of quantum mechanics, we refrain here from a discussion of the foundations or even the various interpretations of quantum mechanics. The reader interested in these aspects may consult [41–44] or the more recent explorations [45–47].

Physical implementations of quantum computers   Neither do we touch on the myriad ways currently attempted to implement quantum gates, or even circuits and ultimately quantum computers. This has now become a fast moving field and presenting even a few of these efforts in a meaningful way would require a lot of additional material from various branches of quantum physics, such as nuclear, atomic, molecular, solid state physics or quantum optics. At the time of writing no comprehensive review of the many ways currently explored to physically implement a quantum computer was available and the reader interested in this is best advised to search the internet.

Complexity theory   A thorough exposition of the information theoretic fundamentals and questions around complexity theory would also exceed the intended scope of this book. To learn more about this, the reader may consult, for example, 'Quantum Information Science' in [48].

Topological quantum computer   This approach is both from the physical as well as the mathematical viewpoint very exciting and challenging. However, precisely because of the latter, the mathematical know-how required is quite extensive and would probably necessitate a (or even more) volume(s) on its own. A relatively recent introductory survey was given by NAYAK et al. [49].

These and other aspects are certainly important and interesting. But they would be misplaced in a book, which claims to be an *introduction to the mathematics* of quantum computing.

1.4   Notation and References

A detailed list of most of the symbols used in this book is given in the list of symbols preceding this chapter. In the following we give some additional general remarks about the notation used here.

General HILBERT spaces are denoted by the symbol $\mathbb{H}$. For the two-dimensional HILBERT space of the qubits we use the symbol $\mathbb{\rlap{\textbar}H}$. The n-fold tensor products are denoted by $\mathbb{H}^{\otimes n}$, resp. $\mathbb{\rlap{\textbar}H}^{\otimes n}$.

For vectors in HILBERT spaces, we initially use the symbols $\psi, \varphi, \ldots$ After the concept of the dual space has been introduced, we use from thereon the DIRAC bra- and ket-notation $|\psi\rangle, |\varphi\rangle, \ldots$ The symbols $|\Psi\rangle, |\Phi\rangle$ mostly denote vectors in composite multi-particle HILBERT spaces. For non-negative integers x and y less than 2^n, the vectors $|x\rangle, |y\rangle, \ldots$ denote elements of the so-called computational basis in $\mathbb{\rlap{\textbar}H}^{\otimes n}$.

Generally, capital letters like A, B, C, D, F etc. are used for operators on HILBERT spaces. Exceptions are: I, which denotes index sets, J and N, which denote natural numbers and L, which mostly stands for the bit-length of a natural number such as N.

The letters i, j, k, l are mostly used for indices. Except, of course, for the imaginary unit i. For symbols with two or more indices such as, for example, a matrix A_{jk} we may insert a comma to improve readability and clarity. This does not alter their meaning. Whilst A_{ij} is just as clear as $A_{i,j}$, the comma in $A_{l-3,l-2}$ is necessary to avoid almost inevitable misunderstandings if we were to write A_{l-3l-2} instead.

We use both ways $e^A = \exp(A)$ to denote the application of the exponential function to A and choose whichever makes a formula less cluttered.

In this book we use natural physical units, such that $\hbar = 1$, where $\hbar = \frac{h}{2\pi}$ denotes the PLANCK constant divided by 2π. This is why $\hbar$ does not appear in many expressions, where the pre-ordained reader might have expected it, such as the HEISENBERG uncertainty relation or the SCHRÖDINGER equation.

Very often we provide a 'justifying' reference for a relation by making use of the following display method.

$$L \underbrace{=}_{\text{(N.nn)}} R.$$

This is to state that the reason L equals R can be found in equation (or a similarly referenced item) with number N.nn.

Concerning the literature, Sect. 1.1 attempts to do justice to important historical contributions and quotes references accordingly. In the remainder of this book, however, references will be given rather sparingly. This is not done with the intention to deny the many original contributors a mentioning of their part in developing the subject. Rather, the intention is not to overload the reader of this introductory text with too many references. The more so, since all necessary material will be presented here.

Chapter 2
Basic Notions of Quantum Mechanics

2.1 Generalities

Quantum mechanics is a theory making predictions about the statistics of micro-scopic objects (such as electrons, protons, atoms, etc.) often with implications for macroscopic phenomena. On such objects measurements of certain quantities can be performed the outcome of which are real numbers. Measurements with equally prepared objects show that the measured values occur with a **relative frequency** and are distributed around a **mean value**. Here the relative frequency is defined as

$$\text{relative frequency of measurement result } a := \frac{\text{number of measurements with result } a}{\text{total number } N \text{ of all measurements}}$$

and the mean value as

$$\text{mean value} := \sum_{a \in \text{measurements}} a \times \binom{\text{relative frequency of}}{\text{measurement result } a}.$$

In such measurements all observed objects have to be prepared in the same way. The following steps are then performed in an experiment:

$$\text{Preparation} \longrightarrow \text{Measurement} \longrightarrow \begin{array}{l}\text{Calculation, such as of relative}\\\text{frequency and mean value.}\end{array}$$

Quantum mechanics is a theory which provides a mathematical model for these steps and makes predictions about relative frequencies and mean values possible. In this context the following notions are used in quantum mechanics.

© Springer Nature Switzerland AG 2019
W. Scherer, *Mathematics of Quantum Computing*,
https://doi.org/10.1007/978-3-030-12358-1_2

- A measurable physical quantity is called an **observable**.
- The quantum mechanical prediction for the relative frequency of a measurement result is called the **probability** of the result.
- The quantum mechanical prediction for the mean value of an observable in a sequence of measurements is called **expectation value**.
- The preparation of objects that yields a statistical ensemble which results in distributions of measurement results and mean values for observables is described by a **state**.

A particular class of such preparations—so-called pure states—can mathematically be described by a vector in a HILBERT space. In their most general form—for so-called mixed states—the preparations are described by operators on a HILBERT space, which are positive, self-adjoint and have trace 1. Observables of the prepared objects are mathematically represented by self-adjoint operators on that HILBERT space. Together with the states (describing the ensemble of objects) the operators (describing the observable physical quantities) then provide a prescription for calculating probabilities and expectation values.

In Sect. 2.2 we will thus first study the tool-set of this theory and exhibit the necessary mathematical objects and notions.

In Sect. 2.3 we then turn to the physical applications of the mathematical objects in quantum mechanics. In doing so, we begin with a description of pure states in Sect. 2.3.1 before we cover the more general case of mixed states in Sect. 2.3.2.

2.2 Mathematical Notions: HILBERT Spaces and Operators

Definition 2.1 A HILBERT space $\mathbb{H}$ is a

(i) complete complex vector space, that is,

$$\psi, \varphi \in \mathbb{H} \text{ and } a, b \in \mathbb{C} \;\Rightarrow\; a\psi + b\varphi \in \mathbb{H},$$

(ii) with a (positive-definite) **scalar product**

$$\langle \cdot | \cdot \rangle : \mathbb{H} \times \mathbb{H} \longrightarrow \mathbb{C}$$
$$(\psi, \varphi) \longmapsto \langle \psi | \varphi \rangle$$

such that for all $\varphi, \psi, \varphi_1, \varphi_2 \in \mathbb{H}$ and $a, b \in \mathbb{C}$

$$\langle \psi | \varphi \rangle = \overline{\langle \varphi | \psi \rangle} \tag{2.1}$$

$$\langle \psi | \psi \rangle \geq 0 \tag{2.2}$$

$$\langle \psi | \psi \rangle = 0 \Leftrightarrow \psi = 0 \tag{2.3}$$

$$\langle \psi | a\varphi_1 + b\varphi_2 \rangle = a\langle \psi | \varphi_1 \rangle + b\langle \psi | \varphi_2 \rangle \tag{2.4}$$

and this scalar product induces a **norm**

$$\begin{aligned} ||\cdot|| : \mathbb{H} &\longrightarrow \mathbb{R} \\ \psi &\longmapsto \sqrt{\langle \psi | \psi \rangle} \end{aligned} \tag{2.5}$$

in which $\mathbb{H}$ is complete.

A subset $\mathbb{H}_{sub} \subset \mathbb{H}$ which is a vector space and inherits the scalar product and the norm from $\mathbb{H}$ is called a sub HILBERT space or simply a **subspace** of $\mathbb{H}$.

In the definition given here the scalar product is linear in the second argument and *anti-linear* (see Exercise 2.1) in the first argument. In some books the opposite convention is used.

Because of (2.1) one has $\langle \psi | \psi \rangle \in \mathbb{R}$, and due to (2.2) the norm is thus well-defined.

Exercise 2.1 Prove the following statements for the scalar product defined in Definition 2.1.

(i) For all $a \in \mathbb{C}$ and $\psi, \varphi \in \mathbb{H}$

$$\langle a\psi | \varphi \rangle = \overline{a}\langle \psi | \varphi \rangle \tag{2.6}$$

$$||a\varphi|| = |a| \, ||\varphi|| \tag{2.7}$$

(ii) For $\psi \in \mathbb{H}$

$$\langle \psi | \varphi \rangle = 0 \quad \forall \varphi \in \mathbb{H} \qquad \Leftrightarrow \qquad \psi = 0 \tag{2.8}$$

(iii) For all $\psi, \varphi \in \mathbb{H}$

$$\langle \psi | \varphi \rangle = \frac{1}{4}\left(||\psi + \varphi||^2 - ||\psi - \varphi||^2 + i\,||\psi - i\varphi||^2 - i\,||\psi + i\varphi||^2 \right). \tag{2.9}$$

For a solution see Solution 2.1.

Completeness of $\mathbb{H}$ in the norm $||\cdot||$ means, that every in $\mathbb{H}$ CAUCHY-convergent[1] sequence $(\varphi_n)_{n\in\mathbb{N}} \subset \mathbb{H}$ has a limit $\lim_{n\to\infty}\varphi_n = \varphi \in \mathbb{H}$ that also lies in $\mathbb{H}$. Finite-dimensional vector spaces, which are the only cases relevant for us in this book, are always complete.

Definition 2.2 A vector $\psi \in \mathbb{H}$ is called **normed** or a **unit vector** if $||\psi||=1$. Two vectors $\psi, \varphi \in \mathbb{H}$ are called **orthogonal** to each other if $\langle\psi|\varphi\rangle = 0$. The subspace in $\mathbb{H}$ of vectors orthogonal to ψ is denoted by

$$\mathbb{H}_{\psi^\perp} := \{\varphi \in \mathbb{H} \,|\, \langle\psi|\varphi\rangle = 0\}.$$

Exercise 2.2 Let $\psi, \varphi \in \mathbb{H}$ with $||\psi|| \neq 0$. Show that

$$\varphi - \frac{\langle\psi|\varphi\rangle}{||\psi||^2}\psi \in \mathbb{H}_{\psi^\perp}$$

and illustrate this graphically.

For a solution see Solution 2.2.

Definition 2.3 Let $\mathbb{H}$ be a HILBERT space and I an index set. A set $\{\varphi_j \,|\, j \in I\} \subset \mathbb{H}$ of vectors is called **linearly independent** if for every finite subset $\{\varphi_1, \varphi_2, \ldots, \varphi_n\}$ and $a_k \in \mathbb{C}$ with $k = 1, \ldots, n$

$$a_1\varphi_1 + a_2\varphi_2 + \cdots + a_n\varphi_n = 0$$

holds only if $a_1 = a_2 = \cdots = a_n = 0$.

A HILBERT space $\mathbb{H}$ is called **finite-dimensional** if $\mathbb{H}$ contains at most $n = \dim\mathbb{H} < \infty$ linearly independent vectors; otherwise $\mathbb{H}$ is called *infinite-dimensional* ($\dim\mathbb{H} = \infty$).

A set of vectors $\{\varphi_j \,|\, j \in I\} \subset \mathbb{H}$ is said to **span** $\mathbb{H}$ if for every vector $\varphi \in \mathbb{H}$ there are $a_j \in \mathbb{C}$ with $j \in I$ such that

$$\varphi = \sum_{j\in I} a_j\varphi_j,$$

---

[1]A sequence $(\varphi_j)_{j\in I}$ is called CAUCHY-convergent, if for every, $\varepsilon > 0$ there exists an $N(\varepsilon)$ such that for all $m,n \geq N(\varepsilon)$ one has $||\varphi_m - \varphi_n|| < \varepsilon$.

which is expressed by writing

$$\mathbb{H} = \text{Span}\left\{\varphi_j \mid j \in I\right\}.$$

A linearly independent set $\{\varphi_j \mid j \in I\}$ of vectors that spans $\mathbb{H}$ is called a **basis** of $\mathbb{H}$ and the vectors φ_j of such a set are called basis vectors. A basis $\{e_j \mid j \in I\} \subset \mathbb{H}$ whose vectors satisfy

$$\langle e_j | e_k \rangle = \delta_{jk} := \begin{cases} 0 \text{ if } j \neq k \\ 1 \text{ if } j = k, \end{cases} \tag{2.10}$$

is called an **orthonormal basis (ONB)**. The HILBERT space $\mathbb{H}$ is called separable if it has a *countable* basis.

Example 2.4

$$\mathbb{H} = \mathbb{C}^n := \left\{ z = \begin{pmatrix} z_1 \\ \vdots \\ z_n \end{pmatrix} \mid z_j \in \mathbb{C} \right\}$$

with the usual scalar product

$$\langle z | w \rangle := \sum_{j=1}^{n} \overline{z}_j w_j$$

is a HILBERT space of dimension n. An ONB for it is given by the countable set of basis vectors

$$\left\{ e_1 = \begin{pmatrix} 1 \\ 0 \\ 0 \\ \vdots \\ 0 \end{pmatrix}, e_2 = \begin{pmatrix} 0 \\ 1 \\ 0 \\ \vdots \\ 0 \end{pmatrix}, \ldots, e_n = \begin{pmatrix} 0 \\ 0 \\ 0 \\ \vdots \\ 1 \end{pmatrix} \right\}.$$

We will encounter HILBERT spaces of this type time and again in the context of quantum computing.

In general quantum mechanics makes use of infinite-dimensional HILBERT spaces. If the position or the momentum of a particle that has to be described by quantum mechanics were our observables of interest, we would have to use the infinite-dimensional HILBERT space of Example 2.5.

Example 2.5 Let $d^3\mathbf{x}$ denote the LEBESGUE measure [50] in $\mathbb{R}^3$. Then

$$L^2(\mathbb{R}^3) := \left\{ \psi : \mathbb{R}^3 \to \mathbb{C} \,\Big|\, \int_{\mathbb{R}^3} |\psi(\mathbf{x})|^2 \, d^3\mathbf{x} < \infty \right\}$$

with the usual scalar product

$$\langle \psi_1 | \psi_2 \rangle := \int_{\mathbb{R}^3} \overline{\psi_1(\mathbf{x})} \, \psi_2(\mathbf{x}) d^3\mathbf{x}$$

is an infinite-dimensional HILBERT space, which is used to describe position and momentum of a particle in three-dimensional space.

In general, however, the position or the momentum of a particle are not observables that are relevant in the context of quantum computing, which is why we will not consider $L^2(\mathbb{R}^3)$ any further.

Rather than position or momentum of a particle, the quantum computational process generally observes and manipulates intrinsic quantum observables such as spin or (photon) polarization. In order to understand the aspects of quantum mechanics relevant for computing it is thus sufficient to consider only finite-dimensional HILBERT spaces. These are necessarily separable. Where possible we shall introduce further notions in the most general form regardless of the dimension of the underlying space $\mathbb{H}$. But in doing so we will mostly ignore the additional mathematical detail required in the infinite-dimensional case, such as convergence for infinite sums or densely defined domains of operators. Including all of these would overload the presentation and unnecessarily distract from the essential features of quantum computing. In particular, we will only consider separable HILBERT spaces.

Every vector $\psi \in \mathbb{H}$ can be expressed with the help of a basis $\{e_j\}$ and complex numbers $\{a_j\}$

$$\psi = \sum_j a_j e_j \,.$$

For a given basis this so-called **basis expansion** ψ is unique because $\psi = \sum_j b_j e_j$ implies that $\sum_j (a_j - b_j) e_j = 0$ and due to the linear independence of the e_j it then follows from Definition 2.3 that we must have $a_j = b_j$.

If $\{e_j\}$ is an ONB, then we have $a_j = \langle e_j | \psi \rangle$, that is,

$$\psi = \sum_j \langle e_j | \psi \rangle e_j \tag{2.11}$$

and

$$||\psi||^2 = \sum_j |\langle e_j | \psi \rangle|^2 \,. \tag{2.12}$$

These claims are to be shown in Exercise 2.3.

Exercise 2.3 Let $\psi, \varphi \in \mathbb{H}$ and $\{e_j\}$ be an ONB. Moreover, let $\psi_j = \langle e_j | \psi \rangle$ and similarly $\varphi_j = \langle e_j | \varphi \rangle$. Show that

(i)

$$\psi = \sum_j \langle e_j | \psi \rangle e_j = \sum_j \psi_j e_j$$

(ii)

$$\langle \varphi | \psi \rangle = \sum_j \overline{\langle e_j | \varphi \rangle} \langle e_j | \psi \rangle = \sum_j \langle \varphi | e_j \rangle \langle e_j | \psi \rangle = \sum_j \overline{\varphi_j} \psi_j \qquad (2.13)$$

(iii)

$$\|\psi\|^2 = \sum_j |\langle e_j | \psi \rangle|^2 = \sum_j |\psi_j|^2 \qquad (2.14)$$

(iv) If $\varphi \in \mathbb{H}_{\psi^\perp}$, then

$$\|\varphi + \psi\|^2 = \|\varphi\|^2 + \|\psi\|^2. \qquad (2.15)$$

This is a generalized version of the **Theorem of PYTHAGORAS**.

For a solution see Solution 2.3.

Another useful relation is the **SCHWARZ-inequality**

$$|\langle \psi | \varphi \rangle| \leq \|\psi\| \, \|\varphi\|, \qquad (2.16)$$

which is to be proven in Exercise 2.4.

Exercise 2.4 Show that for any $\varphi, \psi \in \mathbb{H}$

$$|\langle \psi | \varphi \rangle| \leq \|\psi\| \, \|\varphi\|. \qquad (2.17)$$

First consider the case $\psi = 0$ or $\varphi = 0$. In the case $\psi \neq 0 \neq \varphi$ use Exercises 2.2 and 2.3 and make a suitable estimate. With the help of (2.17) show that also for any $\varphi, \psi \in \mathbb{H}$

$$\|\psi + \varphi\| \leq \|\psi\| + \|\varphi\|. \qquad (2.18)$$

For a solution see Solution 2.4.

With the help of the scalar product every vector $\psi \in \mathbb{H}$ defines a *linear map* from $\mathbb{H}$ to $\mathbb{C}$, which we denote by $\langle \psi |$

$$\langle\psi| : \mathbb{H} \longrightarrow \mathbb{C}$$
$$\varphi \longmapsto \langle\psi|\varphi\rangle \qquad . \tag{2.19}$$

Exercise 2.5 Show that the map defined in (2.19) is continuous.

For a solution see Solution 2.5.

Conversely, it can be shown (see **Theorem of RIESZ** [51]) that every linear and continuous[2] map from $\mathbb{H}$ to $\mathbb{C}$ can be expressed with a $\psi \in \mathbb{H}$ in the form given in (2.19) as $\langle\psi|$. This means that there is a bijection between $\mathbb{H}$ and its **dual space**

$$\mathbb{H}^* := \{f : \mathbb{H} \to \mathbb{C} \mid f \text{ linear and continuous}\} .$$

Essentially, this bijection states that every linear and continuous map from $\mathbb{H}$ to $\mathbb{C}$ is uniquely represented as a scalar product with a suitable vector in $\mathbb{H}$.

The dual space $\mathbb{H}^*$ of a separable HILBERT space $\mathbb{H}$ is also a vector space with the same dimension as $\mathbb{H}$. This identification[3] of $\mathbb{H}$ with $\mathbb{H}^*$ motivates the *'bra'* and *'ket'* notation derived from the word bracket and introduced by DIRAC. **Bra-vectors** are elements of $\mathbb{H}^*$ and are written as $\langle\varphi|$. **Ket-vectors** are elements of $\mathbb{H}$ and are written as $|\psi\rangle$. Because of the above-mentioned bijection between the HILBERT space $\mathbb{H}$ and its dual space $\mathbb{H}^*$ each vector $|\varphi\rangle \in \mathbb{H}$ corresponds to a vector in $\mathbb{H}^*$, which is then denoted as $\langle\varphi|$.

The application of the bra (the linear map) $\langle\varphi|$ on the ket (the vector) $|\psi\rangle$ as the argument of the linear map is then the 'bracket' $\langle\varphi|\psi\rangle \in \mathbb{C}$. One writes (2.11) in the form

$$|\psi\rangle = \sum_j |e_j\rangle\langle e_j|\psi\rangle . \tag{2.20}$$

With (2.20) and the notation $A|\psi\rangle = |A\psi\rangle$ one then has

$$A|\psi\rangle = |A\psi\rangle = \sum_j |e_j\rangle\langle e_j|A\psi\rangle \underset{(2.20)}{=} \sum_j |e_j\rangle\langle e_j|A \sum_k |e_k\rangle\langle e_k|\psi\rangle\rangle$$
$$\underset{(2.4)}{=} \sum_{j,k} |e_j\rangle\langle e_j|Ae_k\rangle\langle e_k|\psi\rangle .$$

Therefore, we can express A in the form

$$A = \sum_{j,k} |e_j\rangle\langle e_j|Ae_k\rangle\langle e_k| = \sum_{j,k} |e_j\rangle A_{jk} \langle e_k| , \tag{2.21}$$

---

[2]Continuity needs to be mentioned separately only in the infinite-dimensional case. In finite-dimensional spaces every linear map is necessarily continuous.

[3]Identified with each other are the sets, but not the linear structures of the vector spaces, since the bijection $\mathbb{H} \ni |\varphi\rangle \to \langle\varphi| \in \mathbb{H}^*$ is *anti-linear*.

where $A_{jk} := \langle e_j | A e_k \rangle$. This motivates the following definition.

Definition 2.6 For an operator A on a HILBERT space $\mathbb{H}$ and an ONB $\{|e_j\rangle\}$ in $\mathbb{H}$ one defines

$$A_{jk} := \langle e_j | A e_k \rangle \tag{2.22}$$

as the (j,k) **matrix element** of A in the basis $\{|e_j\rangle\}$. The matrix $(A_{jk})_{j,k=1,...,\dim \mathbb{H}}$ is called the **matrix representation** or simply the **matrix** of the operator A in the basis $\{|e_j\rangle\}$. The same symbol A is used to denote the operator and its matrix.

For any finite-dimensional HILBERT space $\mathbb{H}$ with $\dim \mathbb{H} = n$ and a given ONB $\{|e_j\rangle\} \subset \mathbb{H}$ we can define an isomorphism $\mathbb{H} \cong \mathbb{C}^n$ by identifying the given basis with the standard basis in $\mathbb{C}^n$, in other words, we make the identification[4]

$$|e_1\rangle = \begin{pmatrix} 1 \\ 0 \\ \vdots \\ 0 \end{pmatrix}, \ldots, |e_n\rangle = \begin{pmatrix} 0 \\ \vdots \\ 0 \\ 1 \end{pmatrix}. \tag{2.23}$$

Likewise, we have for the dual basis[5]

$$\langle e_1| = \begin{pmatrix} 1 & 0 & \ldots & 0 \end{pmatrix}, \ldots, \langle e_n| = \begin{pmatrix} 0 & \ldots & 0 & 1 \end{pmatrix}. \tag{2.24}$$

In (2.23) the right side of each equation is a column vector, which may be considered a complex $n \times 1$ matrix, whereas in (2.24) it is a row vector, which may be viewed as a complex $1 \times n$ matrix. The operator $|e_j\rangle\langle e_k|$ is the product of an $n \times 1$ matrix with an $1 \times n$ matrix, which turns out to be an $n \times n$ matrix. In the basis $\{|e_j\rangle\}$ this matrix is then found to be

---

[4]Strictly speaking, these are not equalities between the vectors $|e_a\rangle \in \mathbb{H}$ and the standard basis in $\mathbb{C}^n$. Rather, we have a linear map, that is, an isomorphism of HILBERT spaces $\iota : \mathbb{H} \to \mathbb{C}^n$ and,

for example, $\iota(|e_1\rangle) = \begin{pmatrix} 1 \\ 0 \\ \vdots \\ 0 \end{pmatrix}$. But the agreed convention, which we have adopted here, is to state

equality without writing out ι explicitly.

[5]This is actually a basis $\{\langle u_a|\}$ of the dual space $\mathbb{H}^*$ satisfying $\langle u_a | e_{a'}\rangle = \delta_{a,a'}$. But as remarked before, we can identify $\mathbb{H}^*$ with $\mathbb{H}$ and thus $\{\langle u_a|\} = \{\langle e_a|\}$.

$$
|e_j\rangle\langle e_k| = \begin{matrix} 1 \\ \vdots \\ 0 \\ j \\ 0 \\ \vdots \\ n \end{matrix} \begin{pmatrix} 0 \\ \vdots \\ 0 \\ 1 \\ 0 \\ \vdots \\ 0 \end{pmatrix} \quad \begin{matrix} 1 & \ldots & k & \ldots & n \end{matrix} \\ \begin{pmatrix} 0 & \ldots & 0 & 1 & 0 & \ldots & 0 \end{pmatrix}
$$

$$
= \begin{matrix} \\ 1 \\ \vdots \\ j \\ \vdots \\ n \end{matrix} \quad \begin{matrix} 1 & \ldots & k & \ldots & n \end{matrix} \\ \begin{pmatrix} & & | & & \\ & & | & & \\ -- & -- & -- & 1 & \\ & & & & \\ & & & & \end{pmatrix} , \tag{2.25}
$$

where the row above a matrix shows the column indices and the column to the left indicates the row indices. *Only the non-zero element* in the matrix is shown, in other words, all other matrix elements apart from the j,k matrix element are zero. In particular, the matrix of $|e_j\rangle\langle e_j|$ would be zero everywhere, except for the j-th entry on the diagonal, which would have the value 1.

As can be seen from (2.23) and (2.24) the vector in $\mathbb{C}^n$ identified with $\langle e_j|$ is just the transpose of the vector in $\mathbb{C}^n$ identified with $|e_j\rangle$. This holds generally and we have that

$$
|\psi\rangle = \sum_j b_j |e_j\rangle = \begin{pmatrix} b_1 \\ \vdots \\ b_n \end{pmatrix} \quad \underset{(2.33)}{\Leftrightarrow} \quad \langle\psi| = \sum_j \overline{b_j}\langle e_j| = \left(\overline{b_1} \ldots \overline{b_n}\right), \tag{2.26}
$$

such that with a

$$
|\varphi\rangle = \sum_j a_j |e_j\rangle = \begin{pmatrix} a_1 \\ \vdots \\ a_n \end{pmatrix}
$$

we then obtain

$$
|\varphi\rangle\langle\psi| = \begin{pmatrix} a_1 \\ \vdots \\ a_n \end{pmatrix} \left(\overline{b_1} \ldots \overline{b_n}\right) = \begin{pmatrix} a_1\overline{b_1} & \ldots & a_1\overline{b_n} \\ \vdots & & \vdots \\ a_n\overline{b_1} & \ldots & a_n\overline{b_n} \end{pmatrix} . \tag{2.27}
$$

When the operator $|\varphi\rangle\langle\psi|$ is multiplied by a number $z \in \mathbb{C}$ we also use the notation $|\varphi\rangle z\langle\psi| := z|\varphi\rangle\langle\psi|$, such that

$$|\varphi\rangle z\langle\psi| = z \begin{pmatrix} a_1\overline{b_1} & \cdots & a_1\overline{b_n} \\ \vdots & & \vdots \\ a_n\overline{b_1} & \cdots & a_n\overline{b_n} \end{pmatrix}.$$

Example 2.7 Let $\mathbb{H} \cong \mathbb{C}^2$ be a two-dimensional HILBERT space with an ONB $\{|0\rangle,|1\rangle\}$, which we identify with the standard basis in $\mathbb{C}^2$ by

$$|0\rangle = \begin{pmatrix} 1 \\ 0 \end{pmatrix}, \qquad |1\rangle = \begin{pmatrix} 0 \\ 1 \end{pmatrix}.$$

Likewise, we have for the dual basis $\{\langle e_a|\} \subset \mathbb{H}^* \cong \mathbb{H}$ the identification

$$\langle 0| = \begin{pmatrix} 1 & 0 \end{pmatrix}, \qquad \langle 1| = \begin{pmatrix} 0 & 1 \end{pmatrix}$$

in $\mathbb{H}^* \cong \mathbb{H} \cong \mathbb{C}^2$. For the operators $|x\rangle\langle y| : \mathbb{H} \to \mathbb{H}$ with $x,y \in \{0,1\}$ we then find from (2.27) that

$$
\begin{aligned}
&|0\rangle\langle 0| = \begin{pmatrix} 1 \\ 0 \end{pmatrix}\begin{pmatrix} 1 & 0 \end{pmatrix} = \begin{pmatrix} 1 & 0 \\ 0 & 0 \end{pmatrix} && |0\rangle\langle 1| = \begin{pmatrix} 1 \\ 0 \end{pmatrix}\begin{pmatrix} 0 & 1 \end{pmatrix} = \begin{pmatrix} 0 & 1 \\ 0 & 0 \end{pmatrix} \\
&|1\rangle\langle 0| = \begin{pmatrix} 0 \\ 1 \end{pmatrix}\begin{pmatrix} 1 & 0 \end{pmatrix} = \begin{pmatrix} 0 & 0 \\ 1 & 0 \end{pmatrix} && |1\rangle\langle 1| = \begin{pmatrix} 0 \\ 1 \end{pmatrix}\begin{pmatrix} 0 & 1 \end{pmatrix} = \begin{pmatrix} 0 & 0 \\ 0 & 1 \end{pmatrix},
\end{aligned}
\tag{2.28}
$$

where the matrices are in the standard basis of $\mathbb{C}^2$. With

$$|\varphi\rangle = a|0\rangle + b|1\rangle, \qquad |\psi\rangle = c|0\rangle + d|1\rangle$$

we thus obtain for a general $|\varphi\rangle\langle\psi| : \mathbb{H} \to \mathbb{H}$

$$
\begin{aligned}
|\varphi\rangle\langle\psi| &= \big(a|0\rangle + b|1\rangle\big)\big(\overline{c}\langle 0| + \overline{d}\langle 1|\big) \\
&= a\overline{c}|0\rangle\langle 0| + a\overline{d}|0\rangle\langle 1| + b\overline{c}|1\rangle\langle 0| + b\overline{d}|1\rangle\langle 1| \\
&\underset{(2.28)}{=} \begin{pmatrix} a\overline{c} & a\overline{d} \\ b\overline{c} & b\overline{d} \end{pmatrix}.
\end{aligned}
\tag{2.29}
$$

From now on we shall use the bra-ket-notation for vectors $|\varphi\rangle \in \mathbb{H}$ and their corresponding elements $\langle\varphi| \in \mathbb{H}^*$ in the dual space. However, in order not to overload the notation we may drop the bra-ket-notation, if the vector appears as an argument of a function. For example, instead of $|\,\|\psi\rangle\|$ we shall simply write $\|\psi\|$, where, of course, ψ and $|\psi\rangle$ denote the same vector in $\mathbb{H}$.

Definition 2.8 A linear map $A : \mathbb{H} \to \mathbb{H}$ is called an **operator** on the HILBERT space $\mathbb{H}$. The set of all operators on $\mathbb{H}$ is denoted by $\mathrm{L}(\mathbb{H})$. A linear map $T : \mathrm{L}(\mathbb{H}) \to \mathrm{L}(\mathbb{H})$, that is, an operator acting on operators, is called a **super-operator**.

The operator $A^* : \mathbb{H} \to \mathbb{H}$ that satisfies

$$\langle A^* \psi | \varphi \rangle = \langle \psi | A \varphi \rangle \quad \forall | \psi \rangle, | \varphi \rangle \in \mathbb{H} \tag{2.30}$$

is called the **adjoint** operator to A. If $A^* = A$ then A is called **self-adjoint**.

In the infinite-dimensional case this means that A and A^* are densely defined and have the same domain on which they coincide. In the finite-dimensional case self-adjoint is the same as **hermitian**. To be precise, A^* is actually a map $A^* : \mathbb{H}^* \to \mathbb{H}^*$ but as mentioned before we can identify $\mathbb{H}^*$ with $\mathbb{H}$.

In Exercise 2.6 it is to be shown that A is self-adjoint if and only if its matrix elements satisfy $\overline{A_{kj}} = A_{jk}$.

Exercise 2.6 Show that

(i)
$$(A^*)^* = A \tag{2.31}$$

(ii) For any $c \in \mathbb{C}$
$$(cA)^* = \bar{c} A^* \tag{2.32}$$

(iii)
$$\langle A \psi | = \langle \psi | A^*, \tag{2.33}$$

where the right side is understood as the map $\mathbb{H} \xrightarrow{A^*} \mathbb{H} \xrightarrow{\langle \psi |} \mathbb{C}$ and $\langle \psi |$ and $\langle A \psi |$ are as given in (2.19).

(iv)
$$A_{jk}^* = \overline{A_{kj}} \tag{2.34}$$

such that
$$A^* = A \quad \Leftrightarrow \quad \overline{A_{kj}} = A_{jk}. \tag{2.35}$$

(v) For any $| \psi \rangle, | \varphi \rangle \in \mathbb{H}$
$$\big(| \varphi \rangle \langle \psi | \big)^* = | \psi \rangle \langle \varphi | \tag{2.36}$$

For a solution see Solution 2.6.

Definition 2.9 An operator U on $\mathbb{H}$ is called **unitary** if

$$\langle U\psi | U\varphi \rangle = \langle \psi | \varphi \rangle \quad \forall | \psi \rangle, | \varphi \rangle \in \mathbb{H}.$$

The set of all unitary operators on $\mathbb{H}$ is denoted by $\mathcal{U}(\mathbb{H})$.

Unitary operators have their adjoint operator as their inverse and do not change the norm.

Exercise 2.7 Show that

$$U \in \mathcal{U}(\mathbb{H}) \quad \Leftrightarrow \quad U^* U = \mathbf{1} \quad \Leftrightarrow \quad ||U\psi|| = ||\psi|| \quad \forall | \psi \rangle \in \mathbb{H}, \qquad (2.37)$$

where $\mathbf{1}$ is the identity operator on $\mathbb{H}$.

For a solution see Solution 2.7.

Definition 2.10 Let A be an operator on a HILBERT space $\mathbb{H}$. A vector $| \psi \rangle \in \mathbb{H} \smallsetminus \{0\}$ is called **eigenvector** of A with **eigenvalue** $\lambda \in \mathbb{C}$ if

$$A | \psi \rangle = \lambda | \psi \rangle.$$

The linear subspace that is spanned by all eigenvectors for a given eigenvalue λ of an operator A is called **eigenspace** of λ and denoted by $\mathrm{Eig}(A, \lambda)$. An eigenvalue λ is called **non-degenerate** if its eigenspace is one-dimensional. Otherwise, λ is called **degenerate**. The set

$$\sigma(A) := \{ \lambda \in \mathbb{C} \,|\, (A - \lambda\mathbf{1})^{-1} \text{does not exist} \}$$

is called the **spectrum** of the operator A.

Eigenvalues of an operator A are thus per definition contained in the spectrum of A. In infinite-dimensional HILBERT spaces the spectrum of an operator may—in addition to the eigenvalues—also contain a so-called continuous part. Since in this book we are dealing exclusively with finite-dimensional HILBERT spaces, we may identify the spectrum of an operator with the set of its eigenvalues for all operators we are dealing with here.

The eigenvalues of self-adjoint operators are always real and the eigenvalues of unitary operators always have absolute value 1.

Exercise 2.8 Let $A|\psi\rangle = \lambda|\psi\rangle$. Show that:

(i)
$$\langle\psi|A^* = \overline{\lambda}\langle\psi|.$$

(ii) The eigenvalues of a self-adjoint operators are always real.
(iii) The eigenvalues of a unitary operators have absolute value 1.

For a solution see Solution 2.8.

Self-adjoint operators are **diagonalizable**, that is, for every self-adjoint operator A there is an ONB consisting of eigenvectors $\{|e_{j,\alpha}\rangle\}$ of A such that

$$A|e_{j,\alpha}\rangle = \lambda_j|e_{j,\alpha}\rangle.$$

The matrix elements in this basis have the form $A_{j,\alpha;k,\beta} = \lambda_j\delta_{j,k}\delta_{\alpha,\beta}$ and thus (2.21) becomes

$$A = \sum_{j,k,\alpha,\beta} |e_{j,\alpha}\rangle\lambda_j\delta_{j,k}\delta_{\alpha,\beta}\langle e_{k,\beta}| = \sum_{j,\alpha}\lambda_j|e_{j,\alpha}\rangle\langle e_{j,\alpha}|. \tag{2.38}$$

Expressing a self-adjoint operator with the help of its eigenvalues and eigenvectors as in (2.38) is referred to as writing the operator in its **diagonal form**.

In the finite-dimensional case it is fairly straightforward to show that the smallest and largest eigenvalues of a self-adjoint operator A serve as lower and upper bound of the scalar product of $\langle\psi|A\psi\rangle$ for $|\psi\rangle$ normalized to 1.

Exercise 2.9 Let $\mathbb{H}$ be a HILBERT space with $\dim\mathbb{H} = d < \infty$ and let A be a self-adjoint operator with eigenvalues $\{\lambda_j \mid j \in \{1,\ldots,d\}\}$ satisfying

$$\lambda_1 \leq \lambda_2 \leq \cdots \leq \lambda_d. \tag{2.39}$$

Show that then for any $|\psi\rangle \in \mathbb{H}$ with $\|\psi\| = 1$

$$\lambda_1 \leq \langle\psi|A\psi\rangle \leq \lambda_d \tag{2.40}$$

holds.

For a solution see Solution 2.9.

An important and special type of operator is what is called projection or projector.

Definition 2.11 Let $\mathbb{H}$ be a HILBERT space. An operator $P \in L(\mathbb{H})$ satisfying $P^2 = P$ is called a **projection** or projector. If in addition $P^* = P$, then P is called an **orthogonal projection**.

Let $\mathbb{H}_{sub}$ be a subspace of $\mathbb{H}$. If P_{sub} is an orthogonal projection and satisfies $P_{sub}|\psi\rangle = |\psi\rangle$ for all $|\psi\rangle \in \mathbb{H}_{sub}$ we call P_{sub} the projection onto this subspace.

Exercise 2.10 As in (2.38) let the $|e_{j,\alpha}\rangle$ with $\alpha \in \{1,\ldots,d_j\}$ be orthonormal eigenvectors for the possibly d_j-fold degenerate eigenvalues λ_j of a self-adjoint operator A. Show that then

$$P_j = \sum_{\alpha=1}^{d_j} |e_{j,\alpha}\rangle\langle e_{j,\alpha}| \tag{2.41}$$

is the projection onto the eigenspace $\mathrm{Eig}(A,\lambda_j)$ for the eigenvalue λ_j and $d_j = \dim\mathrm{Eig}(A,\lambda_j)$.

For a solution see Solution 2.10.

With P_j thus defined, we can write any self-adjoint A in the form

$$A \underbrace{=}_{(2.38)} \sum_{j,\alpha} \lambda_j |e_{j,\alpha}\rangle\langle e_{j,\alpha}| \underbrace{=}_{(2.41)} \sum_j \lambda_j P_j. \tag{2.42}$$

Note that any ONB $\{|e_j\rangle\}$ constitutes an ONB of eigenvectors of the identity operator $\mathbf{1}$ for the eigenvalue 1. Hence, (2.42) implies that for any ONB $\{|e_j\rangle\}$

$$\mathbf{1} \underbrace{=}_{(2.21)} \sum_{j,k} |e_j\rangle\langle e_j|e_k\rangle\langle e_k| \underbrace{=}_{(2.10)} \sum_{j,k} \delta_{jk}|e_j\rangle\langle e_k| = \sum_j |e_j\rangle\langle e_j| = \sum_j P_j, \tag{2.43}$$

where $P_j = |e_j\rangle\langle e_j|$ denotes the projection onto the subspace spanned by $|e_j\rangle$. Due to the $|e_j\rangle$ being orthonormal these projections furthermore satisfy

$$P_j P_k = |e_j\rangle\langle e_j|e_k\rangle\langle e_k| \underbrace{=}_{(2.10)} \delta_{jk}|e_j\rangle\langle e_k| = \delta_{jk}|e_j\rangle\langle e_j| = \delta_{jk}P_j. \tag{2.44}$$

Generally, for each orthogonal projection P there is a set of orthonormal vectors $\{|\psi_j\rangle\}$ such that

$$P = \sum_j |\psi_j\rangle\langle\psi_j|.$$

In case this set consists of one normed vector $|\psi\rangle$ only, P is called a projection onto $|\psi\rangle$ and it is denoted as P_ψ. In other words, for a $|\psi\rangle$ with $||\psi|| = 1$ we have

$$P_\psi := |\psi\rangle\langle\psi|$$

as the orthogonal projection onto $|\psi\rangle$.

Exercise 2.11 Let P be an orthogonal projection. Show that then a set of orthonormal vectors $\{|\psi_j\rangle\} \subset \mathbb{H}$ exists such that

$$P = \sum_j |\psi_j\rangle\langle\psi_j|.$$

Hint: From $P^2 = P = P^*$ deduce the possible eigenvalues of P and use (2.42).

For a solution see Solution 2.11.

Definition 2.12 Let $\mathbb{H}$ be a HILBERT space. An operator $A \in L(\mathbb{H})$ is called **bounded** if

$$||A|| := \sup\{||A\psi|| \mid |\psi\rangle \in \mathbb{H} \text{ and } ||\psi|| = 1\} < \infty \qquad (2.45)$$

and in this case $||A||$ is called the **norm of the operator** A. We denote the set of bounded operators on $\mathbb{H}$ by $B(\mathbb{H})$. The set of bounded self-adjoint operators on $\mathbb{H}$ is denoted by $B_{sa}(\mathbb{H})$.

A self-adjoint operator A is called **positive** if for all $|\psi\rangle \in \mathbb{H}$

$$\langle\psi|A\psi\rangle \geq 0,$$

which is written as $A \geq 0$. A self-adjoint operator A is called **strictly positive** if $\langle\psi|A\psi\rangle > 0$ for all $|\psi\rangle \in \mathbb{H} \smallsetminus \{0\}$ and this is denoted by $A > 0$. For two operators A and B the statement $A \geq B$ is defined as $A - B \geq 0$ and, likewise, $A > B$ is defined as $A - B > 0$.

Furthermore, one defines the **commutator** of two operators A and B as

$$[A,B] := AB - BA. \qquad (2.46)$$

We say A and B *commute* if their commutator vanishes, that is, if $[A,B] = 0$.

If $\mathbb{H}$ is infinite-dimensional, these definitions have to be slightly amended. However, since in this book we deal only with finite-dimensional $\mathbb{H}$, Definition 2.12

is sufficient as is for our purposes. For the same reason we will only encounter bounded operators in this book.

Exercise 2.12 Let $\mathbb{H}$ be a HILBERT space. Show that for $A, B \in \mathrm{B}(\mathbb{H})$ one has

$$(AB)^* = B^* A^* \tag{2.47}$$

and that if $A, B \in \mathrm{B}_{\mathrm{sa}}(\mathbb{H})$ then

$$(AB)^* = AB \qquad \Leftrightarrow \qquad [A, B] = 0. \tag{2.48}$$

Furthermore, show that for any $c \geq 0$

$$A^* A \leq c B^* B \qquad \Leftrightarrow \qquad ||A|| \leq \sqrt{c}\, ||B||, \tag{2.49}$$

such that

$$A^* A \leq c\mathbf{1} \qquad \Leftrightarrow \qquad ||A|| \leq \sqrt{c}.$$

For a solution see Solution 2.12.

Using the result of Exercise 2.9 allows us to show that the norm of a bounded self-adjoint operator is given by the absolute value of its largest eigenvalue.

Exercise 2.13 Let $\mathbb{H}$ be a HILBERT space with $\dim \mathbb{H} = d < \infty$ and let $A \in \mathrm{B}_{\mathrm{sa}}(\mathbb{H})$ with a set of eigenvalues $\sigma(A) = \left\{ \lambda_j \mid j \in \{1, \ldots, d\} \right\}$ such that

$$\lambda_1 \leq \lambda_2 \leq \cdots \leq \lambda_d.$$

Show that then

$$||A|| = |\lambda_d| \tag{2.50}$$

holds.

For a solution see Solution 2.13.

Further relations for operator norms, which will be useful for us, are to be shown as Exercise 2.14.

Exercise 2.14 Show that for all operators $A, B \in \mathrm{B}(\mathbb{H})$, vectors $|\psi\rangle \in \mathbb{H}$ and $a \in \mathbb{C}$ one has

$$||A\psi|| \leq ||A|| \, ||\psi|| \tag{2.51}$$

$$||AB|| \leq ||A|| \, ||B|| \tag{2.52}$$

$$||A+B|| \leq ||A|| + ||B|| \tag{2.53}$$

$$||aA|| = |a| \, ||A|| \tag{2.54}$$

and that every projection P on $\mathbb{H}$ and every unitary operator $U \in \mathcal{U}(\mathbb{H})$ satisfy

$$||P|| = 1 = ||U|| \, . \tag{2.55}$$

For a solution see Solution 2.14.

It turns out that the sum of the diagonal matrix elements of an operator A does not depend on the basis chosen to calculate it, that is, every two ONBs $\{|e_j\rangle\}, \{|\tilde{e}_j\rangle\}$ satisfy

$$\sum_j \langle e_j | A e_j \rangle = \sum_k \langle \tilde{e}_k | A \tilde{e}_k \rangle \, ,$$

which is to be shown as Exercise 2.15.

Exercise 2.15 Let $\{|e_j\rangle\}$ be an ONB in a HILBERT space $\mathbb{H}$ and $A, U \in L(\mathbb{H})$ Show that then:

(i)

$$\left\{ |\tilde{e}_j\rangle = U|e_j\rangle \right\} \text{ is ONB in } \mathbb{H} \qquad \Leftrightarrow \qquad U \in \mathcal{U}(\mathbb{H}) \, . \tag{2.56}$$

(ii)

$$\sum_j \langle e_j | A e_j \rangle = \sum_j \langle \tilde{e}_j | A \tilde{e}_j \rangle . $$

For a solution see Solution 2.15.

This invariance property of the sum of diagonal matrix elements allows us to define an important map $\mathrm{tr} : L(\mathbb{H}) \to \mathbb{C}$ known as the trace.

Definition 2.13 Let $\{|e_j\rangle\}$ be an ONB in a HILBERT space $\mathbb{H}$. The **trace** is defined as the map

$$\begin{aligned} \mathrm{tr} : L(\mathbb{H}) &\longrightarrow \mathbb{C} \\ A &\longmapsto \mathrm{tr}(A) := \sum_j \langle e_j | A e_j \rangle \underbrace{=}_{(2.22)} \sum_j A_{jj} \, . \end{aligned} \tag{2.57}$$

For any $A \in L(\mathbb{H})$ the expression $\mathrm{tr}(A)$ is called the trace of A.

The trace is a linear map since for any $A, B \in L(\mathbb{H})$ we have $(A+B)_{jj} = A_{jj} + B_{jj}$. Two of its further properties are to be shown in Exercise 2.16.

Exercise 2.16 Show that the trace has the following properties:

(i) For all $A, B \in L(\mathbb{H})$
$$\mathrm{tr}\,(AB) = \mathrm{tr}\,(BA)\,. \tag{2.58}$$

(ii) For $B \in L(\mathbb{H})$
$$\mathrm{tr}\,(AB) = 0 \quad \forall A \in L(\mathbb{H}) \quad \Leftrightarrow \quad B = 0. \tag{2.59}$$

For a solution see Solution 2.16.

2.3   Physical Notions: States and Observables

As mentioned before, quantum mechanics is a theory, which in general only allows statements about the statistics of the system described. The statement 'a system is in a given state' thus means that the system has been prepared as a member of a statistical ensemble and the observable statistics of this ensemble can be calculated with the help of the mathematical object representing this state. For the selection of the mathematical object to describe the ensemble one distinguishes between so-called *pure states* and the more general case of *mixed states*, which we shall discuss in Sect. 2.3.2.

To begin with, the specification of the mathematical objects used to describe certain physical entities will be given in the form of five *postulates*. Further mathematical objects, which may be related to physical quantities, will then still be introduced in the form of definitions.

2.3.1   Pure States

Postulate 1 (Observables and Pure States) *An **observable**, that is, a physically measurable quantity of a quantum system is represented by a self-adjoint operator on a HILBERT space $\mathbb{H}$. If the preparation of a statistical ensemble is such that for any observable represented by its self-adjoint operator A the mean value of the observable can be calculated with the help of a vector $|\psi\rangle \in \mathbb{H}$ satisfying $\||\psi\rangle\| = 1$ as*

$$\langle A \rangle_\psi := \langle \psi | A \psi \rangle\,, \tag{2.60}$$

> *then the preparation is said to be described by a **pure state** represented by the*
> *vector* $|\psi\rangle \in \mathbb{H}$. *One calls* $|\psi\rangle$ *the **state vector** or simply the state, and* $\langle A \rangle_\psi$
> *is called the (quantum mechanical) **expectation value** of the observable A in*
> *the pure state* $|\psi\rangle$.
>   *The space* $\mathbb{H}$ *is said to be the* HILBERT *space of the quantum system.*

We will often refer to a quantum system by its HILBERT space. In other words, if a system designated by S is described by states in a HILBERT space $\mathbb{H}^S$, we will simply speak of the 'system $\mathbb{H}^S$'. Likewise, given that every observable is represented by a self-adjoint operator, we shall from now on use the same symbol to denote the observable as well as its associated operator.

It is somewhat intuitive that we require

$$\langle \mathbf{1} \rangle_\psi = ||\psi||^2 = 1 \tag{2.61}$$

for any state vector $|\psi\rangle$. This is because the operator $\mathbf{1}$ can be interpreted as the observable 'is there anything present.' For systems containing any quantum mechanical object such an observable should always have the expectation value 1.

Using the diagonal representation of any self-adjoint operator A in terms of its eigenbasis, the expectation value of the observable represented by A becomes

$$\langle A \rangle_\psi \underbrace{=}_{(2.60)} \langle \psi | A \psi \rangle \underbrace{=}_{(2.38)} \langle \psi | \sum_j \lambda_j |e_j\rangle \langle e_j | \psi \rangle \underbrace{=}_{(2.4)} \sum_j \lambda_j \langle \psi | e_j \rangle \langle e_j | \psi \rangle$$

$$\underbrace{=}_{(2.1)} \sum_j \lambda_j |\langle \psi | e_j \rangle|^2 \,.$$

Indeed, in measurements one always observes an element of the spectrum (see Definition 2.10) of the associated operator. In the case where the system is described by states in an infinite-dimensional HILBERT space, these observations may also include elements of the so-called continuous spectrum of the operator. As already mentioned a couple of times, we restrict ourselves here exclusively to finite-dimensional systems. For our purposes we can thus identify the eigenvalues $\{\lambda_j\}$ of a self-adjoint operator A as the possible measurement results of the associated observable. In the case of a purely non-degenerate spectrum the positive numbers $|\langle e_j | \psi \rangle|^2$ are interpreted as the probabilities with which the respective value λ_j is observed. This is formalized more generally in the following postulate.

Postulate 2 (Measurement Probability) *In a quantum system with* HILBERT *space* $\mathbb{H}$ *the possible measurement values of an observable are given by the spectrum* $\sigma(A)$ *(see Definition 2.10) of the operator* $A \in \mathrm{B}_{sa}(\mathbb{H})$ *associated* *with the observable. The probability $\mathbf{P}_\psi(\lambda)$ that for a quantum system in the pure state $|\psi\rangle \in \mathbb{H}$ a measurement of the observable yields the eigenvalue λ of A is given with the help of the projection P_λ onto the eigenspace $\mathrm{Eig}(A, \lambda)$ of λ as*

$$\mathbf{P}_\psi(\lambda) = ||P_\lambda|\psi\rangle||^2 . \tag{2.62}$$

That (2.62) indeed defines a probability measure (see Definition A.2) on the spectrum of A requires in the general case a technically demanding proof [50]. Here we provide the following plausibility argument for the case of a purely discrete spectrum $\sigma(A) = \{\lambda_j \mid j \in I\}$ with eigenvalue degeneracies $d_j = \dim \mathrm{Eig}(A, \lambda_j)$ and an ONB of eigenvectors $\{|e_{j,\alpha}\rangle \mid j \in I, \alpha \in \{1, \ldots, d_j\}\}$. In this case we have

$$\mathbf{P}_\psi(\lambda_j) \underset{(2.62)}{=} \left|\left|P_{\lambda_j}|\psi\rangle\right|\right|^2 \underset{(2.41)}{=} \left|\left|\sum_{\alpha=1}^{d_j} |e_{j,\alpha}\rangle\langle e_{j,\alpha}|\psi\rangle\right|\right|^2 \underset{(2.12)}{=} \sum_{\alpha=1}^{d_j} |\langle e_{j,\alpha}|\psi\rangle|^2 \geq 0.$$

$$\tag{2.63}$$

That these terms add up to 1 then follows from the requirement made in Postulate 1 that states $|\psi\rangle$ be normalized to 1.

$$\sum_{j \in I} \mathbf{P}_\psi(\lambda_j) \underset{(2.62)}{=} \sum_{j \in I} \left|\left|P_{\lambda_j}|\psi\rangle\right|\right|^2 \underset{(2.63)}{=} \sum_{j \in I} \sum_{\alpha=1}^{d_j} |\langle e_{j,\alpha}|\psi\rangle|^2 \underset{(2.12)}{=} ||\psi||^2 = 1.$$

Hence, the map $\mathbf{P}_\psi(\cdot) : \sigma(A) \to [0, 1]$ can be viewed as a probability measure on $\sigma(A)$.

As a consequence of (2.60) one also finds that for any observable A and complex numbers of the form $\mathrm{e}^{\mathrm{i}\alpha} \in \mathbb{C}$ with $\alpha \in \mathbb{R}$ one has

$$\langle A \rangle_{\mathrm{e}^{\mathrm{i}\alpha}\psi} \underset{(2.60)}{=} \langle \mathrm{e}^{\mathrm{i}\alpha}\psi | A \mathrm{e}^{\mathrm{i}\alpha}\psi\rangle \underset{(2.4),(2.6)}{=} \left|\mathrm{e}^{\mathrm{i}\alpha}\right| \langle \psi | A\psi\rangle = \langle \psi | A\psi\rangle = \langle A \rangle_\psi ,$$

that is, the expectation values of any observable A in the state $\mathrm{e}^{\mathrm{i}\alpha}|\psi\rangle$ and in the state $|\psi\rangle$ are the same. Since

$$\left|\langle \mathrm{e}^{\mathrm{i}\alpha}\psi | e_j\rangle\right|^2 \underset{(2.6)}{=} \left|\mathrm{e}^{-\mathrm{i}\alpha}\langle \psi | e_j\rangle\right|^2 = |\langle \psi | e_j\rangle|^2$$

the measurement probabilities in the two states are also the same. This means that physically the state $\mathrm{e}^{\mathrm{i}\alpha}|\psi\rangle \in \mathbb{H}$ and the state $|\psi\rangle \in \mathbb{H}$ are indistinguishable. In other words, they describe the same state.

Definition 2.14 For every $|\psi\rangle \in \mathbb{H}$ with $\||\psi\|| = 1$ the set

$$S_\psi := \{e^{i\alpha}|\psi\rangle \mid \alpha \in \mathbb{R}\}$$

is called a **ray** in $\mathbb{H}$ with $|\psi\rangle$ as a representative.

Every element of a ray S_ψ describes the same physical situation. The phase $\alpha \in \mathbb{R}$ in $e^{i\alpha}$ can be arbitrarily chosen. More precisely, pure states are thus described by a representative $|\psi\rangle$ of a ray S_ψ in the HILBERT space. In the designation of a state one uses only the symbol $|\psi\rangle$ of a representative of the ray, keeping in mind that $|\psi\rangle$ and $e^{i\alpha}|\psi\rangle$ are physically indistinguishable. We shall use this fact explicitly on several occasions.

Conversely, every unit vector in a HILBERT space $\mathbb{H}$ corresponds to a physical state, in other words, describes the statistics of a quantum mechanical system. If $|\varphi\rangle, |\psi\rangle \in \mathbb{H}$ are states, then $a|\varphi\rangle + b|\psi\rangle \in \mathbb{H}$ for $a, b \in \mathbb{C}$ with $\|a\varphi + b\psi\| = 1$ is a state as well. This is the quantum mechanical **superposition principle**: any normalized linear combination of states is again a state and thus (in principle) a physically realizable preparation.

A word of caution, though: whereas the *global* phase of a linear combination is physically irrelevant, this is no longer true for the *relative* phases in the linear combination. More precisely, let $|\varphi\rangle, |\psi\rangle \in \mathbb{H}$ be two states satisfying $\langle\varphi|\psi\rangle = 0$. Then $\frac{1}{\sqrt{2}}(|\varphi\rangle + |\psi\rangle)$ as well as $\frac{1}{\sqrt{2}}(|\varphi\rangle + e^{i\alpha}|\psi\rangle)$ are normalized state vectors. However, while $|\psi\rangle$ and $e^{i\alpha}|\psi\rangle$ represent the same state, that is, describe the identical physical situation, the states $\frac{1}{\sqrt{2}}(|\varphi\rangle + |\psi\rangle)$ and $\frac{1}{\sqrt{2}}(|\varphi\rangle + e^{i\alpha}|\psi\rangle)$ do differ and correspond to different physical situations. This is because for any observable A we have

$$\langle A\rangle_{(|\varphi\rangle+|\psi\rangle)/\sqrt{2}} \underbrace{=}_{(2.60)} \frac{1}{2}\left(\langle\varphi+\psi|A(\varphi+\psi)\rangle\right)$$

$$\underbrace{=}_{(2.4)} \frac{1}{2}\left(\langle\varphi|A\varphi\rangle + \langle\psi|A\psi\rangle + \langle\varphi|A\psi\rangle + \langle\psi|A\varphi\rangle\right)$$

$$\underbrace{=}_{(2.30),\, A^*=A} \frac{1}{2}\left(\langle\varphi|A\varphi\rangle + \langle\psi|A\psi\rangle + \langle\varphi|A\psi\rangle + \langle A\psi|\varphi\rangle\right)$$

$$\underbrace{=}_{(2.60),(2.1)} \frac{1}{2}\left(\langle A\rangle_\varphi + \langle A\rangle_\psi\right) + \mathrm{Re}\left(\langle\varphi|A\psi\rangle\right),$$

where the term with the real part $\mathrm{Re}\left(\langle\varphi|A\psi\rangle\right)$ contains the so-called **interference term**. Exactly this term is different in the state $\frac{1}{\sqrt{2}}(|\varphi\rangle + e^{i\alpha}|\psi\rangle)$ because, similarly,

$$\langle A \rangle_{(|\varphi\rangle + e^{i\alpha}|\psi\rangle)/\sqrt{2}} = \frac{1}{2} \langle \varphi + e^{i\alpha}\psi | A(\varphi + e^{i\alpha}\psi) \rangle$$

$$= \frac{1}{2} \left(\langle \varphi | A\varphi \rangle + \langle e^{i\alpha}\psi | A e^{i\alpha}\psi \rangle + e^{i\alpha} \langle \varphi | A\psi \rangle + e^{-i\alpha} \langle \psi | A\varphi \rangle \right)$$

$$= \frac{1}{2} \left(\langle A \rangle_\varphi + \langle A \rangle_\psi \right) + \mathrm{Re} \left(e^{i\alpha} \langle \varphi | A\psi \rangle \right),$$

and for $\langle \varphi | A\psi \rangle \neq 0$ the real part of $\langle \varphi | A\psi \rangle$ and of $e^{i\alpha} \langle \varphi | A\psi \rangle$ differ. In other words, while changing $\alpha \in \mathbb{R}$ in $e^{i\alpha}|\psi\rangle$ does not change the state, this is no longer true for $\frac{1}{\sqrt{2}} (|\varphi\rangle + e^{i\alpha}|\psi\rangle)$.

If a system is prepared in a state $|\psi\rangle$, how likely is it that a measurement reveals it to be in a state $|\varphi\rangle$? This is answered by the following proposition.

Proposition 2.15 *Let the states of a quantum system be described by the rays in a HILBERT space $\mathbb{H}$. If the system has been prepared in the state $|\psi\rangle \in \mathbb{H}$, then the probability to observe it in the state $|\varphi\rangle \in \mathbb{H}$ is given by*

$$\mathbf{P} \left\{ \begin{matrix} \text{System prepared in state } |\psi\rangle \\ \text{observed in state } |\varphi\rangle \end{matrix} \right\} = |\langle \varphi | \psi \rangle|^2 . \tag{2.64}$$

Proof Let $|\psi\rangle, |\varphi\rangle \in \mathbb{H}$ with $||\psi|| = 1 = ||\varphi||$. The observable we measure when querying if the system is in the state $|\varphi\rangle$ is the orthogonal projection $P_\varphi = |\varphi\rangle\langle\varphi|$ onto that state. This observable has the eigenvalues 0 and 1. The eigenvalue $\lambda = 1$ is non-degenerate and its eigenspace is spanned by $|\varphi\rangle$, hence the projection onto the eigenspace for eigenvalue $\lambda = 1$ is also given by P_φ and (2.62) of Postulate 2 becomes

$$\mathbf{P}_\psi(\lambda = 1) = ||P_1|\psi\rangle||^2 = ||P_\varphi|\psi\rangle||^2 = |||\varphi\rangle\langle\varphi|\psi\rangle||^2$$

$$\underbrace{=}_{(2.7)} |\langle\varphi|\psi\rangle|^2 \underbrace{||\varphi||^2}_{=1} = |\langle\varphi|\psi\rangle|^2 . \qquad \square$$

How widely around its expectation value are the measurement results distributed? A statement about that is given by the so-called uncertainty or standard deviation defined similarly to the same notion in standard probability theory (see Appendix A).

Definition 2.16 The **uncertainty** of an observable A in the state $|\psi\rangle$ is defined as

$$\Delta_\psi(A) := \sqrt{\langle \psi | (A - \langle A \rangle_\psi \mathbf{1})^2 \psi \rangle} = \sqrt{\left\langle (A - \langle A \rangle_\psi \mathbf{1})^2 \right\rangle_\psi}. \qquad (2.65)$$

If the uncertainty vanishes, that is, if $\Delta_\psi(A) = 0$, one says that the value of the observable A in the state $|\psi\rangle$ is **sharp**.

A sharp value of an observable A in a state $|\psi\rangle$ means that all measurements of A on systems in the state $|\psi\rangle$ always yield the same result. This is the case if and only if $|\psi\rangle$ is an eigenvector of A as stated in the following proposition.

Proposition 2.17 *For any observable A and state $|\psi\rangle$ the following equivalence holds*

$$\Delta_\psi(A) = 0 \quad \Leftrightarrow \quad A|\psi\rangle = \langle A \rangle_\psi |\psi\rangle.$$

Proof Since as an observable A is self-adjoint, that is, $A^* = A$, we have

$$\langle A \rangle_\psi \underbrace{=}_{(2.60)} \langle \psi | A \psi \rangle \underbrace{=}_{(2.30)} \langle A^* \psi | \psi \rangle = \langle A \psi | \psi \rangle \underbrace{=}_{(2.1)} \overline{\langle \psi | A \psi \rangle} \underbrace{=}_{(2.60)} \overline{\langle A \rangle_\psi}$$

and it follows that $\langle A \rangle_\psi \in \mathbb{R}$. Consequently,

$$\left(A - \langle A \rangle_\psi \mathbf{1} \right)^* = A - \langle A \rangle_\psi \mathbf{1} \qquad (2.66)$$

as well and thus

$$\left(\Delta_\psi(A) \right)^2 \underbrace{=}_{(2.65)} \langle \psi | (A - \langle A \rangle_\psi \mathbf{1})^2 \psi \rangle \underbrace{=}_{(2.30)} \langle (A - \langle A \rangle_\psi \mathbf{1}) \psi | (A - \langle A \rangle_\psi \mathbf{1}) \psi \rangle$$

$$\underbrace{=}_{(2.5)} \left\| (A - \langle A \rangle_\psi \mathbf{1}) \psi \right\|^2, \qquad (2.67)$$

such that

$$\Delta_\psi(A) = 0 \underbrace{\Leftrightarrow}_{(2.5),(2.3)} A|\psi\rangle = \langle A \rangle_\psi |\psi\rangle$$

that is, the value of the observable A is sharp if and only if $|\psi\rangle$ is an eigenvector of A with eigenvalue $\langle A \rangle_\psi$. $\qquad \Box$

A state which is an eigenvector of an operator associated to an observable is also called an **eigenstate** of that operator or observable.

A preparation in an eigenstate of A thus implies that all measurements of A in that state always yield the corresponding eigenvalue. The converse is also true: If for a given preparation the uncertainty of A vanishes, then the preparation is described by an eigenstate of A.

Definition 2.18 Two observables A and B are called **compatible** if the associated operators commute, that is, if $[A,B] = 0$. If $[A,B] \neq 0$, they are called **incompatible**.

A result from linear algebra tells us that A and B self-adjoint and $[A,B] = 0$ implies that there is an ONB $\{|e_j\rangle\}$, in which A and B are diagonal, that is,

$$A = \sum_j a_j |e_j\rangle\langle e_j| \quad \text{and} \quad B = \sum_j b_j |e_j\rangle\langle e_j|.$$

A system in the state $|e_k\rangle$ is then in an eigenstate of A *and* B. Hence, measurements of compatible observables A *and* B in this state yield sharp results (here a_k and b_k) for both these observables and *do not exhibit uncertainty*.

However, the product of the uncertainties of incompatible observables is bounded from below as the following proposition shows.

Proposition 2.19 *For any observables $A, B \in \mathrm{B}_{sa}(\mathbb{H})$ and state $|\psi\rangle \in \mathbb{H}$ the following **uncertainty relation** holds*

$$\Delta_\psi(A)\Delta_\psi(B) \geq \left| \left\langle \frac{1}{2i}[A,B] \right\rangle_\psi \right|. \tag{2.68}$$

Proof The relation (2.68) is a consequence of the following estimates

$$\left(\Delta_\psi(A)\right)^2 \left(\Delta_\psi(B)\right)^2$$

$$\underset{(2.67)}{=} \left\| \left(A - \langle A\rangle_\psi \mathbf{1}\right)\psi \right\|^2 \left\| \left(B - \langle B\rangle_\psi \mathbf{1}\right)\psi \right\|^2$$

$$\underset{(2.16)}{\geq} \left| \langle (A - \langle A\rangle_\psi \mathbf{1})\psi | (B - \langle B\rangle_\psi \mathbf{1})\psi\rangle \right|^2$$

$$\geq \left(\mathrm{Im}\left(\langle (A - \langle A\rangle_\psi \mathbf{1})\psi | (B - \langle B\rangle_\psi \mathbf{1})\psi\rangle \right) \right)^2$$

$$= \left(\frac{1}{2i}\langle (A - \langle A\rangle_\psi \mathbf{1})\psi | (B - \langle B\rangle_\psi \mathbf{1})\psi\rangle - \frac{1}{2i}\overline{\langle (A - \langle A\rangle_\psi \mathbf{1})\psi | (B - \langle B\rangle_\psi \mathbf{1})\psi\rangle} \right)^2$$

$$\underbrace{=}_{(2.1)} \left(\frac{1}{2i} \langle (A - \langle A \rangle_\psi 1) \psi | (B - \langle B \rangle_\psi 1) \psi \rangle - \frac{1}{2i} \langle (B - \langle B \rangle_\psi 1) \psi | (A - \langle A \rangle_\psi 1) \psi \rangle \right)^2$$

$$\underbrace{=}_{(2.66)} \left(\frac{1}{2i} \langle (A - \langle A \rangle_\psi 1)^* \psi | (B - \langle B \rangle_\psi 1) \psi \rangle - \frac{1}{2i} \langle (B - \langle B \rangle_\psi 1)^* \psi | (A - \langle A \rangle_\psi 1) \psi \rangle \right)^2$$

$$\underbrace{=}_{(2.30)} \left(\frac{1}{2i} \langle \psi | (A - \langle A \rangle_\psi 1)(B - \langle B \rangle_\psi 1) \psi \rangle - \frac{1}{2i} \langle \psi | (B - \langle B \rangle_\psi 1)(A - \langle A \rangle_\psi 1) \psi \rangle \right)^2$$

$$= \left(\frac{1}{2i} \left\langle [A - \langle A \rangle_\psi 1, B - \langle B \rangle_\psi 1] \right\rangle_\psi \right)^2$$

$$= \left(\left\langle \frac{1}{2i} [A, B] \right\rangle_\psi \right)^2 .$$

$\square$

From (2.68) we see that if for a state $|\psi\rangle$ we have $|\langle [A,B] \rangle_\psi| > 0$ then the product of the uncertainties of the observables A and B in the state $|\psi\rangle$ is bounded from below. The smaller the uncertainty of A the bigger that of B and vice versa.

Example 2.20 The **HEISENBERG uncertainty relation** is a special case of (2.68), where $\mathbb{H} = L^2(\mathbb{R}^3)$ (see Example 2.5) and A is given by one of the position operators Q_j and B by one of the corresponding momentum operators P_j for the three spatial dimensions $j \in \{1, 2, 3\}$. For these two operators one has the following action on states[6] $|\psi\rangle \in \mathbb{H} = L^2(\mathbb{R}^3)$

$$|Q_j \psi\rangle(\mathbf{x}) = |x_j \psi\rangle(\mathbf{x})$$
$$|P_j \psi\rangle(\mathbf{x}) = |-i \frac{\partial}{\partial x_j} \psi\rangle(\mathbf{x}),$$

such that

$$[Q_j, P_k] |\psi\rangle(\mathbf{x}) = |-i x_j \frac{\partial}{\partial x_k} \psi\rangle(\mathbf{x}) - |-i \frac{\partial}{\partial x_k} (x_j \psi\rangle(\mathbf{x})) = i \delta_{jk} |\psi\rangle(\mathbf{x}),$$

that is, $[Q_j, P_k] = i \delta_{jk} 1$. Consequently, in this case

$$\Delta_\psi(Q_j) \Delta_\psi(P_k) \geq \frac{1}{2} \delta_{jk}.$$

A measurement of an observable $A = \sum_j \lambda_j |e_j\rangle \langle e_j|$ on an object prepared in the state $|\psi\rangle = \sum_j |e_j\rangle \langle e_j |\psi\rangle$ yields an eigenvalue $\lambda_k \in \sigma(A)$. Out of the possible measurement values $\sigma(A)$ for A a value λ_k has been observed. On an object for which λ_k has been measured and which has not been influenced by any outside interaction,

---

[6]As always in this book, here the system of units with $\hbar = 1$ is used, since otherwise one would have for the momentum operators $P_j = -i\hbar \frac{\partial}{\partial x_j}$.

another measurement of A thereafter always yields the value λ_k again. The set of all such prepared objects, in other words, objects originally in $|\psi\rangle$ that yield the value λ_k upon measurement of A, then constitutes a preparation in which A has the sharp value λ_k. Such a state is to be described by the eigenvector $|e_k\rangle$ of A. Accordingly, a measurement of an observable A that yielded the value $\lambda_k \in \sigma(A)$ can be viewed as a preparation of the object in the state $|e_k\rangle \in \mathbb{H}$. One says that with the probability $|\langle e_k|\psi\rangle|^2$ the measurement 'forces' or 'projects' the object that was originally in the state $|\psi\rangle$ into the eigenstate $|e_k\rangle$ of the measured observable. Selecting all objects with the measurement result λ_k, we have thus prepared an ensemble that is described by vectors $|e_k\rangle$ in the eigenspace of λ_k. This physical phenomenon is formulated as the Projection Postulate.

> **Postulate 3** (Projection Postulate) *If a measurement of the observable A on a quantum mechanical system in the pure state $|\psi\rangle \in \mathbb{H}$ yields the eigenvalue λ, then the measurement has effected the following state transition*
>
> $$|\psi\rangle = \begin{matrix} state \quad before \\ measurement \end{matrix} \xrightarrow{\;measurement\;} \frac{P_\lambda|\psi\rangle}{||P_\lambda|\psi\rangle||} = \begin{matrix} state \quad after \\ measurement, \end{matrix}$$
>
> *where P_λ is the projection onto the eigenspace of λ.*

Historically, the state $|\psi\rangle$ of a quantum mechanical system has also been called **wave function**. For this reason the Projection Postulate is also known as **collapse of the wave function**.

A state can also change without a measurement being performed on it. The time evolution of a state not caused by measurements is given by a unitary operator obtained as a solution of an operator initial value problem.

Exercise 2.17 Let $\mathbb{H}$ be a HILBERT space and for $t \geq t_0$ let $t \mapsto U(t,t_0) \in L(\mathbb{H})$ be a solution of the initial value problem

$$i\frac{d}{dt}U(t,t_0) = H(t)U(t,t_0)$$
$$U(t_0,t_0) = \mathbf{1}, \tag{2.69}$$

where $H(t) \in B_{sa}(\mathbb{H})$. Show that then $U(t,t_0)$ is unitary and unique.

For a solution see Solution 2.17.

We will not concern ourselves here with the notoriously difficult and technical aspects of the existence of solutions $t \mapsto U(t,t_0)$ to (2.69), but will always assume that $H(t)$ is such that a solution exists and is unique.

Postulate 4 (Time Evolution) *In a quantum system with* HILBERT *space* $\mathbb{H}$ *every change of a pure state over time*

$$|\psi(t_0)\rangle = state\ at\ time\ t_0 \qquad \overset{no\ measurement}{\longrightarrow} \qquad |\psi(t)\rangle = state\ at\ time\ t$$

that has not been caused by a measurement is described by the time evolution operator $U(t,t_0) \in \mathfrak{U}(\mathbb{H})$. *The time-evolved state* $|\psi(t)\rangle$ *originating from* $|\psi_0\rangle$ *is then given by*

$$|\psi(t)\rangle = U(t,t_0)|\psi(t_0)\rangle. \tag{2.70}$$

The time evolution operator $U(t,t_0)$ *is the solution of the initial value problem*

$$\begin{aligned} i\frac{d}{dt}U(t,t_0) &= \mathsf{H}(t)U(t,t_0) \\ U(t_0,t_0) &= \mathbf{1}, \end{aligned} \tag{2.71}$$

where $\mathsf{H}(t)$ *is the self-adjoint* HAMILTON *operator (aka* **Hamiltonian***), which is said to generate the time evolution of the quantum system.*

The operator version of time evolution given in Postulate 4 is completely equivalent to the well-known SCHRÖDINGER **equation**[7]

$$i\frac{d}{dt}|\psi(t)\rangle = \mathsf{H}(t)|\psi(t)\rangle, \tag{2.72}$$

which describes the time evolution of pure states as expressed by its effect on the the state vectors. This is because application of (2.71) to (2.70) results in the SCHRÖDINGER equation (2.72), and, conversely, any solution of the SCHRÖDINGER equation for arbitrary initial states $|\psi(t_0)\rangle$ yields a solution for $U(t,t_0)$. The formulation of the time evolution making use of the time evolution operator $U(t,t_0)$ given in Postulate 4 has the advantage over the SCHRÖDINGER equation that it can be used for mixed states (see Postulate 5) as well.

The operator $\mathsf{H}(t)$ corresponds to the observable energy of the quantum system. Hence, the expectation value $\langle \mathsf{H}(t)\rangle_\psi$ (see Postulate 1) of the Hamiltonian gives the expectation value for the energy of the system in the state $|\psi\rangle$. If H is time-independent, that is, $\frac{d}{dt}\mathsf{H}(t) = 0$, then the energy of the system is constant and is given by the eigenvalues $\{E_j \mid j \in I\}$ of H. The fact that these eigenvalues are discrete for certain Hamiltonians is at the heart of the designation 'quantum'. It was PLANCK's assumption that the energy of a black body can only be integer multiples of a fixed quantum of energy, which helped him derive the correct radiation formula. But the origins of this assumptions were not understood at the time. Only quantum

---

[7]We remind the reader here once more that in this book we use natural physical units, such that $\hbar = 1$, which is why this constant does not appear as a factor on the left sides of (2.71) and (2.72).

mechanics subsequently provided a theoretical and mathematical theory delivering a proof for discrete energy levels.

Exercise 2.18 Let $U(t,t_0)$ be a time evolution operator satisfying (2.71) for a Hamiltonian $H(t)$.

Show that then

$$i\frac{d}{dt}U(t,t_0)^* = -U(t,t_0)^* H(t)$$

$$U(t_0,t_0)^* = \mathbf{1}.$$

(2.73)

For a solution see Solution 2.18.

The HAMILTON operator $H(t)$ not only corresponds to the energy observable of the system, but also—as is evident from (2.71)—determines the time evolution of the system. The specific form of the operator $H(t)$ is determined by the internal and external interactions to which the quantum system is exposed. As we shall see in Chap. 5, circuits in quantum computers are built up from elementary gates that act as unitary operators V on the states. In order to implement such gates one then tries to create HAMILTON operators that generate a time evolution $U(t,t_0)$ implementing the desired gate, that is, one attempts to find $H(t)$ and t such that $V = U(t,t_0)$.

That time evolution acts as a linear transformation on the space of states is a result of the superposition principle. That it ought to be unitary results from the requirement to preserve the norm (see Exercise 2.7) of the time-evolved state, which in turn is a requirement originating from the probability interpretation (see (2.61)).

As an example of other observables of eminent importance for quantum computing we consider the internal angular momentum of an electron, its so-called **spin**. It consists of three observables S_x, S_y, S_z, which are grouped together in the internal angular momentum vector $\mathbf{S} = (S_x, S_y, S_z)$. Since we are only interested in the spin and not in the position or momentum of the electron, the HILBERT space we need to consider is two-dimensional $\mathbb{H} \cong \mathbb{C}^2$. The operators on this two-dimensional HILBERT space for the spin observable $\mathbf{S}$ are[8]

$$S_j = \frac{1}{2}\sigma_j \qquad \text{for } j \in \{x,y,z\},$$

where the σ_j are the so-called PAULI matrices defined as follows.

Definition 2.21 The matrices $\sigma_j \in \mathrm{Mat}(2 \times 2, \mathbb{C})$ indexed by either $j \in \{1,2,3\}$ or $j \in \{x,y,z\}$ and defined as

---

[8]In non-natural units $\hbar$ would appear as a factor on the right side.

$$\sigma_x := \sigma_1 := \begin{pmatrix} 0 & 1 \\ 1 & 0 \end{pmatrix}, \quad \sigma_y := \sigma_2 := \begin{pmatrix} 0 & -i \\ i & 0 \end{pmatrix}, \quad \sigma_z := \sigma_3 := \begin{pmatrix} 1 & 0 \\ 0 & -1 \end{pmatrix} \quad (2.74)$$

are called **PAULI matrices**. With $\sigma_0 := \begin{pmatrix} 1 & 0 \\ 0 & 1 \end{pmatrix}$ denoting the 2×2 unit matrix we define the enlarged set

$$\left\{ \sigma_\alpha \,\middle|\, \alpha \in \{0,\ldots,3\} \right\} = \left\{ \sigma_0, \sigma_1, \sigma_2, \sigma_3 \right\} \quad (2.75)$$

by using the extended notation with Greek subscripts.

For a two-dimensional HILBERT space $\mathbb{H}$ with a designated ONB we also use the symbols

$$\sigma_0 = \mathbf{1}, \quad X = \sigma_1 = \sigma_x, \quad Y = \sigma_2 = \sigma_y, \quad Z = \sigma_3 = \sigma_z,$$

to denote the operators in $L(\mathbb{H})$ that have the corresponding matrix in the designated ONB.

Exercise 2.19 For $j,k,l \in \{1,2,3\}$ let ε_{jkl} denote the *completely anti-symmetric tensor* with

$$\varepsilon_{123} = \varepsilon_{231} = \varepsilon_{312} = 1 = -\varepsilon_{213} = -\varepsilon_{132} = -\varepsilon_{321}$$

and $\varepsilon_{jkl} = 0$ otherwise. Verify the following properties of the PAULI matrices:

(i)

$$\sigma_j \sigma_k = \delta_{jk} \mathbf{1} + i \varepsilon_{jkl} \sigma_l \quad (2.76)$$

(ii) The commutation relations

$$[\sigma_j, \sigma_k] = \sigma_j \sigma_k - \sigma_k \sigma_j = 2i\varepsilon_{jkl} \sigma_l \quad (2.77)$$

(iii) The anti-commutation relations

$$\{\sigma_j, \sigma_k\} := \sigma_j \sigma_k + \sigma_k \sigma_j = 2\delta_{jk} \mathbf{1}$$

(iv)

$$\sigma_j^* = \sigma_j \quad \text{and} \quad \sigma_j^* \sigma_j = \mathbf{1},$$

that is, the σ_j are self-adjoint and unitary.

For a solution see Solution 2.19.

For the states

$$| \uparrow_{\hat{z}} \rangle := |0\rangle := \begin{pmatrix} 1 \\ 0 \end{pmatrix}, \qquad | \downarrow_{\hat{z}} \rangle := |1\rangle := \begin{pmatrix} 0 \\ 1 \end{pmatrix} \qquad (2.78)$$

one finds

$$S_z | \uparrow_{\hat{z}} \rangle = \frac{1}{2} | \uparrow_{\hat{z}} \rangle, \qquad S_z | \downarrow_{\hat{z}} \rangle = -\frac{1}{2} | \downarrow_{\hat{z}} \rangle,$$

that is, S_z has the eigenvalues $\{\pm \frac{1}{2}\}$ with the eigenvectors $\{| \uparrow_{\hat{z}} \rangle, | \downarrow_{\hat{z}} \rangle\}$, which are known as the up, resp. down state for the spin in the z-direction. The S_j are physical observables which can be measured in a STERN–GERLACH experiment. For simplicity we shall, however, use $\sigma_j = 2S_j$ as observables in order to avoid the unwieldy factor $\frac{1}{2}$.

The reason to denote the eigenvectors $| \uparrow_{\hat{z}} \rangle$ and $| \downarrow_{\hat{z}} \rangle$ by $|0\rangle$ and $|1\rangle$ in (2.78) is the identification of these states with the classical bit values 0 and 1. Denoting the eigenvectors of σ_z by $|0\rangle$ and $|1\rangle$ has become standard in quantum computing and we shall use this notation henceforth. We note already here that $a|0\rangle + b|1\rangle$ with $|a|^2 + |b|^2 = 1$ is a possible state (more about this in Sect. 2.4). In contrast, a classical bit-value $a0 + b1$ is meaningless. To avoid misunderstandings, note that $|0\rangle$ *is not the null-vector* in HILBERT space. The null-vector in HILBERT space is denoted by the same symbol 0, which we use throughout for the null in sets like $\mathbb{N}_0$, fields like $\mathbb{R}$ and $\mathbb{C}$ and vector spaces like $\mathbb{H}$.

The observable σ_z thus has the eigenvalues ± 1 and the eigenvectors $|0\rangle = | \uparrow_{\hat{z}} \rangle$ and $|1\rangle = | \downarrow_{\hat{z}} \rangle$ as well as the expectation values

$$\langle \sigma_z \rangle_{|0\rangle} = \langle 0|\sigma_z|0\rangle = \langle 0|0\rangle = +1, \qquad \langle \sigma_z \rangle_{|1\rangle} = \langle 1|\sigma_z|1\rangle = -\langle 1|1\rangle = -1.$$

As an illustration we show that in the state $|0\rangle$ indeed the uncertainty vanishes. First, one has

$$\sigma_z - \langle \sigma_z \rangle_{|0\rangle} \mathbf{1} = \begin{pmatrix} 1 & 0 \\ 0 & -1 \end{pmatrix} - \begin{pmatrix} 1 & 0 \\ 0 & 1 \end{pmatrix} = \begin{pmatrix} 0 & 0 \\ 0 & -2 \end{pmatrix}$$

and thus

$$\langle 0| \left(\sigma_z - \langle \sigma_z \rangle_{|0\rangle} \mathbf{1} \right)^2 |0\rangle = (1 \; 0) \begin{pmatrix} 0 & 0 \\ 0 & 4 \end{pmatrix} \begin{pmatrix} 1 \\ 0 \end{pmatrix} = 0,$$

which implies

$$\Delta_{|0\rangle}(\sigma_z) \underbrace{=}_{(2.65)} 0.$$

Similarly, one shows that $\Delta_{|1\rangle}(\sigma_z) = 0$, which follows from the general theory, since $|0\rangle$ and $|1\rangle$ are eigenstates of σ_z and therefore the measurement of the observable σ_z does not show any uncertainty.

On the other hand, σ_x and σ_z are incompatible since

$$[\sigma_x, \sigma_z] \underbrace{=}_{(2.77)} -2i\sigma_y \neq 0,$$

and one finds

$$\langle\sigma_x\rangle_{|0\rangle} = (1\ 0)\begin{pmatrix}0 & 1\\ 1 & 0\end{pmatrix}\begin{pmatrix}1\\ 0\end{pmatrix} = 0$$

$$\sigma_x - \langle\sigma_x\rangle_{|0\rangle}\,\mathbf{1} = \begin{pmatrix}0 & 1\\ 1 & 0\end{pmatrix}$$

$$\langle 0|\left(\sigma_x - \langle\sigma_x\rangle_{|0\rangle}\,\mathbf{1}\right)^2|0\rangle = (1\ 0)\begin{pmatrix}1 & 0\\ 0 & 1\end{pmatrix}\begin{pmatrix}1\\ 0\end{pmatrix} = 1\,.$$

This implies that

$$\Delta_{|0\rangle}(\sigma_x)\underbrace{=}_{(2.65)}1\,, \tag{2.79}$$

that is, a measurement of σ_x in the state $|0\rangle$ *cannot be sharp*. Similarly, one shows $\Delta_{|1\rangle}(\sigma_x)=1$. Consequently, σ_z and σ_x can never be measured in the same state with vanishing uncertainty. The same holds true for σ_z and σ_y as well as the pair σ_x and σ_y.

Exercise 2.20 Find the eigenvalues and normalized eigenstates $|\uparrow_{\hat{\mathbf{x}}}\rangle$ and $|\downarrow_{\hat{\mathbf{x}}}\rangle$ of σ_x as a linear combination of $|0\rangle$ and $|1\rangle$, and calculate the probabilities $|\langle\uparrow_{\hat{\mathbf{x}}}|0\rangle|^2$ and $|\langle\downarrow_{\hat{\mathbf{x}}}|0\rangle|^2$ to measure the eigenvalues of σ_x in the state $|0\rangle = |\uparrow_{\hat{\mathbf{z}}}\rangle$.

For a solution see Solution 2.20.

The following example provides a simple illustration of the diagonal form (2.38) of an operator and the projector onto a state.

Example 2.22 As an illustration of its diagonal form σ_z can be expressed with the help of its eigenvectors and eigenvalues:

$$\sigma_z = (+1)|0\rangle\langle 0| + (-1)|1\rangle\langle 1| = \begin{pmatrix}1\\ 0\end{pmatrix}(1\ 0) - \begin{pmatrix}0\\ 1\end{pmatrix}(0\ 1)$$

$$= \begin{pmatrix}1 & 0\\ 0 & 0\end{pmatrix} - \begin{pmatrix}0 & 0\\ 0 & 1\end{pmatrix} = \begin{pmatrix}1 & 0\\ 0 & -1\end{pmatrix}\,.$$

For $|\psi\rangle = a_0|0\rangle + a_1|1\rangle \in \mathbb{H}$ with $|a_0|^2 + |a_1|^2 = 1$ we find for the orthogonal projection

$$P_\psi = |\psi\rangle\langle\psi| = \begin{pmatrix}a_0\\ a_1\end{pmatrix}(\overline{a_0}\ \overline{a_1}) = \begin{pmatrix}|a_0|^2 & a_0\overline{a_1}\\ \overline{a_0}a_1 & |a_1|^2\end{pmatrix}\,.$$

Similar to Example 2.22, Exercise 2.21 provides a further illustration of the diagonal form (2.38) of an operator.

Exercise 2.21 Verify the diagonal form of

$$\sigma_x = \sum_j \lambda_j |e_j\rangle\langle e_j|,$$

that is, with the results of Exercise 2.20 calculate the right side of this equation.

For a solution see Solution 2.21.

We can use the results of the previous exercises to illustrate the content of the Projection Postulate as follows. Let an electron be prepared in a state $|0\rangle$ and a measurement of σ_x be performed on it. From Exercise 2.20 we know that the value $+1$ or -1 will be observed, each with a probability of $\frac{1}{2}$. Those electrons for which $+1$ has been measured then constitute an ensemble, which is described by the eigenstate $|\uparrow_{\hat{x}}\rangle$ of σ_x for the eigenvalue $+1$. The selection after the measurement of only those electrons for which $+1$ has been observed is akin to a preparation of the state $|\uparrow_{\hat{x}}\rangle$.

2.3.2 Mixed States

Pure states, however, are not the most general form in which quantum systems can appear. Quantum systems can also exist in so-called mixed states, which include pure states as a special case.

Loosely speaking, a quantum particle needs to be described by a mixed state if it is in one of a set of states $\{|\psi_j\rangle\}$, in other words, its statistics is to be described by one of the $|\psi_j\rangle$, but we do not know by which one. All we know is the probability p_j with which it is in one of the states $|\psi_j\rangle$. The statistical properties of the ensemble of particles thus produced are described by a mixed state.

This is to be distinguished from the system being in the pure state $|\psi\rangle = \sum_j \sqrt{p_j}|\psi_j\rangle$, in which case the quantum statistics is described by $|\psi\rangle$ and not by one of the $|\psi_j\rangle$ alone as in the case of a mixed state.

The mathematically all-encompassing description of quantum mechanical systems covering pure and mixed states is given by so-called density operators.

Postulate 5 (Mixed States) *In general a quantum mechanical system is described mathematically by an operator ρ acting on a* HILBERT *space* $\mathbb{H}$ *with ρ having the properties:*

(i) ρ is self-adjoint

$$\rho^* = \rho. \tag{2.80}$$

(ii) ρ is positive

$$\rho \geq 0. \tag{2.81}$$

(iii) ρ has trace 1

$$\mathrm{tr}(\rho) = 1. \tag{2.82}$$

*The operator ρ is called **density operator** and we denote the **set of density operators** on a* HILBERT *space* $\mathbb{H}$ *by* $\mathrm{D}(\mathbb{H})$*, that is,*

$$\mathrm{D}(\mathbb{H}) := \left\{ \rho \in \mathrm{L}(\mathbb{H}) \,\middle|\, \rho^* = \rho,\, \rho \geq 0,\, \mathrm{tr}(\rho) = 1 \right\}. \tag{2.83}$$

The quantum states described by density operators in $\mathrm{D}(\mathbb{H})$ *are called **mixed states**.*

In general, the sum of two density operators is no longer a density operator, but the set of density operators—and hence mixed states—is convex[9] in $\mathrm{L}(\mathbb{H})$, as is to be shown in Exercise 2.22.

Exercise 2.22 Let $\mathbb{H}$ be a HILBERT space and $\rho_i \in \mathrm{D}(\mathbb{H})$ with $i \in \{1,2\}$. Show that then for any $u \in [0,1] \subset \mathbb{R}$

$$u\rho_1 + (1-u)\rho_2 \in \mathrm{D}(\mathbb{H}).$$

For a solution see Solution 2.22.

For any $U \in \mathcal{U}(\mathbb{H})$ transformations of the type $\rho \mapsto U\rho U^*$ on any mixed state described by a density operator ρ produce again a mixed state described by the density operator $U\rho U^*$. This claim is to be shown in Exercise 2.23.

Exercise 2.23 Let $\mathbb{H}$ be a HILBERT space and $U \in \mathcal{U}(\mathbb{H})$. Show that then

$$\rho \in \mathrm{D}(\mathbb{H}) \qquad \Rightarrow \qquad U\rho U^* \in \mathrm{D}(\mathbb{H}). \tag{2.84}$$

For a solution see Solution 2.23.

---

[9]A set K of a linear space is called convex if for every two elements $x, y \in K$ their connecting line is in K as well, that is, if $x, y \in K$ implies $\{ux + (1-u)y \mid u \in [0,1]\} \subset K$.

This result is used in the generalization of the postulates to mixed states.

Postulate 6 *For mixed states the Postulates 1–4 are generalized as follows:*

Postulate 1 (Observables and States)   *The quantum mechanical **expectation value** of an observable A in a mixed state ρ is given by*

$$\langle A \rangle_\rho := \operatorname{tr}(\rho A). \tag{2.85}$$

Postulate 2 (Measurement Probability)   *If the quantum system is in a state ρ, λ is an eigenvalue of A and P_λ the projection onto the eigenspace of λ, then the probability $\mathbf{P}_\rho(\lambda)$ that a measurement of A yields the value λ is given by*

$$\mathbf{P}_\rho(\lambda) = \operatorname{tr}(\rho P_\lambda). \tag{2.86}$$

Postulate 3 (Projection Postulate)   *If the quantum system is initially described by the state ρ, and then the measurement of the observable A yields the eigenvalue λ of A, then this measurement has effected the following change of state*

$$\rho = \begin{matrix} state \ before \\ measurement \end{matrix} \xrightarrow{measurement} \frac{P_\lambda \rho P_\lambda}{\operatorname{tr}(\rho P_\lambda)} = \begin{matrix} state \quad after \\ measurement \end{matrix} \tag{2.87}$$

where P_λ is the projection onto the eigenspace of λ.

Postulate 4 (Time Evolution)   *Any time evolution of a quantum system that is not caused by a measurement is described as an evolution of states*

$$\rho(t_0) = state \ at \ time \ t_0 \xrightarrow{no \ measurement} \rho(t) = state \ at \ time \ t$$

given by a unitary time evolution operator $U(t,t_0)$ acting on the density operator as

$$\rho(t) = U(t,t_0)\rho(t_0)U(t,t_0)^*. \tag{2.88}$$

Here, as in the case of pure states, the time evolution operator $U(t,t_0)$ is a solution of the initial value problem (2.71).

The uncertainty is given analogously to (2.65) as

$$\Delta_\rho(A) := \sqrt{\left\langle \left(A - \langle A \rangle_\rho \mathbf{1} \right)^2 \right\rangle_\rho}.$$

Pure states $|\psi\rangle \in \mathbb{H}$ are given by special density operators of the form

$$\rho_\psi := |\psi\rangle\langle\psi| = P_\psi. \tag{2.89}$$

Note that, deviating slightly from the definition given above, some authors reserve the term 'mixed state' for the truly non-pure states. Here we shall use the term in the general sense defined above, which includes pure states as a special case, and refer to truly non-pure states as 'true mixtures'. The fact that the generalizations given in Postulate 6 in the case of a pure state $\rho = |\psi\rangle\langle\psi|$ coincide with the Postulates 1-4 given earlier for pure states is to be shown as Exercise 2.24.

Exercise 2.24 Verify that the generalizations for the expectation value, measurement probability, projection onto the state after measurement and time evolution given in Postulate 6 for the case $\rho_\psi = |\psi\rangle\langle\psi|$ coincide with the statements made for a pure state $|\psi\rangle$ in the Postulates 1–4.

For a solution see Solution 2.24.

The reason that in general states can be described by positive, self-adjoint operators with trace 1 lies in the theorem of GLEASON [52], which we will touch upon here briefly. Since a measurement of an observable always yields an eigenvalue of the corresponding operator, we can interpret the observable corresponding to orthogonal projections ($P^* = P = P^2$) as a *yes-no* measurement. This is because orthogonal projections only have the eigenvalues 0 and 1. A mathematical description of such systems should then provide a map

$$\mathbf{P}: \{\text{Projections onto } \mathbb{H}\} \longrightarrow [0,1]$$
$$P \longmapsto \mathbf{P}(P) \tag{2.90}$$

in which we want to interpret $\mathbf{P}(P)$ as the probability to measure the value 1, and which should also have the following properties:

$$\mathbf{P}(0) = 0$$
$$\mathbf{P}(1) = 1 \tag{2.91}$$
$$P_1 P_2 = 0 \Rightarrow \mathbf{P}(P_1 + P_2) = \mathbf{P}(P_1) + \mathbf{P}(P_2).$$

The properties (2.90)–(2.91) are basic requirements for a probability function $\mathbf{P}$ for quantum mechanical systems. The following theorem of GLEASON, which we state her without proof, then tells us that the set of self-adjoint, positive operators with trace 1 is sufficient to construct above mentioned probability function.

Theorem 2.23 (GLEASON [52]) *For a* HILBERT *space* $\mathbb{H}$ *with* $3 \leq \dim \mathbb{H} < \infty$ *any map* $\mathbf{P}$ *with the properties* (2.90)–(2.91) *can be represented with the help of a positive, self-adjoint operator* ρ *with* $\mathrm{tr}(\rho) = 1$ *such that* $\mathbf{P}$ *is given by*

$$\mathbf{P}(P) = \mathrm{tr}(\rho P).$$

With suitable modifications this statement is also valid if $\dim \mathbb{H} = \infty$ and thus applies to quantum mechanics in general.

The following theorem provides some additional properties of density operators.

Theorem 2.24 *A density operator ρ on a* HILBERT *space* $\mathbb{H}$ *has the following properties:*

(i) There exist $p_j \in \mathbb{R}$ with $j \in I \subset \mathbb{N}$ that satisfy

$$p_j \geq 0 \tag{2.92}$$

$$\sum_{j \in I} p_j = 1 \tag{2.93}$$

and an ONB $\{|\psi_j\rangle \mid j \in I\}$ in $\mathbb{H}$ such that

$$\rho = \sum_{j \in I} p_j |\psi_j\rangle\langle\psi_j| = \sum_{j \in I} p_j P_{\psi_j} \tag{2.94}$$

(ii)

$$0 \leq \rho^2 \leq \rho \tag{2.95}$$

(iii)

$$\|\rho\| \leq 1. \tag{2.96}$$

Proof We first show (2.94). Since ρ as a density operator is per definition self-adjoint, its eigenvalues are real and there exists an ONB $\{|\psi_j\rangle\}$ in which ρ has the diagonal form (2.94). Another defining property of ρ is its positivity, which implies for every vector $|\psi_i\rangle$ of the ONB

$$0 \leq \langle\psi_i|\rho\,\psi_i\rangle = \sum_j p_j \langle\psi_i|\psi_j\rangle \underbrace{\langle\psi_j|\psi_i\rangle}_{=\delta_{ji}} = p_i, \tag{2.97}$$

proving (2.92). Lastly, we have, again per definition, also $\mathrm{tr}\,(\rho) = 1$ and thus

$$1 = \mathrm{tr}\,(\rho) \underbrace{=}_{(2.57)} \sum_i \langle\psi_i|\rho\,\psi_i\rangle = \sum_{i,j} p_j \langle\psi_i|\psi_j\rangle\langle\psi_j|\psi_i\rangle$$

$$= \sum_{i,j} \delta_{ij} p_j = \sum_i p_i,$$

which implies (2.93).

The positivity of ρ^2 follows from the fact that for any $|\psi\rangle \in \mathbb{H}$

$$\langle\psi|\rho^2\psi\rangle \underbrace{=}_{(2.30)} \langle\rho^*\psi|\rho\psi\rangle \underbrace{=}_{(2.80)} \langle\rho\psi|\rho\psi\rangle \underbrace{=}_{(2.5)} \|\rho\psi\|^2 \geq 0.$$

The p_j in (2.94) are such that $0 \leq p_j \leq 1 = \sum_j p_j$, hence $p_j^2 \leq p_j$. One then has

$$\rho^2 = \left(\sum_j p_j|\psi_j\rangle\langle\psi_j|\right)^2 = \sum_{j,k} p_j p_k |\psi_j\rangle \underbrace{\langle\psi_j|\psi_k\rangle}_{\delta_{jk}} \langle\psi_k| = \sum_j p_j^2 |\psi_j\rangle\langle\psi_j| \qquad (2.98)$$

and thus for any $|\psi\rangle \in \mathbb{H}$

$$\langle\psi|(\rho-\rho^2)\psi\rangle \underbrace{=}_{(2.94),(2.98)} \langle\psi|\sum_j (p_j-p_j^2)|\psi_j\rangle\langle\psi_j|\psi\rangle = \sum_j (p_j-p_j^2)\langle\psi|\psi_j\rangle\langle\psi_j|\psi\rangle$$

$$\underbrace{=}_{(2.1)} \sum_j \underbrace{(p_j-p_j^2)}_{\geq 0}\underbrace{|\langle\psi_j|\psi\rangle|^2}_{\geq 0} \geq 0, \qquad (2.99)$$

that is, $\rho - \rho^2 \geq 0$, which proves (2.95). From this it follows in turn that

$$\|\rho\psi\|^2 \underbrace{=}_{(2.5)} \langle\rho\psi|\rho\psi\rangle \underbrace{=}_{(2.80)} \langle\rho^*\psi|\rho\psi\rangle \underbrace{=}_{(2.30)} \langle\psi|\rho^2\psi\rangle \underbrace{\leq}_{(2.99)} \langle\psi|\rho\psi\rangle$$

$$\underbrace{\leq}_{(2.16)} \|\psi\|\|\rho\psi\|,$$

which implies

$$\frac{\|\rho\psi\|}{\|\psi\|} \leq 1$$

and thus, because of Definition 2.12 of the operator norm, we obtain (2.96). □

The representation (2.94) of a density operator ρ given in Theorem 2.24 allows us to view particles described by a mixed state as a statistical ensemble of particles constructed as follows. Suppose we have a device that, when a switch is set to j, produces a particle in the state $|\psi_j\rangle$. Moreover, suppose we have a random number generator that produces the switch setting j with probability p_j. We then run the random number generator many a times and each time use its output j to set the switch of the device generating a particle in the state $|\psi_j\rangle$. The statistics of the ensemble of particles thus produced is described by the density operator ρ. We shall see in Proposition 2.26 that if ρ is a true mixture, the statistics of this ensemble cannot be described by a pure state.

Exercise 2.25 Let $\{|\psi_j\rangle \mid j \in I\}$ be an ONB in a HILBERT space $\mathbb{H}$ and for $j \in I$ let $p_j \in [0,1]$ be such that $\sum_{j \in I} p_j = 1$. Show that then with

$$|\psi\rangle = \sum_{j \in I} \sqrt{p_j} |\psi_j\rangle, \quad \rho = \sum_{j \in I} p_j |\psi_j\rangle \langle \psi_j| \qquad (2.100)$$

we have for any $A \in B_{\mathrm{sa}}(\mathbb{H})$

$$\langle A \rangle_\psi = \langle A \rangle_\rho + \sum_{j,k \in I : j \neq k} \sqrt{p_j p_k} \langle \psi_j | A \psi_k \rangle .$$

For a solution see Solution 2.25.

In a mixed state $\rho = \sum_j p_j |\psi_j\rangle \langle \psi_j|$ the probability that a measurement of an observable $A = \sum_i \lambda_i |e_i\rangle \langle e_i|$ yields an eigenvalue λ_i corresponding to the eigenstate $|e_i\rangle$ is given by

$$\langle P_{e_i} \rangle_\rho \underbrace{=}_{(2.86)} \mathrm{tr}\left(\rho |e_i\rangle \langle e_i|\right) \underbrace{=}_{(2.57)} \sum_{k,j} p_j \langle e_k | \psi_j \rangle \langle \psi_j | e_i \rangle \underbrace{\langle e_i | e_k \rangle}_{\delta_{ik}} = \sum_j p_j |\langle e_i | \psi_j \rangle|^2 ,$$

where P_{e_i} denotes the orthogonal projection onto the state $|e_i\rangle$.

One advantage of the description of pure states with the help of a density operator (2.89) is that it becomes obvious that the physical information, that is, the state, does not depend on the overall phase of the state vector since for any $\alpha \in \mathbb{R}$

$$\rho_{e^{i\alpha}\psi} = |e^{i\alpha}\psi\rangle \langle e^{i\alpha}\psi| \underbrace{=}_{(2.26)} e^{i\alpha} |\psi\rangle \langle \psi| e^{-i\alpha} = \rho_\psi .$$

For the probability to measure an eigenvalue λ_i of $A = \sum_i |e_i\rangle \lambda_i \langle e_i|$ one then finds

$$\langle P_{e_i} \rangle_{\rho_\psi} \underbrace{=}_{(2.86)} \mathrm{tr}\left(\rho_\psi P_{e_i}\right) = \mathrm{tr}\left(|\psi\rangle \langle \psi| e_i\rangle \langle e_i|\right) \underbrace{=}_{(2.57)} \sum_k \langle e_k | \psi \rangle \langle \psi | e_i \rangle \underbrace{\langle e_i | e_k \rangle}_{=\delta_{ik}}$$

$$\underbrace{=}_{(2.1)} |\langle e_i | \psi \rangle|^2 \qquad (2.101)$$

and for the expectation value

$$\langle A \rangle_{\rho_\psi} \underbrace{=}_{(2.86)} \mathrm{tr}\left(\rho_\psi A\right) = \mathrm{tr}\left(|\psi\rangle \langle \psi| A\right) \underbrace{=}_{(2.57)} \sum_{k,i} \langle e_k | \psi \rangle \langle \psi | e_i \rangle \lambda_i \langle e_i | e_k \rangle$$

$$\underbrace{=}_{(2.1)} \sum_i \lambda_i |\langle e_i | \psi \rangle|^2 ,$$

exactly as stated in (2.62) for pure states.

Whether a given density operator ρ is a true mixture or a pure state can be decided with the help of the following proposition.

Proposition 2.25 *A density operator ρ describes a pure state if and only if $\rho^2 = \rho$, that is, for a density operator ρ the following equivalence holds*

$$\exists |\psi\rangle \in \mathbb{H} : \rho = |\psi\rangle\langle\psi| \qquad \Leftrightarrow \qquad \rho = \rho^2. \tag{2.102}$$

Proof We first show $\Rightarrow$: From the left side in (2.102) it necessarily follows that $||\psi|| = 1$ since, per definition $\mathrm{tr}\,(\rho) = 1$, and thus

$$1 = \mathrm{tr}\,(\rho) = \mathrm{tr}\,(|\psi\rangle\langle\psi|) \underbrace{=}_{(2.57)} \sum_k |\langle e_k|\psi\rangle|^2 \underbrace{=}_{(2.12)} ||\psi||^2 .$$

From $\rho = |\psi\rangle\langle\psi|$ with $||\psi|| = 1$ it follows immediately that

$$\rho^2 = |\psi\rangle\langle\psi|\psi\rangle\langle\psi| = |\psi\rangle\langle\psi| = \rho .$$

Now for $\Leftarrow$: From (2.94) in Theorem 2.24 we know that there exists an ONB $\{|\psi_j\rangle\}$ and real-valued p_j such that $\rho = \sum_j p_j |\psi_j\rangle\langle\psi_j|$. Because of $\rho^2 = \rho$ one has for all k thus

$$0 = \langle\psi_k|(\rho^2 - \rho)\psi_k\rangle = \langle\psi_k|\sum_j (p_j^2 - p_j)|\psi_j\rangle \underbrace{\langle\psi_j|\psi_k\rangle}_{=\delta_{jk}} = p_k^2 - p_k .$$

Consequently, the p_j can take only the values 0 or 1, and it follows that

$$\rho = \sum_{j:p_j=1} |\psi_j\rangle\langle\psi_j| .$$

Calculating $\mathrm{tr}\,(\rho) = 1$ in the ONB $\{|\psi_j\rangle\}$, we find

$$1 = \mathrm{tr}\,(\rho) = \sum_i \langle\psi_i| \sum_{j:p_j=1} |\psi_j\rangle \underbrace{\langle\psi_j|\psi_i\rangle}_{=\delta_{ji}} = \sum_{j:p_j=1} 1 .$$

Hence, $p_j = 1$ for exactly one $\breve{j}$ and $p_i = 0$ for all $i \neq \breve{j}$. Finally, with $|\psi\rangle = |\psi_{\breve{j}}\rangle$ then $\rho = |\psi\rangle\langle\psi|$ has the claimed form. $\square$

Density operators that satisfy $\rho^2 < \rho$ thus describe *true mixtures*, in other words, the statistics of such preparations cannot be described by a pure state. We formalize this in the following proposition.

> **Proposition 2.26** *Let ρ be a density operator on a* HILBERT *space* $\mathbb{H}$*. Then the following equivalence holds*
>
> $$\rho^2 \neq \rho \quad \Leftrightarrow \quad \nexists |\psi\rangle \in \mathbb{H} \text{ such that for every } A \in B_{sa}(\mathbb{H}) : \langle A \rangle_\psi = \langle A \rangle_\rho .$$

Proof We actually show the contrapositive

$$\rho^2 = \rho \quad \Leftrightarrow \quad \exists |\psi\rangle \in \mathbb{H} \text{ such that for every } A \in B_{sa}(\mathbb{H}) : \langle A \rangle_\psi = \langle A \rangle_\rho .$$
(2.103)

First, we show $\Rightarrow$ in (2.103). Let $\rho^2 = \rho$ and $\{|e_j\rangle\}$ be an ONB in $\mathbb{H}$. From (2.102) we know that then there exists a $|\psi\rangle \in \mathbb{H}$ such that $\rho = |\psi\rangle\langle\psi|$, where

$$|\psi\rangle \underset{(2.11)}{=} \sum_j \langle e_j|\psi\rangle |e_j\rangle .$$
(2.104)

It then follows that for any $A \in B_{sa}(\mathbb{H})$

$$\langle A \rangle_\rho \underset{(2.85)}{=} \mathrm{tr}(\rho A) = \mathrm{tr}(|\psi\rangle\langle\psi|A) \underset{(2.57)}{=} \sum_j \langle e_j|\psi\rangle\langle\psi|Ae_j\rangle$$
$$= \langle\psi|A \sum_j \langle e_j|\psi\rangle e_j\rangle \underset{(2.104)}{=} \langle\psi|A\psi\rangle$$
$$\underset{(2.60)}{=} \langle A \rangle_\psi .$$

Next, we show $\Leftarrow$ in (2.103). For this let

$$\rho = \sum_{j_1,j_2} \rho_{j_1 j_2} |e_{j_1}\rangle\langle e_{j_2}|$$
(2.105)

be a given density operator and suppose $|\psi\rangle = \sum_j \psi_j |e_j\rangle \in \mathbb{H}$ is such that for all $A \in B_{sa}(\mathbb{H})$

$$\langle A \rangle_\rho = \langle A \rangle_\psi .$$
(2.106)

For the left side of (2.106) we have

$$\langle A \rangle_\rho \underbrace{=}_{(2.85)} \mathrm{tr}\,(\rho A) \underbrace{=}_{(2.57)} \sum_j \langle e_j | \rho A e_j \rangle \underbrace{=}_{(2.105)} \sum_j \langle e_j | \sum_{j_1,j_2} \rho_{j_1 j_2} | e_{j_1} \rangle \langle e_{j_2} | A e_{j_1} \rangle$$

$$\underbrace{=}_{(2.22)} \sum_{j_1,j_2} \rho_{j_1 j_2} \sum_j \underbrace{\langle e_j | e_{j_1} \rangle}_{=\delta_{j j_1}} A_{j_2 j} = \sum_{j_1,j_2} \rho_{j_1 j_2} A_{j_2 j_1}\,. \tag{2.107}$$

Whereas for the right side of (2.106) we find

$$\langle A \rangle_\psi \underbrace{=}_{(2.60)} \langle \psi | A \psi \rangle \underbrace{=}_{(2.104)} \langle \sum_{j_2} \psi_{j_2} e_{j_2} | A \sum_{j_1} \psi_{j_1} e_{j_1} \rangle$$

$$\underbrace{=}_{(2.4),(2.6)} \sum_{j_1,j_2} \overline{\psi_{j_2}}\, \psi_{j_1} \langle e_{j_2} | A e_{j_1} \rangle \underbrace{=}_{(2.22)} \sum_{j_1,j_2} \overline{\psi_{j_2}}\, \psi_{j_1} A_{j_2 j_1}\,. \tag{2.108}$$

Using (2.107) and (2.108) in (2.106) and the fact that (2.106) is assumed to hold for all $A \in B_{sa}(\mathbb{H})$, shows that the matrix elements of ρ must be of the form $\rho_{j_1 j_2} = \psi_{j_1} \overline{\psi_{j_2}}$, and thus

$$\rho \underbrace{=}_{(2.105)} \sum_{j_1,j_2} \psi_{j_1} \overline{\psi_{j_2}} | e_{j_1} \rangle \langle e_{j_2} | = | \psi \rangle \langle \psi |\,.$$

From (2.102) we know that this is equivalent to $\rho^2 = \rho$ and this completes the proof of (2.103). $\qquad\qquad\qquad\qquad\qquad\qquad\qquad\qquad\qquad\qquad\qquad\qquad\qquad\quad\square$

Proposition 2.26 states that if there is a pure state which can replicate the statistics of ρ for all observables then ρ itself has to be the density operator of a pure state and vice-versa. Consequently, if ρ is a true mixture, no pure state is able to reproduce all the expectation values of ρ.

Density operators contain all measurable information about the described system. The diagonal representation of a density operator given in (2.94) in terms of its eigenvalues and eigenvectors is unique up to basis changes in degenerate eigenspaces. However, other not necessary diagonal representations are possible as the following proposition shows.

Proposition 2.27 *Let* $\mathbb{H}$ *be a finite-dimensional* HILBERT *space and* $\rho \in D(\mathbb{H})$ *be a density operator with diagonal form*

$$\rho = \sum_{j=1}^n p_j | \psi_j \rangle \langle \psi_j |, \tag{2.109}$$

where the $p_j \in]0,1]$ *for* $j \in \{1,\dots,n\}$ *with* $n \leq \dim \mathbb{H}$ *are the non-zero eigenvalues of* ρ *and* $\{ | \psi_j \rangle \mid j \in \{1,\dots,\dim \mathbb{H}\} \}$ *is an ONB of its eigenvectors. Moreover, for* $i \in \{1,\dots,m\}$ *with* $m \leq \dim \mathbb{H}$ *let* $q_i \in]0,1]$ *be such that* $\sum_{i=1}^m q_i = 1$ *and* $| \varphi_i \rangle \in \mathbb{H}$ *such that* $|| \varphi_i || = 1$. *Then we have the following equivalence*

$$\rho = \sum_{i=1}^{m} q_i |\varphi_i\rangle\langle\varphi_i| \quad (2.110) \quad \Leftrightarrow \quad \begin{cases} m \geq n \text{ and there is a } U \in U(m) \text{ such that} \\[2em] \sqrt{q_i}|\varphi_i\rangle = \sum_{j=1}^{n} U_{ij}\sqrt{p_j}|\psi_j\rangle \quad (2.111) \\[2em] \text{for } i \in \{1,\dots,m\}. \end{cases}$$

Proof To begin with, we note that the definition (2.83) of $D(\mathbb{H})$ implies that for $\rho \in D(\mathbb{H})$ we have $\rho = \rho^*$ and $\rho \geq 0$. It follows that ρ has real, non-negative eigenvalues r_j, its eigenvectors $\{|\psi_j\rangle\}$ form an ONB and it can be written as given in (2.109).

We prove $\Rightarrow$ first. Let $|\varphi_i\rangle$ be such that

$$\rho = \sum_{i=1}^{m} q_i |\varphi_i\rangle\langle\varphi_i|. \tag{2.112}$$

Since the $|\psi_j\rangle$ are vectors of an ONB, and thus linearly independent, we have for the image $\rho\{\mathbb{H}\}$ of ρ

$$n = \dim \mathrm{Span}\left\{|\psi_i\rangle \,\middle|\, i \in \{1,\dots,n\}\right\} \underbrace{=}_{(2.27)} \dim\rho\{\mathbb{H}\}.$$

On the other hand, $\rho\{\mathbb{H}\} \subset \mathrm{Span}\left\{|\varphi_i\rangle \,\middle|\, i \in \{1,\dots,m\}\right\}$ such that

$$\dim\rho\{\mathbb{H}\} \leq \dim \mathrm{Span}\left\{|\varphi_i\rangle \,\middle|\, i \in \{1,\dots,m\}\right\} \leq m,$$

and it follows that $n \leq m$. By assumption, the eigenvectors of ρ for the eigenvalue 0 are given by $|\psi_k\rangle$, where $k \in \{n+1,\dots,\dim\mathbb{H}\}$. They satisfy

$$0 = \langle\psi_k|\rho\psi_k\rangle \underbrace{=}_{(2.112)} \sum_{i=1}^{m} \underbrace{q_i}_{>0}\,\underbrace{|\langle\psi_k|\varphi_i\rangle|^2}_{\geq 0},$$

which implies $\langle\psi_k|\varphi_i\rangle = 0$ for all $k \in \{n+1,\dots,\dim\mathbb{H}\}$ and $i \in \{1,\dots,m\}$. Since the $\{|\psi_j\rangle\}$ are an ONB of $\mathbb{H}$, therefore

$$|\varphi_i\rangle \underbrace{=}_{(2.11)} \sum_{k=1}^{\dim \mathbb{H}} \langle \psi_k | \varphi_i \rangle | \psi_k \rangle = \sum_{k=1}^{n} \langle \psi_k | \varphi_i \rangle | \psi_k \rangle . \qquad (2.113)$$

Next, we define the matrix $V \in \mathrm{Mat}(m \times n, \mathbb{C})$ by

$$V_{ij} = \sqrt{\frac{q_i}{p_j}} \langle \psi_j | \varphi_i \rangle \qquad (2.114)$$

for $i \in \{1, \dots, m\}$ and $j \in \{1, \dots, n\}$, such that

$$\sum_{j=1}^{n} V_{ij} \sqrt{p_j} | \psi_j \rangle \underbrace{=}_{(2.114)} \sum_{j=1}^{n} \sqrt{q_i} \langle \psi_j | \varphi_i \rangle | \psi_j \rangle \underbrace{=}_{(2.113)} \sqrt{q_i} | \varphi_i \rangle . \qquad (2.115)$$

We proceed to show that we can extend $V \in \mathrm{Mat}(m \times n, \mathbb{C})$ to a $U \in \mathrm{U}(m)$. For this we first note that

$$V_{ji}^* \underbrace{=}_{(2.34)} \overline{V_{ij}} \underbrace{=}_{(2.114)} \sqrt{\frac{q_i}{p_j}} \overline{\langle \psi_j | \varphi_i \rangle} \underbrace{=}_{(2.1)} \sqrt{\frac{q_i}{p_j}} \langle \varphi_i | \psi_j \rangle \qquad (2.116)$$

and thus for $j, k \in \{1, \dots, n\}$

$$\sum_{i=1}^{m} V_{ji}^* V_{ik} \underbrace{=}_{(2.114),(2.116)} \frac{1}{\sqrt{p_j p_k}} \sum_{i=1}^{m} q_i \langle \psi_j | \varphi_i \rangle \langle \varphi_i | \psi_k \rangle \underbrace{=}_{(2.112)} \frac{1}{\sqrt{p_j p_k}} \langle \psi_j | \rho \, \psi_k \rangle$$

$$\underbrace{=}_{(2.27)} \frac{1}{\sqrt{p_j p_k}} \langle \psi_j | \sum_{l=1}^{n} p_l | \psi_l \rangle \langle \psi_l | \psi_k \rangle = \sum_{l=1}^{n} \frac{p_l \delta_{jl} \delta_{lk}}{\sqrt{p_j p_k}} = \delta_{jk} .$$

Hence, we have specified $V \in \mathrm{Mat}(m \times n, \mathbb{C})$, where $n \leq m$, such that $V^* V = \mathbf{1} \in \mathrm{Mat}(n \times n, \mathbb{C})$. This means that the n column vectors of V viewed as vectors in $\mathbb{C}^m$ are mutually orthogonal and normalized to 1. Applying the standard orthogonalization procedure, we can thus add $m - n$ more column vectors such that all m column vectors of the resulting matrix $U \in \mathrm{Mat}(m \times m, \mathbb{C})$ form a basis in $\mathbb{C}^m$. By this procedure we extend V to form a matrix $U \in \mathrm{U}(m)$ such that $U_{ij} = V_{ij}$ for $j \in \{1, \dots, n\}$ and (2.115) establishes the claim (2.111).

To prove $\Leftarrow$, let $m \geq n$ and $U \in \mathrm{U}(m)$ such that for $i \in \{1, \dots, m\}$

$$\sqrt{q_i} | \varphi_i \rangle = \sum_{j=1}^{n} U_{ij} \sqrt{p_j} | \psi_j \rangle .$$

This implies

$$\sum_{i=1}^{m} q_i |\varphi_i\rangle\langle\varphi_i| = \sum_{i=1}^{m} q_i \left(\frac{1}{\sqrt{q_i}} \sum_{j=1}^{n} U_{ij} \sqrt{p_j} |\psi_j\rangle \right) \left(\frac{1}{\sqrt{q_i}} \sum_{k=1}^{n} \overline{U_{ik}} \sqrt{p_k} \langle\psi_k| \right)$$

$$= \sum_{j,k=1}^{n} \sqrt{p_j p_k} \left(\sum_{i=1}^{m} U_{ki}^* U_{ij} \right) |\psi_j\rangle\langle\psi_k|$$

$$= \sum_{j,k=1}^{n} \sqrt{p_j p_k} \underbrace{\left((U^*U)_{ij} \right)}_{=\delta_{ij}} |\psi_j\rangle\langle\psi_k| = \sum_{j=1}^{n} p_i |\psi_j\rangle\langle\psi_j| \underbrace{=}_{} \rho, \qquad (2.109)$$

verifying (2.110). $\qquad\qquad\square$

Note that, while the $|\varphi_i\rangle$ in (2.110) are normalized, they are not necessarily orthogonal. The following Exercise 2.26 serves as an illustration of the non-uniqueness of the representation of a density operator.

Exercise 2.26 In $\mathbb{H} \cong \mathbb{C}^2$ let $\rho = \sum_{i=1}^{2} q_i |\varphi_i\rangle\langle\varphi_i|$ be given by

$$q_1 = \frac{2}{5}, \quad q_2 = \frac{3}{5} \quad \text{and} \quad |\varphi_1\rangle = |\uparrow_{\hat{x}}\rangle, \quad |\varphi_2\rangle = |0\rangle.$$

Then one has $\||\varphi_1\|| = 1 = \||\varphi_2\||$ and $\langle\varphi_1|\varphi_2\rangle = \frac{1}{\sqrt{2}}$. Verify that $\mathrm{tr}(\rho) = 1$, determine the eigenvalues p_1, p_2 and the (orthonormal) eigenvectors $|\psi_1\rangle, |\psi_2\rangle$ of ρ. With these, give the alternative diagonal form

$$\rho = \sum_{j=1}^{2} p_j |\psi_j\rangle\langle\psi_j|,$$

and verify that $\rho^2 < \rho$.

For a solution see Solution 2.26.

In a mixture described by $\rho = \sum_j p_j |\psi_j\rangle\langle\psi_j|$ the relative phases of the $|\psi_j\rangle$ are not physically observable. This is because for $\alpha_j \in \mathbb{R}$ one has

$$\sum_j p_j |e^{i\alpha_j}\psi_j\rangle\langle e^{i\alpha_j}\psi_j| \underbrace{=}_{(2.32),(2.33)} \sum_j p_j e^{i\alpha_j} |\psi_j\rangle\langle\psi_j| e^{-i\alpha_j} = \sum_j p_j |\psi_j\rangle\langle\psi_j| = \rho$$

such that the states $e^{i\alpha_j}|\psi_j\rangle$ generate the same mixture as the $|\psi_j\rangle$. There is no interference. Thus, one speaks of **incoherent** superposition as opposed to **coherent** superposition in the case of pure states where the relative phases are observable (see the discussion about interference preceding Proposition 2.15).

Exercise 2.27 Show that for $|\varphi\rangle, |\psi\rangle \in \mathbb{H} \cong \mathbb{C}^2$ and $\alpha \in \mathbb{R}$ the density operator $\rho_{\varphi+\psi}$ generally differs from the density operator $\rho_{\varphi+e^{i\alpha}\psi}$.

For a solution see Solution 2.27.

Interactions of a quantum system with its environment can change pure states into true mixtures. This is called **decoherence**. One of the most challenging problems in the practical implementation of the theory of quantum computing is the avoidance of decoherence for sufficiently long times.

We emphasize here once more that the knowledge of a state ρ or $|\psi\rangle$ only allows statements about the *statistics* of the ensemble described by the state. In general it is not possible to predict the behavior of single objects of the ensemble with certainty. This is how the phrase '*a particle or object is in the state ρ (or $|\psi\rangle$)*' has to be understood.

Exercise 2.28 Calculate the probability to measure the value $+1$ for the observable σ_z

  (i)   in the state $|\uparrow_{\hat{x}}\rangle$
  (ii)   in the state $|\downarrow_{\hat{x}}\rangle$
  (iii)   in the state $\frac{1}{\sqrt{2}}(|\uparrow_{\hat{x}}\rangle + |\downarrow_{\hat{x}}\rangle)$
  (iv)   in the state $\rho = \frac{1}{2}(|\uparrow_{\hat{x}}\rangle\langle\uparrow_{\hat{x}}| + |\downarrow_{\hat{x}}\rangle\langle\downarrow_{\hat{x}}|)$.

For a solution see Solution 2.28.

2.4 Qubits

A classical bit is the smallest possible unit of information. The information given by this unit consists of the selection of binary alternatives which are usually denoted by 0 and 1 or *Yes* and *No* or *True* and *False*. The classical bit is realized physically by assigning the alternatives to two different states of a physical system, such as opposite magnetization in a well defined space on a hard disk.

With the help of quantum mechanics we can represent the binary alternatives by two basis vectors in a *two-dimensional* quantum mechanical state space. Generally, however, the quantum mechanical state space of a microscopic object is an infinite-dimensional HILBERT space. The physical realization of a two-dimensional state space is then mostly accomplished by restricting the preparation to two-dimensional eigenspaces of a suitably chosen observable. Examples of such quantum systems and observables with two-dimensional eigenspaces are:

- *Electrons and their spin*
  Ignoring the position or momentum of the electron, we only measure their spin state and can map the binary alternatives
  - to the vectors
  $$|0\rangle = |\uparrow_{\hat{z}}\rangle \quad \text{and} \quad |1\rangle = |\downarrow_{\hat{z}}\rangle$$

  as ONB, which consists of eigenstates of σ_z
  - or to the vectors
  $$|+\rangle := |\uparrow_{\hat{x}}\rangle = \frac{1}{\sqrt{2}}(|\uparrow_{\hat{z}}\rangle + |\downarrow_{\hat{z}}\rangle) \quad \text{and} \quad |-\rangle := |\downarrow_{\hat{x}}\rangle = \frac{1}{\sqrt{2}}(|\uparrow_{\hat{z}}\rangle - |\downarrow_{\hat{z}}\rangle)$$

  as the ONB, which consists of eigenstates of σ_x
  - or to the vectors
  $$|\uparrow_{\hat{y}}\rangle = \frac{1}{\sqrt{2}}(|\uparrow_{\hat{z}}\rangle + i|\downarrow_{\hat{z}}\rangle) \quad \text{and} \quad |\downarrow_{\hat{y}}\rangle = \frac{1}{\sqrt{2}}(i|\uparrow_{\hat{z}}\rangle + |\downarrow_{\hat{z}}\rangle)$$

  as the ONB, which consists of eigenstates of σ_y.

- *Photons (Light) and their polarization*
  For photons with a given direction of propagation the polarization is described by a two-dimensional complex vector, the so-called polarization-vector. The state space is thus $\mathbb{H} \cong \mathbb{C}^2$ and we can map the binary alternatives to
  - the vectors
  $$|0\rangle = |H\rangle = \begin{pmatrix} 1 \\ 0 \end{pmatrix} \text{ representing horizontal polarization} \quad \text{and} \quad |1\rangle = |V\rangle = \begin{pmatrix} 0 \\ 1 \end{pmatrix} \text{ representing vertical polarization}$$

  as ONB, which consists of eigenstates of the operator $\sigma_z = |H\rangle\langle H| - |V\rangle\langle V|$; the orthogonal projections $|H\rangle\langle H|$ and $|V\rangle\langle V|$ are called horizontal and vertical polarizors
  - or to the vectors
  $$|+\rangle = \frac{1}{\sqrt{2}}(|H\rangle + |V\rangle) \quad \text{and} \quad |-\rangle = \frac{1}{\sqrt{2}}(|H\rangle - |V\rangle)$$

  as ONB, which consists of eigenstates of polarizors $|+\rangle\langle+|$ and $|-\rangle\langle-|$, which are rotated by 45° relative to the horizontal and vertical polarizors,
  - or to the vectors
  $$|R\rangle = \frac{1}{\sqrt{2}}(|H\rangle + i|V\rangle) \quad \text{and} \quad |L\rangle = \frac{1}{\sqrt{2}}(i|H\rangle + |V\rangle)$$

as ONB, which consists of eigenstates of the so-called left- resp. right-circular polarizors.

For the representation of the classical bit values 0 and 1 with an electron for example, we can prepare it in an eigenstate of σ_z, such as $|0\rangle$ for 0 and $|1\rangle$ for 1. If we then isolate the electron from any interactions—that is, maintain its state— and afterwards measure σ_z, we know that then the eigenvalue corresponding to the prepared eigenstate $|0\rangle$ or $|1\rangle$ will be measured. In other words, the electron stores the value of the binary alternative 0 or 1. A measurement of σ_z corresponds to a reading of the stored information.

In order to maintain the stored bit unaltered it is crucial that the electron is not disturbed by interactions that could change its state. With information storage in a classical computer, such as on a hard disk, this is relatively easily accomplished since most external disturbances such as light or heat do not alter the stored bit. In that case it suffices to avoid exposing the hard disk to strong magnetic fields. In quantum mechanical systems, however, it is far more difficult to isolate the system from state-changing interactions with its environment. This is one of the major challenges in a realization of quantum computers currently being addressed in numerous ways.

The classical bit can thus be represented by an ONB in a two-dimensional HILBERT space. The choice of the ONB is arbitrary, as long as we have a suitable observable which has the two vectors of this ONB as its eigenvectors. Measurement of this observable then serves as a read-out of the stored bit. As candidates for a physical realization we mentioned above electrons and their spin or photons and their polarization. But any other quantum system with a suitable two-dimensional subspace may be chosen. Mathematically we can always identify the two-dimensional HILBERT space $\mathbb{H}$ with $\mathbb{C}^2$ by choosing an ONB .

Quantum mechanics, however, also allows states of the form $a|0\rangle + b|1\rangle$ with $a, b \in \mathbb{C}$ and the normalization $|a|^2 + |b|^2 = 1$ for two-dimensional systems. These linear combinations of the states $|0\rangle$ and $|1\rangle$ have no analogue in the world of classical bits. They do not occur in classical computing. The information that can be stored in a two-dimensional quantum system is thus much greater than what can be stored in a classical bit. Writing, reading or transforming such information in a quantum system also requires special care. All this motivates the new notion of a **qubit**, denoting two-dimensional quantum systems with a view towards their information content.

> **Definition 2.28** A **qubit** is a quantum mechanical system described by a two-dimensional HILBERT space denoted by $\mathbb{H}$ and called **qubit space**. The information stored in a qubit is contained in the **qubit state** in $\mathbb{H}$ in which the system is, and it is manipulated and read according to the postulates of quantum mechanics. In $\mathbb{H}$ we select an ONB $\{|0\rangle, |1\rangle\}$ and an observable represented

by a self-adjoint operator denoted by σ_z that has the normalized eigenvector $|0\rangle$ with eigenvalue $+1$ and $|1\rangle$ with eigenvalue -1.

The terminology here is in accordance with that of classical computing: a classical bit is the elementary information container and the classical information content is given by the bit values $\{0, 1\}$. The elementary information container in quantum computing is the two-dimensional quantum system described by $\mathbb{H}$. The 'value' of the quantum information is the state $|\psi\rangle \in \mathbb{H}$ in which that system is.

As a consequence of the Projection Postulate 3 we thus have the following corollary.

Corollary 2.29 *A measurement of σ_z on a qubit yields either $+1$ or -1 as observed value and projects the qubit in the eigenstate $|0\rangle$ or $|1\rangle$ corresponding to the observed value.*

The orthonormal eigenvectors $|0\rangle, |1\rangle$ of σ_z form a standard basis in $\mathbb{H}$, with which the qubit-HILBERT space $\mathbb{H}$ can be identified with $\mathbb{C}^2$. From now on we will also use these states to represent the classical bit values 0 and 1. This representation of classical bit values by the eigenstates of σ_z is as shown in Table 2.1, and is to be understood as follows: a measurement of σ_z in the qubit state yielding the value $+1$ represents the classical bit value 0. The Projection Postulate 3 furthermore tells us that then the qubit is in the state $|0\rangle$. Conversely, if we want to represent the classical bit value 0 in a qubit, we prepare the state $|0\rangle$. An analogous statement holds for the eigenvalue -1, the eigenvector $|1\rangle$, and the classical bit value 1.

This way each classical bit value is mapped to a qubit state. However, not every qubit state can be mapped to a classical bit value. This is because a general qubit state is of the form

$$|\psi\rangle = a|0\rangle + b|1\rangle \tag{2.117}$$

with $a, b \in \mathbb{C}$ and $|a|^2 + |b|^2 = 1$. In case $ab \neq 0$, this is a superposition of $|0\rangle$ and $|1\rangle$ for which there is *no corresponding classical bit value*. As we shall see later in the presentation of some quantum algorithms in Chap. 6, it is the presence of superpositions having no classical equivalents that contributes to the gains in efficiency compared to classical algorithms.

Table 2.1 Representation of classical bits by qubits

Observed value of σ_z	qubit state	Represented classical bit value
$+1$	$|0\rangle = |\uparrow_{\hat{z}}\rangle$	0
-1	$|1\rangle = |\downarrow_{\hat{z}}\rangle$	1

How can we suitably parametrize a pure qubit state of the form (2.117)? Because $|a|^2 + |b|^2 = 1$ we can find $\alpha, \beta, \theta \in \mathbb{R}$ such that $a = e^{i\alpha} \cos \frac{\theta}{2}$ and $b = e^{i\beta} \sin \frac{\theta}{2}$. Thus, a qubit state has the general form

$$|\psi\rangle = e^{i\alpha} \cos \frac{\theta}{2} |0\rangle + e^{i\beta} \sin \frac{\theta}{2} |1\rangle . \tag{2.118}$$

Physically equivalent to $|\psi\rangle$—and thus representing the same qubit state (see discussion around Definition 2.14)—is the following member of its ray

$$e^{-i\frac{\alpha+\beta}{2}} |\psi\rangle = e^{i\frac{\alpha-\beta}{2}} \cos \frac{\theta}{2} |0\rangle + e^{i\frac{\beta-\alpha}{2}} \sin \frac{\theta}{2} |1\rangle .$$

With $\phi = \beta - \alpha$ this leads to

$$|\psi(\theta, \phi)\rangle := e^{-i\frac{\phi}{2}} \cos \frac{\theta}{2} |0\rangle + e^{i\frac{\phi}{2}} \sin \frac{\theta}{2} |1\rangle = \begin{pmatrix} e^{-i\frac{\phi}{2}} \cos \frac{\theta}{2} \\ e^{i\frac{\phi}{2}} \sin \frac{\theta}{2} \end{pmatrix} . \tag{2.119}$$

An observable that has this state as an eigenstate can be constructed as follows. For a vector $\mathbf{a} = \begin{pmatrix} a_1 \\ a_2 \\ a_3 \end{pmatrix} \in \mathbb{R}^3$ one defines the 2×2 *matrix*

$$\mathbf{a} \cdot \sigma := \sum_{j=1}^{3} a_j \sigma_j = \begin{pmatrix} a_3 & a_1 - i a_2 \\ a_1 + i a_2 & -a_3 \end{pmatrix} . \tag{2.120}$$

The notation $\mathbf{a} \cdot \sigma$ introduced in (2.120) is standard in physics. Confusion with the normal scalar product $\mathbf{a} \cdot \mathbf{b}$ for $\mathbf{a}, \mathbf{b} \in \mathbb{R}^n$ can be avoided by keeping in mind that whenever the second factor is σ then the expression denotes the 2×2 matrix given by (2.120).

Exercise 2.29 Show that for $\mathbf{a}, \mathbf{b} \in \mathbb{R}^3$

$$(\mathbf{a} \cdot \sigma)(\mathbf{b} \cdot \sigma) = (\mathbf{a} \cdot \mathbf{b})\mathbf{1} + i(\mathbf{a} \times \mathbf{b}) \cdot \sigma , \tag{2.121}$$

where $\mathbf{a} \cdot \mathbf{b} = \sum_{j=1}^{3} a_j b_j \in \mathbb{R}$ is the usual scalar product and

$$\mathbf{a} \times \mathbf{b} = \begin{pmatrix} a_2 b_3 - a_3 b_2 \\ a_3 b_1 - a_1 b_3 \\ a_1 b_2 - a_2 b_1 \end{pmatrix} \in \mathbb{R}^3$$

is the usual vector (or cross) product.

For a solution see Solution 2.29.

With the unit vector

$$\hat{\mathbf{n}} = \hat{\mathbf{n}}(\theta, \phi) := \begin{pmatrix} \sin\theta\cos\phi \\ \sin\theta\sin\phi \\ \cos\theta \end{pmatrix} \in \mathbb{R}^3 \tag{2.122}$$

it follows that

$$\hat{\mathbf{n}} \cdot \sigma = \begin{pmatrix} \cos\theta & e^{-i\phi}\sin\theta \\ e^{i\phi}\sin\theta & -\cos\theta \end{pmatrix}. \tag{2.123}$$

This is the operator for the observable *spin in direction* $\hat{\mathbf{n}}$, and one has

$$\begin{aligned}
\hat{\mathbf{n}}(\theta, \phi) \cdot \sigma \,|\psi(\theta, \phi)\rangle &= \begin{pmatrix} \cos\theta & e^{-i\phi}\sin\theta \\ e^{i\phi}\sin\theta & -\cos\theta \end{pmatrix} \begin{pmatrix} e^{-i\frac{\phi}{2}}\cos\frac{\theta}{2} \\ e^{i\frac{\phi}{2}}\sin\frac{\theta}{2} \end{pmatrix} \\
&= \begin{pmatrix} e^{-i\frac{\phi}{2}}\left(\cos\theta\cos\frac{\theta}{2} + \sin\theta\sin\frac{\theta}{2}\right) \\ e^{i\frac{\phi}{2}}\left(\sin\theta\cos\frac{\theta}{2} - \cos\theta\sin\frac{\theta}{2}\right) \end{pmatrix} \\
&= \begin{pmatrix} e^{-i\frac{\phi}{2}}\cos\frac{\theta}{2} \\ e^{i\frac{\phi}{2}}\sin\frac{\theta}{2} \end{pmatrix} \\
&= |\psi(\theta, \phi)\rangle. \tag{2.124}
\end{aligned}$$

The state $|\psi(\theta, \phi)\rangle$ is thus the spin-up-state $|\uparrow_{\hat{\mathbf{n}}}\rangle$ for spin in the direction $\hat{\mathbf{n}}$:

$$|\uparrow_{\hat{\mathbf{n}}}\rangle := \begin{pmatrix} e^{-i\frac{\phi}{2}}\cos\frac{\theta}{2} \\ e^{i\frac{\phi}{2}}\sin\frac{\theta}{2} \end{pmatrix}. \tag{2.125}$$

Analogously, one finds for

$$|\downarrow_{\hat{\mathbf{n}}}\rangle := \begin{pmatrix} -e^{-i\frac{\phi}{2}}\sin\frac{\theta}{2} \\ e^{i\frac{\phi}{2}}\cos\frac{\theta}{2} \end{pmatrix} \tag{2.126}$$

then

$$\hat{\mathbf{n}} \cdot \sigma |\downarrow_{\hat{\mathbf{n}}}\rangle = -|\downarrow_{\hat{\mathbf{n}}}\rangle.$$

In particular, it follows from (2.123) and (2.119) that

$$\hat{\mathbf{n}}(0,0) \cdot \sigma = \sigma_z \qquad \text{and} \qquad |\uparrow_{\hat{\mathbf{n}}(0,0)}\rangle = |\uparrow_{\hat{\mathbf{z}}}\rangle$$

or

$$\hat{\mathbf{n}}\left(\frac{\pi}{2}, 0\right) \cdot \sigma = \sigma_x \qquad \text{and} \qquad |\uparrow_{\hat{\mathbf{n}}(\frac{\pi}{2},0)}\rangle = |\uparrow_{\hat{\mathbf{x}}}\rangle.$$

The state $|\psi(\theta,\phi)\rangle = |\uparrow_{\hat{\mathbf{n}}}\rangle$ parametrized by θ,ϕ thus represents an *arbitrary qubit state*, and the operator $\hat{\mathbf{n}} \cdot \boldsymbol{\sigma}$ represents the observable that has this qubit state as an eigenstate with eigenvalue $+1$.

How can we then parametrize *mixtures of qubits*? For this consider the complex 2×2 matrix of a density operator

$$\rho = \begin{pmatrix} a & b \\ c & d \end{pmatrix}.$$

Since $\rho^* = \rho$, it follows that $a,d \in \mathbb{R}$ and $b = \bar{c}$, and since $\text{tr}(\rho) = 1$, that $a + d = 1$. Hence, we can write a and d with the help of a real number x_3 in the form $a = \frac{1+x_3}{2}$ and $d = \frac{1-x_3}{2}$. Moreover, introducing the real numbers $x_1 = 2\,\text{Re}(c)$ and $x_2 = 2\,\text{Im}(c)$, and making use of (2.120) we can thus write an arbitrary qubit-density operator in the form

$$\rho_{\mathbf{x}} = \frac{1}{2}\begin{pmatrix} 1+x_3 & x_1 - ix_2 \\ x_1 + ix_2 & 1-x_3 \end{pmatrix} = \frac{1}{2}\left(1 + \mathbf{x} \cdot \boldsymbol{\sigma}\right). \tag{2.127}$$

Exercise 2.30 Let $\rho_{\mathbf{x}}$ be given as in (2.127). Show that then

$$x_j = \text{tr}(\rho_{\mathbf{x}}\sigma_j) \quad \text{for } j \in \{1,2,3\},$$

which can be succinctly written as $\mathbf{x} = \text{tr}(\rho\boldsymbol{\sigma})$.

For a solution see Solution 2.30.

So far we have only used $\rho^* = \rho$ and $\text{tr}(\rho) = 1$ from the defining properties of a density operator. A further defining requirement is that it has to be positive, that is, it has to satisfy $\rho \geq 0$. We know from (2.97) that the positivity of ρ is equivalent to the requirement that all eigenvalues of ρ be non-negative. The eigenvalues q_1, q_2 of $\rho_{\mathbf{x}}$ are found from (2.127) as

$$q_{1,2} = \frac{1 \pm \sqrt{|\mathbf{x}|^2}}{2} = \frac{1 \pm |\mathbf{x}|}{2} \tag{2.128}$$

such that the requirement $\rho \geq 0$ is satisfied only for $|\mathbf{x}| \leq 1$. Thus we have shown: the density operators for mixtures of qubits can be parametrized by vectors $\mathbf{x}$ in the *closed unit ball* $B^1_{\mathbb{R}^3}$ of $\mathbb{R}^3$. This parametrization is called the **BLOCH representation** and shown in Fig. 2.1. From the fact that every $\mathbf{x} \in B^1_{\mathbb{R}^3}$ represents a mixed state we can see graphically that mixed states form a convex set, as was already shown in Exercise 2.22. Moreover, the *edge points* $|\mathbf{x}| = 1$ correspond exactly to *pure* states, as can be seen as follows. With the help of (2.121) and (2.128) we obtain

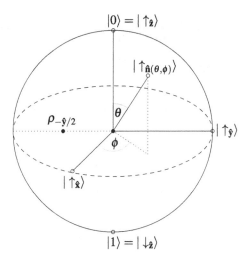

Fig. 2.1 Representing qubits with the BLOCH representation. The boundary sphere $S^1_{\mathbb{R}^3}$ of pure states is called the BLOCH sphere. Note that whereas the whole of the closed ball $B^1_{\mathbb{R}^3}$ (including its interior as mixed states) represents all possible states of a qubit, the only two classical bit values correspond to the north and south pole. The circles ∘ show the representatives of the pure states $|\uparrow_{\hat{z}}\rangle, |\uparrow_{\hat{y}}\rangle, |\uparrow_{\hat{x}}\rangle, |\downarrow_{\hat{z}}\rangle$ and $|\uparrow_{\hat{n}(\theta,\phi)}\rangle$. The black dots • show the mixed states $\rho_{-\frac{1}{2}\hat{y}} = \frac{1}{4}\begin{pmatrix} 2 & i \\ -i & 2 \end{pmatrix}$ and at the center $\rho_0 = \frac{1}{2}\mathbf{1}$

$$\rho^2_{\mathbf{x}} = \frac{1}{4}\left(\mathbf{1} + \mathbf{x}\cdot\boldsymbol{\sigma}\right)^2 = \frac{1}{4}\left(\mathbf{1}(1 + |\mathbf{x}|^2) + 2\mathbf{x}\cdot\boldsymbol{\sigma}\right)$$

such that $\rho^2_{\mathbf{x}} = \rho_{\mathbf{x}}$ if and only if $|\mathbf{x}| = 1$, and as shown in Proposition 2.25, the property $\rho^2 = \rho$ is equivalent to ρ describing a pure state.

From Exercise 2.22 we know already that density operators—and hence mixed states—form a convex set. A particular property of two-dimensional systems, and thus qubits, is the fact that *all* edge points correspond to pure states. If the quantum system is described by states in a HILBERT space with dimension larger than two, only certain edge points correspond to pure states.

2.5   Operators on Qubits

By Definition 2.28 qubits are two-dimensional quantum mechanical systems described by vectors in a two-dimensional HILBERT space $\mathbb{H}$. With the exception of measurements, their time evolution is described by unitary operators. Thus, apart from measurements, all transformations we wish to apply to a qubit have to be unitary as well. This also applies to the action of elementary components of qubit circuits (so-called quantum gates), which we shall discuss extensively in Sect. 5.2.

In the following we thus present a number of results relating to the construction of unitary operators on $\mathbb{H}$, which will be particularly helpful in the context of aforementioned quantum gates. We begin with a general result about exponentials of certain operators in Exercise 2.31.

Exercise 2.31 Let A be an operator on a vector space satisfying $A^2 = \mathbf{1}$. Show that then for any $\alpha \in \mathbb{R}$

$$e^{i\alpha A} := \sum_{n=0}^{\infty} \frac{(i\alpha)^n}{n!} A^n = \cos\alpha \mathbf{1} + i\sin\alpha A. \tag{2.129}$$

For a solution see Solution 2.31.

Throughout this book we will use both notations $e^A = \exp(A)$ to denote the exponential function of an operator A, the choice being made depending on which of the two provides more transparent expressions.

The following Definition 2.30 is motivated by the theory of representations of the group SO(3), that is, the norm- and orientation-preserving linear maps of $\mathbb{R}^3$ (see Example F.8).

Definition 2.30 Let $\hat{\mathbf{n}} \in S^1_{\mathbb{R}^3}$ be a unit vector in $\mathbb{R}^3$ and $\alpha \in \mathbb{R}$. The action of a rotation around the vector $\hat{\mathbf{n}}$ by the angle α on the qubit space $\mathbb{H}$ (also called **spin rotation**) is defined as the operator

$$D_{\hat{\mathbf{n}}}(\alpha) := e^{-i\frac{\alpha}{2}\hat{\mathbf{n}}\cdot\sigma}.$$

The spin-rotation $D_{\hat{\mathbf{n}}}(\alpha)$ is an operator on $\mathbb{H}$. It is the action in the qubit space that represents the action of a rotation in the position space $\mathbb{R}^3$ around $\hat{\mathbf{n}}$ by the angle α.

Lemma 2.31 Let $\hat{\mathbf{n}} \in B^1_{\mathbb{R}^3}$ and $\alpha \in \mathbb{R}$. Then one has

$$D_{\hat{\mathbf{n}}}(\alpha) = \cos\frac{\alpha}{2}\mathbf{1} - i\sin\frac{\alpha}{2}\hat{\mathbf{n}}\cdot\sigma \tag{2.130}$$

$$D_{\hat{\mathbf{n}}}(\alpha)^* = D_{\hat{\mathbf{n}}}(-\alpha) = \cos\frac{\alpha}{2}\mathbf{1} + i\sin\frac{\alpha}{2}\hat{\mathbf{n}}\cdot\sigma \tag{2.131}$$

$$D_{\hat{\mathbf{n}}}(\alpha)D_{\hat{\mathbf{n}}}(\alpha)^* = \mathbf{1},$$

that is, $D_{\hat{\mathbf{n}}}(\alpha)$ is a unitary operator on $\mathbb{H}$.

Proof From (2.121) and because of $\hat{\mathbf{n}} \cdot \hat{\mathbf{n}} = 1$ it follows that $(\hat{\mathbf{n}} \cdot \boldsymbol{\sigma})^2 = \mathbf{1}$. With this (2.130) follows immediately from the result (2.129) in Exercise 2.31.

From $(-\mathrm{i}\hat{\mathbf{n}} \cdot \boldsymbol{\sigma})^* = \mathrm{i}\hat{\mathbf{n}} \cdot \boldsymbol{\sigma}$ and (2.130) we obtain then (2.131). Finally, one has

$$
\begin{aligned}
D_{\hat{\mathbf{n}}}(\alpha) D_{\hat{\mathbf{n}}}(\alpha)^* \underset{(2.130),(2.131)}{=} & \left(\cos\frac{\alpha}{2}\mathbf{1} - \mathrm{i}\sin\frac{\alpha}{2}\hat{\mathbf{n}} \cdot \boldsymbol{\sigma}\right)\left(\cos\frac{\alpha}{2}\mathbf{1} + \mathrm{i}\sin\frac{\alpha}{2}\hat{\mathbf{n}} \cdot \boldsymbol{\sigma}\right) \\
= & \left(\cos\frac{\alpha}{2}\mathbf{1}\right)^2 - \left(\mathrm{i}\sin\frac{\alpha}{2}\hat{\mathbf{n}} \cdot \boldsymbol{\sigma}\right)^2 = \left(\cos^2\frac{\alpha}{2} + \sin^2\frac{\alpha}{2}\right)\mathbf{1} \\
= & \ \mathbf{1}.
\end{aligned}
$$

$\square$

The result $D_{\hat{\mathbf{n}}}(\alpha) D_{\hat{\mathbf{n}}}(-\alpha) = \mathbf{1}$ has a generalization, as is to be shown in Exercise 2.32.

Exercise 2.32 Show that for any $\hat{\mathbf{n}} \in S^1_{\mathbb{R}^3}$ and $\alpha, \beta \in \mathbb{R}$

$$
D_{\hat{\mathbf{n}}}(\alpha) D_{\hat{\mathbf{n}}}(\beta) = D_{\hat{\mathbf{n}}}(\alpha + \beta). \tag{2.132}
$$

For a solution see Solution 2.32.

Before we show in Exercise 2.33 that every unitary operator on $\mathbb{H}$ can be constructed by multiplication of suitably chosen spin-rotations $D_{\hat{\mathbf{n}}}(\cdot)$ and a phase-factor $\mathrm{e}^{\mathrm{i}\alpha}$, it is helpful to prove the following Lemma 2.32 as an intermediate step.

Lemma 2.32 *Let U be a unitary operator on $\mathbb{H}$. Then there are $\alpha, \beta, \delta, \gamma \in \mathbb{R}$ such that the matrix of U in the standard basis $\{|0\rangle, |1\rangle\}$ is given by*

$$
U = \mathrm{e}^{\mathrm{i}\alpha}\begin{pmatrix} \mathrm{e}^{-\mathrm{i}\frac{\beta+\delta}{2}}\cos\frac{\gamma}{2} & -\mathrm{e}^{\mathrm{i}\frac{\delta-\beta}{2}}\sin\frac{\gamma}{2} \\ \mathrm{e}^{\mathrm{i}\frac{\beta-\delta}{2}}\sin\frac{\gamma}{2} & \mathrm{e}^{\mathrm{i}\frac{\beta+\delta}{2}}\cos\frac{\gamma}{2} \end{pmatrix}. \tag{2.133}
$$

Proof Let the matrix of U in the standard basis $\{|0\rangle, |1\rangle\}$ be given by

$$
U = \begin{pmatrix} a & b \\ c & d \end{pmatrix} \tag{2.134}
$$

with $a, b, c, d \in \mathbb{C}$. Because of

$$
U^* = \begin{pmatrix} \bar{a} & \bar{c} \\ \bar{b} & \bar{d} \end{pmatrix} \tag{2.135}
$$

and $UU^* = \mathbb{1}$, it follows that

$$|a|^2 + |b|^2 = 1 = |c|^2 + |d|^2 \tag{2.136}$$

$$a\bar{c} + b\bar{d} = 0. \tag{2.137}$$

If $c = 0$, then $|d| = 1$ follows necessarily, which implies $b = 0$ and thus $|a| = 1$ as well. In this case U is of the form

$$U = \begin{pmatrix} e^{i\xi} & 0 \\ 0 & e^{i\eta} \end{pmatrix}, \tag{2.138}$$

and with $\alpha = \frac{\xi+\eta}{2}, \beta = \eta - \xi$ and $\delta = \gamma = 0$ can thus be written in the form (2.133). Similarly, it follows in the case $a = 0$ that U is of the form

$$U = \begin{pmatrix} 0 & e^{i\omega} \\ e^{i\tau} & 0 \end{pmatrix} \tag{2.139}$$

and with $\alpha = \frac{\omega+\tau+\pi}{2}, \delta = \omega + \pi - \tau$ and $\beta = 0, \gamma = \pi$ can be written in the form (2.133).

Now let $a\bar{c} \neq 0$. Then

$$a = -b\frac{\bar{d}}{\bar{c}}$$

$$\Rightarrow \quad |a|^2 = |b|^2 \frac{|d|^2}{|c|^2}$$

$$\Rightarrow \quad 1 = |a|^2 + |b|^2 = |b|^2 \left(1 + \frac{|d|^2}{|c|^2}\right) = |b|^2 \frac{|c|^2 + |d|^2}{|c|^2} \underbrace{=}_{(2.136)} \frac{|b|^2}{|c|^2},$$

which implies

$$|b| = |c| \quad \text{and} \quad |a| = |d|. \tag{2.140}$$

Hence, there are $\xi, \eta, \gamma \in \mathbb{R}$ such that

$$a = e^{i\xi} \cos\frac{\gamma}{2}, \qquad d = e^{i\eta} \cos\frac{\gamma}{2}, \tag{2.141}$$

from which in turn it follows that

$$|c|^2 \underbrace{=}_{(2.140)} |b|^2 \underbrace{=}_{(2.136)} 1 - |a|^2 = \sin^2\frac{\gamma}{2}. \tag{2.142}$$

Thus, there are $\omega, \tau \in \mathbb{R}$ such that

$$b = -e^{i\omega} \sin\frac{\gamma}{2}, \qquad c = e^{i\tau} \sin\frac{\gamma}{2}. \tag{2.143}$$

Because of $a\bar{c} = -b\bar{d} \neq 0$, we have

$$e^{i(\xi-\tau)} \sin\frac{\gamma}{2} \cos\frac{\gamma}{2} = e^{i(\omega-\eta)} \sin\frac{\gamma}{2} \cos\frac{\gamma}{2} \tag{2.144}$$

and thus $\xi - \tau = \omega - \eta + 2k\pi$. We choose $\eta = \omega + \tau - \xi$ such that

$$U = \begin{pmatrix} e^{i\xi} \cos\frac{\gamma}{2} & -e^{i\omega} \sin\frac{\gamma}{2} \\ e^{i\tau} \sin\frac{\gamma}{2} & e^{i(\omega+\tau-\xi)} \cos\frac{\gamma}{2} \end{pmatrix}. \tag{2.145}$$

With the change of variables

$$\alpha := \frac{\omega+\tau}{2}, \qquad \beta := \tau - \xi, \qquad \delta := \omega - \xi \tag{2.146}$$

we obtain in (2.145)

$$\begin{aligned} \xi &= \alpha - \frac{\beta+\delta}{2}, & \omega &= \alpha + \frac{\delta-\beta}{2}, \\ \tau &= \alpha + \frac{\beta-\delta}{2}, & \omega+\tau-\xi &= \alpha + \frac{\beta+\delta}{2}, \end{aligned} \tag{2.147}$$

that is, U has the form claimed in (2.133). $\qquad\qquad\square$

With the help of Lemma 2.32 it can be shown that any unitary operator on qubits can be expressed as a product of suitable spin-rotations around the $\hat{\mathbf{z}}$- and $\hat{\mathbf{y}}$-axis.

Exercise 2.33 Let U be a unitary operator on $\mathbb{H}$. Show that there exist $\alpha, \beta, \delta, \gamma \in \mathbb{R}$ such that

$$U = e^{i\alpha} D_{\hat{\mathbf{z}}}(\beta) D_{\hat{\mathbf{y}}}(\gamma) D_{\hat{\mathbf{z}}}(\delta).$$

For a solution see Solution 2.33.

A consequence of the claim shown in Exercise 2.33 is that a device which can execute spin-rotations around the $\hat{\mathbf{z}}$- and $\hat{\mathbf{y}}$-axis would be sufficient to implement any unitary transformation of qubits.

Example 2.33 One has, for example,

$$\begin{aligned} e^{i\frac{\alpha}{2}} D_{\hat{\mathbf{z}}}(\alpha) D_{\hat{\mathbf{y}}}(0) D_{\hat{\mathbf{z}}}(0) &= e^{i\frac{\alpha}{2}} \left(\cos\frac{\alpha}{2} \mathbf{1} - i\sin\frac{\alpha}{2} \hat{\mathbf{z}} \cdot \boldsymbol{\sigma} \right) \\ &= e^{i\frac{\alpha}{2}} \left(\cos\frac{\alpha}{2} \mathbf{1} - i\sin\frac{\alpha}{2} \sigma_z \right) \end{aligned}$$

$$= e^{i\frac{\alpha}{2}} \begin{pmatrix} \cos\frac{\alpha}{2} - i\sin\frac{\alpha}{2} & 0 \\ 0 & \cos\frac{\alpha}{2} + i\sin\frac{\alpha}{2} \end{pmatrix}$$

$$= \begin{pmatrix} 1 & 0 \\ 0 & e^{i\alpha} \end{pmatrix}, \tag{2.148}$$

or

$$e^{i\frac{\pi}{2}} D_{\hat{z}}(\beta) D_{\hat{y}}(\pi) D_{\hat{z}}(\beta + \pi) = e^{i\frac{\pi}{2}} \begin{pmatrix} e^{-i(\beta+\frac{\pi}{2})}\cos\frac{\pi}{2} & -e^{i\frac{\pi}{2}}\sin\frac{\pi}{2} \\ e^{-i\frac{\pi}{2}}\sin\frac{\pi}{2} & e^{i(\beta+\frac{\pi}{2})}\cos\frac{\pi}{2} \end{pmatrix}$$

$$= \begin{pmatrix} 0 & 1 \\ 1 & 0 \end{pmatrix}$$

$$= \sigma_x, \tag{2.149}$$

or else

$$e^{i\frac{3\pi}{2}} D_{\hat{z}}(0) D_{\hat{y}}\left(\frac{\pi}{2}\right) D_{\hat{z}}(-\pi) = -i\left(\cos\frac{\pi}{4}\mathbf{1} - i\sin\frac{\pi}{4}\sigma_y\right)\left(\cos\frac{-\pi}{2}\mathbf{1} - i\sin\frac{-\pi}{2}\sigma_z\right)$$

$$= -i\left(\frac{1}{\sqrt{2}}\mathbf{1} - \frac{i}{\sqrt{2}}\sigma_y\right)i\sigma_z = \frac{1}{\sqrt{2}}\Big(\sigma_z - \underbrace{i\sigma_y\sigma_z}_{=i\sigma_x}\Big)$$

$$= \frac{\sigma_z + \sigma_x}{\sqrt{2}}. \tag{2.150}$$

Inverses of spin-rotations around the $\hat{z}$- and $\hat{y}$-axis can be obtained by left- and right-multiplication with σ_x.

Exercise 2.34 Show that

$$\sigma_x D_{\hat{y}}(\eta)\sigma_x = D_{\hat{y}}(-\eta)$$
$$\sigma_x D_{\hat{z}}(\eta)\sigma_x = D_{\hat{z}}(-\eta). \tag{2.151}$$

For a solution see Solution 2.34.

The claim in the following Lemma 2.34 plays an important role in connection with quantum gates, which we shall consider in more depth in Sect. 5.2.

Lemma 2.34 *For every unitary operator U on $\mathbb{H}$ there exist operators A, B and C on $\mathbb{H}$ and $\alpha \in \mathbb{R}$ such that*

$$ABC = 1$$
$$U = e^{i\alpha} A\sigma_x B\sigma_x C.$$

Proof From Exercise 2.33 we know that there exist $\alpha, \beta, \gamma, \delta \in \mathbb{R}$ such that

$$U = e^{i\alpha} D_{\hat{z}}(\beta) D_{\hat{y}}(\gamma) D_{\hat{z}}(\delta). \tag{2.152}$$

We thus set

$$
\begin{aligned}
A &:= D_{\hat{z}}(\beta) D_{\hat{y}}\left(\frac{\gamma}{2}\right) \\
B &:= D_{\hat{y}}\left(-\frac{\gamma}{2}\right) D_{\hat{z}}\left(-\frac{\delta + \beta}{2}\right) \\
C &:= D_{\hat{z}}\left(\frac{\delta - \beta}{2}\right).
\end{aligned}
\tag{2.153}
$$

Then it follows that

$$
\begin{aligned}
ABC \underset{(2.153)}{=} \; & D_{\hat{z}}(\beta) \underbrace{D_{\hat{y}}\left(\frac{\gamma}{2}\right) D_{\hat{y}}\left(-\frac{\gamma}{2}\right)}_{=D_{\hat{y}}(0)=1} \underbrace{D_{\hat{z}}\left(-\frac{\delta+\beta}{2}\right) D_{\hat{z}}\left(\frac{\delta-\beta}{2}\right)}_{=D_{\hat{z}}\left(-\frac{\delta+\beta}{2}+\frac{\delta-\beta}{2}\right)=D_{\hat{z}}(-\beta)} \\
= \; & D_{\hat{z}}(\beta) D_{\hat{z}}(-\beta) \\
= \; & 1
\end{aligned}
$$

and

$$
\begin{aligned}
& e^{i\alpha} A \sigma_x B \sigma_x C \\
\underset{(2.153)}{=} \; & e^{i\alpha} D_{\hat{z}}(\beta) D_{\hat{y}}\left(\frac{\gamma}{2}\right) \sigma_x D_{\hat{y}}\left(-\frac{\gamma}{2}\right) D_{\hat{z}}\left(-\frac{\delta+\beta}{2}\right) \sigma_x D_{\hat{z}}\left(\frac{\delta-\beta}{2}\right) \\
= \; & e^{i\alpha} D_{\hat{z}}(\beta) D_{\hat{y}}\left(\frac{\gamma}{2}\right) \sigma_x D_{\hat{y}}\left(-\frac{\gamma}{2}\right) \overbrace{\sigma_x \sigma_x}^{=1} D_{\hat{z}}\left(-\frac{\delta+\beta}{2}\right) \sigma_x D_{\hat{z}}\left(\frac{\delta-\beta}{2}\right) \\
= \; & e^{i\alpha} D_{\hat{z}}(\beta) D_{\hat{y}}\left(\frac{\gamma}{2}\right) \underbrace{\sigma_x D_{\hat{y}}\left(-\frac{\gamma}{2}\right) \sigma_x}_{=D_{\hat{y}}\left(\frac{\gamma}{2}\right)} \underbrace{\sigma_x D_{\hat{z}}\left(-\frac{\delta+\beta}{2}\right) \sigma_x}_{=D_{\hat{z}}\left(\frac{\delta+\beta}{2}\right)} D_{\hat{z}}\left(\frac{\delta-\beta}{2}\right) \\
\underset{(2.151)}{=} \; & e^{i\alpha} D_{\hat{z}}(\beta) \underbrace{D_{\hat{y}}\left(\frac{\gamma}{2}\right) D_{\hat{y}}\left(\frac{\gamma}{2}\right)}_{=D_{\hat{y}}(\gamma)} \underbrace{D_{\hat{z}}\left(\frac{\delta+\beta}{2}\right) D_{\hat{z}}\left(\frac{\delta-\beta}{2}\right)}_{=D_{\hat{z}}(\delta)} \\
\underset{(2.132)}{=} \; & e^{i\alpha} D_{\hat{z}}(\beta) D_{\hat{y}}(\gamma) D_{\hat{z}}(\delta) \\
\underset{(2.152)}{=} \; & U.
\end{aligned}
$$

□

Indeed, for every unitary operator U on $\mathbb{H}$ we can always find a suitable unit vector $\hat{\mathbf{n}} \in S^1_{\mathbb{R}^3}$ and angles $\alpha, \xi \in \mathbb{R}$ such that U can be written as a product of a phase factor $e^{i\alpha}$ and a spin-rotation with angle ξ around $\hat{\mathbf{n}}$. This is shown in the following Lemma 2.35.

Lemma 2.35 *Let U be a unitary operator on $\mathbb{H}$. Then there exist $\alpha, \xi \in \mathbb{R}$ and $\hat{\mathbf{n}} \in S^1_{\mathbb{R}^3}$ such that*

$$U = e^{i\alpha} D_{\hat{\mathbf{n}}}(\xi).$$

Proof From Lemma 2.32 we know that there exist $\alpha, \beta, \delta, \gamma \in \mathbb{R}$ such that in the standard basis $\{|0\rangle, |1\rangle\}$ the matrix of U is given by

$$
\begin{aligned}
U &= e^{i\alpha}
\begin{pmatrix}
e^{-i\frac{\beta+\delta}{2}} \cos\frac{\gamma}{2} & -e^{i\frac{\delta-\beta}{2}} \sin\frac{\gamma}{2} \\
\exp i\frac{\beta-\delta}{2} \sin\frac{\gamma}{2} & \exp i\frac{\beta+\delta}{2} \cos\frac{\gamma}{2}
\end{pmatrix} \\
&= e^{i\alpha}
\begin{pmatrix}
(\cos\frac{\beta+\delta}{2} - i\sin\frac{\beta+\delta}{2})\cos\frac{\gamma}{2} & -(\cos\frac{\delta-\beta}{2} + i\sin\frac{\delta-\beta}{2})\sin\frac{\gamma}{2} \\
(\cos\frac{\beta-\delta}{2} + i\sin\frac{\beta-\delta}{2})\sin\frac{\gamma}{2} & (\cos\frac{\beta+\delta}{2} + i\sin\frac{\beta+\delta}{2})\cos\frac{\gamma}{2}
\end{pmatrix} \quad (2.154) \\
&= e^{i\alpha}\Bigg(\cos\frac{\beta+\delta}{2} \cos\frac{\gamma}{2} \mathbf{1} \\
&\quad - i\left[\sin\frac{\beta+\delta}{2} \cos\frac{\gamma}{2}\sigma_z + \cos\frac{\delta-\beta}{2} \sin\frac{\gamma}{2}\sigma_y + \sin\frac{\delta-\beta}{2} \sin\frac{\gamma}{2}\sigma_x \right] \Bigg).
\end{aligned}
$$

We now find θ and ϕ in

$$
\hat{\mathbf{n}} = \hat{\mathbf{n}}(\theta, \phi) = \begin{pmatrix} \sin\theta\cos\phi \\ \sin\theta\sin\phi \\ \cos\theta \end{pmatrix},
$$

and ξ such that $e^{-i\alpha} U = D_{\hat{\mathbf{n}}(\theta,\phi)}(\xi)$. For this we first choose a $\tilde{\xi}$ such that

$$
\cos\frac{\tilde{\xi}}{2} = \cos\frac{\delta+\beta}{2} \cos\frac{\gamma}{2}. \quad (2.155)
$$

Then we have

$$
\begin{aligned}
\left| \sin\frac{\tilde{\xi}}{2} \right| &= \sqrt{1 - \cos^2\frac{\tilde{\xi}}{2}} = \sqrt{1 - \cos^2\frac{\delta+\beta}{2}\cos^2\frac{\gamma}{2}} \\
&\geq \sqrt{1 - \cos^2\frac{\gamma}{2}} = \left| \sin\frac{\gamma}{2} \right|.
\end{aligned}
$$

We choose $\xi = \tilde{\xi}$ if $\sin\frac{\tilde{\xi}}{2}$ and $\sin\frac{\gamma}{2}$ have the same sign and $\xi = -\tilde{\xi}$ otherwise. Then there exist $\theta_1 \in [0, \frac{\pi}{2}]$ and $\theta_2 = \pi - \theta_1 \in [\frac{\pi}{2}, \pi]$ such that

$$\sin\theta_j \sin\frac{\xi}{2} = \sin\frac{\gamma}{2}, \qquad j \in \{1,2\}.$$

With this choice of ξ then (2.155) holds for ξ as well, and so far we have altogether

$$\cos\frac{\xi}{2} = \cos\frac{\delta+\beta}{2}\cos\frac{\gamma}{2} \tag{2.156}$$

$$\sin\theta_j \sin\frac{\xi}{2} = \sin\frac{\gamma}{2}. \tag{2.157}$$

From (2.157) it follows that

$$(1 - \cos^2\theta_j)\sin^2\frac{\xi}{2} = 1 - \cos^2\frac{\gamma}{2},$$

and this in turn implies

$$
\begin{aligned}
\cos^2\theta_j \sin^2\frac{\xi}{2} &= \cos^2\frac{\gamma}{2} + \sin^2\frac{\xi}{2} - 1 = \cos^2\frac{\gamma}{2} - \cos^2\frac{\xi}{2} \\
&\underbrace{=}_{(2.156)} \left(1 - \cos^2\frac{\delta+\beta}{2}\right)\cos^2\frac{\gamma}{2} \\
&= \sin^2\frac{\delta+\beta}{2}\cos^2\frac{\gamma}{2}.
\end{aligned}
$$

Thus, we have

$$\left|\cos\theta_j \sin\frac{\xi}{2}\right| = \left|\sin\frac{\delta+\beta}{2}\cos\frac{\gamma}{2}\right|.$$

If $\sin\frac{\xi}{2}$ and $\sin\frac{\delta+\beta}{2}\cos\frac{\gamma}{2}$ have the same sign, we set $\theta = \theta_1$, otherwise $\theta = \theta_2$, such that in every case

$$\cos\theta \sin\frac{\xi}{2} = \sin\frac{\delta+\beta}{2}\cos\frac{\gamma}{2}.$$

We now set $\phi := \frac{\beta-\delta+\pi}{2}$ such that

$$\sin\phi = \sin\frac{\beta-\delta+\pi}{2} = \cos\frac{\beta-\delta}{2}$$

$$\cos\phi = \cos\frac{\beta-\delta+\pi}{2} = -\sin\frac{\beta-\delta}{2} = \sin\frac{\delta-\beta}{2}.$$

Altogether, we thus have in (2.155)

$$\cos\frac{\beta+\delta}{2}\cos\frac{\gamma}{2}=\cos\frac{\xi}{2}$$

$$\sin\frac{\beta+\delta}{2}\cos\frac{\gamma}{2}=\sin\frac{\xi}{2}\cos\theta=\sin\frac{\xi}{2}\hat{n}_z$$

$$\cos\frac{\beta-\delta}{2}\sin\frac{\gamma}{2}=\sin\frac{\xi}{2}\sin\theta\sin\phi=\sin\frac{\xi}{2}\hat{n}_y$$

$$\sin\frac{\delta-\beta}{2}\sin\frac{\gamma}{2}=\sin\frac{\xi}{2}\sin\theta\cos\phi=\sin\frac{\xi}{2}\hat{n}_x,$$

and finally

$$\mathrm{e}^{-\mathrm{i}\alpha}U=\cos\frac{\xi}{2}\mathbf{1}-\mathrm{i}\sin\frac{\xi}{2}\hat{\mathbf{n}}\cdot\boldsymbol{\sigma}$$
$$=D_{\hat{\mathbf{n}}}(\xi).\qquad\square$$

Lemma 2.35 implies that any $U\in\mathcal{U}(\mathbb{H})$ can be expressed as a linear combination of the unit matrix and the PAULI matrices as is to be shown in the following exercise.

Exercise 2.35 Show that every $A\in L(\mathbb{H})$ can be written in the form

$$A=z_0\mathbf{1}+\mathbf{z}\cdot\boldsymbol{\sigma}=\sum_{\alpha=0}^{3}z_\alpha\sigma_\alpha,$$

where $z_0\in\mathbb{C}$ and $\mathbf{z}\in\mathbb{C}^3$, and in the last equation we used the notation (2.75) with $\sigma_0=\mathbf{1}$.

Moreover, show that if $A\in\mathcal{U}(\mathbb{H})$ then the z_α have to satisfy

$$|z_0|^2+|\mathbf{z}|^2=1.$$

For a solution see Solution 2.35.

From Lemma 2.35 it also follows as a corollary that every unitary operator U on $\mathbb{H}$ has an—albeit not necessarily unique—root.

Corollary 2.36 *Every $U\in\mathcal{U}(\mathbb{H})$ has a root, that is, there exists an operator $\sqrt{U}\in\mathcal{U}(\mathbb{H})$ such that*
$$\left(\sqrt{U}\right)^2=U.$$

Proof From Lemma 2.35 we know that there exist $\alpha, \xi \in \mathbb{R}$ and $\hat{\mathbf{n}} \in S^1_{\mathbb{R}^3}$ such that

$$U = \mathrm{e}^{\mathrm{i}\alpha} D_{\hat{\mathbf{n}}}(\xi)\,.$$

With this we choose

$$\sqrt{U} = \mathrm{e}^{\mathrm{i}\frac{\alpha}{2}} D_{\hat{\mathbf{n}}}\left(\frac{\xi}{2}\right)\,.$$

From Lemma 2.31 we know that $D_{\hat{\mathbf{n}}}\left(\frac{\xi}{2}\right) \in \mathcal{U}(\mathbb{H})$, and since $\mathrm{e}^{\mathrm{i}\frac{\alpha}{2}} \in \mathcal{U}(\mathbb{H})$, it follows that $\sqrt{U} \in \mathcal{U}(\mathbb{H})$ as well. Moreover, we find

$$\left(\sqrt{U}\right)^2 = \mathrm{e}^{\mathrm{i}\frac{\alpha}{2}} D_{\hat{\mathbf{n}}}\left(\frac{\xi}{2}\right) \mathrm{e}^{\mathrm{i}\frac{\alpha}{2}} D_{\hat{\mathbf{n}}}\left(\frac{\xi}{2}\right) \underbrace{=}_{(2.132)} \mathrm{e}^{\mathrm{i}\alpha} D_{\hat{\mathbf{n}}}(\xi) = U\,. \qquad \square$$

Example 2.37 Consider, for example,

$$\mathrm{e}^{\mathrm{i}\frac{\pi}{2}} D_{\hat{\mathbf{x}}}(\pi) = \mathrm{i}\left(\cos\frac{\pi}{2}\mathbf{1} - \mathrm{i}\sin\frac{\pi}{2}\hat{\mathbf{x}}\cdot\boldsymbol{\sigma}\right) = \hat{\mathbf{x}}\cdot\boldsymbol{\sigma} = \sigma_x \qquad (2.158)$$

such that

$$\sqrt{\sigma_x} = \mathrm{e}^{\mathrm{i}\frac{\pi}{4}} D_{\hat{\mathbf{x}}}\left(\frac{\pi}{2}\right) \underbrace{=}_{(2.130)} \frac{1+\mathrm{i}}{\sqrt{2}}\left(\cos\frac{\pi}{4}\mathbf{1} - \mathrm{i}\sin\frac{\pi}{4}\hat{\mathbf{x}}\cdot\boldsymbol{\sigma}\right)$$

$$= \frac{1+\mathrm{i}}{\sqrt{2}}\left(\frac{1}{\sqrt{2}}\mathbf{1} - \mathrm{i}\frac{1}{\sqrt{2}}\sigma_x\right) = \frac{1+\mathrm{i}}{2}(\mathbf{1} - \mathrm{i}\sigma_x)$$

$$= \frac{1+\mathrm{i}}{2}\begin{pmatrix} 1 & -\mathrm{i} \\ -\mathrm{i} & 1 \end{pmatrix}$$

and one verifies

$$\left(\frac{1+\mathrm{i}}{2}(\mathbf{1} - \mathrm{i}\sigma_x)\right)^2 = \frac{1+2\mathrm{i}+\mathrm{i}^2}{4}(\mathbf{1} - 2\mathrm{i}\sigma_x + \mathrm{i}^2\underbrace{\sigma_x^2}_{=1}) = \frac{2\mathrm{i}}{4}(-2\mathrm{i}\sigma_x) = \sigma_x\,.$$

On the other hand, we also have

$$\left(\frac{1-\mathrm{i}}{2}(\mathbf{1} + \mathrm{i}\sigma_x)\right)^2 = \frac{1-2\mathrm{i}+\mathrm{i}^2}{4}(\mathbf{1} + 2\mathrm{i}\sigma_x + \mathrm{i}^2\underbrace{\sigma_x^2}_{=1}) = \frac{-2\mathrm{i}}{4}(2\mathrm{i}\sigma_x) = \sigma_x\,.$$

Another widely used operator on $\mathbb{H}$ is the HADAMARD transformation, which is also known as the WALSH–HADAMARD transformation.

Definition 2.38 The HADAMARD **transformation** is defined as

$$H := \frac{\sigma_x + \sigma_z}{\sqrt{2}} : \mathbb{H} \to \mathbb{H}.$$

A few useful properties of the HADAMARD transformation are collected in the following lemma.

Lemma 2.39 *In the basis* $\{|0\rangle, |1\rangle\}$ *the* HADAMARD *transformation has the matrix*

$$H = \frac{1}{\sqrt{2}} \begin{pmatrix} 1 & 1 \\ 1 & -1 \end{pmatrix} \tag{2.159}$$

and satisfies

$$H|0\rangle = \frac{|0\rangle + |1\rangle}{\sqrt{2}} \tag{2.160}$$

$$H|1\rangle = \frac{|0\rangle - |1\rangle}{\sqrt{2}} \tag{2.161}$$

$$H|x_j\rangle = \frac{|0\rangle + e^{\pi i x_j}|1\rangle}{\sqrt{2}} \tag{2.162}$$

$$H^2 = 1, \tag{2.163}$$

as well as

$$H = e^{i\frac{3\pi}{2}} D_{\hat{z}}(0) D_{\hat{y}}\left(\frac{\pi}{2}\right) D_{\hat{z}}(-\pi). \tag{2.164}$$

Proof The claim (2.159) follows immediately from the Definition 2.38 of H and the PAULI matrices in (2.74). This implies (2.160) and (2.161), and because of $x_j \in \{0, 1\}$ these in turn imply (2.162). Equation (2.163) is easily verified by taking the square of the matrix in (2.159).

The representation (2.164) of the HADAMARD transformation by a multiplication of a phase-factor and a spin-rotation was already shown in (2.150).   □

2.6 Further Reading

For the reader who would like to get more background on functional analysis in general and HILBERT spaces in particular a good entry level exposition can be found in the book by RYNNE and YOUNGSON [53]. The book by KREYSZIG [54] introduces the reader to the most important elements of functional analysis at the same level. More advanced is the first tome in the series by REED and SIMON [50], which also contains extensive chapters on topological and convex spaces as well as unbounded operators. An excellent reference way beyond functional analysis and HILBERT spaces is the book by CHOQUET-BRUHAT and DEWITT-MORETTE [55]. This collects a lot of material in mathematical physics and presents it in a concise, yet rigorous, fashion, albeit most without proofs.

As for quantum mechanics, there are, of course, a great number of books. The two volumes by GALINDO and PASCUAL [37] contain sufficient mathematical rigor and cover a great breadth of topics. Similarly modern in style and broad in coverage of topics is the two-volume set by COHEN-TANNOUDJI, DIU and LALOE [39]. Neither of them, however, specializes in quantum computing. Rather, both present quantum mechanics from its historical origins and postulates through to the myriad of special topics, such as the hydrogen atom, symmetry transformations, angular momentum, perturbation theory, etc.

Chapter 3
Tensor Products and Composite Systems

3.1 Towards Qbytes

Classically, information is represented by finite chunks of bits—such as *bytes*—and multiples thereof. These are essentially words $(x_1, x_2, x_3, \ldots, x_n)$ built from the alphabet $\{0,1\} \ni x_l$, where $l \in \{1, \ldots, n\}$. Hence, we need 2^n classical storage configurations in order to represent all such words.

A classical two-bit word (x_1, x_2) is an element of the set $\{0,1\} \times \{0,1\} = \{0,1\}^2$, and classically we can represent the words 00, 01, 10, 11 by storing the first letter x_1 (the first bit) and the second letter x_2 (the second bit) accordingly. If we represent each of these bits quantum mechanically by qubits, we are dealing with a two-qubit quantum system composed of two quantum mechanical sub-systems.

Many quantum mechanical systems are composed of several parts, each of which is again a quantum mechanical system. The hydrogen atom, for example, consists of a proton and an electron. Let the states of the proton be given by elements of a HILBERT space $\mathbb{H}^P$ and those of the electron by $\mathbb{H}^E$. What, then, is the HILBERT space of the hydrogen atom? The answer is: the *tensor product* $\mathbb{H}^P \otimes \mathbb{H}^E$ of the HILBERT spaces of the sub-systems.[1] The tensor product $\mathbb{H}^A \otimes \mathbb{H}^B$ of two HILBERT spaces $\mathbb{H}^A$ and $\mathbb{H}^B$ is again a HILBERT space and provides the state-space for the quantum mechanical description of the total system composed of the sub-systems $\mathbb{H}^A$ and $\mathbb{H}^B$. Before we turn our attention to multi-qubit systems, we shall thus first review tensor products of HILBERT spaces.

---

[1] Quite often the proton is viewed as an object (fixed at a place in space), which exerts a COULOMB force on the electron. In this approximation the state of the proton remains unchanged and one considers only the effects on the electron such that $\mathbb{H}^E$ suffices for the description of the total system. A more precise description includes the reaction of the electron on the proton and uses center-of-gravity and relative coordinates. For isolated systems the center of gravity changes trivially (in other words, maintains constant velocity), and the corresponding HILBERT space is then ignored in this description as well.

© Springer Nature Switzerland AG 2019
W. Scherer, *Mathematics of Quantum Computing*,
https://doi.org/10.1007/978-3-030-12358-1_3

3.2 Tensor Products of HILBERT Spaces

3.2.1 Definition

Here we give a more informal definition of the tensor product of two finite-dimensional HILBERT spaces. This is sufficient for our purposes. For a strict and generally valid version that includes the infinite-dimensional case the reader is referred to [50]. More important than the most general definition, however, it is for us here that we can give the calculation-rules for tensor products, such as for the calculation of the scalar product, with the help of the known rules for the sub-systems.

Let $\mathbb{H}^A$ and $\mathbb{H}^B$ be HILBERT spaces, $|\varphi\rangle \in \mathbb{H}^A$ and $|\psi\rangle \in \mathbb{H}^B$ vectors in these, and define

$$
\begin{aligned}
|\varphi\rangle \otimes |\psi\rangle : \mathbb{H}^A \times \mathbb{H}^B &\longrightarrow \mathbb{C} \\
(\xi, \eta) &\longmapsto \langle \xi | \varphi \rangle^{\mathbb{H}^A} \langle \eta | \psi \rangle^{\mathbb{H}^B} \cdot
\end{aligned}
\tag{3.1}
$$

This map is anti-linear in ξ and η and continuous. Define the set of all such maps and denote it by

$$
\mathbb{H}^A \otimes \mathbb{H}^B := \{ \Psi : \mathbb{H}^A \times \mathbb{H}^B \to \mathbb{C} \,|\, \text{anti-linear and continuous} \} .
\tag{3.2}
$$

This is a vector space over $\mathbb{C}$ since for $\Psi_1, \Psi_2 \in \mathbb{H}^A \otimes \mathbb{H}^B$ and $a, b \in \mathbb{C}$ the map defined by

$$
\left(a\Psi_1 + b\Psi_2 \right)(\xi, \eta) := a\Psi_1(\xi, \eta) + b\Psi_2(\xi, \eta)
\tag{3.3}
$$

is also in $\mathbb{H}^A \otimes \mathbb{H}^B$. The null-map is the null-vector, and $-\Psi$ is the additive-inverse vector for Ψ. According to (3.1) then $|\varphi\rangle \otimes |\psi\rangle$ is a vector in the vector space of the anti-linear and continuous maps $\mathbb{H}^A \otimes \mathbb{H}^B$ from $\mathbb{H}^A \times \mathbb{H}^B$ to $\mathbb{C}$ as defined in (3.2).

Exercise 3.36 Let $|\varphi\rangle \in \mathbb{H}^A$ and $|\psi\rangle \in \mathbb{H}^B$ and $a, b \in \mathbb{C}$. Verify the following identities.

$$
\begin{aligned}
(a|\varphi\rangle) \otimes |\psi\rangle &= |\varphi\rangle \otimes (a|\psi\rangle) = a(|\varphi\rangle \otimes |\psi\rangle) \\
a(|\varphi\rangle \otimes |\psi\rangle) + b(|\varphi\rangle \otimes |\psi\rangle) &= (a + b)\,|\varphi\rangle \otimes |\psi\rangle \\
(|\varphi_1\rangle + |\varphi_2\rangle) \otimes |\psi\rangle &= |\varphi_1\rangle \otimes |\psi\rangle + |\varphi_2\rangle \otimes |\psi\rangle \\
|\varphi\rangle \otimes (|\psi_1\rangle + |\psi_2\rangle) &= |\varphi\rangle \otimes |\psi_1\rangle + |\varphi\rangle \otimes |\psi_2\rangle .
\end{aligned}
$$

For a solution see Solution 3.36.

In order to simplify the notation we shall also write

$$
|\varphi \otimes \psi\rangle := |\varphi\rangle \otimes |\psi\rangle .
$$

For vectors $|\varphi_k\rangle \otimes |\psi_k\rangle \in \mathbb{H}^A \otimes \mathbb{H}^B$ with $k \in \{1,2\}$ and $|\varphi_k\rangle \in \mathbb{H}^A$, $|\psi_k\rangle \in \mathbb{H}^B$ we define

$$\langle \varphi_1 \otimes \psi_1 | \varphi_2 \otimes \psi_2 \rangle := \langle \varphi_1 | \varphi_2 \rangle^{\mathbb{H}^A} \langle \psi_1 | \psi_2 \rangle^{\mathbb{H}^B}, \tag{3.4}$$

where in the following we shall often omit the superscripts, which indicate in which HILBERT space the scalar product is to be calculated. With (3.4) we have a scalar product for vectors of the form $|\varphi\rangle \otimes |\psi\rangle$ in $\mathbb{H}^A \otimes \mathbb{H}^B$. In order to define a scalar product for all $\Psi \in \mathbb{H}^A \otimes \mathbb{H}^B$ we consider ONBs in the subspaces. Let $\{|e_a\rangle\} \subset \mathbb{H}^A$ be an ONB in $\mathbb{H}^A$ and $\{|f_b\rangle\} \subset \mathbb{H}^B$ be an ONB in $\mathbb{H}^B$. The set $\{|e_a\rangle \otimes |f_b\rangle\} \subset \mathbb{H}^A \otimes \mathbb{H}^B$ is then orthonormal since

$$\langle e_{a_1} \otimes f_{b_1} | e_{a_2} \otimes f_{b_2} \rangle \underset{(3.4)}{=} \langle e_{a_1} | e_{a_2} \rangle \langle f_{b_1} | f_{b_2} \rangle \underset{(2.10)}{=} \delta_{a_1 a_2} \delta_{b_1 b_2}. \tag{3.5}$$

Considering an arbitrary vector $\Psi \in \mathbb{H}^A \otimes \mathbb{H}^B$, one has for this anti-linear map

$$\begin{aligned}
\Psi(\xi, \eta) &= \Psi\left(\sum_a |e_a\rangle\langle e_a|\xi\rangle, \sum_b |f_b\rangle\langle f_b|\eta\rangle \right) \\
&= \sum_{a,b} \underbrace{\Psi(|e_a\rangle, |f_b\rangle)}_{=:\Psi_{ab} \in \mathbb{C}} \langle \xi | e_a \rangle \langle \eta | f_b \rangle \\
&= \sum_{a,b} \Psi_{ab}(|e_a\rangle \otimes |f_b\rangle)(\xi, \eta) \\
&= \sum_{a,b} \Psi_{ab} | e_a \otimes f_b \rangle(\xi, \eta).
\end{aligned}$$

This proves that every vector $|\Psi\rangle \in \mathbb{H}^A \otimes \mathbb{H}^B$ can be written as a linear combination of the form[2]

$$|\Psi\rangle = \sum_{a,b} \Psi_{ab} |e_a \otimes f_b\rangle. \tag{3.6}$$

Exercise 3.37 Let $\{|e_a\rangle\} \subset \mathbb{H}^A$ and $\{|f_b\rangle\} \subset \mathbb{H}^B$ be ONBs. Show that the set $\{|e_a \otimes f_b\rangle\}$ is linearly independent in $\mathbb{H}^A \otimes \mathbb{H}^B$.

For a solution see Solution 3.37.

The scalar product of $|\Psi\rangle$ in (3.6) and a vector

$$|\Phi\rangle = \sum_{a,b} \Phi_{ab} |e_a \otimes f_b\rangle$$

---

[2]With possibly an infinite number of terms in the infinite-dimensional case.

is then defined with (3.5) as

$$\langle \Psi | \Phi \rangle = \sum_{a_1,b_1} \sum_{a_2,b_2} \overline{\Psi_{a_1 b_1}} \Phi_{a_2 b_2} \langle e_{a_1} \otimes f_{b_1} | e_{a_2} \otimes f_{b_2} \rangle$$
$$= \sum_{a,b} \overline{\Psi_{ab}} \Phi_{ab} . \tag{3.7}$$

The scalar product thus defined on all of $\mathbb{H}^A \otimes \mathbb{H}^B$ is positive-definite and independent of the choice of the ONBs, as is to be shown in Exercise 3.38.

Exercise 3.38 Show that $\langle \Psi | \Phi \rangle$ as defined in (3.7) is positive-definite and does not depend on the choice of the ONBs $\{|e_a\rangle\} \subset \mathbb{H}^A$ and $\{|f_b\rangle\} \subset \mathbb{H}^B$.

For a solution see Solution 3.38.

The bra-vector belonging to $|\Psi\rangle$ in (3.6) is then

$$\langle \Psi | = \sum_{a,b} \overline{\Psi_{ab}} \langle e_a \otimes f_b | \tag{3.8}$$

and acts as in (3.7) on a $|\Phi\rangle \in \mathbb{H}^A \otimes \mathbb{H}^B$.

The norm of $|\Psi\rangle \in \mathbb{H}^A \otimes \mathbb{H}^B$ is calculated as

$$||\Psi||^2 = \langle \Psi | \Psi \rangle \underbrace{=}_{(3.7)} \sum_{a,b} |\Psi_{ab}|^2 , \tag{3.9}$$

and for any $|\varphi\rangle \in \mathbb{H}^A$ and $|\psi\rangle \in \mathbb{H}^B$ we have

$$||\varphi \otimes \psi|| \underbrace{=}_{(2.5)} \sqrt{\langle \varphi \otimes \psi | \varphi \otimes \psi \rangle} \underbrace{=}_{(3.4)} \sqrt{\langle \varphi | \varphi \rangle \langle \psi | \psi \rangle} \underbrace{=}_{(2.5)} ||\varphi|| \, ||\psi|| . \tag{3.10}$$

Hence, $\mathbb{H}^A \otimes \mathbb{H}^B$ is a complex vector space with scalar product (3.7), which induces a norm (3.9). For finite-dimensional subspaces then $\mathbb{H}^A \otimes \mathbb{H}^B$ is complete in this norm and thus according to Definition 2.1 a HILBERT space.[3] For our purposes it suffices to view $\mathbb{H}^A \otimes \mathbb{H}^B$ as the set of linear combinations of the form (3.6) with $\sum_{a,b} |\Psi_{ab}|^2 < \infty$ and the calculation rules (3.7) and (3.9).

---

[3]Only in the case of infinite-dimensional subspaces the space $\mathbb{H}^A \otimes \mathbb{H}^B$ needs to be completed in this norm (see [50]) for it to become a HILBERT space.

Definition 3.1 The HILBERT space $\mathbb{H}^A \otimes \mathbb{H}^B$ with the scalar product (3.7) is called the **tensor product** of the HILBERT spaces $\mathbb{H}^A$ and $\mathbb{H}^B$.

Proposition 3.2 *Let* $\{|e_a\rangle\} \subset \mathbb{H}^A$ *be an ONB in* $\mathbb{H}^A$ *and* $\{|f_b\rangle\} \subset \mathbb{H}^B$ *be an ONB in* $\mathbb{H}^B$. *The set* $\{|e_a \otimes f_b\rangle\} = \{|e_a\rangle \otimes |f_b\rangle\}$ *forms an ONB in* $\mathbb{H}^A \otimes \mathbb{H}^B$, *and for finite-dimensional* $\mathbb{H}^A$ *and* $\mathbb{H}^B$ *one has*

$$\dim\left(\mathbb{H}^A \otimes \mathbb{H}^B\right) = \dim \mathbb{H}^A \dim \mathbb{H}^B.$$

Proof From Exercise 3.37 we know that the set $\{|e_a \otimes f_b\rangle\}$ is linearly independent and from (3.6) that every $|\Psi\rangle \in \mathbb{H}^A \otimes \mathbb{H}^B$ can be written as a linear combination of vectors from this set. Orthonormality of this set follows from (3.5).

The statement about dimensionality follows immediately from counting the elements in the set. $\qquad\square$

For several tensor products, such as $\mathbb{H}^A \otimes \mathbb{H}^B \otimes \mathbb{H}^C$, *associativity* holds

$$\left(\mathbb{H}^A \otimes \mathbb{H}^B\right) \otimes \mathbb{H}^C = \mathbb{H}^A \otimes \left(\mathbb{H}^B \otimes \mathbb{H}^C\right) = \mathbb{H}^A \otimes \mathbb{H}^B \otimes \mathbb{H}^C$$

and, accordingly,

$$\langle \varphi_1 \otimes \psi_1 \otimes \chi_1 | \varphi_2 \otimes \psi_2 \otimes \chi_2 \rangle = \langle \varphi_1 | \varphi_2 \rangle \langle \psi_1 | \psi_2 \rangle \langle \chi_1 | \chi_2 \rangle.$$

Likewise, with the ONBs $\{|e_a\rangle\} \subset \mathbb{H}^A$, $\{|f_b\rangle\} \subset \mathbb{H}^B$ and $\{|g_c\rangle\} \subset \mathbb{H}^C$ one has

$$|\Psi\rangle \in \mathbb{H}^A \otimes \mathbb{H}^B \otimes \mathbb{H}^C \qquad \Leftrightarrow \qquad |\Psi\rangle = \sum_{a,b,c} \Psi_{abc} |e_a \otimes f_b \otimes g_c\rangle$$

with the $\Psi_{abc} \in \mathbb{C}$ such that $\sum_{a,b,c} |\Psi_{abc}|^2 < \infty$.

Recall that in the case $\dim \mathbb{H} = n < \infty$ we identified a given basis $\{|e_j\rangle\} \subset \mathbb{H}$ in (2.23) with the standard basis in $\mathbb{C}^n$. As shown in (2.26) this then allowed us to express any vector in $\mathbb{H}$ with the help of the standard basis in $\mathbb{C}^n$. How does such a construction look like in a tensor product $\mathbb{H}^A \otimes \mathbb{H}^B$ with $\dim \mathbb{H}^X = n_X < \infty$ where $X \in \{A, B\}$ and ONBs $\{|e_a\rangle\} \subset \mathbb{H}^A$ and $\{|f_b\rangle\} \subset \mathbb{H}^B$? For this we assume that the isomorphisms $\mathbb{H}^X \cong \mathbb{C}^{n_X}$ with $X \in \{A, B\}$ are established by (2.23). We then establish the isomorphism $\mathbb{H}^A \otimes \mathbb{H}^B \cong \mathbb{C}^{n_A n_B}$ by identifying the basis $\{|e_a \otimes f_b\rangle\} \subset \mathbb{H}^A \otimes \mathbb{H}^B$ with the standard basis in $\mathbb{C}^{n_A n_B}$ as follows:

$$
|e_1 \otimes f_1\rangle =
\begin{matrix} 1 \\ 2 \\ \vdots \\ \\ n_A n_B \end{matrix}
\begin{pmatrix} 1 \\ 0 \\ \vdots \\ \vdots \\ 0 \end{pmatrix}, \quad
|e_1 \otimes f_2\rangle =
\begin{matrix} 1 \\ 2 \\ \vdots \\ \\ n_A n_B \end{matrix}
\begin{pmatrix} 0 \\ 1 \\ 0 \\ \vdots \\ 0 \end{pmatrix}, \quad \ldots,
$$

$$
|e_a \otimes f_b\rangle =
\begin{matrix} 1 \\ \vdots \\ \vdots \\ \\ (a-1)n_B + b \\ \\ \vdots \\ \vdots \\ n_A n_B \end{matrix}
\begin{pmatrix} 0 \\ \vdots \\ 0 \\ 1 \\ 0 \\ \vdots \\ 0 \end{pmatrix}, \quad \ldots, \quad
|e_{n_A} \otimes f_{n_B}\rangle =
\begin{matrix} 1 \\ 2 \\ \vdots \\ \\ n_A n_B \end{matrix}
\begin{pmatrix} 0 \\ 0 \\ \vdots \\ 0 \\ 1 \end{pmatrix}, \quad (3.11)
$$

where the columns to the left of the parenthesis show the row numbers for illustration and are not to be considered part of the equations. As in (2.23) and (2.24) the representation of $\langle e_a \otimes f_b|$ in $\mathbb{C}^{n_A n_B}$ is the transpose of the right side of (3.11). Note that our way of arranging the isomorphism means that we are essentially dividing the $n_A n_B$ rows of a vector in $\mathbb{C}^{n_A n_B}$ into n_A row-blocks of n_B rows. In other words, the first n_B vectors of the standard basis in $\mathbb{C}^{n_A n_B}$ are identified with

$$
|e_1 \otimes f_1\rangle, |e_1 \otimes f_2\rangle, \ldots, |e_1 \otimes f_{n_B}\rangle \, .
$$

The next n_B vectors of the standard basis in $\mathbb{C}^{n_A n_B}$ are identified with

$$
|e_2 \otimes f_1\rangle, |e_2 \otimes f_2\rangle, \ldots, |e_2 \otimes f_{n_B}\rangle
$$

and so on until the last n_B vectors of the standard basis in $\mathbb{C}^{n_A n_B}$ are identified with

$$
|e_{n_A} \otimes f_1\rangle, |e_{n_A} \otimes f_2\rangle, \ldots, |e_{n_A} \otimes f_{n_B}\rangle \, .
$$

For a general vector $|\Psi\rangle \in \mathbb{H}^A \otimes \mathbb{H}^B$ we thus have

$$
|\Psi\rangle = \sum_{a=1}^{n_A} \sum_{b=1}^{n_B} \Psi_{ab} |e_a \otimes f_b\rangle =
\begin{matrix} 1 \\ \vdots \\ \\ (a-1)n_B + b \\ \\ \vdots \\ n_A n_B \end{matrix}
\begin{pmatrix} \Psi_{11} \\ \vdots \\ \\ \Psi_{ab} \\ \\ \vdots \\ \Psi_{n_A n_B} \end{pmatrix} .
$$

Example 3.3 Consider the case $\mathbb{H}^A = \mathbb{H}^B = {}^1\!\mathbb{H} \cong \mathbb{C}^2$ with the ONBs

$$\{|e_a\rangle\} = \{|f_b\rangle\} = \{|0\rangle, |1\rangle\} = \left\{ \begin{pmatrix} 1 \\ 0 \end{pmatrix}, \begin{pmatrix} 0 \\ 1 \end{pmatrix} \right\},$$

where the rightmost set denotes the standard basis in $\mathbb{C}^2$. For $\mathbb{H}^A \otimes \mathbb{H}^B \cong \mathbb{C}^4$ we then have the ONB

$$\{|e_a \otimes f_b\rangle\} = \{|00\rangle, |01\rangle, |10\rangle, |11\rangle\} = \left\{ \begin{pmatrix} 1 \\ 0 \\ 0 \\ 0 \end{pmatrix}, \begin{pmatrix} 0 \\ 1 \\ 0 \\ 0 \end{pmatrix}, \begin{pmatrix} 0 \\ 0 \\ 1 \\ 0 \end{pmatrix}, \begin{pmatrix} 0 \\ 0 \\ 0 \\ 1 \end{pmatrix} \right\}, \quad (3.12)$$

where the rightmost set denotes the standard basis in $\mathbb{C}^4$. Moreover, for $j \in \{1, 2\}$ let $a_j, b_j \in \mathbb{C}$ and

$$|\varphi_1\rangle = a_1|0\rangle + b_1|1\rangle = \begin{pmatrix} a_1 \\ b_1 \end{pmatrix}, \qquad |\varphi_2\rangle = a_2|0\rangle + b_2|1\rangle = \begin{pmatrix} a_2 \\ b_2 \end{pmatrix}.$$

Then we have

$$\begin{aligned}
|\varphi_1\rangle \otimes |\varphi_2\rangle &= \left(a_1|0\rangle + b_1|1\rangle\right) \otimes \left(a_2|0\rangle + b_2|1\rangle\right) \\
&= a_1 a_2 |0\rangle \otimes |0\rangle + a_1 b_2 |0\rangle \otimes |1\rangle + b_1 a_2 |1\rangle \otimes |0\rangle + b_1 b_2 |1\rangle \otimes |1\rangle \\
&= a_1 a_2 |00\rangle + a_1 b_2 |01\rangle + b_1 a_2 |10\rangle + b_1 b_2 |11\rangle \\
&\underset{(3.12)}{=} \begin{pmatrix} a_1 a_2 \\ a_1 b_2 \\ b_1 a_2 \\ b_1 b_2 \end{pmatrix}.
\end{aligned} \qquad (3.13)$$

With a further $\mathbb{H}^C = {}^1\!\mathbb{H} \cong \mathbb{C}^2$ with the ONB.

$$\{|g_a\rangle\} = \{|0\rangle, |1\rangle\} = \left\{ \begin{pmatrix} 1 \\ 0 \end{pmatrix}, \begin{pmatrix} 0 \\ 1 \end{pmatrix} \right\}$$

we then find for $\mathbb{H}^A \otimes \mathbb{H}^B \otimes \mathbb{H}^C \cong \mathbb{C}^8$ the ONB

$$\{|e_a \otimes f_b \otimes g_a\rangle\} = \{|000\rangle, |001\rangle, |010\rangle, |011\rangle, |100\rangle, |101\rangle, |110\rangle, |111\rangle\}$$

$$= \left\{ \begin{pmatrix} 1 \\ 0 \\ 0 \\ 0 \\ 0 \\ 0 \\ 0 \\ 0 \end{pmatrix}, \begin{pmatrix} 0 \\ 1 \\ 0 \\ 0 \\ 0 \\ 0 \\ 0 \\ 0 \end{pmatrix}, \begin{pmatrix} 0 \\ 0 \\ 1 \\ 0 \\ 0 \\ 0 \\ 0 \\ 0 \end{pmatrix}, \begin{pmatrix} 0 \\ 0 \\ 0 \\ 1 \\ 0 \\ 0 \\ 0 \\ 0 \end{pmatrix}, \begin{pmatrix} 0 \\ 0 \\ 0 \\ 0 \\ 1 \\ 0 \\ 0 \\ 0 \end{pmatrix}, \begin{pmatrix} 0 \\ 0 \\ 0 \\ 0 \\ 0 \\ 1 \\ 0 \\ 0 \end{pmatrix}, \begin{pmatrix} 0 \\ 0 \\ 0 \\ 0 \\ 0 \\ 0 \\ 1 \\ 0 \end{pmatrix}, \begin{pmatrix} 0 \\ 0 \\ 0 \\ 0 \\ 0 \\ 0 \\ 0 \\ 1 \end{pmatrix} \right\},$$

where the last set now denotes the standard basis in $\mathbb{C}^8$. Also, let $a_3, b_3 \in \mathbb{C}$ and

$$|\varphi_3\rangle = a_3|0\rangle + b_3|1\rangle .$$

Then we have

$$\begin{aligned}
|\varphi_1\rangle \otimes |\varphi_2\rangle \otimes |\varphi_3\rangle &= (a_1|0\rangle + b_1|1\rangle) \otimes (a_2|0\rangle + b_2|1\rangle) \otimes (a_3|0\rangle + b_3|1\rangle) \\
&= a_1 a_2 a_3 |000\rangle + a_1 a_2 b_3 |001\rangle + a_1 b_2 a_3 |010\rangle + a_1 b_2 b_3 |011\rangle \\
&\quad + b_1 a_2 a_3 |100\rangle + b_1 a_2 b_3 |101\rangle + b_1 b_2 a_3 |110\rangle + b_1 b_2 b_3 |111\rangle \\
&= \begin{pmatrix} a_1 a_2 a_3 \\ a_1 a_2 b_3 \\ a_1 b_2 a_3 \\ a_1 b_2 b_3 \\ b_1 a_2 a_3 \\ b_1 a_2 b_3 \\ b_1 b_2 a_3 \\ b_1 b_2 b_3 \end{pmatrix} .
\end{aligned}$$

Concerning the tensor product of bra-vectors, note that because of

$$\langle \varphi_1 \otimes \psi_1 | \left(|\varphi_2\rangle \otimes |\psi_2\rangle \right) = \langle \varphi_1 \otimes \psi_1 | \varphi_2 \otimes \psi_2 \rangle \underset{(3.4)}{=} \langle \varphi_1 | \varphi_2 \rangle \langle \psi_1 | \psi_2 \rangle , \qquad (3.14)$$

we can also write

$$\langle \varphi \otimes \psi | = \langle \varphi | \otimes \langle \psi | . \qquad (3.15)$$

Example 3.4 Consider again the case $\mathbb{H}^A = \mathbb{H}^B = \P\mathbb{H} \cong \mathbb{C}^2$ with the dual[4] ONBs

$$\{\langle e_a |\} = \{\langle f_b |\} = \{\langle 0|, \langle 1| \} = \{ (1\ 0), (0\ 1) \} ,$$

---

[4] These are actually basis $\{\langle u_a |\}$, $\{\langle v_b |\}$ of the dual spaces $(\mathbb{H}^A)^*$, $(\mathbb{H}^B)^*$ satisfying $\langle u_a | e_{a'} \rangle = \delta_{a,a'}$ and $\langle v_b | f_b \rangle = \delta_{b,b'}$. But as remarked before, we can identify $(\mathbb{H}^A)^*$ with $\mathbb{H}^A$ and $(\mathbb{H}^B)^*$ with $\mathbb{H}^B$ and thus $\{\langle u_a |\}$, $\{\langle v_b |\} = \{\langle e_a |\}$, $\{\langle f_b |\}$.

where the last set denotes the standard basis in the dual space $(\mathbb{C}^2)^* \cong \mathbb{C}^2$. For $\mathbb{H}^A \otimes \mathbb{H}^B \cong \mathbb{C}^4$ we then obtain the ONB

$$\{\langle e_a \otimes f_b|\} = \{\langle 00|, \langle 01|, \langle 10|, \langle 11|\}$$
$$= \{(1\ 0\ 0\ 0), (0\ 1\ 0\ 0), (0\ 0\ 1\ 0), (0\ 0\ 0\ 1)\}\ , \quad (3.16)$$

where the last set denotes the standard basis in the dual space $(\mathbb{C}^4)^* \cong \mathbb{C}^4$. Moreover, for $j \in \{1, 2\}$ let $c_j, d_j \in \mathbb{C}$ and

$$\langle \psi_1| = c_1 \langle 0| + d_1 \langle 1| = (c_1\ d_1)\ , \qquad \langle \psi_2| = c_2 \langle 0| + d_2 \langle 1| = (c_2\ d_2)\ .$$

In the basis $\{|e_a \otimes f_b\rangle\}$ we then have for the matrix of

$$\begin{aligned}
\langle \psi_1| \otimes \langle \psi_2| &= (c_1 \langle 0| + d_1 \langle 1|) \otimes (c_2 \langle 0| + d_2 \langle 1|) \\
&= c_1 c_2 \langle 0| \otimes \langle 0| + c_1 d_2 \langle 0| \otimes \langle 1| + d_1 c_2 \langle 1| \otimes \langle 0| + d_1 d_2 \langle 1| \otimes \langle 1| \\
&= c_1 c_2 \langle 00| + c_1 d_2 \langle 01| + d_1 c_2 \langle 10| + d_1 d_2 \langle 11| \\
&\underbrace{=}_{(3.4)} (c_1 c_2\ \ c_1 d_2\ \ d_1 c_2\ \ d_1 d_2)\ .
\end{aligned} \qquad (3.17)$$

Using the standard basis of $\mathbb{H}$ to build basis of higher tensor powers $\mathbb{H}^{\otimes n}$ of $\mathbb{H}$ as in the Examples 3.3 and 3.4 can be generalized and leads to a natural one-to-one correspondence between these basis vectors in $\mathbb{H}^{\otimes n}$ and natural numbers less than 2^n. This is what the computational basis, covered in the next section, is about.

3.2.2 Computational Basis

Definition 3.5 The n-fold tensor product of qubit spaces is defined as

$$\mathbb{H}^{\otimes n} := \underbrace{\mathbb{H} \otimes \cdots \otimes \mathbb{H}}_{n \text{ factors}}\ .$$

We denote the $j + 1$-th factor space *counting from the right* in $\mathbb{H}^{\otimes n}$ by $\mathbb{H}_j$. In other words, we define

$$\mathbb{H}^{\otimes n} = \mathbb{H}_{n-1} \otimes \cdots \otimes \overbrace{\mathbb{H}_j}^{j+1\text{-th factor}} \otimes \cdots \otimes \mathbb{H}_0\ . \qquad (3.18)$$

The HILBERT space $\mathbb{H}^{\otimes n}$ is 2^n-dimensional. The reason to count the factor spaces from the right will become evident further below when we define the very useful computational basis. Every number $x \in \mathbb{N}_0$ with $x < 2^n$ can be expressed in the form

$$x = \sum_{j=0}^{n-1} x_j 2^j \qquad \text{with } x_j \in \{0, 1\},$$

which results in the usual **binary representation**

$$(x)_{\text{Basis } 2} = x_{n-1} \ldots x_1 x_0 {}_2 \qquad \text{with } x_j \in \{0, 1\}. \tag{3.19}$$

For example, $5 = 101_2$. All possible combinations of $x_0, \ldots, x_{n-1}$ thus yield all integers from 0 to $2^n - 1$. Conversely, every natural number x less than 2^n corresponds uniquely to an n-tuple $x_0, \ldots, x_{n-1} \in \{0, 1\}^n$ and thus to a vector $|x_{n-1}\rangle \otimes \cdots \otimes |x_1\rangle \otimes |x_0\rangle \in {}^\P\mathbb{H}^{\otimes n}$.

Definition 3.6 Let $x \in \mathbb{N}_0$ with $x < 2^n$ and $x_0, \ldots, x_{n-1} \in \{0, 1\}^n$ be the coefficients of the binary representation

$$x = \sum_{j=0}^{n-1} x_j 2^j$$

of x. For each such x we define a vector $|x\rangle \in {}^\P\mathbb{H}^{\otimes n}$ as

$$|x\rangle^n := |x\rangle := |x_{n-1} \ldots x_1 x_0\rangle$$

$$:= |x_{n-1}\rangle \otimes \cdots \otimes |x_1\rangle \otimes |x_0\rangle = \bigotimes_{j=n-1}^{0} |x_j\rangle. \tag{3.20}$$

If it is clear in which product space ${}^\P\mathbb{H}^{\otimes n}$ the vector $|x\rangle^n$ lies, we will also simply write $|x\rangle$ instead of $|x\rangle^n$.

Note that in (3.20) in accordance with the usual binary representation (3.19), the counting of indices in $|x\rangle = |x_{n-1} \ldots x_1 x_0\rangle$ starts from the right. We also express this by the bounds on the index j in $\bigotimes_{j=n-1}^{0}$. The way in which the $|x\rangle$ in Definition 3.6 are defined explains the counting of the factor spaces in (3.18). This is because with $|x_j\rangle \in {}^\P\mathbb{H}_j$ for $j \in \{0, \ldots, n-1\}$, one then has

$$
{}^\P\mathbb{H}^{\otimes n} = \quad {}^\P\mathbb{H}_{n-1} \otimes \cdots \otimes \overbrace{{}^\P\mathbb{H}_j}^{j+1\text{-th factor}} \otimes \cdots \otimes {}^\P\mathbb{H}_0 . \tag{3.21}
$$
$$
\ni |x_{n-1}\rangle \otimes \cdots \otimes \quad |x_j\rangle \quad \otimes \cdots \otimes |x_0\rangle
$$

For the smallest and largest in ${}^\P\mathbb{H}^{\otimes n}$ representable numbers 0 and $2^n - 1$ we have

$$|2^n - 1\rangle^n = |11\ldots 1\rangle = \bigotimes_{j=0}^{n-1} |1\rangle \in \mathbb{H}^{\otimes n} \qquad (3.22)$$

$$|0\rangle^n = |00\ldots 0\rangle = \bigotimes_{j=0}^{n-1} |0\rangle \in \mathbb{H}^{\otimes n}. \qquad (3.23)$$

Since the factors in the tensor products in (3.22) and (3.23) are all equal, the sequence of indexing does not matter in these special cases.

Lemma 3.7 *The set of vectors* $\{|x\rangle \in \mathbb{H}^{\otimes n} \mid x \in \mathbb{N}_0, x < 2^n\}$ *forms an ONB in* $\mathbb{H}^{\otimes n}$.

Proof For $|x\rangle, |y\rangle \in \mathbb{H}^{\otimes n}$ one has

$$
\begin{aligned}
\langle x|y\rangle &= \langle x_{n-1}\ldots x_0 | y_{n-1}\ldots y_0\rangle \\
&\underset{(3.4)}{=} \prod_{j=0}^{n-1} \langle x_j | y_j\rangle = \begin{cases} 1 & \text{if } x_j = y_j \quad \forall j \\ 0 & \text{else} \end{cases} \\
&= \delta_{xy}.
\end{aligned} \qquad (3.24)
$$

Hence, the $\{|x\rangle \mid x \in \mathbb{N}_0 \text{ and } x < 2^n\}$ form a set of $2^n = \dim \mathbb{H}^{\otimes n}$ orthonormal vectors in $\mathbb{H}^{\otimes n}$. As the number of orthonormal vectors in this set is equal to the dimension of $\mathbb{H}^{\otimes n}$, the set constitutes an ONB in this HILBERT space. $\square$

The ONB in $\mathbb{H}^{\otimes n}$ defined by the numbers $x \in \mathbb{N}_0$ with $x < 2^n$ is very useful and thus has its own name.

Definition 3.8 The ONB in $\mathbb{H}^{\otimes n}$ defined for $x \in \{0, 1, \ldots, 2^n - 1\}$ by $|x\rangle = |x_{n-1}\ldots x_0\rangle$ is called **computational basis**.

Example 3.9 In $\mathbb{H}$ the computational basis is identical with the standard basis:

$$|0\rangle^1 = |0\rangle = \begin{pmatrix} 1 \\ 0 \end{pmatrix}, \qquad |1\rangle^1 = |1\rangle = \begin{pmatrix} 0 \\ 1 \end{pmatrix},$$

where the rightmost equalities show the identification with the standard basis in $\mathbb{C}^2 \cong \mathbb{H}$. The four basis vectors of the computational basis in $\mathbb{H}^{\otimes 2} \cong \mathbb{C}^4$ are

$$|0\rangle^2 = |00\rangle = |0\rangle \otimes |0\rangle = \begin{pmatrix} 1 \\ 0 \\ 0 \\ 0 \end{pmatrix}$$

$$|1\rangle^2 = |01\rangle = |0\rangle \otimes |1\rangle = \ \vdots \qquad\qquad\qquad (3.25)$$

$$|2\rangle^2 = |10\rangle = |1\rangle \otimes |0\rangle = \ \vdots$$

$$|3\rangle^2 = |11\rangle = |1\rangle \otimes |1\rangle = \begin{pmatrix} 0 \\ 0 \\ 0 \\ 1 \end{pmatrix}.$$

Whereas in $\mathbb{H}^{\otimes 3} \cong \mathbb{C}^8$ one has the computational basis

$$|0\rangle^3 = |000\rangle = |0\rangle \otimes |0\rangle \otimes |0\rangle = \begin{pmatrix} 1 \\ 0 \\ 0 \\ 0 \\ 0 \\ 0 \\ 0 \\ 0 \end{pmatrix}$$

$$|1\rangle^3 = |001\rangle = \ \vdots$$
$$|2\rangle^3 = |010\rangle$$
$$|3\rangle^3 = |011\rangle$$
$$|4\rangle^3 = |100\rangle$$
$$|5\rangle^3 = |101\rangle$$
$$|6\rangle^3 = |110\rangle$$
$$|7\rangle^3 = |111\rangle.$$

For example, in $\mathbb{H}$ we may consider

$$|\varphi_1\rangle = \frac{|0\rangle + |1\rangle}{\sqrt{2}} = \frac{1}{\sqrt{2}} \begin{pmatrix} 1 \\ 1 \end{pmatrix}$$

$$|\varphi_2\rangle = \frac{|0\rangle - |1\rangle}{\sqrt{2}} = \frac{1}{\sqrt{2}} \begin{pmatrix} 1 \\ -1 \end{pmatrix}$$

$$|\psi_1\rangle = |0\rangle = \begin{pmatrix} 1 \\ 0 \end{pmatrix}$$

$$|\psi_2\rangle = |1\rangle = \begin{pmatrix} 0 \\ 1 \end{pmatrix}.$$

With these one finds in $\mathbb{H}^{\otimes 2}$

$$|\varphi_1 \otimes \varphi_2\rangle = |\varphi_1\rangle \otimes |\varphi_2\rangle = \frac{1}{\sqrt{2}} \begin{pmatrix} 1 \\ 1 \end{pmatrix} \otimes \frac{1}{\sqrt{2}} \begin{pmatrix} 1 \\ -1 \end{pmatrix} \underset{(3.13)}{=} \frac{1}{2} \begin{pmatrix} 1 \\ -1 \\ 1 \\ -1 \end{pmatrix}$$

$$\langle \psi_1 \otimes \psi_2| = \langle \psi_1| \otimes \langle \psi_2| = \begin{pmatrix} 1 & 0 \end{pmatrix} \otimes \begin{pmatrix} 0 & 1 \end{pmatrix} \underset{(3.17)}{=} \begin{pmatrix} 0 & 1 & 0 & 0 \end{pmatrix} ,$$

where the rightmost vectors are now expressed in the basis given in (3.25) and its dual. Using this, we find that in this basis the matrix of $|\varphi_1 \otimes \varphi_2\rangle\langle \psi_1 \otimes \psi_2|$ is given by

$$|\varphi_1 \otimes \varphi_2\rangle\langle \psi_1 \otimes \psi_2| = \frac{1}{2} \begin{pmatrix} 1 \\ -1 \\ 1 \\ -1 \end{pmatrix} \begin{pmatrix} 0 & 1 & 0 & 0 \end{pmatrix} \underset{(2.27)}{=} \frac{1}{2} \begin{pmatrix} 0 & 1 & 0 & 0 \\ 0 & -1 & 0 & 0 \\ 0 & 1 & 0 & 0 \\ 0 & -1 & 0 & 0 \end{pmatrix} . \quad (3.26)$$

On the other hand, we have

$$|\varphi_1\rangle\langle \psi_1| = \frac{1}{\sqrt{2}} \begin{pmatrix} 1 \\ 1 \end{pmatrix} \begin{pmatrix} 1 & 0 \end{pmatrix} \underset{(2.27)}{=} \frac{1}{\sqrt{2}} \begin{pmatrix} 1 & 0 \\ 1 & 0 \end{pmatrix}$$

$$|\varphi_2\rangle\langle \psi_2| = \frac{1}{\sqrt{2}} \begin{pmatrix} 1 \\ -1 \end{pmatrix} \begin{pmatrix} 0 & 1 \end{pmatrix} \underset{(2.27)}{=} \frac{1}{\sqrt{2}} \begin{pmatrix} 0 & 1 \\ 0 & -1 \end{pmatrix} . \quad (3.27)$$

The fact that the vectors of the computational basis are identifiable by numbers in $\mathbb{N}_0$ makes this basis play an important role in many areas of quantum computing, such as quantum gates (see Chap. 5) or algorithms (see Sects. 6.5 and 6.9).

The computational basis consists of so-called *separable* (or product-) states (see Definition 4.1). They are called this way because in each of these states of the composite system the sub-systems are in pure states as well. For example, in the state $|01\rangle$ of the computational basis (3.25) of the composite system $\mathbb{H}^{\otimes 2}$ the first sub-system is in the pure state $|0\rangle$. An observer of this sub-system, who measures σ_z in his sub-system, will always observe the value $+1$. At the same time, the second sub-system is in the pure state $|1\rangle$, in other words, an observer of the second sub-system, who measures σ_z in his sub-system, will always observe the value -1. But the four-dimensional space $\mathbb{H}^{\otimes 2}$ also admits other ONBs. One such basis is the BELL basis.

Definition 3.10 The BELL basis in the four-dimensional space $\mathbb{H}^{\otimes 2}$ consists of the basis vectors

$$
\begin{aligned}
|\Phi^{\pm}\rangle &:= \tfrac{1}{\sqrt{2}}\Big(|00\rangle \pm |11\rangle\Big) \\
|\Psi^{\pm}\rangle &:= \tfrac{1}{\sqrt{2}}\Big(|01\rangle \pm |10\rangle\Big).
\end{aligned}
\tag{3.28}
$$

Exercise 3.39 Show that the BELL basis is orthonormal.

For a solution see Solution 3.39.

As we shall see later, the BELL basis does not consist of separable, but *entangled* states (see Definition 4.1). From (3.55) and the results of Exercise 3.44 it even follows that the BELL basis vectors are *maximally entangled* (see Definition 4.4).

As the remark after (3.55) shows, this implies that in the *pure* state $|\Phi^+\rangle \in {}^\P\mathbb{H}^{\otimes 2}$ of the composite system the first sub-system is not in a pure state, but in a *true mixture*. This can perhaps be formulated as: a *qubit-word* (= a state in ${}^\P\mathbb{H}^{\otimes 2} = {}^\P\mathbb{H} \otimes {}^\P\mathbb{H}$) in general does *not consist* of *pure* qubit letters (= pure states in ${}^\P\mathbb{H}$). We shall look at this in much more detail in Sect. 3.3.

3.3 States and Observables for Composite Systems

Quantum systems, which may be described as separate systems with their respective HILBERT spaces, can—and sometimes have to—be combined to form a larger composite system. Although heuristic arguments for the construction of the HILBERT space of such composite systems from those of the sub-systems can be given, there does not seem to be a rigorous derivation, and this construction is best given in 'axiomatic' fashion in the form of another postulate.

Postulate 7 (Composite Systems) *The* HILBERT *space of a composite system that consists of the sub-systems* $\mathbb{H}^A$ *and* $\mathbb{H}^B$ *is the tensor product* $\mathbb{H}^A \otimes \mathbb{H}^B$.

It follows from Postulate 5 that the states of the composite system are thus generally represented by density operators ρ on $\mathbb{H}^A \otimes \mathbb{H}^B$. As shown in Theorem 2.24, these can be written in the form

$$
\rho = \sum_{j\in I} p_j |\Psi_j\rangle\langle\Psi_j|,
$$

where the $\{|\Psi_j\rangle \mid j \in I\}$ are an ONB in $\mathbb{H}^A \otimes \mathbb{H}^B$ and the $p_j \in [0,1]$ satisfy $\sum_{j\in I} p_j = 1$.

Combining first two systems $\mathbb{H}^A$ and $\mathbb{H}^B$ to a composite system with HILBERT space $\mathbb{H}^A \otimes \mathbb{H}^B$, we may then combine this with a third system $\mathbb{H}^C$ and we see that the total composite system of all three sub-systems has the HILBERT space $\mathbb{H}^A \otimes \mathbb{H}^B \otimes \mathbb{H}^C$. Continuing in this fashion we see that the HILBERT space of a composite system formed of n sub-systems $\mathbb{H}^{A_j}$, where $j \in \{1, \ldots, n\}$, is given by $\bigotimes_{j=1}^{n} \mathbb{H}^{A_j} = \mathbb{H}^{A_1} \otimes \cdots \otimes \mathbb{H}^{A_n}$.

Example 3.11 We know from Example 2.5 that the HILBERT space for a single particle in three-dimensional space is $\mathbb{H} = L^2(\mathbb{R}^3)$. The HILBERT space for the composite system formed by n such particles is $\mathbb{H}^{\text{comp}} = L^2(\mathbb{R}^{3n})$. In this case it is indeed a mathematical property of $L^2(\mathbb{R}^{3n})$, that

$$\mathbb{H}^{\text{comp}} = L^2(\mathbb{R}^{3n}) = \left(L^2(\mathbb{R}^3)\right)^{\otimes n} = \mathbb{H}^{\otimes n} .$$

Applying a terminology widely used in quantum computing, we shall always assume that system A is controlled (in other words, can be read and operated on) by **Alice** and that system B is controlled by **Bob**. The association of sub-systems with persons is indeed helpful when describing the systems. For example, instead of saying 'an observer of sub-system A observes' it is simpler to state 'Alice observes'; or instead of saying 'a state is prepared in sub-system B' it is more concise to state simply 'Bob prepares the state'.

From observables of the sub-systems we can build observables of the composite system.[5] For example, let $M^X : \mathbb{H}^X \to \mathbb{H}^X$ be self-adjoint operators of observables in the respective sub-systems $X \in \{A, B\}$. We can then form the operator $M^A \otimes M^B$, which acts factor-wise on tensor products $|\varphi \otimes \psi\rangle = |\varphi\rangle \otimes |\psi\rangle$, that is,

$$\left(M^A \otimes M^B\right) |\varphi \otimes \psi\rangle = \underbrace{\left(M^A |\varphi\rangle\right)}_{\in \mathbb{H}^A} \otimes \underbrace{\left(M^B |\psi\rangle\right)}_{\in \mathbb{H}^B} . \tag{3.29}$$

Using linearity, the operator then acts on an arbitrary vector

$$|\Phi\rangle = \sum_{a,b} \Phi_{ab} |e_a\rangle \otimes |f_b\rangle \in \mathbb{H}^A \otimes \mathbb{H}^B$$

as follows

$$\left(M^A \otimes M^B\right) |\Phi\rangle = \sum_{a,b} \Phi_{ab} \left(M^A |e_a\rangle\right) \otimes \left(M^B |f_b\rangle\right) \quad \in \mathbb{H}^A \otimes \mathbb{H}^B . \tag{3.30}$$

---

[5]There are, of course, observables of composite system which cannot be built from those of the sub-systems.

Example 3.12 As an example we consider the j-th component L_j of the total angular momentum of a non-relativistic electron, which with $j \in \{1,2,3\}$ constitutes a vector-valued observable in $\mathbb{R}^3$. It is built from the orbital angular momentum operator J_j and the intrinsic angular momentum (spin) operator S_j as

$$L_j = J_j \otimes \mathbf{1} + \mathbf{1} \otimes S_j .$$

The adjoint of a tensor product of operators is the tensor product of the adjoint operators as Exercise 3.40 shows.

Exercise 3.40 Show that for operators $M^A : \mathbb{H}^A \to \mathbb{H}^A$, $M^B : \mathbb{H}^B \to \mathbb{H}^B$ and $M^A \otimes M^B : \mathbb{H}^A \otimes \mathbb{H}^B \to \mathbb{H}^A \otimes \mathbb{H}^B$ one has

$$\left(M^A \otimes M^B\right)^* = (M^A)^* \otimes (M^B)^* \tag{3.31}$$

and thus

$$(M^X)^* = M^X \text{ for } X \in \{A,B\} \qquad \Rightarrow \qquad \left(M^A \otimes M^B\right)^* = M^A \otimes M^B , \tag{3.32}$$

that is, the tensor product of self-adjoint operators is self-adjoint.

For a solution see Solution 3.40.

Suppose that for $X \in \{A,B\}$ the operators $M^X : \mathbb{H}^X \to \mathbb{H}^X$ in the respective basis $\{|e_a\rangle\} \in \mathbb{H}^A$ and $\{|f_b\rangle\} \in \mathbb{H}^B$ have the matrices

$$M^X = \begin{pmatrix} M^X_{11} & \cdots & M^X_{1n_X} \\ \vdots & & \vdots \\ M^X_{n_X 1} & \cdots & M^X_{n_X n_X} \end{pmatrix} .$$

What is then the matrix of $M^A \otimes M^B$ in the basis $\{|e_a \otimes f_b\rangle\} \in \mathbb{H}^A \otimes \mathbb{H}^B$? To answer this, we first note that

$$\begin{aligned} M^A \otimes M^B &= \sum_{a,a'=1}^{n_A} \sum_{b,b'=1}^{n_B} |e_a \otimes f_b\rangle\langle e_a \otimes f_b|(M^A \otimes M^B)e_{a'} \otimes f_{b'}\rangle\langle e_{a'} \otimes f_{b'}| \\ &= \sum_{a,a'=1}^{n_A} \sum_{b,b'=1}^{n_B} |e_a \otimes f_b\rangle\langle e_a \otimes f_b|M^A e_{a'} \otimes M^B f_{b'}\rangle\langle e_{a'} \otimes f_{b'}| \\ &= \sum_{a,a'=1}^{n_A} \sum_{b,b'=1}^{n_B} |e_a \otimes f_b\rangle\langle e_a|M^A e_{a'}\rangle\langle f_b|M^B f_{b'}\rangle\langle e_{a'} \otimes f_{b'}| \end{aligned}$$

$$= \sum_{a,a'=1}^{n_A} \sum_{b,b'=1}^{n_B} M_{aa'}^A M_{bb'}^B |e_a \otimes f_b\rangle \langle e_{a'} \otimes f_{b'}|. \tag{3.33}$$

From (2.25) we infer that

$$M_{aa'}^A M_{bb'}^B |e_a \otimes f_b\rangle \langle e_{a'} \otimes f_{b'}| = \begin{matrix} & & 1 & \cdots & & k & \cdots & n \\ 1 & \\ \vdots & \\ j & \\ \vdots & \\ n & \end{matrix} \begin{pmatrix} & & & & | & & \\ & & & & | & & \\ -- & -- & -- & & M_{aa'}^A M_{bb'}^B & & \\ & & & & & & \\ & & & & & & \end{pmatrix},$$

where now in accordance with (2.25) and (3.11) we have $j = (a-1)n_B + b$ and $k = (a'-1)n_B + b'$, and we have written out the only non-zero matrix element. Inserting this into (3.33) we find

$$M^A \otimes M^B = \tag{3.34}$$

$$\begin{matrix} & 1 & \cdots & n_B & n_B+1 & \cdots & 2n_B & \cdots & n_A n_B \\ 1 & \begin{pmatrix} M_{11}^A M_{11}^B & \cdots & M_{11}^A M_{1n_B}^B & M_{12}^A M_{11}^B & \cdots & M_{12}^A M_{1n_B}^B & \cdots & M_{1n_A}^A M_{1n_B}^B \\ \vdots & & \vdots & \vdots & & \vdots & & \vdots \\ n_B & M_{11}^A M_{n_B 1}^B & \cdots & M_{11}^A M_{n_B n_B}^B & M_{12}^A M_{n_B 1}^B & \cdots & M_{12}^A M_{n_B n_B}^B & \cdots & M_{1n_A}^A M_{n_B n_B}^B \\ n_B+1 & M_{21}^A M_{11}^B & \cdots & M_{21}^A M_{1n_B}^B & M_{22}^A M_{11}^B & \cdots & M_{22}^A M_{1n_B}^B & \cdots & M_{2n_A}^A M_{1n_B}^B \\ \vdots & & \vdots & \vdots & & \vdots & & \vdots \\ 2n_B & M_{21}^A M_{n_B 1}^B & \cdots & M_{21}^A M_{n_B n_B}^B & M_{22}^A M_{n_B 1}^B & \cdots & M_{22}^A M_{n_B n_B}^B & \cdots & M_{2n_A}^A M_{n_B n_B}^B \\ \vdots & & \vdots & \vdots & & \vdots & & \vdots \\ n_A n_B & M_{n_A 1}^A M_{n_B 1}^B & \cdots & M_{n_A 1}^A M_{n_B n_B}^B & M_{n_A 2}^A M_{n_B 1}^B & \cdots & M_{n_A 2}^A M_{n_B n_B}^B & \cdots & M_{n_A n_A}^A M_{n_B n_B}^B \end{pmatrix} \end{matrix},$$

A closer inspection of (3.34) reveals that the matrix is comprised of blocks consisting of the matrix for M^B multiplied by matrix elements $M_{aa'}^A$. In other words, the matrix of $M^A \otimes M^B$ in the basis $\{|e_a \otimes f_b\rangle\}$ is given by

$$M^A \otimes M^B = \begin{pmatrix} M_{11}^A \begin{pmatrix} M_{11}^B & \cdots & M_{1n_B}^B \\ \vdots & & \vdots \\ M_{n_B 1}^B & \cdots & M_{n_B n_B}^B \end{pmatrix} & \cdots & M_{1n_A}^A \begin{pmatrix} M_{11}^B & \cdots & M_{1n_B}^B \\ \vdots & & \vdots \\ M_{n_B 1}^B & \cdots & M_{n_B n_B}^B \end{pmatrix} \\ \vdots & & \vdots \\ M_{n_A 1}^A \begin{pmatrix} M_{11}^B & \cdots & M_{1n_B}^B \\ \vdots & & \vdots \\ M_{n_B 1}^B & \cdots & M_{n_B n_B}^B \end{pmatrix} & \cdots & M_{n_A n_A}^A \begin{pmatrix} M_{11}^B & \cdots & M_{1n_B}^B \\ \vdots & & \vdots \\ M_{n_B 1}^B & \cdots & M_{n_B n_B}^B \end{pmatrix} \end{pmatrix}. \tag{3.35}$$

As an example of a tensor product of operators, we show that the projection onto a tensor product of states is equal to the tensor product of projections onto the factor states.

Lemma 3.13 *For arbitrary* $|\varphi_1\rangle, |\varphi_2\rangle \in \mathbb{H}^A$ *and* $|\psi_1\rangle, |\psi_2\rangle \in \mathbb{H}^B$ *one has*

$$|\varphi_1 \otimes \psi_1\rangle\langle\varphi_2 \otimes \psi_2| = |\varphi_1\rangle\langle\varphi_2| \otimes |\psi_1\rangle\langle\psi_2| \,. \tag{3.36}$$

Proof For any $|\xi_1\rangle, |\xi_2\rangle \in \mathbb{H}^A$ and $|\zeta_1\rangle, |\zeta_2\rangle \in \mathbb{H}^B$ we have

$$
\begin{aligned}
\langle\xi_1 \otimes \zeta_1| \Big(|\varphi_1 \otimes \psi_1\rangle\langle\varphi_2 \otimes \psi_2| \Big) \xi_2 \otimes \zeta_2\rangle &= \langle\xi_1 \otimes \zeta_1|\varphi_1 \otimes \psi_1\rangle\langle\varphi_2 \otimes \psi_2|\xi_2 \otimes \zeta_2\rangle \\
&\underbrace{=}_{(3.14)} \langle\xi_1|\varphi_1\rangle\langle\zeta_1|\psi_1\rangle\langle\varphi_2|\xi_2\rangle\langle\psi_2|\zeta_2\rangle \\
&= \langle\xi_1|\varphi_1\rangle\langle\varphi_2|\xi_2\rangle\langle\zeta_1|\psi_1\rangle\langle\psi_2|\zeta_2\rangle \\
&\underbrace{=}_{(3.14)} \langle\xi_1 \otimes \zeta_1| \Big(|\varphi_1\rangle\langle\varphi_2| \otimes |\psi_1\rangle\langle\psi_2| \Big) \xi_2 \otimes \zeta_2\rangle \,.
\end{aligned}
$$

$\square$

Example 3.14 For $j \in \{1, 2\}$ let $|\varphi_j\rangle, |\psi_j\rangle$ be as in Example 3.9. From (3.26) we know that then in the basis (3.25)

$$|\varphi_1 \otimes \varphi_2\rangle\langle\psi_1 \otimes \psi_2| = \frac{1}{2} \begin{pmatrix} 0 & 1 & 0 & 0 \\ 0 & -1 & 0 & 0 \\ 0 & 1 & 0 & 0 \\ 0 & -1 & 0 & 0 \end{pmatrix} . \tag{3.37}$$

On the other hand, we have in the same basis

$$|\varphi_1\rangle\langle\psi_1| \otimes |\varphi_2\rangle\langle\psi_2| \underbrace{=}_{(3.27)} \frac{1}{\sqrt{2}} \begin{pmatrix} 1 & 0 \\ 1 & 0 \end{pmatrix} \otimes \frac{1}{\sqrt{2}} \begin{pmatrix} 0 & 1 \\ 0 & -1 \end{pmatrix}$$

$$\underbrace{=}_{(3.35)} \frac{1}{2} \begin{pmatrix} 0 & 1 & 0 & 0 \\ 0 & -1 & 0 & 0 \\ 0 & 1 & 0 & 0 \\ 0 & -1 & 0 & 0 \end{pmatrix} . \tag{3.38}$$

Together (3.37) and (3.38) verify (3.36) for this particular example.

The operator $M^A \otimes M^B$ thus represents an observable of the composite system. As an example for the action of an observable in the tensor product $\P\mathbb{H}^{AB} = \P\mathbb{H} \otimes \P\mathbb{H}$ of two qubit spaces we show here how $M^A \otimes M^B = \sigma_z^A \otimes \sigma_z^B$ acts on the BELL basis (3.28) of the composite system $\P\mathbb{H}^{AB}$.

$$
\begin{aligned}
\left(\sigma_z \otimes \sigma_z \right) |\Phi^\pm\rangle &= \left(\sigma_z \otimes \sigma_z \right) \frac{1}{\sqrt{2}} \left(|00\rangle \pm |11\rangle \right) \\
&= \frac{1}{\sqrt{2}} \left(\sigma_z \otimes \sigma_z \right) \left(|0\rangle \otimes |0\rangle \pm |1\rangle \otimes |1\rangle \right) \\
&= \frac{1}{\sqrt{2}} \left\{ \left(\sigma_z |0\rangle \right) \otimes \left(\sigma_z |0\rangle \right) \pm \left(\sigma_z |1\rangle \right) \otimes \left(\sigma_z |1\rangle \right) \right\} \\
&= \frac{1}{\sqrt{2}} \left\{ |0\rangle \otimes |0\rangle \pm \left(-|1\rangle \right) \otimes \left(-|1\rangle \right) \right\} \\
&= \frac{1}{\sqrt{2}} \left\{ |0\rangle \otimes |0\rangle \pm |1\rangle \otimes |1\rangle \right\} = \frac{1}{\sqrt{2}} \left(|00\rangle \pm |11\rangle \right) \\
&= |\Phi^\pm\rangle .
\end{aligned}
\tag{3.39}
$$

Analogously, one shows

$$
\begin{aligned}
\left(\sigma_z \otimes \sigma_z \right) |\Psi^\pm\rangle &= -|\Psi^\pm\rangle \\
\left(\sigma_x \otimes \sigma_x \right) |\Phi^\pm\rangle &= \pm|\Phi^\pm\rangle \\
\left(\sigma_x \otimes \sigma_x \right) |\Psi^\pm\rangle &= \pm|\Psi^\pm\rangle .
\end{aligned}
\tag{3.40}
$$

In the BELL basis $\{|\Phi^+\rangle, |\Phi^-\rangle, |\Psi^+\rangle, |\Psi^-\rangle\}$ the operators $\sigma_z \otimes \sigma_z$ and $\sigma_x \otimes \sigma_x$ thus have the matrix (see Definition 2.6)

$$
\sigma_z \otimes \sigma_z|_{\text{in BELL-Basis}} = \begin{pmatrix} 1 & 0 & 0 & 0 \\ 0 & 1 & 0 & 0 \\ 0 & 0 & -1 & 0 \\ 0 & 0 & 0 & -1 \end{pmatrix}
$$

$$
\sigma_x \otimes \sigma_x|_{\text{in BELL-Basis}} = \begin{pmatrix} 1 & 0 & 0 & 0 \\ 0 & -1 & 0 & 0 \\ 0 & 0 & 1 & 0 \\ 0 & 0 & 0 & -1 \end{pmatrix} ,
$$

from which it is evident that these operators commute

$$
\left[\sigma_z \otimes \sigma_z, \sigma_x \otimes \sigma_x \right] = 0 ,
\tag{3.41}
$$

Table 3.1 State determination via joint measurement of $\sigma_z \otimes \sigma_z$ and $\sigma_x \otimes \sigma_x$

Measured value of		State after measurement
$\sigma_z \otimes \sigma_z$	$\sigma_x \otimes \sigma_x$	
$+1$	$+1$	$\lvert \Phi^+ \rangle$
$+1$	-1	$\lvert \Phi^- \rangle$
-1	$+1$	$\lvert \Psi^+ \rangle$
-1	-1	$\lvert \Psi^- \rangle$

and the corresponding observables are compatible. In particular, they have—as can be seen immediately from (3.39)–(3.40)—common eigenvectors, and it is possible to measure these two observables sharply, in other words, without uncertainty. The combination of the measured values of these observables thus reveals in which state the system is after the measurement as shown in Table 3.1. The state determination shown in Table 3.1 will play a role again in the context of teleportation.

Carrying out a measurement on a sub-system—an observable of system A with the operator M^A, say—is a measurement of an observable of the composite system with the operator $M^A \otimes \mathbf{1}^B$. Analogously, measurements on the sub-system B are represented by operators of the form $\mathbf{1}^A \otimes M^B$. Consider, for example, the *pure state of the composite system*

$$\lvert \Psi \rangle = \sum_{a,b} \Psi_{ab} \lvert e_a \rangle \otimes \lvert f_b \rangle \, ,$$

in which the observable M^A of the sub-system A is measured. The expectation value in this state is calculated in accordance with (2.60) as

$$
\begin{aligned}
\langle M^A \otimes \mathbf{1}^B \rangle_\Psi &= \langle \Psi \lvert M^A \otimes \mathbf{1}^B \Psi \rangle \underbrace{=}_{(3.30)} \sum_{a_1,b_1} \sum_{a_2,b_2} \overline{\Psi_{a_2 b_2}} \Psi_{a_1 b_1} \langle e_{a_2} \otimes f_{b_2} \lvert M^A e_{a_1} \otimes f_{b_1} \rangle \\
&\underbrace{=}_{(3.4)} \sum_{a_1,b_1} \sum_{a_2,b_2} \overline{\Psi_{a_2 b_2}} \Psi_{a_1 b_1} \langle e_{a_2} \lvert M^A e_{a_1} \rangle \underbrace{\langle f_{b_2} \lvert f_{b_1} \rangle}_{= \delta_{b_2 b_1}} \\
&= \sum_{a_2,a_1,b} \overline{\Psi_{a_2 b}} \Psi_{a_1 b} \langle e_{a_2} \lvert M^A e_{a_1} \rangle \, .
\end{aligned}
\tag{3.42}
$$

This is the expectation value of the observable M^A, which Alice will find with measurements on her sub-system. She would find exactly the *same expectation value* if her sub-system alone were in the following state:

$$\rho^A(\Psi) := \sum_{a_2,a_1,b} \overline{\Psi_{a_2 b}} \Psi_{a_1 b} \lvert e_{a_1} \rangle \langle e_{a_2} \rvert. \tag{3.43}$$

As we will show below $\rho^A(\Psi)$ satisfies all defining properties of a density operator. Hence, $\rho^A(\Psi)$ describes a mixed state for the sub-system A, which depends on the composite state $|\Psi\rangle$. For observables of the form M^A the state $\rho^A(\Psi)$ reproduces the expectation values of $M^A \otimes \mathbf{1}^B$ in the composite state $|\Psi\rangle$. We now show that $\rho^A(\Psi)$ has all properties of a density operator, in other words, that $\rho^A(\Psi)$ is self-adjoint, positive, and has trace 1. First, we have

$$
\left(\rho^A(\Psi)\right)^* = \sum_{a_1,a_2,b} \overline{\Psi_{a_2b}\Psi_{a_1b}} \underbrace{\left(|e_{a_1}\rangle\langle e_{a_2}|\right)^*}_{=|e_{a_2}\rangle\langle e_{a_1}|} = \sum_{a_1,a_2,b} \overline{\Psi_{a_1b}\Psi_{a_2b}}|e_{a_2}\rangle\langle e_{a_1}|
$$

$$
= \rho^A(\Psi)\,,
$$

proving that $\rho^A(\Psi)$ is self-adjoint. That it is positive follows from

$$
\langle\varphi|\rho^A(\Psi)\varphi\rangle = \sum_{a_1,a_2,b} \overline{\Psi_{a_2b}\Psi_{a_1b}}\langle\varphi|e_{a_1}\rangle\langle e_{a_2}|\varphi\rangle
$$

$$
= \sum_b \left(\sum_{a_1}\Psi_{a_1b}\langle\varphi|e_{a_1}\rangle\right)\overline{\left(\sum_{a_2}\Psi_{a_2b}\langle\varphi|e_{a_2}\rangle\right)}
$$

$$
= \sum_b \left|\sum_a \Psi_{ab}\langle\varphi|e_a\rangle\right|^2
$$

$$
\geq 0\,,
$$

and its trace property is verified by

$$
\mathrm{tr}\left(\rho^A(\Psi)\right) = \sum_{a_3,a_1,a_2,b} \overline{\Psi_{a_2b}\Psi_{a_1b}}\langle e_{a_3}|e_{a_1}\rangle\langle e_{a_2}|e_{a_3}\rangle = \sum_{a,b}|\Psi_{ab}|^2 \underset{(3.9)}{=} ||\Psi||^2
$$

$$
= 1.
$$

Next, we prove the claimed equality of the expectation values of the states $|\Psi\rangle$ and $\rho^A(\Psi)$ for observables of the sub-system A. From (3.43) it follows that

$$
\langle M^A\rangle_{\rho^A(\Psi)} = \mathrm{tr}\left(\rho^A(\Psi)M^A\right) = \sum_a \langle e_a|\rho^A(\Psi)M^A e_a\rangle
$$

$$
= \sum_{a,a_1,a_2,b} \overline{\Psi_{a_2b}\Psi_{a_1b}}\langle e_a|e_{a_1}\rangle\langle e_{a_2}|M^A e_a\rangle = \sum_{a_1,a_2,b} \overline{\Psi_{a_2b}\Psi_{a_1b}}\langle e_{a_2}|M^A e_{a_1}\rangle
$$

$$
\underset{(3.42)}{=} \langle M^A \otimes \mathbf{1}^B\rangle_\Psi\,.
$$

For Alice all measurements on her sub-system, which is part of a composite system in the state $|\Psi\rangle$, indicate that her system is in the mixed state $\rho^A(\Psi)$. This means that in a composite system that is in the pure state $|\Psi\rangle \in \mathbb{H}^A \otimes \mathbb{H}^B$ and thus described by the density operator $\rho = |\Psi\rangle\langle\Psi|$ on $\mathbb{H}^A \otimes \mathbb{H}^B$ the sub-system in $\mathbb{H}^A$ is described by the density operator

$$\rho^A(\Psi) = \sum_{a_1,a_2,b} \overline{\Psi_{a_2 b}} \Psi_{a_1 b} |e_{a_1}\rangle\langle e_{a_2}|. \tag{3.44}$$

Analogously, in the state $|\Psi\rangle$ of the composite system the expectation values of observables M^B for the sub-system B only are given by

$$\langle \mathbf{1}^A \otimes M^B \rangle_\Psi = \sum_{b_1,b_2,a} \Psi_{a b_1} \overline{\Psi_{a b_2}} \langle f_{b_2}|M^B f_{b_1}\rangle,$$

and with

$$\rho^B(\Psi) = \sum_{b_1,b_2,a} \overline{\Psi_{a b_2}} \Psi_{a b_1} |f_{b_1}\rangle\langle f_{b_2}| \tag{3.45}$$

one has accordingly

$$\langle M^B \rangle_{\rho^B(\Psi)} = \langle \mathbf{1}^A \otimes M^B \rangle_\Psi.$$

Loosely speaking, the expression for $\rho^A(\Psi)$ can be viewed as if the trace over $\{|f_b\rangle\}$ has been taken and that for $\rho^B(\Psi)$ as if the trace over $\{|e_a\rangle\}$. This can indeed be more generally and formally defined, but before we do so, we prove a result about existence and uniqueness of what will become known as the partial trace.

Theorem 3.15 *Let* $\mathbb{H}^A$ *and* $\mathbb{H}^B$ *be* HILBERT *spaces with respective ONBs* $\{|e_a\rangle\}$ *and* $\{|f_b\rangle\}$. *Moreover, let* $M \in \mathrm{L}(\mathbb{H}^A \otimes \mathbb{H}^B)$ *and let* $M_{a_1 b_1, a_2 b_2}$ *be the matrix of* M *in the ONB* $\{|e_a \otimes f_b\rangle\}$ *of* $\mathbb{H}^A \otimes \mathbb{H}^B$ *and let the operators* $\mathrm{tr}^B(M) \in \mathrm{L}(\mathbb{H}^A)$ *and* $\mathrm{tr}^A(M) \in \mathrm{L}(\mathbb{H}^B)$ *be given by*

$$\mathrm{tr}^B(M) = \sum_{a_1 a_2 b} M_{a_1 b, a_2 b} |e_{a_1}\rangle\langle e_{a_2}|$$
$$\mathrm{tr}^A(M) = \sum_{b_1 b_2 a} M_{a b_1, a b_2} |f_{b_1}\rangle\langle f_{b_2}|. \tag{3.46}$$

Then $\mathrm{tr}^B(M)$ *and* $\mathrm{tr}^A(M)$ *as given in* (3.46) *do not depend on the choice of the ONBs* $\{|e_a\rangle\}$ *and* $\{|f_b\rangle\}$ *and are the unique operators satisfying*

$$\forall M^A \in \mathrm{L}(\mathbb{H}^A): \quad \mathrm{tr}(M^A \, \mathrm{tr}^B(M)) = \mathrm{tr}((M^A \otimes \mathbf{1}^B)M)$$
$$\forall M^B \in \mathrm{L}(\mathbb{H}^B): \quad \mathrm{tr}(M^B \, \mathrm{tr}^A(M)) = \mathrm{tr}((\mathbf{1}^B \otimes M^B)M). \tag{3.47}$$

Proof We only prove the result for $\mathrm{tr}^B(M)$. The proof for $\mathrm{tr}^A(M)$ is, of course, similar. The proof of the independence of the choice of ONBs is left as an exercise.

Exercise 3.41 Show that the operator $\text{tr}^B(M)$ as given in (3.46) does not depend on the choice of the ONBs $\{|e_a\rangle\}$ and $\{|f_b\rangle\}$.

For a solution see Solution 3.41.

Next, we verify that $\text{tr}^B(M)$ as given in (3.46) does indeed satisfy the first equation in (3.47). For this let $\{|e_a\rangle\}$ be an ONB in $\mathbb{H}^A$, $\{|f_b\rangle\}$ an ONB in $\mathbb{H}^B$ and $M \in L(\mathbb{H}^A \otimes \mathbb{H}^B)$ be given by

$$M = \sum_{a_1,a_2,b_1,b_2} M_{a_1 b_1,a_2 b_2} |e_{a_1} \otimes f_{b_1}\rangle\langle e_{a_2} \otimes f_{b_2}| \,.$$

Furthermore, let

$$M^A = \sum_{a_1,a_2} M^A_{a_1 a_2} |e_{a_1}\rangle\langle e_{a_2}|$$

be an arbitrary operator in $L(\mathbb{H}^A)$. Then one finds that

$$
\begin{aligned}
&\text{tr}\left((M^A \otimes \mathbf{1}^B)M\right) \\
&\underbrace{=}_{(2.57)} \sum_{a_3,b_3} \langle e_{a_3} \otimes f_{b_3}| (M^A \otimes \mathbf{1}^B) \sum_{a_1,a_2,b_1,b_2} |e_{a_1} \otimes f_{b_1}\rangle M_{a_1 b_1,a_2 b_2} \underbrace{\langle e_{a_2} \otimes f_{b_2}|e_{a_3} \otimes f_{b_3}\rangle}_{=\delta_{a_2 a_3}\delta_{b_2 b_3}} \\
&= \sum_{a_1,a_2,b_1,b_2} \langle e_{a_2} \otimes f_{b_2}|(M^A e_{a_1}) \otimes f_{b_1}\rangle M_{a_1 b_1,a_2 b_2} \\
&= \sum_{a_1,a_2,b_1,b_2} \langle e_{a_2}|M^A e_{a_1}\rangle \underbrace{\langle f_{b_2}|f_{b_1}\rangle}_{=\delta_{b_1 b_2}} M_{a_1 b_1,a_2 b_2} = \sum_{a_1,a_2,b} M^A_{a_2 a_1} M_{a_1 b,a_2 b} \\
&\underbrace{=}_{(3.15)} \sum_{a_1,a_2} M^A_{a_2 a_1}\, \text{tr}^B(M)_{a_1 a_2} = \sum_{a_2} \left(M^A\, \text{tr}^B(M)\right)_{a_2 a_2} \\
&\underbrace{=}_{(2.57)} \text{tr}\left(M^A\, \text{tr}^B(M)\right) \,,
\end{aligned}
$$

verifying that $\text{tr}^B(M)$ as given in (3.46) indeed satisfies the first equation in (3.47).

Lastly, we show uniqueness. Let $\widetilde{\text{tr}^B}(M)$ be another operator on $\mathbb{H}^A$, which satisfies the first equation in (3.41). Then for any $M^A \in L(\mathbb{H}^A)$ one finds

$$
\begin{aligned}
\text{tr}\left(M^A\left(\widetilde{\text{tr}^B}(M) - \text{tr}^B(M)\right)\right) &= \text{tr}\left(M^A \widetilde{\text{tr}^B}(M)\right) - \text{tr}\left(M^A\, \text{tr}^B(M)\right) \\
&\underbrace{=}_{(3.47)} \text{tr}\left((M^A \otimes \mathbf{1}^B)M\right) - \text{tr}\left((M^A \otimes \mathbf{1}^B)M\right) \\
&= 0 \,,
\end{aligned}
$$

and because of (2.59) thus $\widetilde{\text{tr}^B}(M) = \text{tr}^B(M)$. $\qquad\square$

Note that $\mathrm{tr}^B(M)$ is an operator in $\mathrm{L}(\mathbb{H}^A)$. Hence, $M^A \mathrm{tr}^B(M) \in \mathrm{L}(\mathbb{H}^A)$ and the trace on the left side of the first equation in (3.47) is a complex number obtained from taking the trace on an operator in $\mathrm{L}(\mathbb{H}^A)$. The trace on the right side of the first equation in (3.47) is a complex number obtained from taking the trace on the operator $(M^A \otimes \mathbf{1}^B)M \in \mathrm{L}(\mathbb{H}^A \otimes \mathbb{H}^B)$. Theorem 3.15 states that there is a unique operator $\mathrm{tr}^B(M)$ such that for every $M^A \in \mathrm{L}(\mathbb{H}^A)$ these two complex numbers coincide and that this operator is given as in (3.46). Likewise, the theorem makes analogous statements for $\mathrm{tr}^A(M)$. These results of Theorem 3.15 thus allow the following definition.

Definition 3.16 Let $\mathbb{H}^A$ and $\mathbb{H}^B$ be two HILBERT spaces. The **partial trace** over $\mathbb{H}^B$ is defined as the map

$$\mathrm{tr}^B : \mathrm{L}(\mathbb{H}^A \otimes \mathbb{H}^B) \longrightarrow \mathrm{L}(\mathbb{H}^A)$$
$$M \longmapsto \mathrm{tr}^B(M) \, ,$$

where $\mathrm{tr}^B(M) \in \mathrm{L}(\mathbb{H}^A)$ is the unique operator that satisfies

$$\forall M^A \in \mathrm{L}(\mathbb{H}^A) : \quad \mathrm{tr}\left(M^A \mathrm{tr}^B(M)\right) = \mathrm{tr}\left((M^A \otimes \mathbf{1}^B)M\right) . \tag{3.48}$$

Similarly, the partial trace tr^A over $\mathbb{H}^A$ is defined as

$$\mathrm{tr}^A : \mathrm{L}(\mathbb{H}^A \otimes \mathbb{H}^B) \longrightarrow \mathrm{L}(\mathbb{H}^B)$$
$$M \longmapsto \mathrm{tr}^A(M) \, ,$$

where $\mathrm{tr}^A(M) \in \mathrm{L}(\mathbb{H}^B)$ is the unique operator that satisfies

$$\forall M^B \in \mathrm{L}(\mathbb{H}^B) : \quad \mathrm{tr}\left(M^B \mathrm{tr}^A(M)\right) = \mathrm{tr}\left((\mathbf{1}^A \otimes M^B)M\right) .$$

The standard terminology 'partial trace' has the potential to mislead. This is because whereas the trace on operators of a HILBERT space $\mathbb{H}$ is defined in Definition 2.13 as a linear map

$$\mathrm{tr} : \mathrm{L}(\mathbb{H}) \to \mathbb{C} \, ,$$

such that evaluating it results in a *complex number*, the partial trace is a linear map

$$\mathrm{tr}^B : \mathrm{L}(\mathbb{H}^A \otimes \mathbb{H}^B) \to \mathrm{L}(\mathbb{H}^A) \, ,$$

and evaluating the partial trace on an operator results in an *operator*.

However, taking the trace of a partial trace gives the same number as taking the trace of the original operator as is to be shown in Exercise 3.42.

Exercise 3.42 Show that for any $M \in L(\mathbb{H}^A \otimes \mathbb{H}^B)$ we have

$$\mathrm{tr}\left(\mathrm{tr}^B\left(M\right)\right) = \mathrm{tr}\left(M\right) = \mathrm{tr}\left(\mathrm{tr}^A\left(M\right)\right) . \tag{3.49}$$

For a solution see Solution 3.42.

With the help of the partial trace we can define from the state ρ of the composite system an operator ρ^A, which has the properties of a density operator and describes the state of the sub-system A when observed alone.

Theorem 3.17 *Let $\rho \in D(\mathbb{H}^A \otimes \mathbb{H}^B)$ be the density operator describing the state of a composite system $\mathbb{H}^A \otimes \mathbb{H}^B$. Then*

$$\rho^A(\rho) := \mathrm{tr}^B\left(\rho\right) \tag{3.50}$$

is the uniquely determined density operator on $\mathbb{H}^A$, which describes the state if only the sub-system A is observed. For any observable $M^A \in B_{sa}(\mathbb{H}^A)$ it satisfies

$$\left\langle M^A \right\rangle_{\rho^A(\rho)} = \left\langle M^A \otimes \mathbf{1}^B \right\rangle_\rho . \tag{3.51}$$

Let furthermore $\{|e_a\rangle\}$ be an ONB in $\mathbb{H}^A$ and $\{|f_b\rangle\}$ an ONB in $\mathbb{H}^B$ as well as $\rho_{a_1 b_1, a_2 b_2}$ be the matrix of ρ in the ONB $\{|e_a \otimes f_b\rangle\}$ in $\mathbb{H}^A \otimes \mathbb{H}^B$. Then the matrix of $\rho^A(\rho)$ in the ONB $\{|e_a\rangle\}$ is given by

$$\rho^A(\rho)_{a_1 a_2} = \sum_b \rho_{a_1 b, a_2 b} . \tag{3.52}$$

Proof That $\mathrm{tr}^B\left(\rho\right)$ exists and is unique was shown in Theorem 3.15. There we also showed with (3.46) that $\rho^A(\rho) = \mathrm{tr}^B\left(\rho\right)$ has the matrix given in (3.52).

Observation of an observable M^A of the sub-system A of a composite system described by states in $D(\mathbb{H}^A \otimes \mathbb{H}^B)$ is akin to observation of the observable $M^A \otimes \mathbf{1}^B$ in the composite system. As a consequence of Definition 3.16 we have that $\rho^A(\rho) = \mathrm{tr}^B\left(\rho\right)$ satisfies

$$\left\langle M^A \otimes \mathbf{1}^B \right\rangle_\rho \underset{(2.85)}{=} \mathrm{tr}\left((M^A \otimes \mathbf{1}^B)\rho\right) \underset{(3.48)}{=} \mathrm{tr}\left(M^A\, \mathrm{tr}^B\left(\rho\right)\right) \underset{(3.50)}{=} \mathrm{tr}\left(M^A \rho^A(\rho)\right)$$

$$\underset{(2.85)}{=} \left\langle M^A \right\rangle_{\rho^A(\rho)} , \tag{3.53}$$

verifying (3.51).

That ρ^A is a density operator on $\mathbb{H}^A$ follows from the fact that it satisfies all defining properties of a density operator as we now show.

ρ^A is self-adjoint: to prove this it suffices to show $\rho^A(\rho)^*_{a_1 a_2} = \rho^A(\rho)_{a_1 a_2}$ in an arbitrary ONB $\{|e_a\rangle\} \subset \mathbb{H}^A$:

$$
\begin{aligned}
\rho^A(\rho)^*_{a_1 a_2} &= \overline{\rho^A(\rho)_{a_2 a_1}} \\
&\underset{(3.52)}{=} \overline{\sum_b \rho_{a_2 b, a_1 b}} = \sum_b \overline{\rho_{a_2 b, a_1 b}} \underset{\rho^*=\rho}{=} \sum_b \rho_{a_1 b, a_2 b} \\
&\underset{(3.52)}{=} \rho^A(\rho)_{a_1 a_2}
\end{aligned}
$$

ρ^A is positive: let $\{|f_b\rangle\}$ be an ONB in $\mathbb{H}^B$ and $|\varphi\rangle \in \mathbb{H}^A$ be arbitrary. Then it follows that

$$
\begin{aligned}
\langle \varphi | \rho^A(\rho) \varphi \rangle &= \sum_{a_1, a_2} \overline{\varphi_{a_1}} \rho^A(\rho)_{a_1 a_2} \varphi_{a_2} \\
&\underset{(3.52)}{=} \sum_{a_1, a_2} \overline{\varphi_{a_1}} \sum_b \rho_{a_1 b, a_2 b} \varphi_{a_2} = \sum_{a_1, a_2, b} \overline{\varphi_{a_1}} \rho_{a_1 b, a_2 b} \varphi_{a_2} \\
&= \sum_b \underbrace{\langle \varphi \otimes f_b | \rho (\varphi \otimes f_b) \rangle}_{\geq 0 \text{ since } \rho \geq 0} \\
&\geq 0 .
\end{aligned}
$$

ρ^A has trace 1:

$$
\begin{aligned}
\mathrm{tr}\left(\rho^A(\rho)\right) &= \sum_a \rho^A(\rho)_{aa} \\
&\underset{(3.52)}{=} \sum_a \sum_b \rho_{ab, ab} = \mathrm{tr}(\rho) \\
&\underset{(2.82)}{=} 1 .
\end{aligned}
$$

This shows that ρ^A is a density operator on $\mathbb{H}^A$ and thus it describes a state in the sub-system A. Observing only the sub-system A of a composite system means measuring only observables of the form $M^A \otimes \mathbf{1}^B$ in the composite system. Because of (3.53) the expectation values of such observables in the composite state ρ are identical to the expectation values of M^A in the state ρ^A. Consequently, ρ^A is the state that describes the physical situation when observations are restricted to the sub-system A. $\square$

Regarding the notation, note that $\rho^A(\rho)$ is the state, which describes the physics if one only observes sub-system A. It is obtained from the state ρ of the composite system by taking the partial trace over the sub-system B, that is, $\rho^A(\rho) = \mathrm{tr}^B(\rho)$.

Definition 3.18 For a density operator ρ on $\mathbb{H}^A \otimes \mathbb{H}^B$ the **reduced density operator** on $\mathbb{H}^A$ is defined as

$$\rho^A(\rho) := \mathrm{tr}^B(\rho)$$

and the reduced density operator on $\mathbb{H}^B$ as

$$\rho^B(\rho) := \mathrm{tr}^A(\rho).$$

Example 3.19 As an example with qubit spaces we determine $\rho^A(\Phi^+)$ for the BELL basis vector

$$|\Phi^+\rangle = \frac{1}{\sqrt{2}}\left(|00\rangle + |11\rangle\right) = \underbrace{\frac{1}{\sqrt{2}}|0\rangle \otimes |0\rangle}_{=\Phi_{00}^+} + \underbrace{\frac{1}{\sqrt{2}}|1\rangle \otimes |1\rangle}_{=\Phi_{11}^+}. \tag{3.54}$$

Then we obtain

$$\begin{aligned}
\rho^A(\Phi^+) \underbrace{=}_{(3.44)} & \sum_{a_1,a_2,b} \overline{\Phi_{a_2 b}^+} \Phi_{a_1 b}^+ |e_{a_1}\rangle\langle e_{a_2}| \\
= & \sum_{a_1,a_2} \left(\overline{\Phi_{a_2 0}^+}\Phi_{a_1 0}^+ + \overline{\Phi_{a_2 1}^+}\Phi_{a_1 1}^+\right)|e_{a_1}\rangle\langle e_{a_2}| \\
\underbrace{=}_{(3.54)} & \frac{1}{2}|0\rangle\langle 0| + \frac{1}{2}|1\rangle\langle 1| \\
= & \frac{1}{2}\mathbf{1}^A. \tag{3.55}
\end{aligned}$$

Since $\left(\rho^A(\Phi^+)\right)^2 = \frac{1}{4}\mathbf{1}^A < \frac{1}{2}\mathbf{1}^A = \rho^A(\Phi^+)$, Alice indeed observes a true mixture in her sub-system even though the composite system is in a pure state $|\Phi^+\rangle$.

Corollary 3.20 *Similarly, we have for a state of the composite system given by the density operator $\rho \in \mathrm{D}(\mathbb{H}^A \otimes \mathbb{H}^B)$, that the sub-system in $\mathbb{H}^B$ is* *described by the reduced density operator*

$$\rho^B(\rho) = \text{tr}^A(\rho) \, . \tag{3.56}$$

For all observables M^B one then has

$$\left\langle M^B \right\rangle_{\rho^B(\rho)} = \left\langle \mathbf{1}^A \otimes M^B \right\rangle_\rho \, ,$$

where now the state $\rho^B(\rho) = \text{tr}^A(\rho)$ in B is obtained by calculating the partial trace over A of the state ρ of the composite system. Accordingly, we have for the matrix elements

$$\rho^B(\rho)_{b_1 b_2} = \sum_a \rho_{ab_1, ab_2} \, .$$

Proof The proof is very much the same as that for Theorem 3.17.   □

Exercise 3.43 For $X \in \{A, B\}$ let $M^X \in \text{L}(\mathbb{H}^X)$. Show that then $M^A \otimes M^B \in \text{L}(\mathbb{H}^A \otimes \mathbb{H}^B)$ satisfies

$$\begin{aligned}
\text{tr}\left(M^A \otimes M^B\right) &= \text{tr}\left(M^A\right)\text{tr}\left(M^B\right) \\
\text{tr}^B\left(M^A \otimes M^B\right) &= \text{tr}\left(M^B\right)M^A \\
\text{tr}^A\left(M^A \otimes M^B\right) &= \text{tr}\left(M^A\right)M^B \, .
\end{aligned} \tag{3.57}$$

For a solution see Solution 3.43.

As an example for reduced density operators we compute those arising from considering one sub-system when the composite two-qubit system is in one of the BELL basis states.

Exercise 3.44 Determine $\rho^A(\Phi^-), \rho^A(\Psi^\pm), \rho^B(\Phi^\pm)$ and $\rho^B(\Psi^\pm)$ for the vectors $|\Phi^\pm\rangle, |\Psi^\pm\rangle$ of the BELL basis (3.28).

For a solution see Solution 3.44.

From Postulate 7 it follows that a system comprised of n qubits is described by the HILBERT space $\mathbb{H}^{\otimes n}$. In general any observable of such a system or its operator thus acts on n qubits. As Exercise 3.45 shows, any such operator can be expressed as a suitable linear combination of n-fold tensor products of operators on one qubit.

Exercise 3.45 Let $\mathbb{V}$ be a finite-dimensional vector space over a field $\mathbb{F}$. Show that then for any $n \in \mathbb{N}$

$$\mathrm{L}\big(\mathbb{V}^{\otimes n}\big) = \mathrm{L}(\mathbb{V})^{\otimes n} . \tag{3.58}$$

For a solution see Solution 3.45.

Note that (3.58) does not mean that every $A \in \mathrm{L}(\mathbb{V}^{\otimes n})$ is of the form $A = A_1 \otimes \cdots \otimes A_n$ for some $A_1, \ldots, A_n \in \mathrm{L}(\mathbb{V})$. Rather, it means that every $A \in \mathrm{L}(\mathbb{V}^{\otimes})$ is of the form

$$A = \sum_j a_{j_1 \ldots j_n} A_{j_1} \otimes \cdots \otimes A_{j_n}$$

for some $a_{j_1 \ldots j_n} \in \mathbb{F}$ and $A_{j_1}, \ldots, A_{j_n} \in \mathrm{L}(\mathbb{V})$.

A special notion has been reserved for operators which can be written as a sum of operators, each of which acts non-trivially on no more than $k \leq n$ qubits.

Definition 3.21 An operator $A \in \mathrm{L}(\text{'}\mathbb{H}^{\otimes n})$ is said to be k-**local** if it is of the form

$$A = \sum_{j \in I} a_{j_{n-1}, \ldots, j_0} A_{j_{n-1}} \otimes \cdots \otimes A_{j_0} ,$$

where $I \subset \mathbb{N}$ and for each $j \in I$ and $l \in \{0, \ldots, n-1\}$ we have $a_{j_l} \in \mathbb{C}$ and $A_{j_l} \in \mathrm{L}(\text{'}\mathbb{H})$, and the index sets

$$I_j = \big\{ l \in \{0, \ldots, n-1\} \,\big|\, A_{j_l} \neq \mathbf{1} \big\}$$

of qubits on which the $A_{j_{n-1}} \otimes \cdots \otimes A_{j_0}$ act non-trivially satisfy $|I_j| \leq k$.

As any physical implementation of well controlled interactions of many qubits is very challenging one is clearly interested in cases where $k \ll n$.

Example 3.22 For $j \in \{0, \ldots, n-1\}$ let

$$\Sigma_z^j = \mathbf{1}^{\otimes n-1-j} \otimes \sigma_z \otimes \mathbf{1}^{\otimes j}$$

such that $\Sigma_z^j \in \mathrm{L}(\text{'}\mathbb{H}^{\otimes n})$ and with $a(t), b(t), K_j, J_{jl} \in \mathbb{R}$ let

$$\mathsf{H}(t) = a(t) \sum_{j=0}^{n-1} K_j \Sigma_z^j + b(t) \sum_{j,l=0, j \neq l}^{n-1} J_{jl} \Sigma_z^j \Sigma_z^l$$

be a Hamiltonian on $\P\mathbb{H}^{\otimes n}$ of ISING-type [56]. Then $\mathsf{H}(t)$ is 2-local.

3.4 SCHMIDT Decomposition

For pure states $|\varPsi\rangle \in \mathbb{H}^A \otimes \mathbb{H}^B$ in composite systems one can utilize the eigenvectors of the reduced density operators to obtain ONBs in $\mathbb{H}^A$ and $\mathbb{H}^B$, which allow a lean and useful representation of $|\varPsi\rangle$. In the following we briefly present this construction, which is known as the SCHMIDT decomposition.

Let

$$|\varPsi\rangle = \sum_{a,b} \varPsi_{ab} |e_a \otimes f_b\rangle$$

be a pure state in $\mathbb{H}^A \otimes \mathbb{H}^B$ and let

$$\rho^A(\varPsi) = \sum_{a_1,a_2,b} \varPsi_{a_1 b} \overline{\varPsi_{a_2 b}} |e_{a_1}\rangle \langle e_{a_2}|$$

be the corresponding density operator. Since $\rho^A(\varPsi)$ is a self-adjoint and positive operator on $\mathbb{H}^A$ there exists an ONB $\{|\widetilde{e_a}\rangle\}$ in $\mathbb{H}^A$ consisting of eigenvectors of $\rho^A(\varPsi)$, such that

$$\rho^A(\varPsi) = \sum_a q_a |\widetilde{e_a}\rangle \langle \widetilde{e_a}| \,, \tag{3.59}$$

where the $q_a \geq 0$ are the eigenvalues. From (2.56) we know that the ONBs $\{|\widetilde{e_a}\rangle\}$ and $\{|e_a\rangle\}$ are mapped into each other by a unitary operator $U \in \mathcal{U}(\mathbb{H}^A)$:

$$|\widetilde{e_a}\rangle = U|e_a\rangle = \sum_{a_1} |e_{a_1}\rangle \underbrace{\langle e_{a_1}|U e_a\rangle}_{=:U_{a_1 a}} \,.$$

With

$$\widetilde{\varPsi_{ab}} := \sum_{a_1} U_{aa_1}^* \varPsi_{a_1 b}$$

one finds

$$|\varPsi\rangle = \sum_{a,b} \widetilde{\varPsi_{ab}} |\widetilde{e_a} \otimes f_b\rangle \,,$$

which, according to Definition (3.44) of the reduced density operator, implies

$$\rho^A(\Psi) = \sum_{a_1,a_2,b} \widetilde{\Psi_{a_1b}} \overline{\widetilde{\Psi_{a_2b}}} |\widetilde{e_{a_1}}\rangle\langle\widetilde{e_{a_2}}| . \tag{3.60}$$

Comparison of (3.59) with (3.60) shows that

$$\sum_b \widetilde{\Psi_{a_1b}} \overline{\widetilde{\Psi_{a_2b}}} = \delta_{a_1a_2} q_{a_2} \tag{3.61}$$

has to hold. In particular, we thus have

$$q_a = 0 \quad \Leftrightarrow \quad \widetilde{\Psi_{ab}} = 0 \quad \forall b . \tag{3.62}$$

For $q_a > 0$ we define the vectors

$$|\widetilde{f_a}\rangle := \frac{1}{\sqrt{q_a}} \sum_b \widetilde{\Psi_{ab}} |f_b\rangle \quad \in \mathbb{H}^B . \tag{3.63}$$

The set of such defined $|\widetilde{f_a}\rangle$ is orthonormal since

$$\langle\widetilde{f_{a_1}}|\widetilde{f_{a_2}}\rangle = \frac{1}{\sqrt{q_{a_1}q_{a_2}}} \sum_{b_1,b_2} \overline{\widetilde{\Psi_{a_1b_1}}} \widetilde{\Psi_{a_2b_2}} \underbrace{\langle f_{b_1}|f_{b_2}\rangle}_{=\delta_{b_1b_2}} = \frac{1}{\sqrt{q_{a_1}q_{a_2}}} \sum_b \overline{\widetilde{\Psi_{a_1b}}} \widetilde{\Psi_{a_2b}}$$

$$\underbrace{=}_{(3.61)} \delta_{a_1a_2} .$$

With this we obtain

$$|\Psi\rangle = \sum_{a,b} \widetilde{\Psi_{ab}} |\widetilde{e_a} \otimes f_b\rangle$$

$$= \sum_{q_a\neq 0} |\widetilde{e_a}\rangle \otimes \underbrace{\sum_b \widetilde{\Psi_{ab}} |f_b\rangle}_{=\sqrt{q_a}|\widetilde{f_a}\rangle} + \sum_{q_a=0} \sum_b \underbrace{\widetilde{\Psi_{ab}}}_{=0} |\widetilde{e_a} \otimes f_b\rangle$$

$$= \sum_{q_a\neq 0} \sqrt{q_a} |\widetilde{e_a} \otimes \widetilde{f_a}\rangle ,$$

where in the second equation we used Definition (3.63) in the first sum and the relation (3.62) in the second sum. Lastly, we can drop the restriction $q_a \neq 0$ in the last equation, since the corresponding terms do not contribute anything. One can then also extend the set of orthonormal vectors $|\widetilde{f_a}\rangle$ to an ONB in $\mathbb{H}^B$ by supplementing the set with suitable vectors. Then basis vectors $|\widetilde{f_a}\rangle$ would also be defined for $q_a = 0$. The result is the **SCHMIDT decomposition** of $|\Psi\rangle \in \mathbb{H}^A \otimes \mathbb{H}^B$:

$$|\Psi\rangle = \sum_a \sqrt{q_a}|\widetilde{e}_a \otimes \widetilde{f}_a\rangle .$$ (3.64)

Note that the ONB $\{|\widetilde{e}_a\rangle\}$ and the $\{|\widetilde{f}_a\rangle\}$ *depend on* $|\Psi\rangle$, that is, for other vectors $|\Phi\rangle \in \mathbb{H}^A \otimes \mathbb{H}^B$ one in general obtains different ONBs $\{|\widetilde{e}_a\rangle\}$ and $\{\widetilde{f}_a\}$. With the help of Definition (3.43) of the reduced density operator it follows immediately from the SCHMIDT decomposition that

$$\rho^A(\Psi) = \sum_a q_a|\widetilde{e}_a\rangle\langle\widetilde{e}_a| ,$$

which is a necessary consequence of (3.59) since this was the starting point of the construction. From (3.64) and (3.45) it also follows that

$$\rho^B(\Psi) = \sum_b q_b|\widetilde{f}_b\rangle\langle\widetilde{f}_b| .$$

The ONBs $\{|\widetilde{e}_a\rangle\}$ and $\{|\widetilde{f}_b\rangle\}$ in the SCHMIDT decomposition are only *unique* in case all non-vanishing eigenvalues of $\rho^A(\Psi)$—and thus according to the above also those of $\rho^B(\Psi)$—are non-degenerate. In case a non-zero eigenvalue of $\rho^A(\Psi)$ is degenerate, the ONB in the corresponding eigenspace is not uniquely determined. Let $d_{\bar{a}} > 1$ denote the dimension of the eigenspace of the degenerate $\bar{a}$-th eigenvalue $q_{\bar{a}} \neq 0$ of $\rho^A(\psi)$. Moreover, for $k \in \{1, \ldots, d_{\bar{a}}\}$ let $|\widetilde{e_{\bar{a},k}}\rangle$ be the eigenvectors belonging to $q_{\bar{a}}$. Then one has

$$\rho^A(\Psi) = \sum_{a\neq\bar{a}} q_a|\widetilde{e}_a\rangle\langle\widetilde{e}_a| + q_{\bar{a}} \sum_{k=1}^{d_{\bar{a}}} |\widetilde{e_{\bar{a},k}}\rangle\langle\widetilde{e_{\bar{a},k}}|$$

$$= \sum_{a\neq\bar{a}} q_a|\widetilde{e}_a\rangle\langle\widetilde{e}_a| + q_{\bar{a}} \sum_{k=1}^{d_{\bar{a}}} |\widetilde{\widetilde{e_{\bar{a},k}}}\rangle\langle\widetilde{\widetilde{e_{\bar{a},k}}}|$$

with

$$|\widetilde{\widetilde{e_{\bar{a}k}}}\rangle = \sum_{l=1}^{d_{\bar{a}}} U_{kl}^{\bar{a}}|\widetilde{e_{\bar{a}l}}\rangle ,$$

where $U_{kl}^{\bar{a}}$ is the matrix of an arbitrary unitary transformation in the eigenspace for $q_{\bar{a}}$. Then one has for the SCHMIDT decomposition the following options

$$|\Psi\rangle = \sum_{q_a\neq q_{\bar{a}}} \sqrt{q_a}|\widetilde{e}_a \otimes \widetilde{f}_a\rangle + \sqrt{q_{\bar{a}}} \sum_{k=1}^{d_{\bar{a}}} |\widetilde{e}_{\bar{a}} \otimes \widetilde{f}_{\bar{a}}\rangle$$

$$= \sum_{q_a\neq q_{\bar{a}}} \sqrt{q_a}|\widetilde{e}_a \otimes \widetilde{f}_a\rangle + \sqrt{q_{\bar{a}}} \sum_{k=1}^{d_{\bar{a}}} |\widetilde{\widetilde{e}}_{\bar{a}} \otimes \widetilde{\widetilde{f}}_{\bar{a}}\rangle ,$$

demonstrating the non-uniqueness of the ONBs in the case of degenerate eigenvalues.

3.5   Quantum Operations

According to the Projection and Time Evolution specifications in Postulate 6 there
are two ways a quantum system can change: by unitary time evolution generated by
some Hamiltonian or by a state transformation effected by a measurement.

Yet another way to generate a state transformation is to combine our system of
interest with another system to a composite system. Then time evolution or measurements may be performed on the combined system after which the second system is
discarded and only the first system is retained. This process of combining our system of interest with another system to a composite system on which some action is
performed and then ignoring the added sub-system can come about in two ways:

On purpose   That is, we want to make use of the added system as a computational
   resource. This is the case when we build circuits with ancillas (see Sect. 5.3). This
   includes the case where we might want to observe the added system, in other
   words, the ancillas, and where such observation has an effect on our principal
   system A.
By error   This happens when we cannot shield our system of interest and it interacts with the environment. This is the case we will need to deal with when we
   consider quantum error correction (see Sect. 7.3). Here, too, we include the case
   where the environment is observed and such observation affects the principle
   system.

The notion of a quantum operation will be very useful to describe these state
transformations in a rather compact way. As a start for preparing its definition in
more detail we recap the stages of the state transformations alluded to above.

1. We begin by preparing the principle system A of interest to us in the state

$$\rho^A \in D(\mathbb{H}^A) .$$

2. We then proceed to combine system A in state ρ^A with a system B in the state
   $\rho^B \in D(\mathbb{H}^B)$. System B will be the ancillas in case we are looking at circuits
   or the environment in case we are considering errors. We assume that the two
   systems A and B are initially separable. Mathematically we can describe this
   enlargement to a composite system as the embedding

$$\iota_{\rho^B} : D(\mathbb{H}^A) \longrightarrow D(\mathbb{H}^A \otimes \mathbb{H}^B)$$
$$\rho^A \longmapsto \rho^A \otimes \rho^B \quad .$$

3. The time evolution $U \in \mathcal{U}(\mathbb{H}^A \otimes \mathbb{H}^B)$ of the combined system transforms the
   combined state as

$$U : D(\mathbb{H}^A \otimes \mathbb{H}^B) \longrightarrow D(\mathbb{H}^A \otimes \mathbb{H}^B)$$
$$\rho^A \otimes \rho^B \longmapsto U(\rho^A \otimes \rho^B)U^* \quad .$$

In general this will result in an entangled state of the combined system.

4. Let P^B be the projector onto the eigenspace of an observable of system B. Suppose upon measuring this observable we obtain the eigenvalue corresponding to the eigenspace onto which P^B projects. According to (2.87) in the Projection Postulate for mixed states this results in the state transformation

$$U(\rho^A \otimes \rho^B)U^* \to \frac{(\mathbf{1}^A \otimes P^B)U(\rho^A \otimes \rho^B)U^*(\mathbf{1}^A \otimes P^B)}{\mathrm{tr}\left((\mathbf{1}^A \otimes P^B)U(\rho^A \otimes \rho^B)U^*\right)}\,.$$

If we do not measure system B, we set $P^B = \mathbf{1}^B$. Noting that then

$$\mathrm{tr}\left((\mathbf{1}^A \otimes P^B)U(\rho^A \otimes \rho^B)U^*\right) = \mathrm{tr}\left(U(\rho^A \otimes \rho^B)U^*\right) \underset{(2.58)}{=} \mathrm{tr}\left((\rho^A \otimes \rho^B)U^*U\right)$$

$$\underset{(2.7)}{=} \mathrm{tr}\left(\rho^A \otimes \rho^B\right) \underset{(3.57)}{=} \mathrm{tr}\left(\rho^A\right)\mathrm{tr}\left(\rho^B\right)$$

$$\underset{(2.83)}{=} 1$$

we see that in this case the state of the combined system remains $U(\rho^A \otimes \rho^B)U^*$.

5. Discarding or ignoring system B we can still obtain a description of system A only by taking the partial trace over B:

$$\frac{(\mathbf{1}^A \otimes P^B)U(\rho^A \otimes \rho^B)U^*(\mathbf{1}^A \otimes P^B)}{\mathrm{tr}\left((\mathbf{1}^A \otimes P^B)U(\rho^A \otimes \rho^B)U^*\right)} \to \frac{\mathrm{tr}^B\left((\mathbf{1}^A \otimes P^B)U(\rho^A \otimes \rho^B)U^*(\mathbf{1}^A \otimes P^B)\right)}{\mathrm{tr}\left((\mathbf{1}^A \otimes P^B)U(\rho^A \otimes \rho^B)U^*\right)}\,,$$

where in the description of the final state on the right we have used the linearity of the partial trace to pull the denominator $\mathrm{tr}(\cdots)$ out of tr^B.

Altogether the initial state ρ^A of the principal system A of interest is thus transformed by the steps 1–5 as

$$\begin{aligned}\mathrm{D}\big(\mathbb{H}^A\big) &\longrightarrow \mathrm{D}\big(\mathbb{H}^A\big)\\ \rho^A &\longmapsto \frac{\mathrm{tr}^B\big((\mathbf{1}^A \otimes P^B)U(\rho^A \otimes \rho^B)U^*(\mathbf{1}^A \otimes P^B)\big)}{\mathrm{tr}\big((\mathbf{1}^A \otimes P^B)U(\rho^A \otimes \rho^B)U^*\big)}\,.\end{aligned} \tag{3.65}$$

It turns out that this transformation can be written in terms of suitable operators on $\mathbb{H}^A$ only. Before we can establish this we need the following preparatory lemma.

Lemma 3.23 *Let $\mathbb{H}$ be a* Hilbert *space and for $l \in \{1,\dots,m\}$ let $K_l \in \mathrm{L}(\mathbb{H})$. Then for any $\rho \in \mathrm{D}(\mathbb{H})$ the operator*

$$K(\rho) = \sum_{l=1}^m K_l \rho K_l^* \tag{3.66}$$

satisfies $K(\rho)^* = K(\rho)$ and $0 \le K(\rho)$. Moreover, for any $\kappa \in {]0,1]}$ we have

$$\sum_{l=1}^{m} K_l^* K_l \le \kappa \mathbf{1} \quad \Leftrightarrow \quad \operatorname{tr}(K(\rho)) \le \kappa \quad \forall \rho \in D(\mathbb{H}) \qquad (3.67)$$

and equality in one side of (3.67) implies equality on the other.

Proof To begin with, recall from (2.83) that $\rho \in D(\mathbb{H})$ implies $\rho^* = \rho$, $\rho \ge 0$ and $\operatorname{tr}(\rho) = 1$. Then we have

$$K(\rho)^* \underset{(3.23)}{=} \left(\sum_{l=1}^{m} K_l \rho K_l^*\right)^* = \sum_{l=1}^{m} (K_l \rho K_l^*)^* \underset{(2.47)}{=} \sum_{l=1}^{m} (K_l^*)^* \rho^* K_l^* \underset{(2.31),(2.83)}{=} \sum_{l=1}^{m} K_l \rho K_l^*$$

$$\underset{(3.23)}{=} K(\rho).$$

and likewise

$$\left(\sum_{l=1}^{m} K_l^* K_l\right)^* = \sum_{l=1}^{m} (K_l^* K_l)^* \underset{(2.47)}{=} \sum_{l=1}^{m} K_l^* (K_l^*)^* \underset{(2.31)}{=} \sum_{l=1}^{m} K_l^* K_l. \qquad (3.68)$$

Moreover, for any $|\psi\rangle \in \mathbb{H}$

$$\langle\psi|K(\rho)\psi\rangle = \langle\psi|\sum_{l=1}^{m} K_l \rho K_l^* \psi\rangle = \sum_{l=1}^{m}\langle\psi|K_l\rho K_l^*\psi\rangle \underset{(2.30)}{=} \sum_{l=1}^{m}\underbrace{\langle K_l^*\psi|\rho K_l^*\psi\rangle}_{\ge 0 \text{ since } \rho \ge 0} \ge 0.$$

So far, we have shown $K(\rho)^* = K(\rho)$ and $0 \le K(\rho)$. Next, we show $\Rightarrow$ in (3.67). From (3.68) we know that $\sum_{l=1}^{m} K_l^* K_l$ is self-adjoint. Thus, there exists an ONB $\{|e_a\rangle\} \in \mathbb{H}$ of its eigenvectors for its eigenvalues $\{\lambda_a\}$ such that

$$\sum_{l=1}^{m} K_l^* K_l \underset{(2.38)}{=} \sum_{a} \lambda_a |e_a\rangle\langle e_a|. \qquad (3.69)$$

Consequently, we find for every eigenvalue λ_a that

$$\lambda_a = \langle e_a|\left(\sum_{a'}\lambda_{a'}|e_{a'}\rangle\langle e_{a'}|\right)e_a\rangle \underset{(3.69)}{=} \langle e_a|\left(\sum_{l=1}^{m} K_l^* K_l\right)e_a\rangle$$

$$\underset{(3.67)}{\le} \langle e_a|\kappa e_a\rangle = \kappa. \qquad (3.70)$$

Therefore, we obtain for any $\rho \in D(\mathbb{H})$

$$\text{tr}\,(K(\rho)) \underbrace{=}_{(3.66)} \text{tr}\left(\sum_{l=1}^{m} K_l \rho K_l^*\right) = \sum_{l=1}^{m} \text{tr}\,(K_l \rho K_l^*) \underbrace{=}_{(2.58)} \sum_{l=1}^{m} \text{tr}\,(\rho K_l^* K_l)$$

$$= \text{tr}\left(\rho \sum_{l=1}^{m} K_l^* K_l\right) \underbrace{=}_{(3.69)} \text{tr}\left(\rho \sum_a \lambda_a |e_a\rangle\langle e_a|\right) = \sum_a \lambda_a \,\text{tr}\,(\rho|e_a\rangle\langle e_a|)$$

$$\underbrace{\leq}_{(3.70)} \kappa \sum_a \text{tr}\,(\rho|e_a\rangle\langle e_a|) \underbrace{=}_{(2.57)} \kappa \sum_{a,a'} \langle e_{a'}|\rho e_a\rangle \underbrace{\langle e_a|e_{a'}\rangle}_{=\delta_{aa'}} = \kappa \sum_a \langle e_a|\rho e_a\rangle$$

$$\underbrace{=}_{(2.57)} \kappa\,\text{tr}\,(\rho) \underbrace{=}_{(2.83)} \kappa \tag{3.71}$$

proving

$$\sum_{l=1}^{m} K_l^* K_l \leq \kappa\mathbf{1} \quad \Rightarrow \quad \text{tr}\,(K(\rho)) \leq \kappa \quad \forall \rho \in D(\mathbb{H})\,.$$

From the second line in (3.71) we also see that $\sum_{l=1}^{m} K_l^* K_l = \kappa\mathbf{1}$ implies

$$\text{tr}\,(K(\rho)) \underbrace{=}_{(3.71)} \text{tr}\,(\rho\kappa) = \kappa\,\text{tr}\,(\rho) \underbrace{=}_{(3.69)} \kappa \quad \forall \rho \in D(\mathbb{H})\,.$$

To show $\Leftarrow$ in (3.67) note that the second line in (3.71) also shows that $\text{tr}\,(K(\rho)) \leq \kappa$ implies for every $\rho \in D(\mathbb{H})$

$$\kappa \geq \text{tr}\left(\rho \sum_{l=1}^{m} K_l^* K_l\right) \underbrace{=}_{(3.69)} \sum_a \lambda_a \,\text{tr}\,(\rho|e_a\rangle\langle e_a|)\,.$$

Choosing $\rho = |e_{a'}\rangle\langle e_{a'}|$, we have $\text{tr}\,(\rho|e_a\rangle\langle e_a|) = \delta_{aa'}$. It follows that $\lambda_a \leq \kappa$ for all a and thus from (3.69) that $\sum_{l=1}^{m} K_l^* K_l \leq \kappa\mathbf{1}$. Similarly, $\text{tr}\,(K(\rho)) = \kappa$ implies $\sum_{l=1}^{m} K_l^* K_l = \kappa\mathbf{1}$. $\qquad\square$

Event though the constituent maps $\iota_{\rho^B}, U(\cdot)U^*, \mathbf{1}^A \otimes P^B$ and $\text{tr}^B(\cdot)$ in (3.65) operate on the composite state space $D(\mathbb{H}^A \otimes \mathbb{H}^B)$ the resulting state transformation is a map from the state space $D(\mathbb{H}^A)$ onto itself. This raises the question if such a transformation can be expressed with the help of operators acting on $\mathbb{H}^A$ only. The following theorem provides the results to answer this question affirmatively.

Theorem 3.24 *([57]) Let $\mathbb{H}^A$ be a finite-dimensional* HILBERT *space and let*

$$K : D(\mathbb{H}^A) \longrightarrow D_\le(\mathbb{H}^A)$$
$$\rho^A \longmapsto K(\rho^A) \qquad (3.72)$$

be a convex-linear map where

$$D_\le(\mathbb{H}^A) := \left\{ \rho \in \mathbb{H}^A \mid \rho^* = \rho,\ \rho \ge 0,\ \mathrm{tr}\,(\rho) \le 1 \right\}.$$

Then the following equivalence holds for any $\kappa \in]0,1]$.

$$\left. \begin{array}{c} \exists\ \text{HILBERT } space\ \mathbb{H}^B,\ \dim \mathbb{H}^B < \infty, \\ V \in B(\mathbb{H}^A \otimes \mathbb{H}^B),\ \rho^B \in D(\mathbb{H}^B)\ such\ that \\[2mm] V^* V \le \kappa \mathbf{1}^{AB} \qquad (3.73) \\[2mm] and \\[2mm] K(\rho^A) = \mathrm{tr}^B \left(V(\rho^A \otimes \rho^B) V^* \right) \quad (3.74) \end{array} \right\} \Leftrightarrow \left\{ \begin{array}{c} \exists K_l \in L(\mathbb{H}^A)\ for\ l \in \{1,\dots,m\} \\ with\ m \le (\dim \mathbb{H}^B)^2\ such\ that \\[2mm] \displaystyle\sum_{l=1}^m K_l^* K_l \le \kappa \mathbf{1}^A \qquad (3.75) \\[2mm] and \\[2mm] K(\rho^A) = \displaystyle\sum_{l=1}^m K_l \rho^A K_l^* . \qquad (3.76) \end{array} \right.$$

Within this equivalence we have the special case

$$V^* V = \kappa \mathbf{1}^{AB} \quad \Leftrightarrow \quad \sum_{l=1}^m K_l^* K_l = \kappa \mathbf{1}^A,$$

Proof From Lemma 3.23 we know already that any superoperator of the form (3.76) where the K_l satisfy (3.75) maps into $D_\le(\mathbb{H}^A)$, that is, satisfies (3.72). Hence, it remains to prove the equivalence.

We prove $\Rightarrow$ first. Let $V \in L(\mathbb{H}^A \otimes \mathbb{H}^B)$ be such that it satisfies (3.73) and (3.74) and let $\rho^B \in D(\mathbb{H}^B)$. Moreover, let $\{|e_a\rangle\}$ be an ONB in $\mathbb{H}^A$. From Theorem 2.24 we know that there exists an ONB $\{|f_b\rangle\} \subset \mathbb{H}^B$ and a set of $q_b \ge 0$ such that

$$\sum_b q_b = 1 \qquad (3.77)$$

and

$$\rho^B = \sum_b q_b |f_b\rangle\langle f_b| . \qquad (3.78)$$

Using this, we also define

$$\sqrt{\rho^B} = \sum_b \sqrt{q_b} |f_b\rangle\langle f_b| , \qquad (3.79)$$

which satisfies

$$(\sqrt{\rho^B})^2 \underbrace{=}_{(3.79)} \sum_{b_1,b_2} \sqrt{q_{b_1}q_{b_2}}|f_{b_1}\rangle \underbrace{\langle f_{b_1}|f_{b_2}\rangle}_{=\delta_{b_1,b_2}} \langle f_{b_2}| = \sum_b q_b |f_b\rangle\langle f_b| \underbrace{=}_{(3.78)} \rho^B \qquad (3.80)$$

$$(\sqrt{\rho^B})^* \underbrace{=}_{(3.79)} \left(\sum_b \sqrt{q_b}|f_b\rangle\langle f_b|\right)^* \underbrace{=}_{(2.32)} \sum_b \sqrt{q_b}(|f_b\rangle\langle f_b|)^* \underbrace{=}_{(2.36)} \sum_b \sqrt{q_b}|f_b\rangle\langle f_b|$$

$$\underbrace{=}_{(3.79)} \sqrt{\rho^B} . \qquad (3.81)$$

For $b_1, b_2 \in \{1,\ldots,\dim \mathbb{H}^B\}$ we then define the operators $K_{(b_1,b_2)}$ by specifying their matrix elements $(K_{(b_1,b_2)})_{a_1a_2}$ in the ONB $\{|e_a\rangle\}$ as

$$(K_{(b_1,b_2)})_{a_1a_2} = \left(V(\mathbf{1}^A \otimes \sqrt{\rho^B})\right)_{a_1b_1,a_2b_2} . \qquad (3.82)$$

Then we have

$$(K^*_{(b_1,b_2)})_{a_1a_2} \underbrace{=}_{(2.34)} \overline{(K_{(b_1,b_2)})_{a_2a_1}} \underbrace{=}_{(3.82)} \overline{\left(V(\mathbf{1}^A \otimes \sqrt{\rho^B})\right)_{a_2b_1,a_1b_2}} \qquad (3.83)$$

$$\underbrace{=}_{(2.34)} \left(V(\mathbf{1}^A \otimes \sqrt{\rho^B})\right)^*_{a_1b_2,a_2b_1} \underbrace{=}_{(2.47)} \left((\mathbf{1}^A \otimes \sqrt{\rho^B})^* V^*\right)_{a_1b_2,a_2b_1}$$

$$\underbrace{=}_{(3.31)} \left((\mathbf{1}^A \otimes \sqrt{\rho^{B^*}})V^*\right)_{a_1b_2,a_2b_1} \underbrace{=}_{(3.81)} \left((\mathbf{1}^A \otimes \sqrt{\rho^B})V^*\right)_{a_1b_2,a_2b_1}$$

and thus

$$\left(\sum_{b_1,b_2} K_{(b_1,b_2)}\rho^A K^*_{(b_1,b_2)}\right)_{a_1a_2} = \sum_{b_1,b_2}\sum_{a_3,a_4} (K_{(b_1,b_2)})_{a_1a_3}\rho^A_{a_3a_4}(K^*_{(b_1,b_2)})_{a_4a_2}$$

$$\underbrace{=}_{(3.82),(3.83)} \sum_{b_1,b_2}\sum_{a_3,a_4} \left(V(\mathbf{1}^A \otimes \sqrt{\rho^B})\right)_{a_1b_1,a_3b_2}\rho^A_{a_3a_4}\left((\mathbf{1}^A \otimes \sqrt{\rho^B})V^*\right)_{a_4b_2,a_2b_1}$$

$$= \sum_{b_1,b_2,b_3}\sum_{a_3,a_4} \left(V(\mathbf{1}^A \otimes \sqrt{\rho^B})\right)_{a_1b_1,a_3b_2}\rho^A_{a_3a_4}\delta_{b_2,b_3}\left((\mathbf{1}^A \otimes \sqrt{\rho^B})V^*\right)_{a_4b_3,a_2b_1}$$

$$= \sum_{b_1,b_2,b_3}\sum_{a_3,a_4} \left(V(\mathbf{1}^A \otimes \sqrt{\rho^B})\right)_{a_1b_1,a_3b_2} \overbrace{(\rho^A \otimes \mathbf{1}^B)_{a_3b_2,a_4b_3}}^{=\rho^A_{a_3a_4}\delta_{b_2,b_3}}$$

$$\left((\mathbf{1}^A \otimes \sqrt{\rho^B})V^*\right)_{a_4b_3,a_2b_1}$$

$$= \sum_b \left(V(\mathbf{1}^A \otimes \sqrt{\rho^B})(\rho^A \otimes \mathbf{1}^B)(\mathbf{1}^A \otimes \sqrt{\rho^B})V^*\right)_{a_1b,a_2b}$$

$$\underbrace{=}_{(3.80)} \sum_b \left(V(\rho^A \otimes \rho^B)V^*\right)_{a_1b,a_2b} \underbrace{=}_{(3.52)} \mathrm{tr}^B \left(V(\rho^A \otimes \rho^B)V^*\right)_{a_1a_2}$$

$$\underbrace{=}_{(3.74)} K(\rho^A)_{a_1a_2}$$

proving (3.76). The sums $\sum_{b_1,b_2}$ run over the index sets indexing an ONB of $\mathbb{H}^B$ and thus the total number m of operators K_{b_1,b_2} cannot exceed $(\dim \mathbb{H}^B)^2$. To show that (3.73) implies (3.75) we note first that

$$\sum_{b_1,b_2} \left(K^*_{(b_1,b_2)}K_{(b_1,b_2)}\right)_{a_1a_2} = \sum_{b_1,b_2,a_3} \left(K^*_{(b_1,b_2)}\right)_{a_1a_3}\left(K_{(b_1,b_2)}\right)_{a_3a_2}$$

$$\underbrace{=}_{(3.82),(3.83)} \sum_{b_1,b_2,a_3} \left((1^A \otimes \sqrt{\rho^B})V^*\right)_{a_1b_2,a_3b_1}\left(V(1^A \otimes \sqrt{\rho^B})\right)_{a_3b_1,a_2b_2}$$

$$= \sum_b \left((1^A \otimes \sqrt{\rho^B})V^*V(1^A \otimes \sqrt{\rho^B})\right)_{a_1b,a_2b}$$

$$\underbrace{=}_{(3.52)} \mathrm{tr}^B \left((1^A \otimes \sqrt{\rho^B})V^*V(1^A \otimes \sqrt{\rho^B})\right)_{a_1a_2} ,$$

such that

$$\sum_{b_1,b_2} K^*_{(b_1,b_2)}K_{(b_1,b_2)} = \mathrm{tr}^B \left((1^A \otimes \sqrt{\rho^B})V^*V(1^A \otimes \sqrt{\rho^B})\right) . \tag{3.84}$$

Exercise 3.46 Show that (3.84) implies for any $|\psi\rangle \in \mathbb{H}^A \smallsetminus \{0\}$ that

$$\langle\psi| \sum_{b_1,b_2} K^*_{(b_1,b_2)}K_{(b_1,b_2)} \psi\rangle = \frac{1}{||\psi||^2} \mathrm{tr} \left((|\psi\rangle\langle\psi| \otimes \sqrt{\rho^B})V^*V(|\psi\rangle\langle\psi| \otimes \sqrt{\rho^B})\right) . \tag{3.85}$$

For a solution see Solution 3.46.

With the help of (3.73) we can establish an upper bound for trace on the right side of (3.85).

Exercise 3.47 Show that (3.73) implies for any $|\psi\rangle \in \mathbb{H}^A$ that

$$\mathrm{tr} \left((|\psi\rangle\langle\psi| \otimes \sqrt{\rho^B})V^*V(|\psi\rangle\langle\psi| \otimes \sqrt{\rho^B})\right) \le \kappa \, ||\psi||^4 . \tag{3.86}$$

For a solution see Solution 3.47.

Together (3.85) and (3.86) imply that for any $|\psi\rangle \in \mathbb{H}^A$

$$\langle\psi| \sum_{b_1,b_2} K^*_{(b_1,b_2)}K_{(b_1,b_2)}\psi\rangle \leq \kappa \, ||\psi||^2 = \langle\psi|\kappa\mathbf{1}^A\psi\rangle \, ,$$

which by Definition 2.12 is equivalent to (3.75).

For the special case $V^*V = \kappa\mathbf{1}^{AB}$ we have

$$\sum_{b_1,b_2} K^*_{(b_1,b_2)}K_{(b_1,b_2)} \underbrace{=}_{(3.84)} \mathrm{tr}^B\left((\mathbf{1}^A \otimes \sqrt{\rho^B}) \underbrace{V^*V}_{=\kappa\mathbf{1}^{AB}} (\mathbf{1}^A \otimes \sqrt{\rho^B})\right)$$

$$= \kappa\,\mathrm{tr}^B\left((\mathbf{1}^A \otimes \sqrt{\rho^B})(\mathbf{1}^A \otimes \sqrt{\rho^B})\right)$$

$$\underbrace{=}_{(3.80)} \kappa\,\mathrm{tr}^B\left((\mathbf{1}^A \otimes \rho^B)\right) \underbrace{=}_{(3.57)} \kappa\,\mathrm{tr}\left(\rho^B\right)\mathbf{1}^A$$

$$\underbrace{=}_{(2.83)} \kappa\mathbf{1}^A \, .$$

To prove $\Leftarrow$ let $K_l \in \mathrm{L}(\mathbb{H}^A)$ for $l \in \{1,\ldots,m\}$ be such that they satisfy (3.75) and (3.76). Furthermore, let $\mathbb{H}^B$ be a HILBERT space with an ONB $\{|f_b\rangle \mid b \in \{1,\ldots,m\}\}$. We embed $\mathbb{H}^A$ in $\mathbb{H}^A \otimes \mathbb{H}^B$ by

$$\iota: \mathbb{H}^A \longrightarrow \mathbb{H}^A \otimes \mathbb{H}^B$$
$$|\psi\rangle \longmapsto |\psi \otimes f_1\rangle = |\psi\rangle \otimes |f_1\rangle$$

and define

$$\check{V}: \iota\{\mathbb{H}^A\} \longrightarrow \mathbb{H}^A \otimes \mathbb{H}^B$$
$$|\psi \otimes f_1\rangle \longmapsto \sum_{l=1}^m K_l|\psi\rangle \otimes |f_l\rangle \tag{3.87}$$

Exercise 3.48 Show that for a $\check{V}$ defined as in (3.87) one has

$$\langle\psi \otimes f_1|\check{V}^* = \sum_{l=1}^m \langle\psi|K_l^* \otimes \langle f_l| \, . \tag{3.88}$$

For a solution see Solution 3.48.

For any $|\psi \otimes f_1\rangle \in \iota\{\mathbb{H}^A\}$ the linear operator $\check{V}$ satisfies

$$\langle\psi \otimes f_1|\check{V}^*\check{V}(\psi \otimes f_1)\rangle \underbrace{=}_{(2.30),(2.31)} \langle\check{V}(\psi \otimes f_1)|\check{V}(\psi \otimes f_1)\rangle$$

$$\underbrace{=}_{(3.87),(2.4)} \sum_{l_1,l_2} \langle K_{l_1}\psi \otimes f_{l_1}|K_{l_2}\psi \otimes f_{l_2}\rangle$$

$$\underbrace{=}_{(3.4)} \sum_{l_1,l_2} \langle K_{l_1}\psi|K_{l_2}\psi\rangle \underbrace{\langle f_{l_1}|f_{l_2}\rangle}_{=\delta_{l_1 l_2}} = \sum_l \langle K_l\psi|K_l\psi\rangle$$

$$\underbrace{=}_{(2.30),(2.4)} \langle\psi|\sum_l K_l^* K_l \psi\rangle . \tag{3.89}$$

We thus have

$$\sum_l K_l^* K_l \leq \kappa\mathbf{1}^A$$

$$\underbrace{\Rightarrow}_{\text{Def. 2.12}} \langle\psi|\sum_l K_l^* K_l\psi\rangle \leq \kappa\,||\psi||^2 = \kappa\,||\psi||^2\,||f_1||^2 \quad \forall|\psi\rangle \in \mathbb{H}^A$$

$$\underbrace{\Rightarrow}_{(2.89),(3.10)} \langle\psi\otimes f_1|\check{V}^*\check{V}(\psi\otimes f_1)\rangle \leq \kappa\,||\psi\otimes f_1||^2 \quad \forall|\psi\rangle\otimes f_1 \in \iota\{\mathbb{H}^A\}$$

$$\underbrace{\Rightarrow}_{\text{Def. 2.12}} \check{V}^*\check{V} \leq \kappa\mathbf{1}^{\iota\{\mathbb{H}^A\}} ,$$

where equality implies equality in each step. The operator $\check{V}$ is defined on the $\dim \mathbb{H}^A$-dimensional subspace $\iota\{\mathbb{H}^A\} \subset \mathbb{H}^A \otimes \mathbb{H}^B$ of the HILBERT space $\mathbb{H}^A \otimes \mathbb{H}^B$ of dimension $(\dim \mathbb{H}^A)(\dim \mathbb{H}^B)$. To extend $\check{V}$ to an operator V on all of $\mathbb{H}^A \otimes \mathbb{H}^B$ we use the result of Exercise 3.49.

Exercise 3.49 Let $m, n \in \mathbb{N}$ with $n > m$ and $A \in \text{Mat}(n \times m, \mathbb{C})$ as well as $c \in]0, 1]$ be given. Show that then we can always find $B \in \text{Mat}(n \times (n-m), \mathbb{C})$ such that

$$V = \begin{pmatrix} A & B \end{pmatrix} \in \text{Mat}(n \times n, \mathbb{C})$$

satisfies

$$V^*V = \begin{pmatrix} A^*A & 0_{m\times(n-m)} \\ 0_{(n-m)\times m} & c\mathbf{1}_{(n-m)\times(n-m)} \end{pmatrix} \in \text{Mat}(n \times n, \mathbb{C}), \tag{3.90}$$

where $0_{k\times l}, \mathbf{1}_{k\times l}$ denote the zero resp. unit matrix in $\text{Mat}(k \times l, \mathbb{C})$

For a solution see Solution 3.49.

By choosing $c = \kappa$ in Exercise 3.49 we can extend $\check{V}$ to an operator V on all of $\mathbb{H}^A \otimes \mathbb{H}^B$ such that $V^*V \leq \kappa\mathbf{1}^{AB}$ if $\sum_l K_l^* K_l \leq \kappa\mathbf{1}^A$ or $V^*V = \kappa\mathbf{1}^{AB}$ if $\sum_l K_l^* K_l = \kappa\mathbf{1}^A$. Thus, we have shown that (3.75) implies (3.73) including the special case of equality.

To show (3.74), we set

$$\rho^B = |f_1\rangle\langle f_1| \in D(\mathbb{H}^B) \tag{3.91}$$

and recall that any $\rho^A \in D(\mathbb{H}^A)$ can be written in the form

$$\rho^A = \sum_a p_a |e_a\rangle\langle e_a| , \tag{3.92}$$

where the $p_a \in [0,1]$ satisfy

$$\sum_a p_a = 1 ,$$

and the $\{|e_a\rangle\}$ form an ONB in $\mathbb{H}^A$. Therefore, we obtain

$$V(\rho^A \otimes \rho^B)V^* \underbrace{=}_{(3.91),(3.92)} \sum_a p_a V(|e_a\rangle\langle e_a| \otimes |f_1\rangle\langle f_1|)V^*$$

$$\underbrace{=}_{(3.36)} \sum_a p_a V|e_a \otimes f_1\rangle\langle e_a \otimes f_1|V^* , \tag{3.93}$$

where

$$V|e_a \otimes f_1\rangle = \check{V}|e_a \otimes f_1\rangle \underbrace{=}_{(3.87)} \sum_l K_l|e_a\rangle \otimes |f_l\rangle$$

$$\langle e_a \otimes f_1|V^* = \langle e_a \otimes f_1|\check{V}^* \underbrace{=}_{(3.88)} \sum_l \langle e_a|K_l^* \otimes \langle f_l| . \tag{3.94}$$

Using (3.94) in (3.93) yields

$$V(\rho^A \otimes \rho^B)V^* = \sum_{a,l_1,l_2} p_a\big(K_{l_1}|e_a\rangle \otimes |f_{l_1}\rangle\big)\big(\langle e_a|K_{l_2}^* \otimes \langle f_{l_2}|\big)$$

$$\underbrace{=}_{(3.36)} \sum_{a,l_1,l_2} p_a K_{l_1}|e_a\rangle\langle e_a|K_{l_2}^* \otimes |f_{l_1}\rangle\langle f_{l_2}|$$

$$= \sum_{l_1,l_2} K_{l_1}\left(\sum_a p_a|e_a\rangle\langle e_a|\right) K_{l_2}^* \otimes |f_{l_1}\rangle\langle f_{l_2}|$$

$$\underbrace{=}_{(3.92)} \sum_{l_1,l_2} K_{l_1}\rho^A K_{l_2}^* \otimes |f_{l_1}\rangle\langle f_{l_2}| . \tag{3.95}$$

Taking the partial trace over $\mathbb{H}^B$ it follows that

$$\mathrm{tr}^B\left(V(\rho^A \otimes \rho^B)V^*\right) \underbrace{=}_{(3.95)} \mathrm{tr}^B\left(\sum_{l_1,l_2} K_{l_1}\rho^A K_{l_2}^* \otimes |f_{l_1}\rangle\langle f_{l_2}|\right)$$

$$= \sum_{l_1,l_2} \mathrm{tr}^B\left(K_{l_1}\rho^A K_{l_2}^* \otimes |f_{l_1}\rangle\langle f_{l_2}|\right)$$

$$\underbrace{=}_{(3.57)} \sum_{l_1,l_2} \mathrm{tr}\left(|f_{l_1}\rangle\langle f_{l_2}|\right) K_{l_1}\rho^A K_{l_2}^*, \qquad (3.96)$$

where we can use

$$\mathrm{tr}\left(|f_{l_1}\rangle\langle f_{l_2}|\right) \underbrace{=}_{(2.57)} \sum_b \langle f_b|f_{l_1}\rangle\langle f_{l_2}|f_b\rangle = \sum_b \delta_{bl_1}\delta_{l_2 b} = \delta_{l_1 l_2}$$

such that (3.96) becomes

$$\mathrm{tr}^B\left(V(\rho^A \otimes \rho^B)V^*\right) = \sum_l K_l\rho^A K_l^* \underbrace{=}_{(3.76)} K(\rho^A)$$

verifying (3.74) and completing the proof of $\Leftarrow$. $\qquad\qquad\qquad\qquad\square$

Note that Theorem 3.24 does not assert that every convex-linear map as in (3.72) does have one of the equivalent forms (3.74) or (3.75). It merely states that if one exists then the other does, too.

However, it can be shown that any completely positive linear map $\Phi : \rho \mapsto \Phi\rho$ which satisfies $(\Phi\rho)^* = \Phi\rho$ and $\mathrm{tr}\,(\Phi\rho) \leq \mathrm{tr}\,(\rho)$ is indeed of the form given by (3.76) with (3.75) [58, 59]. The infinite-dimensional version of this statement is known as the STINESPRING factorization theorem [60]. We shall not pursue this generalization here, as the results stated in Theorem 3.24 are sufficient for our purposes.

Before we give a formal definition of a quantum operation it is useful to exhibit the relations between inequalities for the operators V^*V or $\sum_l K_l^* K_l$ and the trace of $K(\rho^A)$.

Corollary 3.25 *Let $\mathbb{H}^A$ and $\mathbb{H}^B$ be finite-dimensional* HILBERT *spaces and let $K : \mathrm{D}(\mathbb{H}^A) \rightarrow \mathrm{D}_{\leq}(\mathbb{H}^A)$ have the equivalent representations given in Theorem 3.24 with $V \in \mathrm{L}(\mathbb{H}^A \otimes \mathbb{H}^B)$, $\rho^B \in \mathrm{D}(\mathbb{H}^B)$ and $K_l \in \mathrm{L}(\mathbb{H}^A)$ for $l \in \{1,\dots,m\}$. For any $\kappa \in]0,1]$ we then have*

$$V^*V \leq \kappa\mathbf{1}^{AB} \quad \Leftrightarrow \quad \sum_l K_l^* K_l \leq \kappa\mathbf{1}^A \quad \Leftrightarrow \quad \mathrm{tr}\left(K(\rho^A)\right) \leq \kappa \quad \forall \rho^A \in \mathrm{D}(\mathbb{H}^A),$$

and equality in one relation is equivalent to equality in the other two relations. In particular, for equality with $\kappa = 1$ we have

$$V \in \mathcal{U}\left(\mathbb{H}^A \otimes \mathbb{H}^B\right) \quad \Leftrightarrow \quad \sum_l K_l^* K_l = \mathbf{1}^A \quad \Leftrightarrow \quad \mathrm{tr}\left(K(\rho^A)\right) = 1 \quad \forall \rho^A \in \mathrm{D}\left(\mathbb{H}^A\right).$$

Proof From Theorem 3.24 we know already that

$$V^* V \leq \kappa \mathbf{1}^{AB} \quad \Leftrightarrow \quad \sum_l K_l^* K_l \leq \kappa \mathbf{1}^A$$

with equality on one side implying equality on the other and from Lemma 3.23 we know that

$$\sum_l K_l^* K_l \leq \kappa \mathbf{1}^A \quad \Leftrightarrow \quad \mathrm{tr}\left(K(\rho^A)\right) \leq \kappa \quad \forall \rho^A \in \mathrm{D}(\mathbb{H}^A),$$

where again equality on one side implies equality on the other.   □

We are now in a position to define quantum operations. For our purposes it is sufficient to consider the finite-dimensional case and to make do without the general axiomatic approach utilizing the notion of complete positivity.

Definition 3.26 Let $\mathbb{H}$ be a finite-dimensional HILBERT space. A **quantum operation** is a convex-linear map

$$\begin{aligned} K : \mathrm{D}(\mathbb{H}) &\longrightarrow \mathrm{D}_{\leq}(\mathbb{H}) \\ \rho &\longmapsto K(\rho) \end{aligned}$$

that can be expressed in the two equivalent forms (3.74) and (3.76) given in Theorem 3.24. The representation (3.76) of the form

$$K(\rho) = \sum_{l=1}^m K_l \rho K_l^*, \tag{3.97}$$

where the $K_l \in \mathrm{L}(\mathbb{H})$ for $l \in \{1, \ldots, m\}$ satisfy

$$\sum_{l=1}^m K_l^* K_l \leq \mathbf{1},$$

is called **operator-sum representation** of the quantum operation K. The K_l are called KRAUS **operators** or **operation elements** of the quantum operation. The representation (3.74) of the form

$$K(\rho) = \mathrm{tr}^B \left(V(\rho \otimes \rho^B)V^*\right)$$

using an additional HILBERT space $\mathbb{H}^B$, where $V \in B(\mathbb{H} \otimes \mathbb{H}^B)$ satisfies

$$V^*V \leq 1,$$

is called **environmental representation** of K.

   If

$$\sum_{l=1}^{m} K_l^* K_l = 1$$

(and thus equivalently $\mathrm{tr}\,(K(\rho)) = 1 = \mathrm{tr}\,(\rho)$ for all $\rho \in D(\mathbb{H})$), then the quantum operation is called **trace-preserving** or **quantum channel** and constitutes a map $K : D(\mathbb{H}) \to D(\mathbb{H})$.

The K_l in the operator sum representation (3.76) of K depend on the ρ^B that is used in (3.74) to build K. The construction of the K_l in (3.82) in the proof of Theorem 3.24 used the representation of ρ^B given in (3.78). However, from Proposition 2.27 we know that such a decomposition of a given density operator is not unique. This non-uniqueness carries over to the KRAUS operators, which are thus not unique either.

Corollary 3.27 *Let $K : D(\mathbb{H}) \to D_{\leq}(\mathbb{H})$ be a quantum operation with KRAUS operators $K_l \in L(\mathbb{H})$, where $l \in \{1, \ldots, m\}$. Moreover, let $\widetilde{m} \geq m$ and $U \in U(\widetilde{m})$. Then the $\widetilde{K}_j \in L(\mathbb{H})$ with $j \in \{1, \ldots, \widetilde{m}\}$ given by*

$$\widetilde{K}_j = \sum_{l=1}^{m} U_{jl} K_l \tag{3.98}$$

are KRAUS operators for K as well.

Proof To begin with, note that

$$\widetilde{K}_j^* \underbrace{=}_{(2.32),(3.98)} \sum_{l=1}^{m} \overline{U_{jl}} K_l^* \underbrace{=}_{(2.34)} \sum_{l=1}^{m} U_{lj}^* K_l^*, \tag{3.99}$$

such that

$$\sum_{j=1}^{\widetilde{m}} \widetilde{K}_j^* \widetilde{K}_j \underbrace{=}_{(3.98),(3.99)} \sum_{j=1}^{\widetilde{m}} \left(\sum_{l=1}^{m} U_{lj}^* K_l^* \right) \left(\sum_{k=1}^{m} U_{jk} K_k \right) = \sum_{l,k=1}^{m} \left(\sum_{j=1}^{\widetilde{m}} U_{lj}^* U_{jk} \right) K_l^* K_k$$

$$= \sum_{l,k=1}^{m} \underbrace{(U^*U)_{lk}}_{=\delta_{lk}} K_l^* K_k = \sum_{l=1}^{m} K_l^* K_l \leq \kappa \mathbf{1} \,,$$

and the $\widetilde{K}_j$ satisfy (3.75) because the K_l do. The proof of (3.76) is almost identical since for any $\rho \in D(\mathbb{H})$ we have

$$\sum_{j=1}^{\widetilde{m}} \widetilde{K}_j^* \rho \widetilde{K}_j \underbrace{=}_{(3.98),(3.99)} \sum_{j=1}^{\widetilde{m}} \left(\sum_{l=1}^{m} U_{lj}^* K_l^* \right) \rho \left(\sum_{k=1}^{m} U_{jk} K_k \right) = \sum_{l,k=1}^{m} \left(\sum_{j=1}^{\widetilde{m}} U_{lj}^* U_{jk} \right) K_l^* \rho K_k$$

$$= \sum_{l,k=1}^{m} \underbrace{(U^*U)_{lk}}_{=\delta_{lk}} K_l^* \rho K_k = \sum_{l=1}^{m} K_l^* \rho K_l = K(\rho)$$

and the $\widetilde{K}_j$ also satisfy (3.76) because the K_l do.   □

We return to the motivating considerations for quantum operations given at the beginning of this section and show how quantum operations provide a compact means for the description when a system $\mathbb{H}^A$ is combined with another system $\mathbb{H}^B$, which, after some interaction, is subsequently ignored. Recalling the resulting state transformation (3.65) from such process steps 1–5 discussed at the beginning of this section, we see from the results in Theorem 3.24 that this state transformation can be formulated with the help of the quantum operation

$$K(\rho^A) = \operatorname{tr}^B \left(V(\rho^A \otimes \rho^B) V^* \right) \tag{3.100}$$

in its environmental representation (hence this name), where

$$V = (\mathbf{1}^A \otimes P^B) U \tag{3.101}$$

with $U \in \mathcal{U}(\mathbb{H}^A \otimes \mathbb{H}^B)$.

Exercise 3.50 Let a quantum operation K be given in the environmental representation (3.100) with V as in (3.101). Show that then

$$\operatorname{tr} \left((\mathbf{1}^A \otimes P^B) U (\rho^A \otimes \rho^B) U^* \right) = \operatorname{tr} \left(K(\rho^A) \right) \,. \tag{3.102}$$

For a solution see Solution 3.50.

Using (3.100) with (3.101) and (3.102), we see that in the total state transformation (3.65) from the process steps 1–5 discussed at the beginning of this section we have

$$\frac{\operatorname{tr}^B \left((\mathbf{1}^A \otimes P^B) U (\rho^A \otimes \rho^B) U^* (\mathbf{1}^A \otimes P^B) \right)}{\operatorname{tr} \left((\mathbf{1}^A \otimes P^B) U (\rho^A \otimes \rho^B) U^* \right)} = \frac{K(\rho^A)}{\operatorname{tr} \left(K(\rho^A) \right)} \,.$$

Consequently, the total state transformation obtained when our system is

1. initially prepared in state $\rho^A \in D(\mathbb{H}^A)$
2. combined with another system
3. subject to the time evolution of the combined system
4. subject to possible measurements on the newly added system
5. viewed in isolation after discarding the intermittently added system

can be expressed succinctly with the help of a quantum operation K in the form

$$
\begin{aligned}
D(\mathbb{H}^A) &\longrightarrow D(\mathbb{H}^A) \\
\rho^A &\longmapsto \frac{K(\rho^A)}{\mathrm{tr}(K(\rho^A))} \ .
\end{aligned}
$$

This form to represent the state transformation will be used in the context of quantum error correction in Sect. 7.3.

In case the time evolution $U \in \mathcal{U}(\mathbb{H}^A \otimes \mathbb{H}^B)$ of the combined system acts separably on the two sub-systems $\mathbb{H}^A$ and $\mathbb{H}^B$ in the sense that it commutes with the measurement projections $\mathbf{1}^A \otimes P^B$ on sub-system $\mathbb{H}^B$, the trace of the overall quantum operation can be determined from the product $\rho^B P^B$ as the following corollary shows.

Corollary 3.28 *Let $K : D(\mathbb{H}^A) \to D_{\leq}(\mathbb{H}^A)$ be a quantum operation in the environmental representation given by*

$$
K(\rho^A) = \mathrm{tr}^B \left((\mathbf{1}^A \otimes P^B) U (\rho^A \otimes \rho^B) U^* (\mathbf{1}^A \otimes P^B) \right) , \tag{3.103}
$$

where $P^B \in L(\mathbb{H}^B)$ is an orthogonal projection, $U \in \mathcal{U}(\mathbb{H}^A \otimes \mathbb{H}^B)$ and $\rho^B \in D(\mathbb{H}^B)$. Then we have

$$
\left[\mathbf{1}^A \otimes P^B, U\right] = 0 \quad \Rightarrow \quad \mathrm{tr}\left(K(\rho^A)\right) = \mathrm{tr}\left(\rho^B P^B\right) \quad \forall \rho^A \in D(\mathbb{H}^A)
$$

and in particular for $P^B = \mathbf{1}^B$ it follows that then

$$
\mathrm{tr}\left(K(\rho^A)\right) = 1 \quad \forall \rho^A \in D(\mathbb{H}^A) .
$$

Proof The quantum operation K in (3.103) is in environmental representation with $V = (\mathbf{1}^A \otimes P^B) U$ such that

$$
V^* = \underbrace{\left((\mathbf{1}^A \otimes P^B) U \right)^*}_{} \underbrace{= U^* (\mathbf{1}^A \otimes P^B)^*}_{(2.47)} \underbrace{= U^* (\mathbf{1}^A \otimes (P^B)^*)}_{(3.31)}
$$

$$
\underbrace{= U^* (\mathbf{1}^A \otimes P^B)}_{\text{Def. 2.11}} \tag{3.104}
$$

and thus

$$
\begin{aligned}
V^*V &\underbrace{=}_{(3.104)} U^*(\mathbf{1}^A \otimes P^B)(\mathbf{1}^A \otimes P^B)U = U^*(\mathbf{1}^A \otimes (P^B)^2)U \\
&\underbrace{=}_{\text{Def. 2.11}} U^*(\mathbf{1}^A \otimes P^B)U \underbrace{=}_{[\mathbf{1}^A \otimes P^B, U]=0 \text{ and } (2.11)} U^*U(\mathbf{1}^A \otimes P^B) \\
&\underbrace{=}_{(2.37)} \mathbf{1}^A \otimes P^B .
\end{aligned}
\tag{3.105}
$$

From (3.84) in the proof of Theorem 3.24 we recall that we have a set of KRAUS operators K_l for K such that

$$
\begin{aligned}
\sum_l K_l^* K_l &\underbrace{=}_{(3.84)} \mathrm{tr}^B\left((\mathbf{1}^A \otimes \sqrt{\rho^B})V^*V(\mathbf{1}^A \otimes \sqrt{\rho^B})\right) \\
&\underbrace{=}_{(3.105)} \mathrm{tr}^B\left((\mathbf{1}^A \otimes \sqrt{\rho^B})(\mathbf{1}^A \otimes P^B)(\mathbf{1}^A \otimes \sqrt{\rho^B})\right) \\
&= \mathrm{tr}^B\left(\mathbf{1}^A \otimes \sqrt{\rho^B}P^B\sqrt{\rho^B}\right) \\
&\underbrace{=}_{(3.57)} \mathrm{tr}\left(\sqrt{\rho^B}P^B\sqrt{\rho^B}\right)\mathbf{1}^A \underbrace{=}_{(2.58)} \mathrm{tr}\left((\sqrt{\rho^B})^2 P^B\right)\mathbf{1}^A \\
&\underbrace{=}_{(3.81)} \mathrm{tr}\left(\rho^B P^B\right)\mathbf{1}^A .
\end{aligned}
$$

Applying Corollary 3.25 with $\kappa = \mathrm{tr}\left(\rho^B P^B\right)$ then implies that $\mathrm{tr}\left(K(\rho)\right) = \mathrm{tr}\left(\rho^B P^B\right)$. If $P^B = \mathbf{1}^B$, then $\mathrm{tr}\left(K(\rho)\right) = \mathrm{tr}\left(\rho^B\right) = 1$ since $\rho^B \in \mathrm{D}(\mathbb{H}^B)$. $\square$

Finally, note that the domain $\mathrm{D}(\mathbb{H})$ of a quantum operation K is a convex set, that is, for each $\rho_1, \rho_2 \in \mathrm{D}(\mathbb{H})$ and $\mu \in [0, 1]$ we have

$$
\mu\rho_1 + (1 - \mu)\rho_2 \in \mathrm{D}(\mathbb{H}),
$$

and every quantum operation K is convex-linear, which means that for any $\rho_1, \rho_2 \in \mathrm{D}(\mathbb{H})$ and $\mu \in [0, 1]$ it satisfies

$$
K\left(\mu\rho_1 + (1 - \mu)\rho_2\right) = \mu K(\rho_1) + (1 - \mu)K(\rho_2).
\tag{3.106}
$$

When the system consists only of a single qubit we have $\mathbb{H} = {}^{\P}\mathbb{H}$. In this case we know already from (2.127) that every $\rho \in \mathrm{D}({}^{\P}\mathbb{H})$ can be described by an $\mathbf{x} \in B_{\mathbb{R}^3}^1$ in the form

$$
\rho_{\mathbf{x}} = \frac{1}{2}\left(\mathbf{1} + \mathbf{x} \cdot \sigma\right).
\tag{3.107}
$$

Hence, any image of a trace-preserving quantum operation $K : D(\P\H) \to D(\P\H)$ on a single qubit has to be of the same form

$$K(\rho_{\mathbf{x}}) = \frac{1}{2}\left(\mathbf{1} + \mathbf{y}(\mathbf{x}) \cdot \boldsymbol{\sigma}\right),$$

and from Exercise 2.30 we know that $\mathbf{y}(\mathbf{x}) = \mathrm{tr}\,(K(\rho_{\mathbf{x}})\boldsymbol{\sigma})$. As a consequence, every trace-preserving quantum operation K on a single qubit defines a map

$$
\begin{aligned}
\widehat{K} : B^1_{\mathbb{R}^3} &\longrightarrow B^1_{\mathbb{R}^3} \\
\mathbf{x} &\longmapsto \mathrm{tr}\,(K(\rho_{\mathbf{x}})\boldsymbol{\sigma})
\end{aligned}
\tag{3.108}
$$

In other words, every trace-preserving quantum operation K induces a map $\widehat{K}$ of the BLOCH ball $B^1_{\mathbb{R}^3}$ onto itself.

Exercise 3.51 Show that $\widehat{K}$ given by (3.108) is convex-linear, in other words, that it satisfies (3.106).

For a solution see Solution 3.51.

Different types of trace-preserving quantum operations K on qubits can thus be visualized as deformations of the BLOCH ball $B^1_{\mathbb{R}^3}$ effected by $\widehat{K}$ [59, 61].

3.6 Further Reading

Material on tensor products of HILBERT spaces (and a lot more on operators, in particular in the infinite-dimensional setting) can be found in the first book of the multi-volume series by REED and SIMON [50]. Chapter 2 of the book by PARTHASARATHY [62] contains a very detailed but rather advanced exposition of many aspects around states and observables in tensor products of HILBERT spaces.

For a condensed and modern coverage of quantum operations including physical aspects thereof the reader may consult the treatise by NIELSEN and CHUANG [61] or for a geometrical view the book by BENGTSON and ŻYCZKOWSKI [59].

Chapter 4
Entanglement

4.1 Generalities

The notion of *entanglement* goes back to SCHRÖDINGER [7]. The existence of entangled states is arguably the most important difference between classical and quantum computing. Indeed, the existence of entangled states allows *new effects* like teleportation and *new algorithms* like SHOR's algorithm, which performs prime factorization much faster than with a classical computer. Before we concern ourselves with these in Chap. 6, we first want to look at entanglement and some of its resultant effects, which are at odds with our intuition, in this chapter.

We begin in Sect. 4.2 with a mathematical definition of entanglement and present a handy criterion to test, if a pure state is entangled.

In Sect. 4.3 we then show, how entangled states can be generated even though the sub-systems have not interacted before. This effect has become known as 'entanglement swapping'.

The second essential difference between classical and quantum computing is the existence of incompatible observables and the fact that they cannot be measured sharply (see Sect. 2.3.1). That entanglement together with the inability to measure incompatible observables sharply lead to effects which contradict our intuitive understanding of *reality* and *causality* has been exhibited by EINSTEIN, PODOLSKY, and ROSEN [4]. This has since gained prominence as the EPR-paradox and was meant by the aforementioned authors to show that quantum mechanics *does not give a complete description* of reality. We shall examine this line of arguments in Sect. 4.4.

The supposed incompleteness of quantum mechanics initially lead to the concept of additional hidden variables, which are not captured by quantum mechanics. Such variables were assumed to determine the outcome of experiments, but the observer's ignorance of them leads to the observed statistical character of measurement results. Assuming that such local[1] variables exist, BELL [63] derived an

---

[1]Local means here that the variables of one system do not depend on those of another space-like separated system.

© Springer Nature Switzerland AG 2019
W. Scherer, *Mathematics of Quantum Computing*,
https://doi.org/10.1007/978-3-030-12358-1_4

inequality for correlations of various measurable observables. The existence of variables not captured by quantum mechanics would thus resolve the EPR-contradiction to our intuitive understanding of causality and reality and at the same time imply the BELL inequality of correlations. Experiments have shown, however, that certain quantum mechanical systems *violate* the BELL inequality [9]. If these systems had been describable by hidden local variables this should not have happened. Faced with the EPR-alternative

(1) quantum mechanics provides a complete description of a system
(2) our usual intuitive understanding of reality and causality applies to all systems

nature thus obviously voted against (2). These questions in connection with the EPR paradox and the BELL inequality will be treated in more detail in Sect. 4.4.

At first glance the properties mentioned in the context of the EPR paradox may lead to the belief that they could be used for transmitting signals with a speed greater than the speed of light. However, such a device, which has been coined a BELL telephone, does not exist as we shall show in Sect. 4.6.1. In Sect. 4.6.2 we shall consider another impossible device by showing that no apparatus can be built that copies arbitrary unknown qubits.

4.2 Definition and Characterization

We begin by establishing the following result about combining density operators of sub-systems to form a density operator of the composite system.

Exercise 4.52 Let $\mathbb{H}^A$ and $\mathbb{H}^B$ be HILBERT spaces. Show that

$$\rho^X \in D(\mathbb{H}^X) \quad \text{for } X \in \{A, B\} \quad \Rightarrow \quad \rho^A \otimes \rho^B \in D(\mathbb{H}^A \otimes \mathbb{H}^B). \qquad (4.1)$$

For a solution see Solution 4.52.

With (4.1) we can give a general definition for entanglement, which also applies for mixed states.

Definition 4.1 ([64]) A state $\rho \in D(\mathbb{H}^A \otimes \mathbb{H}^B)$ in a composite system $\mathbb{H}^A \otimes \mathbb{H}^B$, which is composed of the sub-systems $\mathbb{H}^A$ and $\mathbb{H}^B$ is called **separable** or **product-state** with respect to the sub-systems $\mathbb{H}^A$ and $\mathbb{H}^B$, if there exist states $\rho_j^A \in D(\mathbb{H}^A)$ in sub-system $\mathbb{H}^A$ and $\rho_j^B \in D(\mathbb{H}^B)$ in sub-system $\mathbb{H}^B$ indexed by $j \in I \subset \mathbb{N}$ together with positive real numbers p_j satisfying

$$\sum_{j \in I} p_j = 1,$$

such that

$$\rho = \sum_{j \in I} p_j \rho_j^A \otimes \rho_j^B. \tag{4.2}$$

Otherwise, ρ is called **entangled**.

In general, the defining properties in this definition are not easy to ascertain such that this definition alone does not provide a practical criterion to determine if a given state is entangled or not. The search for alternative characterizations of entanglement for truly mixed states is still the subject of ongoing research. For our purposes it suffices to restrict our considerations to pure states only. The following theorem thus provides an alternative criterion for separability of pure states. Indeed, the criterion given there is often stated as defining criterion of separability for pure states.

Theorem 4.2 *A pure state $|\Psi\rangle \in \mathbb{H}^A \otimes \mathbb{H}^B$ is **separable**, if and only if there exist pure states $|\varphi\rangle \in \mathbb{H}^A$ and $|\psi\rangle \in \mathbb{H}^B$ such that*

$$|\Psi\rangle = |\varphi\rangle \otimes |\psi\rangle. \tag{4.3}$$

Otherwise, $|\Psi\rangle$ is entangled.

Proof First, we show that (4.3) is sufficient for separability. Suppose we have $|\Psi\rangle = |\varphi\rangle \otimes |\psi\rangle \in \mathbb{H}^A \otimes \mathbb{H}^B$. Then it follows that

$$\rho(\Psi) \underbrace{=}_{(2.89)} |\Psi\rangle\langle\Psi| = |\varphi \otimes \psi\rangle\langle\varphi \otimes \psi| \underbrace{=}_{(3.36)} |\varphi\rangle\langle\varphi| \otimes |\psi\rangle\langle\psi|.$$

Setting $\rho^A = |\varphi\rangle\langle\varphi|$ and $\rho^B = |\psi\rangle\langle\psi|$, this amounts to (4.2).

To show that (4.3) is also necessary, let ρ be a pure and separable state. Hence, there exist ρ_j^A, ρ_j^B and p_j for $j \in I$ as in Definition 4.1 and $|\Psi\rangle \in \mathbb{H}^A \otimes \mathbb{H}^B$, such that

$$\rho = \sum_{j \in I} p_j \rho_j^A \otimes \rho_j^B$$

and simultaneously

$$\rho = |\Psi\rangle\langle\Psi|. \tag{4.4}$$

We now show, that then there exist $|\varphi\rangle \in \mathbb{H}^A$ and $|\psi\rangle \in \mathbb{H}^B$ such that $|\Psi\rangle = |\varphi\rangle \otimes |\psi\rangle$. For all $j \in I$ we set

$$\rho_j = \rho_j^A \otimes \rho_j^B. \tag{4.5}$$

From (4.1) we know that every ρ_j defined as in (4.5) is a density operator, in other words, every ρ_j is self-adjoint, positive and has trace 1. From 1. in Theorem 2.24 we know that then for every $j \in I$ there exist $p_{j,k} \in]0,1]$ with $k \in I_j \subset \mathbb{N}$ satisfying $\sum_{k \in I_j} p_{j,k} = 1$ and an ONB $\{|\Omega_{j,k}\rangle \mid k \in \{1,\ldots,\dim \mathbb{H}^A \otimes \mathbb{H}^B\}\}$ of $\mathbb{H}^A \otimes \mathbb{H}^B$ such that

$$\rho_j = \sum_{k \in I_j} p_{j,k} |\Omega_{j,k}\rangle \langle \Omega_{j,k}| \, . \tag{4.6}$$

We extend $|\Psi\rangle$ to another ONB $\{|\Psi\rangle, |\Psi_l\rangle \mid l \in \{1,\ldots,\dim \mathbb{H}^A \otimes \mathbb{H}^B - 1\}\}$ of $\mathbb{H}^A \otimes \mathbb{H}^B$. It follows that

$$0 = |\langle \Psi_l | \Psi \rangle|^2 = \langle \Psi_l | \rho \Psi_l \rangle = \sum_{j \in I} p_j \langle \Psi_l | \rho_j \Psi_l \rangle$$

and, since $p_j > 0$, we must have $\langle \Psi_l | \rho_j \Psi_l \rangle = 0$ for all $j \in I$ and $l \in \{1,\ldots,\dim \mathbb{H}^A \otimes \mathbb{H}^B - 1\}$. Together with (4.6) this implies

$$\sum_{k \in I_j} p_{j,k} \left| \langle \Psi_l | \Omega_{j,k} \rangle \right|^2 = 0$$

and thus, again because $p_{j,k} > 0$, for all $l \in \{1,\ldots,\dim \mathbb{H}^A \otimes \mathbb{H}^B - 1\}$, $j \in I$ and $k \in I_j$

$$\langle \Psi_l | \Omega_{j,k} \rangle = 0 \, .$$

Hence, for every $j \in I$ and $k \in I_j$ the basis vector $|\Omega_{j,k}\rangle$ is orthogonal to all $|\Psi_l\rangle$ from the ONB $\{|\Psi\rangle, |\Psi_l\rangle \mid l \in \{1,\ldots,\dim \mathbb{H}^A \otimes \mathbb{H}^B - 1\}\}$. Consequently, every $|\Omega_{j,k}\rangle$ is in the ray (see Definition 2.14) of $|\Psi\rangle$ and there exist $\alpha_{j,k} \in \mathbb{R}$ such that

$$|\Omega_{j,k}\rangle = e^{i\alpha_{j,k}} |\Psi\rangle \, .$$

This implies

$$\rho_j^A \otimes \rho_j^B \underbrace{=}_{(4.5)} \rho_j \underbrace{=}_{(4.6)} \sum_{k \in I_j} p_{j,k} e^{i\alpha_{j,k}} |\Psi\rangle \langle \Psi| e^{-i\alpha_{j,k}} = \underbrace{\sum_{k \in I_j} p_{j,k}}_{=1} |\Psi\rangle \langle \Psi| = |\Psi\rangle \langle \Psi|$$

$$\underbrace{=}_{(4.4)} \rho$$

and thus there are $\rho^A \in \mathrm{D}(\mathbb{H}^A)$ and $\rho^B \in \mathrm{D}(\mathbb{H}^B)$ such that for all $j \in I$

$$\rho_j = \rho^A \otimes \rho^B = \rho \, . \tag{4.7}$$

With the help of the SCHMIDT decomposition (see Sect. 3.4) we can find $q_a \in]0,1]$ and ONBs $\{|e_a\rangle \mid a \in \{1,\ldots,\dim \mathbb{H}^A\}\}$ and $\{|f_b\rangle \mid b \in \{1,\ldots,\dim \mathbb{H}^B\}\}$ such that we can write

$$|\Psi\rangle = \sum_a \sqrt{q_a}|e_a\rangle \otimes |f_b\rangle. \tag{4.8}$$

This implies

$$\underbrace{\rho^A \otimes \rho^B}_{(4.7)} = \underbrace{\rho}_{(4.4)} = |\Psi\rangle\langle\Psi| = \sum_{a,b} \sqrt{q_a q_b}|e_a\rangle\langle e_a| \otimes |f_b\rangle\langle f_b| \tag{4.9}$$

such that in the ONB $\{|e_a\rangle \otimes |f_b\rangle\}$ of $\mathbb{H}^A \otimes \mathbb{H}^B$ we see that ρ has the matrix

$$(\rho^A)_{a_1 a_2}(\rho^B)_{b_1 b_2} = \rho_{a_1 b_1, a_2 b_2} = \sqrt{q_{a_1} q_{a_2}}\,\delta_{a_1 b_1}\delta_{a_2 b_2}\,.$$

From this it follows that

$$\begin{aligned}
\rho^A &= \mathrm{tr}^B(\rho) = \sum_a q_a|e_a\rangle\langle e_a| \\
\rho^B &= \mathrm{tr}^A(\rho) = \sum_b q_b|f_b\rangle\langle f_b|\,.
\end{aligned} \tag{4.10}$$

Now,

$$\begin{aligned}
1 \underset{(2.82)}{=} \mathrm{tr}(\rho) &\underset{(4.4)}{=} \mathrm{tr}(\rho^2) \underset{(4.9)}{=} \mathrm{tr}\big((\rho^A)^2 \otimes (\rho^B)^2\big) \\
&\underset{(4.10)}{=} \sum_{c,d}\langle e_c \otimes f_d|\sum_{a,b} q_a^2 q_b^2|e_a \otimes f_b\rangle\langle e_a \otimes f_b|e_c \otimes f_d\rangle \\
&= \sum_{a,b,c,d} q_a^2 q_b^2 \delta_{ca}\delta_{db} = \sum_{a,b} q_a^2 q_b^2 \\
&= \left(\sum_a q_a^2\right)^2,
\end{aligned} \tag{4.11}$$

where $q_a \in [0,1]$ for all a. On the other hand, it follows from (4.8) that

$$\sum_a q_a = ||\Psi|| = 1\,. \tag{4.12}$$

Together (4.11) and (4.12) imply that there can be only one $\hat{a}$ with $q_{\hat{a}} = 1$ and else $q_a = 0$ for all $a \neq \hat{a}$ has to hold. Consequently, (4.10) becomes

$$\rho^A = |e_{\hat{a}}\rangle\langle e_{\hat{a}}| \quad \text{and} \quad \rho^B = |f_{\hat{a}}\rangle\langle f_{\hat{a}}|\,,$$

and (4.8) implies

$$|\Psi\rangle = |e_{\hat{a}}\rangle \otimes |f_{\hat{a}}\rangle\,. \qquad \Box$$

Even the statement in Theorem 4.2 does not suit itself to an easy way to test if a pure state is separable or entangled. For example, consider the state

$$|\Psi\rangle = \frac{1}{2}\left(|00\rangle + |01\rangle + |10\rangle + |11\rangle\right),$$

for which it is not obvious that it is a separable state, which, however, it is since

$$|\Psi\rangle = \frac{|0\rangle + |1\rangle}{\sqrt{2}} \otimes \frac{|0\rangle + |1\rangle}{\sqrt{2}}.$$

How do we then find for a given $|\Psi\rangle$ a $|\varphi\rangle \in \mathbb{H}^A$ and a $|\psi\rangle \in \mathbb{H}^B$ such that $|\Psi\rangle = |\varphi\rangle \otimes |\psi\rangle$ or how do we exclude that there exist such vectors $|\varphi\rangle$ and $|\psi\rangle$? In other words, how does one verify separability or entanglement? For pure states the following theorem provides a helpful criterion for that query.

Theorem 4.3 *For pure states $|\Psi\rangle \in \mathbb{H}^A \otimes \mathbb{H}^B$ the following equivalence holds*

$$|\Psi\rangle \text{ is separable } \quad \Leftrightarrow \quad \rho^X(\Psi) \text{ is pure for all } X \in \{A,B\}.$$

or, equivalently,

$$|\Psi\rangle \text{ is entangled } \quad \Leftrightarrow \quad \rho^X(\Psi) \text{ is a true mixture for any } X \in \{A,B\}.$$

Proof The two statements are, of course, equivalent. It is thus sufficient to prove only the first statement. We show $\Rightarrow$ first. Let $|\Psi\rangle$ be separable. Then we know from Theorem 4.2 that there exist $|\varphi\rangle \in \mathbb{H}^A$ and $|\psi\rangle \in \mathbb{H}^B$ with $|\Psi\rangle = |\varphi\rangle \otimes |\psi\rangle$. Because of

$$1 = ||\Psi|| = \sqrt{\langle\Psi|\Psi\rangle} \underbrace{=}_{(3.4)} \sqrt{\langle\varphi|\varphi\rangle}\sqrt{\langle\psi|\psi\rangle} \underbrace{=}_{(2.5)} ||\varphi||\,||\psi||$$

we must have $||\varphi|| \neq 0 \neq ||\psi||$. We define the unit vectors $|e_0\rangle := \frac{|\varphi\rangle}{||\varphi||}$ and $|f_0\rangle := \frac{|\psi\rangle}{||\psi||}$ and augment them by suitable vectors $|e_1\rangle, |e_2\rangle, \ldots$ and $|f_1\rangle, |f_2\rangle, \ldots$ in order to form the ONBs

$$\left\{|e_0\rangle := \frac{|\varphi\rangle}{||\varphi||}, |e_1\rangle, |e_2\rangle, \ldots\right\} \subset \mathbb{H}^A \qquad \text{and} \qquad \left\{|f_0\rangle := \frac{|\psi\rangle}{||\psi||}, |f_1\rangle, |f_2\rangle, \ldots\right\} \subset \mathbb{H}^B,$$

such that

$$|\Psi\rangle = |\varphi\rangle \otimes |\psi\rangle = ||\varphi||\,||\psi||\,|e_0\rangle \otimes |f_0\rangle = \sum_{a,b}\Psi_{ab}|e_a\rangle \otimes |f_b\rangle,$$

where

$$\Psi_{ab} = \begin{cases} ||\varphi||\,||\psi|| = 1 & \text{if } a = 0 = b \\ 0 & \text{else.} \end{cases}$$

Thus, we have

$$\rho^A(\Psi) = \sum_{a_1,a_2,b} \Psi_{a_1b}\overline{\Psi_{a_2b}}|e_{a_2}\rangle\langle e_{a_1}| = ||\varphi||\,||\psi||\,|e_0\rangle\langle e_0|$$
$$= |e_0\rangle\langle e_0|,$$

which as a projection onto a one-dimensional subspace is a pure state. Consequently, it satisfies $\left(\rho^A(\Psi)\right)^2 = |e_0\rangle\underbrace{\langle e_0|e_0\rangle}_{=1}\langle e_0| = \rho^A(\Psi)$. Similarly, one shows $\rho^B(\Psi) = |f_0\rangle\langle f_0|$, proving that $\rho^B(\Psi)$ is a pure state as well.

We proceed to prove the reverse implication $\Leftarrow$. Let $\rho^A(\Psi)$ be a pure state. Hence, there exists a unit vector $|\varphi\rangle \in \mathbb{H}^A$ such that $\rho^A(\Psi) = |\varphi\rangle\langle\varphi|$. This density operator $\rho^A(\Psi)$ has exactly one eigenvector with eigenvalue 1 and a degenerate eigenvalue 0. According to the SCHMIDT decomposition (3.64) the vector $|\Psi\rangle$ then has the form $|\Psi\rangle = |\varphi\rangle \otimes |\psi\rangle$ with unit vectors $|\varphi\rangle \in \mathbb{H}^A$ and $|\psi\rangle \in \mathbb{H}^B$. The same arguments apply if $\rho^B(\Psi)$ is assumed as a pure state. $\square$

Definition 4.4 A pure state $|\Psi\rangle$ in the tensor product of identical HILBERT spaces $\mathbb{H}^A$ is said to be **maximally entangled** if

$$\rho^A(\Psi) = \lambda\mathbf{1}$$

with $0 < \lambda < 1$.

From (3.55) and the result shown in Exercise 3.44 we see that the vectors $|\Phi^\pm\rangle, |\Psi^\pm\rangle$ of the BELL basis are maximally entangled.

4.3   Entanglement Swapping

As we shall see in Sect. 4.4, entanglement leads to phenomena, which EINSTEIN called 'spooky action at a distance' and which contributed considerably to his doubts about quantum mechanics. It may thus seem even more spooky that systems can be entangled even if they have not interacted with each other. This phenomenon, which has become known as **entanglement swapping** [18, 65, 66], has indeed been performed experimentally [67]. It comes about as follows.

Suppose a four-qubit state $|\Phi\rangle^{ABCD} \in \mathbb{H}^A \otimes \mathbb{H}^B \otimes \mathbb{H}^C \otimes \mathbb{H}^D =: \mathbb{H}^{ABCD}$ has been prepared as a separable product-state of two entangled two-qubit BELL states $|\Psi^-\rangle^{AB} \in \mathbb{H}^A \otimes \mathbb{H}^B =: \mathbb{H}^{AB}$ and $|\Psi^-\rangle^{CD} \in \mathbb{H}^C \otimes \mathbb{H}^D =: \mathbb{H}^{CD}$ such that

$$
\begin{aligned}
|\Phi\rangle^{ABCD} &= |\Psi^-\rangle^{AB} \otimes |\Psi^-\rangle^{CD} \\
&= \frac{1}{2}\left(|0101\rangle - |0110\rangle - |1001\rangle + |1010\rangle\right) \\
&= \frac{1}{2}\Big(|\Psi^+\rangle^{AD} \otimes |\Psi^+\rangle^{BC} - |\Psi^-\rangle^{AD} \otimes |\Psi^-\rangle^{BC} \\
&\quad - |\Phi^+\rangle^{AD} \otimes |\Phi^+\rangle^{BC} + |\Phi^-\rangle^{AD} \otimes |\Phi^-\rangle^{BC}\Big),
\end{aligned}
\tag{4.13}
$$

where, for example,

$$
\begin{aligned}
|\Psi^+\rangle^{AD} \otimes |\Psi^+\rangle^{BC} &= \frac{1}{2}\Big(|0\rangle^A \otimes \left[|0\rangle^B \otimes |1\rangle^C + |1\rangle^B \otimes |0\rangle^C\right] \otimes |1\rangle^D \\
&\quad + |1\rangle^A \otimes \left[|0\rangle^B \otimes |1\rangle^C + |1\rangle^B \otimes |0\rangle^C\right] \otimes |0\rangle^D\Big) \\
&= \frac{1}{2}\Big(|0011\rangle + |0101\rangle + |1010\rangle + |1100\rangle\Big).
\end{aligned}
$$

Systems A and B may have interacted in some way to form the entangled state $|\Psi^-\rangle^{AB}$. Likewise, systems C and D may have interacted to form the entangled state $|\Psi^-\rangle^{CD}$. However, we can prepare the entangled states $|\Psi^-\rangle^{AB}$ and $|\Psi^-\rangle^{CD}$ such that system A has never interacted with either C or D or be influenced in any way by these systems. Nevertheless, we will now show that by suitable measurements in the state $|\Phi\rangle^{ABCD}$ of the total composite system it is possible to create entangled states in the system AD composed of the sub-systems A and D.

From (3.41) we see that the operators

$$
\begin{aligned}
\Sigma_z^{BC} &:= \mathbf{1} \otimes \sigma_z \otimes \sigma_z \otimes \mathbf{1} \\
\Sigma_x^{BC} &:= \mathbf{1} \otimes \sigma_x \otimes \sigma_x \otimes \mathbf{1}
\end{aligned}
$$

commute. Hence, the corresponding observables BC-spin in the z-direction and BC-spin in the x-direction can both be measured sharply in a given state. The measurement of the observables defined by Σ_z^{BC} and Σ_x^{BC} in the state $|\Phi\rangle^{ABCD}$ collapses the state of the qubit-pair BC to one of the states $|\Psi^\pm\rangle^{BC}$ or $|\Phi^\pm\rangle^{BC}$ depending on which values have been observed. In Table 3.1 we can read off, which BC-state corresponds to which pair of measured values. If, for example, for $(\Sigma_z^{BC}, \Sigma_x^{BC})$ the values $(-1, +1)$ have been observed, then the particle-pair BC is in the state $|\Psi^+\rangle^{BC}$. The middle column in Table 4.1 lists, in which state in $\mathbb{H}^{ABCD}$ the composite system is after measurement, given the observed values of Σ_z^{BC} and Σ_x^{BC}.

With regard to the sub-systems AD and BC the composite system is thus *after the measurement* always in a state separable in the BELL basis vectors in $\mathbb{H}^{AD}$ and $\mathbb{H}^{BC}$. If we only consider the sub-system AD the reduced density operators ρ^{AD} of the

Table 4.1   Determination of post-measurement state by measurement of Σ_z^{BC} and Σ_x^{BC} on $|\Phi\rangle^{ABCD}$

Measured value of		Composite state after measurement of Σ_z^{BC} and Σ_x^{BC}	State of sub-system AD after measurement of Σ_z^{BC} and Σ_x^{BC} on $|\Phi\rangle^{ABCD}$
Σ_z^{BC}	Σ_x^{BC}		
$+1$	$+1$	$|\Phi^+\rangle^{AD} \otimes |\Phi^+\rangle^{BC}$	$|\Phi^+\rangle^{AD}$
$+1$	-1	$|\Phi^-\rangle^{AD} \otimes |\Phi^-\rangle^{BC}$	$|\Phi^-\rangle^{AD}$
-1	$+1$	$|\Psi^+\rangle^{AD} \otimes |\Psi^+\rangle^{BC}$	$|\Psi^+\rangle^{AD}$
-1	-1	$|\Psi^-\rangle^{AD} \otimes |\Psi^-\rangle^{BC}$	$|\Psi^-\rangle^{AD}$

states in the middle column of Table 4.1 represent pure states as given in the right column of Table 4.1. Thus, after the measurement of Σ_z^{BC} and Σ_x^{BC} the qubit-pair AD is in the state that in (4.13) is paired with the observed BC-state. This observed BC-state is given by the pair of measured values for Σ_z^{BC} and Σ_x^{BC}. Consequently, after the measurement the qubits A and D are entangled even though they have not interacted with each other at all.

4.4   EINSTEIN–PODOLSKY–ROSEN-Paradox

We begin by exhibiting a slightly modified version of the chain of arguments given in the original article of EINSTEIN, PODOLSKY and ROSEN (EPR) [4]. The origin of this article was EINSTEIN's dissatisfaction—or even rejection—of quantum mechanics, which he considered 'incomplete.' The goal of the arguments given by EPR is thus to show that the following statement is *wrong*.

EPR Claim 1   *The quantum mechanical description of a system by its state vector is* complete.

For simplicity we shall abbreviate EPR Claim 1 as:

Quantum mechanics
is complete.

Accordingly, the negation of this statement will be abbreviated as 'Quantum mechanics is *in*complete.'

EPR then begin by considering what constitutes a *complete* description of the reality of a system by a physical theory. Their minimal requirement for a complete theory of a system is that *every element of the physical reality of the system must have a corresponding element in the physical theory.* What then are elements of reality of a system? For the arguments of EPR it suffices that certain physical quantities constitute elements of reality. In their definition a *physical quantity is an element of reality of a system if the value of this quantity can be predicted with certainty, that is, with probability equal* 1 *without having to interact with the system.* For example, our experience tells us that a pencil resting on a table exposed only to the gravitational pull of the earth and the opposite neutralizing force from the table-top will remain at the same place. We can thus predict the physical quantity 'position' of the system 'pencil' with certainty *without looking.* Consequently, for the system pencil the physical quantity position constitutes an element of its reality. Now, consider a qubit described by the state $|0\rangle = |\uparrow_{\hat{z}}\rangle$. Since $|0\rangle$ is the eigenvector for the eigenvalue $+1$ of the observable spin in the z-direction, we know without measuring that the value of the physical quantity 'spin in z-direction' is $+1$. For qubits described by the state $|0\rangle$ the spin in z-direction thus constitutes an element of their reality. On the other hand, we *cannot predict with certainty* the value of the physical quantity 'spin in x-direction' for a qubit described by the state $|0\rangle$ since $|0\rangle$ is not an eigenvector of σ_x, and one finds for the uncertainty (see (2.79)) in this case

$$\Delta_{|0\rangle}(\sigma_x) = 1.$$

Hence, for a system prepared in the state $|0\rangle$ spin in x-direction does not constitute an element of its reality. In general, quantum mechanical observables M_1 and M_2 of a system *cannot be jointly* elements of reality *if they do not commute,* that is, if $M_1 M_2 \neq M_2 M_1$. This is because in this case not all eigenvectors of M_1 can also be eigenvectors of M_2. But the value an observable reveals, when measured, can be predicted with certainty (in other words, with vanishing uncertainty) only if the system is in an eigenstate of the observable. Consequently, the values of M_1 and M_2 cannot be predicted jointly with certainty if $M_1 M_2 \neq M_2 M_1$. We formulate this as

EPR Claim 2 *The physical quantities of a system belonging to two incompatible observables cannot be* jointly *elements of reality for that system.*

We abbreviate EPR Claim 2 as

The values of *incompatible observables* are *not jointly real.*

The negation of this statement is abbreviated accordingly as 'The values of incompatible observables are jointly real.'

EPR show then that the completeness of quantum mechanics (EPR Claim 1) implies EPR Claim 2. To do this they apply the implications of quantum mechanics about sharp measurement results and uncertainty as follows: suppose, the negation of EPR Claim 2 were true, in other words, the physical quantities corresponding to two incompatible observables of a system were both elements of reality of that system and thus could both be predicted with certainty. If the quantum mechanical description were complete, that is, if EPR Claim 1 were true, then the state vector should provide a prediction of the values with certainty. But it does not, since the observables are assumed incompatible. Hence, we have the implication

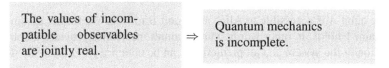

Contraposition of this implication yields

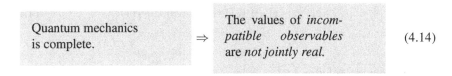

(4.14)

EPR then proceed to prove with the help of entangled states and a 'reasonable definition of reality' that *apparently*

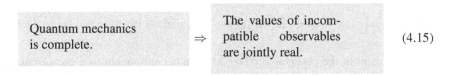

(4.15)

holds.

This is the **EPR paradox**: the implications in (4.14) and (4.15) cannot be simultaneously true. EPR conclude from that, that then

$$\boxed{\text{Quantum mechanics is complete.}} \quad \text{is} \quad \text{FALSE}$$

has to hold, and to show that was the goal of EPR in [4]. There EPR demonstrate the supposed validity of (4.15) with a line of arguments, at the end of which they make use of a reasonable definition of reality, which we shall look at in the following. That (4.15) does not hold after all, comes from the fact that the reality of the considered systems, is surprisingly 'unreasonable' to a degree, which EPR in [4] did not believe possible. However, this counter-intuitive reality of quantum mechanical systems has since repeatedly been confirmed in experiments.

Let us then have a look at the chain of arguments, suitably modified for our know-how that EPR provide for the proof of (4.15). For this we consider BOHM's version of the EPR thought-experiment, which should be readily digestible with the material presented in previous chapters. Consider the preparation of two qubits, the composite system of which is described by the BELL state

$$|\Phi^+\rangle = \frac{1}{\sqrt{2}}\Big(|00\rangle + |11\rangle\Big) = \frac{1}{\sqrt{2}}\left(|\uparrow_{\hat{z}}\rangle \otimes |\uparrow_{\hat{z}}\rangle + |\downarrow_{\hat{z}}\rangle \otimes |\downarrow_{\hat{z}}\rangle\right). \qquad (4.16)$$

Of these qubit A is accessible to Alice and qubit B to Bob. We further assume that EPR Claim 1 holds, in other words, that quantum mechanics provides a complete description of the system and all predictions can be obtained from the state $|\Phi^+\rangle$.

Exercise 4.53 Show that for the eigenvectors $|\uparrow_{\hat{x}}\rangle$ and $|\downarrow_{\hat{x}}\rangle$ of σ_x for the eigenvalues ± 1

$$|\uparrow_{\hat{x}}\rangle \otimes |\uparrow_{\hat{x}}\rangle + |\downarrow_{\hat{x}}\rangle \otimes |\downarrow_{\hat{x}}\rangle = |00\rangle + |11\rangle \qquad (4.17)$$

holds.

For a solution see Solution 4.53.

From (4.16) and (4.17) it follows that

$$|\Phi^+\rangle = \frac{1}{\sqrt{2}}\left(|\uparrow_{\hat{z}}\rangle \otimes |\uparrow_{\hat{z}}\rangle + |\downarrow_{\hat{z}}\rangle \otimes |\downarrow_{\hat{z}}\rangle\right) = \frac{1}{\sqrt{2}}\left(|\uparrow_{\hat{x}}\rangle \otimes |\uparrow_{\hat{x}}\rangle + |\downarrow_{\hat{x}}\rangle \otimes |\downarrow_{\hat{x}}\rangle\right). \quad (4.18)$$

A measurement of the observable σ_z by Alice in her sub-system is a measurement of $\sigma_z \otimes \mathbf{1}$ in the composite system. The eigenvalues of this composite observable are ± 1 and are degenerate. The eigenspaces for the eigenvalues ± 1 are

$$\mathrm{Eig}(\sigma_z, +1) = \mathrm{Span}\left\{|\uparrow_{\hat{z}}\rangle \otimes |\psi\rangle \,\big|\, |\psi\rangle \in \mathbb{H}^B\right\}$$
$$\mathrm{Eig}(\sigma_z, -1) = \mathrm{Span}\left\{|\downarrow_{\hat{z}}\rangle \otimes |\psi\rangle \,\big|\, |\psi\rangle \in \mathbb{H}^B\right\}.$$

The projections onto these eigenspaces are

$$P_{z,+1} = |\uparrow_{\hat{z}}\rangle\langle\uparrow_{\hat{z}}| \otimes \mathbf{1} \qquad \text{and} \qquad P_{z,-1} = |\downarrow_{\hat{z}}\rangle\langle\downarrow_{\hat{z}}| \otimes \mathbf{1}. \qquad (4.19)$$

They satisfy

$$
P_{z,+1}|\Phi^+\rangle \underbrace{=}_{(4.18),(4.19)} \frac{1}{\sqrt{2}}\left(\left(|\uparrow_{\hat{z}}\rangle\langle\uparrow_{\hat{z}}|\otimes\mathbf{1}\right)\left(|\uparrow_{\hat{z}}\rangle\otimes|\uparrow_{\hat{z}}\rangle+|\downarrow_{\hat{z}}\rangle\otimes|\downarrow_{\hat{z}}\rangle\right)\right)
$$

$$
= \frac{1}{\sqrt{2}}\left(|\uparrow_{\hat{z}}\rangle\underbrace{\langle\uparrow_{\hat{z}}|\uparrow_{\hat{z}}\rangle}_{=1}\otimes|\uparrow_{\hat{z}}\rangle+|\uparrow_{\hat{z}}\rangle\underbrace{\langle\uparrow_{\hat{z}}|\downarrow_{\hat{z}}\rangle}_{=0}\otimes|\downarrow_{\hat{z}}\rangle\right)
$$

$$
= \frac{1}{\sqrt{2}}|\uparrow_{\hat{z}}\rangle\otimes|\uparrow_{\hat{z}}\rangle \tag{4.20}
$$

and

$$
P_{z,-1}|\Phi^+\rangle = \frac{1}{\sqrt{2}}|\downarrow_{\hat{z}}\rangle\otimes|\downarrow_{\hat{z}}\rangle
$$

$$
\left|\left|P_{z,\pm1}|\Phi^+\rangle\right|\right| = \frac{1}{\sqrt{2}}. \tag{4.21}
$$

If Alice measures the observable σ_z in her sub-system and detects the value $+1$, then the composite system is—in accordance with the Projection Postulate 3 in Sect. 2.3.1—after that measurement in the normalized composite state

$$
|\Psi_{z,+1}\rangle := \frac{P_{z,+1}|\Phi^+\rangle}{||P_{z,+1}|\Phi^+\rangle||} = \frac{\left(|\uparrow_{\hat{z}}\rangle\langle\uparrow_{\hat{z}}|\otimes\mathbf{1}\right)|\Phi^+\rangle}{\left|\left|\left(|\uparrow_{\hat{z}}\rangle\langle\uparrow_{\hat{z}}|\otimes\mathbf{1}\right)|\Phi^+\rangle\right|\right|} \underbrace{=}_{(4.20),(4.21)} |\uparrow_{\hat{z}}\rangle\otimes|\uparrow_{\hat{z}}\rangle. \tag{4.22}
$$

This means that Bob's system will be described by

$$
\rho^B\left(|\Psi_{z,+1}\rangle\langle\Psi_{z,+1}|\right) \underbrace{=}_{(3.56)} \mathrm{tr}^A\left(|\Psi_{z,+1}\rangle\langle\Psi_{z,+1}|\right) \underbrace{=}_{(4.22)} \mathrm{tr}^A\left(\left(|\uparrow_{\hat{z}}\rangle\otimes|\uparrow_{\hat{z}}\rangle\right)\left(\langle\uparrow_{\hat{z}}|\otimes\langle\uparrow_{\hat{z}}|\right)\right)
$$

$$
\underbrace{=}_{(3.36)} \mathrm{tr}^A\left(|\uparrow_{\hat{z}}\rangle\langle\uparrow_{\hat{z}}|\otimes|\uparrow_{\hat{z}}\rangle\langle\uparrow_{\hat{z}}|\right) \underbrace{=}_{(3.57)} \underbrace{\mathrm{tr}\left(|\uparrow_{\hat{z}}\rangle\langle\uparrow_{\hat{z}}|\right)}_{=1}|\uparrow_{\hat{z}}\rangle\langle\uparrow_{\hat{z}}|
$$

$$
= |\uparrow_{\hat{z}}\rangle\langle\uparrow_{\hat{z}}|,
$$

which is the density operator of the pure state $|\uparrow_{\hat{z}}\rangle$. Hence, after a measurement of σ_z, in which Alice observes the value $+1$, Bob's system *has to be* in the state $|\uparrow_{\hat{z}}\rangle$. The value which would be observed if σ_z were measured by Bob on system B can then be predicted *with certainty* to be $+1$ without actually having to measure it. Analogously, if Alice measures σ_z on her qubit and observes the value -1, then the composite system becomes

$$
\frac{P_{z,-1}|\Phi^+\rangle}{||P_{z,-1}|\Phi^+\rangle||} = \frac{\left(|\downarrow_{\hat{z}}\rangle\langle\downarrow_{\hat{z}}|\otimes\mathbf{1}\right)|\Phi^+\rangle}{\left|\left|\left(|\downarrow_{\hat{z}}\rangle\langle\downarrow_{\hat{z}}|\otimes\mathbf{1}\right)|\Phi^+\rangle\right|\right|} \underbrace{=}_{(4.21)} |\downarrow_{\hat{z}}\rangle\otimes|\downarrow_{\hat{z}}\rangle.
$$

In this case Bob's system has to be in the state $|\downarrow_{\hat{z}}\rangle$ and the value which would be observed if σ_z were measured by Bob on system B can be predicted *with certainty* to be -1 without actually having to measure it. Consequently, the spin in z-direction is an element of reality for Bob's qubit.

If, however, Alice chooses instead to measure σ_x (rather than σ_z) and observes the value $+1$, then according to the Projection Postulate 3 in Sect. 2.3.1 the composite system is after that measurement in the normalized composite state

$$\frac{P_{x,+1}|\Phi^+\rangle}{||P_{x,+1}|\Phi^+\rangle||} = \frac{\left(|\uparrow_{\hat{x}}\rangle\langle\uparrow_{\hat{x}}|\otimes\mathbf{1}\right)|\Phi^+\rangle}{\left\|\left(|\uparrow_{\hat{x}}\rangle\langle\uparrow_{\hat{x}}|\otimes\mathbf{1}\right)|\Phi^+\rangle\right\|} \underset{(4.18)}{=} |\uparrow_{\hat{x}}\rangle\otimes|\uparrow_{\hat{x}}\rangle.$$

In this case Bob's qubit will be in the state $|\uparrow_{\hat{x}}\rangle$, and the value of the spin in x-direction can be predicted with certainty to be $+1$ without the need to have it measured. Similarly, if Alice measures σ_x and observes the value -1, then Bob's qubit after a measurement will be in the state $|\downarrow_{\hat{x}}\rangle$, and, likewise, we can predict with certainty that any measurement of the spin in x-direction on Bob's qubit will reveal the value -1. Thus, if Alice measures σ_x on her qubit, then for Bob's qubit the spin in x-direction is an element of reality.

Regardless, in which direction Alice measures the spin of her qubit, the spin of Bob's qubit in the same direction can always be predicted with certainty without the need to measure it. This means that Alice's choice of the direction z or x for a spin-measurement on her qubit determines whether for Bob's qubit the spin in z- or x-direction is an element of its reality.

This also holds when Alice and Bob are separated by such a distance and perform their measurements in a way that no signal from Alice traveling at the speed of light can reach Bob before he would perform his measurement. Since Alice is free to choose σ_z or σ_x, and Bob's qubit cannot 'know' which direction Alice has chosen, both spin in z-direction *and* spin in x-direction are elements of reality for Bob's qubit and this is *despite* $\sigma_x\sigma_z \neq \sigma_z\sigma_x$.

The objection that Alice cannot measure σ_x and σ_z *jointly sharply*, but only ever one of them and thus for Bob's system only the spin in that same direction can be an element of reality and not both simultaneously is refuted by EPR with the argument that then the possible elements of reality of Bob's qubit are determined by Alice's choice even though no signal from Alice can reach Bob's qubit in time to communicate that choice. Regarding this EPR state: 'no reasonable definition of reality can admit that' [4].

If one accepts this last argument, then (4.15) would be proven and it were shown that quantum mechanics is not a complete description of the systems. One possibility could be that there are further variables that determine the behavior of the systems, but which are not revealed by the quantum mechanical description of the system by a state vector. These were called (local) *hidden variables*.

But the incompleteness of quantum mechanics, that is, the existence of hidden variables, is not the only way out. Rather, all experiments to this date exhibit that reality is—in the sense of EPR—'unreasonable'. In our example this means that (local) measurements by Alice are indeed shown to have an immediate (in other words, faster than light) impact on Bob's system even though no detectable signal has been sent. This is commonly called quantum mechanical *non-locality*.

The central role occupied in this context by BELL's inequality is due to the fact that on the one hand this inequality is based on the assumption of hidden variables, whereas on the other hand a violation of this inequality is predicted by quantum mechanics for certain states. This opened up the possibility of an experimental test for the existence of hidden variables.

4.5 BELL Inequality

In an article, which at the time drew rather little attention, BELL [63] considered a pair of qubits in a composite entangled state $|\Psi^-\rangle$ and assumed that there exist variables that determine the spin-observables of the qubits in any direction. From that he derived an inequality for the expectation values of the spin-observables in various directions. That is the BELL *inequality*, which we will first derive in the form originally given by BELL [63] and then in a more general form derived later by CLAUSER, HORNE, SHIMONY and HOLT (CHSH) [2]. Experiments have since shown that nature violates the BELL inequality [9]. More precisely, there exist entangled states, in which the expectation values of products of spin-observables (aka 'spin-correlations') in certain directions violate the BELL inequality. This implies that the behavior of such systems is not determined by hidden local variables, since their existence is the starting assumption for the derivation of the BELL inequality.

4.5.1 Original BELL Inequality

Assuming the existence of hidden variables, which determine the observed values of spin-observables completely, is equivalent to assuming that the measured values have a joint distribution (see Appendix A). Essentially, the BELL inequality thus follows from the assumption, that the results of spin-measurements on two qubits can be represented as discrete random variables of a joint distribution. As we shall see, the BELL inequality is violated in certain entangled states. Consequently, the assumption of a joint distribution for spin values of two qubits in certain entangled states is invalid and thus also the equivalent assumption of hidden variables.

The derivation originally given by BELL [63] for the inequality which now bears his name runs as follows. Consider a pair of qubits that have been prepared in the entangled BELL state

$$|\Psi^-\rangle \underbrace{=}_{(3.28)} \frac{1}{\sqrt{2}}\Big(|01\rangle - |10\rangle\Big) \in \mathbb{H}^A \otimes \mathbb{H}^B,$$

of which qubit A is sent to Alice and qubit B to Bob. Alice can perform a spin-measurement, in which she can arbitrarily select the direction in which she measures the spin of qubit A. This direction is represented by a unit vector in $\mathbb{R}^3$ as defined in (2.122) and given as

$$\hat{\mathbf{n}} = \hat{\mathbf{n}}(\theta,\phi) = \begin{pmatrix} \sin\theta\cos\phi \\ \sin\theta\sin\phi \\ \cos\theta \end{pmatrix} \in \mathbb{R}^3.$$

On her qubit A she thus measures the observable $\hat{\mathbf{n}}^A \cdot \boldsymbol{\sigma}$, that is, spin in the direction determined by the unit vector $\hat{\mathbf{n}}^A$. We denote this observable by

$$\Sigma^A_{\hat{\mathbf{n}}^A} = \hat{\mathbf{n}}^A \cdot \boldsymbol{\sigma} \tag{4.23}$$

and the values observed when measuring it by $s^A_{\hat{\mathbf{n}}^A}$. The $s^A_{\hat{\mathbf{n}}^A}$ thus constitute a set of discrete random variables (see Appendix A) parametrized by $\hat{\mathbf{n}}^A$ that can only take values in $\{\pm 1\}$. Similarly, Bob can measure the spin of his qubit B in a direction $\hat{\mathbf{n}}^B$, which he can select arbitrarily and independently of Alice. We denote his spin-observable by

$$\Sigma^B_{\hat{\mathbf{n}}^B} = \hat{\mathbf{n}}^B \cdot \boldsymbol{\sigma}. \tag{4.24}$$

Likewise, the values he observes when measuring $\Sigma^B_{\hat{\mathbf{n}}^B}$ constitute discrete random variables, which we denote by $s^B_{\hat{\mathbf{n}}^B}$. These are parametrized by $\hat{\mathbf{n}}^B$ and can only take values in $\{\pm 1\}$.

Exercise 4.54 Let $|\uparrow_{\hat{\mathbf{n}}}\rangle$ and $|\downarrow_{\hat{\mathbf{n}}}\rangle$ be defined as in (2.125) and (2.126). Show that then

$$|\Psi^-\rangle = \frac{1}{\sqrt{2}}\big(|\uparrow_{\hat{\mathbf{n}}}\rangle \otimes |\downarrow_{\hat{\mathbf{n}}}\rangle - |\downarrow_{\hat{\mathbf{n}}}\rangle \otimes |\uparrow_{\hat{\mathbf{n}}}\rangle\big). \tag{4.25}$$

For a solution see Solution 4.54

The quantum mechanical expectation value of the observable $\Sigma^A_{\hat{\mathbf{n}}^A} \otimes \Sigma^B_{\hat{\mathbf{n}}^B}$ in the state $|\Psi^-\rangle$ is given by the negative of the cosine of the angle between $\hat{\mathbf{n}}^A$ and $\hat{\mathbf{n}}^B$. This is to be shown in Exercise 4.55.

Exercise 4.55 Show that for $\Sigma^A_{\hat{\mathbf{n}}^A}$ and $\Sigma^B_{\hat{\mathbf{n}}^B}$ as defined in (4.23) and (4.24) one has

$$\left\langle \Sigma^A_{\hat{\mathbf{n}}^A} \otimes \Sigma^B_{\hat{\mathbf{n}}^B} \right\rangle_{\Psi^-} = -\hat{\mathbf{n}}^A \cdot \hat{\mathbf{n}}^B. \tag{4.26}$$

For a solution see Solution 4.55

In particular, it follows from (4.26) for measurements that Alice and Bob perform in the same direction $\hat{\mathbf{n}}^A = \hat{\mathbf{n}} = \hat{\mathbf{n}}^B$, that

$$\left\langle \Sigma_{\hat{\mathbf{n}}}^A \otimes \Sigma_{\hat{\mathbf{n}}}^B \right\rangle_{\Psi^-} = -1. \tag{4.27}$$

Suppose now that the properties of each qubit are determined by a variable ω, the value of which we do not know, in other words, which is 'hidden'. We may, however, assume that ω lies in a suitable set Ω. Furthermore, we assume that the description of the qubits by $\omega \in \Omega$ is complete in the sense that the measured spin values $s_{\hat{\mathbf{n}}^A}^A(\omega)$ and $s_{\hat{\mathbf{n}}^B}^B(\omega)$ in any directions $\hat{\mathbf{n}}^A$ and $\hat{\mathbf{n}}^B$ are completely determined by the variable ω. If Alice were to know for her qubit the value of ω, she would be able to determine the function $s_{\hat{\mathbf{n}}^A}^A(\omega)$ by sufficiently many measurements. The knowledge of that function would then enable her to predict the result of spin measurements on her qubit, given the knowledge of ω. The same applies for Bob. But the value of ω for a given qubit is not known, which is why they are called *hidden* variables. We can only assume that every value of $\omega \in \Omega$ occurs with a certain probability $0 \leq \mathbf{P}(\omega) \leq 1$ with the property

$$\mathbf{P}(\Omega) = \int_\Omega d\mathbf{P}(\omega) = 1.$$

Altogether this means that we assume that the observable spin measurement values $s_{\hat{\mathbf{n}}^A}^A$ and $s_{\hat{\mathbf{n}}^B}^B$ constitute discrete random variables on a probability space $(\Omega, A, \mathbf{P})$ (see Appendix A), which are parametrized by unit vectors $\hat{\mathbf{n}}^A$ and $\hat{\mathbf{n}}^B$. These random variables do depend on the state in which the particles are prepared. For our considerations this is the state $|\Psi^-\rangle$. Since we consider $s_{\hat{\mathbf{n}}^A}^A$ and $s_{\hat{\mathbf{n}}^B}^B$ as the values of observed spins for the particles in state $|\Psi^-\rangle$, we also require that they satisfy the equivalent of (4.27), which means that for arbitrary $\hat{\mathbf{n}}$

$$\mathbf{E}\left[s_{\hat{\mathbf{n}}}^A s_{\hat{\mathbf{n}}}^B\right] \underbrace{=}_{(A.3)} \sum_{(s_1,s_2)\in\{\pm 1, \pm 1\}} s_1 s_2 \mathbf{P}\left\{s_{\hat{\mathbf{n}}}^A = s_1 \text{ and } s_{\hat{\mathbf{n}}}^B = s_2\right\}$$

satisfies

$$\mathbf{E}\left[s_{\hat{\mathbf{n}}}^A s_{\hat{\mathbf{n}}}^B\right] = -1.$$

With these assumptions BELL then proves the following theorem.

Theorem 4.5 *Let $s_{\hat{\mathbf{n}}}^A$ and $s_{\hat{\mathbf{n}}}^B$ be two discrete random variables on a probability space $(\Omega, A, \mathbf{P})$ that are parametrized by unit vectors $\hat{\mathbf{n}} \in \mathbb{R}^3$ and take values in $\{\pm 1\}$, that is,*

$$s^X : \begin{matrix} S^1_{\mathbb{R}^3} \times \Omega \longrightarrow \{\pm 1\} \\ (\hat{\mathbf{n}}, \omega) \longmapsto s^X_{\hat{\mathbf{n}}}(\omega) \end{matrix}, \qquad \text{for } X \in \{A, B\}, \tag{4.28}$$

and which, in addition, satisfy for all $\hat{\mathbf{n}} \in S^1_{\mathbb{R}^3}$

$$\mathbf{E}\left[s^A_{\hat{\mathbf{n}}} s^B_{\hat{\mathbf{n}}}\right] = -1. \tag{4.29}$$

Then for arbitrary unit vectors $\hat{\mathbf{n}}^i$ with $i \in \{1, 2, 3\}$ the **BELL** inequality

$$\left| \mathbf{E}\left[s^A_{\hat{\mathbf{n}}^1} s^B_{\hat{\mathbf{n}}^2}\right] - \mathbf{E}\left[s^A_{\hat{\mathbf{n}}^1} s^B_{\hat{\mathbf{n}}^3}\right] \right| - \mathbf{E}\left[s^A_{\hat{\mathbf{n}}^2} s^B_{\hat{\mathbf{n}}^3}\right] \leq 1 \tag{4.30}$$

holds.

Proof From (4.28), (4.29) and (A.3) it follows that

$$-1 = \mathbf{E}\left[s^A_{\hat{\mathbf{n}}} s^B_{\hat{\mathbf{n}}}\right] = \underbrace{\mathbf{P}\left\{s^A_{\hat{\mathbf{n}}} = s^B_{\hat{\mathbf{n}}}\right\}}_{=1-\mathbf{P}\left\{s^A_{\hat{\mathbf{n}}} = -s^B_{\hat{\mathbf{n}}}\right\}} - \mathbf{P}\left\{s^A_{\hat{\mathbf{n}}} = -s^B_{\hat{\mathbf{n}}}\right\} = 1 - 2\mathbf{P}\left\{s^A_{\hat{\mathbf{n}}} = -s^B_{\hat{\mathbf{n}}}\right\}$$

and thus

$$\mathbf{P}\left\{s^A_{\hat{\mathbf{n}}} = -s^B_{\hat{\mathbf{n}}}\right\} = 1 \tag{4.31}$$

for arbitrary directions $\hat{\mathbf{n}}$. Furthermore, we have then

$$
\begin{aligned}
\mathbf{E}\left[s^A_{\hat{\mathbf{n}}^1} s^B_{\hat{\mathbf{n}}^2}\right] - \mathbf{E}\left[s^A_{\hat{\mathbf{n}}^1} s^B_{\hat{\mathbf{n}}^3}\right] &\underbrace{=}_{(4.31)} -\mathbf{E}\left[s^A_{\hat{\mathbf{n}}^1} s^A_{\hat{\mathbf{n}}^2}\right] + \mathbf{E}\left[s^A_{\hat{\mathbf{n}}^1} s^A_{\hat{\mathbf{n}}^3}\right] = \mathbf{E}\left[s^A_{\hat{\mathbf{n}}^1}\left(s^A_{\hat{\mathbf{n}}^3} - s^A_{\hat{\mathbf{n}}^2}\right)\right] \\
&\underbrace{=}_{\left(s^A_{\hat{\mathbf{n}}^2}\right)^2 = 1} \mathbf{E}\left[s^A_{\hat{\mathbf{n}}^1}\left(\left(s^A_{\hat{\mathbf{n}}^2}\right)^2 s^A_{\hat{\mathbf{n}}^3} - s^A_{\hat{\mathbf{n}}^2}\right)\right] \\
&= \mathbf{E}\left[s^A_{\hat{\mathbf{n}}^1} s^A_{\hat{\mathbf{n}}^2}\left(s^A_{\hat{\mathbf{n}}^2} s^A_{\hat{\mathbf{n}}^3} - 1\right)\right].
\end{aligned}
$$

This implies the claimed inequality as follows:

$$
\begin{aligned}
\left| \mathbf{E}\left[s^A_{\hat{\mathbf{n}}^1} s^B_{\hat{\mathbf{n}}^2}\right] - \mathbf{E}\left[s^A_{\hat{\mathbf{n}}^1} s^B_{\hat{\mathbf{n}}^3}\right] \right| &= \left| \mathbf{E}\left[s^A_{\hat{\mathbf{n}}^1} s^A_{\hat{\mathbf{n}}^2}\left(s^A_{\hat{\mathbf{n}}^2} s^A_{\hat{\mathbf{n}}^3} - 1\right)\right] \right| \\
&\leq \mathbf{E}\left[\left| s^A_{\hat{\mathbf{n}}^1} s^A_{\hat{\mathbf{n}}^2}\left(s^A_{\hat{\mathbf{n}}^2} s^A_{\hat{\mathbf{n}}^3} - 1\right)\right|\right] \\
&= \mathbf{E}\left[\left| s^A_{\hat{\mathbf{n}}^1} s^A_{\hat{\mathbf{n}}^2}\right| \left| s^A_{\hat{\mathbf{n}}^2} s^A_{\hat{\mathbf{n}}^3} - 1\right|\right] \\
&\underbrace{=}_{\left| s^A_{\hat{\mathbf{n}}^1} s^A_{\hat{\mathbf{n}}^2}\right| = 1} \mathbf{E}\left[\left| 1 - s^A_{\hat{\mathbf{n}}^2} s^A_{\hat{\mathbf{n}}^3}\right|\right] \underbrace{=}_{s^A_{\hat{\mathbf{n}}^2} s^A_{\hat{\mathbf{n}}^3} \leq 1} \mathbf{E}\left[1 - s^A_{\hat{\mathbf{n}}^2} s^A_{\hat{\mathbf{n}}^3}\right]
\end{aligned}
$$

$$= \quad 1 - \mathbf{E}\left[s^A_{\hat{\mathbf{n}}^2} s^A_{\hat{\mathbf{n}}^3}\right]$$

$$\underbrace{=}_{(4.31)} \quad 1 + \mathbf{E}\left[s^A_{\hat{\mathbf{n}}^2} s^B_{\hat{\mathbf{n}}^3}\right]. \qquad\qquad \Box$$

Here we point out once more, that the the assumption, that there exist hidden variables, which determine the values of spin-observables in arbitrary directions, is equivalent to the assumption, that $s^A_{\hat{\mathbf{n}}^A}$ and $s^B_{\hat{\mathbf{n}}^B}$ for arbitrary $\hat{\mathbf{n}}^A$ and $\hat{\mathbf{n}}^B$ are random variables on the same probability space $(\Omega, A, \mathbf{P})$ with a *joint distribution*. This means, in particular, that the spin-observables are determined by $\omega \in \Omega$, and that sets of the form $\{\omega \in \Omega \mid s^A_{\hat{\mathbf{n}}^1}(\omega) = a \text{ and } s^A_{\hat{\mathbf{n}}^2}(\omega) = b\}$ with $(a,b) \in \{\pm 1, \pm 1\}$ for arbitrary directions $\hat{\mathbf{n}}^1$ and $\hat{\mathbf{n}}^2$ are measurable with the probability measure $\mathbf{P}$. This latter property was essential for the proof as it has been used in deriving (4.29). Thus, if there exist hidden variables ω that determine the values of the spins Alice and Bob observe (and that make them simultaneous elements of reality for their particles), then the BELL inequality (4.30) has to hold.

What then do we get, if we insert for the left side of the BELL inequality the quantum mechanical expectation values? With (4.26) and the choice

$$\hat{\mathbf{n}}^1 = \begin{pmatrix} 1 \\ 0 \\ 0 \end{pmatrix}, \quad \hat{\mathbf{n}}^2 = \begin{pmatrix} \frac{1}{\sqrt{2}} \\ 0 \\ \frac{1}{\sqrt{2}} \end{pmatrix}, \quad \hat{\mathbf{n}}^3 = \begin{pmatrix} 0 \\ 0 \\ 1 \end{pmatrix} \qquad (4.32)$$

we obtain

$$\left| \left\langle \Sigma^A_{\hat{\mathbf{n}}^1} \otimes \Sigma^B_{\hat{\mathbf{n}}^2} \right\rangle_{\Psi^-} - \left\langle \Sigma^A_{\hat{\mathbf{n}}^1} \otimes \Sigma^B_{\hat{\mathbf{n}}^3} \right\rangle_{\Psi^-} \right| - \left\langle \Sigma^A_{\hat{\mathbf{n}}^2} \otimes \Sigma^B_{\hat{\mathbf{n}}^3} \right\rangle_{\Psi^-}$$

$$= \left| \hat{\mathbf{n}}^1 \cdot \hat{\mathbf{n}}^3 - \hat{\mathbf{n}}^1 \cdot \hat{\mathbf{n}}^2 \right| + \hat{\mathbf{n}}^2 \cdot \hat{\mathbf{n}}^3$$

$$= \left| -\frac{1}{\sqrt{2}} \right| + \frac{1}{\sqrt{2}} = \sqrt{2} > 1, \qquad (4.33)$$

which means that the quantum mechanical description predicts for the state $|\Psi^-\rangle$ and the choice (4.32) of directions the *violation of the* BELL *inequality*.

Which of the two possibilities (4.30) or (4.33) is then chosen by nature? The answer to that question was given by an experiment performed by ASPECT, DALIBARD and ROGER [9], which, however, used the CHSH-generalization of the BELL inequality. We shall thus first derive that generalization in Sect. 4.5.2 before we discuss the experiment.

But the answer to the question may be given here already: *nature behaves in accordance with the quantum mechanical prediction. It violates the* BELL *inequality in states for which quantum mechanics predicts the violation.*

The fact that in connection with the BELL inequality one often speaks of *correlations* is due to the following. The correlation of two random variables Z_1 and Z_2 is by Definition A.7 given as

$$\mathbf{cor}[Z_1, Z_2] = \frac{\mathbf{E}[Z_1 Z_2] - \mathbf{E}[Z_1] \mathbf{E}[Z_2]}{\sqrt{\left(\mathbf{E}[Z_1^2] - \mathbf{E}[Z_1]^2\right)\left(\mathbf{E}[Z_2^2] - \mathbf{E}[Z_2]^2\right)}} .$$

For random variables Z_i with $i \in \{1, 2\}$ and the properties

$$\begin{aligned} \mathbf{E}[Z_i] &= 0 \\ Z_i^2 &= 1 \end{aligned} \tag{4.34}$$

it thus follows that

$$\mathbf{cor}[Z_1, Z_2] = \mathbf{E}[Z_1 Z_2] .$$

Exercise 4.56 Show that for arbitrary $\hat{\mathbf{n}}^A$ and $\hat{\mathbf{n}}^B$

$$\left\langle \Sigma_{\hat{\mathbf{n}}^A}^A \otimes \mathbf{1}^B \right\rangle_{\Psi^-} = 0 = \left\langle \mathbf{1}^A \otimes \Sigma_{\hat{\mathbf{n}}^B}^B \right\rangle_{\Psi^-} . \tag{4.35}$$

For a solution see Solution 4.56

If we then require for $s_{\hat{\mathbf{n}}^A}^A$ and $s_{\hat{\mathbf{n}}^B}^B$ the equivalent of (4.35), that is,

$$\mathbf{E}\left[s_{\hat{\mathbf{n}}^A}^A\right] = 0 = \mathbf{E}\left[s_{\hat{\mathbf{n}}^B}^B\right] ,$$

then (4.34) is satisfied for the random variables $Z_1 = s_{\hat{\mathbf{n}}^A}^A$ and $Z_2 = s_{\hat{\mathbf{n}}^B}^B$, and we find indeed that

$$\mathbf{cor}[s_{\hat{\mathbf{n}}^A}^A, s_{\hat{\mathbf{n}}^B}^B] = \mathbf{E}\left[s_{\hat{\mathbf{n}}^A}^A s_{\hat{\mathbf{n}}^B}^B\right] .$$

In view of (4.33) compared to (4.30) it is thus often said that 'quantum correlations are stronger than classical correlations'. The correlations generated by entangled states are often called **EPR-correlations**.

4.5.2   CHSH Generalization of the BELL Inequality

Just as the original BELL inequality, the generalization derived by CLAUSER, HORNE, SIMONY and HOLT (CHSH) [2] also considers a pair of particles on which individual measurements yielding possible values in $\{\pm 1\}$ can be performed. The

CHSH generalization also provides an upper bound for expectation values of products of observable single-particle measurements. The generalization is that, unlike in BELL's original derivation, no requirement of the form (4.29) needs to be made.

The CHSH variant of the inequality is based on a surprisingly simple result, which we prove as the following lemma.

Lemma 4.6 *Let s_i with $i \in \{1, \ldots, 4\}$ be four discrete random variables on a probability space (Ω, A, P) that can take only the values in $\{\pm 1\}$, that is,*

$$s_i : \Omega \longrightarrow \{\pm 1\} \atop \omega \longmapsto s_i(\omega) , \qquad for\ i \in \{1, \ldots, 4\}.$$

Then the following inequality holds

$$|\mathbf{E}[s_1 s_2] - \mathbf{E}[s_1 s_3] + \mathbf{E}[s_2 s_4] + \mathbf{E}[s_3 s_4]| \leq 2. \tag{4.36}$$

Proof Because we have for all $\omega \in \Omega$ and $i \in \{1, \ldots, 4\}$

$$s_i(\omega) \in \{\pm 1\},$$

it follows that *either*

$$s_2(\omega) - s_3(\omega) = 0 \quad \Rightarrow \quad s_2(\omega) + s_3(\omega) = \pm 2$$

or

$$s_2(\omega) + s_3(\omega) = 0 \quad \Rightarrow \quad s_2(\omega) - s_3(\omega) = \pm 2,$$

and, using again $s_1(\omega), s_4(\omega) \in \{\pm 1\}$, thus,

$$s_1(\omega)\big(s_2(\omega) - s_3(\omega)\big) + s_4(\omega)\big(s_2(\omega) + s_3(\omega)\big) = \pm 2.$$

This implies

$$\big|s_1(\omega)\big(s_2(\omega) - s_3(\omega)\big) + s_4(\omega)\big(s_2(\omega) + s_3(\omega)\big)\big| \leq 2$$

and it follows from Lemma A.6 that

$$|\mathbf{E}[s_1 s_2] - \mathbf{E}[s_1 s_3] + \mathbf{E}[s_2 s_4] + \mathbf{E}[s_3 s_4]| = |\mathbf{E}[s_1(s_2 - s_3) + s_4(s_2 + s_3)]| \leq 2.$$

$\square$

We apply Lemma 4.6 to random variables given by the results of measurements to obtain a generalization of the BELL inequality as follows. We consider again pairs of particles of which one is accessible to Alice and the other to Bob. The observables they can measure on their respective particles are no longer restricted to observables identified by a direction $\hat{\mathbf{n}}$ in $\mathbb{R}^3$. Rather, more generally, Alice can perform a measurement in which she can select the measurement device by selecting a (possibly multi-dimensional) parameter $\mathbf{p}^A$ from a set P of device-parameters.[2] This means she measures an observable denoted by $S_{\mathbf{p}^A}^A$, the measured value of which we denote by $s_{\mathbf{p}^A}^A$ and which can take only values in $\{\pm 1\}$. Similarly, Bob has at his disposal a measurement device, which he can adjust at his own choosing. The state of the measurement device is described by parameters $\mathbf{p}^B \in P$, and the observable thus measured on his particle is denoted by $S_{\mathbf{p}^B}^B$. This observable can take only values in $\{\pm 1\}$ and the observed values are likewise denoted $s_{\mathbf{p}^B}^B$. Suppose now that each particle is described completely by a variable $\omega \in \Omega$, which we do not know, in other words, which is hidden. Completeness of description means that the variable $\omega \in \Omega$ determines $s_{\mathbf{p}^A}^A$ and $s_{\mathbf{p}^B}^B$. In other words, the values $s_{\mathbf{p}^A}^A$ and $s_{\mathbf{p}^B}^B$ measured for their respective device settings $\mathbf{p}^A$ and $\mathbf{p}^B$ are assumed to be parametrized random variables with a joint distribution on a probability space $(\Omega, A, \mathbf{P})$. CHSH then prove the following Theorem.

Theorem 4.7 *Let $s_{\mathbf{p}}^A$ and $s_{\mathbf{p}}^B$ be two discrete random variables on a probability space $(\Omega, A, \mathbf{P})$ that can take only values in $\{\pm 1\}$ and that are parametrized by a (possibly multi-dimensional) parameter $\mathbf{p}$ in a parameter set P, that is,*

$$s^X : P \times \Omega \longrightarrow \{\pm 1\} \atop (\mathbf{p}, \omega) \longmapsto s_{\mathbf{p}}^X(\omega), \qquad for\ X \in \{A, B\}.$$

Then for arbitrary parameters $\mathbf{p}_1, \ldots, \mathbf{p}_4 \in P$ the following generalization of the BELL *inequality given by* CLAUSER, HORNE, SHIMONY *and* HOLT *(CHSH) holds:*

$$\left| \mathbf{E}\left[s_{\mathbf{p}_1}^A s_{\mathbf{p}_2}^B \right] - \mathbf{E}\left[s_{\mathbf{p}_1}^A s_{\mathbf{p}_3}^B \right] + \mathbf{E}\left[s_{\mathbf{p}_4}^A s_{\mathbf{p}_2}^B \right] + \mathbf{E}\left[s_{\mathbf{p}_4}^A s_{\mathbf{p}_3}^B \right] \right| \leq 2. \qquad (4.37)$$

Proof The claim (4.37) follows immediately from Lemma 4.6 by setting in (4.36) $s_i = s_{\mathbf{p}_i}^A$ for $i \in \{1, 4\}$ and $s_i = s_{\mathbf{p}_i}^B$ for $i \in \{2, 3\}$. $\qquad\qquad\square$

In the derivation of (4.37) the EPR-implication has been used that all observables $S_{\mathbf{p}_i}^X$ for $X \in \{A, B\}$ and $i \in \{1, \ldots, 4\}$ are jointly elements of reality, in other words, that these observables for the considered particles always have one of the values from $\{\pm 1\}$ determined uniquely by a hidden variable ω, the value of which

---

[2]This includes, but is not restricted to, a spin-measurement in which she can select the direction in which the spin is measured.

Fig. 4.1 Choice of directions $\hat{\mathbf{n}}^i$ for $i \in \{1,\ldots,4\}$ in the (x,z)-plane with (4.40) for spin-measurements used to test the CHSH variant of the BELL inequality

is unknown to the observer. What then does quantum mechanics predict for the left side of (4.37)? For this we consider again two particles that are parts of an entangled BELL state $|\Psi^-\rangle$ and as observables $S_{\mathbf{p}_i}^X$ the spin-observables defined in (4.23) and (4.24). In this case the parameters are again given by directions, that is, $\mathbf{p}_i = \hat{\mathbf{n}}^i$ and we have

$$S_{\mathbf{p}_i}^X = \Sigma_{\hat{\mathbf{n}}^i}^X \qquad \text{for } X \in \{A,B\} \text{ and } i \in \{1,\ldots,4\}.$$

We choose directions in the (x,z)-plane

$$\hat{\mathbf{n}}^i = \begin{pmatrix} \cos v_i \\ 0 \\ \sin v_i \end{pmatrix} \in S_{\mathbb{R}^3}^1 \quad \text{for } i \in \{1,\ldots,4\}, \tag{4.38}$$

with angles v_i yet to be selected. With the result (4.26) from Exercise 4.55 it then follows that

$$\left\langle \Sigma_{\hat{\mathbf{n}}^1}^A \otimes \Sigma_{\hat{\mathbf{n}}^2}^B \right\rangle_{\Psi^-} - \left\langle \Sigma_{\hat{\mathbf{n}}^1}^A \otimes \Sigma_{\hat{\mathbf{n}}^3}^B \right\rangle_{\Psi^-} + \left\langle \Sigma_{\hat{\mathbf{n}}^4}^A \otimes \Sigma_{\hat{\mathbf{n}}^2}^B \right\rangle_{\Psi^-} + \left\langle \Sigma_{\hat{\mathbf{n}}^4}^A \otimes \Sigma_{\hat{\mathbf{n}}^3}^B \right\rangle_{\Psi^-}$$
$$= -\cos(v_1 - v_2) + \cos(v_1 - v_3) - \cos(v_4 - v_2) - \cos(v_4 - v_3). \tag{4.39}$$

With the choice of directions shown in Fig. 4.1

$$v_1 = \frac{3\pi}{4} \qquad v_2 = \frac{\pi}{2} \qquad v_3 = 0 \qquad v_4 = \frac{\pi}{4} \tag{4.40}$$

for the spin-measurements, this implies

$$\left\langle \Sigma_{\hat{\mathbf{n}}^1}^A \otimes \Sigma_{\hat{\mathbf{n}}^2}^B \right\rangle_{\Psi^-} - \left\langle \Sigma_{\hat{\mathbf{n}}^1}^A \otimes \Sigma_{\hat{\mathbf{n}}^3}^B \right\rangle_{\Psi^-} + \left\langle \Sigma_{\hat{\mathbf{n}}^4}^A \otimes \Sigma_{\hat{\mathbf{n}}^2}^B \right\rangle_{\Psi^-} + \left\langle \Sigma_{\hat{\mathbf{n}}^4}^A \otimes \Sigma_{\hat{\mathbf{n}}^3}^B \right\rangle_{\Psi^-} = -2\sqrt{2},$$
$$\tag{4.41}$$

which is evidently contradicting (4.37)!

Which of the two exclusive options (4.37) or (4.41) is then realized in nature? The answer to this question was given by the experiment conducted with photons by ASPECT, DALIBARD and ROGER [9], which is graphically summarized in Fig. 4.2. The answer is: *nature behaves in accordance with* (4.41) *and violates the CHSH variant of the* BELL *inequality* (4.37). In that experiment a source emits, by means of two successive transitions (aka cascade), two photons in an entangled state one of which is sent to Alice and the other to Bob. The time from emission to arrival at

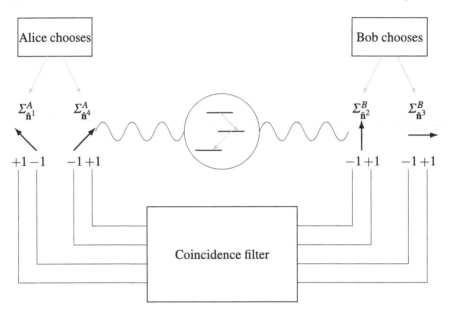

Fig. 4.2 Schematic summary of the experiment by ASPECT, DALIBARD and ROGERS

either of them is 40 ns. During the time of travel of the photons Alice selects either the observable $\Sigma^A_{\hat{\mathbf{n}}^1}$ or $\Sigma^A_{\hat{\mathbf{n}}^4}$ to be measured. The time needed to switch from one to the other is no more than 10 ns. Likewise, Bob selects, during the time the photons travel and *independently from* Alice, either the observable $\Sigma^B_{\hat{\mathbf{n}}^2}$ or $\Sigma^B_{\hat{\mathbf{n}}^3}$. Thus, the observables measured by Alice and Bob are determined *after* the photons have left the source. A coincidence filter is used to select only photons originating from the same cascade. This filter ensures that the photons selected and arriving at Alice and Bob each form part of the same entangled state. Finally, detectors register one of the two possible measurement values from $\{\pm 1\}$ for each of these photons.

The measurement of many photons thus yields, for example, the fictitious results shown in Table 4.2. Let $M^{A,B}_{i,j}$ for $i,j \in \{1,\ldots,4\}$ be the set of measurements, in which $\Sigma^A_{\hat{\mathbf{n}}^i}$ *and* $\Sigma^B_{\hat{\mathbf{n}}^j}$ have been measured. Furthermore, let $N^{A,B}_{i,j}$ be the number of such measurements and let $s^X_{\hat{\mathbf{n}}^i}(l)$ for $X \in \{A,B\}$ be the values observed in measurement $l \in M^{A,B}_{i,j}$. We can then calculate the experimentally observed expectation values denoted by $\overline{\Sigma^A_{\hat{\mathbf{n}}^i}\Sigma^B_{\hat{\mathbf{n}}^j}}$ as

$$\overline{\Sigma^A_{\hat{\mathbf{n}}^i}\Sigma^B_{\hat{\mathbf{n}}^j}} = \frac{1}{N^{A,B}_{i,j}} \sum_{l \in M^{A,B}_{i,j}} s^A_{\hat{\mathbf{n}}^i}(l) s^B_{\hat{\mathbf{n}}^j}(l). \tag{4.42}$$

Table 4.2 Fictitious results of a run of measurements of the experiment shown in Fig. 4.2 in which Alice measures one of the spins $\Sigma^A_{\hat{n}^1}$ or $\Sigma^A_{\hat{n}^4}$ and Bob measures one of the spins $\Sigma^B_{\hat{n}^2}$ or $\Sigma^B_{\hat{n}^3}$ with the choice (4.38) and (4.40) of $\hat{n}^i$ for $i \in \{1,\ldots,4\}$. For these fictitious results the left side of the CHSH inequality (4.37) yields a value close to $-2.8 < -2$. In other words, the measurement results are in good agreement with the quantum mechanical prediction (4.41). The gray cells indicate that the value of these respective observables for that pair of photons is not known and, in accordance with the rule of quantum mechanics, cannot be known sharply since $\Sigma^A_{\hat{n}^1}$ and $\Sigma^A_{\hat{n}^4}$ with the choice of $\hat{n}^i$ are incompatible observables

Particle-pair No.	Alice observes $s^A_{\hat{n}^1}$	Alice observes $s^A_{\hat{n}^4}$	Bob observes $s^B_{\hat{n}^2}$	Bob observes $s^B_{\hat{n}^3}$
1	+1		+1	
2	−1		−1	
3	+1		+1	
4	+1		−1	
5	−1		+1	
6	+1		−1	
7	+1		−1	
8	−1		+1	
9	−1		+1	
10	−1		+1	
11	−1		+1	
12	−1		+1	
13	+1		−1	
14	−1		+1	
15	+1		−1	
16	+1		−1	
17	−1		+1	
18	+1		−1	
19	+1		−1	
20	−1		+1	
21	−1			−1
22	−1			−1
23	−1			−1
24	+1			+1
25	+1			+1
26	+1			+1
27	−1			−1
28	+1			+1
29	−1			−1
30	−1			+1
31	+1			+1
32	−1			−1
33	−1			+1
34	+1			+1
35	+1			+1
36	−1			−1
37	−1			−1
38	+1			+1
39	+1			+1
40	+1			−1

Particle-pair No.	Alice observes $s^A_{\hat{n}^1}$	Alice observes $s^A_{\hat{n}^4}$	Bob observes $s^B_{\hat{n}^2}$	Bob observes $s^B_{\hat{n}^3}$
41		+1	−1	
42		+1	−1	
43		+1	−1	
44		+1	−1	
45		−1	+1	
46		+1	+1	
47		−1	+1	
48		+1	−1	
49		−1	+1	
50		+1	−1	
51		+1	+1	
52		+1	−1	
53		+1	−1	
54		+1	−1	
55		−1	−1	
56		−1	+1	
57		−1	+1	
58		+1	−1	
59		−1	+1	
60		−1	+1	
61		+1		−1
62		+1		−1
63		−1		+1
64		+1		−1
65		−1		+1
66		−1		+1
67		−1		+1
68		+1		−1
69		−1		−1
70		−1		+1
71		−1		−1
72		+1		−1
73		−1		+1
74		+1		−1
75		+1		−1
76		−1		+1
77		−1		+1
78		+1		−1
79		−1		−1
80		−1		+1

Inserting the expectation values $\overline{\Sigma_{\hat{\mathbf{n}}^i}^A \Sigma_{\hat{\mathbf{n}}^j}^B}$ thus calculated as approximation of the quantum mechanical expectation values $\left\langle \Sigma_{\hat{\mathbf{n}}^i}^A \otimes \Sigma_{\hat{\mathbf{n}}^j}^B \right\rangle_{\Psi^-}$ in the left side of (4.41) (approximately) confirms this equation and thus the quantum mechanical prediction. On the other hand, inserting the expectation values $\overline{\Sigma_{\hat{\mathbf{n}}^i}^A \Sigma_{\hat{\mathbf{n}}^j}^B}$ thus calculated in place of the classical expectation values $\mathbf{E}\left[s_{\hat{\mathbf{n}}^i}^A s_{\hat{\mathbf{n}}^j}^B \right]$ (which assume a joint distribution) in the left side of (4.37) shows that this BELL inequality is violated. The reader may check with the help of (4.42) that the left side of the CHSH variant of the BELL inequality (4.37) for the fictitious measurement values shown in Table 4.2 does indeed yield approximately the value $-2.8 < -2$, in other words, that it agrees well with (4.41).

The EPR-implication made by excluding 'unreasonable behavior of reality' that $\Sigma_{\hat{\mathbf{n}}^1}^A$ and $\Sigma_{\hat{\mathbf{n}}^4}^A$ jointly constitute elements of reality would have as a consequence that one can fill in the gray cells in Table 4.2 of the measurement values with the only possible values from $\{\pm 1\}$. But regardless of how we apply our own metaphorical 'gray cells' to fill in the literal gray cells in Table 4.2 with values $+1$ or -1, we always find that the BELL inequality (4.37) is satisfied.

Let us spell out the really baffling aspect once more: every measurement of $\Sigma_{\hat{\mathbf{n}}^i}^X$ for $X \in \{A, B\}$ and $i \in \{1, \dots, 4\}$ yields a value in $\{\pm 1\}$. Measurements of these observables never reveal a different value. It is thus 'reasonable' to assume that these observables always would have the value $+1$ or -1. Consequently, every pair of observables $\left(\Sigma_{\hat{\mathbf{n}}^i}^A, \Sigma_{\hat{\mathbf{n}}^j}^B \right)$ for $i, j \in \{1, \dots, 4\}$ would always have a pair of values in $\{\pm 1, \pm 1\}$. Precisely this, however, necessarily implies the validity of the CHSH version (4.37) of the BELL inequality. But, as we have shown in (4.41), this inequality is violated by quantum mechanics. It is thus impossible, that $\Sigma_{\hat{\mathbf{n}}^i}^A$ and $\Sigma_{\hat{\mathbf{n}}^j}^B$ jointly assume one of their possible values, which we always observe if we measure each particle alone. In other words: even though each of these observables can be measured individually and each measurement yields a value in $\{\pm 1\}$, both together cannot have these values at the same time.

Finally, it is worth remarking that quantum mechanics predicts the violation of the BELL inequality only for entangled states and even then only for certain spin-directions. If, for example, in the state $|\Psi^-\rangle$ we choose to measure spins in the directions $\hat{\mathbf{n}}^2 = \hat{\mathbf{n}}^3$, then the quantum mechanical prediction for the left side of (4.39) gives the value $-\sqrt{2}$, in other words, satisfies the CHSH version (4.37) of the BELL inequality. The quantum mechanical prediction of expectation values for spin-observables in separable (that is, non-entangled) states also always satisfies this inequality, as is shown in the following proposition.

Proposition 4.8 *In any separable state* $|\varphi\rangle \otimes |\psi\rangle \in \mathbb{H}^A \otimes \mathbb{H}^B$ *the expectation values of spin-observables* $\Sigma_{\hat{\mathbf{n}}^i}^A \otimes \Sigma_{\hat{\mathbf{n}}^i}^B$ *in arbitrary spin-directions* $\hat{\mathbf{n}}^i$ *with* $i \in \{1, \dots, 4\}$ *satisfy the CHSH variant of the* BELL *inequality, that is,*

$$\left| \left\langle \Sigma_{\hat{\mathbf{n}}^1}^A \otimes \Sigma_{\hat{\mathbf{n}}^2}^B \right\rangle_{\varphi \otimes \psi} - \left\langle \Sigma_{\hat{\mathbf{n}}^1}^A \otimes \Sigma_{\hat{\mathbf{n}}^3}^B \right\rangle_{\varphi \otimes \psi} + \left\langle \Sigma_{\hat{\mathbf{n}}^4}^A \otimes \Sigma_{\hat{\mathbf{n}}^2}^B \right\rangle_{\varphi \otimes \psi} + \left\langle \Sigma_{\hat{\mathbf{n}}^4}^A \otimes \Sigma_{\hat{\mathbf{n}}^3}^B \right\rangle_{\varphi \otimes \psi} \right| \leq 2$$

(4.43)

holds.

Proof Generally, the expectation values of products of observables $M^A \otimes M^B$ factorize in separable states $|\varphi\rangle \otimes |\psi\rangle \in \mathbb{H}^A \otimes \mathbb{H}^B$, that is, one has

$$
\begin{aligned}
\left\langle M^A \otimes M^B \right\rangle_{\varphi \otimes \psi} &= \langle \varphi \otimes \psi | M^A \otimes M^B (\varphi \otimes \psi) \rangle = \langle \varphi \otimes \psi | M^A \varphi \otimes M^B \psi \rangle \\
&\underset{(3.4)}{=} \langle \varphi | M^A \varphi \rangle \langle \psi | M^B \psi \rangle \\
&= \left\langle M^A \right\rangle_\varphi \left\langle M^B \right\rangle_\psi .
\end{aligned}
$$

(4.44)

From (2.118) we know that an arbitrary state $|\varphi\rangle \in \mathbb{H}^A$ can be given in the form

$$|\varphi\rangle = e^{i\alpha} \cos\beta |0\rangle + e^{i\gamma} \sin\beta |1\rangle .$$

Moreover, from (2.125) we know that a spin-up state for a spin in the direction of $\hat{\mathbf{n}}(\theta, \phi)$ is given by

$$|\uparrow_{\hat{\mathbf{n}}(\theta,\phi)}\rangle = e^{-i\frac{\phi}{2}} \cos\frac{\theta}{2} |0\rangle + e^{i\frac{\phi}{2}} \sin\frac{\theta}{2} |1\rangle$$

such that we can write $|\varphi\rangle$ with the help of a unit vector $\hat{\mathbf{n}}^\varphi := \hat{\mathbf{n}}(2\beta, \frac{\gamma-\alpha}{2})$ in the form

$$|\varphi\rangle = e^{i\frac{\alpha+\gamma}{2}} |\uparrow_{\hat{\mathbf{n}}(2\beta, \frac{\gamma-\alpha}{2})}\rangle = e^{i\frac{\alpha+\gamma}{2}} |\uparrow_{\hat{\mathbf{n}}^\varphi}\rangle .$$

The same holds for $|\psi\rangle = e^{i\delta} |\uparrow_{\hat{\mathbf{n}}^\psi}\rangle$ with suitably chosen δ and $\hat{\mathbf{n}}^\psi$.

Exercise 4.57 Show that

$$\langle \Sigma_{\hat{\mathbf{n}}} \rangle_{|\uparrow_{\hat{\mathbf{m}}}\rangle} = \hat{\mathbf{n}} \cdot \hat{\mathbf{m}} .$$

(4.45)

For a solution see Solution 4.57

Combining (4.44) with (4.45) and the fact that the complex phase-factor is irrelevant (see Definition 2.14 and subsequent paragraph), then yields,

$$\left\langle \Sigma_{\hat{\mathbf{n}}^i}^A \otimes \Sigma_{\hat{\mathbf{n}}^j}^B \right\rangle_{\varphi \otimes \psi} \underset{(4.44)}{=} \left\langle \Sigma_{\hat{\mathbf{n}}^i}^A \right\rangle_\varphi \left\langle \Sigma_{\hat{\mathbf{n}}^j}^B \right\rangle_\psi \underset{(4.45)}{=} \left(\hat{\mathbf{n}}^i \cdot \hat{\mathbf{n}}^\varphi \right) \left(\hat{\mathbf{n}}^j \cdot \hat{\mathbf{n}}^\psi \right)$$

and thus

$$
\begin{aligned}
&\left|\left\langle \Sigma^A_{\hat{\mathbf{n}}^1} \otimes \Sigma^B_{\hat{\mathbf{n}}^2}\right\rangle_{\varphi\otimes\psi} - \left\langle \Sigma^A_{\hat{\mathbf{n}}^1} \otimes \Sigma^B_{\hat{\mathbf{n}}^3}\right\rangle_{\varphi\otimes\psi} + \left\langle \Sigma^A_{\hat{\mathbf{n}}^4} \otimes \Sigma^B_{\hat{\mathbf{n}}^2}\right\rangle_{\varphi\otimes\psi} + \left\langle \Sigma^A_{\hat{\mathbf{n}}^4} \otimes \Sigma^B_{\hat{\mathbf{n}}^3}\right\rangle_{\varphi\otimes\psi}\right| \\
&= \left|\hat{\mathbf{n}}^1 \cdot \hat{\mathbf{n}}^\varphi \left(\hat{\mathbf{n}}^2 \cdot \hat{\mathbf{n}}^\psi - \hat{\mathbf{n}}^3 \cdot \hat{\mathbf{n}}^\psi\right) + \hat{\mathbf{n}}^4 \cdot \hat{\mathbf{n}}^\varphi \left(\hat{\mathbf{n}}^2 \cdot \hat{\mathbf{n}}^\psi + \hat{\mathbf{n}}^3 \cdot \hat{\mathbf{n}}^\psi\right)\right| \\
&\le \left|\hat{\mathbf{n}}^1 \cdot \hat{\mathbf{n}}^\varphi\right|\left|\hat{\mathbf{n}}^2 \cdot \hat{\mathbf{n}}^\psi - \hat{\mathbf{n}}^3 \cdot \hat{\mathbf{n}}^\psi\right| + \left|\hat{\mathbf{n}}^4 \cdot \hat{\mathbf{n}}^\varphi\right|\left|\hat{\mathbf{n}}^2 \cdot \hat{\mathbf{n}}^\psi + \hat{\mathbf{n}}^3 \cdot \hat{\mathbf{n}}^\psi\right| \\
&\le \left|\left(\hat{\mathbf{n}}^2 - \hat{\mathbf{n}}^3\right)\cdot\hat{\mathbf{n}}^\psi\right| + \left|\left(\hat{\mathbf{n}}^2 + \hat{\mathbf{n}}^3\right)\cdot\hat{\mathbf{n}}^\psi\right| \quad (4.46)
\end{aligned}
$$

For arbitrary $x,y \in \mathbb{R}$ one has

$$
|x| + |y| = \begin{cases} |x+y| & \text{if } xy \ge 0 \\ |x-y| & \text{if } xy < 0 \end{cases}
$$

and thus

$$
\left|\left(\hat{\mathbf{n}}^2 - \hat{\mathbf{n}}^3\right)\cdot\hat{\mathbf{n}}^\psi\right| + \left|\left(\hat{\mathbf{n}}^2 + \hat{\mathbf{n}}^3\right)\cdot\hat{\mathbf{n}}^\psi\right| = 2\max\left\{\left|\hat{\mathbf{n}}^2 \cdot \hat{\mathbf{n}}^\psi\right|, \left|\hat{\mathbf{n}}^3 \cdot \hat{\mathbf{n}}^\psi\right|\right\} \le 2. \quad (4.47)
$$

Inserting (4.47) into (4.46) then yields (4.43). $\qquad\square$

4.6 Two Impossible Devices

4.6.1 BELL *Telephone*

The—according to EPR 'unreasonable'—behavior of quantum mechanics, in other words, the instantaneous effect on the reality of Bob's particle by measurements performed by Alice, has tempted some people to attempt to construct a means of super-luminal communication between Alice and Bob. However, as we now show, such a device, which has been termed **BELL telephone**, cannot be used to transmit information at all, not even slower than the speed of light.

The BELL telephone is supposed to function as follows. Suppose Alice and Bob each have a particle which together are in the BELL state $|\Phi^+\rangle$. As shown in Sect. 4.4 after (4.22), Alice can then, by measuring σ_z on her particle, project Bob's particle into $|0\rangle = |\uparrow_{\hat{\mathbf{z}}}\rangle$ or $|1\rangle = |\downarrow_{\hat{\mathbf{z}}}\rangle$. If, however, she measures σ_x, she then projects Bob's particle into the states $|+\rangle = |\uparrow_{\hat{\mathbf{x}}}\rangle$ or $|-\rangle = |\downarrow_{\hat{\mathbf{x}}}\rangle$. Alice thus tries to send a message to Bob by using the protocol shown in Table 4.3. Depending on whether Bob's particle is in a state from $\{|0\rangle,|1\rangle\}$ or from $\{|+\rangle,|-\rangle\}$, he is supposed to read out 0 or 1. As we now show, this attempt to transmit information does not work because after Alice's measurement Bob's particle is in a mixed state. This mixed state can be described using either $|0\rangle$ and $|1\rangle$ or $|+\rangle$ and $|-\rangle$, but, regardless

Table 4.3 Protocol for the BELL telephone

Agreed bit value	Alice measures	Bob's qubit in the state
0	σ_z	$|0\rangle$ or $|1\rangle$
1	σ_x	$|+\rangle$ or $|-\rangle$

of which observable Alice measures and which pair we use for the description, the
mixed state is always the same, and Bob cannot read what Alice wrote.

Our considerations will be general in the sense that we do not restrict ourselves
to the case of the observables σ_z and σ_x or a single qubit system only. We assume
that Alice has control over a sub-system $\mathbb{H}^A$ and Bob over a sub-system $\mathbb{H}^B$, each of
which forms part of a composite system $\mathbb{H}^A \otimes \mathbb{H}^B$. Furthermore, we assume Alice
has two distinct observables M^A and $\widetilde{M}^A$ with purely discrete spectrum in her sub-
system at her disposal, which she can choose to measure and with which she encodes
the classical bits 0 and 1 that she wants to send to Bob. For example, the agreed
communication protocol could be such that she encodes 0 by measuring M^A and
encodes 1 by measuring $\widetilde{M}^A$.

To keep the notation manageable, we assume that the eigenvalues $\lambda_a \in \sigma(M^A)$
and $\widetilde{\lambda}_a \in \sigma(\widetilde{M}^A)$ are all non-degenerate in $\mathbb{H}^A$. The following line of arguments will
also be valid in case there are degenerate eigenvalues, only the notation will become
unnecessarily cumbersome.

There exists an ONB $\{|e_a\rangle\}$ of $\mathbb{H}^A$ consisting of eigenvectors $|e_a\rangle \in \mathrm{Eig}(M^A, \lambda_a)$
as well as an ONB $\{|\widetilde{e}_a\rangle\} \subset \mathbb{H}^A$ consisting of eigenvectors $|\widetilde{e}_a\rangle \in \mathrm{Eig}(\widetilde{M}^A, \widetilde{\lambda}_a)$. From
Exercise 2.15 we know that these ONBs are necessarily related to each other by a
unitary transformation such that

$$|\widetilde{e}_a\rangle = U|e_a\rangle = \sum_{a_1}\langle e_{a_1}|Ue_a\rangle|e_{a_1}\rangle = \sum_{a_1} U_{a_1 a}|e_{a_1}\rangle, \qquad (4.48)$$

where $U \in \mathcal{U}(\mathbb{H}^A)$.

For Bob's system let $\{|f_b\rangle\} \in \mathbb{H}^B$ be an ONB in $\mathbb{H}^B$. From Proposition 3.2 we
know that then the set of vectors $\{|e_a \otimes f_b\rangle\}$ as well as the set $\{|\widetilde{e}_a\rangle \otimes |f_b\rangle\}$ each
constitute an ONB in the HILBERT space $\mathbb{H}^A \otimes \mathbb{H}^B$ of the composite system. In
$\mathbb{H}^A \otimes \mathbb{H}^B$ the vectors $|e_a \otimes f_b\rangle$ are eigenvectors of the observable $M^A \otimes \mathbf{1}^B$ and the
vectors $|\widetilde{e}_a\rangle \otimes |f_b\rangle$ of the observable $\widetilde{M}^A \otimes \mathbf{1}^B$, that is, we have

$$\left(M^A \otimes \mathbf{1}^B\right)|e_a \otimes f_b\rangle = \lambda_a|e_a \otimes f_b\rangle$$
$$\left(\widetilde{M}^A \otimes \mathbf{1}^B\right)|\widetilde{e}_a \otimes f_b\rangle = \widetilde{\lambda}_a|\widetilde{e}_a \otimes f_b\rangle$$

such that $\sigma(M^A \otimes \mathbf{1}^B) = \sigma(M^A)$ as well as $\sigma(\widetilde{M}^A \otimes \mathbf{1}^B) = \sigma(\widetilde{M}^A)$. As an eigenvalue
of the observable $M^A \otimes \mathbf{1}^B \in \mathrm{B_{sa}}(\mathbb{H}^A \otimes \mathbb{H}^B)$ each of these eigenvalues is $\left(\dim \mathbb{H}^B\right)$-
fold degenerate, that is, $\dim \mathrm{Eig}(M^A \otimes \mathbf{1}^B, \lambda_a) = \dim \mathbb{H}^B$. A general eigenstate of
$M^A \otimes \mathbf{1}^B$ is of the form

$$|e_a \otimes \varphi\rangle = \sum_b \varphi_b |e_a \otimes f_b\rangle,$$

and similar statements hold for $\widetilde{M}^A \otimes \mathbf{1}^B$.

Let the composite system initially be prepared in the pure state

$$|\Psi\rangle = \sum_{a,b} \Psi_{ab} |e_a \otimes f_b\rangle = \sum_{a,b} \widetilde{\Psi}_{ab} |\widetilde{e}_a \otimes f_b\rangle \tag{4.49}$$

and the particles of sub-system $\mathbb{H}^A$ be distributed to Alice and those of sub-system $\mathbb{H}^B$ to Bob. Alice would like to exploit the fact that each of them has a sub-system of the same composite system in order to send the classical bit 0 to Bob. To accomplish that, she measures the observable M^A. For the composite system this is a measurement of the observable $M^A \otimes \mathbf{1}^B$. If λ_a is the observed value of this measurement, then (2.87) of the Projection Postulate tells us that after the measurement the composite system is in the state

$$\rho_{\lambda_a} := \frac{P_{\lambda_a} \rho_\Psi P_{\lambda_a}}{\mathrm{tr}\left(\rho P_{\lambda_a}\right)}, \tag{4.50}$$

where $P_{\lambda_a} = |e_a\rangle\langle e_a| \otimes \mathbf{1}^B$ is the projector onto the eigenspace $\mathrm{Eig}(M^A \otimes \mathbf{1}^B, \lambda_a)$ and $\rho_\Psi = |\Psi\rangle\langle\Psi|$ is the density operator of the original pure state (4.49).

From (2.86) we see that he probability to observe λ_a, and thus to end up in the state ρ_{λ_a}, is given by $\mathrm{tr}\left(\rho P_{\lambda_a}\right)$. For all who do not know the measured eigenvalue— and Bob is one of them—the composite system after Alice's measurement is then described by the mixed state ρ which is a statistical ensemble of states ρ_{λ_a} each occurring with a probability $\mathrm{tr}\left(\rho P_{\lambda_a}\right)$, that is,

$$\rho = \sum_a \mathbf{P}\{\text{To observe } \lambda_a\} \rho_{\lambda_a} \underbrace{=}_{(2.86)} \sum_a \mathrm{tr}\left(\rho P_{\lambda_a}\right) \rho_{\lambda_a} \underbrace{=}_{(4.50)} \sum_a P_{\lambda_a} \rho_\Psi P_{\lambda_a}. \tag{4.51}$$

The mixed state, which describes Bob's sub-system after Alice's measurement of the observable M^A, is given by the reduced density operator $\rho^B(\rho)$.

Exercise 4.58 Show that the partial trace of ρ over $\mathbb{H}^A$, that is, the reduced density operator $\rho^B(\rho)$ describing the sub-system B, is given by

$$\rho^B(\rho) = \sum_{b_1,b_2} \sum_a \Psi_{ab_1} \overline{\Psi_{ab_2}} |f_{b_1}\rangle\langle f_{b_2}|. \tag{4.52}$$

For a solution see Solution 4.58

Now, in order to send the classical bit 1 to Bob, Alice measures the observable $\widetilde{M}^A$. With the same line of arguments as for M^A, Bob's sub-system is then in the mixed state

$$\rho^B(\widetilde{\rho}) = \sum_{b_1,b_2} \sum_a \widetilde{\Psi}_{ab_1} \overline{\widetilde{\Psi}_{ab_2}} |f_{b_1}\rangle\langle f_{b_2}|. \tag{4.53}$$

From (4.48) and (4.49) it follows that

$$\Psi_{ab} = \sum_{a_1} U_{aa_1} \widetilde{\Psi}_{a_1 b} \tag{4.54}$$

and thus

$$
\begin{aligned}
\sum_a \Psi_{ab_1} \overline{\Psi_{ab_2}} &\underset{(4.54)}{=} \sum_{a,a_1,a_2} U_{aa_1} \widetilde{\Psi}_{a_1 b_1} \overline{U_{aa_2} \widetilde{\Psi}_{a_2 b_2}} \underset{(2.34)}{=} \sum_{a,a_1,a_2} U_{aa_1} U_{a_2a}^* \widetilde{\Psi}_{a_1 b_1} \overline{\widetilde{\Psi}_{a_2 b_2}} \\
&= \sum_{a_1,a_2} \underbrace{(U^*U)_{a_2 a_1}}_{=\delta_{a_2 a_1}} \widetilde{\Psi}_{a_1 b_1} \overline{\widetilde{\Psi}_{a_2 b_2}} \\
&= \sum_a \widetilde{\Psi}_{ab_1} \overline{\widetilde{\Psi}_{ab_2}}, \tag{4.55}
\end{aligned}
$$

where the last equation follows from the unitarity of U. From (4.52) and (4.53) together with (4.55) it finally follows that

$$\rho^B(\rho) = \rho^B(\widetilde{\rho}),$$

that is, Bob's sub-system is always in the same mixed state regardless of which observable Alice measures. This means that Bob cannot detect the difference between Alice's choice of M^A in order to communicate 0 or $\widetilde{M}^A$ in order to send 1. This constitutes proof of the following statement.

Corollary 4.9 *There is no* BELL *telephone.*

We illustrate this once more using the protocol given in Table 4.3 with the composite state $|\Phi^+\rangle$. How is Bob to read the message? He has to determine whether his particle is described by the states from $\{|0\rangle, |1\rangle\}$ or from $\{|+\rangle, |-\rangle\}$. He can attempt to find this out by measuring σ_z or σ_x on his particle. Suppose he measures on his particle the observable σ_z and observes the value $+1$. Can he deduce from that, that his particle was in the state $|0\rangle = |\uparrow_{\hat{z}}\rangle$? Obviously *not*, since the probability, to observe the value $+1$ when measuring σ_z, is also different from zero in the states $|+\rangle$ and $|-\rangle$:

$$|\langle 0|+\rangle|^2 = \frac{1}{2} = |\langle 0|-\rangle|^2.$$

Consequently, performing a measurement on his particle does not reveal to Bob in which state his particle was and which bit-value was sent by Alice.

This conclusion would be invalid, if Bob were able to *copy* the state of his particle, in other words, if he were able to do the following: from a particle given to him, which is in a state unknown to him, he is able to prepare many (at least two) particles in the same state. To see how this would work, suppose Alice has measured σ_z. Then Bob's qubit is either in the state $|0\rangle$ or $|1\rangle$. Suppose it is in the state $|0\rangle$. Bob then makes several copies of this state unknown to him. He measures σ_z in his copied states. Each time he observes the value $+1$. Now suppose Alice had measured σ_x. Then Bob's qubit will be either in the state $|+\rangle$ or $|-\rangle$. Suppose it is in $|+\rangle$. Bob again makes multiple copies of his state. In each of his copied states he measures σ_z. But now half of the results will be $+1$ and half will be -1. This is in contrast to the case where Alice had measured σ_z in which case Bob's observations of σ_z always yield exclusively $+1$ or exclusively -1. Hence, if Bob were able to copy the state unknown to him, he could deduce from the results of his measurements on the copies he has produced in which state his original particle was, and thus, which classical bit (see Table 4.3) Alice encoded. Consequently, a quantum-copier would allow the construction of a BELL telephone [68].

However, such a device to copy an unknown quantum state, which is called quantum-copier or cloner, cannot exist either, as we show in the next section.

4.6.2 Perfect Quantum Copier

The fact that a quantum-copier does not exist, or as formulated alternatively, that 'qubits cannot be cloned' [15], is due to the linear structure of the HILBERT space containing the state vectors. A quantum-copier for a system with state vectors in $\mathbb{H}$ is defined as follows.

Definition 4.10 Given

(i) an arbitrary state (the 'original') $|\psi\rangle \in \mathbb{H}$ to be copied and
(ii) a state (the 'white-page') $|\omega\rangle \in \mathbb{H}$ to emerge as a copy,

a **quantum-copier** K is a linear transformation that leaves the original state $|\psi\rangle$ unchanged and transforms the white-page-state $|\omega\rangle$ such that it becomes the original state $|\psi\rangle$, that is, K is an operator that satisfies

$$K : |\psi\rangle \otimes |\omega\rangle \mapsto |\psi\rangle \otimes |\psi\rangle \qquad (4.56)$$

for arbitrary $|\psi\rangle \in \mathbb{H}$ and a given fixed $|\omega\rangle \in \mathbb{H}$.

An arbitrary number of copies can then be produced by multiple application of the copier. As we shall see, it is, however, straightforward to prove the following proposition, which states, that no such quantum-copier exists. This statement is also known as the **Quantum No-Cloning Theorem**.

Proposition 4.11 (Quantum No-Cloning Theorem [15]) *A quantum-copier cannot exist.*

Proof It suffices to consider qubits, that is, to consider the case $\mathbb{H} = {}^{\P}\mathbb{H}$ and the action of a quantum-copier on the qubit-states $|0\rangle, |1\rangle$ and $\frac{1}{\sqrt{2}}(|1\rangle + |0\rangle)$. Per definition K has to satisfy

$$K(|0\rangle \otimes |\omega\rangle) \underbrace{=}_{(4.56)} |0\rangle \otimes |0\rangle \tag{4.57}$$

$$K(|1\rangle \otimes |\omega\rangle) \underbrace{=}_{(4.56)} |1\rangle \otimes |1\rangle \tag{4.58}$$

$$K\left(\frac{|1\rangle + |0\rangle}{\sqrt{2}} \otimes |\omega\rangle\right) \underbrace{=}_{(4.56)} \frac{|1\rangle + |0\rangle}{\sqrt{2}} \otimes \frac{|1\rangle + |0\rangle}{\sqrt{2}}. \tag{4.59}$$

As K is supposed to be linear, we find that in place of (4.59) instead K satisfies

$$K\left(\frac{|1\rangle + |0\rangle}{\sqrt{2}} \otimes |\omega\rangle\right) = K\left(\frac{1}{\sqrt{2}}(|1\rangle \otimes |\omega\rangle) + \frac{1}{\sqrt{2}}(|0\rangle \otimes |\omega\rangle)\right)$$

$$\underbrace{=}_{(4.56)} \frac{1}{\sqrt{2}}\left(K(|1\rangle \otimes |\omega\rangle) + K(|0\rangle \otimes |\omega\rangle)\right)$$

$$\underbrace{=}_{(4.57),(4.58)} \frac{1}{\sqrt{2}}\left(|1\rangle \otimes |1\rangle + |0\rangle \otimes |0\rangle\right)$$

$$\neq \frac{|1\rangle + |0\rangle}{\sqrt{2}} \otimes \frac{|1\rangle + |0\rangle}{\sqrt{2}}.$$

From this it follows that there is no device that can copy *arbitrary* qubits. Since qubits are particular quantum systems, the general statement in Proposition 4.11 holds. □

It is worth noting that there can be devices that copy *particular* states as specified in Definition 4.10. For example, the controlled NOT $\Lambda^1(X)$ defined in Fig. 5.5 satisfies the requirements in (4.57) and (4.58), that is, clones the states $|0\rangle$ and $|1\rangle$. The Quantum No-Cloning Theorem only makes the statement that there is no device which does that for *all* states.

4.7 Further Reading

The more historically minded reader may want to consult the original article by
EINSTEIN, PODOLSKY and ROSEN [4], where the the paradox with the same name
was first spelled out. Similarly, the original articles by BELL [8, 63] are a worthwhile read for those who want to get to the origins of the inequalities bearing his
name. They can also be found in the collection by BELL [44], which, in addition,
covers a wide range of fundamental questions of quantum theory.

The book by AUDRETSCH [69] serves as an introductory text on entanglement,
which also includes many of its aspects related to quantum computing and quantum
information. A highly mathematical and more geometrically inspired perspective
on entanglement is given in the book by BENGTSSON and ZYCZKOWSKI [59]. The
book by LALOË [45] offers a very good combination of addressing wider philo-
sophical questions without abandoning a mathematical (formulaic) description, as
so many other works in this context do. Apart from chapters on interpretations
of quantum mechanics, EPR and BELL inequalities, it also covers entanglement
together with its current role in quantum information.

Chapter 5
Quantum Gates and Circuits
for Elementary Calculations

5.1 Classical Gates

Before we turn to quantum gates, in other words, gates for qubits, we first consider the usual 'classical' gates. In a classical computer the processor essentially performs nothing more than a sequence of transformations of a classical state into another one:

$$\begin{aligned} f : \{0,1\}^n &\longrightarrow \{0,1\}^m \\ x &\longmapsto f(x) \end{aligned} \tag{5.1}$$

This is what we will refer to as the **classical computational process**, which is realized with a concatenation of classical gates and circuits.[1]

> **Definition 5.1** An (elementary) **classical (logical-)gate** g is defined as a map
>
> $$\begin{aligned} g : \{0,1\}^n &\longrightarrow \{0,1\} \\ (x_1,\ldots,x_n) &\longmapsto g(x_1,\ldots,x_n) \end{aligned}$$
>
> We define an extended classical logical gate g as a map
>
> $$\begin{aligned} g : \{0,1\}^n &\longrightarrow \{0,1\}^m \\ (x_1,\ldots,x_n) &\longmapsto (g_1(x_1,\ldots,x_n),\ldots,g_m(x_1,\ldots,x_n)) \end{aligned}$$
>
> where each g_j is an elementary gate. A classical gate g is called **reversible** if it is a bijection and thus invertible.

---

[1]The underlying model of computation that we use throughout this book is the sequential model based on a TURING Machine [70].

© Springer Nature Switzerland AG 2019
W. Scherer, *Mathematics of Quantum Computing*,
https://doi.org/10.1007/978-3-030-12358-1_5

Fig. 5.1 Graphical
representation of a generic
classical gate

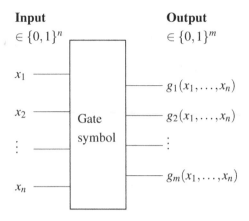

Any operation of a classical computer is of the form (5.1), in other words, a transformation of classical bits from one state into another. Hence, a classical processor is nothing more than a physical implementation of an array of suitable universal classical logical gates or circuits.

Usually, gates are represented by special graphical symbols of the form shown in Fig. 5.1.

For the description of classical gates it is helpful to use the binary addition defined as follows.

Definition 5.2 For $u, v \in \{0, 1\}$ we define the **binary addition** $u \overset{2}{\oplus} v$ as

$$u \overset{2}{\oplus} v := (u + v) \mod 2, \qquad (5.2)$$

where the expression $a \mod n$ is defined in Appendix D.

The most prominent examples of classical gates are:

Classical **NOT-gate**

$$\begin{aligned} \text{NOT} : \{0, 1\} &\longrightarrow \{0, 1\} \\ x_1 &\longmapsto \text{NOT}(x_1) := 1 \overset{2}{\oplus} x_1 \end{aligned} \qquad (5.3)$$

The name arises from the usual association $0 = No = False$ and $1 = Yes = True$ and the effect of NOT as 'negation'.

Classical **AND-gate**

$$\begin{aligned} \text{AND} : \{0, 1\}^2 &\longrightarrow \{0, 1\} \\ (x_1, x_2) &\longmapsto \text{AND}(x_1, x_2) := x_1 x_2 \end{aligned} \qquad (5.4)$$

Classical **OR-gate**

$$\text{OR} : \{0,1\}^2 \longrightarrow \{0,1\}$$
$$(x_1,x_2) \longmapsto \text{OR}(x_1,x_2) := x_1 \overset{2}{\oplus} x_2 \overset{2}{\oplus} x_1 x_2$$

Classical (exclusive Or) **XOR-gate**

$$\text{XOR} : \{0,1\}^2 \longrightarrow \{0,1\}$$
$$(x_1,x_2) \longmapsto \text{XOR}(x_1,x_2) := x_1 \overset{2}{\oplus} x_2 \tag{5.5}$$

Classical **TOFFOLI-gate**

$$\text{TOF} : \ \{0,1\}^3 \ \longrightarrow \{0,1\}^3$$
$$(x_1,x_2,x_3) \longmapsto \text{TOF}(x_1,x_2,x_3) := (x_1,x_2,x_1 x_2 \overset{2}{\oplus} x_3) \tag{5.6}$$

In the graphical representation of the TOFFOLI-gate in Fig. 5.2 large dots have been introduced as symbols for the conditional application of the connected operator. In general, in gate representations these large dots symbolize that the operators connected to them will only be applied if the value of the bit flowing through the dot is 1. Indeed, one can see from (5.6) or Fig. 5.2 that x_3 in the third channel changes only if $x_1 = 1$ as well as $x_2 = 1$ holds.

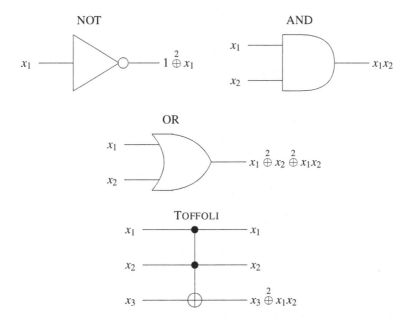

Fig. 5.2 Symbolic representation of the classical NOT-, AND-, OR- and TOFFOLI- gates

For the sake of completeness we also list here the 'gates'

$$
\begin{aligned}
\mathrm{ID} : \{0,1\} &\longrightarrow \{0,1\} \\
(x_1) &\longmapsto \mathrm{ID}(x_1) := x_1 \\
\mathrm{FALSE} : \{0,1\} &\longrightarrow \{0,1\} \\
(x_1) &\longmapsto \mathrm{FALSE}(x_1) := 0 \\
\mathrm{TRUE} : \{0,1\} &\longrightarrow \{0,1\} \\
(x_1) &\longmapsto \mathrm{TRUE}(x_1) := 1 \\
\mathrm{COPY}^{(1)} : \{0,1\} &\longrightarrow \{0,1\}^2 \\
(x_1) &\longmapsto \mathrm{COPY}(x_1) := (x_1,x_1)
\end{aligned}
\tag{5.7}
$$

However, these are often not shown explicitly in listings of logical gates. Altogether there are four elementary gates of the form $g : \{0,1\} \to \{0,1\}$, namely, ID, NOT, FALSE and TRUE. Elementary gates are combined in various ways to form logical circuits.

Definition 5.3 We denote by $\mathcal{F}[g_1,\ldots,g_K]$ the set of all gates which can be constructed from $g_1,\ldots,g_K$. This set is defined by the following construction rules:

(i) the $g_1,\ldots,g_K$ are elements of this set, that is,

$$
g_1,\ldots,g_K \in \mathcal{F}[g_1,\ldots,g_K]
$$

(ii) **padding** operations of the form

$$
p^{(n)}_{y_1,\ldots,y_l;j_1,\ldots,j_l} : \{0,1\}^n \longrightarrow \{0,1\}^{n+l} \\
(x_1,\ldots,x_n) \longmapsto (x_1,\ldots,x_{j_1-1},y_{j_1},x_{j_1+1},\ldots,x_n)
$$

which insert pre-determined bit values $y_1,\ldots,y_l \in \{0,1\}$ at pre-determined slots $j_1,\ldots,j_l \in \{1,\ldots,n+l\}$ are elements of the set, that is, for any $l,n \in \mathbb{N}$, $y_1,\ldots,y_l \in \{0,1\}$ and pairwise distinct $j_1,\ldots,j_l \in \{1,\ldots,n+l\}$

$$
p^{(n)}_{y_1,\ldots,y_l;j_1,\ldots,j_l} \in \mathcal{F}[g_1,\ldots,g_K]
$$

(iii) **restriction and/or re-ordering** operations

$$
r^{(n)}_{j_1,\ldots,j_l} : \{0,1\}^n \longrightarrow \{0,1\}^l \\
(x_1,\ldots,x_n) \longmapsto (x_{j_1},\ldots,x_{j_l})
\tag{5.8}
$$

are elements of the set, that is, for any $l,n \in \mathbb{N}$, and pairwise distinct $j_1,\ldots,j_l \in \{1,\ldots,l\}$

$$r_{j_1,\ldots,j_l}^{(n)} \in \mathcal{F}[g_1,\ldots,g_K]$$

(iv) **compositions** of elements of the set belong to the set, that is, for any $h_1 : \{0,1\}^n \to \{0,1\}^m$ and $h_2 : \{0,1\}^l \to \{0,1\}^n$ we have that

$$h_1, h_2 \in \mathcal{F}[g_1,\ldots,g_K] \quad \Rightarrow \quad h_1 \circ h_2 \in \mathcal{F}[g_1,\ldots,g_K]$$

(v) **cartesian products** of elements of the set belong to the set, that is, for any $h : \{0,1\}^n \to \{0,1\}^m$ and $k : \{0,1\}^p \to \{0,1\}^q$ we have that

$$h, k \in \mathcal{F}[g_1,\ldots,g_K] \quad \Rightarrow \quad h \times k \in \mathcal{F}[g_1,\ldots,g_K],$$

where $h \times k : \{0,1\}^{n+p} \to \{0,1\}^{m+q}$ with

$$h \times k(x_1,\ldots,x_{n+p})$$
$$= (h(x_1,\ldots,x_n)_1,\ldots,h(x_1,\ldots,x_n)_m, k(x_{n+1},\ldots,x_{n+p})_1,\ldots,k(x_{n+1},\ldots,x_{n+p})_q).$$

A set $G = \{g_1,\ldots,g_J\}$ of classical gates is called **universal** if any gate g can be constructed with gates from G, that is, if for every gate g

$$g \in \mathcal{F}[g_1,\ldots,g_J] \quad \text{for } g_1,\ldots,g_J \in G.$$

When many (sometimes many millions) of gates are connected the resulting arrays are called classical digital **logical circuits**.

If gates are built from another set, then any gates gates that can be built from the former can also be built from the latter.

Lemma 5.4 *For gates $h_1,\ldots,h_L, g_1,\ldots,g_K$ we have*

$$h_1,\ldots,h_L \in \mathcal{F}[g_1,\ldots,g_K] \quad \Rightarrow \quad \mathcal{F}[h_1,\ldots,h_L] \subset \mathcal{F}[g_1,\ldots,g_K]$$

such that, in particular,

$$\mathcal{F}[\mathcal{F}[g_1,\ldots,g_K]] \subset \mathcal{F}[g_1,\ldots,g_K].$$

Proof The stated inclusion is a direct consequence of Definition 5.3 since the operations to build any element in $\mathcal{F}[h_1,\ldots,h_L]$ from the $h_1,\ldots,h_L$ are the same as to build any element in $\mathcal{F}[g_1,\ldots,g_K]$ from its elements and since the $h_1,\ldots,h_L$ are members of this set. $\square$

Example 5.5 For $(x_1, x_2, x_3) \in \{0,1\}^3$ we have

$$\left(\text{ID} \times \text{ID} \times \text{XOR}\right) \circ \left(\text{ID} \times \text{ID} \times \text{AND} \times \text{ID}\right) \circ r_{1,3,2,4,5}^{(5)}$$
$$\circ \left(\text{COPY} \times \text{COPY} \times \text{ID}\right)(x_1, x_2, x_3)$$

$$\underbrace{=}_{(5.7)} \left(\text{ID} \times \text{ID} \times \text{XOR}\right) \circ \left(\text{ID} \times \text{ID} \times \text{AND} \times \text{ID}\right) \circ r_{1,3,2,4,5}^{(5)}(x_1, x_1, x_2, x_2, x_3)$$

$$\underbrace{=}_{(5.8)} \left(\text{ID} \times \text{ID} \times \text{XOR}\right) \circ \left(\text{ID} \times \text{ID} \times \text{AND} \times \text{ID}\right)(x_1, x_2, x_1, x_2, x_3)$$

$$\underbrace{=}_{(5.4)} \left(\text{ID} \times \text{ID} \times \text{XOR}\right)(x_1, x_2, x_1 x_2, x_3)$$

$$\underbrace{=}_{(5.5)} (x_1, x_2, x_1 x_2 \overset{2}{\oplus} x_3)$$

$$\underbrace{=}_{(5.6)} \text{TOF}(x_1, x_2, x_3).$$

Hence

$$\text{TOF} = \left(\text{ID} \times \text{ID} \times \text{XOR}\right) \circ \left(\text{ID} \times \text{ID} \times \text{AND} \times \text{ID}\right) \circ r_{1,3,2,4,5}^{(5)} \circ \left(\text{COPY} \times \text{COPY} \times \text{ID}\right)$$

and thus Definition 5.3 implies

$$\text{TOF} \in \mathcal{F}[\text{ID}, \text{AND}, \text{XOR}, \text{COPY}].$$

Theorem 5.6 *The classical* TOFFOLI-*gate is universal and reversible.*

Proof Since every gate $g : \{0,1\}^n \to \{0,1\}^m$ can be decomposed in m gates $g_j :$ $\{0,1\}^n \to \{0,1\}$, where $j \in \{1, \ldots, m\}$, it suffices to show universality only for a gate of the form $f : \{0,1\}^n \to \{0,1\}$, which we shall do by induction in n.

We begin with the initialization of the induction at $n = 1$. With the help of their gate definitions (5.3)–(5.7) one sees that ID, FALSE, TRUE and NOT can be replicated by the various channels of TOF as follows:

$$\text{ID}(x_1) = x_1 = \text{TOF}_1(x_1, 1, 1) = r_1^{(3)} \circ \text{TOF} \circ p_{1,1;2,3}^{(1)}(x_1)$$

$$\text{FALSE}(x_1) = 0 = \text{TOF}_1(0,0,0) = r_1^{(3)} \circ \text{TOF} \circ p_{0,0,0;1,2,3}^{(0)}(x_1)$$

$$\text{TRUE}(x_1) = 1 = \text{TOF}_1(1,0,0) = r_1^{(3)} \circ \text{TOF} \circ p_{1,0,0;12,3}^{(1)}(x_1) \qquad (5.9)$$

$$\text{NOT}(x_1) = 1 \overset{2}{\oplus} x_1 = \text{TOF}_3(1,1,x_1) = r_3^{(3)} \circ \text{TOF} \circ p_{1,1;1,2}^{(1)}(x_1).$$

Consequently, every gate $f : \{0,1\} \to \{0,1\}$ can be constructed with TOF. Moreover, the following shows that we can also build $\mathrm{AND}, \mathrm{XOR}, \mathrm{COPY}^{(1)}$ and $\mathrm{COPY}^{(n)}$ with TOF, that is, we have

$$\mathrm{AND}(x_1,x_2) = x_1 x_2 = \mathrm{TOF}_3(x_1,x_2,0) = r_3^{(3)} \circ \mathrm{TOF} \circ p_{0;3}^{(2)}(x_1,x_2)$$

$$\mathrm{XOR}(x_1,x_2) = x_1 \overset{2}{\oplus} x_2 = \mathrm{TOF}_3(1,x_1,x_2) = r_3^{(3)} \circ \mathrm{TOF} \circ p_{1;1}^{(2)}(x_1,x_2)$$

(5.10)

and

$$\mathrm{COPY}^{(1)}(x_1) = (x_1,x_1) = \big(\mathrm{TOF}_1(x_1,1,0), \mathrm{TOF}_3(x_1,1,0)\big)$$
$$= r_{1,3}^{(3)} \circ \mathrm{TOF} \circ p_{1,0;2,3}^{(1)}(x_1)$$
$$\mathrm{COPY}^{(n)}(x_1,\ldots x_n) = (x_1 \ldots,x_n,x_1,\ldots,x_n)$$
$$= r_{1,3,5,\ldots,2n-1,2,4,\ldots,2n}^{(2n)} \circ \mathrm{COPY}^{(1)} \times \cdots \times \mathrm{COPY}^{(1)}(x_1,\ldots x_n).$$

Hence, according to Definition 5.3, it follows that

$$\mathrm{ID, FALSE, TRUE}, \mathrm{NOT}, \mathrm{AND}, \mathrm{XOR}, \mathrm{COPY}^{(n)} \in \mathcal{F}[\mathrm{TOF}]. \qquad (5.11)$$

Turning to the inductive step from $n-1$ to n, we suppose that TOF is universal for gates of the form $g : \{0,1\}^{n-1} \to \{0,1\}$.

Let $f : \{0,1\}^n \to \{0,1\}$ be arbitrary. Then, for $x_n \in \{0,1\}$ define

$$g_{x_n}(x_1,\ldots,x_{n-1}) := f(x_1,\ldots,x_{n-1},x_n)$$

and

$$h(x_1,\ldots,x_n) := \mathrm{XOR}\Big(\mathrm{AND}\big(g_0(x_1,\ldots,x_{n-1}), \mathrm{NOT}(x_n) \big),$$
$$\mathrm{AND}\big(g_1(x_1,\ldots,x_{n-1}), \quad x_n \quad \big) \Big). \qquad (5.12)$$

Due to the induction-assumption, g_0 and g_1 can be built with TOF, and because of (5.9) and (5.10), we know that NOT, AND and XOR can be constructed with TOF. Altogether thus h in (5.12) can be built with TOF. At the same time we have $h = f$, since

$$h(x_1,\ldots,x_{n-1},0) = \mathrm{XOR}\Big(\mathrm{AND}\big(g_0(x_1,\ldots,x_{n-1}), \mathrm{NOT}(0) \big),$$
$$\mathrm{AND}\big(g_1(x_1,\ldots,x_{n-1}), \quad 0 \quad \big) \Big)$$

$$= \mathrm{XOR}\big(\quad g_0(x_1,\ldots,x_{n-1}), \quad 0 \quad \big)$$

$$= \quad g_0(x_1,\ldots,x_{n-1})$$

$$= \quad f(x_1,\ldots,x_{n-1},0)$$

$$h(x_1,\ldots,x_{n-1},1) = \text{XOR}\Big(\ \text{AND}\big(\ g_0(x_1,\ldots,x_{n-1}), \ \text{NOT}(1) \ \big),$$
$$\text{AND}\big(\ g_1(x_1,\ldots,x_{n-1}), \quad 1 \quad \big) \ \Big)$$

$$= \text{XOR}\Big(\qquad\qquad 0,$$
$$g_1(x_1,\ldots,x_{n-1}) \qquad\qquad\qquad \Big)$$
$$= \qquad\qquad g_1(x_1,\ldots,x_{n-1})$$
$$= \qquad\qquad f(x_1,\ldots,x_{n-1},1).$$

In other words, we have

$$f = \text{XOR} \circ \big(\text{AND} \times \text{AND}\big) \circ \big((g_0 \times \text{NOT}) \times (g_1 \times \text{ID})\big) \circ \text{COPY}^{(n)},$$

and because of (5.11) and the induction-assumption $g_0, g_1 \in \mathcal{F}[\text{TOF}]$, it follows from Definition 5.3 that $f \in \mathcal{F}[\text{TOF}]$. Consequently, TOF is universal. The invertibility of TOF follows from

$$\text{TOF} \circ \text{TOF}(x_1,x_2,x_3) \underbrace{=}_{(5.6)} \text{TOF}(x_1,x_2,x_3 \overset{2}{\oplus} x_1 x_2)$$

$$\underbrace{=}_{(5.6)} \big(x_1,x_2,x_3 \overset{2}{\oplus} \underbrace{x_1 x_2 \overset{2}{\oplus} x_1 x_2}_{=0}\big)$$

$$= \big(x_1,x_2,x_3\big),$$

which means that TOF is its own inverse gate and thus reversible. □

Theoretically, it would thus suffice to only build physical realizations of the TOF-FOLI gate. From them all possible classical gates can be constructed by suitable combination. But the physical realization with the help of TOFFOLI gates is not always the most efficient. Depending on the application, it can be more efficient to use special gates especially built for the intended purpose.

5.2   Quantum Gates

The underlying model of computation that we subscribe to for a quantum computer is akin to the classical sequential model based on a TURING Machine [70]. We will not go into the details of such a computational model here. For our purposes it suffices that, quite analogously to the classical computational process (5.1), we consider the quantum mechanical computational process to be a transformation of a state of n qubits to another such state. In the quantum mechanical case the states are no longer represented by elements in $\{0,1\}^n$, but in the case of pure states are described by vectors in the HILBERT space $\mathbb{H}^{I/O} = \mathbb{H}^{\otimes n}$ or, more generally, by density operators $\rho \in \text{D}(\mathbb{H}^{I/O})$. In the computational process the linear structure of the state space as well as the norm of the original state should be preserved. The latter so that total probability is conserved. From (2.37) we thus know that the purely **quantum mechanical computational process** without any measurement necessarily has to be a unitary transformation, that is,

$$U : \mathbb{H}^{\otimes n} \longrightarrow \mathbb{H}^{\otimes n} \quad \in \mathcal{U}\big(\mathbb{H}^{\otimes n}\big) . \tag{5.13}$$

Physically, such unitary transformations are obtained by subjecting the system to a suitable Hamiltonian for an appropriate period of time, thus generating U as a solution to the initial value problem (2.71).

The quantum computational process forms the core part of any quantum algorithm. However, in order to read out the result, a measurement of one or several observables becomes necessary. With measurement as part of the total process, the transition from initially prepared quantum state to the final measured stated ceases to be unitary.

The qubits on which a quantum processor performs computational operations are called **quantum register** or simply **q-register**. Analogous to classical gates, we then consider quantum gates as maps on the state space of several qubits that preserve the linear structure (superposition) and the normalization to 1 (conservation of probability). We thus define them as follows.

Definition 5.7 A **quantum n-gate** is a unitary operator

$$U : \mathbb{H}^{\otimes n} \to \mathbb{H}^{\otimes n} .$$

For $n = 1$ a gate U is called a **unary quantum gate** and for $n = 2$ a **binary quantum gate**.

It should be noted that quantum gates are *linear* transformations on the state space. As such they are completely defined on the full space by specifying their action on a basis, and in such a basis they can be represented by a matrix. As usual in quantum computing, we choose for such a matrix representation the computational basis of the given spaces.

Before we turn to quantum n-gates, we first consider the simpler cases of unary and binary quantum gates. For the general case of n-gates we will then show that these can be built from elementary unary and binary gates.

5.2.1 Unary Quantum Gates

According to Definition 5.7, unary quantum gates are unitary operators $V : \mathbb{H} \to \mathbb{H}$. These can be represented in the standard basis $\{|0\rangle, |1\rangle\}$ by unitary 2×2 matrices. In

Name	Symbol	Operator	Matrix in basis $\{\lvert 0\rangle, \lvert 1\rangle\}$
Identity	——————	$\mathbf{1}$	$\begin{pmatrix} 1 & 0 \\ 0 & 1 \end{pmatrix}$
Phase-factor	$M(\alpha)$	$M(\alpha) := \mathrm{e}^{\mathrm{i}\alpha}\mathbf{1}$	$\begin{pmatrix} \mathrm{e}^{\mathrm{i}\alpha} & 0 \\ 0 & \mathrm{e}^{\mathrm{i}\alpha} \end{pmatrix}$
Phase-shift	$P(\alpha)$	$P(\alpha) := \\ \lvert 0\rangle\langle 0\rvert + \mathrm{e}^{\mathrm{i}\alpha}\lvert 1\rangle\langle 1\rvert$	$\begin{pmatrix} 1 & 0 \\ 0 & \mathrm{e}^{\mathrm{i}\alpha} \end{pmatrix}$
PAULI-X or Q-NOT	X	$X := \sigma_x$	$\begin{pmatrix} 0 & 1 \\ 1 & 0 \end{pmatrix}$
PAULI-Y	Y	$Y := \sigma_y$	$\begin{pmatrix} 0 & -\mathrm{i} \\ \mathrm{i} & 0 \end{pmatrix}$
PAULI-Z	Z	$Z := \sigma_z$	$\begin{pmatrix} 1 & 0 \\ 0 & -1 \end{pmatrix}$
HADAMARD	H	$H := \frac{\sigma_x + \sigma_z}{\sqrt{2}}$	$\frac{1}{\sqrt{2}}\begin{pmatrix} 1 & 1 \\ 1 & -1 \end{pmatrix}$
Spin-rotation by angle α around $\hat{\mathbf{n}}$	$D_{\hat{\mathbf{n}}}(\alpha)$	$D_{\hat{\mathbf{n}}}(\alpha)$	$\begin{pmatrix} \cos\frac{\alpha}{2} - \mathrm{i}\sin\frac{\alpha}{2}n_z & -\mathrm{i}\sin\frac{\alpha}{2}(n_x - \mathrm{i}n_y) \\ -\mathrm{i}\sin\frac{\alpha}{2}(n_x + \mathrm{i}n_y) & \cos\frac{\alpha}{2} + \mathrm{i}\sin\frac{\alpha}{2}n_z \end{pmatrix}$
Arbitrary unary gate	V	V unitary	$\begin{pmatrix} v_{00} & v_{01} \\ v_{10} & v_{11} \end{pmatrix}$
Measurement of observable A	A λ	$\begin{bmatrix}$ Not a gate, but a non-unitary transformation of the input-state to an eigenstate of A and delivery of measured value λ $\end{bmatrix}$	

Fig. 5.3 Unary quantum gates

Fig. 5.3 we show a list of the most common unary quantum gates. The most prominent unary gates are summarized once more in the following separate itemization:

Quantum-NOT-gate   This is the well known PAULI matrix

$$X := \sigma_x.$$

In the literature on quantum computing the usage of the symbol X in place of σ_x has become common. We shall thus adopt this convention from now on. Because of $\sigma_x = \sigma_x^*$ and $\sigma_x^* \sigma_x = (\sigma_x)^2 = \mathbf{1}$, we know that X is unitary, and because of

$$\sigma_x|0\rangle = \begin{pmatrix} 0 & 1 \\ 1 & 0 \end{pmatrix} \begin{pmatrix} 1 \\ 0 \end{pmatrix} = \begin{pmatrix} 0 \\ 1 \end{pmatrix} = |1\rangle$$

$$\sigma_x|1\rangle = \begin{pmatrix} 0 & 1 \\ 1 & 0 \end{pmatrix} \begin{pmatrix} 0 \\ 1 \end{pmatrix} = \begin{pmatrix} 1 \\ 0 \end{pmatrix} = |0\rangle,$$

it is considered the analogue of the classical negation and thus termed as the quantum NOT gate.

HADAMARD gate   We have already come across the HADAMARD gate

$$H = \frac{\sigma_x + \sigma_z}{\sqrt{2}}$$

in Definition 2.38 as HADAMARD transformation. Some of its properties have been given in in Lemma 2.39.

Representation of a rotation in $\mathbb{R}^3$ by n̂ as spin-rotation   These operators are also known to us from Sect. 2.5.

$$D_{\hat{\mathbf{n}}}(\alpha) = \exp\left(-i\frac{\alpha}{2}\hat{\mathbf{n}} \cdot \sigma\right) = \begin{pmatrix} \cos\frac{\alpha}{2} - i\sin\frac{\alpha}{2}n_z & -i\sin\frac{\alpha}{2}(n_x - in_y) \\ -i\sin\frac{\alpha}{2}(n_x + in_y) & \cos\frac{\alpha}{2} + i\sin\frac{\alpha}{2}n_z \end{pmatrix}.$$

We remind the reader here once more that—as shown in Lemma 2.35—spin-rotations generate all unitary operators on $\mathbb{H}$.

Measurement   As formulated in the Projection Postulate 3, a measurement of an observable A transforms a pure state $|\psi\rangle$ into an eigenstate of A belonging to the eigenvalue λ that has been observed as a result of the measurement. This is an irreversible and thus non-unitary transformation. Hence, a measurement cannot be a gate as defined in Definition 5.7. Nevertheless, we have included measurements here because they are part of some circuits or protocols, such as dense quantum coding (see Sect. 6.2) or teleportation (see Sect. 6.3).

5.2.2   Binary Quantum Gates

Binary quantum gates are unitary operators $U : \mathbb{H}^{\otimes 2} \rightarrow \mathbb{H}^{\otimes 2}$. In the computational basis $\{|0\rangle^2, |1\rangle^2, |2\rangle^2, |3\rangle^2\}$ (see Definition 3.8 and Example 3.9) these are represented by unitary 4×4 matrices. In Figs. 5.4 and 5.5 we show the most important

binary quantum gates. Apart form the matrix representations, we also see there that
a binary gate U in general can be built in various ways from unary and other binary
gates. The function of the gate, that is, the operator U, is, of course, always the
same. But it my be that one of the possible ways to build the gate may be easier to
implement physically or have other advantages in a given application.

These figures also show the symbols commonly used in quantum computing for
the respective gate. The fat dots • and circles ○ used in these figures symbolize the
conditional application of the operators in different channels connected to them by
lines. Those operators in a gate that are connected to a fat dot • will only be applied
if the qubit going through the channel with the dot is in the state $|1\rangle$. The qubit in
the channel with the dot is unaltered. In case the channel with the dot is traversed
by a qubit in the state $|0\rangle$, the operator connected to the dot will not be applied, in
other words, in this case nothing happens. Conversely, those operators which are
connected to a circle ○ are only applied if the qubit traversing the channel is in the
state $|0\rangle$. Nothing happens if the channel with the circle is traversed by a qubit in
the state $|1\rangle$. On linear combinations $|\psi\rangle = \psi_0|0\rangle + \psi_1|1\rangle$ these gates act by linear
continuation of their behavior on $|0\rangle$ and $|1\rangle$.

A further notation used in the graphic representations is the usage of $\oplus$ for NOT,
which is given by the operator $X = \sigma_x$. The notation $\Lambda^.(\cdot)$ used in Figs. 5.4 and 5.5
for the so-called controlled gates will be defined in Definition 5.10.

It should be noted that in the graphic representation of gates the transformation
effected by the gates is assumed to happen by the qubits traversing the gate from *left
to right*, that is, the initial state enters on the left and the transformed state leaves
on the right. This means that—in contrast to operator-products—the *leftmost operator shown in the graphic representation acts first* and the *rightmost operator in the
graphic representation acts last* on the traversing qubit. For example, in the case
of the controlled U gate $\Lambda^1(U)$ the sequence of symbols in the graphic representation $\boxed{C} - \boxed{X} - \boxed{B} - \boxed{X} - \boxed{A}$ of this gate is exactly the *reversed* sequence of
the operator-product $\left(P(\alpha) \otimes A\right) \Lambda^1(X) \left(\mathbf{1} \otimes B\right) \Lambda^1(X) \left(\mathbf{1} \otimes C\right)$ representing the very
same gate.

Every one of the three ways to represent a gate—by its graphical symbol, by
its operator or by its matrix—has its advantages. Sometimes it can be helpful to
represent a gate graphically in order to assist analysis and understanding of a circuit.
Quite often the operator-representation is most suited for general proofs, whereas
the matrix representation may be useful to elucidate proofs in special cases.

5.2.3 General Quantum Gates

As with classical gates, we combine elementary quantum gates to build ever larger
quantum gates and eventually quantum circuits.

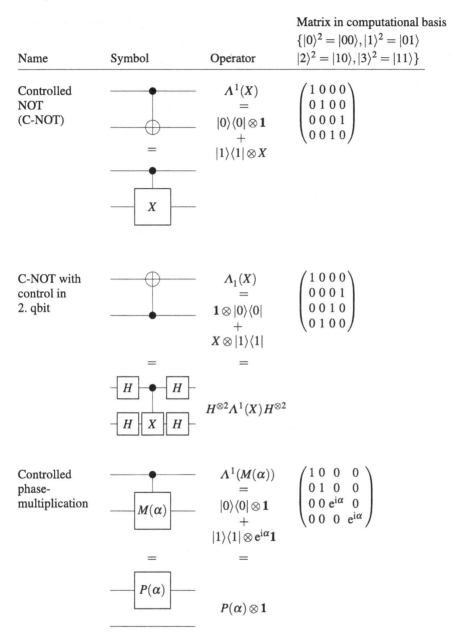

| | | | Matrix in computational basis $\{|0\rangle^2 = |00\rangle, |1\rangle^2 = |01\rangle$ |
| Name | Symbol | Operator | $|2\rangle^2 = |10\rangle, |3\rangle^2 = |11\rangle\}$ |

Fig. 5.4 Binary quantum gates (1/2). Some of the symbols are explained in more detail in Sect. 5.2.2

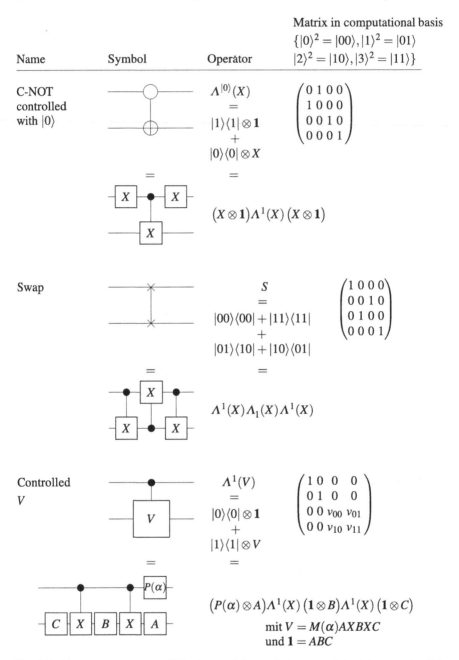

Fig. 5.5 Binary quantum gates (2/2). Some of the symbols are explained in more detail in Sect. 5.2.2

Definition 5.8 For $U_j \in \mathcal{U}(\mathbb{H}^{\otimes n_j})$ with $j \in \{1,\dots,K\}$ we denote by $\mathcal{F}[U_1,\dots,U_K]$ the set of gates which can be constructed with the $U_1,\dots,U_K$. This set is defined by the following rules:

(i)
$$U_1,\dots,U_K \in \mathcal{F}[U_1,\dots,U_K]$$

(ii) for any $n \in \mathbb{N}$
$$\mathbf{1}^{\otimes n} \in \mathcal{F}[U_1,\dots,U_K]$$

(iii) for any $V_1, V_2 \in \mathcal{U}(\mathbb{H}^{\otimes n})$ we have
$$V_1, V_2 \in \mathcal{F}[U_1,\dots,U_K] \quad \Rightarrow \quad V_1 V_2 \in \mathcal{F}[U_1,\dots,U_K]$$

(iv) for any $V_i \in \mathcal{U}(\mathbb{H}^{\otimes n_i})$ with $i \in \{1,2\}$ we have
$$V_1, V_2 \in \mathcal{F}[U_1,\dots,U_K] \quad \Rightarrow \quad V_1 \otimes V_2 \in \mathcal{F}[U_1,\dots,U_K].$$

A set of quantum gates $\mathsf{U} = \{U_1,\dots,U_J\}$ is called **universal** if any quantum gate U can be constructed with gates from U, that is, if for every quantum gate U
$$U \in \mathcal{F}[U_1,\dots,U_J] \quad \text{for } U_1,\dots,U_J \in \mathsf{U}.$$

When acting on a system in the state $\rho \in D(\mathbb{H})$ the gate U transforms it to a new state $U\rho U^*$.

Similarly to Lemma 5.4 in the classical case, we have the following obvious inclusion.

Lemma 5.9 *For gates $V_1,\dots,V_L,U_1,\dots,U_K$ we have*

$$V_1,\dots,V_L \in \mathcal{F}[U_1,\dots,U_K] \quad \Rightarrow \quad \mathcal{F}[V_1,\dots,V_L] \subset \mathcal{F}[U_1,\dots,U_K]$$

such that, in particular,

$$\mathcal{F}[\mathcal{F}[U_1,\dots,U_K]] \subset \mathcal{F}[U_1,\dots,U_K]. \tag{5.14}$$

Proof The stated inclusion is a direct consequence of Definition 5.7 since the operations to construct any element in $\mathcal{F}[V_1,\dots,V_L]$ from the $V_1,\dots,V_L$ are the same as to construct any element in $\mathcal{F}[U_1,\dots,U_K]$ from its elements and since the $V_1,\dots,V_L$ are members of this set. $\qquad\square$

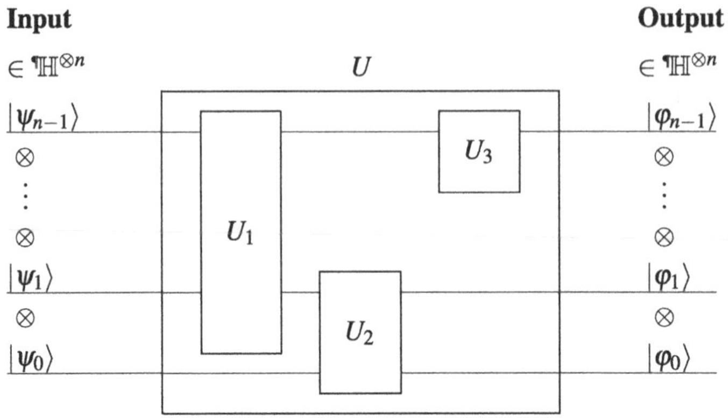

Input   **Output**

$\in \mathbb{IH}^{\otimes n}$ U $\in \mathbb{IH}^{\otimes n}$

$|\psi_{n-1}\rangle$ $|\varphi_{n-1}\rangle$

$\otimes$ U_3 $\otimes$

$\vdots$ $\vdots$

$\otimes$ U_1 $\otimes$

$|\psi_1\rangle$ $|\varphi_1\rangle$

$\otimes$ U_2 $\otimes$

$|\psi_0\rangle$ $|\varphi_0\rangle$

Fig. 5.6 Graphical representation of a generic quantum gate U, which can be built from the gates U_1, U_2, U_3, that is, for which $U \in \mathcal{F}[U_1, U_2, U_3]$ holds

Figure 5.6 shows a schematic representation of a quantum gate which is built from three smaller gates.

The result of the following Exercise 5.59 will allow us to define more general controlled gates.

Exercise 5.59 Let $n, n_a, n_b \in \mathbb{N}_0$ with $n = n_a + n_b$ and let $|a\rangle \in \mathbb{IH}^{\otimes n_a}$ and $|b\rangle \in \mathbb{IH}^{\otimes n_b}$ be vectors in the respective computational basis. Moreover, let $V \in \mathcal{U}(\mathbb{IH})$. Show that then

$$\mathbf{1}^{\otimes n+1} + |a\rangle\langle a| \otimes (V - \mathbf{1}) \otimes |b\rangle\langle b| \in \mathcal{U}(\mathbb{IH}^{n+1}).$$

For a solution see Solution 5.59.

Definition 5.10 Let $n, n_a, n_b \in \mathbb{N}_0$ with $n = n_a + n_b$ and let $|a\rangle \in \mathbb{IH}^{\otimes n_a}$ and $|b\rangle \in \mathbb{IH}^{\otimes n_b}$ be vectors in the respective computational basis. Moreover, let $V \in \mathcal{U}(\mathbb{IH})$. We denote the $(|a\rangle, |b\rangle)$-**controlled** V by $\Lambda_{|b\rangle}^{|a\rangle}(V)$ and define it as the $n+1$-gate

$$\Lambda_{|b\rangle}^{|a\rangle}(V) := \mathbf{1}^{\otimes n+1} + |a\rangle\langle a| \otimes (V - \mathbf{1}) \otimes |b\rangle\langle b|$$

$$= \mathbf{1}^{\otimes n+1} + \bigotimes_{j=n_a-1}^{0} |a_j\rangle\langle a_j| \otimes (V - \mathbf{1}) \otimes \bigotimes_{j=n_b-1}^{0} |b_j\rangle\langle b_j|.$$

The qubit on which V acts is called **target-qubit**. In the special case $a = 2^{n_a} - 1$ and $b = 2^{n_b} - 1$ one has $|a\rangle = |1\dots1\rangle^{n_a}$ and $|b\rangle = |1\dots1\rangle^{n_b}$, and we define the abbreviating notation

$$\Lambda_{n_b}^{n_a}(V) := \Lambda_{|2^{n_b}-1\rangle}^{|2^{n_a}-1\rangle}(V)$$

as well as in the case $n_a = n$ and $a = 2^n - 1$

$$\Lambda^n(V) := \Lambda^{|2^n-1\rangle}(V).$$

Likewise, in the case $n_b = n$ and $b = 2^n - 1$ we define

$$\Lambda_n(V) := \Lambda_{|2^n-1\rangle}(V).$$

In the case $n = 0$ we define

$$\Lambda^0(V) := V =: \Lambda_0(V).$$

Exercise 5.60 Let $V \in \mathcal{U}(\mathbb{H})$ and $\alpha \in \mathbb{R}$. Show that

$$\Lambda^1(V) = |0\rangle\langle 0| \otimes \mathbf{1} + |1\rangle\langle 1| \otimes V \qquad (5.15)$$
$$\Lambda_1(X) = H^{\otimes 2} \Lambda^1(X) H^{\otimes 2} \qquad (5.16)$$
$$\Lambda^1(M(\alpha)) = P(\alpha) \otimes \mathbf{1}. \qquad (5.17)$$

For a solution see Solution 5.60.

Theorem 5.11 *For an arbitrary unitary operator* $V : \mathbb{H} \to \mathbb{H}$ *the following holds*

$$V \in \mathcal{F}[M, D_{\hat{\mathbf{y}}}, D_{\hat{\mathbf{z}}}] \qquad (5.18)$$
$$\Lambda^1(V), \Lambda_1(V) \in \mathcal{F}[M, D_{\hat{\mathbf{y}}}, D_{\hat{\mathbf{z}}}, \Lambda^1(X)], \qquad (5.19)$$

that is, any unitary $V : \mathbb{H} \to \mathbb{H}$ *can be generated from phase-multiplications* M *and spin-rotations around* $\hat{\mathbf{y}}$ *and* $\hat{\mathbf{z}}$. *In order to generate the controlled gates* $\Lambda^1(V)$ *and* $\Lambda_1(V)$ *one needs, in addition, the controlled NOT* $\Lambda^1(X)$.

Proof From Lemma 2.34 and the accompanying proof we know already that for any unitary operator V on $\mathbb{H}$ there exist angles $\alpha, \beta, \gamma, \delta$, so that the operators

$$A := D_{\hat{z}}(\beta) D_{\hat{y}}\left(\frac{\gamma}{2}\right)$$

$$B := D_{\hat{y}}\left(-\frac{\gamma}{2}\right) D_{\hat{z}}\left(-\frac{\delta+\beta}{2}\right)$$

$$C := D_{\hat{z}}\left(\frac{\delta-\beta}{2}\right).$$

(5.20)

satisfy

$$ABC = \mathbf{1} \tag{5.21}$$

$$V = e^{i\alpha} A \sigma_x B \sigma_x C. \tag{5.22}$$

In this we evidently have on account of (5.20) that

$$A, B, C \in \mathcal{F}[D_{\hat{y}}, D_{\hat{z}}], \tag{5.23}$$

and from (2.149) in Example 2.33 we see that

$$X = \sigma_x \in \mathcal{F}[M, D_{\hat{y}}, D_{\hat{z}}]. \tag{5.24}$$

Together (5.22)–(5.24) thus imply

$$V \in \mathcal{F}[M, D_{\hat{y}}, D_{\hat{z}}].$$

From (2.148) in Example 2.33 we also see that

$$P(\alpha) \in \mathcal{F}[M, D_{\hat{y}}, D_{\hat{z}}]. \tag{5.25}$$

From (5.23) and (5.25) it follows thus that

$$\left(P(\alpha)\otimes A\right)\Lambda^1(X)\left(\mathbf{1}\otimes B\right)\Lambda^1(X)\left(\mathbf{1}\otimes C\right) \in \mathcal{F}[M, D_{\hat{y}}, D_{\hat{z}}, \Lambda^1(X)]. \tag{5.26}$$

Finally, one has

$$\left(P(\alpha)\otimes A\right)\Lambda^1(X)\left(\mathbf{1}\otimes B\right)\Lambda^1(X)\left(\mathbf{1}\otimes C\right)$$

$$\underset{(5.15)}{=} \left(P(\alpha)\otimes A\right)\Lambda^1(X)\left(\mathbf{1}\otimes B\right)\left[|0\rangle\langle 0|\otimes \mathbf{1}+|1\rangle\langle 1|\otimes X\right]\left(\mathbf{1}\otimes C\right)$$

$$= \left(P(\alpha)\otimes A\right)\Lambda^1(X)\left[|0\rangle\langle 0|\otimes BC+|1\rangle\langle 1|\otimes BXC\right]$$

$$\underset{(5.15)}{=} \left(P(\alpha)\otimes A\right)\left[|0\rangle\langle 0|\otimes \mathbf{1}+|1\rangle\langle 1|\otimes X\right]\left[|0\rangle\langle 0|\otimes BC+|1\rangle\langle 1|\otimes BXC\right]$$

$$= \left(P(\alpha)\otimes A\right)\left[|0\rangle\langle 0|\otimes BC+|1\rangle\langle 1|\otimes XBXC\right]$$

$$
= \underbrace{P(\alpha)|0\rangle\langle 0|}_{=|0\rangle\langle 0|} \otimes \underbrace{ABC}_{\substack{=\mathbf{1}\\(5.21)}} + \underbrace{P(\alpha)|1\rangle\langle 1|}_{=\mathrm{e}^{\mathrm{i}\alpha}|1\rangle\langle 1|} \otimes AXBXC
$$

$$
= |0\rangle\langle 0|\otimes \mathbf{1} + |1\rangle\langle 1|\otimes \underbrace{\mathrm{e}^{\mathrm{i}\alpha}AXBXC}_{\substack{=V\\(5.22)}} = |0\rangle\langle 0|\otimes \mathbf{1} + |1\rangle\langle 1|\otimes V
$$

$$
\underset{(5.15)}{=} \Lambda^{1}(V) , \tag{5.27}
$$

and with (5.26) the claim (5.19) follows for $\Lambda^{1}(V)$. In order to proof it for $\Lambda_{1}(V)$, one exploits that from (2.150) in Example 2.33 it also follows that

$$
H \in \mathcal{F}[M, D_{\hat{\mathbf{y}}}, D_{\hat{\mathbf{z}}}].
$$

Because of (5.16), thus,

$$
\Lambda_{1}(X) \in \mathcal{F}[M, D_{\hat{\mathbf{y}}}, D_{\hat{\mathbf{z}}}, \Lambda^{1}(X)]
$$

holds and one verifies

$$
\Lambda_{1}(V) = \big(A\otimes P(\alpha)\big)\Lambda_{1}(X)\big(B\otimes \mathbf{1}\big)\Lambda_{1}(X)\big(C\otimes \mathbf{1}\big)
$$

analogously to (5.27). □

We now show that $\Lambda^{n}(V)$, too, can be built from phase-multiplication, spin-rotation and controlled NOT.

Lemma 5.12 *For any operator $V \in \mathcal{U}(\mathbb{H})$ and number $n \in \mathbb{N}_{0}$ we have*

$$
\Lambda^{n}(V) \in \mathcal{F}[M, D_{\hat{\mathbf{y}}}, D_{\hat{\mathbf{z}}}, \Lambda^{1}(X)]. \tag{5.28}
$$

Proof We show this by induction, which we start at $n = 1$. For $n = 0$ or $n = 1$ the claim is true on account of Theorem 5.11. For the inductive step from $n-1$ to n suppose that $\Lambda^{n-1}(V) \in \mathcal{F}[M, D_{\hat{\mathbf{y}}}, D_{\hat{\mathbf{z}}}, \Lambda^{1}(X)]$ holds for arbitrary $V \in \mathcal{U}(\mathbb{H})$.

First, we consider the gates A, B, C, D shown in Fig. 5.7. These satisfy $A, C, D \in \mathcal{F}[\Lambda^{n-1}(W)]$ with $W = \sqrt{V}, \sqrt{V^{*}}$ and $B \in \mathcal{F}[\Lambda^{1}(X)]$. According to the inductive assumption one then also has

$$
A, B, C, D \in \mathcal{F}[M, D_{\hat{\mathbf{y}}}, D_{\hat{\mathbf{z}}}, \Lambda^{1}(X)]. \tag{5.29}
$$

The action of these gates and that of $\Lambda^{n}(V)$ can be described with the help of the computational basis $|x\rangle = \bigotimes_{j=n}^{0}|x_{j}\rangle$ in $\mathbb{H}^{\otimes n+1}$ as follows

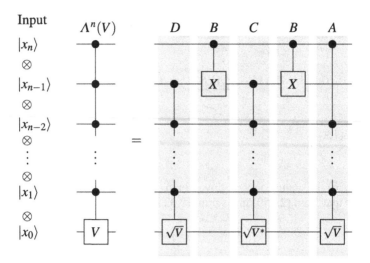

Fig. 5.7 Generation of the controlled V gate $\Lambda^n(V)$ by $A, C, D \in \mathcal{F}[\Lambda^{n-1}(\sqrt{V}), \Lambda^{n-1}(\sqrt{V^*})]$ and $B \in \mathcal{F}[\Lambda^1(X)]$. Note that $\Lambda^n(V)$ acts on $\mathbb{H}^{\otimes n+1}$. We point out once more here that in the graphical representation the leftmost gate is applied first and the rightmost last, so that the operator D of the leftmost gate appears on the right of the operator product and the operator A of the rightmost gate appears on the left of the operator product, that is, $\Lambda^n(V) = ABCBD$ holds

$$D|x\rangle = |x_n \dots x_1\rangle \otimes V^{\frac{1}{2}\prod_{j=1}^{n-1} x_j}|x_0\rangle$$

$$B|x\rangle = |x_n(x_n \overset{2}{\oplus} x_{n-1})x_{n-2}\dots x_1\rangle \otimes |x_0\rangle$$

$$C|x\rangle = |x_n \dots x_1\rangle \otimes V^{-\frac{1}{2}\prod_{j=1}^{n-1} x_j}|x_0\rangle$$

$$A|x\rangle = |x_n \dots x_1\rangle \otimes V^{x_n \frac{1}{2}\prod_{j=1}^{n-2} x_j}|x_0\rangle$$

$$\Lambda^n(V)|x\rangle = |x_n \dots x_1\rangle \otimes V^{\prod_{j=1}^{n} x_j}|x_0\rangle .$$

This implies

$$ABCBD|x\rangle = ABCB|x_n \dots x_1\rangle \otimes V^{\frac{1}{2}\prod_{j=1}^{n-1} x_j}|x_0\rangle$$

$$= ABC|x_n(x_n \overset{2}{\oplus} x_{n-1})x_{n-2}\dots x_1\rangle \otimes V^{\frac{1}{2}\prod_{j=1}^{n-1} x_j}|x_0\rangle$$

$$= AB|x_n(x_n \overset{2}{\oplus} x_{n-1})x_{n-2}\dots x_1\rangle \otimes V^{\frac{x_{n-1}-(x_n \overset{2}{\oplus} x_{n-1})}{2}\prod_{j=1}^{n-2} x_j}|x_0\rangle$$

$$= A|x_n \underbrace{(x_n \overset{2}{\oplus} (x_n \overset{2}{\oplus} x_{n-1}))}_{=x_{n-1}}x_{n-2}\dots x_1\rangle \otimes V^{\frac{x_{n-1}-(x_n \overset{2}{\oplus} x_{n-1})}{2}\prod_{j=1}^{n-2} x_j}|x_0\rangle$$

$$= |x_n \dots x_1\rangle \otimes V^{\frac{\overbrace{x_n + x_{n-1} - (x_n \overset{2}{\oplus} x_{n-1})}^{=x_n x_{n-1}}}{2}\prod_{j=1}^{n-2} x_j}|x_0\rangle$$

$$= |x_n \dots x_1\rangle \otimes V^{\Pi_{j=1}^n x_j} |x_0\rangle$$
$$= \Lambda^n(V) |x\rangle$$

and thus because of (5.29)

$$\Lambda^n(V) = ABCBD \in \mathcal{F}[M, D_{\hat{y}}, D_{\hat{z}}, \Lambda^1(X)]. \qquad \square$$

We also need the following generalization of the swap gate S.

Definition 5.13 For $n \in \mathbb{N}$ and $j, k \in \mathbb{N}_0$ with $k < j \leq n-1$ we define on $\mathbb{H}^{\otimes n}$

$$S_{jk}^{(n)} := \mathbf{1}^{\otimes n-1-j} \otimes |0\rangle\langle 0| \otimes \mathbf{1}^{\otimes j-k-1} \otimes |0\rangle\langle 0| \otimes \mathbf{1}^{\otimes k}$$
$$+ \mathbf{1}^{\otimes n-1-j} \otimes |1\rangle\langle 1| \otimes \mathbf{1}^{\otimes j-k-1} \otimes |1\rangle\langle 1| \otimes \mathbf{1}^{\otimes k}$$
$$+ \mathbf{1}^{\otimes n-1-j} \otimes |0\rangle\langle 1| \otimes \mathbf{1}^{\otimes j-k-1} \otimes |1\rangle\langle 0| \otimes \mathbf{1}^{\otimes k}$$
$$+ \mathbf{1}^{\otimes n-1-j} \otimes |1\rangle\langle 0| \otimes \mathbf{1}^{\otimes j-k-1} \otimes |0\rangle\langle 1| \otimes \mathbf{1}^{\otimes k}.$$

It is useful to define as well $S_{jj}^{(n)} = \mathbf{1}^{\otimes n}$. The **global swap** or **exchange operator** $S^{(n)}$ on $\mathbb{H}^{\otimes n}$ is defined as

$$S^{(n)} := \prod_{j=0}^{\lfloor \frac{n}{2} \rfloor - 1} S_{n-1-j,j}^{(n)}. \qquad (5.30)$$

With $S_{jk}^{(n)}$ the qubits in the factor-spaces $\mathbb{H}_j$ and $\mathbb{H}_k$ inside the tensor products $\mathbb{H}^{\otimes n}$ are swapped, that is, exchanged. With $S^{(n)}$ the sequence of factors in the tensor product is completely reversed.

Exercise 5.61 Suppose $n \in \mathbb{N}$ and $j, k \in \mathbb{N}_0$ with $k < j \leq n-1$ as well as $\bigotimes_{l=n-1}^0 |\psi_l\rangle \in \mathbb{H}^{\otimes n}$. Show that

$$S_{jk}^{(n)} \bigotimes_{l=n-1}^0 |\psi_l\rangle = |\psi_{n-1} \dots \psi_{j+1} \psi_k \psi_{j-1} \dots \psi_{k+1} \psi_j \psi_{k-1} \dots \psi_0\rangle \qquad (5.31)$$

$$\left(S_{jk}^{(n)}\right)^2 = \mathbf{1}^{\otimes n} \qquad (5.32)$$

$$\left[S_{jk}^{(n)}, S_{lm}^{(n)}\right] = 0 \quad \text{for } j, k \notin \{l, m\} \qquad (5.33)$$

$$S^{(n)} \bigotimes_{l=n-1}^0 |\psi_l\rangle = \bigotimes_{l=0}^{n-1} |\psi_l\rangle = |\psi_0 \psi_1 \dots \psi_{n-2} \psi_{n-1}\rangle. \qquad (5.34)$$

For a solution see Solution 5.61.

Example 5.14 As an example for swap operators we consider the case $n = 3, j = 2$ and $k = 0$. Then one has $\left\lfloor \frac{n}{2} \right\rfloor - 1 = 0$ and $S^{(3)} = S_{20}^{(3)}$. To illustrate the swap action, we apply $S_{20}^{(3)}$ on $|\psi\rangle \otimes |\xi\rangle \otimes |\varphi\rangle \in \mathbb{H}^{\otimes 3}$. For such a state we have

$$
\begin{aligned}
|\psi\rangle \otimes |\xi\rangle \otimes |\varphi\rangle &= \big(\psi_0|0\rangle + \psi_1|1\rangle\big) \otimes \big(\xi_0|0\rangle + \xi_1|1\rangle\big) \otimes \big(\varphi_0|0\rangle + \varphi_1|1\rangle\big) \\
&= \psi_0\xi_0\varphi_0|000\rangle + \psi_0\xi_0\varphi_1|001\rangle + \psi_0\xi_1\varphi_0|010\rangle + \psi_0\xi_1\varphi_1|011\rangle \\
&\quad + \psi_1\xi_0\varphi_0|100\rangle + \psi_1\xi_0\varphi_1|101\rangle + \psi_1\xi_1\varphi_0|110\rangle + \psi_1\xi_1\varphi_1|111\rangle
\end{aligned}
$$

and for the swap operator $S_{20}^{(3)}$ we find

$$
\begin{aligned}
S_{20}^{(3)} &= |0\rangle\langle 0| \otimes \mathbf{1} \otimes |0\rangle\langle 0| + |1\rangle\langle 1| \otimes \mathbf{1} \otimes |1\rangle\langle 1| \\
&\quad + |0\rangle\langle 1| \otimes \mathbf{1} \otimes |1\rangle\langle 0| + |1\rangle\langle 0| \otimes \mathbf{1} \otimes |0\rangle\langle 1|.
\end{aligned}
$$

This yields

$$
\begin{aligned}
S_{20}^{(3)}|\psi\rangle \otimes |\xi\rangle \otimes |\varphi\rangle &= \psi_0\xi_0\varphi_0|000\rangle + \psi_0\xi_0\varphi_1|100\rangle + \psi_0\xi_1\varphi_0|010\rangle + \psi_0\xi_1\varphi_1|110\rangle \\
&\quad + \psi_1\xi_0\varphi_0|001\rangle + \psi_1\xi_0\varphi_1|101\rangle + \psi_1\xi_1\varphi_0|011\rangle + \psi_1\xi_1\varphi_1|111\rangle \\
&= \big(\varphi_0|0\rangle + \varphi_1|1\rangle\big) \otimes \big(\xi_0|0\rangle + \xi_1|1\rangle\big) \otimes \big(\psi_0|0\rangle + \psi_1|1\rangle\big) \\
&= |\varphi\rangle \otimes |\xi\rangle \otimes |\psi\rangle.
\end{aligned}
$$

Next, we show that the gate $\Lambda_{n_b}^{n_a}(V)$ can be built from gates of the form $\Lambda_1(X)$, $\Lambda^1(X)$ and $\Lambda^{n_a+n_b}(V)$.

Lemma 5.15 *For any $V \in \mathcal{U}(\mathbb{H})$ and $n_a, n_b \in \mathbb{N}_0$ we have*

$$
\Lambda_{n_b}^{n_a}(V) = S_{n_b 0}^{(n_a+n_b+1)} \Lambda^{n_a+n_b}(V) S_{n_b 0}^{(n_a+n_b+1)} \tag{5.35}
$$

and thus

$$
\Lambda_{n_b}^{n_a}(V) \in \mathcal{F}[\Lambda^1(X), \Lambda_1(X), \Lambda^{n_a+n_b}(V)]. \tag{5.36}
$$

Proof The identity claimed in (5.35) is illustrated graphically in Fig. 5.8.

We use the abbreviation $n = n_a + n_b$ in this proof. Per Definition 5.10 one has

$$
\begin{aligned}
\Lambda_{n_b}^{n_a}(V) &= \mathbf{1}^{\otimes n+1} + |2^{n_a} - 1\rangle\langle 2^{n_a} - 1| \otimes (V - \mathbf{1}) \otimes |2^{n_b} - 1\rangle\langle 2^{n_b} - 1| \\
\Lambda^n(V) &= \mathbf{1}^{\otimes n+1} + |2^n - 1\rangle\langle 2^n - 1| \otimes (V - \mathbf{1}).
\end{aligned}
$$

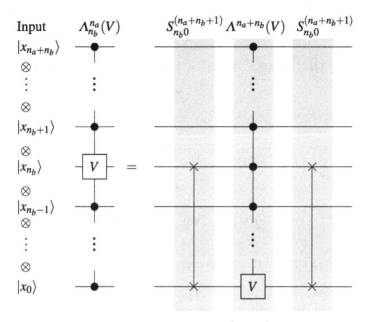

Fig. 5.8 Generation of a controlled V gate $\Lambda_{n_b}^{n_a}(V)$ from $S_{n_b 0}^{(n_a+n_b+1)}$ and $\Lambda^{n_a+n_b}(V)$

Because of (5.32), this implies

$$S_{n_b 0}^{(n+1)} \Lambda^n(V) S_{n_b 0}^{(n+1)} = \mathbf{1}^{\otimes n+1} + S_{n_b 0}^{(n+1)} \left[|2^n - 1\rangle\langle 2^n - 1| \otimes (V - \mathbf{1}) \right] S_{n_b 0}^{(n+1)},$$

and to prove (5.35) it suffices to show that

$$|2^{n_a} - 1\rangle\langle 2^{n_a} - 1| \otimes (V - \mathbf{1}) \otimes |2^{n_b} - 1\rangle\langle 2^{n_b} - 1|$$
$$= S_{n_b 0}^{(n+1)} \left[|2^n - 1\rangle\langle 2^n - 1| \otimes (V - \mathbf{1}) \right] S_{n_b 0}^{(n+1)}. \tag{5.37}$$

For this we consider an arbitrary vector

$$\overset{0}{\underset{j=n}{\bigotimes}} |\psi_j\rangle = |\psi_n \dots \psi_0\rangle \in \P\mathbb{H}^{\otimes n+1}.$$

Then it follows that

$$\left[|2^{n_a} - 1\rangle\langle 2^{n_a} - 1| \otimes (V - \mathbf{1}) \otimes |2^{n_b} - 1\rangle\langle 2^{n_b} - 1| \right] |\psi_n \dots \psi_0\rangle$$
$$= |2^{n_a} - 1\rangle\langle 2^{n_a} - 1| \psi_n \dots \psi_{n-n_a+1}\rangle \tag{5.38}$$
$$\otimes (V - \mathbf{1}) |\psi_{n_b}\rangle \otimes |2^{n_b} - 1\rangle\langle 2^{n_b} - 1| \psi_{n_b-1} \dots \psi_0\rangle.$$

Here we have $n - n_a = n_b$ and

$$|2^{n_a} - 1\rangle\langle 2^{n_a} - 1|\psi_n \ldots \psi_{n_b+1}\rangle = \underbrace{\left[\bigotimes_{l=0}^{n_a-1} |1\rangle\right]}_{\in \mathbb{H}^{\otimes n_a}} \underbrace{\langle 1 \ldots 1|\psi_n \ldots \psi_{n_b+1}\rangle}_{\in \mathbb{C}}$$

$$= \left[\prod_{j=n_b+1}^{n} \langle 1|\psi_j\rangle\right]\left[\bigotimes_{l=0}^{n_a-1} |1\rangle\right] \qquad (5.39)$$

and, analogously,

$$|2^{n_b} - 1\rangle\langle 2^{n_b} - 1|\psi_{n_b-1} \ldots \psi_0\rangle = \left[\prod_{j=0}^{n_b-1} \langle 1|\psi_j\rangle\right]\left[\bigotimes_{l=0}^{n_b-1} |1\rangle\right]. \qquad (5.40)$$

With (5.39) and (5.40) then (5.38) becomes

$$\left[|2^{n_a}-1\rangle\langle 2^{n_a}-1| \otimes (V-\mathbf{1}) \otimes |2^{n_b}-1\rangle\langle 2^{n_b}-1|\right]|\psi_n \ldots \psi_0\rangle$$

$$= \left[\prod_{\substack{j=0\\j\neq n_b}}^{n} \langle 1|\psi_j\rangle\right]\left[\bigotimes_{l=0}^{n_a-1} |1\rangle\right] \otimes (V-\mathbf{1})|\psi_{n_b}\rangle \otimes \left[\bigotimes_{l=0}^{n_b-1} |1\rangle\right]. \qquad (5.41)$$

On the other hand, we have

$$S_{n_b0}^{(n+1)}\left[|2^n-1\rangle\langle 2^n-1| \otimes (V-\mathbf{1})\right] S_{n_b0}^{(n+1)}|\psi_n \ldots \psi_0\rangle$$

$$= S_{n_b0}^{(n+1)}\left[|2^n-1\rangle\langle 2^n-1| \otimes (V-\mathbf{1})\right]|\psi_n \ldots \psi_{n_b+1}\psi_0\psi_{n_b-1} \ldots \psi_1\psi_{n_b}\rangle$$

$$= S_{n_b0}^{(n+1)}\left[|2^n-1\rangle\underbrace{\langle 2^n-1|\psi_n \ldots \psi_{n_b+1}\psi_0\psi_{n_b-1} \ldots \psi_1\rangle}_{=\Pi_{\substack{j=0\\j\neq n_b}}^{n}\langle 1|\psi_j\rangle} \otimes(V-\mathbf{1})|\psi_{n_b}\rangle\right]$$

$$= \left[\prod_{\substack{j=0\\j\neq n_b}}^{n} \langle 1|\psi_j\rangle\right] S_{n_b0}^{(n+1)}\left[\bigotimes_{l=0}^{n-1} |1\rangle \otimes (V-\mathbf{1})|\psi_{n_b}\rangle\right]$$

$$= \left[\prod_{\substack{j=0\\j\neq n_b}}^{n} \langle 1|\psi_j\rangle\right]\left[\bigotimes_{l=n_b+1}^{n} |1\rangle \otimes (V-\mathbf{1})|\psi_{n_b}\rangle \otimes \bigotimes_{l=0}^{n_b-1} |1\rangle\right]. \qquad (5.42)$$

From (5.41) and (5.42) follows (5.37) and thus the claim (5.35). From this in turn it follows that

$$\Lambda_{n_b}^{n_a}(V) \in \mathcal{F}[S_{n_b0}^{(n+1)}, \Lambda^n(V)].$$

Since $S_{n_b 0}^{(n+1)} \in \mathcal{F}[S]$ and S can be built from $\Lambda^1(X)$ and $\Lambda_1(X)$ (see Fig. 5.5), the claim (5.36) follows. □

Definition 5.16 Let A be an operator on $\mathbb{H}$. For vectors $|b\rangle$ of the computational basis of $\mathbb{H}^{\otimes n}$ we define

$$A^{\otimes |b\rangle} := A^{b_{n-1}} \otimes \cdots \otimes A^{b_0}$$

as well as

$$|\neg b\rangle := |\neg b_{n-1} \ldots \neg b_0\rangle = X|b_{n-1}\rangle \otimes \cdots \otimes X|b_0\rangle,$$

where $X = \sigma_x$ is the NOT-operator and $\neg b_j := 1 \overset{2}{\oplus} b_j$ is the classical negation.

The general $(|a\rangle, |b\rangle)$-controlled $n+1$ gate $\Lambda_{|b\rangle}^{|a\rangle}(V)$ can be built with the help of X as a function of the controlled gate $\Lambda^n(V)$.

Lemma 5.17 *Let* $n_a, n_b \in \mathbb{N}_0$ *and* $|a\rangle \in \mathbb{H}^{\otimes n_a}$ *and* $|b\rangle \in \mathbb{H}^{\otimes n_b}$ *vectors of the respective computational basis as well as* V *a unitary operator on* $\mathbb{H}$. *Then the following holds*

$$\Lambda_{|b\rangle}^{|a\rangle}(V) = \left(X^{\otimes |\neg a\rangle} \otimes \mathbf{1} \otimes X^{\otimes |\neg b\rangle}\right) \Lambda_{n_b}^{n_a}(V) \left(X^{\otimes |\neg a\rangle} \otimes \mathbf{1} \otimes X^{\otimes |\neg b\rangle}\right)$$

and thus

$$\Lambda_{|b\rangle}^{|a\rangle}(V) \in \mathcal{F}[X, \Lambda_{n_b}^{n_a}(V)]. \tag{5.43}$$

Proof For $c_j \in \{0,1\}$ one has in general $X^{\neg c_j}|c_j\rangle = |1\rangle$ as well as $\left(X^{\neg c_j}\right)^2 = X^{2 \neg c_j} = \mathbf{1}$. Thus, one has for $c \in \{a, b\}$ the following identities:

$$X^{\otimes |\neg c\rangle}|c\rangle = \left(X^{\neg c_{n_c-1}} \otimes \cdots \otimes X^{\neg c_0}\right)|c_{n_c-1}\rangle \otimes \cdots \otimes |c_0\rangle$$

$$= X^{\neg c_{n_c-1}}|c_{n_c-1}\rangle \otimes \cdots \otimes X^{\neg c_0}|c_0\rangle$$

$$= \overset{0}{\underset{j=n_c-1}{\bigotimes}} |1\rangle = |2^{n_c} - 1\rangle$$

$$\left(X^{\otimes |\neg c\rangle}\right)^2 = \mathbf{1}^{\otimes n_c}$$

$$X^{\otimes |\neg c\rangle}|2^{n_c} - 1\rangle = |c\rangle.$$

With these we obtain

$$
\left(X^{\otimes|\neg a\rangle}\otimes\mathbf{1}\otimes X^{\otimes|\neg b\rangle}\right)\Lambda_{n_b}^{n_a}(V)\left(X^{\otimes|\neg a\rangle}\otimes\mathbf{1}\otimes X^{\otimes|\neg b\rangle}\right)
$$
$$
=\left(X^{\otimes|\neg a\rangle}\otimes\mathbf{1}\otimes X^{\otimes|\neg b\rangle}\right)
$$
$$
\left(\mathbf{1}^{\otimes n_a+n_b+1}+|2^{n_a}-1\rangle\langle2^{n_a}-1|\otimes(V-\mathbf{1})\otimes|2^{n_b}-1\rangle\langle2^{n_b}-1|\right)
$$
$$
\left(X^{\otimes|\neg a\rangle}\otimes\mathbf{1}\otimes X^{\otimes|\neg b\rangle}\right)
$$
$$
=\underbrace{\left(X^{\otimes|\neg a\rangle}\right)^2}_{=\mathbf{1}^{\otimes n_a}}\otimes\mathbf{1}\otimes\underbrace{\left(X^{\otimes|\neg b\rangle}\right)^2}_{=\mathbf{1}^{\otimes n_b}}
$$
$$
+\underbrace{X^{\otimes|\neg a\rangle}|2^{n_a}-1\rangle\langle2^{n_a}-1|X^{\otimes|\neg a\rangle}}_{=|a\rangle\langle a|}
$$
$$
\otimes(V-\mathbf{1})
$$
$$
\otimes\overbrace{X^{\otimes|\neg b\rangle}|2^{n_b}-1\rangle\langle2^{n_b}-1|X^{\otimes|\neg b\rangle}}^{=|b\rangle\langle b|}
$$
$$
=\mathbf{1}^{\otimes n_a+n_b+1}+|a\rangle\langle a|\otimes(V-\mathbf{1})\otimes|b\rangle\langle b|
$$
$$
=\Lambda_{|b\rangle}^{|a\rangle}(V),
$$

from which it follows that

$$
\Lambda_{|b\rangle}^{|a\rangle}(V)\in\mathcal{F}[X,\Lambda_{n_b}^{n_a}(V)].
$$

$\square$

Next, we show that every unitary operator U on $\mathbb{H}^{\otimes n}$ can be written as a product of suitably embedded unitary operators V on $\mathbb{H}$. In order to do so we first define the necessary embedding operators T.

Definition 5.18 Let $n,x,y\in\mathbb{N}_0$ with $0\leq x<y<2^n$ as well as V be a unitary operator on $\mathbb{H}$ with the matrix representation

$$
V=\begin{pmatrix}v_{00}&v_{01}\\v_{10}&v_{11}\end{pmatrix}
$$

in the basis $\{|0\rangle,|1\rangle\}$. With the help of the computational basis in $\mathbb{H}^{\otimes n}$ we define the operator $T_{|x\rangle|y\rangle}(V):\mathbb{H}^{\otimes n}\to\mathbb{H}^{\otimes n}$ as follows

$$T_{|x\rangle|y\rangle}(V)$$

$$:= \sum_{\substack{z=0 \\ z \neq x,y}}^{2^n-1} |z\rangle\langle z| + v_{00}|x\rangle\langle x| + v_{01}|x\rangle\langle y| + v_{10}|y\rangle\langle x| + v_{11}|y\rangle\langle y| \qquad (5.44)$$

$$= \mathbf{1}^{\otimes n} + (v_{00}-1)|x\rangle\langle x| + v_{01}|x\rangle\langle y| + v_{10}|y\rangle\langle x| + (v_{11}-1)|y\rangle\langle y|.$$

In the computational basis $T_{|x\rangle|y\rangle}(V)$ has the matrix representation

$$(5.45)$$

where only the entries different from zero have been shown along with the bra-and ket-vectors to indicate the rows and columns of these non-zero elements in the matrix.

Exercise 5.62 Let n, x, y and $T_{|x\rangle|y\rangle}(\cdot)$ be as defined in Definition 5.18. Show that for unitary operators V, W on $\mathbb{H}$ the following holds

$$T_{|x\rangle|y\rangle}(V) T_{|x\rangle|y\rangle}(W) = T_{|x\rangle|y\rangle}(VW) \qquad (5.46)$$

$$T_{|x\rangle|y\rangle}(V)^* = T_{|x\rangle|y\rangle}(V^*) \qquad (5.47)$$

$$T_{|x\rangle|y\rangle}(V) T_{|x\rangle|y\rangle}(V)^* = \mathbf{1}^{\otimes n}, \qquad (5.48)$$

that is, $T_{|x\rangle|y\rangle}(V)$ is unitary.

For a solution see Solution 5.62.

Before we come to the aforementioned representation of a unitary operator U on $\mathbb{H}^{\otimes n}$ with the help of suitably embedded operators on $\mathbb{H}$, we prove the following helpful intermediate result.

Lemma 5.19 *Let $n \in \mathbb{N}$ and $N = 2^n - 1$ and $U \in \mathcal{U}(\mathbb{H}^{\otimes n})$. Then there exist $V^{(0)}, \ldots, V^{(N-1)} \in \mathcal{U}(\mathbb{H})$ such that the operator*

$$U^{(N)} := UT_{|N-1\rangle|N\rangle}\left(V^{(N-1)}\right) \ldots T_{|0\rangle|N\rangle}\left(V^{(0)}\right) \quad \in \mathcal{U}(\mathbb{H}^{\otimes n})$$

in the computational basis of $\mathbb{H}^{\otimes n}$ has the matrix

$$U^{(N)} = \begin{pmatrix} & & 0 \\ A^{(N)} & & \vdots \\ & & 0 \\ 0 & \cdots & 0\ 1 \end{pmatrix}, \tag{5.49}$$

where $A^{(N)}$ is a unitary $N \times N$ matrix.

Proof In general one has

$$UT_{|x\rangle|y\rangle}(V) = \left(\sum_{a,b=0}^{N} U_{ab}|a\rangle\langle b|\right)$$

$$\left(\sum_{\substack{z=0 \\ z \neq x,y}}^{N} |z\rangle\langle z| + v_{00}|x\rangle\langle x| + v_{01}|x\rangle\langle y| + v_{10}|y\rangle\langle x| + v_{11}|y\rangle\langle y|\right)$$

$$= \sum_{\substack{a,z=0 \\ z \neq x,y}}^{N} U_{az}|a\rangle\langle z| \tag{5.50}$$

$$+ \sum_{a=0}^{N}\left(U_{ax}v_{00}|a\rangle\langle x| + U_{ax}v_{01}|a\rangle\langle y| + U_{ay}v_{10}|a\rangle\langle x| + U_{ay}v_{11}|a\rangle\langle y|\right).$$

We now consider $x = N - j$ and $y = N$ and define

$$\widetilde{U}^{(0)} := U \tag{5.51}$$

$$\widetilde{U}^{(j)} := \widetilde{U}^{(j-1)}T_{|N-j\rangle|N\rangle}\left(V^{(N-j)}\right),$$

where we shall make a suitable choice for the operators $V^{(N-j)}$. For this we consider the Nth row of the matrix $\widetilde{U}^{(j)}$ for which it follows from (5.50) that

$$\widetilde{U}^{(j)} = \widetilde{U}^{(j-1)} T_{|N-j\rangle|N\rangle}\left(V^{(N-j)}\right)$$

$$= \sum_{a=0}^{N} \sum_{\substack{b=0 \\ b \neq N-j}}^{N-1} \widetilde{U}_{ab}^{(j-1)} |a\rangle\langle b|$$

$$+ \sum_{a=0}^{N} \left(\widetilde{U}_{aN-j}^{(j-1)} v_{00}^{(N-j)} + \widetilde{U}_{aN}^{(j-1)} v_{10}^{(N-j)}\right) |a\rangle\langle N-j|$$

$$+ \sum_{a=0}^{N} \left(\widetilde{U}_{aN-j}^{(j-1)} v_{01}^{(N-j)} + \widetilde{U}_{aN}^{(j-1)} v_{11}^{(N-j)}\right) |a\rangle\langle N|.$$

Hence, the matrix elements of $\widetilde{U}^{(j)}$ are

$$\begin{aligned}
\widetilde{U}_{Nb}^{(j)} &= \widetilde{U}_{Nb}^{(j-1)} && \text{if } b \notin \{N-j, N\} \\
\widetilde{U}_{NN-j}^{(j)} &= \widetilde{U}_{NN-j}^{(j-1)} v_{00}^{(N-j)} + \widetilde{U}_{NN}^{(j-1)} v_{10}^{(N-j)} && (5.52) \\
\widetilde{U}_{NN}^{(j)} &= \widetilde{U}_{NN-j}^{(j-1)} v_{01}^{(N-j)} + \widetilde{U}_{NN}^{(j-1)} v_{11}^{(N-j)}.
\end{aligned}$$

To choose $V^{(N-j)}$ suitably, we distinguish two cases:

1. If $\widetilde{U}_{NN-j}^{(j-1)}$ and $\widetilde{U}_{NN}^{(j-1)}$ both vanish, then due to (5.52) also $\widetilde{U}_{NN-j}^{(j)}$ and $\widetilde{U}_{NN}^{(j)}$ vanish, and we choose $V^{(N-j)} = \mathbf{1}$.
2. Otherwise, we set

$$V^{(N-j)} = \frac{1}{\sqrt{\left|\widetilde{U}_{NN-j}^{(j-1)}\right|^2 + \left|\widetilde{U}_{NN}^{(j-1)}\right|^2}} \begin{pmatrix} \widetilde{U}_{NN}^{(j-1)} & \overline{\widetilde{U}_{NN-j}^{(j-1)}} \\ -\widetilde{U}_{NN-j}^{(j-1)} & \widetilde{U}_{NN}^{(j-1)} \end{pmatrix}.$$

Then $V^{(N-j)}$ is unitary, and one has

$$\begin{aligned}
\widetilde{U}_{NN-j}^{(j)} &= \widetilde{U}_{NN-j}^{(j-1)} v_{00}^{(N-j)} + \widetilde{U}_{NN}^{(j-1)} v_{10}^{(N-j)} = 0 \\
\widetilde{U}_{NN}^{(j)} &= \widetilde{U}_{NN-j}^{(j-1)} v_{01}^{(N-j)} + \widetilde{U}_{NN}^{(j-1)} v_{11}^{(N-j)} = \sqrt{\left|\widetilde{U}_{NN-j}^{(j-1)}\right|^2 + \left|\widetilde{U}_{NN}^{(j-1)}\right|^2}.
\end{aligned} \qquad (5.53)$$

Starting with $j = 1$ in either case we thus obtain successively

$$\widetilde{U}_{NN-j}^{(j)} = 0 \qquad \text{for } j \in \{1, \ldots, N\}. \qquad (5.54)$$

For $b \in \{0, \ldots, N-j-1\}$ we have, because of (5.52) and (5.51), that $\widetilde{U}_{Nb}^{(j)} = U_{Nb}$. With this and (5.53) it follows that

$$\widetilde{U}_{NN}^{(j)} = \sqrt{\sum_{l=0}^{j}\left|\widetilde{U}_{NN-l}^{(0)}\right|^2} = \sqrt{\sum_{l=0}^{j}|U_{NN-l}|^2}.$$

Since U is assumed to be unitary, the squares of the absolute values in each row have to add up to one. Thus, we obtain finally

$$\widetilde{U}_{NN}^{(N)} = \sqrt{\sum_{l=0}^{N}|U_{NN-l}|^2} = 1. \tag{5.55}$$

Because of (5.54), (5.55) and (5.52) then $\widetilde{U}^{(N)}$ has the matrix representation

$$\widetilde{U}^{(N)} = \begin{pmatrix} & & & b_0 \\ & A^{(N)} & & \vdots \\ & & & b_{N-1} \\ 0 & \cdots & 0 & 1 \end{pmatrix}.$$

Since $\widetilde{U}^{(N)}$ as a product of unitary operators has to be unitary, it follows that

$$\widetilde{U}^{(N)}\widetilde{U}^{(N)*} = \mathbf{1}^{\otimes n}$$

has to hold. This implies $b_0 = \cdots = b_{N-1} = 0$, and thus that $A^{(N)}$ is a $2^n - 1 = N$-dimensional unitary matrix. Consequently, the matrix representation of $U^{(N)} = \widetilde{U}^{(N)}$ is of the claimed form (5.49). $\qquad\square$

The claim of the following theorem appears to have been utilized in the context of quantum computing for the first time in [71].

Theorem 5.20 *Let $n \in \mathbb{N}$ and U be a unitary operator on $\mathbb{H}^{\otimes n}$. Then there exist $2^{n-1}(2^n - 1)$ unitary operators $W^{(k,k-j)}$ on $\mathbb{H}$ with $k \in \{1,\ldots,2^n - 1\}$ and $j \in \{1,\ldots,k\}$ such that*

$$U = \prod_{k=1}^{2^n-1}\left(\prod_{j=1}^{k} T_{|j-1\rangle|k\rangle}\left(W^{(k,k-j)}\right)\right) \tag{5.56}$$

and thus

$$U \in \mathcal{F}[T_{|x\rangle|y\rangle}(V)] \tag{5.57}$$

for a suitably chosen V.

Proof Let $N = 2^n - 1$. From Lemma 5.19 we know that there exist unitary operators $V^{(N,j)} \in \mathcal{U}(\mathbb{H})$ such that

$$U^{(N)} = U \prod_{j=N}^{1} T_{|j-1\rangle|N\rangle} \left(V^{(N,j-1)} \right)$$

has the matrix representation

$$U^{(N)} = \begin{pmatrix} & & & 0 \\ A^{(N)} & & \vdots \\ & & & 0 \\ 0 & \cdots & 0 & 1 \end{pmatrix}.$$

We can now multiply $U^{(N)}$ with $T_{|N-2\rangle|N-1\rangle} \left(V^{(N-1,N-2)} \right) \ldots T_{|0\rangle|N-1\rangle} \left(V^{(N-1,0)} \right)$ from the right and choose the $V^{(N-1,N-2)}, \ldots, V^{(N-1,0)}$ according to the construction in the proof of Lemma 5.19 such that

$$U^{(N-1)} = U^{(N)} \prod_{j=N-1}^{1} T_{|j-1\rangle|N-1\rangle} \left(V^{(N-1,j-1)} \right)$$

has the matrix representation

$$U^{(N-1)} = \begin{pmatrix} & & & 0 & 0 \\ A^{(N-1)} & & \vdots & \vdots \\ & & & 0 & 0 \\ 0 & \cdots & 0 & 1 & 0 \\ 0 & \cdots & 0 & 0 & 1 \end{pmatrix}, \tag{5.58}$$

where $A^{(N-1)}$ is a unitary $N-1 \times N-1$ matrix. In arriving at (5.58) it has also been used that—as can be seen from the matrix representation (5.45) of the $T_{|l\rangle|N-1\rangle}(\cdot)$— the multiplication of these with $U^{(N)}$ leaves the last row and column of $U^{(N)}$ unchanged. We continue these multiplications and, starting with $l = N$ and counting down until $l = 2$, successively build the sequence of operators

$$U^{(l)} = U^{(l+1)} \prod_{j=l}^{1} T_{|j-1\rangle|l\rangle} \left(V^{(l,j-1)} \right),$$

which have the matrix representations

$$U^{(l)} = \begin{pmatrix} & & 0 \cdots 0 \\ & A^{(l)} & \vdots & \vdots \\ 0 & \cdots & 0 & 1 \\ \vdots & & \vdots & & \ddots \\ 0 & \cdots & & 0 & 1 \end{pmatrix}.$$

(5.59)

The $A^{(l)}$ in (5.59) are always unitary $l \times l$ matrices. Consequently, $A^{(2)}$ in $U^{(2)}$ is a 2×2 matrix. In order to calculate $U^{(1)}$ we thus set $V^{(1,0)} = A^{(2)*}$. Then we have

$$\mathbf{1}^{\otimes n} = U^{(1)} = U^{(2)} T_{|0\rangle|1\rangle} \left(V^{(1,0)} \right) = \cdots = U \prod_{l=N}^{1} \left(\prod_{j=l}^{1} T_{|j-1\rangle|l\rangle} \left(V^{(l,j-1)} \right) \right). \quad (5.60)$$

Solving (5.60) for U yields

$$
\begin{aligned}
U &= \left(\prod_{l=N}^{1} \left(\prod_{j=l}^{1} T_{|j-1\rangle|l\rangle} \left(V^{(l,j-1)} \right) \right) \right)^{-1} = \left(\prod_{l=N}^{1} \left(\prod_{j=l}^{1} T_{|j-1\rangle|l\rangle} \left(V^{(l,j-1)} \right) \right) \right)^{*} \\
&= \prod_{l=1}^{N} \left(\prod_{j=1}^{l} T_{|j-1\rangle|l\rangle} \left(V^{(l,j-1)*} \right) \right),
\end{aligned}
$$

which is the representation claimed in (5.56) The number of factors is

$$n_F = \sum_{l=1}^{N} l = \frac{(N+1)N}{2} = \frac{2^n (2^n - 1)}{2} = 2^{n-1} (2^n - 1).$$

□

Example 5.21 As an example we consider the unitary operator $U \in \mathbb{H}^{\otimes 2}$ with the following matrix representation in the computational basis

$$U = \frac{1}{2} \begin{pmatrix} 1 & 1 & 1 & 1 \\ 1 & i & -1 & -i \\ 1 & -1 & 1 & -1 \\ 1 & -i & -1 & i \end{pmatrix}.$$

This yields the following operators $W^{(k,k-j)}$:

$$W^{(3,2)} = \begin{pmatrix} -\frac{i}{\sqrt{2}} & \frac{1}{\sqrt{2}} \\ -\frac{1}{\sqrt{2}} & \frac{i}{\sqrt{2}} \end{pmatrix}, \qquad W^{(3,1)} = \begin{pmatrix} \sqrt{\frac{2}{3}} & -\frac{i}{\sqrt{3}} \\ -\frac{i}{\sqrt{3}} & \sqrt{\frac{2}{3}} \end{pmatrix}, \qquad W^{(3,0)} = \begin{pmatrix} \frac{\sqrt{3}}{2} & -\frac{1}{2} \\ \frac{1}{2} & \frac{\sqrt{3}}{2} \end{pmatrix}$$

$$W^{(2,1)} = \begin{pmatrix} -\frac{i+1}{4}\sqrt{3} & \frac{3-i}{4} \\ -\frac{3+i}{4} & \frac{i-1}{4}\sqrt{3} \end{pmatrix}, \qquad W^{(2,0)} = \begin{pmatrix} \sqrt{\frac{2}{3}} & -\frac{1}{\sqrt{3}} \\ \frac{1}{\sqrt{3}} & \sqrt{\frac{2}{3}} \end{pmatrix}$$

$$W^{(1,0)} = \begin{pmatrix} \frac{1}{\sqrt{2}} & \frac{1}{\sqrt{2}} \\ -\frac{i}{\sqrt{2}} & \frac{i}{\sqrt{2}} \end{pmatrix}.$$

The claim (5.56) can then be verified by explicit calculation, which is left to the reader.

Next, we show that any $T_{|x\rangle|y\rangle}(V)$ can be built from gates of the form $\Lambda_{|b\rangle}^{|a\rangle}(W)$. For this we require the construction of a sequence, which is based on the so-called GRAY-Code. This is a sequence of vectors in $\mathbb{H}^{\otimes n}$ the consecutive elements of which differ only in one qubit. We formalize this in the following definition.

Definition 5.22 Let $n \in \mathbb{N}$ and $x, y \in \mathbb{N}_0$ with $0 \le x < y < 2^n$. Moreover, let $|x\rangle$ and $|y\rangle$ be the corresponding vectors of the computational basis of $\mathbb{H}^{\otimes n}$. A **GRAY-coded** transition from $|x\rangle$ to $|y\rangle$ is defined as a finite sequence of vectors $|g^0\rangle, \ldots, |g^{K+1}\rangle$ of the computational basis having the following properties.

(i)

$$|g^0\rangle = |x\rangle$$
$$|g^{K+1}\rangle = |y\rangle.$$

(ii) For all $l \in \{1, \ldots, K+1\}$ there exist $n_{a^l}, n_{b^l} \in \mathbb{N}_0$ with $n_{a^l} + n_{b^l} + 1 = n$ and $n_{b^l} \ne n_{b^j}$ for all $l \ne j$ such that

$$|g^l\rangle = \mathbf{1}^{\otimes n_{a^l}} \otimes X \otimes \mathbf{1}^{\otimes n_{b^l}} |g^{l-1}\rangle$$

and

$$
\begin{aligned}
(g^K)_{n_{bK+1}} &= 0 \\
(g^{K+1})_{n_{bK+1}} &= 1.
\end{aligned}
\tag{5.61}
$$

With the help of $|g^{l-1}\rangle$ we also define for $l \in \{1, \ldots, K+1\}$

$$
\begin{aligned}
|a^l\rangle &:= |g_{n-1}^{l-1} \cdots g_{n_{b^l}+1}^{l-1}\rangle \in \mathbb{H}^{\otimes n_{a^l}} \\
|b^l\rangle &:= |g_{n_{b^l}-1}^{l-1} \cdots g_0^{l-1}\rangle \in \mathbb{H}^{\otimes n_{b^l}}.
\end{aligned}
\tag{5.62}
$$

In the GRAY-coded transition thus defined two consecutive elements $|g^{l-1}\rangle$ and $|g^l\rangle$ only differ in the qubit in the factor-space $\mathbb{H}_{n_{b^l}}$ (see (3.21)) of $\mathbb{H}^{\otimes n}$

$$
\begin{aligned}
|g^l\rangle &= \mathbf{1}^{\otimes n_{a^l}} \otimes X \otimes \mathbf{1}^{\otimes n_{b^l}} |g^{l-1}\rangle \\
&= |(g^{l-1})_{n-1} \cdots (g^{l-1})_{n_{b^l}+1} \neg (g^{l-1})_{n_{b^l}} (g^{l-1})_{n_{b^l}-1} \cdots (g^{l-1})_0\rangle .
\end{aligned}
$$

Moreover, from $n_{b^l} \neq n_{b^j}$ for all $l \neq j$ it follows that $|g^{l+k}\rangle \neq |g^l\rangle$ if $k \geq 1$.

GRAY-coded transitions are not unique. Between two vectors $|x\rangle$ and $|y\rangle$ there can be several such transitions.

Example 5.23 We consider the case $n = 3$ and $x = 1$ and $y = 6$. Then one possible GRAY-coded transition is

$$
\begin{aligned}
|x\rangle &= |1\rangle^3 = |001\rangle \\
|g^1\rangle &= |5\rangle^3 = |101\rangle \\
|g^2\rangle &= |4\rangle^3 = |100\rangle \\
|y\rangle &= |6\rangle^3 = |110\rangle ,
\end{aligned}
\tag{5.63}
$$

that is, here one has $n_{b^1} = 2$, $n_{b^2} = 0$ and $n_{b^3} = 1$.

An alternative GRAY-coded transition is

$$
\begin{aligned}
|x\rangle &= |1\rangle^3 = |001\rangle \\
|g^1\rangle &= |0\rangle^3 = |000\rangle \\
|g^2\rangle &= |4\rangle^3 = |100\rangle \\
|y\rangle &= |6\rangle^3 = |110\rangle ,
\end{aligned}
$$

where one has $n_{b^1} = 0$, $n_{b^2} = 2$ and $n_{b^3} = 1$.

The following transition

$$
\begin{aligned}
|x\rangle &= |1\rangle^3 = |001\rangle \\
|g^1\rangle &= |5\rangle^3 = |101\rangle \\
|g^2\rangle &= |7\rangle^3 = |111\rangle \\
|y\rangle &= |7\rangle^3 = |110\rangle
\end{aligned}
$$

also changes in only one qubit in the step from from $|g^{l-1}\rangle$ to $|g^l\rangle$, but the step from $|g^2\rangle$ to $|y\rangle$ does not satisfy the condition (5.61). As we shall see, this condition simplifies the generation of the $T_{|x\rangle|y\rangle}(V)$ with the help of controlled gates of the form $\Lambda_{|b\rangle}^{|a\rangle}(V)$.

In Exercise 5.63 we show that for any $0 \leq x < y < 2^n$ there always exists a GRAY-coded transition from $|x\rangle$ to $|y\rangle$.

Exercise 5.63 Let $n \in \mathbb{N}$ and $x, y \in \mathbb{N}_0$ with $0 \leq x < y < 2^n$. Show that there exists a GRAY-coded transition from $|x\rangle$ to $|y\rangle$.

For a solution see Solution 5.63.

Before we can prove the universality of the set $\mathrm{U} = \{M, D_{\hat{y}}, D_{\hat{z}}, \Lambda^1(X)\}$ of gates we still need the following intermediate result.

Theorem 5.24 *Let $n \in \mathbb{N}$ and $x, y \in \mathbb{N}_0$ with $0 \leq x < y < 2^n$ and let $|x\rangle$ and $|y\rangle$ be the corresponding vectors of the computational basis in $\mathbb{H}^{\otimes n}$. Moreover, let V be a unitary operator on $\mathbb{H}$. Then every GRAY-coded transition $|g^l\rangle$ with $l \in \{0, \ldots, K+1\}$ from $|x\rangle$ to $|y\rangle$ satisfies*

$$\Lambda_{|b^l\rangle}^{|a^l\rangle}(X) = \sum_{\substack{z=0 \\ z \neq g^{l-1}, g^l}} |z\rangle\langle z| + |g^{l-1}\rangle\langle g^l| + |g^l\rangle\langle g^{l-1}| \qquad (5.64)$$

$$T_{|g^K\rangle|y\rangle}(V) = \Lambda_{|b^{K+1}\rangle}^{|a^{K+1}\rangle}(V) \qquad (5.65)$$

$$T_{|g^{l-1}\rangle|y\rangle}(V) = \Lambda_{|b^l\rangle}^{|a^l\rangle}(X) \, T_{|g^l\rangle|y\rangle}(V) \, \Lambda_{|b^l\rangle}^{|a^l\rangle}(X) \qquad (5.66)$$

$$T_{|x\rangle|y\rangle}(V) = \left(\prod_{l=1}^{K} \Lambda_{|b^l\rangle}^{|a^l\rangle}(X) \right) \Lambda_{|b^{K+1}\rangle}^{|a^{K+1}\rangle}(V) \left(\prod_{j=K}^{1} \Lambda_{|b^j\rangle}^{|a^j\rangle}(X) \right). \qquad (5.67)$$

Proof We start with the proof of (5.64). Per Definition 5.10 we have

$$\Lambda_{|b^l\rangle}^{|a^l\rangle}(X) = \mathbf{1}^{\otimes n_{a^l} + n_{b^l} + 1} + |a^l\rangle\langle a^l| \otimes (X - \mathbf{1}) \otimes |b^l\rangle\langle b^l|.$$

With $n = n_{a^l} + n_{b^l} + 1$ and

$$X - \mathbf{1} = |0\rangle\langle 1| + |1\rangle\langle 0| - |0\rangle\langle 0| - |1\rangle\langle 1|$$

one has

$$\Lambda_{|b^l\rangle}^{|a^l\rangle}(X) = \mathbf{1}^{\otimes n} + \underbrace{|a^l\rangle\langle a^l| \otimes \left(|0\rangle\langle 1| + |1\rangle\langle 0|\right) \otimes |b^l\rangle\langle b^l|}_{= |g^{l-1}\rangle\langle g^l| + |g^l\rangle\langle g^{l-1}|}$$

$$\underbrace{- |a^l\rangle\langle a^l| \otimes \left(|0\rangle\langle 0| + |1\rangle\langle 1|\right) \otimes |b^l\rangle\langle b^l|}_{= -|g^{l-1}\rangle\langle g^{l-1}| - |g^l\rangle\langle g^l|}$$

$$= \sum_{\substack{z=0 \\ z \neq g^{l-1}, g^l}}^{2^n - 1} |z\rangle\langle z| + |g^{l-1}\rangle\langle g^l| + |g^l\rangle\langle g^{l-1}|.$$

For the proof of (5.65) one exploits that from (5.61)–(5.62) it follows that

$$|g^K\rangle = |a^{K+1}\rangle \otimes |0\rangle \otimes |b^{K+1}\rangle$$
$$|g^{K+1}\rangle = |a^{K+1}\rangle \otimes |1\rangle \otimes |b^{K+1}\rangle = |y\rangle.$$

With Definition 5.18 we then find

$$T_{|g^K\rangle|y\rangle}(V) = \mathbf{1}^{\otimes n} + (v_{00} - 1)|g^K\rangle\langle g^K| + v_{01}|g^K\rangle\langle y| + v_{10}|y\rangle\langle g^K| + (v_{11} - 1)|y\rangle\langle y|$$

$$= \mathbf{1}^{\otimes n}$$
$$+ (v_{00} - 1)\left(|a^{K+1}\rangle \otimes |0\rangle \otimes |b^{K+1}\rangle\right)\left(\langle a^{K+1}| \otimes \langle 0| \otimes \langle b^{K+1}|\right)$$
$$+ v_{01}\left(|a^{K+1}\rangle \otimes |0\rangle \otimes |b^{K+1}\rangle\right)\left(\langle a^{K+1}| \otimes \langle 1| \otimes \langle b^{K+1}|\right)$$
$$+ v_{10}\left(|a^{K+1}\rangle \otimes |1\rangle \otimes |b^{K+1}\rangle\right)\left(\langle a^{K+1}| \otimes \langle 0| \otimes \langle b^{K+1}|\right)$$
$$+ (v_{11} - 1)\left(|a^{K+1}\rangle \otimes |1\rangle \otimes |b^{K+1}\rangle\right)\left(\langle a^{K+1}| \otimes \langle 1| \otimes \langle b^{K+1}|\right)$$

$$\underbrace{=}_{(3.36)} \mathbf{1}^{\otimes n}$$
$$+ (v_{00} - 1)|a^{K+1}\rangle\langle a^{K+1}| \otimes |0\rangle\langle 0| \otimes |b^{K+1}\rangle\langle b^{K+1}|$$
$$+ v_{01}|a^{K+1}\rangle\langle a^{K+1}| \otimes |0\rangle\langle 1| \otimes |b^{K+1}\rangle\langle b^{K+1}|$$
$$+ v_{10}|a^{K+1}\rangle\langle a^{K+1}| \otimes |1\rangle\langle 0| \otimes |b^{K+1}\rangle\langle b^{K+1}|$$
$$+ (v_{11} - 1)|a^{K+1}\rangle\langle a^{K+1}| \otimes |1\rangle\langle 1| \otimes |b^{K+1}\rangle\langle b^{K+1}|$$

$$= \mathbf{1}^{\otimes n}$$
$$+ |a^{K+1}\rangle\langle a^{K+1}| \otimes \Big((v_{00} - 1)|0\rangle\langle 0| + v_{01}|0\rangle\langle 1|$$
$$+ v_{10}|1\rangle\langle 0| + (v_{11} - 1)|1\rangle\langle 1|\Big) \otimes |b^{K+1}\rangle\langle b^{K+1}|$$

$$= \mathbf{1}^{\otimes n} + |a^{K+1}\rangle\langle a^{K+1}| \otimes (V - \mathbf{1}) \otimes |b^{K+1}\rangle\langle b^{K+1}|$$

$$= \Lambda_{|b^{K+1}\rangle}^{|a^{K+1}\rangle}(V).$$

In order to prove (5.66), we use (5.64), and with Definition 5.18 one then obtains

$$T_{|g^l\rangle|y\rangle}(V)\Lambda_{|b^l\rangle}^{|a^l\rangle}(X)$$

$$= \left(\sum_{\substack{z=0 \\ z\neq g^l, y}} |z\rangle\langle z| + v_{00}|g^l\rangle\langle g^l| + v_{01}|g^l\rangle\langle y| + v_{10}|y\rangle\langle g^l| + v_{11}|y\rangle\langle y| \right)$$

$$\times \left(\sum_{\substack{r=0 \\ r\neq g^{l-1}, g^l}} |r\rangle\langle r| + |g^{l-1}\rangle\langle g^l| + |g^l\rangle\langle g^{l-1}| \right)$$

$$= \sum_{\substack{z=0 \\ z\neq g^{l-1}, g^l, y}} |z\rangle\langle z| + |g^{l-1}\rangle\langle g^l|$$

$$+ v_{00}|g^l\rangle\langle g^{l-1}| + v_{01}|g^l\rangle\langle y| + v_{10}|y\rangle\langle g^{l-1}| + v_{11}|y\rangle\langle y|$$

and thus

$$\Lambda_{|b^l\rangle}^{|a^l\rangle}(X)\, T_{|g^l\rangle|y\rangle}(V)\, \Lambda_{|b^l\rangle}^{|a^l\rangle}(X)$$

$$= \left(\sum_{\substack{r=0 \\ r\neq g^{l-1}, g^l}} |r\rangle\langle r| + |g^{l-1}\rangle\langle g^l| + |g^l\rangle\langle g^{l-1}| \right)$$

$$\times \left(\sum_{\substack{z=0 \\ z\neq g^{l-1}, g^l, y}} |z\rangle\langle z| + |g^{l-1}\rangle\langle g^l| \right.$$

$$\left. + v_{00}|g^l\rangle\langle g^{l-1}| + v_{01}|g^l\rangle\langle y| + v_{10}|y\rangle\langle g^{l-1}| + v_{11}|y\rangle\langle y| \right)$$

$$= \sum_{\substack{z=0 \\ z\neq g^{l-1}, y}} |z\rangle\langle z| + v_{00}|g^{l-1}\rangle\langle g^{l-1}| + v_{01}|g^{l-1}\rangle\langle y| + v_{10}|y\rangle\langle g^{l-1}| + v_{11}|y\rangle\langle y|$$

$$\underbrace{=}_{(5.44)} T_{|g^{l-1}\rangle|y\rangle}(V) .$$

Lastly, we turn to the proof of (5.67). This is accomplished with the help of (5.65) and (5.66) as follows:

$$\prod_{l=1}^{K} \Lambda_{|b^l\rangle}^{|a^l\rangle}(X)\, \Lambda_{|b^{K+1}\rangle}^{|a^{K+1}\rangle}(V) \prod_{j=K}^{1} \Lambda_{|b^j\rangle}^{|a^j\rangle}(X) \underbrace{=}_{(5.65)} \prod_{l=1}^{K} \Lambda_{|b^l\rangle}^{|a^l\rangle}(X)\, T_{|g^K\rangle|y\rangle}(V) \prod_{j=K}^{1} \Lambda_{|b^j\rangle}^{|a^j\rangle}(X)$$

$$\underbrace{=}_{(5.66)} \prod_{l=1}^{K-1} \Lambda_{|b^l\rangle}^{|a^l\rangle}(X)\, T_{|g^{K-1}\rangle|y\rangle}(V) \prod_{j=K-1}^{1} \Lambda_{|b^j\rangle}^{|a^j\rangle}(X)$$

$$\vdots$$

$$= \Lambda_{|b^1\rangle}^{|a^1\rangle}(X)\, T_{|g^1\rangle|y\rangle}(V)\, \Lambda_{|b^1\rangle}^{|a^1\rangle}(X)$$

$$\underbrace{=}_{(5.66)} T_{|g^0\rangle|y\rangle}(V) = T_{|x\rangle|y\rangle}(V) ,$$

where we used in the last equation that $|x\rangle = |g^0\rangle$ holds. $\qquad\square$

Example 5.25 As in Example 5.23 we consider the case $n = 3$ and $x = 1$ and $y = 3$ with the GRAY-coded transition (5.63), that is, $K = 2$ and

$$\begin{aligned}
|x\rangle = |g^0\rangle = |1\rangle^3 = |001\rangle &= |0\rangle \otimes |01\rangle \\
|g^1\rangle = |5\rangle^3 = |101\rangle &= |1\rangle \otimes \underbrace{|01\rangle}_{=|b^1\rangle}
\end{aligned}$$

$$|g^2\rangle = |4\rangle^3 = |100\rangle = \underbrace{|10\rangle}_{=|a^2\rangle} \otimes |0\rangle$$

$$|y\rangle = |g^3\rangle = |6\rangle^3 = |110\rangle = \underbrace{|1\rangle}_{=|a^3\rangle} \otimes |1\rangle \otimes \underbrace{|0\rangle}_{=|b^3\rangle} .$$

Then one has at first

$$T_{|1\rangle|6\rangle}(V) = \begin{array}{c} \\ |0\rangle \\ |1\rangle \\ |2\rangle \\ |3\rangle \\ |4\rangle \\ |5\rangle \\ |6\rangle \\ |7\rangle \end{array} \begin{array}{c} \langle 0| \ \ \langle 1| \ \ \langle 2| \ \ \langle 3| \ \ \langle 4| \ \ \langle 5| \ \ \langle 6| \ \ \langle 7| \\ \left(\begin{array}{cccccccc} 1 & & & & & & & \\ & v_{00} & & & & & v_{01} & \\ & & 1 & & & & & \\ & & & 1 & & & & \\ & & & & 1 & & & \\ & & & & & 1 & & \\ & v_{10} & & & & & v_{11} & \\ & & & & & & & 1 \end{array} \right) \end{array},$$

where—in order to improve readability—we have indicated once more, by showing $|a\rangle = |a\rangle^3$ in the rows and $\langle b| = {}^3\langle b|$ in the columns, to which $|a\rangle\langle b|$ the matrix elements belong. Furthermore, one has

$$\begin{aligned}
\Lambda_{|b^1\rangle}(X) &= \mathbf{1}^{\otimes 3} + (X - 1) \otimes |0\rangle\langle 0| \otimes |1\rangle\langle 1| \\
&= \mathbf{1}^{\otimes 3} + \begin{pmatrix} -1 & 1 \\ 1 & -1 \end{pmatrix} \otimes \begin{pmatrix} 1 \\ 0 \end{pmatrix} (1\ 0) \otimes \begin{pmatrix} 0 \\ 1 \end{pmatrix} (0\ 1) \\
&\underset{(3.35)}{=} \mathbf{1}^{\otimes 3} + \begin{pmatrix} -1 & 1 \\ 1 & -1 \end{pmatrix} \otimes \begin{pmatrix} 1 & 0 \\ 0 & 0 \end{pmatrix} \otimes \begin{pmatrix} 0 & 0 \\ 0 & 1 \end{pmatrix} \\
&\underset{(3.35)}{=} \mathbf{1}^{\otimes 3} + \begin{pmatrix} -1 & 1 \\ 1 & -1 \end{pmatrix} \otimes \begin{pmatrix} 0 & 0 & 0 & 0 \\ 0 & 1 & 0 & 0 \\ 0 & 0 & 0 & 0 \\ 0 & 0 & 0 & 0 \end{pmatrix}
\end{aligned}$$

$$\underbrace{=}_{(3.35)} \mathbf{1}^{\otimes 3} +$$

	$\langle 0|$	$\langle 1|$	$\langle 2|$	$\langle 3|$	$\langle 4|$	$\langle 5|$	$\langle 6|$	$\langle 7|$
$|0\rangle$								
$|1\rangle$		-1				1		
$|2\rangle$								
$|3\rangle$								
$|4\rangle$								
$|5\rangle$		1				-1		
$|6\rangle$								
$|7\rangle$								

$$=$$

	$\langle 0|$	$\langle 1|$	$\langle 2|$	$\langle 3|$	$\langle 4|$	$\langle 5|$	$\langle 6|$	$\langle 7|$
$|0\rangle$	1							
$|1\rangle$						1		
$|2\rangle$			1					
$|3\rangle$				1				
$|4\rangle$					1			
$|5\rangle$		1						
$|6\rangle$							1	
$|7\rangle$								1

Analogously, one finds

$$\Lambda^{|a^2\rangle}(X) =$$

	$\langle 0|$	$\langle 1|$	$\langle 2|$	$\langle 3|$	$\langle 4|$	$\langle 5|$	$\langle 6|$	$\langle 7|$
$|0\rangle$	1							
$|1\rangle$		1						
$|2\rangle$			1					
$|3\rangle$				1				
$|4\rangle$						1		
$|5\rangle$					1			
$|6\rangle$							1	
$|7\rangle$								1

$$\Lambda^{|a^3\rangle}_{|b^3\rangle}(V) =$$

	$\langle 0|$	$\langle 1|$	$\langle 2|$	$\langle 3|$	$\langle 4|$	$\langle 5|$	$\langle 6|$	$\langle 7|$
$|0\rangle$	1							
$|1\rangle$		1						
$|2\rangle$			1					
$|3\rangle$				1				
$|4\rangle$					v_{00}		v_{01}	
$|5\rangle$						1		
$|6\rangle$					v_{10}		v_{11}	
$|7\rangle$								1

The reader is invited to verify (5.67) for

$$T_{|1\rangle|6\rangle}(V) = \Lambda_{|b^1\rangle}(X)\,\Lambda^{|a^2\rangle}(X)\,\Lambda^{|a^3\rangle}_{|b^3\rangle}(V)\,\Lambda^{|a^2\rangle}(X)\,\Lambda_{|b^1\rangle}(X)$$

by explicit multiplication.

With this latest result (5.67) from Theorem 5.24 we can now finally prove the previously announced universality of phase-multiplication, spin-rotation, and controlled NOT.

Theorem 5.26 *The set of quantum gates* $U = \{M, D_{\hat{y}}, D_{\hat{z}}, \Lambda^1(X)\}$ *is universal, that is, for any* $n \in \mathbb{N}$ *and* $U \in \mathcal{U}(\mathbb{H}^{\otimes n})$

$$ U \in \mathcal{F}[M, D_{\hat{y}}, D_{\hat{z}}, \Lambda^1(X)], $$

meaning that any quantum gate $U \in \mathcal{U}(\mathbb{H}^{\otimes n})$ *can be built with elements from* **U**.

Proof We prove the claim with the help of the preceding results

$$ U \underbrace{\in}_{(5.57)} \mathcal{F}[T_{|x\rangle|y\rangle}(V)] \tag{5.68} $$

$$ T_{|x\rangle|y\rangle}(V) \underbrace{\in}_{(5.67)} \mathcal{F}[\Lambda_{|b\rangle}^{|a\rangle}(V)] \tag{5.69} $$

$$ \Lambda_{|b\rangle}^{|a\rangle}(V) \underbrace{\in}_{(5.43)} \mathcal{F}[X, \Lambda_{n_b}^{n_a}(V)] $$

$$ \Lambda_{n_b}^{n_a}(V) \underbrace{\in}_{(5.36)} \mathcal{F}[\Lambda_1(X), \Lambda^1(X), \Lambda^{n_a+n_b}(V)] $$

$$ X, V, \Lambda_1(V), \Lambda^m(V) \underbrace{\in}_{(5.18),(5.19),(5.28)} \mathcal{F}[M, D_{\hat{y}}, D_{\hat{z}}, \Lambda^1(X)]. \tag{5.70} $$

With these one has

$$ U \underbrace{\in}_{(5.68)} \mathcal{F}[T_{|x\rangle|y\rangle}(V)] $$

$$ \underbrace{\in}_{(5.69)} \mathcal{F}[\mathcal{F}[\Lambda_{|b\rangle}^{|a\rangle}(V)]] $$

$$ \vdots $$

$$ \underbrace{\in}_{(5.70)} \mathcal{F}[\mathcal{F}[\mathcal{F}[\mathcal{F}[M, D_{\hat{y}}, D_{\hat{z}}, \Lambda^1(X)]]]]] $$

$$ \underbrace{\in}_{(5.14)} \mathcal{F}[M, D_{\hat{y}}, D_{\hat{z}}, \Lambda^1(X)]. $$

$\square$

The importance of Theorem 5.26 lies in the fact that in principle only the gates $M, D_{\hat{y}}, D_{\hat{z}}$ and $\Lambda^1(X)$ need to be implemented physically in sufficient numbers. All other gates can then be built by suitably combining them. Such constructions of general gates, however, are not necessarily the most efficient ones. Theorem 5.26 only states that the unary gates phase-multiplication and spin-rotation and the one binary gate controlled NOT suffice to build any gate of arbitrary dimension.

5.3   Quantum Circuits

Gates perform elementary transformations. In order to execute more sophisticated applications one has to connect a large number of such elementary gates. Such constructions are called circuits. Similarly to the classical case, we thus denote a combination of quantum gates in order to perform certain transformations on an input/output register $\mathbb{H}^{I/O} = \mathbb{H}^{\otimes n}$ as a quantum circuit. We will use the following three labels for the respective types of circuits:

Plain circuits   These are just compositions of quantum gates.

Circuits with ancillas   In these the input/output register $\mathbb{H}^{I/O}$ is first enlarged to a bigger composite system by adding an auxiliary quantum system (the 'ancilla'). Then a plain circuit is applied to the enlarged system and at the end the ancilla is discarded and only the original system is processed further. Unlike in the case of classical circuits, we have to ensure that before discarding the ancilla any possible entanglement with it has to be reversed, that is, disentangled.

Circuits with classical in/output and/or measurements   In these the quantum system is manipulated depending on classical input, possibly subject to a measurement and delivering partly classical output. In general, however, these are non-reversible transformations of the system.

In what follows we will give more formal definitions of the first two types of circuits. To define circuits with classical in- and output or measurements in a general and formal fashion is quite cumbersome and we shall refrain from doing so here.

Definition 5.27   (*Plain quantum circuit*) Let $n, L \in \mathbb{N}$ and $U_1, \ldots, U_L \in \mathcal{U}(\mathbb{H}^{\otimes n})$ be a set of quantum n-gates as defined in Definition 5.7. We call

$$\dot{U} = U_L \ldots U_1 \quad \in \mathcal{U}(\mathbb{H}^{\otimes n})$$

a **plain quantum circuit** constructed from the gates $U_1, \ldots, U_L$ and $L \in \mathbb{N}$ the **length or depth of the circuit** relative to the gate set $\{U_1, \ldots, U_L\}$.

When acting on a system in the state $\rho \in D(\mathbb{H})$ the plain circuit U transforms it to a new state $U \rho U^*$.

Note that with our Definition 5.7 of gates we need to specify the set of gates in order to have a meaningful notion of circuit length. This is because we may just as well declare $\tilde{U}_1 = U_2 U_1$ as a gate and thereby reduce the length. Hence, whenever we speak of the length of a circuit, it is understood with respect to a given gate set.

Because of (2.89) the action of a plain circuit U on a pure state $|\Psi\rangle$ is, of course, just $|\Psi\rangle \rightarrow U|\Psi\rangle$.

Before we can define quantum circuits with ancillas, we need to prove certain properties of that construction involving the input/output and auxiliary HILBERT space. We formalize this in the following theorem.

Theorem 5.28 *Let $\mathbb{H}^{I/O}$ and $\mathbb{H}^W$ be HILBERT spaces and let $|\omega_i\rangle, |\omega_f\rangle \in \mathbb{H}^W$ be such that $\|\omega_i\| = 1 = \|\omega_f\|$. Moreover, let $\hat{U} \in \mathcal{U}(\mathbb{H}^{I/O} \otimes \mathbb{H}^W)$ be such that for all $|\Psi\rangle \in \mathbb{H}^{I/O}$*

$$\hat{U}|\Psi \otimes \omega_i\rangle = (U|\Psi\rangle) \otimes |\omega_f\rangle \qquad (5.71)$$

and let $\rho_{\omega_i} = |\omega_i\rangle\langle\omega_i|$ be the density operator of the pure state $|\omega_i\rangle \in \mathbb{H}^W$.

Then $U \in \mathcal{U}(\mathbb{H}^{I/O})$ and we have for any density operator $\rho \in \mathrm{D}(\mathbb{H}^{I/O})$ that

$$\rho \otimes \rho_{\omega_i} \in \mathrm{D}(\mathbb{H}^{I/O} \otimes \mathbb{H}^W) \qquad (5.72)$$

as well as

$$\mathrm{tr}^W\left(\hat{U}(\rho \otimes \rho_{\omega_i})\hat{U}^*\right) = U\rho U^*, \qquad (5.73)$$

that is, U is unitary and in a state $\hat{U}(\rho \otimes \rho_{\omega_i})\hat{U}^*$ of the composite system the sub-system I/O is described by the state $U\rho U^*$.

Proof We show the unitarity of U first. Since $|\omega_i\rangle$ and $|\omega_f\rangle$ are normalized to 1, one has for arbitrary $|\Psi\rangle \in \mathbb{H}^{I/O}$

$$\|U\Psi\|^2 \underbrace{=}_{(2.5)} \langle U\Psi | U\Psi \rangle = \langle U\Psi | U\Psi \rangle \underbrace{\langle \omega_f | \omega_f \rangle}_{=1}$$

$$\underbrace{=}_{(3.4)} \langle (U\Psi) \otimes \omega_f | (U\Psi) \otimes \omega_f \rangle \underbrace{=}_{(5.71)} \langle \hat{U}|\Psi \otimes \omega_i\rangle | \hat{U}|\Psi \otimes \omega_i\rangle \rangle$$

$$\underbrace{=}_{(2.5)} \left\|\hat{U}|\Psi \otimes \omega_i\rangle\right\|^2 \underbrace{=}_{(2.37)} \|\Psi \otimes \omega_i\|^2 \underbrace{=}_{(2.5),(3.4)} \|\Psi\|^2 \underbrace{\|\omega_i\|}_{=1}$$

$$= \|\Psi\|^2 .$$

Hence, for all $|\Psi\rangle \in \mathbb{H}^{I/O}$ then $\|U\Psi\| = \|\Psi\|$ holds and from (2.37) it follows that $U \in \mathcal{U}(\mathbb{H}^{I/O})$.

Let now $\rho \in D(\mathbb{H}^{I/O})$ and $\rho_{\omega_i} = |\omega_i\rangle\langle\omega_i| \in D(\mathbb{H}^W)$. Then (4.1) implies (5.72). Moreover, we know from Theorem 2.24 that there exist $p_j \in [0,1]$ and an ONB $|\Psi_j\rangle$ in $\mathbb{H}^{I/O}$ such that

$$\rho = \sum_j p_j |\Psi_j\rangle\langle\Psi_j|. \tag{5.74}$$

Consequently, we have

$$
\begin{aligned}
\hat{U}(\rho \otimes \rho_{\omega_i})\hat{U}^* &\underset{(5.74)}{=} \hat{U}\left(\sum_j p_j |\Psi_j\rangle\langle\Psi_j| \otimes |\omega_i\rangle\langle\omega_i|\right)\hat{U}^* \\
&= \sum_j p_j \hat{U}\left(|\Psi_j\rangle\langle\Psi_j| \otimes |\omega_i\rangle\langle\omega_i|\right)\hat{U}^* \\
&\underset{(3.36)}{=} \sum_j p_j \hat{U}|\Psi_j \otimes \omega_i\rangle\langle\Psi_j \otimes \omega_i|\hat{U}^*. \tag{5.75}
\end{aligned}
$$

Using

$$
\begin{aligned}
\langle\Psi_j \otimes \omega_i|\hat{U}^* &\underset{(2.33)}{=} \langle\hat{U}(\Psi_j \otimes \omega_i)| \underset{(5.71)}{=} \langle(U\Psi_j) \otimes \omega_f| \underset{(3.15)}{=} \langle U\Psi_j| \otimes \langle\omega_f| \\
&\underset{(2.33)}{=} \langle\Psi_j|U^* \otimes \langle\omega_f| \tag{5.76}
\end{aligned}
$$

and (5.71) in (5.75) we obtain

$$
\begin{aligned}
\hat{U}(\rho \otimes \rho_{\omega_i})\hat{U}^* &\underset{(5.71),(5.76)}{=} \sum_j p_j \left(U|\Psi_j\rangle \otimes |\omega_f\rangle\right)\left(\langle\Psi_j|U^* \otimes \langle\omega_f|\right) \\
&\underset{(3.36)}{=} \sum_j p_j U|\Psi_j\rangle\langle\Psi_j|U^* \otimes |\omega_f\rangle\langle\omega_f| \\
&= U\left(\sum_j p_j |\Psi_j\rangle\langle\Psi_j|\right)U^* \otimes |\omega_f\rangle\langle\omega_f| \\
&\underset{(5.77)}{=} U\rho U^* \otimes |\omega_f\rangle\langle\omega_f| \tag{5.77}
\end{aligned}
$$

such that finally

$$
\begin{aligned}
\mathrm{tr}^W\left(\hat{U}(\rho \otimes \rho_{\omega_i})\hat{U}^*\right) &\underset{(5.77)}{=} \mathrm{tr}^W\left(U\rho U^* \otimes |\omega_f\rangle\langle\omega_f|\right) \\
&\underset{(3.57)}{=} \mathrm{tr}\left(|\omega_f\rangle\langle\omega_f|\right)U\rho U^* = U\rho U^*,
\end{aligned}
$$

where we used that

$$\mathrm{tr}\left(|\omega_f\rangle\langle\omega_f|\right)\underbrace{=}_{(2.57)}\sum_j\langle e_j|\omega_f\rangle\langle\omega_f|e_j\rangle\underbrace{=}_{(2.1)}\sum_j|\langle e_j|\omega_f\rangle|^2\underbrace{=}_{(2.12)}||\omega_f||^2=1$$

in the last equation.   □

Note that the left side of (5.73) is the environmental representation of a trace-preserving quantum operation (see Definition 3.26). In some instances we utilize the 'environment' in the form of an additional quantum register, which helps us to build a circuit that is supposed to implement a given $U\in\mathcal{U}(\mathbb{H}^{I/O})$. The auxiliary register is what has become known as ancilla and we extend our definition of quantum circuits to include such constructions.

> **Definition 5.29** (*Quantum circuit with ancilla*) Let $\mathbb{H}^{I/O}=\mathbb{H}^{\otimes n}$ and $\mathbb{H}^W=\mathbb{H}^{\otimes w}$. A circuit U on $\mathbb{H}^{I/O}$ is said to be **a circuit with ancilla** or said to be implemented with **ancilla** $|\omega_i\rangle$ in an **auxiliary (or ancilla) register** $\mathbb{H}^W$ if there exist states $|\omega_i\rangle,|\omega_f\rangle\in\mathbb{H}^W$ and a plain circuit $\hat{U}\in\mathcal{U}(\mathbb{H}^{\otimes n+w})$ on the composite system $\mathbb{H}^{I/O}\otimes\mathbb{H}^W$ such that for all $|\Psi\rangle\in\mathbb{H}^{I/O}$
>
> $$\hat{U}|\Psi\otimes\omega_i\rangle=(U|\Psi\rangle)\otimes|\omega_f\rangle \tag{5.78}$$
>
> holds. The length of U with respect to a given gate set is defined as the length of the plain circuit $\hat{U}$ defined in Definition 5.27.
> When acting on a system in the state $\rho\in D(\mathbb{H}^{I/O})$ the circuit U transforms it to a new state $U\rho U^*=\mathrm{tr}^W\left(\hat{U}(\rho\otimes\rho_{\omega_i})\hat{U}^*\right)$.

Any plain circuit $U\in\mathcal{U}(\mathbb{H}^{I/O})$ as defined in Definition 5.27 can be implemented with ancilla by simply taking $\hat{U}=U\otimes\mathbf{1}\in\mathcal{U}(\mathbb{H}^{I/O}\otimes\mathbb{H}^W)$. In this sense circuits with ancillas are a superset of plain circuits. We will, however, reserve the term 'circuit with ancilla' for those circuits that satisfy (5.78) but are not of the simple form $\hat{U}=U\otimes\mathbf{1}$.

Ancilla registers are, as their name indicates, registers $\mathbb{H}^W$ in which intermittent information is stored within a circuit, recalled and processed, but is not read at the end of the computation. The computation, that is, the circuit, may be performed by applying intermediate gates subsequently. During this intermediate computational process, in general, the states in the ancilla register become entangled with the states in the input or output register. A measurement of the ancilla register at that intermediate stage would thus affect the state in the input or output register. Such a measurement of the ancilla register would be necessary to reset it to a known initial state.[2] In order to avoid that resetting the ancilla register affects the output register, any entanglement between them has to be removed by suitable transformations. In doing so, the desired effect of the circuit should not be altered. In other words, we

---

[2] For example, in order to reset the ancilla register in the state $|0\rangle^W$ one would measure σ_z on each qubit and apply X, if the observed value was -1.

have to disentangle the ancilla from the input/output register without changing the latter. The circuit for the quantum adder discussed in Sect. 5.5.1 is a first example of such a construction.

In Definition 5.29 the ancilla state $|\omega_i\rangle$ is a fixed initial state and $|\omega_f\rangle$ is a fixed final state of the auxiliary register. Often they are both chosen to be $|0\rangle^w$, but they do not have to be identical. They could always be made to coincide by a suitable transformation on $|\omega_f\rangle$.

What is crucial, however, is that—as is evidenced from the right side of (5.78)—the result of the action of $\hat{U}$ decomposes into factors in $\mathbb{H}^{I/O}$ and $\mathbb{H}^W$. As shown in the proof of Theorem 5.28, this required factorization guarantees that a state $\rho \otimes \rho_{\omega_i}$ in the composite system $\mathbb{H}^{I/O} \otimes \mathbb{H}^W$ is then transformed by $\hat{U}$ to a state $(U\rho U^*) \otimes \rho_{\omega_f}$ and taking the partial trace over $\mathbb{H}^W$ leaves us with the sub-system $\mathbb{H}^{I/O}$ in the state $U\rho U^*$. Any measurement or observation of the sub-system $\mathbb{H}^{I/O}$ is thus solely determined by $U\rho U^*$ and does not depend on which state the ancilla register $\mathbb{H}^W$ is in. That is why the ancilla register can safely be ignored after the use of the circuit $\hat{U}$ for the implementation of U. We say that **the ancilla can be discarded**. Figure 5.9 illustrates a generic circuit with ancilla graphically.

For a circuit with initial ancilla state $|\omega_i\rangle$ and final ancilla state $|\omega_f\rangle$ the reverse circuit starts with the ancilla state $|\omega_f\rangle$ and terminates with the ancilla state $|\omega_i\rangle$.

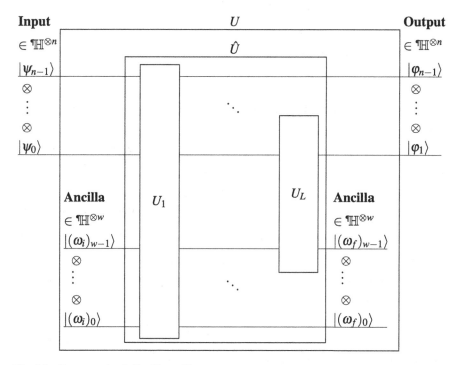

Fig. 5.9 Quantum circuit U with ancilla

Corollary 5.30 *A circuit U that has been implemented with the help of a unitary $\hat{U}$ and initial and final states $|\omega_i\rangle$ and $|\omega_f\rangle$ in the auxiliary register satisfies*

$$\hat{U}^*|\Psi \otimes \omega_f\rangle = (U^*|\Psi\rangle) \otimes |\omega_i\rangle, \tag{5.79}$$

that is, U^* is implemented with the help of $\hat{U}^*$, and the roles of $|\omega_i\rangle$ and $|\omega_f\rangle$ are interchanged.

Proof From (5.71) and the unitarity of U shown in Theorem 5.28 it follows that

$$\hat{U}\left(U^*|\Psi\rangle \otimes |\omega_i\rangle\right) = \left(UU^*|\Psi\rangle\right) \otimes |\omega_f\rangle = |\Psi\rangle \otimes |\omega_f\rangle$$

and thus (5.79). □

Since we have defined circuits as particular instances of quantum operations, it follows from Definition 3.26 and Theorem 3.24 that every circuit can be represented by a suitable set $\{K_j \mid j \in I\}$ of KRAUS operators on $\mathbb{H}^{I/O}$.

Plain circuits and those with ancillas were defined as reversible devices and thus trace-preserving quantum operations. Circuits with measurements are generally not reversible and their formal definition is quite elaborate and will not be attempted here. Suffice it to say that, loosely speaking, they may be viewed as non-trace-preserving quantum operations.

5.4   On the Process of Quantum Algorithms

For algorithms and computations as part of them it is necessary that we can suitably represent the action of functions of the type $f : S_{\text{ini}} \to S_{\text{fin}}$ on quantum registers, where S_{ini} and S_{fin} are finite subsets of $\mathbb{N}_0$. Moreover, we want to implement such functions physically with quantum circuits that perform these mappings. This can be achieved with a construction that makes use of the binary addition per factor. The latter is defined as follows.

Definition 5.31 With the help of the binary addition given in Definition 5.2 we define for vectors $|a\rangle$ and $|b\rangle$ of the computational basis in $\mathbb{H}^{\otimes m}$ the **factor-wise binary addition** $\boxplus$ as

$$\boxplus : \mathbb{H}^{\otimes m} \otimes \mathbb{H}^{\otimes m} \longrightarrow \mathbb{H}^{\otimes m}$$
$$|a\rangle \otimes |b\rangle \longmapsto |a\rangle \boxplus |b\rangle := \bigotimes_{j=m-1}^{0} |a_j \overset{2}{\oplus} b_j\rangle . \tag{5.80}$$

Instead of $|a\rangle \boxplus |b\rangle$ we will also write this abbreviatingly as $|a \boxplus b\rangle$, that is, we use the notation

$$|a \boxplus b\rangle := \bigotimes_{j=m-1}^{0} |a_j \overset{2}{\oplus} b_j\rangle . \tag{5.81}$$

Moreover, let $f : \mathbb{N}_0 \to \mathbb{N}_0$ and $n, m \in \mathbb{N}$ as well as $\mathbb{H}^A := \P\mathbb{H}^{\otimes n}$ and $\mathbb{H}^B := \P\mathbb{H}^{\otimes m}$. We say that a circuit described by the **operator** U_f **implements the function** f on $\mathbb{H}^A \otimes \mathbb{H}^B$ if

$$\begin{aligned} U_f : \mathbb{H}^A \otimes \mathbb{H}^B &\longrightarrow \mathbb{H}^A \otimes \mathbb{H}^B \\ |x\rangle \otimes |y\rangle &\longmapsto |x\rangle \otimes |y \boxplus f(x)\rangle \end{aligned} . \tag{5.82}$$

Exercise 5.64 Show that U_f as defined in (5.82) is unitary.

For a solution see Solution 5.64.

The implementation of a function as given in (5.82) is an important ingredient in the sequence of steps in a quantum algorithm or computational protocol. Generally, these consist of the following stages:

1. Preparation of the input register
2. Implementation of classical functions f by means of quantum circuits U_f on a suitable quantum register
3. Transformation of the quantum register by means of suitable quantum gates or circuits
4. Reading (observing) the result in the output register.

In what follows, we first consider in Sect. 5.4.1 the first and in Sect. 5.4.3 the fourth stage, which are quite similar in most algorithms. Some general aspects of the second stage will be treated in Sect. 5.4.2. The special form of f and thus U_f in the third stage is more particular to a given algorithm. In Sect. 5.5 we thus consider various quantum circuits that are required for the execution of elementary computational operations in the factorization algorithm of SHOR (see Sect. 6.5).

5.4.1   Preparation of Input and Use of Auxiliary Registers

Quite often the starting point of an algorithm is the state in the input register $\mathbb{H}^{I/O} := \P\mathbb{H}^{\otimes n}$ that is an equally weighted linear combination of all vectors of the computational basis. That is, the algorithm starts with the initial state

$$|\Psi_0\rangle = \frac{1}{2^{\frac{n}{2}}} \sum_{x=0}^{2^n-1} |x\rangle^n \in \mathbb{H}^{I/O}.$$

This is indeed the case in the SHOR algorithm for factorizing large numbers (see Sect. 6.5) as well as in the GROVER search algorithm (see Sect. 2.38). With the help of the HADAMARD transformation (see Definition 2.38) such a state $|\Psi_0\rangle$ can be generated as follows. Because of

$$H|0\rangle = \frac{|0\rangle + |1\rangle}{\sqrt{2}},$$

the application of the n-fold tensor product of H on $|0\rangle^n \in \mathbb{H}^{I/O}$ yields

$$
\begin{aligned}
H^{\otimes n}|0\rangle^n = H^{\otimes n}\Big(|0\rangle \otimes |0\rangle \otimes \cdots \otimes |0\rangle\Big) &= \bigotimes_{j=n-1}^{0} H|0\rangle = \bigotimes_{j=n-1}^{0} \frac{|0\rangle + |1\rangle}{\sqrt{2}} \\
&= \frac{1}{2^{\frac{n}{2}}} (|0\rangle + |1\rangle) \otimes \cdots \otimes (|0\rangle + |1\rangle) \\
&= \frac{1}{2^{\frac{n}{2}}} \Big(\underbrace{|0\dots0\rangle}_{=|0\rangle^n} + \underbrace{|0\ldots1\rangle}_{=|1\rangle^n} + \cdots + \underbrace{|1\dots1\rangle}_{=|2^n-1\rangle^n}\Big) \\
&= \frac{1}{2^{\frac{n}{2}}} \sum_{x=0}^{2^n-1} |x\rangle^n,
\end{aligned}
\tag{5.83}
$$

which is the desired initial state.

5.4.2 Implementation of Functions and Quantum Parallelism

From (5.13) we know that the representation of functions of the type $f : \mathbb{N}_0 \to \mathbb{N}_0$ on a quantum register has to be implemented as a unitary transformation U_f. This can be achieved with a construction, which makes use of the binary addition per factor previously defined, as follows.

Definition 5.32 With the help of the factor-wise binary addition given in Definition 5.31 we define the operator

$$
\begin{aligned}
U_{\boxplus} : \mathbb{H}^{\otimes m} \otimes \mathbb{H}^{\otimes m} &\longrightarrow \mathbb{H}^{\otimes m} \otimes \mathbb{H}^{\otimes m} \\
|a\rangle \otimes |b\rangle &\longmapsto |a\rangle \otimes |a \boxplus b\rangle
\end{aligned}
\tag{5.84}
$$

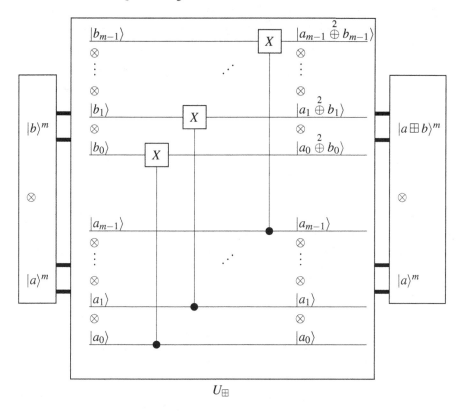

Fig. 5.10 Quantum circuit to implement the operator $U_{\boxplus}$ for the binary addition of two vectors $|a\rangle, |b\rangle \in \mathbb{H}^{\otimes m}$. As before, the thin lines represent the channels (aka 'wires') for single qubits. The pairs of thick lines represent channels for several (here m) qubits the states of which are vectors in tensor products of qubit spaces (here in $\mathbb{H}^{\otimes m}$)

Since $\sum_{j=0}^{m-1} (a_j \overset{2}{\oplus} b_j) 2^j < 2^m$ holds, it follows that $|a \boxplus b\rangle$ is also a vector of the computational basis in $\mathbb{H}^{\otimes m}$. As one can see in Fig. 5.10, the operator $U_{\boxplus}$ can simply be implemented with m controlled NOTs $\Lambda^1(X)$. Moreover, it is unitary.

Lemma 5.33 $U_{\boxplus}$ *defined as in (5.84) is unitary.*

Proof We show that $U_{\boxplus}^2 = \mathbf{1}$ first. To prove this, it suffices to show that it holds for any basis vector $|a\rangle \otimes |b\rangle \in \mathbb{H}^{\otimes m} \otimes \mathbb{H}^{\otimes m}$. Applying the definition of $U_{\boxplus}$ twice yields

$$U_{\boxplus}^2\big(|a\rangle \otimes |b\rangle\big) \underbrace{=}_{(5.84)} U_{\boxplus}\big(|a\rangle \otimes |a \boxplus b\rangle\big) \underbrace{=}_{(5.84)} |a\rangle \otimes |a \boxplus (a \boxplus b)\rangle$$

$$= |a\rangle \otimes \bigotimes_{j=m-1}^{0} |a_j \overset{2}{\oplus} \underbrace{(a \boxplus b)_j}_{=a_j \overset{2}{\oplus} b_j}\rangle \underbrace{=}_{(5.80)} |a\rangle \otimes \bigotimes_{j=m-1}^{0} |a_j \overset{2}{\oplus} \underbrace{(a_j \overset{2}{\oplus} b_j)}_{=b_j}\rangle$$

$$= |a\rangle \otimes |b\rangle .$$

Hence, $U_{\boxplus}$ is invertible and thus maps the basis $|a\rangle \otimes |b\rangle$ in $\mathbb{H}^{\otimes m} \otimes \mathbb{H}^{\otimes m}$ onto itself. Using the result of Exercise 2.15 then shows that $U_{\boxplus}$ is unitary. $\qquad\square$

Next, we show a general construction how to build unitary circuits that implement functions of the form $f : \mathbb{N}_0 \to \mathbb{N}_0$. Precondition for that is the existence of two circuits A_f and B_f, which already implement f in a certain form. The importance of the following construction is that it allows to implement a unitary operator U_f even when f is not bijective. We will see in Sect. 5.5.4 how A_f and B_f can be built in the case of the SHOR algorithm.

Theorem 5.34 *Let $f : \mathbb{N}_0 \to \mathbb{N}_0$ and $n,m \in \mathbb{N}$ as well as $\mathbb{H}^A := \mathbb{H}^{\otimes n}$ and $\mathbb{H}^B := \mathbb{H}^{\otimes m}$. Moreover, let A_f and B_f be circuits on $\mathbb{H}^A \otimes \mathbb{H}^B$ such that there exist states $|\omega_i\rangle, |\omega_f\rangle \in \mathbb{H}^B$ and for any vector of the computational basis $|x\rangle \in \mathbb{H}^A$ there is a state $|\psi(x)\rangle \in \mathbb{H}^A$ such that*

$$A_f\big(|x\rangle \otimes |\omega_i\rangle\big) = |\psi(x)\rangle \otimes |f(x)\rangle \qquad (5.85)$$
$$B_f\big(|\psi(x)\rangle \otimes |f(x)\rangle\big) = |x\rangle \otimes |\omega_f\rangle \qquad (5.86)$$

holds. Then we define on $\mathbb{H}^A \otimes \mathbb{H}^B \otimes \mathbb{H}^B$

$$\hat{U}_f := \big(\mathbf{1}^A \otimes S^{B,B}\big)\big(B_f \otimes \mathbf{1}^B\big)\big(\mathbf{1}^A \otimes U_{\boxplus}\big)\big(A_f \otimes \mathbf{1}^B\big)\big(\mathbf{1}^A \otimes S^{B,B}\big), \qquad (5.87)$$

where $S^{B,B} : |b_1\rangle \otimes |b_2\rangle \mapsto |b_2\rangle \otimes |b_1\rangle$ is the swap operator on $\mathbb{H}^B \otimes \mathbb{H}^B$, and this $\hat{U}_f$ satisfies

$$\hat{U}_f\big(|x\rangle \otimes |y\rangle \otimes |\omega_i\rangle\big) = |x\rangle \otimes |y \boxplus f(x)\rangle \otimes |\omega_f\rangle .$$

With $\hat{U}_f$ we can implement U_f with the help of an auxiliary register and the states $|\omega_i\rangle$ and $|\omega_f\rangle$, and one has

$$U_f : \mathbb{H}^A \otimes \mathbb{H}^B \longrightarrow \mathbb{H}^A \otimes \mathbb{H}^B$$
$$|x\rangle \otimes |y\rangle \longmapsto |x\rangle \otimes |y \boxplus f(x)\rangle . \qquad (5.88)$$

Proof From the definition in (5.87) it follows that

$$
\begin{aligned}
&\hat{U}_f\left(|x\rangle \otimes |y\rangle \otimes |\omega_i\rangle\right)\\
&= \left(\mathbf{1}^A \otimes S^{B,B}\right)\left(B_f \otimes \mathbf{1}^B\right)\left(\mathbf{1}^A \otimes U_{\boxplus}\right)\left(A_f \otimes \mathbf{1}^B\right)\left(|x\rangle \otimes |\omega_i\rangle \otimes |y\rangle\right)\\
&\underset{(5.85)}{=} \left(\mathbf{1}^A \otimes S^{B,B}\right)\left(B_f \otimes \mathbf{1}^B\right)\left(\mathbf{1}^A \otimes U_{\boxplus}\right)\left(|\psi(x)\rangle \otimes |f(x)\rangle \otimes |y\rangle\right)\\
&\underset{(5.84)}{=} \left(\mathbf{1}^A \otimes S^{B,B}\right)\left(B_f \otimes \mathbf{1}^B\right)\left(|\psi(x)\rangle \otimes |f(x)\rangle \otimes |y \boxplus f(x)\rangle\right)\\
&\underset{(5.86)}{=} \left(\mathbf{1}^A \otimes S^{B,B}\right)\left(|x\rangle \otimes |\omega_f\rangle \otimes |y \boxplus f(x)\rangle\right)\\
&= |x\rangle \otimes |y \boxplus f(x)\rangle \otimes |\omega_f\rangle .
\end{aligned}
\tag{5.89}
$$

The claim (5.88) about U_f then follows from (5.89) and the Definition 5.29. □

The swap operator $S^{B,B}$ used here acts on $|a\rangle \otimes |b\rangle \in \mathbb{H}^B \otimes \mathbb{H}^B$ by exchanging the factors and can be implemented with the help of m simple swaps as represented in Fig. 5.5. The circuit to implement U_f is shown in Fig. 5.11.

We define U_f for vectors $|\Phi\rangle$ on the whole of $\mathbb{H}^A \otimes \mathbb{H}^B$ by linear continuation

$$
U_f|\Phi\rangle := \sum_{x=0}^{2^n-1}\sum_{y=0}^{2^m-1} \Phi_{xy}|x\rangle \otimes |y \boxplus f(x)\rangle .
$$

If we apply U_f to $|\Psi_0\rangle := (H^n|0\rangle^n) \otimes |0\rangle^m \in \mathbb{H}^A \otimes \mathbb{H}^B$, we obtain

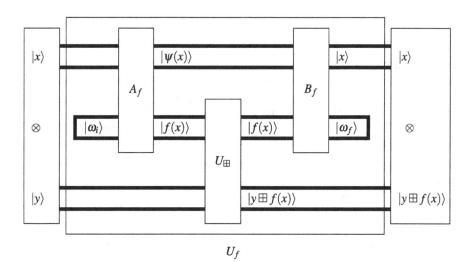

Fig. 5.11 Circuit for implementation of the operator U_f

$$U_f |\Psi_0\rangle = U_f\left((H^n|0\rangle^n)\otimes|0\rangle^m\right) \underset{(5.83)}{=} \frac{1}{2^{\frac{n}{2}}}\sum_{x=0}^{2^n-1} U_f\left(|x\rangle^n\otimes|0\rangle^m\right)$$

$$\underset{(5.88)}{=} \frac{1}{2^{\frac{n}{2}}}\sum_{x=0}^{2^n-1} \underbrace{|x\rangle\otimes|f(x)\rangle}_{\in\mathbb{H}^A\otimes\mathbb{H}^B}. \tag{5.90}$$

As we see in (5.90), applying U_f *once* to $|\Psi_0\rangle$ results in a state, which is given by a linear combination of *all* 2^n states of the form $|x\rangle\otimes|f(x)\rangle$ for $x\in\{0,\ldots,2^n-1\}$. Intuitively, this can be seen as amounting to a simultaneous evaluation of the function f on its total domain $\{0,\ldots,2^n-1\}$ in one step, and is thus called **massive quantum parallelism**. This interpretation seems to originate from the fact that the appearance of all terms of the form $|x\rangle\otimes|f(x)\rangle$ in $U_f|\Psi_0\rangle$ is similar to a complete evaluation-table $(x,f(x))_{x=0,\ldots,2^n-1}$ of the function f. In general the production of such a table would require 2^n evaluations of the function f. This would require $O(2^n)$ computational steps, whereas the evaluation of all $|x\rangle\otimes|f(x)\rangle$ in (5.90) only requires one application of U_f. However, even though one application of U_f yields a superposition of all possible $|x\rangle\otimes|f(x)\rangle$ at once, it is not possible to read the values $f(x)$ for each x separately from the state $U_f|\Psi_0\rangle$. In order to access information encoded in the linear combination of all $|x\rangle\otimes|f(x)\rangle$ in $U_f|\Psi_0\rangle$, we need to apply further transformations that exploit particular properties of the function f. For example, in the case of the SHOR algorithm (see Sect. 6.5) one applies the quantum FOURIER transform (see Sect. 5.5.5) to $U_f|\Psi_0\rangle$ and makes use of the periodicity of the function f.

5.4.3   Reading the Output Register

According to Definition 2.28 of qubits, there exists an observable σ_z the measurement of which yields one of the values in $\{\pm 1\}$ and according to Corollary 2.29 projects the qubit onto the corresponding eigenstate $|0\rangle$ or $|1\rangle$. In a composite system comprised of n qubits, which is described by states in $\mathbb{H}^{\otimes n}$, such measurements can be performed on each qubit, that is, for observables operating on each factor space $\mathbb{H}_j$ in $\mathbb{H}^{\otimes n}$, where $j\in\{0,\ldots,n-1\}$. Each such measurement of σ_z on a factor space corresponds to a measurement of the observable $\Sigma_z^j = \mathbf{1}^{\otimes n-1-j}\otimes\sigma_z\otimes\mathbf{1}^{\otimes j}$ on the composite quantum system $\mathbb{H}^{\otimes n}$. Since Σ_z^j only acts non-trivially on the factor space $\mathbb{H}_j$, one has $\Sigma_z^j\Sigma_z^k = \Sigma_z^k\Sigma_z^j$ for all j and k. The Σ_z^j are thus compatible and can all be measured sharply.

Definition 5.35 Let $n\in\mathbb{N}$ and for $j\in\{0,\ldots,n-1\}$ and $\alpha\in\{0,\ldots,3\}$ (or, equivalently, $\alpha\in\{0,x,y,z\}$) define

$$\Sigma_\alpha^j := \mathbf{1}^{\otimes n-1-j} \otimes \sigma_\alpha \otimes \mathbf{1}^{\otimes j} \quad \in \mathrm{B}_{\mathrm{sa}}\big(\mathbb{H}^{\otimes n}\big),$$

where the σ_α are as in Definition 2.21. The **observation of a state in the quantum register** $\mathbb{H}^{\otimes n}$ is defined as the measurement of all compatible observables

$$\Sigma_z^j = \mathbf{1}^{\otimes n-1-j} \otimes \sigma_z \otimes \mathbf{1}^{\otimes j}$$

for $j \in \{0,\ldots,n-1\}$ in the state of the quantum register. Such an observation is also called **read-out** or **measurement** of the register.

The read-out of the register $\mathbb{H}^{\otimes n}$ yields n observed values $\big(s_{n-1},\ldots,s_0\big) \in \{\pm 1\}^n$ after measuring $\Sigma_z^{n-1},\ldots,\Sigma_z^0$. We identify these observed values with classical bit values x_j as shown in Table 2.1, and use these classical bit values $(x_{n-1},\ldots,x_0)$ for the binary representation $x = \sum_{j=0}^{n-1} x_j 2^j$ of a non-negative integer $x < 2^n$. The measurement of the observables Σ_z^j projects the state in factor space $\mathbb{H}_j$ onto the eigenstate $|0\rangle$ or $|1\rangle$ corresponding to the observed value s_j. Altogether the read-out of the register $\mathbb{H}^{\otimes n}$ thus reveals a non-negative integer $x < 2^n$ and leaves the register in the computational basis state $|x\rangle$.

5.5 Circuits for Elementary Arithmetic Operations

In the following section we first consider a quantum circuit that implements the addition of two non-negative integers [72]. Building on that we look at further circuits that implement additional elementary arithmetic operations. These will finally allow us to present a quantum circuit that implements the modular exponentiation $x \mapsto b^x \mod N$, which is required for the SHOR factorization algorithm.

5.5.1 Quantum Adder

In the following we show how, with the help of elementary quantum gates, one can build a circuit that implements the addition of two numbers $a,b \in \mathbb{N}_0$ [72]. In doing so, we make use of the results about elementary algorithms for addition and subtraction in binary form presented in Appendix B.

We begin with the implementation of the sum bit s_j from Corollary B.2 by a gate U_s. For this we define the following operators on $\mathbb{H}^{\otimes 3}$.

$$
\begin{aligned}
A &:= \mathbf{1}^{\otimes 3} + (X-\mathbf{1}) \otimes |1\rangle\langle 1| \otimes \quad \mathbf{1} \\
B &:= \mathbf{1}^{\otimes 3} + (X-\mathbf{1}) \otimes \quad \mathbf{1} \quad \otimes |1\rangle\langle 1| \\
U_s &:= BA
\end{aligned}
\tag{5.91}
$$

Fig. 5.12 Gate U_s for binary sum used in addition

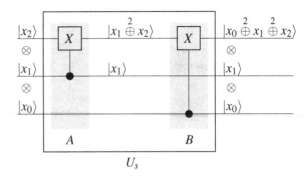

Because of $|1\rangle\langle 1|^* = |1\rangle\langle 1| = (|1\rangle\langle 1|)^2$, $X^* = X$, $X^2 = \mathbf{1}$ and thus $2(X-\mathbf{1}) + (X-\mathbf{1})^2 = 0$, it follows that A and B are self-adjoint and unitary. As can be seen from (5.91), one also has $AB = BA$. Then it follows that U_s is also unitary since

$$U_s^* = (BA)^* = A^*B^* = AB = BA = U_s$$

as well as

$$(U_s)^2 = ABAB = BAAB = B^2 = \mathbf{1}.$$

On vectors of the computational basis $|x\rangle^3 = |x_2\rangle \otimes |x_1\rangle \otimes |x_0\rangle$ in $\mathbb{H}^{\otimes 3}$ the operators A, B and U_s act as follows:

$$A\left(|x_2\rangle \otimes |x_1\rangle \otimes |x_0\rangle\right) = |x_1 \overset{2}{\oplus} x_2\rangle \otimes |x_1\rangle \otimes |x_0\rangle$$

$$B\left(|x_2\rangle \otimes |x_1\rangle \otimes |x_0\rangle\right) = |x_0 \overset{2}{\oplus} x_2\rangle \otimes |x_1\rangle \otimes |x_0\rangle$$

$$U_s\left(|x_2\rangle \otimes |x_1\rangle \otimes |x_0\rangle\right) = B\left(|x_1 \overset{2}{\oplus} x_2\rangle \otimes |x_1\rangle \otimes |x_0\rangle\right) \qquad (5.92)$$

$$= |x_0 \overset{2}{\oplus} x_1 \overset{2}{\oplus} x_2\rangle \otimes |x_1\rangle \otimes |x_0\rangle .$$

In Fig. 5.12 we show U_s graphically as a gate.

From Corollary B.2 we know that the sum of two numbers $a, b \in \mathbb{N}_0$ with $a, b < 2^n$ and the binary representations

$$a = \sum_{j=0}^{n-1} a_j 2^j , \qquad b = \sum_{j=0}^{n-1} b_j 2^j$$

is given by

$$a + b = \sum_{j=0}^{n-1} s_j 2^j + c_n^+ 2^n , \qquad (5.93)$$

where $a_j, b_j \in \{0, 1\}$ and $c_0^+ := 0$ as well as

$$c_j^+ := a_{j-1}b_{j-1} \overset{2}{\oplus} a_{j-1}c_{j-1}^+ \overset{2}{\oplus} b_{j-1}c_{j-1}^+ \qquad \text{for } j \in \{1, \ldots, n\} \qquad (5.94)$$

$$s_j := a_j \overset{2}{\oplus} b_j \overset{2}{\oplus} c_j^+ \qquad \text{for } j \in \{0, \ldots, n-1\} \qquad (5.95)$$

holds. From (5.92) and (5.95) we thus obtain

$$U_s\left(|b_j\rangle \otimes |a_j\rangle \otimes |c_j^+\rangle\right) = |s_j\rangle \otimes |a_j\rangle \otimes |c_j^+\rangle. \qquad (5.96)$$

By repeated application of U_s we can then generate the qubits $|s_j\rangle$ of the sum bits defined in (5.95) and needed in (5.93) if we have the qubits $|c_j^+\rangle$ of the carry terms c_j^+ available. In order to calculate these, we build a gate U_c with the help of the following four operators on $\mathbb{H}^{\otimes 4}$:

$$\begin{aligned}
C &:= \mathbf{1}^{\otimes 4} + (X - \mathbf{1}) \otimes \; |1\rangle\langle 1| \; \otimes |1\rangle\langle 1| \otimes \quad \mathbf{1} \\
D &:= \mathbf{1}^{\otimes 4} + \quad \mathbf{1} \quad \otimes (X - \mathbf{1}) \otimes |1\rangle\langle 1| \otimes \quad \mathbf{1} \\
E &:= \mathbf{1}^{\otimes 4} + (X - \mathbf{1}) \otimes \; |1\rangle\langle 1| \; \otimes \quad \mathbf{1} \quad \otimes |1\rangle\langle 1| \\
U_c &:= EDC .
\end{aligned}$$

For the action on a vector $|x\rangle^4 = |x_3\rangle \otimes |x_2\rangle \otimes |x_1\rangle \otimes |x_0\rangle$ of the computational basis of $\mathbb{H}^{\otimes 4}$ we obtain for these operators

$$C\left(|x_3\rangle \otimes |x_2\rangle \otimes |x_1\rangle \otimes |x_0\rangle\right) = |x_1x_2 \overset{2}{\oplus} x_3\rangle \otimes |x_2\rangle \otimes |x_1\rangle \otimes |x_0\rangle$$

$$D\left(|x_3\rangle \otimes |x_2\rangle \otimes |x_1\rangle \otimes |x_0\rangle\right) = |x_3\rangle \otimes |x_1 \overset{2}{\oplus} x_2\rangle \otimes |x_1\rangle \otimes |x_0\rangle$$

$$E\left(|x_3\rangle \otimes |x_2\rangle \otimes |x_1\rangle \otimes |x_0\rangle\right) = |x_0x_2 \overset{2}{\oplus} x_3\rangle \otimes |x_2\rangle \otimes |x_1\rangle \otimes |x_0\rangle \qquad (5.97)$$

$$\begin{aligned}
U_c\left(|x_3\rangle \otimes |x_2\rangle \otimes |x_1\rangle \otimes |x_0\rangle\right) &= ED\left(|x_1x_2 \overset{2}{\oplus} x_3\rangle \otimes |x_2\rangle \otimes |x_1\rangle \otimes |x_0\rangle\right) \\
&= E\left(|x_1x_2 \overset{2}{\oplus} x_3\rangle \otimes |x_1 \overset{2}{\oplus} x_2\rangle \otimes |x_1\rangle \otimes |x_0\rangle\right) \\
&= |x_0(x_1 \overset{2}{\oplus} x_2) \overset{2}{\oplus} x_1x_2 \overset{2}{\oplus} x_3\rangle \otimes |x_1 \overset{2}{\oplus} x_2\rangle \otimes |x_1\rangle \otimes |x_0\rangle.
\end{aligned}$$

From (5.97) and (5.94) one finds then

$$U_c\left(|0\rangle \otimes |b_{j-1}\rangle \otimes |a_{j-1}\rangle \otimes |c_{j-1}^+\rangle\right)$$

$$= |c_j^+\rangle \otimes |b_{j-1} \overset{2}{\oplus} a_{j-1}\rangle \otimes |a_{j-1}\rangle \otimes |c_{j-1}^+\rangle. \qquad (5.98)$$

Hence, we can generate the qubit $|c_j^+\rangle$ of the carry term c_j^+ by suitably repeated application of U_c. The gate U_c is represented graphically in Fig. 5.13

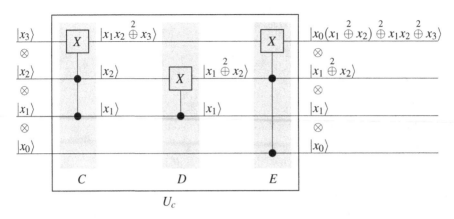

Fig. 5.13 Gate U_c for carry in addition

For the same reasons as for A and B (see discussion after (5.91)) it also follows that C, D and E are all self-adjoint and unitary. However, even though U_c as a product of unitary operators is also unitary, it is no longer self-adjoint since one has

$$U_c^* = (EDC)^* = C^* D^* E^* = CDE \neq EDC .$$

For U_c^* we find, instead of (5.97), for the action on a vector of the computational basis of $\mathbb{H}^{\otimes 4}$

$$
\begin{aligned}
U_c^* \left(|x_3\rangle \otimes |x_2\rangle \otimes |x_1\rangle \otimes |x_0\rangle \right) &= CD \left(|x_0 x_2 \overset{2}{\oplus} x_3\rangle \otimes |x_2\rangle \otimes |x_1\rangle \otimes |x_0\rangle \right) \\
&= C \left(|x_0 x_2 \overset{2}{\oplus} x_3\rangle \otimes |x_1 \overset{2}{\oplus} x_2\rangle \otimes |x_1\rangle \otimes |x_0\rangle \right) \qquad (5.99) \\
&= |x_1 (x_1 \overset{2}{\oplus} x_2) \overset{2}{\oplus} x_0 x_2 \overset{2}{\oplus} x_3\rangle \otimes |x_1 \overset{2}{\oplus} x_2\rangle \otimes |x_1\rangle \otimes |x_0\rangle \\
&= |(x_0 \overset{2}{\oplus} x_1) x_2 \overset{2}{\oplus} x_1 \overset{2}{\oplus} x_3\rangle \otimes |x_1 \overset{2}{\oplus} x_2\rangle \otimes |x_1\rangle \otimes |x_0\rangle .
\end{aligned}
$$

Exercise 5.65 Show that $U_c^* U_c = \mathbf{1}$.

For a solution see Solution 5.65.

By suitably combining U_s, U_c and U_c^* we will build a quantum circuit that implements the addition of two numbers $a, b \in \mathbb{N}_0$. In order to formalize the statement about such a quantum adder we still need a few more definitions.

Definition 5.36 Let $n \in \mathbb{N}$ and

$$\mathbb{H}^B := \mathbb{H}^{\otimes n+1}, \qquad \mathbb{H}^A := \mathbb{H}^{\otimes n}, \qquad \mathbb{H}^W := \mathbb{H}^{\otimes n}.$$

For vectors of the computational basis $|b\rangle \otimes |a\rangle \otimes |w\rangle \in \mathbb{H}^B \otimes \mathbb{H}^A \otimes \mathbb{H}^W$ we define U_0 and $|\Psi[b,a,w]\rangle \in \mathbb{H}^B \otimes \mathbb{H}^A \otimes \mathbb{H}^W$ by

$$U_0\Big(|b\rangle \otimes |a\rangle \otimes |w\rangle\Big) := |b_n\rangle \otimes \overset{0}{\underset{l=n-1}{\bigotimes}} \big(|b_l\rangle \otimes |a_l\rangle \otimes |w_l\rangle\big)$$

$$=: |\Psi[b,a,w]\rangle \tag{5.100}$$

and on all of $\mathbb{H}^B \otimes \mathbb{H}^A \otimes \mathbb{H}^W$ by means of linear continuation.
Furthermore, we define on $\mathbb{H}^B \otimes \mathbb{H}^A \otimes \mathbb{H}^W$ the operators

$$U_1 := \prod_{l=1}^{n-1} \Big(\mathbf{1}^{\otimes 3l} \otimes U_c \otimes \mathbf{1}^{\otimes 3(n-1-l)}\Big)$$

$$U_2 := \Big[\big(\mathbf{1} \otimes U_s\big)\big(\mathbf{1} \otimes \Lambda_{|1\rangle^1}(X) \otimes \mathbf{1}\big)U_c\Big] \otimes \mathbf{1}^{\otimes 3(n-1)}$$

$$U_3 := \prod_{l=n-1}^{1} \Big(\mathbf{1}^{\otimes 3l} \otimes \big(\mathbf{1} \otimes U_s\big)U_c^* \otimes \mathbf{1}^{\otimes 3(n-1-l)}\Big)$$

$$\hat{U}_+ := U_0^* U_3 U_2 U_1 U_0.$$

Note that $\mathbb{H}^B$ has one qubit more than $\mathbb{H}^A$ and $\mathbb{H}^W$. This additional qubit is always zero for $b < 2^n$. It is necessary, however, for the addition $b + a$ in which it will be set equal to the highest carry qubit $|c_n^+\rangle$. We refer the reader to Appendix B for the definitions and roles of the carry and sum bits c_j^+ and s_j in the addition $b + a$.

In the formal Definition 5.36 alone, the construction of the operators $U_0, \ldots, U_3$ is rather obscure. They are easier to understand if we display their constructions and their roles in the addition graphically. Figure 5.14 shows such a graphical representation of the operators $U_0, \ldots, U_3, \hat{U}_+, U_+$ as well as $|\Psi[b,a,0]\rangle$.

Lemma 5.37 *The operators $U_0, \ldots, U_3$ and $\hat{U}_+$ defined in Definition 5.36 are unitary.*

Proof As can be seen in (5.100), the operator U_0 maps each vector of the computational basis bijectively to a vector of the computational basis. According to the first statement in Exercise 2.15, it is thus unitary.

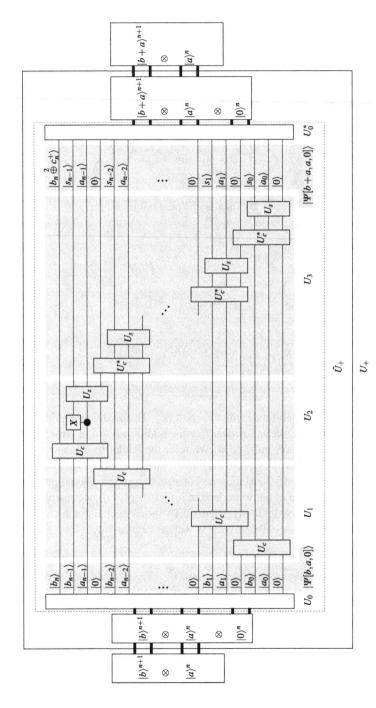

Fig. 5.14 Circuit for the quantum adder U_+ to compute $a + b$ for $a, b \in \mathbb{N}_0$ with $a, b < 2^n$

From Exercise 5.65 we know that U_c is unitary and since for each $l \in \{1, \ldots, n - 1\}$

$$\left(\mathbf{1}^{\otimes 3l} \otimes U_c^* \otimes \mathbf{1}^{\otimes 3(n-1-l)}\right)\left(\mathbf{1}^{\otimes 3l} \otimes U_c \otimes \mathbf{1}^{\otimes 3(n-1-l)}\right) = \mathbf{1}^{\otimes 3l} \otimes U_c^* U_c \otimes \mathbf{1}^{\otimes 3(n-1-l)}$$
$$= \mathbf{1}^{\otimes n}$$

then U_1 as a product of unitary operators is itself unitary. The proof that U_3 is unitary is similar.

For U_2 we have

$$U_2^* = \left[U_c^* \Big(\mathbf{1} \otimes \underbrace{\Lambda_{|1\rangle^1}(X)^*}_{=\Lambda_{|1\rangle^1}(X)} \otimes \mathbf{1}\Big)\Big(\mathbf{1} \otimes \underbrace{U_s^*}_{=U_s}\Big) \right] \otimes \mathbf{1}^{\otimes 3(n-1)}$$

and thus

$$U_2^* U_2 = \left[U_c^* \big(\mathbf{1} \otimes \Lambda_{|1\rangle^1}(X) \otimes \mathbf{1}\big) \underbrace{(\mathbf{1} \otimes U_s)^2}_{=\mathbf{1}^{\otimes 4}} \big(\mathbf{1} \otimes \Lambda_{|1\rangle^1}(X) \otimes \mathbf{1}\big) U_c \right] \otimes \mathbf{1}^{\otimes 3(n-1)}$$

$$= U_c^* \underbrace{\big(\mathbf{1} \otimes \Lambda_{|1\rangle^1}(X) \otimes \mathbf{1}\big)^2}_{=\mathbf{1}^{\otimes 4}} U_c \otimes \mathbf{1}^{\otimes 3(n-1)}$$

$$= \mathbf{1}^{\otimes 3n+1}.$$

Finally, $\hat{U}_+$ being a product of unitary operators is again unitary. $\qquad\square$

Theorem 5.38 *There exists a circuit U_+ on $\mathbb{H}^{I/O} = \mathbb{H}^B \otimes \mathbb{H}^A$, which can be implemented with the help of the auxiliary register $\mathbb{H}^W$ by $\hat{U}_+$, that is, there exists a $\hat{U}_+ \in \mathcal{U}\big(\mathbb{H}^{I/O} \otimes \mathbb{H}^W\big)$ such that for arbitrary $|\Phi\rangle \in \mathbb{H}^{I/O}$ one has*

$$\hat{U}_+\big(|\Phi\rangle \otimes |0\rangle^n\big) = \big(U_+|\Phi\rangle\big) \otimes |0\rangle^n. \tag{5.101}$$

Furthermore, for $a, b \in \mathbb{N}_0$ with $a, b < 2^n$ we have that

$$U_3 U_2 U_1 |\Psi[b, a, 0]\rangle = |\Psi[b + a, a, 0]\rangle \tag{5.102}$$

and thus

$$U_+\big(|b\rangle \otimes |a\rangle\big) = |b + a\rangle \otimes |a\rangle. \tag{5.103}$$

Proof First we show (5.102). The proof of this claim by means of the operator definitions and a sequence of equations is laborious and not very instructive. Much more

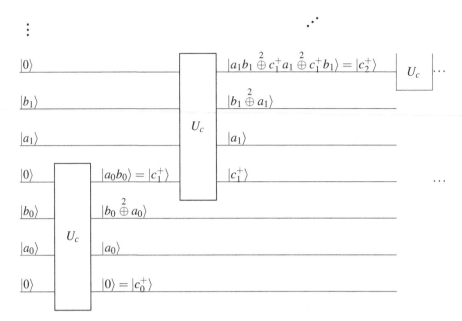

Fig. 5.15 Sub-circuit U_1 of the quantum adder

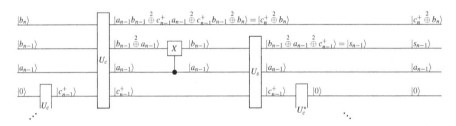

Fig. 5.16 Sub-circuit U_2 of the quantum adder

illuminating and just as valid is a proof with the help of the graphical representations of the individual operators or parts thereof.

From (5.98) and Fig. 5.15 we see that the sequence of the U_c in U_1 delivers the carry qubits $|c_j^+\rangle$ (see Corollary B.2) of the addition of a and b in the uppermost fourth channel, starting with $|c_1^+\rangle$ and then successively up to $|c_{n-1}^+\rangle$. The third channels of U_c in U_1 always deliver $|b_{j-1} \overset{2}{\oplus} a_{j-1}\rangle$, while in the first and second channels the input passes through unaltered.

Similarly, one sees from (5.96), (5.98) and Fig. 5.16, that U_2 delivers in the fourth channel $|b_n \overset{2}{\oplus} c_n^+\rangle$ and in the third the sum-qubit $|s_{n-1}\rangle$ of the addition of $b + a$ (see Corollary B.2). Hence, in the case $b < 2^n$ the most significant carry qubit $|c_n^+\rangle$ of the addition $b + a$ is delivered in the topmost channel of U_2.

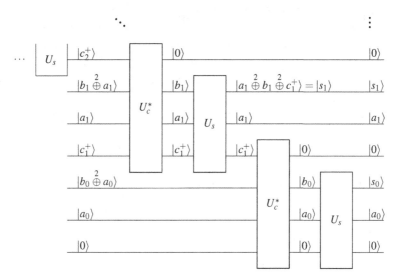

Fig. 5.17 Sub-circuit U_3 of the quantum adder

Finally, one obtains from (5.99) and Fig. 5.17, that U_3 reverts the channels with the carry qubits $|c_{n-1}^+\rangle, \dots, |c_1^+\rangle$ to $|0\rangle$ and delivers in the third channels the sum qubits $|s_{n-1}\rangle, \dots, |s_0\rangle$. Furthermore, U_3 delivers the $|a_{n-1}\rangle, \dots, |a_0\rangle$ unchanged.

Altogether thus

$$|\Psi[b,a,0]\rangle = |0\rangle \otimes \bigotimes_{l=n-1}^{0} \big(|b_l\rangle \otimes |a_l\rangle \otimes |0\rangle\big) \tag{5.104}$$

is transformed by $U_3 U_2 U_1$ into

$$U_3 U_2 U_1 |\Psi[b,a,0]\rangle = |c_n^+\rangle \otimes \bigotimes_{l=n-1}^{0} \big(|s_l\rangle \otimes |a_l\rangle \otimes |0\rangle\big) = |\Psi[b+a,a,0]\rangle. \tag{5.105}$$

This completes the proof of (5.102).

In order to prove (5.101), we note that, because of

$$|\Phi\rangle = \sum_{b=0}^{2^{n+1}-1} \sum_{a=0}^{2^n-1} \Phi_{ba} |b\rangle \otimes |a\rangle,$$

it suffices to prove the claim for an arbitrary vector $|b\rangle \otimes |a\rangle$ of the computational basis of $\mathbb{H}^B \otimes \mathbb{H}^A$. For these we have

$$\hat{U}_+\left(|b\rangle \otimes |a\rangle \otimes |0\rangle^n\right) = U_0^* U_3 U_2 U_1 U_0\left(|b\rangle \otimes |a\rangle \otimes |0\rangle^n\right)$$

$$= U_0^* U_3 U_2 U_1 \left(|b_n\rangle \otimes \bigotimes_{l=n-1}^{0} \left(|b_l\rangle \otimes |a_l\rangle \otimes |0\rangle\right)\right)$$

$$= U_0^* U_3 U_2 U_1 |\Psi[b,a,0]\rangle . \tag{5.106}$$

The only difference between the argument of $U_0^* U_3 U_2 U_1$ in (5.106) and the right side of (5.104) is that b_n in (5.106) can be different from zero. But this changes only the output of the most significant qubit in $\mathbb{H}^B$. From (5.97) and Fig. 5.16 we see that U_2 for this most significant qubit delivers $|b_n \overset{2}{\oplus} c_n^+\rangle$, which is the sum qubit $|s_n\rangle$ since $a_n = 0$ holds. All other qubits are transformed by $U_3 U_2 U_1$ exactly as in (5.105). However, the qubit in the carry state $|c_{n+1}^+\rangle$ from $b + a$ will be lost. Hence, the number $b + a - c_{n+1}^+ 2^{n+1}$ is generated in $\mathbb{H}^B$. Consequently, for a and b such that $0 \le a < 2^n$ and $0 \le b < 2^{n+1}$ we have

$$\hat{U}_+\left(|b\rangle \otimes |a\rangle \otimes |0\rangle^n\right) = U_0^* U_3 U_2 U_1 |\Psi[b,a,0]\rangle$$

$$= U_0^* |\Psi[b + a - c_{n+1}^+ 2^{n+1}, a, 0]\rangle$$

$$= |b + a - c_{n+1}^+ 2^{n+1}\rangle \otimes |a\rangle \otimes |0\rangle^n$$

$$= U_+\left(|b\rangle \otimes |a\rangle\right) \otimes |0\rangle^n .$$

This proves (5.101). For $a, b < 2^n$ one has $c_{n+1}^+ = 0$, and thus (5.103) follows as well.   □

From Theorem 5.28 it follows that U_+ is unitary and thus invertible. Indeed, the inverse of U_+ is a circuit, which implements the algorithm of the binary subtraction $b - a$ formalized in Corollary B.5.

Corollary 5.39 *There exists a circuit U_- on $\mathbb{H}^{I/O} = \mathbb{H}^B \otimes \mathbb{H}^A$, which is implemented with the help of the auxiliary register $\mathbb{H}^W$ by $\hat{U}_+^* = \hat{U}_+^{-1}$, that is, for arbitrary $|\Phi\rangle \in \mathbb{H}^{I/O}$ one has*

$$\hat{U}_+^*\left(|\Phi\rangle \otimes |0\rangle^n\right) = \left(U_- |\Phi\rangle\right) \otimes |0\rangle^n , \tag{5.107}$$

where also $U_- = U_+^* = U_+^{-1}$ holds. Furthermore, for $a, b \in \mathbb{N}_0$ with $a, b < 2^n$ we have that

$$U_1^* U_2^* U_3^* |\Psi[b,a,0]\rangle = |\Psi[c_n^- 2^{n+1} + b - a, a, 0]\rangle \tag{5.108}$$

and thus

$$U_-\left(|b\rangle \otimes |a\rangle\right) = |c_n^- 2^{n+1} + b - a\rangle \otimes |a\rangle = \begin{cases} |b-a\rangle \otimes |a\rangle & \text{if } b \geq a \\ |2^{n+1} + b - a\rangle \otimes |a\rangle & \text{if } b < a \end{cases}. \tag{5.109}$$

Proof From Corollary 5.30 we know that for arbitrary $|\Phi\rangle \in \mathbb{H}^{I/O}$

$$\hat{U}_+^*\left(|\Phi\rangle \otimes |0\rangle^n\right) = \left(U_+^*|\Phi\rangle\right) \otimes |0\rangle^n$$

holds. With $U_- = U_+^*$ this implies (5.107).

The proof of (5.108) is similar to the proof of Theorem 5.38 in that we consider the respective actions of U_3^*, U_2^* and U_1^*. From (5.92) as well as Fig. 5.18 we see that the U_s leave the first two input channels unaltered. In the third output channel the U_s deliver the qubit $|d_j\rangle$ of the difference bit of the subtraction $b - a$ as defined in (B.12). Thereafter, the U_c act in that they also deliver the first two inputs $|c_j^-\rangle$ and $|a_j\rangle$ unaltered, but in the third channel they deliver $|b_j \overset{2}{\oplus} c_j^-\rangle$ (see (5.97)). Moreover, we see from Fig. 5.18 as well as (5.97) that the U_c deliver in the fourth channel the qubits $|c_j^-\rangle$ of the carry term of the subtraction $b - a$ defined in Corollary B.5. This is because

$$c_{j-1}^-(a_{j-1} \overset{2}{\oplus} d_{j-1}) \overset{2}{\oplus} a_{j-1} d_{j-1}$$

$$\underset{\text{(B.12)}}{=} c_{j-1}^-(a_{j-1} \overset{2}{\oplus} a_{j-1} \overset{2}{\oplus} b_{j-1} \overset{2}{\oplus} c_{j-1}^-) \overset{2}{\oplus} a_{j-1}(a_{j-1} \overset{2}{\oplus} b_{j-1} \overset{2}{\oplus} c_{j-1}^-)$$

$$= c_{j-1}^- b_{j-1} \overset{2}{\oplus} c_{j-1}^- \overset{2}{\oplus} a_{j-1} \overset{2}{\oplus} a_{j-1} b_{j-1} \overset{2}{\oplus} a_{j-1} c_{j-1}^-$$

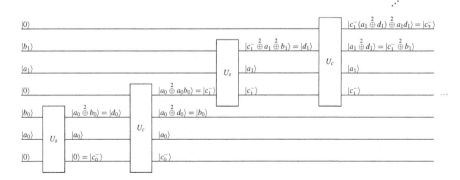

Fig. 5.18 Sub-circuit U_3^* of the quantum subtractor

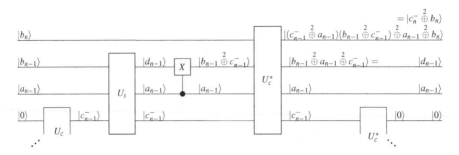

Fig. 5.19 Sub-circuit U_2^* of the quantum subtractor

$$= (1 \overset{2}{\oplus} b_{j-1})(a_{j-1} \overset{2}{\oplus} c_{j-1}^-) \overset{2}{\oplus} a_{j-1}c_{j-1}^-$$

$$\underbrace{=}_{(B.11)} c_j^-$$

holds.

From (5.99) and Fig. 5.19 one sees that U_2^* delivers in the first channel the qubit $|c_{n-1}^-\rangle$ of the carry, in the second channel $|a_{n-1}\rangle$, in the third the qubit in the state $|d_{n-1}\rangle$ corresponding to the difference term and in the fourth $|b_n \overset{2}{\oplus} c_n^-\rangle$. In the case $b < 2^n$ one has $b_n = 0$, and in this case U_2^* delivers in its topmost channel the most significant qubit $|c_n^-\rangle$ of the carry of the subtraction $b - a$.

The fact that, as shown in Fig. 5.20, every $|c_{n-1}^-\rangle, \ldots, |c_0^-\rangle$ is transformed by U_1^* to $|0\rangle$ can be seen as follows:

$$(c_{j-1}^- \overset{2}{\oplus} a_{j-1})(c_{j-1}^- \overset{2}{\oplus} b_{j-1}) \overset{2}{\oplus} a_{j-1} \overset{2}{\oplus} c_j^-$$

$$= c_{j-1}^- \overset{2}{\oplus} c_{j-1}^- b_{1j-} \overset{2}{\oplus} a_{j-1}c_{j-1}^- \overset{2}{\oplus} a_{j-1}b_{j-1} \overset{2}{\oplus} a_{j-1} \overset{2}{\oplus} c_j^-$$

$$\underbrace{=}_{(B.11)} c_{j-1}^- \overset{2}{\oplus} c_{j-1}^- b_{j-1} \overset{2}{\oplus} a_{j-1}c_{j-1}^- \overset{2}{\oplus} a_{j-1}b_{j-1} \overset{2}{\oplus} a_{j-1}$$

$$\overset{2}{\oplus} \underbrace{(1 \overset{2}{\oplus} b_{j-1})(a_{j-1} \overset{2}{\oplus} c_{j-1}^-) \overset{2}{\oplus} a_{j-1}c_{j-1}^-}_{=c_j^-}$$

$$= c_{j-1}^- \overset{2}{\oplus} c_{1j-}^- b_{j-1} \overset{2}{\oplus} a_{j-1}c_{j-1}^- \overset{2}{\oplus} a_{j-1}b_{j-1} \overset{2}{\oplus} a_{j-1} \overset{2}{\oplus} a_{j-1}$$

$$\overset{2}{\oplus} c_{j-1}^- \overset{2}{\oplus} a_{j-1}b_{j-1} \overset{2}{\oplus} c_{j-1}^- b_{j-1} \overset{2}{\oplus} a_{j-1}c_{j-1}^-$$

$$= 0.$$

Finally, the U_c^* in U_1^* invert the action of the U_c in U_3^* also in the third channels and thus deliver there $|d_j\rangle$.

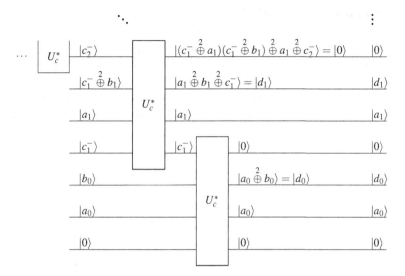

Fig. 5.20 Sub-circuit U_1^* of the quantum subtractor

Altogether, one has thus for $a, b < 2^n$

$$U_1^* U_2^* U_3^* |\Psi[b,a,0]\rangle \underset{(5.100)}{=} U_1^* U_2^* U_3^* \left(|b_n\rangle \otimes \overset{0}{\underset{l=n-1}{\bigotimes}} \left(|b_l\rangle \otimes |a_l\rangle \otimes |0\rangle \right) \right)$$

$$= |c_n^-\rangle \otimes \overset{0}{\underset{l=n-1}{\bigotimes}} \left(|d_l\rangle \otimes |a_l\rangle \otimes |0\rangle \right)$$

$$\underset{(5.100)}{=} \left| \Psi \left[c_n^- 2^n + \sum_{l=0}^{n-1} d_l 2^l, a, 0 \right] \right\rangle. \tag{5.110}$$

On the other hand, we know from Corollary B.5 that

$$\sum_{j=0}^{n-1} d_j 2^j = c_n^- 2^n + b - a \tag{5.111}$$

holds, where

$$c_n^- = \begin{cases} 0 & \text{if } b \geq a \\ 1 & \text{if } b < a. \end{cases} \tag{5.112}$$

Hence, (5.108) follows from (5.110) and (5.111). From (5.108) and (5.112), in turn, follows (5.109). $\qquad \square$

5.5.2   Quantum Adder Modulo N

With the help of the quantum adder U_+ and subtractor U_- we can now build a quantum adder modulo $N \in \mathbb{N}$, which we denote by $U_{+\%N}$. In general one has $(b + a) \mod N \in \{0, \ldots, N - 1\}$. On the other hand, it is not necessarily the case that $N = 2^n$, such that the image under $\mod N$ does not coincide with a total space $\mathbb{H}^{\otimes n}$. Since $U_{+\%N}$ ought to be unitary, we need to suitably restrict the HILBERT space on which the operator $U_{+\%N}$ acts.

Definition 5.40 For $N \in \mathbb{N}$ with $N < 2^n$ we define $\mathbb{H}^{<N}$ as the linear subspace of $\mathbb{H}^{\otimes n}$ spanned by the basis vectors $|0\rangle^n, \ldots, |N-1\rangle^n$

$$\mathbb{H}^{<N} := \mathrm{Span}\{|0\rangle^n, \ldots, |N-1\rangle^n\}$$
$$= \left\{ |\Phi\rangle \in \mathbb{H}^{\otimes n} \,\middle|\, |\Phi\rangle = \sum_{a=0}^{N-1} \Phi_a |a\rangle^n \right\}.$$

It is laborious and not very instructive to define the operator $U_{+\%N}$ with the help of formulas. Instead, we use the simpler and clearer graphical representation shown in Fig. 5.21 as definition.

Definition 5.41 The **quantum adder modulo** N is the operator $U_{+\%N}$ on $\mathbb{H}^{I/O} = \mathbb{H}^{<N} \otimes \mathbb{H}^{<N}$ representing the circuit shown in Fig. 5.21, which is implemented with the help of the state $|\omega_i\rangle = |N\rangle^n \otimes |0\rangle^1 = |\omega_f\rangle$ in an auxiliary register $\mathbb{H}^W = \mathbb{H}^{\otimes n+1}$.

Theorem 5.42 Let n and N be natural numbers satisfying $N < 2^n$. The operator $U_{+\%N}$ shown in Fig. 5.21 satisfies

$$U_{+\%N} : \mathbb{H}^{<N} \otimes \mathbb{H}^{<N} \longrightarrow \mathbb{H}^{<N} \otimes \mathbb{H}^{<N}$$
$$|b\rangle \otimes |a\rangle \longmapsto |(b+a) \mod N\rangle \otimes |a\rangle \quad . \tag{5.113}$$

Furthermore, $U_{+\%N}$ is unitary and we have

$$U_{-\%N} := U_{+\%N}^* : \mathbb{H}^{<N} \otimes \mathbb{H}^{<N} \longrightarrow \mathbb{H}^{<N} \otimes \mathbb{H}^{<N}$$
$$|b\rangle \otimes |a\rangle \longmapsto |(b-a) \mod N\rangle \otimes |a\rangle \quad . \tag{5.114}$$

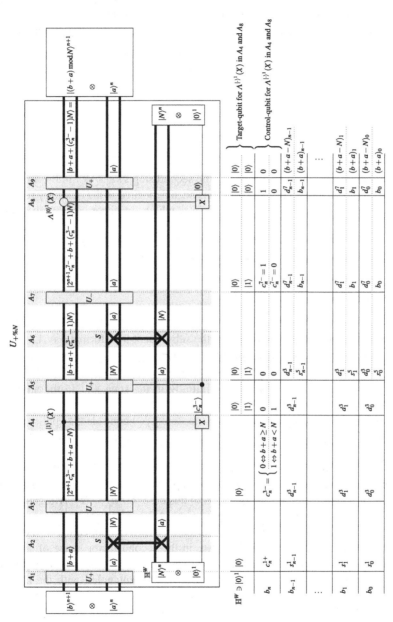

Fig. 5.21 Circuit for the quantum adder $U_{+\%N}$ modulo N applied to $a, b < N$. The table shows how the qubits for $|0\rangle^1$ in the auxiliary register and for b in the input and output register change when stepping through the gates of the circuit

Proof Since $N < 2^n$ is assumed, $\mathbb{H}^{<N}$ is a subspace of $\mathbb{TH}^{\otimes n}$ and $\mathbb{TH}^{\otimes n+1}$ and can be embedded in these. In Fig. 5.21 we thus consider the arguments $|b\rangle \otimes |a\rangle \in \mathbb{H}^{I/O} = \mathbb{H}^{<N} \otimes \mathbb{H}^{<N}$ as vectors in $\mathbb{TH}^{\otimes n+1} \otimes \mathbb{TH}^{\otimes n}$ on which, according to Definition 5.36, Theorem 5.38 and Corollary 5.39 the adder U_+ and the subtractor $U_- = U_+^{-1}$ are defined. The auxiliary register $\mathbb{H}^W = \mathbb{TH}^{\otimes n+1}$ is pre-set with $|N\rangle^n \otimes |0\rangle^1$. One has

$$U_{+\%N} = \prod_{l=9}^{1} A_l ,$$

and for the proof we consider successively the results of the transformations $A_1, \ldots, A_9$ defined in Fig. 5.21.

To begin with, U_+ in A_1 is applied to $|b\rangle \otimes |a\rangle$, which, according to Theorem 5.38, yields $|b+a\rangle \otimes |a\rangle$.

Application of the swap operator S in the second step swaps $|N\rangle$ in the auxiliary register with $|a\rangle$ such that afterwards $|a\rangle$ has been deposited in the auxiliary register.

In A_3 then U_- is applied to $|b+a\rangle \otimes |N\rangle$, which, in accordance with Corollary 5.39, yields $|c_n^{3-} 2^{n+1} + b + a - N\rangle \otimes |N\rangle$. Here we have indexed the carry bit c_n^{3-} with the superscript $3-$ in order to distinguish it from the carry bit of a later subtraction. From Corollary 5.39 we also know that

$$c_n^{3-} = \begin{cases} 0 \Leftrightarrow b+a \geq N \\ 1 \Leftrightarrow b+a < N \end{cases}$$

holds. The value of c_n^{3-} will then serve in the subsequent transformations $A_4, \ldots, A_9$ within $U_{+\%N}$ as a distinguishing indicator for the cases $b+a \geq N$ or $b+a < N$.

In A_4 the state $|c_n^{3-}\rangle$ of the carry qubit is written by means of a controlled NOT $\Lambda^{|1\rangle^1}(X)$ in the target-qubit, which was initially in the state $|0\rangle^1$ in the auxiliary register.

In the fifth step in A_5 the target-qubit in the state $|c_n^{3-}\rangle$ controls the application of the addition U_+ on $|c_n^{3-} 2^{n+1} + b + a - N\rangle \otimes |N\rangle$. In the case $c_n^{3-} = 0$ no addition is performed. In this case the result of the fifth step is $|c_n^{3-} 2^{n+1} + b + a - N\rangle \otimes |N\rangle = |b+a-N\rangle \otimes |N\rangle$. In the case $c_n^{3-} = 1$ the addition will be performed. This addition is the inverse of the previous subtraction in step three. Thus, the state prior to that subtraction is recovered. The result of A_5 is in this case $|b+a\rangle \otimes |N\rangle$. Altogether the result of A_5 can thus be written as $|b + a + (c_n^{3-} - 1)N\rangle \otimes |N\rangle$.

In A_6 the swap of the second step is inverted by a further application of the swap operator S. After that $|a\rangle$ is again the state in the second factor space of $\mathbb{H}^{I/O} = \mathbb{H}^{<N} \otimes \mathbb{H}^{<N} \subset \mathbb{TH}^{\otimes n+1} \otimes \mathbb{TH}^{\otimes n}$, and $|N\rangle$ becomes the state in the auxiliary register.

But the target-qubit in the auxiliary register is still entangled with the state in $\mathbb{TH}^{\otimes n+1} \otimes \mathbb{TH}^{\otimes n}$. In order to disentangle these states (see remarks before and after Definition 5.29), we subtract a with U_- in a seventh step from $b + a + (c_n^{3-} - 1)N$. If $c_n^{3-} = 0$, the result of this subtraction is $b - N < 0$, and thus the state of the carry qubit becomes $|c_n^{7-}\rangle = |1\rangle$. If, on the other hand, $|c_n^{3-}\rangle = |1\rangle$, then the subtraction results in $b \geq 0$, and the state of the carry qubit becomes $|c_n^{7-}\rangle = |0\rangle$.

The value of the carry bit c_n^{7-} then controls in A_8 the re-setting of the target-qubit in the auxiliary register to $|0\rangle$. Finally, in A_9 the subtraction of A_7 is inverted. Because of $a, b < N$, the final result in the first factor space $\mathbb{H}^{<N}$ of $\mathbb{H}^{I/O}$ is thus

$$|b+a+(c_n^{3-}-1)N\rangle = \begin{cases} |b+a-N\rangle & \text{if } b+a \geq N \\ |b+a\rangle & \text{if } b+a < N \end{cases} = |(b+a) \mod N\rangle.$$
(5.115)

For $U_{+\%N}^*$ one has

$$U_{+\%N}^* = \prod_{l=1}^{9} A_l^*.$$

Here it should be noted that in A_1^*, A_5^* and A_9^* then $U_+^* = U_-$ holds and in A_3 and A_7 conversely $U_-^* = U_+$ holds. With exactly the same arguments as in the derivation of (5.115) one obtains that for $a, b < N$

$$U_{+\%N}^*\left(|b\rangle \otimes |a\rangle\right) = \begin{cases} |b-a\rangle \otimes |a\rangle & \text{if } b \geq a \\ |b-a+N\rangle \otimes |a\rangle & \text{if } b < a \end{cases} = |(b-a) \mod N\rangle \otimes |a\rangle.$$

With this and (5.115) it follows that for $a, b < N$

$$\begin{aligned} U_{+\%N}^* U_{+\%N}\left(|b\rangle \otimes |a\rangle\right) &= U_{+\%N}^*\left(|(b+a) \mod N\rangle \otimes |a\rangle\right) \\ &= |((b+a) \mod N - \underbrace{a}_{=a \mod N}) \mod N\rangle \otimes |a\rangle \\ &\underbrace{=}_{(D.23)} |b \mod N\rangle \otimes |a\rangle \\ &= |b\rangle \otimes |a\rangle \end{aligned}$$

holds. Consequently, $U_{+\%N}$ is unitary. $\qquad\square$

5.5.3 Quantum Multiplier Modulo N

With the help of the quantum adder we now define the multiplication modulo N with a number $c \in \mathbb{N}_0$.

Definition 5.43 For $c \in \mathbb{N}_0$ and $n, N \in \mathbb{N}$ we define $U_{\times c\%N}$ as the **quantum multiplier modulo** N on $\mathbb{H}^{I/O} = \mathbb{H}^{<N} \otimes \mathbb{H}^{\otimes n}$ as the operator representing the circuit shown in Fig. 5.22, which is implemented with the help of the states $|\omega_i\rangle$ and $|\omega_f\rangle$ in an auxiliary register $\left(\mathbb{H}^{<N}\right)^{\otimes n+1}$.

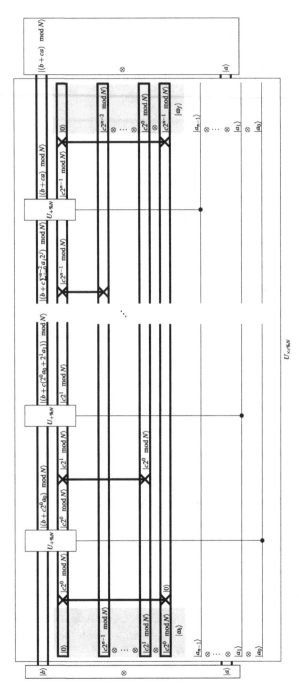

Fig. 5.22   Circuit for the quantum multiplier $U_{\times c \% N}$ modulo N for $a, N < 2^n$ and $b < N$

As one can see in Fig. 5.22, the operator $U_{\times c\%N}$ is implemented with the help of an auxiliary register $\mathbb{H}^W = \left(\mathbb{H}^{<N}\right)^{\otimes n+1}$. In this auxiliary register the initial state is set to

$$|\omega_i\rangle = |0\rangle \otimes |c2^{n-1} \mod N\rangle \otimes \cdots \otimes |c2^0 \mod N\rangle.$$

To prepare the initial state, one calculates $c2^{n-1} \mod N, \ldots, c2^0 \mod N$ with the help of a classical computer and prepares $|\omega_i\rangle$ in the auxiliary register accordingly.

The final state in the auxiliary register is given by

$$|\omega_f\rangle = |0\rangle \otimes |c2^{n-2} \mod N\rangle \otimes \cdots \otimes |c2^0 \mod N\rangle \otimes |c2^{n-1} \mod N\rangle$$

and differs from the initial state $|\omega_i\rangle$, but is always the same, independent of $|b\rangle \otimes |a\rangle$. This means that states in the auxiliary register remain separable from those in the input/output register (see discussion around Definition 5.29). One could, of course, transform $|\omega_f\rangle$ via suitable swap operations to $|\omega_i\rangle$, but to keep things as simple as possible, we have refrained from doing this here.

> **Theorem 5.44** *For any operator $U_{\times c\%N}$ defined as in Definition 5.43 we have*
>
> $$U_{\times c\%N}\left(|b\rangle \otimes |a\rangle\right) = |(b+ca) \mod N\rangle \otimes |a\rangle \qquad (5.116)$$
>
> *as well as*
>
> $$U_{\times c\%N}^*\left(|b\rangle \otimes |a\rangle\right) = |(b-ca) \mod N\rangle \otimes |a\rangle, \qquad (5.117)$$
>
> *and $U_{\times c\%N} : \mathbb{H}^{I/O} \to \mathbb{H}^{I/O}$ is unitary.*

Proof As we see in Fig. 5.22, the operator $U_{\times c\%N}$ consists of repeated additions $U_{+\%N}$ controlled by $|a_k\rangle$ with $k \in \{0, \ldots, n-1\}$, where, before each of these controlled additions, the state $|c2^k \mod N\rangle$ is swapped from the prepared auxiliary register to the entry register for the second summand. In the first step one has, because of $b < N$, after the addition controlled by $|a_0\rangle$ in $\mathbb{H}^{I/O}$ the state

$$\left(U_{+\%N}\right)^{a_0}\left(|b\rangle \otimes |c2^0 \mod N\rangle\right)$$
$$= \left(U_{+\%N}\right)^{a_0}\left(|b \mod N\rangle \otimes |c2^0 \mod N\rangle\right)$$
$$\underset{(5.113)}{=} |(b \mod N + a_0 c2^0 \mod N) \mod N\rangle \otimes |c2^0 \mod N\rangle$$
$$\underset{(D.23)}{=} |(b + a_0 c2^0) \mod N\rangle \otimes |c2^0 \mod N\rangle. \qquad (5.118)$$

After that, $|c2^1 \mod N\rangle$ is swapped into the input for the second summand, and the addition controlled by $|a_1\rangle$ is executed. In the kth step one has, analogously,

$$(U_{+\%N})^{a_k} \left(|(b + c \sum_{j=0}^{k-1} a_j 2^j) \mod N\rangle \otimes |c2^k \mod N\rangle \right)$$

$$= |\left((b + c \sum_{j=0}^{k-1} a_j 2^j) \mod N + a_k c2^k \mod N \right) \mod N\rangle \otimes |c2^k \mod N\rangle$$

$$= |(b + c \sum_{j=0}^{k} a_j 2^j) \mod N\rangle \otimes |c2^k \mod N\rangle.$$

After the last addition, the first channel of the adder controlled by $|a_{n-1}\rangle$ thus delivers the state

$$|(b + c \sum_{j=0}^{n-1} a_j 2^j) \mod N\rangle = |(b + ca) \mod N\rangle$$

in its output. In the second channel the state $|c2^{n-1} \mod N\rangle$ is delivered, which is swapped with $|0\rangle$. This last swap is not really necessary since even without it the auxiliary register is separable from the input/output register. This completes the proof of (5.116).

For the proof of (5.117) one notes that $U^*_{\times c\%N}$ corresponds to a reverse run through the circuit shown in Fig. 5.22 from right to left. This amounts to a circuit in which the steps of $U_{\times c\%N}$ are traversed in reverse order and where the initial state in the auxiliary register is now $|\omega_f\rangle$, and the final state is $|\omega_i\rangle$ as well as where $U_{+\%N}$ has to be replaced by $U^*_{+\%N} = U_{-\%N}$. Analogous to (5.118), the first step is then a subtraction of $c2^{n-1} \mod N$ controlled by $|a_{n-1}\rangle$. These controlled subtractions are continued until the last subtraction of $c2^0 \mod N$ controlled by $|a_0\rangle$. Altogether, the input $|b\rangle \otimes |a\rangle$ is transformed by $U^*_{\times c\%N}$ into $|(b - ca) \mod N\rangle \otimes |a\rangle$ as claimed in (5.117).

We thus have

$$
\begin{aligned}
U^*_{\times c\%N} U_{\times c\%N} (|b\rangle \otimes |a\rangle) &= U^*_{\times c\%N} (|(b + ca) \mod N\rangle \otimes |a\rangle) \\
&= |((b + ca) \mod N - ca) \mod N\rangle \otimes |a\rangle \\
&\underbrace{=}_{\text{(D.23)}} |((b + ca) \mod N - ca \mod N) \mod N\rangle \otimes |a\rangle \\
&\underbrace{=}_{\text{(D.23)}} |b \mod N\rangle \otimes |a\rangle \\
&= |b\rangle \otimes |a\rangle
\end{aligned}
$$

proving unitarity of $U_{\times c\%N}$ on $\mathbb{H}^{I/O} = \mathbb{H}^{<N} \otimes \mathbb{H}^{\otimes n}$. $\square$

Its name as a multiplier deserves $U_{\times c \% N}$ because

$$U_{\times c \% N}\left(|0\rangle \otimes |a\rangle\right) = |ca \mod N\rangle \otimes |a\rangle$$

holds. With the help of the construction in Theorem 5.34, the function $a \mapsto ca \mod N$ can then be implemented.

5.5.4 Quantum Circuit for Exponentiation Modulo N

At long last we are now in a position to present a way to implement the function $f_{b,N}(x) = b^x \mod N$ by means of a quantum circuit.

Definition 5.45 For $b, n, N \in \mathbb{N}$ we define $A_{f_{b,N}}$ on $\P\mathbb{H}^{\otimes n} \otimes \mathbb{H}^{<N}$ as the circuit shown in Fig. 5.23 with the state $|\omega_i\rangle = \left(\bigotimes_{l=0}^{n-2} |0\rangle\right) \otimes |1\rangle = |\omega_f\rangle$ in the auxiliary register $\mathbb{H}^W = \left(\mathbb{H}^{<N}\right)^{\otimes n}$.

The implementation of $A_{f_{b,N}}$ is essentially a version of the fast or so-called **binary exponentiation** with quantum circuits for a given b. In this construction the numbers $\beta_0 := b^{2^0} \mod N, \ldots, \beta_{n-1} := b^{2^{n-1}} \mod N$ are pre-calculated with a classical computer and then the quantum multipliers $U_{\times \beta_j \% N}$ are prepared.

In Definition 5.45 we have $A_{f_{b,N}}$ restricted in the second argument to $\mathbb{H}^{<N}$, but for $N < 2^m$ we can view $\mathbb{H}^{<N}$ as a subspace of $\P\mathbb{H}^{\otimes m}$. We use this in Theorem 5.46.

Theorem 5.46 *Let $b, n, N, m \in \mathbb{N}$ with $N < 2^m$ and $f_{b,N}(x) := b^x \mod N$. Then for any $x \in \mathbb{N}_0$ with $x < 2^n$*

$$A_{f_{b,N}}\left(|x\rangle^n \otimes |0\rangle^m\right) = |x\rangle^n \otimes |f_{b,N}(x)\rangle^m \qquad (5.119)$$

holds as well as

$$A_{f_{b,N}}^*\left(|x\rangle^n \otimes |f_{b,N}(x)\rangle^m\right) = |x\rangle^n \otimes |0\rangle^m . \qquad (5.120)$$

Proof In Fig. 5.23 we use the abbreviating notation $\beta_j = b^{2^j} \mod N$. We see there that the first part of $A_{f_{b,N}}$ consists of successive applications of $U_{\times \beta_j \% N}$ and then $U_{+\% N}$. Each of these multiplications is controlled by a $|x_j\rangle$. For these one has, in general, for $s \in \{0,1\}, c \in \mathbb{N}_0$ and $|a\rangle \in \mathbb{H}^{<N}$ that

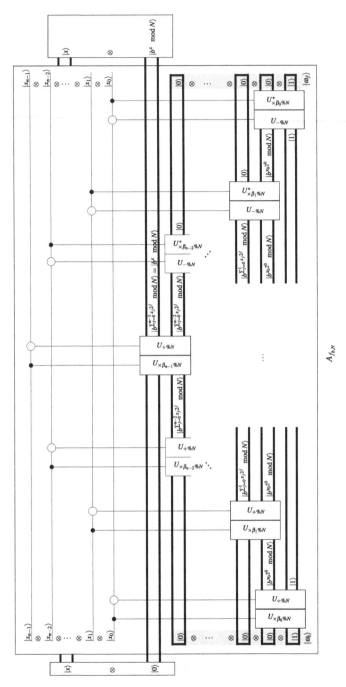

Fig. 5.23 Quantum circuit $A_{f_{b,N}}$ to implement the function $f_{b,N}(x) = b^x \mod N$; here we use the abbreviating notation $\beta_j = b^{2^j} \mod N$

$$\left(U_{+\%N}\right)^{1-s}\left(U_{\times c\%N}\right)^{s}\left(|0\rangle\otimes|a\rangle\right) \underset{(5.116)}{=} \begin{cases} \left(U_{+\%N}\right)\left(|0\rangle\otimes|a\rangle\right) & \text{if } s=0 \\ |ca \mod N\rangle\otimes|a\rangle & \text{if } s=1 \end{cases}$$

$$\underset{(5.113)}{=} \begin{cases} |a \mod N\rangle\otimes|a\rangle & \text{if } s=0 \\ |ca \mod N\rangle\otimes|a\rangle & \text{if } s=1 \end{cases}$$

$$= |c^{s}a \mod N\rangle\otimes|a\rangle. \qquad (5.121)$$

Beginning with the first two factors of the auxiliary register, one has then

$$\left(U_{+\%N}\right)^{1-x_0}\left(U_{\times \beta_0\%N}\right)^{x_0}\left(|0\rangle\otimes|1\rangle\right) \underset{(5.121)}{=} |\beta_0^{x_0} \mod N\rangle\otimes|1\rangle$$

$$= |\left(b^{2^0} \mod N\right)^{x_0} \mod N\rangle\otimes|1\rangle$$

$$\underset{(D.22)}{=} |b^{x_0 2^0} \mod N\rangle\otimes|1\rangle.$$

In the kth step this becomes

$$\left(U_{+\%N}\right)^{1-x_k}\left(U_{\times\beta_k\%N}\right)^{x_k}\left(|0\rangle\otimes|b^{\sum_{j=0}^{k-1}x_j 2^j} \mod N\rangle\right)$$

$$\underset{(5.121)}{=} |\left(\beta_k^{x_k}\left(b^{\sum_{j=0}^{k-1}x_j 2^j} \mod N\right)\right) \mod N\rangle\otimes|b^{\sum_{j=0}^{k-1}x_j 2^j} \mod N\rangle$$

$$= |\left(\left(b^{2^k} \mod N\right)^{x_k}\left(b^{\sum_{j=0}^{k-1}x_j 2^j} \mod N\right)\right) \mod N\rangle\otimes|b^{\sum_{j=0}^{k-1}x_j 2^j} \mod N\rangle$$

$$\underset{(D.21),(D.22)}{=} |b^{\sum_{j=0}^{k}x_j 2^j} \mod N\rangle\otimes|b^{\sum_{j=0}^{k-1}x_j 2^j} \mod N\rangle.$$

After the application of $\left(U_{+\%N}\right)^{1-x_{n-1}}\left(U_{\times\beta_{n-1}\%N}\right)^{x_{n-1}}$ one then has in the second input/-output channel as desired

$$|b^{\sum_{j=0}^{n-1}x_j 2^j} \mod N\rangle = |b^x \mod N\rangle.$$

This remains unchanged by subsequent applications of the

$$\left(U^*_{\times\beta_k\%N}\right)^{x_k}\left(U^*_{+\%N}\right)^{1-x_k} = \left(U^*_{\times\beta_k\%N}\right)^{x_k}\left(U_{-\%N}\right)^{1-x_k}$$

for $k \in \{n-2,\dots,0\}$. These conditional operators disentangle the auxiliary register from the input/output register as can be seen as follows. First, we have, in analogy to (5.121), for $s \in \{0,1\}, c \in \mathbb{N}_0$ and $|u\rangle,|v\rangle \in \mathbb{H}^{<N}$ that

$$\left(U^*_{\times c\%N}\right)^{s}\left(U_{-\%N}\right)^{1-s}\left(|u\rangle\otimes|v\rangle\right) \underset{(5.114)}{=} \begin{cases} |(u-v) \mod N\rangle\otimes|v\rangle & \text{if } s=0 \\ U^*_{\times c\%N}\left(|u\rangle\otimes|v\rangle\right) & \text{if } s=1 \end{cases}$$

$$\underbrace{=}_{(5.117)} \begin{cases} |(u-v) \mod N\rangle \otimes |v\rangle & \text{if } s=0 \\ |(u-cv) \mod N\rangle \otimes |v\rangle & \text{if } s=1 \end{cases}$$

$$= |(u-c^s v) \mod N\rangle \otimes |v\rangle. \tag{5.122}$$

In the second part of $A_{f_b,N}$ one obtains thus with the kth step

$$\left(U^*_{\times\beta_k \%N}\right)^{x_k} \left(U_{-\%N}\right)^{1-x_k} \left(|b^{\sum_{j=0}^{k} x_j 2^j} \mod N\rangle \otimes |b^{\sum_{j=0}^{k-1} x_j 2^j} \mod N\rangle\right)$$

$$\underbrace{=}_{(5.122)} |\left(b^{\sum_{j=0}^{k} x_j 2^j} \mod N - \left(b^{2^k} \mod N\right)^{x_k} \left(b^{\sum_{j=0}^{k-1} x_j 2^j} \mod N\right)\right) \mod N\rangle$$

$$\otimes |b^{\sum_{j=0}^{k-1} x_j 2^j} \mod N\rangle$$

$$\underbrace{=}_{(D.21)-(D.23)} |0\rangle \otimes |b^{\sum_{j=0}^{k-1} x_j 2^j} \mod N\rangle. \tag{5.123}$$

In particular, for $k = 0$ we have

$$\left(U^*_{\times\beta_0 \%N}\right)^{x_0} \left(U_{-\%N}\right)^{1-x_0} \left(|b^{x_0 2^0} \mod N\rangle \otimes |1\rangle\right)$$

$$\underbrace{=}_{(5.122)} |\left(b^{x_0 2^0} \mod N - \left(b^{2^0} \mod N\right)^{x_0}\right) \mod N\rangle \otimes |1\rangle$$

$$= |0\rangle \otimes |1\rangle.$$

This proves (5.119).

In order to show (5.120), consider that $A^*_{f_b,N}$ emerges from the circuit of $A_{f_b,N}$ by the replacements

$$U_{+\%N} U_{\times\beta_k \%N} \rightarrow U^*_{\times\beta_k \%N} U_{-\%N} \quad \text{for } k \in \{0,\dots,n-1\}$$
$$U^*_{\times\beta_k \%N} U_{-\%N} \rightarrow U_{+\%N} U_{\times\beta_k \%N} \quad \text{for } k \in \{n-2,\dots,0\}.$$

Due to the symmetry of the circuit, this means that the circuit of $A^*_{f_b,N}$ only differs from the one for $A_{f_b,N}$ in that in place of $(U_{+\%N})^{1-x_{n-1}} (U_{\times\beta_{n-1} \%N})^{x_{n-1}}$ then $(U^*_{\times\beta_{n-1} \%N})^{x_{n-1}} (U_{-\%N})^{1-x_{n-1}}$ appears. This implies that

$$\left(U^*_{\times\beta_{n-1} \%N}\right)^{x_{n-1}} \left(U_{-\%N}\right)^{1-x_{n-1}} \left(|b^x \mod N\rangle \otimes |b^{\sum_{j=0}^{n-2} x_j 2^j} \mod N\rangle\right)$$

$$= \left(U^*_{\times\beta_{n-1} \%N}\right)^{x_{n-1}} \left(U_{-\%N}\right)^{1-x_{n-1}} \left(|b^{\sum_{j=0}^{n-1} x_j 2^j} \mod N\rangle \otimes |b^{\sum_{j=0}^{n-2} x_j 2^j} \mod N\rangle\right)$$

$$\underbrace{=}_{(5.123)} |0\rangle \otimes |b^{\sum_{j=0}^{n-2} x_j 2^j} \mod N\rangle$$

holds and (5.120) follows. $\square$

With the help of the construction in Theorem 5.34 we can then implement the function $x \mapsto b^x \mod N$.

Corollary 5.47 *Let $b, n, N, m \in \mathbb{N}$ with $N < 2^m$ and $\mathbb{H}^A = \mathbb{H}^{\otimes n}$ as well as $\mathbb{H}^B = \mathbb{H}^{\otimes m}$ and $f_{b,N}(x) = b^x \mod N$. With the help of the states $|\omega_i\rangle = |0\rangle^m = |\omega_f\rangle$ in the auxiliary register $\mathbb{H}^B$ we can implement $U_{f_{b,N}}$ for which*

$$U_{f_{b,N}} : \mathbb{H}^A \otimes \mathbb{H}^B \longrightarrow \mathbb{H}^A \otimes \mathbb{H}^B$$
$$|x\rangle \otimes |y\rangle \longmapsto |x\rangle \otimes |y \boxplus f_{b,N}(x)\rangle$$

holds. In particular, one has

$$U_{f_{b,N}}(|x\rangle \otimes |0\rangle) = |x\rangle \otimes |f_{b,N}(x)\rangle.$$

Proof The claim follows from Theorem 5.34 by setting there $A_f = A_{f_{b,N}}$ and $B_f = A^*_{f_{b,N}}$ from Theorem 5.46. □

The statement in Corollary 5.47 is essential for the SHOR factorization algorithm in Sect. 6.5. This also applies to the quantum FOURIER transform, which is covered in the following section.

5.5.5 Quantum FOURIER Transform

The quantum FOURIER transform [73] is an important part of several algorithms and serves as a further example how elementary gates can be used to build a unitary transformation. It is defined as an operator on the tensor product of the qubit space $\mathbb{H}$ as follows.

Definition 5.48 The **quantum FOURIER transform** F on $\mathbb{H}^{\otimes n}$ is defined as the operator

$$F := \frac{1}{\sqrt{2^n}} \sum_{x,y=0}^{2^n-1} \exp\left(2\pi i \frac{xy}{2^n}\right) |x\rangle\langle y|,$$

where $|x\rangle$ and $|y\rangle$ denote vectors of the computational basis of $\mathbb{H}^{\otimes n}$.

In fact, as shown in Example F.52, the definition given above is a special case of the quantum FOURIER transform on groups given in Definition F.51.

With the help of

$$\omega_n := \exp\left(\frac{2\pi\mathrm{i}}{2^n}\right)$$

the matrix representation of F in the computational basis can be given as

$$F = \frac{1}{\sqrt{2^n}}\begin{pmatrix} 1 & 1 & \cdots & 1 \\ 1 & \omega_n & \cdots & \omega_n^{2^n-1} \\ \vdots & \vdots & \ddots & \vdots \\ 1 & \omega_n^{2^n-1} & \cdots & \omega_n^{(2^n-1)^2} \end{pmatrix}.$$

Exercise 5.66 Show that F is unitary.

For a solution see Solution 5.66.

There is a connection between the quantum FOURIER transform and the **discrete FOURIER transform** used in signal-processing.

Definition 5.49 Let $N \in \mathbb{N}$. The discrete FOURIER transform is a linear map

$$F_{dis} : \mathbb{C}^N \longrightarrow \mathbb{C}^N$$
$$\mathbf{c} \longmapsto F_{dis}(\mathbf{c})$$

defined component-wise by

$$F_{dis}(\mathbf{c})_k = \frac{1}{\sqrt{N}}\sum_{l=0}^{N-1}\exp\left(\frac{2\pi\mathrm{i}}{N}kl\right)c_l. \qquad (5.124)$$

For an arbitrary vector $|\Psi\rangle \in \mathbb{H}^{\otimes n}$ it then follows that the components of the quantum FOURIER transformed vector $(F|\Psi\rangle)_x = \langle x|F\Psi\rangle$ in the computational basis are given by the discrete FOURIER transform $F_{dis}(\mathbf{c})_x$ of a vector $\mathbf{c} \in \mathbb{C}^N$ that has the components $c_x = \langle x|\Psi\rangle$.

Lemma 5.50 *Let $n, N \in \mathbb{N}$ with $N = 2^n$ and*

$$|\Psi\rangle = \sum_{x=0}^{2^n-1}\Psi_x|x\rangle \quad \in \mathbb{H}^{\otimes n}.$$

Moreover, let $\mathbf{c} \in \mathbb{C}^N$ be the vector with components $c_x = \Psi_x = \langle x | \Psi \rangle$ for $x \in \{0, \ldots, N-1\}$. Then we have

$$\langle k | F\Psi \rangle = F_{dis}(\mathbf{c})_k .$$

Proof One has

$$
\begin{aligned}
F|\Psi\rangle &= \sum_{x=0}^{2^n-1} \Psi_x \underbrace{F|x\rangle}_{(5.124)} = \sum_{x=0}^{2^n-1} \Psi_x \frac{1}{\sqrt{2^n}} \sum_{z,y=0}^{2^n-1} \exp\left(2\pi i \frac{zy}{2^n}\right) |z\rangle \underbrace{\langle y|x\rangle}_{=\delta_{xy}} \\
&= \frac{1}{\sqrt{2^n}} \sum_{x,z=0}^{2^n-1} \Psi_x \exp\left(\frac{2\pi i}{2^n} zx\right) |z\rangle \\
&\underset{\substack{N=2^n, c_x = \Psi_x}}{=} \frac{1}{\sqrt{N}} \sum_{x,z=0}^{N-1} c_x \exp\left(\frac{2\pi i}{N} zx\right) |z\rangle
\end{aligned}
$$

and thus

$$
\begin{aligned}
\langle k | F\Psi \rangle &= \frac{1}{\sqrt{N}} \sum_{x,z=0}^{N-1} c_x \exp\left(\frac{2\pi i}{N} zx\right) \underbrace{\langle k|z\rangle}_{=\delta_{kz}} = \frac{1}{\sqrt{N}} \sum_{x=0}^{N-1} c_x \exp\left(\frac{2\pi i}{N} kx\right) \\
&\underbrace{=}_{(5.124)} F_{dis}(\mathbf{c})_z .
\end{aligned}
$$

$\square$

We further introduce the following notation for **binary fractions**:

Definition 5.51 For $a_1, \ldots, a_m \in \{0, 1\}$ we define

$$0.a_1 a_2 \ldots a_m := \frac{a_1}{2} + \frac{a_2}{4} + \cdots \frac{a_m}{2^m} = \sum_{l=1}^{m} a_l 2^{-l} . \qquad (5.125)$$

With the help of this notation for binary fractions the quantum FOURIER transform can be represented as follows.

Lemma 5.52 *Let $n \in \mathbb{N}$ and*

$$x = \sum_{j=0}^{n-1} x_j 2^j, \tag{5.126}$$

where $x_j \in \{0,1\}$ for $j \in \{0,\ldots,n-1\}$.

Then the action of the quantum FOURIER *transform F on any vector $|x\rangle$ of the computational basis of $\mathbb{H}^n$ can be written as*

$$F|x\rangle = \frac{1}{\sqrt{2^n}} \bigotimes_{j=0}^{n-1} \left[|0\rangle + e^{2\pi i 0.x_j \ldots x_0} |1\rangle \right]. \tag{5.127}$$

Proof According to Definition 5.48 one has

$$
\begin{aligned}
F|x\rangle &= \frac{1}{\sqrt{2^n}} \sum_{y=0}^{2^n-1} \exp\left(\frac{2\pi i}{2^n} xy \right) |y\rangle \\
&= \frac{1}{\sqrt{2^n}} \sum_{y=0}^{2^n-1} \exp\left(\frac{2\pi i}{2^n} x \sum_{j=0}^{n-1} y_j 2^j \right) |y_{n-1} \ldots y_0\rangle \\
&= \frac{1}{\sqrt{2^n}} \sum_{y=0}^{2^n-1} \prod_{j=0}^{n-1} \exp\left(\frac{2\pi i}{2^n} x y_j 2^j \right) |y_{n-1} \ldots y_0\rangle \\
&= \frac{1}{\sqrt{2^n}} \sum_{y_0 \ldots y_{n-1} \in \{0,1\}} \prod_{j=0}^{n-1} \exp\left(\frac{2\pi i}{2^n} x y_j 2^j \right) \bigotimes_{k=n-1}^{0} |y_k\rangle \\
&= \frac{1}{\sqrt{2^n}} \sum_{y_0 \ldots y_{n-1} \in \{0,1\}} \bigotimes_{k=n-1}^{0} \exp\left(\frac{2\pi i}{2^n} x y_k 2^k \right) |y_k\rangle \\
&= \frac{1}{\sqrt{2^n}} \bigotimes_{k=n-1}^{0} \sum_{y_k \in \{0,1\}} \exp\left(\frac{2\pi i}{2^n} x y_k 2^k \right) |y_k\rangle \\
&= \frac{1}{\sqrt{2^n}} \bigotimes_{k=n-1}^{0} \left[|0\rangle + \exp\left(\frac{2\pi i}{2^n} x 2^k \right) |1\rangle \right]. \tag{5.128}
\end{aligned}
$$

In the last equation we further use (5.126) and the notation for binary fractions given in (5.125) in order to obtain

$$\exp\left(\frac{2\pi i}{2^n}x2^k\right) = \exp\left(2\pi i \sum_{l=0}^{n-1} x_l 2^{l+k-n}\right)$$

$$= \exp\left(2\pi i \left[\sum_{l=0}^{n-k-1} x_l 2^{l+k-n} + \underbrace{\sum_{l=n-k}^{n-1} x_l 2^{l+k-n}}_{\in\mathbb{N}}\right]\right)$$

$$= \exp\left(2\pi i \sum_{l=0}^{n-k-1} x_l 2^{l+k-n}\right)$$

$$= \exp\left(2\pi i \left[\frac{x_0}{2^{n-k}} + \frac{x_1}{2^{n-(k+1)}} + \cdots + \frac{x_{n-1-k}}{2}\right]\right)$$

$$= e^{2\pi i 0.x_{n-1-k}\cdots x_0}. \tag{5.129}$$

Insertion of (5.129) into (5.128) then yields

$$F|x\rangle = \frac{1}{\sqrt{2^n}} \bigotimes_{k=n-1}^{0} \left[|0\rangle + e^{2\pi i 0.x_{n-1-k}\cdots x_0}|1\rangle\right] = \frac{1}{\sqrt{2^n}} \bigotimes_{j=0}^{n-1} \left[|0\rangle + e^{2\pi i 0.x_j\cdots x_0}|1\rangle\right].$$

□

We still need the following result for the HADAMARD transform if we want to express the quantum FOURIER transform with the help of elementary gates.

Lemma 5.53 *Let $n \in \mathbb{N}$ and $j \in \mathbb{N}_0$ with $j < n$ and let $|x\rangle$ be a vector in the computational basis in $\mathbb{H}^{\otimes n}$. Then*

$$H|x_j\rangle = \frac{|0\rangle + e^{2\pi i 0.x_j}|1\rangle}{\sqrt{2}}, \tag{5.130}$$

holds, and with

$$H_j := \mathbf{1}^{\otimes(n-1-j)} \otimes H \otimes \mathbf{1}^{\otimes j} \tag{5.131}$$

for $j \in \{0,\ldots,n-1\}$ one has

$$H_j|x\rangle = |x_{n-1}\rangle \otimes \cdots \otimes |x_{j+1}\rangle \otimes \frac{|0\rangle + e^{2\pi i 0.x_j}|1\rangle}{\sqrt{2}} \otimes |x_{j-1}\rangle \otimes \cdots \otimes |x_0\rangle. \tag{5.132}$$

Proof From (2.162) in Lemma 2.39 we know that

$$H|x_j\rangle = \frac{|0\rangle + e^{\pi i x_j}|1\rangle}{\sqrt{2}}$$

holds. Then (5.130) follows from the Definition 5.51 of the binary fraction.

The action of H_j as claimed in (5.132) follows directly from definition (5.131) and (5.130). □

A further transformation needed to build the quantum FOURIER transform is the conditional phase shift.

Definition 5.54 Let $j, k \in \{0, \ldots, n-1\}$ with $j > k$ and define $\theta_{jk} := \frac{\pi}{2^{j-k}}$. The **conditional phase shift** is a linear transformation on $\mathbb{H}^{\otimes n}$ defined as

$$P_{jk} := \mathbf{1}^{\otimes(n-1-k)} \otimes |0\rangle\langle 0| \otimes \mathbf{1}^{\otimes k}$$
$$+ \mathbf{1}^{\otimes(n-1-j)} \otimes \left[|0\rangle\langle 0| + e^{i\theta_{jk}}|1\rangle\langle 1|\right] \otimes \mathbf{1}^{j-k-1} \otimes |1\rangle\langle 1| \otimes \mathbf{1}^{\otimes k}.$$

The action of P_{jk} is an application of $\Lambda_{|1\rangle^1}\left(P(\theta_{jk})\right)$ on the $(k+1)$-th and $(j+1)$-th factor space in $\mathbb{H}^{\otimes n}$ (see Figs. 5.4 and 5.5 for the definition of $\Lambda_{|1\rangle}(V)$ and $P(\alpha)$). This can be illustrated if we consider the restriction onto the respective subspaces. Let $\mathbb{H}_k$ denote the $(k+1)$-th factor space counted from the right and $\mathbb{H}_j$ the $(j+1)$-th factor space in $\mathbb{H}^{\otimes n}$ (see (3.18)). Furthermore, let

$$|0\rangle_j \otimes |0\rangle_k, \quad |1\rangle_j \otimes |0\rangle_k, \quad |0\rangle_j \otimes |1\rangle_k, \quad |1\rangle_j \otimes |1\rangle_k$$

be the four vectors of the computational basis in $\mathbb{H}_j \otimes \mathbb{H}_k$. Then the matrix representation of the restriction to these two factor spaces is given by

$$P_{jk}\Big|_{\mathbb{H}_j \otimes \mathbb{H}_k} = \begin{pmatrix} 1 & 0 & 0 & 0 \\ 0 & 1 & 0 & 0 \\ 0 & 0 & 1 & 0 \\ 0 & 0 & 0 & e^{i\theta_{jk}} \end{pmatrix} = \Lambda_{|1\rangle^1}\left(P(\theta_{jk})\right).$$

The action of P_{jk} is thus only non-trivial if both states in the $(j+1)$-th as well as the $(k+1)$-th factor space each have a non-vanishing component along $|1\rangle$ in which case it consists of a multiplication with a phase factor $e^{i\theta_{jk}}$. Otherwise, P_{jk} leaves the total state unchanged, in other words, acts as the identity.

Lemma 5.55 Let $j, k \in \{0, \ldots, n-1\}$ with $j > k$ and let $l \in \{j+1, \ldots, n-1\}$. Moreover, let $|\psi_l\rangle \in \mathbb{H}$ and $\psi_{0j}, \psi_{1j} \in \mathbb{C}$ as well as $x_0, \ldots, x_{j-1} \in \{0, 1\}$. Then we have

$$P_{jk}|\psi_{n-1}\rangle \otimes \cdots \otimes |\psi_{j+1}\rangle \otimes \big[\psi_{0j}|0\rangle + \psi_{1j}|1\rangle\big] \otimes |x_{j-1}\rangle \otimes \cdots \otimes |x_0\rangle$$

$$= |\psi_{n-1}\rangle \otimes \cdots \otimes |\psi_{j+1}\rangle \otimes \big[\psi_{0j}|0\rangle + \psi_{1j}e^{i\pi\frac{x_k}{2^{j-k}}}|1\rangle\big] \otimes |x_{j-1}\rangle \otimes \cdots \otimes |x_0\rangle.$$

Proof With $|x_k\rangle = (1-x_k)|0\rangle + x_k|1\rangle$ and $\theta_{jk} = \frac{\pi}{2^{j-k}}$ one has

$$P_{jk}|\psi_{n-1}\rangle \otimes \cdots \otimes |\psi_{j+1}\rangle \otimes \big[\psi_{0j}|0\rangle + \psi_{1j}|1\rangle\big] \otimes |x_{j-1}\rangle \otimes \cdots \otimes |x_0\rangle$$

$$= (1-x_k)|\psi_{n-1}\rangle \otimes \cdots \otimes |\psi_{j+1}\rangle \otimes \big[\psi_{0j}|0\rangle + \psi_{1j}|1\rangle\big] \otimes |x_{j-1}\rangle \otimes \cdots \otimes |x_0\rangle$$

$$+ x_k|\psi_{n-1}\rangle \otimes \cdots \otimes |\psi_{j+1}\rangle \otimes \big[\psi_{0j}|0\rangle + \psi_{1j}e^{i\theta_{jk}}|1\rangle\big] \otimes |x_{j-1}\rangle \otimes \cdots \otimes |x_0\rangle$$

$$= |\psi_{n-1}\rangle \otimes \cdots \otimes |\psi_{j+1}\rangle \otimes \big[\psi_{0j}|0\rangle + \psi_{1j}e^{i\pi\frac{x_k}{2^{j-k}}}|1\rangle\big] \otimes |x_{j-1}\rangle \otimes \cdots \otimes |x_0\rangle.$$

$\square$

Up to a re-ordering, that is, a reversal of the factors in the nth fold tensor product $\mathbb{H}^{\otimes n}$, the quantum FOURIER transform can be built as a product of HADAMARD transforms and conditional phase shifts.

Theorem 5.56 *The quantum* FOURIER *transform F can be built from the swap operator $S^{(n)}$ defined in (5.30),* HADAMARD *transforms and conditional phase shifts as follows:*

$$F = S^{(n)} \prod_{j=0}^{n-1} \left(\left[\prod_{k=0}^{j-1} P_{jk} \right] H_j \right) \tag{5.133}$$

$$= S^{(n)} H_0 P_{1,0} H_1 P_{2,0} P_{2,1} H_2 \ldots P_{n-1,0} \ldots P_{n-1,n-2} H_{n-1}.$$

Proof With (5.132) we have at first

$$H_{n-1}|x\rangle = \frac{|0\rangle + e^{2\pi i 0.x_{n-1}}|1\rangle}{\sqrt{2}} \otimes |x_{n-2}\rangle \otimes \cdots \otimes |x_0\rangle.$$

According to Lemma 5.55, this implies

$$P_{n-1,n-2}H_{n-1}|x\rangle = \frac{|0\rangle + e^{2\pi i 0.x_{n-1}+i\pi\frac{x_{n-2}}{2}}|1\rangle}{\sqrt{2}} \otimes |x_{n-2}\rangle \otimes \cdots \otimes |x_0\rangle$$

$$\underset{(5.125)}{=} \frac{|0\rangle + e^{2\pi i 0.x_{n-1}x_{n-2}}|1\rangle}{\sqrt{2}} \otimes |x_{n-2}\rangle \otimes \cdots \otimes |x_0\rangle$$

and

$$P_{n-1,0} P_{n-1,1} \ldots P_{n-1,n-2} H_{n-1} |x\rangle$$

$$= \frac{|0\rangle + e^{2\pi i 0.x_{n-1}\ldots x_0}|1\rangle}{\sqrt{2}} \otimes |x_{n-2}\rangle \otimes \cdots \otimes |x_0\rangle \,.$$

Furthermore,

$$H_{n-2} P_{n-1,0} P_{n-1,1} \ldots P_{n-1,n-2} H_{n-1} |x\rangle$$

$$= \frac{|0\rangle + e^{2\pi i 0.x_{n-1}\ldots x_0}|1\rangle}{\sqrt{2}} \otimes \frac{|0\rangle + e^{2\pi i 0.x_{n-2}}|1\rangle}{\sqrt{2}} \otimes |x_{n-3}\rangle \otimes \cdots \otimes |x_0\rangle$$

and

$$P_{n-2,0} P_{n-2,1} \ldots P_{n-2,n-3} H_{n-2} P_{n-1,0} P_{n-1,1} \ldots P_{n-1,n-2} H_{n-1} |x\rangle$$

$$= \frac{|0\rangle + e^{2\pi i 0.x_{n-1}\ldots x_0}|1\rangle}{\sqrt{2}} \otimes \frac{|0\rangle + e^{2\pi i 0.x_{n-2}\ldots x_0}|1\rangle}{\sqrt{2}} \otimes |x_{n-3}\rangle \otimes \cdots \otimes |x_0\rangle \,.$$

Similarly, a repeated application to the remaining tensor products $|x_{n-3}\rangle \otimes \cdots \otimes |x_0\rangle$ then yields

$$\prod_{j=0}^{n-1} \left(\prod_{k=0}^{j-1} (P_{jk}) H_j \right) |x\rangle = \frac{1}{\sqrt{2^n}} \bigotimes_{k=n-1}^{0} \left[|0\rangle + e^{2\pi i 0.x_k \ldots x_0}|1\rangle \right] \,. \qquad (5.134)$$

This is $F|x\rangle$ up to an inversion of the sequence of the factor spaces. Thus, we have finally

$$S^{(n)} \prod_{j=0}^{n-1} \left(\prod_{k=0}^{j-1} (P_{jk}) H_j \right) |x\rangle \underset{(5.134)}{=} \frac{1}{\sqrt{2^n}} S^{(n)} \bigotimes_{k=n-1}^{0} \left[|0\rangle + e^{2\pi i 0.x_k \ldots x_0}|1\rangle \right]$$

$$\underset{(5.34)}{=} \frac{1}{\sqrt{2^n}} \bigotimes_{k=0}^{n-1} \left[|0\rangle + e^{2\pi i 0.x_k \ldots x_0}|1\rangle \right]$$

$$\underset{(5.127)}{=} F|x\rangle \,.$$

$\square$

Figure 5.24 shows the circuit comprised of gates, which generate the quantum FOURIER transform.

The representation (5.133) of $F|x\rangle$ with the help of H_j, P_{jk} and $S^{(n)}$ allows us, to provide bounds for the growth rate of the computational steps required for the quantum FOURIER transform.

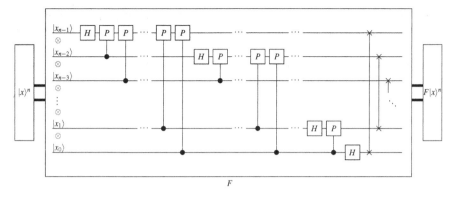

Fig. 5.24 Quantum circuit to build the quantum FOURIER transform with the help of HADAMARD gates, conditional phase shifts and swap gates

Corollary 5.57 *Let F be the quantum* FOURIER *transform on* $\mathbb{H}^{\otimes n}$. *Then the number of required computational steps* S_F *to perform F as a function of n satisfies*

$$S_F(n) \in O(n^2) \text{ for } n \to \infty.$$

Proof The application of H_j and P_{jk} each require a fixed number of computational steps independent of n: $S_{H_j}(n), S_{P_{jk}}(n) \in O(1)$. The application of $S^{(n)}$ requires instead $S_{S^{(n)}}(n) \in O(n)$ steps for $n \to \infty$. Because of (5.133) from Theorem 5.56 we can perform F by

- an n-fold application of H_j for $j \in \{0,\dots,n-1\}$, with $S_H(n) \in O(n)$,
- \+ an $\frac{n(n-1)}{2}$-fold application of P_{jk} for $j \in \{0,\dots,n-1\}$ and $k \in \{0,\dots,j-1\}$ with $S_P(n) \in O(n^2)$
- \+ a single application of $S^{(n)}$ with $S_{S^{(n)}}(n) \in O(n)$.

Hence, we find for the number of computational steps S_F to perform F

$$S_F(n) = S_H(n) + S_P(n) + \underbrace{S_{S^{(n)}}(n) \in O(n^2)}_{C.3}$$

as claimed.   □

5.6   Further Reading

One of the earlier introductions to quantum gates can be found in the paper by DEUTSCH [17]. This paper already addresses the question of universality which is discussed in more detail in the paper by BARENCO et al. [30]. Gates for explicit arithmetic operations up to modular exponentiation were presented by VEDRAL et al. [72].

Many quantum gates and circuits for various applications can found in the introductory text by MERMIN [74] as well as in the comprehensive and wide ranging book by NIELSEN and CHUANG [61].

Chapter 6
On the Use of Entanglement

6.1 Early Promise: DEUTSCH–JOZSA Algorithm

In 1992 DEUTSCH and JOZSA [75] devised a problem and a quantum algorithm
for solving it, which showed how dramatically more efficient than their classical
counterparts quantum algorithms could be. This problem has since become known
as DEUTSCH's problem[1] and the quantum algorithm to solve it efficiently as the
DEUTSCH–JOZSA algorithm.

Definition 6.1 (*Deutsch's Problem*) Let $n \in \mathbb{N}$ and $f : \{0,1\}^n \to \{0,1\}$ be a
function of which we know that it is

either constant,   that is,

$$f(x) = c \in \{0,1\} \qquad \text{for all } x \in \{0,1\}^n$$

or balanced,   that is,

$$f(x) = \begin{cases} 0 & \text{for one half of the } x \in \{0,1\}^n \\ 1 & \text{for the other half of the } x \in \{0,1\}^n \end{cases}$$

DEUTSCH'S Problem is to find the most efficient way to decide with certainty
whether f is constant or balanced.

We measure efficiency of a method by the number of function calls, that is, by the
number of evaluations of f or related objects that have to be made to gain absolute
certainty. We may visualize this by being able to submit queries to f via a keyboard.

---

[1] Named after him because DEUTSCH [16] had first considered a version of it already in 1985.

© Springer Nature Switzerland AG 2019
W. Scherer, *Mathematics of Quantum Computing*,
https://doi.org/10.1007/978-3-030-12358-1_6

In order to ascertain whether f is constant or balanced, any classical method will require at least $2^{n-1} + 1$ queries. Only then can we be convinced about f being constant or balanced.

In contrast, the quantum algorithm devised by DEUTSCH and JOZSA [75] requires just *one* application of a suitably implemented unitary operator U_f of the form (5.84). In other words, if the keyboard were connected to a quantum computer that has a circuit implementing U_f, we would need to enter only one query (that is, one execution of U_f) to ascertain which type f is.

Proposition 6.2 *Let f be as in Definition 6.1 and*

$$U_f : \mathbb{H}^{\otimes n} \otimes \mathbb{H} \longrightarrow \mathbb{H}^{\otimes n} \otimes \mathbb{H}$$
$$|x\rangle \otimes |y\rangle \longmapsto |x\rangle \otimes |y \overset{2}{\oplus} f(x)\rangle . \tag{6.1}$$

Then there is a quantum algorithm which uses only one application of U_f and solves DEUTSCH's problem.

Proof We devise an algorithm that solves the problem with only one application of U_f as follows. Recall from (5.83) that the HADAMARD transform satisfies

$$H^{\otimes n}|0\rangle = \frac{1}{2^{\frac{n}{2}}} \sum_{x=0}^{2^n - 1} |x\rangle . \tag{6.2}$$

Moreover, for any computational basis vector $|x\rangle$ the action of $H^{\otimes n}$ can be expressed as given in Exercise 6.67.

Exercise 6.67 Show that

$$H^{\otimes n}|x\rangle = \frac{1}{2^{\frac{n}{2}}} \sum_{y=0}^{2^n - 1} (-1)^{x \overset{2}{\odot} y} |y\rangle , \tag{6.3}$$

where we set for the computational basis vectors $|x\rangle$ and $|y\rangle$

$$x \overset{2}{\odot} y := x_{n-1} y_{n-1} \overset{2}{\oplus} \ldots \overset{2}{\oplus} x_0 y_0 .$$

For a solution see Solution 6.67

The algorithm initially operates on the composite system $\mathbb{H}^A \otimes \mathbb{H}^B$, where $\mathbb{H}^A = \mathbb{H}^{\otimes n}$ and $\mathbb{H}^B = \mathbb{H}$ and starts with the initial state

$$|\Psi_0\rangle = |0\rangle^n \otimes \frac{|0\rangle - |1\rangle}{\sqrt{2}} \in \mathbb{H}^A \otimes \mathbb{H}^B.$$

In the next step in the algorithm we apply $\left(H^{\otimes n} \otimes \mathbf{1}\right) U_f \left(H^{\otimes n} \otimes \mathbf{1}\right)$ to obtain

$$
\begin{aligned}
|\Psi_1\rangle &= \left(H^{\otimes n} \otimes \mathbf{1}\right) U_f \left(H^{\otimes n} \otimes \mathbf{1}\right) |\Psi_0\rangle \\
&= \left(H^{\otimes n} \otimes \mathbf{1}\right) U_f \left(H^{\otimes n}|0\rangle \otimes \frac{|0\rangle - |1\rangle}{\sqrt{2}}\right) \\
&\underset{(6.2)}{=} \left(H^{\otimes n} \otimes \mathbf{1}\right) U_f \left(\frac{1}{2^{\frac{n}{2}}} \sum_{x=0}^{2^n-1} |x\rangle \otimes \frac{|0\rangle - |1\rangle}{\sqrt{2}}\right) \\
&= \left(H^{\otimes n} \otimes \mathbf{1}\right) \frac{1}{2^{\frac{n+1}{2}}} \sum_{x=0}^{2^n-1} \left(U_f\left(|x\rangle \otimes |0\rangle - |x\rangle \otimes |1\rangle\right)\right) \\
&\underset{(6.1)}{=} \left(H^{\otimes n} \otimes \mathbf{1}\right) \frac{1}{2^{\frac{n+1}{2}}} \sum_{x=0}^{2^n-1} |x\rangle \otimes \left(|f(x)\rangle - |1 \overset{2}{\oplus} f(x)\rangle\right) \\
&= \left(H^{\otimes n} \otimes \mathbf{1}\right) \frac{1}{2^{\frac{n+1}{2}}} \sum_{x=0}^{2^n-1} |x\rangle \otimes (-1)^{f(x)}\left(|0\rangle - |1\rangle\right) \\
&= \frac{1}{2^{\frac{n}{2}}} \sum_{x=0}^{2^n-1} (-1)^{f(x)} H^{\otimes n}|x\rangle \otimes \frac{|0\rangle - |1\rangle}{\sqrt{2}} \\
&\underset{(6.3)}{=} \frac{1}{2^n} \sum_{y,x=0}^{2^n-1} (-1)^{f(x)+x\overset{2}{\odot}y}|y\rangle \otimes \frac{|0\rangle - |1\rangle}{\sqrt{2}}.
\end{aligned}
$$

Note that $\frac{|0\rangle - |1\rangle}{\sqrt{2}} = |\downarrow_{\hat{\mathbf{x}}}\rangle$ is one of the two basis states $|\uparrow_{\hat{\mathbf{x}}}\rangle$ and $|\downarrow_{\hat{\mathbf{x}}}\rangle$ in $\mathbb{H}$ so that $|\Psi_1\rangle$ can be written as

$$|\Psi_1\rangle = |\Psi_1^A\rangle \otimes |\downarrow_{\hat{\mathbf{x}}}\rangle,$$

where

$$|\Psi_1^A\rangle = \frac{1}{2^n} \sum_{y,x=0}^{2^n-1} (-1)^{f(x)+x\overset{2}{\odot}y}|y\rangle.$$

Consequently, the density operator of the complete system is

$$\rho_{\Psi_1} \underset{(2.89)}{=} |\Psi_1\rangle\langle\Psi_1| = |\Psi_1^A \otimes \downarrow_{\hat{\mathbf{x}}}\rangle\langle\Psi_1^A \otimes \downarrow_{\hat{\mathbf{x}}}| \underset{(3.36)}{=} |\Psi_1^A\rangle\langle\Psi_1^A| \otimes |\downarrow_{\hat{\mathbf{x}}}\rangle\langle\downarrow_{\hat{\mathbf{x}}}|$$

and using (3.57) with $\mathrm{tr}\left(|\downarrow_{\hat{x}}\rangle\langle\downarrow_{\hat{x}}|\right)=1$ we have

$$\rho^A\left(\rho_{\Psi_1}\right)\underbrace{=}_{(3.50)}\mathrm{tr}^B\left(\rho_{\Psi_1}\right)\underbrace{=}_{(3.57)}|\Psi_1^A\rangle\langle\Psi_1^A|\underbrace{=}_{(2.89)}\rho_{\Psi_1^A}\,.$$

This shows that when now considering the sub-system A alone, it is described by the pure state

$$|\Psi_1^A\rangle=\frac{1}{2^n}\sum_{x=0}^{2^n-1}(-1)^{f(x)}|0\rangle+\frac{1}{2^n}\sum_{y=1,x=0}^{2^n-1}(-1)^{f(x)+x\overset{2}{\odot}y}|y\rangle\,. \tag{6.4}$$

The probability to find system A in the state $|0\rangle\in\mathbb{H}^A$ is given by

$$\mathbf{P}\left\{\begin{array}{c}\text{System }A\text{ is in}\\\text{the state }|0\rangle\end{array}\right\}\underbrace{=}_{(2.64)}\left|\langle0|\Psi_1^A\rangle\right|^2\underbrace{=}_{(6.4)}\left(\frac{1}{2^n}\sum_{x=0}^{2^n-1}(-1)^{f(x)}\right)^2 \tag{6.5}$$

$$=\begin{cases}\left(\frac{1}{2^n}\sum_{x=0}^{2^n-1}(-1)^c\right)^2&=1\quad\text{if }f\text{ is constant}\\\left(\frac{1}{2^n}\left(2^{n-1}-2^{n-1}\right)\right)^2&=0\quad\text{if }f\text{ is balanced}\end{cases}$$

Hence, measuring $P_{|0\rangle}=|0\rangle\langle0|$ in the state $|\Psi_1^A\rangle$, which we have produced by only one application of U_f, will reveal if f is constant or balanced. $\square$

On a step-by-step basis the DEUTSCH–JOZSA algorithm then runs essentially as follows:

DEUTSCH–JOZSA algorithm

Input: A function $f:\{0,1\}^n\rightarrow\{0,1\}$ that is either constant or balanced and an associated U_f acting as

$$U_f:\mathbb{H}^{\otimes n}\otimes\mathbb{H}\longrightarrow\mathbb{H}^{\otimes n}\otimes\mathbb{H}$$
$$|x\rangle\otimes|y\rangle\longmapsto|x\rangle\otimes|y\overset{2}{\oplus}f(x)\rangle$$

and we set $\mathbb{H}^A=\mathbb{H}^{\otimes n}$ and $\mathbb{H}^B=\mathbb{H}$.

Step 1:   Prepare the initial state

$$|\Psi_0\rangle=|0\rangle^n\otimes\frac{|0\rangle-|1\rangle}{\sqrt{2}}\in\mathbb{H}^A\otimes\mathbb{H}^B$$

Step 2:   Apply $\left(H^{\otimes n}\otimes\mathbf{1}\right)U_f\left(H^{\otimes n}\otimes\mathbf{1}\right)$ to $|\Psi_0\rangle$ to obtain

$$|\Psi_1\rangle=\left(H^{\otimes n}\otimes\mathbf{1}\right)U_f\left(H^{\otimes n}\otimes\mathbf{1}\right)|\Psi_0\rangle\,.$$

Step 3:   Ignoring sub-system B, the sub-system A is then in the pure state $|\Psi_1^A\rangle \in \mathbb{H}^A$ given in (6.4). Query the sub-system A for the presence of $|0\rangle \in \mathbb{H}^A$ by measuring the observable $P_{|0\rangle} = |0\rangle\langle 0|$ in the state $|\Psi_1^A\rangle$ and use the observed value together with (6.5) to ascertain

$$\langle P_{|0\rangle}\rangle_{\Psi_1^A} = |\langle 0|\Psi_1^A\rangle|^2 = \begin{cases} 1 & \text{then } f \text{ is constant} \\ 0 & \text{then } f \text{ is balanced} \end{cases}$$

Output: Determination whether f is constant or balanced.

Hence, the DEUTSCH–JOZSA algorithm reduces the $2^{n-1} + 1$ evaluations of the function f, which are required by a classical way to solve DEUTSCH's problem with certainty, to just one application of U_f.

However, for the quantum algorithm to be more efficient than the many classical evaluations of f, we need to be able to build and run the quantum algorithm in much less than $2^n + 1$ steps. In other words, for efficiency gains U_f has to be such that it can be build and run in much less than $2^n + 1$ steps. In particular, this excludes that the knowledge of all $f(x)$ would be required to build U_f. After all, there would be no point in running any algorithm if we knew all values of f already.

Despite these caveats, the DEUTSCH–JOZSA algorithm clearly signaled—albeit on what some might consider a 'toy problem—that quantum algorithms had the potential to vastly outperform classical algorithms. Before we turn to quantum algorithms solving problems more pertinent to 'everyday life,' we present two applications of entanglement which are both curious and interesting in their own right.

6.2   Dense Quantum Coding

If Alice and Bob each have at their disposal[2] qubits that constitute the parts of a total system that was prepared in an entangled state, they can use this to *transmit two classical bits* by only *sending one qubit*. This procedure of transmitting two classical bits by means of sending one qubit is called **dense quantum coding** [76].

Suppose then Alice and Bob each have a qubit of a two-qubit composite system that is in the entangled BELL state

$$|\Phi^+\rangle = \frac{1}{\sqrt{2}}\left(|00\rangle + |11\rangle\right).$$

Depending on which of the four bit-pairs $x_1x_0 \in \{00, 01, 10, 11\}$ Alice wants to send to Bob, she performs on her qubit a unitary transformation $U^A = U^A(x_1x_0)$ according to the following assignment

---

[2] Having a system at their disposal means that they can apply transformations or measurements to it.

Table 6.1 Protocol of dense quantum coding

Alice wants to send the classical bits x_1x_0	So she applies $U^A(x_1x_0)$	The state of the total system becomes $\left(U^A \otimes \mathbf{1}\right)\lvert\Phi^+\rangle$	on which Bob measures $\sigma_z^A \otimes \sigma_z^B$ and $\sigma_x^A \otimes \sigma_x^B$ and observes the values
00	$\mathbf{1}^A$	$\lvert\Phi^+\rangle$	$+1, +1$
01	σ_z^A	$\lvert\Phi^-\rangle$	$+1, -1$
10	σ_x^A	$\lvert\Psi^+\rangle$	$-1, +1$
11	$\sigma_z^A \sigma_x^A$	$\lvert\Psi^-\rangle$	$-1, -1$

$$\begin{aligned}
U^A(00) &= \mathbf{1}^A \\
U^A(01) &= \sigma_z^A \\
U^A(10) &= \sigma_x^A \\
U^A(11) &= \sigma_z^A \sigma_x^A \,.
\end{aligned} \tag{6.6}$$

Table 6.1 shows in the first two columns which of the four possible choices of U^A listed in (6.6) corresponds to which pair of bits. In applying U^A, she transforms the total state $\lvert\Phi^+\rangle$ to one of the four BELL states $\lvert\Phi^\pm\rangle$ and $\lvert\Psi^\pm\rangle$. For example, if she wants to transmit the bit-pair 01, she would apply σ_z^A to her qubit and thus transform the total state $\lvert\Phi^+\rangle$ to

$$\left(\sigma_z^A \otimes \mathbf{1}^B\right)\lvert\Phi^+\rangle = \frac{1}{\sqrt{2}}\left((\sigma_z^A\lvert0\rangle)\otimes\lvert0\rangle + (\sigma_z^A\lvert1\rangle)\otimes\lvert1\rangle\right) = \frac{1}{\sqrt{2}}\left(\lvert00\rangle - \lvert11\rangle\right) = \lvert\Phi^-\rangle\,.$$

Exercise 6.68 Show that $U^A \otimes \mathbf{1}^B$ with $U^A \in \{\mathbf{1}^A, \sigma_z^A, \sigma_x^A, \sigma_z^A\sigma_x^A\}$ is unitary and verify

$$\begin{aligned}
\left(\sigma_x^A \otimes \mathbf{1}^B\right)\lvert\Phi^+\rangle &= \lvert\Psi^+\rangle \\
\left(\sigma_z^A \sigma_x^A \otimes \mathbf{1}^B\right)\lvert\Phi^+\rangle &= \lvert\Psi^-\rangle\,.
\end{aligned}$$

For a solution see Solution 6.68

Alice thus encodes *two classical bits* in her selection of U^A and by application of the selected transformation to her qubit. She then sends her *(one) qubit* to Bob, who then has both qubits at his disposal. On these, he measures the compatible observables $\sigma_z^A \otimes \sigma_z^B$ and $\sigma_x^A \otimes \sigma_x^B$ (see (3.41)). Reading off the observed value-pair out of $\{\pm1, \pm1\}$ and using the correspondence given in Table 2.1, he can thus uniquely determine the bit-pair Alice has sent.

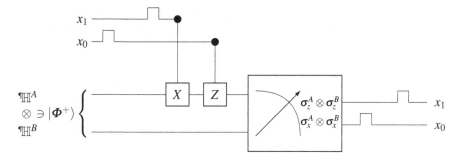

Fig. 6.1 Graphical illustration of the process of dense quantum coding; the inputs x_0 and x_1 are classical bits, which are graphically represented as lines with rectangular humps. They control the application of X and Z on $\mathbb{H}^A$. Here, X will only be applied if $x_1 = 1$. Analogously, Z will only be applied if $x_0 = 1$. This is akin to the application of $X^{x_1} Z^{x_0}$ on $\mathbb{H}^A$. The values observed by measuring $\sigma_z^A \otimes \sigma_z^B$ and $\sigma_x^A \otimes \sigma_x^B$ are translated into the classical bits x_1, x_0 according to Table 2.1

This protocol is illustrated in Table 6.1, where we make use of Tables 2.1 and 3.1. Altogether Bob has read two classical bits, even though he only received one qubit from Alice. A further illustration of this process in form of a circuit diagram is given in Fig. 6.1.

It should be noted, however, that this condensed transmission of classical information presupposes that Alice and Bob each already have received qubits which constitute the parts of a total system in an entangled state. This is in addition to the qubit actually sent from Alice to Bob. In a sense they thus share some form of joint 'information', that is, each already has a qubit of an entangled state, prior to the qubit sent from Alice to Bob.

6.3   Teleportation

In what has become known as **teleportation** [18] we have essentially the inverse of dense quantum coding: if Alice and Bob each have a qubit at their disposal that is part of a total quantum system in an entangled state, then *one qubit can be transferred* (aka 'teleported') from Alice to Bob by *sending two classical bits*. Again, this procedure uses entanglement as the most important ingredient and runs as follows.

Suppose Alice wants to teleport the qubit $|\psi\rangle^S = a|0\rangle + b|1\rangle \in \mathbb{H}^S$ to Bob, and she has another qubit that is part of an entangled BELL state

$$|\Phi^+\rangle^{AB} = \frac{1}{\sqrt{2}}\left(|0\rangle^A \otimes |0\rangle^B + |1\rangle^A \otimes |1\rangle^B\right) \qquad \in \mathbb{H}^A \otimes \mathbb{H}^B.$$

Moreover, suppose the other qubit of this BELL state is with Bob. She then combines the qubit to be teleported with the one that is part of the BELL state and performs measurements on the system in the HILBERT space $\mathbb{H}^S \otimes \mathbb{H}^A$.

For the total system, the addition of the sub-system to be teleported means that a new total system has been formed, which is described by vectors in $\mathbb{H}^S \otimes \mathbb{H}^A \otimes \mathbb{H}^B$ and which is in the state

$$|\psi\rangle^S \otimes |\Phi^+\rangle^{AB} = \left(a|0\rangle^S + b|1\rangle^S\right) \otimes \frac{1}{\sqrt{2}}\left(|0\rangle^A \otimes |0\rangle^B + |1\rangle^A \otimes |1\rangle^B\right)$$

$$= \frac{1}{\sqrt{2}}\left[a \underbrace{|0\rangle^S \otimes |0\rangle^A}_{=\frac{1}{\sqrt{2}}\left(|\Phi^+\rangle^{SA} + |\Phi^-\rangle^{SA}\right)} \otimes |0\rangle^B + a \underbrace{|0\rangle^S \otimes |1\rangle^A}_{=\frac{1}{\sqrt{2}}\left(|\Psi^+\rangle^{SA} + |\Psi^-\rangle^{SA}\right)} \otimes |1\rangle^B \right.$$

$$\left. + b \underbrace{|1\rangle^S \otimes |0\rangle^A}_{=\frac{1}{\sqrt{2}}\left(|\Psi^+\rangle^{SA} - |\Psi^-\rangle^{SA}\right)} \otimes |0\rangle^B + b \underbrace{|1\rangle^S \otimes |1\rangle^A}_{=\frac{1}{\sqrt{2}}\left(|\Phi^+\rangle^{SA} - |\Phi^-\rangle^{SA}\right)} \otimes |1\rangle^B \right]$$

$$= \frac{1}{2}\left[|\Phi^+\rangle^{SA} \otimes \left(a|0\rangle^B + b|1\rangle^B\right) + |\Psi^+\rangle^{SA} \otimes \left(a|1\rangle^B + b|0\rangle^B\right) \right.$$

$$\left. + |\Phi^-\rangle^{SA} \otimes \left(a|0\rangle^B - b|1\rangle^B\right) + |\Psi^-\rangle^{SA} \otimes \left(a|1\rangle^B - b|0\rangle^B\right)\right]$$

$$= \frac{1}{2}\left[|\Phi^+\rangle^{SA} \otimes |\psi\rangle^B + |\Psi^+\rangle^{SA} \otimes \left(\sigma_x^B|\psi\rangle^B\right) \right. \tag{6.7}$$

$$\left. + |\Phi^-\rangle^{SA} \otimes \left(\sigma_z^B|\psi\rangle^B\right) + |\Psi^-\rangle^{SA} \otimes \left(\sigma_x^B\sigma_z^B|\psi\rangle^B\right)\right].$$

Alice now measures the compatible observables $\sigma_z^S \otimes \sigma_z^A$ and $\sigma_x^S \otimes \sigma_x^A$ (see (3.41)) on the composite system $\mathbb{H}^S \otimes \mathbb{H}^A$ formed by the qubit to be teleported and her qubit, which is part of the composite system in the entangled BELL state. According to the Projection Postulate 3 in Sect. 2.3.1, the measurement of these observables projects onto one of the four eigenstates $|\Phi^+\rangle^{SA}, |\Phi^-\rangle^{SA}, |\Psi^+\rangle^{SA}$ and $|\Psi^-\rangle^{SA}$. From the observed values for $\sigma_z^S \otimes \sigma_z^A$ and $\sigma_x^S \otimes \sigma_x^A$, Alice can thus read off in which of the four states appearing in (6.7) the total system is after the measurement (see Table 3.1). Depending on which values she has observed, she sends two classical bits x_1x_0 to Bob, which instruct him to perform on his qubit a unitary transformation $U^B = U^B(x_1x_0)$ according to the following assignment

$$U^B(00) = \mathbf{1}^B$$
$$U^B(01) = \sigma_z^B$$
$$U^B(10) = \sigma_x^B$$
$$U^B(11) = \sigma_z^B\sigma_x^B\,,$$

which is identical to (6.6) used by Alice in the dense coding protocol. Because of $\mathbf{1}^2 = (\sigma_x)^2 = (\sigma_z)^2 = \sigma_z\sigma_x\sigma_x\sigma_z = \mathbf{1}$, this transforms his qubit to the state $|\psi\rangle$. This procedure is illustrated in Table 6.2. The quantum state $|\psi\rangle$ is thus 'teleported' from

Table 6.2 Protocol for teleportation

Alice measures $\sigma_z^S \otimes \sigma_z^A$ and $\sigma_x^S \otimes \sigma_x^A$ to observe	The three-qubit total state after measurement is then: $\lvert\Psi\rangle^{SA} \otimes$ Bob's qubit-state	From bits received Bob determines to apply U^B	The state of Bob's qubit then becomes
$+1, +1$	$\lvert\Phi^+\rangle \otimes \lvert\psi\rangle$	$\mathbf{1}^B$	$(\mathbf{1}^B)^2\lvert\psi\rangle = \lvert\psi\rangle$
$+1, -1$	$\lvert\Phi^-\rangle \otimes \sigma_z^B\lvert\psi\rangle$	σ_z^B	$(\sigma_z^B)^2\lvert\psi\rangle = \lvert\psi\rangle$
$-1, +1$	$\lvert\Psi^+\rangle \otimes \sigma_x^B\lvert\psi\rangle$	σ_x^B	$(\sigma_x^B)^2\lvert\psi\rangle = \lvert\psi\rangle$
$-1, -1$	$\lvert\Psi^-\rangle \otimes \sigma_x^B\sigma_z^B\lvert\psi\rangle$	$\sigma_z^B\sigma_x^B$	$\sigma_z^B\sigma_x^B\sigma_x^B\sigma_z^B\lvert\psi\rangle = \lvert\psi\rangle$

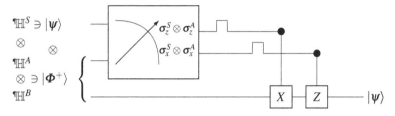

Fig. 6.2 Graphical illustration of the process of teleportation; the results of the measurements are classical bits, which again are graphically represented as lines with rectangular humps. They are used to determine the respective application of X or Z in the sense that X is applied to $\mathbb{H}^B$ only if the measurement of $\sigma_z^S \otimes \sigma_z^A$ yields the value $+1$, which according to Table 2.1 corresponds to the classical bit value 0. Analogously, Z is applied to $\mathbb{H}^B$ only if the measurement of $\sigma_x^S \otimes \sigma_x^A$ yields the value $+1$

Alice to Bob, even though only classical information has been sent from Alice to Bob. However, just as in the case of dense coding, here too, the requirement for this to work is that Alice and Bob already share some kind of joint information in the form of a supply of qubits that are sub-systems of a total system that is in an entangled state. Moreover, it has to be emphasized, that the state $\lvert\psi\rangle$ *is not cloned*, which would contradict the Quantum No-Cloning Theorem 4.11. Instead, this state is destroyed by Alice's measurement, before it is re-generated by Bob through application of a suitable U^B.

Figure 6.2 shows a circuit diagram of the process of teleportation.

6.4 Quantum Cryptography

6.4.1 Ciphers in Cryptography

Cryptography is the science of encrypting and decrypting messages $m \in M$, where M is a finite set of possible messages. Encrypting a plaintext message m means transforming it into an alternative form $c \in C$ called ciphertext that does not allow to retrieve the plaintext m without additional knowledge of some sort. This additional knowledge is what we call a key $k \in K$.

Definition 6.3 Let M, C, K be finite sets. A **cipher** is a cryptographic protocol encrypting plaintexts $m \in M$ with a key $k \in K$ and consists of an **encryption**

$$e : K \times M \longrightarrow C$$
$$(k, m) \longmapsto e(k, m)$$

mapping the key-plaintext pair (k, m) to some ciphertext $e(k, m)$ and a **decryption**

$$d : K \times C \longrightarrow M$$
$$(k, c) \longmapsto d(k, c)$$

mapping the key-ciphertext pair (k, c) to some plaintext $d(k, c)$ such that for all $m \in M$ and $k \in K$

$$d(k, e(k, m)) = m. \tag{6.8}$$

If encryption and decryption use the same key k, the cipher is called **symmetric**. If one part k_{pub} of the the key $k = (k_{\text{priv}}, k_{\text{pub}})$ is used for encryption $m \mapsto e(k_{\text{pub}}, m)$ and another part k_{priv} is used for decryption $c \mapsto d(k_{\text{priv}}, c)$, then the cipher is said to be **asymmetric**.

Effectively, d is the inverse of e. More precisely, if we look at

$$e_k(\cdot) := e(k, \cdot) : M \to C \quad \text{and} \quad d_k(\cdot) := d(k, \cdot) : C \to M$$

then (6.8) implies that

$$d_k \circ e_k = \text{id}_M .$$

As our digital world amply demonstrates, there is no loss of generality if we assume each message to be a binary string $(m_{n_M - 1} \ldots m_0)$ of a fixed length n_M with $m_j \in \{0, 1\}$. Using the m_j as coefficients in a binary expansion, we can thus consider each message to be an integer m satisfying

$$0 \le m = \sum_{j=0}^{n_M - 1} m_j 2^j \le 2^{n_M} - 1 .$$

Likewise, we may assume the ciphertext to be a binary string of length n_C and a natural number c satisfying $0 \le c \le 2^{n_C} - 1$. Sometimes—but not always—the key k is taken to be of the same form with length n_K and $0 \le k \le 2^{n_K} - 1$.

The ciphertext, that is, the encrypted message $c = e(k,m)$, may be transmitted on public channels. A secure cipher would make it impossible to retrieve m from c without knowing k.

Example 6.4 One cipher, which meets this criteria, is the VERNAM cipher also used by CHE GUEVARA and FIDEL CASTRO [77]. This is a *symmetric* cipher where plaintext, key and ciphertext all have the same length $n_M = n_K = n_C = n$. The key k is a random bit-sequence

$$k = (k_{n-1}, \ldots, k_0).$$

Encryption is accomplished by addition modulo 2 of the key-bits k_j to the plaintext-bits m_j. With the help of the notation introduced in Definition 5.2 we can thus write the encryption map as

$$e(k,m) = (k_{n-1} \overset{2}{\oplus} m_{n-1}, \ldots, k_0 \overset{2}{\oplus} m_0).$$

The ciphertext $c = e(k,m)$ is then a purely random sequence of bits and does not allow to obtain any information of the original plaintext m. In order to decrypt the ciphertext, one requires the key k. With the key decryption is accomplished by once again adding modulo 2 to the key-bits to the ciphertext-bits:

$$\begin{aligned}
d\big(k, e(k,m)\big) &= (k_{n-1} \overset{2}{\oplus} e(k,m)_{n-1}, \ldots, k_0 \overset{2}{\oplus} e(k,m)_0) \\
&= (k_{n-1} \overset{2}{\oplus} k_{n-1} \overset{2}{\oplus} m_{n-1}, \ldots, k_0 \overset{2}{\oplus} k_0 \overset{2}{\oplus} m_0) \\
&= (m_{n-1}, \ldots, m_0)
\end{aligned}$$

since $k_j \overset{2}{\oplus} k_j = 0$ for all $k_j \in \{0,1\}$. This is illustrated with a small example in Table 6.3.

The VERNAM cipher is absolutely secure, but has the disadvantage that, in order to maintain its security for every message, we need to use a new key for each new message. Hence, sender and receiver either need to have a large common supply of random key-bits or need to exchange such bits every time they want to exchange a message. The former requires to keep this large supply safe, whereas the latter requires a secure transmission channel, something we are trying to establish in the first place.

Quantum mechanics, however, does offer the possibility to generate a random key-bit-sequence, to transmit it and to check if the transmission has been compromised in the sense that eavesdropping occurred. Thus, even though we cannot prevent eavesdropping, we can at least detect it. In case we detect eavesdropping, we do not use the bit-sequence as a random key. This is called quantum (cryptographic) key distribution.

Table 6.3 Sample en- and decryption with the VERNAM cipher

Encryption	Message	$m = 0$	0	1	0	1	1	0	1
		$\overset{2}{\oplus}$	$\overset{2}{\oplus}$	$\overset{2}{\oplus}$	$\overset{2}{\oplus}$	$\overset{2}{\oplus}$	$\overset{2}{\oplus}$	$\overset{2}{\oplus}$	$\overset{2}{\oplus}$
	Key	$k = 1$	0	0	1	1	0	0	0
	Ciphertext	$c = e(k,m) = 1$	0	1	1	0	1	0	1
Decryption	Ciphertext	$c = 1$	0	1	1	0	1	0	1
		$\overset{2}{\oplus}$	$\overset{2}{\oplus}$	$\overset{2}{\oplus}$	$\overset{2}{\oplus}$	$\overset{2}{\oplus}$	$\overset{2}{\oplus}$	$\overset{2}{\oplus}$	$\overset{2}{\oplus}$
	Key	$k = 1$	0	0	1	1	0	0	0
	Message	$m = 0$	0	1	0	1	1	0	1

There are several protocols of quantum key distribution. In Sect. 6.4.2 we first consider a procedure that—contrary to the title of this chapter—does not use entanglement. Instead, it is based on the quantum mechanical phenomenon that in general an observation, which can only result from a measurement, changes the state of the observed system. In Sect. 6.4.3 we then exhibit a protocol which does utilize entanglement in that it makes use of the fact that the EPR correlations of BELL states cannot be generated by classical random variables. These protocols are two examples of what has become known as quantum cryptography.

The currently most widespread classical cryptographic protocol is the *asymmetric* cipher developed in 1978 by RIVEST, SHAMIR and ADLEMAN [78, 79], which works with one part of the key being publicly known. This has since become known as the RSA protocol, and we study it in more detail in Sect. 6.4.4. Its security relies on the fact that, so far, it is too time-consuming to find the *prime factors p* and *q* of a large number $N = pq$.

However, in Sect. 6.5 we present a quantum algorithm devised by SHOR, which—if quantum computers are realized—allows to find prime factors much faster than hitherto possible. Hence, quantum mechanics on the one hand would allow to compromise the security of RSA, while on the other hand permits to create key distribution protocols such as those presented in Sects. 6.4.2 and 6.4.3, which allow at least the detection of eavesdropping.

6.4.2 Quantum Key Distribution without Entanglement

The following method to distribute a random bit sequence, which can be used as an encryption key for the VERNAM-Code, was proposed in 1984 by BENNETT and BRASSARD [1]. The protocol is thus denoted by the acronym **BB84**. It does not use entanglement but instead relies on the fact that in quantum mechanics measurements alter the state in order to detect eavesdropping.

The objective of the method is to generate a random bit sequence that is only known to Alice and Bob and for which they can check if the transmission has been listened to. This is achieved as follows.

Suppose Alice has a large number of qubits at her disposal. For each of them, she randomly selects to measure either σ_z or σ_x. For each qubit, she records the measurement result and then sends the qubit to Bob. He, too, randomly selects to measure either σ_z or σ_x on the qubit received and also records the measurement result. Alice and Bob then communicate via public channels for which of the qubits they happened to have measured the same observable, that is, either both σ_z or both σ_x. For those qubits, the measurement results have to be the same. This is because, due to the Projection Postulate 3, after the measurement of Alice, the qubit is in an eigenstate of the measured observable corresponding to the observed eigenvalue. A measurement of the same observable by Bob on the qubit in that eigenstate will result in the same eigenvalue to be observed as a result of the measurement. From this set of qubits, where they have measured the same observable, Alice and Bob select a subset and publicly compare their measurement results. We shall show below that, if these measurement results all agree, they can be (almost) certain that no third party has gained knowledge of these measured values. In this case they can use the measured values of those remaining qubits for which they have measured the same observable, but have not not compared the measurement results, as a random bit sequence only known to them. Table 6.4 illustrates the protocol under the assumption that no attempt to intercept the information (aka eavesdropping) has been made.

What then, changes if Eve intercepts the qubits on the way from Alice to Bob and attempts to listen in? The information that Eve wants to get is the random bit-sequence that Alice and Bob have generated as outlined above. The only way for Eve to get that information is to actually measure one[3] of the incompatible observables σ_z or σ_x. Since Alice randomly selects one of the incompatible observables σ_z or σ_x, and Eve does not know which she has selected, Eve will not always happen to measure the same observable that Alice measured.

For those qubits where Eve happens to measure the same observable as Alice, Eve will indeed observe the same value as Alice since the qubit arrives in an eigenstate of the observable. Moreover, this state will not be altered by Eve's measurement.

However, in those cases where Eve happens to measure a different observable from the one Alice measured, the measurements by Eve will change the state of some qubit because the two observables in question are incompatible. Hence, in these cases the qubit arriving at Bob is in a different state from that which Alice prepared. Consequently, the comparison by Alice and Bob of the values measured in case they chose the same observable will reveal disagreeing values for a number of qubits beyond what might reasonably be expected due to transmission errors. Alice and Bob thus conclude that somebody has intercepted the qubits. Hence, they do not use them but start all over again. Table 6.5 illustrates the protocol BB84 *with eavesdropping*.

---

[3]Measuring both does not yield useful information for Eve because the observables are incompatible.

Table 6.4 Key distribution without eavesdropping according to BB84

For qubit no.	1	2	3	4	5	6	7	8	9	10	11	12	...

Alice chooses randomly one of the observables σ_z or σ_x. Suppose she measures

| Alice's observable | σ_x | σ_x | σ_x | σ_z | σ_x | σ_x | σ_x | σ_z | σ_z | σ_x | σ_x | σ_z | ... |

and observes

| Alice's value | +1 | −1 | −1 | +1 | +1 | −1 | +1 | −1 | +1 | +1 | +1 | +1 | ... |

The qubit is then in the

| qubit state | $\lvert\uparrow_{\hat{x}}\rangle$ | $\lvert\downarrow_{\hat{x}}\rangle$ | $\lvert\downarrow_{\hat{x}}\rangle$ | $\lvert\uparrow_{\hat{z}}\rangle$ | $\lvert\uparrow_{\hat{x}}\rangle$ | $\lvert\downarrow_{\hat{x}}\rangle$ | $\lvert\uparrow_{\hat{x}}\rangle$ | $\lvert\downarrow_{\hat{z}}\rangle$ | $\lvert\uparrow_{\hat{z}}\rangle$ | $\lvert\uparrow_{\hat{x}}\rangle$ | $\lvert\uparrow_{\hat{x}}\rangle$ | $\lvert\uparrow_{\hat{z}}\rangle$ | ... |

Alice sends the qubits thus prepared to Bob. He chooses for each qubit randomly one of the observables σ_z or σ_x. Suppose he measures

| Bob's observable | σ_z | σ_x | σ_z | σ_z | σ_z | σ_x | σ_z | σ_z | σ_x | σ_x | σ_z | σ_z | ... |

and observes

| Bob's value | −1 | −1 | −1 | +1 | +1 | −1 | +1 | −1 | +1 | +1 | −1 | +1 | ... |

Alice and Bob publicly compare for which qubit they have measured which observable. But they do not reveal the outcome of the measurement, that is, the observed value. They thus divide the qubits into a set where they chose either the same observable or different ones. Measured

| observables were | ≠ | = | ≠ | = | ≠ | = | ≠ | = | ≠ | = | ≠ | = | ... |

If they have chosen the same observable, their measured values have to agree. As a control they compare publicly every second of the observed values where they measured the same observable:

| control-value Alice | | | | +1 | | | | −1 | | | | +1 | ... |
| control-value Bob | | | | +1 | | | | −1 | | | | +1 | ... |

100% agreement in comparison of control values implies: with a probability increasing with the number of control values the qubits have *not been read* between the measurements by Alice and Bob. Use the observed values in the remaining cases where both measured the same observable as *joint, secret*, and *random*

| *bit sequence:* | | **−1** | | | | **−1** | | | | **+1** | | | ... |

Table 6.5 Key distribution with eavesdropping according to BB84

For qubit no.	1	2	3	4	5	6	7	8	9	10	11	12	...

Alice randomly selects one of the observables σ_z or σ_x and measures

| Alice's observable | σ_x | σ_x | σ_x | σ_z | σ_x | σ_x | σ_x | σ_z | σ_z | σ_x | σ_x | σ_z | ... |

and observes

| Alice's value | +1 | −1 | −1 | +1 | +1 | −1 | +1 | −1 | +1 | +1 | +1 | +1 | ... |

The qubit is then in the

| qubit state | $|\uparrow_{\hat{x}}\rangle$ | $|\downarrow_{\hat{x}}\rangle$ | $|\downarrow_{\hat{x}}\rangle$ | $|\uparrow_{\hat{z}}\rangle$ | $|\uparrow_{\hat{x}}\rangle$ | $|\downarrow_{\hat{x}}\rangle$ | $|\uparrow_{\hat{x}}\rangle$ | $|\downarrow_{\hat{z}}\rangle$ | $|\uparrow_{\hat{z}}\rangle$ | $|\uparrow_{\hat{x}}\rangle$ | $|\uparrow_{\hat{x}}\rangle$ | $|\uparrow_{\hat{z}}\rangle$ | ... |

Alice sends the thus prepared qubits to Bob. Eve intercepts the qubit, but does not know, in which state it is. Thus, Eve measures

| Eve's observable | σ_z | σ_x | σ_x | σ_z | σ_z | σ_x | σ_z | σ_x | σ_x | σ_x | σ_z | σ_z | ... |

and observes

| Eve's value | −1 | −1 | −1 | +1 | −1 | −1 | +1 | −1 | +1 | +1 | +1 | +1 | ... |

The qubit is then in the

| qubit state | $|\downarrow_{\hat{z}}\rangle$ | $|\downarrow_{\hat{x}}\rangle$ | $|\downarrow_{\hat{x}}\rangle$ | $|\uparrow_{\hat{z}}\rangle$ | $|\downarrow_{\hat{z}}\rangle$ | $|\downarrow_{\hat{x}}\rangle$ | $|\uparrow_{\hat{z}}\rangle$ | $|\downarrow_{\hat{x}}\rangle$ | $|\uparrow_{\hat{x}}\rangle$ | $|\uparrow_{\hat{x}}\rangle$ | $|\uparrow_{\hat{z}}\rangle$ | $|\uparrow_{\hat{z}}\rangle$ | ... |

and is passed on from Eve to Bob. He randomly selects one of the observables σ_z or σ_x and measures

| Bob's observable | σ_z | σ_x | σ_z | σ_z | σ_z | σ_x | σ_z | σ_z | σ_x | σ_x | σ_z | σ_z | ... |

He observes

| Bob's value | −1 | −1 | −1 | +1 | −1 | −1 | +1 | +1 | +1 | +1 | +1 | +1 | ... |

Alice and Bob publicly compare for which qubit they have measured which observable. But they do not reveal the outcome of the measurement, that is, the observed value. They thus divide the qubits into a set where they chose either the same observable or different ones. Measured

| observables were | ≠ | = | ≠ | = | ≠ | = | ≠ | = | ≠ | = | ≠ | = | ... |

If they have chosen the same observable, their measured values have to agree and as a control they compare publicly every second of the observed values where they measured the same observable:

| control-value Alice | | | | +1 | | | | −1 | | | | +1 | ... |
| control-value Bob | | | | +1 | | | | +1 | | | | +1 | ... |

33% disagreement in the control values implies eavesdropping. Discard all qubits sent and start all over with a new sequence.

Protocol BB84 does not allow to prevent eavesdropping per se, but it has a built-in strategy to detect it and thus avoid it.[4] The security against eavesdropping in protocol BB84 rests on the fact that in quantum mechanics the measurement of an observable changes the state, and that, in general, it is impossible to restore the state prior to measurement unless it was known (in which case eavesdropping would have unnecessary). This is in contrast to classical bits, which can be read without altering them irreversibly.

The communication of the key, that is, the bit sequence, in protocol BB84 requires that qubits be sent from Alice to Bob without their state being altered since modification to the state is taken as proof of eavesdropping. Such an undisturbed delivery of qubits can be quite difficult in practice as it is notoriously difficult to isolate qubits from external disturbances in the environment and for the time required in everyday applications. The alternative presented in Sect. 6.4.3 uses entanglement and avoids the delivery of qubits during the generation of the key.

6.4.3   Quantum Key Distribution with Entanglement

In the following we describe how entanglement can be utilized to distribute keys in such a way that no qubits need to be sent, provided Alice and Bob each have a supply of single qubits that are part of a composite system in an entangled state. This protocol exploits the CHSH version of the BELL inequality to detect eavesdropping. It was first proposed in 1991 by EKERT [3], which is why we denote it by the acronym **EK91**.

Suppose Alice and Bob each have qubits that are part of a set of composite two-qubit systems that are prepared in the entangled BELL state

$$|\Psi^-\rangle = \frac{1}{\sqrt{2}}\left(|0\rangle\otimes|1\rangle - |1\rangle\otimes|0\rangle\right)\underbrace{=}_{(4.25)}\frac{1}{\sqrt{2}}\left(|\uparrow_{\hat{\mathbf{n}}}\rangle\otimes|\downarrow_{\hat{\mathbf{n}}}\rangle - |\downarrow_{\hat{\mathbf{n}}}\rangle\otimes|\uparrow_{\hat{\mathbf{n}}}\rangle\right),$$

where, due to the result (4.25), any direction $\hat{\mathbf{n}} \in S^1_{\mathbb{R}^3}$ can be chosen to describe the state. To avoid sending qubits, Alice and Bob could each pick up their qubit at a time of their choosing from the source, which produces the composite system in the state $|\Psi^-\rangle$. The protocol then proceeds as follows.

Alice measures $\Sigma^A_{\hat{\mathbf{n}}^A} = \hat{\mathbf{n}}^A \cdot \boldsymbol{\sigma}$ on her qubits by randomly selecting for each qubit a direction $\hat{\mathbf{n}}^A$ from one of the three directions in $\{\hat{\mathbf{n}}^1, \hat{\mathbf{n}}^2, \hat{\mathbf{n}}^4\}$, where the $\hat{\mathbf{n}}^i$ are as defined in (4.38) together with (4.40) and shown in Fig. 4.1. For each qubit, she randomly selects one of these three directions to measure $\Sigma^A_{\hat{\mathbf{n}}^A}$. For each of his qubits, Bob also randomly selects a direction $\hat{\mathbf{n}}^B$ from one of the three directions in $\{\hat{\mathbf{n}}^2, \hat{\mathbf{n}}^3, \hat{\mathbf{n}}^4\}$ and measures $\Sigma^B_{\hat{\mathbf{n}}^B}$ on each of his qubits. Table 6.6 shows a fictitious set of selected directions together with measurement results. For each pair of qubits

---

[4]Although permanent eavesdropping would prevent an exchange of keys.

Table 6.6 Fictitious measurements by Alice and Bob in the protocol EK91 without eavesdropping. The cells in *light gray* show the results for qubits, where Alice and Bob have measured $\Sigma^A_{\hat{n}^A}$ and $\Sigma^B_{\hat{n}^B}$ *in different directions* $\hat{n}^A \neq \hat{n}^B$ and for which they have publicly announced their measurement results. With these results the left side of (4.41) can be calculated as given in (4.42). The values shown here yield $\approx -2\sqrt{2}$, which rules out eavesdropping. The cells in *white*, on the other hand, show measurement results, where they have measured $\Sigma^A_{\hat{n}^A}$ and $\Sigma^B_{\hat{n}^B}$ *in the same directions* $\hat{n}^A = \hat{n}^B$. These results are only known to Alice and Bob, and can be used as a random bit-sequence

qubit-pair no.	Alice measures $\Sigma^A_{\hat{n}^A}$ in direction $\hat{n}^A =$			Bob measures $\Sigma^A_{\hat{n}^B}$ in direction $\hat{n}^B =$		
	$\hat{n}^1$	$\hat{n}^2$	$\hat{n}^4$	$\hat{n}^3$	$\hat{n}^2$	$\hat{n}^4$
1		+1				−1
2	−1			+1		
3	+1				−1	
4		−1			−1	
5		+1			−1	
6	−1					−1
7	+1				−1	
8		+1				−1
9	−1					+1
10		−1				+1
11		+1		+1		
12	−1			−1		
13	−1				+1	
14		+1				−1
15	+1				−1	
16	−1			−1		
17	−1				+1	
18	+1				−1	
19	+1				+1	
20	−1			−1		
21		−1			+1	
22		+1		−1		
23		+1				−1
24		−1		−1		
25	−1					+1
26			−1	+1		
27			+1	−1		
28		−1				−1
29			−1	+1		
30	+1			+1		
31		−1				+1
32			−1			+1

qubit-pair no.	Alice measures $\Sigma^A_{\hat{n}^A}$ in direction $\hat{n}^A =$			Bob measures $\Sigma^A_{\hat{n}^B}$ in direction $\hat{n}^B =$		
	$\hat{n}^1$	$\hat{n}^2$	$\hat{n}^4$	$\hat{n}^3$	$\hat{n}^2$	$\hat{n}^4$
33		+1		−1		
34	−1			+1		
35	+1			+1		
36	−1					−1
37		−1		+1		
38	+1				−1	
39		−1		+1		
40		−1		−1		
41		+1				−1
42	+1				−1	
43		−1		+1		
44		+1		−1		
45	+1					−1
46	−1					−1
47		−1				+1
48		+1		+1		
49		−1				+1
50		+1			−1	
51	+1					−1
52	−1				+1	
53		−1				+1
54	+1				−1	
55	+1				−1	
56	−1					+1
57		−1		−1		
58		+1			−1	
59		−1		+1		
60	+1					−1
61	+1					−1
62			+1			−1
63	+1			+1		

belonging to the same entangled state, Alice and Bob then inform each other via a public channel in which directions $\hat{\mathbf{n}}^A$ and $\hat{\mathbf{n}}^B$ they have measured $\Sigma^A_{\hat{\mathbf{n}}^A}$ and $\Sigma^B_{\hat{\mathbf{n}}^B}$ for each of their respective qubits in an entangled pair. However, they keep the observed values, that is, the measurement results, secret. They divide the measurements (or qubit-pairs) into two disjoint sets:

• one set, where they happened to have measured in the *same direction*

$$\hat{\mathbf{n}}^A = \hat{\mathbf{n}}^2 = \hat{\mathbf{n}}^B \qquad \text{or} \qquad \hat{\mathbf{n}}^A = \hat{\mathbf{n}}^4 = \hat{\mathbf{n}}^B$$

• a separate set, given by those measurement-pairs for which they happened to have measured in *different directions* $\hat{\mathbf{n}}^A \neq \hat{\mathbf{n}}^B$.

For the set where they have measured in the same direction, Alice and Bob will always have measured different values as their qubits are parts of the composite system in the state $|\Psi^-\rangle$. In other words, when Alice has observed the value $+1$ for her qubit, Bob will have observed -1 for his and vice versa. To see this, suppose Alice measures the spin in the direction $\hat{\mathbf{n}}^2$ and observes the value $+1$ ('spin-up'). The projector onto the corresponding eigenspace in the composite system is

$$P_{\hat{\mathbf{n}}^2,+1} = |\uparrow_{\hat{\mathbf{n}}^2}\rangle\langle\uparrow_{\hat{\mathbf{n}}^2}| \otimes \mathbf{1}.$$

From the Projection Postulate 3 we know that Alice's measurement projects the original state

$$|\Psi^-\rangle \underset{(4.25)}{=} \frac{1}{\sqrt{2}}\left(|\uparrow_{\hat{\mathbf{n}}^2}\rangle \otimes |\downarrow_{\hat{\mathbf{n}}^2}\rangle - |\downarrow_{\hat{\mathbf{n}}^2}\rangle \otimes |\uparrow_{\hat{\mathbf{n}}^2}\rangle\right) \tag{6.9}$$

onto

$$|\Psi_{\hat{\mathbf{n}}^2,+1}\rangle := \frac{P_{\hat{\mathbf{n}}^2,+1}|\Psi^-\rangle}{\left\|P_{\hat{\mathbf{n}}^2,+1}|\Psi^-\rangle\right\|} = \frac{\left(|\uparrow_{\hat{\mathbf{n}}^2}\rangle\langle\uparrow_{\hat{\mathbf{n}}^2}| \otimes \mathbf{1}\right)|\Psi^-\rangle}{\left\|\left(|\uparrow_{\hat{\mathbf{n}}^2}\rangle\langle\uparrow_{\hat{\mathbf{n}}^2}| \otimes \mathbf{1}\right)\Psi^-\right\|} = |\uparrow_{\hat{\mathbf{n}}^2}\rangle \otimes |\downarrow_{\hat{\mathbf{n}}^2}\rangle,$$

where we used that

$$\left(|\uparrow_{\hat{\mathbf{n}}^2}\rangle\langle\uparrow_{\hat{\mathbf{n}}^2}| \otimes \mathbf{1}\right)|\Psi^-\rangle$$
$$\underset{(6.9)}{=} \frac{1}{\sqrt{2}}\left(|\uparrow_{\hat{\mathbf{n}}^2}\rangle \underbrace{\langle\uparrow_{\hat{\mathbf{n}}^2}|\uparrow_{\hat{\mathbf{n}}^2}\rangle}_{=1} \otimes |\downarrow_{\hat{\mathbf{n}}^2}\rangle - |\downarrow_{\hat{\mathbf{n}}^2}\rangle \underbrace{\langle\uparrow_{\hat{\mathbf{n}}^2}|\downarrow_{\hat{\mathbf{n}}^2}\rangle}_{=0} \otimes |\uparrow_{\hat{\mathbf{n}}^2}\rangle\right)$$
$$= \frac{1}{\sqrt{2}}|\uparrow_{\hat{\mathbf{n}}^2}\rangle \otimes |\downarrow_{\hat{\mathbf{n}}^2}\rangle.$$

This means that Bob's system will be described by the reduced density operator

$$
\rho^B\big(|\Psi_{\hat{\mathbf{n}}^2,+1}\rangle\langle\Psi_{\hat{\mathbf{n}}^2,+1}|\big) \underbrace{=}_{(3.57)} \mathrm{tr}^A\left(|\Psi_{\hat{\mathbf{n}}^2,+1}\rangle\langle\Psi_{\hat{\mathbf{n}}^2,+1}|\right)
$$

$$
\underbrace{=}_{(4.22)} \mathrm{tr}^A\left(\big(|\uparrow_{\hat{\mathbf{n}}^2}\rangle\otimes|\downarrow_{\hat{\mathbf{n}}^2}\rangle\big)\big(\langle\uparrow_{\hat{\mathbf{n}}^2}|\otimes\langle\downarrow_{\hat{\mathbf{n}}^2}|\big)\right)
$$

$$
\underbrace{=}_{(3.36)} \mathrm{tr}^A\left(|\uparrow_{\hat{\mathbf{n}}^2}\rangle\langle\uparrow_{\hat{\mathbf{n}}^2}|\otimes|\downarrow_{\hat{\mathbf{n}}^2}\rangle\langle\downarrow_{\hat{\mathbf{n}}^2}|\right)
$$

$$
\underbrace{=}_{(3.57)} \underbrace{\mathrm{tr}\left(|\uparrow_{\hat{\mathbf{n}}^2}\rangle\langle\uparrow_{\hat{\mathbf{n}}^2}|\right)}_{=1} |\downarrow_{\hat{\mathbf{n}}^2}\rangle\langle\downarrow_{\hat{\mathbf{n}}^2}|
$$

$$
= |\downarrow_{\hat{\mathbf{n}}^2}\rangle\langle\downarrow_{\hat{\mathbf{n}}^2}|,
$$

which is the density operator of the pure state $|\downarrow_{\hat{\mathbf{n}}^2}\rangle$. If Bob measures in the same direction $\hat{\mathbf{n}}^2$, he will observe the value -1, since the state $|\downarrow_{\hat{\mathbf{n}}^2}\rangle$ is an eigenvector of the observable $\Sigma^B_{\hat{\mathbf{n}}^2}$ with eigenvalue -1. Analogously, we find that Bob will always observe the value $+1$ for his qubit if Alice has measured the value -1 on her qubit. Hence, in the set of identical measurement directions the measurement results of Alice and Bob are always (with certainty, that is) opposite to each other. Since the measurement results are only known to them, they can thus use this set of measurement results as a random and secret bit-sequence.

How could an eavesdropper called Eve possibly compromise the security of this exchange and listen in? There are two ways for her to do this: the first consists of performing measurements on at least one of the qubits of the entangled pair distributed to Alice and Bob; the second consists of manipulating the source of the qubit-pair in such a way that she knows the states in which the qubits sent to Alice and Bob are in. In the following we look at these possibilities in turn.

Alice and Bob do not exchange any qubits, hence, Eve cannot intercept anything. Before Alice or Bob have performed measurements, the pair of qubits is in the entangled state $|\Psi^-\rangle$. The joint information, which Alice and Bob share, comes only into being after their measurements. It is not available before that. We can assume that, after their measurements, Alice and Bob dispose of their qubits so that no further measurements can be performed on them. Hence, the only way for Eve to attempt to access any information by a measurement is to observe, that is, to measure, the qubit of Alice or Bob (or even both) before they perform their measurements. But after any measurement by Eve, the qubits are no longer in an entangled state $|\Psi^-\rangle$, just as is the case after any measurements by Alice or Bob. Instead, after any measurement by Eve, the qubits are in a separable state of the form $|\uparrow_{\hat{\mathbf{n}}^{\varphi_1}}\rangle\otimes|\downarrow_{\hat{\mathbf{n}}^{\psi_1}}\rangle =: |\varphi_1\otimes\psi_1\rangle$ for some directions $\hat{\mathbf{n}}^{\varphi_1}$ and $\hat{\mathbf{n}}^{\psi_1}$ that Eve has chosen.

If Eve would want to avoid making measurements, the only other way for her to compromise the secrecy of the bit-sequence, would be for her to manipulate the original composite state. However, as long as that composite state is entangled, the information which Alice and Bob share, is only created when they measure their qubits. To access information created this way, Eve would have to measure one of the qubits and we are back in the first type of attack discussed above. Eve can only avoid falling back into this line of attack if she prepares the original composite state as a separable state, which we denote by $|\varphi_2 \otimes \psi_2\rangle$. This way, she knows for each pair in which state the respective qubits of Alice and Bob are in.

Hence, in both possible types of eavesdropping attacks that Eve can attempt, the composite system will be in a separable state $|\varphi \otimes \psi\rangle$ before Alice and Bob perform their measurements. From Proposition 4.8 we know, however, that then

$$\left| \left\langle \Sigma_{\hat{\mathbf{n}}^1}^A \otimes \Sigma_{\hat{\mathbf{n}}^2}^B \right\rangle_{\varphi \otimes \psi} - \left\langle \Sigma_{\hat{\mathbf{n}}^1}^A \otimes \Sigma_{\hat{\mathbf{n}}^3}^B \right\rangle_{\varphi \otimes \psi} + \left\langle \Sigma_{\hat{\mathbf{n}}^4}^A \otimes \Sigma_{\hat{\mathbf{n}}^2}^B \right\rangle_{\varphi \otimes \psi} + \left\langle \Sigma_{\hat{\mathbf{n}}^4}^A \otimes \Sigma_{\hat{\mathbf{n}}^3}^B \right\rangle_{\varphi \otimes \psi} \right| \leq 2 \tag{6.10}$$

holds. This can be used by Alice and Bob to detect eavesdropping as follows. As they communicate the directions of their measurements to each other, they can determine the set of qubits where they have measured in different directions. For those they also announce the results of the measurements to each other. With these results $s_{\hat{\mathbf{n}}^i}^X$, where $X \in \{A, B\}$ and $i \in \{1, \ldots, 4\}$, they calculate the empirical expectation values $\overline{\Sigma_{\hat{\mathbf{n}}^i}^A \Sigma_{\hat{\mathbf{n}}^j}^B}$ according to (4.42). From Sect. 4.52 they know that in the state $|\Psi^-\rangle$ for the directions $\hat{\mathbf{n}}^i$ with $i \in \{1, \ldots, 4\}$ given in (4.38) with (4.40) then

$$\left| \left\langle \Sigma_{\hat{\mathbf{n}}^1}^A \otimes \Sigma_{\hat{\mathbf{n}}^2}^B \right\rangle_{\Psi^-} - \left\langle \Sigma_{\hat{\mathbf{n}}^1}^A \otimes \Sigma_{\hat{\mathbf{n}}^3}^B \right\rangle_{\Psi^-} + \left\langle \Sigma_{\hat{\mathbf{n}}^4}^A \otimes \Sigma_{\hat{\mathbf{n}}^2}^B \right\rangle_{\Psi^-} + \left\langle \Sigma_{\hat{\mathbf{n}}^4}^A \otimes \Sigma_{\hat{\mathbf{n}}^3}^B \right\rangle_{\Psi^-} \right| = 2\sqrt{2} \tag{6.11}$$

has to hold. But they will only find this outcome (6.11) if the original composite state and their qubits have not been tampered with. In case eavesdropping has been attempted, they would find (6.10) instead. Consequently, Alice and Bob can conclude if eavesdropping has occurred as follows:

$$\text{If } \left| \overline{\Sigma_{\hat{\mathbf{n}}^1}^A \Sigma_{\hat{\mathbf{n}}^2}^B} - \overline{\Sigma_{\hat{\mathbf{n}}^1}^A \Sigma_{\hat{\mathbf{n}}^3}^B} + \overline{\Sigma_{\hat{\mathbf{n}}^4}^A \Sigma_{\hat{\mathbf{n}}^2}^B} + \overline{\Sigma_{\hat{\mathbf{n}}^4}^A \Sigma_{\hat{\mathbf{n}}^3}^B} \right| \approx 2\sqrt{2} \quad \Rightarrow \quad \text{exchange is secure,}$$

$$\text{if } \left| \overline{\Sigma_{\hat{\mathbf{n}}^1}^A \Sigma_{\hat{\mathbf{n}}^2}^B} - \overline{\Sigma_{\hat{\mathbf{n}}^1}^A \Sigma_{\hat{\mathbf{n}}^3}^B} + \overline{\Sigma_{\hat{\mathbf{n}}^4}^A \Sigma_{\hat{\mathbf{n}}^2}^B} + \overline{\Sigma_{\hat{\mathbf{n}}^4}^A \Sigma_{\hat{\mathbf{n}}^3}^B} \right| \leq 2 \quad \Rightarrow \quad \begin{array}{l} \text{eavesdropping} \\ \text{has occurred.} \end{array}$$

In the former case they use the bit-sequence from the set of qubit-pairs where they happened to have measured in the same direction as a secret random bit-sequence for a VERNAM cipher (see Example 6.4). In the latter case they have to repeat the exchange, possibly within a new set-up, or abandon the key distribution.

6.4.4   RSA Public Key Distribution

Before we turn to SHOR's prime factorization algorithm, let us have a closer look at the widespread cipher devised by RIVEST, SHAMIR and ADLEMAN (RSA) [78, 79] using a combination of private and public keys. As we shall see and mentioned before, the security of this protocol depends crucially on the fact that hitherto it is too time-consuming to determine the prime factors p and q of a publicly known large number[5] $N = pq$. It allows one party (the Receiver) to publish a part k_{pub} of the key $k = (k_{\text{priv}}, k_{\text{pub}})$ with which anybody who wants to send an encrypted message (the Sender(s)) can encrypt his plaintext message, and only the receiver can decrypt the messages. Despite the public knowing one part of the key only, the Receiver is able to decrypt the ciphertext.

As outlined after Definition 6.3, we assume that the plaintext m is an integer satisfying

$$m = \sum_{j=0}^{n_{\text{M}}-1} s_j 2^j \in \{0, \dots, N_{\text{M}}\} \subset \mathbb{N}_0,$$

where n_{M} is the bit-length of messages in M and $N_{\text{M}} = 2^{n_{\text{M}}} - 1$. The RSA protocol then runs essentially as follows:

The Receiver

- picks two primes $p \neq q$ with $p, q > N_{\text{M}}$,
- finds an $a \in \mathbb{N}$ with the property

$$\gcd(a, (p-1)(q-1)) = 1 \tag{6.12}$$

- calculates

$$N := pq \in \mathbb{N}$$

- publishes the *public key* $k_{\text{pub}} = (a, N)$.

Any Sender

- encrypts his plaintext $m \leq N_{\text{M}} < N$ by calculating

$$e(k_{\text{pub}}, m) := m^a \bmod N \tag{6.13}$$

- and sends the ciphertext $c = e(k_{\text{pub}}, m)$ on public channels to the Receiver.

The Receiver

- finds a $b \in \mathbb{N}$ such that

$$ab \bmod (p-1)(q-1) = 1 \tag{6.14}$$

(we will show in Lemma 6.5 below, that given (6.12) one can always find such a b)
- takes $k_{\text{priv}} = (b, N)$ as the *private key*
- and decrypts the ciphertext c by calculating

---

[5] Numbers which are the products of only two primes are also called half-primes.

268   6   On the Use of Entanglement

$$d(k_{\text{priv}}, c) := c^b \bmod N. \tag{6.15}$$

In Theorem 6.6 we will show that then indeed

$$d(k_{\text{priv}}, e(k_{\text{pub}}, m)) = m.$$

Hence, anyone who knows the public key k_{pub} can encrypt a plaintext m by means of (6.13), but only a person who knows a b satisfying (6.14) can decrypt the message by applying (6.15). To find a b satisfying (6.14) requires the knowledge of p and q. Certainly, the Receiver has this knowledge and can decrypt. But so does anyone who can find p and q from the knowledge of $N = pq$, that is, who can factorize N. Consequently, the security of the protocol depends crucially on factorization being so hard that it takes an amount of resources (such as time and/or computing power) normally not available.

With the help of some results from modular arithmetic, we show in Theorem 6.6 below that the decryption map (6.15) indeed reproduces the original plaintext m if the ciphertext is used as an input. Before we prove this, however, we first show that (6.12) guarantees that (6.14) has a solution $b \in \mathbb{N}$.

Lemma 6.5 *Let $a \in \mathbb{N}$ and p and q be two primes such that*

$$\gcd(a, (p-1)(q-1)) = 1. \tag{6.16}$$

Then there exists a $b \in \mathbb{N}$ that satisfies

$$ab \bmod (p-1)(q-1) = 1. \tag{6.17}$$

Proof Applying (D.12) in Theorem D.4 to a and $(p-1)(q-1)$ and (6.16) yields the existence of $x, y \in \mathbb{Z}$ with

$$ax + (p-1)(q-1)y = 1, \tag{6.18}$$

which implies

$$ax = 1 - (p-1)(q-1)y$$

and

$$\left\lfloor \frac{ax}{(p-1)(q-1)} \right\rfloor = \left\lfloor \frac{1}{(p-1)(q-1)} - y \right\rfloor = -y.$$

From this it follows that there exists an $x \in \mathbb{Z}$ satisfying

$$ax - \left\lfloor \frac{ax}{(p-1)(q-1)} \right\rfloor (p-1)(q-1) = 1 - y(p-1)(q-1) - (-y)(p-1)(q-1) = 1$$

and thus

$$ax \bmod (p-1)(q-1) = 1.$$

If $x > 0$, we set $b = x$, if $x < 0$ ($x = 0$ is excluded due to (6.18)), we choose an $l \in \mathbb{N}$ such that $x + l(p-1)(q-1) > 0$ and set $b = x + l(p-1)(q-1)$. In either case then

$$ab \bmod (p-1)(q-1) = 1$$

and we have shown that the assumption (6.16) guarantees the existence of a solution b in (6.17).   □

By construction, we have $N = pq$ and $e(k_{\text{pub}}, m) = m^a \bmod N$ so that the decryption of the ciphertext $c = e(k_{\text{pub}}, m)$ is given as

$$d\big(k_{\text{priv}}, e(k_{\text{pub}}, m)\big) = (m^a \bmod pq)^b \bmod pq.$$

With the result (D.22) of Exercise D.120, this becomes

$$d\big(k_{\text{priv}}, e(k_{\text{pub}}, m)\big) = m^{ab} \bmod pq.$$

Hence, to verify that the decryption (6.15) transforms a ciphertext $c = e(k_{\text{pub}}, m)$ into the original plaintext m, it remains to show that for a b that satisfies (6.14) then

$$m^{ab} \bmod pq = m, \tag{6.19}$$

holds. This is shown in the following theorem, which makes use of results from Appendix D.

Theorem 6.6 *Let p and q be two different primes and let $m \in \mathbb{N}$ with $m < \min\{q, p\}$. Moreover, let $a, b \in \mathbb{N}$ be such that*

$$ab \bmod (p-1)(q-1) = 1.$$

Then we have

$$m^{ab} \bmod pq = m.$$

Proof To begin with, we note that because of $ab \bmod (p-1)(q-1) = 1$, there exists a $k \in \mathbb{N}$ such that $ab = 1 + k(p-1)(q-1)$ and thus

$$m^{ab} = mm^{k(p-1)(q-1)} = m\left(m^{k(p-1)}\right)^{q-1} = m\left(m^{k(q-1)}\right)^{p-1}. \tag{6.20}$$

Because $m < \min\{q,p\}$, the prime factorization of m consists of primes that are smaller than q and p. The same holds for $m^{k(p-1)}$ and $m^{k(q-1)}$, which then cannot have a common divisor with q or p. From Corollary D.19 it then follows that

$$\left(m^{k(p-1)}\right)^{q-1} \bmod q = 1 = \left(m^{k(q-1)}\right)^{p-1} \bmod p,$$

and there exist $r,s \in \mathbb{Z}$ with $1 + rq = m^{k(p-1)(q-1)} = 1 + sp$. This implies

$$\exists r,s \in \mathbb{Z}: \quad rq = sp.$$

The prime factorization of rq thus contains p, that is, there is an $l \in \mathbb{Z}$ with

$$rq = lpq$$

and thus $m^{k(p-1)(q-1)} = 1 + rq = 1 + lpq$ from which it follows that

$$m^{k(p-1)(q-1)} \bmod pq = 1. \tag{6.21}$$

Together with (6.20), this implies

$$m^{ab} \bmod pq \underbrace{=}_{(6.20)} m m^{k(p-1)(q-1)} \bmod pq$$

$$\underbrace{=}_{(D.21)} m(\underbrace{m^{k(p-1)(q-1)} \bmod pq}_{=1}) \bmod pq$$

$$\underbrace{=}_{(6.21)} m \bmod pq = m,$$

where the last equation is a consequence of $m < \min\{p,q\}$. $\square$

As shown in the discussion preceding (6.19), the statement in Theorem 6.6 assures that the decryption of an encrypted plaintext does indeed reproduce the plaintext.

As mentioned before, the security of RSA depends on it being very time-consuming to find the prime factors p and q of a sufficiently large number $N = pq$. In order to verify this, RSA Laboratories, which is nowadays a part of the EMC Corporation, posed a public challenge back in 1991 and offered a reward for the factorization of half-primes, which are defined as the product of two primes [80]. One of these was the number RSA-768 with 232 decimals, which was successfully factorized in 2009 [81, 82]:

RSA-768 = 12301866845301177551304949583849627207728535695
           95334792197322452151726400507263657518745202199
           78646939895647494277406384592519255732630345373
           15482685079170261221429134616704292143116022212

$$4047927473779408066535141959745985690214341 3$$
$$= 334780716989568987860441698482126908177047 94983$$
$$713768568912431388982883793878002287614711652 53$$
$$174308773781446799948 9$$
$$\times\ 367460436667995904282446337996279526322279158 164$$
$$343087642676032283815739666511279233373417143 39$$
$$681027009279873630891 7\ .$$

The method used in this instance, which is believed to be currently the *best classical method* to factorize an arbitrary but very large number $N \in \mathbb{N}$, is the *(General) Number Field Sieve* (NFS) [83]. A heuristic estimate of the growth rate of the computational steps $S_{\text{NFS}}(N)$ required in this method yields for $N \to \infty$ [84, 85]

$$S_{\text{NFS}}(N) \in O\left(\exp\left[\left(\frac{64}{9} + o(1) \right)^{\frac{1}{3}} \left(\log_2 N \right)^{\frac{1}{3}} \left(\log_2 \log_2 N \right)^{\frac{2}{3}} \right] \right),$$

where we have used the LANDAU symbols defined in Definition C.1.

This factorization of RSA-768 required, even with at times several hundred computers, almost three years in real time. With a single 2.2 GHz 2 GB RAM Opteron processor it is estimated to take around 2000 CPU-years [81]. Typically banks use numbers N with around 250 decimals for their implementations of RSA. Breaking it with a single PC would in general take more than 1500 years, which is why it is considered secure.

As we shall show below in Sect. 6.5, however, SHOR in 1994 devised a quantum algorithm that would allow the factorization of a large number N in a number of computational steps which only grows *polynomially* in $\log_2 N$. If this algorithm can be implemented, RSA will become insecure.

Hence, quantum mechanics on the one hand provides tools to break existing ciphers. On the other hand, as we have shown in Sects. 6.4.2 and 6.4.3, it also provides methods to exchange keys in such a fashion that undetected eavesdropping is impossible without violating the laws of (quantum) nature.

6.5 SHOR Factorization Algorithm

6.5.1 Generalities

In 1994 SHOR [19] showed that an algorithm using a quantum computer could factor a number N that has at least two distinct prime factors in a number of computational steps which grow only polynomially with the input length[6] $\log_2 N$. More precisely, the number of computational steps $S_{\text{SHOR}}(N)$ in SHOR's algorithm to factorize N satisfies

$$S_{\text{SHOR}}(N) \in O\big((\log_2 N)^3 \log_2 \log_2 N\big) \qquad \text{for } N \to \infty. \qquad (6.22)$$

The factorization of a number of the order of 10^{1000} with a quantum computer would thus require a number of steps of the order of 10^9.

SHOR's algorithm rests on the facts

- that the factorization of a number N is equivalent to finding the period (see Definition 6.7) of a given function and
- that finding this period can be accelerated with the help of a quantum algorithm.

In the following Sect. 6.5.2 we thus show first how the factorization is accomplished by finding the period, provided certain conditions are satisfied. This re-formulation of the factorization as period-finding problem is based purely on known results from number theory and does not make use of any quantum mechanical process or property.

In Sect. 6.5.4 we present the proper quantum algorithm, which substantially accelerates the finding of the period with the help of quantum mechanical phenomena. The additional properties of the period required for the factorization to succeed are not guaranteed, however, and they only occur with a probability bounded from below for which we need to establish the lower bound. This also implies that the function may have to be changed and the quantum algorithm may have to be repeated sufficiently often. SHOR's algorithm is thus a probabilistic method and the claim (6.22) about the asymptotic efficiency of the algorithm contains the estimated number of repetitions to assure a success-probability close to 1.

However, only if N *contains at least two distinct prime factors*, can we establish adequate bounds on the relevant probabilities so that we can attain a success-probability (of finding any of these prime factors) sufficiently larger than 0. In other words, if N is the power of a prime $N = p^{v_p}$ with $v_p \in \mathbb{N}$, and, in particular, if it is just a prime number $N = p$, then the SHOR algorithm cannot deliver with a sufficiently large enough probability the prime factor p or the statement that N is prime.

Moreover, we can determine with at most $\log_2 N$ divisions by 2 whether N contains a prime power 2^{v_2} of 2, where $v_2 \in \mathbb{N}$. Divisors of N of this type can thus be found efficiently in no more than $\log_2 N$ computational steps. Hence, we assume that a search for divisors of N of the form 2^{v_2} with $v_2 \in \mathbb{N}$ has been performed and any such divisors have been divided out, that is, we assume N to be odd.

Consequently, the claim (6.22) about the efficiency of the SHOR algorithm thus only holds for *odd $N \in \mathbb{N}$ with at least two distinct prime factors*.

---

[6]Since $\log_2 N$ is indicative of the digits in the binary representation of N it is called input length or just length of the number N.

6.5.2 The Algorithm

Let $N \in \mathbb{N}$ be an odd number with at least two distinct prime factors. Our goal is to find a divisor of N. To begin with, we select a natural number $b < N$ and apply the EUCLID algorithm presented in Theorem D.4 to find out if N and b have common divisors. As we can see from Theorem D.4, it is possible to do this in maximally b computational steps. If indeed we happen to find common divisors of N and $b < N$, then the task to find a divisor of N is completed, and we are done. Otherwise, we move on to the next step and determine the period r of the function

$$
\begin{aligned}
f_{b,N} : \mathbb{N}_0 &\longrightarrow \mathbb{N}_0 \\
n &\longmapsto f_{b,N}(n) := b^n \bmod N \,,
\end{aligned}
\tag{6.23}
$$

where the period of a function $f : \mathbb{N}_0 \to \mathbb{N}_0$ is defined as follows.

Definition 6.7 The period r of a function $f : \mathbb{N}_0 \to \mathbb{N}_0$ is defined as

$$
r := \min\{m \in \mathbb{N} \mid f(n+m) = f(n) \quad \forall n \in \mathbb{N}_0\} \,.
$$

For the function $f_{b,N}$ defined in (6.23) the period coincides with the order of b modulo N as defined in Definition D.20.

Exercise 6.69 Let $b, N \in \mathbb{N}$ with $b < N$ and $\gcd(b,N) = 1$. Furthermore, let r be the period of the function $f_{b,N}$ defined in (6.23). Show that then

$$
r = \mathrm{ord}_N(b) \,.
$$

For a solution see Solution 6.69

Furthermore, it follows from EULER's Theorem D.17 that in the case of $\gcd(N,b) = 1$ functions as in (6.23) always have a finite period since $r \leq \phi(N) < \infty$ holds.

With the help of the quantum algorithm of SHOR, which we will describe in more detail below, we can determine this period in a number of computational steps, which for $N \to \infty$ grows asymptotically as $O\big((\log_2 N)^3\big)$. If the period is odd, we choose a different b with $\gcd(b,N) = 1$ and again determine the period of $f_{b,N}$. This is repeated until a b is found such that $f_{b,N}$ has an even period $r \in \mathbb{N}$. How probable it is to find such a b that gives an even period will be determined in Theorem 6.11.

Suppose then r is an even period of $f_{b,N}$. According to the result of Exercise 6.69, we then have $r = \mathrm{ord}_N(b)$, which, because of Definition D.20, leads to $b^r \bmod N = 1$. Using $1 \bmod N = 1$ and (D.2), this is equivalent to $(b^r - 1) \bmod N = 0$ and thus

$$(b^{\frac{r}{2}}+1)(b^{\frac{r}{2}}-1)\bmod N = 0. \tag{6.24}$$

From (6.24) and Lemma D.11 we conclude that N and $b^{\frac{r}{2}}+1$ or $b^{\frac{r}{2}}-1$ have common divisors. Hence, we can again apply EUCLID's algorithm to N and $b^{\frac{r}{2}}+1$ or $b^{\frac{r}{2}}-1$ and thus obtain a factor of N. Without further restrictions, however, this factor may be the trivial factor N itself. In (6.24) we can exclude the case

$$(b^{\frac{r}{2}}-1)\bmod N = 0$$

as a possible solution, since this would imply $b^{\frac{r}{2}}\bmod N = 1$ and thus $\frac{r}{2}$ would be a period of $f_{b,N}$, which contradicts the assumption that r as the period is the smallest such number. If, however, (6.24) holds because

$$(b^{\frac{r}{2}}+1)\bmod N = 0,$$

then it follows that $N \,|\, b^{\frac{r}{2}}+1$ and

$$\gcd(b^{\frac{r}{2}}+1),N) = N,$$

that is, the greatest common divisor of $b^{\frac{r}{2}}+1$ and N yields the trivial factor N. We obtain a *non-trivial* factor of N as a consequence of (6.24) only in case we have chosen $b \in \mathbb{N}$ with $b < N$ so that the event

$$\mathfrak{e}_1 := \left\{ \left[r \text{ even} \right] \text{ and } \left[(b^{\frac{r}{2}}+1)\bmod N \neq 0 \right] \right\} \tag{6.25}$$

has occurred. In this case it follows from (6.24) that N divides the product $(b^{\frac{r}{2}}+1)(b^{\frac{r}{2}}-1)$ but none of the factors $(b^{\frac{r}{2}}\pm 1)$. Consequently, N must have non-trivial common divisors with every factor $(b^{\frac{r}{2}}\pm 1)$.

In order to determine a non-trivial factor for an odd $N \in \mathbb{N}$ that is not a prime power, we can thus execute the following algorithm.

SHOR Factorization Algorithm

Input: An odd natural number N that has at least two distinct prime factors

Step 1:   Choose $b \in \mathbb{N}$ with $b < N$ and determine $\gcd(b,N)$.
            If

   $\gcd(b,N) > 1$, then $\gcd(b,N)$ is a non-trivial factor of N and
                  we are done. Go to Output and show
                  $\gcd(b,N)$ and $\frac{N}{\gcd(b,N)}$
   $\gcd(b,N) = 1$, then   go to Step 2

Step 2:   Determine the period r of the function

$$f_{b,N} : \mathbb{N}_0 \longrightarrow \mathbb{N}_0$$
$$n \longmapsto f_{b,N}(n) := b^n \bmod N.$$

If

r is odd, then   start anew with Step 1
r is even, then   go to Step 3

Step 3: Determine $\gcd(b^{\frac{r}{2}} + 1, N)$.
  If

$\gcd(b^{\frac{r}{2}} + 1, N) = N$, then   start anew with Step 1
$\gcd(b^{\frac{r}{2}} + 1, N) < N$, then   with $\gcd(b^{\frac{r}{2}} + 1, N)$ we have found a non-trivial factor of N. Calculate $\gcd(b^{\frac{r}{2}} - 1, N)$ as a further factor of N. Go to Output and show $\gcd(b^{\frac{r}{2}} \pm 1, N)$

Output: Two non-trivial factors of N

In the following sections we will present the details of the algorithm and, in particular, the computational effort required, that is, the growth rate of computational steps necessary to carry out the algorithm as $N \to \infty$.

In Sect. 6.5.3 we exhibit the effort for Step 1 to check if b and N are coprime, in other words, have no common divisor.

As SHOR [19] has shown, the determination of the period of the function $f_{b,N}$ in Step 2 can be accomplished with the help of quantum mechanics in a number of steps that grows significantly slower for $N \to \infty$ as hitherto known classical methods. We will present this quantum mechanical method and its associated computational effort in Sect. 6.5.4.

In Sect. 6.5.5 we show that for numbers with more than one prime factor the probability to find a b so that the event $\mathfrak{e}_1$, which is defined in (6.25) and which is necessary for the algorithm to succeed, occurs, quickly approaches 1 with increasing number of searches.

Finally, in Sect. 6.5.6 we aggregate the computational effort of all steps in the algorithm to arrive at the claim (6.22) about the efficiency of the SHOR algorithm.

6.5.3 Step 1: Selection of b and Calculation of $\gcd(b, N)$

For a given N we select a natural number b less than N. In order to calculate $\gcd(b, N)$, we can apply the EUCLID algorithm described in Theorem D.4. We established bounds on the growth of the number of computational steps required for this algorithm as $N \to \infty$ in (D.19). Hence, one has for the number of computational steps $S_{\text{SHOR1}}(N)$ in Step 1 that

$$S_{\text{SHOR1}}(N) \in O\big((\log_2 N)^3\big) \qquad \text{for } N \to \infty.$$

6.5.4   Step 2: Determining the Period with a Quantum Computer

From Sect. 6.5.2 we know already that we can determine a factor of N if we manage to find a suitable b and the period r of the function $f_{b,N}(n) = b^n \bmod N$ such that b and N are coprime, $f_{b,N}$ has an even period r and that $(b^{\frac{r}{2}} + 1) \bmod N \neq 0$ holds. As we show below in Theorem 6.8, Step 2 in SHOR's algorithm delivers the period r of a function $f : \mathbb{N}_0 \to \mathbb{N}_0$ with a probability of at least $\frac{const}{\log_2(\log_2 N)}$, and the number of computational steps $S_{SHOR2}(N)$ in a quantum computer required for the determination of the period grows at most with $(\log_2 N)^3$ for $N \to \infty$, that is,

$$S_{SHOR2}(N) \in O\big((\log_2 N)^3\big) \qquad \text{for } N \to \infty. \tag{6.26}$$

The claim about the efficiency of Step 2 to find the period of a function $f : \mathbb{N}_0 \to \mathbb{N}_0$ can be formulated in a slightly more general way (than in our special case $f = f_{b,N}$) for periodic functions f that satisfy the following conditions:

1. The function f can be implemented as a unitary transformation U_f on a suitable HILBERT space so that the growth in the number of computational steps S_{U_f} required to execute U_f is suitably bounded.
2. An upper bound of the period r in the form

$$r < 2^{\frac{L}{2}} \tag{6.27}$$

   exists, where $L \in \mathbb{N}$ is known.
3. The function is injective within a period.

Regarding condition 1, we will show in Proposition 6.12 that for $f = f_{b,N}$ indeed $S_{U_f} \in O\big((\log_2 N)^3\big)$ holds. For our purposes, that is, for $f = f_{b,N}$ with $f_{b,N}(n) = b^n \bmod N$ and $\gcd(b,N) = 1$, condition 2 is satisfied since we know already that $r < N$. Hence, we choose L such that $N^2 \leq 2^L \leq 2N^2$ by setting $L = \lfloor 2\log_2 N \rfloor + 1$. With the help of the result from Exercise 6.69 and the Definition D.20 of the order, we can also verify that condition 3 is satisfied in our case.

The slightly generalized statement about the efficiency of Step 2 in SHOR's algorithm to find the period of a function can be formulated as follows.

Theorem 6.8 *Let* $r, L \in \mathbb{N}$ *with* $19 \leq r < 2^{\frac{L}{2}}$ *and let* r *be the period of a function* $f : \mathbb{N}_0 \to \mathbb{N}_0$ *that is injective within one period and bounded by* 2^K. *Furthermore, let* U_f *be a unitary transformation that implements* f *as follows*

$$\begin{aligned} U_f : \mathbb{H}^{\otimes L} \otimes \mathbb{H}^{\otimes K} &\longrightarrow \mathbb{H}^{\otimes L} \otimes \mathbb{H}^{\otimes K} \\ |x\rangle \otimes |y\rangle &\longmapsto |x\rangle \otimes |y \boxplus f(x)\rangle \end{aligned} \tag{6.28}$$

(where ⊞ is defined in Definition 5.31) and requires a number of computational steps $S_{U_f}(L)$, which satisfies:

$$S_{U_f}(L) \in O\left(L^{K_f}\right) \qquad \text{for } L \to \infty$$

for a $K_f \in \mathbb{N}$. Then there exists a quantum mechanical algorithm A with which we can find the period r with a probability of at least $\frac{1}{10\ln L}$. The number of computational steps of this algorithm $S_A(L)$ satisfies

$$S_A(L) \in O\left(L^{\max\{K_f,3\}}\right) \qquad \text{for } L \to \infty. \tag{6.29}$$

In the case $f = f_{b,N}$ of interest to us we have $L = \lfloor 2\log_2 N \rfloor + 1$ and, as we show in Proposition 6.12, we have $K_f = 3$. Consequently, (6.26) follows from (6.29).

Proof To make things more digestible, we partition the presentation of the algorithm and the proof of Theorem 6.8 in the following paragraphs:

1. Preparation of the input register and the initial state
2. Exploiting massive quantum parallelism
3. Application of the quantum FOURIER transform
4. Probability in measurement of the input register
5. Probability to find r as the denominator in the continued fraction approximation
6. Aggregation of the number of computational steps

In what follows, we consider each of these items in more detail.

1. Preparation of the input register and the initial state

Let $M := \max\left\{f(x) \mid x \in \{0,\ldots,2^L - 1\}\right\}$ and $K \in \mathbb{N}$ with $M < 2^K$ and let $\mathbb{H}$ be the usual qubit HILBERT space with basis $\{|0\rangle, |1\rangle\}$. With these we build the input register

$$\mathbb{H}^A := \mathbb{H}^{\otimes L}.$$

Analogously, we build

$$\mathbb{H}^B := \mathbb{H}^{\otimes K}.$$

As the initial state, we define the state $|\Psi_0\rangle$ in the product space $\mathbb{H}^A \otimes \mathbb{H}^B$

$$|\Psi_0\rangle := |0\rangle^A \otimes |0\rangle^B = \underbrace{|0\rangle \otimes \cdots \otimes |0\rangle}_{L-\text{times}} \otimes \underbrace{|0\rangle \otimes \cdots \otimes |0\rangle}_{K-\text{times}}. \tag{6.30}$$

We then apply the L-fold tensor product of the HADAMARD transform (see Definition 2.38) to the part of the initial state $|\Psi_0\rangle$ in $\mathbb{H}^A$ to obtain

$$|\Psi_1\rangle := H^{\otimes L} \otimes \mathbf{1}^B |\Psi_0\rangle \underbrace{=}_{(5.83)} \frac{1}{2^{\frac{L}{2}}} \sum_{x=0}^{2^L-1} |x\rangle^A \otimes |0\rangle^B. \tag{6.31}$$

This transformation of $|\Psi_0\rangle$ to $|\Psi_1\rangle$ can be performed in a number of computational steps (consisting of the respective application of the HADAMARD transform H) proportional to L. Hence, as a function of L, the number of computational steps in the preparation $S_{\text{Prep}}(L)$ satisfies

$$S_{\text{Prep}}(L) \in O(L) \qquad \text{for } L \to \infty. \tag{6.32}$$

2. Exploiting massive quantum parallelism

By assumption, there exists a unitary transformation U_f on $\mathbb{H}^A \otimes \mathbb{H}^B$ that implements the function f in the form

$$U_f\left(|x\rangle^A \otimes |y\rangle^B\right) \underbrace{=}_{(6.28)} |x\rangle^A \otimes |y \boxplus f(x)\rangle^B \underbrace{=}_{(5.81)} |x\rangle^A \otimes \bigotimes_{j=K-1}^{0} |y_j \overset{2}{\oplus} f(x)_j\rangle, \tag{6.33}$$

where the number of computational steps required $S_{U_f}(L)$ satisfies:

$$S_{U_f}(L) \in O\left(L^{K_f}\right) \qquad \text{for } L \to \infty. \tag{6.34}$$

As we shall show in Proposition 6.12, these assumptions are indeed satisfied for $f(x) = b^x \bmod N$ with $K_f = 3$.

Application of U_f to $|\Psi_1\rangle$ yields

$$|\Psi_2\rangle := U_f |\Psi_1\rangle \underbrace{=}_{(6.31)} U_f \left(\frac{1}{2^{\frac{L}{2}}} \sum_{x=0}^{2^L-1} |x\rangle^A \otimes |0\rangle^B \right) \underbrace{=}_{(6.33)} \frac{1}{2^{\frac{L}{2}}} \sum_{x=0}^{2^L-1} |x\rangle^A \otimes |f(x)\rangle^B. \tag{6.35}$$

In this step we exploit the massive quantum parallelism that makes use of the quantum mechanical superposition in order to generate, by a *a single application* of U_f on a state $|\Psi_1\rangle$, in one go *a superposition of all* 2^L states of the form $|x\rangle^A \otimes |f(x)\rangle^B$. As already discussed in Sect. 5.4.2, this is often viewed as a simultaneous evaluation of the function f on the domain $\{0, \ldots, 2^L - 1\}$. But we cannot simply read off the values $f(x)$ from $|\Psi_2\rangle$ in (6.35) in order to determine the period. Instead, we will apply the quantum FOURIER transform to $|\Psi_2\rangle$ and exploit the periodicity of f to accomplish this. Before we do this, we use the periodicity of f to re-write $|\Psi_2\rangle$ in a form that will make the subsequent steps more transparent. For this we use L and the period r of f to define

$$J := \left\lfloor \frac{2^L - 1}{r} \right\rfloor \tag{6.36}$$

$$R := (2^L - 1) \bmod r. \tag{6.37}$$

Then we have

$$|\Psi_2\rangle =$$

$$
\frac{1}{2^{\frac{L}{2}}} \Bigg[\quad |0\rangle^A \quad \otimes \quad |f(0)\rangle^B \quad + \cdots + \quad |r-1\rangle^A \otimes |f(r-1)\rangle^B
$$
$$
+ \quad |r\rangle^A \quad \otimes \quad \underbrace{|f(r)\rangle^B}_{=|f(0)\rangle^B} \quad + \cdots + \quad |2r-1\rangle^A \otimes \underbrace{|f(2r-1)\rangle^B}_{=|f(r-1)\rangle^B}
$$

$$
+ \quad \vdots \qquad \vdots \qquad \vdots \qquad\qquad \vdots \qquad \vdots \qquad \vdots
$$

$$
+ |(J-1)r\rangle^A \otimes \underbrace{|f((J-1)r)\rangle^B}_{=|f(0)\rangle^B} + \cdots + |Jr-1\rangle^A \otimes \underbrace{|f(Jr-1)\rangle^B}_{=|f(r-1)\rangle^B}
$$

$$
+ \quad |Jr\rangle^A \quad \otimes \quad \underbrace{|f(Jr)\rangle^B}_{=|f(0)\rangle^B} \quad \cdots + |Jr+R\rangle^A \otimes \underbrace{|f(Jr+R)\rangle^B}_{=|f(R)\rangle^B} \Bigg]
$$

$$
= \frac{1}{2^{\frac{L}{2}}} \left[\sum_{j=0}^{J-1} \sum_{k=0}^{r-1} |jr+k\rangle^A \otimes |f(k)\rangle^B + \sum_{k=0}^{R} |Jr+k\rangle^A \otimes |f(k)\rangle^B \right].
$$

Furthermore, defining for $k \in \mathbb{N}_0$

$$
J_k := \begin{cases} J & \text{if } k \leq R \\ J-1 & \text{if } k > R, \end{cases} \tag{6.38}
$$

allows us to write

$$
|\Psi_2\rangle = \frac{1}{2^{\frac{L}{2}}} \sum_{k=0}^{r-1} \sum_{j=0}^{J_k} |jr+k\rangle^A \otimes |f(k)\rangle^B. \tag{6.39}
$$

3. Application of the quantum FOURIER transform

We now apply the quantum FOURIER transform defined in Definition 5.48 to the input register $\mathbb{H}^A$

$$
\begin{aligned}
F : \mathbb{H}^A &\longrightarrow \mathbb{H}^A \\
|x\rangle &\longmapsto F|x\rangle = \frac{1}{2^{\frac{L}{2}}} \sum_{y=0}^{2^L-1} \exp\left(2\pi i \frac{xy}{2^L}\right) |y\rangle.
\end{aligned} \tag{6.40}
$$

This transforms $|\Psi_2\rangle \in \mathbb{H}^A \otimes \mathbb{H}^B$ to the state

$$
\begin{aligned}
|\Psi_3\rangle &:= \left(F \otimes \mathbf{1}^B \right) |\Psi_2\rangle \\
&= \left(F \otimes \mathbf{1}^B \right) \left(\frac{1}{2^{\frac{L}{2}}} \sum_{k=0}^{r-1} \sum_{j=0}^{J_k} |jr+k\rangle^A \otimes |f(k)\rangle^B \right) \\
&\underset{(6.39)}{=} \frac{1}{2^{\frac{L}{2}}} \sum_{k=0}^{r-1} \sum_{j=0}^{J_k} \left(F|jr+k\rangle^A \right) \otimes |f(k)\rangle^B \\
&\underset{(6.40)}{=} \frac{1}{2^L} \sum_{k=0}^{r-1} \sum_{j=0}^{J_k} \sum_{l=0}^{2^L-1} \exp\left(2\pi i \frac{l}{2^L}(jr+k) \right) |l\rangle^A \otimes |f(k)\rangle^B. \quad (6.41)
\end{aligned}
$$

As to the number of computational steps $S_{\text{FOURIER}}(L)$ required for the quantum FOURIER transform, we know from Corollary 5.57 that

$$
S_{\text{FOURIER}}(L) \in O\left(L^2\right) \qquad \text{for } L \to \infty. \quad (6.42)
$$

4. Probability in measurement of the input register

In the next step we observe the input register $\mathbb{H}^A$ (see Definition 5.35) ignoring the system $\mathbb{H}^B$. Through this measurement we project the superposition of the states in $|\Psi_3\rangle$ onto a computational basis state $|z\rangle \in \mathbb{H}^A$, which we can determine as the result of the measurement. The composite system is in the pure state $|\Psi_3\rangle$, which, expressed as a density operator, is given by $\rho_{\Psi_3} = |\Psi_3\rangle\langle\Psi_3|$. When observing the system $\mathbb{H}^A$ alone that system is then in the state given by the partial trace

$$
\rho^A(\rho_{\Psi_3}) \underset{(3.50)}{=} \text{tr}^B \left(|\Psi_3\rangle\langle\Psi_3| \right). \quad (6.43)
$$

We denote probability to observe the state $|z\rangle$ in $\mathbb{H}^A$ for a given $z \in \{0, \ldots, 2^L - 1\}$ when measuring $|\Psi_3\rangle$ by

$$
W(z) := \mathbf{P}\{\text{State } |z\rangle \text{ detected after measurement of input register}\}
$$

It is given by the probability to observe the value 1 when measuring the observable $|z\rangle\langle z|$ in the state $\rho^A(\rho_{\Psi_3})$, that is,

$$
\begin{aligned}
W(z) &\underset{(2.86)}{=} \text{tr}\left(\rho^A(\rho_{\Psi_3})|z\rangle\langle z| \right) \underset{(6.43)}{=} \text{tr}\left(\text{tr}^B \left(|\Psi_3\rangle\langle\Psi_3| \right) |z\rangle\langle z| \right) \\
&\underset{(3.48)}{=} \text{tr}\left(|\Psi_3\rangle\langle\Psi_3| \left(|z\rangle\langle z| \otimes \mathbf{1}^B \right) \right). \quad (6.44)
\end{aligned}
$$

Exercise 6.70 Show that

$$\mathrm{tr}\left(|\Psi_3\rangle\langle\Psi_3|\left(|z\rangle\langle z|\otimes \mathbf{1}^B\right)\right) = \left|\left|\left(|z\rangle\langle z|\otimes \mathbf{1}^B\right)|\Psi_3\rangle\right|\right|^2. \tag{6.45}$$

For a solution see Solution 6.70

Using (6.45) in (6.44), we obtain

$$
\begin{aligned}
W(z) &= \left|\left|\left(|z\rangle\langle z|\otimes \mathbf{1}^B\right)|\Psi_3\rangle\right|\right|^2 \\
&= \left|\left| \frac{1}{2^L}\sum_{k=0}^{r-1}\sum_{j=0}^{J_k}\exp\left(2\pi i \frac{z}{2^L}(jr+k)\right)|z\rangle^A\otimes |f(k)\rangle^B \right|\right|^2 \\
&= \underbrace{\frac{1}{2^{2L}}\sum_{k_1,k_2=0}^{r-1}\sum_{j_1,j_2=0}^{J_{k_1},J_{k_2}}\exp\left(2\pi i \frac{z}{2^L}((j_2-j_1)r+k_2-k_1)\right)\underbrace{{}^A\langle z|z\rangle^A}_{=1}{}^B\underbrace{\langle f(k_1)|f(k_2)\rangle^B}_{=\delta_{k_1,k_2}\text{ since }0\le k_i<r}}_{(6.41)} \\
&= \frac{1}{2^{2L}}\sum_{k=0}^{r-1}\left|\sum_{j=0}^{J_k}\exp\left(2\pi i \frac{zrj}{2^L}\right)\right|^2.
\end{aligned}
$$

Here the injectivity of f within a period has been used to conclude $\langle f(k_1)|f(k_2)\rangle = \delta_{k_1 k_2}$. Using that for $a\in\mathbb{C}$ one has

$$\sum_{j=0}^{D} a^j = \begin{cases} D+1 & \text{if } a=1 \\ \frac{1-a^{D+1}}{1-a} & \text{else,} \end{cases}$$

we obtain

$$\sum_{j=0}^{J_k}\exp\left(2\pi i\frac{zrj}{2^L}\right) = \begin{cases} J_k+1 & \text{if } \frac{zr}{2^L}\in\mathbb{N}_0 \\ \dfrac{1-\exp\left(2\pi i\frac{zr(J_k+1)}{2^L}\right)}{1-\exp\left(2\pi i\frac{zr}{2^L}\right)} & \text{else} \end{cases}$$

and thus

$$W(z) = \begin{cases} W_1(z) := \frac{1}{2^{2L}}\sum_{k=0}^{r-1}(J_k+1)^2 & \text{if } \frac{zr}{2^L}\in\mathbb{N}_0 \\ W_2(z) := \frac{1}{2^{2L}}\sum_{k=0}^{r-1}\left|\dfrac{1-\exp\left(2\pi i\frac{zr}{2^L}(J_k+1)\right)}{1-\exp\left(2\pi i\frac{zr}{2^L}\right)}\right|^2 & \text{else.} \end{cases} \tag{6.46}$$

For the next step in the factorization algorithm it is necessary that the measurement of the input register resulted in finding a z for which there exists an $l\in\mathbb{N}_0$ such that

$$\left|zr - l2^L\right| \le \frac{r}{2} \tag{6.47}$$

holds. Condition (6.47) is essential in order to exploit a property of continued fraction approximations, which is used in the last step to determine r and which we present in the next section.

The objective of the following considerations is to find a suitable lower bound for $W(z)$ for a z that satisfies (6.47). From such a lower bound we can then determine how often we may have to repeat starting again with the preparation of the input register until we measure a z that satisfies (6.47).

In order to determine the lower bound, we first consider the case $\frac{zr}{2^L} \in \mathbb{N}_0$ for which (6.47) obviously holds. If, in addition $\frac{2^L}{r} \in \mathbb{N}$ holds, we obtain the following result.

Exercise 6.71 Show that if $\frac{2^L}{r} =: m \in \mathbb{N}$, then

$$W(z) = \begin{cases} \frac{1}{r} & \text{if } \frac{z}{m} \in \mathbb{N} \\ 0 & \text{else.} \end{cases}$$

For a solution see Solution 6.71

Now we consider the case $\frac{zr}{2^L} \in \mathbb{N}_0$ (but not necessarily $\frac{2^L}{r} \in \mathbb{N}$). With the definition (6.38) of the J_k, one obtains

$$\frac{1}{2^{2L}} \sum_{k=0}^{r-1} (J_k+1)^2 = \frac{1}{2^{2L}} \left(\sum_{k=0}^{R} (J_k+1)^2 + \sum_{k=R+1}^{r-1} (J_k+1)^2 \right)$$

$$= \frac{1}{2^{2L}} \left((R+1) \left(\left\lfloor \frac{2^L-1}{r} \right\rfloor + 1 \right)^2 + (r-1-R) \left\lfloor \frac{2^L-1}{r} \right\rfloor^2 \right)$$

$$\geq \frac{1}{r} \left(\frac{r}{2^L} \left\lfloor \frac{2^L-1}{r} \right\rfloor \right)^2.$$

Note that

$$r-1 \geq (2^L-1) \bmod r = (2^L-1) - \left\lfloor \frac{2^L-1}{r} \right\rfloor r$$

implies that

$$\frac{r}{2^L} \left\lfloor \frac{2^L-1}{r} \right\rfloor = 1 - \frac{1+(2^L-1)\bmod r}{2^L} \geq 1 - \frac{r}{2^L} > 1 - \frac{1}{2^{\frac{L}{2}}}, \qquad (6.48)$$

where we used assumption (6.27) in the last inequality. From (6.46), (6.48) and (6.48) we thus have for the case $\frac{zr}{2^L} \in \mathbb{Z}$ that

$$W_1(z) \geq \frac{1}{r} \left(1 - \frac{1}{2^{\frac{L}{2}}} \right)^2 > \frac{1}{r} \left(1 - \frac{1}{2^{\frac{L}{2}-1}} \right).$$

Lastly, we look for a similar estimate in the case that (6.47) holds, but at the same time $\frac{zr}{2^L} \notin \mathbb{N}$. Given such a z, we know already from (6.46) that

$$
\begin{aligned}
W_2(z) &= \frac{1}{2^{2L}} \sum_{k=0}^{r-1} \left| \frac{1 - \exp\left(2\pi i \frac{zr}{2^L}(J_k+1)\right)}{1 - \exp\left(2\pi i \frac{zr}{2^L}\right)} \right|^2 \\
&= \frac{1}{2^{2L}} \sum_{k=0}^{r-1} \left| \frac{1 - \exp\left(2\pi i \frac{zr-l2^L}{2^L}(J_k+1)\right)}{1 - \exp\left(2\pi i \frac{zr}{2^L}\right)} \right|^2 \\
&= \frac{1}{2^{2L}} \sum_{k=0}^{r-1} \left(\frac{\sin\left(\pi \frac{zr-l2^L}{2^L}(J_k+1)\right)}{\sin\left(\pi \frac{zr}{2^L}\right)} \right)^2 \\
&= \frac{1}{2^{2L}} \sum_{k=0}^{r-1} s(\alpha)^2, \qquad\qquad (6.49)
\end{aligned}
$$

where

$$
s(\alpha) := \frac{\sin\left(\alpha \tilde{J}_k\right)}{\sin(\alpha)}
$$

$$
\alpha := \pi \frac{zr - l2^L}{2^L} \qquad\qquad (6.50)
$$

$$
\tilde{J}_k := J_k + 1
$$

and α and $\tilde{J}_k$ satisfy

$$
|\alpha| = \frac{\pi}{2^L}\left(zr - l2^L\right) \underset{(6.47)}{\le} \frac{\pi}{2^L}\frac{r}{2} \underset{(6.27)}{<} \frac{\pi}{2^L}2^{\frac{L}{2}-1} = \frac{\pi}{2^{\frac{L}{2}+1}} \ll \frac{\pi}{2} \qquad (6.51)
$$

$$
|\tilde{J}_k| = J_k + 1 \underset{(6.38)}{\le} J + 1 \underset{(6.36)}{=} \left\lfloor \frac{2^L-1}{r} \right\rfloor + 1 \le \frac{2^L-1}{r} + 1 < \frac{2^L}{r} + 1
$$

as well as

$$
|\alpha \tilde{J}_k| < \frac{\pi}{2^L}\frac{r}{2}\left(\frac{2^L}{r}+1\right) \underset{(6.27)}{\le} \frac{\pi}{2}\left(1 + \frac{1}{2^{\frac{L}{2}}}\right). \qquad (6.52)
$$

To obtain a lower bound for the probability W_2, we determine a lower bound of the function $s(\alpha)$ defined in (6.50) for a suitable interval of α.

Exercise 6.72 Show that for $|\alpha| \le \alpha_{min}$ with $\alpha_{min} = \frac{\pi r}{2^{L+1}}$ and $s(\cdot)$ as defined in (6.50) one has

$$
s(\alpha)^2 \ge s(\alpha_{min})^2.
$$

For a solution see Solution 6.72

From Exercise 6.72 it follows that

$$s(\alpha)^2 \geq \frac{\sin^2\left(\frac{\pi r}{2^{L+1}}(J_k+1)\right)}{\sin^2\left(\frac{\pi r}{2^{L+1}}\right)}$$

and because of $\sin^2 x \leq x^2$ thus also

$$s(\alpha)^2 \geq \left(\frac{2^{L+1}}{\pi r}\right)^2 \sin^2\left(\frac{\pi r}{2^{L+1}}(J_k+1)\right).$$

The definitions (6.36), (6.37) and (6.38) of J, R and J_k imply

$$\left\lfloor \frac{2^L-1}{r} \right\rfloor \leq J_k + 1$$

$$\Rightarrow \underbrace{\frac{r}{2^L}\left\lfloor \frac{2^L-1}{r} \right\rfloor}_{=1-\frac{R+1}{2^L}} \leq \frac{r}{2^L}(J_k+1)$$

$$\Rightarrow \quad 1 - \frac{R+1}{2^L} \leq \frac{r}{2^L}(J_k+1)$$

$$\Rightarrow \quad 1 - \frac{r}{2^L} \leq \frac{r}{2^L}(J_k+1),$$

such that

$$s(\alpha)^2 \geq \frac{2^{2L+2}}{\pi^2 r^2}\sin^2\left(\frac{\pi}{2}\left(1-\frac{r}{2^L}\right)\right).$$

Moreover, one has

$$\sin\left(\frac{\pi}{2}(1+x)\right) = \cos\left(\frac{\pi x}{2}\right) = \sum_{j=0}^{\infty}\frac{(-1)^j}{(2j)!}\left(\frac{\pi x}{2}\right)^{2j} \geq 1 - \frac{1}{2}\left(\frac{\pi x}{2}\right)^2$$

and thus

$$s(\alpha)^2 \geq \frac{2^{2L+2}}{\pi^2 r^2}\left(1-\frac{1}{2}\left(\frac{\pi}{2}\frac{r}{2^L}\right)^2\right)^2 \geq \frac{2^{2L+2}}{\pi^2 r^2}\left(1-\left(\frac{\pi}{2}\frac{r}{2^L}\right)^2\right)$$

$$\geq \frac{2^{2L+2}}{\pi^2 r^2}\left(1-\left(\frac{\pi}{2}\frac{1}{2^{\frac{L}{2}}}\right)^2\right) = \frac{2^{2L+2}}{\pi^2 r^2}\left(1-\frac{\pi^2}{2^{L+2}}\right),$$

where in the last inequality, we used that according to the assumption (6.27) $r < 2^{\frac{L}{2}}$ holds. With (6.49) we thus obtain

$$W_2(z) \geq \frac{1}{2^{2L}} \sum_{k=0}^{r-1} \frac{2^{2L+2}}{\pi^2 r^2} \left(1 - \frac{\pi^2}{2^{L+2}}\right) = \frac{r}{2^{2L}} \frac{2^{2L+2}}{\pi^2 r^2} \left(1 - \frac{\pi^2}{2^{L+2}}\right)$$

$$= \frac{4}{\pi^2 r} \left(1 - \frac{\pi^2}{2^{L+2}}\right).$$

For $L \geq 4$ the lower bound for the probability to observe a z in the input register that satisfies (6.47) is $\frac{4}{\pi^2 r}\left(1 - \frac{\pi^2}{2^{L+2}}\right)$. This is because, if $L \geq 4$, it follows from $\frac{1}{2^{\frac{L}{2}-1}} \leq \frac{1}{2} < \frac{5}{9} < 1 - \frac{4}{\pi^2}$ that $\frac{1}{2^{\frac{L}{2}-1}} - \frac{1}{2^{2L}} < 1 - \frac{4}{\pi^2}$ and thus

$$W_{\min} := \frac{4}{\pi^2 r}\left(1 - \frac{\pi^2}{2^{L+2}}\right) \leq W_2(z) < W_1(z) < \frac{1}{r}. \tag{6.53}$$

For each $z \in \{0,\ldots,2^L - 1\}$ there exists either none or exactly one $l \in \mathbb{N}_0$ that satisfies (6.47). This is because for $l_1 \neq l_2$ the distance between $l_1 2^L$ and $l_2 2^L$ is at least $2^L > r$, but they cannot be further apart than r if they are to satisfy (6.47) for the same z. This is illustrated graphically in Fig. 6.3. There we also see that there are exactly r such $l \in \mathbb{N}_0$ (namely $l \in \{0,1,2,\ldots,r-1\}$) so that a $z \in \{0,1,2,\ldots,2^L-1\}$ with the property $|zr - l2^L| \leq \frac{r}{2}$ can be found.

A measurement of the input register thus results in a $z \in \{0,\ldots,2^L-1\}$ for which there is

- either no $l \in \mathbb{N}_0$ such that (6.47) is satisfied or
- exactly one $l_z \in \{0,\ldots,r-1\}$ such that (6.47) holds.

Let l_z denote the element of $\{0,\ldots,r-1\}$ uniquely defined by $z \in \{0,\ldots,2^L-1\}$ such that (6.47) holds. It then follows from the assumption (6.27) that

$$\left|\frac{z}{2^L} - \frac{l_z}{r}\right| < \frac{1}{2r^2}. \tag{6.54}$$

We see from Fig. 6.3 that for any $j \in \{0,\ldots,r-1\}$ there is exactly one $z \in \{0,\ldots,2^L-1\}$ such that $|zr - j2^L| \leq \frac{r}{2}$, that is, such that $j = l_z$. Hence, the probability that for a given $j \in \{0,\ldots,r-1\}$ a z exists such that $l_z = j$ is equal to the

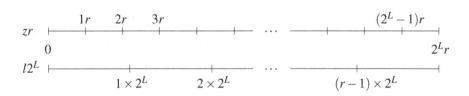

Fig. 6.3 Unique assignment of $z \in \{0,1,\ldots,2^L-1\}$ to $l \in \mathbb{N}$ so that $|zr - l2^L| \leq \frac{r}{2}$ holds. By assumption, we have $r < 2^{\frac{L}{2}}$. We see that, for example, for $z = 1$, no l can be found such that $|zr - l2^L| \leq \frac{r}{2}$, whereas for $z = 2$ exactly one $l_z = 1$ exists that satisfies this condition

probability that (6.47) can be satisfied. This latter probability is bounded from below by $W_{\min}$ such that we have for all $j \in \{0, \ldots, r-1\}$

$$\mathbf{P}\left\{\exists z \in \{0, \ldots, 2^L - 1\} : l_z = j\right\} \geq \frac{4}{\pi^2 r}\left(1 - \frac{\pi^2}{2^{L+2}}\right). \tag{6.55}$$

5. Probability to find r as the denominator in the continued fraction approximation

In (6.54) the value of z as the result of the measurement of the input register and the value of 2^L used in the construction of the register are known and r is to be determined. In order to achieve this, we shall use results from the theory of continued fractions, which are presented in detail in Appendix E. More precisely, we apply the claim in Theorem E.9 to (6.54), which states that $\frac{l_z}{r}$ has to be a partial continued fraction of $\frac{z}{2^L}$.

From Theorem E.4 we know that the defining sequence $(a_j)_{j \in \mathbb{N}_0}$ for the continued fraction of $\frac{z}{2^L}$ is finite, that is

$$\frac{z}{2^L} = a_0 + \cfrac{1}{a_1 + \cfrac{1}{\ddots + \frac{1}{a_n}}} = [a_0; a_1, \ldots, a_n]. \tag{6.56}$$

Since we know z and 2^L, we can efficiently compute the elements a_j of the defining sequence with the continued fraction algorithm presented in Appendix E as follows. Define $r_{-1} := z$ and $r_0 := 2^L$ and for $j \in \mathbb{N}$ as long as $r_{j-1} > 0$ define

$$r_j := r_{j-2} \bmod r_{j-1}.$$

From (E.5) and (E.13) we infer that then the a_j for $j \in \{1, \ldots, n\}$ are given by

$$a_j := \left\lfloor \frac{r_{j-1}}{r_j} \right\rfloor.$$

According to (D.4), the number of computational steps required for this calculation is of the order $O\left((\log_2 \max\{r_{j-2}, r_{j-1}\})^2\right)$. Since $r_{-1}, r_0 \leq 2^L$ and the r_j decrease with increasing j, the number of computational steps required to compute one r_j is of the order $O(L^2)$. The number of r_j, which we have to compute, growths, according to Corollary E.5, as a function of L as $2\min\{\log_2 2^L, \log_2 z\} + 1 \leq 2L+1$. The number of computational steps required to compute all r_j is thus of the order $O(L^3)$. For the calculation of all a_j we have to perform $O(L)$ divisions, each of which requires $O(L)$ steps. Hence, the a_j can be calculated in $O(L^2)$ steps, given the r_j have been computed before.

With the help of the a_j the *partial* continued fractions of $\frac{z}{L}$

$$\frac{p_j}{q_j} := a_0 + \cfrac{1}{a_1 + \cfrac{1}{\ddots + \frac{1}{a_j}}} \qquad \text{for } j \in \{0,\dots,n\}$$

can be calculated. For this we have to carry out $O(L)$ divisions, each of which requires $O(L)$ computational steps. We denote the set of the partial continued fractions by

$$T\left(\frac{z}{2^L}\right) := \left\{\frac{p_j}{q_j} \,\middle|\, j \in \{0,\dots,n\}\right\}.$$

Given the a_j, the computation of all elements in $T\left(\frac{z}{2^L}\right)$ thus requires a number of steps of the order $O(L^2)$. Altogether, the growth of the number of computational steps needed to calculate the partial continued fractions $S_{\text{Part-CF}}(L)$ as a function of L is of the order

$$S_{\text{Part-CF}}(L) \in O(L^3) \qquad \text{for } L \to \infty. \tag{6.57}$$

According to Theorem E.9, one of these partial continued fractions satisfies

$$\frac{p_j}{q_j} = \frac{l_z}{r}. \tag{6.58}$$

We now try to use this and our knowledge of every $\frac{p_j}{q_j} \in T\left(\frac{z}{2^L}\right)$ to find r. Because of (E.23) in Corollary E.8, the p_j and q_j in the left side of (6.58) satisfy $\gcd(q_j, p_j) = 1$. We then check for all $\frac{p_j}{q_j} \in T\left(\frac{z}{2^L}\right)$ if q_j is a period of f. For this we successively compute $f(q_j)$ for all $\frac{p_j}{q_j} \in T\left(\frac{z}{2^L}\right)$.

If we find $f(q_j) = 1$ for a q_j, then $q_j = vr$ has to hold for a $v \in \mathbb{N}$, and (6.58) implies $p_j = vl_z$. Because of $\gcd(q_j, p_j) = 1$, it follows that $v = 1$ and $q_j = r$, yielding the period we have been after all along.

If, instead, we find $f(q_j) \neq 1$ for every $\frac{p_j}{q_j} \in T\left(\frac{z}{2^L}\right)$, we have either found a z in our measurement of the input register that does not satisfy (6.54) for any $l \in \mathbb{N}_0$ or such an l_z that

$$\gcd(l_z, r) > 1.$$

In the case $f(q_j) \neq 1$ we thus have to start anew with the initial state $|\Psi_0\rangle$ in (6.30), measure again the input register to find a new z, determine the partial continued fractions of $\frac{z}{2^L}$, and check if now one of the new q_j is the period of f. The necessity for this repetition increases with the probability that $\gcd(l_z, r) > 1$ for all possible $l_z \in \{0,\dots,r-1\}$. The event

$$\mathfrak{e}_2 := \left\{ \begin{array}{l} \text{To measure a } z \in \{0, 1, \dots, 2^L - 1\} \text{ in the} \\ \text{input register such that there exists } l_z \in \mathbb{N}_0 \\ \text{with } \left| \frac{z}{2^L} - \frac{l_z}{r} \right| < \frac{1}{2r^2} \text{ and } \gcd(l_z, r) = 1 \end{array} \right\}$$

thus guarantees that r can be found from a partial continued fraction $\frac{p_j}{q_j} \in T\left(\frac{z}{2^L}\right)$. Using that, according to its definition (D.29), the EULER function ϕ gives the number $\phi(r)$ of numbers $l \in \{0, \dots, r-1\}$ with the property $\gcd(l, r) = 1$, we can estimate a lower bound for the probability of this event as follows.

$$\begin{aligned} \mathbf{P}\{\mathfrak{e}_2\} &= \sum_{\substack{l \in \{0,\dots,r-1\} \\ \gcd(l,r)=1}} \mathbf{P}\{\exists z \in \{0,\dots,2^L-1\} : l_z = l\} \\ &\underbrace{\geq}_{(6.55)} \sum_{\substack{l \in \{0,\dots,r-1\} \\ \gcd(l,r)=1}} \frac{4}{\pi^2 r} \left(1 - \frac{\pi^2}{2^{L+2}}\right) \\ &= \frac{4}{\pi^2 r} \left(1 - \frac{\pi^2}{2^{L+2}}\right) \underbrace{\sum_{\substack{l \in \{0,\dots,r-1\} \\ \gcd(l,r)=1}} 1}_{=\phi(r)} \\ &= \frac{\phi(r)}{r} \frac{4}{\pi^2} \left(1 - \frac{\pi^2}{2^{L+2}}\right). \end{aligned} \qquad (6.59)$$

In (6.59) we find separate lower bounds for the terms $\frac{\phi}{r}$ and $\frac{4}{\pi^2}\left(1 - \frac{\pi^2}{2^{L+2}}\right)$.

In order to estimate $\frac{\phi}{r}$ we use the following theorem by ROSSER and SCHOENFELD [86], which we quote without proof.

Theorem 6.9 *For $r \geq 3$ the following inequality holds*

$$\frac{r}{\phi(r)} < \exp(\gamma) \ln \ln r + \frac{2.50637}{\ln \ln r},$$

where $\gamma := 0.5772156649\dots$ denotes EULER's constant.

In

$$g(r) := \exp(\gamma) \ln \ln r + \frac{2.50637}{\ln \ln r} = \underbrace{\left(\exp(\gamma) + \frac{2.50637}{(\ln \ln r)^2}\right)}_{=:h(r)} \ln \ln r,$$

the function $h(r)$ is a decreasing function of r. For $r \geq 19$ we have already $h(r) < 4$ and thus for $r \geq 19$

$$\frac{r}{\phi(r)} < g(r) < 4\ln\ln r.$$

The period r we are looking for satisfies $r < 2^{\frac{L}{2}}$ by assumption. Hence, we have for $r \geq 19$

$$\frac{r}{\phi(r)} < g(r) < 4\ln\ln 2^{\frac{L}{2}} < 4\ln L, \tag{6.60}$$

which implies for $r \geq 19$ that

$$\frac{\phi(r)}{r} > \frac{1}{4\ln L}.$$

In order to estimate $\frac{4}{\pi^2}\left(1 - \frac{\pi^2}{2^{L+2}}\right)$ in (6.59) note that $\frac{4}{\pi^2} > \frac{2}{5}$ and for $L \geq 15$ also

$$\frac{4}{\pi^2}\left(1 - \frac{\pi^2}{2^{L+2}}\right) \geq \frac{2}{5} = 40\%. \tag{6.61}$$

Since we are only interested in the asymptotic behavior for $L \to \infty$, the restriction $L \geq 15$ is immaterial for us, and the estimate (6.61) suffices for our purposes.

For $L \geq 19$ we obtain from (6.59), (6.60) and (6.61) for the success-probability of measuring a z, such that the period r can be found as denominator of a partial continued fraction, as

$$\mathbf{P}\{\mathfrak{e}_2\} > \frac{2}{5}\frac{1}{4\ln L} = \frac{1}{10\ln L}. \tag{6.62}$$

This completes the proof of the success-probability claimed in Theorem 6.8.

6. Aggregation of the number of computational steps

From Exercise C.117 we see that in general

$$O(L^{K_1}) + O(L^{K_2}) \in O\left(L^{\max\{K_1, K_2\}}\right) \qquad \text{for } L \to \infty. \tag{6.63}$$

The Steps 1 to 3 are executed successively and from (6.32), (6.34), (6.42) and (6.57) we see that, because of (6.63), the total number $S_A(L)$ of computational steps for a successful execution of the algorithm A satisfies

$$\begin{aligned}
S_A(L) &\in S_{\text{Prep}}(L) + S_{U_f}(L) + S_{\text{FOURIER}}(L) + S_{\text{Part-CF}}(L) \\
&\in O(L) + O(L^{K_f}) + O(L^2) + O(L^3) \\
&\in O\left(L^{\max\{K_f, 3\}}\right) \qquad \text{for } L \to \infty.
\end{aligned}$$

This completes the proof of Theorem 6.8. $\qquad\qquad\qquad\qquad\qquad\qquad\square$

6.5.5 Step 3: Probability of Selecting a Suitable b

We now show that already after a few repetitions it is highly likely that the choice of a $b < N$ will result in the event $\mathfrak{e}_1$ defined in (6.25), which is essential for a successful run of the algorithm. To show this, we first prove the following lemma.

Lemma 6.10 *Let p be an odd prime, $k \in \mathbb{N}$ and $s \in \mathbb{N}_0$ and let b be selected randomly from $\left\{ c \in \{1,\ldots,p^k - 1\} \mid \gcd(p^k, c) = 1 \right\}$ with equally distributed probability $\frac{1}{\phi(p^k)}$. Then the probability that for a given triple (p, k, s) we find* $\mathrm{ord}_{p^k}(b) = 2^s t$ *with an odd t satisfies*

$$\mathbf{P}\left\{ \mathrm{ord}_{p^k}(b) = 2^s t \text{ with } 2 \nmid t \right\} \leq \frac{1}{2}.$$

Proof Let p, k and s be given. By Definition D.12 of the EULER function ϕ, the number of elements in $\left\{ c \in \{1,\ldots,p^k - 1\} \mid \gcd(p^k, c) = 1 \right\}$ is given by $\phi(p^k)$. Furthermore, there exist uniquely determined $u, v \in \mathbb{N}$ with v odd such that

$$\underbrace{\phi(p^k)}_{(\text{D.30})} = p^{k-1}(p-1) = 2^u v.$$

From Theorems D.25 and D.27 it follows that there exists a primitive root $a \in \mathbb{N}$ for p^k and from Theorem D.22 it follows that

$$\left\{ b \in \{1,\ldots,p^k - 1\} \mid \gcd(p^k, b) = 1 \right\} = \left\{ a^j \bmod p^k \mid j \in \{1,\ldots,\phi(p^k)\} \right\}.$$

Hence, via the identification
$$b = a^j \bmod p^k,$$

the random selection of one of the equally distributed b is the same as the random selection of an equally distributed $j \in \{1,\ldots,\phi(p^k)\}$. Moreover, we know from Theorem D.22 that

$$\mathrm{ord}_{p^k}(b) = \frac{\phi(p^k)}{\gcd(j, \phi(p^k))}, \tag{6.64}$$

which means that the event $\mathrm{ord}_{p^k}(b) = 2^s t$ is the same as

$$2^s t = \frac{2^u v}{\gcd(j, 2^u v)}. \tag{6.65}$$

From (6.65) we can deduce that the case $s > u$ cannot occur because in that case we would have

$$v = 2^{s-u}t \gcd(j, 2^u v)$$

and thus $2\,|\,v$, which would contradict the assumption of an odd v in $\phi(p^k) = 2^u v$. Thus, one has

$$\mathbf{P}\left\{\mathrm{ord}_{p^k}(b) = 2^s t \text{ with } 2 \nmid t \text{ and } s > u\right\} = 0. \tag{6.66}$$

For the case that $s \le u$, we can deduce that j has to be of the form $j = 2^{u-s}x$, where x is odd, as follows. Let

$$n = \prod_{p \in \mathrm{Pri}} p^{v_p} \quad \text{and} \quad m = \prod_{p \in \mathrm{Pri}} p^{\mu_p}$$

be the prime factorizations of $n, m \in \mathbb{N}$. Then

$$\gcd(n,m) = \prod_{p \in \mathrm{Pri}} p^{\min\{v_p, \mu_p\}} \tag{6.67}$$

holds. Suppose $j = 2^w x$ with x odd. It then follows from (6.67) that

$$\gcd(j, 2^u v) = 2^{\min\{w,u\}} \prod_{p \in \mathrm{Pri} \setminus \{2\}} p^{\kappa_p} \tag{6.68}$$

with suitably chosen κ_p. In order to have $\mathrm{ord}_{p^k}(b) = 2^s t$, then (6.64) and (6.65) require

$$\gcd(j, 2^u v) = 2^{u-s}\frac{v}{t}. \tag{6.69}$$

Since v and t are assumed odd, it follows that then $\frac{v}{t}$ has to be odd, too. From (6.68) and (6.69) it follows that $\min\{w,u\} = u - s$ and thus $w = u - s$. Because of this, j has to be of the form $j = 2^{u-s}x$ with an odd x and belong to $\{1,\ldots,\phi(p^k) = 2^u v\}$. In this set there exist $2^s v$ multiples of 2^{u-s}, namely

$$\{2^{u-s} \times 1, 2^{u-s} \times 2, \ldots, 2^{u-s} \times 2^s v\}.$$

Of these $2^s v$ multiples of 2^{u-s} only half are of the form $j = 2^{u-s}x$ with an odd x. The fact that all j are chosen with the same probability implies

$$\mathbf{P}\left\{\mathrm{ord}_{p^k}(b) = 2^s t \text{ with } 2 \nmid t \text{ and } s \le u\right\}$$

$$= \frac{\text{Number of possible } j \text{ of the form } j = 2^{u-s}x2 \text{ with } x \text{ odd}}{\text{Number of possible } j}$$

$$= \frac{\frac{1}{2}2^s v}{2^u v} = 2^{s-u-1} \le \frac{1}{2}$$

since $s \le u$. Together with (6.66), this yields

$$\mathbf{P}\left\{\mathrm{ord}_{p^k}(b) = 2^s t \text{ with } t \text{ odd}\right\}$$

$$= \mathbf{P}\left\{\mathrm{ord}_{p^k}(b) = 2^s t \text{ with } s > u \text{ and } t \text{ odd}\right\}$$

$$+\mathbf{P}\left\{\mathrm{ord}_{p^k}(b) = 2^s t \text{ with } s \leq u \text{ and } t \text{ odd}\right\}$$

$$\leq 0 + \frac{1}{2} = \frac{1}{2}. \qquad \square$$

Lastly, in order to estimate the probability that our choice of b does *not* satisfy the criteria in (6.25) and we have to select a new b, we need the following result.

Theorem 6.11 *Let $N \in \mathbb{N}$ be odd with prime factorization $N = \prod_{j=1}^{J} p_j^{v_j}$ consisting of prime powers of J different prime factors $p_1, \ldots, p_J$ and let $b \in \{c \in \{0,1,\ldots,N-1\} \mid \gcd(c,N) = 1\}$ be randomly chosen. Then*

$$\mathbf{P}\left\{\left[\mathrm{ord}_N(b) \text{ even}\right] \text{ and } \left[(b^{\frac{\mathrm{ord}_N(b)}{2}} + 1) \bmod N \neq 0\right]\right\} \geq 1 - \frac{1}{2^{J-1}}$$

holds.

Proof Since by assumption N is odd, all its prime factors $p_1, \ldots, p_J$ have to be odd as well, and we can apply Lemma 6.10 for their powers $p_j^{v_j}$. To abbreviate, we set $r := \mathrm{ord}_N(b)$ and show first

$$\mathbf{P}\left\{\left[r \text{ odd}\right] \text{ or } \left[(b^{\frac{r}{2}} + 1) \bmod N = 0\right]\right\} \leq \frac{1}{2^{J-1}}.$$

From Theorem D.28 we know that every $b \in \{1, \ldots, N-1\}$ with $\gcd(b,N) = 1$ corresponds uniquely to a set of $b_j := b \bmod p_j^{v_j} \in \{1, \ldots, p_j^{v_j} - 1\}$ with $\gcd(b_j, p_j^{v_j}) = 1$ for $j \in \{1, \ldots, J\}$ and vice versa. An arbitrary selection of b is thus equivalent to an arbitrary selection of the tuple $(b_1 = b \bmod p_1^{v_1}, \ldots, b_J = b \bmod p_J^{v_J})$.

According to Definition D.20, $r = \mathrm{ord}_N(b)$ satisfies

$$b^r \bmod N = 1. \qquad (6.70)$$

From (6.70) it follows that there exists a $z \in \mathbb{Z}$ such that $b^r = 1 + zN = 1 + z\prod_{j=1}^{J} p_j^{v_j}$ and thus also

$$b^r \bmod p_j^{v_j} = 1. \qquad (6.71)$$

Furthermore, we set $r_j := \mathrm{ord}_{p_j^{v_j}}(b_j)$ for every $j \in \{1, \ldots, J\}$. Then

$$1 \underbrace{=}_{\text{Def. } r_j} b_j^{r_j} \bmod p_j^{v_j} \underbrace{=}_{\text{Def. } b_j} \left(b \bmod p_j^{v_j}\right)^{r_j} \bmod p_j^{v_j}$$

$$\underbrace{=}_{\text{(D.22)}} b^{r_j} \bmod p_j^{v_j} \tag{6.72}$$

holds, and we have for every $j \in \{1,\ldots,J\}$ in addition to (6.71) also

$$b^{r_j} \bmod p_j^{v_j} = 1. \tag{6.73}$$

Since each r_j is, by its definition, the smallest positive number satisfying the first line in (6.72), it follows that it also is the smallest number satisfying (6.73). Together with (6.71), this implies that for each $j \in \{1,\ldots,J\}$ there exists a $k_j \in \mathbb{N}$ with $r = k_j r_j$. Conversely, every common multiple k of the r_j satisfies $b^k \bmod N = 1$ since

$$\frac{b^k - 1}{p_j^{v_j}} \in \mathbb{Z} \qquad \forall j \in \{1,\ldots,J\}$$

implies, because of $\gcd(p_i, p_j) = 1$ for $i \neq j$, that also

$$\frac{b^k - 1}{\prod_{j=1}^{J} p_j^{v_j}} \in \mathbb{Z}$$

such that $b^k \bmod N = 1$. Since r is, by its definition, the smallest number satisfying $b^r \bmod N = 1$, it follows that it is the smallest common multiple (see Definition D.3) of the r_j, that is,

$$r = \mathrm{scm}(r_1,\ldots,r_J). \tag{6.74}$$

Now let $r = 2^s t$ and $r_j = 2^{s_j} t_j$ with $s, s_j \in \mathbb{N}_0$ and t and t_j odd. Because of (6.74), r is odd (which is the same as $s = 0$) if and only if all r_j are odd, which is the same as $s_j = 0$ for every $j \in \{1,\ldots,J\}$. Consequently, we have

$$r \text{ odd} \quad \Leftrightarrow \quad s_j = 0 \quad \forall j \in \{1,\ldots,J\}. \tag{6.75}$$

Furthermore, because of (6.74) we have for all $j \in \{1,\ldots,J\}$

$$s_j \leq s.$$

Let us now consider the case where r is even and $(b^{\frac{r}{2}} + 1) \bmod N = 0$, that is, where there is an $l \in \mathbb{N}$ such that

$$b^{\frac{r}{2}} + 1 = lN. \tag{6.76}$$

Because of $N = \prod_{j=1}^{J} p_j^{v_j}$, it also follows that for every j there is an $l_j = l \frac{N}{p_j^{v_j}} \in \mathbb{N}$ such that

$$b^{\frac{r}{2}} + 1 = l_j p_j^{v_j}. \tag{6.77}$$

We know already that we must have $s_j \le s$ and now show that (6.76) furthermore implies $s_j = s$. To see this, suppose there is a j with $s_j < s$, then it follows from

$$2^s t = r = k_j r_j = k_j 2^{s_j} t_j$$

that

$$k_j = 2^{s-s_j} \frac{t}{t_j} \in \mathbb{N}$$

and thus

$$\frac{r}{2} = \underbrace{2^{s-s_j-1} \frac{t}{t_j}}_{:=z_j \in \mathbb{N}} r_j$$

since we are considering the case when r is even. Hence, for a j with $s_j < s$ there is a $z_j \in \mathbb{N}$ satisfying

$$\frac{r}{2} = z_j r_j. \tag{6.78}$$

Together with (6.72), this implies

$$b^{\frac{r}{2}} \bmod p_j^{v_j} \underset{(6.78)}{=} b^{z_j r_j} \bmod p_j^{v_j} \underset{(D.22)}{=} \left(b^{r_j} \bmod p_j^{v_j}\right)^{z_j} \bmod p_j^{v_j} \underset{(6.73)}{=} 1 \bmod p_j^{v_j}$$

$$= 1.$$

But this contradicts (6.77). Consequently, one has

$$(b^{\frac{r}{2}} + 1) \bmod N = 0 \quad \Rightarrow \quad s_j = s \quad \forall j \in \{1,\dots,J\}. \tag{6.79}$$

Note that with our choice of notations we have for every $j \in \{1,\dots,J\}$

$$\mathrm{ord}_{p_j^{v_j}}\left(b \bmod p_j^{v_j}\right) = r_j = 2^{s_j} t_j,$$

where the t_j are odd. Hence, a random selection of b entails a random selection of the s_j, and for the set of events under consideration the statements (6.75) and (6.79) imply

$$\{r \text{ odd }\} \subset \{s_j = 0 \quad \forall j\}$$
$$\{[r \text{ even}] \text{ and } [(b^{\frac{r}{2}} + 1) \bmod N = 0]\} \subset \{s_j = s \in \mathbb{N} \quad \forall j\}$$

and thus

$$\left\{ \left[r \text{ odd} \right] \text{ or } \left[\left[r \text{ even} \right] \text{ and } \left[(b^{\frac{r}{2}} + 1) \bmod N = 0 \right] \right] \right\} \subset \left\{ s_j = s \in \mathbb{N}_0 \quad \forall j \right\}.$$

Since we can consider the choice of the s_j as independent, we obtain for the probability that r is odd or r is even and at the same time $(b^{\frac{r}{2}} + 1) \bmod N = 0$ holds

$$\mathbf{P}\left\{ \left[r \text{ odd} \right] \text{ or } \left[\left[r \text{ even} \right] \text{ and } \left[(b^{\frac{r}{2}} + 1) \bmod N = 0 \right] \right] \right\}$$

$$\leq \mathbf{P}\{ s_j = s \in \mathbb{N}_0 \quad \forall j \} = \sum_{s \in \mathbb{N}_0} \mathbf{P}\{ s_j = s \quad \forall j \} = \sum_{s \in \mathbb{N}_0} \prod_{j=1}^{J} \mathbf{P}\{ s_j = s \}$$

$$= \sum_{s \in \mathbb{N}_0} \mathbf{P}\{ s_1 = s \} \prod_{j=2}^{J} \mathbf{P}\{ s_j = s \}$$

$$= \sum_{s \in \mathbb{N}_0} \mathbf{P}\{ s_1 = s \} \prod_{j=2}^{J} \underbrace{\mathbf{P}\{ r_j = 2^s t \text{ with } 2 \nmid t \}}_{\leq \frac{1}{2} \text{ from Lemma 6.10}}$$

$$\leq \underbrace{\sum_{s \in \mathbb{N}_0} \mathbf{P}\{ s_1 = s \}}_{=1} \frac{1}{2^{J-1}} = \frac{1}{2^{J-1}}.$$

From this we finally arrive at the probability for the event $\mathfrak{e}_1$ defined in (6.25) as

$$\mathbf{P}\{\mathfrak{e}_1\} = \mathbf{P}\left\{ \left[r \text{ even} \right] \text{ and } \left[(b^{\frac{r}{2}} + 1) \bmod N \neq 0 \right] \right\}$$

$$= 1 - \mathbf{P}\left\{ \left[r \text{ odd} \right] \text{ or } \left[\left[r \text{ even} \right] \text{ and } \left[(b^{\frac{r}{2}} + 1) \bmod N = 0 \right] \right] \right\}$$

$$\geq 1 - \frac{1}{2^{J-1}}. \tag{6.80}$$

$\square$

For a number N with more than one prime factor, that is, for the case $J \geq 2$, this implies that the probability that with one selection we have found a b such that $\mathrm{ord}_N(b)$ is even and $(b^{\frac{\mathrm{ord}_N(b)}{2}} + 1) \bmod N \neq 0$ holds, is no less than $\frac{1}{2}$. For example, after ten trials the probability to find such a b is already bounded from below by $1 - \frac{1}{2^{10}} = 0.999$.

If, however, N is a prime power, that is, if $J = 1$, we cannot derive any non-trivial statements from (6.80) about the success-probability to find a b with the property that r is even and $(b^{\frac{r}{2}} + 1) \bmod N \neq 0$ holds as well.

6.5.6   Balance Sheet of Steps

Proposition 6.12 *For $f_{b,N}(x) = b^x \bmod N$ there exists a unitary operator $U_{f_{b,N}}$ on $\mathbb{H}^A \otimes \mathbb{H}^B$ satisfying*

$$U_{f_{b,N}}\left(|x\rangle^A \otimes |0\rangle^B\right) = |x\rangle^A \otimes |f_{b,N}(x)\rangle^B \qquad (6.81)$$

and the number of computational steps $S_{U_{f_{b,N}}}$ required for $U_{f_{b,N}}$ grows as a function of $L = \lfloor 2\log_2 N \rfloor + 1$ as

$$S_{U_{f_{b,N}}}(L) \in O\left(L^3\right) \qquad \text{for } L \to \infty. \qquad (6.82)$$

Proof The statement claiming (6.81) has been proven already in Corollary 5.47. To show (6.82), we only need to gather the computational efforts required for the operations which are used to build $U_{f_{b,N}}$.

From Fig. 5.14 we see that for a quantum adder U_+ to add two numbers $a, b < 2^L$, we need to execute U_s, U_c and U_c^* each $O(L)$ times. Hence, the number of computational steps scales for $L \to \infty$ as $S_{U_+}(L) \in O(L)$. The same holds for the quantum subtractor as the inverse of the adder.

From Fig. 5.21 we see that the quantum adder modulo N requires a fixed number of quantum adders U_+ and subtractors U_- that is independent of a, b and N. Consequently, the number of computational steps for the quantum adder modulo N grows for $L \to \infty$ as $S_{U_{+\%N}}(L) \in O(L)$.

For $a, b, c, N < 2^L$ one has to perform $O(L)$ additions U_+ to execute the quantum multiplier $U_{\times c\%N}$ modulo N defined in Definition 5.43. Thus, the number of computational steps grows as $S_{U_{\times c\%N}}(L) \in O\left(L^2\right)$ for $L \to \infty$.

In Fig. 5.23 we see that $A_{f_{b,N}}$ for $x < 2^L$ is performed by executing $O(L)$ quantum multipliers $U_{\times \beta_j \%N}$. The calculation of the $\beta_j = b^{2^j} \bmod N$ for $j \in \{0, \dots, L-1\}$ needed in the multipliers can be implemented classically in an efficient manner as follows. Because of

$$b^{2^j} \bmod N \underbrace{=}_{\text{(D.22)}} \underbrace{\left(b^{2^{j-1}} \bmod N\right)^2}_{<N^2} \bmod N$$

the calculation of the $b^{2^0} \bmod N, \dots, b^{2^{L-1}} \bmod N$ only requires L times the computation of expressions of the form $a \bmod N$ in which $a < N^2$. According to Lemma D.2, we need for each of these expressions $O\left((\log_2 \max\{a, N\})^2\right) \in O\left(L^2\right)$ steps. The number of computational steps for $A_{f_{b,N}}$ thus scales altogether for $L \to \infty$ as $S_{A_{f_{b,N}}}(L) \in O\left(L^3\right)$.

Lastly, one sees from Fig. 5.11, that the construction used in Corollary 5.47 for $U_{f_{b,N}}$ requires the computation of $A_{f_{b,N}}$ and $A_{f_{b,N}}^*$ a fixed number of times, which is independent of N. We thus have also $S_{U_{f_{b,N}}}(L) \in O(L^3)$ for $x, N < 2^L$ and $L \to \infty$. $\square$

With these preparations we can now formulate the claim about the efficiency of the SHOR algorithm as follows.

Theorem 6.13 *For the factorization of an odd number $N \in \mathbb{N}$ that has at least two distinct prime factors with the SHOR algorithm, the number of required computational steps $S_{\mathrm{SHOR}}(N)$ satisfies*

$$S_{\mathrm{SHOR}}(N) \in O\big((\log_2 N)^3 \log_2 \log_2 N\big) \qquad N \to \infty.$$

Proof First, we adapt the estimate (6.62) to a statement using N instead of L. Because of $L = \lfloor 2\log_2 N \rfloor + 1$, we have

$$2\log_2 N < L \leq 2\log_2 N + 1 \tag{6.83}$$

and thus

$$\frac{1}{\ln L} \geq \frac{1}{\ln(2\log_2 N + 1)}. \tag{6.84}$$

Furthermore, because of (6.83) for $L \geq 15$, we have at least $\log_2 N \geq 7$. Such N satisfy $(\log_2 N)^{\frac{17}{12}} \geq 2\log_2 N + 1$ and we have

$$\frac{1}{\ln(2\log_2 N + 1)} \geq \frac{1}{\frac{17}{12}\ln\log_2 N} = \frac{1}{\frac{17}{12}\ln 2\log_2 \log_2 N} > \frac{1}{\log_2 \log_2 N},$$

where in the last inequality we have used $\frac{17}{12}\ln 2 < 1$. Together with (6.84), this implies for $L \geq 15$

$$\frac{1}{\ln L} > \frac{1}{\log_2 \log_2 N},$$

and with (6.62), we have for $L \geq 15$ that

$$\mathbf{P}\{\mathfrak{e}_2\} > \frac{1}{10\log_2 \log_2 N}.$$

The events relevant for the SHOR algorithm and their probabilities are collected once more in Table 6.7. The algorithm together with the required computational effort is then as follows:

Table 6.7 Relevant events in the SHOR factorization algorithm and their respective probabilities

In step	Event	Description	Probability
1	$\mathfrak{e}_1$	b has been chosen such that r is even and $(b^{\frac{r}{2}}+1)\bmod N \neq 0$ holds	$\mathbf{P}\{\mathfrak{e}_1\} \geq \frac{1}{2}$
2	$\mathfrak{e}_2$	In the measurement of the input register $\mathbb{H}^A$ the state $|z\rangle$ for such a $z \in \{0,\ldots,2^L-1\}$ has been measured, that $\exists l_z \in \{0,\ldots,r-1\}$ with $\left|\frac{z}{2^L}-\frac{l_z}{r}\right| < \frac{1}{2r^2}$ and $\gcd(l_z,r)=1$	$\mathbf{P}\{\mathfrak{e}_2\} > \frac{1}{10\log_2\log_2 N}$ for $r \geq 19$
2	Follows from $\mathfrak{e}_2$	A partial continued fraction $\frac{p_j}{q_j}$ of $\frac{z}{2^L}$ has a denominator $q_j = r$ which is a period of $f_{b,N}$	$\mathbf{P}\{\mathfrak{e}_2\} > \frac{1}{10\log_2\log_2 N}$ for $r \geq 19$
3	$\mathfrak{e}_1 \cap \mathfrak{e}_2$	A $b < N$ has been chosen such that an $r = q_j$ has been found that is an even period of $f_{b,N}$ and for which $(b^{\frac{r}{2}}+1)\bmod N \neq 0$ holds	$\mathbf{P}\{\mathfrak{e}_1 \cap \mathfrak{e}_2\} > \frac{1}{20\log_2\log_2 N}$ for $r \geq 19$

SHOR Factorization Algorithm with Computational Effort

Input: An odd natural number N with at least two distinct prime factors

Step 1: Choose a $b \in \mathbb{N}$ with $b < N$ and calculate $\gcd(b,N)$. The number of computational steps required for this satisfies, according to (D.19),

$$S_{\text{SHOR1}}(N) \in O\big((\log_2 N)^3\big) \qquad \text{for } N \to \infty.$$

If

$\gcd(b,N) > 1$, then $\gcd(b,N)$ is a non-trivial factor of N and we are done. Go to Output and show $\gcd(b,N)$ and $\frac{N}{\gcd(b,N)}$

$\gcd(b,N) = 1$, then   go to Step 2

Step 2: Determine the period r of the function

$$\begin{aligned} f_{b,N} : \mathbb{N}_0 &\longrightarrow \mathbb{N}_0 \\ n &\longmapsto f_{b,N}(n) := b^n \bmod N. \end{aligned}$$

For this we first calculate $f_{b,N}$ 20 times by direct computation. According to (D.4), the number of computational steps required for this is of the order $O\big((\log_2 N)^2\big)$. In case this evaluation reveals a period $r < 19$, we continue with the case distinction following (6.85). Otherwise, we use the quantum algorithm given in Theorem 6.8. The number of computational steps required for this is, according to (6.29) together with Proposition 6.12, of the order $O\big((\log_2 N)^3\big)$. Altogether, we have thus

$$S_{\text{SHOR2}}(N) \in O\big((\log_2 N)^3\big) \qquad \text{for } N \to \infty. \tag{6.85}$$

If

r is odd, then   start anew with Step 1
r is even, then   go to Step 3

Step 3:   Calculate $\gcd(b^{\frac{r}{2}} + 1, N)$. The number of computational steps required for this satisfies, according to (D.19),

$$S_{\text{SHOR3}}(N) \in O\big((\log_2 N)^3\big) \qquad \text{for } N \to \infty.$$

If

$\gcd(b^{\frac{r}{2}} + 1, N) = N$, then   start anew with Step 1
$\gcd(b^{\frac{r}{2}} + 1, N) < N$, then   with $\gcd(b^{\frac{r}{2}} + 1, N)$ we have found a non-trivial factor of N. Calculate $\gcd(b^{\frac{r}{2}} - 1, N)$ as a further non-trivial factor of N. Go to Output and show $\gcd(b^{\frac{r}{2}} \pm 1)$.

Output: Two non-trivial factors of N

As we see from Table 6.7, the probability that b has been chosen such that the event $\mathfrak{e}_1 \cap \mathfrak{e}_2$ occurs and thus factors of N can be determined satisfies

$$\mathbf{P}\Big\{\mathfrak{e}_1 \cap \mathfrak{e}_2\Big\} > \frac{1}{20 \log_2 \log_2 N}.$$

To find suitable b and r with a probability close to 1, it thus suffices to repeat Steps 1 to 3 approximately $20 \log_2 \log_2 N$ times.

Altogether, the number of computational steps needed to factorize N with a success-probability close to 1 growths as a function of N as

$$
\begin{aligned}
S_{\text{SHOR}}(N) \;\in\;& \big(S_{\text{SHOR1}}(N) + S_{\text{SHOR2}}(N) + S_{\text{SHOR3}}(N)\big) O(\log_2 \log_2 N) \\
\in\;& \big(O((\log_2 N)^3) + O((\log_2 N)^3) + O((\log_2 N)^3)\big) O(\log_2 \log_2 N) \\
\underbrace{\in\;}_{(\text{C.2})}& O\big((\log_2 N)^3 \log_2 \log_2 N\big) \qquad \text{for } N \to \infty.
\end{aligned}
$$

$\square$

Example 6.14 **for the SHOR algorithm**

Input: Given $N = 143$
Step 1:   We choose $b = 7$ and find $\gcd(b, N) = 1$. Hence, we proceed to Step 2.
Step 2:   The evaluation of $f_{b,N}(x) = 7^x \bmod 143$ for $x \in \{0, \dots 20\}$ shows that the period r of $f_{b,N}$ is greater than 20. We set $L = \lfloor 2 \log_2 N \rfloor + 1 = 15$.
In case we had a quantum computer, we would prepare the initial state in $\mathbb{H}^{\otimes L}$, apply $U_{f_{b,N}}$ and F and then observe the input register to read a z. The probability distribution for observing a z has the form shown in Fig. 6.4. We first find $z = 7646$ with a probability (6.46). The continued fraction representation (6.56) of $\frac{z}{2^L} = \frac{7646}{2^{15}}$ is found to be

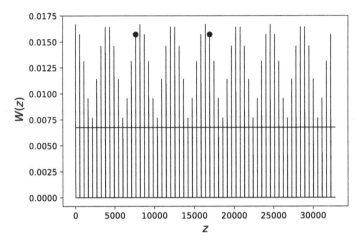

Fig. 6.4 Probability $W(z)$, to observe a $z \in \{0, \ldots, 2^L - 1 = 32767\}$ in the input register. The horizontal line shows the limit-probability $W_{\min}$ from (6.53). The vertical lines show the isolated z-values, for which the observation-probability is greater than $W_{\min}$. For all other $z \in \{0, \ldots, 32767\}$ the probabilities are smaller than 10^{-6}. The two dots show the z-values 7646 and 16930 that we have observed in the input register in this fictitious run of the algorithm

$$\frac{7646}{2^{15}} = [0; 4, 3, 1, 1, 272, 2].$$

From the partial continued fractions (6.58) we thus find as possible candidates $\frac{l_z}{r}$

$$T\left(\frac{7646}{2^{15}}\right) = \left\{\frac{p_1}{q_1}, \ldots, \frac{p_6}{q_6}\right\} = \left\{\frac{1}{4}, \frac{3}{13}, \frac{4}{17}, \frac{7}{30}, \frac{1908}{8177}, \frac{3823}{16384}\right\}. \tag{6.86}$$

Applying $f_{b,N}$ to the q_j from (6.86) yields $f_{b,N}(q_j) \neq 1$ for all these q_j, in other words, none of the q_j is the period of $f_{b,N}$.

Hence, we prepare anew the initial state, apply $U_{f_{b,N}}$ and F and measure the input register once more. This time we observe $z = 16930$. The continued fraction representation (6.56) of $\frac{z}{2^L} = \frac{16930}{2^{15}}$ is now given by

$$\frac{16930}{2^{15}} = [0; 1, 1, 14, 1, 1, 67, 1, 3].$$

For the partial continued fractions from (6.58) we now find as possible candidates for $\frac{l_z}{r}$

$$T\left(\frac{16930}{2^{15}}\right) = \left\{\frac{p_1}{q_1}, \ldots, \frac{p_8}{q_8}\right\} \tag{6.87}$$

$$= \left\{ \frac{1}{1}, \frac{1}{2}, \frac{15}{29}, \frac{16}{31}, \frac{31}{60}, \frac{2093}{4051}, \frac{2124}{4111}, \frac{8465}{16384} \right\}.$$

Applying $f_{b,N}$ to the q_j from (6.87) shows that $f_{b,N}(60) = 1$. Consequently, we have found an even period $r = q_5 = 60$ and proceed to Step 3.

Step 3: In Step 2 we identified 60 as the period of $f_{b,N}$. We thus calculate

$$\gcd(7^{30} + 1, 143) = 13 \quad \text{and} \quad \gcd(7^{30} - 1, 143) = 11.$$

These are non-trivial factors of 143. Indeed, one finds $143 = 13 \times 11$.

Output: The factors 11 and 13 of 143

6.6 Generalizing: The Abelian Hidden Subgroup Problem

It turns out that SHOR's factorization algorithm is a special case of a wider class of quantum algorithms which solve what is known as the Hidden Subgroup Problem. In the following we present the algorithm to solve this more general problem with a quantum computer in the case of a finite abelian group.

We begin with a general definition of a function hiding a subgroup before we turn to the special case of a finite abelian group. The group theoretical notions and results required for all material in this section are presented in Appendix F, and the reader unfamiliar with group theory is urged to read it prior to this section.

Definition 6.15 Let $\mathcal{H}$ be a subgroup of the group $\mathcal{G}$ and let S be a finite set. We say that a function $f : \mathcal{G} \to S$ **hides the subgroup** $\mathcal{H}$ if for any $g_1, g_2 \in \mathcal{G}$

$$f(g_1) = f(g_2) \quad \Leftrightarrow \quad g_1^{-1} g_2 \in \mathcal{H}.$$

We can re-formulate the condition $g_1^{-1} g_2 \in \mathcal{H}$ in terms of left cosets of $\mathcal{H}$.

Exercise 6.73 Let $\mathcal{H}$ be a subgroup of the group $\mathcal{G}$ and $f : \mathcal{G} \to S$, where S is a finite set. Show that then

$$f \text{ hides } \mathcal{H} \quad \Leftrightarrow \quad \forall g_1, g_2 \in \mathcal{G} \quad f(g_1) = f(g_2) \Leftrightarrow g_1 \mathcal{H} = g_2 \mathcal{H}. \tag{6.88}$$

For a solution see Solution 6.73

Hence, f hides $\mathcal{H}$ if and only if it is constant on any given left coset and takes different values on distinct left cosets of $\mathcal{H}$. Consequently, if f hides a normal subgroup $\mathcal{H}$, then it can be seen as an injective function on the quotient group $\mathcal{G}/\mathcal{H}$ (see Definition F.23).

The Hidden Subgroup Problem (HSP) is defined as the problem to identify $\mathcal{H}$ as efficiently as possible with the help of f, that is, with as few evaluations of f as possible.

> **Definition 6.16** Let f hide the subgroup $\mathcal{H}$ of the group $\mathcal{G}$. The problem to identify $\mathcal{H}$ with the help of f is called **Hidden Subgroup Problem (HSP)**.
>
> In case $\mathcal{G}$ is a finite abelian group it is called the **Abelian Hidden Subgroup Problem (AHSP)**.

In solving an HSP we would like to be as efficient as possible in the sense that we require the smallest number of queries (or evaluations) of the function f in order to determine $\mathcal{H}$. Quantum algorithms have been developed to solve non-abelian HSPs efficiently (see [87] for a review), but here we will restrict our presentation to the AHSP, in which case

$$\mathcal{G} = \{g_1, \ldots, g_{|\mathcal{G}|}\},$$

where $|\mathcal{G}|$ denotes the order, that is, the number of elements of $\mathcal{G}$. We define $n := \lceil \log_2 |\mathcal{G}| \rceil$ and identify each element g_l of the group $\mathcal{G}$ with a vector $|g_l\rangle$ of the computational basis in a suitable HILBERT space $\mathbb{H}^{\otimes n}$, in other words, we choose a subset of the computational basis states the elements of which are labeled by group elements

$$\{|g_1\rangle, \ldots, |g_{|\mathcal{G}|}\rangle\} \subset \{|x\rangle \mid x \in \{0, \ldots, 2^n - 1\}\} \subset \mathbb{H}^{\otimes n},$$

such that

$$\langle g_l | g_k \rangle = \delta_{lk}. \tag{6.89}$$

With the $|g_l\rangle$ chosen this way we define

$$\mathbb{H}^A := \mathrm{Span}\{|g_1\rangle, \ldots, |g_{|\mathcal{G}|}\rangle\} \subset \mathbb{H}^{\otimes n} \tag{6.90}$$

such that the $\{|g_l\rangle \mid l \in \{1, \ldots, |\mathcal{G}|\}\}$ form an ONB in $\mathbb{H}^A$.

Furthermore, we define $m := |S|$ for the discrete and finite set S used for hiding the subgroup $\mathcal{H} < \mathcal{G}$ by a function $f : \mathcal{G} \to S$. We also assume that the set S can be ordered and that we can identify each element $s_j \in S = \{s_0, \ldots, s_{|S|-1}\}$ with its index. We denote this identification by $\tilde{\ }$, that is, we have the bijection

$$\tilde{\ } : S \longrightarrow \{0, \ldots, |S| - 1\}$$
$$s_j \longmapsto j$$

which we use to identify each element $s_j \in S$ with a computational basis vector

$$|\widetilde{s_j}\rangle = |j\rangle \in \mathbb{\Psi H}^{\otimes m} =: \mathbb{H}^B .$$

Hence, for any $g \in \mathcal{G}$ we have $0 \leq \widetilde{f(g)} < 2^m$ and $|\widetilde{f(g)}\rangle \in \{|y\rangle \mid y \in \{0, \ldots, 2^m - 1\}\}$.

The first step in the quantum algorithm to solve the AHSP efficiently is then to prepare the initial state

$$|\Psi_0\rangle := \frac{1}{\sqrt{|\mathcal{G}|}} \sum_{g \in \mathcal{G}} |g\rangle^A \otimes |0\rangle^B \in \mathbb{H}^A \otimes \mathbb{H}^B . \tag{6.91}$$

The precise computational effort required to prepare $|\Psi_0\rangle$ depends on the group $\mathcal{G}$ in question, but for the efficiency of our quantum algorithm it suffices to assume a bound on the growths of the number of computational steps as follows.

AHSP Assumption 1 *We assume that the number of computational steps* S_1 *required to prepare* $|\Psi_0\rangle$ *as given in* (6.91) *satisfies*

$$S_1(|\mathcal{G}|) \in \text{poly}\,(\log_2(|\mathcal{G}|)) \quad \textit{for } |\mathcal{G}| \to \infty .$$

In the second step of our quantum algorithm to solve the AHSP we want to exploit massive quantum parallelism in order to evaluate f on all of $\mathcal{G}$ in one step. Hence, we make the following assumption.

AHSP Assumption 2 *For* $f : \mathcal{G} \to S$ *there exists an implementation of a unitary* U_f *defined by its action on the ONB*

$$\{|g\rangle \otimes |y\rangle \mid g \in \mathcal{G}, \, 0 \leq y < 2^m\} \subset \mathbb{H}^A \otimes \mathbb{H}^B \tag{6.92}$$

as

$$\begin{aligned} U_f : \mathbb{H}^A \otimes \mathbb{H}^B &\longrightarrow \mathbb{H}^A \otimes \mathbb{H}^B \\ |g\rangle \otimes |y\rangle &\longmapsto |g\rangle \otimes |y \boxplus \widetilde{f(g)}\rangle \end{aligned} , \tag{6.93}$$

and the number of computational steps S_2 *for the application of* U_f *satisfies*

$$S_2(|\mathcal{G}|) \in \text{poly}\,(\log_2(|\mathcal{G}|)) \quad \textit{for } |\mathcal{G}| \to \infty .$$

The second step of our quantum AHSP algorithm is then the application of U_f to $|\Psi_0\rangle$ to produce

$$|\Psi_1\rangle := U_f|\Psi_0\rangle = \frac{1}{\sqrt{|\mathcal{G}|}} \sum_{g \in \mathcal{G}} U_f(|g\rangle \otimes |0\rangle)$$

$$\underset{(6.92)}{=} \frac{1}{\sqrt{|\mathcal{G}|}} \sum_{g \in \mathcal{G}} |g\rangle \otimes |\widetilde{f(g)}\rangle \in \mathbb{H}^A \otimes \mathbb{H}^B. \tag{6.94}$$

After the first two steps in the algorithm the composite system is in the pure state $\rho = |\Psi_1\rangle\langle\Psi_1|$. We now ignore the sub-system $\mathbb{H}^B$ and only observe the sub-system $\mathbb{H}^A$. Theorem 3.17 tells us that we then have to describe sub-system $\mathbb{H}^A$ by the mixed state

$$\rho^A \underset{(3.50)}{=} \mathrm{tr}^B(\rho) = \mathrm{tr}^B(|\Psi_1\rangle\langle\Psi_1|),$$

where we have

$$|\Psi_1\rangle\langle\Psi_1| \underset{(6.94)}{=} \frac{1}{|\mathcal{G}|} \sum_{g,k \in \mathcal{G}} \left(|g\rangle \otimes |\widetilde{f(g)}\rangle\right)\left(\langle k| \otimes \langle\widetilde{f(k)}|\right)$$

$$\underset{(3.36)}{=} \frac{1}{|\mathcal{G}|} \sum_{g,k \in \mathcal{G}} |g\rangle\langle k| \otimes |\widetilde{f(g)}\rangle\langle\widetilde{f(k)}|. \tag{6.95}$$

The matrix elements of $\rho = |\Psi_1\rangle\langle\Psi_1|$ in the ONB of $\mathbb{H}^A \otimes \mathbb{H}^B$ given in (6.92) thus are

$$\rho_{gy,kz} \underset{(2.22)}{=} \langle g \otimes y|\Psi_1\rangle\langle\Psi_1|k \otimes z\rangle \underset{(6.95)}{=} \frac{1}{|\mathcal{G}|} \langle y|\widetilde{f(g)}\rangle\langle\widetilde{f(k)}|z\rangle. \tag{6.96}$$

The matrix of ρ^A in the ONB $\{g \mid g \in \mathcal{G}\}$ of $\mathbb{H}^A$ is then given by

$$\rho^A_{gk} \underset{(3.52)}{=} \sum_y \rho_{gy,ky} \underset{(6.96)}{=} \frac{1}{|\mathcal{G}|} \sum_{y=0}^{2^m-1} \langle y|\widetilde{f(g)}\rangle\langle\widetilde{f(k)}|y\rangle$$

$$= \frac{1}{|\mathcal{G}|} \langle\widetilde{f(k)}| \sum_{y=0}^{2^m-1} |y\rangle\langle y|\widetilde{f(g)}\rangle \underset{(2.20)}{=} \frac{1}{|\mathcal{G}|} \langle\widetilde{f(k)}|\widetilde{f(g)}\rangle. \tag{6.97}$$

Since $|\widetilde{f(g)}\rangle$ and $|\widetilde{f(k)}\rangle$ are vectors of the computational basis in $\mathbb{H}^B$, they satisfy

$$\langle\widetilde{f(k)}|\widetilde{f(g)}\rangle = \begin{cases} 1 & \text{if } f(k) = f(g) \\ 0 & \text{else.} \end{cases}$$

We know from Exercise 6.73 that, because f hides $\mathcal{H}$, we have that $f(k) = f(g)$ if and only if k and g belong to the same left coset, that is, there is a $\breve{g} \in \mathcal{G}$ such that $k, g \in \breve{g}\mathcal{H}$. Since $\mathcal{G}$ is assumed abelian, left and right cosets coincide and we use the notation $[\breve{g}]_{\mathcal{H}}$ to denote the coset which contains $\breve{g} \in \mathcal{G}$. Hence,

$$\langle \widetilde{f(k)} | \widetilde{f(g)} \rangle = \begin{cases} 1 & \text{if } \exists \breve{g} \in \mathcal{G} : \ g, k \in [\breve{g}]_{\mathcal{H}} \\ 0 & \text{else,} \end{cases}$$

and since such cosets are elements of the quotient group $\mathcal{G}/\mathcal{H}$, (see Definition F.23) the matrix element ρ^A_{gk} in (6.97) then becomes

$$\rho^A_{gk} = \frac{1}{|\mathcal{G}|} \sum_{\substack{[\breve{g}]_{\mathcal{H}} \in \mathcal{G}/\mathcal{H} \\ \text{s.th.:} \ g,k \in [\breve{g}]_{\mathcal{H}}}} 1, \tag{6.98}$$

where the sum is over all cosets that contain g and k. The sub-system $\mathbb{H}^A$ is then described by the mixed state

$$\rho^A \underbrace{=}_{(2.21)} \sum_{g,k \in \mathcal{G}} |g\rangle \rho^A_{gk} \langle k| \underbrace{=}_{(6.98)} \sum_{g,k \in \mathcal{G}} \frac{1}{|\mathcal{G}|} \sum_{\substack{[\breve{g}]_{\mathcal{H}} \in \mathcal{G}/\mathcal{H} \\ \text{s.th.:} \ g,k \in [\breve{g}]_{\mathcal{H}}}} |g\rangle \langle k|$$

$$= \frac{|\mathcal{H}|}{|\mathcal{G}|} \sum_{[\breve{g}]_{\mathcal{H}} \in \mathcal{G}/\mathcal{H}} \left(\frac{1}{\sqrt{|\mathcal{H}|}} \sum_{g \in [\breve{g}]_{\mathcal{H}}} |g\rangle \right) \left(\frac{1}{\sqrt{|\mathcal{H}|}} \sum_{k \in [\breve{g}]_{\mathcal{H}}} \langle k| \right).$$

We can express ρ^A more succinctly by defining for any $g \in \mathcal{G}$ what we might call the coset state

$$|\Psi^A_{[g]_{\mathcal{H}}}\rangle := \frac{1}{\sqrt{|\mathcal{H}|}} \sum_{k \in [g]_{\mathcal{H}}} |k\rangle \tag{6.99}$$

such that

$$\rho^A = \frac{|\mathcal{H}|}{|\mathcal{G}|} \sum_{[g]_{\mathcal{H}} \in \mathcal{G}/\mathcal{H}} |\Psi^A_{[g]_{\mathcal{H}}}\rangle \langle \Psi^A_{[g]_{\mathcal{H}}}|. \tag{6.100}$$

Exercise 6.74 Show that $|\Psi^A_{[g]_{\mathcal{H}}}\rangle$ as defined in (6.99) satisfies

$$\langle \Psi^A_{[g_1]_{\mathcal{H}}} | \Psi^A_{[g_2]_{\mathcal{H}}} \rangle = \begin{cases} 1 & \text{if } [g_1]_{\mathcal{H}} = [g_2]_{\mathcal{H}} \\ 0 & \text{else.} \end{cases} \tag{6.101}$$

For a solution see Solution 6.74

The next step in the algorithm is the application of the FOURIER transform as defined in Definition F.51

$$F_{\mathcal{G}} = \frac{1}{\sqrt{|\mathcal{G}|}} \sum_{g \in \mathcal{G}} \sum_{\chi \in \widehat{\mathcal{G}}} \chi(g)|\chi\rangle\langle g|, \qquad (6.102)$$

where we make use of

$$\dim \mathbb{H}^A \underbrace{=}_{(6.90)} |\mathcal{G}| \underbrace{=}_{(F.70)} |\widehat{\mathcal{G}}|$$

and assume just as for the group $\mathcal{G}$ that we can also identify each element of the dual group $\{\chi_l \mid l \in \{0, \ldots, |\mathcal{G}|\}\} = \widehat{\mathcal{G}}$ with a vector of the computational basis, such that

$$\{|\chi_1\rangle, \ldots, |\chi_{|\widehat{\mathcal{G}}|}\rangle\} = \{|g_1\rangle, \ldots, |g_{|\mathcal{G}|}\rangle\} \subset \{|x\rangle \mid x \in \{0, \ldots, 2^n - 1\}\} \subset \mathbb{H}^{\otimes n}, \qquad (6.103)$$

and we have $F_{\mathcal{G}} : \mathbb{H}^A \to \mathbb{H}^A$.

Applying the FOURIER transform (6.102) to the sub-system $\mathbb{H}^A$ transforms the state of this sub-system as

$$\rho^A \mapsto F_{\mathcal{G}} \rho^A F_{\mathcal{G}}^*.$$

Once more we make a generic assumption on the computational effort for this state transformation.

AHSP Assumption 3 *The number S_3 of computational steps required to perform the state transformation $\rho^A \mapsto F_{\mathcal{G}} \rho^A F_{\mathcal{G}}^*$ of sub-system $\mathbb{H}^A$ effected by the FOURIER transform $F_{\mathcal{G}}$ given in (6.102) satisfies*

$$S_3(|G|) \in \mathrm{poly}\left(\log_2(|\mathcal{G}|)\right) \quad for \ |\mathcal{G}| \to \infty.$$

The state of sub-system $\mathbb{H}^A$ after applying the FOURIER transform is thus

$$F_{\mathcal{G}} \rho^A F_{\mathcal{G}}^* \underbrace{=}_{(6.100)} \frac{|\mathcal{H}|}{|\mathcal{G}|} \sum_{[g]_{\mathcal{H}} \in \mathcal{G}/\mathcal{H}} F_{\mathcal{G}}|\Psi_{[g]_{\mathcal{H}}}^A\rangle\langle\Psi_{[g]_{\mathcal{H}}}^A|F_{\mathcal{G}}^*,$$

where

$$F_{\mathcal{G}}|\Psi_{[g]_{\mathcal{H}}}^A\rangle \underbrace{=}_{(6.102)} \frac{1}{\sqrt{|\mathcal{G}|}} \sum_{\check{g} \in \mathcal{G}} \sum_{\chi \in \widehat{\mathcal{G}}} \chi(\check{g})|\chi\rangle\langle\check{g}|\Psi_{[g]_{\mathcal{H}}}^A\rangle$$

$$\underbrace{=}_{(6.99)} \frac{1}{\sqrt{|\mathcal{G}||\mathcal{H}|}} \sum_{\check{g} \in \mathcal{G}} \sum_{\chi \in \widehat{\mathcal{G}}} \sum_{k \in [g]_{\mathcal{H}}} \chi(\check{g})|\chi\rangle\langle\check{g}|k\rangle$$

$$\underbrace{=}_{(6.89)} \frac{1}{\sqrt{|\mathcal{G}||\mathcal{H}|}} \sum_{\check{g} \in \mathcal{G}} \sum_{\chi \in \widehat{\mathcal{G}}} \sum_{k \in [g]_{\mathcal{H}}} \chi(\check{g})|\chi\rangle\delta_{\check{g},k}$$

$$= \underbrace{\frac{1}{\sqrt{|\mathcal{G}||\mathcal{H}|}} \sum_{\chi \in \hat{\mathcal{G}}} \sum_{k \in [g]_{\mathcal{H}}} \chi(k)|\chi\rangle}_{}$$

$$\underbrace{=}_{(\text{F.22})} \frac{1}{\sqrt{|\mathcal{G}||\mathcal{H}|}} \sum_{\chi \in \hat{\mathcal{G}}} \sum_{h \in \mathcal{H}} \chi(gh)|\chi\rangle \underbrace{=}_{(\text{F.43})} \frac{1}{\sqrt{|\mathcal{G}||\mathcal{H}|}} \sum_{\chi \in \hat{\mathcal{G}}} \sum_{h \in \mathcal{H}} \chi(g)\chi(h)|\chi\rangle$$

$$\underbrace{=}_{(\text{F.75})} \sqrt{\frac{|\mathcal{H}|}{|\mathcal{G}|}} \sum_{\chi \in \mathcal{H}^\perp} \chi(g)|\chi\rangle , \tag{6.104}$$

such that the state of the sub-system $\mathbb{H}^A$ is eventually given by

$$F_{\mathcal{G}} \rho^A F_{\mathcal{G}}^* = \frac{|\mathcal{H}|}{|\mathcal{G}|} \sum_{[g]_{\mathcal{H}} \in \mathcal{G}/\mathcal{H}} \left(\sum_{\chi \in \mathcal{H}^\perp} \chi(g)|\chi\rangle \right) \left(\sum_{\xi \in \mathcal{H}^\perp} \overline{\xi(g)}\langle\xi| \right) . \tag{6.105}$$

Now, let $\zeta \in \mathcal{H}^\perp$ and let $|\zeta\rangle$ be the corresponding basis state in $\mathbb{H}^A$. Then $P_\zeta = |\zeta\rangle\langle\zeta|$ is the orthogonal projector onto that state. According to (2.86), the probability to detect the state $|\zeta\rangle$ when observing system $\mathbb{H}^A$, which has been prepared in the state $F_{\mathcal{G}} \rho^A F_{\mathcal{G}}^*$, is

$$\mathbf{P} \left\{ \begin{array}{l} \text{To observe } |\zeta\rangle \text{ for a } \zeta \in \mathcal{H}^\perp \text{ in system } \mathbb{H}^A \text{ when} \\ \text{it has been prepared in the state } F_{\mathcal{G}} \rho^A F_{\mathcal{G}}^* \end{array} \right\}$$

$$\underbrace{=}_{(2.86)} \text{tr} \left(|\zeta\rangle\langle\zeta| F_{\mathcal{G}} \rho^A F_{\mathcal{G}}^* \right)$$

$$\underbrace{=}_{(2.57)} \sum_a \langle e_a|\zeta\rangle\langle\zeta| F_{\mathcal{G}} \rho^A F_{\mathcal{G}}^* e_a\rangle ,$$

where we have used an ONB $\{e_a\}$ in $\mathbb{H}^A$. Using (6.105), we thus obtain

$$\mathbf{P} \left\{ \begin{array}{l} \text{To observe } |\zeta\rangle \text{ for a } \zeta \in \mathcal{H}^\perp \text{ in system } \mathbb{H}^A \text{ when} \\ \text{it has been prepared in the state } F_{\mathcal{G}} \rho^A F_{\mathcal{G}}^* \end{array} \right\}$$

$$= \sum_a \frac{|\mathcal{H}|}{|\mathcal{G}|} \sum_{[g]_{\mathcal{H}} \in \mathcal{G}/\mathcal{H}} \left(\sum_{\chi \in \mathcal{H}^\perp} \chi(g)\langle e_a|\zeta\rangle \underbrace{\langle\zeta|\chi\rangle}_{=\delta_{\zeta\chi}} \right) \left(\sum_{\xi \in \mathcal{H}^\perp} \overline{\xi(g)}\langle\xi|e_a\rangle \right)$$

$$= \frac{|\mathcal{H}|}{|\mathcal{G}|} \sum_{[g]_{\mathcal{H}} \in \mathcal{G}/\mathcal{H}} \sum_{\xi \in \mathcal{H}^\perp} \left(\sum_a \langle e_a|\zeta\rangle\langle\xi|e_a\rangle \right) \zeta(g)\overline{\xi(g)}$$

$$\underbrace{=}_{(2.13)} \frac{|\mathcal{H}|}{|\mathcal{G}|} \sum_{[g]_{\mathcal{H}} \in \mathcal{G}/\mathcal{H}} \sum_{\xi \in \mathcal{H}^\perp} \underbrace{\langle\xi|\zeta\rangle}_{=\delta_{\xi\zeta}} \zeta(g)\overline{\xi(g)} = \sum_{[g]_{\mathcal{H}} \in \mathcal{G}/\mathcal{H}} \zeta(g)\overline{\zeta(g)}$$

$$\underbrace{=}_{(F.58)} \frac{|\mathcal{H}|}{|\mathcal{G}|} \sum_{[g]_{\mathcal{H}} \in \mathcal{G}/\mathcal{H}} 1 \underbrace{=}_{(F.36)} 1 . \qquad (6.106)$$

Consequently, when observing the sub-system $\mathbb{H}^A$ after the third step of our algorithm we will always find a state $|\zeta\rangle$ that corresponds to a character $\zeta \in \mathcal{H}^{\perp}$.

Re-running the three steps of the algorithm L times, we can then apply Corollary F.50 to our observations of $\zeta_1, \ldots, \zeta_L \in \mathcal{H}^{\perp}$ to assert that

$$\mathbf{P}\left\{\langle \zeta_1, \ldots, \zeta_L \rangle = \mathcal{H}^{\perp}\right\} \underbrace{\geq}_{(F.103)} 1 - \frac{|\mathcal{H}^{\perp}|}{2^L} \underbrace{=}_{(F.83)} 1 - \frac{|\mathcal{G}|}{2^L |\mathcal{H}|} .$$

Hence, after

$$L \geq \log_2 \left(\frac{|\mathcal{G}|}{\varepsilon |\mathcal{H}|} \right)$$

repetitions and observations, the probability that we have observed a generating set of $\mathcal{H}^{\perp}$ is no less than $1 - \varepsilon$, that is,

$$\mathbf{P}\left\{\langle \zeta_1, \ldots, \zeta_L \rangle = \mathcal{H}^{\perp}\right\} \geq 1 - \varepsilon .$$

Finally, it follows from Theorem F.44 that we can find the desired $\mathcal{H}$ from the observed generating set by

$$\mathcal{H} = \bigcap_{l=1}^{L} \text{Ker}(\zeta_l) .$$

Let us summarize the complete algorithm to solve the AHSP.

Abelian Hidden Subgroup Problem algorithm with computational effort

Input: A finite abelian group $\mathcal{G}$ and a function $f : \mathcal{G} \to S$ that hides a subgroup $\mathcal{H} \leq \mathcal{G}$

Step 1:   In $\mathbb{H}^A \otimes \mathbb{H}^B$, where

$$\mathbb{H}^A = \text{Span}\{|g_1\rangle, \ldots, |g_{|\mathcal{G}|}\rangle\} \subset \mathbb{H}^{\otimes n}$$

with $n = \lceil \log_2 |\mathcal{G}| \rceil$ and

$$\mathbb{H}^B = \mathbb{H}^{\otimes m} ,$$

with $m = \lceil \log_2 |S| \rceil$, prepare the initial state

$$|\Psi_0\rangle = \frac{1}{\sqrt{|\mathcal{G}|}} \sum_{g \in \mathcal{G}} |g\rangle \otimes |0\rangle \in \mathbb{H}^A \otimes \mathbb{H}^B .$$

The computational effort required for this is assumed to satisfy

$$S_{\text{AHSP Step 1}}(|\mathcal{G}|) \in \text{poly}\left(\log_2 |\mathcal{G}|\right) \qquad \text{for } |\mathcal{G}| \to \infty$$

Step 2:   Apply U_f of (6.93) to $|\Psi_0\rangle$ to produce

$$|\Psi_1\rangle = U_f |\Psi_0\rangle = \frac{1}{\sqrt{|\mathcal{G}|}} \sum_{g \in \mathcal{G}} |g\rangle \otimes |\widetilde{f(g)}\rangle \in \mathbb{H}^A \otimes \mathbb{H}^B .$$

The computational effort required for this is assumed to satisfy

$$S_{\text{AHSP Step 2}}(|\mathcal{G}|) \in \text{poly}\left(\log_2 |\mathcal{G}|\right) \qquad \text{for } |\mathcal{G}| \to \infty$$

Step 3:   Consider only sub-system $\mathbb{H}^A$, which when disregarding sub-system $\mathbb{H}^B$ will be in the mixed state

$$\rho^A = \frac{|\mathcal{H}|}{|\mathcal{G}|} \sum_{[g]_{\mathcal{H}} \in \mathcal{G}/\mathcal{H}} |\Psi^A_{[g]_{\mathcal{H}}}\rangle \langle \Psi^A_{[g]_{\mathcal{H}}}| ,$$

where

$$|\Psi^A_{[g]_{\mathcal{H}}}\rangle := \frac{1}{\sqrt{|\mathcal{H}|}} \sum_{h \in \mathcal{H}} |gh\rangle .$$

On this sub-system perform the quantum FOURIER transform $F_{\mathcal{G}}$ to transform sub-system $\mathbb{H}^A$ into the state

$$F_{\mathcal{G}} \rho^A F_{\mathcal{G}}^* = \sqrt{\frac{|\mathcal{H}|}{|\mathcal{G}|}} \sum_{[g]_{\mathcal{H}} \in \mathcal{G}/\mathcal{H}} \left(\sum_{\chi \in \mathcal{H}^{\perp}} \chi(g) |\chi\rangle \right) \left(\sum_{\xi \in \mathcal{H}^{\perp}} \overline{\xi(g)} \langle \xi| \right) .$$

The computational effort required for this satisfies

$$S_{\text{AHSP Step 3}}(|\mathcal{G}|) \in \text{poly}\left(\log_2 |\mathcal{G}|\right) \qquad \text{for } |\mathcal{G}| \to \infty$$

Step 4:   Observe the sub-system $\mathbb{H}^A$ to detect a $\zeta \in \mathcal{H}^{\perp}$ with certainty. The computational effort required for this is

$$S_{\text{AHSP Step 4}}(|\mathcal{G}|) \in \text{poly}\left(\log_2 |\mathcal{G}|\right) \qquad \text{for } |\mathcal{G}| \to \infty$$

Step 5:   Repeat Steps 1–4 for

$$L \geq \log_2 \left(\frac{|\mathcal{G}|}{\varepsilon |\mathcal{H}|} \right)$$

times to determine $\zeta_1,\ldots,\zeta_L \in \mathcal{H}^{\perp}$ and from this

$$\bigcap_{l=1}^{L} \text{Ker}(\zeta_l).\tag{6.107}$$

The computational effort required for this is assumed to satisfy

$$S_{\text{AHSP Step 5}}(|\mathcal{G}|) \in \text{poly}\,(\log_2 |\mathcal{G}|) \qquad \text{for } |\mathcal{G}| \to \infty$$

Output:

$$\mathcal{H} = \bigcap_{l=1}^{L} \text{Ker}(\zeta_l)$$

with a probability no less than $1 - \varepsilon$

The total computational effort for the AHSP algorithm grows as function of $|G|$ as

$$S_{\text{AHSP}}(|\mathcal{G}|) = \sum_{j=1}^{5} S_{\text{AHSP Step }j}(|\mathcal{G}|) \in \text{poly}\,(\log_2 |\mathcal{G}|) \qquad \text{for } |\mathcal{G}| \to \infty.$$

Quite a number of problems can be re-formulated as AHSPs [61, 87, 88]. Many of those play a role in cryptography, and finding a discrete logarithm in a group is one of them.

6.7   Finding the Discrete Logarithm as a Hidden Subgroup Problem

As we shall see below, the discrete logarithm plays an essential role in some of today's most advanced cryptographic protocols. It is defined as follows.

Definition 6.17 Let $\mathcal{G}$ be a group and $g, h \in \mathcal{G}$ such that there exists a $d \in \mathbb{N}_0$ such that

$$h = g^d.\tag{6.108}$$

Then d is called the **discrete logarithm** of h to base g, and this is expressed by the notation $d = \text{dlog}_g(h)$.

The task to find $d = \text{dlog}_g(h)$, when only g and and h satisfying (6.108) are known, is called the **Discrete Logarithm Problem (DLP)**.

The use of discrete logarithms in cryptography is owed to the fact that, given $g \in \mathcal{G}$ and $d \in \mathbb{N}_0$, it is quite easy to calculate $h = g^d$ simply by taking the group product of g with itself d times, but that it is generally extremely difficult to find d when given only g and h. Hence, g and h can be published without revealing $\mathrm{dlog}_g(h)$. This is what is used in the DIFFIE–HELLMAN [89] public key cryptographic protocol, which can be illustrated as follows.

DIFFIE–HELLMAN Cryptographic Protocol		
Alice	Bob	Public knows
mutually agree a $g \in \mathcal{G}$		g
selects $a \in \mathbb{N}$, calculates $A = g^a$ and sends A to Bob	selects $b \in \mathbb{N}$, calculates $B = g^b$ and sends B to Alice	A, B
calculates the shared but secret key $K = B^a = g^{ab}$	calculates the shared but secret key $K = A^b = g^{ab}$	

At the end of the above protocol Alice and Bob share a secret key K. If, however, calculating the discrete logarithm were feasible for Eve, her knowledge of A, B and g would allow her to compute $a = \mathrm{dlog}_g(A)$ and thus the key $K = B^a$ as well.

As we now show, the discrete logarithm problem can be formulated as an AHSP for a suitably chosen group, set and function in the AHSP. Hence, a working quantum computer on which the AHSP algorithm is implemented would potentially render the DIFFIE–HELLMAN cryptographic protocols insecure.

The DLP that we solve with the help of the AHSP algorithm is as follows:

Given: (i) a group $\mathcal{G}_{\mathrm{DLP}}$ and an element $g \in \mathcal{G}_{\mathrm{DLP}}$ that has order $N = \mathrm{ord}(g)$, that is $N \in \mathbb{N}$ is the smallest number satisfying

$$g^N = e_{\mathcal{G}_{\mathrm{DLP}}} \tag{6.109}$$

(ii) an $h \in \mathcal{G}_{\mathrm{DLP}}$ such that for some unknown $d \in \mathbb{N}$

$$h = g^d \tag{6.110}$$

Find: $d = \mathrm{dlog}_g(h)$

To find d, we set up a suitable AHSP and then execute the algorithm to solve it. As before, all group theoretical notions used here are defined and explained in Appendix F.

The group $\mathcal{G}$ of the AHSP is not the same as the group $\mathcal{G}_{\mathrm{DLP}}$ in which we want to find the discrete logarithm. Rather, it is given as the direct product group of the quotient group $\mathbb{Z}_N$, that is, we set

$$\mathcal{G} := \mathbb{Z}_N \times \mathbb{Z}_N \,,$$

where N is as given in (6.109). From Lemma F.5 it follows that $|\mathbb{Z}_N| = N$ and from (F.38) that then in the case at hand

$$|\mathcal{G}| = N^2. \tag{6.111}$$

Any element $g \in \mathcal{G}$ is thus of the form $g = ([x]_{N\mathbb{Z}}, [y]_{N\mathbb{Z}})$, where $[x]_{N\mathbb{Z}}$ and $[y]_{N\mathbb{Z}}$ are cosets in $\mathbb{Z}_N = \mathbb{Z}/N\mathbb{Z}$. From (F.24) we know that for these cosets $[x]_{N\mathbb{Z}} = [x \bmod N]_{N\mathbb{Z}}$. We use this to represent the elements of $\mathcal{G}$ as vectors in the HILBERT space $\mathbb{H}^A$ given in (6.90) in the form

$$|g\rangle = |([x]_{N\mathbb{Z}}, [y]_{N\mathbb{Z}})\rangle = |x \bmod N\rangle \otimes |y \bmod N\rangle \tag{6.112}$$

and thus

$$\mathbb{H}^A = \text{Span}\left\{ |u\rangle \otimes |v\rangle \mid u, v \in \{0, \ldots, N-1\} \right\}. \tag{6.113}$$

As the set S in the AHSP, we choose

$$S := \langle g \rangle \le \mathcal{G}_{\text{DLP}},$$

which in itself is a cyclic group of order N.

For the function f we choose

$$f : \begin{array}{ccc} \mathcal{G} & \longrightarrow & S \\ ([x]_{N\mathbb{Z}}, [y]_{N\mathbb{Z}}) & \longmapsto & h^x g^y \end{array}, \tag{6.114}$$

where $[x]_{N\mathbb{Z}}, [y]_{N\mathbb{Z}}$ denote the cosets in the quotient group $\mathbb{Z}_N = \mathbb{Z}/N\mathbb{Z}$ and $h, g \in \mathcal{G}_{\text{DLP}}$ are as given in (6.110). Since $h = g^d$, we find

$$f([x]_{N\mathbb{Z}}, [y]_{N\mathbb{Z}}) = (g^d)^x g^y = g^{dx+y} \in \langle g \rangle.$$

To determine the subgroup hidden by f, note that we have for any $([x]_{N\mathbb{Z}}, [y]_{N\mathbb{Z}}) \in \mathcal{G}$ and $([u]_{N\mathbb{Z}}, [v]_{N\mathbb{Z}}) \in \mathcal{G}$

$$f\big(([x]_{N\mathbb{Z}}, [y]_{N\mathbb{Z}}) +_{\mathcal{G}} ([u]_{N\mathbb{Z}}, [v]_{N\mathbb{Z}})\big) = f([x]_{N\mathbb{Z}}, [y]_{N\mathbb{Z}})$$

$$\underset{(\text{F.35})}{\Leftrightarrow} \qquad f([x+u]_{N\mathbb{Z}}, [y+v]_{N\mathbb{Z}}) = f([x]_{N\mathbb{Z}}, [y]_{N\mathbb{Z}})$$

$$\underset{(6.110)}{\Leftrightarrow} \qquad g^{d(x+u)+y+v} = g^{dx+y}$$

$$\Leftrightarrow \qquad g^{du+v} = e_{\mathcal{G}}$$

$$\underset{(\text{F.24})}{\Leftrightarrow} \qquad (du+v) \bmod N = 0$$

$$\underset{(\text{F.24})}{\Leftrightarrow} \qquad [du+v]_{N\mathbb{Z}} = [0]_{N\mathbb{Z}}$$

$$\Leftrightarrow \qquad [v]_{N\mathbb{Z}} = [-du]_{N\mathbb{Z}}.$$

Exercise 6.75 Show that

$$\mathcal{H} = \left\{ ([u]_{N\mathbb{Z}}, [-du]_{N\mathbb{Z}}) \,\middle|\, [u]_{N\mathbb{Z}} \in \mathbb{Z}_N \right\} \tag{6.115}$$

is a subgroup of $\mathcal{G} = \mathbb{Z}_N \times \mathbb{Z}_N$.

For a solution see Solution 6.75

Hence, $\mathcal{H} \le \mathcal{G}$ as given in (6.115) is the subgroup hidden by the function f defined in (6.114). For any $g = ([x]_{N\mathbb{Z}}, [y]_{N\mathbb{Z}}) \in \mathcal{G}$ its coset $[g]_{\mathcal{H}}$ with $\mathcal{H}$ is then given by

$$[g]_{\mathcal{H}} = \left\{ ([x+u]_{N\mathbb{Z}}, [y-du]_{N\mathbb{Z}}) \,\middle|\, [u]_{N\mathbb{Z}} \in \mathbb{Z}_N \right\}.$$

Moreover, the number of elements in this $\mathcal{H}$ is equal to the number of elements in $\mathbb{Z}_N$, and we have here

$$|\mathcal{H}| = N. \tag{6.116}$$

With this the coset state for a $[g]_{\mathcal{H}}$ is given by

$$|\Psi^A_{[g]_{\mathcal{H}}}\rangle \underbrace{=}_{(6.99),(6.112)} \frac{1}{\sqrt{N}} \sum_{[u]_{N\mathbb{Z}} \in \mathbb{Z}_N} |(x+u) \bmod N\rangle \otimes |(y-du) \bmod N\rangle.$$

Using (F.59) and (F.71), we find that the characters of $\mathcal{G}$ are given by

$$\begin{aligned} \chi_{m,n} : \quad \mathcal{G} &\longrightarrow \mathrm{U}(1) \\ ([x]_{N\mathbb{Z}}, [y]_{N\mathbb{Z}}) &\longmapsto e^{2\pi i \frac{mx+ny}{N}}, \end{aligned} \tag{6.117}$$

where $m, n \in \{0, \ldots, N-1\}$. As assumed in (6.103), we use this to represent the elements of $\widehat{\mathcal{G}}$ as vectors in $\mathbb{H}^A$ in the form

$$|\chi_{m,n}\rangle = |m\rangle \otimes |n\rangle \qquad \text{for } m, n \in \{0, \ldots, N-1\}. \tag{6.118}$$

Using once more (6.99) and the fact that $|\mathcal{G}| = N^2$, the FOURIER transform (6.102) then becomes

$$F_{\mathcal{G}} = \frac{1}{N} \sum_{\substack{m,n,v,w \\ \in \{0,\ldots,N-1\}}} e^{2\pi i \frac{mv+nw}{N}} |m\rangle \otimes |n\rangle\langle v| \otimes \langle w|$$

Recall from Lemma F.39 that $\mathcal{H}^\perp = \left\{ \chi \in \widehat{\mathcal{H}} \,\middle|\, \mathcal{H} \subset \mathrm{Ker}(\chi) \right\}$, and in the case at hand $\mathcal{H}$ is given by (6.115) such that for $m, n \in \{0, \ldots, N-1\}$ we have

$$\chi_{m,n} \in \mathcal{H}^\perp \quad \Leftrightarrow \quad \chi_{m,n}([u]_{N\mathbb{Z}}, [-du]_{N\mathbb{Z}}) = 1 \quad \forall\, [u]_{N\mathbb{Z}} \in \mathbb{Z}_N$$

$$\underset{(6.117)}{\Leftrightarrow} \quad e^{2\pi i \frac{mu - ndu}{N}} = 1 \quad \forall u \in \{0, \dots, N-1\}$$

$$\Leftrightarrow \quad m = dn \bmod N.$$

Consequently, we obtain

$$\mathcal{H}^\perp = \left\{ \chi_{dn \bmod N, n} \;\middle|\; [n]_{N\mathbb{Z}} \in \mathbb{Z}_N \right\}. \tag{6.119}$$

Moreover, $\mathcal{H}^\perp$ is actually a cyclic group generated by $\chi_{d,1}$ as is shown in the following exercise.

Exercise 6.76 Show that

$$\mathcal{H}^\perp = \langle \chi_{d,1} \rangle, \tag{6.120}$$

where $d = \mathrm{dlog}_g(h) < N$.

For a solution see Solution 6.76

We illustrate these constructions with a simple example to be worked out in Exercise 6.77.

Exercise 6.77 Write down $\mathcal{H}$ and $\mathcal{H}^\perp$ for the case $N = 6$ and $d = 3$ as well as $\mathrm{Ker}(\chi_{m,n})$ for $\chi_{m,n} \in \mathcal{H}^\perp$, and verify (6.120) in this particular case.

For a solution see Solution 6.77

The FOURIER transform of a coset state $|\Psi_{[g]\mathcal{H}}^A\rangle$ given in (6.104) for the general AHSP algorithm now becomes for the discrete logarithm problem at hand for a $g = ([x]_{N\mathbb{Z}}, [y]_{N\mathbb{Z}})$

$$F_g |\Psi_{[g]\mathcal{H}}^A\rangle \underset{(6.104)}{=} \frac{1}{\sqrt{N}} \sum_{[n]_{N\mathbb{Z}} \in \mathbb{Z}_N} \chi_{dn \bmod N, n}([x]_{N\mathbb{Z}}, [y]_{N\mathbb{Z}}) |dn \bmod N\rangle \otimes |n\rangle$$

$$\underset{(6.117)}{=} \frac{1}{\sqrt{N}} \sum_{[n]_{N\mathbb{Z}} \in \mathbb{Z}_N} e^{2\pi i \frac{dnx + ny}{N}} |dn \bmod N\rangle \otimes |n\rangle. \tag{6.121}$$

For the mixed state (6.105) in which the sub-system $\mathbb{H}^A$ is, we obtain

$$
\begin{aligned}
F_{\mathcal{G}}\rho^A F_{\mathcal{G}}^* &= \underbrace{\frac{|\mathcal{H}|}{|\mathcal{G}|} \sum_{[g]_{\mathcal{H}} \in \mathcal{G}/\mathcal{H}} F_{\mathcal{G}} |\Psi_{[g]_{\mathcal{H}}}^A\rangle \langle \Psi_{[g]_{\mathcal{H}}}^A| F_{\mathcal{G}}^*}_{(6.111),(6.116)} \\
&= \underbrace{\frac{1}{N} \sum_{[g]_{\mathcal{H}} \in \mathcal{G}/\mathcal{H}} F_{\mathcal{G}} |\Psi_{[g]_{\mathcal{H}}}^A\rangle \langle \Psi_{[g]_{\mathcal{H}}}^A| F_{\mathcal{G}}^*}_{} \\
&= \underbrace{\frac{1}{N^2} \sum_{\substack{[g]_{\mathcal{H}} \in \mathcal{G}/\mathcal{H} \\ [n]_{N\mathbb{Z}},[m]_{N\mathbb{Z}} \in \mathbb{Z}_N}} e^{2\pi i \frac{dx+y}{N}(n-m)} |dn \bmod N\rangle \langle dm \bmod N| \otimes |n\rangle\langle m|}_{(6.121)},
\end{aligned}
\tag{6.122}
$$

where, as before, $g = ([x]_{N\mathbb{Z}}, [y]_{N\mathbb{Z}}) \in \mathcal{G}$.

Exercise 6.78 Let $u,v \in \{0,\dots,N-1\}$ be such that

$$
P_{u,v} = \big(|u\rangle \otimes |v\rangle\big)\big(\langle u| \otimes \langle v|\big) \underbrace{=}_{(3.36)} |u\rangle\langle u| \otimes |v\rangle\langle v|,
\tag{6.123}
$$

in accordance with (6.113), is an orthogonal projector onto a state $|u\rangle \otimes |v\rangle \in \mathbb{H}^A$. Show that then

$$
\mathrm{tr}\left(P_{u,v} F_{\mathcal{G}} \rho^A F_{\mathcal{G}}^*\right) = \frac{|\langle u|dv \bmod N\rangle|^2}{N}.
\tag{6.124}
$$

For a solution see Solution 6.78

From (2.86) we know that the probability to measure the state $|u\rangle \otimes |v\rangle$ when observing the system $\mathbb{H}^A$ when it is described by $F_{\mathcal{G}}\rho^A F_{\mathcal{G}}^*$ is given by the left side of (6.124). From (6.118) and (6.119) we know that the states which represent characters in $\mathcal{H}^\perp$ are of the form $|dn \bmod N\rangle \otimes |n\rangle$, where $n \in \{0,\dots,N-1\}$. Consequently,

$$
\mathbf{P}\left\{\begin{array}{l} \text{When system } \mathbb{H}^A \text{ is prepared in} \\ F_{\mathcal{G}}\rho^A F_{\mathcal{G}}^* \text{ and measured, a state cor-} \\ \text{responding to a } \chi \in \mathcal{H}^\perp \text{ is detected} \end{array}\right\} = \sum_{n=0}^{N-1} \mathrm{tr}\left(P_{dn \bmod N, n} F_{\mathcal{G}} \rho^A F_{\mathcal{G}}^*\right)
$$

$$
\underbrace{=}_{(6.124)} \sum_{n=1}^{N-1} \frac{|\langle dn \bmod N | dn \bmod N\rangle|^2}{N}
$$

$$
= 1,
$$

which tells us, as already ascertained in (6.106) for a general AHSP, that we will always find a state corresponding to an element in $\mathcal{H}^\perp$.

Rather than employing (6.107) to find $\mathcal{H}$ and then $d = \mathrm{dlog}_g(h)$, we can determine d more directly in the DLP version of the AHSP as follows. From (6.119) we infer that finding a state corresponding to an element in $\mathcal{H}^\perp$ means that we have found states of the computational basis of the form $|dn \bmod N\rangle \otimes |n\rangle \in \mathbb{H}^A$ where $n \in \{0,\dots,N-1\}$. Hence, after one execution of the AHSP algorithm for the DLP we know a pair of numbers $(dn \bmod N, n)$ with $n \in \{0,\dots,N-1\}$.

Now, suppose $\gcd(n, N) = 1$. From (D.19) we know that the computational effort to calculate $\gcd(n, N)$ scales as $\text{poly}(\log_2 |\mathcal{G}|)$. The probability for $\gcd(n, N) = 1$ to occur can be bounded from below away from zero as can be seen from the following theorem, which we quote without proof.

Theorem 6.18 ([90] **Theorem 332**) *Let $m \in \mathbb{N}$, and let p, q be selected independently and with uniform probability from $\{0, \dots, m-1\} \subset \mathbb{N}$. Then we have*

$$\lim_{m \to \infty} \mathbf{P}\{\gcd(p, q) = 1\} = \frac{6}{\pi^2} > 0.6.$$

From Lemma D.9 we then know that we can employ the extended EUCLID algorithm to find the inverse of n modulo N, that is, the number denoted $n^{-1} \bmod N$ that satisfies $((n(n^{-1} \bmod N)) \bmod N = 1$. From (D.11) we can infer that the computational effort for this is of order $\text{poly}(\log_2 |\mathcal{G}|)$. Having obtained $n^{-1} \bmod N$, we can use it to compute $d \, \text{dlog}_g(h)$ as follows

$$((dn \bmod N)(n^{-1} \bmod N)) \bmod N \underbrace{=}_{(D.20)} (dnn^{-1} \bmod N) \bmod N \underbrace{=}_{(D.8)} d \bmod N$$

$$= d,$$

where in the last step we used that from (6.109) we can infer that $d < N$.

If, however, $\gcd(n, N) > 1$, we repeat the algorithm to find a second pair $(dm \bmod N, m)$ and determine $\gcd(n, m)$. If $\gcd(n, m) > 1$, we repeat the algorithm again and again until we find a pair, such that $\gcd(n, m) = 1$. As can be seen once more from Theorem 6.18, the probability for this to happen is greater than $\frac{3}{5}$, and the number of potential repetitions of the algorithm does not alter the overall computational effort from $\text{poly}(\log_2 |\mathcal{G}|)$.

Assuming then that we have a second pair $(dm \bmod N, m)$ such that $\gcd(n, m) = 1$, we can employ the extended EUCLID algorithm of Theorem D.4 to find integers a, b such that (D.12) ensures

$$an + bm = \gcd(n, m) = 1. \tag{6.125}$$

Consequently, we have

$$a(dn \bmod N) + b(dm \bmod N) \underbrace{=}_{(D.1)} d(an + bm) - N\left(a\left\lfloor \frac{dn}{N} \right\rfloor + b\left\lfloor \frac{dm}{N} \right\rfloor\right)$$

$$\underbrace{=}_{(6.125)} d - N\left(a\left\lfloor \frac{dn}{N} \right\rfloor + b\left\lfloor \frac{dm}{N} \right\rfloor\right)$$

and thus, since $0 < d < N$, we can use the known integers $dn \bmod N, dm \bmod N, a, b$ to obtain

$$\left(a(dn \bmod N) + b(dm \bmod N)\right) \bmod N \underbrace{}_{(D.1)} = d.$$

This shows that we can indeed use the AHSP algorithm to solve the DLP. We now turn to show how such a solution of the DLP could potentially render the bitcoin transaction signature protocol unsafe.

6.8 Breaking Bitcoin Signatures

In today's digital world it becomes increasingly important to be able to sign documents digitally, in other words, to add a digital signature that cannot be forged but which can be verified to a document. This is accomplished by so-called *digital signature algorithms* (DSA), which constitute cryptographic protocols by which a signer can authenticate a publicly known document and provide the public with a means to verify the authentication. The scheme of such protocols is as follows.

Digital Signature Algorithm (DSA) Protocol	
Signer	**Public knows**
	algorithm parameters $\mathcal{A}$
	verification statement v
chooses a private key k	
creates a public *verification key* **by**	
computing a $V = V(k, \mathcal{A})$	
and publishing it	verification key V
signs document by	
taking document d,	document d
computing a *signature* $s(d, \mathcal{A})$	
and publishing it	signature s
	and can verify by
	checking the verification statement $v(s, d, V, \mathcal{A}) = \text{TRUE}$?

The security of this protocol relies on the difficulty with which the secret private key k can be obtained from the knowledge of the public verification key $V = V(k, \mathcal{A})$ and the publicly known algorithm parameters $\mathcal{A}$.

If k could be found by a fraudster, he could publish altered or new documents with a valid signature by the signer and the public would be made to believe that these documents were authenticated by the signer.

A widely used version of such a DSA is based on the difficulty to find discrete logarithms for elements of elliptic curves (see Definition F.56). This is known as the Elliptic Curve Digital Signature Algorithm (ECDSA), and a version of it is used for creating the digital signature of transactions conducted with bitcoins [79, 91].

With the help of a suitably chosen elliptic curve the ECDSA provides a document authentication protocol for a document in the public domain. Prior to input into the ECDSA these documents are standardized by a so-called **hash function**.   Such hash functions deterministically map input of arbitrary length to output of fixed bit-string length. For cryptographic purposes it is desirable that small changes in input produce significant changes in output and that it is extremely improbable that two distinct inputs produce identical output.

Example 6.19 An example of a hash function provided by the NSA is the Secure Hashing Algorithm SHA256 which converts any ASCII into a 64 digit hexadecimal string. As an example consider the following text.
The SHA256 hash output of the text in this line in hexadecimal form displayed across two lines is:

$$A3C431026DDD514C6D0C7E5EB253D424$$

$$B6A4AF20EC00A8C4CBE8E57239BBB848$$

Such a 64 digit hexadecimal string can be interpreted as a 256-bit natural number d, which in our example would be (given in binary format first)

d

$$=(1010001111000100001100010000001001101101110111010101010000000$$
$$0000000000000000000000000000000000000000000000000000000000000000$$
$$0000000000000000000000000000000000000000000000000000000000000000$$
$$0000000000000000000000000000000000000000000000000000000000000000$$
$$00000000000000000000)_2$$
$$=7.407363459482995\cdots \times 10^{76} < 2^{256}.$$

The hash functions used for pre-processing the documents are part of the algorithm specification of the ECDSA in the public domain, and we may thus assume that the document to be signed is given as a positive integer not exceeding a known upper bound N, such as $N = 2^{256}$ in the case of SHA256 in Example 6.19.

ECDSAs are usually based on elliptic curves $E(\mathbb{F}_p)$ for which p is a large prime. From Corollary F.59 we know that for a prime p the elliptic curve $E(\mathbb{F}_p)$ over the finite field $\mathbb{F}_p = \mathbb{Z}/p\mathbb{Z}$ together with the addition $+_E$ given in Theorem F.58 forms a finite abelian group such that we can define addition of two points $P, Q \in E(\mathbb{F}_p)$. In particular, we can add P to itself $k \in \mathbb{N}$ times to obtain

$$kP = \underbrace{P +_E P +_E \cdots +_E P}_{k \text{ times}}.$$

In addition to the hash function, the protocol for ECDSA then requires the specification of five publicly known parameters $\mathcal{A} = (p,A,B,P,q)$ consisting of the following items.

ECDSA Parameters (p,A,B,P,q) in the Public Domain

1. A prime p specifying the finite field $\mathbb{F}_p$
2. Two elements $A,B \in \mathbb{F}_p$ specifying the WEIERSTRASS equation

$$y^2 = x^3 + Ax + B$$

   of the elliptic curve $E(\mathbb{F}_p)$. This is an equation in the finite field $\mathbb{F}_p$. The underlying set of $\mathbb{F}_p$ consists of cosets in $\mathbb{Z}/p\mathbb{Z} \cong \mathbb{Z}_p$. From Lemma F.5 and Example F.19 we know that any such coset (or equivalently element in $\mathbb{Z}_p$) can be uniquely identified with a number in $\{0,\dots,p-1\}$. Hence, we consider A and B and the components x and y of elements $P = (x,y) \in E(\mathbb{F}_p) \smallsetminus \{0_E\}$ as elements of the set $\{0,\dots,p-1\}$
3. An element

$$P = (x_P, y_P) \in E(\mathbb{F}_p) \smallsetminus \{0_E\} \subset \mathbb{F}_p \times \mathbb{F}_p,$$

   which is often called the base point of the ECDSA
4. The element P is chosen such that it has prime order, that is,

$$q = \mathrm{ord}(P) := \min\{n \in \mathbb{N} \mid nP = 0_E \in E(\mathbb{F}_p)\} \qquad (6.126)$$

   is a publicly known prime

Given a document d in the appropriate format, the process steps of ECDSA can then be divided into the three sections Public Key Generation, Signature Generation and Verification as follows [92].

ECDSA Public Key Generation

1. Select a private key

$$k \in \{1,\dots,q-1\} \subset \mathbb{N}$$

2. Compute the *verification key*

$$V = kP \in E(\mathbb{F}_p) \smallsetminus \{0_E\}. \qquad (6.127)$$

   Note that $V \neq 0_E$ since $k < q$, and q is the smallest number such that $qP = 0_E$
3. Publish the verification key $V \in E(\mathbb{F}_p) \smallsetminus \{0_E\}$

ECDSA Signature Generation

1. Select a natural number

$$a \in \{1, \ldots, q-1\} \tag{6.128}$$

2. Compute

$$aP = (x_{aP}, y_{aP}) \in E(\mathbb{F}_p) \setminus \{0_E\}, \tag{6.129}$$

   where, as above, we are guaranteed $aP \neq 0_E$ since $a < q$, and we consider $x_{aP} \in \mathbb{F}_p$ to be represented by a number in $\{0, \ldots, p-1\}$

3. Compute

$$s_1 = x_{aP} \bmod q \in \{0, \ldots, q-1\} \tag{6.130}$$

4. If $s_1 = 0$, go back to Step 1 of the signature generation and select a new $a \in \{1, \ldots, q-1\}$.
   If $s_1 \neq 0$, calculate the multiplicative inverse of a modulo q

$$\widehat{a} = a^{-1} \bmod q \in \{1, \ldots, q-1\} \tag{6.131}$$

   defined in Definition D.8, that is, the number $\widehat{a}$ such that $a\widehat{a} \bmod q = 1$. Note that since $a \in \{0, \ldots, q-1\}$ and q is a prime, we always have $\gcd(a, q) = 1$ and the multiplicative inverse exists.
   With $\widehat{a}$ compute

$$s_2 = \big((d + ks_1)\widehat{a}\big) \bmod q \in \{0, \ldots, q-1\}$$

5. If $s_2 = 0$, go back to Step 1 of the signature generation and select a new $a \in \{1, \ldots, q-1\}$.
   Else, set the *signature* as

$$(s_1, s_2) \in \{1, \ldots, q-1\} \times \{1, \ldots, q-1\}$$

6. Publish the signature (s_1, s_2)

Before we turn to the verification procedure, we first show that the statement which will be tested in the verification query is indeed true. This is done in the following proposition.

Proposition 6.20 *Let (p, A, B, P, q) be parameters of an ECDSA with elliptic curve $E(\mathbb{F}_p)$, let $d \in \mathbb{N}$ be the document to be signed, and let $k, V, a, \widehat{a}, s_1, s_2$ be as given in the ECDSA public key and signature generation. With*

$$\widehat{s_2} = s_2^{-1} \bmod q$$
$$u_1 = d\widehat{s_2} \bmod q \tag{6.132}$$
$$u_2 = s_1\widehat{s_2} \bmod q$$

it then follows that

$$u_1 P + u_2 V = aP$$

and thus for $(x,y) = u_1 P + u_2 V$ *that*

$$x \bmod q = s_1. \tag{6.133}$$

Proof To begin with, we note that from (D.1) we know that for any $b \in \mathbb{Z}$ there is a $z \in \mathbb{Z}$ such that $b = b \bmod q + zq$ and thus

$$bP = (b \bmod q)P + zqP \underbrace{=}_{(6.126)} (b \bmod q)P. \tag{6.134}$$

Next, we have

$$u_1 P + u_2 V \underbrace{=}_{(6.127)} u_1 P + u_2 kP = (u_1 + u_2 k)P$$
$$\underbrace{=}_{(6.134)} \big((u_1 + u_2 k) \bmod q\big)P. \tag{6.135}$$

The coefficient of P in the last equation of (6.135) can be evaluated as follows

$$(u_1 + u_2 k) \bmod q \underbrace{=}_{(6.132)} \big(d\widehat{s_2} \bmod q + (s_1\widehat{s_2} \bmod q)k\big) \bmod q$$
$$\underbrace{=}_{(D.20)-(D.23)} \big((d + s_1 k)\widehat{s_2}\big) \bmod q$$
$$\underbrace{=}_{(6.132)} \big((d + s_1 k)((d + s_1 k)\widehat{a})^{-1} \bmod q\big) \bmod q$$
$$\underbrace{=}_{(D.8)} \widehat{a}^{-1} \bmod q \underbrace{=}_{(6.131)} a \bmod q$$
$$\underbrace{=}_{(6.128)} a,$$

which, inserted into (6.135), shows that $u_1 P + u_2 V = aP$. This in turn implies that

$$(x_{aP}, y_{aP}) \underbrace{=}_{(6.129)} aP = u_1 P + u_2 V = (x,y)$$

such that the claim (6.133) follows immediately from (6.130).   □

Note that the left side of (6.133) can be calculated from the publicly available information $p, A, B, P, q, V, d, s_1, s_2$ and the right side s_1 is also publicly known. Hence, the veracity of (6.133) can be verified with the help of publicly available information and it constitutes the verification statement.

ECDSA Verification

1. Compute

$$\widehat{s_2} = s_2^{-1} \bmod q$$
$$u_1 = d\widehat{s_2} \bmod q$$
$$u_2 = s_1\widehat{s_2} \bmod q$$

and with these calculate

$$(x, y) = u_1 P + u_2 V$$

2. Check if

$$x \bmod q \stackrel{?}{=} s_1$$

is true. If it is, then (s_1, s_2) constitutes a valid signature of the document d. Otherwise, it does not

An ECDSA protocol can be summarized as follows [79].

Elliptic Curve Digital Signature (ECDSA) Protocol	
Signer	**Public knows**
	algorithm parameters $\mathcal{A}$: large prime p elliptic curve $E(\mathbb{F}_p)$ public point $P \in E(\mathbb{F}_p) \smallsetminus \{0_E\}$ with a large prime order q
creates key by choosing a *secret signing key* $k \in \mathbb{N}$ with $1 < k < q$, computing the *verification key* $V = kP$ and publishing it	verification key V
signs document by taking document d and a random $a \in \mathbb{N}$ with $a < q$, computing $aP \in E(\mathbb{F}_p) \smallsetminus \{0_E\}$ $s_1 = x_{aP} \bmod q$ $s_2 = ((d + ss_1)(a^{-1} \bmod q)) \bmod q$ and publishing the *signature* (s_1, s_2)	document d signature (s_1, s_2)
	and verifies by computing $u_1 = (d(s_2^{-1} \bmod q)) \bmod q$ $u_2 = (s_1(s_2^{-1} \bmod q)) \bmod q$ $(x, y) = u_1 P +_E u_2 V \in E(\mathbb{F}_p) \smallsetminus \{0_E\}$ and checking the verification statement **is** $x \bmod q = s_1$ **TRUE?**

In Definition 6.17 we defined for any $V, P \in E(\mathbb{F}_p)$ such that $V = kP$

$$k = \mathrm{dlog}_P(V)$$

as the **discrete logarithm** in $E(\mathbb{F}_p)$ of V to base P. The security of ECDSA depends on the fact that it is very hard to calculate the discrete logarithm for this group. If it were feasible to calculate the discrete logarithm in a relatively short time span, then ECDSA would become insecure. This is because V and P are in the public domain, and anyone who can calculate the elliptic curve discrete logarithm from the publicly known V and P would obtain the the *secret signing key k*. With this key any fraudster could forge the signature of the signer. In other words, they could publish an alternative document $\widetilde{d}$, perform the signing procedure on this document $\widetilde{d}$, publish the new signature $(\widetilde{s}_1, \widetilde{s}_2)$, and the public would believe it were from the signer who has published V. For example, in the context of bitcoins a fraudster could submit a new transaction to the blockchain claiming that the signer has transferred bitcoins to him.

Example 6.21 Bitcoins use the `secp256k1` ECDSA [93] protocol with the WEIER-STRASS equation defined by $A = 0$ and $B = 7$, that is,

$$y^2 = x^3 + 7,$$

the prime

$$p = 2^{256} - 2^{32} - 2^9 - 2^8 - 2^7 - 2^6 - 2^4 - 1 \tag{6.136}$$

and the public point $P = (x_P, y_P)$ given by

$x_P = 5506626302227734366957871889516853432625060345377759417550018$
7360389116729240

$y_P = 3267051002075881697808308513050704318447127338065924327593890$
4335757337482424 .

The best known classical method to calculate $k = \mathrm{dlog}_P(V)$ for $E(\mathbb{F}_p)$ requires $O(\sqrt{p})$ computational steps and thus for the bitcoin ECDSA of the order of $O(10^{77})$ computational steps. In contrast, a quantum computer could potentially calculate $k = \mathrm{dlog}_P(V)$ for $E(\mathbb{F}_p)$ requiring only

$$O(\text{polynomial in } \log_2(p)) \underbrace{=}_{(6.136)} O(\text{polynomial in } 256)$$

computational steps and thus render the bitcoin signature insecure.

6.9 GROVER Search Algorithm

The search algorithm developed in 1996 by GROVER [21] describes a method in which known objects ('needles') can be found in a large unordered set ('haystack') of N objects. The method achieves a success-probability greater than 50% in $O(\sqrt{N})$ steps. Previously known methods required $O(\frac{N}{2})$ steps to find the object with such a probability.

For example, if we wanted to find the name belonging to a given phone number in a phone book of four million entries, the GROVER search algorithm would have a success-probability greater than 50% after $O(2 \times 10^3)$ steps, whereas the previously known methods would require $O(2 \times 10^6)$ steps to achieve the same success-probability.

The GROVER search algorithm begins by representing the objects as quantum states, in other words, normalized vectors in a suitable HILBERT space. The vectors of the objects which we try to find span a subspace in this HILBERT space and the algorithm constructs operators that successively transform (or 'rotate') a given initial state into a state which has a maximal component in the subspace of desired objects. This implies that, when measuring the rotated states, we have a greater probability of detecting a state which lies in the subspace of desired objects. This method of rotating the initial state into the solution space is also used in other quantum algorithms and has become known as **amplitude amplification**.

In the following we first present this method in Sect. 6.9.1 for the case where the number of objects we are searching for (aka 'solutions') is known. In case the number of possible solutions is not known, a modification of the algorithm allows to perform a search with a success-probability of at least 25%. This is discussed in more detail in Sect. 6.9.2.

6.9.1 Search Algorithm for Known Number of Objects

We assume that the objects of the unordered list in which we search can be identified by numbers in $\{0, 1, \ldots, |L|\}$. In case the cardinality $|L|$ of the list is less than 2^n, we supplement the list with $2^n - |L|$ placeholder objects such that, without loss of generality, we may consider a search in the set $\{0, \ldots, 2^n - 1\}$ with cardinality $N := 2^n$. Since every number $x \in \{0, \ldots, 2^n - 1\}$ can be uniquely associated with a vector in the computational basis of $\mathbb{H}^{\otimes n}$, we can execute the search in $\mathbb{H}^{\otimes n}$ if we succeed to suitably identify the objects we are looking for in this HILBERT space.

We denote the set of the m objects we are looking for by S. By allowing $m \geq 1$ we also permit the case where there is more than one solution.

Definition 6.22 Let S denote the set of objects we are searching for, and let $m \geq 1$ be the cardinality of this set. The set S is called solution set, and we call its elements solutions. For the algorithm to search an $x \in S \subset \{0, \ldots, N-1\}$, where $N = 2^n$, we define the input and output register as $\mathbb{H}^{I/O} = \P\mathbb{H}^{\otimes n}$. Furthermore, we denote the set of objects that are not a solution by

$$S^{\perp} := \{0, \ldots N - 1\} \smallsetminus S$$

and define the subspaces

$$\mathbb{H}_S := \mathrm{Span}\left\{|x\rangle \,\middle|\, x \in S\right\} \subset \mathbb{H}^{I/O}$$
$$\mathbb{H}_{S^{\perp}} := \mathrm{Span}\left\{|x\rangle \,\middle|\, x \in S^{\perp}\right\} \subset \mathbb{H}^{I/O}$$

and the operators

$$P_S := \sum_{x \in S} |x\rangle\langle x|$$
$$P_{S^{\perp}} := \sum_{x \in S^{\perp}} |x\rangle\langle x| = \mathbf{1}^{\otimes n} - P_S \tag{6.137}$$

on $\mathbb{H}^{I/O}$ as well as the vectors

$$|\Psi_S\rangle := \frac{1}{\sqrt{m}} \sum_{x \in S} |x\rangle$$
$$|\Psi_{S^{\perp}}\rangle := \frac{1}{\sqrt{N-m}} \sum_{x \in S^{\perp}} |x\rangle \tag{6.138}$$

in $\mathbb{H}^{I/O}$.

The subspace $\mathbb{H}_S$ is spanned by vectors corresponding to objects in the solution set S. The operator P_S is a projection onto this subspace because $P_S|\Psi\rangle = |\Psi\rangle$ for any $|\Psi\rangle \in \mathbb{H}_S$, and it satisfies the requirements of an orthogonal projection given in Definition (2.11), that is,

$$P_S^* = \left(\sum_{x \in S} |x\rangle\langle x|\right)^* = \sum_{x \in S} \left(|x\rangle\langle x|\right)^* \underbrace{=}_{(2.36)} \sum_{x \in S} |x\rangle\langle x| = P_S$$

as well as

$$P_S^2 = \sum_{x,y \in S} |x\rangle\langle x|y\rangle\langle y| \underbrace{=}_{(3.24)} \sum_{x,y \in S} |x\rangle \delta_{xy} \langle y| = \sum_{x \in S} |x\rangle\langle x| = P_S.$$

Analogously, $P_{S^\perp}$ is a projection onto the subspace $\mathbb{H}_{S^\perp}$ spanned by vectors corresponding to objects which are outside of the solution set S.

The state $|\Psi_S\rangle$ is an equally weighted linear combination of those vectors $|x\rangle$ of the computational basis that are solutions $x \in S$. An observation (see Definition 5.35) of this state will reveal a solution with certainty.

Because of $S \cup S^\perp = \{0,\ldots,2^n-1\}$, every state $|\Psi\rangle \in \mathbb{H}^{I/O}$ can be expanded in the computational basis as follows

$$|\Psi\rangle = (P_{S^\perp} + P_S)|\Psi\rangle = \sum_{x\in S^\perp} \Psi_x|x\rangle + \sum_{x\in S} \Psi_x|x\rangle, \tag{6.139}$$

where $\Psi_x \in \mathbb{C}$. An observation of the input/output register $\mathbb{H}^{I/O}$ in a state $|\Psi\rangle$, that is, a measurement of the observable Σ_z^j defined in Definition 5.35, projects this state onto a $|x\rangle$ and yields an observed x. It is the goal of the algorithm to create states $|\Psi\rangle$ for which the probability to observe an $x \in S$ is maximized. This is accomplished by starting from an initial state $|\Psi_0\rangle$ and by applying suitable transformations (aka 'rotations') which increase the component in $\mathbb{H}_S$. The probability to observe an $x \in S$ when observing $|\Psi\rangle$ is given by

$$\mathbf{P}\left\{\begin{array}{l}\text{Observation of }|\Psi\rangle\text{ projects}\\\text{onto a state }|x\rangle\text{ with }x\in S\end{array}\right\} \underset{(2.62)}{=} ||P_S|\Psi\rangle||^2 \underset{(6.137)}{=} \left\|\sum_{x\in S}\Psi_x|x\rangle\right\|^2$$

$$\underset{(2.14)}{=} \sum_{x\in S}|\Psi_x|^2. \tag{6.140}$$

The search algorithm consists of a construction of a sequence of transformations that generate a $|\Psi\rangle$ from an initial $|\Psi_0\rangle$ such as to maximize the right side of (6.140).

Before we can discuss this in more detail, we need to introduce a method to identify solutions. For this we assume that we can decide in a limited number of computational steps, which is independent of N, whether an $x \in \{0,\ldots,N-1\}$ is a solution. We assume that this decision is accomplished by the application of a function g which yields $g(x) = 1$ if x is a solution and $g(x) = 0$ otherwise. Furthermore, we assume that this function can be implemented on a suitable HILBERT space with the help of an operator which also can be executed with a limited number of computational steps that is independent of N.

Definition 6.23 Let S be the set of solutions with cardinality $m \geq 1$. For the algorithm to search for an $x \in S \subset \{0,\ldots,N-1\}$, where $N = 2^n$, we define the function

$$g : \{0,\ldots,N-1\} \longrightarrow \{0,1\}$$
$$x \longmapsto g(x) := \begin{cases} 0 & \text{if } x \in S^\perp \\ 1 & \text{if } x \in S \end{cases} \tag{6.141}$$

as the **oracle-function** of the search problem. With the help of the auxiliary register $\mathbb{H}^W = {}^{\mathfrak{I}}\mathbb{H}$ we define on $\mathbb{H}^{I/O} \otimes \mathbb{H}^W$ the **oracle** $\widehat{U}_g$ via the following action on the computational basis

$$\widehat{U}_g\left(|x\rangle \otimes |y\rangle\right) := |x\rangle \otimes |y \boxplus g(x)\rangle. \tag{6.142}$$

By linear continuation, $\widehat{U}_g$ is then defined for all vectors in $\mathbb{H}^{I/O} \otimes \mathbb{H}^W$.

As mentioned above, we assume that the oracle $\widehat{U}_g$ can be efficiently executed, that is, with a finite number of computational steps independent of N. According to Definition 5.29 and Theorem 5.28 we can implement with $\widehat{U}_g$ an operator on the sub-system $\mathbb{H}^{I/O}$.

Lemma 6.24 *For the oracle $\widehat{U}_g$ and the state*

$$|\omega_i\rangle = |\omega_f\rangle = |-\rangle := \frac{|0\rangle - |1\rangle}{\sqrt{2}}$$

in the auxiliary register $\mathbb{H}^W$ one has for arbitrary $|\Psi\rangle \in \mathbb{H}^{I/O}$

$$\widehat{U}_g\left(|\Psi\rangle \otimes |-\rangle\right) = \left(R_{\mathrm{S}^\perp}|\Psi\rangle\right) \otimes |-\rangle, \tag{6.143}$$

where

$$
\begin{aligned}
R_{\mathrm{S}^\perp}|\Psi\rangle &= \sum_{x \in \mathrm{S}^\perp} \Psi_x |x\rangle - \sum_{x \in \mathrm{S}} \Psi_x |x\rangle \\
&= (\mathbf{1}^{\otimes n} - 2P_{\mathrm{S}})|\Psi\rangle.
\end{aligned} \tag{6.144}
$$

Proof To begin with, we have

$$
\begin{aligned}
\widehat{U}_g\left(|x\rangle \otimes |0\rangle\right) &\underset{(6.142),(5.81)}{=} |x\rangle \otimes |g(x)\rangle \underset{(6.141)}{=} \begin{cases} |x\rangle \otimes |0\rangle, & \text{if } x \in \mathrm{S}^\perp \\ |x\rangle \otimes |1\rangle, & \text{if } x \in \mathrm{S} \end{cases} \\
\widehat{U}_g\left(|x\rangle \otimes |1\rangle\right) &\underset{(6.142),(5.81)}{=} |x\rangle \otimes |1 \overset{2}{\oplus} g(x)\rangle \underset{(6.141)}{=} \begin{cases} |x\rangle \otimes |1\rangle, & \text{if } x \in \mathrm{S}^\perp \\ |x\rangle \otimes |0\rangle, & \text{if } x \in \mathrm{S}. \end{cases}
\end{aligned} \tag{6.145}
$$

The oracle $\widehat{U}_g$ then acts on the states

$$|x\rangle \otimes |-\rangle = |x\rangle \otimes \left(\frac{|0\rangle - |1\rangle}{\sqrt{2}}\right)$$

in $\mathbb{H}^{I/O} \otimes \mathbb{H}^W$ as follows

$$\widehat{U}_g\left(|x\rangle \otimes |-\rangle\right) \underset{(6.145)}{=} \begin{cases} |x\rangle \otimes |-\rangle, & \text{if } x \in S^\perp \\ -|x\rangle \otimes |-\rangle, & \text{if } x \in S \end{cases}$$

$$= (-1)^{g(x)}|x\rangle \otimes |-\rangle. \tag{6.146}$$

Together with (6.139) it thus follows for a $|\Psi\rangle \in \mathbb{H}^{I/O}$ that

$$\widehat{U}_g\left(|\Psi\rangle \otimes |-\rangle\right) \underset{(6.139),(6.146)}{=} \left(\sum_{x \in S^\perp} \Psi_x|x\rangle - \sum_{x \in S} \Psi_x|x\rangle\right) \otimes |-\rangle$$

$$\underset{(6.137)}{=} \left((P_{S^\perp} - P_S)|\Psi\rangle\right) \otimes |-\rangle.$$

Because of $P_{S^\perp} + P_S = \mathbf{1}^{\otimes n}$ this implies (6.144). $\qquad \square$

Note that $R_{S^\perp}$ reverses the component $P_S|\Psi\rangle$ of $|\Psi\rangle$ in the space $\text{Span}\left\{|x\rangle \mid x \in S\right\}$ in the opposite direction $-P_S|\Psi\rangle$. This can be viewed as a reflection about $\mathbb{H}_{S^\perp}$, where a reflection about a subspace is generally defined as follows.

Definition 6.25 Let $\mathbb{H}_{\text{sub}}$ be a subspace of the HILBERT space $\mathbb{H}$, and let P_{sub} be the projection onto this subspace. The **reflection about the subspace** $\mathbb{H}_{\text{sub}}$ is defined as the operator

$$R_{\text{sub}} := 2P_{\text{sub}} - \mathbf{1}. \tag{6.147}$$

If the subspace is one-dimensional and spanned by a $|\Psi\rangle \in \mathbb{H}$, we simply write R_Ψ and call this a reflection about $|\Psi\rangle$.

Note that the projection onto the orthogonal complement $\mathbb{H}_{\text{sub}^\perp}$ of $\mathbb{H}_{\text{sub}}$ is given by $P_{\text{sub}^\perp} = \mathbf{1} - P_{\text{sub}}$ such that the reflection $R_{\text{sub}^\perp}$ about $\mathbb{H}_{\text{sub}^\perp}$ becomes

$$R_{\text{sub}^\perp} \underset{(6.147)}{=} 2P_{\text{sub}^\perp} - \mathbf{1} = 2\left(\mathbf{1} - P_{\text{sub}}\right) - \mathbf{1} = \mathbf{1} - 2P_{\text{sub}} \underset{(6.147)}{=} -R_{\text{sub}}.$$

Figure 6.5 illustrates the geometry of these constructions.

For the start of the search algorithm we have to initialize the input/output register.

Definition 6.26 Let S be the solution set with cardinality $m \geq 1$. For the algorithm to search for an $x \in S \subset \{0, \dots, N-1\}$, where $N = 2^n$, we define the initial state in the input/output register as

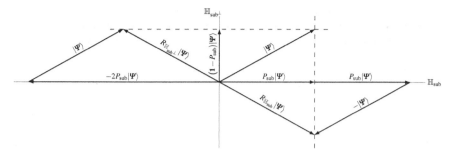

Fig. 6.5 Geometry of the reflections of a vector $|\Psi\rangle \in \mathbb{H}$ about a subspace $\mathbb{H}_{\text{sub}}$ and its orthogonal complement $\mathbb{H}_{\text{sub}^\perp}$

$$|\Psi_0\rangle := \frac{1}{\sqrt{N}} \sum_{x=0}^{N-1} |x\rangle \in \mathbb{H}^{I/O} . \qquad (6.148)$$

Moreover, we define the angle

$$\theta_0 := \arcsin\left(\sqrt{\frac{m}{N}}\right) \in \left[0, \frac{\pi}{2}\right] , \qquad (6.149)$$

and with the help of $|\Psi_0\rangle$ we define the operator

$$R_{\Psi_0} = 2|\Psi_0\rangle\langle\Psi_0| - \mathbf{1}^{\otimes n} \qquad (6.150)$$

on $\mathbb{H}^{I/O}$ as well as the initial state in the composite system

$$|\widehat{\Psi_0}\rangle := |\Psi_0\rangle \otimes |-\rangle \in \mathbb{H}^{I/O} \otimes \mathbb{H}^W . \qquad (6.151)$$

The initialization of the input/output register in the state $|\Psi_0\rangle$ can be performed as described in Sect. 5.4.1 with the help of the HADAMARD transform as shown in (5.83).

Exercise 6.79 Show that

$$|\Psi_0\rangle = \cos\theta_0 |\Psi_{S\perp}\rangle + \sin\theta_0 |\Psi_S\rangle \qquad (6.152)$$

holds and that R_{Ψ_0} is a *reflection about* $|\Psi_0\rangle$.

For a solution see Solution 6.79

The transformation that increases the component of $|\Psi_0\rangle$ in $\mathbb{H}_S$ is defined as follows.

Definition 6.27 The GROVER **iteration** is defined as the operator

$$\widehat{G} := (R_{\Psi_0} \otimes 1)\, \widehat{U}_g$$

on $\mathbb{H}^{I/O} \otimes \mathbb{H}^W$.

As we will now show, the GROVER iteration $\widehat{G}$ transforms separable states in $\mathbb{H}^{I/O} \otimes \mathbb{H}^W$ of the form $|\widehat{\Psi}_j\rangle = |\Psi_j\rangle \otimes |-\rangle$ to separable states $|\widehat{\Psi}_{j+1}\rangle = |\Psi_{j+1}\rangle \otimes |-\rangle$ of a similar form. We will see that in the input/output register $\mathbb{H}^{I/O}$ an application of $\widehat{G}$ can then be viewed as *a rotation of* $2\theta_0$ *in* $\mathbb{H}^{I/O}$ *in the direction of* $|\Psi_S\rangle$. Consequently, repeated application of $\widehat{G}$ increases the component along $|\Psi_S\rangle$ in the resulting state and thus the probability to find a solution upon observation of the input/output register.

Proposition 6.28 *For $j \in \mathbb{N}_0$ let*

$$|\widehat{\Psi}_j\rangle := \widehat{G}^j |\widehat{\Psi}_0\rangle\,.$$

Then we have for all $j \in \mathbb{N}_0$

$$|\widehat{\Psi}_j\rangle = |\Psi_j\rangle \otimes |-\rangle \tag{6.153}$$

with $|\Psi_j\rangle \in \mathbb{H}^{I/O}$ and

$$|\Psi_j\rangle = \cos\theta_j |\Psi_{S\perp}\rangle + \sin\theta_j |\Psi_S\rangle\,, \tag{6.154}$$

where

$$\theta_j = (2j+1)\theta_0\,. \tag{6.155}$$

Proof We show this by induction in j, which we start at $j = 0$. From the definition of $|\widehat{\Psi}_0\rangle$ in (6.151) and the result (6.152) we know that (6.153)–(6.155) are already satisfied for $j = 0$.

For the inductive step from j to $j+1$ assume that for a $j \in \mathbb{N}_0$

$$|\widehat{\Psi}_j\rangle = \big(\cos\theta_j |\Psi_{S\perp}\rangle + \sin\theta_j |\Psi_S\rangle\big) \otimes |-\rangle$$

holds with $\theta_j = (2j+1)\theta_0$. Then we have

$$\begin{aligned}
|\widehat{\Psi}_{j+1}\rangle &= \widehat{G}|\widehat{\Psi}_j\rangle \\
&= (R_{\Psi_0} \otimes 1)\widehat{U}_g\Big[\big(\cos\theta_j|\Psi_{S\perp}\rangle + \sin\theta_j|\Psi_S\rangle\big) \otimes |-\rangle\Big] \\
&\underbrace{=}_{(6.143)} (R_{\Psi_0} \otimes 1)\Big[\big(\cos\theta_j|\Psi_{S\perp}\rangle - \sin\theta_j|\Psi_S\rangle\big) \otimes |-\rangle\Big] \\
&\underbrace{=}_{(6.150)} \Big(\big(2|\Psi_0\rangle\langle\Psi_0| - \mathbf{1}^{\otimes n}\big)\big(\cos\theta_j|\Psi_{S\perp}\rangle - \sin\theta_j|\Psi_S\rangle\big)\Big) \otimes |-\rangle \\
&= \Big(\cos\theta_j\big(2|\Psi_0\rangle\langle\Psi_0|\Psi_{S\perp}\rangle - |\Psi_{\Psi_{S\perp}}\rangle\big) \\
&\qquad - \sin\theta_j\big(2|\Psi_0\rangle\langle\Psi_0|\Psi_S\rangle - |\Psi_S\rangle\big)\Big) \otimes |-\rangle \\
&= |\Psi_{j+1}\rangle \otimes |-\rangle
\end{aligned}$$

with

$$\begin{aligned}
|\Psi_{j+1}\rangle &= \cos\theta_j\big(2|\Psi_0\rangle\underbrace{\langle\Psi_0|\Psi_{S\perp}\rangle}_{=\cos\theta_0} - |\Psi_{S\perp}\rangle\big) - \sin\theta_j\big(2|\Psi_0\rangle\underbrace{\langle\Psi_0|\Psi_S\rangle}_{=\sin\theta_0} - |\Psi_S\rangle\big) \\
&\underbrace{=}_{(6.152)} \cos\theta_j\Big(2\big(\cos\theta_0|\Psi_{S\perp}\rangle + \sin\theta_0|\Psi_S\rangle\big)\cos\theta_0 - |\Psi_{S\perp}\rangle\Big) \\
&\qquad - \sin\theta_j\Big(2\big(\cos\theta_0|\Psi_{S\perp}\rangle + \sin\theta_0|\Psi_S\rangle\big)\sin\theta_0 - |\Psi_S\rangle\Big) \\
&= \Big(\cos\theta_j\big(\underbrace{2\cos^2\theta_0 - 1}_{=\cos 2\theta_0}\big) - \sin\theta_j\underbrace{2\cos\theta_0\sin\theta_0}_{=\sin 2\theta_0}\Big)|\Psi_{S\perp}\rangle \\
&\qquad + \Big(\cos\theta_j\underbrace{2\cos\theta_0\sin\theta_0}_{=\sin 2\theta_0} + \sin\theta_j\big(\underbrace{1 - 2\sin^2\theta_0}_{=\cos 2\theta_0}\big)\Big)|\Psi_S\rangle \\
&= \big(\cos\theta_j\cos 2\theta_0 - \sin\theta_j\sin 2\theta_0\big)|\Psi_{S\perp}\rangle \\
&\qquad + \big(\cos\theta_j\sin 2\theta_0 + \sin\theta_j\cos 2\theta_0\big)|\Psi_S\rangle \\
&= \cos(\theta_j + 2\theta_0)|\Psi_{S\perp}\rangle + \sin(\theta_j + 2\theta_0)|\Psi_S\rangle \\
&= \cos\theta_{j+1}|\Psi_{S\perp}\rangle + \sin\theta_{j+1}|\Psi_S\rangle
\end{aligned}$$

and thus finally

$$\theta_{j+1} = \theta_j + 2\theta_0 = (2j+1)\theta_0 + 2\theta_0 = \Big(2(j+1) + 1\Big)\theta_0.$$

Hence, (6.153)–(6.155) also hold for $j+1$. □

The geometry of the sequence of the $|\Psi_j\rangle$ is graphically illustrated for an example in Fig. 6.6.

As shown in Proposition 6.28, the j GROVER iterations transform the composite system $\mathbb{H}^{I/O} \otimes \mathbb{H}^W$ from the initial state $|\widehat{\Psi}_0\rangle = |\Psi_0\rangle \otimes |-\rangle$ to the separable state $|\widehat{\Psi}_j\rangle = |\Psi_j\rangle \otimes |-\rangle$. From Theorem 3.17 we know that then the sub-system in the

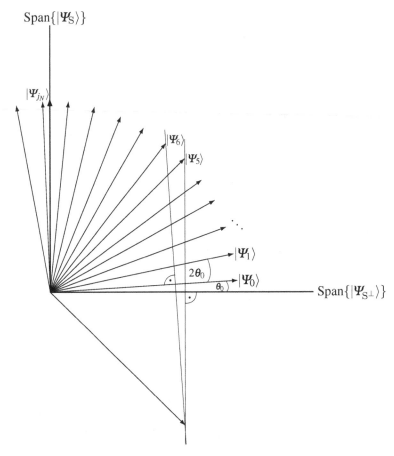

Fig. 6.6 Geometry of the GROVER iteration in the input/output register with $m = 5, N = 2^{10}$ and $j_N = 11$. In the two-dimensional subspace $\mathrm{Span}\{|\Psi_{S\perp}\rangle, |\Psi_S\rangle\}$ the initial state $|\Psi_0\rangle$ is rotated towards $|\Psi_S\rangle$. The illustrated transition from $|\Psi_5\rangle$ to $|\Psi_6\rangle$ shows that $\widehat{G}$ in the sub-system I/O first performs a reflection about $|\Psi_{S\perp}\rangle$ and then a reflection about $|\Psi_0\rangle$. The vector immediately to the right of $|\Psi_{j_N}\rangle$ is $|\Psi_S\rangle$ and is a state in the subspace $\mathbb{H}_S$ of the solution set. We can see that $|\Psi_{j_N}\rangle$ comes close to that

input/output register is described by the reduced density operator

$$
\rho^{I/O} \underbrace{=}_{(3.50)} \mathrm{tr}^W\left(|\widehat{\Psi}_j\rangle\langle\widehat{\Psi}_j|\right) = \mathrm{tr}^W\left(|\Psi_j\rangle\otimes|-\rangle\langle\Psi_j|\otimes\langle-|\right)
$$

$$
\underbrace{=}_{(3.36)} \mathrm{tr}^W\left(|\Psi_j\rangle\langle\Psi_j|\otimes|-\rangle\langle-|\right) \underbrace{=}_{(3.57)} \mathrm{tr}\left(|-\rangle\langle-|\right)|\Psi_j\rangle\langle\Psi_j|
$$

$$
= |\Psi_j\rangle\langle\Psi_j|, \tag{6.156}
$$

where in the last equation we used that $\{|\pm\rangle\}$ is an ONB in $\mathbb{H}^W = \mathbb{H}$ and thus

$$\text{tr}\left(|-\rangle\langle-|\right) \underbrace{=}_{(2.57)} \langle-|-\rangle\langle-|-\rangle + \langle+|-\rangle\langle-|+\rangle = 1 .$$

From (6.156) we see that the sub-system in the input/output register $\mathbb{H}^{I/O}$ after j GROVER iterations is described by the pure states $|\Psi_{j+1}\rangle$. We can thus ignore the auxiliary register $\mathbb{H}^{W}$ and restrict our considerations to the input/output register.

When observing the input/output register in state $|\Psi_j\rangle$ the probability to find a state $|x\rangle$ for which x is a solution is determined by the projection of $|\Psi_j\rangle$ onto the sub-space $\mathbb{H}_S$ spanned by vectors corresponding to the solution set. Using (6.154) we obtain

$$\mathbf{P}\left\{\begin{array}{l} \text{Observation of } |\Psi_j\rangle \text{ projects} \\ \text{onto a state } |x\rangle \text{ with } x \in S \end{array}\right\} = ||P_S|\Psi_j\rangle||^2 \underbrace{=}_{\substack{(6.137),(6.138),\\(6.154)}} \sin^2\theta_j . \quad (6.157)$$

To optimize the success-probability of the search, we need to determine the number j of applications of $\widehat{G}$ on $|\widehat{\Psi_0}\rangle$ that maximizes $\sin^2\theta_j$, which, as shown in (6.157), is the probability to find a solution. The following lemma gives a lower bound for this probability if we determine j such that θ_j is as close as possible to $\frac{\pi}{2}$.

Lemma 6.29 *Let S be the solution set with cardinality $m \geq 1$, and let $N = 2^n$ be the number of objects in which we search for solutions. If we apply the* GROVER *iteration $\widehat{G}$*

$$j_N := \left\lfloor \frac{\pi}{4\arcsin\left(\sqrt{\frac{m}{N}}\right)} \right\rfloor \quad (6.158)$$

times to $|\widehat{\Psi_0}\rangle$ and observe the state $|\Psi_{j_N}\rangle$ in the input/output register, then the probability to observe in the sub-system $\mathbb{H}^{I/O}$ a state $|x\rangle$ with $x \in S$ satisfies

$$\mathbf{P}\left\{\begin{array}{l} \text{Observation of } |\Psi_j\rangle \text{ projects} \\ \text{onto a state } |x\rangle \text{ with } x \in S \end{array}\right\} \geq 1 - \frac{m}{N} . \quad (6.159)$$

Proof Because of $\theta_0 = \arcsin\left(\sqrt{\frac{m}{N}}\right)$, the choice (6.158) is equivalent to

$$j_N = \left\lfloor \frac{\pi}{4\theta_0} \right\rfloor .$$

With this choice we have

$$j_N \leq \frac{\pi}{4\theta_0} < j_N + 1$$

and thus

$$-\theta_0 \le \frac{\pi}{2} - (2j_N + 1)\theta_0 < \theta_0,$$

which implies

$$\frac{\pi}{2} - \theta_0 < \theta_{j_N} \le \frac{\pi}{2} + \theta_0. \tag{6.160}$$

For the probability that the observation of $|\Psi_j\rangle$ finds an $x \in S$ it follows that

$$
\begin{aligned}
\left\| P_S |\Psi_{j_N}\rangle \right\|^2 &= \sin^2 \theta_{j_N} \\
&\underset{(6.160)}{\ge} \sin^2 \left(\frac{\pi}{2} + \theta_0 \right) = 1 - \cos^2 \left(\frac{\pi}{2} + \theta_0 \right) \\
&= 1 - \sin^2 \theta_0 \underset{(6.149)}{=} 1 - \frac{m}{N}.
\end{aligned}
$$

$\square$

From (6.158) we see that the number j_N of GROVER iterations decreases with an increase in the number m of solutions. However, at first sight it might seem strange that also the lower bound in (6.159) of the probability to find a solution decreases with an increase in the number m of possible solutions. The reason for both is that θ_0 is an increasing function of m and thus an increase in m means that, while we need fewer GROVER iterations to reach θ_{j_N}, the possible distance of θ_{j_N} to $\frac{\pi}{2}$ increases and the lower bound of $\sin^2 \theta_{j_N}$ decreases. But for m much smaller than N this latter effect is almost negligible.

To see how the number of GROVER iterations grows with a large and increasing cardinality N of the search set, we use that the TAYLOR expansion of $\arcsin(y)$ around $y = 0$ is given by

$$\arcsin(y) = y + \sum_{k=1}^{\infty} \frac{\prod_{l=1}^{k}(2l-1)}{\prod_{j=1}^{k}(2j)} \frac{y^{2k+1}}{2k+1},$$

which implies that

$$\lim_{N \to \infty} \sqrt{N} \arcsin\left(\sqrt{\frac{m}{N}} \right) = \sqrt{m}$$

and thus

$$\lim_{N \to \infty} \frac{j_N}{\sqrt{N}} = \frac{\pi}{4\sqrt{m}}.$$

Consequently, we have

$$j_N = O\left(\sqrt{\frac{N}{m}} \right) \quad \text{for } N \to \infty,$$

that is, the optimal number j_N of iterations scales for $N \to \infty$ with $O\left(\sqrt{\frac{N}{m}}\right)$.

In the case where the number m of possible solutions in S is known, the GROVER search algorithm to find an $x \in$ S in a total set of $N = 2^n$ objects consists of the following steps.

GROVER Search Algorithm for Known Number of Solutions

Input: A set $\{0, \ldots, N-1\}$ of $N = 2^n$ objects that contains a subset S of $m \geq 1$ objects to be searched for and an oracle-function $g : \{0, \ldots, N-1\} \to \{0,1\}$ that takes the value 1 in S and the value 0 elsewhere

Step 1: In $\mathbb{H}^{I/O} \otimes \mathbb{H}^W = \mathbb{H}^{\otimes n} \otimes \mathbb{H}$ prepare the composite system in the state $|\widehat{\Psi}_0\rangle = |\Psi_0\rangle \otimes |-\rangle$ with

$$|\Psi_0\rangle = \frac{1}{\sqrt{N}} \sum_{x=0}^{N-1} |x\rangle.$$

The number of computational steps required for Step 1 scales for $N \to \infty$ with

$$S_{\text{GROVER1}}(N) \in O(1)$$

Step 2: With $\theta_0 = \arcsin\left(\sqrt{\frac{m}{N}}\right)$ apply the transform $\widehat{G} = (R_{\Psi_0} \otimes 1)\widehat{U}_g$

$$j_N = \left\lfloor \frac{\pi}{4\theta_0} \right\rfloor$$

times to $|\widehat{\Psi}_0\rangle$ in order to transform the composite system to the state

$$|\widehat{\Psi}_{j_N}\rangle = \widehat{G}^{j_N} |\widehat{\Psi}_0\rangle.$$

The number of computational steps required for Step 2 scales for $N \to \infty$ with

$$S_{\text{GROVER2}}(N) \in O\left(\sqrt{\frac{N}{m}}\right)$$

Step 3: Observe the sub-system $\mathbb{H}^{I/O}$ and infer from the observed state $|x\rangle$ the value $x \in \{0, \ldots, N-1\}$. The number of computational steps required for Step 3 scales for $N \to \infty$ with

$$S_{\text{GROVER3}}(N) \in O(1)$$

Step 4: By evaluating $g(x)$, check if $x \in$ S. The number of computational steps required for Step 4 scales for $N \to \infty$ with

$$S_{\text{GROVER4}}(N) \in O(1)$$

Output: A solution $x \in S$ with probability no less than $1 - \frac{m}{N}$

Altogether, the number of computational steps required for one run of the algorithm, which finds a solution $x \in S$ with a probability of at least $1 - \frac{m}{N}$, satisfies

$$S_{\text{GROVER}}(N) = \sum_{i=1}^{4} S_{\text{GROVER}i}(N) \in O\left(\sqrt{\frac{N}{m}}\right) \quad \text{for } N \to \infty.$$

We collect these results in the following theorem.

> **Theorem 6.30** *Let $S \subset \{0, \ldots, N-1\}$ be the solution set with cardinality $m \geq 1$, and let $N = 2^n$ be the number of objects in which we search for solutions. Let the elements of S be identified with an oracle-function g as given in (6.141), and let the corresponding oracle $\widehat{U}_g$ be as given in (6.142) such that g and $\widehat{U}_g$ require a limited number of computational steps independent of N. Then the success-probability of a search in $\{0, \ldots, N-1\}$ with the GROVER search algorithm to find a solution in S satisfies*
>
> $$\mathbf{P}\{\text{The algorithm finds an } x \in S\} \geq 1 - \frac{m}{N}.$$
>
> *The number of computational steps required for the algorithm scales as a function of N as*
>
> $$S_{\text{GROVER}}(N) \in O\left(\sqrt{\frac{N}{m}}\right) \quad \text{for } N \to \infty. \tag{6.161}$$

In particular, in the case $\frac{m}{N} < \frac{1}{2}$ we have for the success-probability

$$\mathbf{P}\{\text{The algorithm finds an } x \in S\} > \frac{1}{2}.$$

On the other hand, the case $\frac{m}{N} \geq \frac{1}{2}$ does not require any algorithm since a plain search of $m \geq \frac{N}{2}$ solutions in N objects will yield a solution with probability greater than $\frac{1}{2}$.

Note that he determination of the optimal number j_N of GROVER iterations $\widehat{G}$ requires knowledge of N as well as m, in other words, we not only need to know in how many objects N we have to search, but also how many solutions m are among them.

6.9.2 Search Algorithm for Unknown Number of Objects

There is, however, an extended version [94] of the algorithm that does not require prior knowledge of the number m of solutions present among the N objects. We need a few intermediate results before we can introduce this version.

Exercise 6.80 Let $J \in \mathbb{N}$ and $\alpha \in \mathbb{R}$. Show that

$$\sum_{j=0}^{J-1} \cos((2j+1)\alpha) = \frac{\sin(2J\alpha)}{2\sin\alpha}. \tag{6.162}$$

For a solution see Solution 6.80

Lemma 6.31 *Let $N \in \mathbb{N}$ be the number of objects, among which an unknown —but non-zero—number $m \in \mathbb{N}$ of solutions exists, and let $\theta_0 \in [0, \frac{\pi}{2}]$ be such that $\sin^2\theta_0 = \frac{m}{N}$. For a $J \in \mathbb{N}$ let j be selected randomly from $\{0, \ldots, J-1\}$ with equal probability $\frac{1}{J}$ and let $|\widehat{\Psi}_j\rangle$ be the state obtained after j GROVER iterations have been applied to $|\widehat{\Psi}_0\rangle$ as described in Proposition 6.28. Then the probability for the event*

$$\mathfrak{e}_3 := \left\{ \begin{array}{l} \text{Observation of the input/output reg-} \\ \text{ister yields an } x \in S \end{array} \right\}$$

is given as

$$\mathbf{P}\{\mathfrak{e}_3\} = \frac{1}{2} - \frac{\sin(4J\theta_0)}{4J\sin(2\theta_0)}. \tag{6.163}$$

In particular, in the case $J \geq \frac{1}{\sin(2\theta_0)}$ we have

$$\mathbf{P}\{\mathfrak{e}_3\} \geq \frac{1}{4}. \tag{6.164}$$

Proof The probability of the event $\mathfrak{e}_3$ can be decomposed as

$$\mathbf{P}\{\mathfrak{e}_3\} = \sum_{j=0}^{J-1} \mathbf{P}\left\{ \begin{array}{l} j \text{ has been} \\ \text{selected} \end{array} \right\} \mathbf{P}\left\{ \begin{array}{l} \text{Observation of } |\Psi_j\rangle \text{ in } \mathbb{H}^{I/O} \\ \text{yields a solution } x \in S \end{array} \right\}. \tag{6.165}$$

Because of the assumed distribution of the j we have

$$\mathbf{P}\left\{ \begin{array}{l} j \text{ has been} \\ \text{selected} \end{array} \right\} = \frac{1}{J}, \tag{6.166}$$

and using the results of Proposition 6.28, we find

$$\mathbf{P}\left\{ \begin{array}{l} \text{Observation of } |\Psi_j\rangle \text{ in } \mathbb{H}^{I/O} \\ \text{yields a solution } x \in S \end{array} \right\} \underset{(6.157)}{=} \sin^2\theta_j \underset{(6.155)}{=} \sin^2((2j+1)\theta_0)$$

$$= \frac{1-\cos(2(2j+1)\theta_0)}{2}. \tag{6.167}$$

Inserting (6.166) and (6.167) into (6.165) yields

$$\mathbf{P}\{\mathfrak{e}_3\} = \sum_{j=0}^{J-1} \frac{1}{J}\left(\frac{1-\cos(2(2j+1)\theta_0)}{2}\right)$$

$$= \frac{1}{2} - \frac{1}{2J}\sum_{j=0}^{J-1}\cos(2(2j+1)\theta_0)$$

$$\underset{(6.162)}{=} \frac{1}{2} - \frac{\sin(4J\theta_0)}{4J\sin(2\theta_0)}, \tag{6.168}$$

which proves (6.163).

If $J \geq \frac{1}{\sin(2\theta_0)}$ holds, it follows that

$$\sin(4J\theta_0) \leq 1 \leq J\sin(2\theta_0)$$

and thus

$$\frac{\sin(4J\theta_0)}{4J\sin(2\theta_0)} \leq \frac{1}{4},$$

which, together with (6.168), implies the claim (6.164).   □

In case we know that at least one solution $x \in S \neq \emptyset$ exists but do not know the cardinality of the solutions set S, the following variant of the GROVER search algorithm provides a method to find a solution with a probability of at least $\frac{1}{4}$. As before, we assume that $\{0,\ldots,2^n-1\}$ is the set in which we search and that $N = 2^n$ is the cardinality of this set.

Modified GROVER Algorithm for Unknown Cardinality of $S \neq \emptyset$

Input: A set $\{0,\ldots,N-1\}$ of $N = 2^n$ objects that contains a subset S of $m \geq 0$ objects to be searched for and an oracle-function $g : \{0,\ldots,N-1\} \rightarrow \{0,1\}$ that takes the value 1 in S and the value 0 elsewhere

Step 1: Randomly select an $x \in \{0,\ldots,N-1\}$ and check if $x \in S$. If it is not a solution, go to Step 2. Otherwise, we have already found a solution. The number of computational steps required for this scales for $N \rightarrow \infty$ as

$$S_{\widetilde{\text{GROVER1}}}(N) \in O(1)$$

Step 2:   In $\mathbb{H}^{I/O} \otimes \mathbb{H}^W = \mathbb{TH}^{\otimes n} \otimes \mathbb{TH}$ prepare the composite system in the state $|\widehat{\Psi_0}\rangle = |\Psi_0\rangle \otimes |-\rangle$ with

$$|\Psi_0\rangle = \frac{1}{\sqrt{N}} \sum_{x=0}^{N-1} |x\rangle .$$

The number of computational steps required for Step 2 scales for $N \to \infty$ as

$$S_{\widetilde{\text{GROVER2}}}(N) \in O(1)$$

Step 3:   Set $J := \lfloor \sqrt{N} \rfloor + 1$ and randomly select an integer $j \in \{0,\ldots,J-1\}$ with equal probability $\frac{1}{J}$. Apply j times the GROVER iteration $\widehat{G} = (R_{\Psi_0} \otimes 1)\widehat{U}_g$ to $|\widehat{\Psi_0}\rangle$ to transform the composite system to the state

$$|\widehat{\Psi_j}\rangle = \widehat{G}^j |\widehat{\Psi_0}\rangle .$$

Because of $j \leq \lfloor \sqrt{N} \rfloor$, the number of computational steps required for Step 3 scales for $N \to \infty$ as

$$S_{\widetilde{\text{GROVER3}}}(N) \in O\left(\sqrt{N}\right)$$

Step 4:   Observe the input/output register $\mathbb{TH}^{I/O}$ and read off the observed $x \in \{0,\ldots,N-1\}$. The number of computational steps required for Step 4 scales for $N \to \infty$ as

$$S_{\widetilde{\text{GROVER4}}}(N) \in O(1)$$

Step 5:   Check if $x \in S$ by evaluating $g(x)$. The number of computational steps required for Step 5 scales for $N \to \infty$ as

$$S_{\widetilde{\text{GROVER5}}}(N) \in O(1) .$$

Output: In case $S \neq \emptyset$, a solution $x \in S$ with probability no less than $\frac{1}{4}$

In case at least one solution exists in $\{0,\ldots,N-1\}$, we can formulate the efficiency and success-probability of the modified algorithm as follows.

Theorem 6.32 *Let $N = 2^n$, and let the solution set $S \subset \{0,\ldots,N-1\}$ be nonempty. Let the elements of S be identified with the help of an oracle-function g as given in (6.141), and let the corresponding oracle be $\widehat{U}_g$ as in (6.142) such* *that g and $\widehat{U}_g$ require a finite number of computational steps independent of N. Then the search in $\{0,\ldots,N-1\}$ with the modified* GROVER *search algorithm finds a solution with a probability*

$$\mathbf{P}\{\text{The algorithm finds an } x \in S\} \geq \frac{1}{4}.$$

The number of computational steps required for this algorithm scales as a function of N as

$$S_{\widetilde{\text{GROVER}}}(N) \in O\left(\sqrt{N}\right) \quad \text{for } N \to \infty.$$

Proof We show the claim about the success-probability first. Let $m \in \mathbb{N}$ be the unknown number of solutions in $\{0,1,\ldots,N-1\}$. We distinguish two cases:

if $m > \frac{3N}{4}$,   the purely random choice of an $x \in \{0,\ldots,N-1\}$ in Step 1 of the modified algorithm has already a success-probability of at least $\frac{3}{4} > \frac{1}{4}$;

if $1 \leq m \leq \frac{3N}{4}$,   we have for θ_0 with $\sin^2 \theta_0 = \frac{m}{N} > 0$ at first

$$\frac{1}{\sin(2\theta_0)} = \frac{1}{2\sin\theta_0\cos\theta_0} = \frac{1}{2\sqrt{\sin^2\theta_0(1-\sin^2\theta_0)}}$$

$$= \frac{N}{2\sqrt{m(N-m)}}. \tag{6.169}$$

The case assumption $1 \leq m \leq \frac{3N}{4}$ implies $\frac{1}{4m} \leq \frac{1}{4}$ as well as $\frac{4}{3} \leq \frac{N}{m}$. From the former it follows that $1 - \frac{1}{4m} \geq \frac{3}{4}$ and thus

$$1 \leq \left(1 - \tfrac{1}{4m}\right)\tfrac{N}{m}$$
$$\Rightarrow \quad 4m^2 \leq (4m-1)N$$
$$\Rightarrow \quad N \leq 4mN - 4m^2 = 4m(N-m)$$
$$\Rightarrow \quad \frac{1}{2\sqrt{m(N-m)}} \leq \frac{1}{\sqrt{N}}$$
$$\Rightarrow \quad \frac{N}{2\sqrt{m(N-m)}} \leq \sqrt{N} \leq \lfloor\sqrt{N}\rfloor + 1 = J$$
$$\underset{(6.169)}{\Rightarrow} \quad \frac{1}{\sin(2\theta_0)} \leq J$$

holds. Hence, the number $J = \lfloor\sqrt{N}\rfloor + 1$ used in Step 3 of the modified algorithm satisfies the assumptions for (6.164) in Lemma 6.31, and the success-probability for a run through the Steps 1 to 5 of the algorithm is at least $\frac{1}{4}$ in the case $1 \leq m \leq \frac{3N}{4}$ as well.

The total number of computational steps required for a run of the modified GROVER algorithm satisfies

$$S_{\widetilde{\text{GROVER}}}(N) = \sum_{i=1}^{5} S_{\widetilde{\text{GROVER}i}}(N) = O\left(\sqrt{N}\right) \quad \text{for } N \to \infty.$$

$\square$

In the case $m \geq 1$, Theorem 6.32 tells us that the probability of failure in one run through the search algorithm is less than $\frac{3}{4}$. After s searches it is less than $\left(\frac{3}{4}\right)^s$ and thus after 20 searches already smaller than 0.32%.

If it is not known if any solution exists at all, that is, if $m = 0$ is possible, and after s searches we have not found a solution, we can only say that with a probability of $1 - \left(\frac{3}{4}\right)^s$ no solution exists.

6.10   Further Reading

A rather comprehensive and up-to-date list of quantum algorithms of all types, including adiabatic ones, is maintained online by JORDAN in the Quantum Algorithm Zoo [95]. Each of the more than 60 entries contains a brief description of the algorithm along with extensive references.

A more detailed albeit more selective overview on algorithms for algebraic problems, including the non-abelian hidden subgroup problem, is given by CHILDS and VAN DAM [87]. A more condensed version of this can be found in the paper by VAN DAM and SASAKI [96].

A slightly more comprehensive overview of quantum algorithms, which contains nice summaries of each algorithm, is given by MOSCA [88].

An accessible and modern introduction to cryptography, covering most of the current methods, is given by HOFFSTEIN, PIPHER and SILVERMAN [79].

For a treatise on elliptic curves the reader may consult the book by WASHINGTON [97]. Details about settings of the elliptic curve digital signature can be found online at [93].

Chapter 7
Error Correction

7.1 What Can Go Wrong?

Already the classical computational process is not free of errors and extensive methods and protocols have been developed to detect and correct them. Ideally, these methods should enable us to recover the non-erroneous state of the process. We introduce the basics of classical error correcting codes in Sect. 7.2.

As any physical implementation of quantum computation will in most cases be based on devices which are very sensitive to disturbances, the possibility of states being corrupted by errors is even bigger than in classical computation. Hence, the need for error correcting methods without which quantum computation will be infeasible. The sources or causes of errors in quantum computation can be divided roughly into the following categories [98].

Decoherence   By far the most common and important cause of errors is that the system interacts with its environment in a way that is not in accordance with the planned computational process. This can happen if the system is not perfectly isolated and will introduce an error in that it changes the state in an uncontrolled way. To ensure the integrity of the computation, it is necessary that any such errors be corrected and the uncorrupted state be restored as faithfully as possible.

Coherent Errors   We expect gates to perform certain unitary transformations of the states. Unlike the transformation of a classical state, which is of binary nature, unitary operators form a continuous set. Hence, it is possible that instead of performing the desired U, the implementation of the gate actually performs U' which, even though being close to U, is nonetheless different.

Corrupt Input   The initial state of a computation may not have been prepared as required. Rather than starting from the desired state $|\psi_{\text{ini}}\rangle$, our preparation mechanism may not work with perfect precision, and the state $|\psi'_{\text{ini}}\rangle$ is prepared instead.

© Springer Nature Switzerland AG 2019
W. Scherer, *Mathematics of Quantum Computing*,
https://doi.org/10.1007/978-3-030-12358-1_7

Leakage   In most cases our qubit H<small>ILBERT</small> space $\mathbb{H}$ will be a subspace of a larger
H<small>ILBERT</small> space $\mathbb{H}$, and we assume that once a system is in a state $|\psi\rangle \in \mathbb{H}$, all
our interactions are such that it remains in that subspace $\mathbb{H}$, in other words,
remains a qubit state. For example, our qubit system may be comprised of the
two lowest energy eigenstates of an atom, and the next higher energy eigenstate
is at a substantially higher energy level than those two. If none of our interactions
with the system imparts sufficient energy to make the system transition to the
higher energy eigenstates, we are effectively dealing with a two state system
as desired. However, uncontrolled (by us) interactions of the system with the
environment or insufficient control of our interactions may cause the system to
transition to a higher energy level state and thus leave the qubit H<small>ILBERT</small> space.
This phenomenon is called leakage.

To model these errors, we will utilize quantum operations (see Sect. 3.5), that is, we
assume that any error can be described by a suitable (possibly non-trace-preserving)
quantum operation.

However, in order to detect if an error has occurred, we face the obstacle that any
measurement of a quantum system can change it in an irreversible way as described
in the Projection Postulate 3. This prevents us from applying a simple measurement
to find out if an error has corrupted the state.

A second obstacle particular to quantum systems is that we cannot copy an
unknown quantum state as the Quantum No-Cloning Theorem 4.11 tells us. In other
words, we cannot produce multiple copies of our state which we could use to detect
and correct errors.

Fortunately, both these obstacles can be overcome by building in suitable redundancy in our description. This is similar to the classical approach to error correction
where a single bit is encoded into several bits. We briefly review the basics of classical error correction in Sect. 7.2.

In quantum error correction redundancy will result in representing a single qubit
by several qubits. This will enable us to perform measurements on this enlarged
representation to detect potential errors but without obtaining information about the
state of the original single qubit. It is to be emphasized that building in redundancy
does not mean duplication and is no violation of the Theorem 4.11.

Quantum error correction is a large field within quantum computing and it is
impossible to present all aspects in an introductory text. Nevertheless, the selection
of topics presented in this chapter should provide the reader with a good understanding of the basic issues and methods of the field.

For the group theoretic notions used in this chapter, and in particular for the
stabilizer formalism, the reader is advised to consult Appendix F in general and
Sect. F.5 on the P<small>AULI</small> group in particular.

7.2 Classical Error Correction

Here we briefly review how errors are detected and corrected in classical computing before we present error detection and correction in the quantum setting in the following section.

The principle means to devise systems allowing for the detection and correction of errors is to build in redundancy. In our treatment of classical error correction we always work in the binary alphabet consisting of the finite field $\mathbb{F}_2$ (see Corollary F.55). This means that the letters of the alphabet $\mathbb{F}_2$ are the by now familiar bit values 0 and 1 and that we can perform addition and multiplication on these letters in accordance with the rules for the finite field $\mathbb{F}_2$, namely, we have for any letters $a, b \in \mathbb{F}_2$ in our alphabet

$$
\begin{aligned}
a +_{\mathbb{F}_2} b &= a \overset{2}{\oplus} b \underbrace{=}_{(5.2)} (a+b) \mod 2 \\
a \cdot_{\mathbb{F}_2} b &= ab = (ab) \mod 2 \\
a \overset{2}{\oplus} b &= 0 \quad \Leftrightarrow \quad a = b.
\end{aligned}
\tag{7.1}
$$

As in normal language, letters in $\mathbb{F}_2$ can be strung together to form words.

Definition 7.1 A word of length $k \in \mathbb{N}$ in $\mathbb{F}_2$ is defined as the vector

$$
\mathbf{w} = \begin{pmatrix} w_1 \\ \vdots \\ w_k \end{pmatrix} \in \mathbb{F}_2^k,
$$

where $\mathbb{F}_2^k = \mathbb{F}_2 \times \cdots \mathbb{F}_2$ denotes the vector space over $\mathbb{F}_2$ comprised of the k-fold cartesian product of the field $\mathbb{F}_2$.

The binary sum of two vectors $\mathbf{u}, \mathbf{v} \in \mathbb{F}_2^k$ is defined as

$$
\mathbf{u} \overset{2}{\oplus} \mathbf{v} = \begin{pmatrix} u_1 \overset{2}{\oplus} v_1 \\ \vdots \\ u_k \overset{2}{\oplus} v_k \end{pmatrix} \in \mathbb{F}_2^k,
\tag{7.2}
$$

where $a \overset{2}{\oplus} b$ is as defined in (5.2).

Hence, words of length k are bit-strings of length k which we can add or multiply with 0 or 1. Note that the vector space $\mathbb{F}_2^k$ is comprised of a a finite set of vectors.

Linear independence in this vector space and its dimension are defined in the usual way.

Definition 7.2 Let $k \in \mathbb{N}$ and I be a finite subset of $\mathbb{N}$. A set of vectors $\{\mathbf{w}_i \mid i \in I\} \subset \mathbb{F}_2^k$ is said to be linearly independent if for any set $\{a_i \mid i \in I\} \subset \mathbb{F}_2$

$$\sum_{i \in I} a_i \mathbf{w}_i = 0 \quad \Rightarrow \quad a_i = 0 \quad \forall i \in I.$$

For any set of vectors $\{\mathbf{w}_i \mid i \in I\} \subset \mathbb{F}_2^k$ the dimension of the linear space

$$\text{Span}\{\mathbf{w}_i \mid i \in I\} = \Big\{ \sum_{i \in I} a_i \mathbf{w}_i \,\Big|\, a_i \in \mathbb{F}_2 \Big\}$$

is defined as the maximal number of linearly independent vectors in this space. Any such set of vectors is called a basis in this space.

Using the usual basis vectors

$$\mathbf{e}_1 = \begin{pmatrix} 1 \\ 0 \\ \vdots \\ 0 \end{pmatrix}, \ldots, \mathbf{e}_k = \begin{pmatrix} 0 \\ \vdots \\ 0 \\ 1 \end{pmatrix} \in \mathbb{F}_2^k,$$

we see immediately that they are linearly independent and form a basis in $\mathbb{F}_2^k$ such that $\dim \mathbb{F}_2^k = k$. A word of caution, though: not every set of vectors linearly independent over $\mathbb{R}$ is linearly independent over $\mathbb{F}_2$ as the following example shows.

Example 7.3 Consider the vectors

$$\mathbf{a} = \begin{pmatrix} 1 \\ 0 \\ 1 \end{pmatrix}, \mathbf{b} = \begin{pmatrix} 0 \\ 1 \\ 1 \end{pmatrix}, \mathbf{c} = \begin{pmatrix} 1 \\ 1 \\ 0 \end{pmatrix}$$

in $\mathbb{R}^3$. Then any $a, b, c \in \mathbb{R}$ such that

$$a\mathbf{a} + b\mathbf{b} + c\mathbf{c} = \begin{pmatrix} a + c \\ b + c \\ a + b \end{pmatrix} = \begin{pmatrix} 0 \\ 0 \\ 0 \end{pmatrix}$$

satisfy $a = b = c = 0$. Hence, $\mathbf{a}, \mathbf{b}, \mathbf{c}$ when viewed as vectors of $\mathbb{R}^3$, are linearly independent. However, when viewed as elements of $\mathbb{F}_2^3$ we see that for $a = b = c = 1$

we have

$$\mathbf{a} \overset{2}{\oplus} \mathbf{b} \overset{2}{\oplus} \mathbf{c} = \begin{pmatrix} 1 \overset{2}{\oplus} 1 \\ 1 \overset{2}{\oplus} 1 \\ 1 \overset{2}{\oplus} 1 \end{pmatrix} = \begin{pmatrix} 0 \\ 0 \\ 0 \end{pmatrix},$$

and thus $\mathbf{a}, \mathbf{b}, \mathbf{c}$ are not linearly independent as vectors in $\mathbb{F}_2^3$.

The vector space $\mathbb{F}_2^k$ can be made into a metric space by endowing it with a suitable distance function.

Definition 7.4 A **metric space** is a set M for which there exists a mapping $d : M \times M \to \mathbb{R}$ satisfying

(i)
$$d(u, v) \geq 0$$

(ii)
$$d(u, v) = 0 \quad \Leftrightarrow \quad u = v$$

(iii)
$$d(u, v) = d(v, u)$$

(iv)
$$d(u, v) \leq d(u, v) + d(v, w). \tag{7.3}$$

The function d is called the **distance** function of the metric space.

Exercise 7.81 Let $k \in \mathbb{N}$. Show that the function

$$d_H : \mathbb{F}_2^k \times \mathbb{F}_2^k \longrightarrow \mathbb{N}_0$$
$$(\mathbf{u}, \mathbf{v}) \longmapsto \textstyle\sum_{j=1}^{k} u_j \overset{2}{\oplus} v_j$$

satisfies the defining properties of a distance function given in Definition 7.4 and makes $\mathbb{F}_2^k$ a metric space.

For a solution see Solution 7.81.

Definition 7.5 Let $k \in \mathbb{N}$. The distance function

$$d_H : \mathbb{F}_2^k \times \mathbb{F}_2^k \longrightarrow \mathbb{N}_0$$
$$(\mathbf{u}, \mathbf{v}) \longmapsto \sum_{j=1}^{k} u_j \overset{2}{\oplus} v_j \qquad (7.4)$$

is called the HAMMING **distance** between the words $\mathbf{u}$ and $\mathbf{v}$ in the alphabet $\mathbb{F}_2$. The function

$$w_H : \mathbb{F}_2^k \longrightarrow \mathbb{N}_0$$
$$\mathbf{u} \longmapsto d_H(\mathbf{u}, \mathbf{0}) \qquad (7.5)$$

is called the HAMMING **weight** for words of length k in the alphabet $\mathbb{F}_2$.

Obviously,

$$w_H(\mathbf{u}) = d_H(\mathbf{u}, \mathbf{0}) = \sum_{j=1}^{k} u_j$$

is just the number of bits with the value 1 in the bit-string (word) $\mathbf{u} \in \mathbb{F}_2^k$. The HAMMING distance of two words $\mathbf{u}$ and $\mathbf{v}$ is equal to the weight of their binary sum since

$$d_H(\mathbf{u}, \mathbf{v}) \underset{(7.4)}{=} \sum_{j=1}^{k} u_j \overset{2}{\oplus} v_j \underset{(7.2)}{=} \sum_{j=1}^{k} (\mathbf{u} \overset{2}{\oplus} \mathbf{v})_j \underset{(7.4)}{=} d_H(\mathbf{u} \overset{2}{\oplus} \mathbf{v}, \mathbf{0})$$

$$\underset{(7.5)}{=} w_H(\mathbf{u} \overset{2}{\oplus} \mathbf{v}) . \qquad (7.6)$$

Altogether, there are 2^k words we can write in $\mathbb{F}_2^k$. In order to enable redundancy, the words are embedded in a larger space $\mathbb{F}_2^n$ with $n > k$. This is called encoding the words and formally given by a map

$$C_c : \mathbb{F}_2^k \longrightarrow \mathbb{F}_2^n$$
$$\mathbf{w} \longmapsto C_c(\mathbf{w}) .$$

Encoding does not mean copying the word but only adding redundancy to it as we encode k bits with $n > k$ bits. One straightforward—and perhaps computationally most efficient—way is a linear encoding which is determined by specifying a matrix $G \in \text{Mat}(n \times k, \mathbb{F}_2)$ such that $C_c(\mathbf{w}) = G \mathbf{w}$. We want the encoding to be injective since otherwise the original word $\mathbf{w}$ could no longer be retrieved (decoded) uniquely from the encoded word $C_c(\mathbf{w})$. The following definition formalizes these requirements.

Definition 7.6 Let $n, k \in \mathbb{N}$ with $n > k$. An injective map

$$C_c : \mathbb{F}_2^k \longrightarrow \mathbb{F}_2^n$$
$$\mathbf{w} \longmapsto C_c(\mathbf{w})$$

is called a **classical $[\![n, k]\!]_c$ code** or **encoding** C_c with alphabet $\mathbb{F}_2$. The images $C_c(\mathbf{w})$ are called (classical) **codewords**.

If the map C_c is linear and specified by a matrix $G \in \text{Mat}(n \times k, \mathbb{F}_2)$ of maximal rank k such that $C_c(\mathbf{w}) = G\,\mathbf{w}$, then the code is called linear and the matrix G is called the **generator of the code**. The **distance of the code** C_c is defined as

$$d_H(C_c) := \min \left\{ d_H(\mathbf{u}, \mathbf{v}) \mid \mathbf{u}, \mathbf{v} \in G\{\mathbb{F}_2^k\}, \mathbf{u} \neq \mathbf{v} \right\}. \tag{7.7}$$

Note that because of (7.6), the distance of a linear $[\![n, k]\!]_c$ code C_c can be given in the alternative form

$$d_H(C_c) \underset{(7.6)}{=} \min \left\{ w_H(\mathbf{u} \overset{2}{\oplus} \mathbf{v}) \mid \mathbf{u}, \mathbf{v} \in G\{\mathbb{F}_2^k\}, \mathbf{u} \neq \mathbf{v} \right\}$$
$$= \min \left\{ w_H(\mathbf{u}) \mid \mathbf{u} \in G\{\mathbb{F}_2^k\}, \mathbf{u} \neq \mathbf{0} \right\}.$$

For a linear $[\![n, k]\!]_c$ code with alphabet $\mathbb{F}_2$ the set of all codewords $\{G\,\mathbf{w} \mid \mathbf{w} \in \mathbb{F}_2^k\}$ forms a linear subspace

$$G\{\mathbb{F}_2^k\} = \text{Span}\left\{ G\,\mathbf{w} \mid \mathbf{w} \in \mathbb{F}_2^k \right\}$$

of $\mathbb{F}_2^n$ of dimension k. Since $\mathbb{F}_2^n$ is of dimension $n > k$, the elements of $G\{\mathbb{F}_2^k\}$ have to satisfy $n - k$ independent linear equations

$$\sum_{l=1}^{n} H_{jl}(G\,\mathbf{w})_l = 0 \qquad \text{for } j \in \{1, \ldots, n-k\}.$$

These equations are called parity check equations. Independence of the equations means that the the matrix $H \in \text{Mat}\left((n-k) \times n, \mathbb{F}_2\right)$ is of maximal rank.

Definition 7.7 Let G be the generator of a linear $[\![n, k]\!]_c$ code with alphabet $\mathbb{F}_2$. A matrix $H \in \text{Mat}\left((n-k) \times n, \mathbb{F}_2\right)$ with the property

$$\text{Ker}(H) = G\{\mathbb{F}_2^k\} \tag{7.8}$$

is called a **parity check matrix** of the code.

Any vector $\mathbf{u} \in \mathbb{F}_2^n$ satisfying $H\mathbf{u} = \mathbf{0}$ is a valid codeword, namely, it is the image $\mathbf{u} = G\mathbf{w}$ of some word $\mathbf{w} \in \mathbb{F}_2^k$. Vectors $\mathbf{v} \in \mathbb{F}_2^n$ for which $H\mathbf{v} \neq \mathbf{0}$ do not lie in the image of the encoding G and are thus corrupted by some error.

As is to be shown in the following exercise, the parity check matrix is necessarily of rank $n - k$ but not unique.

Exercise 7.82 Let H be a parity check matrix of a linear $[\![n, k]\!]_c$ code with generator G. Show that then

(i)
$$HG = 0$$

(ii)
$$\dim H\{\mathbb{F}_2^n\} = n - k \tag{7.9}$$

(iii) H is not unique.

For a solution see Solution 7.82.

Note, however, that even though parity check matrices of a code are not unique, it follows from (7.8) that their kernels have to coincide.

Example 7.8 Consider the linear $[\![7, 4]\!]_c$ code with generator

$$G = \begin{pmatrix} 1\,0\,0\,1 \\ 0\,1\,0\,1 \\ 1\,1\,0\,1 \\ 0\,0\,1\,1 \\ 1\,1\,1\,0 \\ 0\,0\,0\,1 \\ 1\,0\,1\,0 \end{pmatrix} = \begin{pmatrix} \mathbf{r}_1^T \\ \vdots \\ \mathbf{r}_7^T \end{pmatrix}.$$

To show that $\mathbf{r}_1, \ldots, \mathbf{r}_4$ are linearly independent let $a_1, \ldots, a_4 \in \mathbb{F}_2$ be such that

$$\sum_{j=1}^{4} a_j \mathbf{r}_j = 0,$$

which is equivalent to

$$a_1 \begin{pmatrix} 1 \\ 0 \\ 0 \\ 1 \end{pmatrix} \overset{2}{\oplus} a_2 \begin{pmatrix} 0 \\ 1 \\ 0 \\ 1 \end{pmatrix} \overset{2}{\oplus} a_3 \begin{pmatrix} 1 \\ 1 \\ 0 \\ 1 \end{pmatrix} \overset{2}{\oplus} a_4 \begin{pmatrix} 0 \\ 0 \\ 1 \\ 1 \end{pmatrix} = \begin{pmatrix} a_1 \overset{2}{\oplus} a_3 \\ a_2 \overset{2}{\oplus} a_3 \\ a_4 \\ a_1 \overset{2}{\oplus} a_2 \overset{2}{\oplus} a_3 \overset{2}{\oplus} a_4 \end{pmatrix} = \begin{pmatrix} 0 \\ 0 \\ 0 \\ 0 \end{pmatrix}$$

and results in $a_1 = \cdots = a_4 = 0$. Hence, G is of maximal rank 4. A parity check matrix for this code is given by

$$H = \begin{pmatrix} 0\ 1\ 1\ 1\ 0\ 1\ 1 \\ 0\ 1\ 0\ 0\ 1\ 1\ 1 \\ 1\ 0\ 0\ 1\ 0\ 0\ 1 \end{pmatrix} = \begin{pmatrix} \mathbf{h}_1^T \\ \mathbf{h}_2^T \\ \mathbf{h}_3^T \end{pmatrix},$$

and, as above for $\mathbf{r}_1, \ldots, \mathbf{r}_4$, it can be verified for $\mathbf{h}_1, \mathbf{h}_2, \mathbf{h}_3$ that they are linearly independent. It follows that H is of maximal rank 3. We also have

$$H\,G = \begin{pmatrix} 0\ 1\ 1\ 1\ 0\ 1\ 1 \\ 0\ 1\ 0\ 0\ 1\ 1\ 1 \\ 1\ 0\ 0\ 1\ 0\ 0\ 1 \end{pmatrix} \begin{pmatrix} 1\ 0\ 0\ 1 \\ 0\ 1\ 0\ 1 \\ 1\ 1\ 0\ 1 \\ 0\ 0\ 1\ 1 \\ 1\ 1\ 1\ 0 \\ 0\ 0\ 0\ 1 \\ 1\ 0\ 1\ 0 \end{pmatrix} = \begin{pmatrix} 1 \overset{2}{\oplus} 1 & 1 \overset{2}{\oplus} 1 & 1 \overset{2}{\oplus} 1 & 1 \overset{2}{\oplus} 1 & 1 \overset{2}{\oplus} 1 & 1 \overset{2}{\oplus} 1 \\ 1 \overset{2}{\oplus} 1 & 1 \overset{2}{\oplus} 1 & 1 \overset{2}{\oplus} 1 & & 1 \overset{2}{\oplus} 1 \\ 1 \overset{2}{\oplus} 1 & 0 & 1 \overset{2}{\oplus} 1 & & 1 \overset{2}{\oplus} 1 \end{pmatrix}$$

$$= \begin{pmatrix} 0\ 0\ 0\ 0 \\ 0\ 0\ 0\ 0 \\ 0\ 0\ 0\ 0 \end{pmatrix}.$$

Finally, the reader may verify in a similar manner that

$$\widetilde{H} = \begin{pmatrix} 0\ 0\ 1\ 1\ 1\ 0\ 0 \\ 0\ 1\ 0\ 0\ 1\ 1\ 1 \\ 1\ 0\ 0\ 1\ 0\ 0\ 1 \end{pmatrix}$$

also satisfies $\dim \widetilde{H}\{\mathbb{F}_2^7\} = 3$ and $\widetilde{H}\,G = 0$.

A parity check matrix is a means to detect errors in the encoded words $\mathbf{c} = G\,\mathbf{w}$ as follows. Any encoded word $\mathbf{c} = G\,\mathbf{w}$ has to satisfy

$$H\,\mathbf{c} = H\,G\,\underbrace{\mathbf{w} = }_{(7.8)}\mathbf{0}. \tag{7.10}$$

Suppose the codeword $\mathbf{c}$ has been corrupted in transmission or while stored by an error ε such that the received or retrieved codeword is

$$\mathbf{c}' = \mathbf{c} \overset{2}{\oplus} \varepsilon \in \mathbb{F}_2^n.$$

This implies

$$H\,\mathbf{c}' = H\,\mathbf{c} \overset{2}{\oplus} H\,\varepsilon \underbrace{=}_{(7.10)} H\,\varepsilon.$$

Comparing this with (7.10) shows that $H\mathbf{c}' \neq \mathbf{0}$ indicates that the codeword $\mathbf{c}'$ has been corrupted, in other words, it is not the image $G\mathbf{w}$ of a word $\mathbf{w} \in \mathbb{F}_2^k$. This motivates the following definition.

Definition 7.9 Let H be a parity check matrix of a linear $[\![n, k]\!]_c$ code with alphabet $\mathbb{F}_2$. The map

$$\text{syn}_c : \mathbb{F}_2^n \longrightarrow \mathbb{F}_2^{n-k} \qquad (7.11)$$
$$\mathbf{a} \longmapsto H\mathbf{a}$$

is called a **syndrome mapping** of the code. The vector $H\mathbf{a}$ is called a syndrome of $\mathbf{a}$.

It follows from (7.11) that uncorrupted codewords $\mathbf{c} \in G\{\mathbb{F}_2^k\} = \text{Ker}(H)$ have vanishing syndrome $\text{syn}_c(\mathbf{c}) = H\mathbf{c} = \mathbf{0}$, namely that

$$\text{Ker}(H) = \text{Ker}(\text{syn}_c). \qquad (7.12)$$

Exercise 7.83 Show that syn_c, as defined in Definition 7.9, satisfies for any $\mathbf{a}, \mathbf{b} \in \mathbb{F}_2^n$

$$\text{syn}_c(\mathbf{a} \overset{2}{\oplus} \mathbf{b}) = \text{syn}_c(\mathbf{a}) \overset{2}{\oplus} \text{syn}_c(\mathbf{b})$$

such that syn_c is a homomorphism, that is,

$$\text{syn}_c \in \text{Hom}(\mathbb{F}_2^n, \mathbb{F}_2^{n-k}).$$

For a solution see Solution 7.83.

For $\mathbf{c}' = \mathbf{c} \overset{2}{\oplus} \varepsilon$, where $\mathbf{c} \in G\{\mathbb{F}_2^k\} = \text{Ker}(H)$, we find that

$$\text{syn}_c(\mathbf{c}') = H(\mathbf{c} \overset{2}{\oplus} \varepsilon) = H(\varepsilon) = \text{syn}_c(\varepsilon),$$

and $\text{syn}_c(\mathbf{c}') = \text{syn}_c(\varepsilon) \neq \mathbf{0}$ tells us that $\mathbf{c}'$ has been corrupted by an error. Hence, a non-vanishing syndrome is sufficient for error detection. However, in general the knowledge of $\mathbf{c}'$ or $\text{syn}_c(\mathbf{c}')$ is not sufficient for unmistakable error correction since it does not suffice to uniquely determine ε and thus the original uncorrupted codeword $\mathbf{c}$. This is because

$$\text{syn}_c(\mathbf{a} \overset{2}{\oplus} \mathbf{h}) = \text{syn}_c(\mathbf{a}) \quad \Leftrightarrow \quad \mathbf{h} \in \text{Ker}(H)$$

such that the knowledge of $\text{syn}_c(\varepsilon)$ only determines the coset (see Definition F.18)

$$[\varepsilon] = [\varepsilon]_{\mathrm{Ker(H)}} = \left\{ \varepsilon \overset{2}{\oplus} \mathbf{h} \,\middle|\, \mathbf{h} \in \mathrm{Ker(H)} \right\}.$$

From Lemma F.30 we know that $\mathrm{Ker(H)} = \mathrm{Ker(syn}_c)$ is a normal subgroup of $\mathbb{F}_2^n$, and thus the quotient group $\mathbb{F}_2^n/\mathrm{Ker(syn}_c)$ defined in Definition F.23 exists. The syndrome mapping provides an isomorphism between the quotient group $\mathbb{F}_2^n/\mathrm{Ker(syn}_c)$ and $\mathbb{F}_2^{n-k}$.

Proposition 7.10 *For a linear $[\![n,k]\!]_c$ code with parity check matrix* H *and syndrome* syn_c *the map*

$$\widehat{\mathrm{syn}_c} : \mathbb{F}_2^n/\mathrm{Ker(H)} \longrightarrow \mathbb{F}_2^{n-k}$$
$$[\mathbf{a}] \longmapsto \mathrm{syn}_c(\mathbf{a})$$

is an isomorphism.

Proof From (7.9) we know that H has maximal rank. This implies that for any $\mathbf{b} \in \mathbb{F}_2^{n-k}$ there exists an $\mathbf{a} \in \mathbb{F}_2^n$ such that $\mathbf{b} = \mathrm{H}\,\mathbf{a} = \mathrm{syn}_c(\mathbf{a})$ proving that $\mathbb{F}_2^{n-k} = \mathrm{syn}_c\{\mathbb{F}_2^n\}$. Since $\mathrm{Ker(H)} = \mathrm{Ker(syn}_c)$, the claim then follows immediately from the First Group Isomorphism Theorem F.31. $\qquad\square$

Proposition 7.10 tells us that there is a bijection between syndromes in $\mathbb{F}_2^{n-k}$ and cosets in $\mathbb{F}_2^n/\mathrm{Ker(H)}$ such that $\widehat{\mathrm{syn}_c}^{-1}$ uniquely identifies a coset from a given syndrome. Let $\mathbf{c} \in \mathrm{Ker(H)} = \mathrm{G}\{\mathbb{F}_2^k\}$ be a valid codeword sent or stored, and let $\tilde{\mathbf{c}} = \mathbf{c} \overset{2}{\oplus} \varepsilon$ be what has been received or retrieved. Therefore, any potential error ε satisfies

$$\varepsilon = \varepsilon \overset{2}{\oplus} \underbrace{\mathbf{c} \overset{2}{\oplus} \mathbf{c}}_{\mathbf{c} \in \mathrm{Ker(H)}} = \tilde{\mathbf{c}} \overset{2}{\oplus} \mathbf{c} \underbrace{\in}_{} [\tilde{\mathbf{c}}]_{\mathrm{Ker(H)}} \qquad (7.13)$$

such that

$$\mathrm{d}_H(\tilde{\mathbf{c}}, \mathbf{c}) \underbrace{=}_{(7.6)} \mathrm{w}_H(\tilde{\mathbf{c}} \overset{2}{\oplus} \mathbf{c}) \underbrace{=}_{(7.13)} \mathrm{w}_H(\varepsilon). \qquad (7.14)$$

How do we then go about to recover the uncorrupted codeword $\mathbf{c}$ from the corrupted version $\tilde{\mathbf{c}}$? The codeword known to us is $\tilde{\mathbf{c}}$. From this we can infer $\mathrm{syn}_c(\tilde{\mathbf{c}})$ and thus $[\tilde{\mathbf{c}}]_{\mathrm{Ker(H)}}$ by applying $\widehat{\mathrm{syn}}^{-1}$. We also know from (7.13) that $\varepsilon \in [\tilde{\mathbf{c}}]_{\mathrm{Ker(H)}}$, in other words, the error which corrupted $\mathbf{c}$ is an element of the coset determined by $\tilde{\mathbf{c}}$. Our error correction strategy is then to find the $\mathbf{r} \in [\tilde{\mathbf{c}}]_{\mathrm{Ker(H)}}$ with the smallest weight and to attempt to recover $\mathbf{c}$ by adding $\mathbf{r}$ to $\tilde{\mathbf{c}}$. Choosing $\mathbf{r} \in [\tilde{\mathbf{c}}]_{\mathrm{Ker(H)}}$ with the smallest weight is based on the assumption that in the coset $[\tilde{\mathbf{c}}]$ the most likely error which corrupted $\mathbf{c}$ is the one which changed the least number of bits. From (7.14) we see that for a given syndrome this would mean the smallest distance between observed $\tilde{\mathbf{c}}$

and the presumed original codeword $\mathbf{c}$. Our classical error detection and correction protocol is thus as follows:

Classical Error Detection and Correction Protocol

1. Determine $\mathrm{syn}_c(\tilde{\mathbf{c}}) = \mathrm{H}\,\tilde{\mathbf{c}}$ by applying the parity check matrix H to $\tilde{\mathbf{c}}$
2. Determine $[\tilde{\mathbf{c}}]_{\mathrm{Ker(H)}} = \widehat{\mathrm{syn}_c}^{-1}\left(\mathrm{syn}_c(\tilde{\mathbf{c}})\right)$
3. Determine the set of elements in this coset with minimal weight

$$R_c^{\min}[\tilde{\mathbf{c}}]_{\mathrm{Ker(H)}} := \left\{\mathbf{a} \in [\tilde{\mathbf{c}}]_{\mathrm{Ker(H)}} \mid \mathrm{w}_H(\mathbf{a}) \leq \mathrm{w}_H(\mathbf{b})\ \forall \mathbf{b} \in [\tilde{\mathbf{c}}]_{\mathrm{Ker(H)}}\right\}$$

4. Select a random element

$$\mathbf{r} \in R_c^{\min}[\tilde{\mathbf{c}}]_{\mathrm{Ker(H)}}$$

5. Calculate $\tilde{\mathbf{c}} \overset{2}{\oplus} \mathbf{r}$ and proceed with this as the presumed original codeword for $\mathbf{c}$.

Note that the case where we leave $\tilde{\mathbf{c}}$ unchanged is included in this protocol as the case where $\mathrm{syn}_c(\tilde{\mathbf{c}}) = \mathbf{0}$. In this case $[\tilde{\mathbf{c}}]_{\mathrm{Ker(H)}} = \mathrm{Ker(H)}$ and $R_c^{\min}[\tilde{\mathbf{c}}]_{\mathrm{Ker(H)}} = \{\mathbf{0}\}$.

In general, however, the set $R_c^{\min}[\tilde{\mathbf{c}}]_{\mathrm{Ker(H)}}$ may contain more than one element. In that case it is not possible to guarantee that the protocol will always recover the original codeword $\mathbf{c}$ with certainty. This is why this type of recovery method is called **maximum-likelihood** or nearest neighbor **decoding**. The name originates from the assumption that the most likely error is the one that corrupts the fewest number of bits.

However, if the error does not change too many bits relative to the distance of the code, then the original codeword can be recovered with certainty as the following proposition shows.

Proposition 7.11 *Let* C_c *be a linear* $[\![n,k]\!]_c$ *code, let* $\mathbf{c} \in \mathrm{Ker(H)} = \mathrm{G}\{\mathbb{F}_2^k\}$ *be a valid codeword sent or stored, and let* $\tilde{\mathbf{c}} = \mathbf{c} \overset{2}{\oplus} \varepsilon$ *be the received or retrieved word corrupted by* ε. *Then* $\mathbf{c}$ *can be correctly recovered from* $\tilde{\mathbf{c}}$ *with certainty if*

$$\mathrm{w}_H(\varepsilon) \leq \left\lfloor \frac{\mathrm{d}_H(C_c) - 1}{2} \right\rfloor, \tag{7.15}$$

where $\lfloor b \rfloor$ *denotes the integer part of* $b \in \mathbb{R}$.

Proof Let

$$B_{\mathbb{F}_2^n}^r(\mathbf{a}) := \left\{\mathbf{v} \in \mathbb{F}_2^n \mid \mathrm{d}_H(\mathbf{a}, \mathbf{v}) \leq r\right\} \tag{7.16}$$

denote the ball of radius $r \geq 0$ centered at $\mathbf{a}$ in $\mathbb{F}_2^n$. We will show that the intersection of all codewords $\mathrm{G}\{\mathbb{F}_2^k\}$ with the ball of radius $\frac{\mathrm{d}_H(C_c)}{2}$ centered at $\tilde{\mathbf{c}}$ only contains $\mathbf{c}$. To begin with, we have

$$d_H(\tilde{\mathbf{c}}, \mathbf{c}) \underbrace{=}_{(7.14)} w_H(\varepsilon) \underbrace{<}_{(7.15)} \frac{d_H(C_c)}{2}$$

such that (7.16) implies

$$\mathbf{c} \in B_{\mathbb{F}_2^n}^{d_H(C_c)/2}(\tilde{\mathbf{c}}). \tag{7.17}$$

Next, let $\mathbf{a} \in G\{\mathbb{F}_2^k\} \smallsetminus \{\mathbf{c}\}$. Then we find

$$d_H(C_c) \underbrace{\leq}_{(7.7)} d_H(\mathbf{c}, \mathbf{a}) \underbrace{\leq}_{(7.3)} d_H(\mathbf{c}, \tilde{\mathbf{c}}) + d_H(\tilde{\mathbf{c}}, \mathbf{a})$$

$$\underbrace{=}_{(7.14)} w_H(\varepsilon) + d_H(\tilde{\mathbf{c}}, \mathbf{a}) \underbrace{\leq}_{(7.15)} \left\lfloor \frac{d_H(C_c) - 1}{2} \right\rfloor + d_H(\tilde{\mathbf{c}}, \mathbf{a})$$

$$< \frac{d_H(C_c)}{2} + d_H(\tilde{\mathbf{c}}, \mathbf{a})$$

such that $\frac{d_H(C_c)}{2} < d_H(\tilde{\mathbf{c}}, \mathbf{a})$ for any $\mathbf{a} \in G\{\mathbb{F}_2^k\} \smallsetminus \{\mathbf{c}\}$, which implies

$$\mathbf{a} \notin B_{\mathbb{F}_2^n}^{d_H(C_c)/2}(\tilde{\mathbf{c}}) \tag{7.18}$$

for any such $\mathbf{a}$. From (7.17) and (7.18) it follows that

$$G\{\mathbb{F}_2^k\} \cap B_{\mathbb{F}_2^n}^{d_H(C_c)/2}(\tilde{\mathbf{c}}) = \{\mathbf{c}\}.$$

Certain recovery of $\mathbf{c}$ from the knowledge of $\tilde{\mathbf{c}}$ is then accomplished by finding the unique element of $G\{\mathbb{F}_2^k\}$ in a ball of radius $d_H(C_c)/2$ centered at $\tilde{\mathbf{c}}$. $\qquad\square$

Proposition 7.11 is often stated by saying that a linear code C_c can correct any errors ε satisfying (7.15). We will obtain similar results for quantum error correction to which we turn in the next section.

7.3   Quantum Error Correction

7.3.1   Correctable Errors

Any quantum system is quite easily changed by interactions with its environment, and in general it is very difficult to avoid this. In the context of quantum computation any such unwanted interaction is likely to introduce errors which need correcting if the computation is to succeed.

Just as in the classical case, the first ingredient for quantum error correction is a means to create redundancy. This is achieved by what is generally described as a quantum code.

Definition 7.12 A **quantum error correcting code** (QECC) is specified with the help of an injective and norm-preserving operator

$$C_q : \mathbb{H}^K \to \mathbb{H}^N$$

called **encoding** between HILBERT spaces $\mathbb{H}^K$ and $\mathbb{H}^N$ with $\dim \mathbb{H}^K <$ $\dim \mathbb{H}^N < \infty$. The space $\mathbb{H}^N$ is called the **quantum encoding space**, and the image of the encoding

$$\mathbb{H}^{C_q} := C_q \left\{ \mathbb{H}^K \right\} \subsetneq \mathbb{H}^N$$

is called a **quantum code** (or quantum code space). The elements of $\mathbb{H}^{C_q}$ are called **codewords**. A convex-linear map

$$D_q : D\left(\mathbb{H}^{C_q} \right) \to D\left(\mathbb{H}^K \right)$$

satisfying

$$D_q \left(C_q \left| \psi \right\rangle \left\langle \psi \right| C_q^* \right) = \left| \psi \right\rangle \left\langle \psi \right| \quad \forall \left| \psi \right\rangle \in \mathbb{H}^K$$

is called **decoding**.

If $\mathbb{H}^K = \mathbb{H}^{\otimes k}$ and $\mathbb{H}^N = \mathbb{H}^{\otimes n}$ with $k < n$, then the QECC is called an $[\![n, k]\!]_q$ QECC and the images

$$\left| \Psi_x \right\rangle := C_q \left| x \right\rangle \in \mathbb{H}^{C_q}$$

of the computational basis states $\left| x \right\rangle \in \mathbb{H}^K$ are called **basis codewords**.

In the case $k = 1$ of a single qubit $\mathbb{H}^K = \mathbb{H}$, the basis codewords $C_q \left| 0 \right\rangle$ and $C_q \left| 1 \right\rangle$ are called **logical qubits**, and the single qubit-factors $\mathbb{H}$ in $\mathbb{H}^N = \mathbb{H}^{\otimes n}$ are referred to as **physical qubits**.

We will often use the symbol C_q of the encoding operator to denote the QECC.

Note that just like in the classical case, the code is a subspace $\mathbb{H}^{C_q}$ of dimension 2^k of the larger encoding space $\mathbb{H}^N$, which has dimension 2^n. The redundancy for potential error correction is provided by the $2^n - 2^k$ extra dimensions in $\mathbb{H}^N$. Enlarging the original system in this way, means that we are combining it with a system of ancillas (see Figs. 7.1 or 7.3 for examples) to a larger composite system. Decoding is then accomplished by discarding, that is, tracing out, these ancillas. This is why the decoding map has to be formulated on the density operators.

Lemma 7.13 *Let* C_q *be an* $[\![n, k]\!]_q$ *QECC. Then the set of its basis codewords* $\{|\Psi_x\rangle \mid x \in \{0, \ldots, 2^k - 1\}\}$ *is an ONB in* $\mathbb{H}^{C_q}$.

Proof To begin with, note that since C_q is by definition injective, it follows that $|\varphi_1\rangle - |\varphi_2\rangle \neq 0$ for any $|\varphi_1\rangle, |\varphi_2\rangle \in \mathbb{H}^K$ implies $C_q(|\varphi_1\rangle - |\varphi_2\rangle) \neq 0$. Hence, we have

$$C_q|\varphi\rangle = 0 \quad \Leftrightarrow \quad |\varphi\rangle = 0. \tag{7.19}$$

Next, we show that the $|\Psi_x\rangle = C_q|x\rangle$ are linearly independent. For $x \in \{0, \ldots, 2^k - 1\}$ let $a_x \in \mathbb{C}$. Then we have the following chain of implications

$$\sum_{x=0}^{2^k-1} a_x|\Psi_x\rangle = 0 \implies \sum_{x=0}^{2^k-1} a_x C_q|x\rangle = C_q\left(\sum_{x=0}^{2^k-1} a_x|x\rangle\right) = 0$$

$$\underset{(7.19)}{\implies} \sum_{x=0}^{2^k-1} a_x|x\rangle = 0$$

$$\implies a_x = 0 \quad \forall x \in \{0, \ldots, 2^k - 1\},$$

where for the last implication we used that the computational basis $\{|x\rangle\}$ is an ONB in $\mathbb{H}^K$. This shows that the $|\Psi_x\rangle$ are linearly independent.

Now, for every $|\Phi\rangle \in \mathbb{H}^{C_q} = C_q\{\mathbb{H}^K\}$ there exists a $|\varphi\rangle \in \mathbb{H}^K$ such that $|\Phi\rangle = C_q|\varphi\rangle$. Using again that that the $\{|x\rangle\}$ form an ONB in $\mathbb{H}^K$, we thus find

$$|\Phi\rangle = C_q|\varphi\rangle = C_q\left(\sum_{x=0}^{2^k-1} \varphi_x|x\rangle\right) = \sum_{x=0}^{2^k-1} \varphi_x C_q|x\rangle = \sum_{x=0}^{2^k-1} \varphi_x|\Psi_x\rangle,$$

which shows that every $|\Phi\rangle \in \mathbb{H}^{C_q}$ can be expressed as a linear combination of the $|\Psi_x\rangle$.

Finally, to show that the $|\Psi_x\rangle = C_q|x\rangle$ are orthonormal, we use that by definition C_q satisfies $||C_q|\varphi\rangle|| = ||\varphi||$ for all $|\varphi\rangle \in \mathbb{H}^K$ to obtain

$$\langle\Psi_x|\Psi_y\rangle \underset{(2.9)}{=} \frac{1}{4}\Big(\big|\big| |\Psi_x\rangle + |\Psi_y\rangle \big|\big|^2 - \big|\big| |\Psi_x\rangle - |\Psi_y\rangle \big|\big|^2$$

$$+ \mathrm{i} \big|\big| |\Psi_x\rangle - \mathrm{i}|\Psi_y\rangle \big|\big|^2 - \mathrm{i} \big|\big| |\Psi_x\rangle + \mathrm{i}|\Psi_y\rangle \big|\big|^2 \Big)$$

$$= \frac{1}{4}\Big(\big|\big| C_q(|x\rangle + |y\rangle) \big|\big|^2 - \big|\big| C_q(|x\rangle - |y\rangle) \big|\big|^2$$

$$+ \mathrm{i} \big|\big| C_q(|x\rangle - \mathrm{i}|y\rangle) \big|\big|^2 - \mathrm{i} \big|\big| C_q(|x\rangle + \mathrm{i}|y\rangle) \big|\big|^2 \Big)$$

$$= \frac{1}{4} \Big(\| \, |x\rangle + |y\rangle \, \|^2 - \| \, |x\rangle - |y\rangle \, \|^2 + i \| \, |x\rangle - i|y\rangle \, \|^2 - i \| \, |x\rangle + i|y\rangle \, \|^2 \Big)$$

$$\underbrace{=}_{(2.9)} \langle x|y\rangle \underbrace{=}_{(3.24)} \delta_{xy} \, . \qquad\qquad\qquad\qquad \square$$

Before we consider our first example, we remind the reader that here as throughout this book we use all three notations

$$\sigma_0 = \mathbf{1}, \quad \sigma_1 = \sigma_x = X, \quad \sigma_2 = \sigma_y = Y, \quad \sigma_3 = \sigma_z = Z \, .$$

for the PAULI matrices/operators.

Example 7.14 As first examples, we consider two $[\![3, 1]\!]_q$ QECCs in which one logical qubit is encoded with three physical qubits. The first QECC, which we denote by $[\![3, 1]\!]_{q_1}$, is given by the encoding map

$$C_{q_1} : \mathbb{H} \overset{\iota}{\rightarrow} \mathbb{H}^{\otimes 3} \overset{A}{\rightarrow} \mathbb{H}^{\otimes 3} \, , \qquad\qquad (7.20)$$

where

$$\iota : \mathbb{H} \longrightarrow \mathbb{H}^{\otimes 3}$$
$$|\psi\rangle \longmapsto |\psi\rangle \otimes |0\rangle \otimes |0\rangle$$

and

$$A = |1\rangle\langle 1| \otimes X \otimes X + |0\rangle\langle 0| \otimes \mathbf{1} \otimes \mathbf{1} \, . \qquad\qquad (7.21)$$

The circuit for this encoding is given in Fig. 7.1

For the $[\![3, 1]\!]_{q_1}$ basis codewords $|\Psi_0\rangle, |\Psi_1\rangle \in \mathbb{H}^{\otimes 3}$ we thus find

$$\begin{aligned} |\Psi_0\rangle &= C_{q_1} |0\rangle = A|000\rangle = |000\rangle \\ |\Psi_1\rangle &= C_{q_1} |1\rangle = A|100\rangle = |1\rangle \otimes X|0\rangle \otimes X|0\rangle = |111\rangle \, , \end{aligned} \qquad (7.22)$$

Fig. 7.1 Circuit for the encoding of the first $[\![3, 1]\!]_{q_1}$ QECC given in (7.20)

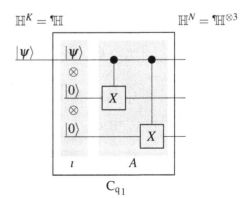

such that the code is

$$\mathbb{H}^{C_{q_1}} = \mathrm{Span}\{|000\rangle, |111\rangle\} \subset \mathbb{H}^{\otimes 3} . \tag{7.23}$$

Note that

$$A^2 \underbrace{=}_{(7.21)} |1\rangle\langle 1| \otimes X^2 \otimes X^2 + |0\rangle\langle 0| \otimes \mathbf{1} \otimes \mathbf{1} = |1\rangle\langle 1| \otimes \mathbf{1} \otimes \mathbf{1} + |0\rangle\langle 0| \otimes \mathbf{1} \otimes \mathbf{1}$$

$$= \mathbf{1}^{\otimes 3} . \tag{7.24}$$

Hence, the decoding of any

$$|\Psi\rangle = C_{q_1} |\psi\rangle = A\big(|\psi\rangle \otimes |0\rangle \otimes |0\rangle\big) = A\big(|\psi\rangle \otimes |00\rangle\big) \in \mathbb{H}^{\otimes 3} \tag{7.25}$$

can be given as a map on its density matrix by

$$D_{q_1} : \quad \begin{array}{ccccc} D\big(\mathbb{H}^{\otimes 3}\big) & \xrightarrow{A(\cdot)A^*} & D\big(\mathbb{H}^{\otimes 3}\big) & \xrightarrow{\mathrm{tr}^2(\cdot)} & D(\mathbb{H}) \\ \rho_{|\Psi\rangle} = |\Psi\rangle\langle\Psi| & \longmapsto & A|\Psi\rangle\langle\Psi|A^* & \longmapsto & |\psi\rangle\langle\psi| \end{array} ,$$

where $\mathrm{tr}^2(\cdot)$ denotes tracing out the two ancilla qubits introduced by ι. This is because

$$D_{q_1} \rho_{|\Psi\rangle} = \mathrm{tr}^2\big(A|\Psi\rangle\langle\Psi|A^*\big) \underbrace{=}_{(7.25)} \mathrm{tr}^2\big(AA\big(|\psi\rangle \otimes |00\rangle\langle\psi| \otimes \langle 00|\big)A^*A^*\big)$$

$$\underbrace{=}_{(7.24),(3.36)} \mathrm{tr}^2\big(|\psi\rangle\langle\psi| \otimes |00\rangle\langle 00|\big) \underbrace{=}_{(3.57)} \underbrace{\mathrm{tr}\big(|00\rangle\langle 00|\big)}_{=1} |\psi\rangle\langle\psi|$$

$$= |\psi\rangle\langle\psi| .$$

The decoding circuit is depicted in Fig. 7.2.

Fig. 7.2 Circuit for the decoding of the first $[3, 1]_{q_1}$ QECC given in (7.20)

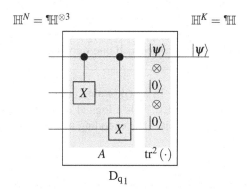

The second QECC, which we denote by $[\![3,1]\!]_{q2}$, is given by the encoding map

$$C_{q_1} : \mathbb{H} \overset{\iota}{\to} \mathbb{H}^{\otimes 3} \overset{A}{\to} \mathbb{H}^{\otimes 3} \overset{H^{\otimes 3}}{\to} \mathbb{H}^{\otimes 3} ,$$

where H denotes the HADAMARD transformation defined in Definition 2.38. It has the property

$$H|0\rangle \underset{(2.160)}{=} \frac{|0\rangle + |1\rangle}{\sqrt{2}} = |+\rangle \quad \text{and} \quad H|1\rangle \underset{(2.161)}{=} \frac{|0\rangle - |1\rangle}{\sqrt{2}} = |-\rangle . \qquad (7.26)$$

For the $[\![3,1]\!]_{q2}$ basis codewords $|\Phi_0\rangle, |\Phi_1\rangle \in \mathbb{H}^{\otimes 3}$ we thus find

$$|\Phi_0\rangle = C_{q2}|0\rangle = H^{\otimes 3}A|000\rangle \underset{(7.22)}{=} H^{\otimes 3}|000\rangle \underset{(7.26)}{=} |++\,+\rangle$$

$$|\Phi_1\rangle = C_{q2}|1\rangle = H^{\otimes 3}A|100\rangle \underset{(7.22)}{=} H^{\otimes 3}|111\rangle \underset{(7.26)}{=} |-\,--\rangle \qquad (7.27)$$

such that the code is

$$\mathbb{H}^{C_{q2}} = \text{Span}\{|++\,+\rangle, |-\,--\rangle\} \subset \mathbb{H}^{\otimes 3} . \qquad (7.28)$$

To illustrate the point that the encoding creates redundancy and not duplication, we consider the $[\![3,1]\!]_{q1}$ code of Example 7.14. There we have

$$C_{q_1}(a|0\rangle + b|1\rangle) \underset{(7.22)}{=} a|000\rangle + b|111\rangle \neq (a|0\rangle + b|1\rangle)^{\otimes 3} ,$$

where the state to the right of the inequality shows what a three-fold copy of the initial state would look like. Hence, a quantum encoding does not constitute a violation of the Quantum No-Cloning Theorem 4.11.

Another code with more redundancy than those of Example 7.14 is due to SHOR [99]. It requires nine physical qubits to encode one logical qubit. As we will show later, it corrects all one-qubit errors.

Example 7.15 In SHOR's nine qubit $[\![9,1]\!]_q$ QECC, which we denote by C_{qS}, one qubit is encoded with a logical qubit consisting of a tensor product state of nine (physical) qubits. The encoding map is given by

$$C_{qS} : \mathbb{H} \overset{\iota}{\to} \mathbb{H}^{\otimes 9} \overset{A}{\to} \mathbb{H}^{\otimes 9} , \qquad (7.29)$$

Fig. 7.3 Circuit for the encoding of SHOR's $[\![9,1]\!]_q$ QECC given in (7.29) as $C_{q_S} = A_3 A_2 A_1 \iota$

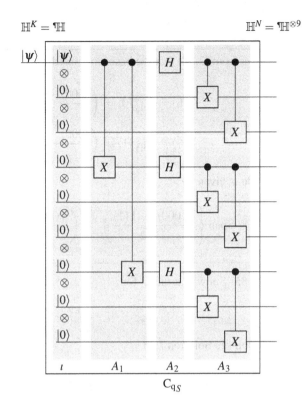

where

$$\iota : \quad \mathbb{H} \longrightarrow \mathbb{H}^{\otimes 9}$$
$$|\psi\rangle \longmapsto |\psi\rangle \otimes |0\rangle^8 \tag{7.30}$$

and $A = A_3 A_2 A_1$ with

$$A_1 = |1\rangle\langle 1| \otimes \mathbf{1}^{\otimes 2} \otimes X \otimes \mathbf{1}^{\otimes 2} \otimes X \otimes \mathbf{1}^{\otimes 2} + |0\rangle\langle 0| \otimes \mathbf{1}^{\otimes 8}$$
$$A_2 = \left(H \otimes \mathbf{1}^{\otimes 2}\right)^{\otimes 3} \tag{7.31}$$
$$A_3 = \left(|1\rangle\langle 1| \otimes X \otimes X + |0\rangle\langle 0| \otimes \mathbf{1}^{\otimes 2}\right)^{\otimes 3}$$

and, as usual $X = \sigma_x$, and H denotes the HADAMARD operator given in Definition 2.38. The circuit for this encoding is given in Fig. 7.3.

Exercise 7.84 Show that

$$C_{q_S}\left(a|0\rangle + b|1\rangle\right) = a\left(\frac{|000\rangle + |111\rangle}{\sqrt{2}}\right)^{\otimes 3} + b\left(\frac{|000\rangle - |111\rangle}{\sqrt{2}}\right)^{\otimes 3}. \tag{7.32}$$

For a solution see Solution 7.84.

It follows from (7.32) that the basis codewords are

$$|\Psi_0\rangle = C_{qS}|0\rangle = \left(\frac{|000\rangle + |111\rangle}{\sqrt{2}}\right)^{\otimes 3}$$

$$|\Psi_1\rangle = C_{qS}|1\rangle = \left(\frac{|000\rangle - |111\rangle}{\sqrt{2}}\right)^{\otimes 3} , \tag{7.33}$$

and the code is given by

$$\mathbb{H}^{C_{qS}} = \mathrm{Span}\left\{\left(\frac{|000\rangle + |111\rangle}{\sqrt{2}}\right)^{\otimes 3} , \left(\frac{|000\rangle - |111\rangle}{\sqrt{2}}\right)^{\otimes 3}\right\} \subset \mathbb{H}^{\otimes 9} .$$

Exercise 7.85 Show that the A_i for $i \in \{1, 2, 3\}$ defined in (7.31) satisfy

$$A_i^2 = \mathbf{1}^{\otimes 9} . \tag{7.34}$$

For a solution see Solution 7.85.

The statement (7.34) of Exercise 7.85 implies that

$$A_1 A_2 A_3 A_3 A_2 A_1 = \mathbf{1}^{\otimes 9} \tag{7.35}$$

and thus

$$A_1 A_2 A_3\, C_{qS}|\psi\rangle \underbrace{=}_{(7.29)} A_1 A_2 A_3 A_3 A_2 A_1 \iota|\psi\rangle \underbrace{=}_{(7.35)} \iota|\psi\rangle \underbrace{=}_{(7.30)} |\psi\rangle \otimes |0\rangle^8 .$$

The decoding $D_{qS} : D(\mathbb{H}^{\otimes 9}) \to D(\mathbb{H})$ is then given by

$$D_{qS}(\rho) = \mathrm{tr}^8\left(A_1 A_2 A_3 \rho A_3 A_2 A_1\right) ,$$

where we have made use of the fact that $A_i^* = A_i$ for $i \in \{1, 2, 3\}$. The circuit for this decoding is shown in Fig. 7.4.

Each basis codeword $|\Psi_x\rangle \in \mathbb{H}^{C_q}$ is the image of a unique un-encoded word $|x\rangle \in \mathbb{H}^K$, which can be recovered by the decoding D_q. Generally, the more redundancy the encoding provides, the more errors will be detectable. On the other hand, the more redundancy is encoded, the more computational resources are required.

We wish to create a mechanism by which errors corrupting the codewords can be corrected. In doing so, we need to specify which type of errors we consider acting in $\mathbb{H}^{C_q}$ and how we can construct recovery mechanisms for these errors.

Fig. 7.4 Circuit for the decoding of SHOR's $[\![9, 1]\!]_q$ QECC

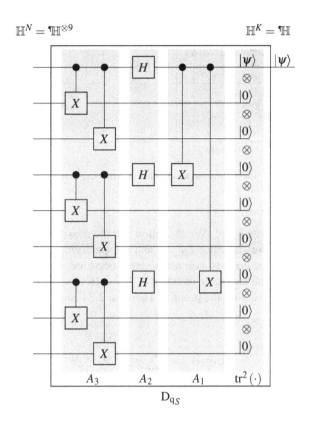

Of course, the encoding C_q and the decoding D_q themselves could be prone to error, that is, the methods to correct errors can suffer from errors. The treatment of such combined errors belongs to the realm of fault tolerant quantum computation [100, 101], which we will not address here. In our discussion we will only consider errors corrupting (encoded) codewords, in other words, we assume errors occur after encoding and before decoding. This is illustrated in Fig. 7.5 with the $[\![3, 1]\!]_{q1}$ code of Example 7.14.

The situation we are trying to describe by modeling the occurrence of errors in quantum systems is then as follows. We first consider the case of pure states in the code space $\mathbb{H}^{C_q}$ and then the case of mixed states in $\mathrm{D}(\mathbb{H}^{C_q})$. Let us assume our system is initially in the uncorrupted pure state $|\Psi\rangle \in \mathbb{H}^{C_q} \subset \mathbb{H}^N$. We then need to specify which possible errors may occur. We do this by specifying so-called error operators $\hat{\mathcal{E}}_a \in \mathrm{L}(\mathbb{H}^N)$, where $a \in I \subset \mathbb{N}_0$ and I is a set indexing the errors covered by our error model. The original uncorrupted states are in $\mathbb{H}^{C_q}$, but the effect of an error will in general take them out of the code space. Therefore, the error operators are elements of $\mathrm{L}(\mathbb{H}^N)$.

If the error indexed by a occurs, then the system ends up in a state $\hat{\mathcal{E}}_a|\Psi\rangle \in \mathbb{H}$. Since this is a random event, we need to assign it a probability of occurring, which

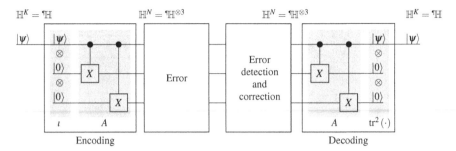

Fig. 7.5 Illustrating our assumption of error occurrence between en- and decoding for the $[[3,1]]_{q_1}$ code

we denote by p_a. We also allow for the possibility of no error by always including the 'error' operator $\hat{\mathcal{E}}_0 = \mathbf{1}$ and by specifying the probability of no error with p_0. After the error is assumed to have occurred, our computational state is given by a statistical ensemble of states $\hat{\mathcal{E}}_a|\Psi\rangle$ each of which occurs with a probability p_a. As discussed after the proof of Theorem 2.24, such a system is described by the mixed state

$$
\begin{aligned}
\mathcal{E}(\rho_\Psi) &= \sum_{a\in I} p_a \hat{\mathcal{E}}_a |\Psi\rangle\langle\hat{\mathcal{E}}_a\Psi| \underset{(2.33)}{=} \sum_{a\in I} p_a \hat{\mathcal{E}}_a |\Psi\rangle\langle\Psi|\hat{\mathcal{E}}_a^* \\
&\underset{(2.89)}{=} \sum_{a\in I} p_a \hat{\mathcal{E}}_a \rho_\Psi \hat{\mathcal{E}}_a^* = \sum_{a\in I} \mathcal{E}_a \, \rho_\Psi \, \mathcal{E}_a^* \,,
\end{aligned}
\tag{7.36}
$$

where we set $\mathcal{E}_a = \sqrt{p_a}\hat{\mathcal{E}}_a$ in the last equation.

The very same arguments apply if our original uncorrupted system is in a truly mixed state $\rho \in \mathrm{D}(\mathbb{H}^{C_q})$. In this case each error a produces a new mixed state $\hat{\mathcal{E}}_a \rho \hat{\mathcal{E}}_a^*$ with probability p_a. The resulting statistical ensemble of mixed state is described as in (7.36) with ρ_Ψ replaced by ρ.

From Definition 3.26 and Theorem 3.24 we know that any quantum operation K can be expressed in terms of its operation elements (aka KRAUS operators) K_a by

$$
K(\rho) = \sum_a K_a \rho K_a^* \,,
$$

which is precisely of the form (7.36). Moreover, most errors are caused by interactions of $\mathbb{H}^N$ with the environment, and quantum operations have an environmental representation reflecting this. These considerations lead us to model the effect of errors occurring in a system $\mathbb{H}^N$ with the help of a quantum operation (see Sect. 3.5)

$$
\begin{aligned}
\mathcal{E} : \mathrm{D}(\mathbb{H}^N) &\longrightarrow \mathrm{D}_{\leq}(\mathbb{H}^N) \\
\rho &\longmapsto \mathcal{E}(\rho) = \textstyle\sum_{a\in I} \mathcal{E}_a \rho \, \mathcal{E}_a^* \,,
\end{aligned}
\tag{7.37}
$$

where the operation elements $\mathcal{E}_a$ are of the form $\mathcal{E}_a = \sqrt{p_a}\hat{\mathcal{E}}_a$ and satisfy

$$\sum_{a \in I} \mathcal{E}_a^* \mathcal{E}_a \leq \kappa \mathbf{1} \tag{7.38}$$

for some $\kappa \in]0, 1]$. From Corollary 3.25 we know that equality in (7.38) is equivalent to $\mathrm{tr}(\mathcal{E}(\rho)) = \kappa$ for all $\rho \in \mathrm{D}(\mathbb{H}^N)$. For reasons which will become obvious once we have introduced recovery operations correcting errors, we will have to restrict quantum operations classified as errors to this category of constant trace.

While the quantum operation $\mathcal{E}$ in (7.37) is defined on all of $\mathbb{H}^N$, we are only interested in the effect of $\mathcal{E}$ on $\rho \in \mathrm{D}(\mathbb{H}^{C_q})$. However, since $\mathbb{H}^{C_q} \subsetneq \mathbb{H}^N$ there is a natural embedding $\mathrm{D}(\mathbb{H}^{C_q}) \hookrightarrow \mathrm{D}(\mathbb{H}^N)$ given as follows. From Theorem 2.24 we know that for any $\rho \in \mathrm{D}(\mathbb{H}^{C_q})$ there exists an ONB $\{|\Theta_j\rangle\}$ in $\mathbb{H}^{C_q}$ and $q_j \in [0, 1]$ with $\sum_j q_j = 1$ such that

$$\rho = \sum_j q_j |\Theta_j\rangle\langle\Theta_j| . \tag{7.39}$$

Any ONB in $\mathbb{H}^{C_q} \subsetneq \mathbb{H}^N$ can be extended to an ONB in $\mathbb{H}^N$ and thus ρ in (7.39) can be viewed as an element of $\mathrm{D}(\mathbb{H}^N)$.

We finally arrive at the following definition of an error operation and related notions.

Definition 7.16 Let $\mathbb{H}^N$ be the encoding space of a QECC with code $\mathbb{H}^{C_q}$. An **error operation in the code** $\mathbb{H}^{C_q}$ is defined as a quantum operation

$$\begin{aligned} \mathcal{E} : \mathrm{D}(\mathbb{H}^N) &\longrightarrow \mathrm{D}_{\leq}(\mathbb{H}^N) \\ \rho &\longmapsto \sum_{a \in I} \mathcal{E}_a \rho \, \mathcal{E}_a^* \end{aligned} \tag{7.40}$$

with operation elements $\mathcal{E}_a \in \mathrm{L}(\mathbb{H}^N)$, where $a \in I$ and I is some index set.

An error operation in an $[\![n, k]\!]_q$ QECC is called an m**-qubit error** operation if all its operation elements are m-local (see Definition 3.21).

To model a set of errors indexed by $a \in I$ which change the states by acting with $\hat{\mathcal{E}}_a \in \mathrm{L}(\mathbb{H}^N)$ and which occur with probabilities $p_a \in [0, 1]$, we choose operation elements in the form

$$\mathcal{E}_a = \sqrt{p_a}\hat{\mathcal{E}}_a ,$$

where the p_a satisfy $\sum_{a \in I} p_a = 1$. The operators $\hat{\mathcal{E}}_a \in \mathrm{L}(\mathbb{H}^N)$ are called **error operators**. They include the 'no error' operator $\hat{\mathcal{E}}_0 = \mathbf{1}$.

We will often write $\mathcal{E} = \{\mathcal{E}_a\}$ to denote the error operation and its operation elements at the same time. Just as we saw in Corollary 3.27 that the operation elements for a given quantum operation are not unique, the operation elements for a given error operation are not unique either as is to be re-confirmed in the following exercise.

Exercise 7.86 Let $\mathcal{E} = \{\mathcal{E}_a\}$ be an error operation, $U \in \mathrm{U}(m)$ a unitary $m \times m$ matrix with matrix elements U_{ab} and

$$\tilde{\mathcal{E}}_a = \sum_{b=1}^{m} U_{ab} \, \mathcal{E}_b \ . \tag{7.41}$$

Show that then $\tilde{\mathcal{E}} = \{\tilde{\mathcal{E}}_a\}$ generates the same error operation, that is,

$$\tilde{\mathcal{E}}(\rho) = \sum_a \tilde{\mathcal{E}}_a \rho \tilde{\mathcal{E}}_a^* = \sum_b \mathcal{E}_b \, \rho \, \mathcal{E}_b^* = \mathcal{E}(\rho) \quad \forall \rho \in \mathrm{D}\big(\mathbb{H}^N\big)$$

and

$$\sum_a \tilde{\mathcal{E}}_a^* \tilde{\mathcal{E}}_a = \sum_b \mathcal{E}_b^* \, \mathcal{E}_b \ .$$

For a solution see Solution 7.86.

To illustrate the naming convention used in connection with certain error operators, let us consider for a moment the case of a single qubit-space $\mathbb{H}$ only by assuming $n = 1$. In $\mathbb{H}$ consider an error operation $\mathcal{E}$ with the error operators $\hat{\mathcal{E}}_\alpha = \sigma_\alpha$ and the following operation elements

$$\mathcal{E}_\alpha = \sqrt{p_\alpha} \sigma_\alpha \underbrace{=}_{(2.74)} \sqrt{p_\alpha} \sigma_\alpha^* = \mathcal{E}_\alpha^* \quad \text{for } \alpha \in \{0, \ldots, 3\}, \tag{7.42}$$

where we require $p_\alpha \geq 0$. Then we find

$$\sum_{\alpha=0}^{3} \mathcal{E}_\alpha^* \, \mathcal{E}_\alpha \underbrace{=}_{(7.42)} \sum_{\alpha=0}^{3} p_\alpha \sigma_\alpha^2 \underbrace{=}_{(2.76)} \left(\sum_{\alpha=0}^{3} p_\alpha \right) \mathbf{1} \ .$$

It follows from Corollary 3.25 that

$$\sum_\alpha p_\alpha = 1 \quad \Leftrightarrow \quad \mathrm{tr}\,(\mathcal{E}(\rho)) = 1 \quad \forall \rho \in \mathrm{D}(\mathbb{H}) \ .$$

The error operation $\mathcal{E}$ generated by these error operators transforms the density operator of a pure state $\rho_\psi = |\psi\rangle\langle\psi| \in \mathrm{D}(\mathbb{H})$ to the corrupted density operator

$$\mathcal{E}(\rho_\psi) = \sum_{\alpha=0}^{3} \mathcal{E}_\alpha \, \rho_\psi \, \mathcal{E}_\alpha^* \underbrace{=}_{(7.42)} \sum_{\alpha=0}^{3} p_\alpha \sigma_\alpha |\psi\rangle\langle\psi| \sigma_\alpha^* .$$

Therefore, the error operation $\{\mathcal{E}_\alpha = \sqrt{p_\alpha}\sigma_\alpha \mid \alpha \in \{0, \ldots, 3\}\}$, leaves the state $|\psi\rangle$ unchanged (no error) with probability p_0 and transforms the state $|\psi\rangle$ to a corrupted state $\sigma_j|\psi\rangle$ for $j \in \{1, 2, 3\}$ with respective probabilities p_j.

Due to their action on single qubits, the error operators σ_j for $j \in \{1, 2, 3\}$ have been given the following intuitive names.

The **bit-flip**   error $X = \sigma_x$ 'flips' the qubits since it acts as

$$X(a|0\rangle + b|1\rangle) = a|1\rangle + b|0\rangle$$

The **phase-flip**   error $Z = \sigma_z$ changes the phase since it acts as

$$Z(a|0\rangle + b|1\rangle) = a|0\rangle - b|1\rangle$$

The **bit-phase-flip**   error $Y = \sigma_y$ flips the qubits and changes their phase since it acts as

$$Y(a|0\rangle + b|1\rangle) = i(a|1\rangle - b|0\rangle) .$$

Returning to the general case of $\mathbb{H}^N = \mathbb{H}^{\otimes n}$ with $n > 1$, a one-qubit error operation $\mathcal{E}$ affecting only qubits in the factor space $\mathbb{H}_j$ can be comprised of the error operators

$$\hat{\mathcal{E}}_{(\alpha,j)} = \Sigma_\alpha^j = 1^{\otimes n-1-j} \otimes \sigma_\alpha \otimes 1^{\otimes j} \quad \text{for } \alpha \in \{0, \ldots, z\} . \qquad (7.43)$$

Here we use the observable Σ_α^j introduced in Definition 5.35, where j denotes the j-th qubit counted from the right starting with $j = 0$ and $\alpha \in \{0, \ldots, z\}$ with $\sigma_0 = 1$. This counting convention is chosen to remain consistent with our previous counting convention (3.18) introduced in Definition 3.5 of the computational basis, where it was adapted to mimic the binary representation of natural numbers.

One-qubit error operations of the form

$$\mathcal{E} = \left\{ \mathcal{E}_{(\alpha,j)} = \sqrt{p_\alpha}\Sigma_\alpha^j \mid \alpha \in \{0, \ldots, z\}\right\} ,$$

transform a state

$$|\Psi\rangle = |\psi_{n-1}\rangle \otimes \cdots \otimes |\psi_{j+1}\rangle \otimes \ |\psi_j\rangle \ \otimes|\psi_{j-1}\rangle \otimes \cdots \otimes |\psi_0\rangle$$

to states

$$\Sigma_\alpha^j|\Psi\rangle = |\psi_{n-1}\rangle \otimes \cdots \otimes |\psi_{j+1}\rangle \otimes \sigma_\alpha|\psi_j\rangle \otimes|\psi_{j-1}\rangle \otimes \cdots \otimes |\psi_0\rangle$$

with respective probabilities p_α. Error operators of the form

$$\hat{\mathcal{E}}_{(\alpha_1,j_1),\ldots,(\alpha_l,j_l)} = \Sigma_{\alpha_1}^{j_1} \cdots \Sigma_{\alpha_l}^{j_l}$$

describe l-qubit error operations. They affect up to l qubits when they occur with a probability $p_{(\alpha_1,j_1),...,(\alpha_l,j_l)}$. An example of a 2 qubit-error operation is given by error operators of the form

$$\hat{\mathcal{E}}_{(\alpha_1,j_1),(\alpha_2,j_2)} = \Sigma_{\alpha_1}^{j_1}\Sigma_{\alpha_2}^{j_2} = \mathbf{1}^{\otimes n-1-j_2} \otimes \sigma_{\alpha_2} \otimes \mathbf{1}^{\otimes j_2-1-j_1} \otimes \sigma_{\alpha_1} \otimes \mathbf{1}^{\otimes j_1} ,$$

where we have assumed $j_1 \leq j_2$ in displaying the rightmost equation. These error operators have non-trivial action with probability $p_{(\alpha_1,j_1),(\alpha_2,j_2)}$ in at most two of the factor spaces $\mathbb{H}_{j_1}$ and $\mathbb{H}_{j_2}$ if $\alpha_1 \neq 0 \neq \alpha_2$.

Example 7.17 Consider the two QECCs $[\![3,1]\!]_{q_1}$ and $[\![3,1]\!]_{q_2}$ from Example 7.14. The $[\![3,1]\!]_{q_1}$ code has the code space

$$\underbrace{\mathbb{H}^{[\![3,1]\!]_{q_1}}}_{(7.23)} = \text{Span}\{|000\rangle, |111\rangle\} = \text{Span}\{|\Psi_0\rangle, |\Psi_1\rangle\} .$$

The one-qubit bit-flip error operation $\mathcal{E}^{\text{bf}} = \left\{ \mathcal{E}^{\text{bf}}_{st} \mid s,t \in \{0,1\} \right\}$ is generated by the error operators

$$
\begin{aligned}
\hat{\mathcal{E}}^{\text{bf}}_{00} &= \mathbf{1} \otimes \mathbf{1} \otimes \mathbf{1} & \hat{\mathcal{E}}^{\text{bf}}_{01} &= \mathbf{1} \otimes \mathbf{1} \otimes \sigma_x \\
\hat{\mathcal{E}}^{\text{bf}}_{10} &= \mathbf{1} \otimes \sigma_x \otimes \mathbf{1} & \hat{\mathcal{E}}^{\text{bf}}_{11} &= \sigma_x \otimes \mathbf{1} \otimes \mathbf{1} .
\end{aligned}
\tag{7.44}
$$

Recalling that $\sigma_x|0\rangle = |1\rangle$ and $\sigma_x|1\rangle = |0\rangle$, we see that these errors transform the $[\![3,1]\!]_{q_1}$-basis codewords as follows

$$
\begin{aligned}
\hat{\mathcal{E}}^{\text{bf}}_{01}|\Psi_0\rangle &= \hat{\mathcal{E}}^{\text{bf}}_{01}|000\rangle = |001\rangle & \hat{\mathcal{E}}^{\text{bf}}_{01}|\Psi_1\rangle &= \hat{\mathcal{E}}^{\text{bf}}_{01}|111\rangle = |110\rangle \\
\hat{\mathcal{E}}^{\text{bf}}_{10}|\Psi_0\rangle &= \hat{\mathcal{E}}^{\text{bf}}_{10}|000\rangle = |010\rangle & \hat{\mathcal{E}}^{\text{bf}}_{10}|\Psi_1\rangle &= \hat{\mathcal{E}}^{\text{bf}}_{10}|111\rangle = |101\rangle \\
\hat{\mathcal{E}}^{\text{bf}}_{11}|\Psi_0\rangle &= \hat{\mathcal{E}}^{\text{bf}}_{11}|000\rangle = |100\rangle & \hat{\mathcal{E}}^{\text{bf}}_{11}|\Psi_1\rangle &= \hat{\mathcal{E}}^{\text{bf}}_{11}|111\rangle = |011\rangle .
\end{aligned}
\tag{7.45}
$$

The one-qubit phase-flip error operation $\mathcal{E}^{\text{pf}} = \left\{ \mathcal{E}^{\text{pf}}_{st} \mid s,t \in \{0,1\} \right\}$ is given by the error operators

$$
\begin{aligned}
\hat{\mathcal{E}}^{\text{pf}}_{00} &= \mathbf{1} \otimes \mathbf{1} \otimes \mathbf{1} & \hat{\mathcal{E}}^{\text{pf}}_{01} &= \mathbf{1} \otimes \mathbf{1} \otimes \sigma_z \\
\hat{\mathcal{E}}^{\text{pf}}_{10} &= \mathbf{1} \otimes \sigma_z \otimes \mathbf{1} & \hat{\mathcal{E}}^{\text{pf}}_{11} &= \sigma_z \otimes \mathbf{1} \otimes \mathbf{1} .
\end{aligned}
\tag{7.46}
$$

Using that $\sigma_z|0\rangle = |0\rangle$ and $\sigma_z|1\rangle = -|1\rangle$, we see that their action on the $[\![3,1]\!]_{q_1}$-basis codewords is quite simple

$$
\begin{aligned}
\hat{\mathcal{E}}^{\text{pf}}_{st}|\Psi_0\rangle &= \hat{\mathcal{E}}^{\text{pf}}_{st}|000\rangle = |000\rangle = |\Psi_0\rangle \\
\hat{\mathcal{E}}^{\text{pf}}_{st}|\Psi_1\rangle &= \hat{\mathcal{E}}^{\text{pf}}_{st}|111\rangle = -|111\rangle = -|\Psi_1\rangle .
\end{aligned}
\tag{7.47}
$$

The $[\![3,1]\!]_{q_2}$ code has the code space

$$\mathbb{H}^{[\![3,1]\!]_{q2}} \underbrace{=}_{(7.28)} \text{Span}\{|+++\rangle, |---\rangle\} = \text{Span}\{|\Phi_0\rangle, |\Phi_1\rangle\}.$$

In the $[\![3,1]\!]_{q2}$ code the situation between bit-flip $\mathcal{E}^{\text{bf}}$ and phase-flip $\mathcal{E}^{\text{pf}}$ is reversed, that is, we have

$$\hat{\mathcal{E}}_{st}^{\text{bf}}|\Phi_0\rangle = \hat{\mathcal{E}}_{st}^{\text{bf}}|+++\rangle = |+++\rangle = |\Phi_0\rangle$$
$$\hat{\mathcal{E}}_{st}^{\text{bf}}|\Phi_1\rangle = \hat{\mathcal{E}}_{st}^{\text{bf}}|---\rangle = -|---\rangle = -|\Phi_1\rangle \tag{7.48}$$

but

$$\hat{\mathcal{E}}_{01}^{\text{pf}}|\Phi_0\rangle = \hat{\mathcal{E}}_{01}^{\text{pf}}|+++\rangle = |++-\rangle \qquad \hat{\mathcal{E}}_{01}^{\text{pf}}|\Phi_1\rangle = \hat{\mathcal{E}}_{01}^{\text{pf}}|---\rangle = |--+\rangle$$
$$\hat{\mathcal{E}}_{10}^{\text{pf}}|\Phi_0\rangle = \hat{\mathcal{E}}_{10}^{\text{pf}}|+++\rangle = |+-+\rangle \qquad \hat{\mathcal{E}}_{10}^{\text{pf}}|\Phi_1\rangle = \hat{\mathcal{E}}_{10}^{\text{pf}}|---\rangle = |-+-\rangle$$
$$\hat{\mathcal{E}}_{11}^{\text{pf}}|\Phi_0\rangle = \hat{\mathcal{E}}_{11}^{\text{pf}}|+++\rangle = |-++\rangle \qquad \hat{\mathcal{E}}_{11}^{\text{pf}}|\Phi_1\rangle = \hat{\mathcal{E}}_{11}^{\text{pf}}|---\rangle = |+--\rangle. \tag{7.49}$$

Two qubit-flips are generated by error operators of the form

$$\hat{\mathcal{E}}_{l,j}^{\text{2bf}} = \Sigma_x^l \Sigma_x^j$$

and we have, for example,

$$\hat{\mathcal{E}}_{2,0}^{\text{2bf}}|\Psi_0\rangle = \hat{\mathcal{E}}_{2,0}^{\text{2bf}}|000\rangle = |101\rangle. \tag{7.50}$$

Now that we have defined the type of possible errors we will consider in our theory of error correction, we need to identify means to rectify them. We begin this by defining the recovery or correction operation.

Definition 7.18 Let $\mathbb{H}^N$ be the encoding space of a QECC with code $\mathbb{H}^{C_q}$ and $\mathcal{E}$ an error operation in the code $\mathbb{H}^{C_q}$. A **recovery operation** is a trace-preserving quantum operation

$$R : D(\mathbb{H}^N) \longrightarrow D(\mathbb{H}^N)$$
$$\rho \longmapsto \Sigma_r R_r \rho R_r^*.$$

The operation elements $R_r \in L(\mathbb{H}^N)$ are called **recovery operators**. A recovery operation R is said to **recover or correct an error operation** $\mathcal{E}$ in $\mathbb{H}^{C_q}$ if

$$R\left(\frac{\mathcal{E}(\rho)}{\text{tr}(\mathcal{E}(\rho))}\right) = \rho \quad \forall \rho \in D(\mathbb{H}^{C_q}). \tag{7.51}$$

An error operation $\mathcal{E}$ in the code $\mathbb{H}^{C_q}$ is said to be **correctable** if a recovery operation for it exists.

Just like for the error operation, we will often write $R = \{R_r\}$ to denote the recovery operation and its recovery operators at the same time.

The condition (7.51) for an error operation $\mathcal{E}$ to be corrected by a recovery operation R is quite intuitive. It states that any state $\rho \in \mathbb{H}^{C_q}$ in the code which is corrupted by $\mathcal{E}$ is recovered by R. It also gives a first condition for an error to be correctable as the next lemma shows.

Lemma 7.19 *For an error operation $\mathcal{E}$ in the code $\mathbb{H}^{C_q}$ of a QECC to be correctable it is necessary that it satisfies*

$$\mathrm{tr}\left(\mathcal{E}(\rho)\right) = const = \kappa_{\mathcal{E}} \in [0,1] \quad \forall \rho \in D\left(\mathbb{H}^{C_q}\right).$$

Proof The claim follows from the result in the following exercise.

Exercise 7.87 Let $\mathbb{H}$ be a HILBERT space and S and T quantum operations on $D(\mathbb{H})$. Show that then

$$T\left(\frac{S(\rho)}{\mathrm{tr}\left(S(\rho)\right)}\right) = \rho \quad \forall \rho \in D(\mathbb{H}) \quad \Rightarrow \quad \mathrm{tr}\left(S(\rho)\right) = const \quad \forall \rho \in D(\mathbb{H}).$$

For a solution see Solution 7.87.

$\square$

As it stands, however, (7.51) is not so useful in determining if an error operation $\mathcal{E}$ is recovered by R since we have to check it for all ρ. Fortunately, there exist equivalent conditions for an error operation to be correctable that can be stated in terms of the error and recovery operators.

The first such characterization given in Theorem 7.20 is in the form of necessary and sufficient conditions which the operation elements $\mathcal{E}_a$ of an error operation $\mathcal{E}$ and recovery operators R_r have to satisfy for R to correct $\mathcal{E}$. The second formulation given in Theorem 7.22 states such necessary and sufficient conditions in terms of the operation elements $\mathcal{E}_a$ and the code alone.

Theorem 7.20 *Let $\mathcal{E} = \left\{\mathcal{E}_a \mid a \in \{1,\ldots,l\}\right\} \subset L\left(\mathbb{H}^N\right)$ be an error operation in the code $\mathbb{H}^{C_q}$ of a QECC, and let $R = \left\{R_r \mid r \in \{1,\ldots,m\}\right\} \subset L\left(\mathbb{H}^N\right)$ be a recovery operation. Then the following equivalence holds.*

$$\left.\begin{array}{l} \textit{The error operation}\\ \mathcal{E} = \{\mathcal{E}_a\} \textit{ in } \mathbb{H}^{\mathbb{C}^q}\\ \textit{is correctable by } R \end{array}\right\} \quad \Leftrightarrow \quad \left\{\begin{array}{l} \exists Z \in \mathrm{Mat}(m \times l, \mathbb{C}) \quad \textit{such} \quad \textit{that}\\ \forall R_r \in R \textit{ and } \mathcal{E}_a \in \mathcal{E}:\\[2mm] \left(R_r \mathcal{E}_a - Z_{ra}\mathbf{1}\right)\big|_{\mathbb{H}^{\mathbb{C}^q}} = 0, \end{array}\right.$$

and, if these statements are true, we also have

$$\mathrm{tr}\left(Z^*Z\right) = \mathrm{tr}\left(\mathcal{E}(\rho)\right) \quad \forall \rho \in D\left(\mathbb{H}^{\mathbb{C}^q}\right). \tag{7.52}$$

Proof To prove $\Rightarrow$, let $\mathcal{E}$ be correctable. Then we know from Lemma 7.19 that

$$\mathrm{tr}\left(\mathcal{E}(\rho)\right) = \kappa_{\mathcal{E}} \quad \forall \rho \in D\left(\mathbb{H}^{\mathbb{C}^q}\right). \tag{7.53}$$

Since R is by definition trace-preserving, that is, $\mathrm{tr}\left(R(\rho)\right) = 1$ for all ρ, Corollary 3.25 implies that

$$\frac{1}{\kappa_{\mathcal{E}}} \sum_a \mathcal{E}_a^* \mathcal{E}_a = \mathbf{1} = \sum_r R_r^* R_r. \tag{7.54}$$

Moreover, using the fact that $\mathcal{E}$ and R are quantum operations with operator-sum representations, we have for any $\rho \in D\left(\mathbb{H}^{\mathbb{C}^q}\right)$

$$\rho \underbrace{=}_{(7.51)} R\left(\frac{\mathcal{E}(\rho)}{\kappa_{\mathcal{E}}}\right) \underbrace{=}_{(3.97)} \sum_r R_r \frac{\mathcal{E}(\rho)}{\kappa_{\mathcal{E}}} R_r^* = \frac{1}{\kappa_{\mathcal{E}}} \sum_r R_r \mathcal{E}(\rho) R_r^* \underbrace{=}_{(3.97)} \frac{1}{\kappa_{\mathcal{E}}} \sum_{r,a} R_r \mathcal{E}_a \rho \mathcal{E}_a^* R_r^*$$

such that

$$\kappa_{\mathcal{E}} \rho = \sum_{r,a} R_r \mathcal{E}_a \rho \mathcal{E}_a^* R_r^* \quad \forall \rho \in D\left(\mathbb{H}^{\mathbb{C}^q}\right).$$

For $\rho = |\Psi\rangle\langle\Psi|$ with $|\Psi\rangle \in \mathbb{H}^{\mathbb{C}^q}$ such that $||\Psi|| = 1$ this becomes

$$\kappa_{\mathcal{E}} |\Psi\rangle\langle\Psi| = \sum_{r,a} R_r \mathcal{E}_a |\Psi\rangle\langle\Psi| \mathcal{E}_a^* R_r^* \underbrace{=}_{(2.47)} \sum_{r,a} R_r \mathcal{E}_a |\Psi\rangle\langle\Psi| (R_r \mathcal{E}_a)^* \tag{7.55}$$

such that

$$\begin{aligned} \kappa_{\mathcal{E}} &= \kappa_{\mathcal{E}} \underbrace{||\Psi||^4}_{=1} \underbrace{=}_{(2.5)} \kappa_{\mathcal{E}} \langle\Psi|\Psi\rangle\langle\Psi|\Psi\rangle \underbrace{=}_{(7.55)} \sum_{r,a} \langle\Psi| R_r \mathcal{E}_a \Psi\rangle\langle\Psi|(R_r \mathcal{E}_a)^* \Psi\rangle\\ &\underbrace{=}_{(2.30),(2.31)} \sum_{r,a} \langle\Psi| R_r \mathcal{E}_a \Psi\rangle\langle R_r \mathcal{E}_a \Psi|\Psi\rangle\\ &\underbrace{=}_{(2.1)} \sum_{r,a} |\langle\Psi| R_r \mathcal{E}_a \Psi\rangle|^2. \end{aligned} \tag{7.56}$$

Exercise 7.88 Show that for any HILBERT space $\mathbb{H}$, operator $A \in B(\mathbb{H})$ and vector $|\psi\rangle \in \mathbb{H}$ with $||\psi|| = 1$ one has

$$\left|\left|\left(A - \langle\psi|A\psi\rangle\right)|\psi\rangle\right|\right|^2 = \langle\psi|A^*A\psi\rangle - |\langle\psi|A\psi\rangle|^2 \,. \tag{7.57}$$

For a solution see Solution 7.88.

Using the result of Exercise 7.88 in our case for $\mathbb{H} = \mathbb{H}^{C_q}$, the operator $A = R_r\,\mathcal{E}_a$ and the vector $|\psi\rangle = |\Psi\rangle \in \mathbb{H}^{C_q}$ we obtain

$$\left|\left|\left(R_r\,\mathcal{E}_a - \langle\Psi|R_r\,\mathcal{E}_a\,\Psi\rangle\right)|\Psi\rangle\right|\right|^2 \underset{(7.57)}{=} \langle\Psi|(R_r\,\mathcal{E}_a)^* R_r\,\mathcal{E}_a\,\Psi\rangle - |\langle\Psi|R_r\,\mathcal{E}_a\,\Psi\rangle|^2$$

$$\underset{(2.47)}{=} \langle\Psi|\,\mathcal{E}_a^* R_r^* R_r\,\mathcal{E}_a\,\Psi\rangle - |\langle\Psi|R_r\,\mathcal{E}_a\,\Psi\rangle|^2$$

such that

$$\sum_{r,a}\left|\left|\left(R_r\,\mathcal{E}_a - \langle\Psi|R_r\,\mathcal{E}_a\,\Psi\rangle\right)|\Psi\rangle\right|\right|^2 = \sum_{r,a}\langle\Psi|\,\mathcal{E}_a^* R_r^* R_r\,\mathcal{E}_a\,\Psi\rangle - \sum_{r,a}|\langle\Psi|R_r\,\mathcal{E}_a\,\Psi\rangle|^2$$

$$\underset{(7.56)}{=} \langle\Psi|\sum_a\mathcal{E}_a^*\left(\sum_r R_r^* R_r\right)\mathcal{E}_a\,\Psi\rangle - \kappa_\mathcal{E}$$

$$\underset{(7.54)}{=} 0\,.$$

As the left side is a sum of non-negative terms, this can only be true if each term vanishes, and it follows that for every $R_r \in R$, $\mathcal{E}_a \in \mathcal{E}$ and $|\Psi\rangle \in \mathbb{H}$ with $||\Psi|| = 1$ we have

$$R_r\,\mathcal{E}_a\,|\Psi\rangle = \langle\Psi|R_r\,\mathcal{E}_a\,\Psi\rangle)|\Psi\rangle\,. \tag{7.58}$$

Exercise 7.89 Let $\mathbb{H}$ be a HILBERT space and $A \in B(\mathbb{H})$. Show that then

$$A|\psi\rangle = \langle\psi|A\psi\rangle|\psi\rangle \quad \forall|\psi\rangle \in S_\mathbb{H}^1 \quad \Leftrightarrow \quad \exists a \in \mathbb{C}: \; A = a\mathbf{1}\,,$$

where $S_\mathbb{H}^1$ denotes the unit sphere (vectors of norm 1) in $\mathbb{H}$.

For a solution see Solution 7.89.

Using (7.58), we see that the statement in Exercise 7.89 then implies for $A = R_r\,\mathcal{E}_a$ that for each $R_r \in R$ and $\mathcal{E}_a \in \mathcal{E}$ there exists a $Z_{ra} \in \mathbb{C}$ such that for all $|\Psi\rangle \in \mathbb{H}^{C_q}$

$$R_r \, \mathcal{E}_a \, |\Psi\rangle = Z_{ra}|\Psi\rangle \, . \tag{7.59}$$

Therefore, we have

$$|Z_{ra}|^2 \, ||\Psi||^2 \underbrace{=}_{(2.7)} \, ||Z_{ra}|\Psi\rangle||^2 \underbrace{=}_{(7.59)} ||R_r \, \mathcal{E}_a \, |\psi\rangle||^2 \underbrace{=}_{(2.5)} \langle R_r \, \mathcal{E}_a \, \Psi | R_r \, \mathcal{E}_a \, \Psi \rangle$$

$$\underbrace{=}_{(2.31),(2.30)} \langle \Psi | \, \mathcal{E}_a^* \, R_r^* \, R_r \, \mathcal{E}_a \, \Psi \rangle \, , \tag{7.60}$$

and in summing over r and a, we obtain for any $|\Psi\rangle \neq 0$

$$\sum_{r=1}^{m} \sum_{a=1}^{l} |Z_{ra}|^2 \underbrace{=}_{(7.60)} \frac{1}{||\Psi||^2} \langle \Psi | \sum_{a=1}^{l} \mathcal{E}_a^* \Big(\sum_{r=1}^{m} R_r^* \, R_r \Big) \mathcal{E}_a \, \Psi \rangle \underbrace{=}_{(7.54)} \kappa_{\mathcal{E}} \, . \tag{7.61}$$

On the other hand, we have

$$\mathrm{tr} \, (Z^* Z) \underbrace{=}_{(2.57)} \sum_a (Z^* Z)_{aa} = \sum_{r,a} Z_{ar}^* Z_{ra} \underbrace{=}_{(2.34)} \sum_{r,a} \overline{Z}_{ra} Z_{ra} = \sum_{r,a} |Z_{ra}|^2 \underbrace{=}_{(7.61)} \kappa_{\mathcal{E}}$$

$$\underbrace{=}_{(7.53)} \mathrm{tr} \, (\mathcal{E}(\rho))$$

for all $\rho \in D(\mathbb{H}^{C_q})$, which completes the proof of $\Rightarrow$ and the claim (7.52).

To show $\Leftarrow$, let each $R_r \in R$ and $\mathcal{E}_a \in \mathcal{E}$ be such that there exists a $Z_{ra} \in \mathbb{C}$ so that for all $|\Psi\rangle \in \mathbb{H}^{C_q}$

$$R_r \, \mathcal{E}_a \, |\Psi\rangle = Z_{ra}|\Psi\rangle \, . \tag{7.62}$$

Then it follows that for any $|\Phi\rangle \in \mathbb{H}^{C_q}$

$$\langle \Phi | \, \mathcal{E}_a^* \, R_r^* \underbrace{=}_{(2.47)} \langle \Phi | (R_r \, \mathcal{E}_a)^* \underbrace{=}_{(2.33)} \langle R_r \, \mathcal{E}_a \, \Phi | \underbrace{=}_{(7.62)} \langle Z_{ra} \Phi | \underbrace{=}_{(2.33),(2.32)} \overline{Z}_{ra} \langle \Phi | \, .$$

$$\tag{7.63}$$

Since R is by definition trace-preserving, Corollary 3.25 implies that $\sum_r R_r^* \, R_r = \mathbf{1}$ and thus

$$\sum_{r,a} |Z_{ra}|^2 \langle \Phi | \Psi \rangle \underbrace{=}_{(7.62),(7.63)} \langle \Phi | \sum_a \mathcal{E}_a^* \Big(\sum_r R_r^* \, R_r \Big) \mathcal{E}_a \, \Psi \rangle \underbrace{=}_{(2.47)} \langle \Phi | \sum_a \mathcal{E}_a^* \, \mathcal{E}_a \, \Psi \rangle \, .$$

As this holds for any $|\Phi\rangle, |\Psi\rangle \in \mathbb{H}^{C_q}$, it follows that $\sum_a \mathcal{E}_a^* \, \mathcal{E}_a = \Big(\sum_{r,a} |Z_{ra}|^2 \Big) \mathbf{1}$, and Corollary 3.25 implies that then

$$\mathrm{tr}\left(\mathcal{E}(\rho)\right) = \Big(\sum_{r,a} |Z_{ra}|^2 \Big) \quad \forall \rho \in \mathrm{D}\big(\mathbb{H}^{C_q}\big). \tag{7.64}$$

Now, let $\rho \in \mathrm{D}\big(\mathbb{H}^{C_q}\big)$. Then we know from Theorem 2.24 that there exist $p_j \in [0,1]$ with $\sum_j p_j = 1$ and an ONB $\{|\Theta_j\rangle\}$ in $\mathbb{H}^{C_q}$ such that

$$\rho = \sum_j p_j |\Theta_j\rangle\langle\Theta_j|. \tag{7.65}$$

Therefore, we have

$$
\begin{aligned}
\mathrm{R}\left(\frac{\mathcal{E}(\rho)}{\mathrm{tr}\left(\mathcal{E}(\rho)\right)}\right) &\underset{(3.97)}{=} \sum_r \mathrm{R}_r \frac{\sum_a \mathcal{E}_a \rho \, \mathcal{E}_a^*}{\mathrm{tr}\left(\mathcal{E}(\rho)\right)} \mathrm{R}_r^* = \frac{1}{\mathrm{tr}\left(\mathcal{E}(\rho)\right)} \sum_{r,a} \mathrm{R}_r \, \mathcal{E}_a \rho \, \mathcal{E}_a^* \, \mathrm{R}_r^* \\
&\underset{(7.65)}{=} \frac{1}{\mathrm{tr}\left(\mathcal{E}(\rho)\right)} \sum_{r,a,j} p_j \, \mathrm{R}_r \, \mathcal{E}_a \, |\Theta_j\rangle\langle\Theta_j| \, \mathcal{E}_a^* \, \mathrm{R}_r^* \\
&\underset{(7.62),(7.63)}{=} \frac{1}{\mathrm{tr}\left(\mathcal{E}(\rho)\right)} \sum_{r,a,j} p_j \, |Z_{ra}|^2 \, |\Theta_j\rangle\langle\Theta_j| = \frac{\sum_{r,a} |Z_{ra}|^2}{\mathrm{tr}\left(\mathcal{E}(\rho)\right)} \sum_j p_j |\Theta_j\rangle\langle\Theta_j| \\
&\underset{(7.64),(7.65)}{=} \rho\,,
\end{aligned}
$$

and $\mathcal{E}$ is correctable. $\qquad\square$

An almost direct consequence of Theorem 7.20 is that if $\mathcal{E} = \{\mathcal{E}_a\}$ is correctable by R, then the error operation generated by any linear combination of the $\mathcal{E}_a$ is also correctable by R.

Corollary 7.21 *Let* $\hat{\mathcal{E}} = \big\{\hat{\mathcal{E}}_a \,\big|\, a \in \{1,\dots,l\}\big\}$ *and* $\check{\mathcal{E}} = \big\{\check{\mathcal{E}}_s \,\big|\, s \in \{1,\dots,t\}\big\}$ *be two error operations in the code* $\mathbb{H}^{C_q}$ *of a QECC which are both correctable by* $\mathrm{R} = \big\{ \mathrm{R}_r \,\big|\, r \in \{1,\dots,m\}\big\}$. *Furthermore, let* $V \in \mathrm{Mat}(v \times l, \mathbb{C})$ *and* $W \in \mathrm{Mat}(v \times t, \mathbb{C})$, *and let* $\mathcal{E} = \big\{ \mathcal{E}_u \,\big|\, u \in \{1,\dots,v\}\big\}$ *be the error operation in* $\mathbb{H}^{C_q}$ *given by operation elements*

$$\mathcal{E}_u = \sum_{a=1}^{l} V_{au}\hat{\mathcal{E}}_a + \sum_{s=1}^{t} W_{su}\check{\mathcal{E}}_s\,. \tag{7.66}$$

Then $\mathcal{E}$ *is also correctable by* R.

Proof From Theorem 7.20 we know that, since $\hat{\mathcal{E}}$ and $\check{\mathcal{E}}$ are correctable by R, there exist $\hat{Z} \in \mathrm{Mat}(m \times l, \mathbb{C})$ and $\check{Z} \in \mathrm{Mat}(m \times t, \mathbb{C})$ such that for any R_r, $\hat{\mathcal{E}}_a$, $\check{\mathcal{E}}_s$ and $|\Psi\rangle \in \mathbb{H}^{C_q}$

$$\mathrm{R}_r \, \hat{\mathcal{E}}_a|\Psi\rangle = \hat{Z}_{ra}|\Psi\rangle \quad \text{and} \quad \mathrm{R}_r \, \check{\mathcal{E}}_s|\Psi\rangle = \check{Z}_{rs}|\Psi\rangle\,. \tag{7.67}$$

Therefore, we have for every $R_r \in R$, $\mathcal{E}_u \in \mathcal{E}$ and $|\Psi\rangle \in \mathbb{H}^{C_q}$

$$R_r \mathcal{E}_u |\Psi\rangle \underset{(7.66)}{=} \sum_a V_{au} R_r \hat{\mathcal{E}}_a |\Psi\rangle + \sum_s W_{su} R_r \check{\mathcal{E}}_s |\Psi\rangle$$

$$\underset{(7.67)}{=} \sum_a V_{au} \hat{Z}_{ra} |\Psi\rangle + \sum_s W_{su} \check{Z}_{rs} |\Psi\rangle = \left(\hat{Z}V + \check{Z}W\right)_{ru} |\Psi\rangle$$

$$= Z_{ru} |\Psi\rangle,$$

and it follows from Theorem 7.20 that $\mathcal{E}$ is correctable by R. $\qquad\square$

Note that since we would like $\mathcal{E}$ to be an error operation, the matrices V and W in (7.66) cannot be completely arbitrary as we still want to ensure that $\text{tr}\,(\mathcal{E}(\rho)) \leq 1$. As long as this is assured, however, any linear combination of operation elements generates an error operation which remains correctable by R.

The result of Theorem 7.20 allows us to determine from the error and recovery operators if the error operation is correctable. However, it suffers from the shortcoming that we need to know the recovery operators R_r in order to determine whether $\mathcal{E}$ is correctable. The next theorem alleviates this in that it provides necessary and sufficient conditions for an error operation $\mathcal{E}$ in $\mathbb{H}^{C_q}$ to be correctable, where the conditions are given in terms of its operation elements $\mathcal{E}_a$ and the code $\mathbb{H}^{C_q}$.

Theorem 7.22 Let $\mathcal{E} = \left\{ \mathcal{E}_a \mid a \in \{1,\dots,l\} \right\} \in L(\mathbb{H}^N)$ be an error operation in the code $\mathbb{H}^{C_q}$ of an $[n,k]_c$ QECC, and let $\left\{ |\Psi_w\rangle \mid w \in \{0,\dots,2^k - 1\} \right\}$ be the ONB of basis codewords in $\mathbb{H}^{C_q}$. Then the following equivalence holds.

$$\left.\begin{array}{l} \textit{The error operation} \\ \mathcal{E} = \{\mathcal{E}_a\} \textit{ in } \mathbb{H}^{C_q} \\ \textit{is correctable} \end{array}\right\} \quad \Leftrightarrow \quad \left\{\begin{array}{l} \exists C \in \text{Mat}(l \times l, \mathbb{C}) \textit{ such that } C^* = C \textit{ and} \\ \forall \mathcal{E}_a, \mathcal{E}_b \in \mathcal{E} \textit{ and } |\Psi_x\rangle, |\Psi_y\rangle \in \{|\Psi_w\rangle\} : \\[4pt] \qquad \langle \mathcal{E}_a \Psi_x | \mathcal{E}_b \Psi_y \rangle = C_{ab} \delta_{xy}, \qquad (7.68) \end{array}\right.$$

and, if these statements are true, we also have

$$\text{tr}\,(C) = \text{tr}\,(\mathcal{E}(\rho)) \quad \forall \rho \in D(\mathbb{H}^{C_q}).$$

Proof To prove $\Rightarrow$, let $\mathcal{E}$ be correctable. Then there exists a recovery operation $R = \left\{ R_r \in L(\mathbb{H}^N) \mid r \in \{1,\dots,m\} \right\}$, which is a trace-preserving quantum operation satisfying

$$\sum_{r=1}^m R_r^* R_r = \mathbf{1} \qquad (7.69)$$

and which corrects $\mathcal{E}$. Theorem 7.20 implies that there exist $Z_{ra} \in \mathbb{C}$ satisfying (7.52) such that for all $|\Psi\rangle \in \mathbb{H}^{C_q}$

$$R_r \, \mathcal{E}_a \, |\Psi\rangle = Z_{ra} |\Psi\rangle \, . \tag{7.70}$$

Hence, we have for any basis codewords $|\Psi_x\rangle, |\Psi_y\rangle \in \mathbb{H}^{C_q}$

$$\langle \mathcal{E}_a \, \Psi_x | \, \mathcal{E}_b \, \Psi_y \rangle \underset{(7.69)}{=} \langle \mathcal{E}_a \, \Psi_x | \Big(\sum_{r=1}^{m} R_r^* R_r \Big) \mathcal{E}_b \, \Psi_y \rangle = \sum_{r=1}^{m} \langle \mathcal{E}_a \, \Psi_x | \, R_r^* R_r \, \mathcal{E}_b \, \Psi_y \rangle$$

$$\underset{(2.30),(2.31)}{=} \sum_{r=1}^{m} \langle R_r \, \mathcal{E}_a \, \Psi_x | \, R_r \, \mathcal{E}_b \, \Psi_y \rangle \underset{(7.70)}{=} \sum_{r=1}^{m} \langle Z_{ra} \Psi_x | Z_{rb} \Psi_y \rangle$$

$$\underset{(2.4),(2.6)}{=} \sum_{r=1}^{m} \overline{Z_{ra}} Z_{rb} \underbrace{\langle \Psi_x | \Psi_y \rangle}_{= \delta_{xy}} = C_{ab} \delta_{xy} \, ,$$

where in the last equation we used Lemma 7.13, and where $C_{ab} = \sum_r \overline{Z_{ra}} Z_{rb}$ is such that

$$C_{ab}^* \underset{(2.34)}{=} \overline{C_{ba}} = \sum_{r=1}^{m} \overline{\overline{Z_{rb}} Z_{ra}} = \sum_{r=1}^{m} \overline{Z_{ra}} Z_{rb} = C_{ab} \, .$$

Moreover, we obtain

$$\text{tr} \, (C) \underset{(2.57)}{=} \sum_{a=1}^{l} C_{aa} = \sum_{a=1}^{l} \sum_{r=1}^{m} \overline{Z_{ra}} Z_{ra} = \sum_{a=1}^{l} \sum_{r=1}^{m} |Z_{ra}|^2 \underset{(7.52)}{=} \text{tr} \, (\mathcal{E}(\rho)) \, . \tag{7.71}$$

To prove $\Leftarrow$, let

$$\langle \mathcal{E}_a \, \Psi_x | \, \mathcal{E}_b \, \Psi_y \rangle = C_{ab} \delta_{xy} \, , \tag{7.72}$$

where $C \in \text{Mat}(l \times l, \mathbb{C})$. Then it follows for any $|\Psi_x\rangle$ that

$$C_{ab}^* \underset{(2.34)}{=} \overline{C_{ba}} \underset{(7.72)}{=} \overline{\langle \mathcal{E}_b \, \Psi_x | \, \mathcal{E}_a \, \Psi_x \rangle} \underset{(2.1)}{=} \langle \mathcal{E}_a \, \Psi_x | \, \mathcal{E}_b \, \Psi_y \rangle \underset{(7.72)}{=} C_{ab} \, .$$

Since C is self-adjoint, it can be diagonalized by a unitary $l \times l$ matrix U such that

$$\sum_{b,c=1}^{l} U_{ab}^* C_{bc} U_{cd} = C_a \delta_{ad} \, , \tag{7.73}$$

where the C_a are the eigenvalues of the matrix C. With the matrix U we define the operators

$$\tilde{\mathcal{E}}_a = \sum_{b=1}^{l} U_{ba} \mathcal{E}_b \in L(\mathbb{H}^N) \, , \tag{7.74}$$

and from Corollary 3.27 we know that the operators $\{\widetilde{\mathcal{E}}_a\}$ generate the same quantum operation as the $\{\mathcal{E}_a\}$, that is, we have

$$\widetilde{\mathcal{E}}(\rho) = \mathcal{E}(\rho) \quad \forall \rho \in \mathrm{D}(\mathbb{H}^N) \,. \tag{7.75}$$

Moreover, we find

$$\langle \widetilde{\mathcal{E}}_a \Psi_x | \widetilde{\mathcal{E}}_b \Psi_y \rangle \underbrace{=}_{(7.74)} \Big\langle \sum_c U_{ca}\, \mathcal{E}_c\, \Psi_x \Big| \sum_d U_{db}\, \mathcal{E}_d\, \Psi_y \Big\rangle \underbrace{=}_{(2.4),(2.6)} \sum_{c,d} \overline{U_{ca}} U_{db} \langle \mathcal{E}_c\, \Psi_x | \mathcal{E}_d\, \Psi_y \rangle$$

$$\underbrace{=}_{(7.72),(2.34)} \sum_{c,d} U^*_{ac} C_{cd} U_{db}\, \delta_{xy} \underbrace{=}_{(7.73)} C_a \delta_{ab} \delta_{xy} \,, \tag{7.76}$$

and we can conclude that

$$C_a \underbrace{=}_{(7.76)} \langle \widetilde{\mathcal{E}}_a \Psi_x | \widetilde{\mathcal{E}}_a \Psi_x \rangle \underbrace{\geq}_{(2.2)} 0 \,.$$

Also note that since the trace does not depend on the basis (see Exercise 2.16), we have

$$\sum_{a=1}^{l} C_a = \mathrm{tr}\,(C) \underbrace{=}_{(7.71)} \mathrm{tr}\,(\mathcal{E}(\rho)) \,. \tag{7.77}$$

We divide the index set $\widetilde{I} = \{1, \dots, l\}$ indexing the operation elements $\widetilde{\mathcal{E}}_a$ into two sets

$$I_0 = \{a \in \widetilde{I} \mid C_a = 0\} \quad \text{and} \quad I_> = \{a \in \widetilde{I} \mid C_a > 0\} \,. \tag{7.78}$$

For $a \in I_0$ we have for any basis codeword $|\Psi_x\rangle \in \mathbb{H}^{C_q}$ that $\left\|\widetilde{\mathcal{E}}_a \Psi\right\|^2 = C_a = 0$ and thus

$$\widetilde{\mathcal{E}}_a\big|_{\mathbb{H}^{C_q}} = 0 \quad \forall a \in I_0 \,. \tag{7.79}$$

For every $a \in I_>$ and basis codeword $|\Psi_x\rangle \in \mathbb{H}^{C_q}$ we define a vector

$$|\Phi_{a,x}\rangle = \frac{1}{\sqrt{C_a}} \widetilde{\mathcal{E}}_a |\Psi_x\rangle \,. \tag{7.80}$$

The $|\Phi_{a,x}\rangle \in \mathbb{H}^{C_q} \subset \mathbb{H}^N$ satisfy

$$\langle \Phi_{a,x} | \Phi_{b,y} \rangle \underbrace{=}_{(7.80)} \frac{1}{\sqrt{C_a C_b}} \langle \widetilde{\mathcal{E}}_a \Psi_x | \widetilde{\mathcal{E}}_b \Psi_y \rangle \underbrace{=}_{(7.76)} \frac{C_a \delta_{ab} \delta_{xy}}{\sqrt{C_a C_b}} = \delta_{ab} \delta_{xy} \,. \tag{7.81}$$

Hence, they form a set of orthonormal vectors in $\mathbb{H}^N$ and we can extend them to an ONB of $\mathbb{H}^N$, namely, we can find $|\Xi_p\rangle \in \mathbb{H}^N$ such that $\{|\Phi_{a,y}\rangle, |\Xi_p\rangle\}$ form an ONB in $\mathbb{H}^N$ satisfying in addition to (7.81)

$$\langle \Phi_{a,x}|\Xi_p\rangle = 0 \quad \text{and} \quad \langle \Xi_p|\Xi_s\rangle = \delta_{ps} \,. \tag{7.82}$$

With the $|\Phi_{a,x}\rangle$ and $|\Xi_p\rangle$ we define recovery operators $R_0, R_a \in L(\mathbb{H}^N)$ as

$$R_0 := \sum_p |\Xi_p\rangle\langle\Xi_p| \quad \text{and} \quad \forall a \in I_> : \quad R_a := \sum_x |\Psi_x\rangle\langle\Phi_{a,x}| \,, \tag{7.83}$$

such that

$$\begin{aligned}
R_0^* &= \sum_p (|\Xi_p\rangle\langle\Xi_p|)^* \underbrace{=}_{(2.36)} \sum_p |\Xi_p\rangle\langle\Xi_p| = R_0 \\
R_a^* &= \sum_x (|\Psi_x\rangle\langle\Phi_{a,x}|)^* \underbrace{=}_{(2.36)} \sum_x |\Phi_{a,x}\rangle\langle\Psi_x|
\end{aligned} \tag{7.84}$$

and thus

$$\begin{aligned}
R_0^* R_0 + \sum_{a \in I_>} R_a^* R_a \underbrace{=}_{(7.83),(7.84)} &\sum_{p,s} |\Xi_p\rangle \underbrace{\langle\Xi_p|\Xi_s\rangle}_{=\delta_{ps}} \langle\Xi_s| + \sum_{a \in I_>} \sum_{x,y} |\Phi_{a,x}\rangle \underbrace{\langle\Psi_x|\Psi_y\rangle}_{=\delta_{xy}} \langle\Phi_{a,y}| \\
= &\sum_p |\Xi_p\rangle\langle\Xi_p| + \sum_{a \in I_>} \sum_x |\Phi_{a,x}\rangle\langle\Phi_{a,x}| \\
= &\quad \mathbf{1}^{\mathbb{H}^N}
\end{aligned}$$

since $\{|\Phi_{a,x}\rangle, |\Xi_p\rangle\}$ is an ONB in $\mathbb{H}^N$. Corollary 3.25 then implies that the R_0 and R_a can be taken as operation elements to construct a trace-preserving quantum operation R. It remains to show that this operation corrects $\mathcal{E}$. Since $\{|\Psi_w\rangle\}$ forms an ONB of the basis codewords in $\mathbb{H}^{C_q}$, we can write any $\rho \in D(\mathbb{H}^{C_q})$ in the form

$$\rho = \sum_{x,y} \rho_{xy} |\Psi_x\rangle\langle\Psi_y| \,. \tag{7.85}$$

such that

$$\widetilde{\mathcal{E}}_a \rho \widetilde{\mathcal{E}}_a^* = \sum_{x,y} \rho_{xy} \widetilde{\mathcal{E}}_a |\Psi_x\rangle\langle\Psi_y| \widetilde{\mathcal{E}}_a^* \underbrace{=}_{(7.79),(7.80)} \begin{cases} 0 & \text{if } a \in I_0 \\ C_a \sum_{x,y} \rho_{xy} |\Phi_{a,x}\rangle\langle\Phi_{a,y}| & \text{if } a \in I_> \end{cases}$$

and thus

$$\mathcal{E}(\rho) \underbrace{=}_{(7.75)} \widetilde{\mathcal{E}}(\rho) \underbrace{=}_{(7.40)} \sum_a \widetilde{\mathcal{E}}_a \rho \widetilde{\mathcal{E}}_a^* = \sum_{a \in I_>} C_a \sum_{x,y} \rho_{x,y} |\Phi_{a,x}\rangle\langle\Phi_{a,y}| \,.$$

With this we find

$$R_0\,\mathcal{E}(\rho)\,R_0^* \underbrace{=}_{(7.84)} \sum_{p,s}\sum_{a\in I_>}\sum_{x,y} C_a\rho_{xy}|\Xi_p\rangle\langle\Xi_p|\Phi_{a,x}\rangle\langle\Phi_{a,y}|\Xi_s\rangle\langle\Xi_s|\underbrace{=}_{(7.82)}0 \qquad (7.86)$$

as well as

$$\sum_{b\in I_>}R_b\,\mathcal{E}(\rho)\,R_b^* \underbrace{=}_{(7.83),(7.84)} \sum_{b\in I_>}\sum_{x',y'}\sum_{a\in I_>}\sum_{x,y}C_a\rho_{ij}|\Psi_{x'}\rangle\langle\Phi_{b,x'}|\Phi_{a,x}\rangle\langle\Phi_{a,y}|\Phi_{b,y'}\rangle\langle\Psi_{y'}|$$

$$\underbrace{=}_{(7.81)} \sum_{a\in I_>}\sum_{x,y}C_a\rho_{xy}|\Psi_x\rangle\langle\Psi_y|\underbrace{=}_{(7.85)}\Big(\sum_{a\in I_>}C_a\Big)\rho\underbrace{=}_{(7.78)}\Big(\sum_{a=1}^{l}C_a\Big)\rho$$

$$\underbrace{=}_{(7.77)} \operatorname{tr}(\mathcal{E}(\rho))\,\rho\,. \qquad (7.87)$$

Setting $I=\{0\}\cup I_>$, we thus have created a recovery operation $R=\{R_\alpha\mid\alpha\in I\}$ that satisfies for any $\rho\in D(\mathbb{H}^{C_q})$

$$R\left(\frac{\mathcal{E}(\rho)}{\operatorname{tr}(\mathcal{E}(\rho))}\right) \underbrace{=}_{(3.97)} \sum_{\alpha\in I}R_\alpha\frac{\mathcal{E}(\rho)}{\operatorname{tr}(\mathcal{E}(\rho))}R_\alpha^* = \frac{1}{\operatorname{tr}(\mathcal{E}(\rho))}\sum_{\alpha\in I}R_\alpha\,\mathcal{E}(\rho)\,R_\alpha^*$$

$$\underbrace{=}_{(7.86),(7.87)} \rho\,,$$

and the R we have constructed corrects $\mathcal{E}$. $\qquad\square$

With Theorem 7.22 we can determine if an error operation $\mathcal{E}=\{\mathcal{E}_a\}$ in a code $\mathbb{H}^{C_q}$ is correctable. This determination can be made with information from the operation elements $\mathcal{E}_a$ and the codeword basis $\{|\Psi_x\rangle\}$ alone. To give a rationale for the condition (7.68), it is worthwhile to consider the equivalent operation elements $\widetilde{\mathcal{E}}_a$ constructed in the proof from the matrix diagonalizing C. The $\widetilde{\mathcal{E}}_a$ generate the same error operation $\mathcal{E}$ as the $\mathcal{E}_a$. If $a\in I_0$, then we have $\widetilde{\mathcal{E}}_a\big|_{\mathbb{H}^{C_q}}=0$. If $a\in I_>$, then the $\widetilde{\mathcal{E}}_a|\Psi_x\rangle$ give rise to orthonormal basis vectors $|\Phi_{a,x}\rangle\in\mathbb{H}^N$, and we have

$$\langle\widetilde{\mathcal{E}}_a\Psi_x|\widetilde{\mathcal{E}}_b\Psi_y\rangle=C_a\delta_{ab}\delta_{xy}\,.$$

This states that the operation elements $\widetilde{\mathcal{E}}_a$ generating $\mathcal{E}$ map orthogonal codewords $|\Psi_x\rangle$ to orthogonal subspaces (spanned by the $|\Phi_{a,x}\rangle$). This means that after different elements $\widetilde{\mathcal{E}}_a\neq\widetilde{\mathcal{E}}_b$ have acted on any pair of codewords, the probability to observe $\widetilde{\mathcal{E}}_a|\Psi_x\rangle$ when the system is in the state $\widetilde{\mathcal{E}}_b|\Psi_y\rangle$ is zero if $a\neq b$ or $x\neq y$. If this were not the case, it would be impossible to uniquely recover any codeword after an error has occurred. Therefore, the condition (7.68) guarantees that after $\mathcal{E}$ has corrupted any codeword, the corrupted state is sufficiently 'distant' from any other corrupted state so that we can recover the uncorrupted codeword uniquely.

Another important feature of the condition (7.68) is that the matrix C does not depend on the basis vectors $|\Psi_x\rangle$ of the code, namely, it implies

$$\langle \mathcal{E}_a \, \Psi_x | \, \mathcal{E}_b \, \Psi_x \rangle = C_{ab} \quad \forall |\Psi_x\rangle \in \{|\Psi_w\rangle\}, \tag{7.88}$$

where C_{ab} does not depend on x. If C_{ab} were to depend on x, it would reveal information about the codeword $|\Psi_x\rangle$. This is akin to observing $|\Psi_x\rangle$ and if the encoded states are part of an entangled state, any such entanglement would be destroyed rendering the computational process useless.

Example 7.23 Consider the two QECCs $[\![3,1]\!]_{q1}$ and $[\![3,1]\!]_{q2}$ from Examples 7.14 and 7.17. The basis codewords in the $[\![3,1]\!]_{q1}$ code were determined in (7.22) to be

$$|\Psi_0\rangle = |000\rangle \quad \text{and} \quad |\Psi_1\rangle = |111\rangle$$

and those for the $[\![3,1]\!]_{q2}$ code were determined in (7.27) as

$$|\Phi_0\rangle = |+++\rangle \quad \text{and} \quad |\Phi_1\rangle = |---\rangle.$$

We find for the bit-flip error operators $\mathcal{E}_{st}^{\mathrm{bf}}$ and phase-flip operators $\mathcal{E}_{st}^{\mathrm{pf}}$ in the $[\![3,1]\!]_{q1}$ code that

$$\langle \hat{\mathcal{E}}_{st}^{\mathrm{bf}} \Psi_u | \hat{\mathcal{E}}_{lm}^{\mathrm{bf}} \Psi_v \rangle \underbrace{=}_{(7.44),(7.45)} C_{st,lm}^{\mathrm{bf}} \delta_{uv} \tag{7.89}$$

$$\langle \hat{\mathcal{E}}_{00}^{\mathrm{pf}} \Psi_0 | \hat{\mathcal{E}}_{01}^{\mathrm{pf}} \Psi_0 \rangle \underbrace{=}_{(7.46),(7.47)} -\langle \hat{\mathcal{E}}_{00}^{\mathrm{pf}} \Psi_1 | \hat{\mathcal{E}}_{01}^{\mathrm{pf}} \Psi_1 \rangle. \tag{7.90}$$

From (7.89) we see that the bit-flip error operation $\mathcal{E}^{\mathrm{bf}}$ satisfies condition (7.68) in Theorem 7.22 in the code $[\![3,1]\!]_{q1}$ and is thus correctable in this code. On the other hand, (7.90) shows that the phase-flip error operation $\mathcal{E}^{\mathrm{pf}}$ does not satisfy (7.68) in the code $[\![3,1]\!]_{q1}$ because it violates (7.88). Therefore, phase-flip errors are *not* correctable in this code.

Likewise, we find in the $[\![3,1]\!]_{q2}$ code that

$$\langle \hat{\mathcal{E}}_{st}^{\mathrm{pf}} \Phi_u | \hat{\mathcal{E}}_{lm}^{\mathrm{pf}} \Phi_v \rangle \underbrace{=}_{(7.46),(7.49)} C_{st,lm}^{\mathrm{pf}} \delta_{uv} \tag{7.91}$$

$$\langle \hat{\mathcal{E}}_{00}^{\mathrm{bf}} \Phi_0 | \hat{\mathcal{E}}_{01}^{\mathrm{bf}} \Phi_0 \rangle \underbrace{=}_{(7.44),(7.48)} -\langle \hat{\mathcal{E}}_{00}^{\mathrm{bf}} \Phi_1 | \hat{\mathcal{E}}_{01}^{\mathrm{bf}} \Phi_1 \rangle. \tag{7.92}$$

Here the situation is reversed, and (7.91) shows that the phase-flip operation $\mathcal{E}^{\mathrm{pf}}$ is correctable in the $[\![3,1]\!]_{q2}$ code, whereas (7.92) shows that the bit-flip operation $\mathcal{E}^{\mathrm{bf}}$ is *not* correctable in this code. In summary:

- one-qubit bit-flip errors are correctable in the $[\![3,1]\!]_{q1}$ code, but not in the $[\![3,1]\!]_{q2}$ code.
- one-qubit phase-flip errors are correctable in the $[\![3,1]\!]_{q2}$ code, but not in the $[\![3,1]\!]_{q1}$ code.

Moreover, we see from (7.45) and (7.50) that

$$\langle \hat{\mathcal{E}}_{01}^{\mathrm{bf}} \Psi_1 | \hat{\mathcal{E}}_{2,0}^{2\mathrm{bf}} \Psi_0 \rangle \neq 0 \,,$$

and it follows from Theorem 7.22 that error operations with more than one-qubit-flip error are not correctable in the $[\![3,1]\!]_{q1}$ code either.

We mentioned already that condition (7.68) ensures that the corrupted states are sufficiently 'distant' from each other so that we can recover the uncorrupted codeword. Similar to the classical world, it is thus useful to define the notion of a distance of a quantum code. Its definition incorporates the condition (7.68) by making use of the fact that $\mathbb{H}^{C_q} \subset \mathbb{H}^N = \mathbb{H}^{\otimes n}$ and that the n-fold PAULI group $\mathcal{P}_n$ (see Sect. F.5) acts on $\mathbb{H}^{\otimes n}$.

Definition 7.24 Let C_q be an $[\![n,k]\!]_q$ QECC, and let $\{|\Psi_w\rangle\}$ be the ONB of basis codewords in $\mathbb{H}^{C_q}$. The **distance** of C_q is defined as

$$d_{\mathcal{P}}(C_q) := \min \left\{ w_{\mathcal{P}}(g) \; \middle| \; \begin{array}{l} \text{all } g \in \mathcal{P}_n \text{ such that there exist } |\Psi_x\rangle, |\Psi_y\rangle \in \\ \{|\Psi_w\rangle\} \quad \text{satisfying} \quad \langle \Psi_x | g \Psi_y \rangle \neq f(g)\delta_{xy} \text{ for} \\ \text{every } f : \mathcal{P}_n \to \mathbb{C} \end{array} \right\} ,$$

where $w_{\mathcal{P}}(g)$ is the weight as defined in Definition F.63. An $[\![n,k]\!]_q$ QECC C_q with distance $d = d_{\mathcal{P}}(C_q)$ is called an $[\![n,k,d]\!]_q$ code.

Exercise 7.90 Show that for the SHOR's $[\![9,1]\!]_q$ code C_{qS} of Example 7.15 there exist $g \in \mathcal{P}_9$ with $w_{\mathcal{P}}(g) = 3$ such that

$$\langle \Psi_x | g \Psi_y \rangle \neq f(g)\delta_{xy}$$

but that for any $h \in \mathcal{P}_9$ with $w_{\mathcal{P}}(h) \leq 2$ we have instead

$$\langle \Psi_x | h \Psi_y \rangle = f(h)\delta_{xy} \,,$$

and conclude that this code has distance three, namely that

$$d_{\mathcal{P}}(C_{q_S}) = 3.\tag{7.93}$$

For a solution see Solution 7.90.

Next, we establish the useful fact that any linear operator on $\mathbb{H}^N = \mathbb{H}^{\otimes n}$ can be written as a linear combination of elements in the PAULI group.

> **Lemma 7.25** *Any $A \in L(\mathbb{H}^{\otimes n})$ can be written as a finite linear combination of the form*
> $$A = \sum_j a_j h_j,$$
> *where $a_j \in \mathbb{C}$ and $h_j \in \mathcal{P}_n$.*

Proof From Exercise 3.45 we know that any $A \in L(\mathbb{H}^{\otimes n})$ can be written in the form

$$A = \sum_{j_{n-1},\dots,j_0} a_{j_{n-1},\dots,j_0} A_{j_{n-1}} \otimes \cdots \otimes A_{j_0},\tag{7.94}$$

where $a_{j_{n-1},\dots,j_0} \in \mathbb{C}$ and $A_{j_l} \in L(\mathbb{H})$, and from Exercise 2.35 we know that any $A_{j_l} \in L(\mathbb{H})$ can be written in the form

$$A_{j_l} = \sum_{\alpha_{j_l}=0}^{3} z_{\alpha_{j_l}} \sigma_{\alpha_{j_l}}.\tag{7.95}$$

Inserting (7.95) into (7.94) yields

$$A = \sum_{j_{n-1},\dots,j_0} a_{j_{n-1},\dots,j_0} \left(\sum_{\alpha_{j_{n-1}}=0}^{3} z_{\alpha_{j_{n-1}}} \sigma_{\alpha_{j_{n-1}}} \right) \otimes \cdots \otimes \left(\sum_{\alpha_{j_0}=0}^{3} z_{\alpha_{j_0}} \sigma_{\alpha_{j_0}} \right)$$
$$= \sum_{j_{n-1},\dots,j_0} \sum_{\alpha_{j_{n-1}},\dots,\alpha_{j_0}=0}^{3} \tilde{a}_{j_{n-1},\dots,j_0} \sigma_{\alpha_{j_{n-1}}} \otimes \cdots \otimes \sigma_{\alpha_{j_0}},$$

where Definition F.63 tells us that $\sigma_{\alpha_{j_{n-1}}} \otimes \cdots \otimes \sigma_{\alpha_{j_0}} \in \mathcal{P}_n$. $\qquad\square$

Since for $[\![n,k]\!]_q$ codes we have $\mathbb{H}^N = \mathbb{H}^{\otimes n}$, the operation elements $\mathcal{E}_a$ of any error operation belong to $L(\mathbb{H}^{\otimes n})$, and it follows from Lemma 7.25 that any operation element can be expressed as a linear combination of elements of the PAULI group. If the elements h_j of the PAULI group used in the linear combination $\mathcal{E}_a = \sum_j a_j h_j$ are correctable by some recovery R, then Corollary 7.21 tells us that the error generated by the $\mathcal{E}_a$ remains correctable by the same recovery. This way it suffices to study errors in the PAULI group: its elements generate all linear operators of $\mathbb{H}^N$ and correctability is preserved under linear combination. This also explains why in the stabilizer formalism to be presented in Sect. 7.3.3 it suffices to consider error correction on the n-fold PAULI group $\mathcal{P}_n$.

The following proposition is in some way a quantum analogue of Proposition 7.11.

Proposition 7.26 *Let C_q be an $[\![n,k]\!]_q$ QECC and $\mathcal{E} = \{\mathcal{E}_a\}$ a u-qubit error in $\mathbb{H}^{C_q}$ such that*

$$u \le \left\lfloor \frac{d_{\mathcal{P}}(C_q) - 1}{2} \right\rfloor. \tag{7.96}$$

Then $\mathcal{E}$ is correctable.

Proof From Lemma 7.25 we know that every operation element $\mathcal{E}_a \in L(\mathbb{H}^N)$ of $\mathcal{E}$ can be written as

$$\mathcal{E}_a = \sum_j a_j h_j \tag{7.97}$$

with $a_j \in \mathbb{C}$ and $h_j \in \mathcal{P}_n$. By assumption $\mathcal{E}$ is a u-qubit error. It follows from Definition 7.16 and 3.21 that each h_j can act non-trivially on at most u qubits, and we must have for any $\mathcal{E}_a \in \mathcal{E}$ and all the PAULI group operators h_j in their expansion (7.97) that

$$w_{\mathcal{P}}(h_j) \le u. \tag{7.98}$$

With $\mathcal{E}_b = \sum_l b_l h_l$ this implies for all $\mathcal{E}_a, \mathcal{E}_b \in \mathcal{E}$ and their PAULI group expansion operators h_j and h_l that

$$w_{\mathcal{P}}\left(h_j^* h_l\right) \underbrace{\le}_{\text{(F.136),(F.135)}} w_{\mathcal{P}}\left(h_j\right) + w_{\mathcal{P}}\left(h_l\right) \underbrace{\le}_{\text{(7.98)}} 2u \underbrace{<}_{\text{(7.96)}} d_{\mathcal{P}}(C_q). \tag{7.99}$$

It follows for any $\mathcal{E}_a, \mathcal{E}_b \in \mathcal{E}$ and any basis codewords $|\Psi_x\rangle, |\Psi_y\rangle \in \mathbb{H}^{C_q}$ that

$$\langle \mathcal{E}_a \Psi_x | \mathcal{E}_b \Psi_y \rangle \underbrace{=}_{\text{(2.31),(2.30)}} \langle \Psi_x | \mathcal{E}_a^* \mathcal{E}_b \Psi_y \rangle = \langle \Psi_x | \left(\sum_j a_j h_j\right)^* \left(\sum_l b_l h_l\right) \Psi_y \rangle$$

$$\underbrace{=}_{\text{(2.32),(2.4)}} \sum_{j,l} \overline{a_j} b_l \langle \Psi_x | h_j^* h_l \Psi_y \rangle \underbrace{=}_{\substack{\text{(7.99),} \\ \text{Def. 7.24}}} \underbrace{\sum_{j,l} \overline{a_j} b_l f(h_j^* h_l)}_{=C_{ab}} \delta_{xy}$$

$$= C_{ab} \delta_{xy}$$

and thus from Theorem 7.22 that $\mathcal{E}$ is correctable. $\qquad\square$

Proposition 7.26 allows us to make a statement about the error correcting capabilities of SHOR's $[\![9,1]\!]_q$ code.

Corollary 7.27 SHOR's $[\![9,1]\!]_q$ code C_{qS} of Example 7.15 corrects all one-qubit errors.

Proof The claim follows immediately from the result (7.93) of Exercise 7.90 and Proposition 7.26. □

7.3.2 Detection and Correction

Having defined error and recovery operations together with their operation elements, we now proceed to establish a process to detect and subsequently correct an error. When trying to correct an error in the code, we cannot apply all recovery operators as this would in general apply more changes than are needed. Rather, we need to take into account which error has occurred and gear our recovery operation accordingly. Finding out which error has actually occurred is called error detection.

Error detection is accomplished with the so-called error syndrome. The syndrome will serve to point to the error operator which has corrupted the code. This information is subsequently used to select the appropriate recovery operator to correct the error identified.

Definition 7.28 Let C_q be an $[\![n,k]\!]_q$ QECC and $\mathcal{E} = \{\mathcal{E}_a \in L(\mathbb{H}^N) \mid a \in I \subset \mathbb{N}_0\}$ be an error operation in $\mathbb{H}^{C_q}$ such that it includes the non-error $\mathcal{E}_0 = \sqrt{p_0}\mathbb{1}^{\otimes n}$ and its operational elements satisfy for all $a \in I$

$$\mathcal{E}_a = \sqrt{p_a}\hat{\mathcal{E}}_a, \quad p_a \in [0,1], \quad \sum_a p_a \le 1, \quad \hat{\mathcal{E}}_a \in \mathcal{U}(\mathbb{H}^N). \quad (7.100)$$

Furthermore, let $\mathbb{H}^S$ be a HILBERT space of dimension $\dim \mathbb{H}^S = |I|$ and $\{|\varphi_a\rangle \mid a \in I\}$ be an ONB in $\mathbb{H}^S$. We define the **syndrome detection operator** for $\mathcal{E}$ as a norm-preserving operator $S_{\mathcal{E}} : \mathbb{H}^N \to \mathbb{H}^N \otimes \mathbb{H}^S$ which satisfies

$$S_{\mathcal{E}}(\hat{\mathcal{E}}_a|\Psi\rangle) = \hat{\mathcal{E}}_a|\Psi\rangle \otimes |\varphi_a\rangle \quad \forall \mathcal{E}_a \in \mathcal{E} \text{ and } |\Psi\rangle \in \mathbb{H}^{C_q}. \quad (7.101)$$

The space $\mathbb{H}^S$ is called a syndrome space for $\mathcal{E}$ and the state $|\varphi_a\rangle$ the syndrome state for $\hat{\mathcal{E}}_a$. Observing $\mathbb{H}^S$ to detect $|\varphi_a\rangle$ and determine a is called **syndrome extraction**, and we call $a = \mathrm{syn}_q(\hat{\mathcal{E}}_a)$ the **syndrome of** $\hat{\mathcal{E}}_a$.

With the help of syndrome detectors we can then formulate a quantum error detection and correction protocol for a given error operation $\mathcal{E}$ in an $[\![n,k]\!]_q$ QECC as follows:

Quantum Error Detection and Correction Protocol

1. Let $|\widetilde{\Psi}\rangle \in \mathbb{H}^N$ be a (possibly corrupted) codeword
2. Apply the syndrome detection gate $\mathrm{S}_\mathcal{E}$ for the error operation $\mathcal{E}$ to $|\widetilde{\Psi}\rangle$ to obtain $\mathrm{S}_\mathcal{E}|\widetilde{\Psi}\rangle$
3. In the syndrome space $\mathbb{H}^S$ measure the mutually commuting observables

$$P_a = |\varphi_a\rangle\langle\varphi_a|$$

for all $a \in I$ in the state

$$\rho^S = \mathrm{tr}^N\left(\mathrm{S}_\mathcal{E}\,|\widetilde{\Psi}\rangle\langle\widetilde{\Psi}|\,\mathrm{S}_\mathcal{E}^*\right) \in \mathrm{D}\big(\mathbb{H}^S\big)$$

until one observation yields the (eigen-)value 1. Let $\widetilde{a} \in I$ be the single value in I for which this is the case
4. Apply the recovery operator

$$\mathrm{R}_{\widetilde{a}} := \hat{\mathcal{E}}_{\widetilde{a}}^* \tag{7.102}$$

to the state

$$\rho^N = \mathrm{tr}^S\left(\mathrm{S}_\mathcal{E}\,|\widetilde{\Psi}\rangle\langle\widetilde{\Psi}|\,\mathrm{S}_\mathcal{E}^*\right) \in \mathrm{D}\big(\mathbb{H}^N\big)$$

to obtain the corrected state

$$\mathrm{R}(\rho^N) = \mathrm{R}_{\widetilde{a}}\,\rho^N\,\mathrm{R}_{\widetilde{a}}^*, \tag{7.103}$$

and proceed with this as the state of the presumed original uncorrupted codeword $|\Psi\rangle$.

The proof of the following proposition shows how the detection and correction protocol indeed corrects errors belonging to the error operation $\mathcal{E}$.

Proposition 7.29 *Let C_q be an $[\![n,k]\!]_q$ QECC and $\mathcal{E} = \{\mathcal{E}_a \mid a \in I\}$ a correctable error operation in $\mathbb{H}^q$ with error operators $\hat{\mathcal{E}}_a$. Moreover, let $|\Psi\rangle \in \mathbb{H}^{\mathrm{C}_q}$ be a valid codeword, and let*

$$|\tilde{\Psi}\rangle = \hat{\mathcal{E}}_{\tilde{a}}|\Psi\rangle \qquad\qquad (7.104)$$

be the state after some error operator $\hat{\mathcal{E}}_{\tilde{a}}$ has corrupted it.
Then the error detection and correction protocol recovers $|\Psi\rangle$ from $|\tilde{\Psi}\rangle$.

Proof After the detection operator $S_{\mathcal{E}}$ is applied to the corrupted state $|\tilde{\Psi}\rangle$ the combined system is in the state (written as a density operator)

$$\rho_{S_{\mathcal{E}}|\tilde{\Psi}\rangle} \underset{(2.89)}{=} S_{\mathcal{E}}\,|\tilde{\Psi}\rangle\langle\tilde{\Psi}|\,S_{\mathcal{E}}^{*} \underset{(7.104),(7.101)}{=} \hat{\mathcal{E}}_{\tilde{a}}|\Psi\rangle\otimes|\varphi_a\rangle\langle\Psi|\hat{\mathcal{E}}_{\tilde{a}}^{*}\otimes\langle\varphi_a|$$

$$\underset{(3.36)}{=} \hat{\mathcal{E}}_{\tilde{a}}|\Psi\rangle\langle\Psi|\hat{\mathcal{E}}_{\tilde{a}}^{*}\otimes|\varphi_a\rangle\langle\varphi_a|\,.$$

From Theorem 3.17 we know that then the sub-system $\mathbb{H}^N$ describing the encoded system is in the state

$$\rho^N \underset{(3.50)}{=} \mathrm{tr}^S\left(\hat{\mathcal{E}}_{\tilde{a}}|\Psi\rangle\langle\Psi|\hat{\mathcal{E}}_{\tilde{a}}^{*}\otimes|\varphi_a\rangle\langle\varphi_a|\right) \underset{(3.57)}{=} \overbrace{\mathrm{tr}\left(|\varphi_{\tilde{a}}\rangle\langle\varphi_{\tilde{a}}|\right)}^{=1}\hat{\mathcal{E}}_{\tilde{a}}|\Psi\rangle\langle\Psi|\hat{\mathcal{E}}_{\tilde{a}}^{*}$$

$$= \hat{\mathcal{E}}_{\tilde{a}}|\Psi\rangle\langle\Psi|\hat{\mathcal{E}}_{\tilde{a}}^{*}\,,$$

where we used that $\{|\varphi_a\rangle\}$ is an ONB in $\mathbb{H}^S$ such that

$$\mathrm{tr}\left(|\varphi_{\tilde{a}}\rangle\langle\varphi_{\tilde{a}}|\right) \underset{(2.57)}{=} \sum_a \underbrace{\langle\varphi_a|\varphi_{\tilde{a}}\rangle}_{=\delta_{a\tilde{a}}}\underbrace{\langle\varphi_{\tilde{a}}|\varphi_a\rangle}_{=\delta_{a\tilde{a}}} = 1\,.$$

Similarly, we know from Corollary 3.20 that the syndrome sub-system $\mathbb{H}^S$ is in the state

$$\rho^S \underset{(3.50)}{=} \mathrm{tr}^N\left(\hat{\mathcal{E}}_{\tilde{a}}|\Psi\rangle\langle\Psi|\hat{\mathcal{E}}_{\tilde{a}}^{*}\otimes|\varphi_a\rangle\langle\varphi_a|\right) \underset{(3.57)}{=} \overbrace{\mathrm{tr}\left(\hat{\mathcal{E}}_{\tilde{a}}|\Psi\rangle\langle\Psi|\hat{\mathcal{E}}_{\tilde{a}}^{*}\right)}^{=1}|\varphi_{\tilde{a}}\rangle\langle\varphi_{\tilde{a}}|$$

$$= |\varphi_{\tilde{a}}\rangle\langle\varphi_{\tilde{a}}|\,, \qquad\qquad (7.105)$$

where we used that by assumption (7.100) we have $\hat{\mathcal{E}}_a \in \mathcal{U}\left(\mathbb{H}^N\right)$ such that

$$\mathrm{tr}\left(\hat{\mathcal{E}}_{\tilde{a}}|\Psi\rangle\langle\Psi|\hat{\mathcal{E}}_{\tilde{a}}^{*}\right) \underset{(2.58)}{=} \mathrm{tr}\left(|\Psi\rangle\langle\Psi|\hat{\mathcal{E}}_{\tilde{a}}^{*}\hat{\mathcal{E}}_{\tilde{a}}\right) \underset{(2.37)}{=} \mathrm{tr}\left(|\Psi\rangle\langle\Psi|\right) \underset{(3.50)}{=} \mathrm{tr}\left(\rho_\Psi\right) = 1\,.$$

From (7.105) we see that with $P_a = |\varphi_a\rangle\langle\varphi_a|$ we find that

$$P_a \rho^S P_a \underbrace{=}_{(7.105)} |\varphi_a\rangle \underbrace{\langle\varphi_a|\varphi_{\tilde{a}}\rangle}_{=\delta_{a\tilde{a}}} \underbrace{\langle\varphi_{\tilde{a}}|\varphi_a\rangle}_{=\delta_{a\tilde{a}}} \langle\varphi_a| = \delta_{a\tilde{a}} P_{\tilde{a}},$$

and only a measurement of $P_{\tilde{a}}$ in this state will yield a value of 1. Therefore, we will find $\tilde{a}$ as the error syndrome of Step 3 of the protocol. Consequently, we obtain for the state corrected by $R_{\tilde{a}} = \hat{\mathcal{E}}_{\tilde{a}}^*$ of Step 4 of the protocol that

$$R(\rho^N) \underbrace{=}_{(7.103)} R_{\tilde{a}} \rho^N R_{\tilde{a}}^* \underbrace{=}_{(7.102)} \hat{\mathcal{E}}_{\tilde{a}}^* \rho^N \hat{\mathcal{E}}_{\tilde{a}} \underbrace{=}_{(7.105)} \hat{\mathcal{E}}_{\tilde{a}}^* \hat{\mathcal{E}}_{\tilde{a}} |\Psi\rangle\langle\Psi| \hat{\mathcal{E}}_{\tilde{a}}^* \hat{\mathcal{E}}_{\tilde{a}} \underbrace{=}_{\hat{\mathcal{E}}_{\tilde{a}} \in \mathcal{U}(\mathbb{H}^N)} |\Psi\rangle\langle\Psi|$$

$$\underbrace{=}_{(2.89)} \rho_\Psi.$$

$\square$

The above mentioned error detection and correction protocol is thus designed to recover from the errors which occur in the error operation $\mathcal{E}$. Errors caused by operators which do not belong to $\mathcal{E}$ will in general not be corrected by this procedure. This is the case when the error is caused by an error operator $\widehat{\xi} \in \mathcal{U}(\mathbb{H}^N)$ which has the same syndrome as an error in $\mathcal{E}$ but is not part of $\mathcal{E}$. In other words, whenever syndrome detection yields $\mathrm{syn}_q(\widehat{\xi}) = \mathrm{syn}_q(\hat{\mathcal{E}}_{\tilde{a}}) = \tilde{a}$ but $\widehat{\xi} \neq z\hat{\mathcal{E}}_a$ for all $z \in \mathbb{C}$ and $\mathcal{E}_a \in \mathcal{E}$, then the correction protocol fails to recover the original codeword $|\Psi\rangle$ from

$$|\widetilde{\Psi}\rangle = \widehat{\xi} |\Psi\rangle.$$

This is because having detected the syndrome $\tilde{a}$, the recovery operation will apply the recovery operator $R_{\tilde{a}} = \hat{\mathcal{E}}_{\tilde{a}}^*$ and produce

$$\hat{\mathcal{E}}_{\tilde{a}}^* |\widetilde{\Psi}\rangle = \underbrace{\hat{\mathcal{E}}_{\tilde{a}}^* \widehat{\xi}}_{\neq \mathbb{1}^{\otimes n}} |\Psi\rangle \neq |\Psi\rangle.$$

Therefore, we ought to collect the most probable errors in $\mathcal{E}$ first and correct them. In this case errors like the aforementioned $\widehat{\xi}$ would be less likely and more of a rare event. A much better strategy, however, is to expand the error operation to include such $\widehat{\xi}$. In general this also requires that we enlarge the code such that the enlarged error operation remains correctable. An example for this is provided by the QECCs $[\![3, 1]\!]_{q_1}$ and $[\![3, 1]\!]_{q_2}$ of Examples 7.23 vs. the SHOR-code $[\![9, 1]\!]_{q_S}$ of Example 7.15. From Example 7.23 we know that in $[\![3, 1]\!]_{q_1}$ all one-qubit-flips are correctable, but not the phase-flips, whereas in $[\![3, 1]\!]_{q_2}$ all one-qubit phase-flips are correctable, but not the bit-flips. On the other hand, we know from Corollary 7.27 that $[\![9, 1]\!]_{q_S}$ corrects all one-qubit errors. But then this code also requires a much larger redundancy.

Example 7.30 We illustrate the error detection and correction protocol for the one-qubit-flip error $\mathcal{E}^{bf}$ in (7.44) with the QECC $[\![3, 1]\!]_{q_1}$ of the previous Examples 7.14, 7.17 and 7.23.

In this case we have $\mathbb{H}^N = \mathbb{H}^{\otimes 3}$ and $\mathcal{E}^{bf}$ contains four operational elements with the four error operators given in (7.44). Hence, we set $\mathbb{H}^S = \mathbb{H}^{\otimes 2}$ as our syndrome space and select the basis $|\varphi_{st}\rangle = |st\rangle$, where $s, t \in \{0, 1\}$. A syndrome detection operator

$$S_{\mathcal{E}^{bf}} : \mathbb{H}^N = \mathbb{H}^{\otimes 3} \rightarrow \mathbb{H}^N \otimes \mathbb{H}^S = \mathbb{H}^{\otimes 3} \otimes \mathbb{H}^{\otimes 2}$$

is given by

$$
\begin{aligned}
S_{\mathcal{E}^{bf}} = \quad & (|111\rangle\langle 111| + |000\rangle\langle 000|) \otimes |00\rangle \\
+ & (|110\rangle\langle 110| + |001\rangle\langle 001|) \otimes |01\rangle \\
+ & (|101\rangle\langle 101| + |010\rangle\langle 010|) \otimes |10\rangle \\
+ & (|100\rangle\langle 100| + |011\rangle\langle 011|) \otimes |11\rangle .
\end{aligned}
\tag{7.106}
$$

Any valid codeword is of the form

$$|\Psi\rangle \underset{(7.23)}{=} a|000\rangle + b|111\rangle \in \mathbb{H}^{\otimes 3} \tag{7.107}$$

with $a, b \in \mathbb{C}$ such that $||\Psi|| = 1$. Suppose then this is corrupted by the error operator

$$\hat{\mathcal{E}}_{10}^{bf} = \mathbf{1} \otimes \sigma_x \otimes \mathbf{1}$$

such that the corrupted state is

$$|\widetilde{\Psi}\rangle = \hat{\mathcal{E}}_{10}^{bf}|\Psi\rangle \underset{(7.107)}{=} a\hat{\mathcal{E}}_{10}^{bf}|000\rangle + b\hat{\mathcal{E}}_{10}^{bf}|111\rangle \underset{(7.45)}{=} a|010\rangle + b|101\rangle . \tag{7.108}$$

Applying the syndrome detection $S_{\mathcal{E}^{bf}}$, we see that the correct syndrome 10 is detected:

$$S_{\mathcal{E}^{bf}}|\widetilde{\Psi}\rangle \underset{(7.108)}{=} a\,S_{\mathcal{E}^{bf}}|010\rangle + b\,S_{\mathcal{E}^{bf}}|101\rangle \underset{(7.106)}{=} (a|010\rangle + b|101\rangle) \otimes |10\rangle$$

such that according to the protocol we apply $R_{10} = \hat{\mathcal{E}}_{10}^*$ to $|\widetilde{\Psi}\rangle = \hat{\mathcal{E}}_{10}^{bf}|\Psi\rangle$ and recover $|\Psi\rangle$ as desired.

Figure 7.6 shows the circuit for detection and correction of one-qubit-flip errors in the $[\![3, 1]\!]_{q_1}$ code.

Fig. 7.6 Illustrating the error detection, syndrome extraction and correction process between en- and decoding for the $[\![3,1]\!]_{q_1}$ code and the bit-flip error operation $\mathcal{E}^{\mathrm{bf}}$. The error operators for this error operation are given in (7.44). Measurement of $\sigma_z \otimes \sigma_z$ on the ancillas reveals the syndrome $(s_1\ s_0)$ which is then used to control which recovery operator $\mathrm{R}_{s_1 s_0}$ is applied

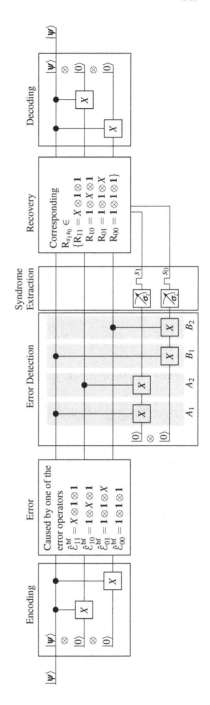

Exercise 7.91 Show that the operator $S = B_2B_1A_2A_1$ in Fig. 7.6 is given by

$$
\begin{aligned}
S = \quad & (|111\rangle\langle 111| + |000\rangle\langle 000|) \otimes \mathbf{1} \otimes \mathbf{1} \\
+ & (|110\rangle\langle 110| + |001\rangle\langle 001|) \otimes \mathbf{1} \otimes X \\
+ & (|101\rangle\langle 101| + |010\rangle\langle 010|) \otimes X \otimes \mathbf{1} \\
+ & (|100\rangle\langle 100| + |011\rangle\langle 011|) \otimes X \otimes X \, .
\end{aligned}
\tag{7.109}
$$

For a solution see Solution 7.91.

To illustrate that errors $\widehat{\xi}$ not in $\mathcal{E}^{\mathrm{bf}}$ will in general not be properly corrected by this code and protocol, let us now assume that the valid codeword as in (7.107) is corrupted by a two qubit-error $\widehat{\xi} = \hat{\mathcal{E}}_{11}^{\mathrm{bf}}\hat{\mathcal{E}}_{10}^{\mathrm{bf}}$ such that the corrupted state is

$$
\begin{aligned}
|\widetilde{\Psi}\rangle &= \hat{\mathcal{E}}_{11}^{\mathrm{bf}}\hat{\mathcal{E}}_{10}^{\mathrm{bf}}|\Psi\rangle \underbrace{=}_{(7.107)} a\hat{\mathcal{E}}_{11}^{\mathrm{bf}}\hat{\mathcal{E}}_{10}^{\mathrm{bf}}|000\rangle + b\hat{\mathcal{E}}_{11}^{\mathrm{bf}}\hat{\mathcal{E}}_{10}^{\mathrm{bf}}|111\rangle \\
&\underbrace{=}_{(7.45)} a\hat{\mathcal{E}}_{11}^{\mathrm{bf}}|010\rangle + b\hat{\mathcal{E}}_{11}^{\mathrm{bf}}|101\rangle \underbrace{=}_{(7.44)} a|110\rangle + b|001\rangle \, .
\end{aligned}
\tag{7.110}
$$

Applying the syndrome detection $S_{\mathcal{E}^{\mathrm{bf}}}$, we see that the syndrome 01 is detected:

$$
S_{\mathcal{E}^{\mathrm{bf}}}|\widetilde{\Psi}\rangle \underbrace{=}_{(7.110)} a\, S_{\mathcal{E}^{\mathrm{bf}}}|110\rangle + b\, S_{\mathcal{E}^{\mathrm{bf}}}|001\rangle \underbrace{=}_{(7.106)} \big(a|110\rangle + b|001\rangle\big) \otimes |01\rangle
$$

such that, according to the protocol, we apply $R_{01} = \hat{\mathcal{E}}_{01}^{*}$ to $|\widetilde{\Psi}\rangle$ and obtain

$$
\hat{\mathcal{E}}_{01}^{*}|\widetilde{\Psi}\rangle \underbrace{=}_{(7.110)} \hat{\mathcal{E}}_{01}^{*}\big(a|110\rangle + b|001\rangle\big) \underbrace{=}_{(7.44)} a|111\rangle + b|000\rangle \underbrace{\neq}_{(7.107)} |\Psi\rangle \, .
$$

7.3.3 Stabilizer Formalism

The stabilizer formalism provides a neat and compact way to describe QECCs with the help of group theory [102, 103]. Due to the fact proven in Corollary 7.21 that correctability is preserved under linear combinations of operation elements, and because any operation element can be expressed as a linear combination of elements of the PAULI group, it is sufficient to consider only operation elements which are elements of the PAULI group.

All necessary group theoretical notions required for this section are provided in Appendix F.

From Definition F.27 we recall that the left action Λ of a group $\mathcal{G}$ on a set M is defined as the map

$$\Lambda : \mathcal{G} \times M \longrightarrow M$$
$$(g, m) \longmapsto g.m$$

satisfying for all $g, h \in \mathcal{G}$ and $m \in M$

$$e_{\mathcal{G}}.m = m$$
$$(gh).m = g.(h.m).$$

and that the stabilizer of any subset $Q \subset M$ is defined as

$$\mathrm{Sta}_{\mathcal{G}}(Q) = \{ g \in \mathcal{G} \mid g.m = m \ \forall m \in Q \}.$$

Moreover, from Exercise F.130 we know that any such stabilizer is a subgroup of $\mathcal{G}$.

In the context of QECCs these notions are applied as follows

$$\mathcal{G} = \mathcal{P}_n$$
$$M = \mathbb{H}^N = \mathbb{H}^{\otimes n}$$
$$Q = \mathbb{H}^{C_q}.$$

The code is then determined by the specification of a suitable subgroup $\mathcal{S} < \mathcal{P}_n$. This subgroup in turn determines the code space. Our first task is to establish that the space determined in this fashion has the right properties to serve as the code space $\mathbb{H}^{C_q}$ of a QECC.

Theorem 7.31 *Let $k, n \in \mathbb{N}$ with $k < n$ and let $\mathcal{S}$ be an abelian subgroup of $\mathcal{P}_n$ generated by independent $\{g_1, \ldots, g_{n-k}\}$ satisfying $-\mathbf{1}^{\otimes n} \notin \mathcal{S}$, that is,*

$$-\mathbf{1}^{\otimes n} \notin \mathcal{S} = \langle g_1, \ldots, g_{n-k} \rangle < \mathcal{P}_n.$$

Then

$$\mathrm{Span}\{ |\Psi\rangle \in \mathbb{H}^{\otimes n} \mid g|\Psi\rangle = |\Psi\rangle \ \ \forall g \in \mathcal{S} \}$$

is a subspace of $\mathbb{H}^{\otimes n}$ of dimension 2^k.

Proof We begin by establishing that the g_l only have the two eigenvalues ± 1. From Proposition F.62 we know that every $g \in \mathcal{P}_n < \mathcal{U}(\mathbb{H}^{\otimes n})$ satisfies $g^2 = \pm \mathbf{1}^{\otimes n}$ and $g^* g = \mathbf{1}^{\otimes n}$. Since by assumption $-\mathbf{1}^{\otimes n} \notin \langle g_1, \ldots, g_{n-k} \rangle$, it follows that every $g_l \in \{g_1, \ldots, g_{n-k}\}$ satisfies $g_l^2 = \mathbf{1}^{\otimes n}$. Furthermore, if we had $g_l^* = -g_l$, it would follow that $-g_l^2 = \mathbf{1}^{\otimes n}$ and thus $g_l^2 = -\mathbf{1}^{\otimes n}$. But this would contradict $-\mathbf{1}^{\otimes n} \notin \langle g_1, \ldots, g_{n-k} \rangle$, and it follows that we must have $g_l^* = g_l$. Consequently, $g_l \in B_{sa}(\mathbb{H}^{\otimes n})$, and the eigenvalues of all g_l are real. Moreover, since $g \in \mathcal{U}(\mathbb{H}^{\otimes n})$, we infer from (iii) in Exercise 2.8 that any such eigenvalue satisfies $|\lambda_l| = 1$. Hence, every g_l has only the two eigenvalues $\lambda_l = \pm 1$.

For each $(n-k)$-tuple $(\lambda_1, \ldots, \lambda_{n-k}) \in \{\pm 1\}^{n-k}$ we define the subspace

$$\mathbb{H}_{(\lambda_1,\ldots,\lambda_{n-k})} := \mathrm{Span}\left\{ |\Psi\rangle \in \mathbb{H}^{\otimes n} \mid g_l|\Psi\rangle = \lambda_l|\Psi\rangle \ \forall l \in \{1,\ldots,n-k\} \right\}, \tag{7.111}$$

which by definition satisfy

$$\mathbb{H}_{(\lambda_1,\ldots,\lambda_{n-k})} \subset \mathrm{Eig}(g_l, \lambda_l) \quad \forall l \in \{1,\ldots,n-k\}. \tag{7.112}$$

It follows from (7.112) that for $(\lambda_1, \ldots, \lambda_{n-k}) \neq (\mu_1, \ldots, \mu_{n-k})$ the subspaces $\mathbb{H}_{(\lambda_1,\ldots,\lambda_{n-k})}$ and $\mathbb{H}_{(\mu_1,\ldots,\mu_{n-k})}$ are orthogonal. This is because there must be at least one l such that $\lambda_l \neq \mu_l$, and since eigenspaces of self-adjoint operators for different eigenvalues are orthogonal, we have

$$\mathbb{H}_{(\lambda_1,\ldots,\lambda_{n-k})} \underbrace{\subset}_{(7.112)} \mathrm{Eig}(g_l, \lambda_l) \perp \mathrm{Eig}(g_l, \mu_l) \underbrace{\supset}_{(7.112)} \mathbb{H}_{(\mu_1,\ldots,\mu_{n-k})}.$$

From Proposition F.72 we know that there exist $h_j \in \mathcal{P}_n$ for $j \in \{1, \ldots, n-k\}$ such that for every $l \in \{1, \ldots, n-k\}$

$$g_l h_j = (-1)^{\delta_{lj}} h_j g_l.$$

Hence, for $|\Psi\rangle \in \mathbb{H}_{(\lambda_1,\ldots,\lambda_{n-k})}$ we have

$$g_l h_j|\Psi\rangle = (-1)^{\delta_{lj}} h_j g_l|\Psi\rangle = (-1)^{\delta_{lj}} \lambda_l h_j|\Psi\rangle = \begin{cases} \lambda_l h_j|\Psi\rangle & \text{if } j \neq l \\ -\lambda_l h_j|\Psi\rangle & \text{if } j = l. \end{cases}$$

such that for all $|\Psi\rangle \in \mathbb{H}_{(\lambda_1,\ldots,\lambda_{n-k})}$

$$h_j|\Psi\rangle \in \mathbb{H}_{(\lambda_1,\ldots,\lambda_{j-1},-\lambda_j,\lambda_{j+1},\ldots,\lambda_{n-k})}. \tag{7.113}$$

Since $h_j \in \mathcal{U}(\mathbb{H}^{\otimes n})$ it follows that

$$\dim \mathbb{H}_{(\lambda_1,\ldots,\lambda_{n-k})} = \dim \mathbb{H}_{(\lambda_1,\ldots,\lambda_{j-1},-\lambda_j,\lambda_{j+1},\ldots,\lambda_{n-k})}. \tag{7.114}$$

Moreover, (7.113) shows that by application of a suitable set of h_j we can unitarily map any $\mathbb{H}_{(\lambda_1,\ldots,\lambda_{n-k})}$ to any $\mathbb{H}_{(\mu_1,\ldots,\mu_{n-k})}$. Together with (7.114) this implies that all these subspaces have the same dimension, namely

$$\dim \mathbb{H}_{(\lambda_1,\ldots,\lambda_{n-k})} = J \quad \forall (\lambda_1, \ldots, \lambda_{n-k}) \in \{\pm 1\}^{n-k}.$$

Altogether we have thus 2^{n-k} mutually orthogonal subspaces of equal dimension J which together comprise $\mathbb{H}^{\otimes n}$. It follows that $J2^{n-k} = \dim \mathbb{H}^{\otimes n} = 2^n$ and thus

$$2^k = J = \dim \mathbb{H}_{(\lambda_1,\dots,\lambda_{n-k})} \qquad \forall(\lambda_1,\dots,\lambda_{n-k}) \in \{\pm 1\}^{n-k}\,.$$

Finally, note that $g = g_1^{a_1} \cdots g_{n-k}^{a_{n-k}}$ for any $g \in \mathcal{S} = \langle g_1,\dots,g_{n-k}\rangle$, where $a_j \in \{0,1\}$ since $g^2 = \mathbf{1}^{\otimes n}$ for any $g \in \mathcal{S}$. Consequently,

$$\mathbb{H}_{(+1,\dots,+1)} \underset{(7.111)}{=} \text{Span}\left\{|\Psi\rangle \in \mathbb{H}^{\otimes n} \mid g_l|\Psi\rangle = |\Psi\rangle \ \forall l \in \{1,\dots,n-k\}\right\}$$
$$= \left\{|\Psi\rangle \in \mathbb{H}^{\otimes n} \mid \forall g \in \mathcal{S}: g|\Psi\rangle = |\Psi\rangle\right\}.$$

$\square$

We thus define the notion of a stabilizer (error correcting) code as follows.

Definition 7.32 An $[\![n,k]\!]_q$ QECC C_q with code $\mathbb{H}^{C_q}$ is called a **stabilizer code** if there exists an abelian subgroup $\mathcal{S}$ of $\mathcal{P}_n$ that is generated by independent $\{g_1,\dots,g_{n-k}\}$ and satisfies $-\mathbf{1}^{\otimes n} \notin \mathcal{S}$ such that the code $\mathbb{H}^{C_q}$ is given by

$$\mathbb{H}^{C_q} = \left\{|\Psi\rangle \in \mathbb{H}^N \mid g|\Psi\rangle = |\Psi\rangle \ \forall g \in \mathcal{S}\right\}. \qquad (7.115)$$

The subgroup

$$\mathcal{S} = \text{Sta}_{\mathcal{P}_n}(\mathbb{H}^{C_q}) = \left\{g \in \mathcal{P}_n \mid g|\Psi\rangle = |\Psi\rangle \ \forall|\Psi\rangle \in \mathbb{H}^{C_q}\right\}$$

is called the **stabilizer of the code**.

The reason $-\mathbf{1}^{\otimes n}$ is excluded from $\mathcal{S}$ is that including it would result in a trivial $\mathbb{H}^{C_q}$. This is because (7.115) would imply for $g = -\mathbf{1}^{\otimes n}$ for all $|\Psi\rangle \in \mathbb{H}^{C_q}$ that $|\Psi\rangle = -|\Psi\rangle$, resulting in $\mathbb{H}^{C_q} = \{0\}$.

In the stabilizer formalism it is the subgroup $\mathcal{S}$ of the n-PAULI group $\mathcal{P}_n$ which determines the code $\mathbb{H}^{C_q}$. In general $\mathcal{S}$ will be specified and $\mathbb{H}^{C_q}$ is then given as the subspace of $\mathbb{H}^N$ consisting of all vectors which are invariant under the action of every $g \in \mathcal{S}$. Hence, any $|\Psi\rangle \in \mathbb{H}^N$ for which there exists at least one $g \in \mathcal{S}$ such that $g|\Psi\rangle \neq |\Psi\rangle$ cannot be in the code $\mathbb{H}^{C_q}$.

Example 7.33 Consider the two $[\![3,1]\!]_q$ codes of Examples 7.14, 7.17 and 7.23. The code $[\![3,1]\!]_{q_1}$, which can correct bit-flip errors, has the basis codewords

$$|\Psi_0\rangle = |000\rangle \quad \text{and} \quad |\Psi_1\rangle = |111\rangle$$

and the bit-flip error operators

$$\hat{\mathcal{E}}_{00}^{\mathrm{bf}} = \mathbf{1} \otimes \mathbf{1} \otimes \mathbf{1}, \qquad \hat{\mathcal{E}}_{01}^{\mathrm{bf}} = \mathbf{1} \otimes \mathbf{1} \otimes \sigma_x, \qquad \hat{\mathcal{E}}_{10}^{\mathrm{bf}} = \mathbf{1} \otimes \sigma_x \otimes \mathbf{1}, \qquad \hat{\mathcal{E}}_{11}^{\mathrm{bf}} = \sigma_x \otimes \mathbf{1} \otimes \mathbf{1}.$$

The stabilizer of this code is $\mathcal{S}_1 = \langle g_1, g_2 \rangle$, where

$$g_1 = \sigma_z \otimes \sigma_z \otimes \mathbf{1} \quad \text{and} \quad g_2 = \sigma_z \otimes \mathbf{1} \otimes \sigma_z.$$

On the other hand, the code $[\![3,1]\!]_{q2}$, which corrects phase-flip errors, has the basis codewords

$$|\Phi_0\rangle = |+++\rangle \quad \text{and} \quad |\Phi_1\rangle = |---\rangle$$

and the phase-flip error operators

$$\hat{\mathcal{E}}_{00}^{\mathrm{pf}} = \mathbf{1} \otimes \mathbf{1} \otimes \mathbf{1}, \qquad \hat{\mathcal{E}}_{01}^{\mathrm{pf}} = \mathbf{1} \otimes \mathbf{1} \otimes \sigma_z, \qquad \hat{\mathcal{E}}_{10}^{\mathrm{pf}} = \mathbf{1} \otimes \sigma_z \otimes \mathbf{1}, \qquad \hat{\mathcal{E}}_{11}^{\mathrm{pf}} = \sigma_z \otimes \mathbf{1} \otimes \mathbf{1}.$$

The stabilizer of this code is $\mathcal{S}_2 = \langle h_1, h_2 \rangle$, where

$$h_1 = \sigma_x \otimes \sigma_x \otimes \mathbf{1} \quad \text{and} \quad h_2 = \sigma_x \otimes \mathbf{1} \otimes \sigma_x.$$

Lemma 7.34 *Let C_q be an $[\![n,k]\!]_q$ QECC with stabilizer $\mathcal{S} < \mathcal{P}_n$ and $N(\mathcal{S})$ the normalizer of $\mathcal{S}$. Furthermore, let $\{|\Psi_w\rangle\}$ be the ONB of basis codewords in $\mathbb{H}^N$, and define*

$$M_{\neq} := \left\{ g \in \mathcal{P}_n \;\middle|\; \begin{array}{l} \text{there exist } |\Psi_x\rangle, |\Psi_y\rangle \in \{|\Psi_w\rangle\} \text{ such that} \\ \langle\Psi_x|g\Psi_y\rangle \neq f(g)\delta_{xy} \text{ for every } f : \mathcal{P}_n \to \mathbb{C} \end{array} \right\}. \quad (7.116)$$

Then we have

$$M_{\neq} \subset N(\mathcal{S}) \smallsetminus \mathcal{S}. \quad (7.117)$$

Proof Recall first that by Definitions F.27 and F.16

$$\mathcal{S} = \{ g \in \mathcal{P}_n \mid g|\Psi\rangle = |\Psi\rangle \;\; \forall|\Psi\rangle \in \mathbb{H}^{C_q} \}$$
$$N(\mathcal{S}) = \{ g \in \mathcal{P}_n \mid g\mathcal{S} = \mathcal{S}g \},$$

and that Exercises F.130 and F.126 showed that they are subgroups, that is, $\mathcal{S}$ is the subgroup of $\mathcal{P}_n$ which leaves every vector in $\mathbb{H}^{C_q}$ unchanged, and the normalizer $N(\mathcal{S})$ is the subgroup of all elements in $\mathcal{P}_n$ which commute with every element of $\mathcal{S}$.

We show the inclusion by proving the contrapositive. First let $g \in \mathcal{S}$. Then it follows for any two basis codewords that

$$\langle \Psi_x | g \Psi_y \rangle \underbrace{=}_{g \in \mathcal{S}} \langle \Psi_x | \Psi_y \rangle = \delta_{xy}$$

and thus $g \notin M_{\neq}$.

Now suppose $g \notin N(\mathcal{S})$ such that $g\mathcal{S} \neq \mathcal{S}g$, which implies that there exists at least one $h \in \mathcal{S}$ such that for all $\tilde{h} \in \mathcal{S}$ we have $gh \neq \tilde{h}g$. This also holds for $\tilde{h} = h$. It follows that for any $g \in N(\mathcal{S}) \smallsetminus \mathcal{S}$ there exists at least one $h \in \mathcal{S}$ that satisfies

$$gh \neq hg \,.$$

But from Proposition F.70 we know that for any $g, h \in \mathcal{P}_n$ either $gh = hg$ or $gh = -hg$. Hence, for the g and h at hand we must have

$$gh = -hg \,, \tag{7.118}$$

and thus any pair $|\Psi_x\rangle, |\Psi_y\rangle \in \mathbb{H}^{C_q}$ of basis codewords satisfies

$$\langle \Psi_x | g \Psi_y \rangle \underbrace{=}_{h \in \mathcal{S}} \langle \Psi_x | g h \Psi_y \rangle \underbrace{=}_{(7.118)} -\langle \Psi_x | h g \Psi_y \rangle \underbrace{=}_{(2.30)} -\langle h^* \Psi_x | g \Psi_y \rangle \,. \tag{7.119}$$

Any $h \in \mathcal{P}_n$ satisfies $h^* h = \mathbf{1}^{\otimes n}$ and thus $h^* = h^{-1}$. From Exercise F.130 we know that $\mathcal{S}$ is a subgroup of $\mathcal{P}_n$, and since $h \in \mathcal{S}$, it follows that $h^* = h^{-1} \in \mathcal{S}$. Hence, h^* leaves every vector in $\mathbb{H}^{C_q}$ invariant. Consequently, we obtain

$$\langle \Psi_x | g \Psi_y \rangle \underbrace{=}_{(7.119)} -\langle h^* \Psi_x | g \Psi_y \rangle \underbrace{=}_{h^* \in \mathcal{S}} -\langle \Psi_x | g \Psi_y \rangle \,,$$

which implies

$$\langle \Psi_x | g \Psi_y \rangle = 0 = f(g) \delta_{xy}$$

with $f(g) \equiv 0$ and thus $g \notin M_{\neq}$. We have shown that

$$g \in \mathcal{S} \quad \Rightarrow \quad g \notin M_{\neq} \qquad \text{and} \qquad g \notin N(\mathcal{S}) \quad \Rightarrow \quad g \notin M_{\neq}$$

and the contrapositive yields $g \in M_{\neq} \Rightarrow g \in N(\mathcal{S}) \smallsetminus \mathcal{S}$, which implies (7.117). $\quad\square$

The following theorem gives a group theoretic criterion for an error to be correctable.

Theorem 7.35 *Let* C_q *be an* $[\![n,k]\!]_q$ *QECC with stabilizer* $\mathcal{S} < \mathcal{P}_n$ *and* $\mathcal{E} = \{\mathcal{E}_a\}$ *an error in the code* $\mathbb{H}^{C_q}$ *with operation elements* $\mathcal{E}_a \in \mathcal{P}_n$. *Moreover, let* $N(\mathcal{S}) = \mathrm{Nor}_{\mathcal{P}_n}(\mathcal{S})$ *denote the normalizer of* $\mathcal{S}$ *in* $\mathcal{P}_n$. *Then we have*

$$\forall \mathcal{E}_a, \mathcal{E}_b \in \mathcal{E}: \; \mathcal{E}_a^* \mathcal{E}_b \notin N(\mathcal{S}) \smallsetminus \mathcal{S} \;\; \Rightarrow \;\; \mathcal{E} = \{\mathcal{E}_a\} \text{ is correctable in } \mathbb{H}^{C_q}.$$

Proof Let $\mathcal{E}_a^* \mathcal{E}_b \notin N(\mathcal{S}) \smallsetminus \mathcal{S}$. Then it follows from Lemma 7.34 that $\mathcal{E}_a^* \mathcal{E}_b \notin M_{\neq}$ and, therefore, that for all basis codewords $|\Psi_x\rangle, |\Psi_y\rangle \in \mathbb{H}^{C_q}$

$$\langle \mathcal{E}_a \Psi_x | \mathcal{E}_b \Psi_y \rangle \underbrace{=}_{(2.30),(2.31)} \langle \Psi_x | \mathcal{E}_a^* \mathcal{E}_b \Psi_y \rangle = f(\mathcal{E}_a^* \mathcal{E}_b)\delta_{xy} = C_{ab}\delta_{xy},$$

and Theorem 7.22 implies that $\mathcal{E} = \{\mathcal{E}_a\}$ is correctable in $\mathbb{H}^{C_q}$. □

From Proposition F.70 we know that any two elements of the n-fold PAULI group either commute or anti-commute. Therefore, the following definition is meaningful.

Definition 7.36 Let C_q be an $[\![n,k]\!]_q$ QECC with stabilizer

$$\mathcal{S} = \langle g_1, \ldots, g_{n-k}\rangle < \mathcal{P}_n. \tag{7.120}$$

We define the **syndrome map of the stabilizer QECC C_q** as

$$\begin{aligned} \mathrm{syn}_q : \mathcal{P}_n &\longrightarrow \mathbb{F}_2^{n-k} \\ g &\longmapsto (l_1(g), \ldots, l_{n-k}(g)) \end{aligned}, \tag{7.121}$$

where

$$l_j(g) = \begin{cases} 0 & \text{if } gg_j = g_jg \\ 1 & \text{if } gg_j = -g_jg \end{cases} \tag{7.122}$$

for $j \in \{1, \ldots, n-k\}$.

Note that the syndrome map depends on C_q since the component functions l_j are defined with the help of the generators of the stabilizer of C_q.

Example 7.37 Consider first the code $[\![3,1]\!]_{q_1}$ with generators g_1 and g_2 given in Example 7.33. From Example 7.23 we know that this code corrects single-qubit bit-flip errors. The bit-flip error operators $\hat{\mathcal{E}}_{st}^{bf}$ for $s,t \in \{0,1\}$ have the syndromes shown in Table 7.1.

From Example 7.23 we also know that the code $[\![3,1]\!]_{q_2}$ with generators h_1 and h_2 given in Example 7.33 corrects phase-flips. The phase-flip error operators $\hat{\mathcal{E}}_{st}^{bf}$ for $s,t \in \{0,1\}$ have the syndromes shown in Table 7.2.

Definition 7.36 is the quantum equivalent of the classical syndrome defined in Definition 7.9. The following lemma establishes the quantum analogue of (7.12).

Lemma 7.38 *Let $\mathcal{S}$ be the stabilizer of an $[\![n, k]\!]_q$ QECC C_q and syn_q the syndrome map of C_q, and let $N(\mathcal{S})$ denote the normalizer of $\mathcal{S}$. Then we have*

$$N(\mathcal{S}) = \mathrm{Ker}(\mathrm{syn}_q). \tag{7.123}$$

Proof Let $h \in N(\mathcal{S})$. By Definition F.16 this implies $h\mathcal{S} = \mathcal{S}h$, which means

$$\forall g \in \mathcal{S} \quad \exists \tilde{g} \in \mathcal{S}: \quad hg = \tilde{g}h. \tag{7.124}$$

On the other hand, we know from Proposition F.70 that for any $h, g \in \mathcal{P}_n$ either $hg = -gh$ or $hg = gh$. Suppose $hg = -gh$. Then it follows from (7.124) that $\tilde{g} = -g$ and thus

$$-\mathbf{1}^{\otimes n} = \tilde{g}g^{-1} \in \mathcal{S}.$$

But $-\mathbf{1}^{\otimes n} \in \mathcal{S}$ contradicts the defining assumption in Definition 7.32 of $\mathcal{S}$ being a stabilizer of the QECC C_q. Therefore we must have $\tilde{g} = g$ and thus

Table 7.1 $[\![3, 1]\!]_{q_1}$-syndromes for one-qubit bit-flip errors

g	$gg_1 = \pm g_1 g$	$gg_2 = \pm g_2 g$	$l_1(g)$	$l_2(g)$
$\hat{\mathcal{E}}_{00}^{\mathrm{bf}}$	$\hat{\mathcal{E}}_{00}^{\mathrm{bf}} g_1 = g_1 \hat{\mathcal{E}}_{00}^{\mathrm{bf}}$	$\hat{\mathcal{E}}_{00}^{\mathrm{bf}} g_2 = g_2 \hat{\mathcal{E}}_{00}^{\mathrm{bf}}$	0	0
$\hat{\mathcal{E}}_{01}^{\mathrm{bf}}$	$\hat{\mathcal{E}}_{01}^{\mathrm{bf}} g_1 = g_1 \hat{\mathcal{E}}_{01}^{\mathrm{bf}}$	$\hat{\mathcal{E}}_{01}^{\mathrm{bf}} g_2 = -g_2 \hat{\mathcal{E}}_{01}^{\mathrm{bf}}$	0	1
$\hat{\mathcal{E}}_{10}^{\mathrm{bf}}$	$\hat{\mathcal{E}}_{10}^{\mathrm{bf}} g_1 = -g_1 \hat{\mathcal{E}}_{10}^{\mathrm{bf}}$	$\hat{\mathcal{E}}_{10}^{\mathrm{bf}} g_2 = g_2 \hat{\mathcal{E}}_{10}^{\mathrm{bf}}$	1	0
$\hat{\mathcal{E}}_{11}^{\mathrm{bf}}$	$\hat{\mathcal{E}}_{11}^{\mathrm{bf}} g_1 = -g_1 \hat{\mathcal{E}}_{11}^{\mathrm{bf}}$	$\hat{\mathcal{E}}_{11}^{\mathrm{bf}} g_2 = -g_2 \hat{\mathcal{E}}_{11}^{\mathrm{bf}}$	1	1

Table 7.2 $[\![3, 1]\!]_{q_2}$-syndromes for the one-qubit phase-flip errors

g	$gh_1 = \pm h_1 g$	$gh_2 = \pm h_2 g$	$l_1(g)$	$l_2(g)$
$\hat{\mathcal{E}}_{00}^{\mathrm{pf}}$	$\hat{\mathcal{E}}_{00}^{\mathrm{pf}} h_1 = h_1 \hat{\mathcal{E}}_{00}^{\mathrm{pf}}$	$\hat{\mathcal{E}}_{00}^{\mathrm{pf}} h_2 = h_2 \hat{\mathcal{E}}_{00}^{\mathrm{pf}}$	0	0
$\hat{\mathcal{E}}_{01}^{\mathrm{pf}}$	$\hat{\mathcal{E}}_{01}^{\mathrm{pf}} h_1 = h_1 \hat{\mathcal{E}}_{01}^{\mathrm{pf}}$	$\hat{\mathcal{E}}_{01}^{\mathrm{pf}} h_2 = -h_2 \hat{\mathcal{E}}_{01}^{\mathrm{pf}}$	0	1
$\hat{\mathcal{E}}_{10}^{\mathrm{pf}}$	$\hat{\mathcal{E}}_{10}^{\mathrm{pf}} h_1 = -h_1 \hat{\mathcal{E}}_{10}^{\mathrm{pf}}$	$\hat{\mathcal{E}}_{10}^{\mathrm{pf}} h_2 = h_2 \hat{\mathcal{E}}_{10}^{\mathrm{pf}}$	1	0
$\hat{\mathcal{E}}_{11}^{\mathrm{pf}}$	$\hat{\mathcal{E}}_{11}^{\mathrm{pf}} h_1 = -h_1 \hat{\mathcal{E}}_{11}^{\mathrm{pf}}$	$\hat{\mathcal{E}}_{11}^{\mathrm{pf}} h_2 = -h_2 \hat{\mathcal{E}}_{11}^{\mathrm{pf}}$	1	1

$$hg = gh \qquad \forall g \in \mathcal{S} = \langle g_1, \ldots, g_{n-k} \rangle$$
$$\underset{(7.120)}{\Leftrightarrow} \qquad hg_j = g_j h \qquad \forall g_j \in \{g_1, \ldots, g_{n-k}\}$$
$$\underset{(7.122)}{\Leftrightarrow} \qquad l_j(h) = 0 \qquad \forall j \in \{1, \ldots, n-k\} \tag{7.125}$$
$$\underset{(7.121)}{\Leftrightarrow} \quad \mathrm{syn}_q(h) = \mathbf{0},$$

which implies $h \in \mathrm{Ker}(\mathrm{syn}_q)$ and thus

$$N(\mathcal{S}) \subset \mathrm{Ker}(\mathrm{syn}_q). \tag{7.126}$$

To prove the converse inclusion, let $h \in \mathrm{Ker}(\mathrm{syn}_q)$ such that $\mathrm{syn}_q(h) = \mathbf{0}$. Then it follows from (7.125) that $h\mathcal{S} = \mathcal{S}h$ and thus $h \in N(\mathcal{S})$. Therefore, $\mathrm{Ker}(\mathrm{syn}_q) \subset N(\mathcal{S})$ and together with (7.126) thus $N(\mathcal{S}) = \mathrm{Ker}(\mathrm{syn}_q)$. $\qquad\square$

The following Exercise 7.92 is the quantum analogue of Exercise 7.83.

Exercise 7.92 Show that syn_q as defined in Definition 7.36 satisfies for any $h_1, h_2 \in \mathcal{P}_n$

$$\mathrm{syn}_q(h_1 h_2) = \mathrm{syn}_q(h_1) \overset{2}{\oplus} \mathrm{syn}_q(h_2)$$

such that syn_q is a homomorphism, that is,

$$\mathrm{syn}_q \in \mathrm{Hom}(\mathcal{P}_n, \mathbb{F}_2^{n-k}). \tag{7.127}$$

For a solution see Solution 7.92.

Lemma 7.39 *Let $\mathcal{S}$ be the stabilizer of an $[\![n,k]\!]_q$ QECC C_q. Then the normalizer $N(\mathcal{S})$ of $\mathcal{S}$ is a normal subgroup of $\mathcal{P}_n$, that is,*

$$N(\mathcal{S}) \trianglelefteq \mathcal{P}_n.$$

Proof From (7.127) and (7.123) we infer that $N(\mathcal{S})$ is the kernel of a homomorphism from $\mathcal{P}_n$ to $\mathbb{F}_2^{n-k}$. The claim then follows immediately from Lemma F.30. $\qquad\square$

Having established that $N(\mathcal{S}) \trianglelefteq \mathcal{P}_n$, it follows from Proposition F.22 that the quotient group $\mathcal{P}_n/N(\mathcal{S})$ exists. This allows us to formulate the following proposition, which is the quantum equivalent of Proposition 7.10 and which will enable us to establish a quantum error correction protocol similar to the one presented after that proposition.

Proposition 7.40 *Let* $\text{syn}_q : \mathcal{P}_n \to \mathbb{F}_2^{n-k}$ *be the syndrome map for an* $[\![n, k]\!]_q$ *QECC* C_q *with stabilizer* $\mathcal{S}$, *and let* $N(\mathcal{S})$ *be the normalizer of* $\mathcal{S}$. *Then*

$$\widehat{\text{syn}_q} : \mathcal{P}_n/N(\mathcal{S}) \longrightarrow \mathbb{F}_2^{n-k}$$
$$[g]_{N(\mathcal{S})} \longmapsto \text{syn}_q(g)$$

is an isomorphism.

Proof From Lemma 7.38 we know that $N(\mathcal{S}) = \text{Ker}(\text{syn}_q)$ and from Exercise 7.92 that $\text{syn}_q \in \text{Hom}(\mathcal{P}_n, \mathbb{F}_2^{n-k})$. If we can show that $\mathbb{F}_2^{n-k} = \text{syn}_q\{\mathcal{P}_n\}$, then the claim follows immediately from the First Group Isomorphism Theorem F.31.

To show that $\mathbb{F}_2^{n-k} = \text{syn}_q\{\mathcal{P}_n\}$, let $\mathbf{a} \in \mathbb{F}_2^{n-k}$. Suppose $a_m = 1$ for some $m \in \{1, \ldots, n-k\}$. From Proposition F.72 we know that there exists an $h_m \in \mathcal{P}_n$ such that for all $j \in \{1, \ldots, n-k\}$

$$g_j h_m = (-1)^{\delta_{jm}} h_m g_j = \begin{cases} (-1)^{a_m} h_m g_j & \text{if } m = j \\ h_m g_j & \text{if } m \neq j. \end{cases} \tag{7.128}$$

For each $m \in \{1, \ldots, n-k\}$ for which $a_m = 1$ let h_m be one of those that satisfy (7.128). For those $m \in \{1, \ldots, n-k\}$ for which $a_m = 0$, set $h_m = \mathbf{1}^{\otimes n}$. Then we have for

$$h = h_1^{a_1} \cdots h_{n-k}^{a_{n-k}} \tag{7.129}$$

that

$$g_j h \underbrace{=}_{(7.129)} g_j h_1^{a_1} \cdots h_{n-k}^{a_{n-k}} \underbrace{=}_{(7.128)} h_1^{a_1} \cdots h_{j-1}^{a_{j-1}} g_j h_j^{a_j} \cdots h_{n-k}^{a_{n-k}}$$

$$\underbrace{=}_{(7.128)} h_1^{a_1} \cdots h_{j-1}^{a_{j-1}} (-1)^{a_j} h_j^{a_j} g_j h_{j+1}^{a_{j+1}} \cdots h_{n-k}^{a_{n-k}} \underbrace{=}_{(7.128)} (-1)^{a_j} h_1^{a_1} \cdots h_{n-k}^{a_{n-k}} g_j$$

$$\underbrace{=}_{(7.129)} (-1)^{a_j} h g_j.$$

Hence, (7.122) implies $l_j(h) = a_j$, and it follows from (7.121) that $\text{syn}_q(h) = \mathbf{a}$, which implies $\text{syn}_q\{\mathbb{F}_2^n\} = \mathbb{F}_2^{n-k}$ and completes the proof. $\square$

Similar to the classical case with syn_c of Definition 7.9, we would like to use the quantum syndrome syn_q for detection of errors and subsequent correction in a quantum code. To see how this can be achieved with a stabilizer QECC $[\![n, k]\!]_q$ with stabilizer $\mathcal{S}$, let $\hat{\mathcal{E}} \in \mathcal{P}_n$ be an error operator having corrupted a codeword $|\Psi\rangle \in \mathbb{H}^{C_q} = \mathbb{H}^{n-k}$ such that

$$|\widetilde{\Psi}\rangle = \hat{\mathcal{E}}|\Psi\rangle \tag{7.130}$$

is the codeword received or retrieved. Then we have for any generator $g_j \in \mathcal{S}$ used to determine the syndrome

$$g_j|\widetilde{\Psi}\rangle \underbrace{=}_{(7.130)} g_j\hat{\mathcal{E}}|\Psi\rangle \underbrace{=}_{(7.122)} (-1)^{l_j(\hat{\mathcal{E}})}\hat{\mathcal{E}}g_j|\Psi\rangle \underbrace{=}_{g_j\in\mathcal{S}} (-1)^{l_j(\hat{\mathcal{E}})}\hat{\mathcal{E}}|\Psi\rangle \underbrace{=}_{(7.130)} (-1)^{l_j(\hat{\mathcal{E}})}|\widetilde{\Psi}\rangle,$$

that is, the corrupted codeword $|\widetilde{\Psi}\rangle$ is an eigenstate of each generator g_j with eigenvalue $(-1)^{l_j(\hat{\mathcal{E}})}$. Therefore, we can measure the the observables g_j, where $j \in \{1,\ldots,n-k\}$, without changing the corrupted state $|\widetilde{\Psi}\rangle$ to obtain the measurement results $(-1)^{l_j(\hat{\mathcal{E}})}$. Taken together, all these measurement results reveal the syndrome of $\hat{\mathcal{E}}$:

$$\mathrm{syn}_q(\mathcal{E}) = \big(l_1(\hat{\mathcal{E}}),\ldots,l_{n-k}(\hat{\mathcal{E}})\big) \in \mathbb{F}_2^{n-k}.$$

But Proposition 7.40 tells us that $\mathrm{syn}_q(\hat{\mathcal{E}})$ does not uniquely identify the error $\mathcal{E}$. It only determines the coset $\left[\hat{\mathcal{E}}\right]_{N(\mathcal{S})}$ to which the error operator $\hat{\mathcal{E}}$ belongs. We can then use the isomorphism provided by $\widehat{\mathrm{syn}_q}$ to identify this coset $\left[\hat{\mathcal{E}}\right]_{N(\mathcal{S})}$. We assume that the error which changes the smallest number of qubits is the most likely to have occurred. Therefore, we pick an element $g \in \left[\hat{\mathcal{E}}\right]_{N(\mathcal{S})}$ which has the minimal weight in the coset. Having chosen $g \in \left[\hat{\mathcal{E}}\right]_{N(\mathcal{S})}$ means that there exists an $h \in N(\mathcal{S})$ such that $g = \hat{\mathcal{E}}h$. We then apply a gate implementing g^* to $|\widetilde{\Psi}\rangle$ to correct the corrupted codeword. This way, we obtain the codeword

$$g^*|\widetilde{\Psi}\rangle = (\hat{\mathcal{E}}h)^*|\widetilde{\Psi}\rangle \underbrace{=}_{(2.48)} h^*\hat{\mathcal{E}}^*|\widetilde{\Psi}\rangle \underbrace{=}_{(7.130)} h^*\hat{\mathcal{E}}^*\hat{\mathcal{E}}|\Psi\rangle \underbrace{=}_{(F.131)} h^*|\Psi\rangle.$$

Exercise 7.93 shows that the result $h^*|\Psi\rangle$ of our correcting transformation $|\widetilde{\Psi}\rangle \mapsto g^*|\widetilde{\Psi}\rangle$ is again a valid codeword in $\mathbb{H}^{C_q}$.

Exercise 7.93 Let $\mathcal{S}$ be the stabilizer of a QECC C_q with code $\mathbb{H}^{C_q}$, and let $N(\mathcal{S})$ be the normalizer of $\mathcal{S}$. Show that then for any $h \in N(\mathcal{S})$ and $|\Psi\rangle \in \mathbb{H}^{C_q}$

$$h|\Psi\rangle \in \mathbb{H}^{C_q}.$$

For a solution see Solution 7.93.

Since $N(\mathcal{S})$ is a subgroup of $\mathcal{P}_n$ and $h \in N(\mathcal{S})$, it follows that $h^* = h^{-1} \in N(\mathcal{S})$ as well. Moreover, since the original codeword $|\Psi\rangle$ was an element of $\mathbb{H}^{C_q}$, the result of Exercise 7.93 tells us that $h^*|\Psi\rangle \in \mathbb{H}^{C_q}$ and is thus a valid codeword. In general, however, we have no guarantee that $h^*|\Psi\rangle = |\Psi\rangle$ and the original codeword is fully recovered. Only if we have been lucky enough to have selected h in $g = \hat{\mathcal{E}}h$ from the stabilizer, namely such that $h \in \mathcal{S} < N(\mathcal{S})$, would we end up with $h^*|\Psi\rangle = |\Psi\rangle$.

In summary, our quantum error detection and correction protocol is thus as follows:

Quantum Error Detection and Correction Protocol for Stabilizer QECC

1. Let $|\widetilde{\Psi}\rangle \in \mathbb{H}^N$ be a (possibly corrupted) codeword
2. Determine $\widetilde{l}_1, \ldots, \widetilde{l}_{n-k}$ by observing all generators g_j of $\mathcal{S} = \langle g_1, \ldots, g_{n-k}\rangle$ in the state $|\widetilde{\Psi}\rangle$, which satisfies

$$g_j|\widetilde{\Psi}\rangle = (-1)^{\widetilde{l}_j}|\widetilde{\Psi}\rangle\,,$$

and set $\mathrm{syn}_q(\hat{\mathcal{E}}) = (\widetilde{l}_1, \ldots, \widetilde{l}_{n-k})$
3. Determine $\left[\hat{\mathcal{E}}\right]_{N(\mathcal{S})} = \widehat{\mathrm{syn}}_q^{-1}\left(\mathrm{syn}_q(\hat{\mathcal{E}})\right)$
4. Determine the set of elements in this coset with minimal weight

$$R_q^{\min}\left[\hat{\mathcal{E}}\right]_{N(\mathcal{S})} := \left\{h \in \left[\hat{\mathcal{E}}\right]_{N(\mathcal{S})} \mid \mathrm{w}_{\mathcal{P}}(h) \leq \mathrm{w}_{\mathcal{P}}(g) \; \forall g \in \left[\hat{\mathcal{E}}\right]_{N(\mathcal{S})}\right\}$$

5. Select a random element

$$g \in R_q^{\min}\left[\hat{\mathcal{E}}\right]_{N(\mathcal{S})}$$

6. Transform $|\widetilde{\Psi}\rangle \mapsto g^*|\widetilde{\Psi}\rangle$ and proceed with $g^*|\widetilde{\Psi}\rangle$ as the presumed original codeword for $|\Psi\rangle$.

Note that, whenever the measurement of all generators g_j of $\mathcal{S} = \langle g_1, \ldots, g_{n-k}\rangle$ reveals $\mathrm{syn}_q(\hat{\mathcal{E}}) = \mathbf{0}$, the identification of the coset by $\widehat{\mathrm{syn}}_q^{-1}$ yields $\left[\hat{\mathcal{E}}\right]_{N(\mathcal{S})} = N(\mathcal{S})$. Since $N(\mathcal{S})$ is a subgroup, it contains $\mathbf{1}^{\otimes n}$. Therefore, we have in this case

$$R_q^{\min}\left[\hat{\mathcal{E}}\right]_{N(\mathcal{S})} = \left\{h \in \left[\hat{\mathcal{E}}\right]_{N(\mathcal{S})} \mid \mathrm{w}_{\mathcal{P}}(h) = 0\right\} \underbrace{=}_{\text{(F.134)}} \mathrm{Ctr}(\mathcal{P}_n)\,,$$

and the action of g^* is at most only an irrelevant phase multiplication of i^a, where $a \in \{0, \ldots, 3\}$, in other words, the error correction procedure leaves the state $|\widetilde{\Psi}\rangle$ unchanged if the syndrome vanishes.

We conclude this chapter on error correction with a result which is a quantum analogue of Proposition 7.11 and a stabilizer version of Proposition 7.26. It gives yet another sufficient condition for correctability, this time in terms of the weight of error operators and the distance of the code.

Corollary 7.41 *Let* C_q *be an* $[\![n, k]\!]_q$ *stabilizer QECC and* $\mathcal{E} = \{\mathcal{E}_a\}$ *an error in* $\mathbb{H}^{C_q}$ *with operation elements* $\mathcal{E}_a \in \mathcal{P}_n$. *Then we have*

$$w_{\mathcal{P}}(\mathcal{E}_a) \leq \left\lfloor \frac{d_{\mathcal{P}}(C_q) - 1}{2} \right\rfloor \quad \forall \mathcal{E}_a \in \mathcal{E} \quad \Rightarrow \quad \mathcal{E} \text{ is correctable in } \mathbb{H}^{C_q}.$$

Proof Let

$$w_{\mathcal{P}}(\mathcal{E}_a) \leq \left\lfloor \frac{d_{\mathcal{P}}(C_q) - 1}{2} \right\rfloor \tag{7.131}$$

hold for all $\mathcal{E}_a \in \mathcal{E}$. Since $\lfloor x \rfloor \leq x$, we obtain for any $\mathcal{E}_a, \mathcal{E}_b \in \mathcal{E}$ that

$$w_{\mathcal{P}}(\mathcal{E}_a^* \mathcal{E}_b) \underset{(F.136)}{\leq} w_{\mathcal{P}}(\mathcal{E}_a^*) + w_{\mathcal{P}}(\mathcal{E}_b) \underset{(F.135)}{\leq} w_{\mathcal{P}}(\mathcal{E}_a) + w_{\mathcal{P}}(\mathcal{E}_b)$$

$$\underset{(7.131)}{\leq} 2 \left\lfloor \frac{d_{\mathcal{P}}(C_q) - 1}{2} \right\rfloor \leq 2 \frac{d_{\mathcal{P}}(C_q) - 1}{2}$$

$$< d_{\mathcal{P}}(C_q) \underset{\text{Def. 7.24, (7.116)}}{=} \min \left\{ w_{\mathcal{P}}(g) \,\middle|\, g \in M_{\neq} \right\}.$$

Therefore, we must have $\mathcal{E}_a^* \mathcal{E}_b \notin M_{\neq}$, and (7.117) implies that then $\mathcal{E}_a^* \mathcal{E}_b \notin N(\mathcal{S}) \smallsetminus \mathcal{S}$. It follows from Theorem 7.35 that $\mathcal{E}$ is correctable in $\mathbb{H}^{C_q}$. $\qquad\square$

7.4 Further Reading

An introduction to classical error correction can be found in the book by PLESS [104].

For a more introductory and less mathematical exposition the reader may consult the review article by STEANE [105] or the more recent one by DEVITT, MUNRO and NEMOTO [98], which also references a great many of the original contributions.

More mathematical, albeit without proofs, but nonetheless eminently readable, is the review by GOTTESMANN [100]. A good comprehensive and mathematical treatise on the subject of quantum error correction is the book by GAITAN [106].

The voluminous collection edited by by LIDAR and BRUN [101] covers a very wide range of topics related to quantum error correction and fault tolerant quantum computing including more recent developments. This book also contains a chapter on topological quantum codes first proposed by KITAEV [36, 107, 108].

Chapter 8
Adiabatic Quantum Computing

8.1 Introduction

In exploring how the laws of quantum mechanics can be utilized to improve algorithms, we have essentially looked at what happens when we replace classical logic gates or circuits with their quantum versions. We call this circuit (or gate) based computation or computational model. Adiabatic quantum computation is an altogether different, albeit equivalent, way to harness quantum effects for the benefit of computation. Even though both the circuit based and adiabatic quantum computational model utilize the laws of quantum mechanics, they differ substantially in their approach.

The circuit based model (see Definition 5.27) uses a finite number L of gates $U_1, \ldots, U_L$ to build a circuit $U = U_L U_{L-1} \cdots U_2 U_1$. For pure states this circuit U then acts on an initial state $|\Psi_{\text{ini}}\rangle$ to produce the desired final state

$$|\Psi_{\text{fin}}\rangle = U|\Psi_{\text{ini}}\rangle = U_L U_{L-1} \cdots U_2 U_1 |\Psi_{\text{ini}}\rangle$$

as the output of the computation.

In contrast, the adiabatic quantum computation uses a time-dependent Hamiltonian $\mathsf{H}(t)$ to generate a time evolution $U(t_{\text{fin}}, t_{\text{ini}})$ that transforms the initial pure state $|\Psi_{\text{ini}}\rangle$ into a state close enough to the desired final state $|\Psi_{\text{fin}}\rangle$, that is,

$$\big\| \, |\Psi_{\text{fin}}\rangle - U(t_{\text{fin}}, t_{\text{ini}})|\Psi_{\text{ini}}\rangle \, \big\| \ll 1$$

such that after $|\Psi_{\text{ini}}\rangle$ has been subject to the time evolution we should find $|\Psi_{\text{fin}}\rangle$ with a high enough probability.

The quality of the adiabatic method then depends on finding a suitable $\mathsf{H}(t)$ that generates a time evolution such that $U(t_{\text{fin}}, t_{\text{ini}})|\Psi_{\text{ini}}\rangle$ is as close as possible to the desired final state $|\Psi_{\text{fin}}\rangle$. However, the transition time $T = t_{\text{fin}} - t_{\text{ini}}$ required for this should not become too large, so as not to suffer from decoherence effects. We shall explore all these aspects in this chapter, which is organized as follows.

© Springer Nature Switzerland AG 2019
W. Scherer, *Mathematics of Quantum Computing*,
https://doi.org/10.1007/978-3-030-12358-1_8

In Sect. 8.2 we state the assumptions underlying the adiabatic method and derive important results about the quality of the adiabatic approximation, which is at the heart of this method. In doing so, we make use of the Quantum Adiabatic Theorem, for which we give a thorough proof in Appendix G.

A generic version of the adiabatic method is presented in Sect. 8.3. As an application of this, we then look in Sect. 8.4 at a search algorithm using the adiabatic method. There we also show that with a suitably adapted algorithm the efficiency of the GROVER search algorithm of Sect. 6.9 can be obtained.

Since both the adiabatic and circuit based method produce the final state, the question arises if one method might be more efficient than the other. In Sects. 8.5 and 8.6 we shall show that the two approaches are indeed equivalent in their efficiency. More precisely, we shall show:

(i) in Sect. 8.5 that adiabatic quantum computation can efficiently approximate any circuit computation, in other words, the final state $U_L \cdots U_1 |\Psi_{ini}\rangle$ of a circuit computation with L gates $\{U_j \mid j \in \{1, \ldots, L\}\}$ can be obtained via an adiabatic quantum computation with a probability $\frac{1-\varepsilon^2}{L+1}$, and a transition time $T \in O\left(\frac{1}{\varepsilon} L^6\right)$ is sufficient to achieve this;

(ii) in Sect. 8.6 that the circuit based approach can efficiently approximate any adiabatic quantum computation, that is, the adiabatic time evolution $U(T + t_{ini}, t_{ini})$ can be approximated up to an error of δ by a circuit of J gates, where $J \in O\left(\left(\frac{T}{\delta}\right)^2\right)$.

Throughout this chapter we often use notions and results from Appendix G, which the reader is encouraged to consult.

8.2 Starting Point and Assumptions

Ever since its first formulation by BORN and FOCK [109] in 1928, the Quantum Adiabatic Theorem has been recast in various guises. Here we follow the approach first developed by KATO [110], which has been refined numerous times, and adopted to the context of quantum computing by JANSEN et al. [111].

All these formulations concern the following situation: we are given a quantum system with

• an initial Hamiltonian H_{ini} at some initial time t_{ini}
• a final Hamiltonian H_{fin} at some final time t_{fin}
• a time dependent Hamiltonian

$$
\begin{aligned}
H : [t_{ini}, t_{fin}] &\longrightarrow B_{sa}(\mathbb{H}) \\
t &\longmapsto H(t)
\end{aligned}
$$

such that

$$H(t_{\text{ini}}) = H_{\text{ini}} \qquad \text{and} \qquad H(t_{\text{fin}}) = H_{\text{fin}}, \tag{8.1}$$

and the time evolution between t_{ini} and t_{fin} is generated by $H(t)$, that is, the time evolution operator $U(t, t_{\text{ini}})$ for the system is given as the solution of the initial value problem

$$\begin{aligned} i\frac{d}{dt}U(t, t_{\text{ini}}) &= H(t)U(t, t_{\text{ini}}) \\ U(t_{\text{ini}}, t_{\text{ini}}) &= \mathbf{1}. \end{aligned} \tag{8.2}$$

Any initial state $|\Psi(t_{\text{ini}})\rangle$ evolves into the state $U(t, t_{\text{ini}})|\Psi(t_{\text{ini}})\rangle$ at a later time $t \geq t_{\text{ini}}$.

Let the system be prepared initially at time t_{ini} in the jth eigenstate $|\Phi_j(t_{\text{ini}})\rangle$ of the initial Hamiltonian H_{ini}, and then let it evolve according to the time evolution generated by $H(t)$ until t_{fin}. The question the Quantum Adiabatic Theorem addresses is the following: how well does this time-evolved state $U(t_{\text{fin}}, t_{\text{ini}})|\Phi_j(t_{\text{ini}})\rangle$ agree with with the jth eigenstate $|\Phi_j(t_{\text{fin}})\rangle$ of the final Hamiltonian H_{fin}? Figure 8.1 illustrates this setup and the question for $j = 0$.

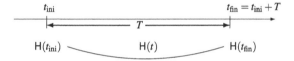

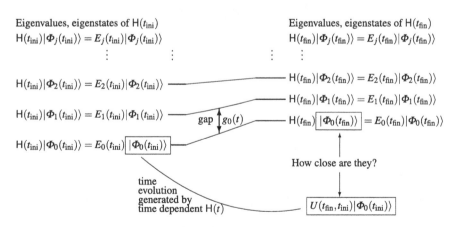

Fig. 8.1 Schematic representation of the question the Quantum Adiabatic Theorem addresses, where we consider the case of the ground state; in this illustration the left side depicts the eigenvalues, in other words, the 'energy levels' of the initial Hamiltonian $H(t_{\text{ini}})$; the right side depicts the energy levels of the final Hamiltonian $H(t_{\text{fin}})$; the curve emanating from $|\Phi_0(t_{\text{ini}})\rangle$ represents the time evolution $U(t_{\text{fin}}, t_{\text{ini}})$ generated by $H(t)$ for this state and attempts to illustrate that it need not coincide with the ground state of $H(t_{\text{fin}})$

The Quantum Adiabatic Theorem tells us that under certain conditions these two states can be made made arbitrarily close to each other. One of these conditions is that the system exhibits no eigenvalue crossing during the time evolution. Another condition is that the transition from H_{ini} to H_{fin} is sufficiently slow.

Before we proceed to make these statements more exact and ponder their relevance for quantum computing, we introduce some further notation. For the formulation of the Quantum Adiabatic Theorem it is helpful to define time-rescaled objects.

Definition 8.1 We denote the time elapsed between t_{ini} and $t_{fin} \geq t_{ini}$ by

$$T := t_{fin} - t_{ini} \tag{8.3}$$

and define the rescaled Hamiltonian by

$$\begin{aligned} H_T : [0,1] &\longrightarrow B_{sa}(\mathbb{H}) \\ s &\longmapsto H_T(s) := H(t_{ini} + sT) \end{aligned} \tag{8.4}$$

Similarly, we introduce the time-rescaled evolution operator

$$\begin{aligned} U_T : [0,1] &\longrightarrow B(\mathbb{H}) \\ s &\longmapsto U_T(s) := U(t_{ini} + sT, t_{ini}) \end{aligned} \tag{8.5}$$

As a consequence of this definition, we have that

$$H_T\left(\frac{t - t_{ini}}{T}\right) \underbrace{=}_{(8.4)} H(t)$$

$$H_T(0) \underbrace{=}_{(8.4)} H(t_{ini}) \underbrace{=}_{(8.1)} H_{ini}$$

$$H_T(1) \underbrace{=}_{(8.4)} H(t_{fin}) \underbrace{=}_{(8.1)} H_{fin} .$$

Lemma 8.2 *The time-rescaled evolution operator* $U_T(\cdot)$ *satisfies*

$$i\frac{d}{ds}U_T(s) = TH_T(s)U_T(s)$$
$$U_T(0) = \mathbf{1} \tag{8.6}$$
$$U_T(1) = U(t_{fin}, t_{ini}) .$$

Proof We have

$$
\begin{aligned}
i\frac{d}{ds}U_T(s) \underbrace{=}_{(8.5)} \ & i\frac{d}{ds}U(t_{\text{ini}} + sT, t_{\text{ini}}) \\
= \ & iT\frac{d}{dt}U(t, t_{\text{ini}})\Big|_{t=t_{\text{ini}}+sT} \\
\underbrace{=}_{(8.2)} \ & T\left[H(t)U(t, t_{\text{ini}})\right]\Big|_{t=t_{\text{ini}}+sT} \\
= \ & TH(t_{\text{ini}} + sT)U(t_{\text{ini}} + sT, t_{\text{ini}}) \\
\underbrace{=}_{(8.4),(8.5)} \ & TH_T(s)U_T(s)\,,
\end{aligned}
$$

which proves (8.6). Moreover,

$$
U_T(0) \underbrace{=}_{(8.5)} U(t_{\text{ini}}, t_{\text{ini}}) \underbrace{=}_{(8.2)} \mathbf{1}
$$

and

$$
U_T(1) \underbrace{=}_{(8.5)} U(t_{\text{ini}} + T, t_{\text{ini}}) \underbrace{=}_{(8.3)} U(t_{\text{fin}}, t_{\text{ini}})\,.
$$

$\square$

We denote the jth eigenvalue of $H_T(s)$ by $E_j(s)$, where the index j belongs to some index set $I \subset \mathbb{N}_0$. Each eigenvalue $E_j(s)$ may have a finite, d_j-fold degeneracy. The corresponding orthonormalized eigenvectors are thus denoted by $|\Phi_{j,\alpha}(s)\rangle$, where $j \in I$ and $\alpha \in \{1, \ldots, d_j\}$. For $j, k \in I$, $\alpha \in \{1, \ldots, d_j\}$ and $\beta \in \{1, \ldots, d_k\}$ they satisfy

$$
\langle \Phi_{j,\alpha}(s) | \Phi_{k,\beta}(s) \rangle = \delta_{j,k}\delta_{\alpha,\beta}
$$

and

$$
H_T(s)|\Phi_{j,\alpha}(s)\rangle = E_j(s)|\Phi_{j,\alpha}(s)\rangle\,. \tag{8.7}
$$

The eigenvalues $E_j(s)$ are often referred to as 'instantaneous' eigenvalues and the eigenvectors $|\Phi_{j,\alpha}(s)\rangle$ as instantaneous eigenvectors. Generally, the instantaneous jth eigenvector $|\Phi_{j,\alpha}(s)\rangle$ does not coincide with the state reached by time evolution from the initial state $|\Phi_{j,\alpha}(0)\rangle$. That is, in general we have that

$$
|\Phi_{j,\alpha}(s)\rangle \neq U_T(s)|\Phi_{j,\alpha}(0)\rangle = U(t_{\text{ini}} + sT, t_{\text{ini}})|\Phi_{j,\alpha}(0)\rangle\,,
$$

and the goal of the adiabatic method is to design H_T such that, at least for a particular j, this difference is as small as possible. In fact, what the Quantum Adiabatic Theorem provides is a bound on their difference.

Whereas t carries the physical dimension of time, the evolution parameter s is a pure number in $[0, 1]$ and is dimensionless. Using $H_T(s)$ and $U_T(s)$ for $s \in [0, 1]$ instead of $H(t)$ and $U(t, t_{\text{ini}})$ for $t \in [t_{\text{ini}}, t_{\text{fin}}]$ is a matter of pure convenience. We can always obtain $H(t)$ and $U(t, t_{\text{ini}})$ from $H_T(s)$ and $U_T(s)$ by the substitution $s = \frac{t - t_{\text{ini}}}{T}$ and vice versa by substituting $t = t_{\text{ini}} + sT$. For example, the $E_j\left(\frac{t-t_{\text{ini}}}{T}\right)$ are the instantaneous eigenvalues of $H(t)$ and $|\Phi_{j,\alpha}\left(\frac{t-t_{\text{ini}}}{T}\right)\rangle$ their respective eigenvectors since (8.4) implies

$$H(t) = H_T\left(\frac{t - t_{\text{ini}}}{T}\right)$$

and thus

$$
\begin{aligned}
H(t)|\Phi_{j,\alpha}\left(\frac{t - t_{\text{ini}}}{T}\right)\rangle &= H_T\left(\frac{t - t_{\text{ini}}}{T}\right)|\Phi_{j,\alpha}\left(\frac{t - t_{\text{ini}}}{T}\right)\rangle \\
&\underset{(8.7)}{=} E_j\left(\frac{t - t_{\text{ini}}}{T}\right)|\Phi_{j,\alpha}\left(\frac{t - t_{\text{ini}}}{T}\right)\rangle .
\end{aligned}
$$

For convenience, we shall mostly use the formulation in terms of s. For any s dependent object $A(s)$ we also use the abbreviation

$$\dot{A} = \dot{A}(s) := \frac{d}{ds}A(s) .$$

Exercise 8.94 Let

$$
\begin{aligned}
A : [0, 1] &\longrightarrow B(\mathbb{H}) \\
s &\longmapsto A(s)
\end{aligned}
$$

be differentiable with respect to s. Show that then

$$\left(\dot{A}(s)\right)^* = \left(\frac{d}{ds}A(s)\right)^* = \frac{d}{ds}\left(A(s)^*\right) = \left(A(s)^*\right)^{\cdot}$$

such that, in particular, if $A(s)^* = A(s)$, then also

$$\dot{A}(s)^* = \dot{A}(s) .$$

For a solution see Solution 8.94.

Throughout this chapter we always assume that during the time evolution between $s = 0$ and $s = 1$ (or, equivalently, between $t = t_{\text{ini}}$ and $t = t_{\text{fin}}$) the instantaneous eigenvalues $E_j(s)$ do not cross. In other words, we make the following assumptions, which we collectively denote by the term Adiabatic Assumption.

Adiabatic Assumption *(AA). We assume that*

$$H_T : [0, 1] \longrightarrow B_{sa}(\mathbb{H})$$
$$s \longmapsto H_T(s)$$

is such that

(i) $H_T(s)$ is at least twice differentiable with respect to every $s \in]0, 1[$

(ii) $H_T(s)$ has purely discrete spectrum $\{E_j(s) \mid j \in I \subset \mathbb{N}_0\}$ for every $s \in [0, 1]$; the eigenvalues may be $d_j(s)$-fold degenerate, where the degeneracies $d_j(s)$ are constant for $s \in]0, 1]$

(iii) the eigenvalues $E_j(s)$ of $H_T(s)$ do not cross, that is, for $s \in]0, 1]$ we have

$$E_0(s) < E_1(s) < \cdots < E_j(s) < E_{j+1}(s) < \cdots .$$

In some cases the Adiabatic Assumption (AA) (in particular item (iii)) may not be satisfied when considering $H_T(s)$ on the whole HILBERT space $\mathbb{H}$. However, it may hold only on a subspace $\mathbb{H}_{\text{sub}} \subset \mathbb{H}$. If this subspace is left invariant by H_T, that is, if $H_T(s)\{\mathbb{H}_{\text{sub}}\} \subset \mathbb{H}_{\text{sub}}$ for all $s \in [0, 1]$, then the Quantum Adiabatic Theorem and its implications still hold for the restriction $H_T|_{\mathbb{H}_{\text{sub}}} : \mathbb{H}_{\text{sub}} \to \mathbb{H}_{\text{sub}}$.

Exercise 8.95 Let $|\Phi\rangle, |\Psi\rangle \in \mathbb{H}$ with $\||\Phi\|| = 1 = \||\Psi\||$ and let $\varepsilon \in \mathbb{R}$ be such that

$$\||\,|\Phi\rangle - |\Psi\rangle\,\|| \leq \varepsilon . \tag{8.8}$$

Show that then

$$|\langle\Phi|\Psi\rangle|^2 \geq 1 - \varepsilon^2 .$$

For a solution see Solution 8.95.

With the help of our version of the Quantum Adiabatic Theorem G.15 we can now make a statement about the probability of finding a system in an eigenstate of H_{fin} if it was initially prepared in the corresponding eigenstate of H_{ini}.

Corollary 8.3 *Let* H_T *be as defined in Definition 8.1 and be such that it sat-isfies the Adiabatic Assumption (AA). Moreover, let* $\varepsilon_j := \frac{C_j(1)}{T}$, *where* $C_j(s)$ *is as defined in (G.45). Let a system initially be prepared in an eigenstate* $|\Xi_j\rangle$ *of* H_{ini} *with eigenvalue* $E_j(0)$ *and then subject to the time evolution generated by* $H_T(s)$ *for* $s \in [0,1]$.

Then the probability to find that system at time t_{fin} *an eigenstate of* H_{fin} *with the eigenvalue* $E_j(1)$ *is given by* $\big||P_j(1)U_T(1)\Xi_j|\big|^2$ *and satisfies*

$$\big||P_j(1)U_T(1)\Xi_j|\big|^2 \geq 1 - \varepsilon_j^2,\qquad(8.9)$$

that is, we have

$$\mathbf{P}\left\{\begin{array}{l}\text{To observe an eigenstate of } H_{fin} \text{ for}\\ \text{the eigenvalue } E_j(1) \text{ at time } t_{fin},\\ \text{if system was in an eigenstate } |\Xi_j\rangle\\ \text{of } H_{ini} \text{ for the eigenvalue } E_j(0) \text{ at}\\ \text{time } t_{ini}\end{array}\right\} = \big||P_j(1)U_T(1)\Xi_j|\big|^2 \geq 1 - \varepsilon_j^2.$$

Proof Note that since $H_T(1) = H_{fin}$, $P_j(1)$ is the projector onto the eigenspace of H_{fin} for the eigenvalue $E_j(1)$. Moreover, by Lemma 8.2 we have

$$U_T(1) = U(t_{fin}, t_{ini}),$$

where $U(t, t_{ini})$ is the time evolution generated by $H(t)$. Starting from an initial state $|\Xi_j\rangle$, the system evolves to the state $U(t_{fin}, t_{ini})|\Xi_j\rangle = U_T(1)|\Xi_j\rangle$ at t_{fin}.

The statement that $\big||P_j(1)U_T(1)\Xi_j|\big|^2$ is the probability to observe an eigenstate of H_{fin} for the eigenvalue $E_j(1)$, when the system is in the time evolved state $U_T(1)|\Xi_j\rangle$, is the content of Postulate 2. It thus remains to show (8.9). For this let

$$\begin{aligned}|\Phi(s)\rangle &= U_{A,j}(s)P_j(0)|\Xi_j\rangle = U_{A,j}(s)|\Xi_j\rangle\\ |\Psi(s)\rangle &= U_T(s)P_j(0)|\Xi_j\rangle = U_T(s)|\Xi_j\rangle,\end{aligned}\qquad(8.10)$$

where $U_{A,j}(s)$ is the adiabatic intertwiner defined in Definition G.9. Consequently, we have

$$\begin{aligned}\big|||\Phi(s)\rangle - |\Psi(s)\rangle|\big| &= \big||(U_{A,j}(s) - U_T(s))P_j(0)\Xi_j|\big|\\ &\underset{(2.51)}{\leq} \big||(U_{A,j}(s) - U_T(s))P_j(0)|\big|\underbrace{\big||\Xi_j|\big|}_{=1}\\ &\underset{(G.44)}{\leq} \frac{C_j(s)}{T} = \varepsilon_j(s).\end{aligned}$$

The result of Exercise 8.95 then implies

$$|\langle \Phi(s)|\Psi(s)\rangle|^2 \geq 1 - \varepsilon_j(s)^2 .\qquad (8.11)$$

With this we can show claim (8.9) as follows. For any $s \in [0, 1]$ we have

$$||P_j(s)U_T(s)\Xi_j||^2 \underset{(2.37)}{=} ||U_{A,j}(s)^* P_j(s)U_T(s)\Xi_j||^2$$

$$\underset{(G.23)}{=} ||P_j(0)U_{A,j}(s)^* U_T(s)\Xi_j||^2 .\qquad (8.12)$$

Since $|\Xi_j\rangle$ is assumed to be an eigenstate of $\mathsf{H}_{\text{ini}} = \mathsf{H}_T(0)$ for the eigenvalue $E_j(0)$, we can form an ONB of the eigenspace for this eigenvalue that contains $|\Xi_j\rangle$ as one basis vector, that is, we form an ONB $\{|\widetilde{\Phi_{j,\alpha}}(0)\rangle \mid \alpha \in \{1, \ldots d_j\}\}$ of $\text{Eig}(\mathsf{H}_T(0), E_j(0))$, where $|\widetilde{\Phi_{j,1}}(0)\rangle = |\Xi_j\rangle$. This means that we can express the projector $P_j(0)$ onto the eigenspace $\text{Eig}(\mathsf{H}_T(0), E_j(0))$ in the form

$$P_j(0) = \sum_{\alpha=1}^{d_j} |\widetilde{\Phi_{j,\alpha}}(0)\rangle\langle\widetilde{\Phi_{j,\alpha}}(0)| = |\Xi_j\rangle\langle\Xi_j| + \sum_{\alpha=2}^{d_j} |\widetilde{\Phi_{j,\alpha}}(0)\rangle\langle\widetilde{\Phi_{j,\alpha}}(0)| .$$

Moreover, as the $\{|\widetilde{\Phi_{j,\alpha}}(0)\rangle \mid \alpha \in \{1, \ldots d_j\}\}$ are mutually orthogonal, it follows for any $|\Omega\rangle \in \mathbb{H}$ that

$$||P_j(0)\Omega||^2 \underset{(2.15)}{=} \sum_{\alpha=1}^{d_j} \left|\langle\widetilde{\Phi_{j,\alpha}}(0)|\Omega\rangle\right|^2 \geq |\langle\Xi_j|\Omega\rangle|^2 .\qquad (8.13)$$

Setting $|\Omega\rangle = U_{A,j}(s)^* U_T(s)|\Xi_j\rangle$, we obtain from (8.12) that

$$||P_j(s)U_T(s)\Xi_j||^2 \underset{(8.13)}{\geq} \left|\langle\Xi_j|U_{A,j}(s)^* U_T(s)\Xi_j\rangle\right|^2$$

$$\underset{(2.30)}{=} \left|\langle U_{A,j}(s)\Xi_j|U_T(s)\Xi_j\rangle\right|^2$$

$$\underset{(8.10)}{=} |\langle \Phi(s)|\Psi(s)\rangle|^2$$

$$\underset{(8.11)}{\geq} 1 - \varepsilon_j(s)^2$$

and the claim (8.9) follows from taking $s = 1$. □

Another way to use the result of Theorem G.15 is to derive a lower bound for the transition time T if we want to achieve a given minimal success-probability to find an eigenstate of $H_{fin} = H_T(1)$ at the end of the transition. In other words, Theorem G.15 tells us how slowly we have to change $H_{ini} = H_T(0)$ into $H_{fin} = H_T(1)$ if we want to find the system in an eigenstate of H_{fin}, given that it was initially prepared in the corresponding eigenstate of H_{ini}. This is made more precise in the following corollary.

Corollary 8.4 *Let H_T be defined as in Definition 8.1 and such that it satisfies the Adiabatic Assumption (AA), and let $C_j(s)$ be defined as in (G.45). Let a system initially be prepared in an eigenstate $|\Xi_j\rangle$ of H_{ini} with eigenvalue $E_j(0)$ and then subject to the time evolution generated by $H_T(s)$ for $s \in [0, 1]$.*
If the time T for the transition from H_{ini} to H_{fin} satisfies

$$T \geq \frac{C_j(1)}{\sqrt{1 - p_{min}}} \tag{8.14}$$

for $p_{min} \in [0, 1[$, then the probability to find that system at time t_{fin} in an eigenstate of H_{fin} with the eigenvalue $E_j(1)$ is bounded from below by p_{min}, that is,

$$\mathbf{P} \left\{ \begin{array}{l} \text{To observe an eigenstate of } H_{fin} \text{ for} \\ \text{the eigenvalue } E_j(1) \text{ at time } t_{fin}, \\ \text{if system was in an eigenstate } |\Xi_j\rangle \\ \text{of } H_{ini} \text{ for the eigenvalue } E_j(0) \text{ at} \\ \text{time } t_{ini} \end{array} \right\} \geq p_{min} \,.$$

Proof This is a direct consequence of Corollary 8.3 since the success-probability in question is given by $\left\| P_j(1)U_T(1)\Xi_j \right\|^2$ and satisfies

$$\left\| P_j(1)U_T(1)\Xi_j \right\|^2 \underbrace{\geq}_{(8.9)} 1 - \left(\frac{C_j(1)}{T} \right)^2 \underbrace{\geq}_{(8.14)} p_{min} \,.$$

$\square$

We recall from (G.45) that $C_j(1)$ is given as

$$C_j(1) = \frac{\left\| \dot{H}_T(1) \right\|}{g_j(1)^2} + \frac{\left\| \dot{H}_T(0) \right\|}{g_j(0)^2} + \int_0^1 \left(\frac{\left\| \ddot{H}_T(u) \right\|}{g_j(u)^2} + 10 \frac{\left\| \dot{H}_T(u) \right\|^2}{g_j(u)^3} \right) du \,. \tag{8.15}$$

In many versions of the Adiabatic Theorem quoted in the literature [33, 112, 113] this expression for $C_j(1)$ is further simplified by using

$$g_{j,\min} := \min_{s \in [0,1]} \{g_j(s)\} \tag{8.16}$$

and

$$\left\| \dot{\mathsf{H}} \right\|_{\max} := \max_{s \in [0,1]} \left\{ \left\| \dot{\mathsf{H}}_T(s) \right\| \right\}$$

to obtain an upper bound on $C_j(1)$. As we shall see when considering the adiabatic quantum search in Sect. 8.4, this may be too generous, and, instead, making use of more explicit expressions for $\left\| \dot{\mathsf{H}}_T(s) \right\|$, $\left\| \ddot{\mathsf{H}}_T(s) \right\|$ and $g_j(s)$ can give much improved bounds for $C_j(1)$.

8.3 Generic Adiabatic Algorithm

How can the results of the previous section be of help in solving problems via quantum computation? To see this, let us look at how an algorithm based on adiabatic time evolution can be utilized to solve a problem, the solution of which is given by some state $|\Psi_s\rangle$ in some HILBERT space $\mathbb{H}$. For example, the solution of our problem may be some natural number $x < 2^n$ for some $n \in \mathbb{N}$. Then it could be obtained by devising a solution-algorithm that finds the corresponding computational basis state $|x\rangle \in \mathbb{H} = \mathbb{H}^{\otimes n}$.

For such problems we can describe a **generic adiabatic algorithm** as follows:

Generic Adiabatic Algorithm

Input: A problem for which the solution can be obtained through knowledge of a state $|\Psi_s\rangle$ in some HILBERT space $\mathbb{H}$

Step 1: Find a Hamiltonian H_{fin} such that the solution state $|\Psi_s\rangle$ is an eigenstate of H_{fin} for some eigenvalue $E_{\text{fin},j}$, that is, $|\Psi_s\rangle \in \text{Eig}\left(\mathsf{H}_{\text{fin}}, E_{\text{fin},j}\right)$. Typically, one chooses $E_{\text{fin},j} = E_{\text{fin},0}$ to be the lowest eigenvalue such that the solution is a ground state of H_{fin}

Step 2: Find a Hamiltonian H_{ini} for which an eigenstate $|\Phi_{\text{ini},j}\rangle$ with the same j as in Step 1 is known and can easily be prepared

Step 3: Find a time dependent Hamiltonian $\mathsf{H}(t)$ such that

$$\mathsf{H}(t_{\text{ini}}) = \mathsf{H}_{\text{ini}}$$
$$\mathsf{H}(t_{\text{fin}}) = \mathsf{H}_{\text{fin}}$$

and $\mathsf{H}_T(s) = \mathsf{H}(t_{\text{ini}} + sT)$, where $T = t_{\text{fin}} - t_{\text{ini}}$, satisfies the Adiabatic Assumption (AA)

Step 4: Prepare the quantum system in the initial state $|\Phi_{\text{ini},j}\rangle$ at time t_{ini}

Step 5: Evolve the system from time t_{ini} for a period of length T until t_{fin} with the time evolution $U(t, t_{\text{ini}})$ generated by $\mathsf{H}(t)$ according to (8.2)

Step 6:   Observe the final state

$$U(t_{\text{fin}}, t_{\text{ini}})|\Phi_{\text{ini},j}\rangle = U_T(1)|\Phi_{\text{ini},j}\rangle \, .$$

From Corollary 8.3 we know that the probability to find a state in Eig $\left(\mathsf{H}_{\text{fin}}, E_{\text{fin},j}\right)$ is bounded from below by $1 - \left(\frac{C_j(1)}{T}\right)^2$

Output: A solution state $|\Psi_s\rangle \in$ Eig $\left(\mathsf{H}_{\text{fin}}, E_{\text{fin},j}\right)$ with a probability of at least $1 - \left(\frac{C_j(1)}{T}\right)^2$

From this algorithm we see that we can increase the probability of finding our desired solution state $|\Psi_s\rangle$ by increasing T, which is akin to performing the time evolution more slowly. One might be tempted to think that by making T large enough we could bring our success-probability arbitrarily close to 1. The reason that this does not work is that we cannot isolate quantum systems an arbitrarily large time from its environment, and the natural decoherence time puts a limit on how large we can make T.

Another way, however, to increase the success-probability is to make $C_j(1)$ as small as possible. From (8.15) we see that one way to contribute to this is by choosing the eigenvalue E_j with the largest gap g_j. Typically, this is the lowest eigenvalue $E_{\text{ini},0}$ of H_{ini}, and the corresponding eigenstate is its ground state.

From (8.15) we also see that another way to decrease $C_j(1)$ consists of finding a time dependent $\mathsf{H}_T(s)$ such that $\left\|\dot{\mathsf{H}}_T(s)\right\|$ and $\left\|\ddot{\mathsf{H}}_T(s)\right\|$ are small. However, slowing down the transition will inevitably increase T.

Definition 8.5 If in an adiabatic algorithm $t \mapsto \mathsf{H}(t)$ is a convex combination of H_{ini} and H_{fin} such that it results in

$$\mathsf{H}_T(s) = \left(1 - f(s)\right)\mathsf{H}_{\text{ini}} + f(s)\mathsf{H}_{\text{fin}} , \qquad (8.17)$$

where $f : [0,1] \to [0,1]$ is at least twice differentiable with $\dot{f} > 0$ and satisfies $f(0) = 0$ as well as $f(1) = 1$, then the function f is called the **adiabatic schedule** of the adiabatic algorithm.

If $\mathsf{H}_T(s)$ is a convex combination of the initial H_{ini} and the final H_{fin}, then reducing $\left\|\dot{\mathsf{H}}_T(s)\right\|$ and $\left\|\ddot{\mathsf{H}}_T(s)\right\|$ means finding an optimal schedule $f(s)$. In Theorem 8.16 in Sect. 8.4 we will encounter an example where the gap $g_j(s)$ is known explicitly and where we can then optimize the schedule $f(s)$ to attain a small $C_j(1)$. The following corollary gives a general expression for $C_j(s)$ in the case of a convex combination.

Corollary 8.6 *Let $\mathsf{H}_T(s)$ be a convex combination of $\mathsf{H}_{ini}, \mathsf{H}_{fin} \in \mathsf{B}_{sa}(\mathbb{H})$ of the form*

$$\mathsf{H}_T(s) = \big(1 - f(s)\big)\mathsf{H}_{ini} + f(s)\mathsf{H}_{fin} \qquad (8.18)$$

with adiabatic schedule $f : [0,1] \to [0,1]$. Then $C_j(s)$ for $s \in [0,1]$ as defined in Theorem G.15 and used in Corollaries 8.3 and 8.4 is given by

$$C_j(s) = \left|\left|\mathsf{H}_{fin} - \mathsf{H}_{ini}\right|\right| \left[\frac{\dot{f}(s)}{g_j(s)^2} + \frac{\dot{f}(0)}{g_j(0)^2} \right. \qquad (8.19)$$

$$\left. + \int_0^s \left(\frac{\left|\ddot{f}(u)\right|}{g_j(u)^2} + 10\left|\left|\mathsf{H}_{fin} - \mathsf{H}_{ini}\right|\right| \frac{(\dot{f}(u))^2}{g_j(u)^3} \right) du \right],$$

and an upper bound for it can be given by replacing $\left|\left|\mathsf{H}_{fin} - \mathsf{H}_{ini}\right|\right|$ in (8.19) by $\left|\left|\mathsf{H}_{fin}\right|\right| + \left|\left|\mathsf{H}_{ini}\right|\right|$.

Proof From (8.18) it follows that

$$\dot{\mathsf{H}}_T(s) = \dot{f}(s)\big(\mathsf{H}_{fin} - \mathsf{H}_{ini}\big) \quad \text{and} \quad \ddot{\mathsf{H}}_T(s) = \ddot{f}(s)\big(\mathsf{H}_{fin} - \mathsf{H}_{ini}\big).$$

Using $\dot{f} > 0$, we thus obtain

$$\left|\left|\dot{\mathsf{H}}_T(s)\right|\right| = \dot{f}(s)\left|\left|\mathsf{H}_{fin} - \mathsf{H}_{ini}\right|\right| \quad \text{and} \quad \left|\left|\ddot{\mathsf{H}}_T(s)\right|\right| = \left|\ddot{f}(s)\right|\left|\left|\mathsf{H}_{fin} - \mathsf{H}_{ini}\right|\right|.$$

The claim (8.19) then follows by inserting these expressions for the norms of the derivatives into (G.45), and the last statement in Corollary 8.6 follows from (2.53). □

Since it is notoriously difficult to obtain the explicit form of $g_j(s)$, it may be necessary to replace it by a lower bound, such as its minimum as defined in (8.16) if at least that can be found.

It may happen that, although we can specify and (hopefully) implement a time dependent Hamiltonian $\mathsf{H}(\cdot)$ starting at H_{ini} and ending at H_{fin}, the eigenvalues $E_j(\cdot)$ along its path do not satisfy the Adiabatic Assumption (AA) on the full HILBERT space $\mathbb{H}$. Hence, the preconditions of the Adiabatic Theorem G.15 would not be satisfied.

If, however, the Adiabatic Assumption (AA) is satisfied on a subspace $\mathbb{H}_{sub} \subset \mathbb{H}$ and if $H(\cdot)$ leaves this subspace invariant, then we can apply Theorem G.15 and its Corollaries 8.3, 8.4 and 8.6 to the reduced setup, in other words, consider only the restriction of $H(\cdot)$ to $\mathbb{H}_{sub}$.

Before we encounter such a situation in the context of the adiabatic search algorithm, we make the above statements more precise with the following lemma, which shows that if $H(\cdot)$ maps vectors of $\mathbb{H}_{sub}$ into $\mathbb{H}_{sub}$, then the time evolution generated by $H(\cdot)$ does too. This lemma will be useful in the construction of an adiabatic version of GROVER's search algorithm to which we turn in Sect. 8.4.

Lemma 8.7 *Let* $H : [t_{ini}, t_{fin}] \rightarrow B_{sa}(\mathbb{H})$ *be a time-dependent Hamiltonian which generates the time evolution U such that*

$$\mathrm{i}\frac{d}{dt}U(t, t_{ini}) = H(t)U(t, t_{ini}) \tag{8.20}$$
$$U(t_{ini}, t_{ini}) = \mathbf{1}.$$

Moreover, let $\mathbb{H}_{sub} \subset \mathbb{H}$ *be a subspace of* $\mathbb{H}$ *and* P_{sub} *the projection onto this subspace. If* P_{sub} *is such that for all* $t \in [t_{ini}, t_{fin}]$ *we have*

$$[H(t), P_{sub}] = 0, \tag{8.21}$$

then

$$U(t, t_{ini})\{\mathbb{H}_{sub}\} \subset \mathbb{H}_{sub}$$

holds for $t \in [t_{ini}, t_{fin}]$.

Proof The claim is proven, if we can show that

$$U(t, t_{ini})^* P_{sub} U(t, t_{ini}) = P_{sub} \tag{8.22}$$

holds for all $t \in [t_{ini}, t_{fin}]$. This is because (8.22) implies $U(t, t_{ini})P_{sub} = P_{sub}U(t, t_{ini})$ and thus for any $|\Psi\rangle \in \mathbb{H}_{sub}$

$$U(t, t_{ini})|\Psi\rangle = U(t, t_{ini})P_{sub}|\Psi\rangle = P_{sub}U(t, t_{ini})|\Psi\rangle \in \mathbb{H}_{sub}.$$

From (8.20) we see that (8.22) is obviously true for $t = t_{ini}$. To show that it holds for every $t \in [t_{ini}, t_{fin}]$, we show that the left side of (8.22) is constant:

$$\frac{d}{dt}\big(U(t,t_{\text{ini}})^{*}P_{\text{sub}}U(t,t_{\text{ini}})\big)$$

$$= \frac{dU(t,t_{\text{ini}})^{*}}{dt}P_{\text{sub}}U(t,t_{\text{ini}}) + U(t,t_{\text{ini}})^{*}P_{\text{sub}}\frac{dU(t,t_{\text{ini}})}{dt}$$

$$\underset{(8.20),(2.73)}{=} i\big(U(t,t_{\text{ini}})^{*}\mathsf{H}(t)P_{\text{sub}}U(t,t_{\text{ini}}) - U(t,t_{\text{ini}})^{*}P_{\text{sub}}\mathsf{H}(t)U(t,t_{\text{ini}})\big)$$

$$\underset{(2.46)}{=} i U(t,t_{\text{ini}})^{*}\big[\mathsf{H}(t),P_{\text{sub}}\big]U(t,t_{\text{ini}})$$

$$\underset{(8.21)}{=} 0,$$

which completes the proof of (8.22). $\qquad\qquad\qquad\qquad\qquad\qquad\square$

Before we turn to the adiabatic search, we illustrate how the adiabatic method may be used to solve commonly occurring quadratic binary optimization problems.

Definition 8.8 (*QUBO*) A **Quadratic Unconstrained Binary Optimization (QUBO)** problem is defined as follows: for $n \in \mathbb{N}$ and a given $Q \in \text{Mat}(n\times,n,\mathbb{R})$ find the extremum of the function

$$B: \quad \{0,1\}^{n} \longrightarrow \mathbb{R}$$
$$(x_0,\ldots,x_{n-1}) \longmapsto \sum_{i,j=0}^{n-1} x_i Q_{ij} x_j .$$

Example 8.9 As an example of how the adiabatic method might be applied to solve a QUBO *minimization* problem for

$$B(x) = B(x_0,\ldots,x_{n-1}) = \sum_{i,j=0}^{n-1} x_i Q_{ij} x_j , \qquad x_i \in \{0,1\} \qquad (8.23)$$

with a given $Q \in \text{Mat}(n \times n, \mathbb{R})$, we consider the Hamiltonian

$$\mathsf{H}_T(s) = \big(1-f(s)\big)\mathsf{H}_{\text{ini}} + f(s)\mathsf{H}_{\text{fin}}$$

on $\mathbb{H}^{\otimes n}$, where $f : [0,1] \to [0,1]$ is an adiabatic schedule as given in Definition 8.5. To find the minimum for B, we choose the initial and final Hamiltonians as

$$H_{ini} = \sum_{j=0}^{n-1} \Sigma_z^j$$

$$H_{fin} = \sum_{j=0}^{n-1} K_j \Sigma_z^j + \sum_{\substack{i,j=0 \\ i \neq j}}^{n-1} J_{ij} \Sigma_z^i \Sigma_z^j + c\mathbf{1}^{\otimes n} \tag{8.24}$$

with Σ_z^j as defined in Definition 5.35 and

$$J_{ij} = \frac{1}{4} Q_{ij} \qquad \text{for } i \neq j$$

$$K_j = -\frac{1}{4} \sum_{\substack{i=0 \\ i \neq j}}^{n-1} (Q_{ij} + Q_{ji}) - \frac{1}{2} Q_{jj} \tag{8.25}$$

$$c = \frac{1}{4} \sum_{\substack{i,j=0 \\ i \neq j}}^{n-1} Q_{ji} + \frac{1}{2} \sum_{j=0}^{n-1} Q_{jj} \, .$$

Recall that $\sigma_z|0\rangle = |0\rangle$ and $\sigma_z|1\rangle = -|1\rangle$, which we can write succinctly as

$$\sigma_z|x_j\rangle = (1 - 2x_j)|x_j\rangle \qquad \text{for } x_j \in \{0, 1\}$$

such that for any computational basis vector $|x\rangle = |x_{n-1} \ldots x_0\rangle \in \mathbb{H}^{\otimes n}$ we have

$$\Sigma_z^j|x\rangle = (1 - 2x_j)|x\rangle \, . \tag{8.26}$$

Exercise 8.96 Show that H_{ini} as given in (8.24) has the eigenvalues

$$E_{ini,l} = -n + 2l \qquad \text{for } l \in \{0, \ldots, n\}$$

with degeneracy $d_l = \binom{n}{l}$ such that its lowest eigenvalue $E_{ini,0} = -n$ is non-degenerate and its ground state is given by

$$|\Phi_0\rangle = |11 \ldots 1\rangle = |2^n - 1\rangle \, .$$

For a solution see Solution 8.96.

The result of Exercise 8.96 shows that the ground state of H_{ini} is easy to prepare. The result of the next exercise shows that the computational basis states $\{|x\rangle \mid x \in \mathbb{N}_0, x < 2^n\}$ of $\mathbb{H}^{\otimes n}$ are the eigenstates of H_{fin} with the eigenvalues given by $B(x)$.

Exercise 8.97 Show that for H_{fin} as given in (8.24) and (8.25) and B as given in (8.23) any computational basis state $|x\rangle = |x_{n-1} \ldots x_0\rangle \in \mathbb{H}^{\otimes n}$ satisfies

$$H_{fin}|x\rangle = B(x)|x\rangle. \qquad (8.27)$$

For a solution see Solution 8.97.

We can thus easily prepare the system in the initial ground state $|\Phi_{ini,0}\rangle = |11 \ldots 1\rangle$ of $H_T(0) = H_{ini}$ and then subject the system to the time evolution generated by $H_T(s)$ for $s = 0 \mapsto s = 1$. Provided that the Adiabatic Assumption (AA) is satisfied, Corollary 8.4 tells us that if T satisfies (8.14), the system will end up in the ground state, that is, the eigenstate with the lowest eigenvalue of $H_T(1) = H_{fin}$, with a probability of at least p_{min}. From (8.27) we see that this is the lowest possible value of $B(x)$. Hence, observing the time-evolved state $U_T(1)|\Phi_0\rangle$ will reveal a solution of the QUBO minimization problem with a probability of at least p_{min}.

8.4   Adiabatic Quantum Search

We illustrate the adiabatic method by using it for the search problem which we dealt with in the GROVER algorithm in Sect. 6.9. This means that we are trying to find one of m elements of S, which is the set of 'solutions', by the adiabatic method [111, 113, 114]. In other words, we are trying to find an

$$x \in S \subset \{0, 1, \ldots, N-1\},$$

where $N = 2^n$. For the adiabatic search we initially prepare the system in the state

$$|\Psi_0\rangle = \frac{1}{\sqrt{N}} \sum_{x=0}^{N-1} |x\rangle \in \mathbb{H}^{\otimes n} =: \mathbb{H},$$

which is the eigenstate for the eigenvalue 0 of the initial Hamiltonian operator

$$H_{ini} = \mathbf{1} - |\Psi_0\rangle\langle\Psi_0|.$$

For all vectors $|\Psi\rangle \in \mathbb{H}_{|\Psi_0\rangle^\perp}$ orthogonal to $|\Psi_0\rangle$ we have

$$H_{\text{ini}}|\Psi\rangle = |\Psi\rangle \,,$$

that is, they are eigenvectors of H_{ini} with eigenvalue 1. Whereas the eigenspace for the eigenvalue 0 is one-dimensional and spanned by $|\Psi_0\rangle$, the eigenspace for the eigenvalue 1 is $(N-1)$-dimensional and given by the orthogonal complement of $|\Psi_0\rangle$.

The final Hamiltonian H_{fin} of our adiabatic search is given by

$$H_{\text{fin}} = \mathbf{1} - P_S \,,$$

where P_S is defined in (6.137) as the projection onto the subspace spanned by basis vectors $|x\rangle$ of the objects $x \in S$ we are searching for. Note that

$$|\Psi_S\rangle = \frac{1}{\sqrt{m}} \sum_{x \in S} |x\rangle$$

as an equal superposition of all solution states is in the subspace onto which P_S projects.

With S and $S^{\perp}$ as defined in Definition 6.22, S is assumed to have m elements, and since

$$H_{\text{fin}}|x\rangle = 0 \qquad \forall x \in S \,,$$

the eigenspace of H_{fin} for the eigenvalue 0 is m-fold degenerate. On the other hand, we have

$$H_{\text{fin}}|x\rangle = |x\rangle - P_S|x\rangle \underbrace{=}_{(6.137)} |x\rangle \qquad \forall x \in S^{\perp} \,,$$

which means that H_{fin} also has 1 as an eigenvalue and that the eigenspace of H_{fin} for this eigenvalue 1 is spanned by the $N-m$ computational basis vectors $|x\rangle$ for which $x \in S^{\perp}$. Hence, the eigenvalue 1 of H_{fin} is $(N-m)$-fold degenerate.

Definition 8.10 Let $N = 2^n - 1$ and $m \in \mathbb{N}$ with $m < N$. An **adiabatic search algorithm** for the search of m objects (aka 'solutions') $x \in S \subset \{0, \ldots, N-1\}$ consists of subjecting a system in $\mathbb{H} := \mathbb{H}^{\otimes n}$ between the times t_{ini} and $t_{\text{fin}} = t_{\text{ini}} + T$ to the time evolution generated by the time dependent Hamiltonian

$$H(t) = H_T\left(\frac{t - t_{\text{ini}}}{T}\right) = \left[1 - f\left(\frac{t - t_{\text{ini}}}{T}\right)\right] H_{\text{ini}} + f\left(\frac{t - t_{\text{ini}}}{T}\right) H_{\text{fin}} \,,$$

where $f[0,1] \to [0,1]$ is an adiabatic schedule of the time evolution. We define

$$H_{\text{ini}} := \mathbf{1} - |\Psi_0\rangle\langle\Psi_0| \tag{8.28}$$

with

$$|\Psi_0\rangle = \frac{1}{\sqrt{N}} \sum_{x=0}^{N-1} |x\rangle\langle x| \in \mathbb{H} \tag{8.29}$$

as the initial and

$$\mathsf{H}_{\mathrm{fin}} := \mathbf{1} - P_{\mathrm{S}} \tag{8.30}$$

with

$$P_{\mathrm{S}} = \sum_{x\in\mathrm{S}} |x\rangle\langle x| \tag{8.31}$$

as the final Hamiltonian of the adiabatic search algorithm.

Since we are only interested in the time evolution between t_{ini} and t_{fin}, it is more convenient to work with the rescaled Hamiltonian as defined in (8.4), that is,

$$\mathsf{H}_T(s) = \mathsf{H}(t_{\mathrm{ini}} + sT) = \big(1 - f(s)\big)\mathsf{H}_{\mathrm{ini}} + f(s)\mathsf{H}_{\mathrm{fin}} , \tag{8.32}$$

where $s \in [0,1]$.

Exercise 8.98 Show that $\mathsf{H}_{\mathrm{ini}}, \mathsf{H}_{\mathrm{fin}}$ and $\mathsf{H}_T(s)$ are all self-adjoint and positive.

For a solution see Solution 8.98.

Theorem 8.11 *The eigenvalues and eigenspaces of the Hamiltonian $\mathsf{H}_T(s)$ of an adiabatic search algorithm with schedule f for the search of m solutions in $N = 2^n$ objects can be characterized as follows:*

(i) $\mathsf{H}_T(0)$ has

   (a) an eigenvalue $E_-(0) = 0$, which is non-degenerate with eigenstate spanned by $|\Psi_0\rangle$
   (b) and an eigenvalue $E_+(0) = E_1(0) = E_2(0) = 1$, which is $N - 1$ fold degenerate

(ii) $\mathsf{H}_T(s)$ for $s \in]0, 1[$ has

   (a) two distinct non-degenerate eigenvalues $E_\pm(s)$
   (b) an eigenvalue $E_1(s)$, which is $(m - 1)$-fold degenerate
   (c) an eigenvalue $E_2(s)$, which is $(N - m - 1)$-fold degenerate

and these eigenvalues satisfy

$$E_-(s) < E_1(s) < E_+(s) < E_2(s)$$

(iii) $\mathsf{H}_T(1)$ *has*

    (a) an eigenvalue $E_-(1) = E_1(1) = 0$, which is m-fold degenerate
    (b) an eigenvalue $E_+(1) = E_2(1) = 1$, which is $(N - m)$-fold degenerate

(iv) The eigenvalues $E_j(\cdot) : [0,1] \to [0,1]$ for $j \in \{+,-,1,2\}$ mentioned in (i)–(iii) are given by

$$E_\pm(s) = \frac{1}{2} \pm \frac{1}{2}\sqrt{\tilde{m} + 4(1-\tilde{m})\left(f(s) - \frac{1}{2}\right)^2} \qquad (8.33)$$

$$E_1(s) = 1 - f(s)$$

$$E_2(s) = 1\,,$$

where $\tilde{m} = \frac{m}{N}$.

Proof To begin with, we have for $\mathsf{H}_T(s)$ that

$$
\mathsf{H}_T(s) \underset{(8.5)}{=} \left(1 - f(s)\right)\mathsf{H}_{\mathrm{ini}} + f(s)\mathsf{H}_{\mathrm{fin}}
$$

$$
\underset{(8.28),(8.30)}{=} \left(1 - f(s)\right)\left(\mathbf{1} - |\Psi_0\rangle\langle\Psi_0|\right) + f(s)\left(\mathbf{1} - P_S\right)
$$

$$
= \mathbf{1} - \left(1 - f(s)\right)|\Psi_0\rangle\langle\Psi_0| - f(s)P_S\,. \qquad (8.34)
$$

As before, let $|\Phi_{j,\alpha}(s)\rangle$, where $\alpha \in \{1,\dots,d_j\}$, denote the eigenvectors of $\mathsf{H}_T(s)$ for the eigenvalue $E_j(s)$, that is,

$$\mathsf{H}_T(s)|\Phi_{j,\alpha}(s)\rangle = E_j(s)|\Phi_{j,\alpha}(s)\rangle\,. \qquad (8.35)$$

These can be expressed in the computational basis $\{|x\rangle \mid x \in \{0,\dots,N-1\}\}$ (see Sect. 3.2.2) in $\mathbb{H}$ with the help of their components $\Phi_{j,\alpha}(s)_x \in \mathbb{C}$ as

$$|\Phi_{j,\alpha}(s)\rangle = \sum_{x=0}^{N-1} \Phi_{j,\alpha}(s)_x |x\rangle\,. \qquad (8.36)$$

Hence,

$$\mathsf{H}_T(s)|\Phi_{j,\alpha}(s)\rangle \underbrace{=}_{(8.34)} |\Phi_{j,\alpha}(s)\rangle - \big(1-f(s)\big)|\Psi_0\rangle\langle\Psi_0|\Phi_{j,\alpha}(s)\rangle$$

$$-f(s)P_S|\Phi_{j,\alpha}(s)\rangle$$

$$\underbrace{=}_{\substack{(8.29),(8.31),\\(8.36)}} \sum_{x=0}^{N-1}\left(\Phi_{j,\alpha}(s)_x - \big(1-f(s)\big)\frac{1}{\sqrt{N}}\langle\Psi_0|\Phi_{j,\alpha}(s)\rangle\right)|x\rangle$$

$$-f(s)\sum_{x\in S}\Phi_{j,\alpha}(s)_x|x\rangle ,$$

and the eigenvalue equation (8.35) implies

$$\big(1-f(s)-E_j(s)\big)\Phi_{j,\alpha}(s)_x = \frac{1-f(s)}{\sqrt{N}}\langle\Psi_0|\Phi_{j,\alpha}(s)\rangle \quad \forall x\in S \qquad (8.37)$$

$$\big(1-E_j(s)\big)\Phi_{j,\alpha}(s)_x = \frac{1-f(s)}{\sqrt{N}}\langle\Psi_0|\Phi_{j,\alpha}(s)\rangle \quad \forall x\in S^\perp. \qquad (8.38)$$

We distinguish the three cases listed in the theorem, and in the case $0 < s < 1$ we consider first the subcase $\langle\Psi_0|\Phi_{j,\alpha}(s)\rangle = 0$ and then $\langle\Psi_0|\Phi_{j,\alpha}(s)\rangle \neq 0$.

(i) First, let $s = 0$. In this case Definition 8.5 of a schedule implies $f(0) = 0$ and thus (8.34) gives

$$\mathsf{H}_T(0) = \mathbf{1} - |\Psi_0\rangle\langle\Psi_0|$$

from which it follows that

$$\mathsf{H}_T(0)|\Psi_0\rangle = 0$$
$$\mathsf{H}_T(0)|\Psi\rangle = |\Psi\rangle \qquad \forall\,|\Psi\rangle \in \mathbb{H}_{|\Psi_0\rangle^\perp} .$$

Hence, $\mathsf{H}_T(0)$ has the eigenvalues 0 and 1. The eigenspace for 0 is one-dimensional and spanned by $|\Psi_0\rangle$. The eigenspace for the eigenvalue 1 consists of the orthogonal complement $\mathbb{H}_{|\Psi_0\rangle^\perp}$ of $|\Psi_0\rangle$. This forms an $(N-1)$-dimensional subspace of $\mathbb{H}$ and the eigenvalue 1 is thus $(N-1)$-fold degenerate.

(ii) Next, let $s \in {]}0,1{[}$. If $\langle\Psi_0|\Phi_{j,\alpha}(s)\rangle = 0$, then (8.37) and (8.38) become

$$\big(1-f(s)-E_j(s)\big)\Phi_{j,\alpha}(s)_x = 0 \quad \forall x\in S \qquad (8.39)$$
$$\big(1-E_j(s)\big)\Phi_{j,\alpha}(s)_x = 0 \quad \forall x\in S^\perp , \qquad (8.40)$$

which allow two solutions for the eigenvalues. The first is

$$E_1(s) = 1 - f(s).$$

Since $f(s) \neq 0$ for $s \in]0, 1[$, it follows that $1 - E_1(s) \neq 0$ and (8.40) implies $\Phi_{1,\alpha}(s)_x = 0$ for all $x \in S^\perp$. Hence, the eigenvectors for the eigenvalue $E_1(s)$ satisfy

$$|\Phi_{1,\alpha}(s)\rangle = \sum_{x \in S} \Phi_{1,\alpha}(s)_x |x\rangle . \qquad (8.41)$$

Since the set S has m elements, the subspace $\mathrm{Span}\,\{|x\rangle \mid x \in S\}$ is m-dimensional. However, the eigenvectors also have to satisfy the starting assumption of this subcase, namely $\langle \Psi_0 | \Phi_{1,\alpha}(s)\rangle = 0$, which together with (8.29) and (8.41) translates to

$$\sum_{x \in S} \Phi_{1,\alpha}(s)_x = 0$$

and reduces the dimension by 1. Consequently, the eigenspace of the eigenvalue $E_1(s)$ is the $(m-1)$-dimensional subspace $\mathrm{Span}\,\{|x\rangle \mid x \in S\} \cap \mathbb{H}_{|\Psi_0\rangle^\perp}$. The other solution for the eigenvalue implied by (8.39) and (8.40) is

$$E_2(s) = 1 .$$

Then it follows from (8.39) and $f(s) \neq 0$ for $s \in]0, 1[$ that $\Phi_{2,\alpha}(s)_x = 0$ for all $x \in S$. Therefore, the eigenvectors for the eigenvalue $E_2(s)$ satisfy

$$|\Phi_{2,\alpha}(s)\rangle = \sum_{x \in S^\perp} \Phi_{2,\alpha}(s)_x |x\rangle .$$

Since the set $S^\perp$ has $N - m$ elements, the subspace $\mathrm{Span}\,\{|x\rangle \mid x \in S^\perp\}$ is $(N - m)$-dimensional. As before, the eigenvectors also have to satisfy the starting assumption of this subcase, that is, $\langle \Psi_0 | \Phi_{2,\alpha}(s)\rangle = 0$, which reduces the dimension by 1. Consequently, the eigenspace of the eigenvalue $E_2(s)$ is the $(N - m - 1)$-dimensional subspace $\mathrm{Span}\,\{|x\rangle \mid x \in S^\perp\} \cap \mathbb{H}_{|\Psi_0\rangle^\perp}$.

If, instead $\langle \Psi_0 | \Phi_{j,\alpha}(0)\rangle \neq 0$, we first observe that since f is injective on $[0, 1]$ and $f(1) = 1$, it follows that $1 - f(s) \neq 0$ for $\in]0, 1[$. Consequently, (8.37) and (8.38) imply

$$\left(1 - f(s) - E_j(s)\right)\left(1 - E_j(s)\right) \neq 0$$

and thus

$$\Phi_{j,\alpha}(s)_x = \frac{1 - f(s)}{\sqrt{N}\left(1 - f(s) - E_j(s)\right)} \langle \Psi_0 | \Phi_{j,\alpha}(s)\rangle \quad \forall x \in S$$

$$\Phi_{j,\alpha}(s)_x = \frac{1 - f(s)}{\sqrt{N}\left(1 - E_j(s)\right)} \langle \Psi_0 | \Phi_{j,\alpha}(s)\rangle \quad \forall x \in S^\perp$$

such that

$$
\begin{aligned}
\langle \Psi_0 | \Phi_{j,\alpha}(s) \rangle \underbrace{=}_{(8.29),(8.36)} \frac{1}{\sqrt{N}} \sum_{x=0}^{N-1} \Phi_{j,\alpha}(s)_x \\
= \frac{(1 - f(s)) \langle \Psi_0 | \Phi_{j,\alpha}(s) \rangle}{N} \left(\sum_{x \in S} \frac{1}{1 - f(s) - E_j(s)} + \sum_{x \in S^{\perp}} \frac{1}{1 - E_j(s)} \right) \\
= \frac{(1 - f(s)) \langle \Psi_0 | \Phi_{j,\alpha}(s) \rangle}{N} \left(\frac{m}{1 - f(s) - E_j(s)} + \frac{N - m}{1 - E_j(s)} \right) .
\end{aligned}
$$

Since the starting assumption in this subcase is $\langle \Psi_0 | \Phi_{j,\alpha}(s) \rangle \neq 0$, it follows that

$$
1 = (1 - f(s)) \left(\frac{\widetilde{m}}{1 - f(s) - E_j(s)} + \frac{1 - \widetilde{m}}{1 - E_j(s)} \right) , \tag{8.42}
$$

where $\widetilde{m} = \frac{m}{N}$. The solutions of (8.42) for $E_j(s)$ are

$$
E_{\pm}(s) = \frac{1}{2} \pm \frac{1}{2} \sqrt{\widetilde{m} + 4(1 - \widetilde{m}) \left(f(s) - \frac{1}{2} \right)^2} . \tag{8.43}
$$

Here we make use of the claim about the ordering of $E_j(s)$ for $j \in \{-, 1, +, 2\}$, which is to be shown in Exercise 8.99.

Exercise 8.99 Show that for $s \in\ ]0, 1[$

$$
E_-(s) < E_1(s) < E_+(s) < E_2(s) .
$$

For a solution see Solution 8.99.

Hence, $E_1(s), E_2(s)$ and $E_{\pm}(s)$ are four distinct eigenvalues and since we know already that

$$
\dim \mathrm{Eig}(H_T(s), E_1(s)) = m - 1
$$
$$
\dim \mathrm{Eig}\left(H_T(s), E_2(s)\right) = N - m - 1 ,
$$

it follows that $E_+(s)$ and $E_-(s)$ are each eigenvalues with one-dimensional eigenspaces. We can thus drop the degeneracy index α, and the components of the associated eigenvectors $| \Phi_{\pm}(s) \rangle$ are given by

$$\Phi_\pm(s)_x = \frac{1-f(s)}{\sqrt{N}\left(1-f(s)-E_\pm(s)\right)} \langle \Psi_0 | \Phi_\pm(s)\rangle \quad \forall x \in S$$

$$\Phi_\pm(s)_x = \frac{1-f(s)}{\sqrt{N}\left(1-E_\pm(s)\right)} \langle \Psi_0 | \Phi_\pm(s)\rangle \quad \forall x \in S^\perp$$

with $E_\pm(s)$ as given in (8.43). With the help of the vectors $|\Psi_S\rangle$ and $|\Psi_{S^\perp}\rangle$ defined in (6.138) we can write the eigenvectors as

$$|\Phi_\pm(s)\rangle = a_\pm(s)|\Psi_{S^\perp}\rangle + b_\pm(s)|\Psi_S\rangle, \tag{8.44}$$

where

$$a_\pm(s) := \frac{\left(1-\sqrt{\widetilde{m}}\right)\left(1-f(s)\right)}{\left(1-E_\pm(s)\right)} \langle \Psi_0 | \Phi_\pm(s)\rangle$$

$$b_\pm(s) := \frac{\sqrt{\widetilde{m}}\left(1-f(s)\right)}{\left(1-f(s)-E_\pm(s)\right)} \langle \Psi_0 | \Phi_\pm(s)\rangle. \tag{8.45}$$

(iii) Finally, let $s = 1$. From Definition 8.5 we know that a schedule function f has to satisfy $f(1) = 1$. Hence, (8.34) gives

$$\mathsf{H}_T(1) = \mathbf{1} - P_S.$$

Consequently, $\mathsf{H}_T(1)$ has the two eigenvalues 0 and 1. The former has the eigenspace $P_S\{\mathbb{H}\}$, which has dimension m. The eigenspace for the eigenvalue 1 is $P_{S^\perp}\{\mathbb{H}\} = \left(P_S\{\mathbb{H}\}\right)^\perp$ and has dimension $N - m$.

$$\square$$

Recall from Proposition 6.28 that in GROVER's search algorithm the states after the jth step

$$|\Psi_j\rangle = \cos\theta_j |\Psi_{S^\perp}\rangle + \sin\theta_j |\Psi_S\rangle$$

are obtained from the initial state

$$|\Psi_0\rangle = \frac{1}{\sqrt{N}} \sum_{x=0}^{N-1} |x\rangle \underset{(6.138)}{=} \sqrt{1-\widetilde{m}}|\Psi_{S^\perp}\rangle + \sqrt{\widetilde{m}}|\Psi_S\rangle \tag{8.46}$$

by rotating in the subspace

$$\mathbb{H}_{\mathrm{sub}} := \mathrm{Span}\left\{|\Psi_{S^\perp}\rangle, |\Psi_S\rangle\right\} \tag{8.47}$$

by an angle $\theta_j = (2j+1)\theta_0$, where

$$\theta_0 = \arcsin\left(\sqrt{\frac{m}{N}}\right) = \arcsin\left(\sqrt{\widetilde{m}}\right).$$

In the adiabatic version of the search algorithm we have a very similar situation as the following corollary shows.

Corollary 8.12 *Let* $\mathsf{H}_T(\cdot)$ *be the time-rescaled adiabatic Hamiltonian of the search algorithm for the search of m solutions* $x \in S \subset \{0, \ldots, N-1\}$ *as defined in Definition 8.10. Moreover, define*

$$\theta_0 := \arcsin\left(\sqrt{\widetilde{m}}\right) \in \left[0, \frac{\pi}{2}\right[\tag{8.48}$$

and

$$\begin{aligned} |\widehat{\Phi}_-(s)\rangle &:= \cos\theta(s)|\Psi_{S\perp}\rangle + \sin\theta(s)|\Psi_S\rangle \\ |\widehat{\Phi}_+(s)\rangle &:= \sin\theta(s)|\Psi_{S\perp}\rangle - \cos\theta(s)|\Psi_S\rangle\,, \end{aligned} \tag{8.49}$$

where

$$\theta(s) = \begin{cases} \arctan\left(\frac{1-E_-(s)}{1-f(s)-E_-(s)}\tan\theta_0\right) \in \left[\theta_0, \frac{\pi}{2}\right[& \text{for } s \in [0,1[\\ \lim_{s\nearrow 1}\theta(s) & \text{for } s = 1\,. \end{cases} \tag{8.50}$$

Then we have $\theta(0) = \theta_0$ *and* $\theta(1) = \frac{\pi}{2}$ *as well as for* $s \in [0,1]$

$$|\widehat{\Phi}_\pm(s)\rangle \in \mathrm{Eig}\left(\mathsf{H}_T(s), E_\pm(s)\right)\,. \tag{8.51}$$

Proof To begin with, note that the definition of $|\Psi_{S\perp}\rangle$ and $|\Psi_S\rangle$ in (6.138) implies $\langle\Psi_{S\perp}|\Psi_S\rangle = 0$. It then follows from (8.49) that also $\langle\widehat{\Phi}_+(s)|\widehat{\Phi}_-(s)\rangle = 0$ for all $s \in [0,1]$.

First, consider $s = 0$ for which we have $f(0) = 0 = E_-(0)$ and thus $\theta(0) = \theta_0$ such that

$$|\widehat{\Phi}_-(0)\rangle \underbrace{=}_{(8.49)} \cos\theta_0|\Psi_{S\perp}\rangle + \sin\theta_0|\Psi_S\rangle \underbrace{=}_{(8.48)} \sqrt{1-\widetilde{m}}|\Psi_{S\perp}\rangle + \sqrt{\widetilde{m}}|\Psi_S\rangle \underbrace{=}_{(8.46)} |\Psi_0\rangle\,,$$

and we know from (i) in Theorem 8.11 that $|\Psi_0\rangle \in \mathrm{Eig}\left(\mathsf{H}_T(0), E_-(0)\right)$. This proves (8.51) for $|\widehat{\Phi}_-(0)\rangle$. Moreover, (i) in Theorem 8.11 also states that $E_-(0)$ is non-degenerate and $E_+(0)$ is $(N-1)$-fold degenerate. Hence, any vector orthogonal to $\mathrm{Eig}\left(\mathsf{H}_T(0), E_-(0)\right)$ is an eigenvector of $E_+(0)$, which proves (8.51) for $|\widehat{\Phi}_+(0)\rangle$.

For $s \in]0, 1[$ we recall from (8.44) that

$$|\Phi_\pm(s)\rangle = a_\pm(s)|\Psi_{S\perp}\rangle + b_\pm(s)|\Psi_S\rangle \tag{8.52}$$

with $a_\pm(s)$ and $b_\pm(s)$ as given in (8.45). The latter shows that $a_\pm(s)$ and $b_\pm(s)$ have the same complex phase, which we denote by $\omega_\pm(s)$. This, together with $\langle \Psi_{S\perp} | \Psi_S \rangle = 0$ and $||\Psi_S||^2 = ||\Psi_{S\perp}||^2 = ||\Phi_\pm(s)||^2 = 1$, implies that we can write $a_\pm(s)$ and $b_\pm(s)$ in the form

$$a_\pm(s) = e^{i\omega_\pm(s)} \cos \theta_\pm(s) \quad \text{and} \quad b_\pm(s) = e^{i\omega_\pm(s)} \sin \theta_\pm(s).$$

Using this and the fact that eigenvectors for different eigenvalues are orthogonal, we obtain

$$
\begin{aligned}
0 &= \langle \Phi_+(s) | \Phi_-(s) \rangle \\
&= e^{i(\omega_-(s) - \omega_+(s))} \left(\cos \theta_+(s) \cos \theta_-(s) + \sin \theta_+(s) \sin \theta_-(s) \right) \\
&= e^{i(\omega_-(s) - \omega_+(s))} \cos \left(\theta_+(s) - \theta_-(s) \right),
\end{aligned}
$$

which requires

$$\theta_+(s) = \theta_-(s) \pm (2k+1) \frac{\pi}{2}$$

for a $k \in \mathbb{Z}$. Setting $\theta(s) = \theta_-(s)$ yields

$$
\begin{aligned}
a_-(s) &= e^{i\omega_-(s)} \cos \theta(s) \\
b_-(s) &= e^{i\omega_-(s)} \sin \theta(s)
\end{aligned}
\tag{8.53}
$$

such that (8.52) gives

$$|\Phi_-(s)\rangle = e^{i\omega_-(s)} \left(\cos \theta(s) | \Psi_{S\perp} \rangle + \sin \theta(s) | \Psi_S \rangle \right) \in \text{Eig} \left(H_T(s), E_-(s) \right).$$

It follows that

$$|\widehat{\Phi}_-(s)\rangle \underset{(8.49)}{=} \cos \theta(s) | \Psi_{S\perp} \rangle + \sin \theta(s) | \Psi_S \rangle = e^{-i\omega_-(s)} |\Phi_-(s)\rangle \in \text{Eig} \left(H_T(s), E_-(s) \right),$$

proving (8.51) for $|\widehat{\Phi}_-(s)\rangle$, where $s \in]0,1[$. We will show below that $\theta(s)$ has the form given in (8.50). Noting that the choice $\theta(s) = \theta_-(s)$ implies

$$
\begin{aligned}
\cos \theta_+(s) &= \cos \left(\theta(s) \pm (2k+1) \frac{\pi}{2} \right) = \mp(-1)^k \sin \theta(s) \\
\sin \theta_+(s) &= \sin \left(\theta(s) \pm (2k+1) \frac{\pi}{2} \right) = \pm(-1)^k \cos \theta(s),
\end{aligned}
$$

we therefore also obtain

$$
\begin{aligned}
a_+(s) &= e^{i\omega_+(s)}\cos\theta_+(s) = \mp(-1)^k e^{i\omega_+(s)}\sin\theta(s)\\
&= e^{i\widetilde{\omega}_+(s)}\sin\theta(s)\\
b_+(s) &= e^{i\omega_+(s)}\sin\theta_+(s) = \pm(-1)^k e^{i\omega_+(s)}\cos\theta(s)\\
&= -e^{i\widetilde{\omega}_+(s)}\cos\theta(s)\,,
\end{aligned}
\tag{8.54}
$$

where $\widetilde{\omega}_+$ is such that $\mp(-1)^k e^{i\omega_+(s)} = e^{i\widetilde{\omega}_+(s)}$. Using (8.54) in (8.52) gives

$$
|\Phi_+(s)\rangle = e^{i\widetilde{\omega}_+(s)}\big(\sin\theta(s)|\Psi_{S\perp}\rangle - \cos\theta(s)|\Psi_S\rangle\big) \in \mathrm{Eig}\big(H_T(s), E_+(s)\big)\,,
$$

which implies

$$
|\widehat{\Phi}_+(s)\rangle \underset{(8.49)}{=} \sin\theta(s)|\Psi_{S\perp}\rangle - \cos\theta(s)|\Psi_S\rangle = e^{-i\widetilde{\omega}_+(s)}|\Phi_+(s)\rangle \in \mathrm{Eig}\big(H_T(s), E_+(s)\big)
$$

and proves (8.51) for $|\widehat{\Phi}_+(s)\rangle$, where $s \in]0,1[$.

Before we consider the case $s = 1$, we prove (8.50). We know already that $\theta(0) = \theta_0$. For $s \in]0,1[$ we have

$$
\begin{aligned}
\tan\theta(s) &\underset{(8.53)}{=} \frac{b_-(s)}{a_-(s)} \underset{(8.45)}{=} \frac{1-E_-(s)}{1-f(s)-E_-(s)}\frac{\sqrt{\widetilde{m}}}{\sqrt{1-\widetilde{m}}}\\
&\underset{(8.48)}{=} \frac{1-E_-(s)}{1-f(s)-E_-(s)}\tan\theta_0\,,
\end{aligned}
\tag{8.55}
$$

and we have thus verified (8.50) for $s \in [0,1[$.

As for $s = 1$, we know that $f(1) = 1$ and from (8.33) that $E_-(1) = 0$. To show that then (8.50) implies $\theta(1) = \frac{\pi}{2}$, it suffices to note that $\tan\theta_0 > 0$ and to use the result of the following exercise when considering the $\lim_{s\nearrow 1}$ of the right side of (8.55).

Exercise 8.100 Show that $0 \le f(s) \le 1$ implies

$$
1 - f(s) - E_-(s) \ge 0\,.
$$

For a solution see Solution 8.100.

It follows that then $\lim_{s \nearrow 1} \tan \theta(s) = +\infty$ and thus $\theta(1) = \frac{\pi}{2}$. With this the claims (8.51) for $s = 1$ take the form

$$|\widehat{\Phi}_-(1)\rangle \underbrace{=}_{(8.49)} |\Psi_S\rangle \in \mathrm{Eig}\left(\mathsf{H}_T(1), E_-(1)\right)$$

$$|\widehat{\Phi}_+(1)\rangle \underbrace{=}_{(8.49)} |\Psi_{S\perp}\rangle \in \mathrm{Eig}\left(\mathsf{H}_T(1), E_+(1)\right),$$

which we know to be true since $|\Psi_S\rangle$ is an eigenvector of $\mathsf{H}_T(1) = \mathsf{H}_{\mathrm{fin}} = \mathbf{1} - P_S$ with eigenvalue $E_-(1) = 0$, and $|\Psi_{S\perp}\rangle$ is an eigenvector for the eigenvalue $E_+(1) = 1$. □

Corollary 8.12 essentially states that the adiabatic time evolution $U_T(\cdot)$ generated by the Hamiltonian $\mathsf{H}_T(\cdot)$ rotates the initial state $|\Psi_0\rangle$ within the two-dimensional subspace $\mathbb{H}_{\mathrm{sub}}$ to the state $|\Psi_S\rangle$, which is an equally weighted superposition of all solution basis states $|x\rangle$, where $x \in S$.

Let us briefly summarize what we have learned of the adiabatic search algorithm so far: among the eigenvalues of $\mathsf{H}_T(s)$ we have $E_\pm(s)$, which satisfy $0 \le E_-(s) < E_+(s) \le 1$, and their eigenspaces are contained in $\mathbb{H}_{\mathrm{sub}} = \mathrm{Span}\left\{|\Psi_S\rangle, |\Psi_{S\perp}\rangle\right\}$.

From Lemma 8.7 we know that if $\mathsf{H}_T(s)$ leaves $\mathbb{H}_{\mathrm{sub}}$ invariant, then this is also true for the time evolution it generates. In this case we can then apply the adiabatic estimates of Corollaries 8.3 and 8.4 to the two-dimensional problem in $\mathbb{H}_{\mathrm{sub}}$. As the following lemma shows, $\mathsf{H}_T(s)$ for the adiabatic search indeed leaves the subspace $\mathbb{H}_{\mathrm{sub}}$ invariant.

Lemma 8.13 *For $\mathsf{H}_T(s)$ as defined in (8.32) and $\mathbb{H}_{\mathrm{sub}}$ as defined in (8.47) we have for all $s \in [0,1]$ that*

$$[\mathsf{H}_T(s), P_{\mathrm{sub}}] = 0. \tag{8.56}$$

Proof As we have already seen in (8.34), $\mathsf{H}_T(s)$ can be written as

$$\mathsf{H}_T(s) = \mathbf{1} - \left(1 - f(s)\right)|\Psi_0\rangle\langle\Psi_0| - f(s)P_S, \tag{8.57}$$

where

$$P_S = \sum_{x \in S} |x\rangle\langle x|. \tag{8.58}$$

From (8.46) we obtain

$$|\Psi_0\rangle\langle\Psi_0| = (1 - \widetilde{m})|\Psi_{S\perp}\rangle\langle\Psi_{S\perp}| + \widetilde{m}|\Psi_S\rangle\langle\Psi_S|$$
$$+ \sqrt{\widetilde{m}(1 - \widetilde{m})}\left(|\Psi_{S\perp}\rangle\langle\Psi_S| + |\Psi_S\rangle\langle\Psi_{S\perp}|\right). \tag{8.59}$$

On the other hand, since $\mathbb{H}_{\text{sub}} = \text{Span}\left\{|\Psi_S\rangle, |\Psi_{S\perp}\rangle\right\}$, the projector onto this subspace is given by

$$P_{\text{sub}} = |\Psi_S\rangle\langle\Psi_S| + |\Psi_{S\perp}\rangle\langle\Psi_{S\perp}| \tag{8.60}$$

such that together with (8.59) we evidently have

$$|\Psi_0\rangle\langle\Psi_0|P_{\text{sub}} = |\Psi_0\rangle\langle\Psi_0| = P_{\text{sub}}|\Psi_0\rangle\langle\Psi_0| \tag{8.61}$$

and thus

$$\left[\mathbf{1} - \left(1 - f(s)\right)|\Psi_0\rangle\langle\Psi_0|, P_{\text{sub}}\right] = 0 \,. \tag{8.62}$$

Moreover, we find

$$
\begin{aligned}
P_S P_{\text{sub}} &\underbrace{=}_{(8.58),(8.60)} \left(\sum_{x \in S} |x\rangle\langle x|\right)\left(|\Psi_S\rangle\langle\Psi_S| + |\Psi_{S\perp}\rangle\langle\Psi_{S\perp}|\right) \\
&= |\Psi_S\rangle\langle\Psi_S| = P_{\text{sub}} P_S \,.
\end{aligned}
\tag{8.63}
$$

Hence,

$$\left[f(s)P_S, P_{\text{sub}}\right] = 0 \,,$$

and together with (8.62) and (8.57) this proves the claim (8.56). $\square$

As a consequence of Lemmas 8.7 and 8.13 we can now restrict all considerations to the subspace $\mathbb{H}_{\text{sub}}$ since it is in this subspace where the initial state $|\Psi_0\rangle$ at t_{ini} resides and where we remain until $t_{\text{fin}} = t_{\text{ini}} + T$. This means that instead of H_T we only need to consider its restriction to the subspace $\mathbb{H}_{\text{sub}}$

$$\mathsf{H}_T(s)\big|_{\mathbb{H}_{\text{sub}}} = P_{\text{sub}}\mathsf{H}_T(s)P_{\text{sub}} \,, \tag{8.64}$$

where P_{sub} is as given in (8.60) and when multiplying $\mathsf{H}_T(s)$ on the right is viewed as an operator $P_{\text{sub}} : \mathbb{H}_{\text{sub}} \to \mathbb{H}$, whereas when multiplying on the left as $P_{\text{sub}} : \mathbb{H} \to \mathbb{H}_{\text{sub}}$. Because of (8.56) and $P_{\text{sub}}^2 = P_{\text{sub}}$ we also have

$$P_{\text{sub}}\mathsf{H}_T(s)P_{\text{sub}} = P_{\text{sub}}\mathsf{H}_T(s) = \mathsf{H}_T(s)P_{\text{sub}} \,,$$

but the right side of (8.64) is more suggestive of the fact that we are dealing with an operator on $\mathbb{H}_{\text{sub}}$ only.

Exercise 8.101 Show that for $s \in [0, 1]$

$$\mathsf{H}_T(s)\big|_{\mathbb{H}_{\text{sub}}}|\widehat{\Phi}_\pm(s)\rangle = E_\pm(s)|\widehat{\Phi}_\pm(s)\rangle$$

such that the spectrum of $\mathsf{H}_T(s)$, when restricted to $\mathbb{H}_{\text{sub}}$, is given by

$$\sigma\left(\mathsf{H}_T(s)\big|_{\mathbb{H}_{\text{sub}}}\right) = \{E_\pm(s)\} . \tag{8.65}$$

For a solution see Solution 8.101.

Note that $\mathbb{H}_{\text{sub}}$ being two-dimensional implies that the two distinct eigenvalues $E_\pm(s)$ of $\mathsf{H}_T(s)\big|_{\mathbb{H}_{\text{sub}}}$ are non-degenerate. Consequently, we only need to consider the eigenvalues $E_\pm(s)$ and the eigenstates $|\widehat{\Phi}_\pm(s)\rangle$ of $\mathsf{H}_T(s)$. From Theorem (8.11) we know that $E_-(s)$ is the lower of these and it is also the eigenvalue which has the initial state $|\Psi_0\rangle$ as its eigenstate at $s = 0$.

Corollary 8.4 tells us that if we want to find the eigenstate $|\widehat{\Phi}_-(1)\rangle = |\Psi_S\rangle$ of $E_-(1)$—and thus a solution searched for—with a minimal probability $p_{\min}$, then the transition time T has to satisfy

$$T \geq \frac{C_-(1)}{\sqrt{1 - p_{\min}}} .$$

In order to know how T grows for $N \to \infty$, or equivalently $\widetilde{m} = \frac{m}{N} \to 0$, we thus need to determine the growth of $C_-(1)$ as a function of $\widetilde{m} \to 0$.

From Corollary 8.6 we see that $C_-(1)$ is determined by $\|\mathsf{H}_{\text{fin}} - \mathsf{H}_{\text{ini}}\|$ and the functions g_- and f. The value of the first of these three is given in Exercise 8.102.

Exercise 8.102 Let H_{ini} and H_{fin} be as defined in Definition 8.10, and let P_{sub} be as in (8.60). Show that then

$$\left\|\left(\mathsf{H}_{\text{fin}} - \mathsf{H}_{\text{ini}}\right)\big|_{\mathbb{H}_{\text{sub}}}\right\| = \sqrt{1 - \widetilde{m}} . \tag{8.66}$$

For a solution see Solution 8.102.

This determines one ingredient of $C_-(1)$. Before we consider the schedule f in more detail below, let us have a look at the gap function $g_-(s)$. From Definition G.4 of the gap function we see that for the eigenvalue $E_-(s)$ of the reduced operator $\mathsf{H}_T(s)\big|_{\mathbb{H}_{\text{sub}}}$ it is given by

$$g_-(s) \underbrace{=}_{\text{(G.8)}} \min\left\{ |E_j(s) - E_-(s)| \;\big|\; E_j \in \sigma\left(\mathsf{H}_T(s)\big|_{\mathbb{H}_{\text{sub}}}\right) \smallsetminus \{E_-(s)\}\right\}$$

$$\underbrace{=}_{\text{(8.65)}} E_+(s) - E_-(s)$$

$$\underbrace{=}_{\text{(8.33)}} \sqrt{\widetilde{m} + 4(1 - \widetilde{m})\left(f(s) - \frac{1}{2}\right)^2} .$$

With the help of the function

$$
\begin{aligned}
g : [0, 1] &\longrightarrow [0, 1] \\
u &\longmapsto g(u) = \sqrt{\tilde{m} + 4(1 - \tilde{m})\left(u - \tfrac{1}{2}\right)^2}
\end{aligned}
\tag{8.67}
$$

we can write g_- in the form

$$
g_-(s) = g\big(f(s)\big) .
$$

The next proposition shows that if we choose a linear schedule $f(s) = s$, then we end up requiring a transition time $T \in O\left(\frac{N}{m}\right)$ for $N \to \infty$ if we want to guarantee to find the solution with a given minimal probability. This does not reflect the quadratic speedup seen in (6.161) for the circuit based version of GROVER's search algorithm. As we shall see later, a more suitable choice of the schedule f allows us to replicate the efficiency of the GROVER's search algorithm in the adiabatic setting.

> **Proposition 8.14** *In an adiabatic search with schedule $f(s) = s$ the success-probability of finding a solution after the time evolution from t_{ini} to $t_{fin} = t_{ini} + T$ can be bounded from below by $p_{min} \in {]}0, 1{[}$ if*
>
> $$
> T \in O\left(\frac{N}{m}\right) \qquad \text{for } N \to \infty.
> \tag{8.68}
> $$

Proof For the linear schedule $f(s) = s$ we have $g_-(s) = g(s)$ with g as defined in (8.67). Inserting this and the result (8.66) of Exercise 8.102 into (8.19) yields

$$
C_-(1) = \sqrt{1 - \tilde{m}} \left(\frac{1}{g(1)^2} + \frac{1}{g(0)^2} + 10\sqrt{1 - \tilde{m}} \int_0^1 \frac{du}{g(u)^3} \right) ,
\tag{8.69}
$$

where we have used that $\dot{f}(s) = 1$ and $\ddot{f}(s) = 0$ for $s \in [0, 1]$. Using

$$
\int \frac{du}{\left(au^2 + bu + c\right)^{3/2}} = \frac{2(2au + b)}{(4ac - b^2)\sqrt{au^2 + bu + c}} ,
$$

we find

$$
\int_0^1 \frac{du}{g(u)^3} = \int_0^1 \frac{du}{\left(\tilde{m} + 4(1 - \tilde{m})\left(u - \tfrac{1}{2}\right)^2\right)^{3/2}} = \frac{1}{\tilde{m}}
$$

With this and $g(1) = 1 = g(0)$ it follows from (8.69) that

$$C_-(1) = 2\sqrt{1 - \widetilde{m}} + +10\frac{1 - \widetilde{m}}{\widetilde{m}} \in O\left(\frac{1}{\widetilde{m}}\right) \quad \text{for } \widetilde{m} \to 0.$$

From Corollary 8.4 we know that in order to ensure that the success-probability is no less than $p_{\min}$, the transition time T must satisfy

$$T \geq \frac{C_-(1)}{\sqrt{1 - p_{\min}}}.$$

Since $\widetilde{m} = \frac{m}{N}$ it follows that for $N \to \infty$

$$T \in O(C_-(1)) \in O\left(\frac{N}{m}\right).$$

$\square$

The result of Proposition 8.14 shows that using a linear schedule leads to a transition time that is of the same order as the usual search, in other words, grows linearly with N. It turns out, however, that by judiciously choosing the schedule f we can improve on (8.68) to the extent that we indeed obtain the quadratic GROVER speedup $T \in O\left(\sqrt{\frac{N}{m}}\right)$. In order to show this, we need the following preparatory lemma [111].

Lemma 8.15 *Let* $0 < \widetilde{m} < 1$ *and* $g : [0, 1] \to [0, 1]$ *be such that*

$$g(u) := \sqrt{\widetilde{m} + 4(1 - \widetilde{m})\left(u - \frac{1}{2}\right)^2}. \tag{8.70}$$

Then the following hold.

(i) For $a \in \mathbb{R}$ *with* $a > 1$ *we have*

$$\int_0^1 g(u)^{-a} du \in O\left(\widetilde{m}^{\frac{1-a}{2}}\right) \quad \text{for } \widetilde{m} \to 0. \tag{8.71}$$

(ii) For $b \in \mathbb{R}$ *with* $1 < b < 2$ *we have*

$$\int_0^1 g(u)^{b-3} |\dot{g}(u)| \, du \in O\left(\widetilde{m}^{\frac{b-2}{2}}\right) \quad \text{for } \widetilde{m} \to 0. \tag{8.72}$$

Proof (i) In

$$\int_0^1 g(u)^{-a} du = \int_0^1 \frac{du}{\left(\tilde{m} + 4(1 - \tilde{m})\left(u - \frac{1}{2}\right)^2\right)^{a/2}}$$

$$= \tilde{m}^{-\frac{a}{2}} \int_0^1 \frac{du}{\left(1 + 4(\frac{1}{\tilde{m}} - 1)\left(u - \frac{1}{2}\right)^2\right)^{a/2}}$$

we substitute $z = 2\sqrt{\frac{1}{\tilde{m}} - 1}\left(u - \frac{1}{2}\right)$ such that

$$\int_0^1 g(u)^{-a} du = \frac{\tilde{m}^{\frac{1-a}{2}}}{2\sqrt{1 - \tilde{m}}} \int_{-\sqrt{\frac{1}{\tilde{m}} - 1}}^{\sqrt{\frac{1}{\tilde{m}} - 1}} \frac{dz}{(1 + z^2)^{a/2}} = \frac{\tilde{m}^{\frac{1-a}{2}}}{\sqrt{1 - \tilde{m}}} \int_0^{\sqrt{\frac{1}{\tilde{m}} - 1}} \frac{dz}{(1 + z^2)^{a/2}}$$

$$\leq \frac{\tilde{m}^{\frac{1-a}{2}}}{\sqrt{1 - \tilde{m}}} \int_0^\infty \frac{dz}{(1 + z^2)^{a/2}} \cdot$$

Here we can use that for $a > 1$

$$\int_0^\infty \frac{dz}{(1 + z^2)^{a/2}} = D < \infty.$$

Hence, we have

$$\int_0^1 g(u)^{-a} du \leq D \frac{\tilde{m}^{\frac{1-a}{2}}}{\sqrt{1 - \tilde{m}}}$$

and thus

$$\int_0^1 g(u)^{-a} du \in O\left(\tilde{m}^{\frac{1-a}{2}}\right) \qquad \text{for } \tilde{m} \to 0$$

as claimed.

(iii) From (8.70) we find

$$\dot{g}(u) = \frac{4(1 - \tilde{m})\left(u - \frac{1}{2}\right)}{g(u)}$$

and since $g(u) > 0$ thus

$$|\dot{g}(u)| = \begin{cases} -\dot{g}(u) & \text{for } u \in [0, \frac{1}{2}] \\ \dot{g}(u) & \text{for } u \in [\frac{1}{2}, 0]. \end{cases}$$

Hence, using again $g(1) = 1 = g(0)$ and $g\left(\frac{1}{2}\right) = \sqrt{\widetilde{m}}$, we obtain

$$
\int_0^1 g(u)^{b-3} \left| \dot{g}(u) \right| du = \int_{\frac{1}{2}}^1 g(u)^{b-3} \dot{g}(u) du - \int_0^{\frac{1}{2}} g(u)^{b-3} \dot{g}(u) du
$$

$$
= \frac{1}{b-2} \left(g(u)^{b-2} \Big|_{\frac{1}{2}}^1 - g(u)^{b-2} \Big|_0^{\frac{1}{2}} \right)
$$

$$
= \frac{2}{2-b} \left(1 - \widetilde{m}^{\frac{2-b}{2}} \right) \widetilde{m}^{\frac{b-2}{2}}
$$

$$
< \frac{2}{2-b} \widetilde{m}^{\frac{b-2}{2}} ,
$$

where we used the assumption $1 < b < 2$ in the last inequality. Consequently,

$$
\int_0^1 g(u)^{b-3} \left| \dot{g}(u) \right| du \in O\left(\widetilde{m}^{\frac{b-2}{2}} \right) \qquad \text{for } \widetilde{m} \to 0
$$

holds as claimed.

□

The next theorem shows that if we suitably adapt the schedule f to the function g, we can indeed achieve $T \in O\left(\sqrt{\frac{N}{m}} \right)$.

Theorem 8.16 ([111]) *Let* $0 < \widetilde{m} = \frac{m}{N} < 1$ *and* g *be defined as in Lemma 8.15. For* $1 < b < 2$ *define*

$$
\kappa_b := \int_0^1 g(u)^{-b} du \qquad\qquad (8.73)
$$

and let $f : [0,1] \to \mathbb{R}$ *be defined as the solution of the initial value problem*

$$
\dot{f}(s) = \kappa_b g\big(f(s)\big)^b
$$
$$
f(0) = 0. \qquad\qquad (8.74)
$$

Then the following hold.

(i) *f is a permissible schedule, in other words, it is a strictly increasing func-tion from $[0,1]$ onto itself satisfying $f(0) = 0$ and $f(1) = 1$.*

(ii) *In the adiabatic search with schedule f we can guarantee to find a solution with a given minimal success-probability if the transition time $T = t_{fin} - t_{ini}$ grows as*

$$
T \in O\left(\sqrt{\frac{N}{m}} \right) \qquad \text{for } N \to \infty. \qquad\qquad (8.75)
$$

Proof Note that existence and uniqueness of a solution to (8.74) is guaranteed because the function $y \mapsto \kappa_b g(y)^b$ is LIPPSCHITZ-continuous [115].

(i) From its definition in (8.70) we see that $g > 0$, and thus (8.73) implies that $\kappa_b > 0$ as well. It then follows from (8.74) that also $\dot{f} > 0$. Hence, f is strictly increasing on $[0, 1]$ and by its defining property (8.74) satisfies $f(0) = 0$. To show that also $f(1) = 1$, consider that in general

$$\int_{f(0)}^{f(1)} g(u)^{-b} du = \int_0^1 g(f(s))^{-b} \dot{f}(s) ds. \tag{8.76}$$

Using the defining properties of f and κ_b we thus obtain

$$\int_0^{f(1)} g(u)^{-b} du \underbrace{=}_{(8.74),(8.76)} \int_0^1 g(f(s))^{-b} \kappa_b g(f(s))^b ds = \kappa_b \underbrace{=}_{(8.73)} \int_0^1 g(u)^{-b} du.$$

Since $g > 0$, we must have $f(1) = 1$, and altogether f is a permissible schedule as it satisfies the requirements of Definition 8.5.

(ii) From Corollary 8.4 we know that the transition time T needed to guarantee a success probability of at least $p_{\min}$ has to satisfy

$$T \geq \frac{C_-(1)}{\sqrt{1 - p_{\min}}}. \tag{8.77}$$

In order to ascertain the growth of T as a function of N, we thus need to examine $C_-(1)$ in this respect. Inserting the result of Exercise 8.102 into (8.19) yields

$$C_-(1) = \sqrt{1 - \tilde{m}} \left(\frac{\dot{f}(1)}{g_-(1)^2} + \frac{\dot{f}(0)}{g_-(0)^2} \right.$$

$$\left. + \int_0^1 \frac{|\ddot{f}(u)|}{g_-(u)^2} du + 10\sqrt{1 - \tilde{m}} \int_0^1 \frac{\dot{f}(u)^2}{g_-(u)^3} du \right). \tag{8.78}$$

We now consider each term in (8.78) in turn. Using first that per definition

$$\dot{f}(s) = \kappa_b g(f(s))^b \quad \text{and} \quad g_-(s) = g(f(s)) \tag{8.79}$$

together with $f(0) = 0, f(1) = 1$ and $g(0) = 1 = g(1)$ yields

$$\dot{f}(1) = \kappa_b g(f(1)) = \kappa_b g(1) = \kappa_b, \quad \dot{f}(0) = \kappa_b g(f(0)) = \kappa_b g(0) = \kappa_b,$$
$$g_-(1) = g(f(1)) = g(1) = 1, \quad g_-(0) = g(f(0)) = g(0) = 1$$

such that

$$\frac{\dot{f}(1)}{g_-(1)^2} + \frac{\dot{f}(0)}{g_-(0)^2} = 2\kappa_b . \tag{8.80}$$

Next, consider that

$$\ddot{f}(u) = \underbrace{\frac{d}{du}\dot{f}(u)}_{(8.74)} = \frac{d}{du}\left(\kappa_b g(f(u))^b\right) = b\kappa_b g(f(u))^{b-1}\dot{g}(f(u))\dot{f}(u)$$

$$\underbrace{=}_{(8.74)} b\kappa_b^2 g(f(u))^{2b-1}\dot{g}(f(u)) .$$

Hence, we obtain

$$\left|\ddot{f}(u)\right| = b\kappa_b^2 g(f(u))^{2b-1}\left|\dot{g}(f(u))\right| , \tag{8.81}$$

and thus

$$\int_0^1 \frac{\left|\ddot{f}(u)\right|}{g_-(u)^2}du \underbrace{=}_{(8.79),(8.81)} b\kappa_b^2 \int_0^1 \frac{g(f(u))^{2b-1}\left|\dot{g}(f(u))\right|}{g(f(u))^2}du$$

$$= b\kappa_b^2 \int_0^1 g(f(u))^{2b-3}\left|\dot{g}(f(u))\right| du . \tag{8.82}$$

In the last integral we make the substitution $z = f(u)$, which implies

$$dz = \dot{f}(u)du \underbrace{=}_{(8.74)} \kappa_b g(f(u))^b du . \tag{8.83}$$

Consequently, we have

$$\int_0^1 \frac{\left|\ddot{f}(u)\right|}{g_-(u)^2}du \underbrace{=}_{(8.82),(8.83)} b\kappa_b^2 \int_0^1 g(z)^{2b-3}\left|\dot{g}(z)\right| \frac{dz}{\kappa_b g(z)^b}$$

$$= b\kappa_b \int_0^1 g(z)^{b-3}\left|\dot{g}(z)\right| dz . \tag{8.84}$$

Lastly, consider

$$\int_0^1 \frac{\dot{f}(u)^2}{g_-(u)^3}du \underbrace{=}_{(8.74)} \int_0^1 \frac{\left(\kappa_b g(f(u))^b\right)^2}{g(f(u))^3}du = \kappa_b^2 \int_0^1 g(f(u))^{2b-3}du .$$

Making again the substitution $z = f(u)$ and using (8.83) results in

$$\int_0^1 \frac{\dot{f}(u)^2}{g_-(u)^3} = \kappa_b^2 \int_0^1 g(z)^{2b-3} \frac{dz}{\kappa_b g(z)^b}$$

$$= \kappa_b \int_0^1 g(z)^{b-3} dz. \qquad (8.85)$$

Inserting (8.80), (8.84) and (8.85) into (8.78) yields

$$C_-(1) = \sqrt{1 - \widetilde{m}} \kappa_b \left(2 + b \int_0^1 g(z)^{b-3} |\dot{g}(z)| \, dz + 10 \sqrt{1 - \widetilde{m}} \int_0^1 g(z)^{b-3} dz \right). \qquad (8.86)$$

From the definition (8.73) of κ_b and Lemma 8.15 (i) we deduce that

$$\kappa_b \in O\left(\widetilde{m}^{\frac{1-b}{2}} \right) \qquad \text{for } \widetilde{m} \to 0$$

and thus

$$\sqrt{1 - \widetilde{m}} \kappa_b \in O\left(\widetilde{m}^{\frac{1-b}{2}} \right) \qquad \text{for } \widetilde{m} \to 0. \qquad (8.87)$$

From Lemma 8.15 (ii) we see that

$$\int_0^1 g(z)^{b-3} |\dot{g}(z)| \, dz \underbrace{\in}_{(8.72)} O\left(\widetilde{m}^{\frac{b-2}{2}} \right) \qquad \text{for } \widetilde{m} \to 0. \qquad (8.88)$$

Setting $a = 3 - b$, we find that then $1 < a < 2$ and that Lemma 8.15 (i) implies

$$\int_0^1 g(z)^{b-3} dz = \int_0^1 g(z)^{-a} dz \underbrace{\in}_{(8.71)} O\left(\widetilde{m}^{\frac{1-a}{2}} \right) = O\left(\widetilde{m}^{\frac{b-2}{2}} \right) \qquad \text{for } \widetilde{m} \to 0$$

and thus

$$\sqrt{1 - \widetilde{m}} \int_0^1 g(z)^{b-3} dz \in O\left(\widetilde{m}^{\frac{b-2}{2}} \right) \qquad \text{for } \widetilde{m} \to 0. \qquad (8.89)$$

Using (8.87)–(8.89) in (8.86) and applying the properties (C.1) and (C.2) of the LANDAU symbols, we finally have

$$C_-(1) \in O\left(\widetilde{m}^{\frac{1-b}{2}} \widetilde{m}^{\frac{b-2}{2}} \right) = O\left(\widetilde{m}^{-\frac{1}{2}} \right) = O\left(\sqrt{\frac{N}{m}} \right) \qquad \text{for } \widetilde{m} \to 0.$$

The claim (8.75) then follows from (8.77).

$\square$

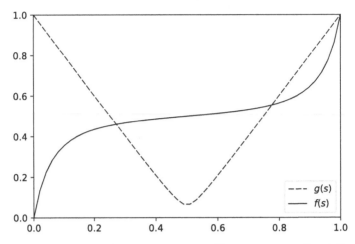

Fig. 8.2 Graph of functions $g(s)$ of (8.70) with $N = 2^{10}, m = 4$ such that $\widetilde{m} = 2^{-8}$ and $f(s)$ of (8.74) for $b = \frac{3}{2}$

Figure 8.2 shows $g(s)$ in the case $\widetilde{m} = 2^{-8}$ together with a numerical solution $f(s)$ of (8.74) for a schedule with $b = \frac{3}{2}$.

With the proof that the efficiency of the gate based search algorithm can be replicated by its adiabatic version we conclude our investigations of quantum adiabatic search algorithms, and we turn to the question of equivalence between gate based and adiabatic algorithms on a general level. First, we show that any gate based computation can be replicated with similar efficiency as an adiabatic computation, before we then show the reverse, namely, that any adiabatic quantum computation can be replicated with similar efficiency by a suitable gate based version.

8.5 Replicating a Circuit Based by an Adiabatic Computation

We consider the situation where we are given a circuit U which acts on n qubits and which is comprised of L gates $U_1, \ldots, U_L$, that is,

$$U = U_L \cdots U_1 : \mathbb{H}^{\otimes n} \to \mathbb{H}^{\otimes n},$$

where L is the length of the circuit U (see Definition 5.27). It helps to simplify the notation in subsequent expressions if we also use the trivial gate

$$U_0 = \mathbf{1}^{\otimes n}.$$

The circuit U is used to perform a quantum computation by transforming a given initial state $|\Psi_{\text{ini}}\rangle$ to a final output state

$$|\Psi_{\text{fin}}\rangle = U|\Psi_{\text{ini}}\rangle .$$

In this section we shall show that given any input state $|\Psi_{\text{ini}}\rangle$, the final output state $|\Psi_{\text{fin}}\rangle$ can also be produced by means of a suitable adiabatic quantum computation with similar efficiency [34, 114]. Comparable efficiency here means that the transition time T in the adiabatic computation grows only polynomially in the length L of the circuit, namely that $T \in \text{poly}(L)$.

How do we then go about constructing $|\Psi_{\text{fin}}\rangle = U|\Psi_{\text{ini}}\rangle$ in the adiabatic computation? As we assume that we have $|\Psi_{\text{ini}}\rangle$ at our disposal, we can construct $\mathsf{H}_{\text{ini}} = \mathbf{1} - |\Psi_{\text{ini}}\rangle\langle\Psi_{\text{ini}}|$ as the initial Hamiltonian that has $|\Psi_{\text{ini}}\rangle$ as its eigenstate. But we cannot do the same to construct the final Hamiltonian, since we do not have $|\Psi_{\text{fin}}\rangle$ directly available to us.[1]

However, we assume that we do have the gates $U_1, \ldots, U_L$ of the circuit U at our disposal. The following Exercise 8.103 serves to illustrate the idea how to utilize the gates U_j to construct $|\Psi_{\text{fin}}\rangle$ in the adiabatic computation.

Exercise 8.103 Let $L \in \mathbb{N}$, $a_1, \ldots, a_L \in \mathbb{R} \setminus \{0\}$ and let $A \in \text{Mat}\big((L+1) \times (L+1), \mathbb{R}\big)$ be given as

$$A = \begin{pmatrix} 1 & -(a_1)^{-1} & 0 & & & \cdots & & & 0 \\ -a_1 & 2 & -(a_2)^{-1} & 0 & & & & & \\ 0 & -a_2 & 2 & -(a_3)^{-1} & & & & & \\ \vdots & & \ddots & \ddots & \ddots & & & & \vdots \\ & & & -a_{j-1} & 2 & & -(a_j)^{-1} & & \\ \vdots & & & & \ddots & \ddots & & & \vdots \\ & & & & -a_{L-2} & 2 & & -(a_{L-1})^{-1} & 0 \\ & & & & & -a_{L-1} & & 2 & -(a_L)^{-1} \\ 0 & & \cdots & & & & 0 & -a_L & 1 \end{pmatrix} .$$

Show that A has the eigenvalue 0 with eigenvector

---

[1] If we had, then there would be no need to perform the adiabatic computation.

$$
\mathbf{e}_0 = \begin{pmatrix} 1 \\ a_1 \\ a_2 a_1 \\ \vdots \\ a_{j-1} a_{j-2} \cdots a_2 a_1 \\ \vdots \\ a_{L-2} a_{L-3} \cdots a_2 a_1 \\ a_{L-1} a_{L-2} \cdots a_2 a_1 \\ a_L a_{L-1} \cdots a_2 a_1 \end{pmatrix} \in \mathbb{R}^{L+1} .
$$

For a solution see Solution 8.103.

In order to illustrate the idea underlying the replication of a gate based computation with an adiabatic one, we use the result of Exercise 8.103 for the following analogy: imagine that in A we replace the a_j with the operators U_j of the gates of our circuit, and in $\mathbf{e}_0$ we replace 1 with $|\Psi_{\text{ini}}\rangle$ as well as a_1 with $U_1|\Psi_{\text{ini}}\rangle$, and the remaining a_j in $\mathbf{e}_0$ are replaced again with the U_j. What the claim of Exercise 8.103 then would tell us is that the eigenstate $\mathbf{e}_0$ of A for the eigenvalue 0 has in its last component $U_L U_{L-1} \cdots U_1 |\Psi_{\text{ini}}\rangle = U|\Psi_{\text{ini}}\rangle$, which is the final state $|\Psi_{\text{fin}}\rangle$ we are trying to construct. Hence, we would have identified the desired state $|\Psi_{\text{fin}}\rangle$ as part of an eigenstate of an operator A which has been built by using the individual gates U_j of our circuit U. This is the idea behind replicating a gate based quantum computation with an adiabatic computation, where the role of A will be taken by a suitable Hamiltonian. The remainder of this section is devoted to making this more precise.

In our analogy the jth component of $\mathbf{e}_0$ is $U_j U_{j-1} \cdots U_1 |\Psi_{\text{ini}}\rangle$ and constitutes that state in the circuit after which the jth gate has been applied. Therefore, we interpret the basis states

$$
\begin{pmatrix} 1 \\ 0 \\ 0 \\ \vdots \\ 0 \end{pmatrix} , \begin{pmatrix} 0 \\ 1 \\ 0 \\ \vdots \\ 0 \end{pmatrix} , \begin{pmatrix} 0 \\ 0 \\ 1 \\ \vdots \\ 0 \end{pmatrix} , \cdots , \begin{pmatrix} 0 \\ 0 \\ 0 \\ \vdots \\ 1 \end{pmatrix}
$$

as the analogues of states of a computer clock advancing with each new gate that is applied in the circuit. A first ingredient required for the adiabatic equivalent is then the 'clock' or 'counter' space and its states.

Definition 8.17 Let $L \in \mathbb{N}$, and for $l \in \{0, \ldots, L\}$ define

$$
x(l) := \begin{cases} 0 & \text{if } l = 0 \\ \sum_{k=L-l}^{L-1} 2^k & \text{if } l > 0. \end{cases}
$$

We define the **clock space** as $\mathbb{H}^C := \mathbb{H}^{\otimes L}$ and the **clock** or **counter states** as the subset of the computational basis states given by

$$
|x(l)\rangle = |x(l)_{L-1} \ldots x(l)_0\rangle \in \mathbb{H}^C ,
$$

where

$$
x(l)_k = \begin{cases} 0 & \text{if } 0 \le k < L - l \\ 1 & \text{if } L - l \le k \le L - 1. \end{cases} \tag{8.90}
$$

Note that for $l \in \{0, \ldots, L\}$ each $|x(l)\rangle$ is a basis vector of the computational basis and that

$$
x(l) = x(m) \quad \Leftrightarrow \quad l = m
$$

as well as

$$
\langle x(l) | x(m) \rangle = \delta_{lm} . \tag{8.91}
$$

Moreover, we have

$$
|x(0)\rangle = |\underbrace{0 \ldots 0}_{L \text{ times}}\rangle , \qquad |x(1)\rangle = |1 \underbrace{0 \ldots 0}_{L-1 \text{ times}}\rangle , \quad \ldots ,
$$
$$
|x(l)\rangle = |\underbrace{1 \ldots 1}_{l \text{ times}} \underbrace{0 \ldots 0}_{L-l \text{ times}}\rangle , \qquad \ldots , \qquad |x(L)\rangle = |\underbrace{1 \ldots 1}_{L \text{ times}}\rangle . \tag{8.92}
$$

Since by Definition 8.17

$$
x(l + 1) = x(l) + 2^{L-l-1} ,
$$

we find that the difference between $|x(l)\rangle$ and $|x(l + 1)\rangle$ is that only $x(l)_{L-l-1} = 0$ changes to $x(l + 1)_{L-l-1} = 1$. Hence, stepping through the finite sequence $(|x(0)\rangle, |x(1)\rangle, \ldots, |x(L)\rangle)$ constitutes a GRAY-coded transition (see Definition 5.22).

The reason that we chose the $|x(l)\rangle$ as the counter states in this form is that each 'time-step', that is, 'advancing the clock' from $|x(l)\rangle$ to $|x(l + 1)\rangle$ remains 3-local (see Definition 3.21). Had we chosen a simple counter, such as the computational basis states $|0\rangle, |1\rangle, |2\rangle, \ldots, |L\rangle$ in $\mathbb{H}^{\otimes J}$ with $J = \lfloor \log_2 L \rfloor + 1$, to count down time, then advancing the clock, for example from

$$|2^{J-1} - 1\rangle = |\sum_{j=0}^{J-2} 2^j\rangle = |0 \underbrace{1\ldots1}_{J-1 \text{ times}}\rangle ,$$

by one step to the clock state

$$|2^{J-1}\rangle = |1 \underbrace{0\ldots0}_{J-1 \text{ times}}\rangle ,$$

would require the change of altogether J qubits. Hence, the clock Hamiltonian would have to be J-local and thus ultimately be $O(\log_2 L)$ local.

Equipped with the clock states $|x(l)\rangle$, we proceed to define various other states which will be used to construct the adiabatic approximation to a given circuit [34].

Definition 8.18 Let U be a plain circuit of length L as defined in Definition 5.27 acting on n qubits

$$U = U_L \cdots U_1 U_0 \quad \in \mathcal{U}(\mathbb{H}^{\otimes n}),$$

where we set $U_0 := \mathbf{1}^{\otimes n}$ and each U_l is 2-local. We call $\mathbb{H}^U := \mathbb{H}^{\otimes n}$ the circuit space. Let $|\Psi_{\text{ini}}\rangle := |0\rangle^n \in \mathbb{H}^U$ be the initial circuit state, and for $l \in \{0,\ldots L\}$ define the states

$$|\Xi(l)\rangle := U_l \cdots U_1 U_0 |\Psi_{\text{ini}}\rangle \in \mathbb{H}^U . \qquad (8.93)$$

In the combined circuit and clock system described by states in the combined circuit and clock space $\mathbb{H}^U \otimes \mathbb{H}^C$ we define

$$|\Gamma(l)\rangle := |\Xi(l)\rangle \otimes |x(l)\rangle \in \mathbb{H}^U \otimes \mathbb{H}^C \qquad (8.94)$$

and

$$|\Gamma\rangle := \frac{1}{\sqrt{L+1}} \sum_{m=0}^{L} |\Gamma(m)\rangle \in \mathbb{H}^U \otimes \mathbb{H}^C . \qquad (8.95)$$

Exercise 8.104 Show that

$$\|\Xi(l)\| = 1 \qquad (8.96)$$
$$\langle\Gamma(l)|\Gamma(m)\rangle = \delta_{lm} . \qquad (8.97)$$

For a solution see Solution 8.104.

From (8.94) and (8.97) we see that the $\{|\Gamma(m)\rangle \mid m \in \{0, \ldots, L\}\}$ are a set of orthonormal vectors in the combined circuit and clock space $\mathbb{H}^U \otimes \mathbb{H}^C$. With their help we define the following subspace

$$\mathbb{H}_{\text{sub}} := \text{Span}\,\{|\Gamma(m)\rangle \mid m \in \{0, \ldots, L\}\} \subset \mathbb{H}^U \otimes \mathbb{H}^C. \tag{8.98}$$

In order to construct an adiabatic quantum computation, we need to specify its initial and final Hamiltonian. This is done in the following rather lengthy definition [34]. The rationale behind the various terms will be given later after some of their properties have been shown.

Definition 8.19 For $l \in \{0, \ldots, L\}$ let U_l be the gates of a circuit U of length L as in Definition 8.18. On the combined circuit and clock space $\mathbb{H}^U \otimes \mathbb{H}^C$ we define the following Hamiltonians

$$H_{\text{ini}} := H_{\text{c-ini}} + H_{\text{input}} + H_{\text{clock}} \tag{8.99}$$

$$H_{\text{fin}} := H_{\text{prop}} + H_{\text{input}} + H_{\text{clock}}, \tag{8.100}$$

where

$$H_{\text{c-ini}} := \mathbf{1}^{\otimes n} \otimes |1\rangle\langle 1| \otimes \mathbf{1}^{\otimes L-1} \tag{8.101}$$

$$H_{\text{input}} := \sum_{j=1}^{n} \mathbf{1}^{\otimes j-1} \otimes |1\rangle\langle 1| \otimes \mathbf{1}^{\otimes n-j} \otimes |0\rangle\langle 0| \otimes \mathbf{1}^{\otimes L-1} \tag{8.102}$$

$$H_{\text{clock}} := \mathbf{1}^{\otimes n} \otimes \sum_{l=0}^{L-2} \mathbf{1}^{\otimes l} \otimes |0\rangle\langle 0| \otimes |1\rangle\langle 1| \otimes \mathbf{1}^{\otimes L-l-2} \tag{8.103}$$

$$H_{\text{prop}} := \frac{1}{2} \sum_{l=1}^{L} H_l. \tag{8.104}$$

The H_l in the propagation Hamiltonian H_{prop} are defined for the cases $l = 1$, $1 < l < L$ and $l = L$ as follows:

$$H_1 := \Big[\mathbf{1}^{\otimes n} \otimes \big(|00\rangle\langle 00| + |10\rangle\langle 10| \big) \tag{8.105}$$
$$- U_1 \otimes |10\rangle\langle 00| - U_1^* \otimes |00\rangle\langle 10| \Big] \otimes \mathbf{1}^{\otimes L-2}$$

$$H_l := \Big[\mathbf{1}^{\otimes n+l-2} \otimes \big(|100\rangle\langle 100| + |110\rangle\langle 110| \big) \tag{8.106}$$
$$- U_l \otimes \mathbf{1}^{\otimes l-2} \otimes |110\rangle\langle 100| - U_l^* \otimes \mathbf{1}^{\otimes l-2} \otimes |100\rangle\langle 110| \Big] \otimes \mathbf{1}^{\otimes L-l-1}$$

$$H_L := \mathbf{1}^{\otimes n+L-2} \otimes \big(|10\rangle\langle 10| + |11\rangle\langle 11| \big) \tag{8.107}$$
$$- U_L \otimes \mathbf{1}^{\otimes L-2} \otimes |11\rangle\langle 10| - U_L^* \otimes \mathbf{1}^{\otimes L-2} \otimes |10\rangle\langle 11|.$$

The reason for the particular form of H_{prop} comes from the illustrative results derived in Exercise 8.103. The discussion subsequent to that exercise shows that with H_{prop} as defined in Definition 8.19, we can expect to find the desired state $U_L \cdots U_1|\Psi_{ini}\rangle$ as a component of the ground state of H_{fin}. This will be confirmed in Theorem 8.23, where we show that $|\Gamma\rangle$, which is the analogue of e_0 in Exercise 8.103, is a ground state of H_{fin}. The rationale for the particular forms of $H_{c\text{-}ini}$, H_{input} and H_{clock} will be easier to explain with the proof of Lemma 8.20 in hand.

It turns out that $|\Gamma(0)\rangle$ is the ground state of H_{ini}, in other words, it is an eigenvector of H_{ini} for the lowest possible eigenvector, which for H_{ini} is zero.

Lemma 8.20 *The operators* $H_{c\text{-}ini}, H_{input}, H_{clock}$ *and* H_{ini} *as defined in Definition 8.19 are self-adjoint and positive.*

Moreover, the lowest eigenvalue of H_{ini} *is* 0. *This eigenvalue is non-degenerate and has* $|\Gamma(0)\rangle = |0\rangle^n \otimes |0\rangle^L$ *as an eigenvector, that is,* $\text{Eig}(H_{ini}, 0) = \text{Span}\{|\Gamma(0)\rangle\}$.

Proof To begin with, we have, for example,

$$H_{clock}^* \underset{(8.103)}{=} \left(\mathbf{1}^{\otimes n} \otimes \sum_{l=0}^{L-2} \mathbf{1}^{\otimes l} \otimes |0\rangle\langle 0| \otimes |1\rangle\langle 1| \otimes \mathbf{1}^{\otimes L-l-2} \right)^*$$

$$\underset{(3.31)}{=} \mathbf{1}^{\otimes n} \otimes \sum_{l=0}^{L-2} \mathbf{1}^{\otimes l} \otimes (|0\rangle\langle 0|)^* \otimes (|1\rangle\langle 1|)^* \otimes \mathbf{1}^{\otimes L-l-2}$$

$$\underset{(2.36)}{=} \mathbf{1}^{\otimes n} \otimes \sum_{l=0}^{L-2} \mathbf{1}^{\otimes l} \otimes |0\rangle\langle 0| \otimes |1\rangle\langle 1| \otimes \mathbf{1}^{\otimes L-l-2}$$

$$\underset{(8.103)}{=} H_{clock} .$$

Similarly, one shows that $H_{c\text{-}ini}$ and H_{input} are self-adjoint. Then H_{ini} as a sum of self-adjoint operators is self-adjoint as well.

To show their positivity, consider the computational basis (see Definition 3.8) vectors

$$|\xi\rangle = |\xi_{n-1} \dots \xi_0\rangle = |\xi_{n-1}\rangle \otimes \cdots \otimes |\xi_0\rangle \in \mathbb{H}^U$$

with $\xi = \sum_{j=0}^{n-1} \xi_j 2^j$, where $\xi_j \in \{0, 1\}$, and

$$|x\rangle = |x_{n-1} \dots x_0\rangle = |x_{n-1}\rangle \otimes \cdots \otimes |x_0\rangle \in \mathbb{H}^C$$

with $x = \sum_{j=0}^{L-1} x_j 2^j$, where $x_j \in \{0, 1\}$. The set of vectors $\{|\xi\rangle \otimes |x\rangle \mid 0 \leq \xi < 2^n,\ 0 \leq x < 2^L\}$ form an ONB in $\mathbb{H}^U \otimes \mathbb{H}^C$, and any vector $|\Psi\rangle \in \mathbb{H}^U \otimes \mathbb{H}^C$ can be written in the form

$$|\Psi\rangle = \sum_{\xi=0}^{2^n-1} \sum_{x=0}^{2^L-1} \Psi_{\xi x} |\xi\rangle \otimes |x\rangle\,.$$

In the following we will often use that

$$\langle 0|\xi_j\rangle = (1 - \xi_j) \quad \text{and} \quad \langle 1|\xi_j\rangle = \xi_j \tag{8.108}$$

and likewise for $|x_j\rangle$. Hence, we can write

$$
\begin{aligned}
H_{\text{c-ini}}\Big(|\xi\rangle \otimes |x\rangle\Big) &\underset{(8.101)}{=} \big(\mathbf{1}^{\otimes n} \otimes |1\rangle\langle 1| \otimes \mathbf{1}^{\otimes L-1}\big)|\xi\rangle \otimes |x\rangle \\
&= |\xi\rangle \otimes |1\rangle \underbrace{\langle 1|x_{L-1}\rangle}_{=x_{L-1}} \otimes |x_{L-2}\rangle \otimes \cdots \otimes |x_0\rangle \\
&= x_{L-1}|\xi\rangle \otimes |1\rangle \otimes |x_{L-2}\rangle \otimes \cdots \otimes |x_0\rangle \\
&= x_{L-1}|\xi\rangle \otimes |x\rangle \tag{8.109}
\end{aligned}
$$

such that for any $|\Psi\rangle \in \mathbb{H}^U \otimes \mathbb{H}^C$

$$
\begin{aligned}
\langle\Psi|H_{\text{c-ini}}\Psi\rangle &= \sum_{\zeta,y}\sum_{\xi,x} \overline{\Psi_{\zeta y}} \Psi_{\xi x} x_{L-1} \underbrace{\langle\zeta|\xi\rangle}_{=\delta_{\zeta\xi}} \underbrace{\langle y_{L-1}|1\rangle}_{=y_{L-1}} \langle y_{L-2}|x_{L-2}\rangle \cdots \langle y_0|x_0\rangle \\
&= \sum_{\xi}\sum_{y,x} y_{L-1} x_{L-1} \overline{\Psi_{\xi y}} \Psi_{\xi x} \underbrace{\langle y_{L-2}|x_{L-2}\rangle}_{=\delta_{y_{L-2},x_{L-2}}} \cdots \underbrace{\langle y_0|x_0\rangle}_{=\delta_{y_0,x_0}} \\
&= \sum_{\xi,x} \delta_{1,x_{L-1}} |\Psi_{\xi x}|^2 \\
&\geq 0\,, \tag{8.110}
\end{aligned}
$$

verifying the positivity of $H_{\text{c-ini}}$. Next, we consider

$$
\begin{aligned}
H_{\text{input}}\Big(|\xi\rangle \otimes |x\rangle\Big) &\underset{(8.102)}{=} \Big(\sum_{j=1}^{n} \mathbf{1}^{\otimes j-1} \otimes |1\rangle\langle 1| \otimes \mathbf{1}^{\otimes n-j} \otimes |0\rangle\langle 0| \otimes \mathbf{1}^{\otimes L-1}\Big)|\xi\rangle \otimes |x\rangle \\
&\underset{(8.108)}{=} \sum_{j=1}^{n} \xi_{n-j}(1 - x_{L-1})|\xi_{n-1}\rangle \otimes \cdots \otimes |\xi_{n-j+1}\rangle \otimes |1\rangle \otimes |\xi_{n-j-1}\rangle \\
&\qquad\qquad \otimes \cdots \otimes |\xi_0\rangle \otimes |0\rangle \otimes |x_{L-2}\rangle \otimes \cdots \otimes |x_0\rangle \tag{8.111}
\end{aligned}
$$

such that for any $|\Psi\rangle \in \mathbb{H}^U \otimes \mathbb{H}^C$

$$
\begin{aligned}
\langle\Psi|\mathsf{H}_{\text{input}}\Psi\rangle = &\sum_{\zeta,\xi,y,x}\sum_{j=1}^{n}\overline{\Psi_{\zeta y}}\Psi_{\xi x}\xi_{n-j}(1-x_{L-1})\langle\zeta_{n-1}|\xi_{n-1}\rangle\cdots\langle\zeta_{n-j+1}|\xi_{n-j+1}\rangle \\
&\times\langle\zeta_{n-j}|1\rangle\langle\zeta_{n-j-1}|\xi_{n-j-1}\rangle\cdots\langle\zeta_0|\xi_0\rangle \\
&\times\langle y_{L-1}|0\rangle\langle y_{L-2}|x_{L-2}\rangle\cdots\langle y_0|x_0\rangle \\
= &\sum_{\xi,x}\sum_{j=1}^{n}|\Psi_{\xi x}|^2\,\delta_{1,\xi_{n-j}}\delta_{0,x_{L-1}} \\
\geq &\,0,
\end{aligned}
$$

showing that $\mathsf{H}_{\text{input}}$ is positive. Finally, we have

$$
\begin{aligned}
\mathsf{H}_{\text{clock}}|\xi\rangle\otimes|x\rangle \underbrace{=}_{(8.103)} &\left(\mathbf{1}^{\otimes n}\otimes\sum_{l=0}^{L-2}\mathbf{1}^{\otimes l}|0\rangle\langle 0|\otimes|1\rangle\langle 1|\otimes\mathbf{1}^{\otimes L-l-2}\right)|\xi\rangle\otimes|x\rangle \\
\underbrace{=}_{(8.108)} &|\xi\rangle\otimes\sum_{l=0}^{L-2}(1-x_{L-l-1})x_{L-l-2}|x_{L-1}\rangle\otimes\cdots\otimes|x_{L-l}\rangle \\
&\otimes|0\rangle\otimes|1\rangle\otimes|x_{L-l-3}\rangle\otimes\cdots\otimes|x_0\rangle
\end{aligned}
\tag{8.112}
$$

such that for any $|\Psi\rangle \in \mathbb{H}^U \otimes \mathbb{H}^C$

$$
\begin{aligned}
\langle\Psi|\mathsf{H}_{\text{clock}}\Psi\rangle = &\sum_{\xi,x}\sum_{l=0}^{L-2}|\Psi_{\xi x}|^2\,\delta_{1,x_{L-l-2}}\delta_{0,x_{L-l-1}} \\
\geq &\,0,
\end{aligned}
\tag{8.113}
$$

which proves the positivity of $\mathsf{H}_{\text{clock}}$. As a sum of positive operators, H_{ini} is thus positive, too, and its lowest possible eigenvalue is zero. Since for $|\Gamma(0)\rangle = |0\rangle^n \otimes |0\rangle^L$ all ξ_j and x_j are zero, it follows from (8.99), (8.109), (8.111) and (8.112) that $\mathsf{H}_{\text{ini}}|\Gamma(0)\rangle = 0$.

To show that this is the only eigenvector for the eigenvalue zero, note that we have from (8.110)–(8.113) that for any $|\Psi\rangle \in \mathbb{H}^U \otimes \mathbb{H}^C$

$$
\langle\Psi|\mathsf{H}_{\text{ini}}\Psi\rangle = \sum_{\xi,x}\Big(\underbrace{\delta_{1,x_{L-1}}}_{\geq 0}+\underbrace{\sum_{j=1}^{n}\delta_{1,\xi_{n-j}}\delta_{0,x_{L-1}}}_{\geq 0}+\underbrace{\sum_{l=0}^{L-2}\delta_{1,x_{L-l-2}}\delta_{0,x_{L-l-1}}}_{\geq 0}\Big)|\Psi_{\xi x}|^2\,.
$$

Hence, for $|\Psi\rangle$ to be an eigenvector of H_{ini} for the eigenvalue zero, each term has to vanish. This means that first only $\Psi_{\xi x}$ for $x_{L-1}=0$ can be non-zero. But then the middle sum implies that only $\Psi_{\xi x}$ with $\xi_j = 0 = x_{L-1}$, where $j \in \{0,\ldots,n-1\}$ can be non-zero, that is, the only possibly non-vanishing $\Psi_{\xi x}$

are of the form $\Psi_{0\ldots0,0x_{L-2}\ldots x_0}$. The last sum only vanishes, if out of these those for which

$$x_{L-l-2} = 1 \quad \text{and} \quad x_{L-l-1} = 0 \quad \text{for } l \in \{0, \ldots, L-2\}$$

vanish. Using this for $l = 0$ implies that the only possibly non-vanishing component of the eigenvector has to be of the form $\Psi_{0\ldots0,00x_{L-3}\ldots x_0}$. Using the same argument successively from $l = 1$ to $l = L - 2$, we find that the only non-vanishing component of an eigenvector $|\Psi\rangle$ of H_{ini} is $\Psi_{0\ldots0,0\ldots0}$, hence, $|\Psi\rangle = e^{i\alpha}|\Gamma(0)\rangle$ as claimed.

□

Our adiabatic replication of the circuit U evolves the ground state of H_{ini} into a ground state of H_{fin} with a probability as described in Corollary 8.4. The rationale behind the form of $H_{c\text{-}ini}$, H_{input} and H_{clock} is to assure that the initial state $|\Psi\rangle = |0\rangle^n \otimes |0\rangle^L = |\Gamma(0)\rangle$ is the ground state of H_{ini} and that the ground state of H_{fin} is akin to e_0 in Exercise 8.103. This means in particular, that we want to start our adiabatic evolution in the starting clock state $|x(0)\rangle = |0\rangle^L$ and then let the ground state evolve only in the subspace of 'legal' clock states

$$\mathbb{H}^C_{leg} = \text{Span}\left\{ |x(l)\rangle \mid l \in \{0, \ldots, L\} \right\}. \tag{8.114}$$

This is achieved by designing $H_{c\text{-}ini}$, H_{input} and H_{clock} in such a way that they vanish on $\mathbb{H}^C_{leg}$ and have eigenvalues no less than 1 on its orthogonal complement. From (8.112) we see that

$$H_{clock}\left(|\xi\rangle \otimes |x\rangle\right) = \begin{pmatrix} \text{count of appearances of sequences} \\ \text{01 in binary form of x} \end{pmatrix} |\xi\rangle \otimes |x\rangle$$

and (8.92) together with (8.114) shows that $H_{clock}\left(|\xi\rangle \otimes |x\rangle\right) = 0$ if and only if $|x\rangle \in \mathbb{H}^C_{leg}$. Similarly, we see from (8.111) that $H_{input}\left(|\xi\rangle \otimes |x\rangle\right) = 0$ if and only if $\xi_j = 0$ for $j \in \{0, \ldots, n-1\}$ or if $x_{L-1} = 1$. Hence, for the initial clock state $|x(0)\rangle = |0\rangle^L$ only $|\xi\rangle = |0\rangle^n$ yields $H_{input}\left(|\xi\rangle \otimes |x\rangle\right) = 0$, in other words, if the clock shows 0, the circuit state must be $|0\rangle^n$ if the combined system is to be in a ground state of H_{ini}. Finally, from (8.109) we see that $H_{c\text{-}ini}\left(|\xi\rangle \otimes |x\rangle\right) = x_{L-1}|\xi\rangle \otimes |x\rangle$ such that it vanishes if and only if $x_{L-1} = 0$. But this together with the earlier requirement $|x\rangle \in \mathbb{H}^C_{leg}$ implies that the ground state of H_{ini} is $|0\rangle^n \otimes |0\rangle^L$. This may suffice as a somewhat heuristic motivation for the constructions of the various Hamiltonians in Definition 8.19. In the remainder of this section we proceed to show in more detail that the way these Hamiltonians are defined does indeed deliver the desired replication.

Now that we have established in Lemma 8.20 that $|\Gamma(0)\rangle$ is indeed the unique (up to a phase, of course) eigenvector of the lowest eigenvalue of H_{ini}, we study the properties of H_{fin}. For this we first need to establish how the various terms in the H_l of H_{prop} act on the clock state vectors $|x(m)\rangle$, where $m \in \{0, \ldots L\}$.

Exercise 8.105 Show that for any clock state $|x(m)\rangle$ as defined in Definition 8.17 and $a, b \in \{0, 1\}$ one has the following identities.

$$\left(|a0\rangle\langle b0| \otimes \mathbf{1}^{\otimes L-2}\right)|x(m)\rangle = \delta_{a,0}\delta_{b,1}\delta_{m,1}|x(m-1)\rangle \tag{8.115}$$
$$+ \left(\delta_{a,0}\delta_{b,0}\delta_{m,0} + \delta_{a,1}\delta_{b,1}\delta_{m,1}\right)|x(m)\rangle$$
$$+ \delta_{a,1}\delta_{b,0}\delta_{m,0}|x(m+1)\rangle$$

$$\left(\mathbf{1}^{\otimes l-2} \otimes |1a0\rangle\langle 1b0|\mathbf{1}^{\otimes L-l-1}\right)|x(m)\rangle = \delta_{a,0}\delta_{b,1}\delta_{m,l}|x(m-1)\rangle \tag{8.116}$$
$$+ \left(\delta_{a,0}\delta_{b,0}\delta_{m,l-1} + \delta_{a,1}\delta_{b,1}\delta_{m,l}\right)|x(m)\rangle$$
$$+ \delta_{a,1}\delta_{b,0}\delta_{m,l-1}|x(m+1)\rangle$$

$$\left(\mathbf{1}^{\otimes L-2} \otimes |1a\rangle\langle 1b|\right)|x(m)\rangle = \delta_{a,0}\delta_{b,1}\delta_{m,L}|x(m-1)\rangle \tag{8.117}$$
$$+ \left(\delta_{a,0}\delta_{b,0}\delta_{m,L-1} + \delta_{a,1}\delta_{b,1}\delta_{m,L}\right)|x(m)\rangle$$
$$+ \delta_{a,1}\delta_{b,0}\delta_{m,L-1}|x(m+1)\rangle .$$

For a solution see Solution 8.105.

Next, we show that $H_{c\text{-ini}}, H_{input}, H_{clock}, H_l$ and thus H_{prop} all leave $\mathbb{H}_{sub}$ as defined in (8.98) invariant.

Lemma 8.21 *For $m \in \{0, \ldots, L\}$ let $|\Gamma(m)\rangle$ be as defined in Definition 8.18 and $H_{c\text{-ini}}, H_{input}, H_{clock}$ and H_l for $l \in \{1, \ldots, L\}$ as defined in Definition 8.19. Then the following hold.*

$$H_{c\text{-ini}}|\Gamma(m)\rangle = (1 - \delta_{m,0})|\Gamma(m)\rangle \tag{8.118}$$
$$H_{input}|\Gamma(m)\rangle = 0 \tag{8.119}$$
$$H_{clock}|\Gamma(m)\rangle = 0 \tag{8.120}$$
$$H_l|\Gamma(m)\rangle = (\delta_{m,l-1} + \delta_{m,l})|\Gamma(m)\rangle \tag{8.121}$$
$$- \delta_{m,l-1}|\Gamma(m+1)\rangle - \delta_{m,l}|\Gamma(m-1)\rangle .$$

Proof We begin with the proof for (8.118). We have

$$H_{c\text{-ini}}|\Gamma(m)\rangle \underset{(8.101)}{=} \mathbf{1}^{\otimes n} \otimes |1\rangle\langle 1| \otimes \mathbf{1}^{\otimes L-1}|\Xi(m)\rangle \otimes |x(m)\rangle$$
$$\underset{(8.109)}{=} x(m)_{L-1}|\Xi(m)\rangle \otimes |1x(m)_{L-2}\ldots x(m)_0\rangle ,$$

where (8.90) implies

$$x(m)_{L-1} = 1 - \delta_{m,0}$$

and

$$|1x(m)_{L-2}\ldots x(m)_0\rangle = |x(m)\rangle \qquad \text{for } m \geq 1.$$

Hence, we obtain

$$H_{\text{c-ini}}|\Gamma(m)\rangle = (1 - \delta_{m,0})|\Xi(m)\rangle \otimes |x(m)\rangle = (1 - \delta_{m,0})|\Gamma(m)\rangle$$

as claimed. Next, we turn to (8.119) for which we have

$$H_{\text{input}}|\Gamma(m)\rangle \underset{(8.102)}{=} \left(\sum_{j=1}^{n} \mathbf{1}^{\otimes j-1} \otimes |1\rangle\langle 1| \otimes \mathbf{1}^{\otimes n-j} \otimes |0\rangle\langle 0| \otimes \mathbf{1}^{\otimes L-1} \right)|\Xi(m)\rangle \otimes |x(m)\rangle$$

$$= \delta_{0,x(m)_{L-1}}\left(\left(\sum_{j=1}^{n} \mathbf{1}^{\otimes j-1} \otimes |1\rangle\langle 1| \otimes \mathbf{1}^{\otimes n-j} \right)|\Xi(m)\rangle \right)$$
$$\otimes |0x(m)_{L-2}\ldots x(m)_0\rangle,$$

where (8.90) implies

$$x(m)_{L-1} = 0 \quad \Leftrightarrow \quad m = 0.$$

Therefore, we obtain

$$H_{\text{input}}|\Gamma(m)\rangle = \delta_{m,0}\left(\sum_{j=1}^{n} \mathbf{1}^{\otimes j-1} \otimes |1\rangle\langle 1| \otimes \mathbf{1}^{\otimes n-j} \right)|0\rangle^{n} \otimes |0\rangle^{L}$$

$$= 0$$

as claimed. To show (8.120), we note that for any $m \in \{0,\ldots,L\}$ we have

$$H_{\text{clock}}|\Gamma(m)\rangle \underset{(8.103)}{=} \left(\mathbf{1}^{\otimes n} \otimes \sum_{l=0}^{L-2} \mathbf{1}^{\otimes l} \otimes |0\rangle\langle 0| \otimes |1\rangle\langle 1| \otimes \mathbf{1}^{\otimes L-l-2} \right)|\Xi(m)\rangle \otimes |x(m)\rangle$$

$$= |\Xi(m)\rangle \otimes$$
$$\sum_{l=0}^{L-2} \delta_{0,x(m)_{L-1-l}}\delta_{1,x(m)_{L-2-l}}|x(m)_{L-1}\ldots x(m)_{L-l}01x(m)_{L-3-l}\ldots x(m)_0\rangle,$$

where (8.90) implies

$$x(m)_{L-1-l} = 0 \quad \Leftrightarrow \quad m < l+1 \quad \text{and} \quad x(m)_{L-2-l} = 1 \quad \Leftrightarrow \quad m \geq l+2,$$

which is impossible and thus $H_{\text{clock}}|\Gamma(m)\rangle = 0$ as claimed.

To show (8.121), we begin with $l = 1$, which yields

$$
\begin{aligned}
\mathsf{H}_1|\Gamma(m)\rangle \underbrace{=}_{(8.105)} & \left(\left[\mathbf{1}^{\otimes n} \otimes (|00\rangle\langle 00| + |10\rangle\langle 10|)\right.\right. \\
& \left.- U_1 \otimes |10\rangle\langle 00| - U_1^* \otimes |00\rangle\langle 10|\right] \otimes \mathbf{1}^{\otimes L-2}\Big) |\Xi(m)\rangle \otimes |x(m)\rangle \\
\underbrace{=}_{(8.115)} & |\Xi(m)\rangle \otimes (\delta_{m,0} + \delta_{m,1})|x(m)\rangle \\
& - U_1|\Xi(m)\rangle \otimes \delta_{m,0}|x(m+1)\rangle - U_1^*|\Xi(m)\rangle \otimes \delta_{m,1}|x(m-1)\rangle \\
\underbrace{=}_{(8.94)} & (\delta_{m,0} + \delta_{m,1})|\Gamma(m)\rangle \\
& - U_1|\Xi(0)\rangle \otimes \delta_{m,0}|x(1)\rangle - U_1^*|\Xi(1)\rangle \otimes \delta_{m,1}|x(0)\rangle .
\end{aligned}
$$

Here we can use that (8.93) implies $U_1|\Xi(0)\rangle = |\Xi(1)\rangle$ and $U_1^*|\Xi(1)\rangle = |\Xi(0)\rangle$. Consequently, we obtain

$$
\begin{aligned}
\mathsf{H}_1|\Gamma(m)\rangle = & (\delta_{m,0} + \delta_{m,1})|\Gamma(m)\rangle - \delta_{m,0}|\Xi(1)\rangle \otimes |x(1)\rangle - \delta_{m,1}|\Xi(0)\rangle \otimes |x(0)\rangle \\
\underbrace{=}_{(8.94)} & (\delta_{m,0} + \delta_{m,1})|\Gamma(m)\rangle - \delta_{m,0}|\Gamma(1)\rangle - \delta_{m,1}|\Gamma(0)\rangle \\
= & (\delta_{m,0} + \delta_{m,1})|\Gamma(m)\rangle - \delta_{m,0}|\Gamma(m+1)\rangle - \delta_{m,1}|\Gamma(m-1)\rangle ,
\end{aligned}
$$

which is (8.121) for $l = 1$. For $1 < l < L$ consider

$$
\begin{aligned}
\mathsf{H}_l|\Gamma(m)\rangle \underbrace{=}_{(8.105)} & \left(\left[\mathbf{1}^{\otimes n+l-2} \otimes (|100\rangle\langle 100| + |110\rangle\langle 110|)\right.\right. \\
& - U_l \otimes \mathbf{1}^{\otimes l-2} \otimes |110\rangle\langle 100| \\
& \left.- U_l^* \otimes \mathbf{1}^{\otimes l-2} \otimes |100\rangle\langle 110|\right] \otimes \mathbf{1}^{\otimes L-l-1}\Big) |\Xi(m)\rangle \otimes |x(m)\rangle \\
\underbrace{=}_{(8.116)} & |\Xi(m)\rangle \otimes (\delta_{m,l-1} + \delta_{m,l})|x(m)\rangle \\
& - U_l|\Xi(m)\rangle \otimes \delta_{m,l-1}|x(m+1)\rangle - U_l^*|\Xi(m)\rangle \otimes \delta_{m,l}|x(m-1)\rangle \\
\underbrace{=}_{(8.94)} & (\delta_{m,l-1} + \delta_{m,l})|\Gamma(m)\rangle \\
& - U_l|\Xi(l-1)\rangle \otimes \delta_{m,l-1}|x(l)\rangle - U_l^*|\Xi(l)\rangle \otimes \delta_{m,l}|x(l-1)\rangle .
\end{aligned}
$$

Here we use that (8.93) implies $U_l|\Xi(l-1)\rangle = |\Xi(l)\rangle$ and $U_l^*|\Xi(l)\rangle = |\Xi(l-1)\rangle$ to obtain

$$
\begin{aligned}
\mathsf{H}_l|\Gamma(m)\rangle &= \big(\delta_{m,l-1} + \delta_{m,l}\big)|\Gamma(m)\rangle \\
&\quad - \delta_{m,l-1}|\Xi(l)\rangle \otimes |x(l)\rangle - \delta_{m,l}|\Xi(l-1)\rangle \otimes |x(l-1)\rangle \\
&\underset{(8.94)}{=} \big(\delta_{m,l-1} + \delta_{m,l}\big)|\Gamma(m)\rangle - \delta_{m,l-1}|\Gamma(l)\rangle - \delta_{m,l}|\Gamma(l-1)\rangle \\
&= \big(\delta_{m,l-1} + \delta_{m,l}\big)|\Gamma(m)\rangle - \delta_{m,l-1}|\Gamma(m+1)\rangle - \delta_{m,l}|\Gamma(m-1)\rangle .
\end{aligned}
$$

Finally, consider the case $l = L$ for which we find

$$
\begin{aligned}
\mathsf{H}_L|\Gamma(m)\rangle &\underset{(8.107)}{=} \Big(\mathbf{1}^{\otimes n+L-2} \otimes \big(|10\rangle\langle 10| + |11\rangle\langle 11|\big) \\
&\quad - U_L \otimes \mathbf{1}^{\otimes L-2} \otimes |11\rangle\langle 10| - U_L^* \otimes \mathbf{1}^{\otimes L-2} \otimes |10\rangle\langle 11|\Big)|\Xi(m)\rangle \otimes |x(m)\rangle \\
&\underset{(8.117)}{=} |\Xi(m)\rangle \otimes \big(\delta_{m,L-1} + \delta_{m,L}\big)|x(m)\rangle \\
&\quad - U_L|\Xi(m)\rangle \otimes \delta_{m,L-1}|x(m+1)\rangle - U_L^*|\Xi(m)\rangle \otimes \delta_{m,L}|x(m-1)\rangle \\
&\underset{(8.94)}{=} \big(\delta_{m,L-1} + \delta_{m,L}\big)|\Gamma(m)\rangle \\
&\quad - U_L|\Xi(L-1)\rangle \otimes \delta_{m,L-1}|x(L)\rangle - U_L^*|\Xi(L)\rangle \otimes \delta_{m,L}|x(L-1)\rangle .
\end{aligned}
$$

Here we use once more that (8.93) implies $U_L|\Xi(L-1)\rangle = |\Xi(L)\rangle$ and $U_L^*|\Xi(L)\rangle = |\Xi(L-1)\rangle$. Therefore,

$$
\begin{aligned}
\mathsf{H}_L|\Gamma(m)\rangle &= \big(\delta_{m,L-1} + \delta_{m,L}\big)|\Gamma(m)\rangle \\
&\quad - \delta_{m,L-1}|\Xi(L)\rangle \otimes |x(L)\rangle - \delta_{m,L}|\Xi(L-1)\rangle \otimes |x(L-1)\rangle \\
&\underset{(8.94)}{=} \big(\delta_{m,L-1} + \delta_{m,L}|\Gamma(m)\rangle - \delta_{m,L-1}|\Gamma(L)\rangle - \delta_{m,L}|\Gamma(L-1)\rangle \\
&= \big(\delta_{m,L-1} + \delta_{m,L}\big)|\Gamma(m)\rangle - \delta_{m,L-1}|\Gamma(m+1)\rangle - \delta_{m,L}|\Gamma(m-1)\rangle ,
\end{aligned}
$$

which is (8.121) for $l = L$. $\qquad\square$

From the results in Lemma 8.21 we see that all the constituents $\mathsf{H}_{\text{c-ini}}$, $\mathsf{H}_{\text{input}}$, $\mathsf{H}_{\text{clock}}$ and H_{prop} of H_{ini} and H_{fin} leave the subspace $\operatorname{Span}\big\{|\Gamma(m)\rangle \,\big|\, m \in \{0,\ldots,L\}\big\}$ invariant. Consequently, this also holds for H_{ini} and H_{fin} and, in particular, for $\mathsf{H}_T(s)$ for all $s \in [0,1]$.

Theorem 8.22 *Let H_{ini} and H_{fin} be defined as in Definition 8.19. Moreover, let*

$$H_T(s) := (1 - s)H_{ini} + sH_{fin} \tag{8.122}$$

for $s \in [0, 1]$ and

$$\mathbb{H}_{sub} := \text{Span}\left\{ |\Gamma(m)\rangle \mid m \in \{0, \ldots, L\} \right\}. \tag{8.123}$$

Then we have

$$H_T(s)\{\mathbb{H}_{sub}\} \subset \mathbb{H}_{sub} \tag{8.124}$$

and $\left\{ |\Gamma(m)\rangle \mid m \in \{0, \ldots, L\} \right\}$ is an ONB of $\mathbb{H}_{sub}$. In this basis the restriction of $H_T(s)$ to $\mathbb{H}_{sub}$ has the matrix

$$H_T(s)\big|_{\mathbb{H}_{sub}} = \begin{pmatrix} \frac{s}{2} & -\frac{s}{2} & 0 & 0 & \cdots & 0 \\ -\frac{s}{2} & 1 & -\frac{s}{2} & 0 & \cdots & 0 \\ 0 & -\frac{s}{2} & 1 & -\frac{s}{2} & 0 & \vdots \\ \vdots & \ddots & \ddots & \ddots & \ddots & \vdots \\ 0 & \cdots & 0 & -\frac{s}{2} & 1 & -\frac{s}{2} \\ 0 & \cdots & 0 & 0 & -\frac{s}{2} & 1-\frac{s}{2} \end{pmatrix}. \tag{8.125}$$

Proof Let

$$|\Psi\rangle = \sum_{m=0}^{L} \Psi_m |\Gamma(m)\rangle$$

be an arbitrary vector of $\mathbb{H}_{sub}$. From (8.118)–(8.121) it follows that then

$$\left\{ H_{c\text{-}ini}|\Psi\rangle, \, H_{input}|\Psi\rangle, \, H_{clock}|\Psi\rangle, \, H_l|\Psi\rangle \right\} \subset \mathbb{H}_{sub}.$$

With Definition 8.19 we thus have $H_{prop}|\Psi\rangle \in \mathbb{H}_{sub}$ as well as

$$\left\{ H_{ini}|\Psi\rangle, \, H_{fin}|\Psi\rangle \right\} \subset \mathbb{H}_{sub},$$

which implies $H_T(s)|\Psi\rangle \in \mathbb{H}_{sub}$ for $H_T(s)$ as defined in (8.122), and the claim (8.124) follows.

To show (8.125), first recall that from (8.97) we know that the $\left\{ |\Gamma(m)\rangle \mid m \in \{0, \ldots, L\} \right\}$ are orthonormal, hence, by definition (8.123) of $\mathbb{H}_{sub}$ they form an ONB of this subspace. From (8.118)–(8.120) in Lemma 8.21 it follows that H_{ini} as defined in Definition 8.19 has the following matrix in the ONB $\left\{ |\Gamma(m)\rangle \mid m\{0, \ldots, L\} \right\}$

$$H_{\text{ini}}\big|_{\mathbb{H}_{\text{sub}}} \underbrace{=}_{(8.99)} H_{\text{c-ini}}\big|_{\mathbb{H}_{\text{sub}}} + H_{\text{input}}\big|_{\mathbb{H}_{\text{sub}}} + H_{\text{clock}}\big|_{\mathbb{H}_{\text{sub}}}$$

$$\underbrace{=}_{(8.119),(8.120)} H_{\text{c-ini}}\big|_{\mathbb{H}_{\text{sub}}} \underbrace{=}_{(8.118)} \begin{pmatrix} 0 & 0 & \cdots & & & 0 \\ 0 & 1 & 0 & \cdots & & 0 \\ 0 & 0 & 1 & 0 & \cdots & 0 \\ \vdots & \ddots & \ddots & \ddots & \ddots & \vdots \\ 0 & & \cdots & 0 & 1 & 0 \\ 0 & & \cdots & & 0 & 1 \end{pmatrix}. \qquad (8.126)$$

Similarly, it follows that

$$H_{\text{fin}}\big|_{\mathbb{H}_{\text{sub}}} \underbrace{=}_{(8.100)} H_{\text{prop}}\big|_{\mathbb{H}_{\text{sub}}} + H_{\text{input}}\big|_{\mathbb{H}_{\text{sub}}} + H_{\text{clock}}\big|_{\mathbb{H}_{\text{sub}}}$$

$$\underbrace{=}_{(8.119),(8.120)} H_{\text{prop}}\big|_{\mathbb{H}_{\text{sub}}}. \qquad (8.127)$$

Moreover, (8.121) implies that H_{prop} as given in (8.104) satisfies

$$H_{\text{prop}}|\Gamma(m)\rangle \underbrace{=}_{(8.104)} \sum_{l=1}^{L} H_l |\Gamma(m)\rangle$$

$$\underbrace{=}_{(8.121)} \frac{1}{2}\sum_{l=1}^{L}(\delta_{m,l-1} + \delta_{m,l})|\Gamma(m)\rangle$$

$$-\frac{1}{2}\sum_{l=1}^{L}\left(\delta_{m,l-1}|\Gamma(m+1)\rangle + \delta_{m,l}|\Gamma(m-1)\rangle\right)$$

$$= \begin{cases} \frac{1}{2}\left(|\Gamma(0)\rangle - |\Gamma(1)\rangle\right) & \text{if } m = 0 \\ |\Gamma(m)\rangle - \frac{1}{2}\left(|\Gamma(m-1)\rangle + |\Gamma(m+1)\rangle\right) & \text{if } 1 \le m \le L-1 \\ \frac{1}{2}\left(|\Gamma(L)\rangle - |\Gamma(L-1)\rangle\right) & \text{if } m = L. \end{cases}$$

The matrix of the restriction of H_{prop} to $\mathbb{H}_{\text{sub}}$ is thus given by

$$H_{\text{prop}}\big|_{\mathbb{H}_{\text{sub}}} = \begin{pmatrix} \frac{1}{2} & -\frac{1}{2} & 0 & \cdots & & & 0 \\ -\frac{1}{2} & 1 & -\frac{1}{2} & 0 & \cdots & & 0 \\ 0 & -\frac{1}{2} & 1 & -\frac{1}{2} & 0 & \cdots & 0 \\ \vdots & \ddots & \ddots & \ddots & \ddots & \ddots & \vdots \\ 0 & \cdots & 0 & -\frac{1}{2} & 1 & -\frac{1}{2} & 0 \\ 0 & & \cdots & 0 & -\frac{1}{2} & 1 & -\frac{1}{2} \\ 0 & & & \cdots & 0 & -\frac{1}{2} & \frac{1}{2} \end{pmatrix}. \qquad (8.128)$$

Consequently, we have

$$H_T(s)\big|_{\mathbb{H}_{sub}} \underbrace{=}_{(8.122)} (1-s)H_{ini}\big|_{\mathbb{H}_{sub}} + sH_{fin}\big|_{\mathbb{H}_{sub}}$$

$$\underbrace{=}_{(8.126),(8.127)} (1-s)H_{c\text{-}ini}\big|_{\mathbb{H}_{sub}} + sH_{prop}\big|_{\mathbb{H}_{sub}},$$

and (8.125) follows from (8.126) and (8.128).   □

Exercise 8.106 Let $H_T(s)$ and $\mathbb{H}_{sub}$ be defined as in Theorem 8.22, and let

$$P_{sub} = \sum_{m=0}^{L} |\Gamma(m)\rangle\langle\Gamma(m)| \qquad (8.129)$$

be the projector onto this subspace. Show that then

$$[H_T(s), P_{sub}] = 0 \qquad (8.130)$$

holds.

For a solution see Solution 8.106.

The result stated in Exercise 8.106 allows us to use Lemma 8.7, which implies that since we start with an initial state $|\Gamma(0)\rangle \in \mathbb{H}_{sub}$, the time evolution $U_T(s)$ generated by $H_T(s)$ does not leave the subspace $\mathbb{H}_{sub}$. Hence, we can restrict all our considerations to that subspace.

Exercise 8.107 Let $\mathbb{H}_{sub}$ be as defined in (8.123) and P_{sub} as given in (8.129). Show that for H_{prop} as defined in Definition 8.19 its restriction[2]

$$H_{prop}\big|_{\mathbb{H}_{sub}} := P_{sub}H_{prop}P_{sub} \qquad (8.131)$$

is positive, that is,

$$H_{prop}\big|_{\mathbb{H}_{sub}} \geq 0. \qquad (8.132)$$

For a solution see Solution 8.107.

---

[2]On the right side P_{sub} on the right of H_{prop} is viewed as a map $P_{sub} : \mathbb{H}_{sub} \to \mathbb{H}^U \otimes \mathbb{H}^C$ and on the left of H_{prop} as a map $P_{sub} : \mathbb{H}^U \otimes \mathbb{H}^C \to \mathbb{H}_{sub}$ such that we can view $H_{prop}\big|_{\mathbb{H}_{sub}}$ as an operator on $\mathbb{H}_{sub}$.

Recall from Lemma 8.20 that the lowest eigenvalue of H_{ini} is zero, non-degenerate and has the eigenvector $|\Gamma(0)\rangle = |0\rangle^n \otimes |0\rangle^L$. In the next theorem we show that $|\Gamma\rangle$ as given in Definition 8.18 is the ground state of H_{fin}, in other words, that it is an eigenvector of H_{fin} for its lowest eigenvalue, which is also zero and non-degenerate [34].

Theorem 8.23 *Let $\mathbb{H}_{sub}$ be as defined in (8.123) and P_{sub} as given in (8.129). Furthermore, let H_{ini} and H_{fin} be as defined in Definition 8.19 and $\mathsf{H}_T(s)$ as in Theorem 8.22. Then for all $s \in [0,1]$ the restriction*

$$\mathsf{H}_T(s)\big|_{\mathbb{H}_{sub}} := P_{sub}\mathsf{H}_T(s)P_{sub} \tag{8.133}$$

is positive, that is,

$$\mathsf{H}_T(s)\big|_{\mathbb{H}_{sub}} \geq 0. \tag{8.134}$$

Moreover, $|\Gamma\rangle$ as defined in Definition 8.18, satisfies

$$\mathsf{H}_{fin}|\Gamma\rangle = 0, \tag{8.135}$$

and the lowest eigenvalue of $\mathsf{H}_{fin}\big|_{\mathbb{H}_{sub}}$ is zero and non-degenerate.

Proof From Lemma 8.20 we know already that $\mathsf{H}_{c\text{-}ini}, \mathsf{H}_{input}, \mathsf{H}_{clock}$ and H_{ini} are positive on $\mathbb{H}^U \otimes \mathbb{H}^C$. Hence, they must be positive on $\mathbb{H}_{sub} \subset \mathbb{H}^U \otimes \mathbb{H}^C$ as well. Moreover, Lemma 8.21 shows that H_{input} and H_{clock} both vanish on $\mathbb{H}_{sub}$. Hence,

$$
\mathsf{H}_T(s)\big|_{\mathbb{H}_{sub}} \underbrace{=}_{(8.122)} (1-s)\mathsf{H}_{ini}\big|_{\mathbb{H}_{sub}} + s\mathsf{H}_{fin}\big|_{\mathbb{H}_{sub}}
$$

$$
\underbrace{=}_{(8.99),(8.100)} (1-s)\mathsf{H}_{c\text{-}ini}\big|_{\mathbb{H}_{sub}} + s\mathsf{H}_{prop}\big|_{\mathbb{H}_{sub}} + \mathsf{H}_{input}\big|_{\mathbb{H}_{sub}} + \mathsf{H}_{clock}\big|_{\mathbb{H}_{sub}}
$$

$$
\underbrace{=}_{(8.119),(8.120)} (1-s)\mathsf{H}_{c\text{-}ini}\big|_{\mathbb{H}_{sub}} + s\mathsf{H}_{prop}\big|_{\mathbb{H}_{sub}}
$$

$$
\underbrace{\geq}_{(8.110),(8.132)} 0,
$$

where we also used $s \in [0,1]$ in the last inequality. This proves (8.134).

To show (8.135), we note that due to (8.119) and (8.120) we have

$$\mathsf{H}_{input}|\Gamma\rangle = 0 = \mathsf{H}_{clock}|\Gamma\rangle.$$

It follows that

$$\mathsf{H}_{\text{fin}}|\Gamma\rangle \underset{(8.100)}{=} \mathsf{H}_{\text{prop}}|\Gamma\rangle + \mathsf{H}_{\text{input}}|\Gamma\rangle + \mathsf{H}_{\text{clock}}|\Gamma\rangle = \mathsf{H}_{\text{prop}}|\Gamma\rangle\,,$$

and it remains to show that

$$\mathsf{H}_{\text{prop}}|\Gamma\rangle = 0\,.$$

To this end, consider

$$
\begin{aligned}
\mathsf{H}_{\text{prop}}|\Gamma\rangle \underset{(8.95),(8.104)}{=} & \frac{1}{2\sqrt{L+1}} \sum_{l=1}^{L}\sum_{m=0}^{L} \mathsf{H}_l|\Gamma(m)\rangle \\
\underset{(8.121)}{=} & \frac{1}{2\sqrt{L+1}} \sum_{l=1}^{L}\sum_{m=0}^{L} \big((\delta_{m,l-1}+\delta_{m,l})|\Gamma(m)\rangle \\
& \qquad\qquad -\delta_{m,l-1}|\Gamma(m+1)\rangle - \delta_{m,l}|\Gamma(m-1)\rangle\big) \\
= & \frac{1}{2\sqrt{L+1}} \sum_{l=1}^{L} \big(|\Gamma(l-1)\rangle + |\Gamma(l)\rangle - |\Gamma(l)\rangle - |\Gamma(l-1)\rangle\big) \\
= & \ 0\,,
\end{aligned}
$$

which completes the proof of (8.135).

Finally, we show that the eigenvalue zero is non-degenerate. For this let

$$|\Phi\rangle = \sum_{m=0}^{L} \Phi_m|\Gamma(m)\rangle \in \mathbb{H}_{\text{sub}} \tag{8.136}$$

be such that $\||\Phi\|| = 1$ and

$$\mathsf{H}_{\text{fin}}\big|_{\mathbb{H}_{\text{sub}}}|\Phi\rangle = 0\,. \tag{8.137}$$

From (8.125) we see that in the ONB $\{|\Gamma(m)\rangle \mid m \in \{0,\dots,L\}\}$ the operator $\mathsf{H}_{\text{fin}}\big|_{\mathbb{H}_{\text{sub}}}$ has the matrix

$$\mathsf{H}_{\text{fin}}\big|_{\mathbb{H}_{\text{sub}}} \underset{(8.122)}{=} \mathsf{H}_T(1)\big|_{\mathbb{H}_{\text{sub}}} \underset{(8.125)}{=}
\begin{pmatrix}
\frac{1}{2} & -\frac{1}{2} & 0 & 0 & \dots & 0 \\
-\frac{1}{2} & 1 & -\frac{1}{2} & 0 & \dots & 0 \\
0 & -\frac{1}{2} & 1 & -\frac{1}{2} & 0 & \vdots \\
\vdots & \ddots & \ddots & \ddots & \ddots & \vdots \\
0 & \dots & 0 & -\frac{1}{2} & 1 & -\frac{1}{2} \\
0 & \dots & 0 & 0 & -\frac{1}{2} & \frac{1}{2}
\end{pmatrix}
\tag{8.138}$$

such that (8.137) implies

$$\frac{1}{2}(\Phi_0 - \Phi_1) = 0$$

$$\Phi_1 - \frac{1}{2}(\Phi_0 + \Phi_2) = 0$$

$$\vdots$$

$$\Phi_m - \frac{1}{2}(\Phi_{m-1} + \Phi_{m+1}) = 0$$

$$\vdots$$

$$\Phi_{L-1} - \frac{1}{2}(\Phi_{L-2} + \Phi_L) = 0$$

$$\frac{1}{2}(\Phi_L - \Phi_{L-1}) = 0\,,$$

which yields

$$\Phi_0 = \Phi_1 = \Phi_2 = \cdots = \Phi_m = \Phi_{m+1} = \cdots = \Phi_{L-1} = \Phi_L\,.$$

Consequently, (8.136) becomes

$$|\Phi\rangle = \Phi_0 \sum_{m=0}^{L} |\Gamma(m)\rangle \underbrace{=}_{(8.95)} \Phi_0\sqrt{L+1}|\Gamma\rangle\,,$$

proving the non-degeneracy of the eigenvalue zero.   □

Recall that our overall purpose in this section is to show that the action of a given circuit $U = U_L \ldots U_1$ composed of L gates U_l can be efficiently replicated (with a given probability) by a suitable adiabatic computation. So far we have identified the initial and final Hamiltonians H_{ini} and H_{fin} as well as $H_T(s)$ for this equivalent adiabatic computation and have established some of their properties. In particular, we have identified $|\Gamma(0)\rangle$ as the unique (up to a phase) ground state of H_{ini} and $|\Gamma\rangle$ as the unique (again, up to a phase) ground state of $H_{fin}|_{H_{sub}}$. The following lemma illuminates how finding the system in the state $|\Gamma\rangle$ is of help in replicating the action of a given circuit $U = U_L \cdots U_1$.

Lemma 8.24 *Observing (see Definition 5.35) the state* $|\Gamma\rangle$ *projects the circuit sub-system into the circuit end state* $U|0\rangle^n \in \mathbb{H}^U$ *with a probability of* $\frac{1}{L+1}$.

Proof Note that by Definition 8.18 we have

$$|\Gamma\rangle \underset{(8.121)}{=} \frac{1}{\sqrt{L+1}} \sum_{m=0}^{L} |\Gamma(m)\rangle$$

$$\underset{(8.94)}{=} \frac{1}{\sqrt{L+1}} \sum_{m=0}^{L} |\Xi(m)\rangle \otimes |x(m)\rangle$$

$$\underset{(8.93)}{=} \frac{1}{\sqrt{L+1}} \sum_{m=0}^{L} U_m \cdots U_0 |0\rangle^n \otimes |x(m)\rangle \qquad (8.139)$$

$$= \frac{1}{\sqrt{L+1}} \left(|0\rangle^n \otimes |x(0)\rangle + U_1 |0\rangle^n \otimes |x(1)\rangle + \cdots + U_L \cdots U_0 |0\rangle^n \otimes |x(L)\rangle \right)$$

$$\underset{(8.92)}{=} \frac{1}{\sqrt{L+1}} \left(|0\rangle^n \otimes |0\rangle^L + U_1 |0\rangle^n \otimes |10\ldots0\rangle^L + \cdots + U |0\rangle^n \otimes |1\ldots1\rangle^L \right),$$

and we see that the circuit end state $U|0\rangle^n$ appears in the tensor product with $|x(L)\rangle = |1\ldots1\rangle^L$ as one of the $L+1$ terms comprising $|\Gamma\rangle$. Since Definition 8.17 implies that $x(m)_0 = 0$ for all $m \in \{0,\ldots,L-1\}$, we find that observing the final state $|\Gamma\rangle$, namely, measuring the observable (see Definition 5.35)

$$\Sigma_z^{U:n,C:L} := \mathbf{1}^{\otimes n} \otimes \mathbf{1}^{\otimes L-1} \otimes \sigma_z$$

in the state $|\Gamma\rangle$, will yield the eigenvalue $+1$ with probability $\frac{L}{L+1}$ (because $\sigma_z|0\rangle = |0\rangle$) and the eigenvalue -1 with probability $\frac{1}{L+1}$ (because $\sigma_z|1\rangle = -|1\rangle$). If we have measured the eigenvalue -1, then, according to the Projection Postulate 3, the system will be in the corresponding eigenstate obtained from projecting $|\Gamma\rangle$ onto the eigenstate. The projector onto the eigenspace $\mathrm{Eig}(\Sigma_z^{U:n,C:L}, -1)$ is given as

$$P_{-1} := \mathbf{1}^{\otimes n} \otimes \mathbf{1}^{\otimes L-1} \otimes |1\rangle\langle 1|,$$

and its application to $|\Gamma\rangle$ yields

$$P_{-1}|\Gamma\rangle \underset{(8.139)}{=} U|0\rangle^n \otimes |1\ldots1\rangle^L \in \mathbb{H}^U \otimes \mathbb{H}^C.$$

This is a separable state with density operator

$$\rho \underset{(2.33)}{=} \left(U|0\rangle^n \otimes |1\ldots1\rangle \right) \left({}^n\langle 0|U^* \otimes \langle 1\ldots1| \right)$$

$$\underset{(3.36)}{=} U|0\rangle^n{}^n\langle 0|U^* \otimes |1\ldots1\rangle\langle 1\ldots1| \qquad (8.140)$$

in the composite circuit and clock system $\mathbb{H}^U \otimes \mathbb{H}^C$. The sub-system in the circuit space is described by the reduced density operator $\rho^U(\rho)$ for which we obtain

$$\rho^U(\rho) \underbrace{=}_{(3.50)} \operatorname{tr}^C(\rho) \underbrace{=}_{(8.140)} \operatorname{tr}^C\left(U|0\rangle^n\langle 0|U^* \otimes |1\ldots1\rangle\langle 1\ldots1|\right)$$

$$\underbrace{=}_{(3.57)} \underbrace{\operatorname{tr}\left(|1\ldots1\rangle\langle 1\ldots1|\right)}_{=1} U|0\rangle^n\langle 0|U^* = U|0\rangle^n\langle 0|U^*$$

$$\underbrace{=}_{(2.89)} \rho_{U|0\rangle^n},$$

that is, the system described by $\mathbb{H}^U$ is then in the pure state $U|0\rangle^n$, which is the end state of the circuit computation. □

The state $|\Gamma(0)\rangle$ will be the initial state of the adiabatic computation. The system will then be subject to the time evolution generated by $\mathsf{H}_T(s)$. Provided $\mathsf{H}_T(s)$ satisfies the Adiabatic Assumption (AA), Theorem G.15 with Corollary 8.3 and Corollary 8.4 yield bounds on the probability to find the system in the ground state $|\Gamma\rangle$ at the end of the adiabatic time evolution generated by $\mathsf{H}_T(s)$. More precisely, Corollary 8.4 tells us how large we have to make the transition period $T = t_{\text{fin}} - t_{\text{ini}}$ (or 'how long we have to wait') for a given spectral gap $g(s)$ in order to achieve a desired lower bound $p_{\min}$ on the probability to find $|\Gamma\rangle$.

Such a replication of the circuit action by means of an adiabatic computation is said to be efficient if the transition period T increases at most polynomially with the number of gates L. In order to verify that our construction of $\mathsf{H}_T(s)$ is sufficient to ensure this efficiency, we thus need to

- ascertain that the Adiabatic Assumption (AA) required for the applicability of the adiabatic Theorem G.15 and its corollaries 8.3 and 8.4 is satisfied;
- determine a lower bound for the spectral gap $g_0(s)$ of the ground state of $\mathsf{H}_T(s)$.

Both items require a more detailed spectral analysis of $\mathsf{H}_T(s)\big|_{\mathbb{H}_{\text{sub}}}$. We begin this by exhibiting the set of equations which the coordinates (in the ONB $\{|\Gamma(m)\rangle \mid m \in \{0,\ldots,L\}\}$) of the eigenvectors of $\mathsf{H}_T(s)\big|_{\mathbb{H}_{\text{sub}}}$ have to satisfy.

Lemma 8.25 *Let $s \in [0,1]$ and*

$$|\Phi(s)\rangle = \sum_{m=0}^{L} \Phi(s)_m |\Gamma(m)\rangle \in \mathbb{H}_{\text{sub}}. \tag{8.141}$$

Then $|\Phi(s)\rangle$ is an eigenvector of $\mathsf{H}_T(s)\big|_{\mathbb{H}_{\text{sub}}}$ with eigenvalue $E(s)$ if and only if the $\Phi(s)_m$ satisfy for $m \in \{0,\ldots,L\}$

$$\Phi(s)_1 = a(s)\Phi(s)_0 \tag{8.142}$$

$$\Phi(s)_m = b(s)\Phi(s)_{m-1} - \Phi(s)_{m-2} \quad for\ m \in \{2,\ldots,L-1\} \tag{8.143}$$

$$\Phi(s)_L = c(s)\Phi(s)_{L-1}, \tag{8.144}$$

where

$$a(s) = a(s, E(s)) = 1 - \frac{2E(s)}{s}$$

$$b(s) = b(s, E(s)) = 2\frac{1 - E(s)}{s} \tag{8.145}$$

$$c(s) = c(s, E(s)) = \frac{s}{2 - 2E(s) - s}.$$

Proof Using the matrix of $H_T(s)\big|_{\mathbb{H}_{sub}}$ in the ONB $\{|\Gamma(m)\rangle \mid m \in \{0,\ldots,L\}\}$ as given in (8.125), together with (8.141) and the eigenvalue equation

$$H_T(s)\big|_{\mathbb{H}_{sub}}|\Phi(s)\rangle = E(s)|\Phi(s)\rangle ,$$

gives

$$\begin{pmatrix} \frac{s}{2} & -\frac{s}{2} & 0 & 0 & \cdots & 0 \\ -\frac{s}{2} & 1 & -\frac{s}{2} & 0 & \cdots & 0 \\ 0 & -\frac{s}{2} & 1 & -\frac{s}{2} & 0 & \vdots \\ \vdots & \ddots & \ddots & \ddots & \ddots & \vdots \\ 0 & \cdots & 0 & -\frac{s}{2} & 1 & -\frac{s}{2} \\ 0 & \cdots & 0 & 0 & -\frac{s}{2} & 1-\frac{s}{2} \end{pmatrix} \begin{pmatrix} \Phi(s)_0 \\ \Phi(s)_1 \\ \Phi(s)_2 \\ \vdots \\ \Phi(s)_{L-1} \\ \Phi(s)_L \end{pmatrix} = E(s) \begin{pmatrix} \Phi(s)_0 \\ \Phi(s)_1 \\ \Phi(s)_2 \\ \vdots \\ \Phi(s)_{L-1} \\ \Phi(s)_L \end{pmatrix} .$$

Evaluating and re-arranging the component equations yields (8.142)–(8.144) with (8.145), showing that these equations are equivalent to the eigenvalue equation for $|\Phi(s)\rangle$ in the ONB $\{|\Gamma(m)\rangle \mid m \in \{0,\ldots,L\}\}$. □

From Theorem 8.23 we know that $H_T(s)\big|_{\mathbb{H}_{sub}} \geq 0$, and it follows that we must have $E(s) \geq 0$ for any of its eigenvalues as well. We continue our investigation of the spectrum of $H_T(s)\big|_{\mathbb{H}_{sub}}$ in a number of exercises until we have enough material to convince ourselves that it satisfies the Adiabatic Assumption (AA).

As we shall see below, we may assume that $0 \leq E(s) < 1 + s$ since we will find that all eigenvalues of $H_T(s)\big|_{\mathbb{H}_{sub}}$ satisfy this property. For further consideration we sub-divide the range $[0, 1 + s[$ for the eigenvalues into two domains. The first domain will contain the lowest eigenvalue $E_0(s)$. The second domain will contain a further L eigenvalues $E_m(s)$ for $m \in \{1,\ldots,L\}$, which will be shown to be non-crossing.

Domain 1: is given by $E \in D_1 = D_1(s) := [0, 1 - s]$, which implies

$$\frac{1 - E}{s} \underbrace{=}_{(8.145)} \frac{b(s, E)}{2} \geq 1$$

Domain 2: is given by $E \in D_2 = D_2(s) :=]1 - s, 1 + s[$, which implies

$$-1 < \frac{1 - E}{s} \underbrace{=}_{(8.145)} \frac{b(s, E)}{2} < 1 .$$

For these domains D_j, where $j \in \{1, 2\}$ we define

$$
\begin{aligned}
\mathrm{Co}_1(u) &:= \cosh u , \quad \mathrm{Si}_1(u) := \sinh u \\
\mathrm{Co}_2(u) &:= \cos u , \quad \mathrm{Si}_2(u) := \sin u \\
\theta_j(s, E) &:= \mathrm{Co}_j^{-1}\left(\frac{1 - E}{s}\right) = \begin{cases} \operatorname{arccosh}\left(\frac{1-E}{s}\right) & \text{for } E \in D_1 \\ \operatorname{arccos}\left(\frac{1-E}{s}\right) & \text{for } E \in D_2 \end{cases}
\end{aligned}
\tag{8.146}
$$

such that for both domains we have

$$b(s, E) = 2 \, \mathrm{Co}_j\left(\theta_j(s, E)\right) . \tag{8.147}$$

Exercise 8.108 For $j \in \{1, 2\}$ and $m \in \{2, \ldots, L - 2\}$ let

$$\Phi(s)_m = A_j \, \mathrm{Co}_j\left(m\theta_j(s, E(s))\right) + B_j \, \mathrm{Si}_j\left(m\theta_j(s, E(s))\right) , \tag{8.148}$$

where $A_j, B_j \in \mathbb{C}$ and Co_j, Si_j and θ_j are as defined in (8.146). Show that the $\Phi(s)_m$ are a solution of the recursion (8.143).

For a solution see Solution 8.108.

The solutions (8.148) of the recursive equations (8.143) for the components of the eigenvectors still contain two free parameters A_j and B_j, which we will determine with the help of the boundary conditions (8.142) and (8.144). These two boundary conditions will yield equations for $\theta_j(s, E)$. Eigenvalues of $H_T(s)\big|_{\mathbb{H}_{\text{sub}}}$ are then found as implicit solutions $E = E(s)$ of these determining equations. Hence, the next step is to establish the equations for $\theta_j(s, E)$ from the boundary conditions.

Exercise 8.109 Using the definitions (8.146) we define furthermore for $j \in \{1, 2\}$

$$\mathrm{Ta}_j(u) := \frac{\mathrm{Si}_j(u)}{\mathrm{Co}_j(u)} = \begin{cases} \tanh u & \text{if } j = 1 \\ \tan u & \text{if } j = 2. \end{cases} \tag{8.149}$$

Let $s \in]0, 1]$ and $j \in \{1, 2\}$. Using the boundary conditions (8.142) and (8.144), show that any eigenvalue $E(s) \in D_j$ of $H_T(s)\big|_{\mathbb{H}_{\text{sub}}}$ has to satisfy

$$\mathrm{Ta}_j\big((L-1)\theta_j(s, E(s))\big) = \frac{(c-a)\sqrt{(-1)^j(4-b^2)}}{b(a+c) - 2ac - 2}, \tag{8.150}$$

where $a = a(s), b = b(s)$ and $c = c(s)$ are defined as in (8.145).

For a solution see Solution 8.109.

Either side of (8.150) can have singularities, but we will ignore this for now and consider them later in the proof of Theorem 8.26, where we use this equation to prove that the eigenvalues of $H_T(s)\big|_{\mathbb{H}_{\text{sub}}}$ are non-crossing for $0 < s \leq 1$.

Note that the right side of (8.150) does not depend on L. It can be written in a form which makes the s and E dependence more explicit and makes it more amenable to further analysis.

Exercise 8.110 Define

$$s_\pm = s_\pm(s) := 1 \pm s$$

$$z_\pm = z_\pm(s) := \frac{1}{2}\left(1 \pm \sqrt{s^2 + s_-^2}\right) = \frac{1}{2}\left(1 \pm \sqrt{1 - 2s + 2s^2}\right)$$

$$p_\pm = p_\pm(s) := \frac{1}{2}\left(s_+ \pm \sqrt{2 - s_+ s_-}\right) = \frac{1}{2}\left(1 + s \pm \sqrt{1 + s^2}\right) \tag{8.151}$$

$$h_j(s, E) := (-1)^j \frac{(E - z_+)(E - z_-)}{(E - p_+)(E - p_-)} \sqrt{\frac{s_+ - E}{(-1)^j(E - s_-)}},$$

where h_j is only defined in the respective domain D_j with $j \in \{1, 2\}$. Show that for each domain D_j

$$\frac{(c-a)\sqrt{(-1)^j(4-b^2)}}{b(a+c) - 2ac - 2} = h_j(s, E(s)). \tag{8.152}$$

For a solution see Solution 8.110.

Combining (8.150) with (8.152) shows that any eigenvalue of $\mathsf{H}_T(s)\big|_{\mathbb{H}_{\text{sub}}}$ has to be a solution $E = E(s)$ of

$$\mathrm{Ta}_j\left((L-1)\theta_j(s,E)\right) = h_j(s,E). \tag{8.153}$$

In the proof of Theorem 8.26 we will show that in $D_1 \cup D_2$ there are exactly $L+1$ such solutions. Since $\dim \mathbb{H}\big|_{\text{sub}} = L+1$, this implies that every solution of (8.153) in $D_1 \cup D_2$ gives an eigenvalue of $\mathsf{H}_T(s)\big|_{\mathbb{H}_{\text{sub}}}$. In order to facilitate the analysis in the aforementioned proof, we first provide some properties of the $s_\pm, z_\pm$ and $p_\pm$ defined in (8.151).

Exercise 8.111 Show that the functions $s_\pm(s), z_\pm(s)$ and $p_{\pm(s)}$ satisfy

$$
\begin{aligned}
& z_-(0) = p_-(0) = 0 < 1 = z_+(0) = p_+(0) = s_\pm(0) \\
& z_-(s) < p_-(s) < s_-(s) < z_+(s) < p_+(s) < s_+(s) && \text{for } 0 < s < \tfrac{3}{4} \\
& z_-(s) < s_-(s) \le p_-(s) < z_+(s) < p_+(s) < s_+(s) && \text{for } \tfrac{3}{4} \le s < 1 \\
& z_-(1) = s_-(1) = 0 < p_-(1) < z_+(1) < p_+(1) < s_+(1).
\end{aligned}
\tag{8.154}
$$

For a solution see Solution 8.111.

Theorem 8.26 *Let $\mathsf{H}_T(s)$ and $\mathbb{H}_{\text{sub}}$ be defined as in Theorem 8.22. Then $\mathsf{H}_T(s)\big|_{\mathbb{H}_{\text{sub}}}$ satisfies the Adiabatic Assumption (AA). In particular, we have for $0 < s \le 1$ that*

$$E_0(s) \le s_-(s) < E_1(s) < \cdots < E_L(s) < s_+(s). \tag{8.155}$$

Proof With $\mathsf{H}_T(s)$ defined as in (8.122) it is obvious that it is twice continuously differentiable, in other words, that item (i) of the Adiabatic Assumption (AA) is satisfied.

That item (ii) is satisfied follows from the fact that $\dim \mathbb{H}_{\text{sub}} = L+1$ and that—as we shall show below—$\mathsf{H}_T(s)\big|_{\mathbb{H}_{\text{sub}}}$ has $L+1$ distinct eigenvalues for $0 < s \le 1$. Consequently, each of the eigenspaces remains one-dimensional for $s > 0$.

As mentioned above, (8.150) with (8.152) shows that any eigenvalue $E = E(s)$ of $\mathsf{H}_T(s)\big|_{\mathbb{H}_{\text{sub}}}$ has to be a solution of

$$\mathrm{Ta}_j\left((L-1)\theta_j(s,E)\right) = h_j(s,E). \tag{8.156}$$

We analyze both sides of this equation in the two domains. As for the right side of (8.156) in the domain D_1, in which $0 \le E \le s_-$, we note from its definition in (8.151) and (8.154) that in D_1 the function $E \mapsto h_1(s,E)$ has a zero at z_- and poles at p_- and s_-, where $p_- < s_-$ for $s \in {]}0, \tfrac{3}{4}{[}$ and $s_- \le p_-$ for $s \in [\tfrac{3}{4}, 1]$.

In the domain D_2 for which $s_- < E < s_+$, we see again from its definition in (8.151) that in this domain the function $E \mapsto h_2(s, E)$ has a zero at z_+ and poles at p_- and p_+, where we know from (8.154) that for $s \in]0, \frac{3}{4}[$ the only pole in the domain is p_+ and for $s \in [\frac{3}{4}, 1]$ both $p_\pm$ are in D_2.

Moreover, we find after some lengthy calculations

$$\frac{\partial}{\partial E} h_j(s, E) = \frac{s^3 \left(2E(s-2) - 2s^2 + s + 1\right)}{(E - (1-s))^2 (2E(E - (1+s)) + s)^2} \sqrt{\frac{(-1)^j (E - s_-)}{s_+ - E}},$$

such that for $s \in]0, 1]$ and $E \in D_1 \cup D_2 \smallsetminus \{p_\pm, s_\pm\}$ we have

$$\frac{\partial}{\partial E} h_j(s, E)\Big|_{\hat{E}} = 0 \quad \Leftrightarrow \quad \hat{E} = \hat{E}(s) := \frac{1 + s - 2s^2}{2(2 - s)} \tag{8.157}$$

and

$$\frac{\partial}{\partial E} h_j(s, E) \gtrless 0 \quad \Leftrightarrow \quad E \lessgtr \hat{E}(s). \tag{8.158}$$

Exercise 8.112   Show that for $s \in]0, 1]$

$$s < \frac{3}{4} \quad \Leftrightarrow \quad p_- < \hat{E} < s_-$$
$$s = \frac{3}{4} \quad \Leftrightarrow \quad p_- = \hat{E} = s_- \tag{8.159}$$
$$s > \frac{3}{4} \quad \Leftrightarrow \quad p_- > \hat{E} > s_-.$$

For a solution see Solution 8.112.

The location of zeros and poles together with the results (8.157)–(8.159) show that the functions h_j in their respective domains have graphs as shown in Fig. 8.3 for $s < \frac{3}{4}$ and in Fig. 8.4 for $s > \frac{3}{4}$.

Let us now turn to the left side of (8.156) in the two domains. First consider this equation in D_1 in which $0 \le E \le s_-$. For any given $s \in]0, 1]$ we take the non-negative solution for $\theta_1(s, E)$ in (8.146). Then it follows that $\theta_1(s, E)$ as a function of E is positive and decreasing for $0 < E \le s_- = 1 - s$ until it becomes zero at s_-. This also holds for $(L - 1)\theta_1(s, E)$ as well as $\mathrm{Ta}_1\left((L - 1)\theta_1(s, E)\right)$.

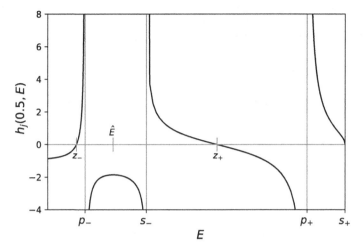

Fig. 8.3 Graph of functions $E \mapsto h_j(\frac{1}{2}, E)$ for both domains D_1 and D_2. The thin vertical lines show the location of the poles. This shape is generic for $s \in]0, \frac{3}{4}[$ and does not depend on L

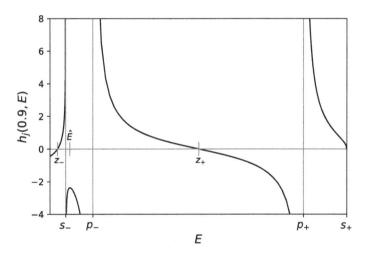

Fig. 8.4 Graph of functions $E \mapsto h_j(\frac{9}{10}, E)$ for both domains D_1 and D_2. The thin vertical lines show the location of the poles. This shape is generic for $s \in]\frac{3}{4}, 1]$ and does not depend on L

In the domain D_2 the function $E \mapsto \mathrm{Ta}_2\left((L-1)\theta_2(s,E)\right)$ has singularities at the points $\overline{E}_{s,q}$ where

$$\theta_2(s, \overline{E}_{s,q}) = \alpha_q := \frac{2q+1}{2(L-1)}\pi \quad \text{for } q \in \{0, 1, \ldots, L-2\} \qquad (8.160)$$

since these are the points where $Co_2\left((L-1)\theta_2(s,\overline{E}_{s,q})\right)=\cos\left((L-1)\theta_2(s,\overline{E}_{s,q})\right)$ $=0$. It follows from (8.146) that

$$\overline{E}_{s,q}=1-s\cos\alpha_q .\tag{8.161}$$

We use these points to divide the domain D_2 for E into the L intervals

$$I_1=I_1(s):=]s_-(s),\overline{E}_{s,0}]$$
$$I_2=I_2(s):=]\overline{E}_{s,0},\overline{E}_{s,1}]$$

$$\vdots\tag{8.162}$$

$$I_{L-1}=I_{L-1}(s):=]\overline{E}_{s,L-3},\overline{E}_{s,L-2}]$$
$$I_L=I_L(s):=]\overline{E}_{s,L-2},s_+(s)[.$$

The functions Ta_j in their respective domains have the graphs of the shapes as shown for $L=7$ in Fig. 8.5 for $s<\frac{3}{4}$ and in Fig. 8.6 for $s>\frac{3}{4}$.

For a given $s\in]0,1]$ the solutions $E=E(s)$ of (8.156) can be characterized as follows:

in $D_1(s)=[0,s_-(s)]$ there is exactly one value $0\leq E_0(s)\leq 1-s$;

in $D_2(s)=]s_-(s),s_+(s)[$ consider first those s where the poles of $E\mapsto Ta_2\left((L-1)\theta_2(s,E)\right)$ and $E\mapsto h_2(s,E)$ do not coincide, that is, where $p_\pm(s)\neq\overline{E}_{s,q}$ for all (s,q).

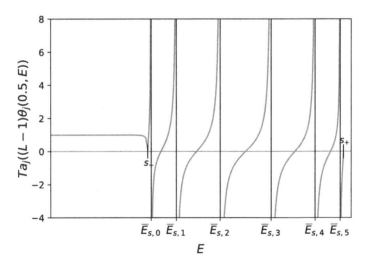

Fig. 8.5 Graph of functions $E\mapsto Ta_j\left((L-1)\theta_j(\frac{1}{2},E)\right)$ for both domains D_1 and D_2 and $L=7$. The thin vertical lines show the location of the poles $\overline{E}_{s,q}$. This shape is generic for $s\in]0,\frac{3}{4}[$

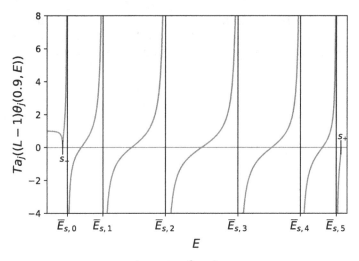

Fig. 8.6 Graph of functions $E \mapsto \mathrm{Ta}_j\left((L-1)\theta_j(\frac{9}{10},E)\right)$ for both domains D_1 and D_2 and $L=7$. The thin vertical lines show the location of the poles $\overline{E}_{s,q}$. This shape is generic for $s \in]\frac{3}{4},1]$

Exercise 8.113 Suppose $s \in]0,1]$ is such that $p_\pm(s) \neq \overline{E}_{s,q}$ for all $q \in \{1,2,\ldots,L\}$. Show that then $p_\pm(s) \in I_{q_\pm}(s)$ for some $q_\pm \in \{1,2,\ldots,L\}$ implies

$$q_- < \frac{L}{2} + 1$$
$$\frac{L}{2} < q_+ \leq L - 1 .$$
(8.163)

For a solution see Solution 8.113.

Note that (8.163) implies that while $p_-(s)$ can fall into $I_1(s)$, none of the poles $p_\pm(s)$ can be in $I_L(s)$. As the following Exercise 8.114 shows, another consequence of (8.163) is that both poles cannot be in the same interval.

Exercise 8.114 Show that for $L > 1$ we cannot have $p_\pm(s) \in I_q$.

For a solution see Solution 8.114.

The locations of the solutions of (8.156) in the intervals of $D_2(s) = I_1(s) \cap \cdots \cap I_L(s)$ are then as follows:

- From (8.163) we see that neither $p_+(s)$ nor $p_-(s)$ can be in the right-most interval $I_L(s)$. Since we have $h_2(s,E) >$ for $E \in]p_+(s), s_+(s)[$ and $\mathrm{Ta}_2\left((L-1)\theta_2(s,E)\right) < 0$ for $E \in]\overline{E}_{s,L-2}, s_+(s)[$, it follows that there is *no solution of* (8.156) *in* $I_L(s)$.

- In the interval $I_{q_+}(s)$ which contains $p_+(s)$, there are *two solutions of* (8.156), one at either side of $p_+(s)$. This is because $h_2(s,E)$ tends to $-\infty$ as $E \nearrow p_+(s)$ and $h_2(s,E) \to +\infty$ as $E \searrow p_+(s)$, and $\mathrm{Ta}_2\left((L-1)\theta_2(s,E)\right)$ is increasing from $-\infty$ at $E \searrow \overline{E}_{s,q_+-2} < p_+(s)$ to $+\infty$ at $E \nearrow \overline{E}_{s,q_+-1} > p_+(s)$.

- Similarly, in any interval $I_{q_-}(s)$, which contains $p_-(s)$ and for which $q_- \neq 1$, there are two solutions of (8.156) separated by $p_-(s)$.

- If $p_-(s) \in I_1(s)$, there is *only one solution* $E \in]p_-(s), \overline{E}_{s,0}[$ of (8.156) since $h_2(s,E) \to +\infty$ as $E \searrow p_-(s)$ and $\mathrm{Ta}_2\left((L-1)\theta_2(s,E)\right) \to +\infty$ as $E \nearrow \overline{E}_{s,0} > p_-(s)$. For $E \in]s_-(s), p_-(s)]$ there is no solution since there $h_2(s,E) < 0$ and $\mathrm{Ta}_2\left((L-1)\theta_2(s,E)\right) \geq 0$.

- In every other interval $I_q(s)$ for $q \in \{2, \ldots, L-1\} \smallsetminus \{q_\pm\}$ that does not contain $p_\pm(s)$ and is neither I_1 nor I_L we find *exactly one solution of* (8.156) since there $h_2(s,E)$ is finite and decreasing and $\mathrm{Ta}_2\left((L-1)\theta_2(s,E)\right)$ covers all of $\mathbb{R}$ as E ranges between $\overline{E}_{s,q-2}$ and $\overline{E}_{s,q-1}$.

- If $p_-(s) \neq I_1(s)$, *there is no solution in* $I_1(s)$ since $h_2(s,E) < 0$ for $E \in s_-(s)$, $p_-(s)[$ and $\mathrm{Ta}_2\left((L-1)\theta_2(s,E)\right) > 0$ for $E \in]s_-(s), \overline{E}_{s,0}[$.

In summary we have in D_2, in the case where $p_\pm(s) \neq \overline{E}_{s,q}$ for $q \in \{1, \ldots, L-2\}$,

- one solution in each $I_q(s), q \in \{1, \ldots, L-1\} \smallsetminus \{q_+\}$ plus two solutions in $I_{q_+}(s)$ if $p_-(s) \in I_1(s)$
- one solution in each $I_q(s), q \in \{2, \ldots, L-1\} \smallsetminus \{q_\pm\}$ plus two solutions in each $I_{q_\pm}(s)$ if $p_-(s) \notin I_1(s)$.

In total we have thus L distinct solutions in $D_2(s)$, which do not cross or coincide as s varies in $]0,1]$, as long as $p_\pm(s) \neq \overline{E}_{s,q}$ for $q \in \{1, \ldots, L-2\}$. Together with the one solution in $D_1(s)$ these give the $L+1$ eigenvalues of $H_T(s)\big|_{\mathbb{H}_{\mathrm{sub}}}$. Generic examples of these solutions are shown for $L = 7$ in Fig. 8.7 for $s < \frac{3}{4}$ and in Fig. 8.8 for $s < \frac{3}{4}$.

Consider now the case $p_+(s) = \overline{E}_{s,r_+}$ for some $r_+ \in \{1, \ldots, L-2\}$. We claim that then $E(s) = \overline{E}_{s,r_+} = p_+(s)$ is an eigenvalue of $H_T(s)\big|_{\mathbb{H}_{\mathrm{sub}}}$. This can be seen as follows.

Recall that our Ansatz (8.148) for the components of the eigenvectors satisfied the recursive part (8.143) of the eigenvalue equations. To obtain a solution for the eigenvalue equation, one still has to impose the boundary conditions. In Exercise 8.109 we showed that this resulted in (8.150). In the proof of this, we found in (G.182) that whenever E is such that

$$(1 + ac - ab)\sin\left((L-1)\theta_2(s,E)\right) = (c-a)\sin\left((L-2)\theta_2(s,E)\right) \quad (8.164)$$

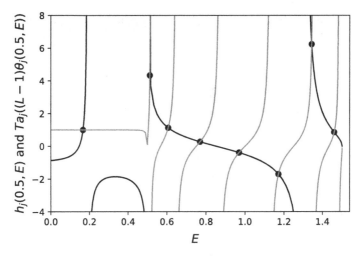

Fig. 8.7 Determination of the eigenvalues of $H_T\left(\frac{1}{2}\right)$ for the case $L = 7$ as intersection of $\mathrm{Ta}_j\left((L-1)\theta_j\left(\frac{1}{2},E\right)\right)$ (gray lines) and $h_j\left(\frac{1}{2},E\right)$ (black lines). Here $\mathrm{Ta}_1\left(\frac{1}{2},E\right)$ and $h_1\left(\frac{1}{2},E\right)$ are shown for $E \in D_1\left(\frac{1}{2}\right)$, whereas $\mathrm{Ta}_2\left(\frac{1}{2},E\right)$ and $h_2\left(\frac{1}{2},E\right)$ are shown for $E \in D_2\left(\frac{1}{2}\right)$

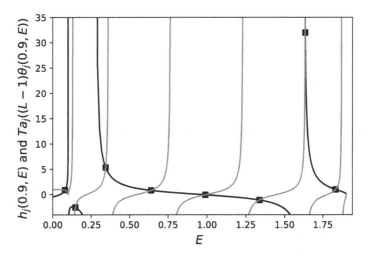

Fig. 8.8 Determination of the eigenvalues of $H_T\left(\frac{9}{10}\right)$ for the case $L = 7$ as intersection of $\mathrm{Ta}_j\left((L-1)\theta_j\left(\frac{9}{10},E\right)\right)$ (gray lines) and $h_j\left(\frac{9}{10},E\right)$ (black lines). Here $\mathrm{Ta}_1\left(\frac{9}{10},E\right)$ and $h_1\left(\frac{9}{10},E\right)$ are shown for $E \in D_1\left(\frac{9}{10}\right)$, whereas $\mathrm{Ta}_2\left(\frac{9}{10},E\right)$ and $h_2\left(\frac{9}{10},E\right)$ are shown for $E \in D_2\left(\frac{9}{10}\right)$

then the boundary conditions are satisfied. Consequently, E is an eigenvalue of $H_T(s)\big|_{\mathbb{H}_{\text{sub}}}$. For $E = \overline{E}_{s,r_+}$ we have

$$\underbrace{\theta_2(s,E)}_{(8.161)} = \alpha_{r_+} = \frac{2r_+ + 1}{2(L-1)}\pi$$

such that $(L-1)\theta_2(s,E) = (2r_+ + 1)\frac{\pi}{2}$ and

$$
\begin{aligned}
\sin\left((L-1)\theta_2(s,E)\right) &= (-1)^{r_+}\\
\cos\left((L-2)\theta_2(s,E)\right) &= 0\,.
\end{aligned}
\tag{8.165}
$$

Moreover, using $\sin(x-y) = \sin x \cos y - \cos x \sin y$ with $x = (L-1)\theta_2$ and $y = \theta_2$, we find

$$\sin\left((L-2)\theta_2(s,E)\right) = (-1)^{r_+}\cos\theta_2 \underbrace{=}_{(8.147)} (-1)^{r_+}\frac{b}{2}\,.
\tag{8.166}$$

Inserting (8.165) and (8.166) into (8.164) gives

$$1 + ac - ab = (c-a)\frac{b}{2}\,,$$

which is equivalent to

$$b(a+c) - 2ac - 2 = 0\,.$$

Using (G.186) (and the fact that $E = p_+(s)$ also precludes $2 - 2E - s = 0$), this in turn is equivalent to

$$(s_- - E)(E - p_+)(E - p_-) = 0,
\tag{8.167}$$

and this holds true since we consider the special case $E = p_+(s) = \overline{E}_{s,r_+}$. Hence, such an E solves (8.164) and has to be an eigenvalue of $H_T(s)\big|_{\mathbb{H}_{\text{sub}}}$.

The very same arguments apply if $E = E(s) = p_-(s) = \overline{E}_{s,r_-}$ since $E - p_-$ also appears in the left side of (8.167), making such E also eigenvalues of $H_T(s)\big|_{\mathbb{H}_{\text{sub}}}$.

With our definition (8.163) of the intervals $I_q = I_q(s)$, we see that $p_+(s) = \overline{E}_{s,r_+}$ or $p_-(s) = \overline{E}_{s,r_-}$ implies $p_+(s) \in I_{r_++2}$ or $p_-(s) \in I_{r_-+2}$ for $r_\pm \in \{0,1,\ldots,L-2\}$. Now, p_- cannot be in I_L. Whenever $p_- \in I_{r_-+2}$ for $r_- \in \{0,\ldots,L-3\}$, we have one eigenvalue given by $E = \overline{E}_{s,r_-} = p_-$ in I_{r_-+2} and a second eigenvalue in I_{r_-+2} as a solution of (8.156). This is because for $E \in]\overline{E}_{s,r_-}, p_+[$, where $\overline{E}_{s,r_-} = p_-$, we know that $h_2(s,E)$ is finite and decreasing, and $\mathrm{Ta}_2\left((L-1)\theta_2(s,E)\right)$ covers all of $\mathbb{R}$ as E ranges between $\overline{E}_{s,r_-}$ and $\overline{E}_{s,r_-+1}$. Hence, every interval which contains p_- also contains two eigenvalues.

The same applies for $p_+ \in I_{r_++2}$ as long as $r_+ + 2 \neq L$. If $r_+ + 2 = L$, then $I_L = [\overline{E}_{s,r_+}, s_+(s)[$ contains exactly the one eigenvalue $E = \overline{E}_{s,r_+}$ and as before $D_2 \smallsetminus I_L$ contains L eigenvalues.

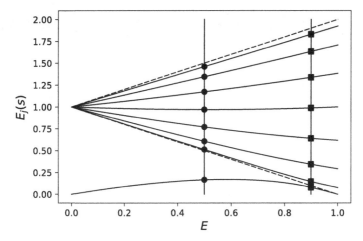

Fig. 8.9 Eigenvalues of $H_T(s)$ as a function of $s \in [0, 1]$ for the case of $L = 7$. Note that then $\dim \mathbb{H}_{\text{sub}} = L + 1 = 8$. The dashed lines show the domain-delimiting lines $s_\pm(s) = 1 \pm s$. The black circles show the eigenvalues at $s = \frac{1}{2}$, whereas the squares show them at $s = \frac{9}{10}$

Our analysis of the solutions $E = E(s)$ of (8.156) thus implies the following: for each $s \in]0, 1]$ we have one eigenvalue $E_0(s) \in D_1(s)$ satisfying $E_0(s) \le s_-(s)$ and L eigenvalues $E_j(s)$ for $j \in \{1, \dots, L\}$ in $D_2(s)$ satisfying $s_-(s) < E_1(s) < \cdots < E_L(s) < s_+(s)$. This is once more illustrated in Fig. 8.9, which shows the eigenvalues of $H_T(s)\big|_{\mathbb{H}_{\text{sub}}}$ for the case $L = 7$ as s varies from 0 to 1. Also shown there more prominently are the eigenvalues for the two sample values $s = \frac{1}{2}$ (as circles) and $s = \frac{9}{10}$ (as squares).   □

Recall that our goal in this section is to show that any quantum circuit $U = U_L \cdots U_1$ comprised of L gates can be simulated efficiently, namely with $T \in$ poly (L), by a suitable adiabatic computation.

So far we have presented a generic method to construct a suitable time dependent Hamiltonian $H_T(s)$. This Hamiltonian is suitable because at the initial time t_{ini} it has the known ground state $|\Gamma(0)\rangle$, and at the final time t_{fin} it has the ground state $|\Gamma\rangle$, which contains the circuit action end state $U|0\rangle^n$ in one of its components. We have also shown that by measuring a certain observable in the state $|\Gamma\rangle$, we will obtain $U|0\rangle^n$ with a probability of $\frac{1}{L+1}$. Moreover, starting from $|\Gamma(0)\rangle$ the time evolution generated by $H_T(s)$ leaves a finite-dimensional subspace $\mathbb{H}_{\text{sub}}$ invariant. On this subspace the Hamiltonian $H_T(s)\big|_{\mathbb{H}_{\text{sub}}}$ satisfies the Adiabatic Assumption (AA) required for the application of the Quantum Adiabatic Theorem G.15, Corollaries 8.3 and 8.4. These tell us that the time-evolved state is suitably close to the desired final ground state $|\Gamma\rangle$, provided $\frac{C_0(1)}{T}$ is sufficiently small. Ultimately, we want to derive an estimate as to how T grows as a function of L if we are to guarantee a given minimal probability to find $|\Gamma\rangle$. Hence, we need to obtain a suitable bound on the growth of $C_0(1)$ as a function of L. Lemma 8.27 provides a first step in that direction.

Lemma 8.27 *Let* $H_T(s)$ *be as defined in (8.22). Then* $C_0(1)$ *as defined in Theorem G.15 satisfies*

$$C_0(1) \leq \frac{3}{g_0(1)^2} + \frac{3}{g_0(0)^2} + 90 \int_0^1 \frac{du}{g_0(u)^3} . \qquad (8.168)$$

Proof Applying the results of Corollary 8.6 to $H_T(s)$ as defined in (8.22) and keeping in mind that we only need to consider the restrictions on $\mathbb{H}_{\text{sub}}$ yields

$$C_0(1) \underbrace{\leq}_{(2.53),(8.19)} \left(\left\| H_{\text{fin}} \big|_{\mathbb{H}_{\text{sub}}} \right\| + \left\| H_{\text{ini}} \big|_{\mathbb{H}_{\text{sub}}} \right\| \right) \left[\frac{1}{g_0(1)^2} + \frac{1}{g_0(0)^2} \right. \qquad (8.169)$$

$$+ 10 \left(\left\| H_{\text{fin}} \big|_{\mathbb{H}_{\text{sub}}} \right\| + \left\| H_{\text{ini}} \big|_{\mathbb{H}_{\text{sub}}} \right\| \right) \left. \int_0^1 \frac{du}{g_0(u)^3} \right] .$$

From (8.126) we see that the largest eigenvalue of $H_{\text{ini}} \big|_{\mathbb{H}_{\text{sub}}}$ is 1 and it follows from (2.50) that

$$\left\| H_{\text{ini}} \big|_{\mathbb{H}_{\text{sub}}} \right\| = 1 . \qquad (8.170)$$

Likewise, (8.155) shows that the eigenvalues of $H_T(s) \big|_{\mathbb{H}_{\text{sub}}}$ are bounded from above by $s_+(s) = 1 + s$. Considering that $H_{\text{fin}} \big|_{\mathbb{H}_{\text{sub}}} = H_T(1) \big|_{\mathbb{H}_{\text{sub}}}$, we thus obtain, again with the help of (2.50), that

$$\left\| H_{\text{fin}} \big|_{\mathbb{H}_{\text{sub}}} \right\| \leq 2 . \qquad (8.171)$$

Inserting (8.170) and (8.171) into (8.169) yields the claim (8.168). $\square$

The one ingredient still missing to obtain a bound on $C_0(1)$ is the behavior of the gap function $g_0(s) = E_1(s) - E_0(s)$ as the number of gates L increases. Before we can exhibit this in more detail, we need a few more auxiliary results. The first of these is given in the following theorem, which contains the so-called **max-min principle**.

Theorem 8.28 *Let* $\mathbb{H}$ *be a* HILBERT *space with* $\dim \mathbb{H} = d < \infty$ *and let* $A \in B_{sd}(\mathbb{H})$ *with a set of eigenvalues* $\sigma(A) = \{ \lambda_j \,|\, j \in \{1, \ldots, d\} \}$ *such that*

$$\lambda_1 \leq \lambda_2 \leq \cdots \leq \lambda_d . \qquad (8.172)$$

Then we have for every $w \in \{1, \ldots, d\}$

$$\lambda_w = \max \left\{ \min \left\{ \langle \Psi | A\Psi \rangle \mid |\Psi\rangle \in \mathbb{H}_w, ||\Psi|| = 1 \right\} \;\middle|\; \begin{array}{l} \mathbb{H}_w \text{ subspace of } \mathbb{H} \text{ with} \\ \dim \mathbb{H}_w = d - w + 1 \end{array} \right\}.$$

$$(8.173)$$

Proof Let $\left\{ |\Phi_j\rangle \mid j \in \{1,\dots,d\} \right\}$ be an ONB of $\mathbb{H}$ consisting of eigenvectors of A and define

$$\tau_w := \max \left\{ \min \left\{ \langle \Psi | A\Psi \rangle \mid |\Psi\rangle \in \mathbb{H}_w, ||\Psi|| = 1 \right\} \;\middle|\; \begin{array}{l} \mathbb{H}_w \text{ subspace of } \mathbb{H} \text{ with} \\ \dim \mathbb{H}_w = d - w + 1 \end{array} \right\}.$$

$$(8.174)$$

In a first step we show that $\tau_w \geq \lambda_w$. For this we choose $\mathbb{H}_w$ as the subspace

$$\mathbb{H}_w = \text{Span}\{|\Phi_w\rangle, \dots, |\Phi_d\rangle\}.$$

Then we have $\dim \mathbb{H}_w = d - w + 1$ and λ_w is the smallest eigenvalue of $A\big|_{\mathbb{H}_w}$. The result (2.40) of Exercise 2.9 then implies that for any $|\Psi\rangle \in \mathbb{H}_w$ with $||\Psi|| = 1$ we must have $\langle \Psi | A\Psi \rangle \geq \lambda_w$. For τ_w as defined in (8.174) it thus follows that

$$\tau_w \geq \lambda_w.$$

$$(8.175)$$

We now proceed to show the reverse inequality. For this let $\mathbb{H}_w$ now be an arbitrary subspace of $\mathbb{H}$ with $\dim \mathbb{H}_w = d - w + 1$, and let $\left\{ |\Omega_k\rangle \mid k \in \{1, \dots, d - w + 1\} \right\}$ be an ONB of this subspace. For any $|\Psi\rangle \in \mathbb{H}_w$ we thus have

$$|\Psi\rangle = \sum_{k=1}^{d-w+1} \Psi_k |\Omega_k\rangle,$$

$$(8.176)$$

where

$$|\Omega_k\rangle = \sum_{j=1}^{d} M_{kj} |\Phi_j\rangle,$$

$$(8.177)$$

since $\left\{ |\Phi_j\rangle \mid j \in \{1,\dots,d\} \right\}$ is an ONB of $\mathbb{H}$. Consequently, we obtain for any $l \in \{1, \dots, d\}$ that

$$\langle \Phi_l | \Psi \rangle \underset{(8.176)}{=} \sum_{k=1}^{d-w+1} \Psi_k \langle \Phi_l | \Omega_k \rangle \underset{(8.177)}{=} \sum_{k=1}^{d-w+1} \Psi_k \sum_{j=1}^{d} M_{kj} \underbrace{\langle \Phi_l | \Phi_j \rangle}_{=\delta_{jl}}$$

$$= \sum_{k=1}^{d-w+1} \Psi_k M_{kl}.$$

$$(8.178)$$

This allows us to find $|\Psi\rangle \in \mathbb{H}_w$ such that

$$|\Psi\rangle \in \big(\mathrm{Span}\{|\Phi_{w+1}\rangle, \ldots, |\Phi_d\rangle\}\big)^\perp = \mathrm{Span}\{|\Phi_1\rangle, \ldots, |\Phi_w\rangle\} \qquad (8.179)$$

as follows. The requirement (8.179) is equivalent to

$$\langle \Phi_l | \Psi\rangle = 0 \quad \text{for } l \in \{w+1, \ldots, d\} \,.$$

From (8.178) we see that this means that $|\Psi\rangle$ has to be such that

$$\sum_{k=1}^{d-w+1} \Psi_k M_{kl} = 0 \quad \text{for } l \in \{w+1, \ldots, d\} \,.$$

This is a set of $d - w$ linear equations for $\dim \mathbb{H}_w = d - w + 1$ unknowns Ψ_k. It can always be solved and the remaining one unknown can be used to normalize $|\Psi\rangle$ to 1. This means that in any subspace $\mathbb{H}_w \subset \mathbb{H}$ of dimension $d - w + 1$ we can always find a $|\Psi\rangle \in \mathrm{Span}\{|\Phi_1\rangle, \ldots, |\Phi_w\rangle\}$ with $||\Psi|| = 1$. Now, the operator A restricted to $\mathrm{Span}\{|\Phi_1\rangle, \ldots, |\Phi_w\rangle\}$ has the eigenvalues $\lambda_1, \ldots, \lambda_w$. Because of the assumption (8.172), the largest of those is λ_w. The result (2.40) of Exercise 2.9 then implies that for any $|\Psi\rangle \in \mathrm{Span}\{|\Phi_1\rangle, \ldots, |\Phi_w\rangle\}$ with $||\Psi|| = 1$ we must have

$$\langle \Psi | A\Psi\rangle \leq \lambda_w \,.$$

Since we have shown above that for any $\mathbb{H}_w \subset \mathbb{H}$ with $\dim \mathbb{H}_w = d - w + 1$ we can always find a $|\Psi\rangle \in \mathbb{H}_w \cap \mathrm{Span}\{|\Phi_1\rangle, \ldots, |\Phi_w\rangle\}$ with $||\Psi|| = 1$, it follows that

$$\min\big\{ \langle \Psi | A\Psi\rangle \mid |\Psi\rangle \in \mathbb{H}_w, ||\Psi|| = 1 \big\} \leq \lambda_w \,.$$

Therefore, due to its definition in (8.174), τ_w also satisfies

$$\tau_w \leq \lambda_w \,,$$

which, together with (8.175), implies $\tau_w = \lambda_w$ and completes the proof of (8.173).   □

For the proof of our desired result about the growth of $g_0(s)$ as a function of L we need one more auxiliary result, which is given in the following lemma.

Lemma 8.29 ([116]) *Let* $|\Gamma(0)\rangle$ *be as defined in (8.94),* H_{fin} *as in (8.100) and* $\mathbb{H}_{sub}$ *as in (8.123). Then the lowest eigenvalue* $\check{E}_0$ *of*

$$\check{\mathsf{H}} := \left(\mathsf{H}_{fin} + \frac{1}{2}|\Gamma(0)\rangle\langle\Gamma(0)| \right)\Big|_{\mathbb{H}_{sub}} \tag{8.180}$$

satisfies for $L \geq 8$

$$\frac{1}{L^2} < \check{E}_0 < 1. \tag{8.181}$$

Proof From (8.138) we see that $\check{\mathsf{H}}$ in the ONB $\{|\Gamma(m)\rangle \mid m \in \{0,\ldots,L\}\}$ has the matrix

$$\check{\mathsf{H}} = \begin{pmatrix} 1 & -\frac{1}{2} & 0 & 0 & \ldots & 0 \\ -\frac{1}{2} & 1 & -\frac{1}{2} & 0 & \ldots & 0 \\ 0 & -\frac{1}{2} & 1 & -\frac{1}{2} & 0 & \vdots \\ \vdots & \ddots & \ddots & \ddots & \ddots & \vdots \\ 0 & \ldots & 0 & -\frac{1}{2} & 1 & -\frac{1}{2} \\ 0 & \ldots & 0 & 0 & -\frac{1}{2} & \frac{1}{2} \end{pmatrix}.$$

Consequently, the eigenvalue equation $\check{\mathsf{H}}|\check{\Phi}\rangle = \check{E}|\check{\Phi}\rangle$ implies for the components $\check{\Phi}_m$ of the eigenvector

$$|\check{\Phi}\rangle = \sum_{m=0}^{L} \check{\Phi}_m|\Gamma(m)\rangle$$

in the basis $\{|\Gamma(m)\rangle \mid m \in \{0,\ldots,L\}\}$ that

$$\check{\Phi}_1 = 2(1 - \check{E})\check{\Phi}_0 \tag{8.182}$$

$$\check{\Phi}_m = 2(1 - \check{E})\check{\Phi}_{m-1} - \check{\Phi}_{m-2} \quad \text{for } m \in \{2,\ldots,L-1\} \tag{8.183}$$

$$\check{\Phi}_L = \frac{1}{1 - 2\check{E}}\check{\Phi}_{L-1}. \tag{8.184}$$

Assuming that

$$\check{E} = 1 - \cos\theta \tag{8.185}$$

for a $\theta \in]0, \pi[$, we make the ansatz

$$\check{\Phi}_m = \sin\big((m+1)\theta\big) \tag{8.186}$$

for $m \in \{-1, \ldots, L\}$. Using

$$2 \cos \alpha \sin \beta = \sin(\alpha + \beta) - \sin(\alpha - \beta), \qquad (8.187)$$

we then find with $\alpha = \theta$ and $\beta = m\theta$ that

$$\check{\Phi}_m \underset{(8.186)}{=} \sin\left((m+1)\theta\right) \underset{(8.187)}{=} 2 \cos \theta \sin(m\theta) - \sin\left((m-1)\theta\right)$$

$$\underset{(8.185),(8.186)}{=} 2(1 - \check{E})\check{\Phi}_{m-1} - \check{\Phi}_{m-2} \qquad (8.188)$$

for $m \in \{2, \ldots, L\}$, which shows that (8.183) is satisfied by our ansatz. But (8.188) also holds for $m = 1$, in which case it shows that (8.185) together with (8.186) also satisfies (8.182).

To satisfy (8.184), we require in addition that

$$(1 - 2\check{E}) \sin\left((L+1)\theta\right) = \sin(L\theta).$$

Using (8.185), this becomes

$$2 \cos \theta \sin\left((L+1)\theta\right) = \sin(L\theta) + \sin\left((L+1)\theta\right),$$

where we can again make use of (8.187) to obtain

$$\sin\left((L+2)\theta\right) = \sin\left((L+1)\theta\right).$$

Since $\sin \alpha = \sin \beta$ implies $\beta + \alpha = (2p+1)\pi$ or $\beta - \alpha = 2q\pi$ for $p, q \in \mathbb{N}_0$, it follows that we must have

$$\theta = \frac{2p+1}{2L+3}\pi \quad \text{or} \quad \theta = 2q\pi \quad \text{for } p, q \in \mathbb{N}_0.$$

The second possibility can be discarded because (8.186) implies that in this case $|\Phi\rangle = 0$, which is not a permissible eigenvector.

The first possibility allows for $L + 1$ distinct eigenvalues of $\check{H}$ given by

$$\check{E}_p = 1 - \cos\left(\frac{2p+1}{2L+3}\pi\right) \quad \text{for } p \in \{0, \ldots, L\},$$

and since (8.123) implies $\dim \mathbb{H}_{\text{sub}} = L + 1$, these are all possible eigenvalues of $\check{H}$. The lowest eigenvalue is

$$\check{E}_0 = 1 - \cos\left(\frac{\pi}{2L+3}\right) < 1 \qquad (8.189)$$

and evidently satisfies the upper bound given in (8.181). In order to obtain the lower bound $\frac{1}{L^2}$ for $\check{E}_0$, we note first that the series expansion

$$\cos x = \sum_{k=0}^{\infty} \frac{(-1)^k x^{2k}}{(2k)!} = 1 - \frac{x^2}{2} + R_4(x)$$

implies

$$1 - \cos x = \frac{x^2}{2} - R_4(x).$$

For $|x| \leq 4$ the remainder

$$R_4(x) = \overbrace{\frac{x^4}{4!} - \frac{x^6}{6!}}^{\geq 0} + \overbrace{\frac{x^8}{8!} - \frac{x^{10}}{10!}}^{\geq 0} + \cdots = \frac{x^4}{4!} - \underbrace{\frac{x^6}{6!} + \frac{x^8}{8!}}_{\leq 0} - \underbrace{\frac{x^{10}}{10!} + \frac{x^{12}}{12!}}_{\leq 0} - \cdots$$

satisfies

$$0 \leq R_4(x) \leq \frac{x^4}{4!}$$

such that

$$1 - \cos x \geq \frac{x^2}{2} - \frac{x^4}{4!}.$$

Applying this to (8.189) yields

$$\check{E}_0 \geq \frac{\pi^2}{2(2L+3)^2} - \frac{\pi^4}{24(2L+3)^4}. \tag{8.190}$$

Here we can use that for $L \geq 8$

$$\frac{\pi^2}{2} - 4 \geq \frac{6}{L} + \frac{9}{L^2} + \frac{\pi^4}{24(2L+3)^2},$$

which implies

$$\frac{\pi^2}{2(2L+3)^2} - \frac{\pi^4}{24(2L+3)^4} \geq \frac{1}{L^2}.$$

Together with (8.190), this gives

$$\check{E} \geq \frac{1}{L^2} \quad \text{for } L \geq 8$$

and completes the proof of (8.181).   □

After all these auxiliary results we are now in a position to give a rigorous lower bound for the ground state eigenvalue gap $g_0(s)$ of $H_T(s)\big|_{\mathbb{H}_{\text{sub}}}$.

Theorem 8.30 ([116]) *Let $H_T(s)\big|_{\mathbb{H}_{\text{sub}}}$ be defined as in (8.133) and let $\{E_j(s)\,|\,j \in \{0,\dots,L\}\}$ be the set of its eigenvalues. Then the ground state energy gap $g_0(s) = E_1(s) - E_0(s)$ satisfies for $L \geq 8$*

$$g_0(s) > \frac{1}{2L^2}\,. \tag{8.191}$$

Proof We know already from Theorem 8.26 that for $s \in\,]0,1]$ the eigenvalues $E_j(s)$ of $H_T(s)\big|_{\mathbb{H}_{\text{sub}}}$ satisfy $E_0(s) < E_1(s) < \cdots < E_L(s)$. From the matrix of $H_T(s)\big|_{\mathbb{H}_{\text{sub}}}$ in the ONB $\{|\Gamma(m)\rangle \,|\, m \in \{0,\dots,L\}\}$ of $\mathbb{H}_{\text{sub}}$ as shown in (8.125) we see that

$$\langle \Gamma(0)|H_T(s)\big|_{\mathbb{H}_{\text{sub}}}\Gamma(0)\rangle = \frac{s}{2}$$

and that for $|\Gamma\rangle$ as defined in (8.95)

$$\langle \Gamma|H_T(s)\big|_{\mathbb{H}_{\text{sub}}}\Gamma\rangle = \frac{L(1-s)}{L+1}\,.$$

The result (2.40) of Exercise 2.9 then implies that

$$E_0(s) \leq \min\left\{\frac{s}{2}, \frac{L(1-s)}{L+1}\right\}\,. \tag{8.192}$$

and the two lines $s \mapsto \frac{s}{2}$ and $s \mapsto \frac{L(1-s)}{L+1}$ cross at

$$s_c := \frac{2L}{3L+1}\,. \tag{8.193}$$

Since we may assume $L > 1$, we have $\frac{L}{2} > \frac{1}{2}$ and thus $2L > \frac{3L+1}{2}$, which implies

$$s_c > \frac{1}{2}\,. \tag{8.194}$$

On the other hand, we have $2L < \frac{2(3L+1)}{3}$ such that

$$s_c < \frac{2}{3}\,. \tag{8.195}$$

These two bounds for s_c will be used later. For now let us turn to $E_1(s)$. From Theorem 8.28 we know that

$$E_1(s) \underset{(8.173)}{=} \max \left\{ \min \left\{ \langle \Psi | H_T(s) \big|_{\mathbb{H}_{\text{sub}}} \Psi \rangle \, \big| \, |\Psi\rangle \in \mathbb{H}_2, ||\Psi|| = 1 \right\} \right.$$

$$\left. \Big| \, \mathbb{H}_2 \text{ subset of } \mathbb{H} \text{ with } \dim \mathbb{H}_2 = L + 1 - 2 + 1 = L \right\}.$$

For the subspace $\mathbb{H}_2 = \text{Span}\{|\Gamma(1)\rangle, \ldots, |\Gamma(L)\rangle\}$ it follows that

$$E_1(s) \geq \min \left\{ \langle \Psi | H_T(s) \big|_{\mathbb{H}_{\text{sub}}} \Psi \rangle \, \big| \, |\Psi\rangle \in \text{Span}\{|\Gamma(1)\rangle, \ldots, |\Gamma(L)\rangle\}, ||\Psi|| = 1 \right\}. \tag{8.196}$$

Now, inserting

$$H_{\text{ini}}(s)\big|_{\mathbb{H}_{\text{sub}}} \underset{(8.126)}{=} \mathbf{1}\big|_{\mathbb{H}_{\text{sub}}} - |\Gamma(0)\rangle\langle\Gamma(0)|$$

$$H_{\text{fin}}(s)\big|_{\mathbb{H}_{\text{sub}}} \underset{(8.180)}{=} \check{H} - \frac{1}{2}|\Gamma(0)\rangle\langle\Gamma(0)|$$

in the right side of

$$H_T(s)\big|_{\mathbb{H}_{\text{sub}}} \underset{(8.122)}{=} (1-s)H_{\text{ini}}(s)\big|_{\mathbb{H}_{\text{sub}}} + sH_{\text{fin}}(s)\big|_{\mathbb{H}_{\text{sub}}}$$

yields

$$H_T(s)\big|_{\mathbb{H}_{\text{sub}}} = (1-s)\mathbf{1}\big|_{\mathbb{H}_{\text{sub}}} + s\check{H} + \left(1 - \frac{s}{2}\right)|\Gamma(0)\rangle\langle\Gamma(0)|.$$

Since for any $|\Psi\rangle \in \mathbb{H}_2 = \text{Span}\{|\Gamma(1)\rangle, \ldots, |\Gamma(L)\rangle\}$ we have $\langle\Gamma(0)|\Psi\rangle = 0$, it follows that for such $|\Psi\rangle \in \mathbb{H}_2$ with $||\Psi|| = 1$

$$\langle \Psi | H_T(s) \big|_{\mathbb{H}_{\text{sub}}} \Psi \rangle = (1-s) + s\langle\Psi|\check{H}\Psi\rangle. \tag{8.197}$$

Denoting the lowest eigenvalue of $\check{H}$ by $\check{E}_0$, the result (2.40) of Exercise 2.9 tells us that $\langle\Psi|\check{H}\Psi\rangle \geq \check{E}_0$, such that (8.196) becomes

$$E_1(s) \geq \left\{ \min \left\{ \langle \Psi | H_T(s) \big|_{\mathbb{H}_{\text{sub}}} \Psi \rangle \, \big| \, |\Psi\rangle \in \text{Span}\{|\Gamma(1)\rangle, \ldots, |\Gamma(L)\rangle\}, ||\Psi|| = 1 \right\}\right.$$

$$\underset{(8.197)}{\geq} 1 - s + \check{E}_0, \tag{8.198}$$

and we obtain

$$E_1(s) - E_s(0) \underbrace{\geq}_{(8.192),(8.198)} 1 - s + s\check{E}_0 - \min\left\{\frac{s}{2}, \frac{L(1-s)}{L+1}\right\}. \qquad (8.199)$$

Here we observe that, with s_c as defined in (8.193), then $s \gtrsim s_c = \frac{2L}{3L+1}$ implies $\frac{s}{2} \gtrsim \frac{L(1-s)}{L+1}$. Hence, we have

$$\min\left\{\frac{s}{2}, \frac{L(1-s)}{L+1}\right\} = \begin{cases} \frac{s}{2} & \text{if } 0 \leq s \leq s_c \\ \frac{L(1-s)}{L+1} & \text{if } s_c < s \leq 1, \end{cases}$$

and thus (8.199) becomes

$$E_1(s) - E_s(0) \geq \begin{cases} 1 - \frac{3s}{2} + s\check{E}_0 & \text{if } 0 \leq s \leq s_c \\ 1 - s + s\check{E}_0 - \frac{L(1-s)}{L+1} & \text{if } s_c < s \leq 1. \end{cases} \qquad (8.200)$$

We consider the two cases separately. For the case $0 \leq s \leq s_c$ we note that

$$\frac{d}{ds}\left(1 - \frac{3s}{2} + s\check{E}_0\right) = \check{E}_0 - \frac{3}{2} \underbrace{<}_{(8.181)} 0$$

such that for $s \in [0, s_c]$ the function $1 - \frac{3s}{2} + s\check{E}_0$ attains its lowest value at the rightmost boundary $s = s_c$. Consequently, we find in the case $0 \leq s \leq s_c$ that

$$E_1(s) - E_2(s) \underbrace{\geq}_{(8.200)} 1 - \frac{3s}{2} + s\check{E}_0 \geq 1 - \frac{3s_c}{2} + s_c\check{E}_0 \underbrace{\geq}_{(8.195)} s_c\check{E}_0. \qquad (8.201)$$

For the case $s \in]s_c, 1]$ we observe that

$$E_1(s) - E_2(s) \underbrace{\geq}_{(8.200)} 1 - s + s\check{E}_0 - \frac{L(1-s)}{L+1} = s\check{E}_0 + \frac{1-s}{L+1} \geq s\check{E}_0 \geq s_c\check{E}_0. \qquad (8.202)$$

Therefore, we finally have

$$g_0(s) = E_1(s) - E_0(s) \underbrace{\geq}_{(8.201),(8.201)} s_c\check{E}_0 \underbrace{\geq}_{(8.194)} \frac{1}{2}\check{E}_0 \underbrace{\geq}_{(8.181)} \frac{1}{2L^2}.$$

□

We can now combine the results of Corollary 8.4, Lemmas 8.24, 8.27 and Theorem 8.30 to prove a statement about how the time T required to approximate the action of the circuit $U = U_L \cdots U_1$ with an adiabatic computation scales as a function of the number of gates L in the circuit.

Theorem 8.31 *Let the circuit U be as in Definition 8.18. To obtain $U|0\rangle^n$ with a probability of at least $\frac{p_{min}}{L+1}$ as a result of a suitable adiabatic computation of duration T and subsequent measurement, it suffices that*

$$T \in O\left(\frac{L^6}{\sqrt{1 - p_{min}}}\right) \qquad for\ L \to \infty. \tag{8.203}$$

Proof The probability to obtain $U|0\rangle^n$ as a result of a suitable adiabatic computation and subsequent measurement satisfies

$$\mathbf{P}\left\{\begin{matrix}\text{To observe } U|0\rangle^n \text{ after} \\ \text{the adiabatic evolution} \\ \text{with } \mathsf{H}_T(s) \text{ in (8.122)}\end{matrix}\right\} = \mathbf{P}\left\{\begin{matrix}\text{To observe} \\ U|0\rangle^n \text{ in the} \\ \text{state } |\Gamma\rangle\end{matrix}\right\} \times \mathbf{P}\left\{\begin{matrix}\text{To produce } |\Gamma\rangle \\ \text{with the adiabatic} \\ \text{evolution } \mathsf{H}_T(s)\end{matrix}\right\}.$$

We know from Theorem 8.23 that $|\Gamma\rangle$ as defined in Definition 8.18 is the ground state of the final Hamiltonian in the adiabatic computation designed to approximate U. From Corollary 8.4 we know that if the time T for the transition from H_{ini} to H_{fin} satisfies

$$T \geq \frac{C_0(1)}{\sqrt{1 - p_{min}}}, \tag{8.204}$$

then the probability to find $|\Gamma\rangle$ at time $t_{fin} = T + t_{ini}$ is bounded from below by p_{min}, that is,

$$\mathbf{P}\left\{\begin{matrix}\text{To produce } |\Gamma\rangle \\ \text{with the adiabatic} \\ \text{evolution } \mathsf{H}_T(s)\end{matrix}\right\} \geq p_{min}.$$

Moreover, Lemma 8.24 tells us that measuring a suitable observable in the state $|\Gamma\rangle$ will project onto the state $U|0\rangle^n$ with a probability of $\frac{1}{L+1}$, that is,

$$\mathbf{P}\left\{\begin{matrix}\text{To observe} \\ U|0\rangle^n \text{ in the} \\ \text{state } |\Gamma\rangle\end{matrix}\right\} = \frac{1}{L+1}.$$

Consequently, (8.204) implies

$$\mathbf{P}\left\{\begin{matrix}\text{To observe } U|0\rangle^n \text{ after} \\ \text{the adiabatic evolution} \\ \text{with } \mathsf{H}_T(s) \text{ in (8.122)}\end{matrix}\right\} \geq \frac{p_{min}}{L+1}.$$

Combining Lemma 8.27 with Theorem 8.30 shows that for $L \geq 8$ in the case at hand

$$
C_0(1) \underbrace{\leq}_{(8.168),(8.191)} 24L^4 + 720L^6 .
\tag{8.205}
$$

Finally, we see that for an adiabatic transition time $T = T(L)$ satisfying

$$
\frac{C_0(1)}{\sqrt{1 - p_{\min}}} \underbrace{\leq}_{(8.205)} \frac{800L^6}{\sqrt{1 - p_{\min}}} \leq T(L) \leq \frac{1000L^6}{\sqrt{1 - p_{\min}}} \in O\left(\frac{L^6}{\sqrt{1 - p_{\min}}}\right)
$$

both (8.203) and (8.204) hold true.   □

From Theorem 8.31 we see that if we repeat the adiabatic computation $O(L)$ times, we will obtain $U|0\rangle^n$ with a probability approximately bounded from below by $p_{\min}$. We can thus replicate the result of the circuit U with a given minimal probability by repeating the adiabatic computation $O(L)$ times such that the total replication has an aggregated running time of $O(L^7)$. Consequently, a circuit based computation can be efficiently replicated by an adiabatic computation.

8.6 Replicating an Adiabatic by a Circuit Based Computation

In this section we show that any adiabatic quantum computation can be efficiently approximated with arbitrary precision by a circuit with sufficiently many gates [33, 114]. More precisely, we assume that we are given an adiabatic quantum computation with initial Hamiltonian H_{ini} at time t_{ini} and final Hamiltonian H_{fin} at time t_{fin} with a linear schedule such that with $T = t_{\mathrm{fin}} - t_{\mathrm{ini}}$ the Hamiltonian

$$
H(t) = \left(1 - \frac{t - t_{\mathrm{ini}}}{T}\right) H_{\mathrm{ini}} + \frac{t - t_{\mathrm{ini}}}{T} H_{\mathrm{fin}}
\tag{8.206}
$$

generates the adiabatic time evolution $U(t, t_{\mathrm{ini}})$ as solution of the initial value problem

$$
\begin{aligned}
i\frac{d}{dt}U(t, t_{\mathrm{ini}}) &= H(t)U(t, t_{\mathrm{ini}}) \\
U(t_{\mathrm{ini}}, t_{\mathrm{ini}}) &= \mathbf{1}
\end{aligned}
\tag{8.207}
$$

for $t \in [t_{\mathrm{ini}}, t_{\mathrm{fin}}]$. To show that $U(t_{\mathrm{fin}}, t_{\mathrm{ini}}) = U_T(1)$ can be obtained by a circuit based computation, we first approximate U by a time evolution $\hat{U}$ generated from a piecewise constant Hamiltonian $\hat{H}$. In a second step we approximate $\hat{U}$ further by a time evolution $\hat{\hat{U}}$ generated by H_{ini} and H_{fin} acting independently.

The approximation of U thus obtained will be shown to be efficient in the sense that the number of gates required to reach a given precision only increases as $O(T^2)$.

Let $J \in \mathbb{N}$ and $\Delta t = \frac{T}{J}$. For $j \in \mathbb{N}_0$ such that $0 \le j \le J$ we define

$$\hat{H}_j := H(j\Delta t) = \left(1 - \frac{j\Delta t}{T}\right) H_{\text{ini}} + \frac{j\Delta t}{T} H_{\text{fin}}$$

$$= \left(1 - \frac{j}{J}\right) H_{\text{ini}} + \frac{j}{J} H_{\text{fin}}. \tag{8.208}$$

Moreover, we define $\kappa : [t_{\text{ini}}, t_{\text{fin}}] \to \mathbb{N}_0$ as

$$\kappa(t) = \left\lceil \frac{J(t - t_{\text{ini}})}{T} \right\rceil = \left\lceil \frac{t - t_{\text{ini}}}{\Delta t} \right\rceil \in \{0, 1, \dots, J\}, \tag{8.209}$$

where

$$\lceil \cdot \rceil : \mathbb{R} \longrightarrow \mathbb{Z}$$
$$x \longmapsto \min\{z \in \mathbb{Z} \mid z \ge x\}$$

such that $\lceil x \rceil$ denotes the smallest integer not less than x. With the help of $\hat{H}_j$ and κ we then define

$$\hat{H}(t) := \hat{H}_{\kappa(t)}. \tag{8.210}$$

Note that since $\kappa(t_{\text{ini}}) = 0$ and $\kappa(t_{\text{fin}}) = J$, we have

$$\hat{H}(t_{\text{ini}}) = \hat{H}_{\kappa(t_{\text{ini}})} = \hat{H}_0 = H_{t_{\text{ini}}}$$
$$\hat{H}(t_{\text{fin}}) = \hat{H}_{\kappa(t_{\text{fin}})} = \hat{H}_J = H_{t_{\text{fin}}}$$

and that $\hat{H}(t)$ is constant whenever $\frac{t - t_{\text{ini}}}{\Delta t} \notin \mathbb{N}_0$.

We begin our effort to approximate the adiabatic time evolution with a lemma giving an estimate of how far two time evolutions can drift apart given a bound on the difference of the Hamiltonians generating them.

Lemma 8.32 *For $t \in [t_{ini}, t_{fin}]$ let $H_A(t), H_B(t) \in B_{sd}(\mathbb{H})$ be two Hamiltonians on a* HILBERT *space $\mathbb{H}$, and for $X \in \{A, B\}$ let $U_X(t, t_{ini})$ be the respective time evolutions they generate, that is, for $X \in \{A, B\}$ and $t \in [t_{ini}, t_{fin}]$ we have*

$$i\frac{d}{dt}U_X(t, t_{ini}) = H_X(t)U_X(t, t_{ini})$$
$$U_X(t_{ini}, t_{ini}) = 1. \tag{8.211}$$

If

$$\|H_A(t) - H_B(t)\| \le \varepsilon \tag{8.212}$$

holds for $t \in [t_{ini}, t_{fin}]$, then we also have for such t that

$$\|U_A(t, t_{ini}) - U_B(t, t_{ini})\| \leq \sqrt{2(t - t_{ini})\varepsilon} \,. \qquad (8.213)$$

Proof To begin with, we have for any $|\psi\rangle \in \mathbb{H}$

$$
\begin{aligned}
& \left\| \left(U_A(t, t_{ini}) - U_B(t, t_{ini}) \right) \psi \right\|^2 \\
\underset{(2.5)}{=} \ & \left\langle \left(U_A(t, t_{ini}) - U_B(t, t_{ini}) \right) \psi \middle| \left(U_A(t, t_{ini}) - U_B(t, t_{ini}) \right) \psi \right\rangle \\
\underset{(2.4),(2.5)}{=} \ & \left\| U_A(t, t_{ini}) \psi \right\|^2 + \left\| U_B(t, t_{ini}) \psi \right\|^2 \qquad\qquad\qquad\qquad (8.214) \\
& - \langle U_A(t, t_{ini}) \psi | U_B(t, t_{ini}) \psi \rangle - \langle U_B(t, t_{ini}) \psi | U_A(t, t_{ini}) \psi \rangle \\
\underset{(2.37)}{=} \ & 2 \|\psi\|^2 - \left(\langle U_A(t, t_{ini}) \psi | U_B(t, t_{ini}) \psi \rangle + \langle U_B(t, t_{ini}) \psi | U_A(t, t_{ini}) \psi \rangle \right).
\end{aligned}
$$

Next, we note that

$$
\begin{aligned}
& \frac{d}{dt} \langle U_A(t, t_{ini}) \psi | U_B(t, t_{ini}) \psi \rangle \\
= \ & \langle \frac{d}{dt} U_A(t, t_{ini}) \psi | U_B(t, t_{ini}) \psi \rangle + \langle U_A(t, t_{ini}) \psi | \frac{d}{dt} U_B(t, t_{ini}) \psi \rangle \\
\underset{(8.211)}{=} \ & \langle -i\mathsf{H}_A(t) U_A(t, t_{ini}) \psi | U_B(t, t_{ini}) \psi \rangle + \langle U_A(t, t_{ini}) \psi | -i\mathsf{H}_B(t) U_B(t, t_{ini}) \psi \rangle \\
= \ & i \langle U_A(t, t_{ini}) \psi | \left(\mathsf{H}_A(t) - \mathsf{H}_B(t) \right) U_B(t, t_{ini}) \psi \rangle \,,
\end{aligned}
$$

where in the last line we have used (2.4), (2.6), and that $\mathsf{H}_A(t)^* = \mathsf{H}_A(t)$. Integrating on both sides then yields

$$
\begin{aligned}
& \langle U_A(t, t_{ini}) \psi | U_B(t, t_{ini}) \psi \rangle \\
= \ & \langle U_A(t_{ini}, t_{ini}) \psi | U_B(t_{ini}, t_{ini}) \psi \rangle \\
& + i \int_{t_{ini}}^{t} \langle U_A(s, t_{ini}) \psi | \left(\mathsf{H}_A(s) - \mathsf{H}_B(s) \right) U_B(s, t_{ini}) \psi \rangle ds \\
\underset{(8.211)}{=} \ & \|\psi\|^2 + i \int_{t_{ini}}^{t} \langle U_A(s, t_{ini}) \psi | \left(\mathsf{H}_A(s) - \mathsf{H}_B(s) \right) U_B(s, t_{ini}) \psi \rangle ds \,.
\end{aligned}
$$

Likewise, we have

$$
\begin{aligned}
& \langle U_B(t, t_{ini}) \psi | U_A(t, t_{ini}) \psi \rangle \\
= \ & \|\psi\|^2 + i \int_{t_{ini}}^{t} \langle U_B(s, t_{ini}) \psi | \left(\mathsf{H}_B(s) - \mathsf{H}_A(s) \right) U_A(s, t_{ini}) \psi \rangle ds
\end{aligned}
$$

such that

$$\langle U_A(t,t_{\text{ini}})\psi|U_B(t,t_{\text{ini}})\psi\rangle + \langle U_B(t,t_{\text{ini}})\psi|U_A(t,t_{\text{ini}})\psi\rangle$$
$$= 2\,||\psi||^2$$
$$+ i\int_{t_{\text{ini}}}^{t}\langle U_A(s,t_{\text{ini}})\psi|(H_A(s)-H_B(s))U_B(s,t_{\text{ini}})\psi\rangle ds$$
$$+ i\int_{t_{\text{ini}}}^{t}\langle U_B(s,t_{\text{ini}})\psi|(H_B(s)-H_A(s))U_A(s,t_{\text{ini}})\psi\rangle ds. \tag{8.215}$$

Inserting (8.215) into (8.214) yields

$$||(U_A(t,t_{\text{ini}})-U_B(t,t_{\text{ini}}))\psi||^2 = \left|||(U_A(t,t_{\text{ini}})-U_B(t,t_{\text{ini}}))\psi||^2\right| \tag{8.216}$$
$$= \left| i\int_{t_{\text{ini}}}^{t}\langle U_A(s,t_{\text{ini}})\psi|(H_A(s)-H_B(s))U_B(s,t_{\text{ini}})\psi\rangle ds \right.$$
$$\left. + i\int_{t_{\text{ini}}}^{t}\langle U_B(s,t_{\text{ini}})\psi|(H_B(s)-H_A(s))U_A(s,t_{\text{ini}})\psi\rangle ds \right|$$
$$\leq \int_{t_{\text{ini}}}^{t}\left|\langle U_A(s,t_{\text{ini}})\psi|(H_A(s)-H_B(s))U_B(s,t_{\text{ini}})\psi\rangle\right| ds$$
$$+ \int_{t_{\text{ini}}}^{t}\left|\langle U_B(s,t_{\text{ini}})\psi|(H_B(s)-H_A(s))U_A(s,t_{\text{ini}})\psi\rangle\right| ds.$$

In the last two lines of (8.216) we can use that, for example,

$$\left|\langle U_A(s,t_{\text{ini}})\psi|(H_A(s)-H_B(s))U_B(s,t_{\text{ini}})\psi\rangle\right|$$
$$\underbrace{\leq}_{(2.16)} ||U_A(s,t_{\text{ini}})\psi||\,||(H_A(s)-H_B(s))U_B(s,t_{\text{ini}})\psi||$$
$$\underbrace{\leq}_{(2.51)} ||U_A(s,t_{\text{ini}})\psi||\,||H_A(s)-H_B(s)||\,||U_B(s,t_{\text{ini}})\psi||$$
$$\underbrace{=}_{(2.37)} ||\psi||^2\,||H_A(s)-H_B(s)||$$
$$\underbrace{\leq}_{(8.212)} ||\psi||^2\,\varepsilon$$

and, likewise, for $\left|\langle U_B(s,t_{\text{ini}})\psi|(H_B(s)-H_A(s))U_A(s,t_{\text{ini}})\psi\rangle\right|$. Using this in (8.216) gives

$$||(U_A(t,t_{\text{ini}})-U_B(t,t_{\text{ini}}))\psi||^2 \leq 2\,||\psi||^2\,\varepsilon\int_{t_{\text{ini}}}^{t} ds = ||\psi||^2\,2(t-t_{\text{ini}})\varepsilon.$$

The claim (8.213) then follows from the definition of the operator norm (2.45). □

In order to apply Lemma 8.32 to our situation, we need an estimate of the difference between $\hat{H}(t)$ and $H(t)$. Exercise 8.115 provides this.

Exercise 8.115 Show that $H(t)$ as defined in (8.206) and $\hat{H}(t)$ as defined in (8.210) satisfy

$$\left|\left|\hat{H}(t) - H(t)\right|\right| \leq \frac{1}{J} \left|\left|H_{\text{fin}} - H_{\text{ini}}\right|\right| . \tag{8.217}$$

For a solution see Solution 8.115.

Whereas the result of Exercise 8.115 shows us how much the time dependent Hamiltonian $H(t)$ of the adiabatic quantum computation differs from the piece-wise constant $\hat{H}(t)$, the statement in Lemma 8.32 tells us how much the time evolutions they generate differ. We combine these two results to obtain a first approximation of $U(t_{\text{fin}}, t_{\text{ini}})$.

Lemma 8.33 *Let $\hat{H}(t)$ be defined for $t \in [t_{ini}, t_{fin}]$ as in (8.210) and let $\hat{U}(t, t_{ini})$ be the time evolution it generates, namely,*

$$i\frac{d}{dt}\hat{U}(t, t_{ini}) = \hat{H}(t)\hat{U}(t, t_{ini}) \tag{8.218}$$

$$\hat{U}(t_{ini}, t_{ini}) = 1 .$$

Moreover, with $J \in \mathbb{N}$ and $\Delta t = \frac{t_{fin} - t_{ini}}{J}$ define

$$\hat{U}_0 := 1$$
$$\hat{U}_j := e^{-i\Delta t \hat{H}_j} \qquad for\ j \in \{1, \ldots, J\} \tag{8.219}$$
$$M := \max\{||H_{ini}||, ||H_{fin}||\}, \tag{8.220}$$

where $\hat{H}_j$ is as defined in (8.208). Then we have

$$\hat{U}(t_{fin}, t_{ini}) = \hat{U}_J\hat{U}_{J-1}\cdots\hat{U}_1\hat{U}_0 \tag{8.221}$$

$$\left|\left|U(t_{fin}, t_{ini}) - \hat{U}(t_{fin}, t_{ini})\right|\right| \leq 2\sqrt{\frac{MT}{J}} = 2\sqrt{M\Delta t}, \tag{8.222}$$

where, as before, $T = t_{fin} - t_{ini}$.

Proof To show (8.221), we note that for $t \in]t_{\mathrm{ini}} + (k-1)\Delta t, t_{\mathrm{ini}} + k\Delta t]$ we have $\kappa(t) = k \in \{1, \ldots, J\}$ and thus $\hat{H}(t) = \hat{H}_{\kappa(t)} = \hat{H}_k$, in other words, inside of each of these time periods of length Δt the Hamiltonian $\hat{H}(t)$ is constant. We define

$$
\begin{aligned}
&\hat{U}(t_{\mathrm{ini}}, t_{\mathrm{ini}}) = \mathbf{1} \\
&\hat{U}(t, t_{\mathrm{ini}}) = e^{-i\left(t - t_{\mathrm{ini}} - (\kappa(t) - 1)\Delta t\right)\hat{H}_{\kappa(t)}} \hat{U}_{\kappa(t)-1} \cdots \hat{U}_0 \quad \text{for } t \in]t_{\mathrm{ini}}, t_{\mathrm{fin}}]
\end{aligned}
\tag{8.223}
$$

and show first that $\hat{U}(t, t_{\mathrm{ini}})$ thus defined is a solution of the initial value problem (8.218). For $t \in]t_{\mathrm{ini}} + (k-1)\Delta t, t_{\mathrm{ini}} + k\Delta t]$ we have

$$
\hat{U}(t, t_{\mathrm{ini}}) = e^{-i\left(t - t_{\mathrm{ini}} - (k-1)\Delta t\right)\hat{H}_k} \hat{U}_{k-1} \cdots \hat{U}_0
\tag{8.224}
$$

and $k = \kappa(t)$ such that

$$
i\frac{d}{dt}\hat{U}(t, t_{\mathrm{ini}}) = \hat{H}_k \hat{U}(t, t_{\mathrm{ini}}) = \hat{H}_{\kappa(t)} \hat{U}(t, t_{\mathrm{ini}}) \underbrace{=}_{(8.210)} \hat{H}(t)\hat{U}(t, t_{\mathrm{ini}}) .
$$

At the discrete points $t_{\mathrm{ini}} + k\Delta t$, where $\in \{1, \ldots, J\}$, we use the left-derivative

$$
\begin{aligned}
i\frac{d}{dt}\hat{U}(t, t_{\mathrm{ini}})|_{t \nearrow t_{\mathrm{ini}} + k\Delta t} &= i \lim_{\delta \searrow 0} \frac{1}{\delta}\left(\hat{U}(t_{\mathrm{ini}} + k\Delta t, t_{\mathrm{ini}}) - \hat{U}(t_{\mathrm{ini}} + k\Delta t - \delta, t_{\mathrm{ini}})\right) \\
&\underbrace{=}_{(8.223)} i \lim_{\delta \searrow 0} \frac{1}{\delta}\left(e^{-i\Delta t \hat{H}_k} - e^{-i(\Delta t - \delta)\hat{H}_k}\right)\hat{U}_{k-1} \cdots \hat{U}_0 \\
&\underbrace{=}_{(8.223)} i \lim_{\delta \searrow 0} \frac{1}{\delta}\left(1 - e^{i\delta \hat{H}_k}\right)\hat{U}_k \cdots \hat{U}_0 \\
&= \hat{H}_k \hat{U}_k \cdots \hat{U}_0 \\
&\underbrace{=}_{\substack{(8.210), \\ (8.223)}} \hat{H}(t_{\mathrm{ini}} + k\Delta t)\hat{U}(t_{\mathrm{ini}} + k\Delta t, t_{\mathrm{ini}}) .
\end{aligned}
$$

Hence, $\hat{U}(t, t_{\mathrm{ini}})$ is a solution of the initial value problem (8.218). Since $\kappa(t_{\mathrm{ini}} + k\Delta t) = k$, we obtain

$$
\hat{U}(t_{\mathrm{ini}} + k\Delta t, t_{\mathrm{ini}}) \underbrace{=}_{(8.224)} e^{-i\Delta t \hat{H}_k}\hat{U}_{k-1} \cdots \hat{U}_1 \underbrace{=}_{(8.219)} \hat{U}_k \cdots \hat{U}_1
$$

and ultimately

$$
\hat{U}(t_{\mathrm{fin}}, t_{\mathrm{ini}}) = \hat{U}_J \cdots \hat{U}_1 .
$$

To show (8.222), we note that using the result (8.217) in Exercise 8.115 and applying (8.213) of Lemma 8.32 to $U(t_{\text{fin}}, t_{\text{ini}})$ and $\hat{U}(t_{\text{fin}}, t_{\text{ini}})$ yields

$$\left\|U(t_{\text{fin}}, t_{\text{ini}}) - \hat{U}(t_{\text{fin}}, t_{\text{ini}})\right\| \underbrace{=}_{(8.217),(8.213)} \sqrt{\frac{2(t_{\text{fin}} - t_{\text{ini}})}{J}} \left\|H_{\text{fin}} - H_{\text{ini}}\right\|. \quad (8.225)$$

Using $T = t_{\text{fin}} - t_{\text{ini}}$ and

$$\|H_{\text{fin}} - H_{\text{ini}}\| \underbrace{\leq}_{(2.53)} \|H_{\text{fin}}\| + \|H_{\text{ini}}\| \underbrace{\leq}_{(8.220)} 2M$$

in (8.225) implies the stated claim (8.222). $\qquad\qquad\qquad\qquad\qquad\qquad\qquad\square$

Having approximated $U(t_{\text{fin}}, t_{\text{ini}})$ by the product $\hat{U}(t_{\text{fin}}, t_{\text{ini}}) = \hat{U}_J \cdots \hat{U}_1$ of the $\hat{U}_j$ defined in (8.219), we now proceed to approximate each $\hat{U}_j$ in turn by

$$\hat{U}_j := e^{-i\Delta t\left(1 - \frac{j}{J}\right)H_{\text{ini}}} e^{-i\Delta t \frac{j}{J} H_{\text{fin}}} \quad (8.226)$$

and $\hat{U}(t_{\text{fin}}, t_{\text{ini}}) = \hat{U}_J \cdots \hat{U}_1$ by

$$\hat{U} := \hat{U}_J \cdots \hat{U}_1. \quad (8.227)$$

The $\hat{U}_j$ will be our gates and $\hat{U}$ will be the circuit with which we approximate the adiabatic evolution. Note that the unitary factors $\hat{U}_j$ are akin to a time evolution generated by first H_{fin} and then by H_{ini} separately each for a period of Δt. More precisely, let U_{H_X} denote the time evolution generated by H_X for $X \in \{\text{ini}, \text{fin}\}$, that is, the U_{H_X} are the solutions of the initial value problem

$$i\frac{d}{dt}U_{H_X}(t) = H_I U_{H_I}(t)$$
$$U_{H_X}(0) = \mathbf{1}.$$

Then we can express the $\hat{U}_j$ of (8.226) as follows

$$\hat{U}_j = U_{H_{\text{ini}}}\left(\Delta t\left(1 - \frac{j}{J}\right)\right) U_{H_{\text{fin}}}\left(\Delta t \frac{j}{J}\right).$$

This shows that we can construct the gate $\hat{U}_j$ by first subjecting the system for a time $\Delta t \frac{j}{J}$ to the time evolution generated by H_{fin} and subsequently for a time $\Delta t\left(1 - \frac{j}{J}\right)$ to the time evolution generated by H_{ini}.

Also note that since H_{ini} and H_{fin} do not necessarily commute,[3] we have in general

$$\hat{U}_j = e^{-i\Delta t\left(1-\frac{j}{J}\right)H_{ini}}e^{-i\Delta t\frac{j}{J}H_{fin}} \neq e^{-i\left(\Delta t\left(1-\frac{j}{J}\right)H_{ini}+\Delta t\frac{j}{J}H_{fin}\right)} = \hat{U}_j.$$

Even though $\hat{U}_j$ and $\hat{U}_j$ differ, we can give bounds on their difference, as the Theorem 8.34 shows.

Theorem 8.34 ([117]) *Let A and B be bounded operators on a* HILBERT *space* $\mathbb{H}$. *Then the following hold:*

$$e^A e^B - e^{A+B} = \int_0^1 e^{uA}[e^{(1-u)(A+B)}, B]e^{uB}du \qquad (8.228)$$

$$\left|\left| [e^A, B] \right|\right| \leq || [A, B] || \, e^{||A||} \qquad (8.229)$$

$$\left|\left| e^{A+B} - e^A e^B \right|\right| \leq \frac{1}{2} ||[A, B]|| \, e^{||A||+||B||} . \qquad (8.230)$$

Proof To begin with, we have

$$e^A e^B - e^{A+B} = e^{uA}e^{(1-u)(A+B)}e^{uB}\Big|_{u=0}^{u=1}, \qquad (8.231)$$

where we used the notation $f(u)\big|_{u=a}^{u=b} := f(b) - f(a)$. It follows that

$$e^A e^B - e^{A+B} \underset{(8.231)}{=} \int_0^1 \frac{d}{du}\left(e^{uA}e^{(1-u)(A+B)}e^{uB}\right)du$$

$$= \int_0^1 \left(Ae^{uA}e^{(1-u)(A+B)}e^{uB} - e^{uA}(A+B)e^{(1-u)(A+B)}e^{uB} + e^{uA}e^{(1-u)(A+B)}Be^{uB}\right)du$$

$$= \int_0^1 e^{uA}\left(Ae^{(1-u)(A+B)} - (A+B)e^{(1-u)(A+B)} + e^{(1-u)(A+B)}B\right)e^{uB}du$$

$$= \int_0^1 e^{uA}\left(e^{(1-u)(A+B)}B - Be^{(1-u)(A+B)}\right)e^{uB}du$$

$$= \int_0^1 e^{uA}[e^{(1-u)(A+B)}, B]e^{uB}du ,$$

---

[3]If they did commute, then there would exist an ONB of eigenvectors for both operators. In this case preparing an eigenstate of H_{ini}, which is the starting point of the adiabatic algorithm, would be the same as preparing an eigenstate of H_{fin}, which is the desired end-state we would like to obtain with the algorithm. Hence, there would be no need for any algorithm.

where we used that $Ae^{uA} = e^{uA}A$ holds for any $u \in \mathbb{C}$ and bounded operator A. This proves (8.228). In order to show (8.229), we use that

$$[e^A, B] = e^{uA}Be^{(1-u)A}\big|_{u=0}^{u=1} \tag{8.232}$$

to write

$$
\begin{aligned}
[e^A, B] \underbrace{=}_{(8.232)} & \int_0^1 \frac{d}{du}(e^{uA}Be^{(1-u)A})du = \int_0^1 \left(Ae^{uA}Be^{(1-u)A} - e^{uA}BAe^{(1-u)A} \right) du \\
= & \int_0^1 \left(e^{uA}ABe^{(1-u)A} - e^{uA}BAe^{(1-u)A} \right) du \\
= & \int_0^1 e^{uA}[A, B]e^{(1-u)A}du .
\end{aligned} \tag{8.233}
$$

From this follows

$$
\begin{aligned}
||[e^A, B]|| \underbrace{=}_{(8.233)} & \left|\left| \int_0^1 e^{uA}[A, B]e^{(1-u)A}du \right|\right| \\
\underbrace{\leq}_{(2.53)} & \int_0^1 \left|\left| e^{uA}[A, B]e^{(1-u)A} \right|\right| du \\
\underbrace{\leq}_{(2.52)} & \int_0^1 \left|\left| e^{uA} \right|\right| \, ||[A, B]|| \left|\left| e^{(1-u)A} \right|\right| du \\
\underbrace{\leq}_{(2.53)} & \int_0^1 e^{u||A||} \, ||[A, B]|| \, e^{(1-u)||A||} du \\
= & \; ||[A, B]|| \, e^{||A||} ,
\end{aligned}
$$

which proves (8.229). Finally, to show (8.230), we first note that

$$
\begin{aligned}
\left|\left| [e^{(1-u)(A+B)}, B] \right|\right| \underbrace{\leq}_{(8.229)} & \; ||[(1-u)(A+B), B]|| \, e^{||(1-u)(A+B)||} \\
\underbrace{\leq}_{(2.54),(2.53)} & \; (1-u) \, ||[A, B]|| \, e^{(1-u)(||A||+||B||)} .
\end{aligned} \tag{8.234}
$$

Finally, we have

$$
\begin{aligned}
||e^{A+B} - e^A e^B|| &\underset{(8.228)}{=} \left\lVert \int_0^1 e^{uA}[e^{(1-u)(A+B)}, B]e^{uB}\,du \right\rVert \\
&\underset{(2.53)}{\leq} \int_0^1 \left\lVert e^{uA}[e^{(1-u)(A+B)}, B]e^{uB} \right\rVert du \\
&\underset{(2.52)}{\leq} \int_0^1 ||e^{uA}|| \, \left\lVert [e^{(1-u)(A+B)}, B] \right\rVert \, ||e^{uB}||\, du \\
&\underset{(2.53),(8.234)}{=} \int_0^1 e^{u||A||}(1-u)\,||\,[A,B]\,||\,e^{(1-u)(||A||+||B||)}e^{u||B||}\,du \\
&= ||\,[A,B]\,||\,e^{||A||+||B||} \int_0^1 (1-u)\,du \\
&= \frac{1}{2}\,||\,[A,B]\,||\,e^{||A||+||B||},
\end{aligned}
$$

proving (8.230). $\qquad\square$

The claim (8.230) of Theorem 8.34 allows us to find bounds on the difference of the $\hat{U}_j$ and $\hat{\hat{U}}_j$ and eventually $\hat{U}$ and $\hat{\hat{U}}_j$.

> **Corollary 8.35** *Let $\hat{U}_j, \hat{U}, J, T, \Delta t$ and M be defined as in Lemma 8.33, and let $\hat{\hat{U}}_j$ and $\hat{\hat{U}}$ be defined as in (8.226) and (8.227). Then we have the following bounds:*
>
> $$\left\lVert \hat{U}_j - \hat{\hat{U}}_j \right\rVert \leq \frac{1}{2}\left(\frac{T}{J}\right)^2 ||\,[\mathsf{H}_{ini}, \mathsf{H}_{fin}]\,||\,e^{\frac{2MT}{J}} \qquad (8.235)$$
>
> $$\left\lVert \hat{U}(t_{fin}, t_{ini}) - \hat{\hat{U}} \right\rVert \leq \frac{T^2}{2J}\,||\,[\mathsf{H}_{ini}, \mathsf{H}_{fin}]\,||\,e^{\frac{2MT}{J}}. \qquad (8.236)$$

Proof To prove (8.235), we apply (8.230) of Theorem 8.34 to

$$A = -\mathrm{i}\Delta t\left(1 - \frac{j}{J}\right)\mathsf{H}_{ini} \quad \text{and} \quad B = -\mathrm{i}\Delta t\frac{j}{J}\mathsf{H}_{fin} \qquad (8.237)$$

such that

$$
e^{A+B} \underbrace{=}_{(8.237)} e^{-i\Delta t\left(1-\frac{j}{J}\right)H_{\text{ini}}+-i\Delta t\frac{j}{J}H_{\text{fin}}} \underbrace{=}_{(8.219),(8.208)} \hat{U}_j
$$

$$
e^{A}e^{B} \underbrace{=}_{(8.237)} e^{-i\Delta t\left(1-\frac{j}{J}\right)H_{\text{ini}}}e^{-i\Delta t\frac{j}{J}H_{\text{fin}}} \underbrace{=}_{(8.226)} \hat{\hat{U}}_j .
$$

Consequently, we find

$$
\left\|\hat{U}_j - \hat{\hat{U}}_j\right\| \underbrace{\leq}_{(8.230)} \frac{1}{2}\left|\Delta t\left(1-\frac{j}{J}\right)\Delta t\frac{j}{J}\right| \,\|\,[H_{\text{ini}}, H_{\text{fin}}]\,\| \, e^{\left|\Delta t\left(1-\frac{j}{J}\right)\right|\|H_{\text{ini}}\|+\left|\Delta t\frac{j}{J}\right|\|H_{\text{fin}}\|}
$$

$$
\leq \frac{1}{2}\Delta t^2 \,\|\,[H_{\text{ini}}, H_{\text{fin}}]\,\| \, e^{\Delta t(\|H_{\text{ini}}\|+\|H_{\text{fin}}\|)}
$$

$$
\leq \frac{1}{2}\left(\frac{T}{J}\right)^2 \,\|\,[H_{\text{ini}}, H_{\text{fin}}]\,\| \, e^{\frac{2MT}{J}} ,
$$

where we used $0 \leq \frac{j}{J} \leq 1$ in the second line and $\Delta t = \frac{T}{J}$ as well as the definition (8.220) of M in the last. This completes the proof of (8.235).

To show (8.236), we first deduce that

$$
\left\|\hat{U}(t_{\text{fin}}, t_{\text{ini}}) - \hat{\hat{U}}\right\| \underbrace{=}_{(8.221),(8.226)} \left\|\hat{U}_J \cdots \hat{U}_1 - \hat{\hat{U}}_J \cdots \hat{\hat{U}}_1\right\|
$$

$$
= \left\|\hat{U}_J \cdots \hat{U}_2(\hat{U}_1 - \hat{\hat{U}}_1 + \hat{\hat{U}}_1) - \hat{\hat{U}}_J \cdots \hat{\hat{U}}_1\right\|
$$

$$
= \left\|\hat{U}_J \cdots \hat{U}_2(\hat{U}_1 - \hat{\hat{U}}_1) + \hat{U}_J \cdots \hat{U}_2\hat{\hat{U}}_1 - \hat{\hat{U}}_J \cdots \hat{\hat{U}}_1\right\|
$$

$$
\underbrace{\leq}_{(2.53)} \left\|\hat{U}_J \cdots \hat{U}_2(\hat{U}_1 - \hat{\hat{U}}_1)\right\| + \left\|(\hat{U}_J \cdots \hat{U}_2 - \hat{\hat{U}}_J \cdots \hat{\hat{U}}_2)\hat{\hat{U}}_1\right\|
$$

$$
\underbrace{\leq}_{(2.52)} \underbrace{\|\hat{U}_J \cdots \hat{U}_2\|}_{=1}\left\|\hat{U}_1 - \hat{\hat{U}}_1\right\| + \left\|\hat{U}_J \cdots \hat{U}_2 - \hat{\hat{U}}_J \cdots \hat{\hat{U}}_2\right\|\underbrace{\|\hat{\hat{U}}_1\|}_{=1}
$$

$$
\underbrace{=}_{(2.55)} \left\|\hat{U}_1 - \hat{\hat{U}}_1\right\| + \left\|\hat{U}_J \cdots \hat{U}_2 - \hat{\hat{U}}_J \cdots \hat{\hat{U}}_2\right\| .
$$

Repeating this with $\left\|\hat{U}_J \cdots \hat{U}_2 - \hat{\hat{U}}_J \cdots \hat{\hat{U}}_2\right\|$ and so on, we obtain

$$
\left\|\hat{U}(t_{\text{fin}}, t_{\text{ini}}) - \hat{\hat{U}}\right\| \leq \sum_{j=1}^{J}\left\|\hat{U}_j - \hat{\hat{U}}_j\right\| . \tag{8.238}
$$

For the terms $\left\| \hat{U}_j - \hat{U}_j \right\|$ on the right side of (8.238) we use the bounds given in (8.235) to arrive at

$$
\left\| \hat{U}(t_{\text{fin}}, t_{\text{ini}}) - \hat{U} \right\| \leq \sum_{j=1}^{J} \frac{1}{2} \left(\frac{T}{J} \right)^2 \| [H_{\text{ini}}, H_{\text{fin}}] \| \, e^{\frac{2MT}{J}}
$$
$$
= \frac{T^2}{2J} \| [H_{\text{ini}}, H_{\text{fin}}] \| \, e^{\frac{2MT}{J}} ,
$$

which completes the proof of (8.236). $\qquad\square$

Recall that $U(t_{\text{fin}}, t_{\text{ini}})$ as defined in (8.207) is the adiabatic time evolution which we want to approximate by a circuit. For the approximating circuit we choose $\hat{U} = \hat{U}_J \cdots \hat{U}_1$ comprised of the J gates $\hat{U}_j$. We can then combine (8.222) in Lemma 8.33 with (8.236) of Corollary 8.35 to obtain a bound on the quality of the approximation of $U(t_{\text{fin}}, t_{\text{ini}})$ by $\hat{U}$. This result will also enable us to state how J has to increase as a function of $T = t_{\text{fin}} - t_{\text{ini}}$ and $\|H_{\text{ini}}\|$ and $\|H_{\text{fin}}\|$ in order to achieve a desired precision of the approximation of $U(t_{\text{fin}}, t_{\text{ini}})$ by $\hat{U}$.

Theorem 8.36 *Let $U(t, t_{ini})$ be defined as in (8.207), and J, T and M as in Lemma 8.33. Moreover, let $\hat{U} = \hat{U}_J \cdots \hat{U}_1$ be the circuit comprised of the J gates $\hat{U}_j$ as given in (8.226). Then we have*

$$
\left\| U(t_{fin}, t_{ini}) - \hat{U} \right\| \leq 2\sqrt{\frac{MT}{J}} + \frac{(MT)^2}{J} e^{\frac{2MT}{J}} . \tag{8.239}
$$

In particular, if we choose $J \in \mathbb{N}$ such that

$$
J \geq \left(\frac{2e}{\delta} \right)^2 \max \left\{ 2MT, (MT)^2 \right\} , \tag{8.240}
$$

where $\delta \in \mathbb{R}$ is such that $0 < \delta < 1$, then we have

$$
\left\| U(t_{fin}, t_{ini}) - \hat{U} \right\| \leq \delta . \tag{8.241}
$$

Proof In order to show (8.239), we note that

$$\left\|\left| U(t_{\mathrm{fin}},t_{\mathrm{ini}}) - \hat{U} \right\|\right| \quad = \quad \left\|\left| U(t_{\mathrm{fin}},t_{\mathrm{ini}}) - \hat{U} + \hat{U} - \hat{U} \right\|\right|$$

$$\underbrace{\leq}_{(2.53)} \quad \left\|\left| U(t_{\mathrm{fin}},t_{\mathrm{ini}}) - \hat{U} \right\|\right| + \left\|\left| \hat{U} - \hat{U} \right\|\right|$$

$$\underbrace{\leq}_{(8.222),(8.236)} \quad 2\sqrt{\frac{MT}{J}} + \frac{T^2}{2J} \left\| [H_{\mathrm{fin}}, H_{\mathrm{ini}}] \right\| \, e^{\frac{2MT}{J}} .$$

Using that

$$\left\| [H_{\mathrm{fin}}, H_{\mathrm{ini}}] \right\| \underbrace{=}_{(2.46)} \left\| H_{\mathrm{fin}} H_{\mathrm{ini}} - H_{\mathrm{ini}} H_{\mathrm{fin}} \right\| \underbrace{\leq}_{(2.53)} \left\| H_{\mathrm{fin}} H_{\mathrm{ini}} \right\| + \left\| H_{\mathrm{ini}} H_{\mathrm{fin}} \right\|$$

$$\underbrace{\leq}_{(2.52)} 2 \left\| H_{\mathrm{fin}} \right\| \, \left\| H_{\mathrm{ini}} \right\| \underbrace{\leq}_{(8.220)} 2M^2$$

then completes the proof of (8.239).

Choosing J as in (8.240) implies

$$\left(\frac{\delta}{2e} \right)^2 \geq \max \left\{ \frac{2MT}{J}, \frac{(MT)^2}{J} \right\}$$

and since $0 < \delta < 1$ thus

$$\frac{2MT}{J} \leq \left(\frac{\delta}{2e} \right)^2 < \frac{\delta}{2e} < 1$$

$$\sqrt{\frac{MT}{J}} < \sqrt{\frac{2MT}{J}} \leq \frac{\delta}{2e} < 1 \qquad\qquad (8.242)$$

$$\frac{(MT)^2}{J} \leq \left(\frac{\delta}{2e} \right)^2 < \frac{\delta}{2e} < 1 .$$

Using (8.242), we can bound the right side of (8.239) as follows

$$\left\|\left| U(t_{\mathrm{fin}},t_{\mathrm{ini}}) - \hat{U} \right\|\right| \underbrace{\leq}_{(8.239)} 2\sqrt{\frac{MT}{J}} + \frac{(MT)^2}{J} e^{\frac{2MT}{J}}$$

$$\underbrace{\leq}_{(8.242)} 2\frac{\delta}{2e} + \frac{\delta}{2e} e = (2 + e) \frac{\delta}{2e}$$

$$< \delta ,$$

which completes the proof of (8.241).   □

Theorem 8.36 tells us that a given adiabatic evolution $U(T + t_{\text{ini}}, t_{\text{ini}})$ can be approximated up to an error of δ by a circuit $\hat{U}$ of J gates, where $J \in O\left(\left(\frac{T}{\delta}\right)^2\right)$. In other words, the number of gates that we need to replicate an adiabatic evolution $U(T + t_{\text{ini}}, t_{\text{ini}})$ up to a desired precision δ growths at most quadratically as a function of $\frac{T}{\delta}$. Therefore, an adiabatic computation can be efficiently replicated by a circuit based computation.

8.7 Further Reading

A rigorous proof of quantum adiabatic theorems with a view towards application in adiabatic quantum computation can be found in the paper by JANSEN et al. [111]. Their proofs apply to infinite dimensions as well and contain the optimization result for the schedule in the adiabatic GROVER search.

Combining quantum adiabatic computation with another field of much recent interest, PUDENZ and LIDAR [118] investigate using the adiabatic method in the context of *machine learning*.

Related to the adiabatic method is the idea of *quantum annealing*, which is described in more detail by DAS and SUZUKI in [119]. This is a method which derives from the classical simulated annealing methodologies. It makes use of quantum tunneling, which is not available in the classical annealing version.

For a most comprehensive, detailed and up to date overview of almost all aspects of adiabatic quantum computing we recommend the recent review by ALBASH and LIDAR [114]. This review also contains a summary of the history as well as an extensive list of further references pertaining to the various sub-branches of adiabatic quantum computing.

Chapter 9
Epilogue

The physical phenomena of quantum mechanics have been extensively investigated and experimentally verified long before the advent of quantum computing. Indeed, it is a perhaps justifiably widely held opinion, that quantum mechanics is the best tested physical theory ever. The example of the EPR paradox, however, shows that a certain discomfort about the apparently counter-intuitive character of quantum mechanics could be found among scientific prominence. Even BOHR has been quoted as saying 'he who is not shocked by quantum mechanics has not understood it.'

Perhaps nowadays it would be too strong to use the term discomfort or even shock, but the opinion that we do not really understand quantum mechanics was said to be shared by FEYNMAN, one of the founders of quantum computing. Such epistemological doubts are viewed by some as unnecessary distractions—a viewpoint often expressed by the slogan 'shut up and calculate.' And calculate they do! Despite it 'being impossible to understand' innumerable applications have emerged from it, and it is hard to overestimate the influence quantum mechanics has had on our daily lives.

It remains to be seen if quantum computing is yet another addition to the success story of quantum mechanics. The theory we have presented here is sound and promising, but much remains to be done before we will see a significant impact of this new paradigm of computing. Certainly, reliable physical realizations still pose a formidable challenge, although the race is on, and many big corporate players have joined it.

But also on the theoretical side work remains to be done. Although after the algorithms by SHOR and GROVER many more have been added such that there is now a well populated zoo [95], it is fair to say their generalizations, such as the hidden subgroup problem and amplitude amplification, make up the bulk of quantum algorithms. For many problems quantum algorithms with significant efficiency gains over hitherto existing classical algorithms have yet to be found. One obstacle for this may be that the practitioners attacking these problems are not aware of the possibilities a quantum computer can offer. Hopefully, this book can make a contribution to alleviate that.

© Springer Nature Switzerland AG 2019   499
W. Scherer, *Mathematics of Quantum Computing*,
https://doi.org/10.1007/978-3-030-12358-1_9

In the previous chapters we have tried to provide the basic mathematical concepts and elements for quantum computing. As we have seen, the theory of quantum computing rests on exploiting phenomena of quantum mechanics combined with suitably adapted results from various branches of mathematics. In fact, one of the appealing aspects of quantum computing is that it uses results from, inter alia, real, complex and functional analysis, linear algebra, group, number, complexity and probability theory and at the same time has at the heart of it physical phenomena that challenge our understanding of reality and reach into the realm of philosophy. This multitude of intellectual challenges, paired with the potential effect any successful implementation can have on our everyday lives, makes it an attractive field of study indeed.

It is the hope of the author that this book will entice the reader to take up the challenges and to contribute to make quantum computing another big chapter in the success story of quantum mechanics.

Appendix A
Elementary Probability Theory

The mathematical concept of probability builds on some basic measure-theoretic notions, which we exhibit first.

Definition A.1 Let Ω be a non-empty set and $A \subset \{B \mid B \subset \Omega\}$ a set of subsets of Ω. The set of subsets A is called σ**-Algebra** on Ω if

$$\Omega \in A$$
$$B \in A \Rightarrow \Omega \smallsetminus B \in A$$
$$\bigcup_{n \in \mathbb{N}} A_n \in A \qquad \text{for every sequence } (A_n)_{n \in \mathbb{N}} \text{ with } A_n \in A$$

holds. The pair (Ω, A) is called a **measurable space**. A **measure** on (Ω, A) is a map μ with the properties

$$\mu : A \to [0, \infty]$$
$$\mu(\emptyset) = 0$$
$$\mu \left(\bigcup_{n \in \mathbb{N}} A_n \right) = \sum_{n \in \mathbb{N}} \mu(A_n) \qquad \begin{array}{l} \text{for every sequence } (A_n)_{n \in \mathbb{N}} \text{ with } A_n \in A \\ \text{and } A_n \cap A_m = \emptyset \text{ for all } n \neq m. \end{array}$$

The triple (Ω, A, μ) is called **measure space**.

For $i \in \{1,2\}$ let (Ω_i, A_i) be two measurable spaces. A function $f : \Omega_1 \to \Omega_2$ is called **measurable** if for every $A \in A_2$ its pre-image $f^{-1}(A) := \{\omega \in \Omega_1 \mid f(\omega) \in A\}$ satisfies $f^{-1}(A) \in A_1$.

By definition every measure takes only non-negative values. If a measure is such that on the complete set Ω it takes the value 1 (and, consequently, takes only values in $[0, 1]$) it becomes a probability measure.

© Springer Nature Switzerland AG 2019
W. Scherer, *Mathematics of Quantum Computing*,
https://doi.org/10.1007/978-3-030-12358-1

Definition A.2 A measure $\mathbf{P}$ on a measurable space (Ω, A) is called **probability measure** if

$$\mathbf{P}(\Omega) = 1. \tag{A.1}$$

The triple $(\Omega, A, \mathbf{P})$ is called **probability space**.

A probability space enables us to define what is called a random variable.

Definition A.3 Let $(\Omega, A, \mathbf{P})$ be a probability space and (M, M) a measurable space. An M-valued **random variable** Z is a measurable map

$$Z : \Omega \to M.$$

The probability measure

$$\mathbf{P}_Z := \mathbf{P} \circ Z^{-1} : M \to [0, 1]$$

on (M, M) is called **probability distribution** (or simply distribution) of Z. If the image of Z is countable, that is, if there exists an index set $I \subset \mathbb{N}$ such that $Z\{\Omega\} = \{m_i \,|\, i \in I\}$, then Z is called a **discrete** random variable. In this case $\mathbf{P}_Z$ becomes a **discrete probability distribution**, which assigns a probability

$$\mathbf{P}_Z(\{m\}) = \mathbf{P} \circ Z^{-1}(\{m\})$$

to every $m \in Z\{\Omega\}$.

For every subset $S \subset M$ satisfying $S \in M$ the real number

$$\mathbf{P}_Z(S) = \mathbf{P}\big(\{\omega \,|\, Z(\omega) \in S\}\big)$$

is called the probability of the 'event' $Z(\omega) \in S$ taking place. To emphasize this, we will often write $\mathbf{P}\{Z \in S\}$ for $\mathbf{P}_Z(S)$ or, in the case of a discrete random variable, $\mathbf{P}\{Z = m\}$ for $\mathbf{P}_Z(\{m\})$.

Example A.4 Let $(\Omega, A, \mathbf{P})$ be a probability space and let $(\mathbb{R}^n, B)$ be the measure space, where B denotes the BOREL sets of $\mathbb{R}^n$. Then any measurable function $Z : \Omega \to \mathbb{R}^n$ constitutes an n-dimensional real-valued random variable. This means that to any BOREL set $B \in B$ we can assign a probability $\mathbf{P}\{\omega \,|\, Z(\omega) \in B\}$.

Definition A.5 Let Z be an n-dimensional real-valued random variable on a probability space (Ω, A, P). The **expectation value** of Z is defined as

$$E[Z] := \int_{\Omega} Z(\omega)dP(\omega) = \int_{\mathbb{R}^n} x\,dP_Z(x), \qquad (A.2)$$

where $dP_Z(x)$ is often (in particular when $n = 1$) written as $P\{Z \in [x, x + dx]\}$. In case Z is an integer-valued discrete random variable with $Z\{\Omega\} = \{x_i \mid i \in I\} \subset \mathbb{Z}$, the expectation value is given by

$$E[Z] := \sum_{i \in I} x_i P\{Z = x_i\}. \qquad (A.3)$$

As can be seen from the right side of (A.2) and (A.3), the knowledge of Ω, P and $Z(\omega)$ is not required to calculate the expectation value as long as one knows the respective probabilities $P\{Z \in [x, x + dx]\}$ or, in case of a discrete random variable, $P\{Z = x_i\}$.

Lemma A.6 *Let Z be an integer-valued random variable on a probability space (Ω, A, P) satisfying*

$$|Z(\omega)| \leq c \qquad (A.4)$$

for all $\omega \in \Omega$ and some non-negative $c \in \mathbb{R}$. Then we have $|E[Z]| \leq c$.

Proof

$$|E[Z]| \underbrace{=}_{(A.3)} \left| \sum_{i \in I} x_i P\{Z = x_i\} \right| \leq \sum_{i \in I} |x_i P\{Z = x_i\}| = \sum_{i \in I} |x_i| P\{Z = x_i\}$$

$$\underbrace{\leq}_{(A.4)} \sum_{i \in I} c P\{Z = x_i\} = c \sum_{i \in I} P\{Z = x_i\} \underbrace{=}_{(A.1)} c$$

□

If Z is an n-dimensional real-valued random variable, then for every BOREL measurable function $f : \mathbb{R}^n \to \mathbb{R}^m$ the function $f \circ Z : \Omega \to \mathbb{R}^m$ is also an m-dimensional real-valued random variable. Its expectation value is given by

$$E[f(Z)] = \int_{\Omega} f(Z(\omega))dP(\omega) = \int_{\mathbb{R}^n} f(x)dP_Z(x). \qquad (A.5)$$

For discrete random variables one has analogously

$$\mathbf{E}\left[f(Z)\right] = \sum_{i \in I} f(x_i)\mathbf{P}\left\{Z = x_i\right\}. \tag{A.6}$$

Definition A.7 Let Z_1 and Z_2 be one-dimensional real- or integer-valued random variables. Their **variance var**$[Z_i]$, **covariance cov**$[Z_1, Z_2]$ and **correlation cor**$[Z_1, Z_2]$ are defined as

$$\mathbf{var}[Z_i] := \mathbf{E}\left[(Z_i - \mathbf{E}\left[Z_i\right])^2\right]$$
$$\mathbf{cov}[Z_1, Z_2] := \mathbf{E}\left[(Z_1 - \mathbf{E}\left[Z_1\right])(Z_2 - \mathbf{E}\left[Z_2\right])\right]$$
$$\mathbf{cor}[Z_1, Z_2] := \frac{\mathbf{cov}[Z_1, Z_2]}{\sqrt{\mathbf{var}[Z_1]\,\mathbf{var}[Z_2]}}.$$

As can be seen from (A.5) and (A.6), the calculation of the expectation value of a function of several random variables $Z_1, Z_2, \ldots$ requires their *joint* distribution $\mathbf{P}\{Z_1 \in [x_1, x_1 + dx], Z_2 \in [x_2, x_2 + dx], \ldots\}$ or $\mathbf{P}\{Z_1 = x_1, Z_2 = x_2, \ldots\}$ for discrete random variables. This holds, in particular, for covariance and correlation of two random variables.

Appendix B
Elementary Arithmetic Operations

The following lemma formalizes the algorithm for the binary representation of the addition of two numbers. It is a binary version of the elementary textbook addition of numbers in decimal representation and is implemented with the quantum adder defined in Sect. 5.5.1. The functions $\left\lfloor \frac{a}{b} \right\rfloor$ and $a \bmod b$ used here are defined in Definition D.1. The binary sum $a \overset{2}{\oplus} b = (a+b) \bmod 2$ was introduced in Definition 5.2.

Lemma B.1 *Let $n \in \mathbb{N}$ and $a, b \in \mathbb{N}_0$ with $a, b < 2^n$ have the binary representations*

$$a = \sum_{j=0}^{n-1} a_j 2^j, \qquad b = \sum_{j=0}^{n-1} b_j 2^j,$$

where $a_j, b_j \in \{0, 1\}$. Furthermore, set $\hat{c}_0^+ := 0$ and

$$\hat{c}_j^+ := \left\lfloor \frac{a_{j-1} + b_{j-1} + \hat{c}_{j-1}^+}{2} \right\rfloor \qquad \text{for } j \in \{1, \ldots, n\} \tag{B.1}$$

$$s_j := a_j \overset{2}{\oplus} b_j \overset{2}{\oplus} \hat{c}_j^+ \qquad \text{for } j \in \{0, \ldots, n-1\}.$$

Then the sum of a and b is given as

$$a + b = \sum_{j=0}^{n-1} s_j 2^j + \hat{c}_n^+ 2^n.$$

Proof We prove the claim by induction in n. For the induction-start we consider first the case $n = 1$. Let $a, b \in \{0, 1\}$ such that $a = a_0$, $b = b_0$ and $a_1 = 0 = b_1$. Then we have

© Springer Nature Switzerland AG 2019
W. Scherer, *Mathematics of Quantum Computing*,
https://doi.org/10.1007/978-3-030-12358-1

$$
\begin{aligned}
a+b &= a_0 + b_0 \underbrace{=}_{(D.1)} (a_0+b_0) \bmod 2 + \left\lfloor \frac{a_0+b_0}{2} \right\rfloor 2 \\
&\underbrace{=}_{(5.2)} a_0 \overset{2}{\oplus} b_0 + \left\lfloor \frac{a_0+b_0}{2} \right\rfloor 2 \\
&= s_0 + \hat{c}_1^+ 2 .
\end{aligned}
$$

This proves the claim for $n = 1$.

For the inductive step from n to $n+1$ suppose that the claim holds for n, that is,

$$
a+b = \sum_{j=0}^{n-1} s_j 2^j + \hat{c}_n^+ 2^n . \tag{B.2}
$$

For $\tilde{a} = a + \tilde{a}_n 2^n$ and $\tilde{b} = a + \tilde{b}_n 2^n$ their binary representatives $\tilde{a}_j$ and $\tilde{b}_j$ satisfy for $j \in \{0,\dots,n-1\}$

$$
\tilde{a}_j = a_j , \qquad \tilde{b}_j = b_j \tag{B.3}
$$

and thus

$$
\begin{aligned}
\tilde{a} + \tilde{b} &= a + b + (\tilde{a}_n + \tilde{b}_n) 2^n \\
&\underbrace{=}_{(B.2)} \sum_{j=0}^{n-1} s_j 2^j + \left(\tilde{a}_n + \tilde{b}_n + \hat{c}_n^+ \right) 2^n \\
&\underbrace{=}_{(D.1)} \sum_{j=0}^{n-1} s_j 2^j + \left((\tilde{a}_n + \tilde{b}_n + \hat{c}_n^+) \bmod 2 + \left\lfloor \frac{\tilde{a}_n + \tilde{b}_n + \hat{c}_n^+}{2} \right\rfloor 2 \right) 2^n \\
&\underbrace{=}_{(5.2)} \sum_{j=0}^{n-1} s_j 2^j + \underbrace{(\tilde{a}_n \overset{2}{\oplus} \tilde{b}_n \overset{2}{\oplus} \hat{c}_n^+)}_{=s_n} 2^n + \underbrace{\left\lfloor \frac{\tilde{a}_n + \tilde{b}_n + \hat{c}_n^+}{2} \right\rfloor 2^{n+1}}_{=\hat{c}_{n+1}^+} \\
&= \sum_{j=0}^{n} s_j 2^j + \hat{c}_{n+1}^+ 2^{n+1} .
\end{aligned}
$$

Because of $\hat{c}_0^+ := 0$ and (B.3), we then have

$$
\hat{c}_j^+ = \left\lfloor \frac{\tilde{a}_{j-1} + \tilde{b}_{j-1} + c_{j-1}^+}{2} \right\rfloor \qquad \text{for } j \in \{1,\dots,n+1\}
$$

$$
s_j = \tilde{a}_j \overset{2}{\oplus} \tilde{b}_j \overset{2}{\oplus} c_j^+ \qquad \text{for } j \in \{0,\dots,n\} .
$$

This shows that the claim also holds true for $n+1$. $\square$

The carry terms $\hat{c}_j^+$ of the addition defined in (B.1) can be written in a form avoiding the explicit use of $\lfloor \ \rfloor$ and more suitable for direct implementation by a quantum adder.

Corollary B.2 (Binary Addition) *Let $n \in \mathbb{N}$ and $a, b \in \mathbb{N}_0$ with $a, b < 2^n$ have the binary representations*

$$a = \sum_{j=0}^{n-1} a_j 2^j, \qquad b = \sum_{j=0}^{n-1} b_j 2^j,$$

where $a_j, b_j \in \{0, 1\}$. Furthermore, set $c_0^+ := 0$ and

$$c_j^+ := a_{j-1} b_{j-1} \overset{2}{\oplus} a_{j-1} c_{j-1}^+ \overset{2}{\oplus} b_{j-1} c_{j-1}^+ \qquad \text{for } j \in \{1, \ldots, n\}$$

$$s_j := a_j \overset{2}{\oplus} b_j \overset{2}{\oplus} c_j^+ \qquad \text{for } j \in \{0, \ldots, n-1\}.$$

Then the sum of a and b can be written as

$$a + b = \sum_{j=0}^{n-1} s_j 2^j + c_n^+ 2^n.$$

Proof Because of Lemma B.1 it suffices for the proof of the claim to show that for $j \in \{0, \ldots, n\}$ we have $c_j^+ = \hat{c}_j^+$. We show this by induction in j. For $j = 0$ it holds by definition. For the inductive step from $j - 1$ to j suppose that $c_{j-1}^+ = \hat{c}_{j-1}^+$. It remains to show that then

$$c_j^+ = a_{j-1} b_{j-1} \overset{2}{\oplus} a_{j-1} c_{j-1}^+ \overset{2}{\oplus} b_{j-1} c_{j-1}^+ = \left\lfloor \frac{a_{j-1} + b_{j-1} + \hat{c}_{j-1}^+}{2} \right\rfloor = \hat{c}_j^+ \qquad (B.4)$$

holds.

The carry terms $\hat{c}_j^+$ can only assume the values 0 or 1, as we begin with $\hat{c}_0^+ = 0$ and then have successively $0 \leq \frac{a_{j-1} + b_{j-1} + \hat{c}_{j-1}^+}{2} \leq \frac{3}{2}$. The proof of (B.4) is given by simply evaluating all possible combinations of the left and right side and verifying equality. This is shown in Table B.1. □

The following lemma formalizes the algorithm for calculating the difference between two numbers in the binary representation.

Table B.1 Value table for the proof of (B.4)

a_{j-1}	b_{j-1}	$\hat{c}^+_{j-1}$	$c^+_j = a_{j-1}b_{j-1} \overset{2}{\oplus} a_{j-1}c^+_{j-1} \overset{2}{\oplus} b_{j-1}c^+_{j-1}$	$\hat{c}^+_j = \left\lfloor \frac{a_{j-1}+b_{j-1}+\hat{c}^+_{j-1}}{2} \right\rfloor$
0	0	0	0	0
0	0	1	0	0
0	1	0	0	0
0	1	1	1	1
1	0	0	0	0
1	0	1	1	1
1	1	0	1	1
1	1	1	1	1

Lemma B.3 *Let $n \in \mathbb{N}$ and $a,b \in \mathbb{N}_0$ with $a,b < 2^n$ have the binary representations as given in Lemma B.1. Furthermore, set $\hat{c}^-_0 := 0$ and*

$$\hat{c}^-_j := \left\lfloor \frac{b_{j-1}-a_{j-1}+\hat{c}^-_{j-1}}{2} \right\rfloor \qquad \text{for } j \in \{1,\ldots,n\}$$

$$\hat{d}_j := \left(b_j - a_j + \hat{c}^-_j\right) \bmod 2 \qquad \text{for } j \in \{0,\ldots,n-1\}.$$

Then the difference between b and a can be written as

$$b - a = \sum_{j=0}^{n-1} \hat{d}_j 2^j + \hat{c}^-_n 2^n. \tag{B.5}$$

Proof We also show this by induction in n. For the induction-start we consider at first the case $n = 1$. Let $a,b \in \{0,1\}$ such that $a = a_0, b = b_0$ and $a_1 = 0 = b_1$. Then we have

$$b - a = b_0 - a_0 = (b_0 - a_0)\bmod 2 + \left\lfloor \frac{b_0 - a_0}{2} \right\rfloor 2$$

$$= \hat{d}_0 + \hat{c}^-_1 2,$$

which proves the claim for $n = 1$.

For the inductive step from n to $n+1$ suppose that the claim holds for n, that is,

$$b - a = \sum_{j=0}^{n-1} \hat{d}_j 2^j + \hat{c}^-_n 2^n. \tag{B.6}$$

For $\tilde{a} = a + \tilde{a}_n 2^n$ and $\tilde{b} = b + \tilde{b}_n 2^n$ it follows for $j \in \{0, \dots, n-1\}$ that

$$\tilde{a}_j = a_j, \qquad \tilde{b}_j = b_j \tag{B.7}$$

and thus

$$
\begin{aligned}
\tilde{b} - \tilde{a} &= b - a + (\tilde{b}_n - \tilde{a}_n) 2^n \\
&\underset{(B.6)}{=} \sum_{j=0}^{n-1} \hat{d}_j 2^j + (\tilde{b}_n - \tilde{a}_n + \hat{c}_n^-) 2^n \\
&\underset{(D.1)}{=} \sum_{j=0}^{n-1} \hat{d}_j 2^j + \left((\tilde{b}_n - \tilde{a}_n + \hat{c}_n^-) \bmod 2 + \left\lfloor \frac{\tilde{b}_n - \tilde{a}_n + \hat{c}_n^-}{2} \right\rfloor 2 \right) 2^n \\
&= \sum_{j=0}^{n-1} \hat{d}_j 2^j + \Bigg(\underbrace{\big((\tilde{b}_n - \tilde{a}_n + \hat{c}_n^-) \bmod 2 \big)}_{=\hat{d}_n} 2^n + \underbrace{\left\lfloor \frac{\tilde{b}_n - \tilde{a}_n + \hat{c}_n^-}{2} \right\rfloor}_{=\hat{c}_{n+1}^-} \Bigg) 2^{n+1} \\
&= \sum_{j=0}^{n} \hat{d}_j 2^j + \hat{c}_{n+1}^- 2^{n+1},
\end{aligned}
$$

where we have $\hat{c}_0^- := 0$ and, because of (B.7), also

$$
\begin{aligned}
\hat{c}_j^- &= \left\lfloor \frac{\tilde{b}_{j-1} - \tilde{a}_{j-1} + \hat{c}_{j-1}^-}{2} \right\rfloor && \text{for } j \in \{1, \dots, n+1\} \\
\hat{d}_j &= (\tilde{b}_j - \tilde{a}_j + \hat{c}_j^-) \bmod 2 && \text{for } j \in \{0, \dots, n\}.
\end{aligned}
$$

This completes the proof of the claim for $n + 1$. □

In contrast to the addition, the carry terms $\hat{c}_j^-$ in the subtraction can become negative. Moreover, for two numbers $a, b < 2^n$ the highest carry $\hat{c}_n^-$ provides information whether $b \geq a$ or $b < a$, as Exercise B.116 shows.

Exercise B.116 Let $n \in \mathbb{N}$ and $a, b \in \mathbb{N}_0$ with $a, b < 2^n$ and $\hat{c}_j^-$ and $\hat{d}_j$ be defined as in Lemma B.3. Then the following hold:

(i)

$$\hat{c}_j^- \in \{0, -1\}$$

and thus $\hat{c}_j^- = -\left| \hat{c}_j^- \right|$.

(ii) In particular, we have that

$$\hat{c}_n^- = \begin{cases} 0 \Leftrightarrow b \geq a \\ -1 \Leftrightarrow b < a. \end{cases} \tag{B.8}$$

For a solution see Solution B.116

The subtraction algorithm can be formulated without negative carries. In preparation for this we first prove the following lemma.

Lemma B.4 *Let $n \in \mathbb{N}$ and $a, b \in \mathbb{N}_0$ as well as $\hat{c}_j^-$ and $\hat{d}_j$ as in Lemma B.3. Then the following holds*

$$\left|\hat{c}_j^-\right| = \left(1 \overset{2}{\oplus} b_{j-1}\right)\left(a_{j-1} \overset{2}{\oplus} \left|\hat{c}_{j-1}^-\right|\right) \overset{2}{\oplus} a_{j-1}\left|\hat{c}_{j-1}^-\right| \in \{0,1\} \tag{B.9}$$

$$\hat{d}_j = a_j \overset{2}{\oplus} b_j \overset{2}{\oplus} \left|\hat{c}_j^-\right|. \tag{B.10}$$

Proof The simplest way to prove both (B.9) and (B.10) simultaneously is by direct evaluation as shown in Table B.2. □

With the result of Lemma B.4 we can formulate the subtraction algorithm in a form which is akin to the inverse quantum adder.

Corollary B.5 (Binary Subtraction) *Let $n \in \mathbb{N}$ and $a, b \in \mathbb{N}_0$ with their binary representations as in Lemma B.1. Furthermore, set $c_0^- := 0$ as well as*

Table B.2 Value table to prove (B.9) and (B.10)

			Proof of (B.9)		Proof of (B.10)	
a_{j-1}	b_{j-1}	$\hat{c}_{j-1}^-$	$\left\|\hat{c}_j^-\right\|$	$(1 \overset{2}{\oplus} b_{j-1})(a_{j-1} \overset{2}{\oplus} \left\|\hat{c}_{j-1}^-\right\|) \overset{2}{\oplus} a_{j-1}\left\|\hat{c}_{j-1}^-\right\|$	$\hat{d}_j$	$a_j \overset{2}{\oplus} b_j \overset{2}{\oplus} \left\|\hat{c}_j^-\right\|$
0	0	0	0	0	0	0
0	0	−1	1	1	1	1
0	1	0	0	0	1	1
0	1	−1	0	0	0	0
1	0	0	1	1	1	1
1	0	−1	1	1	0	0
1	1	0	0	0	0	0
1	1	−1	1	1	1	1

$$c_j^- := \left(1 \overset{2}{\oplus} b_{j-1}\right)\left(a_{j-1} \overset{2}{\oplus} c_{j-1}^-\right) \overset{2}{\oplus} a_{j-1}c_{j-1}^- \quad \text{for } j \in \{1,\dots,n\} \tag{B.11}$$

$$d_j := a_j \overset{2}{\oplus} b_j \overset{2}{\oplus} c_j^- \quad \text{for } j \in \{0,\dots,n-1\}. \tag{B.12}$$

Then we have

$$\sum_{j=0}^{n-1} d_j 2^j = c_n^- 2^n + b - a$$

and

$$c_n^- = \begin{cases} 0 \Leftrightarrow b \geq a \\ 1 \Leftrightarrow b < a. \end{cases} \tag{B.13}$$

Proof From (B.9) and (B.11) we see that $\left|\hat{c}_j^-\right|$ and c_j^- satisfy the same recursion formula. Since both start with the value $\left|\hat{c}_0^-\right| = c_0^- = 0$, it follows that for all $j \in \{1,\dots,n\}$ then

$$c_j^- = \left|\hat{c}_j^-\right| = -\hat{c}_j^- \tag{B.14}$$

holds. From this and from (B.10) together with (B.12), it thus also follows that for all $j \in \{1,\dots,n\}$ then $d_j = \hat{d}_j$ holds. Together with (B.5) and (B.14), this implies

$$b - a = \sum_{j=0}^{n-1} d_j 2^j - c_n^- 2^n.$$

Applying (B.14) once more to (B.8) finally yields (B.13).   □

Appendix C
LANDAU Symbols

The input to many algorithms often contains a natural number N, the increase of which leads to an increase of the computational effort in the algorithm. For example, in the SHOR algorithm N is the number we want to factorize, whereas in the GROVER search algorithm N denotes the cardinality of the set in which we search.

The growth rate of the number of computational steps in the algorithms is usually classified as a function of N with the help of LANDAU symbols. For these we use here the following definitions.

Definition C.1 For functions $f, g : \mathbb{N} \to \mathbb{N}$ and in the limit $N \to \infty$ the **little LANDAU symbol** $o(\cdot)$ is defined as

$$f(N) \in o(g(N)) \text{ for } N \to \infty$$
$$:\Leftrightarrow \forall \varepsilon \in \mathbb{R}_+ \; \exists M \in \mathbb{N} : \; \forall N > M \quad |f(N)| \leq \varepsilon |g(N)| \,,$$

and the **big Landau symbol** $O(\cdot)$ is defined as

$$f(N) \in O(g(N)) \text{ for } N \to \infty$$
$$:\Leftrightarrow \exists C \in \mathbb{R} \text{ and } M \in \mathbb{N} : \; \forall N > M \quad |f(N)| \leq C |g(N)| \,.$$

We say f is of **polynomial order** (or of order $\mathrm{poly}\,(N)$) and write this as

$$f(N) \in \mathrm{poly}\,(N)$$

if

$$f(N) \in O\left(\sum_{j=0}^{k} a_j N^j \right)$$

for a finite $k \in \mathbb{N}_0$ and some $a_j \in \mathbb{R}$.

© Springer Nature Switzerland AG 2019
W. Scherer, *Mathematics of Quantum Computing*,
https://doi.org/10.1007/978-3-030-12358-1

Apart from those given above, slightly modified or generalized definitions of these symbols can be found in the literature, but the above is suitable and sufficient for our purposes.

Example C.2 By application of the L'HOSPITAL rule one can easily show that

$$\lim_{N \to \infty} \left| \frac{\ln N}{N^{\frac{1}{m}}} \right| = 0 = \lim_{N \to \infty} \left| \frac{N^m}{\exp(N)} \right| \qquad \forall m \in \mathbb{N}.$$

This implies for all $m \in \mathbb{N}$:

$$\ln N = o\left(N^{\frac{1}{m}}\right) \quad \text{and} \quad N^m = o(\exp(N)).$$

Exercise C.117 Let $f_i(N) \in O(g_i(N))$ for $i \in \{1,2\}$ and $N \to \infty$. Show that then for $N \to \infty$

(i)
$$f_1(N) + f_2(N) \in O(|g_1(N)| + |g_2(N)|). \tag{C.1}$$

(ii)
$$f_1(N)f_2(N) \in O(g_1(N)g_2(N)). \tag{C.2}$$

(iii) If there exists an $M \in \mathbb{N}$ such that for all $N > M$ we have $|g_1(N)| < |g_2(N)|$, then it follows that
$$f_1(N) + f_2(N) \in O(g_2(N)). \tag{C.3}$$

For a solution see Solution C.117

Appendix D
Modular Arithmetic

Definition D.1 The **integer part** of a real number u is denoted by $\lfloor u \rfloor$ and defined as

$$\lfloor u \rfloor := \max\{z \in \mathbb{Z} \mid z \leq u\}.$$

Likewise, we define

$$\lceil u \rceil := \min\{z \in \mathbb{Z} \mid z \geq u\}.$$

For $a \in \mathbb{Z}$ the **remainder of a after division by** $N \in \mathbb{N}$ is denoted by $a \bmod N$ and is defined as

$$a \bmod N := a - \left\lfloor \frac{a}{N} \right\rfloor N. \tag{D.1}$$

An immediate consequence of (D.1) is that for $a \in \mathbb{Z}$ and $N \in \mathbb{N}$

$$a \bmod N = 0 \quad \Leftrightarrow \quad \exists z \in \mathbb{Z}: \; a = zN,$$

that is, $a \bmod N = 0$ if and only if N divides a.

Exercise D.118 Show that for $a, b \in \mathbb{Z}$ and $N \in \mathbb{N}$

$$a \bmod N = b \bmod N \quad \Leftrightarrow \quad (a-b) \bmod N = 0. \tag{D.2}$$

For a solution see Solution D.118

© Springer Nature Switzerland AG 2019
W. Scherer, *Mathematics of Quantum Computing*,
https://doi.org/10.1007/978-3-030-12358-1

The remainder $a \bmod N$ is bounded from above by $\frac{a}{2}$ and N as is to be shown in Exercise D.119.

Exercise D.119 Let $a, N \in \mathbb{N}$ with $a > N$. Show that then

$$a \bmod N < \min\{\frac{a}{2}, N\}. \tag{D.3}$$

For a solution see Solution D.119

Lemma D.2 *For $a, N \to \infty$ the number of computational steps required for the calculation of $a \bmod N$ scales as*

$$\begin{array}{l} \text{Number of computational steps} \\ \text{required to calculate } a \bmod N \end{array} \in O\big((\log_2 \max\{a, N\})^2\big). \tag{D.4}$$

Proof Because of (D.1), the computational effort to calculate $a \bmod N$ is given by

the number of operations required for dividing a by N and determining $\lfloor \frac{a}{N} \rfloor$, which is of order $O\big((\log_2 \max\{a, N\})^2\big)$
+ the number of operations for the multiplication $\lfloor \frac{a}{N} \rfloor$ with N, which is of order $O\big((\log_2 \max\{a, N\})^2\big)$
+ the number of operations for subtracting $\lfloor \frac{a}{N} \rfloor N$ from a, which is of order $O(\log_2 \max\{a, N\})$.

Using (C.3), it follows that

Number of operations to calculate $a \bmod N \in O\big((\log_2 \max\{a, N\})^2\big)$.

$\square$

Definition D.3 If, for $a, b \in \mathbb{Z}$, there is a $z \in \mathbb{Z}$, such that $b = za$, then a **divides** b (or one also says b is **divisible** by a). If no such z exists, a does not divide b (resp. b is not divisible by a). These two exclusive cases are described with the following notations:

$$\begin{array}{lll} a \mid b & :\Leftrightarrow & \exists z \in \mathbb{Z}: \ b = za \\ a \nmid b & :\Leftrightarrow & \nexists z \in \mathbb{Z}: \ b = za. \end{array}$$

For integers $a_i \in \mathbb{Z}$ with $i \in \{1,\ldots,n\}$ and $\sum_{i=1}^{n} |a_i| \neq 0$ we define the **greatest common divisor** as

$$\gcd(a_1,\ldots,a_n) := \max\{k \in \mathbb{Z} \mid k \mid a_i \quad \forall a_i\}.$$

In case $\prod_{i=1}^{n} |a_i| \neq 0$, we define the **smallest common multiple** as

$$\mathrm{scm}(a_1,\ldots,a_n) := \min\{k \in \mathbb{N} \mid a_i \mid k \quad \forall a_i\}.$$

For $a \neq 0$ one defines $\gcd(a,0) = a$. If a and b have no common divisor other than 1, that is, if

$$\gcd(a,b) = 1,$$

one calls a and b **coprime**.

The extended EUCLID algorithm determines the greatest common divisor $\gcd(a,b)$ of two numbers $a,b \in \mathbb{N}$ and a solution $x,y \in \mathbb{Z}$ of

$$ax + by = \gcd(a,b)$$

in the following way.

Theorem D.4 (Extended EUCLID algorithm) *Let $a,b \in \mathbb{N}$. Define*

$$
\begin{aligned}
r_{-1} &:= \max\{a,b\} \text{ and } r_0 := \min\{a,b\} \\
s_{-1} &:= 1 \qquad\qquad \text{ and } s_0 := 0 \\
t_{-1} &:= 0 \qquad\qquad \text{ and } t_0 := 1
\end{aligned}
\tag{D.5}
$$

and for every $j \in \mathbb{N}$ with $r_{j-1} > 0$

$$r_j := r_{j-2} \bmod r_{j-1} \tag{D.6}$$

$$s_j := s_{j-2} - \left\lfloor \frac{r_{j-2}}{r_{j-1}} \right\rfloor s_{j-1} \tag{D.7}$$

$$t_j := t_{j-2} - \left\lfloor \frac{r_{j-2}}{r_{j-1}} \right\rfloor t_{j-1}. \tag{D.8}$$

Then $r_j < r_{j-1}$ holds and there is an $n \in \mathbb{N}$ after which the sequence terminates, that is, for which

$$r_{n+1} = 0. \tag{D.9}$$

Furthermore, one has

$$r_n = \gcd(a,b) \tag{D.10}$$
$$n \le 2\min\{\log_2 a, \log_2 b\} + 1 \tag{D.11}$$
$$r_{-1}s_n + r_0 t_n = \gcd(a,b). \tag{D.12}$$

Proof Since we know from (D.3) that $u \bmod v < v$, it follows from the definition (D.6) of the r_j that $0 \le r_j < r_{j-1}$, that is, the r_j are strictly decreasing with increasing j. Hence, there must exist an $n \in \mathbb{N}$ with $n \le \min\{a,b\}, r_n > 0$ and $r_{n+1} = 0$. This proves (D.9).

In order to show (D.10), we first prove by descending induction that for all $j \in \{0,\dots,n+1\}$ a $z_{n-j} \in \mathbb{N}$ exists such that

$$r_{n-j} = z_{n-j} r_n. \tag{D.13}$$

For the induction-start let $n \in \mathbb{N}$ be such that

$$r_n > 0 \text{ but } r_{n+1} = 0.$$

It follows that

$$0 = r_{n+1} \underbrace{=}_{(D.6)} r_{n-1} \bmod r_n = r_{n-1} - \left\lfloor \frac{r_{n-1}}{r_n} \right\rfloor r_n.$$

Consequently, there exists a $z_{n-1} := \left\lfloor \frac{r_{n-1}}{r_n} \right\rfloor \in \mathbb{N}$ that satisfies

$$r_{n-1} = z_{n-1} r_n.$$

Furthermore, one has per definition (D.6) that $r_n = r_{n-2} \bmod r_{n-1}$ and thus

$$r_{n-2} = r_n + \left\lfloor \frac{r_{n-2}}{r_{n-1}} \right\rfloor r_{n-1} = \Big(\underbrace{1}_{=:z_n} + \left\lfloor \frac{r_{n-2}}{r_{n-1}} \right\rfloor z_{n-1} \Big) r_n = z_{n-2} r_n$$

for a $z_{n-2} \in \mathbb{N}$. This proves (D.13) for $j \in \{1,2\}$, and the start of the descending induction is established. Next, we turn to the inductive step. We will show that if there exist $z_{n-(j-1)}, z_{n-j} \in \mathbb{N}$, such that

$$r_{n-(j-1)} = z_{n-(j-1)} r_n \tag{D.14}$$
$$r_{n-j} = z_{n-j} r_n, \tag{D.15}$$

then there exists a $z_{n-(j+1)} \in \mathbb{N}$ satisfying

$$r_{n-(j+1)} = z_{n-(j+1)} r_n.$$

From the definition (D.6) of $r_{n-(j-1)}$ and the assumptions (D.14) and (D.15) it follows that

$$r_{n-(j+1)} = r_{n-(j-1)} + \left\lfloor \frac{r_{n-(j+1)}}{r_{n-j}} \right\rfloor r_{n-j} = \underbrace{\left(z_{n-(j-1)} + \left\lfloor \frac{r_{n-(j+1)}}{r_{n-j}} \right\rfloor z_{n-j} \right)}_{:=z_{n-(j+1)} \in \mathbb{N}} r_n$$

$$= z_{n-(j+1)} r_{n-1}.$$

This completes the inductive proof of (D.13). Hence, there exist $z_0, z_{-1} \in \mathbb{N}$, such that

$$\min\{a,b\} = r_0 = z_0 r_n$$
$$\max\{a,b\} = r_{-1} = z_{-1} r_n,$$

and r_n is a common divisor of a and b.

To show that r_n is the greatest such divisor, suppose g is also a common divisor of a and b. Then define $\tilde{a} := \frac{a}{g} \in \mathbb{N}$ and $\tilde{b} := \frac{b}{g} \in \mathbb{N}$. Applying the algorithm to $\tilde{a}, \tilde{b}$ generates $\tilde{r}_j := \frac{r_j}{g}$ and thus $\tilde{r}_n = \frac{r_n}{g} \in \mathbb{N}$, in other words, any common divisor of a and b also divides r_n. Consequently, r_n is the greatest common divisor of a and b. This completes the proof of (D.10).

To prove (D.11), we use that, because of $r_j < r_{j-1}$, we can apply the estimate (D.3) from Exercise D.119 to definition (D.6) of r_j. This implies

$$r_j < \min\{\frac{r_{j-2}}{2}, r_{j-1}\}. \tag{D.16}$$

Repeated application of this yields

$$r_{2k-1} < \frac{r_{2k-3}}{2} < \cdots < \frac{r_{-1}}{2^k} = \frac{\max\{a,b\}}{2^k}$$
$$r_{2k} < \frac{r_{2k-2}}{2} < \cdots < \frac{r_0}{2^k} = \frac{\min\{a,b\}}{2^k}.$$

Because of $r_{2k+1} < r_{2k} < r_{2k-1}$, we then have

$$r_{2k+1} < r_{2k} < \min\{\frac{a}{2^k}, \frac{b}{2^k}\}$$

and thus

$$r_j < \frac{\min\{a,b\}}{2^{\lfloor \frac{j}{2} \rfloor}}.$$

Consequently,

$$\left\lfloor \frac{j}{2} \right\rfloor \geq \min\{\log_2 a, \log_2 b\} \quad \Rightarrow \quad r_j = 0.$$

Since n per definition in (D.9) is the biggest number that still satisfies $r_n > 0$, it follows that

$$\left\lfloor \frac{n}{2} \right\rfloor < \min\{\log_2 a, \log_2 b\}.$$

This in turn implies

$$n < 2\min\{\log_2 a, \log_2 b\} + 1. \tag{D.17}$$

In order to prove (D.12), we show, again by a two-step induction, that

$$r_{-1}s_j + r_0 t_j = r_j. \tag{D.18}$$

Induction start is given by $j \in \{-1, 0\}$ because the defining equations (D.6)–(D.8) imply

$$r_{-1}s_{-1} + r_0 t_{-1} = r_{-1}$$
$$r_{-1}s_0 + r_0 t_0 = r_0.$$

In order to verify the induction-step from j to $j+1$, we suppose that (D.18) holds for j and $j-1$. Then it follows that

$$
\begin{aligned}
r_{-1}s_{j+1} + r_0 t_{j+1} &\underset{(D.7),(D.8)}{=} r_{-1}\left(s_{j-1} - \left\lfloor \frac{r_{j-1}}{r_j} \right\rfloor s_j\right) + r_0\left(t_{j-1} - \left\lfloor \frac{r_{j-1}}{r_j} \right\rfloor t_j\right) \\
&= \underbrace{r_{-1}s_{j-1} + r_0 t_{j-1}}_{=r_{j-1}} - \left\lfloor \frac{r_{j-1}}{r_j} \right\rfloor \underbrace{(r_{-1}s_j + r_0 t_j)}_{=r_j} = r_{j-1} - \left\lfloor \frac{r_{j-1}}{r_j} \right\rfloor r_j \\
&\underset{(D.6)}{=} r_{j+1}.
\end{aligned}
$$

This proves (D.18), and the claim (D.12) follows from the case $j = n$.   □

Example D.5 Table D.1 shows the values obtained in running the extended EUCLID algorithm for $a = 999$ and $b = 351$, yielding $n = 3$ and $\gcd(999, 351) = r_3 = 27$. For $a = 999$ and $b = 352$ the numbers of the algorithm are shown in Table D.2. In this case we obtain $n = 6$ and $\gcd(999, 352) = 1$.

Table D.1 The extended EUCLID algorithm for $a = 999$ and $b = 351$

j	r_j	s_j	t_j	$as_j + bt_j$
-1	$a = 999$	1	0	999
0	$b = 351$	0	1	351
1	$999 \bmod 351 = 297$	1	-2	297
2	$351 \bmod 297 = 54$	-1	3	54
3	$297 \bmod 54 = 27$	6	-17	27
4	$54 \bmod 27 = 0$	-13	37	0

Table D.2 The extended EUCLID algorithm for $a = 999$ and $b = 352$

j	r_j	s_j	t_j	$as_j + bt_j$
-1	$a = 999$	1	0	999
0	$b = 352$	0	1	352
1	$999 \bmod 352 = 295$	1	-2	295
2	$352 \bmod 295 = 57$	-1	3	57
3	$295 \bmod 57 = 10$	6	-17	10
4	$57 \bmod 10 = 7$	-31	88	7
5	$10 \bmod 7 = 3$	37	-105	3
6	$7 \bmod 3 = 1$	-105	-298	1
7	$3 \bmod 1 = 0$	352	-999	0

Lemma D.6 *For $a, b \to \infty$ the number of computational steps required for the calculation of $\gcd(a,b)$ scales as*

$$\text{Number of computational steps required for } \gcd(a,b) \quad \in O\big((\log_2 \min\{a,b\})^3\big). \tag{D.19}$$

Proof From Theorem D.4 we see that for the calculation of $\gcd(a,b)$ with the EUCLID algorithm we need to compute expressions of the form $u \bmod v$, starting with $a \bmod b$ as shown in (D.6). The number of computational steps for each calculation of $a \bmod b$ grows according to Lemma D.2 as $O\big((\log_2 \min\{a,b\})^2\big)$ for $a, b \to \infty$. From (D.11) we infer that in the EUCLID algorithm the number of times we have to calculate expressions of the form $u \bmod v$ grows with $a, b \to \infty$ as $O(\log_2 \min\{a,b\})$. The total effort for the calculation of $\gcd(a,b)$ is thus given by (D.19). $\qquad\square$

Exercise D.120 Show that for $u, v, u_j \in \mathbb{Z}$ and $k, a, N \in \mathbb{N}$ the following hold

(i)

$$u(v \bmod N) \bmod N = uv \bmod N \tag{D.20}$$

(ii)

$$\left(\prod_{j=1}^{k} (u_j \bmod N) \right) \bmod N = \left(\prod_{j=1}^{k} u_j \right) \bmod N \tag{D.21}$$

(iii)

$$(u^a \bmod N)^k \bmod N = u^{ak} \bmod N \tag{D.22}$$

(iv)

$$\left(\sum_{j=1}^{k} (u_j \bmod N) \right) \bmod N = \left(\sum_{j=1}^{k} u_j \right) \bmod N. \tag{D.23}$$

For a solution see Solution D.120

Next, we show the following useful lemma.

Lemma D.7 *Let $a, b, c \in \mathbb{Z}$ and $N \in \mathbb{N}$ with $c \neq 0$ and $\gcd(N, c) = 1$. Then we have*

$$a \bmod N = b \bmod N \qquad \Leftrightarrow \qquad ac \bmod N = bc \bmod N. \tag{D.24}$$

Proof We show $\Rightarrow$ first: By definition, we have

$$a \bmod N = b \bmod N$$

$$\underset{(\text{D.1})}{\Leftrightarrow} \quad a - \left\lfloor \frac{a}{N} \right\rfloor N = b - \left\lfloor \frac{b}{N} \right\rfloor N$$

$$\Leftrightarrow \quad ac = bc + \left(\left\lfloor \frac{a}{N} \right\rfloor - \left\lfloor \frac{b}{N} \right\rfloor \right) Nc \tag{D.25}$$

such that

$$ac \bmod N \underbrace{=}_{(D.1)} ac - \left\lfloor \frac{ac}{N} \right\rfloor N$$

$$\underbrace{=}_{(D.25)} bc + \left(\left\lfloor \frac{a}{N} \right\rfloor - \left\lfloor \frac{b}{N} \right\rfloor \right) Nc - \left\lfloor \frac{bc + \left(\lfloor \frac{a}{N} \rfloor - \lfloor \frac{b}{N} \rfloor \right) Nc}{N} \right\rfloor N$$

$$= bc - \left\lfloor \frac{bc}{N} \right\rfloor N \underbrace{=}_{(D.1)} bc \bmod N.$$

Now, as for $\Leftarrow$: Let $ac \bmod N = bc \bmod N$. Then there exists a $z \in \mathbb{Z}$ such that

$$a - b = \frac{z}{c} N \in \mathbb{Z}.$$

Since c and N are coprime, we must have $\frac{z}{c} \in \mathbb{Z}$. Hence, $(a-b) \bmod N = 0$, and it follows from (D.2) that $a \bmod N = b \bmod N$. $\qquad \Box$

The equivalence (D.24) in Lemma D.7 suggests that there is something akin to a multiplicative inverse. This is indeed the case, and it is defined as follows.

Definition D.8 Let $b, N \in \mathbb{N}$ with $\gcd(b,N) = 1$. The **multiplicative inverse modulo** N of b is denoted by $b^{-1} \bmod N$ and is defined as the number $x \in \{1, \ldots, N-1\}$ that satisfies

$$bx \bmod N = 1.$$

The multiplicative inverse is unique and can easily be determined with the extended EUCLID algorithm.

Lemma D.9 *Let $b, N \in \mathbb{N}$ with $\gcd(b,N) = 1$ and let $x, y \in \mathbb{Z}$ be a solution of*

$$bx + Ny = 1. \tag{D.26}$$

Then $x \bmod N$ is the uniquely determined multiplicative inverse of b modulo N, that is, it satisfies

$$\left(b(x \bmod N) \right) \bmod N = 1. \tag{D.27}$$

Proof We first show uniqueness. Let u and v be two multiplicative inverses for b modulo N, that is, suppose $u, v \in \{1, \ldots, N-1\}$ are such that $bu \bmod N = 1 = bv \bmod N$. Because of $\gcd(b,N) = 1$ and (D.24), then $bu \bmod N = bv \bmod N$ implies that $u \bmod N = v \bmod N$. From (D.2) we know that then N divides $u - v$. Because of $0 < u, v < N$, it thus follows that $u = v$.

As to the existence, we can apply the extended EUCLID algorithm in Theorem D.4 to b and N and use that $\gcd(b, N) = 1$ in (D.12) to find x and y satisfying (D.26). Then we have

$$b(x \bmod N) \bmod N \underbrace{=}_{(D.20)} bx \bmod N \underbrace{=}_{(D.26)} (1 - Ny) \bmod N = 1,$$

and since $0 < x \bmod N < N$, it follows that $x \bmod N$ satisfies all the defining properties of the multiplicative inverse of b modulo N. $\qquad\square$

Example D.10 Consider the extended EUCLID algorithm in Example D.5 for the case $b = 999$ and $N = 352$. We see from Table D.2 that $x = -105$ and $y = 298$ satisfy $bx + Ny = 1$. Hence, $x \bmod N = -105 \bmod 352 = 247$, and we have $b(x \bmod N) \bmod N = 999 \times 247 \bmod 352 = 1$, that is, 247 is the multiplicative inverse of 999 modulo 352.

The following lemma is also helpful in the context of factorization.

Lemma D.11 *Any $a, b, N \in \mathbb{N}$ with $N > 1$ satisfy*

$$ab \bmod N = 0 \quad \Rightarrow \quad \gcd(a, N) \gcd(b, N) > 1.$$

In particular, if N is prime, then

$$ab \bmod N = 0 \quad \Leftrightarrow \quad a \bmod N = 0 \text{ or } b \bmod N = 0 \qquad (D.28)$$

holds.

Proof Let $ab \bmod N = 0$. Then there exists a $q \in \mathbb{N}$ such that $ab = qN$. From the prime decomposition of this equation

$$\overbrace{p_1^{\alpha_1} \cdots p_s^{\alpha_s}}^{=a} \overbrace{p_1^{\beta_1} \cdots p_r^{\beta_r}}^{=b} = \overbrace{p_1^{\kappa_1} \cdots p_v^{\kappa_v}}^{=q} \overbrace{p_1^{\nu_1} \cdots p_u^{\nu_u}}^{=N}$$

one sees that the prime factors of N have to be contained in those of a or b and thus that a or b must have common divisors with N, that is, $\gcd(a, N) > 1$ or $\gcd(b, N) > 1$.

If N is prime and $N\,|\,ab$ holds, then N must be contained in a or b as a prime factor. Conversely, $a\bmod N = 0$ implies $N\,|\,a$ and $b\bmod N = 0$ implies $N\,|\,b$. Either case has $N\,|\,ab$ as a consequence. $\qquad\square$

Definition D.12 The EULER function ϕ is defined as

$$\phi : \mathbb{N} \longrightarrow \mathbb{N}$$
$$n \longmapsto \phi(n) := \Big|\{r \in \{1,\dots,n-1\}\,|\, \gcd(r,n) = 1\}\Big|, \qquad (D.29)$$

that is, $\phi(n)$ is the number of all $r \in \mathbb{N}$ with $1 \leq r < n$ that have no common divisor (are coprime) with n.

Example D.13 For $n = 10$ we have

$$\gcd(1,10) = \gcd(3,10) = \gcd(7,10) = \gcd(9,10) = 1$$

as well as

$$\gcd(2,10),\gcd(4,10),\gcd(5,10),\gcd(6,10),\gcd(8,10) > 1$$

and thus $\phi(10) = 4$.

Generally, it is quite difficult to compute the EULER function. For prime powers, however, it is very easy as shown in the following lemma.

Lemma D.14 *For p prime and $k \in \mathbb{N}$ one has*

$$\phi(p^k) = p^{k-1}(p-1). \qquad (D.30)$$

Proof In the set of the $p^k - 1$ numbers $1,\dots,p^k - 1$ the $p^{k-1} - 1$ multiples $1p,2p,\dots,(p^{k-1}-1)p$ of p are the only numbers that have a non-trivial common divisor with p^k. Consequently, the number $\phi(p^k)$ of those which do not have a common divisor with p^k is given by $\phi(p^k) = p^k - 1 - (p^{k-1}-1) = p^{k-1}(p-1)$. $\qquad\square$

For numbers $N = pq$ that have only two simple prime factors $p,q \in$ Pri, that is, so-called **half-primes**, the knowledge of $\phi(N)$ is equivalent to the knowledge of the prime factors p and q as the following lemma shows.

Lemma D.15 *Let p and q be primes such that $p > q$ and let $N = pq$. Then we have*

$$\phi(N) = (p-1)(q-1)$$

and with

$$S := N + 1 - \phi(N) \tag{D.31}$$

$$D := \sqrt{S^2 - 4N} > 0, \tag{D.32}$$

furthermore,

$$p = \frac{S+D}{2} \tag{D.33}$$

$$q = \frac{S-D}{2}. \tag{D.34}$$

Proof Since p and q are different primes, we find that among the $N - 1$ natural numbers smaller than $N = pq$ only the numbers $1 \times q, 2 \times q, \ldots, (p-1) \times q$ and $1 \times p, 2 \times p, \ldots, (q-1) \times p$ have a common divisor with N. Hence, we have

$$\phi(N) = N - 1 - (p-1) - (q-1) = pq - (p+q) + 1 = (p-1)(q-1).$$

Together with (D.31) and (D.32), this implies

$$S = p + q$$
$$D = p - q,$$

and (D.33) as well as (D.34) follow immediately.   □

Example D.16 In Example D.13 we found for $N = 10$ that $\phi(10) = 4$. Using this in (D.31) and (D.32) yields $S = 7$ and $D = 3$, which in turn gives $p = 5$ and $q = 2$.

The following theorem by EULER is useful for the decryption in the RSA public key encryption method as well as in connection with the prime factorization in Sect. 6.5.2.

Theorem D.17 *(EULER) Any coprime $b, N \in \mathbb{N}$ satisfy*

$$b^{\phi(N)} \bmod N = 1. \tag{D.35}$$

Proof First, we define $a_j := r_j b \bmod N$ for all $r_j \in \mathbb{N}$ with $1 \leq r_j < N$ and $\gcd(r_j, N) = 1$ and set

$$P := \left(\prod_{j=1}^{\phi(N)} a_j \right) \bmod N.$$

From (D.21) follows that then

$$P = \left(b^{\phi(N)} \prod_{j=1}^{\phi(N)} r_j \right) \bmod N. \tag{D.36}$$

For $j \neq k$ we have that $a_j \neq a_k$. To see this, suppose $a_j = a_k$, that is, $r_j b \bmod N = r_k b \bmod N$. Since b and N are coprime, Lemma D.7 implies that then $r_j \bmod N = r_k \bmod N$. Because of the assumption $1 \leq r_j, r_k < N$ it follows that $r_j = r_k$ and thus $j = k$. Hence, $a_j \neq a_k$ for $j \neq k$. Since r_j and N as well as b and N are coprime, it follows that also $r_j b$ and N are coprime, that is,

$$\gcd(r_j b, N) = 1. \tag{D.37}$$

Suppose $a_j = r_j b \bmod N$ has a common divisor $s > 1$ with N such that $a_j = us$ and $N = vs$. Then there exists a $k \in \mathbb{Z}$ such that $us = r_j b + kvs$, which is equivalent to $r_j b = (u - kv)s$. This, however, would imply that $r_j b$ and N have a common divisor $s > 1$, which contradicts (D.37). Consequently, all a_j are coprime with N and there exist $\phi(N)$ distinct a_j with $1 \leq a_j < N$. This means that the set of the a_j is a permutation of the set of the r_j, and thus

$$P = \left(\prod_{j=1}^{\phi(N)} a_j \right) \bmod N = \left(\prod_{j=1}^{\phi(N)} r_j \right) \bmod N,$$

which, together with (D.36), yields

$$\left(b^{\phi(N)} \prod_{j=1}^{\phi(N)} r_j \right) \bmod N = \left(\prod_{j=1}^{\phi(N)} r_j \right) \bmod N. \tag{D.38}$$

Since N and all r_j are coprime, we can apply Lemma D.7 to (D.38) to obtain

$$b^{\phi(N)} \bmod N = 1,$$

which was to be shown. $\square$

Example D.18 As we saw in Example D.13, one has for $b = 7$ and $N = 10$ that $\gcd(7,10) = 1$ and $\phi(10) = 4$. As stated in (D.35) we then find indeed $7^4 \bmod 10 = 2401 \bmod 10 = 1$.

As a corollary to Theorem D.17, we have what sometimes is called FERMAT'S Little Theorem.

Corollary D.19 (FERMAT'S Little Theorem) *Any $b \in \mathbb{N}$ and prime p with the property $p \nmid b$ satisfy*

$$b^{p-1} \bmod p = 1. \qquad (D.39)$$

Proof For a prime p one evidently has $\phi(p) = p - 1$ and that $p \nmid b$ implies $\gcd(p,b) = 1$. Then (D.39) immediately follows from (D.35). $\qquad \Box$

Definition D.20 For $a, N \in \mathbb{N}$ with $\gcd(a,N) = 1$ we define the **order of a modulo N** as

$$\mathrm{ord}_N(a) := \min\{m \in \mathbb{N} \mid a^m \bmod N = 1\}.$$

If

$$\mathrm{ord}_N(a) = \phi(N),$$

then a is called a **primitive root modulo N**.

Example D.21 For $N = 3 \times 5 = 15$ we have $\phi(15) = 2 \times 4 = 8$, and with $a = 7$ we find $\gcd(7,15) = 1$ as well as

$$
\begin{array}{c|l}
m & 1\ 2\ 3\ \ 4\ 5\ 6\ 7\ \ 8\ 9\ 10\ 11\ \ldots \\
\hline
7^m \bmod 15 & 7\ 4\ 13\ 1\ 7\ 4\ 13\ 1\ 7\ \ 4\ \ 13\ \ldots
\end{array}
$$

and thus $\mathrm{ord}_{15}(7) = 4 < \phi(15)$. Whereas for $N = 2 \times 5 = 10$ we find $\phi(10) = 1 \times 4 = 4, \gcd(7,10) = 1$ as well as

$$
\begin{array}{c|l}
m & 1\ 2\ 3\ 4\ 5\ 6\ 7\ 8\ 9\ 10\ 11\ \ldots \\
\hline
7^m \bmod 10 & 7\ 9\ 3\ 1\ 7\ 9\ 3\ 1\ 7\ \ 9\ \ \ 3\ \ldots
\end{array},
$$

that is, $\mathrm{ord}_{10}(7) = 4 = \phi(10)$. Hence, 7 is a primitive root modulo 10.

The following results for orders and primitive roots will be of use for us further on.

Theorem D.22 *Let $a, b, N \in \mathbb{N}$ with $\gcd(a, N) = 1 = \gcd(b, N)$. Then the following hold.*

(i) For all $k \in \mathbb{N}$

$$a^k \bmod N = 1 \quad \Leftrightarrow \quad \mathrm{ord}_N(a) \mid k. \qquad (D.40)$$

(ii)

$$\mathrm{ord}_N(a) \mid \phi(N). \qquad (D.41)$$

(iii) If $\mathrm{ord}_N(a)$ and $\mathrm{ord}_N(b)$ are coprime, then

$$\mathrm{ord}_N(ab) = \mathrm{ord}_N(a)\,\mathrm{ord}_N(b). \qquad (D.42)$$

(iv) If a is a primitive root modulo N, that is, if it also satisfies $\mathrm{ord}_N(a) = \phi(N)$, then we also have

(a)

$$\{d \in \{1, \ldots, N-1\} \mid \gcd(d, N) = 1\} = \{a^j \bmod N \mid j \in \{1, \ldots, \phi(N)\}\}. \qquad (D.43)$$

(b) If $b = a^j \bmod N$ for a $j \in \mathbb{N}$, then

$$\mathrm{ord}_N(b) = \mathrm{ord}_N\left(a^j\right) = \frac{\phi(N)}{\gcd(j, \phi(N))}. \qquad (D.44)$$

Proof Let $a, b, N \in \mathbb{N}$ with $\gcd(a, N) = 1 = \gcd(b, N)$. We first show $\Rightarrow$ in (D.40):

Let k be a natural number satisfying $a^k \bmod N = 1$. Then $k \geq \mathrm{ord}_N(a)$ has to hold since $\mathrm{ord}_N(a)$ is per definition the smallest such number. Now, let $c = k \bmod \mathrm{ord}_N(a)$, that is, $c \in \mathbb{N}_0$ with $c < \mathrm{ord}_N(a)$ and there is an $l \in \mathbb{Z}$ such that $k = \mathrm{ord}_N(a)\, l + c$ and thus $a^k = a^{\mathrm{ord}_N(a)l + c} = \left(a^{\mathrm{ord}_N(a)l}\right)a^c$. This implies

$$
\begin{aligned}
1 &= a^k \bmod N = \left(a^{\mathrm{ord}_N(a)l}\right)a^c \bmod N \\
&\underset{(D.21)}{=} \left(a^{\mathrm{ord}_N(a)l} \bmod N\right)(a^c \bmod N) \bmod N \\
&\underset{(D.22)}{=} \left(\underbrace{a^{\mathrm{ord}_N(a)} \bmod N}_{=1}\right)^l (a^c \bmod N) \bmod N \\
&= (a^c \bmod N) \bmod N \\
&\underset{(D.21)}{=} a^c \bmod N.
\end{aligned}
$$

Per construction, we have $c < \mathrm{ord}_N(a)$, and since $\mathrm{ord}_N(a)$ is per definition the smallest natural number k satisfying $a^k \bmod N = 1$, it follows that c has to vanish and thus $\mathrm{ord}_N(a) \mid k$.

To show $\Leftarrow$ in (D.40), let $\mathrm{ord}_N(a) \mid k$. Hence, there is a natural number l such that $k = \mathrm{ord}_N(a)\, l$ and thus

$$
\begin{aligned}
a^k \bmod N &= \left(a^{\mathrm{ord}_N(a)}\right)^l \bmod N \\
&\underset{(\mathrm{D}.22)}{=} \Big(\underbrace{a^{\mathrm{ord}_N(a)} \bmod N}_{=1}\Big)^l \bmod N = 1 \bmod N \\
&= 1 .
\end{aligned}
$$

This completes the proof of (D.40).

According to Theorem D.17, one has $a^{\phi(N)} \bmod N = 1$, and thus (D.41) follows from (D.40).

To show (D.42), consider first that

$$
\begin{aligned}
&(ab)^{\mathrm{ord}_N(a)\,\mathrm{ord}_N(b)} \bmod N \\
&\underset{(\mathrm{D}.21)}{=} \Big(\big(a^{\mathrm{ord}_N(a)\,\mathrm{ord}_N(b)} \bmod N\big)\big(b^{\mathrm{ord}_N(b)\,\mathrm{ord}_N(a)} \bmod N\big)\Big) \bmod N \\
&\underset{(\mathrm{D}.21)}{=} \Big(\big(\underbrace{a^{\mathrm{ord}_N(a)} \bmod N}_{=1}\big)^{\mathrm{ord}_N(b)} \bmod N \\
&\qquad \times \big(\underbrace{b^{\mathrm{ord}_N(b)} \bmod N}_{=1}\big)^{\mathrm{ord}_N(a)} \bmod N\Big) \bmod N \\
&= 1 \bmod N = 1
\end{aligned}
$$

and thus, because of (D.40),

$$
\mathrm{ord}_N(ab) \mid \mathrm{ord}_N(a)\,\mathrm{ord}_N(b) . \tag{D.45}
$$

Analogously, it follows that

$$
\begin{aligned}
&a^{\mathrm{ord}_N(b)\,\mathrm{ord}_N(ab)} \bmod N \\
&= \Big(a^{\mathrm{ord}_N(b)\,\mathrm{ord}_N(ab)} \bmod N\Big)\Big(\underbrace{b^{\mathrm{ord}_N(b)} \bmod N}_{=1}\Big)^{\mathrm{ord}_N(ab)} \bmod N \\
&\underset{(\mathrm{D}.21)}{=} (ab)^{\mathrm{ord}_N(b)\,\mathrm{ord}_N(ab)} \bmod N \\
&\underset{(\mathrm{D}.21)}{=} \Big(\underbrace{(ab)^{\mathrm{ord}_N(ab)} \bmod N}_{=1}\Big)^{\mathrm{ord}_N(b)} \bmod N \\
&= 1
\end{aligned} \tag{D.46}
$$

and thus, due to (D.40), that

$$\mathrm{ord}_N(a) \mid \mathrm{ord}_N(b)\,\mathrm{ord}_N(ab) \,.$$

However, by assumption $\mathrm{ord}_N(a)$ and $\mathrm{ord}_N(b)$ are coprime, which implies

$$\mathrm{ord}_N(a) \mid \mathrm{ord}_N(ab) \,. \tag{D.47}$$

Similarly, beginning in (D.46) with $b^{\mathrm{ord}_N(a)\,\mathrm{ord}_N(ab)} \bmod N$ yields

$$\mathrm{ord}_N(b) \mid \mathrm{ord}_N(ab) \,. \tag{D.48}$$

Again, as $\mathrm{ord}_N(a)$ and $\mathrm{ord}_N(b)$ are assumed coprime, it follows from (D.47) and (D.48) that

$$\mathrm{ord}_N(a)\,\mathrm{ord}_N(b) \mid \mathrm{ord}_N(ab) \,.$$

This, together with (D.45), yields (D.42).

Suppose now a is a primitive root modulo N. To prove (D.43), we first show the inclusion

$$\left\{ a^j \bmod N \mid j \in \{1,\dots,\phi(N)\} \right\} \subset \left\{ d \in \{1,\dots N-1\} \mid \gcd(d,N) = 1 \right\}. \tag{D.49}$$

Then we will prove that the two sets have the same cardinality and thus are identical.

To verify the inclusion, we show that the elements of $\left\{ a^j \bmod N \mid j \in \{1,\dots,\phi(N)\} \right\}$ have the property $\gcd(a^j \bmod N, N) = 1$. To show this, suppose $l \in \mathbb{N}$ is a common divisor of $a^j \bmod N$ and N, that is, suppose that there are $u,v \in \mathbb{N}$ satisfying

$$a^j \bmod N = lu$$
$$N = lv \,. \tag{D.50}$$

Then it follows that $lu = a^j \bmod N = a^j - \left\lfloor \frac{a^j}{N} \right\rfloor N = a^j - \left\lfloor \frac{a^j}{N} \right\rfloor lv$ and thus $l \mid a^j$. Consequently, every prime factor of l would be a prime factor of a as well. Due to (D.50), such prime factors would then be divisors of a and N. But by assumption $\gcd(a,N) = 1$. Hence, we must have $l = 1$, which implies $\gcd(a^j \bmod N, N) = 1$ and the inclusion (D.49) is proven.

It remains to show that $\left\{ a^j \bmod N \mid j \in \{1,\dots,\phi(N)\} \right\}$ contains indeed $\phi(N)$ distinct elements. Let $i,j \in \mathbb{N}$ be such that $1 \le i < j \le \phi(N)$ and suppose that

$$a^j \bmod N = a^i \bmod N \,. \tag{D.51}$$

With the assumption $\gcd(a,N) = 1$ it follows from (D.51) with Lemma D.7 that

$$a^{j-i} \bmod N = 1,$$

which, together with $0 < j - i < \phi(N)$, contradicts the assumption that a is a primitive root, which means $\mathrm{ord}_N(a) = \phi(N)$. Thus, the set $\{a^j \bmod N \mid j\{1,\ldots,\phi(N)\}$ contains exactly $\phi(N)$ distinct elements each of which is coprime to N. This completes the proof of the equality of the sets in (D.43).

In (D.44) we first show that $b = a^j \bmod N$ implies

$$\mathrm{ord}_N(b) = \mathrm{ord}_N\left(a^j\right). \tag{D.52}$$

We have

$$\begin{aligned}
1 &= b^{\mathrm{ord}_N(b)} \bmod N = \left(a^j \bmod N\right)^{\mathrm{ord}_N(b)} \bmod N \\
&\underbrace{=}_{(\text{D.22})} \left(a^j\right)^{\mathrm{ord}_N(b)} \bmod N,
\end{aligned}$$

which implies $\mathrm{ord}_N\left(a^j\right) \le \mathrm{ord}_N(b)$. Conversely, it follows from

$$1 = \left(a^j\right)^{\mathrm{ord}_N(a^j)} \bmod N \underbrace{=}_{(\text{D.22})} \left(a^j \bmod N\right)^{\mathrm{ord}_N(a^j)} \bmod N$$

$$= (b)^{\mathrm{ord}_N(a^j)} \bmod N$$

that $\mathrm{ord}_N(b) \le \mathrm{ord}_N\left(a^j\right)$ and thus (D.52).

For the right side in (D.44) we know from (D.41) already that

$$\mathrm{ord}_N\left(a^j\right) \mid \phi(N),$$

that is, there exists an $m_1 \in \mathbb{N}$ such that

$$m_1 \, \mathrm{ord}_N\left(a^j\right) = \phi(N). \tag{D.53}$$

Furthermore,
$$1 = \left(a^j\right)^{\mathrm{ord}_N(a^j)} \bmod N = a^{\mathrm{ord}_N(a^j)j} \bmod N$$

implies, because of (D.40), that

$$\mathrm{ord}_N(a) \mid \mathrm{ord}_N\left(a^j\right) j.$$

From the assumption $\mathrm{ord}_N(a) = \phi(N)$ follows the existence of an $m_2 \in \mathbb{N}$ such that

$$m_2 \phi(N) = \mathrm{ord}_N\left(a^j\right) j. \tag{D.54}$$

Insertion of (D.53) in (D.54) yields

$$m_1 \mid j.$$

Altogether, thus with (D.53)

$$\text{ord}_N\left(a^j\right) = \frac{\phi(N)}{m_1},$$

and m_1 divides $\phi(N)$ as well as j. That m_1 is the greatest such divisor can be seen as follows. Suppose that

$$m_1 < \widehat{m} := \gcd(j, \phi(N)).$$

Then we would have for

$$\widehat{r} := \frac{\phi(N)}{\widehat{m}} < \frac{\phi(N)}{m_1} = \text{ord}_N\left(a^j\right),$$

that

$$\left(a^j\right)^{\widehat{r}} \bmod N = \left(a^j\right)^{\frac{\phi(N)}{\widehat{m}}} \bmod N = \left(a^{\phi(N)}\right)^{\frac{j}{\widehat{m}}} \bmod N$$

$$\underbrace{=}_{(D.22)} \left(\underbrace{a^{\phi(N)} \bmod N}_{=1}\right)^{\frac{j}{\widehat{m}}} \bmod N = 1.$$

This contradicts the fact that $\text{ord}_N\left(a^j\right)$ is, by definition, the smallest natural number with the property $(a^j)^r \bmod N = 1$. Consequently, we must have $m_1 = \gcd(j, \phi(N))$, and (D.44) is proven. □

Before we give the proof of the existence of a primitive root for primes, we show two lemmas, which we use in that proof.

Lemma D.23 *Let p be a prime, $k \in \mathbb{N}_0$ and $\{f_j \mid j \in \{0, \dots, k\}\} \subset \mathbb{Z}$ with $p \nmid f_k$ and let f be the polynomial*

$$f : \mathbb{Z} \longrightarrow \mathbb{Z}$$

$$x \longmapsto f(x) := \sum_{j=0}^{k} f_j x^j.$$

Then either

(i) f has at most k distinct zeros modulo p in $\{1, \dots, p-1\} \subset \mathbb{N}$, that is, in $\{1, \dots, p-1\}$ there are no more than k distinct natural numbers n_j that satisfy

$$f(n_j) \bmod p = 0,$$

or

(ii) *f is the zero-polynomial modulo p, that is,*

$$f(x) \bmod p = 0 \quad \forall x \in \mathbb{Z}.$$

Proof We show this by induction in the degree of the polynomial, which we start at $k = 0$: if $f(x) = f_0 \neq 0$ such that $p \nmid f_0$, then it follows that $f_0 \bmod p \neq 0$, and there is no $x \in \mathbb{Z}$ with $f(x) \bmod p = 0$. If $f_0 = 0$, then f is the zero-polynomial.

The inductive step is performed from $k - 1$ to k. Suppose then the claim holds for all polynomials of degree up to $k - 1$ and f is a polynomial of degree k. If f has fewer than k zeros modulo p in $\{1, \ldots, p-1\}$, the claim holds already. Otherwise, let $n_1, \ldots, n_k$ be k arbitrarily selected zeros of f modulo p from the set $\{1, \ldots, p-1\}$. Then

$$g(x) := f(x) - f_k \prod_{j=1}^{k}(x - n_j) = \sum_{l=0}^{k-1} g_l x^l \tag{D.55}$$

is a polynomial of degree not exceeding $k - 1$.

Furthermore, every of the k selected zeros $n_l \in \{n_1, \ldots, n_k\}$ satisfies

$$g(n_l) \bmod p = \left(f(n_l) - f_k \prod_{j=1}^{k}(n_l - n_j) \right) \bmod p = f(n_l) \bmod p = 0. \tag{D.56}$$

Set $m := \max\left\{ l \in \{0, \ldots, k-1\} \mid p \nmid g_l \right\}$ and

$$\tilde{g}(x) := \sum_{l=0}^{m} g_l x^l. \tag{D.57}$$

Since then $p \mid g_l$ for all $l > m$, we have for all $x \in \mathbb{Z}$ that

$$\tilde{g}(x) \bmod p \underset{(D.57)}{=} \left(\sum_{l=0}^{m} g_l x^l \right) \bmod p = \left(\sum_{l=0}^{k-1} g_l x^l \right) \bmod p \underset{(D.55)}{=} g(x) \bmod p, \tag{D.58}$$

and the set of zeros modulo p of $\tilde{g}$ and g coincide. Because of this and (D.56), $\tilde{g}$ has at least k zeros modulo p. At the same time $\tilde{g}$ is a polynomial of degree not exceeding $k - 1$ and thus satisfies the inductive assumption, which then implies that $\tilde{g}$ and, because of (D.58), also g can only be the zero-polynomial modulo p:

$$g(x) \bmod p = 0 \quad \forall x \in \mathbb{Z}.$$

With (D.55) it thus follows that for all $x \in \mathbb{Z}$

$$f(x) \bmod p = f_k \prod_{j=1}^{k} (x - n_j) \bmod p,$$

and for an arbitrary zero z of f modulo p we have

$$0 = f(z) \bmod p = f_k \prod_{j=1}^{k} (z - n_j) \bmod p.$$

Since by assumption $p \nmid f_k$, one of the factors in $\prod_{j=1}^{k} (z - n_j)$ has to satisfy

$$(z - n_j) \bmod p = 0.$$

As we chose the n_j form the set $\{1, \ldots, p-1\}$, it follows that $z \bmod p = n_j$, and z is either one of the k zeros selected from $\{1, \ldots, p-1\}$ or it differs from one of these by a multiple of p and is thus not an element of the set $\{1, \ldots, p-1\}$. $\qquad\square$

Lemma D.24 *Let p be prime, d a natural number satisfying $d \mid p-1$ and let h be the polynomial*

$$h : \mathbb{Z} \longrightarrow \mathbb{Z}$$
$$x \longmapsto h(x) := x^d - 1.$$

Then there are d zeros of h modulo p in $\{1, \ldots, p-1\} \subset \mathbb{N}$, that is, in $\{1, \ldots, p-1\}$ there exist d natural numbers n_j satisfying

$$h(n_j) \bmod p = 0.$$

Proof Let $k \in \mathbb{N}$ be such that $p - 1 = dk$ and set

$$f(x) := \sum_{l=0}^{k-1} \left(x^d \right)^l.$$

Then we have

$$g(x) := h(x) f(x) = \left(x^d - 1 \right) \sum_{l=0}^{k-1} \left(x^d \right)^l = x^{p-1} - 1.$$

Since $p - 1 = \phi(p)$ and, according to the EULER Theorem D.17, $a^{\phi(p)} \bmod p = 1$ for all $a \in \{1, \ldots, p-1\}$, it follows that for all $z \in \{1, \ldots, p-1\}$

$$z^{p-1} \bmod p = 1.$$

Hence, all $p-1 = dk$ integers in $\{1,\ldots,p-1\}$ are zeros modulo p of the polynomial g. Since p is a prime and $g = hf$, each of the dk zeros $n_j \in \{1,\ldots,p-1\}$ of g modulo p has to satisfy

$$h(n_j) \bmod p = 0 \quad \text{or} \quad f(n_j) \bmod p = 0.$$

According to Lemma D.23 the polynomial h has at most d and the polynomial f has at most $d(k-1)$ zeros modulo p in $\{1,\ldots,p-1\}$. Denoting the number of zeros modulo p in $\{1,\ldots,p-1\}$ of the polynomials g,h and f by N_g, N_h and N_f, we have thus

$$dk = N_g \leq N_h + N_f \leq d + d(k-1) = dk.$$

This can only be true if f has exactly $d(k-1)$ and h has exactly d zeros, which was to be shown.   □

Theorem D.25 *For every odd prime p there exists at least one primitive root a modulo p, that is, a natural number a such that*

$$\mathrm{ord}_p(a) = \phi(p).$$

Proof Let q be a prime factor of $p-1$, that is, there exists a $k_q \in \mathbb{N}$ such that $q^{k_q} \mid p - 1$. From Lemma D.24 we know that the polynomial $h(x) := x^{q^{k_q}} - 1$ has exactly q^{k_q} zeros modulo p in $\{1,\ldots,p-1\}$. Let a_q be one of these zeros such that it satisfies

$$\left(a_q^{q^{k_q}} - 1 \right) \bmod p = 0$$

and thus

$$a_q^{q^{k_q}} \bmod p = 1.$$

Since $a_q \in \{1,\ldots,p-1\}$ and $\gcd(a_q,p) = 1$, it follows from (D.40) in Theorem D.22 that

$$\mathrm{ord}_p(a_q) \mid q^{k_q}.$$

If this zero a_q of h has the additional property $\mathrm{ord}_p(a_q) \mid q^j$ for a $j \in \mathbb{N}$ with $j < k_q$, then $\mathrm{ord}_p(a_q) \mid q^{k_q-1}$ holds. This means that there is an $n \in \mathbb{N}$ with $q^{k_q-1} = \mathrm{ord}_p(a_q) n$ and thus according to (D.40) in Theorem D.22

$$a_q^{q^{k_q-1}} \bmod p = 1.$$

Hence, $a_q \in \{1,\ldots,p-1\}$ is a zero modulo p of the polynomial $f(x) := x^{q^{k_q-1}} - 1$. According to Lemma D.24, there are exactly q^{k_q-1} of these. Of the q^{k_q} zeros modulo

p in $\{1,\ldots,p-1\}$ of h at most q^{k_q-1} can be zeros of f as well. This means that of the q^{k_q} zeros a_q of h at most q^{k_q-1} such a_q exist that also satisfy in addition $\operatorname{ord}_p(a_q)\mid q^j$ with $j < k_q$. Consequently, there remain $q^{k_q} - q^{k_q-1}$ zeros $a_q \in \{1,\ldots,p-1\}$ that satisfy

$$\operatorname{ord}_p(a_q)\mid q^{k_q} \quad \text{and} \quad \operatorname{ord}_p(a_q)\nmid q^j \quad \forall j < k_q. \tag{D.59}$$

Since q is assumed prime, it follows for the $q^{k_q} - q^{k_q-1}$ numbers a_q that satisfy (D.59) that

$$q^{k_q} = \operatorname{ord}_p(a_q). \tag{D.60}$$

Now, let

$$p - 1 = \prod_{q\in\operatorname{Pri}(p-1)} q^{k_q}$$

be the prime factorization of $p-1$ and

$$a := \prod_{q\in\operatorname{Pri}(p-1)} a_q. \tag{D.61}$$

For arbitrary $q_1, q_2 \in \operatorname{Pri}(p-1)$ with $q_1 \neq q_2$ we have

$$\gcd(\operatorname{ord}_p(a_{q_1}), \operatorname{ord}_p(a_{q_2})) = \gcd(q_1^{k_{q_1}}, q_2^{k_{q_2}}) = 1.$$

This, together with (D.42) in Theorem D.22, implies

$$\operatorname{ord}_p(a_{q_1} a_{q_2}) = \operatorname{ord}_p(a_{q_1})\operatorname{ord}_p(a_{q_2}) = q_1^{k_{q_1}} q_2^{k_{q_2}} \tag{D.62}$$

and thus, finally,

$$\begin{aligned}
\operatorname{ord}_p(a) \underset{(\text{D.61})}{=} \operatorname{ord}_p\left(\prod_{q\in\operatorname{Pri}(p-1)} a_q\right) &\underset{(\text{D.62})}{=} \prod_{q\in\operatorname{Pri}(p-1)} \operatorname{ord}_p(a_q)\\
&\underset{(\text{D.60})}{=} \prod_{q\in\operatorname{Pri}(p-1)} q^{k_q}\\
&= p - 1\\
&\underset{(\text{D.30})}{=} \phi(p).
\end{aligned}$$

$\square$

This shows that every odd prime p has a primitive root modulo p. To show that also every power of an odd prime has a primitive root, we still need the following lemma.

Lemma D.26 *Suppose an odd prime p and a primitive root a modulo p satisfy*

$$a^{\phi(p)} \bmod p^2 \neq 1. \tag{D.63}$$

Then one has for all $k \in \mathbb{N}$

$$a^{\phi(p^k)} \bmod p^{k+1} \neq 1. \tag{D.64}$$

Proof From Theorem D.17 we have for all $k \in \mathbb{N}$ that

$$a^{\phi(p^k)} \bmod p^k = 1,$$

that is, for every $k \in \mathbb{N}$ there exists an $n_k \in \mathbb{N}$ such that

$$a^{\phi(p^k)} = 1 + n_k p^k \tag{D.65}$$

holds. We prove the claim (D.64) by induction in k. The start of the induction for $k = 1$ is given by (D.63). For the inductive step from k to $k+1$ we suppose that (D.64) holds for k such that for all $m \in \mathbb{N}$ we have

$$a^{\phi(p^k)} \neq 1 + m p^{k+1}. \tag{D.66}$$

From (D.66) it follows for n_k in (D.65) that $p \nmid n_k$. According to Lemma D.14,

$$\phi(p^{k+1}) = p^{k+1} - p^k = p(p^k - p^{k-1}) = p\phi(p^k)$$

holds and thus

$$\begin{aligned} a^{\phi(p^{k+1})} &= a^{p\phi(p^k)} = \left(a^{\phi(p^k)}\right)^p \\ &\underset{(D.65)}{=} (1+n_k p^k)^p = \sum_{l=0}^{p} \binom{p}{l} \left(n_k p^k\right)^l \\ &= 1 + n_k p^{k+1} + \sum_{l=2}^{p} \binom{p}{l} n_k^l p^{kl}, \end{aligned}$$

where in the last equation the binomial coefficient for $l = 1$ contributed a factor p. This implies

$$\frac{a^{\phi(p^{k+1})} - 1}{p^{k+2}} = \underbrace{\frac{n_k}{p}}_{\notin \mathbb{Z}} + \underbrace{\sum_{l=2}^{p} \binom{p}{l} n_k^l p^{k(l-1)-2}}_{\in \mathbb{N}}$$

and thus the claim for $k+1$:

$$a^{\phi(p^{k+1})} \bmod p^{k+2} \neq 1.$$

$\square$

Lastly, we show that every power of an odd prime has a primitive root.

Theorem D.27 *Let p be an odd prime and let a be a primitive root modulo p. Then for all $k \in \mathbb{N}$ either*

$$\operatorname{ord}_{p^k}(a) = \phi(p^k) \tag{D.67}$$

or

$$\operatorname{ord}_{p^k}(a+p) = \phi(p^k),$$

that is, either a or $a+p$ is a primitive root modulo p^k.

Proof We distinguish between two cases.

Case 1:

$$a^{\phi(p)} \bmod p^2 \neq 1. \tag{D.68}$$

Case 2:

$$a^{\phi(p)} \bmod p^2 = 1.$$

Consider first case 1. We show by induction in k that in this case (D.67) holds. The induction-start for $k=1$ is given by the assumption that a is a primitive root modulo p. For the inductive step from k to $k+1$ we suppose that (D.67) holds for k. Per definition D.20 of the order, we have

$$a^{\operatorname{ord}_{p^{k+1}}(a)} \bmod p^{k+1} = 1,$$

that is, there exists an $n \in \mathbb{N}$ such that

$$a^{\operatorname{ord}_{p^{k+1}}(a)} = 1 + np^{k+1} = 1 + npp^k.$$

Then (D.40) in Theorem D.22 implies

$$\operatorname{ord}_{p^k}(a) \mid \operatorname{ord}_{p^{k+1}}(a).$$

By the inductive assumption, $\operatorname{ord}_{p^k}(a) = \phi(p^k) = p^{k-1}(p-1)$ holds and thus

$$p^{k-1}(p-1) \mid \operatorname{ord}_{p^{k+1}}(a). \tag{D.69}$$

Theorem D.22 also implies

$$\text{ord}_{p^{k+1}}(a) \mid \phi(p^{k+1}) = p^k(p-1). \tag{D.70}$$

From (D.69) and (D.70) it follows that there exist a $n_1, n_2 \in \mathbb{N}$ such that

$$n_1 p^{k-1}(p-1) = \text{ord}_{p^{k+1}}(a)$$
$$\text{ord}_{p^{k+1}}(a) n_2 = p^k(p-1),$$

which implies $n_1 n_2 p^{k-1}(p-1) = p^k(p-1)$ and thus $n_1 n_2 = p$. Since p is prime, we can only have either

$$n_1 = 1 \quad \text{and} \quad n_2 = p \tag{D.71}$$

or

$$n_1 = p \quad \text{and} \quad n_2 = 1. \tag{D.72}$$

But the case (D.71) would imply $\text{ord}_{p^{k+1}}(a) = p^{k-1}(p-1) \underbrace{=}_{(\text{D.30})} \phi(p^k)$ as a conse-

quence and thus

$$a^{\phi(p^k)} \bmod p^{k+1} = 1.$$

This, however, is impossible because of the case assumption (D.68) and its implications in Lemma D.26. On the other hand, the case (D.72) implies $\text{ord}_{p^{k+1}}(a) = p^k(p-1) = \phi(p^{k+1})$, which is the claim (D.67) for $k+1$.

Now, let us consider case 2 and suppose

$$a^{\phi(p)} \bmod p^2 = 1. \tag{D.73}$$

We show first that in this case $a + p$ is a primitive root modulo p and then that it satisfies the conditions of case 1. Setting $r := \text{ord}_p(a+p)$, it follows that

$$r \le \phi(p) \tag{D.74}$$

and per definition

$$(a+p)^r \bmod p = 1.$$

The latter means that there exists a natural number m such that

$$\sum_{l=0}^{r} \binom{r}{l} a^{r-l} p^l = 1 + mp.$$

This in turn leads to

$$a^r = 1 + p\left(m - \sum_{l=1}^{r} \binom{r}{l} a^{r-l} p^{l-1}\right)$$

$$\underbrace{\qquad\qquad\qquad\qquad}_{\in \mathbb{N}}$$

and thus

$$a^r \bmod p = 1 \, .$$

Hence, we must have

$$r \geq \mathrm{ord}_p(a) = \phi(p) \, . \tag{D.75}$$

From (D.74), the definition of r and (D.75) it follows that

$$\mathrm{ord}_p(a + p) = \phi(p) \, ,$$

that is, $a + p$ is also (besides a) a primitive root modulo p. From (D.73) it follows that there exists an $n_3 \in \mathbb{N}$ such that $a^{p-1} = 1 + n_3 p^2$. Hence, there also exists an $n_4 \in \mathbb{N}$ such that

$$
\begin{aligned}
(a+p)^{p-1} &= a^{p-1} + (p-1)a^{p-2}p + \sum_{l=2}^{p-1} \binom{p-1}{l} a^{p-1-l} p^l \\
&\underset{(D.73)}{=} 1 + n_3 p^2 + p^2 a^{p-2} - p a^{p-2} + \sum_{l=2}^{p-1} \binom{p-1}{l} a^{p-1-l} p^l \\
&= 1 + n_4 p^2 - p a^{p-2} \, .
\end{aligned}
\tag{D.76}
$$

The assumption $a^{p-1} \bmod p = 1$ implies $p \nmid a^{p-2}$. Hence, it follows from (D.76) that

$$(a+p)^{p-1} \bmod p^2 \neq 1 \, ,$$

implying that $a + p$ is a primitive root modulo p that satisfies the condition (D.68) of case 1. As shown in that case, it is then a primitive root modulo p^k for all $k \in \mathbb{N}$.
□

Finally, we prove one more result, which we need in the context of SHOR's factorization algorithm. To estimate the probability that in this algorithm the selection of b does *not* satisfy the criteria in (6.25), we need the following result.

Theorem D.28 *Let* $N = \prod_{j=1}^{J} n_j$ *with* $n_j \in \mathbb{N}$ *and* $\gcd(n_i, n_j) = 1$ *if* $i \neq j$. *Then there is a bijection between the set*

$$A := \{a \in \{1, \ldots, N-1\} \mid \gcd(a, N) = 1\}$$

and the set

$$B := \{(b_1, \ldots, b_J) \mid \forall j \quad b_j \in \{1, \ldots, n_j - 1\} \text{ and } \gcd(b_j, n_j) = 1\}.$$

This bijection $g : A \to B$ *is defined by*

$$g(a) := (a \bmod n_1, \ldots, a \bmod n_J) =: (g(a)_1, \ldots, g(a)_J).$$

Proof We show first that $g\{A\} \subset B$. By definition, we have $g(a)_j \in \{1, \ldots, n_j - 1\}$. We show that $\gcd(g(a)_j, n_j) = 1$ holds for $a \in A$. Suppose γ is a common divisor of $g(a)_j$ and n_j. Then there exist $l, k \in \mathbb{N}$ such that $\gamma l = g(a)_j = a - \left\lfloor \frac{a}{n_j} \right\rfloor \gamma k$ and thus $\frac{a}{\gamma} = l + \left\lfloor \frac{a}{n_j} \right\rfloor k \in \mathbb{N}$, that is, γ divides a, and since γ per definition also divides n_j, it divides N as well. Hence, γ is a common divisor of a and N, which, because of $a \in A$ and the definition of A, implies $\gamma = 1$. This completes the proof of $g\{A\} \subset B$.

Next, we show that g is injective. Let $a_1, a_2 \in A$ and suppose $a_1 \geq a_2$ and $g(a_1) = g(a_2)$. Then it follows for all $j \in \{1, \ldots, J\}$ that

$$a_1 \bmod n_j = a_2 \bmod n_j$$

and thus

$$a_1 - a_2 = \left(\left\lfloor \frac{a_1}{n_j} \right\rfloor - \left\lfloor \frac{a_2}{n_j} \right\rfloor \right) n_j.$$

Consequently, every n_j divides $a_1 - a_2 \in \mathbb{N}_0$, and since $\gcd(n_i, n_j) = 1$ for $i \neq j$, it follows that then also $N = \prod_{j=1}^{J} n_j$ has to be a divisor of $a_1 - a_2 \geq 0$. This means that there is a $k \in \mathbb{N}_0$ with

$$a_1 = a_2 + kN.$$

Since $a_1, a_2 \in A \subset \{1, \ldots, N-1\}$ we must have $k = 0$ and $a_1 = a_2$, implying that g is injective.

Lastly, we define an $h : B \to A$ and show that $g \circ h = \mathrm{id}_B$. For $\mathbf{b} := (b_1, \ldots, b_J) \in B$ we define $h(\mathbf{b})$ as follows. Let $m_j := \frac{N}{n_j}$. We then have $\gcd(m_j, n_j) = 1$, and because of (D.12) in Theorem D.4, there exist $x_j, y_j \in \mathbb{Z}$ such that $m_j x_j + n_j y_j = 1$. With the m_j we define

$$h(\mathbf{b}) := \left(\sum_{j=1}^{J} m_j x_j b_j \right) \bmod N. \tag{D.77}$$

We proceed to show that h is well defined despite the non-uniqueness of the x_j, y_j. For this suppose that $\tilde{x}_j, \tilde{y}_j \in \mathbb{Z}$, such that also $m_j \tilde{x}_j + n_j \tilde{y}_j = 1$. Then it follows that for all $k = 1, \ldots, J$

$$\frac{1}{n_k} \left(\sum_{j=1}^{J} m_j (x_j - \tilde{x}_j) b_j \right) = \sum_{j \neq k} \underbrace{\frac{m_j}{n_k} (x_j - \tilde{x}_j) b_j}_{\in \mathbb{Z}} + \frac{1 - n_k y_k - (1 - n_k \tilde{y}_k)}{n_k} b_k \in \mathbb{Z}, \tag{D.78}$$

that is, every n_k divides $\sum_{j=1}^{J} m_j x_j b_j - \sum_{j=1}^{J} m_j \tilde{x}_j b_j$ for all $k \in \{1, \ldots, J\}$. Since $\gcd(n_j, n_i) = 1$ if $i \neq j$, thus $N = \prod_{j=1}^{J} n_j$ also divides this difference, and there exists a $z \in \mathbb{Z}$ with

$$\sum_{j=1}^{J} m_j x_j b_j = \sum_{j=1}^{J} m_j \tilde{x}_j b_j + zN,$$

which implies

$$\left(\sum_{j=1}^{J} m_j x_j b_j \right) \bmod N = \left(\sum_{j=1}^{J} m_j \tilde{x}_j b_j \right) \bmod N.$$

This shows that the right side in (D.77) is independent of the choice of the x_j, and $h(\mathbf{b})$ is well defined for all $\mathbf{b} \in B$.

We now show that $h\{B\} \subset A$. Similar to (D.78), one has for all $\mathbf{b} \in B$ and $k \in 1, \ldots, J$ that

$$\frac{1}{n_k} (h(\mathbf{b}) - b_k) = \frac{1}{n_k} \left(\left(\sum_{j=1}^{J} m_j x_j b_j \right) \bmod N - b_k \right)$$

$$\underset{(D.1)}{=} \underbrace{\sum_{j \neq k} \underbrace{\frac{m_j}{n_k} x_j b_j}_{\in \mathbb{Z}}}_{} + \underbrace{\frac{m_k x_k - 1}{n_k} b_k}_{\in \mathbb{Z}} - \left\lfloor \frac{\sum_{j=1}^{J} m_j x_j b_j}{N} \right\rfloor \frac{N}{n_k} \in \mathbb{Z},$$

that is, for every $k \in \{1, \ldots, J\}$ there is a $z_k \in \mathbb{Z}$ with

$$h(\mathbf{b}) = b_k + z_k n_k. \tag{D.79}$$

Therefore, every common divisor v of $h(\mathbf{b})$ and n_k is also a common divisor of b_k and n_k. Because of $\mathbf{b} \in B$, it follows that we must have $v = 1$ and thus

$$\gcd(h(\mathbf{b}), n_k) = 1.$$

Furthermore, according to Definition (D.77), then $h(\mathbf{b}) \in \{0, 1, \ldots, N-1\}$. The case $h(\mathbf{b}) = 0$ can be excluded, since, if there were a $z \in \mathbb{Z}$ with $\sum_{j=1}^{J} m_j x_j b_j = zN$, we would have for every $k \in \{1, \ldots, J\}$ that

$$\frac{1 - n_k y_k}{n_k} b_k = \frac{m_k x_k b_k}{n_k} = z \frac{N}{n_k} - \sum_{j \neq k} \frac{m_j}{n_k} x_j b_j \in \mathbb{Z}.$$

Hence, n_k would be a divisor of b_k, which, however, is excluded for $\mathbf{b} \in B$. Consequently, $h(\mathbf{b}) \in \{1, \ldots, N-1\}$ with $\gcd(h(\mathbf{b}), n_j) = 1$ for all j and thus $h(\mathbf{b}) \in A$.

It also follows from (D.79) and $b_k < n_k$ that

$$g(h(\mathbf{b}))_k = g(b_k + z_k n_k)_k = (b_k + z_k n_k) \bmod n_k = b_k.$$

This shows that $g \circ h = \mathrm{id}_B$, proving that g is also surjective and altogether thus bijective. $\square$

Appendix E
Continued Fractions

Definition E.1 Let a_0 be an integer, and let $(a_1, \ldots, a_n)$ be a finite sequence of natural numbers. The number denoted by $[a_0; a_1, \ldots, a_n]$ and defined as

$$[a_0; a_1, \ldots, a_n] := a_0 + \cfrac{1}{a_1 + \cfrac{1}{\ddots \cfrac{\vdots}{a_{n-1} + \frac{1}{a_n}}}} \qquad (E.1)$$

is called a (regular) **finite continued fraction** generated by $(a_0, \ldots, a_n)$.

For any $j \in \{0, \ldots, n\}$ the number

$$[a_0; a_1, \ldots, a_j] = a_0 + \cfrac{1}{a_1 + \cfrac{1}{\ddots \cfrac{\vdots}{a_{j-1} + \frac{1}{a_j}}}}$$

is called the **jth convergent** of the continued fraction $[a_0; a_1, \ldots, a_n]$. For any sequence $(a_j)_{j \in I \subset \mathbb{N}}$ the sequence $([a_0; a_1, \ldots, a_j])_{j \in I}$ is called the **sequence of convergents**.

Continued fractions can be defined more generally by replacing every 1 in (E.1) by elements b_j of a sequence (b_j) of integers. These would no longer be called 'regular' continued fractions. Since for our purposes we will deal only with the regular continued fractions defined in Definition (E.1), we shall omit the adjective 'regular' henceforth.

© Springer Nature Switzerland AG 2019
W. Scherer, *Mathematics of Quantum Computing*,
https://doi.org/10.1007/978-3-030-12358-1

It can be shown that, as the name suggests, the sequence of convergents *always* converges to a limit

$$[a_0; a_1, \ldots] := \lim_{n \to \infty} [a_0; a_1, \ldots, a_n] = \lim_{n \to \infty} \left(a_0 + \cfrac{1}{a_1 + \cfrac{1}{a_2 + \cfrac{1}{\ddots + \frac{1}{a_n}}}} \right). \qquad (E.2)$$

This is why in the literature a continued fraction is often introduced 'as an expression of the form'

$$[a_0; a_1, \ldots] = a_0 + \cfrac{1}{a_1 + \cfrac{1}{a_2 + \cfrac{1}{\ddots a_j + \cfrac{1}{\ddots}}}}.$$

Moreover, it can be shown that any real number may be expressed as a limit of a suitable sequence of convergents. We will not prove this and the convergence claimed in (E.2) here since it is not required for our purposes. The interested reader can consult the classic treatise by HARDY and WRIGHT [90] or can attempt the proof with the help of Corollary E.8 below from which it is only a small step to establish the 'convergence of the convergents' claimed in (E.2).

However, in the following we show how for a given real number a suitable sequence of convergents can be constructed.

Lemma E.2 *For every $x \in \mathbb{R}$ define sequences (f_j) and (a_j) as follows.*

If $x = 0$, then the sequence (f_j) is empty and the sequence (a_j) consists only of $(a_0 = 0)$.
If $x \neq 0$, then set

$$f_0 := \frac{1}{x} \in \mathbb{R} \quad and \quad a_0 := \left\lfloor \frac{1}{f_0} \right\rfloor = \lfloor x \rfloor \in \mathbb{Z} \qquad (E.3)$$

and for $j \in \mathbb{N}$

if $f_{j-1} \neq 0$, then

$$f_j := \frac{1}{f_{j-1}} - \left\lfloor \frac{1}{f_{j-1}} \right\rfloor \in [0, 1[\quad and \quad a_{j-1} := \left\lfloor \frac{1}{f_{j-1}} \right\rfloor \in \mathbb{N}; \qquad (E.4)$$

*if $f_{j-1} = 0$,   then the sequences (f_j) and (a_j) are finite and (f_j) termi-
nates with the last element given by $f_{j-1} = 0$, whereas (a_j) terminates
with the last element a_{j-2}.*

Then we have

$$x = a_0 + f_1 = a_0 + \cfrac{1}{a_1 + \cfrac{1}{a_2 + \cfrac{1}{\ddots + \cfrac{1}{a_n + f_{n+1}}}}}, \tag{E.5}$$

*where the last equation holds for every $n \in \mathbb{N}$ for which f_{n+1} is defined by the
construction rules for (f_j).*

Proof Note that for the sequences (f_j) and (a_j) defined by an $x < 0$, only a_0 and f_0
are negative. For those f_j and a_j that are defined, we have, per definition given in
(E.4),

$$f_j = \frac{1}{f_{j-1}} - \left\lfloor \frac{1}{f_{j-1}} \right\rfloor \geq 0 \quad \text{and} \quad a_j = \left\lfloor \frac{1}{f_j} \right\rfloor \geq 0.$$

From (E.4) it follows that for any $j \in \mathbb{N}$ for which f_{j+1} is defined, we have

$$f_j = \frac{1}{a_j + f_{j+1}}. \tag{E.6}$$

In particular, we have

$$\frac{1}{x} \underset{(\text{E.3})}{=} f_0 \underset{(\text{E.6})}{=} \frac{1}{a_0 + f_1},$$

which implies

$$x = a_0 + f_1 \tag{E.7}$$

Moreover, we can iterate (E.6) for all j and n such that $j \leq n$ and f_{n+1} exists to
obtain

$$f_j \underset{(\text{E.6})}{=} \frac{1}{a_j + f_{j+1}} \underset{(\text{E.6})}{=} \frac{1}{a_j + \frac{1}{a_{j+1} + f_{j+2}}} = \cdots = \cfrac{1}{a_j + \cfrac{1}{\ddots + \cfrac{1}{a_n + f_{n+1}}}}. \tag{E.8}$$

Inserting (E.8) for $j = 1$ into (E.7) yields the claim (E.5).   □

The sequence (a_j) constructed in Lemma E.2 can then be used to build a sequence
of convergents as defined in Definition E.1.

Example E.3 The sequence of convergents constructed with (a_j) from Lemma E.2 also approximates irrational numbers x relatively quickly. For example, the 6-th convergent of $\sqrt{2}$ satisfies

$$\sqrt{2} = [1;2,2,2,2,2,2,\ldots] = [1;2,2,2,2,2,2] + 1.23789\cdots \times 10^{-05}.$$

Theorem E.4 *For an $x \in \mathbb{R}$ the sequence (a_j) constructed as in Lemma E.2 terminates if and only if x is rational, that is, we have*

$$x = [a_0;a_1,\ldots,a_n] \quad \Leftrightarrow \quad x \in \mathbb{Q}.$$

Proof $\Rightarrow$: Let

$$x = [a_0;a_1,\ldots,a_n] = a_0 + \cfrac{1}{a_1 + \cfrac{1}{a_2 + \cfrac{1}{\ddots + \frac{1}{a_n}}}} \tag{E.9}$$

and for $j \in \{n, n-1, \ldots, 0\}$ define successively

$$p_n := a_n \quad \text{and} \quad q_n := 1 \quad \text{for } j = n$$
$$\frac{p_{j-1}}{q_{j-1}} := \frac{a_{j-1}p_j + q_j}{p_j} \quad \text{and} \quad q_{j-1} := p_j \quad \text{for } j \in \{n-1,\ldots,0\}.$$

Then we have for $j \in \{n,\ldots,0\}$ that $p_j, q_j \in \mathbb{Z}$. Moreover, it follows that

$$\frac{p_{j-1}}{q_{j-1}} = \frac{a_{j-1}p_j + q_j}{p_j} = a_{j-1} + \frac{1}{\frac{p_j}{q_j}} = \cdots = a_{j-1} + \cfrac{1}{a_j + \cfrac{1}{\ddots + \frac{1}{a_n}}}$$

such that finally

$$\mathbb{Q} \ni \frac{p_0}{q_0} = a_0 + \cfrac{1}{a_1 + \cfrac{1}{\ddots + \frac{1}{a_n}}} \underbrace{=}_{(E.9)} x.$$

$\Leftarrow$: Now, let $x = \frac{p}{q} \in \mathbb{Q}$. The case $x = 0$ is trivial. For $x \neq 0$ let (f_j) and (a_j) be the sequences defined in Lemma E.2 starting with $f_0 = \frac{1}{x}$. Define a sequence (r_j) by

$$r_{-1} := p \in \mathbb{Z} \setminus \{0\} \quad \text{and} \quad r_0 := q \in \mathbb{N} \quad \text{and}$$
$$r_{j+1} = r_{j-1} \bmod r_j \quad \text{for } j \in \mathbb{N} \text{ such that } r_j \neq 0. \tag{E.10}$$

Then we have

$$f_0 = \frac{1}{x} = \frac{q}{p} = \frac{r_0}{r_{-1}} \in \mathbb{Q}, \tag{E.11}$$

and we show by induction that for $j \in \mathbb{N}_0$

$$f_j = \frac{r_j}{r_{j-1}} \in \mathbb{Q}. \tag{E.12}$$

The induction-start is given by (E.11). For the inductive step from j to $j+1$ suppose (E.12) is satisfied for a $j \in \mathbb{N}_0$. It follows that

$$f_{j+1} \underset{(E.4)}{=} \frac{1}{f_j} - \left\lfloor \frac{1}{f_j} \right\rfloor \underset{(E.12)}{=} \frac{r_{j-1}}{r_j} - \left\lfloor \frac{r_{j-1}}{r_j} \right\rfloor \underset{(D.1)}{=} \frac{r_{j-1} \bmod r_j}{r_j} \underset{(E.10)}{=} \frac{r_{j+1}}{r_j}, \tag{E.13}$$

proving (E.12) for $j+1$. Since $0 \leq r_{j+1} = r_{j-1} \bmod r_j < r_j$ holds, the sequence (r_j) constitutes a strictly decreasing sequence of non-negative integers. This implies that after a finite number of steps we must find $r_{n+1} = 0$ for some $n \in \mathbb{N}_0$. It follows from (E.12) that $f_{n+1} = 0$, which means that the sequence (a_j) ends with a_n and (E.5) implies that $x = [a_0; a_1, \ldots, a_n]$ holds. $\qquad \square$

Note that the sequence (r_j) defined in (E.10) is similar to the one used in the extended EUCLID algorithm in Theorem D.4. Consequently, the number n of the a_j that have to be calculated for the continued fraction representation of a rational number satisfies the same bound.

Corollary E.5 *Let $p, q \in \mathbb{N}$ be given and let*

$$[a_0; a_1, \ldots, a_n] = \frac{p}{q} \tag{E.14}$$

be the finite continued fraction representation of the rational number $\frac{p}{q}$. Then we have

$$n < 2\min\{\log_2 q, \log_2 p\} + 1.$$

Proof Let (f_j) be constructed from $x = \frac{p}{q}$ as in Lemma E.2. This sequence terminates as soon as $f_j = 0$ for the first time. From (E.12) we see that $f_j = 0$ is equivalent to $r_j = 0$, where the sequence (r_j) is defined as in (E.10) and satisfies the recursion

$$r_j = r_{j-2} \bmod r_{j-1},$$

which is similar to (D.5) in the EUCLID algorithm in Theorem D.4. With the help of (D.3) from Exercise D.119 and just as in (D.16) it follows that

$$r_j < \min\{\frac{r_{j-2}}{2}, r_{j-1}\}.$$

We can thus apply the very same arguments as after (D.16). Since, due to (E.14), n is the largest number for which $r_n > 0$ holds, we infer from (D.17) that

$$n < 2\min\{\log_2 q, \log_2 p\} + 1.$$

$\square$

Example E.6 As an example for a continued fraction expansion of a rational number we have

$$\frac{67}{47} = [1;2,2,1,6] = 1 + \cfrac{1}{2 + \cfrac{1}{2 + \cfrac{1}{1 + \frac{1}{6}}}}.$$

Theorem E.7 *Let $I = \{0,\ldots,n\} \subsetneq \mathbb{N}_0$ or $I = \mathbb{N}_0$, and let $(a_j)_{j\in I}$ be a sequence of numbers, where $a_0 \in \mathbb{Z}$ and $a_j \in \mathbb{N}$ for $j \geq 1$. Then the continued fractions constructed with $(a_j)_{j\in I}$ satisfy:*

(i) *For all $j \in I$ there exist $p_j \in \mathbb{Z}$ and $q_j \in \mathbb{N}$ such that*

$$\frac{p_j}{q_j} = [a_0;\ldots,a_j]. \tag{E.15}$$

(ii) *The p_j and q_j in (E.15) can be obtained by setting*

$$p_{-2} = 0 = q_{-1} \quad and \quad q_{-2} = 1 = p_{-1} \tag{E.16}$$

and for $j \in I$ by defining them through the recursion

$$\begin{aligned} p_j &= a_j p_{j-1} + p_{j-2} \\ q_j &= a_j q_{j-1} + q_{j-2}, \end{aligned} \tag{E.17}$$

that is, the p_j and q_j defined by (E.16) and (E.17) satisfy (E.15).

(iii) *The sequence $(q_j)_{j\in I}$ has only positive elements and growths faster than the FIBONACCI sequence given by $b_0 = 0, b_1 = 1, (b_j = b_{j-1} + b_{j-2})_{j\geq 2}$. In case $a_0 \geq 1$, the same holds for every element in the sequence $(p_j)_{j\in I}$.*

Proof The claim in (i) follows immediately from Theorem E.4 since the continued fraction expansions $[a_0; a_1,\ldots,a_j]$ are finite.

We prove (ii) by induction. Let the p_j and q_j be as defined in (E.16) and (E.17). For the induction-start we show (E.15) for $j \in \{0,1\}$ by using (E.16) and (E.17) as follows:

$$\left.\begin{aligned} p_0 &= a_0 p_{-1} + p_{-2} = a_0 \\ q_0 &= a_0 q_{-1} + q_{-2} = 1 \end{aligned}\right\} \quad \Rightarrow \quad \frac{p_0}{q_0} = a_0 = [a_0;]$$

$$\left.\begin{aligned} p_1 &= a_1 p_0 + p_{-1} = a_1 a_0 + 1 \\ q_1 &= a_1 q_0 + q_{-1} = a_1 \end{aligned}\right\} \quad \Rightarrow \quad \frac{p_1}{q_1} = a_0 + \frac{1}{a_1} = [a_0; a_1]$$

(E.18)

For the inductive step from j to $j+1$ suppose that for a $j \in I$

$$\frac{p_j}{q_j} = [a_0; a_1, \ldots, a_j] \tag{E.19}$$

holds. Define the function $g_j : \mathbb{Q} \setminus \{0\} \to \mathbb{Q}$ by

$$g_j(m) := \frac{m p_{j-1} + p_{j-2}}{m q_{j-1} + q_{j-2}}, \tag{E.20}$$

which satisfies

$$g_j(a_j) \underbrace{=}_{(E.20)} \frac{a_j p_{j-1} + p_{j-2}}{a_j q_{j-1} + q_{j-2}} \underbrace{=}_{(E.17)} \frac{p_j}{q_j} \underbrace{=}_{(E.19)} [a_0; a_1, \ldots, a_j] \underbrace{=}_{(E.1)} a_0 + \cfrac{1}{a_1 + \cfrac{1}{\ddots + \frac{1}{a_j}}}$$

such that

$$g_j(m) \underbrace{=}_{(E.20)} a_0 + \cfrac{1}{a_1 + \cfrac{1}{\ddots + \frac{1}{m}}}. \tag{E.21}$$

From this it follows that

$$\frac{p_{j+1}}{q_{j+1}} \underbrace{=}_{(E.17)} \frac{a_{j+1} p_j + p_{j-1}}{a_{j+1} q_j + q_{j-1}} = \frac{p_j + \frac{1}{a_{j+1}} p_{j-1}}{q_j + \frac{1}{a_{j+1}} q_{j-1}} \underbrace{=}_{(E.17)} \frac{\left(a_j + \frac{1}{a_{j+1}}\right) p_{j-1} + p_{j-2}}{\left(a_j + \frac{1}{a_{j+1}}\right) q_{j-1} + q_{j-2}}$$

$$\underbrace{=}_{(E.20)} g_j\left(a_j + \frac{1}{a_{j+1}}\right) \underbrace{=}_{(E.21)} a_0 + \cfrac{1}{a_1 + \cfrac{1}{\ddots + \cfrac{1}{a_j + \frac{1}{a_{j+1}}}}}$$

$$\underbrace{=}_{(E.1)} [a_0; a_1, \ldots, a_{j-1}, a_j, a_{j+1}]$$

and the inductive step from j to $j+1$ is verified.

To show (iii), note that we have from (E.16) that $q_0 = 1$ and from (E.18) that $q_1 = a_1$ as well as by construction that $a_j \in \mathbb{N}$ for $j \geq 1$. Using this and (E.17), it follows immediately that $q_j = a_j q_{j-1} + q_{j-2} \geq q_{j-1} + q_{j-2}$. In case an a_j is greater than 1, then q_j is strictly greater than the j-th element of the FIBONACCI sequence, and this then also holds for all following elements. Similarly, if $a_0 \geq 1$, we have that $p_0 = a_0 \geq 1, p_1 = a_0 a_1 + 1 > 1$ and with (E.17) that $p_j = a_j p_{j-1} + p_{j-2} \geq p_{j-1} + p_{j-2}$. $\qquad\square$

Corollary E.8 *Let $I = \{0, \ldots, n\} \subsetneq \mathbb{N}_0$ or $I = \mathbb{N}_0$, and let $(a_j)_{j \in I}$ be a sequence of numbers, where $a_0 \in \mathbb{Z}$ and $a_j \in \mathbb{N}$ for $j \geq 1$. Moreover, for all $j \in I$ let*

$$\frac{p_j}{q_j} = [a_0; a_1, \ldots, a_j].$$

Then the following holds:

(i) For all $j \in I \setminus \{0\}$

$$p_j q_{j-1} - q_j p_{j-1} = (-1)^{j-1} \tag{E.22}$$

and

$$\gcd(p_j, q_j) = 1. \tag{E.23}$$

(ii) For $j, k \in I$ such that $j > k \geq 0$

$$\frac{p_k}{q_k} - \frac{p_j}{q_j} = (-1)^j \sum_{l=0}^{j-k-1} \frac{(-1)^l}{q_{j-l} q_{j-l-1}}. \tag{E.24}$$

(iii) For all $k \in I$

$$\frac{p_{2k}}{q_{2k}} < \frac{p_{2k+2}}{q_{2k+2}} \tag{E.25}$$

$$\frac{p_{2k+3}}{q_{2k+3}} < \frac{p_{2k+1}}{q_{2k+1}}. \tag{E.26}$$

(iv) For all $k \in I$

$$\frac{p_0}{q_0} < \frac{p_2}{q_2} < \cdots < \frac{p_{2k}}{q_{2k}} < \cdots < \frac{p_{2k+1}}{q_{2k+1}} < \cdots < \frac{p_3}{q_3} < \frac{p_1}{q_1}. \tag{E.27}$$

Proof To prove (i), note that for $j \in I \setminus \{0\}$ we have

$$
\begin{aligned}
z_j &:= p_j q_{j-1} - q_j p_{j-1} \\
&\underset{(E.17)}{=} (a_j p_{j-1} + p_{j-2}) q_{j-1} - (a_j q_{j-1} + q_{j-2}) p_{j-1} \\
&= p_{j-2} q_{j-1} - q_{j-2} p_{j-1} = -z_{j-1} = \cdots = (-1)^k z_{j-k} = \cdots \\
&= (-1)^{j-1} z_1 = (-1)^{j-1} (p_1 q_0 - q_1 p_0) \\
&\underset{(E.18)}{=} (-1)^{j-1} ((a_1 a_0 + 1) - a_1 a_0) \\
&= (-1)^{j-1},
\end{aligned}
$$

which proves (E.22). This in turn implies (E.23) as follows. Suppose

$$
q_j = \tilde{q}_j \gcd(q_j, p_j) \quad \text{and} \quad p_j = \tilde{p}_j \gcd(q_j, p_j).
$$

With (E.22) it follows that

$$
\mathbb{Z} \ni \tilde{p}_j q_{j+1} - \tilde{q}_j p_{j+1} = \frac{(-1)^{j+1}}{\gcd(q_j, p_j)},
$$

which requires $\gcd(q_j, p_j) = 1$, that is, (E.23).

To show (ii), we note that for $k \in I$

$$
\frac{p_k}{q_k} - \frac{p_{k+1}}{q_{k+1}} = \frac{p_k q_{k+1} - q_k p_{k+1}}{q_k q_{k+1}} \underset{(E.22)}{=} \frac{(-1)^{k+1}}{q_k q_{k+1}} \tag{E.28}
$$

and thus for any $j \in I$ such that $j > k$

$$
\begin{aligned}
\frac{p_k}{q_k} - \frac{p_j}{q_j} &= \frac{p_k}{q_k} - \frac{p_{k+1}}{q_{k+1}} + \frac{p_{k+1}}{q_{k+1}} - \frac{p_{k+2}}{q_{k+2}} + \cdots + \frac{p_{j-1}}{q_{j-1}} - \frac{p_j}{q_j} \\
&\underset{(E.28)}{=} \frac{(-1)^{k+1}}{q_k q_{k+1}} + \frac{(-1)^{k+2}}{q_{k+1} q_{k+2}} + \cdots + \frac{(-1)^j}{q_{j-1} q_j} \\
&= (-1)^j \sum_{l=0}^{j-k-1} \frac{(-1)^l}{q_{j-l} q_{j-l-1}},
\end{aligned}
$$

proving (E.24).

To prove (iii), we observe that

$$\frac{p_{2k}}{q_{2k}} - \frac{p_{2k+2}}{q_{2k+2}} \underset{\text{(E.24)}}{=} (-1)^{2k+2} \left(\frac{1}{q_{2k+2}q_{2k+1}} - \frac{1}{q_{2k+1}q_{2k}} \right)$$

$$= \frac{1}{q_{2k+1}} \underbrace{\left(\frac{1}{q_{2k+2}} - \frac{1}{q_{2k}} \right)}_{<0} < 0,$$

where we used (iii) in Theorem E.7 to ascertain $q_{2k} < q_{2k+2}$. This verifies (E.25). Similarly, one shows (E.26).

In (iv) the inequalities for even indices follow from (E.25) and those for odd indices from (E.26). To prove (E.27), it remains to show that for an arbitrary $k \in I$ the quotient for $2k$ is smaller than that for $2k + 1$. From (iii) in Theorem E.7 follows that $q_{2k}q_{2k+1} > 0$ for arbitrary $k \in I$. With this we obtain the desired inequality as follows

$$\frac{p_{2k+1}}{q_{2k+1}} - \frac{p_{2k}}{q_{2k}} = \frac{p_{2k+1}q_{2k} - p_{2k}q_{2k+1}}{q_{2k}q_{2k+1}} \underset{\text{(E.22)}}{=} \frac{1}{q_{2k}q_{2k+1}} > 0.$$

$\square$

With these auxiliary results we can now show the claim required for the SHOR algorithm, which states that, if a positive rational number is sufficiently close to a second positive rational number, the first rational number has to be a continued fraction of the second.

Theorem E.9 *Let $P, Q \in \mathbb{N}$ be given and let $[a_0; \ldots, a_n]$ be the continued fraction of their quotient, that is*

$$[a_0; a_1, \ldots, a_n] = \frac{P}{Q}. \tag{E.29}$$

If $p, q \in \mathbb{N}$ are such that

$$\left| \frac{P}{Q} - \frac{p}{q} \right| < \frac{1}{2q^2}, \tag{E.30}$$

then $\frac{p}{q}$ is a convergent of the continued fraction of $\frac{P}{Q}$, that is, there exists a $j \in \{0, 1, \ldots, n\}$ such that

$$\frac{p}{q} = [a_0; a_1, \ldots, a_j] = \frac{p_j}{q_j},$$

where p_j and q_j are as constructed in Theorem E.7.

Proof Let $p_0, \ldots, p_n$ and $q_0, \ldots, q_n$ be given by the recursive construction rules (E.16) and (E.17) in Theorem E.7. This implies that for $j \in \{0, \ldots, n\}$ we have $[a_0; a_1, \ldots, a_j] = \frac{p_j}{q_j}$ and, in particular,

$$\frac{p_n}{q_n} = \underbrace{[a_0; a_1, \ldots, a_n]}_{(E.29)} = \frac{P}{Q}. \tag{E.31}$$

First, suppose $q \geq q_n$. Then we have

$$\left| \frac{p_n}{q_n} - \frac{p}{q} \right| \underset{(E.31)}{=} \left| \frac{P}{Q} - \frac{p}{q} \right| \underset{(E.30)}{<} \frac{1}{2q^2},$$

and multiplying both sides by qq_n yields

$$|p_n q - q_n p| < \frac{q_n}{2q} \leq \frac{1}{2}.$$

Since $p_n q - q_n p \in \mathbb{Z}$, we must have $p_n q = q_n p$, which implies $\frac{p}{q} = \frac{p_n}{q_n}$, and the claim holds with $j = n$.

Now, suppose $q < q_n$. From (E.18) we know that $q_0 = 1$. Therefore, in the case $q < q_n$ there must be a $j \in \{0, \ldots, n-1\}$ such that

$$q_j \leq q < q_{j+1}. \tag{E.32}$$

We show that then the following inequality

$$\left| \frac{P}{Q} - \frac{p_j}{q_j} \right| \underset{(E.31)}{=} \left| \frac{p_n}{q_n} - \frac{p_j}{q_j} \right| < \frac{1}{2q_j q}$$

holds. For this we choose $a, b \in \mathbb{Z}$ as follows

$$\begin{aligned} a &= (-1)^{j+1} \left(q_{j+1} p - p_{j+1} q \right) \\ b &= (-1)^{j+1} \left(p_j q - q_j p \right). \end{aligned} \tag{E.33}$$

This implies

$$\begin{aligned} p_j a + p_{j+1} b &= (-1)^{j+1} \left(p_j q_{j+1} p - p_j p_{j+1} q + p_{j+1} p_j q - p_{j+1} q_j p \right) \\ &= (-1)^{j+1} \underbrace{\left(p_j q_{j+1} - p_{j+1} q_j \right)}_{\underset{(E.22)}{=} (-1)^{j+1}} p \\ &= p. \end{aligned} \tag{E.34}$$

Similarly, one shows that

$$q_j a + q_{j+1} b = q. \tag{E.35}$$

From (E.23) in Corollary E.8 we know that

$$\gcd(q_{j+1}, p_{j+1}) = 1. \tag{E.36}$$

This excludes $a = 0$, for, if $a = 0$, it would follow from assumption (E.32) and (E.33) that $p_{j+1} = \frac{q_{j+1}}{q} p > p$ and thus

$$\frac{p_{j+1}}{q_{j+1}} = \frac{p}{q}$$

with $p_{j+1} > p$ and $q_{j+1} > q$, which is impossible because of (E.36). Hence, due to $a \in \mathbb{Z}$ it must be that $|a| \geq 1$ and we obtain

$$\left| q \frac{p_n}{q_n} - p \right| = \left| (a q_j + b q_{j+1}) \frac{p_n}{q_n} - (a p_j + b p_{j+1}) \right|$$
$$= \left| a \underbrace{\left(q_j \frac{p_n}{q_n} - p_j \right)}_{=: c_j} + b \underbrace{\left(q_{j+1} \frac{p_n}{q_n} - p_{j+1} \right)}_{=: c_{j+1}} \right|. \tag{E.37}$$

According to (E.27) in Corollary E.8 one has for even $j \in \{0, \dots, n-1\}$ that

$$\frac{p_j}{q_j} < \frac{p_n}{q_n} \leq \frac{p_{j+1}}{q_{j+1}},$$

whereas for odd j

$$\frac{p_{j+1}}{q_{j+1}} \leq \frac{p_n}{q_n} < \frac{p_j}{q_j}.$$

This implies

$$c_j c_{j+1} \leq 0. \tag{E.38}$$

On the other hand, it follows from (E.34) that

$$b = \frac{p - p_j a}{p_{j+1}}$$

such that

$$a < 0 \quad \Rightarrow \quad b > 0. \tag{E.39}$$

Similarly, it follows from (E.35) that

$$a = \frac{q - q_{j+1}b}{q_j}$$

and with $b \in \mathbb{Z}$ as well as $q < q_{j+1}$, thus,

$$b > 0 \quad \Rightarrow \quad q < bq_{j+1} \quad \Rightarrow \quad a < 0. \tag{E.40}$$

From (E.39) and (E.40) follows $ab \leq 0$ and together with (E.38) that

$$(ac_j)(bc_{j+1}) \geq 0. \tag{E.41}$$

With this we can estimate a lower bound for the left side of (E.37) as follows:

$$\left| q\frac{p_n}{q_n} - p \right| = |ac_j + bc_{j+1}| \underset{(E.41)}{=} |ac_j| + |bc_{j+1}|$$

$$\geq |ac_j| = |a|\,|c_j| \geq |c_j| = \left| q_j\frac{p_n}{q_n} - p_j \right|, \tag{E.42}$$

which leads to

$$\left| \frac{P}{Q} - \frac{p_j}{q_j} \right| \underset{(E.31)}{=} \left| \frac{p_n}{q_n} - \frac{p_j}{q_j} \right| \underset{(E.42)}{\leq} \frac{q}{q_j}\left| \frac{p_n}{q_n} - \frac{p}{q} \right| \underset{(E.31)}{=} \frac{q}{q_j}\left| \frac{P}{Q} - \frac{p}{q} \right| \underset{(E.30)}{<} \frac{q}{q_j}\frac{1}{2q^2} = \frac{1}{2q_j q}. \tag{E.43}$$

Combining (E.30) and (E.43) we then obtain

$$
\begin{aligned}
\left| \frac{p}{q} - \frac{p_j}{q_j} \right| &= \left| \frac{p}{q} - \frac{P}{Q} + \frac{P}{Q} - \frac{p_j}{q_j} \right| \\
&\leq \left| \frac{p}{q} - \frac{P}{Q} \right| + \left| \frac{P}{Q} - \frac{p_j}{q_j} \right| \\
&\underset{(E.30),(E.43)}{<} \frac{1}{2q^2} + \frac{1}{2q_j q}. \tag{E.44}
\end{aligned}
$$

Multiplying both sides in (E.44) with qq_j yields

$$\left| pq_j - qp_j \right| < \frac{q_j}{2q} + \frac{1}{2} \underset{(E.32)}{\leq} 1,$$

that is, $\left| pq_j - qp_j \right| < 1$, which, because of $pq_j - qp_j \in \mathbb{Z}$, finally implies

$$\frac{p}{q} = \frac{p_j}{q_j}.$$

□

Appendix F
Some Group Theory

F.1 Groups, Subgroups and Quotient Groups

Groups play a very important—if not crucial—role in physics in general and in quantum mechanics in particular. They are formally defined as follows.

Definition F.1 A **group** $(\mathcal{G}, \cdot)$ is a set $\mathcal{G}$ together with a binary operation

$$\cdot : \mathcal{G} \times \mathcal{G} \longrightarrow \mathcal{G}$$
$$(g, h) \longmapsto g \cdot h =: gh$$

called (group) multiplication that has the following three properties.

Associativity: the group multiplication is associative, that is, for all $g, h, k \in \mathcal{G}$

$$(g \cdot h) \cdot k = g \cdot (h \cdot k) \tag{F.1}$$

such that $g \cdot h \cdot k$ is well defined since it does not make any difference which of the two multiplications is carried out first.

Existence of a unit or neutral element: there exists an $e \in \mathcal{G}$ such that for every $g \in \mathcal{G}$

$$g \cdot e = g. \tag{F.2}$$

Existence of inverses: for each $g \in \mathcal{G}$ there exists an element $g^{-1} \in \mathcal{G}$ called the inverse of g which has the property

© Springer Nature Switzerland AG 2019
W. Scherer, *Mathematics of Quantum Computing*,
https://doi.org/10.1007/978-3-030-12358-1

$$g \cdot g^{-1} = e. \tag{F.3}$$

A group $\mathcal{G}$ is called **discrete** if the set $\mathcal{G}$ is countable. The group is called **finite** if the set $\mathcal{G}$ is finite. The number of elements in a finite group is called the **order of the group** and denoted by $|\mathcal{G}|$.

If for $g \in \mathcal{G}$ there exists an $n \in \mathbb{N}$ such that $g^n = e$, then the smallest such n is called the **order of the element** g and denoted by $\mathrm{ord}(g)$. If no such n exists, the order is said to be infinite.

A group is called **abelian** if the order of the multiplication does not matter, that is, if $g \cdot h = h \cdot g$ for all $g, h \in \mathcal{G}$.

To emphasize which group multiplication is used, we write at times $\cdot_{\mathcal{G}}$. On the other hand, if there is no danger of confusion, the multiplication sign $\cdot$ is often omitted and one writes $\mathcal{G}$ instead of $(\mathcal{G}, \cdot)$ and gh instead of $g \cdot h$. For abelian groups the multiplication is often denoted by $+$ instead of $\cdot$.

It turns out that the unit element of any group and the inverse of any group element are unique.

Exercise F.121 Show that the unit element of a group $\mathcal{G}$ is unique and that for any $g \in \mathcal{G}$ its inverse g^{-1} is also unique.

For a solution see Solution F.121

Hence, we may speak of *the neutral element* of a group and, likewise, of *the inverse* of a group element.

Exercise F.122 Let $\mathcal{G}$ be a group with unit element e. Show that then

$$e^{-1} = e \tag{F.4}$$

and for all $g, h \in \mathcal{G}$

$$(gh)^{-1} = h^{-1}g^{-1} \tag{F.5}$$

$$\left(g^{-1}\right)^{-1} = g \tag{F.6}$$

$$h = g \quad \Leftrightarrow \quad h^{-1} = g^{-1}, \tag{F.7}$$

where we have already adopted the convention of not writing out the group multiplication sign $\cdot$.

For a solution see Solution F.122

In the definition given above we have defined the unit element as a trivial multiplication from the right and the inverse as a right inverse. It turns out that the unit element also acts trivially from the left, and the right inverse also acts as an inverse from the left.

Exercise F.123 Let $(\mathcal{G}, \cdot)$ be a group with unit element e. Show that then for all $g \in \mathcal{G}$

$$g^{-1}g = e \tag{F.8}$$
$$eg = g \tag{F.9}$$

and that for any $h, k \in \mathcal{G}$

$$gh = gk \quad \Leftrightarrow \quad h = k \quad \Leftrightarrow \quad hg = kg. \tag{F.10}$$

For a solution see Solution F.123

Groups which are not discrete are called continuous groups. If for a continuous group $(\mathcal{G}, \cdot)$ the set $\mathcal{G}$ also has what is called a differentiable structure, such that it is a differentiable manifold, then the group is called a LIE group. Such groups are at the heart of not only quantum mechanics but all of quantum theory such as quantum field theory, elementary particles, string theory, etc. For quantum computing, however, they do not play such a central role, which is why we do not pursue them any further here.

Overall, groups are more ubiquitous than the unknowing reader might be aware of. Hence, it is worthwhile to consider a number of examples.

Example F.2 As a first simple example of an abelian continuous group, we consider $(\mathbb{R}, +)$. Here the unit element is $e = 0$, and the inverse of an $a \in \mathbb{R}$ is $-a$.

A first example of a discrete group is given by the integers.

Example F.3 As a first simple example of an abelian discrete group we consider $(\mathbb{Z}, +)$, that is, the integers with addition as group multiplication. Here again, the unit element is $e = 0$, and the inverse of a $z \in \mathbb{Z}$ is $-z$.

Many sets of maps can also be made into groups if multiplication can be suitably defined.

Example F.4 Let $\mathbb{V}$ be a vector space over a field $\mathbb{F}$ (see Definition F.53), and let

$$\mathrm{L}(\mathbb{V}) := \{M : \mathbb{V} \to \mathbb{V} \mid M \text{ linear}\}$$

denote the set of linear maps of $\mathbb{V}$ onto itself. The set of linear invertible maps

$$\mathrm{GL}(\mathbb{V}) := \left\{ M \in \mathrm{L}(\mathbb{V}) \,\middle|\, M^{-1} \in \mathrm{L}(\mathbb{V}) \right\}$$

forms a group with the multiplication

$$\begin{aligned} \cdot : \mathrm{GL}(\mathbb{V}) \times \mathrm{GL}(\mathbb{V}) &\longrightarrow \mathrm{GL}(\mathbb{V}) \\ (M_1, M_2) &\longmapsto M_1 \cdot M_2 := M_1 M_2 \end{aligned} ,$$

where $M_1 M_2$ denotes the composition of the maps M_1 and M_2. This is because the successive application of linear maps is associative and for any two elements yields another element of the set since the composition of linear maps is again a linear map. The identity map $\mathrm{id}_{\mathbb{V}}$ is the neutral element (also denoted by $\mathbf{1}_{\mathbb{V}}$), and each element of the set has its inverse in the set. This group is called the **general linear group** of $\mathbb{V}$ and denoted by $\mathrm{GL}(\mathbb{V})$. If $\dim \mathbb{V} > 1$, it is non-abelian.

A particular case is $\mathbb{V} = \mathbb{C}^n$ for $n \in \mathbb{N}$ and $\mathbb{F} = \mathbb{C}$, which gives the group of all invertible complex $n \times n$ matrices

$$\mathrm{GL}(n, \mathbb{C}) := \left\{ M \in \mathrm{Mat}(n \times n, \mathbb{C}) \,\middle|\, \det M \neq 0 \right\}.$$

Together with the usual matrix multiplication, this set forms a group which is abelian for $n = 1$ but non-abelian for $n > 1$. This group is denoted by $\mathrm{GL}(n, \mathbb{C})$. It is also a continuous, in other words, a non-discrete group, and since the underlying set is a differentiable manifold, it is also a LIE group.

Likewise,

$$\mathrm{GL}(n, \mathbb{R}) := \left\{ M \in \mathrm{Mat}(n \times n, \mathbb{R}) \,\middle|\, \det M \neq 0 \right\}$$

forms a group.

Lemma F.5 *Let $N \in \mathbb{N}$. Then the set $\mathcal{G} = \{0, 1, \ldots, N-1\}$ with the group multiplication given by addition modulo N, that is,*

$$a +_{\mathbb{Z}_N} b := (a+b) \bmod N \tag{F.11}$$

is a finite abelian group $(\mathcal{G}, +_{\mathbb{Z}_N})$ denoted by $\mathbb{Z}_N$.

Moreover, for a prime p the set $\mathcal{G} \setminus \{0\} = \{1, \ldots, p-1\}$ with group multiplication given by

$$a \cdot_{\mathbb{Z}_p} b := (ab) \bmod p \tag{F.12}$$

also constitutes a finite abelian group $(\mathcal{G} \setminus \{0\}, \cdot_{\mathbb{Z}_p})$, which is denoted by $\mathbb{Z}_p^{\times}$.

Proof First, consider the additive group $\mathbb{Z}_N$. Since $(a+b) \bmod N \in \{0, \ldots, N-1\}$, the group multiplication given in (F.11) is a map $\mathcal{G} \times \mathcal{G} \to \mathcal{G}$ and, due to (D.23), it is associative. The neutral element is $e_+ = 0$ and the inverse of an $a \in \mathcal{G} \setminus \{0\}$ is given by $a^{-1} = N - a \in \mathcal{G}$ since

$$a +_{\mathbb{Z}_N} a^{-1} \underbrace{=}_{(\text{F.11})} (a + N - a) \bmod N = 0 = e_+ . \tag{F.13}$$

Clearly, $\mathcal{G}$ is finite and $a +_{\mathbb{Z}_N} b = b +_{\mathbb{Z}_N} a$.

Now consider $\mathbb{Z}_p^\times$, where p is a prime, and let $a, b \in \{1, \ldots, p-1\}$ such that $a \bmod p \neq 0 \neq b \bmod p$. From (D.28) in Lemma D.11 it then follows that $(ab) \bmod p \neq 0$. Consequently, $(ab) \bmod p \in \{1, \ldots, N-1\}$, and the group multiplication given in (F.12) is a map $\mathcal{G} \setminus \{0\} \times \mathcal{G} \setminus \{0\} \to \mathcal{G} \setminus \{0\}$. Because of (D.20), it is associative. The neutral element of this group is $e_. = 1$ since for any $a \in \{1, \ldots, p-1\}$ we have $a \cdot_{\mathbb{Z}_p} e_. = a \bmod p = a$. Moreover, since for any such a we have $\gcd(a, p) = 1$, we know from (D.12) in the extended EUCLID algorithm given in Theorem D.4 that we can always find $x, y \in \mathbb{Z}$ such that $ax + py = 1$. Lemma D.9 then implies that $x \bmod p = a^{-1} \bmod p \in \mathcal{G} \setminus \{0\}$ such that

$$a \cdot_{\mathbb{Z}_p} (x \bmod p) \underbrace{=}_{(\text{F.12})} \big(a(x \bmod p)\big) \bmod p \underbrace{=}_{(\text{D.27})} 1 .$$

Hence, every $a \in \mathcal{G} \setminus \{0\}$ has an inverse under $\cdot_{\mathbb{Z}_p}$ in $\mathcal{G} \setminus \{0\}$.

Obviously, $a \cdot_{\mathbb{Z}_p} b = (ab) \bmod p = b \cdot_{\mathbb{Z}_p} a$, and the proof that $\mathbb{Z}_p^\times$ is a finite abelian group is complete. $\qquad\square$

Definition F.6 Let $(\mathcal{G}, \cdot)$ be a group with neutral element e. A subset $\mathcal{H} \subset \mathcal{G}$ which satisfies

$$e \in \mathcal{H} \tag{F.14}$$
$$h \in \mathcal{H} \Rightarrow h^{-1} \in \mathcal{H} \tag{F.15}$$
$$h_1, h_2 \in \mathcal{H} \Rightarrow h_1 \cdot h_2 \in \mathcal{H} \tag{F.16}$$

is called a **subgroup** of $\mathcal{G}$, and this is expressed by writing $\mathcal{H} \leq \mathcal{G}$.

A subgroup $\mathcal{H}$ is called a **proper** subgroup of $\mathcal{G}$ if $\mathcal{H} \subsetneqq \mathcal{G}$, and this is expressed by writing $\mathcal{H} < \mathcal{G}$.

A proper subgroup $\mathcal{H}$ of $\mathcal{G}$ is called **maximal** if it is not a proper subgroup of a proper subgroup of $\mathcal{G}$, namely if there is no $\mathcal{K} < \mathcal{G}$ such that $\mathcal{H} < \mathcal{K}$. Likewise, a subgroup $\mathcal{H}$ of $\mathcal{G}$ is called **minimal** if there is no $\mathcal{K} < \mathcal{H}$ such that $\{e\} < \mathcal{K}$.

Example F.7 Consider the group $(\mathcal{G},+) = (\mathbb{Z},+)$ of Example F.3. For $N \in \mathbb{N}$ define

$$N\mathbb{Z} := \{Nk \mid k \in \mathbb{Z}\} = \{0, \pm N, \pm 2N, \pm 3N, \ldots\}.$$

This is clearly a subset of $\mathbb{Z}$ which contains the neutral element 0 of the group $(\mathbb{Z},+)$. Addition of two of its elements $Nk, Nl \in N\mathbb{Z}$ results in $N(k+l) \in N\mathbb{Z}$. Moreover, for any $Nk \in N\mathbb{Z}$, there is $N(-k) \in N\mathbb{Z}$ such that adding them yields the neutral element 0. The case $N = 1$ does not give anything new since $1\mathbb{Z} = \mathbb{Z}$, but for $N > 1$ we have that $N\mathbb{Z}$ is a proper subgroup of $\mathbb{Z}$, that is,

$$N\mathbb{Z} < \mathbb{Z}.$$

Many groups of interest in physics are subgroups of the general linear linear group of $\mathbb{C}^n$ for some $n \in \mathbb{N}$.

Example F.8 Let $n \in \mathbb{N}$. Within the group $\mathrm{GL}(n,\mathbb{C})$ of Example F.4 consider the set of all unitary $n \times n$ matrices

$$\mathrm{U}(n) := \{U \in \mathrm{Mat}(n \times n, \mathbb{C}) \mid UU^* = \mathbf{1}\}.$$

The set $\mathrm{U}(n)$ contains the unit matrix as the neutral element, and since the product of two unitary matrices is again unitary, multiplication of two elements of the set $\mathrm{U}(n)$ yields another element of $\mathrm{U}(n)$. By definition, each element U of the set has its inverse U^* in the set. Hence, $\mathrm{U}(n)$ is a subgroup of $\mathrm{GL}(n,\mathbb{C})$. It is called the **unitary group** in n dimensions and denoted by $\mathrm{U}(n)$.

The case $\mathrm{U}(1)$ is special in the sense that it is an abelian group and we can identify it with the unit circle in $\mathbb{C}$

$$\mathrm{U}(1) = \{z \in \mathrm{Mat}(1 \times 1, \mathbb{C}) = \mathbb{C} \mid z\bar{z} = 1\}.$$

Moreover, within $\mathrm{U}(n)$ the set

$$\mathrm{SU}(n) := \{U \in \mathrm{U}(n) \mid \det U = 1\}$$

also forms a subgroup, which is called the **special unitary group** in n dimensions and denoted by $SU(n)$. It is a subgroup of $\mathrm{U}(n)$, and we have

$$\mathrm{SU}(n) < \mathrm{U}(n) < \mathrm{GL}(n,\mathbb{C}).$$

Similarly, within the group $\mathrm{GL}(n,\mathbb{R})$ of Example F.4 the set of all orthogonal $n \times n$ matrices

$$O(n) := \left\{ M \in \mathrm{Mat}(n \times n, \mathbb{R}) \,\middle|\, MM^T = \mathbf{1} \right\}$$

forms a subgroup of $\mathrm{GL}(n, \mathbb{R})$ called the **orthogonal group**. Within $O(n)$ the set

$$SO(n) := \left\{ M \in O(n) \,\middle|\, \det M = 1 \right\}$$

also forms a subgroup, which is called the **special orthogonal group** in n dimensions and denoted by $SO(n)$.

Intersections of subgroups form another subgroup as is to be shown in Exercise F.124.

Exercise F.124 Let I be an index set and let $\{\mathcal{H}_j \,|\, j \in I\}$ be a set of subgroups of a group $\mathcal{G}$. Show that then

$$\mathcal{H}_\cap := \bigcap_{j \in I} \mathcal{H}_j$$

is a subgroup of $\mathcal{G}$.

For a solution see Solution F.124

Definition F.9 Let K be a nonempty subset of a group $\mathcal{G}$ and

$$S_K := \left\{ \mathcal{H} \leq \mathcal{G} \,\middle|\, K \subset \mathcal{H} \right\}$$

the set of all subgroups that contain K. Then

$$\langle K \rangle := \bigcap_{\mathcal{H} \in S_K} \mathcal{H}$$

is defined as the group **generated** by K.

A group $\mathcal{G}$ is said to be **finitely generated** if there are $g_1, \ldots g_n \in \mathcal{G}$ such that

$$\mathcal{G} = \langle g_1, \ldots, g_n \rangle := \langle \{g_1, \ldots, g_n\} \rangle .$$

A group $\mathcal{G}$ which can be generated from one element $g \in \mathcal{G}$ such that

$$\mathcal{G} = \langle g \rangle$$

is called **cyclic** and in this case g is called the **generator** of $\mathcal{G}$.

Any element of a cyclic group $\mathcal{G} = \langle g \rangle$ is a power of the group generator, that is, for each $\widetilde{g} \in \mathcal{G}$ there exists an $m \in \mathbb{Z}$ such that $\widetilde{g} = g^m$.

The adjective cyclic may mislead to believe that the sequence $(g^m)_{m \in \mathbb{Z}}$ may repeat itself, but this does not have to be the case as the following example shows.

Example F.10 The group $(\mathcal{G}, +) = (\mathbb{Z}, +)$ of Example F.3 is generated by 1 such that as a group $\mathbb{Z} = \langle 1 \rangle$ since, with $g = 1$, we have

$$\mathbb{Z} = \big\{ \pm (\underbrace{1 + \cdots + 1}_{m \text{ times}}) \,\big|\, m \in \mathbb{N}_0 \big\} = \{ g^m \mid m \in \mathbb{Z} \}.$$

However, for a finite cyclic group $\mathcal{G} = \langle g \rangle$ the sequence $(g^m)_{m \in \mathbb{Z}}$ has to repeat itself. In this case there exists an $n \in \mathbb{N}$ such that $g^n = e_{\mathcal{G}}$, where $n = |\mathcal{G}|$ is the number of the elements in $\mathcal{G}$, and in terms of sets we have

$$\mathcal{G} = \langle g \rangle = \{ g^0 = e_{\mathcal{G}}, g^1, \ldots g^{n-1} \}.$$

Cyclic groups appear as minimal subgroups of groups as the following lemma shows.

Lemma F.11 *Every minimal subgroup $\mathcal{H} < \mathcal{G}$ of a group $\mathcal{G}$ is cyclic, that is, it is of the form $\mathcal{H} = \langle g \rangle$ for some $g \in \mathcal{G}$.*

Proof Recall that a subgroup $\mathcal{H} < \mathcal{G}$ is called minimal if there is no subgroup $\mathcal{K} \leq \mathcal{G}$ such that $\langle e_{\mathcal{G}} \rangle < \mathcal{K} < \mathcal{H}$.

Let $\mathcal{H}$ be minimal and $g \in \mathcal{H} \setminus \langle e_{\mathcal{G}} \rangle$. If $\mathcal{H} = \langle g \rangle$, we are done. Otherwise, take a $k \in \mathcal{H} \setminus \langle g \rangle$. But then we have

$$\langle e_{\mathcal{G}} \rangle < \langle g \rangle < \langle g, k \rangle \leq \mathcal{H},$$

which contradicts our assumption that $\mathcal{H}$ is minimal. $\qquad\square$

Another notion that will play a role in our considerations of error correcting codes is that of independence of elements of a group.

Definition F.12 A subset $\{ g_1, \ldots, g_k \} \subset \mathcal{G}$ of a group $\mathcal{G}$ is called **independent** if for any $g_j \in \{ g_1, \ldots, g_k \}$

$$g_j \notin \langle \{ g_1, \ldots, g_k \} \setminus \{ g_j \} \rangle.$$

Let P denote a permutation of the index set $\{1,\ldots,k\}$, that is, the map

$$P: \{1,\ldots,k\} \longrightarrow \{1,\ldots,k\}$$
$$j \longmapsto P(j)$$

is a bijection.

Independence of a subset $\{g_1,\ldots,g_k\} \subset \mathcal{G}$ of a group means that for every $a_1,\ldots,a_k \in \{0,1\}$ and permutation P we must have

$$\prod_{j=1}^{k} g_{P(j)}^{a_j} = e_\mathcal{G} \quad \Rightarrow \quad a_j = 0 \quad \forall j \in \{1,\ldots,k\}.$$

This is because otherwise we would have for some j and P

$$g_{P(j)} = g_{P(j-1)}^{-a_{j-1}} \cdots g_{P(1)}^{-a_1} g_{P(k)}^{-a_k} g_{P(j+1)}^{-a_{j+1}} \in \langle \{g_1,\ldots,g_k\} \setminus \{g_{P(j)}\} \rangle,$$

and since $P(j) \in \{1,\ldots,k\}$ the set $\{g_1,\ldots,g_k\}$ would not be independent.

Definition F.13 Let $\mathcal{G}$ be a group and $g \in \mathcal{G}$. The **centralizer** $\mathrm{Clz}_\mathcal{G}(g)$ of g is the set of elements of $\mathcal{G}$ which commute with g, that is,

$$\mathrm{Clz}_\mathcal{G}(g) := \{h \in \mathcal{G} \mid hg = gh\}.$$

The centralizer of a subset $S \subset \mathcal{G}$ is defined as

$$\mathrm{Clz}_\mathcal{G}(S) := \{h \in \mathcal{G} \mid hg = gh \quad \forall g \in S\}. \tag{F.17}$$

Exercise F.125 Show that the centralizer of any subset $S \subset \mathcal{G}$ of a group $\mathcal{G}$ is a subgroup of $\mathcal{G}$, that is,
$$\mathrm{Clz}_\mathcal{G}(S) \leq \mathcal{G}.$$

For a solution see Solution F.125

Definition F.14 Let $S \subset \mathcal{G}$ be a subset of the group $\mathcal{G}$. The **conjugate** S^g of S by g is defined as the set

$$S^g := gSg^{-1} = \{ghg^{-1} \mid h \in S\}. \tag{F.18}$$

A conjugate of a subgroups is again a subgroup as is to be shown in Exercise F.126.

Exercise F.126 Let $\mathcal{H}$ be a subgroup of the group $\mathcal{G}$. Show that for any $g \in \mathcal{G}$ the conjugate of $\mathcal{H}$ by g

$$\mathcal{H}^g = \left\{ ghg^{-1} \mid h \in \mathcal{H} \right\} \tag{F.19}$$

is a subgroup of $\mathcal{G}$.

For a solution see Solution F.126

Definition F.15 Let $\mathcal{H}$ be a subgroup of the group $\mathcal{G}$. For any $g \in \mathcal{G}$ the set

$$\mathcal{H}^g := \left\{ ghg^{-1} \mid h \in \mathcal{H} \right\}$$

is called a **conjugate subgroup** to $\mathcal{H}$. If for every $g \in \mathcal{G}$

$$\mathcal{H}^g = \mathcal{H},$$

then $\mathcal{H}$ is called a **normal or invariant subgroup** of $\mathcal{G}$, and this is denoted by $\mathcal{H} \trianglelefteq \mathcal{G}$.

For an abelian group any subgroup is normal.

Definition F.16 Let S be a subset of the group $\mathcal{G}$. The **normalizer** $\mathrm{Nor}_{\mathcal{G}}(S)$ of S in $\mathcal{G}$ is defined as

$$\mathrm{Nor}_{\mathcal{G}}(S) := \left\{ g \in \mathcal{G} \mid S^g = S \right\}. \tag{F.20}$$

Note that by definition

$$
\begin{aligned}
g \in \mathrm{Nor}_{\mathcal{G}}(S) \quad &\Leftrightarrow \quad \forall h \in S \quad \exists \widetilde{h} \in S: \; ghg^{-1} = \widetilde{h} \\
&\Leftrightarrow \quad \forall h \in S \quad \exists \widetilde{h} \in S: \; gh = \widetilde{h}g.
\end{aligned}
$$

Exercise F.127 Show that for any subset S of the group $\mathcal{G}$ its normalizer is a subgroup of $\mathcal{G}$, that is,

$$\mathrm{Nor}_{\mathcal{G}}(S) \leq \mathcal{G}.$$

For a solution see Solution F.127

The set of group elements which commute with every element in the group is called the center of the group.

Definition F.17 The **center** of a group $\mathcal{G}$ is defined as

$$\mathrm{Ctr}(\mathcal{G}) := \left\{ h \in \mathcal{G} \,\middle|\, hg = gh \quad \forall g \in \mathcal{G} \right\}. \tag{F.21}$$

The center is actually a normal subgroup as is to be shown in the following exercise.

Exercise F.128 Show that the center of a group $\mathcal{G}$ is a normal subgroup, that is,

$$\mathrm{Ctr}(\mathcal{G}) \trianglelefteq \mathcal{G}.$$

For a solution see Solution F.128

Definition F.18 Let $\mathcal{H}$ be a subgroup of the group $\mathcal{G}$. For any $g \in \mathcal{G}$ the set

$$g\mathcal{H} := \left\{ gh \,\middle|\, h \in \mathcal{H} \right\} \tag{F.22}$$

is called the **left coset** of g, and the set

$$\mathcal{H}g := \left\{ hg \,\middle|\, h \in \mathcal{H} \right\} \tag{F.23}$$

is called the **right coset** of g.

If $\mathcal{H}$ is such that left and right cosets are identical, we only speak of cosets and denote them by $[g]_{\mathcal{H}}$. To simplify notation, we may at times write $[g]$ only if it is clear which is the subgroup for the cosets.

Clearly, for a subgroup $\mathcal{H}$ of an abelian group $\mathcal{G}$ left and right cosets coincide.

Example F.19 Consider the group $(\mathbb{Z}, +)$ of Example F.3 and its subgroup $(N\mathbb{Z}, +)$ of Example F.7 for some $N > 1$. For any element $g \in \mathbb{Z}$ we have the coset

$$[g]_{N\mathbb{Z}} = \{g + Nk \,|\, k \in \mathbb{Z}\}$$
$$= \{g, g \pm N, g \pm 2N, g \pm 3N, \ldots\}$$
$$= \{g \bmod N, g \bmod N \pm N, g \bmod N \pm 2N, g \bmod N \pm 3N, \ldots\}$$
$$= [g \bmod N]_{N\mathbb{Z}}, \qquad (F.24)$$

that is, any coset $[g]_{N\mathbb{Z}} \in N\mathbb{Z}$ is equal to $[m]_{N\mathbb{Z}}$, where $m = g \bmod N \in \{0, 1, \ldots, N - 1\}$.

If $\mathcal{H}$ is a subgroup of the group $\mathcal{G}$, then we have for any $k \in \mathcal{H}$ and any $g \in \mathcal{G}$

$$k\mathcal{H} \underbrace{=}_{(F.22)} \{kh \,|\, h \in \mathcal{H}\} = \{h' \,|\, h' \in \mathcal{H}\} = \mathcal{H}$$

$$gk\mathcal{H} \underbrace{=}_{(F.22)} \{gkh \,|\, h \in \mathcal{H}\} = \{gh' \,|\, h' \in \mathcal{H}\} = g\mathcal{H}. \qquad (F.25)$$

Lemma F.20 *Let $\mathcal{H}$ be a subgroup of the group $\mathcal{G}$. For any two $g_1, g_2 \in \mathcal{G}$ their left cosets $g_1\mathcal{H}$ and $g_2\mathcal{H}$ are either disjoint or they are identical. The same holds for any two right cosets.*

Proof If $g_1\mathcal{H} \cap g_2\mathcal{H} = \emptyset$ they are disjoint, and there is nothing to prove. Suppose then that there is a $g \in g_1\mathcal{H} \cap g_2\mathcal{H}$, namely that there exist $h_1, h_2 \in \mathcal{H}$ such that

$$g_1 h_1 = g = g_2 h_2. \qquad (F.26)$$

Since $h_1, h_2 \in \mathcal{H}$ and $\mathcal{H}$ is a subgroup, we have $h_1 h_2^{-1} \in \mathcal{H}$. Consequently, for any $h \in \mathcal{H}$

$$g_2 h \underbrace{=}_{(F.26)} g_1 \underbrace{h_1 h_2^{-1} h}_{\in \mathcal{H}} \underbrace{\in}_{(F.22)} g_1 \mathcal{H}$$

and thus

$$g_2\mathcal{H} \subset g_1\mathcal{H}. \qquad (F.27)$$

Similarly, we have $h_2 h_1^{-1} \in \mathcal{H}$

$$g_1 \underbrace{=}_{(F.26)} g_2 \underbrace{h_2 h_1^{-1}}_{\in \mathcal{H}} \underbrace{\in}_{(F.22)} g_2 \mathcal{H},$$

which implies

$$g_1\mathcal{H} \subset g_2\mathcal{H}, \qquad (F.28)$$

and it follows from (F.27) and (F.28) that $g_1 \mathcal{H} \cap g_2 \mathcal{H} \neq \emptyset$ implies $g_1 \mathcal{H} = g_2 \mathcal{H}$. $\square$

The previous lemma allows us to prove what in group theory is known as LAGRANGE's Theorem, which states that for a finite group the number of its elements is divisible by the number of elements of any subgroup.

Theorem F.21 *Let $\mathcal{H}$ be a subgroup of the finite group $\mathcal{G}$. Then the number of elements in each left coset $g\mathcal{H}$ is equal to the order $|\mathcal{H}|$ of $\mathcal{H}$, namely the number of elements in $\mathcal{H}$. Moreover, the order of $\mathcal{H}$ divides the order of $\mathcal{G}$ and $\mathcal{G}$ is the disjoint union of*

$$J = \frac{|\mathcal{G}|}{|\mathcal{H}|} \in \mathbb{N}$$

left cosets of $\mathcal{H}$, that is, there are $g_j \in \mathcal{G}$ with $j \in \{1, \dots, J\}$ such that $g_i \mathcal{H} \cap g_j \mathcal{H} = \emptyset$ if $i \neq j$ and

$$\mathcal{G} = \bigcup_{j=1}^{J} g_j \mathcal{H}. \tag{F.29}$$

The same statement holds for the right cosets.

Proof We only prove the statements for left cosets here. The proof for the right cosets is, of course, similar.

First, we prove (F.29). For this we pick any $g \in \mathcal{G}$ and set $g_1 = g$. Then, successively for $j \in \mathbb{N}$ and as long as $\bigcup_{i=1}^{j} g_i \mathcal{H} \neq \mathcal{G}$, we pick any $g \in \mathcal{G} \smallsetminus \bigcup_{i=1}^{j} g_i \mathcal{H}$ and set $g_{j+1} = g$. By construction, we have for any $k \in \{1, \dots, j\}$ that $g_{j+1} \notin g_k \mathcal{H}$ and it follows from Lemma F.20 that $g_{j+1} \mathcal{H}$ is disjoint from all such $g_k \mathcal{H}$. Moreover, since $\mathcal{G}$ is assumed finite, this process terminates for a J that satisfies (F.29).

Next, we prove the statement about the number of elements in cosets $g\mathcal{H}$ for a given $g \in \mathcal{G}$. For any two elements $h_1, h_2 \in \mathcal{H}$ with $h_1 \neq h_2$ it follows that $gh_1 \neq gh_2$. Consequently, the number of elements in $g\mathcal{H} = \{gh \mid h \in \mathcal{H}\}$ is equal to the number of elements in $\mathcal{H}$. Hence, (F.29) implies that $\mathcal{G}$ is the union of J disjoint sets each of which has $|\mathcal{H}|$ elements. Therefore, the number of elements in $\mathcal{G}$ is given by $|\mathcal{G}| = J|\mathcal{H}|$. $\square$

We know already that for abelian groups left and right cosets of any subgroup coincide. This is actually a general property of normal subgroups of any (not necessarily abelian) group.

Exercise F.129 Let $\mathcal{H}$ be a subgroup of the group $\mathcal{G}$. Show that then

$$\mathcal{H} \text{ is normal} \quad \Leftrightarrow \quad g\mathcal{H} = \mathcal{H}g \quad \forall g \in \mathcal{G}.$$

For a solution see Solution F.129

For a normal subgroup we thus do not need to distinguish between left and right cosets. Moreover, the set of cosets of a normal subgroup can be endowed with a multiplication and made into a group itself as the following proposition shows.

Proposition F.22 *Let $\mathcal{G}$ be a group with neutral element e and let $\mathcal{H} \trianglelefteq \mathcal{G}$. Then the set $\{ [g]_{\mathcal{H}} \mid g \in \mathcal{G} \}$ of cosets forms a group with*

multiplication: *for all $g_1, g_2 \in \mathcal{G}$*

$$[g_1]_{\mathcal{H}} \cdot [g_2]_{\mathcal{H}} := [g_1 g_2]_{\mathcal{H}}, \tag{F.30}$$

neutral element:

$$[e]_{\mathcal{H}} = \mathcal{H}, \tag{F.31}$$

inverse: *for each $g \in \mathcal{G}$*

$$([g]_{\mathcal{H}})^{-1} := [g^{-1}]_{\mathcal{H}}. \tag{F.32}$$

Moreover, for any $g_1, g_2 \in \mathcal{G}$ we have

$$[g_1]_{\mathcal{H}} = [g_2]_{\mathcal{H}} \qquad \Leftrightarrow \qquad \exists h \in \mathcal{H}: \ g_1 = g_2 h. \tag{F.33}$$

Proof Since for any $g_1, g_2 \in \mathcal{G}$ we have $g_1 g_2 \in \mathcal{G}$, the multiplication defined in (F.30) is a binary map

$$\cdot : \{ [g]_{\mathcal{H}} \mid g \in \mathcal{G} \} \times \{ [g]_{\mathcal{H}} \mid g \in \mathcal{G} \} \longrightarrow \{ [g]_{\mathcal{H}} \mid g \in \mathcal{G} \} \\ ([g_1]_{\mathcal{H}}, [g_2]_{\mathcal{H}}) \longmapsto [g_1 g_2]_{\mathcal{H}},$$

and associativity of $\cdot$ follows from associativity in $\mathcal{G}$:

$$([g_1]_{\mathcal{H}} \cdot [g_2]_{\mathcal{H}}) \cdot [g_3]_{\mathcal{H}} \underset{(F.30)}{=} [g_1 g_2]_{\mathcal{H}} \cdot [g_3]_{\mathcal{H}} \underset{(F.30)}{=} ((g_1 g_2) g_3]_{\mathcal{H}} \underset{(F.1)}{=} [g_1 g_2 g_3]_{\mathcal{H}}.$$

To show that the product as defined in (F.30) does not depend on the particular g_1 and g_2 chosen to represent the cosets $[g_1]_{\mathcal{H}}$ and $[g_2]_{\mathcal{H}}$, requires the invariance property of $\mathcal{H}$. For this let $i \in \{1, 2\}$ and $\widetilde{g}_i \in \mathcal{G}$ be such that $\widetilde{g}_i \neq g_i$, but $[\widetilde{g}_i]_{\mathcal{H}} = [g_i]_{\mathcal{H}}$. Then there exist $h_i \in \mathcal{H}$ for $i \in \{1, 2\}$ such that $\widetilde{g}_i = g_i h_i$ and thus

$$[\widetilde{g}_1 \widetilde{g}_2]_{\mathcal{H}} = [g_1 h_1 g_2 h_2]_{\mathcal{H}} \underset{(F.25)}{=} [g_1 h_1 g_2]_{\mathcal{H}}. \tag{F.34}$$

Since $\mathcal{H}$ is assumed normal, Definition F.15 implies that for any $\widetilde{h} \in \mathcal{H}$ and $g \in \mathcal{G}$ there exists an $h \in \mathcal{H}$ such that $g\widetilde{h}g^{-1} = h$ and thus $g\widetilde{h} = hg$. Using this for $h = h_1$ and $g = g_2$ in (F.34) gives

$$[\widetilde{g}_1\widetilde{g}_2]_{\mathcal{H}} = [g_1 h_1 g_2]_{\mathcal{H}} = \left[g_1 g_2 \widetilde{h}_1\right]_{\mathcal{H}} \underbrace{=}_{(F.25)} [g_1 g_2]_{\mathcal{H}},$$

which shows that the product of two cosets $[g_1]_{\mathcal{H}} \cdot [g_2]_{\mathcal{H}}$ as defined in (F.30) does not depend on the choice of the g_i to represent the cosets.

For any $g \in \mathcal{G}$ we have

$$[g]_{\mathcal{H}} \cdot [e]_{\mathcal{H}} \underbrace{=}_{(F.30)} [ge]_{\mathcal{H}} \underbrace{=}_{(F.2)} [g]_{\mathcal{H}},$$

which proves that $[e]_{\mathcal{H}}$ is indeed the neutral element.

Finally,

$$[g]_{\mathcal{H}} \cdot ([g]_{\mathcal{H}})^{-1} \underbrace{=}_{(F.32)} [g]_{\mathcal{H}} \cdot [g^{-1}]_{\mathcal{H}} \underbrace{=}_{(F.30)} [gg^{-1}]_{\mathcal{H}} \underbrace{=}_{(F.3)} [e]_{\mathcal{H}},$$

which verifies that every $[g]_{\mathcal{H}}$ has an inverse in $\left\{ [g]_{\mathcal{H}} \mid g \in \mathcal{G} \right\}$.

To prove (F.33), let $g_1, g_2 \in \mathcal{G}$. Then we have

$$[g_1]_{\mathcal{H}} = [g_2]_{\mathcal{H}}$$
$$\underbrace{\Leftrightarrow}_{(F.22)} \quad \{g_1 h_1 \mid h_1 \in \mathcal{H}\} = \{g_2 h_2 \mid h_2 \in \mathcal{H}\}$$
$$\Leftrightarrow \quad \forall h_1 \in \mathcal{H} \quad \exists h_2 \in \mathcal{H} \text{ and } \forall h_2 \in \mathcal{H} \quad \exists h_1 \in \mathcal{H} : \ g_1 h_1 = g_2 h_2$$
$$\Leftrightarrow \quad \forall h_1 \in \mathcal{H} \quad \exists h_2 \in \mathcal{H} \text{ and } \forall h_2 \in \mathcal{H} \quad \exists h_1 \in \mathcal{H} : \ g_1 = g_2 \underbrace{h_2 h_1^{-1}}_{=h \in \mathcal{H}}$$
$$\Leftrightarrow \quad \exists h \in \mathcal{H} : \ g_1 = g_2 h.$$

$\square$

For a normal subgroup, the statements in Proposition F.22 allow us to define a group that consists of cosets. This group formed by cosets is called is the quotient group.

Definition F.23 Let $\mathcal{H}$ be a normal subgroup of the group $\mathcal{G}$. The group $(\{[g]_{\mathcal{H}} \mid g \in \mathcal{G}\}, \cdot)$ given by the cosets of $\mathcal{H}$ in $\mathcal{G}$ with the multiplication, neutral element and inverses as given in (F.30)–(F.32) is called the **quotient group** of $\mathcal{H}$ in $\mathcal{G}$ and denoted by $\mathcal{G}/\mathcal{H}$.

Example F.24 Consider again the group $(\mathbb{Z}, +)$ of Example F.3 and its subgroup $(N\mathbb{Z}, +)$ of Example F.7 for some $N > 1$. The group multiplication of two cosets $[g_1]_{N\mathbb{Z}}, [g_2]_{N\mathbb{Z}} \in \mathbb{Z}/N\mathbb{Z}$, which we write as $+_{\mathbb{Z}/N\mathbb{Z}}$ since we are dealing with an abelian group, is then given by

$$
\begin{aligned}
[g_1]_{N\mathbb{Z}} +_{\mathbb{Z}/N\mathbb{Z}} [g_2]_{N\mathbb{Z}} &\underset{\text{(F.24)}}{=} [g_1 \bmod N]_{N\mathbb{Z}} +_{\mathbb{Z}/N\mathbb{Z}} [g_2 \bmod N]_{N\mathbb{Z}} \\
&\underset{\text{(F.30)}}{=} [g_1 \bmod N + g_2 \bmod N]_{N\mathbb{Z}} \\
&\underset{\text{(F.24)}}{=} [(g_1 \bmod N + g_2 \bmod N) \bmod N]_{N\mathbb{Z}} \\
&\underset{\text{(D.23)}}{=} [(g_1 + g_2) \bmod N]_{N\mathbb{Z}} \\
&\underset{\text{(F.11)}}{=} [(g_1 +_{\mathbb{Z}_N} g_2)]_{N\mathbb{Z}} ,
\end{aligned}
\tag{F.35}
$$

where $+_{\mathbb{Z}_N}$ is the group multiplication in $\mathbb{Z}_N$ of Lemma F.5.

For a finite group, the number of elements in a quotient group of any of its subgroups is indeed the quotient given by the number of elements in the group divided by the number of elements in the subgroup as the following corollary shows.

Corollary F.25 *Let $\mathcal{H}$ be a normal subgroup of the finite group $\mathcal{G}$. Then the order of the quotient group $\mathcal{G}/\mathcal{H}$ is given by the quotient of the orders of $\mathcal{G}$ and $\mathcal{H}$, namely*

$$
|\mathcal{G}/\mathcal{H}| = \frac{|\mathcal{G}|}{|\mathcal{H}|} .
\tag{F.36}
$$

Proof From Theorem F.21 we know that there are exactly $\frac{|\mathcal{G}|}{|\mathcal{H}|}$ distinct cosets in $\{[g]_{\mathcal{H}} \mid g \in \mathcal{G}\}$, which is the set of elements of the group $\mathcal{G}/\mathcal{H}$.   □

Hence, any normal subgroup of a group gives rise to a new group formed by their quotient group. This is one way to construct new groups from existing ones.

Another way to do this is by forming the direct product group of two groups $(\mathcal{G}_1, \cdot_1)$ and $(\mathcal{G}_2, \cdot_2)$. The underlying set of this group is the cartesian product, and multiplication is defined component-wise in each of the groups.

Exercise F.130 Let $(\mathcal{G}_1, \cdot_{\mathcal{G}_1})$ and $(\mathcal{G}_2, \cdot_{\mathcal{G}_2})$ be two groups. Show that the set $\mathcal{G}_1 \times \mathcal{G}_2$ together with the multiplication

$$((g_1,g_2) \cdot_\times (g_1',g_2') := (g_1 \cdot_{\mathcal{G}_1} g_1', g_2 \cdot_{\mathcal{G}_2} g_2') \qquad (\text{F.37})$$

forms a group, and that if $\mathcal{G}_1$ and $\mathcal{G}_2$ are both finite, this group is also finite and satisfies

$$|\mathcal{G}_1 \times \mathcal{G}_2| = |\mathcal{G}_1||\mathcal{G}_2|. \qquad (\text{F.38})$$

For a solution see Solution F.130

As a consequence of the statement in Exercise F.130, we can give the following definition.

> **Definition F.26** Let $(\mathcal{G}_1, \cdot_{\mathcal{G}_1})$ and $(\mathcal{G}_2, \cdot_{\mathcal{G}_2})$ be groups. Their **direct product group** $(\mathcal{G}_1 \times \mathcal{G}_2, \cdot_\times)$ is defined as the set of pairs $(g,k) \in \mathcal{G}_1 \times \mathcal{G}_2$ with component-wise multiplication
>
> $$\begin{aligned} \cdot_\times : (\mathcal{G}_1 \times \mathcal{G}_2) \times (\mathcal{G}_1 \times \mathcal{G}_2) &\longrightarrow \mathcal{G}_1 \times \mathcal{G}_2 \\ ((g_1,g_2),(g_1',g_2')) &\longmapsto (g_1 \cdot_{\mathcal{G}_1} g_1', g_2 \cdot_{\mathcal{G}_2} g_2') \end{aligned}. \qquad (\text{F.39})$$

Before we turn to maps between groups and related concepts, we introduce the notion of a (left) action of a group on a set and that of a stabilizer on a set.

> **Definition F.27** Let $\mathcal{G}$ be a group with neutral element e and let M be a set. A **left action** of $\mathcal{G}$ on M is defined as a map
>
> $$\begin{aligned} \Lambda : \mathcal{G} \times M &\longrightarrow M \\ (g,m) &\longmapsto g.m \end{aligned}$$
>
> that satisfies for all $h,g \in \mathcal{G}$ and $m \in M$
>
> $$e.m = m \qquad (\text{F.40})$$
> $$hg.m = h.(g.m). \qquad (\text{F.41})$$
>
> The **stabilizer** of a subset $Q \subset M$ under the left action is defined as
>
> $$\mathrm{Sta}_{\mathcal{G}}(Q) := \{ g \in \mathcal{G} \mid g.m = m \quad \forall m \in Q \}. \qquad (\text{F.42})$$

Exercise F.131 Let $\mathcal{G}$ be a group which acts by left action on a set M. Show that for any subset $Q \subset M$ its stabilizer is a subgroup of $\mathcal{G}$, that is,

$$\mathrm{Sta}_{\mathcal{G}}(Q) \leq \mathcal{G}.$$

For a solution see Solution F.131

F.2 Homomorphisms, Characters and Dual Groups

Another way to connect two groups is by maps from one group into the other such that the group multiplication in each of the groups is preserved by the map. Such maps are called homomorphisms. If, in addition, they are bijective, they are called isomorphisms.

Definition F.28 A **homomorphism** between two groups $(\mathcal{G}_1, \cdot_{\mathcal{G}_1})$ and $(\mathcal{G}_2, \cdot_{\mathcal{G}_2})$ is a map $\varphi : \mathcal{G}_1 \to \mathcal{G}_2$ that maintains the group multiplication, that is, which satisfies for all $g, h \in \mathcal{G}_1$

$$\varphi(g) \cdot_{\mathcal{G}_2} \varphi(h) = \varphi(g \cdot_{\mathcal{G}_1} h). \tag{F.43}$$

The set of all homomorphisms from a group $\mathcal{G}_1$ to a group $\mathcal{G}_2$ is denoted by $\mathrm{Hom}(\mathcal{G}_1, \mathcal{G}_2)$.

The pre-image in $\mathcal{G}_1$ of the neutral element $e_2 \in \mathcal{G}_2$ under a homomorphism φ, namely the set

$$\mathrm{Ker}(\varphi) := \big\{ g \in \mathcal{G}_1 \,\big|\, \varphi(g) = e_2 \big\}, \tag{F.44}$$

is called the **kernel** of φ.

A map $\varphi : \mathcal{G} \to \mathcal{G}$ is called an **isomorphism** if it is a homomorphism and a bijection. Two groups $\mathcal{G}_1$ and $\mathcal{G}_2$ are said to be isomorphic if there exists an isomorphism between them, and this is expressed by the notation $\mathcal{G}_1 \cong \mathcal{G}_2$.

Example F.29 Consider again the group $(\mathbb{Z}, +)$ of Example F.3 and its subgroup $(N\mathbb{Z}, +)$ of Example F.7 for some $N > 1$. We will show that the quotient group $\mathbb{Z}/N\mathbb{Z}$ is isomorphic to the group $\mathbb{Z}_N$, which was defined in Lemma F.5 and which consists of the integers $\{0, 1, \ldots, N-1\}$ with multiplication given by addition modulo N. More precisely, we will show that the map

$$
\begin{aligned}
\iota : \mathbb{Z}/N\mathbb{Z} &\longrightarrow \mathbb{Z}_N \\
[g]_{N\mathbb{Z}} &\longmapsto g \bmod N
\end{aligned}
\tag{F.45}
$$

constitutes an isomorphism between the two groups $\mathbb{Z}/N\mathbb{Z}$ and $\mathbb{Z}_N$.

To begin with, we show that ι is well defined, in other words, that the image $\iota\big([g]_{N\mathbb{Z}}\big)$ does not depend on the $g \in \mathbb{Z}$ chosen to represent the coset $[g]_{N\mathbb{Z}}$. To see this, let $g_1, g_2 \in \mathbb{Z}$ be such that $[g_1]_{N\mathbb{Z}} = [g_2]_{N\mathbb{Z}}$. Then it follows from (F.24) that $[g_1 \bmod N]_{N\mathbb{Z}} = [g_2 \bmod N]_{N\mathbb{Z}}$. Since $g_i \bmod N \in \{0, 1, \ldots, N-1\}$ for $i \in \{1, 2\}$, this implies $g_1 \bmod N = g_2 \bmod N$, hence, $\iota([g_1]_{N\mathbb{Z}}) = \iota([g_2]_{N\mathbb{Z}})$.

Now, suppose $g_1, g_2 \in \mathbb{Z}$ are such that $[g_1]_{N\mathbb{Z}} \neq [g_2]_{N\mathbb{Z}}$. Then it follows again from (F.24) that $g_1 \bmod N \neq g_2 \bmod N$, since otherwise their cosets would be equal. Consequently, $\iota([g_1]_{N\mathbb{Z}}) \neq \iota([g_2]_{N\mathbb{Z}})$, which means that ι is injective. It is also surjective, since, again using (F.24), every $m \in \{0, 1, \ldots, N-1\}$ uniquely defines a coset $[m]_{N\mathbb{Z}}$, which also satisfies $\iota([m]_{N\mathbb{Z}}) = m$. Therefore, ι is a bijection.

It remains to show that ι is also a homomorphism. For this we apply ι to both sides of (F.35) to obtain

$$\iota\big([g_1]_{N\mathbb{Z}} +_{\mathbb{Z}/N\mathbb{Z}} [g_2]_{N\mathbb{Z}}\big) \underset{(\text{F.35})}{=} \iota\big([(g_1 + g_2) \bmod N]_{N\mathbb{Z}}\big)$$

$$\underset{(\text{F.45})}{=} (g_1 + g_2) \bmod N$$

$$\underset{(\text{F.11})}{=} g_1 \bmod N +_{\mathbb{Z}_N} g_2 \bmod N$$

$$\underset{(\text{F.45})}{=} \iota\big([g_1]_{N\mathbb{Z}}\big) +_{\mathbb{Z}_N} \iota\big([g_2]_{N\mathbb{Z}}\big),$$

which shows that ι satisfies (F.43), hence is also a homomorphism.

Altogether, we have thus shown that

$$\mathbb{Z}/N\mathbb{Z} \cong \mathbb{Z}_N. \tag{F.46}$$

As a result of (F.46) we shall no longer distinguish between $\mathbb{Z}/N\mathbb{Z}$ and $\mathbb{Z}_N$ and also use the notation $[m]_{N\mathbb{Z}}$ to denote an element $m \in \mathbb{Z}_N$.

Lemma F.30 *Let $\varphi \in \mathrm{Hom}(\mathcal{G}_1, \mathcal{G}_2)$ be a homomorphism between the two groups $\mathcal{G}_1$ and $\mathcal{G}_2$. Then $\mathrm{Ker}(\varphi)$ is a normal subgroup of $\mathcal{G}_1$, that is,*

$$\mathrm{Ker}(\varphi) \trianglelefteq \mathcal{G}_1.$$

Proof We first show that $\mathrm{Ker}(\varphi)$ is a subgroup of $\mathcal{G}_1$. For $i \in \{1, 2\}$ let e_i denote the neutral element in $\mathcal{G}_i$. For any $g \in \mathcal{G}_1$ we have

$$\varphi(e_1) \underset{(\text{F.3}),(\text{F.2})}{=} \varphi(e_1)\varphi(g)\big(\varphi(g)\big)^{-1} \underset{(\text{F.43})}{=} \varphi(e_1 g)\big(\varphi(g)\big)^{-1} \underset{(\text{F.9})}{=} \varphi(g)\big(\varphi(g)\big)^{-1}$$

$$\underset{(\text{F.3})}{=} e_2, \tag{F.47}$$

which shows that $e_1 \in \mathrm{Ker}(\varphi)$ and verifies (F.14).

Next, for any $h \in \mathrm{Ker}(\varphi)$ it follows that

$$\varphi(h^{-1}) \underbrace{=}_{(F.9)} e_2\varphi(h^{-1}) \underbrace{=}_{(F.44)} \varphi(h)\varphi(h^{-1}) \underbrace{=}_{(F.43)} \varphi(hh^{-1}) \underbrace{=}_{(F.3)} \varphi(e_1) \underbrace{=}_{(F.47)} e_2,$$

which shows that $h^{-1} \in \mathrm{Ker}(\varphi)$ and verifies (F.15).

Finally, for any $h_1, h_2 \in \mathrm{Ker}(\varphi)$

$$\varphi(h_1 h_2) \underbrace{=}_{(F.43)} \varphi(h_1)\varphi(h_2) \underbrace{=}_{(F.44)} e_2 e_2 = e_2,$$

which shows that $h_1 h_2 \in \mathrm{Ker}(\varphi)$ and verifies (F.16).

Now that we have shown that $\mathrm{Ker}(\varphi)$ is a subgroup, it remains to show that it is normal. For this let $g \in \mathcal{G}_1$ be arbitrary and let $h' \in \mathrm{Ker}(\varphi)^g$, that is, there is an $h \in \mathrm{Ker}(\varphi)$ such that $h' = ghg^{-1}$. Then we have

$$\varphi(h') = \varphi(ghg^{-1}) \underbrace{=}_{(F.43)} \varphi(g)\varphi(h)\varphi(g^{-1}) \underbrace{=}_{(F.44)} \varphi(g)e_2\varphi(g^{-1})$$

$$= \varphi(g)\varphi(g^{-1}) \underbrace{=}_{(F.43)} \varphi(gg^{-1}) = \varphi(e_1)$$

$$= e_2,$$

which shows that for any $g \in \mathcal{G}_1$ we have that $h' \in \mathrm{Ker}(\varphi)^g$ implies $h' \in \mathrm{Ker}(\varphi)$. Hence, we have shown that

$$\mathrm{Ker}(\varphi)^g \subset \mathrm{Ker}(\varphi) \quad \forall g \in \mathcal{G}_1. \tag{F.48}$$

To finally show the reverse inclusion, let $h \in \mathrm{Ker}(\varphi)$ and $g \in \mathcal{G}_1$ be arbitrary. Then

$$k = g^{-1}hg \tag{F.49}$$

satisfies

$$\varphi(k) = \varphi(g^{-1}hg) \underbrace{=}_{(F.43)} \varphi(g^{-1})\varphi(h)\varphi(g) \underbrace{=}_{(F.44)} \varphi(g^{-1})e_2\varphi(g)$$

$$= \varphi(g^{-1})\varphi(g) \underbrace{=}_{(F.43)} \varphi(g^{-1}g) = \varphi(e_1)$$

$$\underbrace{=}_{(F.47)} e_2$$

such that $k \in \mathrm{Ker}(\varphi)$. But then it follows for the arbitrary $h \in \mathrm{Ker}(\varphi)$ that

$$h \underbrace{=}_{(F.49)} gkg^{-1} \in \mathrm{Ker}(\varphi)^g.$$

Consequently, we have

$$\text{Ker}(\varphi) \subset \text{Ker}(\varphi)^g \quad \forall g \in \mathcal{G}_1,$$

which, together with (F.48), finally proves that $\text{Ker}(\varphi) = \text{Ker}(\varphi)^g$ for all $g \in \mathcal{G}_1$, that is, $\text{Ker}(\varphi)$ is a normal subgroup of $\mathcal{G}_1$. $\qquad\square$

Exercise F.132 Show that any homomorphism $\varphi : \mathcal{G}_1 \to \mathcal{G}_2$ between two groups $\mathcal{G}_1$ and $\mathcal{G}_2$ satisfies

$$\varphi(g^{-1}) = \varphi(g)^{-1} \quad \forall g \in \mathcal{G}_1. \tag{F.50}$$

For a solution see Solution F.132

The following theorem is called **First Group Isomorphism Theorem** and is also known as the fundamental homomorphism theorem. It states that the for a homomorphism the quotient group over its kernel can be identified with its image.

Theorem F.31 (First Group Isomorphism) *Let $\mathcal{G}_1$ and $\mathcal{G}_2$ be groups and let $\varphi \in \text{Hom}(\mathcal{G}_1, \mathcal{G}_2)$. Then we have*

$$\mathcal{G}_1 / \text{Ker}(\varphi) \cong \varphi\{\mathcal{G}_1\},$$

where the isomorphism is provided by the map

$$\begin{aligned} \widehat{\varphi} : \mathcal{G}_1 / \text{Ker}(\varphi) &\longrightarrow \varphi\{\mathcal{G}_1\} \\ [g]_{\text{Ker}(\varphi)} &\longmapsto \varphi(g) \end{aligned} . \tag{F.51}$$

Proof From Lemma F.30 we know that $\text{Ker}(\varphi)$ is a normal subgroup of $\mathcal{G}_1$, and we can define the quotient group $\mathcal{G}_1 / \text{Ker}(\varphi)$. To show that $\widehat{\varphi}$ is an isomorphism, we first show that it is well defined. For this let, $g_a, g_b \in \mathcal{G}_1$ and $[g_a]_{\text{Ker}(\varphi)} = [g_b]_{\text{Ker}(\varphi)}$. Then we know from (F.33) that there exists an $h \in \text{Ker}(\varphi)$ such that

$$g_a = g_b h. \tag{F.52}$$

Consequently

$$\widehat{\varphi}\big([g_a]_{\text{Ker}(\varphi)}\big) \underset{(F.51)}{=} \varphi(g_a) \underset{(F.52)}{=} \varphi(g_b h) \underset{(F.43)}{=} \varphi(g_b)\varphi(h) \underset{h \in \text{Ker}(\varphi)}{=} \varphi(g_b) e_2 \underset{(F.3)}{=} \varphi(g_b)$$

$$\underset{(F.51)}{=} \widehat{\varphi}\big([g_b]_{\text{Ker}(\varphi)}\big),$$

proving that $\widehat{\varphi}$ is well defined. Here, as usual, e_2 denotes the neutral element in $\mathcal{G}_2$. Injectivity of $\widehat{\varphi}$ is proven by the following chain of implications for $g_c, g_d \in \mathcal{G}_1$:

$$\widehat{\varphi}\big([g_c]_{\mathrm{Ker}(\varphi)}\big) = \widehat{\varphi}\big([g_d]_{\mathrm{Ker}(\varphi)}\big) \underset{\text{(F.51)}}{\Rightarrow} \varphi(g_c) = \varphi(g_d) \quad \Rightarrow \quad \varphi(g_d)^{-1}\varphi(g_c) = e_2$$

$$\underset{\text{(F.43)}}{\Rightarrow} \varphi(g_d^{-1}g_c) = e_2 \underset{\text{(F.44)}}{\Rightarrow} g_d^{-1}g_c \in \mathrm{Ker}(\varphi)$$

$$\Rightarrow \exists h \in \mathrm{Ker}(\varphi): \; g_c = g_d h$$

$$\underset{\text{(F.22)}}{\Rightarrow} [g_c]_{\mathrm{Ker}(\varphi)} = [g_d]_{\mathrm{Ker}(\varphi)} \;.$$

To show surjectivity, note that $\varphi\{\mathcal{G}_1\} = \{\varphi(g) \,|\, g \in \mathcal{G}_1\}$. Therefore, for any $h \in \varphi\{\mathcal{G}_1\}$ there exists a $g \in \mathcal{G}_1$ such that

$$h = \varphi(g) \underset{\text{(F.51)}}{=} \widehat{\varphi}\big([g]_{\mathrm{Ker}(\varphi)}\big) \,,$$

proving that $\widehat{\varphi}$ is surjective as well. It remains to show that $\widehat{\varphi}$ is a homomorphism. For this consider

$$\widehat{\varphi}\big([g_1]_{\mathrm{Ker}(\varphi)}[g_2]_{\mathrm{Ker}(\varphi)}\big) \underset{\text{(F.30)}}{=} \widehat{\varphi}\big([g_1 g_2]_{\mathrm{Ker}(\varphi)}\big) \underset{\text{(F.51)}}{=} \varphi(g_1 g_2) \underset{\text{(F.43)}}{=} \varphi(g_1)\varphi(g_2)$$

$$\underset{\text{(F.51)}}{=} \widehat{\varphi}\big([g_1]_{\mathrm{Ker}(\varphi)}\big)\widehat{\varphi}\big([g_2]_{\mathrm{Ker}(\varphi)}\big) \,,$$

which shows that indeed $\widehat{\varphi} \in \mathrm{Hom}\big(\mathcal{G}_1 / \mathrm{Ker}(\varphi), \varphi\{\mathcal{G}_1\}\big)$. $\square$

A very useful class of homomorphisms for the study of groups are the so-called characters of a group. They can be defined for any group, but for us the special case of abelian groups suffices.

Definition F.32 A **character of an abelian group** $\mathcal{G}$ is defined as an element $\chi \in \mathrm{Hom}\big(G, \mathrm{U}(1)\big)$, where

$$\mathrm{U}(1) = \{z \in \mathbb{C} \,|\, z\bar{z} = 1\} = \{\mathrm{e}^{\mathrm{i}\alpha} \,|\, \alpha \in \mathbb{R}\} \tag{F.53}$$

is the special unitary group in one dimension.

For any character χ of an abelian group $\mathcal{G}$ we define the **conjugate character** $\overline{\chi}$ as

$$\begin{aligned} \overline{\chi}: \mathcal{G} &\longrightarrow \mathrm{U}(1) \\ g &\longmapsto \overline{\chi(g)} \end{aligned} \,. \tag{F.54}$$

A special character for any abelian group $\mathcal{G}$ is the **trivial character**

$$\begin{aligned} 1_G : \mathcal{G} &\longrightarrow U(1) \\ g &\longmapsto 1 \end{aligned}, \tag{F.55}$$

which maps any group element to $1 \in U(1)$.

Note that by definition any character χ of an abelian group $\mathcal{G}$ being a homomorphism from $\mathcal{G}$ to $U(1)$, it satisfies for any $g_1, g_2 \in \mathcal{G}$

$$\chi(g_1 +_{\mathcal{G}} g_2) \underbrace{=}_{(F.43)} \chi(g_1)\chi(g_2), \tag{F.56}$$

where $+_{\mathcal{G}}$ denotes the group multiplication in the abelian group $\mathcal{G}$, whereas on the right side the product is in $U(1)$, which is just a multiplication of two complex numbers of unit modulus.

As a consequence of (F.56), any character of an abelian group has to map the neutral element e of the group $\mathcal{G}$ to 1, that is, we have

$$\chi(e) \underbrace{=}_{(F.56)} \frac{\chi(g +_{\mathcal{G}} e)}{\chi(g)} = 1 \tag{F.57}$$

since $g +_{\mathcal{G}} e = g$. Actually, this statement already follows from Lemma F.30 since as a subgroup $\mathrm{Ker}(\chi) = \{g \in \mathcal{G} \,|\, \chi(g) = 1\}$ has to contain e.

Moreover, since $\chi(g) \in U(1)$

$$1 \underbrace{=}_{(F.53)} \chi(g)\overline{\chi(g)} \underbrace{=}_{(F.54)} \chi(g)\overline{\chi}(g),$$

we have

$$\overline{\chi}(g) = \chi(g)^{-1} \underbrace{=}_{(F.50)} \chi(g^{-1}). \tag{F.58}$$

Example F.33 For the group $\mathbb{Z}_N$ defined in Lemma F.5 and considered in Example F.29, we have the characters

$$\begin{aligned} \chi_n : \mathbb{Z}_N &\longrightarrow U(1) \\ [g]_{N\mathbb{Z}} &\longmapsto e^{2\pi i \frac{ng}{N}} \end{aligned}, \tag{F.59}$$

where $n \in \{0, 1, \ldots, N-1\}$.

To verify (F.56), we note that for any $[g_1]_{N\mathbb{Z}}, [g_2]_{N\mathbb{Z}} \in \mathbb{Z}_N$

$$\chi_n \big([g_1]_{N\mathbb{Z}} \underbrace{+_{\mathbb{Z}_N} [g_2]_{N\mathbb{Z}} \big) = }_{(\text{F.35})} \chi_n \big([(g_1 + g_2) \bmod N]_{N\mathbb{Z}} \big)$$

$$\underbrace{=}_{(\text{F.59})} e^{2\pi i \frac{n\left((g_1 + g_2) \bmod N \right)}{N}} = e^{2\pi i \frac{n(g_1 + g_2)}{N}}$$

$$\underbrace{=}_{(\text{F.59})} \chi_n \big([g_1]_{N\mathbb{Z}} \big) \chi_n \big([g_2]_{N\mathbb{Z}} \big) .$$

The kernel of a χ_n consists of all cosets $[g]_{N\mathbb{Z}}$ such that $\frac{ng}{N} \in \mathbb{Z}$, that is, for which $ng \bmod N = 0$.

The characters of an abelian group again form a group.

Theorem F.34 *The characters* $\widehat{\mathcal{G}} = \mathrm{Hom}\left(\mathcal{G}, \mathrm{U}(1)\right)$ *of an abelian group* $\mathcal{G}$ *form an abelian group with the the trivial character* $1_{\mathcal{G}}$ *as the neutral element and with the group multiplication*

$$\begin{aligned} \cdot : \ \widehat{\mathcal{G}} \times \widehat{\mathcal{G}} &\longrightarrow \widehat{\mathcal{G}} \\ (\chi_1, \chi_2) &\longmapsto \chi_1 \chi_2 \end{aligned}, \tag{F.60}$$

where $\chi_1 \chi_2$ *is the character*

$$\begin{aligned} \chi_1 \chi_2 : \mathcal{G} &\longrightarrow \mathrm{U}(1) \\ g &\longmapsto \chi_1(g) \chi_2(g) \end{aligned}. \tag{F.61}$$

Proof To show that $\widehat{\mathcal{G}}$ is a group, we need to show that the multiplication of two of its elements as defined in (F.60) gives again an element of $\widehat{\mathcal{G}}$, that there is a neutral element for this multiplication, and that every element has an inverse in $\widehat{\mathcal{G}}$.

Let $\chi_1, \chi_2 \in \widehat{\mathcal{G}}$ and $g_1, g_2 \in \mathcal{G}$. Then we find

$$\underbrace{\chi_1 \chi_2(g_1) \chi_1 \chi_2(g_2) = }_{(\text{F.61})} \chi_1(g_1) \chi_2(g_1) \chi_1(g_2) \chi_2(g_2) \underbrace{=}_{(\text{F.43})} \chi_1(g_1 g_2) \chi_2(g_1 g_2)$$

$$\underbrace{=}_{(\text{F.61})} \chi_1 \chi_2(g_1 g_2)$$

and $\chi_1 \chi_2$ as defined in (F.60) and (F.61) is indeed an element of $\mathrm{Hom}\left(\mathcal{G}, \mathrm{U}(1)\right)$.

For any $\chi \in \widehat{\mathcal{G}}$ and $g \in \mathcal{G}$ we have

$$\underbrace{\chi 1_{\mathcal{G}}(g) = }_{(\text{F.61})} \chi(g) 1_{\mathcal{G}}(g) \underbrace{=}_{(\text{F.55})} \chi(g)$$

such that the multiplication (F.60) gives indeed $\chi 1_{\mathcal{G}} = \chi$ for any $\chi \in \widehat{\mathcal{G}}$ and $1_{\mathcal{G}}$ is the neutral element.

Finally, for any $\chi \in \widehat{\mathcal{G}}$ its inverse is given by its conjugate character $\overline{\chi} \in \widehat{\mathcal{G}}$, since for any $g \in \mathcal{G}$

$$\chi \overline{\chi}(g) \underbrace{=}_{\text{(F.61)}} \chi(g)\overline{\chi}(g) \underbrace{=}_{\text{(F.55)}} \chi(g)\overline{\chi(g)} \underbrace{=}_{\text{(F.53)}} 1. \tag{F.62}$$

$\square$

Definition F.35 Let $\mathcal{G}$ be an abelian group. The group $\widehat{\mathcal{G}} := \mathrm{Hom}\,(\mathcal{G}, \mathrm{U}(1))$ formed by its characters with the group multiplication given in (F.60) and (F.61) is called the **dual** or **character group** of $\mathcal{G}$.

Theorem F.36 *Let $\mathcal{H}$ be a subgroup of the finite abelian group $\mathcal{G}$. Then any character of $\mathcal{H}$ can be extended to a character of $\mathcal{G}$, and the number of such extensions is $\frac{|\mathcal{G}|}{|\mathcal{H}|}$.*

Proof To begin with, we recall that subgroups of abelian groups are always normal, so we can always form quotient groups with them.

If $\mathcal{H} = \mathcal{G}$ we are done. Otherwise, we have $\mathcal{H} < \mathcal{G}$ and can pick a $g_1 \in \mathcal{G} \setminus \mathcal{H}$. Let $\mathcal{H}_1 = \langle \mathcal{H}, g_1 \rangle$ denote the subgroup generated by the set $\mathcal{H} \cup \{g_1\}$ such that

$$\mathcal{H}_1 = \langle \mathcal{H}, g_1 \rangle = \left\{ hg_1^m \mid h \in \mathcal{H}, m \in \mathbb{Z} \right\}$$

and

$$\mathcal{H} < \mathcal{H}_1 \leq \mathcal{G}.$$

Furthermore, we set

$$k := \min \left\{ n \in \mathbb{N} \mid g_1^n \in \mathcal{H} \right\}, \tag{F.63}$$

which exists since at least $g_1^m = e_{\mathcal{G}} \in \mathcal{H}$ for some finite $m \in \mathbb{N}$. Consequently, for any character $\chi_{\mathcal{H}} \in \widehat{\mathcal{H}}$, we have that $\chi(g_1^k) = e^{i\alpha} \in \mathrm{U}(1)$ for some $\alpha \in \mathbb{R}$. There are k k-th roots of $e^{i\alpha}$, which are of the form

$$\beta_{k,l} = \frac{\alpha + 2\pi l}{k} \quad \text{for } l \in \{0, 1, \ldots, k-1\}.$$

These satisfy

$$e^{i\beta_{k,l}} = \left(\chi_{\mathcal{H}}(g_1^k) \right)^{\frac{1}{k}} =: \mu_l(g_1),$$

which implies

$$\mu_l(g_1)^k = \chi_{\mathcal{H}}(g_1^k) \tag{F.64}$$

for each $l \in \{0, 1, \dots, k-1\}$. For each such l we want to define a character $\chi_l \in \widehat{\mathcal{H}_1}$, which acts on a generic element $hg_1^j \in \mathcal{H}_1$ by

$$\chi_l(hg_1^j) := \chi_{\mathcal{H}}(h)\mu_l(g_1)^j.$$

For this to be a meaningful definition, we need to make sure that the right side does not depend on the particular h and g_1^j but only their product hg_1^j. To show this, suppose that $h, \widetilde{h} \in \mathcal{H}$ and $j, \widetilde{j} \in \mathbb{Z}$ are such that $\widetilde{j} \geq j$ and

$$hg_1^j = \widetilde{h}g_1^{\widetilde{j}}.$$

From this it follows that

$$g_1^{\widetilde{j}-j} = h\widetilde{h}^{-1} \in \mathcal{H} \tag{F.65}$$

and since (F.63) implies that k is the smallest natural number such that $g_1^k \in \mathcal{H}$, we must have

$$\widetilde{j} = j + mk \tag{F.66}$$

for some $m \in \mathbb{N}$. Therefore, (F.65) implies

$$h = \widetilde{h}g_1^{mk}. \tag{F.67}$$

Consequently,

$$\chi_{\mathcal{H}}(\widetilde{h})\mu_l(g_1)^{\widetilde{j}} \underset{(\text{F.66})}{=} \chi_{\mathcal{H}}(\widetilde{h})\mu_l(g_1)^{j+mk} = \chi_{\mathcal{H}}(\widetilde{h})\left(\mu_l(g_1)^k\right)^m \mu_l(g_1)^j$$

$$\underset{(\text{F.64})}{=} \chi_{\mathcal{H}}(\widetilde{h})\left(\chi_{\mathcal{H}}(g_1^k)\right)^m \mu_l(g_1)^j \underset{(\text{F.43})}{=} \chi_{\mathcal{H}}(\widetilde{h}g_1^{mk})\mu_l(g_1)^j$$

$$\underset{(\text{F.67})}{=} \chi_{\mathcal{H}}(h)\mu_l(g_1)^j,$$

and

$$\begin{aligned} \chi_l : \mathcal{H}_1 &\longrightarrow U(1) \\ hg_1^j &\longmapsto \chi_{\mathcal{H}}(h)\mu_l(g_1)^j \end{aligned} \tag{F.68}$$

is well defined. Moreover, for any $h_i g_1^{j_i} \in \mathcal{H}_1$, where $i \in \{1, 2\}$, we have

$$\chi_l(h_1 g_1^{j_1} h_2 g_1^{j_2}) \;=\; \underbrace{\chi_l(h_1 h_2 g_1^{j_1+j_2})}_{\text{(F.68)}} \;=\; \chi_{\mathcal{H}}(h_1 h_2)\mu_l(g_1)^{j_1+j_2}$$

$$\underbrace{=}_{\text{(F.43)}} \chi_{\mathcal{H}}(h_1)\mu_l(g_1)^{j_1} \chi_{\mathcal{H}}(h_2)\mu_l(g_1)^{j_2}$$

$$\underbrace{=}_{\text{(F.68)}} \chi_l(h_1 g_1^{j_1})\chi_l(h_2 g_1^{j_2}),$$

showing that $\chi_l \in \mathrm{Hom}(\mathcal{H}_1, U(1))$, in other words, χ_l is a character of $\mathcal{H}_1 = \langle \mathcal{H}, g_1 \rangle$ that satisfies $\chi_l|_{\mathcal{H}} = \chi_{\mathcal{H}}$.

There are k such extensions $\chi_0, \ldots, \chi_{k-1}$ of $\chi_{\mathcal{H}}$ to $\mathcal{H}_1$ and since $g_1, g_1^2, \ldots, g_1^{k-1} \notin \mathcal{H}$, we have

$$\mathcal{H}_1 = \langle \mathcal{H}, g_1 \rangle = [e_{\mathcal{G}}]_{\mathcal{H}} \cup [g_1]_{\mathcal{H}} \cup \cdots \cup \left[g_1^{k-1}\right]_{\mathcal{H}} \qquad (\text{F.69})$$

such that

$$|\mathcal{H}_1/\mathcal{H}| \underbrace{=}_{\text{(F.36)}} \frac{|\mathcal{H}_1|}{|\mathcal{H}|} \underbrace{=}_{\text{(F.69)}} k.$$

We choose any of these extensions and denote it by $\chi_{\mathcal{H}_1}$. We thus have a subgroup $\mathcal{H}_1$ and a character $\chi_{\mathcal{H}_1} \in \widehat{\mathcal{H}_1}$. Now we pick any $g_2 \in \mathcal{G} \setminus \mathcal{H}_1$ and repeat the previous construction with $\mathcal{H}_2 = \langle \mathcal{H}_1, g_2 \rangle$ leading to a character $\chi_{\mathcal{H}_2} \in \widehat{\mathcal{H}_2}$ that satisfies $\chi_{\mathcal{H}_2}|_{\mathcal{H}_1} = \chi_{\mathcal{H}_1}$. We continue with this until there is no more $g_{n+1} \in \mathcal{G} \setminus \mathcal{H}_n$ to be found, which must happen after a finite number of steps since $\mathcal{G}$ is assumed finite.

Altogether, we thus have a sequence of subgroups $\mathcal{H}_r = \langle \mathcal{H}_{r-1}, g_{r-1} \rangle$, elements $g_r \in \mathcal{G} \setminus \mathcal{H}_{r-1}$ and characters $\chi_{\mathcal{H}_r} \in \widehat{\mathcal{H}_r}$ for $r \in \{0, 1, \ldots, n\}$ such that

$$\mathcal{H} = \mathcal{H}_0 < \mathcal{H}_1 < \cdots < \mathcal{H}_n = \mathcal{G}$$

$$\chi_{\mathcal{H}_r}\big|_{\mathcal{H}_{r-1}} = \chi_{\mathcal{H}_{r-1}}$$

$$\chi_{\mathcal{G}}\big|_{\mathcal{H}} = \chi_{\mathcal{H}_n}\big|_{\mathcal{H}} = \left(\chi_{\mathcal{H}_n}\big|_{\mathcal{H}_{n-1}}\right)\big|_{\mathcal{H}} = \chi_{\mathcal{H}_{n-1}}\big|_{\mathcal{H}} = \cdots = \chi_{\mathcal{H}_0}\big|_{\mathcal{H}} = \chi_{\mathcal{H}}.$$

In each step from $r-1$ to r we have $\frac{|\mathcal{H}_r|}{|\mathcal{H}_{r-1}|}$ possible extensions. The number of ways in which $\chi_{\mathcal{H}}$ can be extended from a character on $\mathcal{H} = \mathcal{H}_0$ to one on $\mathcal{G} = \mathcal{H}_n$ is thus

$$\prod_{r=1}^{n} \frac{|\mathcal{H}_r|}{|\mathcal{H}_{r-1}|} = \frac{|\mathcal{H}_n|}{|\mathcal{H}_0|} = \frac{|\mathcal{G}|}{|\mathcal{H}|}.$$

$\square$

A direct consequence of Theorem F.36 is that the dual group of a finite abelian group has the same number of elements as the group itself.

Corollary F.37 *The dual group $\widehat{\mathcal{G}}$ of a finite abelian group $\mathcal{G}$ satisfies*

$$|\widehat{\mathcal{G}}| = |\mathcal{G}|. \tag{F.70}$$

Proof Let e be the neutral element of the group $\mathcal{G}$. The subgroup $\langle e \rangle$ has only one element, that is, $|\langle e \rangle| = 1$. Moreover, its only character is the trivial character $1_{\langle e \rangle}$, which, according to Theorem F.36, can be extended in exactly

$$\frac{|\mathcal{G}|}{|\langle e \rangle|} = |\mathcal{G}|$$

ways. Any character $\chi \in \widehat{\mathcal{G}}$, when restricted to $\langle e \rangle$, is equal to $1_{\langle e \rangle}$. Hence, it must be one of the $|\mathcal{G}|$ extensions constructed from $1_{\langle e \rangle}$ since, otherwise, there would be more than $|\mathcal{G}|$ extensions. Consequently, the number of characters is $|\mathcal{G}|$. $\quad\square$

Exercise F.133 Let $\mathcal{G}_1$ and $\mathcal{G}_2$ be two finite abelian groups and $\widehat{\mathcal{G}_1}$ and $\widehat{\mathcal{G}_2}$ their dual groups. Show that the direct product group satisfies

$$\widehat{\mathcal{G}_1 \times \mathcal{G}_2} = \widehat{\mathcal{G}_1} \times \widehat{\mathcal{G}_2}. \tag{F.71}$$

For a solution see Solution F.133.

Corollary F.38 *Let $\mathcal{G}$ be a finite abelian group with neutral element e and $g \in \mathcal{G} \smallsetminus \langle e \rangle$. Then there exists a character $\chi \in \widehat{\mathcal{G}}$ such that $g \notin \mathrm{Ker}(\chi)$.*

Proof Let $g \in \mathcal{G} \smallsetminus \langle e \rangle$ be given. Since $\mathcal{G}$ is finite, there exists a smallest $n \in \mathbb{N}$ such that $n > 1$ and $g^n = e$. With this n we define

$$\begin{aligned}\mu_g : \langle g \rangle &\longrightarrow \mathrm{U}(1) \\ g^m &\longmapsto e^{2\pi i \frac{m}{n}}\end{aligned}.$$

Since any element $\widetilde{g} \in \langle g \rangle$ can be written in the form $\widetilde{g} = g^m$ for some $m \in \mathbb{Z}$, we have that μ_g is indeed defined on all of $\langle g \rangle$ and for any $g_i = g^{m_i} \in \langle g \rangle$ with $i \in \{1,2\}$, obviously,

$$\mu_g(g_1 g_2) = \mu_g(g^{m_1+m_2}) = e^{2\pi i \frac{m_1+m_2}{n}} = \mu_g(g_1)\mu_g(g_2).$$

Hence, μ_g is a character of the subgroup $\langle g \rangle \leq \mathcal{G}$, and it satisfies $\mu_g(g) = e^{\frac{2\pi i}{n}} \neq 1$.

We then apply Theorem F.36 to the subgroup $\langle g \rangle$ to extend the character μ_g to a character $\chi \in \widehat{\mathcal{G}}$ such that $\chi|_{\langle g \rangle} = \mu_g$ and thus $\chi(g) \neq 1$. $\square$

Lemma F.39 *Let $\mathcal{H}$ be a subgroup of the abelian group $\mathcal{G}$. Then*

$$\mathcal{H}^\perp := \left\{ \chi \in \widehat{\mathcal{G}} \, \middle| \, \mathcal{H} \subset \mathrm{Ker}(\chi) \right\} \tag{F.72}$$

is a subgroup of $\widehat{\mathcal{G}}$.

Proof Let $e_{\widehat{\mathcal{G}}}$ denote the neutral element of the dual group $\widehat{\mathcal{G}}$. Since $e_{\widehat{\mathcal{G}}} = 1_{\mathcal{G}}$ and $\mathrm{Ker}(1_{\mathcal{G}}) = \mathcal{G}$, we have $\mathcal{H} \subset \mathrm{Ker}(e_{\widehat{\mathcal{G}}})$ and thus $e_{\widehat{\mathcal{G}}} \in \mathcal{H}^\perp$, proving (F.14).

Next, we show that $\mathcal{H}^\perp$ is closed under multiplication. For this let $\chi_1, \chi_2 \in \mathcal{H}^\perp$ and $h \in \mathcal{H}$ be arbitrary. Then we have

$$\chi_1 \chi_2(h) \underbrace{=}_{\text{(F.61)}} \chi_1(h)\chi_2(h) \underbrace{=}_{h \in H \subset \mathrm{Ker}(\chi_i)} 1 \,.$$

Hence, $\chi_1 \chi_2 \in \mathcal{H}^\perp$, and (F.16) holds.

It remains to show that for any $\chi \in \mathcal{H}^\perp$ its inverse lies in $\mathcal{H}^\perp$. From (F.62) we know already that the conjugate character $\overline{\chi}$ is the inverse of χ, so we only need to show that it is an element of $\mathcal{H}^\perp$. For this let $h \in \mathcal{H}$ be arbitrary. Then $\chi \in \mathcal{H}^\perp$ implies that $h \in \mathrm{Ker}(\chi)$ and thus $\chi(h) = 1$ from which it follows that

$$1 = \overline{\chi(h)} \underbrace{=}_{\text{(F.54)}} \overline{\chi}(h) \,.$$

Consequently, $\chi^{-1} = \overline{\chi} \in \mathcal{H}^\perp$, and we have verified (F.15), which completes the proof that $\mathcal{H}^\perp$ is a subgroup. $\square$

Example F.40 Let $e_{\mathcal{G}}$ denote the neutral element of the group $\mathcal{G}$ and $e_{\widehat{\mathcal{G}}}$ that of the dual group $\widehat{\mathcal{G}}$. For the trivial subgroup $\langle e_{\mathcal{G}} \rangle < \mathcal{G}$ we find

$$\langle e_{\mathcal{G}} \rangle^\perp = \left\{ \chi \in \widehat{\mathcal{G}} \, \middle| \, e_{\mathcal{G}} \in \mathrm{Ker}(\chi) \right\} = \widehat{\mathcal{G}}, \tag{F.73}$$

whereas for the trivial subgroup $\mathcal{G}$ itself

$$\mathcal{G}^\perp = \left\{ \chi \in \widehat{\mathcal{G}} \, \middle| \, \mathcal{G} \subset \mathrm{Ker}(\chi) \right\} = \langle 1_{\mathcal{G}} \rangle = \langle e_{\widehat{\mathcal{G}}} \rangle \,. \tag{F.74}$$

Exercise F.134 Let $\mathcal{H}$ be a subgroup of the finite abelian group $\mathcal{G}$, and let $\mathcal{H}^\perp$ be as defined in (F.72). Show that then for any $\chi \in \widehat{\mathcal{G}}$

$$\sum_{h \in \mathcal{H}} \chi(h) = \begin{cases} |\mathcal{H}| & \text{if } \chi \in \mathcal{H}^\perp \\ 0 & \text{else.} \end{cases} \tag{F.75}$$

For a solution see Solution F.134

Corollary F.41 *Let $\mathcal{G}$ be a finite abelian group and $\widehat{\mathcal{G}}$ its dual group. For any $\chi_1, \chi_2 \in \widehat{\mathcal{G}}$ we then have*

$$\sum_{g \in \mathcal{G}} \chi_1(g)\chi_2(g) = \begin{cases} |\mathcal{G}| & \text{if } \chi_2 = \chi_1^{-1} \\ 0 & \text{else.} \end{cases}$$

Proof

$$\sum_{g \in \mathcal{G}} \chi_1(g)\chi_2(g) \underset{(F.61)}{=} \sum_{g \in \mathcal{G}} \chi_1\chi_2(g) \underset{(F.75)}{=} \begin{cases} |\mathcal{G}| & \text{if } \chi_1\chi_2 \in \mathcal{G}^\perp \underset{(F.74)}{=} \langle e_{\widehat{\mathcal{G}}} \rangle \\ 0 & \text{else.} \end{cases}$$

$$= \begin{cases} |\mathcal{G}| & \text{if } \chi_2 = \chi_1^{-1} \\ 0 & \text{else.} \end{cases} \tag{F.76}$$

$\square$

Exercise F.135 Let $\mathcal{H}$ be a subgroup of the abelian group $\mathcal{G}$ and let $\mathcal{H}^\perp$ be as defined in (F.72). Show that then

$$\mathcal{H} \leq \bigcap_{\chi \in \mathcal{H}^\perp} \mathrm{Ker}(\chi). \tag{F.77}$$

For a solution see Solution F.135

It turns out that in the case of abelian groups we have that $\mathcal{H}^\perp$ is isomorphic to the dual of the quotient group $\mathcal{G}/\mathcal{H}$.

Theorem F.42 *Let $\mathcal{H}$ be a subgroup of an abelian group $\mathcal{G}$ and $\mathcal{H}^\perp$ as defined in (F.72). Then we have*

$$\mathcal{H}^\perp \cong \widehat{\mathcal{G}/\mathcal{H}}. \tag{F.78}$$

Proof Let $e_\mathcal{G}$ denote the neutral element in the group $\mathcal{G}$ and $e_{\mathcal{G}/\mathcal{H}}$ that in the quotient group $\mathcal{G}/\mathcal{H}$. For any $\Xi \in \widehat{\mathcal{G}/\mathcal{H}}$, we define

$$\iota_\Xi : \mathcal{G} \longrightarrow U(1) \\ g \longmapsto \Xi([g]_\mathcal{H}) \;. \tag{F.79}$$

Then we have for every $h \in \mathcal{H}$ that

$$\iota_\Xi(h) \underset{\text{(F.79)}}{=} \Xi([h]_\mathcal{H}) \underset{\text{(F.25)}}{=} \Xi([e_\mathcal{G}]_\mathcal{H}) \underset{\text{(F.31)}}{=} \Xi(e_{\mathcal{G}/\mathcal{H}}) \underset{\text{(F.57)}}{=} 1$$

such that $\mathcal{H} \subset \mathrm{Ker}(\iota_\Xi)$ and thus

$$\iota_\Xi \underset{\text{(F.72)}}{\in} \mathcal{H}^\perp. \tag{F.80}$$

Moreover, for $\Xi_1, \Xi_2 \in \widehat{\mathcal{G}/\mathcal{H}}$ and any $g \in \mathcal{G}$

$$\iota_{\Xi_1 \Xi_2}(g) \underset{\text{(F.79)}}{=} \Xi_1 \Xi_2([g]_\mathcal{H}) \underset{\text{(F.61)}}{=} \Xi_1([g]_\mathcal{H}) \Xi_2([g]_\mathcal{H}) \underset{\text{(F.79)}}{=} \iota_{\Xi_1}(g) \iota_{\Xi_2}(g),$$

and from this and (F.80), we conclude that the map

$$\iota : \widehat{\mathcal{G}/\mathcal{H}} \longrightarrow \mathcal{H}^\perp \\ \Xi \longmapsto \iota_\Xi$$

satisfies

$$\iota \in \mathrm{Hom}\left(\widehat{\mathcal{G}/\mathcal{H}}, \mathcal{H}^\perp\right). \tag{F.81}$$

Now that we have established that ι is a homomorphism it remains to show that it is also a bijection. To accomplish this, we define for any $\chi \in \mathcal{H}^\perp$

$$j_\chi : \mathcal{G}/\mathcal{H} \longrightarrow U(1) \\ [g]_\mathcal{H} \longmapsto \chi(g) \;. \tag{F.82}$$

This is well defined since $[g_1]_\mathcal{H} = [g_2]_\mathcal{H}$ implies $g_2 = g_1 h$ for some $h \in \mathcal{H}$ and thus

$$\chi(g_2) = \chi(g_1 h) \underset{\text{(F.43)}}{=} \chi(g_1)\chi(h) \underset{h \in \mathcal{H} \subset \mathrm{Ker}(\chi)}{=} \chi(g_1),$$

where in the last equation we used that $\chi \in \mathcal{H}^\perp$ implies $\chi(h) = 1$ for all $h \in \mathcal{H}$. Moreover, for any $[g_1]_{\mathcal{H}}, [g_2]_{\mathcal{H}} \in \mathcal{G}/\mathcal{H}$ we have

$$j_\chi([g_1]_{\mathcal{H}}) j_\chi([g_2]_{\mathcal{H}}) \underbrace{=}_{\text{(F.82)}} \chi(g_1)\chi(g_2) \underbrace{=}_{\text{(F.43)}} \chi(g_1 g_2) \underbrace{=}_{\text{(F.82)}} j_\chi([g_1 g_2]_{\mathcal{H}})$$

$$\underbrace{=}_{\text{(F.30)}} j_\chi([g_1]_{\mathcal{H}} [g_2]_{\mathcal{H}}) .$$

Hence, we have

$$j_\chi \in \widehat{\mathcal{G}/\mathcal{H}},$$

and the map

$$j : \mathcal{H}^\perp \longrightarrow \widehat{\mathcal{G}/\mathcal{H}}$$
$$\chi \longmapsto j_\chi$$

satisfies

$$j \in \mathrm{Hom}\left(\mathcal{H}^\perp, \widehat{\mathcal{G}/\mathcal{H}}\right) .$$

The composition $\iota \circ j : \mathcal{H}^\perp \to \mathcal{H}^\perp$ satisfies

$$\iota \circ j(\chi) : \mathcal{G} \longrightarrow U(1)$$
$$g \longmapsto \iota_{j_\chi}(g) ,$$

where

$$\iota_{j_\chi}(g) \underbrace{=}_{\text{(F.79)}} j_\chi([g]_{\mathcal{H}}) \underbrace{=}_{\text{(F.82)}} \chi(g) ,$$

such that $\iota \circ j(\chi) = \chi$, that is, every $\chi \in \mathcal{H}^\perp$ is an image under ι of a coset $j(\chi) \in \widehat{\mathcal{G}/\mathcal{H}}$, which implies that ι is surjective.

Likewise, the composition $j \circ \iota : \widehat{\mathcal{G}/\mathcal{H}} \to \widehat{\mathcal{G}/\mathcal{H}}$ satisfies

$$j \circ \iota(\Xi) : \mathcal{G}/\mathcal{H} \longrightarrow U(1)$$
$$[g]_{\mathcal{H}} \longmapsto j_{\iota_\Xi}([g]_{\mathcal{H}}) ,$$

where now

$$j_{\iota_\Xi}([g]_{\mathcal{H}}) \underbrace{=}_{\text{(F.82)}} \iota_\Xi(g) \underbrace{=}_{\text{(F.79)}} \Xi([g]_{\mathcal{H}}) ,$$

such that $j \circ \iota(\Xi) = \Xi$. From this it follows that for any $\Xi_1, \Xi_2 \in \widehat{\mathcal{G}/\mathcal{H}}$ satisfying $\Xi_1 \neq \Xi_2$ we must have $j \circ \iota(\Xi_1) \neq j \circ \iota(\Xi_2)$ and thus $\iota(\Xi_1) \neq \iota(\Xi_2)$, proving that ι is injective as well.

Hence, we have established that ι is a bijection and because of (F.81) thus an isomorphism, which completes the proof of $\mathcal{H}^\perp \cong \widehat{\mathcal{G}/\mathcal{H}}$. $\qquad\square$

Corollary F.43 *Let $\mathcal{H}$ be a subgroup of the finite abelian group $\mathcal{G}$ and $\mathcal{H}^\perp$ be defined as in (F.72). Then we have*

$$|\mathcal{H}^\perp| = \frac{|\mathcal{G}|}{|\mathcal{H}|}. \tag{F.83}$$

Proof With Theorem F.42, we have

$$|\mathcal{H}^\perp| \underset{(F.78)}{=} |\widehat{\mathcal{G}/\mathcal{H}}| \underset{(F.70)}{=} |\mathcal{G}/\mathcal{H}| \underset{(F.36)}{=} \frac{|\mathcal{G}|}{|\mathcal{H}|}.$$

$\qquad\square$

Theorem F.44 *Let $\mathcal{H}$ be a subgroup of the finite abelian group $\mathcal{G}$ and $\mathcal{H}^\perp$ be defined as in (F.72). Then for each $g \in \mathcal{G} \setminus \mathcal{H}$ there exists a $\chi \in \mathcal{H}^\perp$ such that $g \notin \mathrm{Ker}(\chi)$, and we have*

$$\mathcal{H} = \bigcap_{\chi \in \mathcal{H}^\perp} \mathrm{Ker}(\chi). \tag{F.84}$$

Moreover, if the $\chi_1, \ldots, \chi_L \in \mathcal{H}^\perp$ are such that $\mathcal{H}^\perp = \langle \chi_1, \ldots, \chi_L \rangle$, then

$$\mathcal{H} = \bigcap_{l=1}^{L} \mathrm{Ker}(\chi_l)$$

holds.

Proof As a subgroup of an abelian group $\mathcal{H}$ is normal and thus Proposition F.22 tells us that there is a quotient group $\mathcal{G}/\mathcal{H}$. Let $e_\mathcal{G}$ denote the neutral element in the group $\mathcal{G}$ and $e_{\mathcal{G}/\mathcal{H}}$ that in the quotient group $\mathcal{G}/\mathcal{H}$. Applying the statement of Corollary F.38 to the group $\mathcal{G}/\mathcal{H}$ means that for each coset $[g]_\mathcal{H} \neq [e_\mathcal{G}]_\mathcal{H} = e_{\mathcal{G}/\mathcal{H}}$ there exists a $\chi_{[g]_\mathcal{H}} \in \widehat{\mathcal{G}/\mathcal{H}}$ such that

$$[g]_\mathcal{H} \notin \mathrm{Ker}(\chi_{[g]_\mathcal{H}}). \tag{F.85}$$

For $g \in \mathcal{G}$, let

$$
\begin{aligned}
\zeta_g : \mathcal{G} &\longrightarrow U(1) \\
\widetilde{g} &\longmapsto \chi_{[g]_{\mathcal{H}}}([\widetilde{g}]_{\mathcal{H}})
\end{aligned}
\tag{F.86}
$$

which satisfies

$$
\zeta_g(g_1)\zeta_g(g_2) \underbrace{=}_{\text{(F.86)}} \chi_{[g]_{\mathcal{H}}}([g_1]_{\mathcal{H}})\chi_{[g]_{\mathcal{H}}}([g_2]_{\mathcal{H}}) \underbrace{=}_{\text{(F.43)}} \chi_{[g]_{\mathcal{H}}}([g_1]_{\mathcal{H}}[g_2]_{\mathcal{H}})
$$

$$
\underbrace{=}_{\text{(F.30)}} \chi_{[g]_{\mathcal{H}}}([g_1g_2]_{\mathcal{H}}) \underbrace{=}_{\text{(F.86)}} \zeta_g(g_1g_2)
\tag{F.87}
$$

as well as for any $h \in \mathcal{H}$

$$
\zeta_g(h) \underbrace{=}_{\text{(F.86)}} \chi_{[g]_{\mathcal{H}}}([h]_{\mathcal{H}}) \underbrace{=}_{\text{(F.25)}} \chi_{[g]_{\mathcal{H}}}(\mathcal{H}) \underbrace{=}_{\text{(F.31)}} \chi_{[g]_{\mathcal{H}}}(e_{\mathcal{G}/\mathcal{H}}) = 1 ,
\tag{F.88}
$$

whereas

$$
\zeta_g(g) \underbrace{=}_{\text{(F.86)}} \chi_{[g]_{\mathcal{H}}}([g]_{\mathcal{H}}) \underbrace{\neq}_{\text{(F.85)}} 1 .
\tag{F.89}
$$

From (F.86) and (F.87) we see that $\zeta_g \in \mathrm{Hom}(\mathcal{G}, U(1)) = \widehat{\mathcal{G}}$ and from (F.88) that $\mathcal{H} \subset \mathrm{Ker}(\zeta_g)$, which together implies $\zeta_g \in \mathcal{H}^{\perp}$, while (F.89) implies $g \notin \mathrm{Ker}(\zeta_g)$, which completes the proof of the first part of the statement in the theorem.

From Exercise F.135 we know already that

$$
\mathcal{H} \leq \bigcap_{\chi \in \mathcal{H}^{\perp}} \mathrm{Ker}(\chi) .
$$

But we have just shown that for any $g \notin \mathcal{H}$ there exists a $\chi \in \mathcal{H}^{\perp}$ such that $g \notin \mathrm{Ker}(\chi)$. Consequently, such a g cannot be in the intersection of all $\chi \in \mathcal{H}^{\perp}$, and we must have

$$
\mathcal{H} = \bigcap_{\chi \in \mathcal{H}^{\perp}} \mathrm{Ker}(\chi) .
$$

Now let $\mathcal{H}^{\perp} = \langle \chi_1, \ldots, \chi_L \rangle$. This means that for any $\chi \in \mathcal{H}^{\perp}$ there exist $m_1, \ldots, m_L \in \mathbb{Z}$ such that

$$
\chi = \chi_1^{m_1} \cdots \chi_L^{m_L} .
$$

Hence, for any $g \in \mathcal{G}$

$$
\chi(g) = \chi_1^{m_1} \cdots \chi_L^{m_L}(g) \underbrace{=}_{\text{(F.61)}} (\chi_1(g))^{m_1} \cdots (\chi_L(g))^{m_L} .
$$

Consequently, $h \in \bigcap_{l=1}^{L} \mathrm{Ker}(\chi_l)$ implies $h \in \mathrm{Ker}(\chi)$ and thus for any $\chi \in \mathcal{H}^{\perp}$

$$\bigcap_{l=1}^{L} \mathrm{Ker}(\chi_l) \subseteq \mathrm{Ker}(\chi)$$

from which it follows that

$$\bigcap_{l=1}^{L} \mathrm{Ker}(\chi_l) \subseteq \bigcap_{\chi \in \mathcal{H}^{\perp}} \mathrm{Ker}(\chi). \tag{F.90}$$

On the other hand, since $\chi_1, \ldots, \chi_L \in \mathcal{H}^{\perp}$, we also have

$$\bigcap_{\chi \in \mathcal{H}^{\perp}} \mathrm{Ker}(\chi) \subseteq \bigcap_{l=1}^{L} \mathrm{Ker}(\chi_l). \tag{F.91}$$

Together, (F.90) and (F.91) imply

$$\bigcap_{l=1}^{L} \mathrm{Ker}(\chi_l) = \bigcap_{\chi \in \mathcal{H}^{\perp}} \mathrm{Ker}(\chi) \underbrace{=}_{(F.84)} \mathcal{H}.$$

$\square$

Finite abelian groups are isomorphic to the dual of their dual group.

Theorem F.45 *Let $\mathcal{G}$ be a finite abelian group and $\widehat{\mathcal{G}}$ its dual group. Then*

$$\begin{aligned} \widehat{}: \mathcal{G} &\longrightarrow \widehat{\widehat{\mathcal{G}}} \\ g &\longmapsto \widehat{\widehat{g}} \end{aligned}, \tag{F.92}$$

where $\widehat{\widehat{g}}$ is defined as

$$\begin{aligned} \widehat{\widehat{g}}: \widehat{\mathcal{G}} &\longrightarrow U(1) \\ \chi &\longmapsto \widehat{\widehat{g}}(\chi) := \chi(g) \end{aligned},$$

is an isomorphism, and we have

$$\widehat{\widehat{\mathcal{G}}} \cong \mathcal{G}.$$

Proof From Theorem F.34 we know that the dual group $\widehat{\mathcal{G}}$ of a finite abelian group $\mathcal{G}$ is a finite abelian group as well and applying the statement of that theorem to $\widehat{\mathcal{G}}$ in turn implies that $\widehat{\widehat{\mathcal{G}}}$ is a finite abelian group, too. From Corollary F.37 we also know that $|\widehat{\mathcal{G}}| = |\mathcal{G}|$ and thus also $|\widehat{\widehat{\mathcal{G}}}| = |\widehat{\mathcal{G}}| = |\mathcal{G}|$, that is the group $\widehat{\widehat{\mathcal{G}}}$ has the same number of elements as the group $\mathcal{G}$.

For any $g_1, g_2 \in \mathcal{G}$ and $\chi \in \widehat{\mathcal{G}}$ we have

$$\widehat{g_1 g_2}(\chi) \underset{(F.45)}{=} \chi(g_1 g_2) \underset{(F.61)}{=} \chi(g_1)\chi(g_2) \underset{(F.45)}{=} \widehat{g_1}(\chi)\widehat{g_2}(\chi).$$

Hence, $\widehat{g_1 g_2} = \widehat{g_1}\widehat{g_2}$, and it follows that $\widehat{} \in \mathrm{Hom}(\mathcal{G}, \widehat{\widehat{\mathcal{G}}})$. Moreover, $\widehat{g_1} = \widehat{g_2}$ implies that for any $\chi \in \widehat{\mathcal{G}}$

$$\chi(g_1) \underset{(F.45)}{=} \widehat{g_1}(\chi) = \widehat{g_2}(\chi) \underset{(F.45)}{=} \chi(g_2) \tag{F.93}$$

and thus

$$\chi(g_1 g_2^{-1}) \underset{(F.61)}{=} \chi(g_1)\chi(g_2^{-1}) \underset{(F.50)}{=} \chi(g_1)\chi(g_2)^{-1} \underset{(F.93)}{=} 1$$

for any $\chi \in \widehat{\mathcal{G}}$ from which it follows that $g_1 g_2^{-1} = e_{\mathcal{G}}$ and thus $g_1 = g_2$. Hence, the map $\widehat{} : \mathcal{G} \to \widehat{\widehat{\mathcal{G}}}$ is also injective, and since $\widehat{\widehat{\mathcal{G}}}$ has the same number of elements as $\mathcal{G}$, it is also a bijection and altogether an isomorphism, which was to be shown. $\square$

Lemma F.46 *Let $\mathcal{H}$ be a subgroup of the finite abelian group $\mathcal{G}$ and $\mathcal{H}^\perp$ as defined in (F.72). Then we have*

$$(\mathcal{H}^\perp)^\perp = \mathcal{H}. \tag{F.94}$$

Proof For any $h \in \mathcal{H}$ and $\chi \in \mathcal{H}^\perp$, we have

$$1 = \chi(h) = \widehat{h}(\chi),$$

where $\widehat{}$ denotes the isomorphism defined in (F.92). Hence,

$$h \in (\mathcal{H}^\perp)^\perp = \{\xi \in \widehat{\widehat{\mathcal{G}}} \mid \mathcal{H}^\perp \subset \mathrm{Ker}(\xi)\},$$

and it follows that

$$\mathcal{H} \subset (\mathcal{H}^\perp)^\perp. \tag{F.95}$$

Applying the result of Corollary F.43 to the subgroup $\mathcal{H}^{\perp} \leq \widehat{\mathcal{G}}$ gives

$$\left|\left(\mathcal{H}^{\perp}\right)^{\perp}\right| \underbrace{=}_{(F.83)} \frac{|\widehat{\mathcal{G}}|}{|\mathcal{H}^{\perp}|} \underbrace{=}_{(F.70),(F.83)} \frac{|\mathcal{G}|}{\frac{|\mathcal{G}|}{|\mathcal{H}|}} = |\mathcal{H}|,$$

which, together with (F.95), implies $\left(\mathcal{H}^{\perp}\right)^{\perp} = \mathcal{H}$. $\square$

Example F.47 Applying (F.94) to the trivial subgroups of Example F.40, we find

$$\langle e_{\mathcal{G}}\rangle \underbrace{=}_{(F.94)} \left(\langle e_{\mathcal{G}}\rangle^{\perp}\right)^{\perp} \underbrace{=}_{(F.73)} \widehat{\mathcal{G}}^{\perp},$$

whereas for the trivial subgroup $\mathcal{G}$ itself

$$\mathcal{G} \underbrace{=}_{(F.94)} \left(\mathcal{G}^{\perp}\right)^{\perp} \underbrace{=}_{(F.74)} \langle e_{\widehat{\mathcal{G}}}\rangle^{\perp}. \tag{F.96}$$

Theorem F.48 *Let $\mathcal{H}_1$ and $\mathcal{H}_2$ be subgroups of the finite abelian group $\mathcal{G}$ and let $\mathcal{H}_1^{\perp}$ and $\mathcal{H}_2^{\perp}$ be as defined in (F.72). Then we have*

$$\mathcal{H}_1 < \mathcal{H}_2 \leq \mathcal{G} \quad \Leftrightarrow \quad \mathcal{H}_2^{\perp} < \mathcal{H}_1^{\perp} \leq \widehat{\mathcal{G}}. \tag{F.97}$$

Proof From Lemma F.39 we know already that $\mathcal{H}_1^{\perp}$ and $\mathcal{H}_2^{\perp}$ are subgroups of $\widehat{\mathcal{G}}$. Now let $\mathcal{H}_1 < \mathcal{H}_2$ and $\chi \in \mathcal{H}_2^{\perp}$, which, on account of (F.72), implies

$$\mathcal{H}_1 \subset \mathcal{H}_2 \subset \text{Ker}(\chi)$$

and thus, using once more (F.72), $\chi \in \mathcal{H}_1^{\perp}$. Consequently, $\mathcal{H}_2^{\perp} \leq \mathcal{H}_1^{\perp}$, but from Theorem F.44 we know that for each $g \in \mathcal{H}_2 \setminus \mathcal{H}_1$ there exists a $\chi \in \mathcal{H}_1^{\perp}$ such that $g \notin \text{Ker}(\chi)$. Hence, $\mathcal{H}_2^{\perp} < \mathcal{H}_1^{\perp}$, and we have shown that

$$\mathcal{H}_1 < \mathcal{H}_2 \leq \mathcal{G} \quad \Rightarrow \quad \mathcal{H}_2^{\perp} < \mathcal{H}_1^{\perp} \leq \widehat{\mathcal{G}}. \tag{F.98}$$

From Corollary F.37 we know that $\widehat{\mathcal{G}}$ is a finite abelian group and from Lemma F.39 that $\mathcal{H}_1^{\perp}$ and $\mathcal{H}_2^{\perp}$ are subgroups of $\widehat{\mathcal{G}}$. Hence, we can apply the result (F.98) to these groups, which gives us

$$\mathcal{H}_2^{\perp} < \mathcal{H}_1^{\perp} \leq \widehat{\mathcal{G}} \quad \Rightarrow \quad \left(\mathcal{H}_1^{\perp}\right)^{\perp} < \left(\mathcal{H}_2^{\perp}\right)^{\perp} \leq \widehat{\widehat{\mathcal{G}}}. \tag{F.99}$$

Theorem F.45 tells us that we can identify $\widehat{\widehat{\mathcal{G}}}$ with $\mathcal{G}$ and Lemma F.46 that $\left(\mathcal{H}_1^\perp\right)^\perp = \mathcal{H}_1$ as well as $\left(\mathcal{H}_2^\perp\right)^\perp = \mathcal{H}_2$. Therefore, (F.99) becomes

$$\mathcal{H}_2^\perp < \mathcal{H}_1^\perp \leq \widehat{\mathcal{G}} \quad \Rightarrow \quad \mathcal{H}_1 < \mathcal{H}_2 \leq \mathcal{G},$$

and the proof of (F.97) is complete.   □

Corollary F.49 *Let $\mathcal{G}$ be a finite abelian group, $\widehat{\mathcal{G}}$ its dual, $\mathcal{H} < \mathcal{G}$ a proper subgroup and $\mathcal{H}^\perp \leq \widehat{\mathcal{G}}$ be defined as in (F.72). Then the following hold*

$$\begin{aligned}\mathcal{H} \text{ maximal} &\quad\Leftrightarrow\quad \mathcal{H}^\perp \text{ minimal} \\ \mathcal{H} \text{ minimal} &\quad\Leftrightarrow\quad \mathcal{H}^\perp \text{ maximal}.\end{aligned} \qquad\text{(F.100)}$$

Proof Let $\mathcal{H} < \mathcal{G}$ be maximal. Hence, there is no $\mathcal{K} < \mathcal{G}$ such that

$$\mathcal{H} < \mathcal{K} < \underbrace{\mathcal{G} = \langle e_{\widehat{\mathcal{G}}} \rangle^\perp}_{\text{(F.96)}}.$$

Theorem F.48 then implies that there is no subgroup $\mathcal{M} < \widehat{\mathcal{G}}$ such that

$$\underbrace{\langle e_{\widehat{\mathcal{G}}} \rangle = \left(\langle e_{\widehat{\mathcal{G}}} \rangle^\perp\right)^\perp}_{\text{(F.94)}} < \mathcal{M} < \mathcal{H}^\perp \qquad\text{(F.101)}$$

because, if there were such an $\mathcal{M}$, then it would follow from (F.97) that the subgroup $\mathcal{M}^\perp$ would satisfy $\mathcal{H} < \mathcal{M}^\perp < \mathcal{G}$, which contradicts the assumption that $\mathcal{H}$ is maximal. Hence, there can be no $\mathcal{M}$ satisfying (F.101) and $\mathcal{H}^\perp$ has to be minimal.

Now let $\mathcal{H}^\perp$ be minimal, which means that there is no subgroup $\mathcal{K}$ satisfying

$$\langle e_{\widehat{\mathcal{G}}} \rangle < \mathcal{K} < \mathcal{H}^\perp.$$

Then again, Theorem F.48 implies that there is no subgroup $\mathcal{M} < \mathcal{G}$ such that

$$\mathcal{H} < \mathcal{M} < \mathcal{G} \qquad\text{(F.102)}$$

because, if there were, then (F.97) would imply

$$\underbrace{\langle e_{\widehat{\mathcal{G}}} \rangle = \mathcal{G}^\perp}_{\text{(F.74)}} < \mathcal{M}^\perp < \mathcal{H}^\perp,$$

which contradicts $\mathcal{H}^\perp$ being minimal. Consequently, there can be no $\mathcal{M}$ satisfying (F.102) and $\mathcal{H}$ has to be maximal.

So far we have shown

$$\mathcal{H} \text{ maximal} \quad \Leftrightarrow \quad \mathcal{H}^\perp \text{ minimal}.$$

Applying this statement to the subgroup $\mathcal{H}^\perp < \widehat{\mathcal{G}}$ yields

$$\mathcal{H}^\perp \text{ maximal} \quad \Leftrightarrow \quad \underbrace{\left(\mathcal{H}^\perp\right)^\perp}_{(F.94)} = \mathcal{H} \text{ minimal}.$$

$\square$

With the previous results we can finally establish a lower bound on the probability that a finite abelian group is generated by a randomly selected set of elements. This result will be useful in the context of the Abelian Hidden Subgroup Problem discussed in detail in Sect. 6.6.

Corollary F.50 *Let $g_1, \ldots, g_L$ be a set of elements of the finite abelian group $\mathcal{G}$ that have been selected at random, independently and with replacement from the uniformly distributed elements of $\mathcal{G}$. Then the probability that the whole group is generated by this set satisfies*

$$\mathbf{P}\{\mathcal{G} = \langle g_1, \ldots, g_L \rangle\} \geq 1 - \frac{|\mathcal{G}|}{2^L}. \tag{F.103}$$

Proof We show the claim by first deriving a bound on the probability of the complementary event $\{\langle g_1, \ldots, g_L \rangle < \mathcal{G}\}$ and then use that

$$\mathbf{P}\{\langle g_1, \ldots, g_L \rangle = \mathcal{G}\} + \mathbf{P}\{\langle g_1, \ldots, g_L \rangle < \mathcal{G}\} = 1$$

If $\langle g_1, \ldots, g_L \rangle \neq \mathcal{G}$, there is a maximal subgroup $\mathcal{H}$ of $\mathcal{G}$ such that

$$\langle g_1, \ldots, g_L \rangle \leq \mathcal{H} < \mathcal{G}.$$

From (F.36) we know that $\frac{|\mathcal{G}|}{|\mathcal{H}|} \in \mathbb{N}$, and since $\mathcal{H} < \mathcal{G}$, we must have

$$\frac{|\mathcal{G}|}{|\mathcal{H}|} \geq 2. \tag{F.104}$$

Hence, when a g_1 is selected at random from the uniformly distributed $|\mathcal{G}|$ elements of $\mathcal{G}$, the probability that it is in $\mathcal{H}$ satisfies

$$\mathbf{P}\{g_1 \in \mathcal{H}\} = \frac{|\mathcal{H}|}{|\mathcal{G}|} \underbrace{\leq}_{(\text{F.104})} \frac{1}{2}.$$

When independently selecting L elements $g_1, \ldots, g_L$ in this fashion, the probability that they are all in $\mathcal{H}$ thus satisfies

$$\mathbf{P}\{g_1, \ldots, g_L \in \mathcal{H}\} \leq \frac{1}{2^L}. \tag{F.105}$$

Hence, the probability that the $g_1, \ldots, g_L$ are all in any maximal subgroup $\mathcal{H} < \mathcal{G}$ satisfies

$$\mathbf{P}\left\{\begin{array}{l} g_1, \ldots, g_L \in \mathcal{H} \quad \text{for} \\ \text{any maximal } \mathcal{H} < \mathcal{G} \end{array}\right\} \leq \sum_{\mathcal{H} < \mathcal{G} \text{ maximal}} \mathbf{P}\{g_1, \ldots, g_L \in \mathcal{H}\}$$

$$\underbrace{=}_{(\text{F.100})} \sum_{\mathcal{H}^\perp \leq \widehat{\mathcal{G}} \text{ minimal}} \mathbf{P}\{g_1, \ldots, g_L \in \mathcal{H}\}$$

$$\underbrace{\leq}_{(\text{F.105})} \frac{\text{number of minimal subgroups of } \widehat{\mathcal{G}}}{2^L}.$$

From Lemma F.11 we know that any minimal subgroup of $\widehat{\mathcal{G}}$ is of the form $\langle \chi \rangle$ for some $\chi \in \widehat{\mathcal{G}}$. Hence,

$$\text{number of minimal subgroups of } \widehat{\mathcal{G}} \leq |\widehat{\mathcal{G}}| \underbrace{=}_{(\text{F.70})} |\mathcal{G}|,$$

such that

$$\mathbf{P}\left\{\begin{array}{l} g_1, \ldots, g_L \in \mathcal{H} \quad \text{for} \\ \text{any maximal } \mathcal{H} < \mathcal{G} \end{array}\right\} \leq \frac{|\mathcal{G}|}{2^L}$$

from which follows that

$$\mathbf{P}\{\langle g_1, \ldots, g_L \rangle < \mathcal{G}\} \leq \frac{|\mathcal{G}|}{2^L}.$$

Consequently,

$$\mathbf{P}\{\langle g_1, \ldots, g_L \rangle = \mathcal{G}\} = 1 - \mathbf{P}\{\langle g_1, \ldots, g_L \rangle < \mathcal{G}\} \geq 1 - \frac{|\mathcal{G}|}{2^L}.$$

$\square$

F.3 Quantum FOURIER Transform on Groups

Definition F.51 Let $\mathcal{G}$ be a finite abelian group and $\widehat{\mathcal{G}}$ its dual group. Moreover, let $\mathbb{H}$ be a HILBERT space with $\dim\mathbb{H} = |\mathcal{G}|$ that has an ONB $\{|g_j\rangle \,|\, j \in \{1,\ldots,|\mathcal{G}|\}\}$ such that we can identify each element g_j of $\mathcal{G}$ with a basis vector of that ONB, and that the same holds for another ONB $\{|\chi_k\rangle \,|\, k \in \{1,\ldots,|\mathcal{G}|\}\}$ of $\mathbb{H}$ and the elements $|\chi_k\rangle$ of $\widehat{\mathcal{G}}$, that is,

$$\mathbb{H} = \mathrm{Span}\{|g_1\rangle,\ldots,|g_{|\mathcal{G}|}\rangle\} = \mathrm{Span}\left\{|g_j\rangle \,\middle|\, g_j \in \mathcal{G}\right\}$$
$$= \mathrm{Span}\{|\chi_1\rangle,\ldots,|\chi_{|\mathcal{G}|}\rangle\} = \mathrm{Span}\left\{|\chi_k\rangle \,\middle|\, \chi_k \in \widehat{\mathcal{G}}\right\} \qquad \text{(F.106)}$$
$$\langle g_j|g_k\rangle = \delta_{jk} = \langle\chi_j|\chi_k\rangle .$$

Then the **quantum FOURIER transform on the group** $\mathcal{G}$ is the operator $F_{\mathcal{G}}$: $\mathbb{H} \to \mathbb{H}$ defined by

$$F_{\mathcal{G}} := \frac{1}{\sqrt{|\mathcal{G}|}} \sum_{\substack{g \in \mathcal{G} \\ \chi \in \widehat{\mathcal{G}}}} \chi(g)|\chi\rangle\langle g| . \qquad \text{(F.107)}$$

Exercise F.136 Show that the quantum FOURIER transform $F_{\mathcal{G}}$ is unitary.

For a solution see Solution F.136

The quantum FOURIER transform introduced in Definition 5.48 is a special case of the quantum FOURIER transform on groups, namely for the group $\mathbb{Z}_{2^n}$.

Example F.52 Consider the group $\mathcal{G} = \mathbb{Z}_N$ of Example F.29 for the special case $N = 2^n$ in which case

$$\mathcal{G} \underbrace{=}_{\text{(F.46)}} \mathbb{Z}_{2^n} = \mathbb{Z}/2^n\mathbb{Z} = \left\{ [x]_{2^n\mathbb{Z}} \,\middle|\, x \in \{0,\ldots,2^n-1\}\right\}$$

$$\widehat{\mathcal{G}} \underbrace{=}_{\text{(F.59)}} \left\{ \chi_y : \begin{array}{rcl} \mathcal{G} & \longrightarrow & \mathrm{U}(1) \\ [x]_{2^n\mathbb{Z}} & \longmapsto & e^{2\pi\mathrm{i}\frac{xy}{2^n}} \end{array} \,\middle|\, y \in \{0,\ldots,2^n-1\}\right\}$$

and $|G| = 2^n = |\widehat{\mathcal{G}}|$. Moreover, we can take $\mathbb{H} = {}^{\P}\mathbb{H}^{\otimes n}$, where ${}^{\P}\mathbb{H} \cong \mathbb{C}^2$ is the qubit HILBERT space defined in Definition 2.28. If we assume that the identification of (F.106) can be written in the form

$$\begin{array}{ll} |\,[x]_{2^n\mathbb{Z}}\rangle = |x\rangle & \forall [x]_{2^n\mathbb{Z}} \in \mathbb{Z}_{2^n} \\ |\chi_y\rangle = |y\rangle & \forall \chi_y \in \widehat{\mathbb{Z}_{2^n}} , \end{array}$$

then $F_{\mathbb{Z}_{2^n}}$ of (F.107) becomes

$$F_{\mathbb{Z}_{2^n}} = \frac{1}{\sqrt{2^n}} \sum_{x,y=0}^{2^n-1} \exp\left(2\pi i\frac{xy}{2^n}\right) |x\rangle\langle y|,$$

which is exactly as it was defined in Definition 5.48.

F.4   Elliptic Curves

Before we can define elliptic curves, we need to introduce the notion of a field.

Definition F.53 A **field** is a set $\mathbb{F}$ together with an operation $+ : \mathbb{F} \times \mathbb{F} \to \mathbb{F}$ called addition and an operation $\cdot : \mathbb{F} \times \mathbb{F} \to \mathbb{F}$ called multiplication such that

(i) $(\mathbb{F}, +)$ is an abelian group with neutral element $e_{+_{\mathbb{F}}}$
(ii) $\left(\mathbb{F} \smallsetminus \{e_{+_{\mathbb{F}}}\}, \cdot\right)$ is an abelian group
(iii) multiplication is distributive over addition, that is, for all $a, b, c \in \mathbb{F}$ we have
$$a \cdot (b+c) = a \cdot b + a \cdot c.$$

Usually, the neutral element of addition $e_{+_{\mathbb{F}}}$ is denoted by 0 and the neutral element of multiplication by 1. If the set $\mathbb{F}$ is finite the field is called finite as well.

Example F.54 The definition of a field formalizes something we are most familiar with since $\mathbb{Q}, \mathbb{R}$ and $\mathbb{C}$ are all fields. But $\mathbb{Z}$ is not a field since any integer other than ± 1 has no integer multiplicative inverse.

Corollary F.55 *For a prime* p

$$\mathbb{F}_p := \mathbb{Z}/p\mathbb{Z} \cong \mathbb{Z}_p$$

is a finite field.

Proof Lemma F.5 states that $\mathbb{Z}_p$ forms a group under addition and that $\mathbb{Z}_p^{\times}$ forms a multiplication group. Moreover, for any $[a]_{p\mathbb{Z}}, [b]_{p\mathbb{Z}}, [c]_{p\mathbb{Z}} \in \mathbb{Z}_p$ we have

$$[a]_{p\mathbb{Z}} \cdot_{\mathbb{Z}_p} \left([b]_{p\mathbb{Z}} +_{\mathbb{Z}_p} [c]_{p\mathbb{Z}}\right) \underset{\text{(F.11),(F.12)}}{=} \left(a\big((b+c)\bmod p\big)\right)\bmod p$$

$$\underset{\text{(D.21),(D.23)}}{=} \big((ab)\bmod p + (ac)\bmod p\big)\bmod p$$

$$\underset{\text{(F.11),(F.12)}}{=} [a]_{p\mathbb{Z}} \cdot_{\mathbb{Z}_p} [b]_{p\mathbb{Z}} +_{\mathbb{Z}_p} [a]_{p\mathbb{Z}} \cdot_{\mathbb{Z}_p} [c]_{p\mathbb{Z}},$$

showing that multiplication is distributive over addition and thus completing the proof that $\mathbb{F}_p$ is a field. From Example F.7 and (F.46) we know that $\mathbb{Z}_p$ is finite. $\square$

Definition F.56 A (non-singular) **elliptic curve** with point at infinity 0_E over a field $\mathbb{F}$ is defined by two elements $A, B \in \mathbb{F}$ as the set

$$E(\mathbb{F}) := \left\{ (x,y) \in \mathbb{F} \times \mathbb{F} \;\middle|\; \begin{array}{l} y^2 = x^3 + Ax + B, \text{ where } A, B \in \mathbb{F} \\ \text{such that } 4A^3 + 27B^2 \neq 0_{\mathbb{F}} \end{array} \right\} \cup \{0_E\}.$$

The cubic equation $y^2 = x^3 + Ax + B$ is called the **WEIERSTRASS equation of the elliptic curve**. The quantity

$$\Delta_E := 4A^3 + 27B^2 \tag{F.108}$$

is called the **discriminant of the elliptic curve**.

In this definition any addition, multiplication and equality are all understood to be in the underlying field $\mathbb{F}$. We shall also use the notation $\frac{1}{x}$ to denote the multiplicative inverse of $x \in \mathbb{F}$.

Strictly speaking—and as the noun 'curve' suggests—an elliptic curve is just the graph $(x,y) \in E(\mathbb{F}) \setminus \{0_E\}$ of the solutions of the WEIERSTRASS equation. We have added the point at infinity 0_E to this set because we want to make the elliptic curve into a group, and for this purpose, the point at infinity will serve as the neutral element of the group.

Note that the elements of $E(\mathbb{F}) \setminus \{0_E\}$ come in pairs, that is, if $(x,y) \in E(\mathbb{F}) \setminus \{0_E\}$ then it follows that $(x,-y) \in E(\mathbb{F}) \setminus \{0_E\}$. This is often stated as the elliptic curve being symmetrical around the x-axis, which is evident in two-dimensional plots for the case $\mathbb{F} = \mathbb{R}$.

Example F.57 As an example we consider the case $\mathbb{F} = \mathbb{R}$ and $A = -2$ and $B = 2$. Figure F.1 then shows the curve $E(\mathbb{R}) \setminus \{0_E\}$.

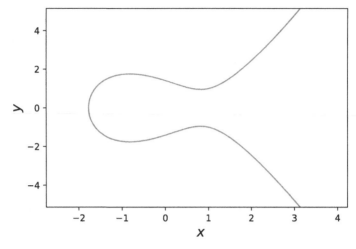

Fig. F.1 Elliptic Curve over $\mathbb{R}$ with $A = -2$ and $B = 2$. The symmetry around the x-axis is evident

The case $\mathbb{F} = \mathbb{R}$ is special in that it allows to display the solutions $(x, y) \in \mathbb{R}^2$ of the WEIERSTRASS equation as an actual curve in the two-dimensional plane. For most other fields $\mathbb{F}$ no such graphical illustration or only a much less intuitive one is possible.

The condition $\Delta_E \neq 0_F$ is what makes the elliptic curve non-singular. To see its relevance, we first note that if the roots $r_1, r_2, r_3 \in \mathbb{F}$ of the cubic on the right side of the WEIERSTRASS equation exist, we obtain

$$
\begin{aligned}
x^3 + Ax + B &= (x - r_1)(x - r_2)(x - r_3) \\
&= x^3 - (r_1 + r_2 + r_3)x^2 + (r_1 r_2 + r_1 r_3 + r_2 r_3)x - r_1 r_2 r_3 \,,
\end{aligned}
$$

which implies

$$r_1 + r_2 + r_3 = 0 \tag{F.109}$$

$$r_1 r_2 + r_1 r_3 + r_2 r_3 = A \tag{F.110}$$

$$-r_1 r_2 r_3 = B \,. \tag{F.111}$$

Exercise F.137 Show that the roots r_1, r_2 and r_3 of the right side of the WEIER-STRASS equation and the discriminant Δ_E of an elliptic curve E satisfy

$$\left((r_1 - r_2)(r_1 - r_3)(r_2 - r_3) \right)^2 = -\Delta_E \,. \tag{F.112}$$

For a solution see Solution F.137

From (F.112) we see that the requirement $\Delta_E \neq 0_{\mathbb{F}}$ in Definition F.56 ensures that all roots of the cubic on the right side of the WEIERSTRASS equation are distinct. This is the only case which is of interest to us.

Theorem F.58 *Let $E(\mathbb{F})$ be a non-singular elliptic curve over $\mathbb{F}$ with WEIERSTRASS equation $y^2 = x^3 + Ax + B$ and with point at infinity 0_E. For any two elements $P, Q \in E(\mathbb{F})$ we define*

$$P +_E Q \in (\mathbb{F} \times \mathbb{F}) \cup \{0_E\}$$

as follows:

if $P = 0_E$, then

$$P +_E Q = Q \tag{F.113}$$

else if $Q = 0_E$, then

$$P +_E Q = P \tag{F.114}$$

else set $P = (x_P, y_P), Q = (x_Q, y_Q)$ and

  if $x_P = x_Q$ and $y_P = -y_Q$, then

$$P +_E Q = 0_E \tag{F.115}$$

  else set

$$m = m(P, Q) = \begin{cases} \frac{3x_P^2 + A}{2y_P} & \text{if } x_P = x_Q \text{ and } y_P = y_Q \neq 0_{\mathbb{F}} \\ \frac{y_Q - y_P}{x_Q - x_P} & \text{if } x_P \neq x_Q \end{cases} \tag{F.116}$$

  and

$$\begin{aligned} x_+ &= m^2 - x_P - x_Q \\ y_+ &= m(x_P - x_+) - y_P \end{aligned} \tag{F.117}$$

  then

$$P +_E Q = (x_+, y_+).$$

Then $(E(\mathbb{F}), +_E)$ is an abelian group with neutral element 0_E and inverse elements $-_E(x, y) = (x, -_{\mathbb{F}} y)$.

Fig. F.2 Graphical
illustration of the addition of
P to itself in $E(\mathbb{R})$ with
$A = -2$ and $B = 2$. The line
at P is the tangent to the
curve. The intersection of
this line with the curve is
$-2P$. Reflecting this point
about the x-axis gives
$2P = P +_E P$

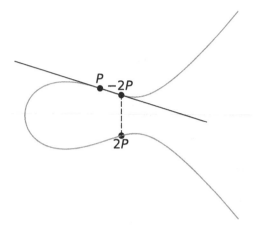

Proof Our first task is to show that $P +_E Q \in E(\mathbb{F})$. From (F.113) and (F.114) we
see that this is indeed the case when $P = 0_E$ or $Q = 0_E$ since we assume $P, Q \in E(\mathbb{F})$
to begin with. Likewise, since $0_E \in E(\mathbb{F})$, it follows from (F.115) that in the case
$P, Q \in E(\mathbb{F}) \smallsetminus \{0_E\}$, where $x_P = x_Q$ and $y_P = -y_Q$, we have $P +_E Q \in E(\mathbb{F})$.

Next, we show that in the case $P, Q \in E(\mathbb{F}) \smallsetminus \{0_E\}$, where $x_P = x_Q$ and $y_P = y_Q \neq 0_{\mathbb{F}}$ or $x_P \neq x_Q$, we have $P +_E Q \in E(\mathbb{F})$. For this we need to show that (x_+, y_+)
as defined in (F.117) satisfies the WEIERSTRASS equation given that (x_P, y_P) and
(x_Q, y_Q) do. To verify this, let $m = m(P, Q) \in \mathbb{F}$ be as given in (F.116), and define
the 'line'

$$l : \mathbb{F} \longrightarrow \mathbb{F}$$
$$x \longmapsto l(x) := m(x - x_P) + y_P \qquad (F.118)$$

Then we have $y_P = l(x_P)$ and $y_Q = l(x_Q)$ as well as

$$y_+ \underbrace{=}_{(F.117),(F.118)} -l(x_+).$$

Moreover, any pair $(x, l(x))$ that satisfies the WEIERSTRASS equation is by definition an element of $E(\mathbb{F})$, that is,

$$\left(x, \pm l(x)\right) \in E(\mathbb{F}) \qquad \Leftrightarrow \qquad \left(l(x)\right)^2 = x^3 + Ax + B.$$

Hence, the point $P +_E Q = (x_+, y_+)$ is the reflection about the x-axis of the intercep-
tion of the line $(x, l(x))$ with the curve $E(\mathbb{F}) \smallsetminus \{0_E\}$. We illustrate this graphically
for $E(\mathbb{R})$ in the case where $P = Q$ in Fig. F.2 and where $P \neq Q$ in Fig. F.3.

Consequently, if we show that $(x_+, l(x_+))$ satisfies the WEIERSTRASS equation,
then if follows that

$$P +_E Q = \left(x_+, y_+\right) = \left(x_+, -l(x_+)\right) \in E(\mathbb{F}).$$

Fig. F.3 Graphical
illustration of the calculation
of $P +_E Q$ in $E(\mathbb{R})$ with
$A = -2$ and $B = 2$. The
intersection of the line
through P and Q with the
curve is $-(P +_E Q)$.
Reflecting this point about
the x-axis gives $P +_E Q$

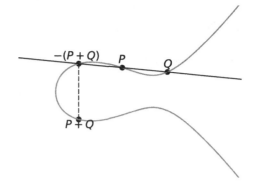

If $x_P = x_Q$ and $y_P = y_Q \neq 0_E$ we show this by direct, albeit cumbersome, calculation.
In this case we have

$$m = \frac{3x_P^2 + A}{2y_P} \qquad (\text{F.119})$$

$$x_+ = m^2 - 2x_P \qquad (\text{F.120})$$

and thus

$$x_+^3 + Ax_+ + B - \left(l(x_+)\right)^2 \underbrace{=}_{(\text{F.120})} 3m^2 x_P^2 + Am^2 + 6mx_P y_P - 2m^3 y_P - 9x_P^3 - 3Ax_P$$

$$= 2y_P \left[m^2 \frac{3x_P^2 + A}{2y_P} - m^3 + 3x_P \left(m - \frac{3x_P^2 + A}{2y_P} \right) \right]$$

$$\underbrace{=}_{(\text{F.119})} 0.$$

Now consider the case $x_P \neq x_Q$. The expression

$$x^3 + Ax + B - \left(l(x)\right)^2 \qquad (\text{F.121})$$
$$= x^3 - m^2 x^2 + (2m^2 x_P - 2my_P + A)x + B - m^2 x_P^2 + 2mx_P y_P - y_P^2$$

is a cubic of which we already know two roots, namely x_P and x_Q since
$P = \left(x_P, y_p = l(x_P)\right)$ and $Q = \left(x_Q, y_Q = l(x_Q)\right)$ are both in $E(\mathbb{F})$. Let x_+ denote
the third root such that

$$x^3 + Ax + B - \left(l(x)\right)^2 = (x - x_P)(x - x_Q)(x - x_+) \qquad (\text{F.122})$$
$$= x^3 - (x_P + x_Q + x_+) + (x_P x_Q + x_P x_+ + x_Q x_+)x - x_P x_Q x_+ .$$

Comparing coefficients of x in (F.121) and (F.122), yields, after some
re-arrangements,

Fig. F.4 Illustration of the associativity of addition for $E(\mathbb{R})$ with $A = -2$ and $B = 2$. Here P is added to Q first, and the result is added to R, resulting in $\left(P +_E Q\right) +_E R$

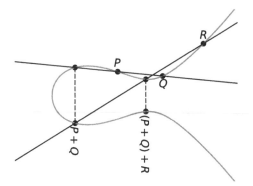

$$x_+ = m^2 - x_P - x_Q \qquad\qquad\qquad (F.123)$$
$$0 = 2my_P - A - 2x_P^2 - x_P x_Q - x_P x_+ + x_Q x_+ \qquad (F.124)$$
$$0 = \left(2my_P - A - 2x_P^2 - x_P x_Q - x_P x_+ + x_Q x_+\right) x_P, \qquad (F.125)$$

and we see that (F.123) gives x_+ as claimed in (F.117). As for the other two equations, suppose first that $x_P = 0$. Then (F.125) is trivially satisfied and only (F.124) remains. Likewise, if $x_P \neq 0$, division by x_P in (F.125) reduces it to (F.124). Using (F.123) in (F.124) and re-arranging it, we find

$$\begin{aligned} m^2(x_Q - x_P)^2 + 2my_P(x_Q - x_P) &= (x_P^2 + x_Q^2 + x_P x_Q + A)(x_Q - x_P) \\ &= x_Q^3 + A x_Q - (x_P^3 + A x_P) \\ &= y_Q^2 - y_P^2, \end{aligned} \qquad (F.126)$$

where in the last equation we used that (x_P, y_P) and (x_Q, y_Q) satisfy the WEIERSTRASS equation. From (F.126) we find that

$$m = \frac{y_Q - y_P}{x_Q - x_P}$$

is one solution for m which guarantees that (x_+, y_+) satisfies the WEIERSTRASS equation, and this completes the proof that $P +_E Q \in E(\mathbb{F})$.

To show that $(E(\mathbb{F}), +_E)$ is indeed a group, we also need to prove associativity of the addition. This can be done by direct computation as well, but is a very cumbersome and tedious calculation, which we do not present here. Instead, we provide a graphical illustration of the associativity of addition in Figs. F.4 and F.5. In the former we display the graphical calculation of $(P +_E Q) +_E R$, whereas in the latter we show the calculation of $P +_E (Q +_E R)$.

Comparing the resulting point $\left(P +_E Q\right) +_E R$ in Fig. F.4 with that of $P +_E \left(Q +_E R\right)$ in Fig. F.5, shows that they are indeed identical, which provides a graphical 'proof' of

Fig. F.5 Illustration of the associativity of addition for $E(\mathbb{R})$ with $A = -2$ and $B = 2$. Here Q is added to R first, and the result is added to P, resulting in $P +_E (Q +_E R)$

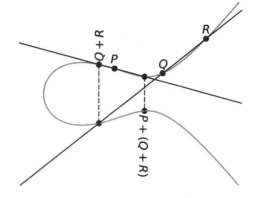

$$\left(P +_E Q\right) +_E R = P +_E \left(Q +_E R\right).$$

It remains to show the statements about the neutral element and inverses. From (F.113) and (F.114) we see that 0_E is the neutral element of $+_E$, and from (F.115) we see that every element $P = (x,y) \in E(\mathbb{F}) \smallsetminus \{0_E\}$ has its inverse given by

$$-_E P = -_E(x,y) = (x, -_{\mathbb{F}} y).$$

Finally, we note that in (F.116) we have $m(P,Q) = m(Q,P)$, making $+_E$ commutative and the group abelian. $\qquad\square$

Corollary F.59 *Let p be prime and $\mathbb{F}_p = \mathbb{Z}/p\mathbb{Z}$. Then the elliptic curve $\left(E(\mathbb{F}_p), +_E\right)$ is a finite abelian group.*

Proof From Corollary F.55 we know that $\mathbb{F}_p$ is a finite field, and from Theorem F.58 it follows that $\left(E(\mathbb{F}_p), +_E\right)$ is then a finite abelian group. $\qquad\square$

Example F.60 For finite fields $\mathbb{F}_p$ the set $E(\mathbb{F}_p) \smallsetminus \{0_E\}$ is no longer given by a curve, but rather a set of points. Moreover, the pairs $\pm_E P \in E(\mathbb{F}_p) \smallsetminus \{0_E\}$ are given by the solution pairs $(x, \pm_{\mathbb{F}_p} y) \in \mathbb{F}_p \times \mathbb{F}_p$ of the WEIERSTRASS equation, where it follows from (F.13) that $-_{\mathbb{F}_p} y = p - y$ for $y \in \{0, \ldots, p-1\}$. This results in the symmetry of the set about the line $p/2$. Figure F.6 shows the elliptic curve $E(\mathbb{F}_p)$ for $p = 541$.

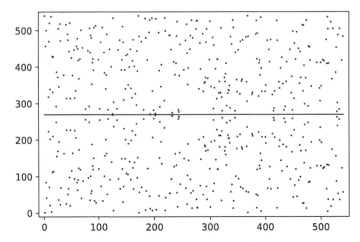

Fig. F.6 Elliptic curve $E(\mathbb{F}_{541})$ over the finite field $\mathbb{F} = \mathbb{F}_{541}$ with $A = -2$ and $B = 2$. The symmetry is made more evident by displaying a 'symmetry axis'. It arises due to the fact that if (x,y) is a solution of the WEIERSTRASS equation, then so is $-_E(x,y) = (x, -_{\mathbb{F}_{541}}y) = (x, 541 - y)$

F.5 The PAULI Group

Exercise F.138 Let σ_j for $j \in \{1,2,3\}$ denote the PAULI matrices defined in (2.74) and let $\sigma_0 := \mathbf{1}$ be the unit 2×2 matrix. Show that then the set

$$\mathcal{P} = \left\{ i^a \sigma_\alpha \mid a, \alpha \in \{0,\ldots,3\} \right\} \subset \mathrm{Mat}(2 \times 2, \mathbb{C})$$

forms a subgroup of $U(2) \cong \mathcal{U}(\mathbb{H})$ with $|\mathcal{P}| = 16$.

For a solution see Solution F.138

By selecting a basis in $\mathbb{H}$, we can identify each matrix in $\mathrm{Mat}(2 \times 2, \mathbb{C})$ bijectively with an element in $L(\mathbb{H})$ such that $\mathrm{Mat}(2 \times 2, \mathbb{C}) \cong L(\mathbb{H})$ and, likewise, $U(2) \cong \mathcal{U}(\mathbb{H})$. Hence, we will consider the elements of the set $\mathcal{P}$ as operators in $\mathcal{U}(\mathbb{H}) \subset L(\mathbb{H})$.

Recall that throughout this book we use the notations

$$\sigma_0 = \mathbf{1}, \quad \sigma_1 = \sigma_x = X, \quad \sigma_2 = \sigma_y = Y, \quad \sigma_3 = \sigma_z = Z$$

to denote PAULI matrices.

Definition F.61 The group $(\mathcal{P}, \cdot)$ with

$$\mathcal{P} := \left\{ i^a \sigma_\alpha \mid a, \alpha \in \{0, \ldots, 3\} \right\} < \mathcal{U}(\mathbb{H})$$

is called the **PAULI group**.

The PAULI group $\mathcal{P}$ is clearly non-abelian. The only elements commuting with every other element, that is, the center $\mathrm{Ctr}(\mathcal{P})$ of the group (see Definition F.17) are powers of $i\mathbf{1}$.

Exercise F.139 Show that

$$\mathcal{P} = \langle \sigma_x, \sigma_y, \sigma_z \rangle \tag{F.127}$$

and

$$\mathrm{Ctr}(\mathcal{P}) = \langle i\sigma_0 \rangle = \langle i\mathbf{1} \rangle = \left\{ i^a \mathbf{1} \mid a \in \{0, \ldots, 3\} \right\} \tag{F.128}$$

such that $|\mathrm{Ctr}(\mathcal{P})| = 4$.

For a solution see Solution F.139

Proposition F.62 *For any $n \in \mathbb{N}$, the set*

$$\mathcal{P}_n = \left\{ i^a \sigma_{\alpha_{n-1}} \otimes \cdots \otimes \sigma_{\alpha_0} \in \mathrm{L}(\mathbb{H}^{\otimes n}) \mid a, \alpha_j \in \{0, \ldots, 3\} \right\}$$

together with operator multiplication forms a group with $\mathbf{1}^{\otimes n} \in \mathrm{L}(\mathbb{H}^{\otimes n})$ as the neutral element and with the following properties for any $g = i^a \sigma_{\alpha_{n-1}} \otimes \cdots \otimes \sigma_{\alpha_0} \in \mathcal{P}_n$:

(i)

$$g^* = \begin{cases} g & \text{if } a \in \{0, 2\} \\ -g & \text{if } a \in \{1, 3\} \end{cases} \tag{F.129}$$

(ii)

$$g^2 = \begin{cases} \mathbf{1}^{\otimes n} & \text{if } a \in \{0, 2\} \\ -\mathbf{1}^{\otimes n} & \text{if } a \in \{1, 3\} \end{cases} \tag{F.130}$$

(iii)

$$g^* g = \mathbf{1}^{\otimes n}, \qquad \text{hence } \mathcal{P}_n < \mathcal{U}(\mathbb{H}^{\otimes n}) \tag{F.131}$$

(iv)

$$\mathrm{Ctr}(\mathcal{P}_n) = \langle i\mathbf{1}^{\otimes n}\rangle \tag{F.132}$$

(v)

$$|\mathcal{P}_n| = 2^{2n+2}.$$

Proof First, we show that $\mathcal{P}_n$ forms a group. Obviously, $\mathbf{1}^{\otimes n}g = g = g\mathbf{1}^{\otimes n}$ for any $g \in \mathcal{P}_n$ such that $\mathbf{1}^{\otimes n}$ is the neutral element. Let

$$g = i^a \sigma_{\alpha_{n-1}} \otimes \cdots \otimes \sigma_{\alpha_0} \qquad \text{and} \qquad h = i^b \sigma_{\beta_{n-1}} \otimes \cdots \otimes \sigma_{\beta_0}$$

such that

$$gh = i^{a+b}\sigma_{\alpha_{n-1}}\sigma_{\beta_{n-1}} \otimes \cdots \otimes \sigma_{\alpha_0}\sigma_{\beta_0}.$$

We know from Exercise F.138 that $\sigma_{\alpha_j}\sigma_{\beta_j} = i^{c_j}\sigma_{\gamma_j} \in \mathcal{P}$ with $c_j, \gamma_j \in \{0,\ldots,3\}$. Using that $i^d = i^{d \bmod 4}$ for any $d \in \mathbb{Z}$, we thus obtain

$$gh = i^{(a+b+\sum_{j=0}^{n-1}c_j)\bmod 4}\sigma_{\gamma_{n-1}} \otimes \cdots \otimes \sigma_{\gamma_0} \in \mathcal{P}_n,$$

showing that the product of two elements of $\mathcal{P}_n$ is again in $\mathcal{P}_n$. That every element of $\mathcal{P}_n$ has an inverse follows from (F.131), which we show below. Assuming this, we have thus shown that $\mathcal{P}_n \subset \mathrm{L}(\mathbb{H}^{\otimes n})$ forms a group, and we proceed with showing the claimed properties.

(i) With $g = i^a \sigma_{\alpha_{n-1}} \otimes \cdots \otimes \sigma_{\alpha_0}$, we have

$$g^* = \overline{i^a}\sigma_{\alpha_{n-1}}^* \otimes \cdots \otimes \sigma_{\alpha_0}^*,$$

where we can use $\overline{i^a} = (-i)^a$ and $\sigma_{\alpha_j}^* = \sigma_{\alpha_j}$ for all $a, \alpha_j \in \{0,\ldots,3\}$ to obtain

$$g^* = (-i)^a \sigma_{\alpha_{n-1}} \otimes \cdots \otimes \sigma_{\alpha_0} = \begin{cases} g & \text{if } a \in \{0,2\} \\ -g & \text{if } a \in \{1,3\} \end{cases} = \pm g.$$

(ii) For any $\alpha_j \in \{0,\ldots,3\}$, we have $\sigma_{\alpha_j}^2 = \mathbf{1}$ and thus

$$g^2 = i^{2a}\sigma_{\alpha_{n-1}}^2 \otimes \cdots \otimes \sigma_{\alpha_0}^2 = (-1)^a\mathbf{1}^{\otimes n} = \begin{cases} \mathbf{1}^{\otimes n} & \text{if } a \in \{0,2\} \\ -\mathbf{1}^{\otimes n} & \text{if } a \in \{1,3\} \end{cases} = \pm\mathbf{1}^{\otimes n}.$$

(iii)

$$g^*g \underbrace{=}_{(\text{F.129})} \begin{cases} g^2 & \text{if } a \in \{0,2\} \\ -g^2 & \text{if } a \in \{1,3\} \end{cases} \underbrace{=}_{(\text{F.130})} \mathbf{1}^{\otimes n}.$$

(iv) Let $h = i^b \sigma_{\beta_{n-1}} \otimes \cdots \otimes \sigma_{\beta_0} \in \mathrm{Ctr}(\mathcal{P}_n)$. Per Definition F.17, this is equivalent to $hg = gh$ for all $g = i^a \sigma_{\alpha_{n-1}} \otimes \cdots \otimes \sigma_{\alpha_0} \in \mathcal{P}_n$. Consequently,

$$\sigma_{\beta_j} \sigma_{\alpha_j} = \sigma_{\alpha_j} \sigma_{\beta_j} \quad \forall \alpha_j \in \{0,\ldots,3\}, j \in \{0,\ldots,n-1\},$$

which means $\sigma_{\beta_j} \in \mathrm{Ctr}(\mathcal{P})$ for every j and thus

$$\mathrm{Ctr}(\mathcal{P}_n) = \big(\mathrm{Ctr}(\mathcal{P})\big)^{\otimes n} \underbrace{=}_{(F.128)} \langle i\mathbf{1}\rangle^{\otimes n} = \langle i\mathbf{1}^{\otimes n}\rangle.$$

(v) In an arbitrary $g = i^a \sigma_{\alpha_{n-1}} \otimes \cdots \otimes \sigma_{\alpha_0} \in \mathcal{P}_n$, there are 4 distinct elements σ_{α_j} with $\alpha_j \in \{0,\ldots,3\}$ possible in each tensor factor $j \in \{0,\ldots,n-1\}$ and 4 overall scalar pre-factors $i^a \in \{\pm 1, \pm i\}$. This gives rise to altogether $4^{n+1} = 2^{2n+2}$ possible distinct elements in $\mathcal{P}_n$.

$\square$

Definition F.63 The group

$$\mathcal{P}_n := \big\{ i^a \sigma_{\alpha_{n-1}} \otimes \cdots \otimes \sigma_{\alpha_0} \in L\big(\mathbb{H}^{\otimes n}\big) \,\big|\, a, \alpha_j \in \{0,\ldots,3\} \big\} < \mathcal{U}\big(\mathbb{H}^{\otimes n}\big)$$

is called the n-**fold PAULI group.**

The **weight** $w_\mathcal{P}$ of an element $i^a \sigma_{\alpha_{n-1}} \otimes \cdots \otimes \sigma_{\alpha_0} \in \mathcal{P}_n$ is defined as the number of tensor factors j for which $\sigma_{\alpha_j} \neq \mathbf{1}$, namely

$$w_\mathcal{P}(i^a \sigma_{\alpha_{n-1}} \otimes \cdots \otimes \sigma_{\alpha_0}) := \big| \{j \in \{0,\ldots,n-1\} \,|\, \alpha_j \neq 0\} \big| \in \mathbb{N}_0. \quad (F.133)$$

Example F.64 For

$$g_1 = \sigma_z \otimes \sigma_z \otimes \mathbf{1} \in \mathcal{P}_3 \quad \text{and} \quad g_2 = \sigma_z \otimes \mathbf{1} \otimes \sigma_z \in \mathcal{P}_3$$

we have

$$w_\mathcal{P}(g_1) = 2 = w_\mathcal{P}(g_2),$$

whereas for

$$h_1 = \sigma_x \otimes \sigma_z \otimes \sigma_z \otimes \sigma_x \in \mathcal{P}_4 \quad \text{and} \quad h_2 = \sigma_y \otimes \sigma_x \otimes \sigma_x \otimes \sigma_y \in \mathcal{P}_4$$

we find

$$w_\mathcal{P}(h_1) = 4 = w_\mathcal{P}(h_2).$$

Before we continue to explore the structure of the PAULI group, we collect a few properties of the weight function $w_{\mathcal{P}}$.

Lemma F.65 *The weight function* $w_{\mathcal{P}} : \mathcal{P}_n \to \mathbb{N}_0$ *satisfies*

$$w_{\mathcal{P}}(g) = 0 \Leftrightarrow g \in \mathrm{Ctr}(\mathcal{P}_n) \tag{F.134}$$

$$w_{\mathcal{P}}(g^{-1}) = w_{\mathcal{P}}(g^*) = w_{\mathcal{P}}(g) \tag{F.135}$$

$$w_{\mathcal{P}}(gh) \leq w_{\mathcal{P}}(g) + w_{\mathcal{P}}(h). \tag{F.136}$$

Proof Let $g = i^a \sigma_{\alpha_{n-1}} \otimes \cdots \otimes \sigma_{\alpha_0} \in \mathcal{P}_n$. Since $w_{\mathcal{P}}(g) = 0$ if and only if $\sigma_{\alpha_j} = \mathbf{1}$ for all $j \in \{0, \ldots, n-1\}$, the claim in (F.134) follows from (F.132).

From (F.131) we have $g^{-1} = g^*$ and thus $w_{\mathcal{P}}(g^{-1}) = w_{\mathcal{P}}(g^*)$. Moreover, since

$$g^* = \left(i^a \sigma_{\alpha_{n-1}} \otimes \cdots \otimes \sigma_{\alpha_0}\right)^* \underbrace{=}_{(2.32),(3.31)} \overline{i^a} \sigma_{\alpha_{n-1}}^* \otimes \cdots \otimes \sigma_{\alpha_0}^* \underbrace{=}_{(2.74)} \overline{i^a} \sigma_{\alpha_{n-1}} \otimes \cdots \otimes \sigma_{\alpha_0},$$

the number of j for which $\alpha_j \neq 0$ is the same for g^* and g, and it follows that $w_{\mathcal{P}}(g^*) = w_{\mathcal{P}}(g)$.

To prove (F.136), let $h = i^b \sigma_{\beta_{n-1}} \otimes \cdots \otimes \sigma_{\beta_0} \in \mathcal{P}_n$ such that

$$gh = i^{a+b} \sigma_{\alpha_{n-1}} \sigma_{\beta_{n-1}} \otimes \cdots \otimes \sigma_{\alpha_0} \sigma_{\beta_0} = g = i^c \sigma_{\gamma_{n-1}} \otimes \cdots \otimes \sigma_{\gamma_0}$$

and observe that

$$\begin{aligned}
&\left\{j \in \{0, \ldots, n-1\} \,\middle|\, \gamma_j \neq 0\right\} \\
&= \left\{j \in \{0, \ldots, n-1\} \,\middle|\, \alpha_j \neq \beta_j\right\} \\
&\subset \left\{j \in \{0, \ldots, n-1\} \,\middle|\, \alpha_j \neq 0\right\} \cup \left\{j \in \{0, \ldots, n-1\} \,\middle|\, \beta_j \neq 0\right\}
\end{aligned}$$

and therefore

$$\begin{aligned}
w_{\mathcal{P}}(gh) \underbrace{=}_{(F.133)} &\left|\left\{j \in \{0, \ldots, n-1\} \,\middle|\, \gamma_j \neq 0\right\}\right| \\
\leq &\left|\left\{j \in \{0, \ldots, n-1\} \,\middle|\, \alpha_j \neq 0\right\}\right| + \left|\left\{j \in \{0, \ldots, n-1\} \,\middle|\, \beta_j \neq 0\right\}\right| \\
\underbrace{=}_{(F.133)} &w_{\mathcal{P}}(g) + w_{\mathcal{P}}(h).
\end{aligned}$$

$\square$

Lemma F.66 *Let $n \in \mathbb{N}$. For $\alpha \in \{x,z\}$ define*

$$\Sigma_\alpha : \qquad \mathbb{F}_2^n \qquad \longrightarrow \mathcal{P}_n$$

$$\mathbf{a} = \begin{pmatrix} a_0 \\ \vdots \\ a_{n-1} \end{pmatrix} \longmapsto \Sigma_\alpha(\mathbf{a}) := \sigma_\alpha^{a_{n-1}} \otimes \cdots \otimes \sigma_\alpha^{a_0} \qquad \text{(F.137)}$$

and for $\mathbf{a}, \mathbf{b} \in \mathbb{F}_2^n$

$$\mathbf{a} \overset{2}{\odot} \mathbf{b} := \left(\sum_{j=0}^{n-1} a_j b_j \right) mod\, 2$$

so that $\overset{2}{\odot} : \mathbb{F}_2^n \times \mathbb{F}_2^n \to \mathbb{F}_2$. Then the following hold.

$$\Sigma_\alpha(\mathbf{a})^* = \Sigma_\alpha(\mathbf{a}) \qquad \text{(F.138)}$$

$$\Sigma_\alpha(\mathbf{a}) \Sigma_\alpha(\mathbf{b}) = \Sigma_\alpha(\mathbf{a} \overset{2}{\oplus} \mathbf{b}) \qquad \text{(F.139)}$$

$$\Sigma_x(\mathbf{a}) \Sigma_z(\mathbf{a}) = (-1)^{\mathbf{a} \overset{2}{\odot} \mathbf{b}} \Sigma_z(\mathbf{b}) \Sigma_x(\mathbf{a}). \qquad \text{(F.140)}$$

Proof To show (F.138), we use that $\sigma_\alpha^* = \sigma_\alpha$ for any $\alpha \in \{0, \ldots, 3\}$ to obtain

$$\Sigma_\alpha(\mathbf{a})^* \underset{\text{(F.137)}}{=} \left(\sigma_\alpha^{a_{n-1}} \otimes \cdots \otimes \sigma_\alpha^{a_0} \right)^* \underset{(3.31)}{=} \left(\sigma_\alpha^{a_{n-1}} \right)^* \otimes \cdots \otimes \left(\sigma_\alpha^{a_0} \right)^* = \sigma_\alpha^{a_{n-1}} \otimes \cdots \otimes \sigma_\alpha^{a_0}$$

$$\underset{\text{(F.137)}}{=} \Sigma_\alpha(\mathbf{a}).$$

To show (F.139), we use that $\sigma_\alpha^2 = \mathbf{1}$ for any $\alpha \in \{0, \ldots, 3\}$ such that $\sigma_\alpha^{a_j} \sigma_\alpha^{b_j} = \sigma_\alpha^{a_j \overset{2}{\oplus} b_j}$ and thus

$$\Sigma_\alpha(\mathbf{a}) \Sigma_\alpha(\mathbf{b}) \underset{\text{(F.137)}}{=} \left(\sigma_\alpha^{a_{n-1}} \otimes \cdots \otimes \sigma_\alpha^{a_0} \right) \left(\sigma_\alpha^{b_{n-1}} \otimes \cdots \otimes \sigma_\alpha^{b_0} \right)$$

$$= \sigma_\alpha^{a_{n-1}} \sigma_\alpha^{b_{n-1}} \otimes \cdots \otimes \sigma_\alpha^{a_0} \sigma_\alpha^{b_0} = \sigma_\alpha^{a_{n-1} \overset{2}{\oplus} b_{n-1}} \otimes \cdots \otimes \sigma_\alpha^{a_0 \overset{2}{\oplus} b_0}$$

$$\underset{\text{(F.137)}}{=} \Sigma_\alpha(\mathbf{a} \overset{2}{\oplus} \mathbf{b}).$$

Lastly, we have

$$\underbrace{\Sigma_x(\mathbf{a})\Sigma_z(\mathbf{b})}_{(\text{F.137})} = \left(\sigma_x^{a_{n-1}} \otimes \cdots \otimes \sigma_x^{a_0}\right)\left(\sigma_z^{b_{n-1}} \otimes \cdots \otimes \sigma_z^{b_0}\right)$$

$$= \sigma_x^{a_{n-1}}\sigma_z^{b_{n-1}} \otimes \cdots \otimes \sigma_x^{a_1}\sigma_z^{b_1} \otimes \sigma_x^{a_0}\sigma_z^{b_0}$$

$$= (-1)^{a_0 b_0}\sigma_x^{a_{n-1}}\sigma_z^{b_{n-1}} \otimes \cdots \otimes \sigma_x^{a_1}\sigma_z^{b_1} \otimes \sigma_z^{b_0}\sigma_x^{a_0}$$

$$\vdots$$

$$= (-1)^{\sum_{j=0}^{n-1} a_j b_j}\sigma_z^{b_{n-1}}\sigma_x^{a_{n-1}} \otimes \cdots \otimes \sigma_z^{b_0}\sigma_x^{a_0}$$

$$\underbrace{=}_{(\text{F.137})} (-1)^{\mathbf{a}\overset{2}{\odot}\mathbf{b}}\Sigma_z(\mathbf{b})\Sigma_x(\mathbf{a}),$$

proving (F.140).   □

Example F.67 For $n = 3$ we have, for example,

$$\Sigma_x\begin{pmatrix}1\\0\\1\end{pmatrix} = \sigma_x \otimes \mathbf{1} \otimes \sigma_x \quad \text{and} \quad \Sigma_z\begin{pmatrix}1\\1\\1\end{pmatrix} = \sigma_z \otimes \sigma_z \otimes \sigma_z$$

such that

$$\Sigma_x\begin{pmatrix}1\\0\\1\end{pmatrix}\Sigma_z\begin{pmatrix}1\\1\\1\end{pmatrix} = \sigma_x\sigma_z \otimes \sigma_z \otimes \sigma_x\sigma_z = (-\mathrm{i}\sigma_y) \otimes \sigma_z \otimes (-\mathrm{i}\sigma_y) = -\sigma_y \otimes \sigma_z \otimes \sigma_z.$$

Lemma F.68 *Let $n \in \mathbb{N}$ and Σ_x and Σ_z be as defined in Lemma F.66. Then any $g \in \mathcal{P}_n$ can be written in the form*

$$g = \mathrm{i}^{c(g)}\Sigma_x\big(\mathbf{x}(g)\big)\Sigma_z\big(\mathbf{z}(g)\big), \tag{F.141}$$

where $c(g) \in \{0,\dots,3\}$ and $\mathbf{x}(g),\mathbf{z}(g) \in \mathbb{F}_2^n$ are uniquely determined by g.

Proof From Definition F.63 of $\mathcal{P}_n$ we know that any $g \in \mathcal{P}_n$ is of the form

$$g = \mathrm{i}^a \sigma_{\alpha_{n-1}} \otimes \cdots \otimes \sigma_{\alpha_0}$$

with $a, \alpha_j \in \{0,\dots,3\}$. On the other hand,

$$\mathrm{i}^c\Sigma_x(\mathbf{x})\Sigma_z(\mathbf{z}) = \mathrm{i}^c\left(\sigma_x^{x_{n-1}} \otimes \cdots \otimes \sigma_x^{x_0}\right)\left(\sigma_z^{z_{n-1}} \otimes \cdots \otimes \sigma_z^{z_0}\right)$$

$$= \mathrm{i}^c \sigma_x^{x_{n-1}}\sigma_z^{z_{n-1}} \otimes \cdots \otimes \sigma_x^{x_0}\sigma_z^{z_0},$$

where for any $j \in \{0,\dots,n-1\}$

$$\sigma_x^{x_j}\sigma_z^{z_j} = \begin{cases} \sigma_0 & \text{if } x_j = 0 = z_j \\ \sigma_x & \text{if } x_j = 1 \text{ and } z_j = 0 \\ \sigma_z & \text{if } x_j = 0 \text{ and } z_j = 1 \\ -i\sigma_y & \text{if } x_j = 1 = z_j. \end{cases}$$

This shows that with a suitable choice of $c = c(g), \mathbf{x} = \mathbf{x}(g), \mathbf{z} = \mathbf{z}(g)$ we obtain

$$g = i^c \Sigma_x(\mathbf{x}) \Sigma_z(\mathbf{z})$$

and have shown (F.141).

Suppose now there are $\tilde{c}, \tilde{\mathbf{x}}, \tilde{\mathbf{z}}$ such that

$$g = i^{\tilde{c}} \Sigma_x(\tilde{\mathbf{x}}) \Sigma_z(\tilde{\mathbf{z}})$$

as well. Then we have

$$g^* = (-i)^{\tilde{c}}\left(\Sigma_x(\tilde{\mathbf{x}})\Sigma_z(\tilde{\mathbf{z}})\right)^* \underbrace{=}_{(2.47)} (-i)^{\tilde{c}}\left(\Sigma_z(\tilde{\mathbf{z}})\right)^*\left(\Sigma_x(\tilde{\mathbf{x}})\right)^*$$

$$\underbrace{=}_{(F.138)} (-i)^{\tilde{c}}\Sigma_z(\tilde{\mathbf{z}})\Sigma_x(\tilde{\mathbf{x}})$$

such that

$$1^{\otimes n} \underbrace{=}_{(F.131)} g^*g = i^c(-i)^{\tilde{c}}\Sigma_z(\tilde{\mathbf{z}})\Sigma_x(\tilde{\mathbf{x}})\Sigma_x(\mathbf{x})\Sigma_z(\mathbf{z})$$

$$\underbrace{=}_{(F.139)} i^{c-\tilde{c}}\Sigma_z(\tilde{\mathbf{z}})\Sigma_x(\mathbf{x}\overset{2}{\oplus}\tilde{\mathbf{x}})\Sigma_z(\mathbf{z}) \underbrace{=}_{(F.140)} i^{c-\tilde{c}}(-1)^{(\mathbf{x}\overset{2}{\oplus}\tilde{\mathbf{x}})\overset{2}{\odot}\tilde{\mathbf{z}}}\Sigma_x(\mathbf{x}\overset{2}{\oplus}\tilde{\mathbf{x}})\Sigma_z(\tilde{\mathbf{z}})\Sigma_z(\mathbf{z})$$

$$\underbrace{=}_{(F.139)} i^{c-\tilde{c}}(-1)^{(\mathbf{x}\overset{2}{\oplus}\tilde{\mathbf{x}})\overset{2}{\odot}\tilde{\mathbf{z}}}\Sigma_x(\mathbf{x}\overset{2}{\oplus}\tilde{\mathbf{x}})\Sigma_z(\tilde{\mathbf{z}}\overset{2}{\oplus}\mathbf{z}).$$

This requires

$$\Sigma_x(\tilde{\mathbf{x}}\overset{2}{\oplus}\mathbf{x}) = 1^{\otimes n} = \Sigma_z(\tilde{\mathbf{z}}\overset{2}{\oplus}\mathbf{z})$$

and thus

$$\tilde{\mathbf{x}}\overset{2}{\oplus}\mathbf{x} = 0 = \tilde{\mathbf{z}}\overset{2}{\oplus}\mathbf{z}.$$

Since $\mathbf{x}, \tilde{\mathbf{x}}, \mathbf{z}, \tilde{\mathbf{z}} \in \mathbb{F}_2^n$, it follows that $\tilde{\mathbf{x}} = \mathbf{x}$ and $\tilde{\mathbf{z}} = \mathbf{z}$ as well as $(-1)^{(\mathbf{x}\overset{2}{\oplus}\tilde{\mathbf{x}})\overset{2}{\odot}\tilde{\mathbf{z}}} = 1$. Thus, we are left with

$$1^{\otimes n} = g^*g = i^{c-\tilde{c}}1^{\otimes n},$$

and since $c, \tilde{c} \in \{0,\dots,3\}$, we must have $c = \tilde{c}$, which completes the proof of the uniqueness of $c = c(g), \mathbf{x} = \mathbf{x}(g)$ and $\mathbf{z} = \mathbf{z}(g)$. $\qquad\square$

Definition F.69 Let $n \in \mathbb{N}$. The mappings

$$(c(\cdot), \mathbf{x}(\cdot), \mathbf{z}(\cdot)) : \mathcal{P}_n \longrightarrow \mathbb{F}_2 \times \mathbb{F}_2^n \times \mathbb{F}_2^n$$
$$g \longmapsto (c(g), \mathbf{x}(g), \mathbf{z}(g))$$

are defined by means of the unique representation of g as

$$g = \mathrm{i}^{c(g)} \Sigma_x (\mathbf{x}(g)) \Sigma_z (\mathbf{z}(g)) . \qquad (\text{F.142})$$

Exercise F.140 Let $c(\cdot), \mathbf{x}(\cdot)$ and $\mathbf{z}(\cdot)$ be as defined in Definition F.69. Show that then for any $g, h \in \mathcal{P}_n$

$$c(gh) = \left(c(g) + c(h) + 2\mathbf{z}(g) \overset{2}{\odot} \mathbf{x}(h) \right) \bmod 4 \qquad (\text{F.143})$$

$$\mathbf{x}(gh) = \mathbf{x}(g) \overset{2}{\oplus} \mathbf{x}(h) \qquad (\text{F.144})$$

$$\mathbf{z}(gh) = \mathbf{z}(g) \overset{2}{\oplus} \mathbf{z}(h) . \qquad (\text{F.145})$$

For a solution see Solution F.140

Proposition F.70 *For any $g, h \in \mathcal{P}_n$ we have*

$$gh = (-1)^{\left(\mathbf{x}(g) \overset{2}{\odot} \mathbf{z}(h) \right) \overset{2}{\oplus} \left(\mathbf{z}(g) \overset{2}{\odot} \mathbf{x}(h) \right)} hg .$$

Proof

$$gh \underset{(\text{F.141})}{=} \mathrm{i}^{c(g)+c(h)} \Sigma_x (\mathbf{x}(g)) \Sigma_z (\mathbf{z}(g)) \Sigma_x (\mathbf{x}(h)) \Sigma_z (\mathbf{z}(h))$$

$$\underset{(\text{F.140})}{=} \mathrm{i}^{c(g)+c(h)} (-1)^{\mathbf{z}(g) \overset{2}{\odot} \mathbf{x}(h)} \Sigma_x (\mathbf{x}(g)) \Sigma_x (\mathbf{x}(h)) \Sigma_z (\mathbf{z}(g)) \Sigma_z (\mathbf{z}(h))$$

$$= (-1)^{\mathbf{z}(g) \overset{2}{\odot} \mathbf{x}(h)} \mathrm{i}^{c(g)+c(h)} \Sigma_x (\mathbf{x}(h)) \Sigma_x (\mathbf{x}(g)) \Sigma_z (\mathbf{z}(h)) \Sigma_z (\mathbf{z}(g))$$

$$\underset{(\text{F.140})}{=} (-1)^{\overset{2}{\oplus} \left(\mathbf{x}(g) \overset{2}{\odot} \mathbf{z}(h) \right) \overset{2}{\oplus} \left(\mathbf{z}(g) \overset{2}{\odot} \mathbf{x}(h) \right)} \mathrm{i}^{c(g)+c(h)} \Sigma_x (\mathbf{x}(h)) \Sigma_z (\mathbf{z}(h)) \Sigma_x (\mathbf{x}(g)) \Sigma_z (\mathbf{z}(g))$$

$$\underset{(\text{F.141})}{=} (-1)^{\overset{2}{\oplus} \left(\mathbf{x}(g) \overset{2}{\odot} \mathbf{z}(h) \right) \overset{2}{\oplus} \left(\mathbf{z}(g) \overset{2}{\odot} \mathbf{x}(h) \right)} hg .$$

$\square$

Lemma F.71 *Let $m, n \in \mathbb{N}$ with $m < n$ and $\{g_1, \ldots, g_m\} \in \mathcal{P}_n$ such that $-\mathbf{1}^{\otimes n} \notin \langle g_1, \ldots, g_m \rangle$. Then we have*

$$\{g_1, \ldots, g_m\} \text{ are independent}$$
$$\Rightarrow \left\{ \begin{pmatrix} \mathbf{x}(g_l) \\ \mathbf{z}(g_l) \end{pmatrix} \,\middle|\, l \in \{1, \ldots, m\} \right\} \subset \mathbb{F}_2^{2n} \text{ are linearly independent.}$$

Proof Let $\{g_1, \ldots, g_m\} \in \mathcal{P}_n$ be such that $-\mathbf{1}^{\otimes n} \notin \langle g_1, \ldots, g_m \rangle$. We show the claim by contraposition. Let the $\left\{ \begin{pmatrix} \mathbf{x}(g_l) \\ \mathbf{z}(g_l) \end{pmatrix} \,\middle|\, l \in \{1, \ldots, m\} \right\} \subset \mathbb{F}_2^{2n}$ be linearly dependent such that there exist $k_1, \ldots, k_m \in \mathbb{F}_2$ with some $k_j = 1$ and

$$\sum_{l=1}^{m} k_l \begin{pmatrix} \mathbf{x}(g_l) \\ \mathbf{z}(g_l) \end{pmatrix} \bmod 2 = \begin{pmatrix} \mathbf{0} \\ \mathbf{0} \end{pmatrix} \in \mathbb{F}_2^{2n}. \tag{F.146}$$

This implies that

$$\mathbf{x}(g_1^{k_1} \cdots g_m^{k_m}) \underset{(\mathrm{F.144})}{=} k_1 \mathbf{x}(g_1) \overset{2}{\oplus} \cdots \overset{2}{\oplus} k_m \mathbf{x}(g_m) = \sum_{l=1}^{m} k_l \mathbf{x}(g_l) \bmod 2 \underset{(\mathrm{F.146})}{=} \mathbf{0}$$

$$\mathbf{z}(g_1^{k_1} \cdots g_m^{k_m}) \underset{(\mathrm{F.145})}{=} k_1 \mathbf{z}(g_1) \overset{2}{\oplus} \cdots \overset{2}{\oplus} k_m \mathbf{z}(g_m) = \sum_{l=1}^{m} k_l \mathbf{z}(g_l) \bmod 2 \underset{(\mathrm{F.146})}{=} \mathbf{0}, \tag{F.147}$$

and thus,

$$g_1^{k_1} \cdots g_m^{k_m} \underset{(\mathrm{F.142})}{=} \mathrm{i}^{c(g_1^{k_1} \cdots g_m^{k_m})} \Sigma_x\big(\mathbf{x}(g_1^{k_1} \cdots g_m^{k_m})\big) \Sigma_z\big(\mathbf{z}(g_1^{k_1} \cdots g_m^{k_m})\big)$$
$$\underset{(\mathrm{F.147})}{=} \mathrm{i}^{c(g_1^{k_1} \cdots g_m^{k_m})} \Sigma_x(\mathbf{0}) \Sigma_z(\mathbf{0}) = \mathrm{i}^{c(g_1^{k_1} \cdots g_m^{k_m})} \mathbf{1}^{\otimes n}. \tag{F.148}$$

We know from (F.130) that any $g \in \mathcal{P}_n$ satisfies $g^2 = \pm \mathbf{1}^{\otimes n}$. Since $-\mathbf{1}^{\otimes n} \notin \langle g_1, \ldots, g_m \rangle$, it follows that

$$g^2 = \mathbf{1}^{\otimes n} \quad \forall g \in \langle g_1, \ldots, g_m \rangle. \tag{F.149}$$

Hence, for any $k_1, \ldots, k_m \in \mathbb{F}_2$ we must have $\left(g_1^{k_1} \cdots g_m^{k_m}\right)^2 = \mathbf{1}^{\otimes n}$, and (F.148) then implies that $c(g_1^{k_1} \cdots g_m^{k_m}) = 0$. By our initial assumption, we have $k_j = 1$ for some j and thus

$$g_1^{k_1} \cdots g_m^{k_m} = \mathbf{1}^{\otimes n}$$
$$\underset{\text{(F.149)}}{\Rightarrow} \quad g_1^{k_2} \cdots g_m^{k_{m-1}} = g_1^{k_1} g_m^{k_m}$$

$$\vdots \qquad\qquad \vdots$$

$$\underset{\text{(F.149)}}{\Rightarrow} \qquad g_j = g_{j-1}^{k_{j-1}} \cdots g_1^{k_1} g_m^{k_m} \cdots g_{j+1}^{k_{j+1}} .$$

Hence, the $\{g_1 \ldots, g_m\}$ are not independent. We have thus shown that

$$\left\{ \begin{pmatrix} \mathbf{x}(g_l) \\ \mathbf{z}(g_l) \end{pmatrix} \;\middle|\; l \in \{1, \ldots, m\} \right\} \subset \mathbb{F}_2^{2n} \text{ are linearly dependent}$$
$$\Rightarrow \quad \{g_1, \ldots, g_m\} \text{ are not independent.}$$

The claim of the lemma then follows by contraposition. $\square$

With the previous lemma we can derive the following proposition, which is useful in the context of the stabilizer formalism in quantum error correction in Sect. 7.3.3.

Proposition F.72 *Let $m, n \in \mathbb{N}$ with $m < n$ and let $\{g_1, \ldots, g_m\} \in \mathcal{P}_n$ be independent and satisfy $-\mathbf{1}^{\otimes n} \notin \langle g_1, \ldots, g_m \rangle$. Then for any $j \in \{1, \ldots, m\}$ there exists an $h \in \mathcal{P}_n$ such that*

$$g_l h = (-1)^{\delta_{lj}} h g_l \quad \forall l \in \{1, \ldots, m\}. \tag{F.150}$$

Proof From Proposition F.70 we know that for any $g, h \in \mathcal{P}_n$

$$gh = (-1)^{\left(\mathbf{x}(g) \overset{2}{\odot} \mathbf{z}(h) \right) \overset{2}{\oplus} \left(\mathbf{z}(g) \overset{2}{\odot} \mathbf{x}(h) \right)} hg .$$

For h to satisfy (F.150), it is thus necessary and sufficient that for all $j, l \in \{1, \ldots, m\}$

$$\left(\mathbf{x}(g_l) \overset{2}{\odot} \mathbf{z}(h) \right) \overset{2}{\oplus} \left(\mathbf{z}(g_l) \overset{2}{\odot} \mathbf{x}(h) \right) = \delta_{lj},$$

which is equivalent to the matrix equation

$$\underbrace{\begin{pmatrix} \mathbf{x}(g_1)^T & \mathbf{z}(g_1)^T \\ \vdots & \vdots \\ \mathbf{x}(g_m)^T & \mathbf{z}(g_m)^T \end{pmatrix}}_{=M \in \mathrm{Mat}(m \times 2n, \mathbb{F}_2)} \begin{pmatrix} \mathbf{z}(h) \\ \mathbf{x}(h) \end{pmatrix} = \begin{pmatrix} \delta_{1j} \\ \vdots \\ \delta_{mj} \end{pmatrix} . \tag{F.151}$$

Since the $\{g_1,\ldots,g_m\}$ are independent, we know from Lemma F.71 that then the $\left\{ \begin{pmatrix} \mathbf{x}(g_l) \\ \mathbf{z}(g_l) \end{pmatrix} \ \middle|\ l \in \{1,\ldots,m\} \right\}$ are linearly independent. Consequently, the matrix M in (F.151) is of maximal rank m and we can always find a solution for $\mathbf{z}(h)$ and $\mathbf{x}(h)$ in (F.151). With this solution

$$h = \Sigma_x\big(\mathbf{x}(h)\big)\,\Sigma_z\big(\mathbf{z}(h)\big) \in \mathcal{P}_n$$

has the desired property (F.150).   □

It turns out that for a subgroup of the PAULI group that does not contain $-\mathbf{1}^{\otimes n}$ its normalizer and centralizer coincide. This is to be shown in Exercise F.141.

Exercise F.141 Let $\mathcal{S}$ be a subgroup of $\mathcal{P}_n$ that satisfies $-\mathbf{1}^{\otimes n} \notin \mathcal{S}$. Show that then

$$\mathrm{Nor}_{\mathcal{P}_n}(\mathcal{S}) = \mathrm{Clz}_{\mathcal{P}_n}(\mathcal{S})\,.$$

For a solution see Solution F.141.

Appendix G
Proof of a Quantum Adiabatic Theorem

G.1 Resolvents and Projections

A first ingredient required for our proof of the Quantum Adiabatic Theorem is the resolvent of an operator.

> **Definition G.1** The **resolvent of an operator** A on a HILBERT space $\mathbb{H}$ is defined as
> $$R_{(\cdot)}(A) : \mathbb{C} \smallsetminus \sigma(A) \longrightarrow L(\mathbb{H})$$
> $$z \longmapsto R_z(A) := (A - z\mathbf{1})^{-1} . \qquad (G.1)$$

Recall from Definition 2.10 that the operator $R_z(A)$ does indeed exist for $z \in \mathbb{C} \smallsetminus \sigma(A)$. Moreover, it is obvious from (G.1) that, by definition,

$$R_z(A)^{-1} = A - z\mathbf{1} .$$

Exercise G.142 Let $A \in B_{sa}(\mathbb{H})$ with purely discrete (and possibly degenerate) spectrum $\sigma(A) = \{\lambda_j \mid j \in I \subset \mathbb{N}_0\}$, and let P_j denote the projections onto the corresponding eigenspaces. Show that $R_z(A)$ can be written as

$$R_z(A) = \sum_{j \in I} \frac{P_j}{\lambda_j - z} . \qquad (G.2)$$

For a solution see Solution G.142.

© Springer Nature Switzerland AG 2019
W. Scherer, *Mathematics of Quantum Computing*,
https://doi.org/10.1007/978-3-030-12358-1

The resolvent is an operator valued analytical function of $z \in \mathbb{C} \setminus \sigma(A)$ in the sense of complex analysis. We will not prove this fact here, but will make use of it when writing the projection onto eigenspaces with the help of the resolvent.

Lemma G.2 *Let $A \in B_{sa}(\mathbb{H})$ with purely discrete (and possibly degenerate) spectrum $\sigma(A) = \{\lambda_l \mid l \in I \subset \mathbb{N}_0\}$ and with resolvent $R_{(\cdot)}(A)$. For any $j \in I$ let γ_j be any closed counter-clockwise curve in $\mathbb{C} \setminus \sigma(A)$ that encloses solely the eigenvalue λ_j from $\sigma(A)$. Then the projection P_j onto the eigenspace of λ_j satisfies*

$$P_j = \frac{-1}{2\pi i} \oint_{\gamma_j} R_z(A) dz. \tag{G.3}$$

Proof Using (G.2) in the right side of (G.3), gives for any $j \in I$

$$\frac{-1}{2\pi i} \oint_{\gamma_j} R_z(A) dz = \sum_{k \in I} \frac{1}{2\pi i} \oint_{\gamma_j} \frac{P_k}{z - \lambda_k} dz. \tag{G.4}$$

Here we can use the operator version of the CAUCHY Theorem from complex analysis, which tells us that for a function $f : \mathbb{C} \to \mathbb{C}$ that is analytic on and inside a curve γ_0 which encloses $z_0 \in \mathbb{C}$ counter-clockwise one has for any $n \in \mathbb{N}_0$

$$\frac{n!}{2\pi i} \oint_{\gamma_0} \frac{f(z)}{(z - z_0)^{n+1}} = \begin{cases} \frac{d^n}{dz^n} f(z_0) & \text{if } \gamma_0 \text{ encloses } z_0 \\ 0 & \text{else.} \end{cases} \tag{G.5}$$

Applying this to the right side of (G.4) gives

$$\frac{1}{2\pi i} \oint_{\gamma_j} \frac{P_k}{z - \lambda_k} dz = \begin{cases} P_k & \text{if } \gamma_j \text{ encloses } \lambda_k \\ 0 & \text{else} \end{cases}$$
$$= \delta_{jk} P_k$$

since, by assumption, γ_j only encloses the eigenvalue λ_j. Inserting this into (G.4) yields the claim (G.3). □

The following transformation, which is defined using the resolvent of operator-valued functions $s \to H_T(s)$ satisfying the Adiabatic Assumption (AA), will be very useful for obtaining bounds on various operators in the context of the proof of the adiabatic theorem.

Definition G.3 Let $\mathbb{H}$ be a HILBERT space and let H_T be an operator-valued function

$$H_T : [0,1] \longrightarrow B_{sa}(\mathbb{H})$$
$$s \longmapsto H_T(s)$$

satisfying the Adiabatic Assumption (AA). Moreover, let $\gamma_j(s)$ be any closed counter-clockwise curve in $\mathbb{C} \smallsetminus \sigma(H_T(s))$ that encloses solely the eigenvalue $E_j(s)$ from $\sigma(H_T(s))$. For any

$$A : [0,1] \longrightarrow B(\mathbb{H})$$
$$s \longmapsto A(s)$$

and $j \in I$ we define

$$X_j[A]: [0,1] \longrightarrow L(\mathbb{H})$$
$$s \longmapsto X_j[A](s)$$

by

$$X_j[A](s) := \frac{1}{2\pi i} \oint_{\gamma_j(s)} R_z(H_T(s))A(s)R_z(H_T(s))dz. \tag{G.6}$$

We shall see later that the $X_j[A](s)$ do not depend on the exact form of the curves $\gamma_j(s)$ as long as these are of the form described in Definition G.3. We will also show later that $X_j[A](s) \in B(\mathbb{H})$.

Exercise G.143 Show that

$$\frac{d}{ds}R_z(H_T(s)) = -R_z(H_T(s))\dot{H}_T(s)R_z(H_T(s)). \tag{G.7}$$

For a solution see Solution G.143.

Definition G.4 Let $H_T : [0,1] \to B_{sa}(\mathbb{H})$ be an operator-valued function satisfying the Adiabatic Assumption (AA). The **gap (or the gap function) of the eigenvalue $E_j(s)$** of $H_T(s)$ is defined as

$$g_j : [0,1] \longrightarrow \mathbb{R}$$
$$s \longmapsto g_j(s) := \min\left\{ |E_k(s) - E_j(s)| \ \middle| \ k \in I \smallsetminus \{j\}\right\}. \tag{G.8}$$

Lemma G.5 *Let* $H_T : [0,1] \to B_{sa}(\mathbb{H})$ *be such that it satisfies the Adiabatic Assumption (AA) and let* $\{P_j(s) \mid j \in I\}$ *denote the projections onto the eigenspaces of* $H_T(s)$. *Then we have for all* $j \in I$ *and* $s \in [0,1]$

$$\dot{P}_j(s) = X_j\left[\dot{H}_T\right](s) \tag{G.9}$$

and

$$\left[H_T(s), X_j\left[\dot{P}_j\right](s)\right] = -\left[\dot{P}_j(s), P_j(s)\right]. \tag{G.10}$$

Proof From (G.3) it follows that

$$P_j(s) = \frac{-1}{2\pi i} \oint_{\gamma_j(s)} R_z(H_T(s)) dz,$$

where we choose the curves $\gamma_j(s)$ to be circles with radius $r = \frac{g_j(s)}{2}$ centered at $E_j(s)$. We then have

$$
\begin{aligned}
\dot{P}_j(s) &= \frac{-1}{2\pi i} \frac{d}{ds} \oint_{\gamma_j(s)} R_z(H_T(s)) dz \\
&= \frac{-1}{2\pi i} \lim_{ds \to 0} \frac{1}{ds} \left(\oint_{\gamma_j(s+ds)} R_z(H_T(s+ds)) dz - \oint_{\gamma_j(s)} R_z(H_T(s)) dz \right) \\
&= \frac{-1}{2\pi i} \lim_{ds \to 0} \frac{1}{ds} \left(\oint_{\gamma_j(s+ds)} \left[R_z(H_T(s)) + \frac{d}{ds} R_z(H_T(s)) ds + o(ds^2) \right] dz \right. \\
&\qquad\qquad\qquad \left. - \oint_{\gamma_j(s)} R_z(H_T(s)) dz \right) \\
&= \frac{-1}{2\pi i} \left[\lim_{ds \to 0} \frac{1}{ds} \left(\oint_{\gamma_j(s+ds)} R_z(H_T(s)) dz - \oint_{\gamma_j(s)} R_z(H_T(s)) dz \right) \right. \\
&\qquad\qquad \left. + \oint_{\gamma_j(s)} \frac{d}{ds} R_z(H_T(s)) dz \right]. \tag{G.11}
\end{aligned}
$$

In (G.11) the limit term comprised of the difference of the integrals along the curves $\gamma_j(s+ds)$ and $\gamma_j(s)$ vanishes. This is because item (iii) in the Adiabatic Assumption (AA) implies that each eigenvalue $E_j(s)$ of $H_T(s)$ is separated by a finite non-zero distance $g_j(s) = 2r$ from all other eigenvalues. Hence, for small enough ds, the only singularity of the function $z \mapsto R_z(H_T(s))$ inside the circle $\gamma_j(s+ds)$ will be at $z_0 = E_j(s)$ making the integrals of $R_z(H_T(s))$ along the two curves $\gamma_j(s+ds)$ and $\gamma_j(s)$ equal. Consequently, (G.11) becomes

$$\dot{P}_j(s) = \frac{-1}{2\pi i} \oint_{\gamma_j(s)} \frac{d}{ds} R_z(H_T(s)) dz$$

$$\underbrace{=}_{(G.7)} \frac{1}{2\pi i} \oint_{\gamma_j(s)} R_z(H_T(s)) \dot{H}_T(s) R_z(H_T(s)) dz$$

$$\underbrace{=}_{(G.6)} X_j\left[\dot{H}_T(s)\right],$$

which proves (G.9).

To show (G.10), note that the definition of $X_j[\cdot]$ in (G.6) implies

$$\left[H_T(s), X_j\left[\dot{P}_j\right](s)\right] = \left[H_T(s), \frac{1}{2\pi i} \oint_{\gamma_j(s)} R_z(H_T(s)) \dot{P}_j(s) R_z(H_T(s)) dz\right]$$

$$= \frac{1}{2\pi i} \oint_{\gamma_j(s)} \left[H_T(s), R_z(H_T(s)) \dot{P}_j(s) R_z(H_T(s)) dz\right]$$

$$= \frac{1}{2\pi i} \oint_{\gamma_j(s)} \Big(H_T(s) R_z(H_T(s)) \dot{P}_j(s) R_z(H_T(s)) \qquad (G.12)$$

$$- R_z(H_T(s)) \dot{P}_j(s) R_z(H_T(s)) H_T(s)\Big) dz,$$

and that by Definition G.142 $R_z(H_T(s))$ satisfies

$$\left(H_T(s) - z\mathbf{1}\right) R_z(H_T(s)) = \mathbf{1} = R_z(H_T(s))\left(H_T(s) - z\mathbf{1}\right)$$

such that

$$H_T(s) R_z(H_T(s)) = \mathbf{1} + z R_z(H_T(s)) = R_z(H_T(s)) H_T(s).$$

Using this in (G.12), we obtain

$$\left[H_T(s), X_j\left[\dot{P}_j\right](s)\right] = \frac{1}{2\pi i} \oint_{\gamma_j(s)} \Big(\left(\mathbf{1} + z R_z(H_T(s))\right) \dot{P}_j(s) R_z(H_T(s))$$

$$- R_z(H_T(s)) \dot{P}_j(s)\left(\mathbf{1} + z R_z(H_T(s))\right)\Big) dz$$

$$= \frac{1}{2\pi i} \oint_{\gamma_j(s)} \left(\dot{P}_j(s) R_z(H_T(s)) - R_z(H_T(s)) \dot{P}_j(s)\right) dz$$

$$= \left[\dot{P}_j(s), \frac{1}{2\pi i} \oint_{\gamma_j(s)} R_z(H_T(s)) dz\right]$$

$$\underbrace{=}_{(G.3)} -\left[\dot{P}_j(s), P_j(s)\right],$$

which completes the proof of (G.10). $\qquad \square$

Let us collect a few more properties of the projections $\{P_j(s) \mid j \in I\}$, which we will need in various proofs later on.

Lemma G.6 *Let $H_T : [0,1] \to B_{sa}(\mathbb{H})$ be such that it satisfies the Adiabatic Assumption (AA) and let $\{E_j(s) \mid j \in I\}$ denote the eigenvalues of $H_T(s)$ each of which may be d_j-fold degenerate. Moreover, let $\{|\Phi_{j,\alpha}(s)\rangle \mid j \in I, \alpha \in \{1,\ldots,d_j\}\}$ denote the ONB of eigenvectors of $H_T(s)$ and let $\{P_j(s) \mid j \in I\}$ denote the projections onto its eigenspaces.*

Then these projections satisfy for all $j, k \in I$ and $s \in [0,1]$

$$\sum_{\alpha=1}^{d_j} |\Phi_{j,\alpha}(s)\rangle\langle\Phi_{j,\alpha}(s)| = P_j(s) \tag{G.13}$$

$$\dot{P}_j(s)P_j(s) + P_j(s)\dot{P}_j(s) = \dot{P}_j(s) \tag{G.14}$$

$$P_k(s)\dot{P}_j(s)P_k(s) = 0 \tag{G.15}$$

$$\sum_{j\in I} \dot{P}_j(s) = 0. \tag{G.16}$$

Proof The claim (G.13) is a direct consequence of (2.41) and the fact that $\{|\Phi_{j,\alpha}(s)\rangle \mid j \in I, \alpha \in \{1,\ldots,d_j\}\}$ is by assumption an ONB of eigenvectors of $H_T(s)$ for each $s \in [0,1]$.

The $P_j(s)$ are orthogonal projections for every $j \in I$ and $s \in [0,1]$. Hence, (2.44) implies $\left(P_j(s)\right)^2 = P_j(s)$. Taking the derivative with respect to s on both sides of this equation yields (G.14).

Multiplying both sides of (G.14) from left and right with $P_k(s)$, we obtain

$$P_k(s)\dot{P}_j(s)P_k(s) \underbrace{=}_{(G.14)} P_k(s)\left(\dot{P}_j(s)P_j(s) + P_j(s)\dot{P}_j(s)\right)P_k(s)$$

$$= P_k(s)\dot{P}_j(s)P_j(s)P_k(s) + P_k(s)P_j(s)\dot{P}_j(s)P_k(s)$$

$$\underbrace{=}_{(2.44)} 2\delta_{jk}P_k(s)\dot{P}_j(s)P_k(s),$$

which implies (G.15).

Finally, (2.43) implies that for all $s \in [0,1]$ we have $\sum_{j\in I} P_j(s) = 1$. Once again, taking the derivative with respect to s on both sides of this equation yields the claim (G.16) $\square$

G.2 Adiabatic Generator and Intertwiner

Definition G.7 Let H_T be as defined in Definition 8.1 and such that it satisfies the Adiabatic Assumption (AA), and let $\{P_j(s) \mid j \in I\}$ be the projections onto the eigenspaces of $H_T(s)$. For each $j \in I$ the **adiabatic generator** $H_{A,j}$ is defined as

$$H_{A,j} : [0,1] \longrightarrow B_{sa}(\mathbb{H})$$
$$s \longmapsto H_{A,j}(s) := T H_T(s) + i[\dot{P}_j(s), P_j(s)]. \tag{G.17}$$

Lemma G.8 *Let $H_{A,j}$ be as defined in Definition G.7, and let $\{P_j(s) \mid j \in I\}$ be the projections onto the eigenspaces of $H_T(s)$. Then we have for all $j \in I$ and $s \in [0,1]$*

$$H_{A,j}(s)^* = H_{A,j}(s) \tag{G.18}$$

and

$$i\dot{P}_j(s) = \left[H_{A,j}(s), P_j(s) \right]. \tag{G.19}$$

Proof From its its definition (8.4) we know already that $H_T(s)$ is self-adjoint and thus the term $T H_T(s)$ in (G.17) is as well. Hence,

$$
\begin{aligned}
H_{A,j}(s)^* \quad &\underset{(G.17)}{=} \quad T H_T(s)^* + \left(i\left[\dot{P}_j(s), P_j(s)\right] \right)^* \\[6pt]
&\underset{(2.32)}{=} \quad T H_T(s) - i\left(\left[\dot{P}_j(s), P_j(s)\right] \right)^* \\[6pt]
&\underset{(2.46)}{=} \quad T H_T(s) - i\left(\dot{P}_j(s)P_j(s) \right)^* + i\left(P_j(s)\dot{P}_j(s) \right)^* \\[6pt]
&\underset{(2.47)}{=} \quad T H_T(s) - iP_j(s)^*\dot{P}_j(s)^* + i\dot{P}_j(s)^*P_j(s)^* \\[6pt]
&\underset{\text{Def. 2.11, Exerc. 8.94}}{=} \quad T H_T(s) - iP_j(s)\dot{P}_j(s) + i\dot{P}_j(s)P_j(s) \\[6pt]
&= \quad T H_T(s) + i\left[\dot{P}_j(s), P_j(s)\right] \\[4pt]
&\underset{(G.17)}{=} \quad H_{A,j}(s),
\end{aligned}
$$

which proves (G.18).

In order to prove (G.19), we observe that, by definition (G.17) of the adiabatic generator, we have

$$\left[\mathsf{H}_{A,j}(s), P_j(s)\right] = T\left[\mathsf{H}_T(s), P_j(s)\right] + \mathrm{i}\left[\left[\dot{P}_j(s), P_j(s)\right], P_j(s)\right]. \qquad (G.20)$$

From (2.42) it follows that $\mathsf{H}_T(s) = \sum_{k \in I} E_k(s) P_k(s)$, which together with (2.44) implies that $\left[\mathsf{H}_T(s), P_j(s)\right] = 0$, that is, $\mathsf{H}_T(s)$ commutes with all its orthogonal projectors $P_j(s)$. Consequently, (G.20) becomes

$$
\begin{aligned}
\left[\mathsf{H}_{A,j}(s), P_j(s)\right] &= \mathrm{i}\left[\left[\dot{P}_j(s), P_j(s)\right], P_j(s)\right] \\
&\underset{(2.47)}{=} \mathrm{i}\left[\dot{P}_j(s) P_j(s) - P_j(s) \dot{P}_j(s), P_j(s)\right] \\
&\underset{(2.47)}{=} \mathrm{i}\left(\dot{P}_j(s)\left(P_j(s)\right)^2 - 2P_j(s)\dot{P}_j(s)P_j(s) + \left(P_j(s)\right)^2 \dot{P}_j(s)\right) \\
&\underset{(2.44),(G.15)}{=} \mathrm{i}\left(\dot{P}_j(s) P_j(s) + P_j(s) \dot{P}_j(s)\right) \\
&\underset{(G.14)}{=} \mathrm{i}\dot{P}_j(s),
\end{aligned}
$$

which completes the proof of (G.19).   □

Definition G.9 Let $\mathsf{H}_{A,j}$ be as defined in Definition G.7. For each $j \in I$ the **adiabatic intertwiner** $U_{A,j} : [0,1] \to \mathcal{U}(\mathbb{H})$ is defined as the solution of the initial value problem

$$
\begin{aligned}
\mathrm{i}\dot{U}_{A,j}(s) &= \mathsf{H}_{A,j}(s) U_{A,j}(s) \\
U_{A,j}(0) &= \mathbf{1}.
\end{aligned} \qquad (G.21)
$$

As usual, we assume that the $\mathsf{H}_{A,j}(s)$ are such that a solution exists and is unique.

Exercise G.144 Show that the adiabatic intertwiners $U_{A,j}$ satisfy

$$
\begin{aligned}
\mathrm{i}\dot{U}_{A,j}(s)^* &= -U_{A,j}(s)^* \mathsf{H}_{A,j}(s) \\
U_{A,j}(0)^* &= \mathbf{1}.
\end{aligned} \qquad (G.22)
$$

For a solution see Solution G.144.

The next lemma shows why the intertwiner has its name.

Lemma G.10 *Let $U_{A,j}$ be defined as in Definition G.9 and $P_j(s)$ as in Definition G.7. For any $j \in I$ and $s \in [0,1]$ we have*

$$U_{A,j}(s)^* P_j(s) U_{A,j}(s) = P_j(0). \tag{G.23}$$

Proof Since $U_{A,j}(0) = \mathbf{1} = U_{A,j}(0)^*$, the statement (G.23) is true for $s = 0$. To prove it for any $s \in [0,1]$, it thus suffices to show that the left side of (G.23) is constant. For this expression we find

$$
\begin{aligned}
\frac{d}{ds}\left(U_{A,j}(s)^* P_j(s) U_{A,j}(s) \right) &= \dot{U}_{A,j}(s)^* P_j(s) U_{A,j}(s) + U_{A,j}(s)^* \dot{P}_j(s) U_{A,j}(s) \\
&\quad + U_{A,j}(s)^* P_j(s) \dot{U}_{A,j}(s) \\
&\overset{\text{(G.21),(G.22)}}{=} \mathrm{i} U_{A,j}(s)^* \mathsf{H}_{A,j}(s) P_j(s) U_{A,j}(s) \\
&\quad + U_{A,j}(s)^* \dot{P}_j(s) U_{A,j}(s) \\
&\quad - \mathrm{i} U_{A,j}(s)^* P_j(s) \mathsf{H}_{A,j}(s) U_{A,j}(s) \\
&= U_{A,j}(s)^* \left(\dot{P}_j(s) + \mathrm{i}\big[\mathsf{H}_{A,j}(s), P_j(s)\big] \right) U_{A,j}(s) \\
&\underset{\text{(G.19)}}{=} 0,
\end{aligned}
$$

which completes the proof of (G.23). $\qquad\square$

G.3 Reduced Resolvent, Bounds on Time Derivatives and an Adiabatic Theorem

Definition G.11 Let $A \in \mathrm{B_{sa}}(\mathbb{H})$ with purely discrete (and possibly degenerate) spectrum $\sigma(A) = \{\lambda_j \mid j \in I \subset \mathbb{N}_0\}$, and let $\{P_j \mid j \in I\}$ denote the projections onto its eigenspaces. For each $j \in I$ the **reduced resolvent** $\check{R}_j(A)$ is defined as

$$\check{R}_j(A) := (\mathbf{1} - P_j)\left((A - \lambda_j \mathbf{1})\big|_{P_j\{\mathbb{H}\}^\perp} \right)^{-1} (\mathbf{1} - P_j) \quad \in \mathrm{L}(\mathbb{H}),$$

where the rightmost factor $\mathbf{1} - P_j$ on the right side is taken as the operator $\mathbf{1} - P_j : \mathbb{H} \to P_j\{\mathbb{H}\}^\perp$, whereas the leftmost factor $\mathbf{1} - P_j$ is to be understood as $\mathbf{1} - P_j : P_j\{\mathbb{H}\}^\perp \to \mathbb{H}$.

Note that we can express $A - \lambda_j \mathbf{1}$ with the help of projectors as

$$A - \lambda_j \mathbf{1} \underbrace{=}_{(2.42),(2.43)} \sum_{k \in I} (\lambda_k - \lambda_j) P_k = \sum_{k \in I \setminus \{j\}} (\lambda_k - \lambda_j) P_k. \tag{G.24}$$

Thus, the reduced resolvent $\check{R}_j(A)$ operates non-trivially only in the orthogonal complement of the eigenspace for the eigenvalue λ_j. The eigenspace is given by $P_j\{\mathbb{H}\}$, and its orthogonal complement is denoted by $P_j\{\mathbb{H}\}^\perp$.

Exercise G.145 Show that the reduced resolvent as defined in Definition G.11 satisfies

$$\check{R}_j(A) = \sum_{k \in I \setminus \{j\}} \frac{P_k}{\lambda_k - \lambda_j} \tag{G.25}$$

$$P_j \check{R}_j(A) = 0 = \check{R}_j(A) P_j \tag{G.26}$$

$$\left(\mathbf{1} - P_j\right) \check{R}_j(A) = \check{R}_j(A) = \check{R}_j(A)\left(\mathbf{1} - P_j\right). \tag{G.27}$$

For a solution see Solution G.145.

The inverse of the gap function defined in Definition G.4 bounds the norm of the reduced resolvent of $\mathsf{H}_T(s)$.

Lemma G.12 *Let* $\mathsf{H}_T : [0,1] \to \mathsf{B}_{sa}(\mathbb{H})$ *be such that it satisfies the Adiabatic Assumption (AA). Then we have*

$$\left\|\check{R}_j\left(\mathsf{H}_T(s)\right)\right\| \leq \frac{1}{g_j(s)}. \tag{G.28}$$

Proof Let $|\Psi\rangle \in \mathbb{H}$. Then we have

$$\left\|\check{R}_j\left(\mathsf{H}_T(s)\right)|\Psi\rangle\right\|^2 \underbrace{=}_{(2.5)} \left\langle \check{R}_j\left(\mathsf{H}_T(s)\right)\Psi \,\middle|\, \check{R}_j\left(\mathsf{H}_T(s)\right)\Psi\right\rangle$$

$$\underbrace{=}_{(G.25)} \sum_{k,l \in I \setminus \{j\}} \frac{\langle P_k(s)\Psi | P_l(s)\Psi\rangle}{\left(E_k(s) - E_j(s)\right)\left(E_l(s) - E_j(s)\right)}$$

$$\underbrace{=}_{P_k^* = P_k} \sum_{k,l \in I \setminus \{j\}} \frac{\langle \Psi | P_k(s) P_l(s)\Psi\rangle}{\left(E_k(s) - E_j(s)\right)\left(E_l(s) - E_j(s)\right)}$$

$$\underbrace{=}_{(2.44)} \sum_{k,l \in I \smallsetminus \{j\}} \delta_{kl} \frac{\langle \Psi | P_k(s) \Psi \rangle}{\left(E_k(s) - E_j(s) \right) \left(E_l(s) - E_j(s) \right)}$$

$$= \sum_{k \in I \smallsetminus \{j\}} \frac{\langle \Psi | P_k(s) \Psi \rangle}{\left(E_k(s) - E_j(s) \right)^2}$$

$$\underbrace{\leq}_{(G.8)} \frac{1}{g_j(s)^2} \sum_{k \in I \smallsetminus \{j\}} \langle \Psi | P_k(s) \Psi \rangle. \qquad (G.29)$$

Definition 2.11 of an orthogonal projection implies

$$\langle \Psi | P_j(s) \Psi \rangle = \langle \Psi | P_j(s)^2 \Psi \rangle = \langle P_j(s) \Psi | P_j(s) \Psi \rangle \underbrace{\geq}_{(2.2)} 0.$$

Using this in (G.29) gives

$$\left\| \check{R}_j \big(\mathsf{H}_T(s) \big) | \Psi \rangle \right\|^2 \leq \frac{1}{g_j(s)^2} \left(\sum_{k \in I \smallsetminus \{j\}} \langle \Psi | P_k(s) \Psi \rangle + \langle \Psi | P_j(s) \Psi \rangle \right)$$

$$= \frac{1}{g_j(s)^2} \sum_{k \in I} \langle \Psi | P_k(s) \Psi \rangle = \frac{1}{g_j(s)^2} \langle \Psi | \sum_{k \in I} P_k(s) \Psi \rangle$$

$$\underbrace{=}_{(2.43)} \frac{1}{g_j(s)^2} \langle \Psi | \Psi \rangle \underbrace{=}_{(2.5)} \frac{\| \Psi \|^2}{g_j(s)^2}.$$

The claim (G.28) then follows from the definition (2.45) of the operator norm. $\qquad \square$

Theorem G.13 *Let* $\mathsf{H}_T : [0,1] \to \mathsf{B}_{sa}(\mathbb{H})$ *be such that it satisfies the Adiabatic Assumption (AA) and let*

$$A : [0,1] \longrightarrow \mathsf{B}_{sa}(\mathbb{H})$$
$$s \longmapsto A(s)$$

be differentiable with respect to s. Moreover, let $X_j[A]$ *be defined as in Definition G.3. Then the following bounds hold.*

(i)

$$\left\| X_j[A](s) \right\| \leq \frac{\| A(s) \|}{g_j(s)} \qquad (G.30)$$

such that $X_j[A]\colon [0,1] \to B(\mathbb{H})$.

(ii)

$$\left\| \frac{d}{ds} X_j[A](s) \right\| \leq \frac{\left\| \dot{A}(s) \right\|}{g_j(s)} + 4 \frac{\left\| \dot{H}_T(s) \right\| \|A(s)\|}{g_j(s)^2}. \tag{G.31}$$

Proof In order to simplify the notation in this proof, we shall omit writing out the s-dependence explicitly and use the following abbreviations

$$
\begin{array}{lll}
P_j = P_j(s) & A = A(s) & \check{R}_j = \check{R}_j\big(H_T(s)\big) \\
E_j = E_j(s) & \gamma_j = \gamma_j(s) & X_j[A] = X_j[A](s) \\
H_T = H_T(s) & H_{A,j} = H_{A,j}(s) & g_j = g_j(s).
\end{array} \tag{G.32}
$$

Beginning with (i), we note that from the expression (G.2) for the resolvent and the Definition G.3 we obtain

$$X_j[A] = \sum_{k,l \in I} \frac{1}{2\pi i} \oint_{\gamma_j} \frac{P_k A P_l}{(E_k - z)(E_l - z)} dz. \tag{G.33}$$

Here we use the CAUCHY-integral formula (G.5), which gives

$$\frac{1}{2\pi i} \oint_{\gamma_j} \frac{P_k A P_l}{(E_k - z)(E_l - z)} dz = \begin{cases} \frac{P_j A P_l}{E_j - E_l} & \text{if } k = j \neq l \\[2mm] \frac{P_k A P_j}{E_j - E_k} & \text{if } k \neq j = l \\[2mm] 0 & \text{else} \end{cases}$$

such that (G.33) becomes

$$X_j[A] = \sum_{l \in I \setminus \{j\}} \frac{P_j A P_l}{E_j - E_l} + \sum_{k \in I \setminus \{j\}} \frac{P_k A P_j}{E_j - E_k} \underbrace{=}_{(\mathrm{G.25})} -\big(P_j A \check{R}_j + \check{R}_j A P_j\big).$$

Note that $P_j A \check{R}_j$ and $\check{R}_j A P_j$ map into orthogonal subspaces, that is, for any $|\Psi\rangle \in \mathbb{H}$ we have

$$\langle P_j A \check{R}_j \Psi | \check{R}_j A P_j \Psi \rangle \underbrace{=}_{P_j^* = P_j} \langle A \check{R}_j \Psi | P_j \check{R}_j A P_j \Psi \rangle \underbrace{=}_{(\mathrm{G.26})} 0. \tag{G.26}$$

Consequently, we can apply (2.15) to obtain

$$
\begin{aligned}
\left|\left|X_j[A]\Psi\right|\right|^2 \underset{(2.15)}{=} &\ \left|\left|\check{R}_j A P_j \Psi\right|\right|^2 + \left|\left|P_j A \check{R}_j \Psi\right|\right|^2 \\
\underset{(G.27)}{=} &\ \left|\left|\check{R}_j A P_j \Psi\right|\right|^2 + \left|\left|P_j A \check{R}_j (1-P_j)\Psi\right|\right|^2 \\
\underset{(2.51),(2.52)}{\leq} &\ \left|\left|\check{R}_j\right|\right|^2 \left|\left|A\right|\right|^2 \left|\left|P_j\Psi\right|\right|^2 + \underbrace{\left|\left|P_j\right|\right|^2}_{=1}\left|\left|A\right|\right|^2 \left|\left|\check{R}_j\right|\right|^2 \left|\left|(1-P_j)\Psi\right|\right|^2 \\
\underset{(G.28)}{\leq} &\ \frac{\left|\left|A\right|\right|^2}{g_j^2}\left(\left|\left|P_j\Psi\right|\right|^2 + \left|\left|(1-P_j)\Psi\right|\right|^2\right).
\end{aligned}
$$
(G.34)

Since $P_j\Psi$ and $(1-P_j)\Psi$ belong to orthogonal subspaces, we can once again apply (2.15) to find that

$$
\left|\left|P_j\Psi\right|\right|^2 + \left|\left|(1-P_j)\Psi\right|\right|^2 = \left|\left|\Psi\right|\right|^2 .
$$

Hence, (G.34) becomes

$$
\left|\left|X_j[A]\Psi\right|\right|^2 \leq \frac{\left|\left|A\right|\right|^2}{g_j^2}\left|\left|\Psi\right|\right|^2 ,
$$

and the claim (G.30) follows from the definition (2.45) of the operator norm. In order to show (G.31) we first note that in taking the derivative with respect to s of $X_j[A](s)$ the same arguments as in the proof of Lemma G.5 apply, in other words, we can ignore the s-dependence of the curve $\gamma_j(s)$ and take the s-derivative under the integral only to obtain

$$
\begin{aligned}
\frac{d}{ds}X_j[A] \underset{(G.6)}{=} &\ \frac{1}{2\pi i}\oint_{\gamma_j}\left(\dot{R}_z A R_z + R_z \dot{A} R_z + R_z A \dot{R}_z\right)dz \\
\underset{(G.6),(G.7)}{=} &\ X_j\left[\dot{A}\right] - \frac{1}{2\pi i}\oint_{\gamma_j}\left(R_z \dot{H}_T R_z A R_z + R_z A R_z \dot{H}_T R_z\right)dz .
\end{aligned}
$$
(G.35)

Consider the integral with the first term, which gives

$$
\begin{aligned}
\frac{1}{2\pi i}\oint_{\gamma_j} R_z \dot{H}_T R_z A R_z dz \underset{(G.2)}{=} &\ \frac{1}{2\pi i}\oint_{\gamma_j}\sum_{k,l,m\in I}\underbrace{\frac{P_k}{E_k - z}\dot{H}_T\frac{P_l}{E_l - z}A\frac{P_m}{E_m - z}}_{=:C_{klm}(z)}dz \\
= &\ \sum_{k,l,m\in I}\frac{1}{2\pi i}\oint_{\gamma_j}C_{klm}(z)dz ,
\end{aligned}
$$
(G.36)

where the CAUCHY integral formula (G.5) can be applied to yield

$$\frac{1}{2\pi i}\oint_{\gamma_j}C_{klm}(z)=\begin{cases}P_j\dot{H}_T\frac{P_l}{E_l-E_j}A\frac{P_m}{E_m-E_j}&\text{if }k=j\neq l,m\\\frac{P_k}{E_k-E_j}\dot{H}_TP_jA\frac{P_m}{E_m-E_j}&\text{if }l=j\neq k,m\\\frac{P_k}{E_k-E_j}\dot{H}_T\frac{P_l}{E_l-E_j}AP_j&\text{if }m=j\neq k,l\\0&\text{else.}\end{cases}$$

Inserting this into (G.36), we obtain

$$
\begin{aligned}
\frac{1}{2\pi i}\oint_{\gamma_j}R_z\dot{H}_TR_zAR_zdz &= \sum_{l,m\in I\smallsetminus\{j\}}P_j\dot{H}_T\frac{P_l}{E_l-E_j}A\frac{P_m}{E_m-E_j}\\
&+\sum_{k,m\in I\smallsetminus\{j\}}\frac{P_k}{E_k-E_j}\dot{H}_TP_jA\frac{P_m}{E_m-E_j}\\
&+\sum_{k,l\in I\smallsetminus\{j\}}\frac{P_k}{E_k-E_j}\dot{H}_T\frac{P_l}{E_l-E_j}AP_j\\
&\underset{(G.25)}{=}P_j\dot{H}_T\check{R}_jA\check{R}_j+\check{R}_j\dot{H}_TP_jA\check{R}_j+\check{R}_j\dot{H}_T\check{R}_jAP_j.
\end{aligned}
\tag{G.37}
$$

Similarly, the second integral term in (G.35) yields

$$\frac{1}{2\pi i}\oint_{\gamma_j}R_zAR_z\dot{H}_TR_zdz=P_jA\check{R}_j\dot{H}_T\check{R}_j+\check{R}_jAP_j\dot{H}_T\check{R}_j+\check{R}_jA\check{R}_j\dot{H}_TP_j.\tag{G.38}$$

With (G.37) and (G.38) the integral term in (G.35) becomes

$$
\begin{aligned}
\frac{1}{2\pi i}\oint_{\gamma_j}\left(R_z\dot{H}_TR_zAR_z+R_zAR_z\dot{H}_TR_z\right)dz &= P_j\left(\dot{H}_T\check{R}_jA+A\check{R}_j\dot{H}_T\right)\check{R}_j\\
&+\check{R}_j\left(\dot{H}_TP_jA+AP_j\dot{H}_T\right)\check{R}_j\\
&+\check{R}_j\left(\dot{H}_T\check{R}_jA+A\check{R}_j\dot{H}_T\right)P_j.
\end{aligned}
$$

Here we can use that (G.26) implies that P_j and $\check{R}_j$ map into orthogonal subspaces such that we can apply (2.15) to obtain for any $|\Psi\rangle\in\mathbb{H}$

$$
\begin{aligned}
&\left\|\frac{1}{2\pi i}\oint_{\gamma_j}\left(R_z\dot{H}_TR_zAR_z+R_zAR_z\dot{H}_TR_z\right)dz|\Psi\rangle\right\|^2\\
&=\left\|P_j\left(\dot{H}_T\check{R}_jA+A\check{R}_j\dot{H}_T\right)\check{R}_j\Psi\right\|^2\\
&+\left\|\check{R}_j\left(\left(\dot{H}_TP_jA+AP_j\dot{H}_T\right)\check{R}_j+\left(\dot{H}_T\check{R}_jA+A\check{R}_j\dot{H}_T\right)P_j\right)\Psi\right\|^2.
\end{aligned}
\tag{G.39}
$$

In (G.39) the first term can be estimated as follows.

$$\left|\left|P_j\left(\dot{\mathsf{H}}_T\check{R}_jA+A\check{R}_j\dot{\mathsf{H}}_T\right)\check{R}_j\Psi\right|\right|^2$$

$$\underbrace{\leq}_{(2.51),(2.52)}\underbrace{||P_j||^2}_{=1}\left(\left|\left|\dot{\mathsf{H}}_T\check{R}_jA+A\check{R}_j\dot{\mathsf{H}}_T\right|\right|\,||\check{R}_j||\right)^2||\Psi||^2$$

$$\underbrace{\leq}_{(2.52),(2.53)}\left(2\left|\left|\dot{\mathsf{H}}_T\right|\right|\,||A||\,||\check{R}_j||^2\right)^2||\Psi||^2\,. \tag{G.40}$$

The second term in (G.39) is estimated similarly.

Exercise G.146 Show that

$$\left|\left|\check{R}_j\left(\left(\dot{\mathsf{H}}_TP_jA+AP_j\dot{\mathsf{H}}_T\right)\check{R}_j+\left(\dot{\mathsf{H}}_T\check{R}_jA+A\check{R}_j\dot{\mathsf{H}}_T\right)P_j\right)\Psi\right|\right|^2$$

$$\leq\left(2\left|\left|\dot{\mathsf{H}}_T\right|\right|\,||A||\,||\check{R}_j||^2\right)^2 3\,||\Psi||^2\,. \tag{G.41}$$

For a solution see Solution G.146

Using (G.40) and (G.41) in (G.39), yields for any $|\Psi\rangle\in\mathbb{H}$

$$\left|\left|\frac{1}{2\pi i}\oint_{\gamma_j}\left(R_z\dot{\mathsf{H}}_TR_zAR_z+R_zAR_z\dot{\mathsf{H}}_TR_z\right)dz|\Psi\rangle\right|\right|^2\leq\left(4\left|\left|\dot{\mathsf{H}}_T\right|\right|\,||A||\,||\check{R}_j||^2\right)^2||\Psi||^2\,,$$

which, together with Definition 2.12 of the operator norm, implies

$$\left|\left|\frac{1}{2\pi i}\oint_{\gamma_j}\left(R_z\dot{\mathsf{H}}_TR_zAR_z+R_zAR_z\dot{\mathsf{H}}_TR_z\right)dz\right|\right|\leq 4\left|\left|\dot{\mathsf{H}}_T\right|\right|\,||A||\,||\check{R}_j||^2$$

$$\underbrace{\leq}_{(G.28)}4\frac{\left|\left|\dot{\mathsf{H}}_T\right|\right|\,||A||}{g_j^2}\,. \tag{G.42}$$

From this and (G.35) it follows that

$$\left|\left|\frac{d}{ds}X_j[A]\right|\right|\underbrace{\leq}_{(2.53)}\left|\left|X_j\left[\dot{A}\right]\right|\right|+\left|\left|\frac{1}{2\pi i}\oint_{\gamma_j}\left(R_z\dot{\mathsf{H}}_TR_zAR_z+R_zAR_z\dot{\mathsf{H}}_TR_z\right)dz\right|\right|$$

$$\underbrace{\leq}_{(G.30),(G.42)}\frac{\left|\left|\dot{A}\right|\right|}{g_j}+4\frac{\left|\left|\dot{\mathsf{H}}_T\right|\right|\,||A||}{g_j^2}\,,$$

completing the proof of (G.31). $\qquad\qquad\square$

Before we can finally formulate and prove our version of the Quantum Adiabatic Theorem, we need one more ingredient.

Corollary G.14 *Let H_T be as defined in Definition 8.1 and such that it satisfies the Adiabatic Assumption (AA), and let $\{P_j(s) \mid j \in I\}$ be the projections onto the eigenspaces of $H_T(s)$. Then we have*

$$\left\|\dot{P}_j(s)\right\| \leq \frac{\left\|\dot{H}_T(s)\right\|}{g_j(s)}. \tag{G.43}$$

Proof From (G.9) in Lemma G.5 we know already that $\dot{P}_j = X_j\left[\dot{H}_T\right]$, therefore,

$$\left\|\dot{P}_j\right\| \underset{(G.9)}{=} \left\|X_j\left[\dot{H}_T\right]\right\| \underset{(G.30)}{\leq} \frac{\left\|\dot{H}_T\right\|}{g_j}$$

as claimed.   □

We can now prove what we might call a precursor to the actual Quantum Adiabatic Theorem. A statement about the probability to find an eigenstate of H_{fin} at t_{fin}, given that we started from a corresponding eigenstate of H_{ini} at t_{ini}, will follow as a corollary from the following theorem.

Theorem G.15 (Quantum Adiabatic Theorem [111]) *Let H_T be as defined in Definition 8.1 and such that it satisfies the Adiabatic Assumption (AA), and let be $\{P_i(s) \mid i \in I\}$ the projections onto the eigenspaces of $H_T(s)$. For a given $j \in I$ let g_j be its gap function as defined in Definition G.4. Moreover, let U_T and $U_{A,j}$ be as defined in Definitions 8.1 and G.9.*
Then we have for all $j \in I$ and $s \in [0,1]$

$$\left\|(U_{A,j}(s) - U_T(s))P_j(0)\right\| \leq \frac{C_j(s)}{T}, \tag{G.44}$$

where

$$C_j(s) := \frac{\left\|\dot{H}_T(s)\right\|}{g_j(s)^2} + \frac{\left\|\dot{H}_T(0)\right\|}{g_j(0)^2} + \int_0^s \left(\frac{\left\|\ddot{H}_T(u)\right\|}{g_j(u)^2} + 10\frac{\left\|\dot{H}_T(u)\right\|^2}{g_j(u)^3} \right) du. \tag{G.45}$$

Proof Once again, for the sake of brevity we often omit writing out the s-dependence explicitly in this proof and abbreviate as shown in (G.32). We consider first

$$
\begin{aligned}
&\frac{d}{ds}\left(U_{A,j}^* U_T\right)\\
=\ &\dot{U}_{A,j}^* U_T + U_{A,j}^* \dot{U}_T\\
\underset{\text{(G.22),(8.6)}}{=}\ &i U_{A,j}^* \mathsf{H}_{A,j} U_T - i U_{A,j}^* T \mathsf{H}_T U_T\\
=\ &i U_{A,j}^* \left(\mathsf{H}_{A,j} - T \mathsf{H}_T\right) U_T\\
\underset{\text{(G.17)}}{=}\ &-U_{A,j}^* [\dot{P}_j, P_j] U_T \underset{\text{(G.10)}}{=} U_{A,j}^* \left[\mathsf{H}_T, X_j\!\left[\dot{P}_j\right]\right] U_T\\
=\ &U_{A,j}^* \left(\mathsf{H}_T X_j\!\left[\dot{P}_j\right] - X_j\!\left[\dot{P}_j\right] \mathsf{H}_T\right) U_T\\
\underset{\text{(G.17)}}{=}\ &U_{A,j}^* \left(\tfrac{1}{T}\left(\mathsf{H}_{A,j} - i[\dot{P}_j, P_j]\right) X_j\!\left[\dot{P}_j\right] - X_j\!\left[\dot{P}_j\right] \mathsf{H}_T\right) U_T\\
=\ &\tfrac{1}{T} U_{A,j}^* \mathsf{H}_{A,j} X_j\!\left[\dot{P}_j\right] U_T - \tfrac{i}{T} U_{A,j}^* [\dot{P}_j, P_j] X_j\!\left[\dot{P}_j\right] U_T - U_{A,j}^* X_j\!\left[\dot{P}_j\right] \mathsf{H}_T U_T\\
\underset{\text{(G.22),(8.6)}}{=}\ &\tfrac{-i}{T} \dot{U}_{A,j}^* X_j\!\left[\dot{P}_j\right] U_T - \tfrac{i}{T} U_{A,j}^* [\dot{P}_j, P_j] X_j\!\left[\dot{P}_j\right] U_T - U_{A,j}^* X_j\!\left[\dot{P}_j\right] \tfrac{i}{T} \dot{U}_T\\
=\ &\tfrac{-i}{T}\left(\frac{d}{ds}\left(U_{A,j}^* X_j\!\left[\dot{P}_j\right] U_T\right) - U_{A,j}^*\left(\frac{d}{ds} X_j\!\left[\dot{P}_j\right]\right) U_T\right)\\
&+ U_{A,j}^* [\dot{P}_j, P_j] X_j\!\left[\dot{P}_j\right] U_T \Big).
\end{aligned}
$$

Integrating both sides and using that $U_{A,j}^*(0) = \mathbf{1} = U_T(0)$, we find

$$
\begin{aligned}
U_{A,j}^*(s) U_T(s) =\ &\mathbf{1} - \frac{i}{T}\left(U_{A,j}^*(s) X_j\!\left[\dot{P}_j\right](s) U_T(s) - X_j\!\left[\dot{P}_j\right](0)\right)\\
&- \frac{i}{T}\int_0^s U_{A,j}^*(u)\left([\dot{P}_j(u), P_j(u)] X_j\!\left[\dot{P}_j\right](u) - \frac{d}{du} X_j\!\left[\dot{P}_j\right](u)\right) U_T(u)\, du.
\end{aligned}
$$

Consequently, we have

$$
\begin{aligned}
U_T(s) =\ &U_{A,j}(s) - \frac{i}{T}\left(X_j\!\left[\dot{P}_j\right](s) U_T(s) - U_{A,j}(s) X_j\!\left[\dot{P}_j\right](0)\right)\\
&- U_{A,j}(s)\frac{i}{T}\int_0^s U_{A,j}^*(u)\left([\dot{P}_j(u), P_j(u)] X_j\!\left[\dot{P}_j\right](u) - \frac{d}{du} X_j\!\left[\dot{P}_j\right](u)\right) U_T(u)\, du.
\end{aligned}
$$

Making use of (2.53) and (2.54), it follows that

$$
\begin{aligned}
&\left\| U_{A,j}(s) - U_T(s)\right\|\\
\leq\ &\frac{1}{T}\left\| X_j\!\left[\dot{P}_j\right](s) U_T(s) - U_{A,j}(s) X_j\!\left[\dot{P}_j\right](0)\right\| \tag{G.46}\\
&+ \frac{1}{T}\left\| U_{A,j}(s)\int_0^s U_{A,j}^*(u)\left([\dot{P}_j(u), P_j(u)] X_j\!\left[\dot{P}_j\right](u) - \frac{d}{du} X_j\!\left[\dot{P}_j\right](u)\right) U_T(u)\, du\right\|.
\end{aligned}
$$

In (G.46) we estimate the two terms separately and begin with

$$
\left\| X_j \left[\dot{P}_j \right](s) U_T(s) - U_{A,j}(s) X_j \left[\dot{P}_j \right](0) \right\|
$$

$$
\underset{(2.52),(2.53)}{\leq} \left\| X_j \left[\dot{P}_j \right](s) \right\| \|U_T(s)\| + \|U_{A,j}(s)\| \left\| X_j \left[\dot{P}_j \right](0) \right\|
$$

$$
\underset{(2.55)}{=} \left\| X_j \left[\dot{P}_j \right](s) \right\| + \left\| X_j \left[\dot{P}_j \right](0) \right\|
$$

$$
\underset{(G.30)}{\leq} \frac{\left\| \dot{P}_j(s) \right\|}{g_j(s)} + \frac{\left\| \dot{P}_j(0) \right\|}{g_j(0)}
$$

$$
\underset{(G.43)}{\leq} \frac{\left\| \dot{H}_T(s) \right\|}{g_j(s)^2} + \frac{\left\| \dot{H}_T(0) \right\|}{g_j(0)^2}. \tag{G.47}
$$

The second term in (G.46) is estimated similarly

$$
\left\| U_{A,j}(s) \int_0^s U_{A,j}^*(u) \left([\dot{P}_j(u), P_j(u)] X_j \left[\dot{P}_j \right](u) - \frac{d}{du} X_j \left[\dot{P}_j \right](u) \right) U_T(u) du \right\|
$$

$$
\underset{(2.52)}{\leq} \|U_{A,j}(s)\| \left\| \int_0^s U_{A,j}^*(u) \left([\dot{P}_j(u), P_j(u)] X_j \left[\dot{P}_j \right](u) - \frac{d}{du} X_j \left[\dot{P}_j \right](u) \right) U_T(u) du \right\|
$$

$$
\underset{(2.55)}{=} \left\| \int_0^s U_{A,j}^*(u) \left([\dot{P}_j(u), P_j(u)] X_j \left[\dot{P}_j \right](u) - \frac{d}{du} X_j \left[\dot{P}_j \right](u) \right) U_T(u) du \right\|
$$

$$
\leq \int_0^s \left\| U_{A,j}^*(u) \left([\dot{P}_j(u), P_j(u)] X_j \left[\dot{P}_j \right](u) - \frac{d}{du} X_j \left[\dot{P}_j \right](u) \right) U_T(u) du \right\|
$$

$$
\underset{(2.52)}{\leq} \int_0^s \left\| U_{A,j}^*(u) \right\| \left\| [\dot{P}_j(u), P_j(u)] X_j \left[\dot{P}_j \right](u) - \frac{d}{du} X_j \left[\dot{P}_j \right](u) \right\| \|U_T(u)\| du
$$

$$
\underset{(2.55)}{=} \int_0^s \left\| [\dot{P}_j(u), P_j(u)] X_j \left[\dot{P}_j \right](u) - \frac{d}{du} X_j \left[\dot{P}_j \right](u) \right\| du
$$

$$
\underset{(2.53)}{=} \int_0^s \left(\left\| [\dot{P}_j(u), P_j(u)] X_j \left[\dot{P}_j \right](u) \right\| + \left\| \frac{d}{du} X_j \left[\dot{P}_j \right](u) \right\| \right) du. \tag{G.48}
$$

In the last equation of (G.48) we estimate the terms under the integral separately. The first term gives

$$
\left\| [\dot{P}_j(u), P_j(u)] X_j \left[\dot{P}_j \right](u) \right\| \underbrace{=}_{(2.46),(2.52),(2.53)} 2 \left\| \dot{P}_j(u) \right\| \left\| P_j(u) \right\| \left\| X_j \left[\dot{P}_j \right](u) \right\|
$$

$$
\underbrace{\leq}_{(2.55),(G.30)} 2 \frac{\left\| \dot{P}_j(u) \right\|^2}{g_j(u)}
$$

$$
\underbrace{\leq}_{(G.43)} 2 \frac{\left\| \dot{H}_T(u) \right\|^2}{g_j(u)^3} . \tag{G.49}
$$

The second term in (G.48) is estimated as follows.

$$
\left\| \frac{d}{du} X_j \left[\dot{P}_j \right](u) \right\| \underbrace{\leq}_{(G.31)} \frac{\left\| \ddot{P}_j(u) \right\|}{g_j(u)} + 4 \frac{\left\| \dot{H}_T(u) \right\| \left\| \dot{P}_j(u) \right\|}{g_j(u)^2}
$$

$$
\underbrace{\leq}_{(G.43)} \frac{\left\| \frac{d}{du} \dot{P}_j(u) \right\|}{g_j(u)} + 4 \frac{\left\| \dot{H}_T(u) \right\|^2}{g_j(u)^3}
$$

$$
\underbrace{=}_{(G.9)} \frac{\left\| \frac{d}{du} X_j \left[\dot{H}_T \right](u) \right\|}{g_j(u)} + 4 \frac{\left\| \dot{H}_T(u) \right\|^2}{g_j(u)^3}
$$

$$
\underbrace{\leq}_{(G.31)} \frac{1}{g_j(u)} \left(\frac{\left\| \ddot{H}_T(u) \right\|}{g_j(u)} + 4 \frac{\left\| \dot{H}_T(u) \right\|^2}{g_j(u)^2} \right) + 4 \frac{\left\| \dot{H}_T(u) \right\|^2}{g_j(u)^3}
$$

$$
= \frac{\left\| \ddot{H}_T(u) \right\|}{g_j(u)^2} + 8 \frac{\left\| \dot{H}_T(u) \right\|^2}{g_j(u)^3} . \tag{G.50}
$$

Using (G.49) and (G.50) in (G.48) gives

$$
\left\| U_{A,j}(s) \int_0^s U_{A,j}^*(u) \left([\dot{P}_j(u), P_j(u)] X_j \left[\dot{P}_j \right](u) - \frac{d}{du} X_j \left[\dot{P}_j \right](u) \right) U_T(u) du \right\|
$$

$$
\leq \int_0^s \left(\frac{\left\| \ddot{H}_T(u) \right\|}{g_j(u)^2} + 10 \frac{\left\| \dot{H}_T(u) \right\|^2}{g_j(u)^3} \right) du . \tag{G.51}
$$

Making use of (G.47) and (G.51) in (G.46), we obtain

$$\left|\left|U_{A,j}(s) - U_T(s)\right|\right| \tag{G.52}$$

$$\leq \frac{1}{T}\left(\frac{\left|\left|\dot{H}_T(s)\right|\right|}{g_j(s)^2} + \frac{\left|\left|\dot{H}_T(0)\right|\right|}{g_j(0)^2} + \int_0^s \left(\frac{\left|\left|\ddot{H}_T(u)\right|\right|}{g_j(u)^2} + 10\frac{\left|\left|\dot{H}_T(u)\right|\right|^2}{g_j(u)^3}\right) du\right).$$

The claim (G.44) then follows from (G.52) and

$$\left|\left|\left(U_{A,j}(s) - U_T(s)\right)P_j(0)\right|\right| \underbrace{\leq}_{(2.52)} \left|\left|U_{A,j}(s) - U_T(s)\right|\right| \left|\left|P_j(0)\right|\right|$$

$$\underbrace{=}_{(2.55)} \left|\left|U_{A,j}(s) - U_T(s)\right|\right|.$$

□

Solutions to Exercises

Solutions to Exercises from Chapter 2

Solution 2.1

(i) The claim (2.6) follows from

$$\langle a\psi|\varphi\rangle \underbrace{=}_{(2.1)} \overline{\langle\varphi|a\psi\rangle} \underbrace{=}_{(2.4)} \overline{a\langle\varphi|\psi\rangle} = \overline{a}\,\overline{\langle\varphi|\psi\rangle} \underbrace{=}_{(2.1)} \overline{a}\langle\psi|\varphi\rangle\,.$$

With $||\varphi||^2 = \langle\varphi|\varphi\rangle$ for all $\varphi \in \mathbb{H}$ it follows that

$$||a\varphi||^2 \underbrace{=}_{(2.5)} \langle a\varphi|a\varphi\rangle \underbrace{=}_{(2.4)} a\langle a\varphi|\varphi\rangle \underbrace{=}_{(2.6)} a\overline{a}\langle\varphi|\varphi\rangle \underbrace{=}_{(2.5)} |a|^2||\varphi||^2$$

and thus upon taking the square root the claim (2.7).

(ii) $\Rightarrow$: $\langle\psi|\varphi\rangle = 0$ for all $\varphi \in \mathbb{H}$ implies $\langle\psi|\psi\rangle = 0$, such that $\psi = 0$ follows from (2.3).
$\Leftarrow$: Let $\psi = 0$ and $\xi, \varphi \in \mathbb{H}$ be arbitrary. Then one has $\psi = 0\xi$ and

$$\langle\psi|\varphi\rangle = \langle 0\xi|\varphi\rangle \underbrace{=}_{(2.4)} 0\langle\xi|\varphi\rangle = 0\,.$$

(iii) Again with $||\psi||^2 = \langle\psi|\psi\rangle$ for all $\psi \in \mathbb{H}$ it follows that

$$\frac{1}{4}\left[||\psi+\varphi||^2 - ||\psi-\varphi||^2 + i||\psi-i\varphi||^2 - i||\psi+i\varphi||^2\right]$$

$$\underbrace{=}_{(2.5)} \frac{1}{4}\left[\langle\psi+\varphi|\psi+\varphi\rangle - \langle\psi-\varphi|\psi-\varphi\rangle\right.$$

$$\left. +i\langle\psi-i\varphi|\psi-i\varphi\rangle - i\langle\psi+i\varphi|\psi+i\varphi\rangle\right]$$

$$\underbrace{=}_{(2.4)} \frac{1}{4}\left[\langle\psi|\psi\rangle + \langle\psi|\varphi\rangle + \langle\varphi|\psi\rangle + \langle\varphi|\varphi\rangle\right.$$

© Springer Nature Switzerland AG 2019
W. Scherer, *Mathematics of Quantum Computing*,
https://doi.org/10.1007/978-3-030-12358-1

Fig. G.1 Graphical representation of $\varphi - \frac{\langle\psi|\varphi\rangle}{||\psi||^2}\psi \in \mathbb{H}_{\psi^\perp}$

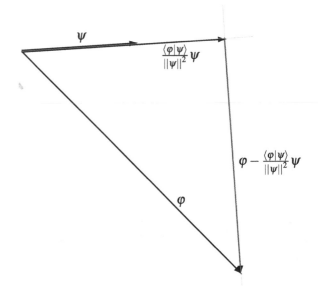

$$-\langle\psi|\psi\rangle + \langle\psi|\varphi\rangle + \langle\varphi|\psi\rangle - \langle\varphi|\varphi\rangle$$
$$+i\langle\psi|\psi\rangle + \langle\psi|\varphi\rangle - \langle\varphi|\psi\rangle + i\langle\varphi|\varphi\rangle$$
$$-i\langle\psi|\psi\rangle + \langle\psi|\varphi\rangle - \langle\varphi|\psi\rangle - i\langle\varphi|\varphi\rangle\Big]$$
$$= \frac{1}{4}\Big[4\langle\psi|\varphi\rangle\Big] = \langle\psi|\varphi\rangle \,.$$

Solution 2.2 Because of

$$\langle\psi|\varphi - \frac{\langle\psi|\varphi\rangle}{||\psi||^2}\psi\rangle \underset{(2.4)}{=} \langle\psi|\varphi\rangle - \frac{\langle\psi|\varphi\rangle}{||\psi||^2}\langle\psi|\psi\rangle \underset{(2.5)}{=} 0$$

it follows from Definition 2.2 that $\varphi - \frac{\langle\psi|\varphi\rangle}{||\psi||^2}\psi \in \mathbb{H}_{\psi^\perp}$. This is represented graphically in Fig. G.1.

Solution 2.3

(i) Let $\{e_j\}$ be an ONB and $\psi = \sum_j a_j e_j$. Then one has

$$\langle e_k|\psi\rangle = \langle e_k|\sum_j a_j e_j\rangle \underset{(2.4)}{=} \sum_j a_j\langle e_k|e_j\rangle \underset{(2.10)}{=} \sum_j a_j\delta_{jk} = a_k$$

and thus

$$\psi = \sum_j \langle e_j|\psi\rangle e_j = \sum_j \psi_j e_j \,. \tag{G.53}$$

(ii) Next,

$$\langle \varphi | \psi \rangle \underset{\text{(G.53)}}{=} \langle \sum_j \varphi_j e_j | \sum_k \psi_k e_k \rangle \underset{\text{(2.4),(2.6)}}{=} \sum_j \sum_k \overline{\varphi_j} \psi_k \langle e_j | e_k \rangle \underset{\text{(2.10)}}{=} \sum_j \sum_k \overline{\varphi_j} \psi_k \delta_{jk}$$

$$= \sum_j \overline{\varphi_j} \psi_j \underset{\varphi_j = \langle e_j | \varphi \rangle}{=} \sum_j \overline{\langle e_j | \varphi \rangle} \langle e_j | \psi \rangle \underset{\text{(2.1)}}{=} \sum_j \langle \varphi | e_j \rangle \langle e_j | \psi \rangle .$$

(iii) With $\psi_j = \langle e_j | \psi \rangle$ we have

$$||\psi||^2 \underset{\text{(2.5)}}{=} \langle \psi | \psi \rangle \underset{\text{(2.13)}}{=} \sum_j \overline{\langle e_j | \psi \rangle} \langle e_j | \psi \rangle = \sum_j \left| \langle e_j | \psi \rangle \right|^2 .$$

(iv) Finally, one has for $\varphi \in \mathbb{H}_{\psi^\perp}$ that

$$\langle \psi | \varphi \rangle \underset{\text{Def. 2.2}}{=} 0 = \overline{\langle \psi | \varphi \rangle} \underset{\text{(2.1)}}{=} \langle \varphi | \psi \rangle ,$$

such that

$$||\varphi + \psi||^2 \underset{\text{(2.5)}}{=} \langle \varphi + \psi | \varphi + \psi \rangle = \langle \varphi | \varphi \rangle + \underbrace{\langle \varphi | \psi \rangle}_{=0} + \underbrace{\langle \psi | \varphi \rangle}_{=0} + \langle \psi | \psi \rangle$$

$$\underset{\text{(2.5)}}{=} ||\varphi||^2 + ||\varphi||^2 .$$

Solution 2.4 For $\psi = 0$ or $\varphi = 0$ both sides of the relation are zero, and thus the relation is true. Consider now the case $\psi \neq 0 \neq \varphi$. Then it follows that

$$0 \leq \left\| \varphi - \frac{\langle \psi | \varphi \rangle}{||\psi||^2} \psi \right\|^2 \underset{\text{(2.5)}}{=} \langle \varphi - \frac{\langle \psi | \varphi \rangle}{||\psi||^2} \psi | \varphi - \frac{\langle \psi | \varphi \rangle}{||\psi||^2} \psi \rangle$$

$$\underset{\text{(2.4),(2.5)}}{=} ||\varphi||^2 - \langle \varphi | \frac{\langle \psi | \varphi \rangle}{||\psi||^2} \psi \rangle - \langle \frac{\langle \psi | \varphi \rangle}{||\psi||^2} \psi | \varphi \rangle + \left\| \frac{\langle \psi | \varphi \rangle}{||\psi||^2} \psi \right\|^2$$

$$\underset{\text{(2.5),(2.6),(2.7)}}{=} ||\varphi||^2 - \frac{\langle \psi | \varphi \rangle \langle \varphi | \psi \rangle}{||\psi||^2} - \frac{\overline{\langle \psi | \varphi \rangle} \langle \psi | \varphi \rangle}{||\psi||^2} + \left| \frac{\langle \psi | \varphi \rangle}{||\psi||^2} \right|^2 ||\psi||^2$$

$$\underset{\text{(2.1)}}{=} ||\varphi||^2 - \frac{|\langle \psi | \varphi \rangle|^2}{||\psi||^2}$$

and thus

$$|\langle \psi | \varphi \rangle|^2 \leq ||\varphi||^2 ||\psi||^2 . \tag{G.54}$$

Next, we have

$$
\begin{aligned}
||\varphi + \psi||^2 &= \underbrace{\left|||\varphi + \psi||^2\right|}_{(2.5)} = |\langle \varphi + \psi | \varphi + \psi \rangle| \\
&\underbrace{=}_{(2.4),(2.5)} \left| ||\varphi||^2 + ||\psi||^2 + \langle \varphi | \psi \rangle + \langle \psi | \varphi \rangle \right| \\
&\leq ||\varphi||^2 + ||\psi||^2 + |\langle \varphi | \psi \rangle| + |\langle \psi | \varphi \rangle| \\
&\underbrace{\leq}_{(G.54)} ||\varphi||^2 + ||\psi||^2 + 2||\varphi|| \, ||\psi|| = \left(||\varphi|| + ||\psi|| \right)^2
\end{aligned}
$$

from which (2.18) follows immediately.

Solution 2.5 For $\psi = 0$ the map is the zero map and as such obviously continuous. Assume then that $\psi \in \mathbb{H} \setminus \{0\}$. For any $\varepsilon > 0$ there is a $\delta(\varepsilon) = \frac{\varepsilon}{||\psi||}$ such that for $\varphi_0, \varphi \in \mathbb{H}$ that satisfy $||\varphi - \varphi_0|| \leq \delta(\varepsilon)$ it follows that

$$
\underbrace{|\langle \psi | \varphi - \varphi_0 \rangle| \leq}_{(2.17)} ||\psi|| \, ||\varphi - \varphi_0|| \leq ||\psi|| \delta(\varepsilon) = \varepsilon . \tag{G.55}
$$

Hence, $\langle \psi| : \varphi \mapsto \langle \psi | \varphi \rangle$ is continuous at φ_0.

Solution 2.6

(i) First we show that $(A^*)^* = A$. For arbitrary $|\psi\rangle, |\varphi\rangle \in \mathbb{H}$ one has

$$
\langle \psi | (A^*)^* \varphi \rangle \underbrace{=}_{(2.1)} \overline{\langle (A^*)^* \varphi | \psi \rangle} \underbrace{=}_{(2.30)} \overline{\langle \varphi | A^* \psi \rangle} \underbrace{=}_{(2.1)} \langle A^* \psi | \varphi \rangle \underbrace{=}_{(2.30)} \langle \psi | A \varphi \rangle .
$$

It follows that $\langle \psi | (A^*)^* \varphi - A \varphi \rangle = 0$ for arbitrary $|\psi\rangle$. With (2.8) this implies $(A^*)^* |\varphi\rangle = A |\varphi\rangle$ for all $|\varphi\rangle$.

(ii) Let $c \in \mathbb{C}$ and $A \in \mathrm{L}(\mathbb{H})$ be arbitrary. Then we have for any $|\psi\rangle, |\varphi\rangle \in \mathbb{H}$

$$
\langle (cA)^* \psi | \varphi \rangle \underbrace{=}_{(2.30)} \langle \psi | cA \varphi \rangle \underbrace{=}_{(2.4)} c \langle \psi | A \varphi \rangle \underbrace{=}_{(2.30)} c \langle A^* \psi | \varphi \rangle \underbrace{=}_{(2.6)} \langle \bar{c} A^* \psi | \varphi \rangle
$$

and thus

$$
\langle \left((cA)^* - \bar{c} A^* \right) \psi | \varphi \rangle = 0 \underbrace{\Rightarrow}_{(2.8)} \left((cA)^* - \bar{c} A^* \right) |\psi\rangle = 0 \quad \forall |\psi\rangle \in \mathbb{H} .
$$

(iii) In the linear maps

$$
\begin{array}{ll}
\langle A\psi| : \mathbb{H} \longrightarrow \mathbb{C} & \qquad \langle \psi | A^* : \mathbb{H} \longrightarrow \mathbb{C} \\
\quad\quad \varphi \longmapsto \langle A\psi | \varphi \rangle \; , & \qquad \quad\quad \varphi \longmapsto \langle \psi | A^* \varphi \rangle
\end{array}
$$

one then has

$$\underbrace{\langle A\psi|\varphi\rangle}_{(2.31)} = \underbrace{\langle (A^*)^*\psi|\varphi\rangle}_{(2.30)} = \langle \psi|A^*\varphi\rangle$$

and thus $\langle A\psi| = \langle \psi|A^*$.

(iv) Let $\{|e_j\rangle\}$ be an ONB in $\mathbb{H}$. The claim (2.35) then follows from

$$\underbrace{A^*_{jk}}_{(2.22)} = \underbrace{\langle e_j|A^*e_k\rangle}_{(2.30)} = \underbrace{\langle (A^*)^*e_j|e_k\rangle}_{(2.31)} = \underbrace{\langle Ae_j|e_k\rangle}_{(2.1)} = \underbrace{\overline{\langle e_k|Ae_j\rangle}}_{(2.22)} = \overline{A_{kj}}.$$

(v) Per Definition 2.8 of the adjoint operator we have for any $|\xi\rangle, |\eta\rangle \in \mathbb{H}$

$$\underbrace{\langle (|\varphi\rangle\langle\psi|)^*\xi|\eta\rangle}_{(2.30)} = \langle \xi|(|\varphi\rangle\langle\psi|)\eta\rangle = \langle \xi|\varphi\rangle\langle\psi|\eta\rangle$$

$$\underbrace{=}_{(2.6)} \langle \psi|\overline{\langle\xi|\varphi\rangle}|\eta\rangle \underbrace{=}_{(2.1)} \langle \psi\langle\varphi|\xi\rangle|\eta\rangle$$

$$= \langle (|\psi\rangle\langle\varphi|)\xi|\eta\rangle \qquad (G.56)$$

and the claim (2.36) follows from the fact that (G.56) holds for any $|\xi\rangle, |\eta\rangle \in \mathbb{H}$.

Solution 2.7 We show first that from the unitarity of U it follows that $U^*U=1$. Then that this implies $||U\psi||=||\psi||$ and finally that this in turn implies the unitarity of U. Let U be unitary. Then one has

$$\langle U\psi|U\varphi\rangle = \langle \psi|\varphi\rangle \quad \forall |\psi\rangle, |\varphi\rangle \in \mathbb{H}$$

$$\underbrace{\Rightarrow}_{(2.30)} \langle \psi|U^*U\varphi\rangle = \langle \psi|\varphi\rangle \quad \forall |\psi\rangle, |\varphi\rangle \in \mathbb{H}$$

$$\Rightarrow \langle \psi|U^*U\varphi - \varphi\rangle = 0 \quad \forall |\psi\rangle, |\varphi\rangle \in \mathbb{H}$$

$$\underbrace{\Rightarrow}_{(2.8)} U^*U|\varphi\rangle - |\varphi\rangle = 0 \quad \forall |\varphi\rangle \in \mathbb{H}$$

$$\Rightarrow U^*U = \mathbf{1}.$$

Now suppose $U^*U = \mathbf{1}$. For any $|\psi\rangle \in \mathbb{H}$ it follows then that

$$\underbrace{||\psi||}_{(2.5)} = \sqrt{\langle \psi|\psi\rangle}$$

$$= \underbrace{\sqrt{\langle \psi|U^*U\psi\rangle}}_{(2.30)} = \underbrace{\sqrt{\langle (U^*)^*\psi|U\psi\rangle}}_{(2.31)} = \sqrt{\langle U\psi|U\psi\rangle}$$

$$\underbrace{=}_{(2.5)} ||U\psi||.$$

Finally, suppose that $||U\psi||=||\psi||$ for all $|\psi\rangle \in \mathbb{H}$. Using (2.9) twice it follows that for any $|\psi\rangle, |\varphi\rangle \in \mathbb{H}$ one has

$$\langle U\psi|U\varphi\rangle$$

$$\underset{(2.9)}{=} \frac{1}{4}\left[||U\psi+U\varphi||^2-||U\psi-U\varphi||^2+\mathrm{i}\,||U\psi-\mathrm{i}U\varphi||^2-\mathrm{i}\,||U\psi+\mathrm{i}U\varphi||^2\right]$$

$$= \frac{1}{4}\left[||U(\psi+\varphi)||^2-||U(\psi-\varphi)||^2+\mathrm{i}\,||U(\psi-\mathrm{i}\varphi)||^2-\mathrm{i}\,||U(\psi+\mathrm{i}\varphi)||^2\right]$$

$$= \frac{1}{4}\left[||\psi+\varphi||^2-||\psi-\varphi||^2+\mathrm{i}\,||\psi-\mathrm{i}\varphi||^2-\mathrm{i}\,||\psi+\mathrm{i}\varphi||^2\right]$$

$$\underset{(2.9)}{=} \langle\psi|\varphi\rangle$$

and U is per Definition 2.9 unitary.

Solution 2.8 Let $A|\psi\rangle=\lambda|\psi\rangle$ for a $|\psi\rangle\neq 0$ in $\mathbb{H}$.

(i) Because of

$$\langle\psi|A^*\varphi\rangle \underset{(2.30)}{=} \langle(A^*)^*\psi|\varphi\rangle \underset{(2.31)}{=} \langle A\psi|\varphi\rangle = \langle\lambda\psi|\varphi\rangle \underset{(2.6)}{=} \overline{\lambda}\langle\psi|\varphi\rangle$$

the linear maps

$$\begin{aligned} \langle\psi|A^* : \mathbb{H} &\longrightarrow \mathbb{C} \\ \varphi &\longmapsto \langle\psi|A^*\varphi\rangle \end{aligned}, \qquad \begin{aligned} \overline{\lambda}\langle\psi| : \mathbb{H} &\longrightarrow \mathbb{C} \\ \varphi &\longmapsto \overline{\lambda}\langle\psi|\varphi\rangle \end{aligned}$$

are identical.

(ii) With $|\psi\rangle\neq 0$ one has

$$\lambda\langle\psi|\psi\rangle \underset{(2.4)}{=} \langle\psi|\lambda\psi\rangle = \langle\psi|A\psi\rangle = \langle A^*\psi|\psi\rangle \underset{A^*=A}{=} \langle A\psi|\psi\rangle = \langle\lambda\psi|\psi\rangle$$

$$\underset{(2.6)}{=} \overline{\lambda}\langle\psi|\psi\rangle$$

and thus $A^*=A$ implies $\overline{\lambda}=\lambda$.

(iii) Let $U|\psi\rangle=\lambda|\psi\rangle$ for a $|\psi\rangle\neq 0$ in $\mathbb{H}$ and let be U unitary, such that

$$||\psi|| \underset{(2.37)}{=} ||U\psi|| = ||\lambda\psi|| \underset{(2.7)}{=} |\lambda|\,||\psi||$$

and thus $|\lambda|=1$.

Solution 2.9 Let $\{|e_{j,\alpha}\rangle \mid j\in\{1,\ldots,d\},\alpha\in\{1,\ldots,d_j\}\}$ be the ONB of eigenstates of A, that is, let

$$A|e_{j,\alpha}\rangle = \lambda_j|e_{j,\alpha}\rangle, \tag{G.57}$$

where we allow each eigenspace to have dimension $d_j \geq 1$. It follows from Definition 2.3 of an ONB that any $|\psi\rangle \in \mathbb{H}$ with $||\psi|| = 1$ can be written in the form

$$|\psi\rangle = \sum_{j=1}^{d} \sum_{\alpha=1}^{d_j} \psi_{j,\alpha} |e_{j,\alpha}\rangle \tag{G.58}$$

with

$$\sum_{j=1}^{d} \sum_{\alpha=1}^{d_j} |\psi_{j,\alpha}|^2 \underbrace{=}_{(2.14)} ||\psi||^2 = 1. \tag{G.59}$$

Then we have

$$\begin{aligned}
\langle \psi | A \psi \rangle \;\underbrace{=}_{(G.58)}\; & \langle \sum_{j=1}^{d} \sum_{\alpha=1}^{d_j} \psi_{j,\alpha} |e_{j,\alpha}\rangle | A \sum_{k=1}^{d} \sum_{\beta=1}^{d_k} \psi_{k,\beta} |e_{k,\beta}\rangle \rangle \\
= \; & \sum_{j,k}^{d} \sum_{\alpha,\beta=1}^{d_j,d_k} \overline{\psi_{j,\alpha}} \psi_{k,\beta} \langle e_{j,\alpha} | A e_{k,\beta} \rangle \\
\underbrace{=}_{(G.57),(2.4)} \; & \sum_{j,k}^{d} \sum_{\alpha,\beta=1}^{d_j,d_k} \overline{\psi_{j,\alpha}} \psi_{k,\beta} \lambda_k \underbrace{\langle e_{j,\alpha} | e_{k,\beta} \rangle}_{=\delta_{j,k}\delta_{\alpha,\beta}} \\
\underbrace{=}_{(2.10)} \; & \sum_{j}^{d} \sum_{\alpha}^{d_j} |\psi_{j,\alpha}|^2 \lambda_j. \tag{G.60}
\end{aligned}$$

Consequently,

$$\begin{aligned}
\lambda_1 \;\underbrace{=}_{(G.59)}\; & \sum_{j=1}^{d} \sum_{\alpha=1}^{d_j} |\psi_{j,\alpha}|^2 \lambda_1 \\
\underbrace{\leq}_{(2.39)} \; & \sum_{j}^{d} \sum_{\alpha}^{d_j} |\psi_{j,\alpha}|^2 \lambda_j \;\underbrace{=}_{(G.60)}\; \langle \psi | A \psi \rangle \\
\underbrace{\leq}_{(2.39)} \; & \sum_{j=1}^{d} \sum_{\alpha=1}^{d_j} |\psi_{j,\alpha}|^2 \lambda_d \\
\underbrace{=}_{(G.59)} \; & \lambda_d.
\end{aligned}$$

Solution 2.10 With (2.41) we have

$$P_j^* \;\underbrace{=}_{(2.41)}\; \left(\sum_{\alpha=1}^{d_j} |e_{j,\alpha}\rangle\langle e_{j,\alpha}| \right)^* = \sum_{\alpha=1}^{d_j} |e_{j,\alpha}\rangle\langle e_{j,\alpha}|^* \;\underbrace{=}_{(2.36)}\; \sum_{\alpha=1}^{d_j} |e_{j,\alpha}\rangle\langle e_{j,\alpha}| \;\underbrace{=}_{(2.41)}\; P_j$$

and

$$
\begin{aligned}
P_j P_k &= \left(\sum_{\alpha=1}^{d_j} |e_{j,\alpha}\rangle\langle e_{j,\alpha}| \right) \left(\sum_{\beta=1}^{d_k} |e_{k,\beta}\rangle\langle e_{k,\beta}| \right) = \sum_{\alpha=1}^{d_j} \sum_{\beta=1}^{d_k} |e_{j,\alpha}\rangle\langle e_{j,\alpha}|e_{k,\beta}\rangle\langle e_{k,\beta}| \\
&\underbrace{=}_{(2.10)} \sum_{\alpha=1}^{d_j} \sum_{\beta=1}^{d_k} |e_{j,\alpha}\rangle \delta_{jk}\delta_{\alpha\beta} \langle e_{k,\beta}| = \delta_{jk} \sum_{\alpha=1}^{d_j} |e_{j,\alpha}\rangle\langle e_{j,\alpha}| = \delta_{jk} P_j ,
\end{aligned}
$$

such that P_j satisfies the defining properties of an orthogonal projection given in Definition 2.11.

As the $\{|e_{j,\alpha}\rangle j \in I, \alpha \in \{1,\dots,d_j\}\}$ are assumed to be an ONB of eigenvectors of A any eigenvector $|\psi\rangle \in \mathrm{Eig}(A,\lambda_j)$ can be written in the form

$$
|\psi\rangle \underbrace{=}_{(2.20)} \sum_{\alpha=1}^{d_j} |e_{j,\alpha}\rangle\langle e_{j,\alpha}|\psi\rangle \underbrace{=}_{(2.41)} P_j|\psi\rangle
$$

showing that P_j is indeed a projection onto $\mathrm{Eig}(A,\lambda_j)$.

Solution 2.11 Let P be a projection and $P|\psi_j\rangle = \lambda_j|\psi_j\rangle$ with $||\psi_j|| = 1$. Since $P^* = P$ all eigenvalues λ_j are real. From $P^2 = P$ it also follows that

$$
\begin{aligned}
\lambda_j^2 &= \lambda_j^2 ||\psi_j||^2 \underbrace{=}_{(2.5)} \lambda_j^2 \langle \psi_j|\psi_j\rangle \underbrace{=}_{(2.4)} \langle \psi_j|\lambda_j^2\psi_j\rangle = \langle\psi_j|P^2\psi_j\rangle \\
&= \langle\psi_j|P\psi\rangle = \langle\psi_j|\lambda_j\psi_j\rangle \underbrace{=}_{(2.4)} \lambda_j\langle\psi_j|\psi_j\rangle \underbrace{=}_{(2.5)} \lambda_j ||\psi_j||^2 = \lambda_j .
\end{aligned}
$$

Hence, we have $\lambda_j = 0$ or 1 and thus

$$
P \underbrace{=}_{(2.42)} \sum_j \lambda_j |\psi_j\rangle\langle\psi_j| = \sum_{j:\lambda_j=1} |\psi_j\rangle\langle\psi_j| .
$$

Solution 2.12 Let $|\varphi\rangle, |\psi\rangle \in \mathbb{H}$ be arbitrary. Then

$$
\langle (AB)^* \psi|\varphi\rangle \underbrace{=}_{(2.30)} \langle\psi|AB\varphi\rangle \underbrace{=}_{(2.30)} \langle A^*\psi|B\varphi\rangle \underbrace{=}_{(2.30)} \langle B^*A^*\psi|\varphi\rangle .
$$

Consequently,

$$
\begin{aligned}
\langle ((AB)^* - B^*A^*)\psi|\varphi\rangle &= 0 \qquad \forall |\psi\rangle, |\varphi\rangle \in \mathbb{H} \\
\underbrace{\Leftrightarrow}_{(2.8)} \quad ((AB)^* - B^*A^*)|\psi\rangle &= 0 \qquad \forall |\psi\rangle \in \mathbb{H},
\end{aligned}
$$

which proves $(AB)^* = B^*A^*$ and thus (2.47). Next, suppose that also

$$A^* = A \qquad \text{and} \qquad B^* = B. \tag{G.61}$$

Then we have

$$[A,B] = 0 \underbrace{\Leftrightarrow}_{(2.46)} \quad AB = BA \underbrace{=}_{(G.61)} B^*A^* \underbrace{=}_{(2.47)} (AB)^* \,,$$

which proves (2.48). To show (2.49) note that

$$A^*A \le cB^*B \underbrace{\Leftrightarrow}_{\text{Def. 2.12}} \quad 0 \le \langle \psi | (cB^*B - A^*A)\psi \rangle \quad \forall |\psi\rangle \in \mathbb{H}$$

$$\underbrace{\Leftrightarrow}_{(2.4)} \quad 0 \le c\langle \psi | B^*B\psi \rangle - \langle \psi | A^*A\psi \rangle \quad \forall |\psi\rangle \in \mathbb{H}$$

$$\underbrace{\Leftrightarrow}_{(2.30),(2.31)} \quad 0 \le c\langle B\psi | B\psi \rangle - \langle A\psi | A\psi \rangle \quad \forall |\psi\rangle \in \mathbb{H}$$

$$\underbrace{\Leftrightarrow}_{(2.5)} \quad 0 \le c||B\psi||^2 - ||A\psi||^2 \quad \forall |\psi\rangle \in \mathbb{H}$$

$$\Leftrightarrow \quad ||A\psi|| \le \sqrt{c}||B\psi|| \quad \forall |\psi\rangle \in \mathbb{H}$$

$$\underbrace{\Leftrightarrow}_{(2.57)} \quad ||A|| \le \sqrt{c}||B|| \,.$$

The last claim of Exercise 2.12 then follows from $||\mathbf{1}|| = 1$.

Solution 2.13 If A is self-adjoint, then so is A^2 and it has the real eigenvalues $\{\lambda_j^2 \mid j \in \{1,\dots,d\}\}$ and the same eigenvectors as A. Applying (2.40) to A^2, we obtain

$$\lambda_d^2 \underbrace{\ge}_{(2.40)} \langle \psi | A^2 \psi \rangle \underbrace{=}_{(2.30)} \langle A\psi | A\psi \rangle \underbrace{=}_{(2.5)} ||A\psi||^2 \,,$$

such that $||A\psi|| \le |\lambda_d|$ for any $|\psi\rangle \in \mathbb{H}$ with $||\psi|| = 1$. It then follows from the definition (2.45) of the operator norm that

$$||A|| \le |\lambda_d| \,. \tag{G.62}$$

On the other hand, any eigenvector $|e_d\rangle$ for the eigenvalue λ_d with $||e_d|| = 1$ satisfies

$$||A|e_d\rangle|| = ||\lambda_d|e_d\rangle|| \underbrace{=}_{(2.7)} |\lambda_d| \,||e_d|| = |\lambda_d| \,.$$

The definition of the operator norm given in (2.45) then implies

$$||A|| \geq |\lambda_d| \, . \tag{G.63}$$

Together (G.62) and (G.63) yield $||A|| = |\lambda_d|$, which is the claim (2.50).

Solution 2.14 Since $A0 = 0$ the statement is trivially true for the zero vector $0 \in \mathbb{H}$. Suppose then that $|\psi\rangle \neq 0$. Thus, we have $||\psi|| \neq 0$ and $\left|\left| \frac{\psi}{||\psi||} \right|\right| = 1$. Consequently,

$$\frac{1}{||\psi||} ||A\psi|| \underset{(2.7)}{=} \left|\left| \frac{1}{||\psi||} A\psi \right|\right| = \left|\left| A \frac{\psi}{||\psi||} \right|\right|$$
$$\leq \sup\left\{ ||A\varphi|| \mid |\varphi\rangle \in \mathbb{H}, ||\varphi|| = 1 \right\} \underset{(2.45)}{=} ||A|| \, .$$

Hence,

$$||A\psi|| \leq ||A|| \, ||\psi|| \tag{G.64}$$

as claimed. According to Definition 2.12

$$||AB|| \underset{(2.45)}{=} \sup\left\{ ||AB\psi|| \mid |\psi\rangle \in \mathbb{H}, ||\psi|| = 1 \right\}$$
$$\underset{(G.64)}{\leq} \sup\left\{ ||A|| \, ||B\psi|| \mid |\psi\rangle \in \mathbb{H}, ||\psi|| = 1 \right\}$$
$$\leq ||A|| \sup\left\{ ||B\psi|| \mid |\psi\rangle \in \mathbb{H}, ||\psi|| = 1 \right\}$$
$$\underset{(2.45)}{=} ||A|| \, ||B|| \, ,$$

which proves (2.52). Again, according to Definition 2.12

$$||A+B|| \underset{(2.45)}{=} \sup\left\{ ||(A+B)\psi|| \mid |\psi\rangle \in \mathbb{H}, ||\psi|| = 1 \right\}$$
$$\underset{(2.18)}{\leq} \sup\left\{ ||A\psi|| + ||B\psi|| \mid |\psi\rangle \in \mathbb{H}, ||\psi|| = 1 \right\}$$
$$\leq \sup\left\{ ||A\psi|| \mid |\psi\rangle \in \mathbb{H}, ||\psi|| = 1 \right\} + \sup\left\{ ||B\psi|| \mid |\psi\rangle \in \mathbb{H}, ||\psi|| = 1 \right\}$$
$$\underset{(2.45)}{=} ||A|| + ||B|| \, .$$

Once more, according to Definition 2.12

$$\underbrace{||aA|| \;=\;}_{(2.45)} \sup \left\{ ||aA\psi|| \;\big|\; |\psi\rangle \in \mathbb{H}, ||\psi|| = 1 \right\}$$

$$\underbrace{=}_{(2.7)} \sup \left\{ |a| \, ||A\psi|| \;\big|\; |\psi\rangle \in \mathbb{H}, ||\psi|| = 1 \right\}$$

$$= |a| \sup \left\{ ||A\psi|| \;\big|\; |\psi\rangle \in \mathbb{H}, ||\psi|| = 1 \right\}$$

$$\underbrace{=}_{(2.45)} |a| \, ||A|| ,$$

proving (2.54).

Next, consider an orthogonal projection P which satisfies

$$||P\psi||^2 \underbrace{=}_{(2.5)} \langle P\psi | P\psi \rangle \underbrace{=}_{(2.30)} \langle \psi | P^* P\psi \rangle \underbrace{=}_{\text{Def. 2.11}} \langle \psi | P^2 \psi \rangle \underbrace{=}_{\text{Def. 2.11}} \langle \psi | P\psi \rangle$$

$$\underbrace{\leq}_{(2.16)} ||\psi|| \, ||P\psi|| .$$

Hence, $||P\psi|| \leq 1$ for any $|\psi\rangle \in \mathbb{H}$ with $||\psi|| = 1$ and (2.45) implies that $||P|| \leq 1$. On the other hand, the results of Exercise 2.11 show that there exist $|\psi_j\rangle \in \mathbb{H}$ with $||\psi_j|| = 1$, such that $P|\psi_j\rangle = |\psi_j\rangle$. Thus $||P\psi_j|| = 1$ and (2.45) then implies that $||P|| = 1$.

Finally, for a unitary operator $U \in \mathfrak{U}(\mathbb{H})$ we know already from (2.37) that $||U\psi|| = ||\psi||$ for any $|\psi\rangle \in \mathbb{H}$. Hence, (2.45) implies that $||U|| = 1$ as claimed.

Solution 2.15

(i) To show $\Rightarrow$, let $\{|\tilde{e}_j\rangle = U|e_j\rangle\}$ be an ONB in $\mathbb{H}$. It follows that $U_{kj} = \langle e_k | U e_j \rangle = \langle e_k | \tilde{e}_j \rangle$ and thus

$$(UU^*)_{kl} = \sum_j U_{kj} U_{jl}^* = \sum_j U_{kj} \overline{U_{lj}} = \sum_j \langle e_k | \tilde{e}_j \rangle \overline{\langle e_l | \tilde{e}_j \rangle}$$

$$= \sum_j \langle e_k | \tilde{e}_j \rangle \langle \tilde{e}_j | e_l \rangle = \langle e_k | \underbrace{\sum_j \tilde{e}_j \langle \tilde{e}_j | e_l \rangle \rangle}_{= |e_l\rangle} = \langle e_k | e_l \rangle$$

$$= \delta_{kl} ,$$

such that $U \in \mathfrak{U}(\mathbb{H})$.

To prove $\Leftarrow$, let $U \in \mathfrak{U}(\mathbb{H})$ and $|\tilde{e}_j\rangle = U|e_j\rangle$. For any $\{a_j\} \subset \mathbb{C}$ it follows that

$$\sum_j a_j |\tilde{e}_j\rangle = 0 \quad \Rightarrow \quad \sum_j a_j U|e_j\rangle = 0 \quad \Rightarrow \quad U \sum_j a_j |e_j\rangle = 0$$

$$\Rightarrow U^* U \sum_j a_j |e_j\rangle = 0 \quad \underbrace{\Rightarrow}_{(2.37)} \quad \sum_j a_j |e_j\rangle = 0 \quad \underbrace{\Rightarrow}_{\text{Def. 2.3}} \quad a_j = 0 \quad \forall j ,$$

such that the $\{|\widetilde{e}_j\rangle\} \subset \mathbb{H}$ are linearly independent. Moreover, we have for any $|\psi\rangle \in \mathbb{H}$

$$U^*|\psi\rangle \underset{(2.11)}{=} \sum_j \langle e_j|U^*\psi\rangle|e_j\rangle \underset{(2.30),(2.31)}{=} \sum_j \langle Ue_j|\psi\rangle|e_j\rangle = \sum_j \langle \widetilde{e}_j|\psi\rangle|e_j\rangle$$

$$\Rightarrow \quad |\psi\rangle \underset{(2.37)}{=} UU^*|\psi\rangle = \sum_j \langle \widetilde{e}_j|\psi\rangle U|e_j\rangle = \sum_j \langle \widetilde{e}_j|\psi\rangle|\widetilde{e}_j\rangle$$

showing that any vector in $\mathbb{H}$ can be written as a linear combination of the $|\widetilde{e}_j\rangle$. Finally, we have

$$\langle \widetilde{e}_j|\widetilde{e}_k\rangle = \langle Ue_j|Ue_k\rangle \underset{(2.30)}{=} \langle e_j|U^*Ue_k\rangle \underset{(2.37)}{=} \langle e_j|e_k\rangle \underset{(2.10)}{=} \delta_{jk}$$

completing the proof that $\{|\widetilde{e}_j\rangle\}$ is an ONB in $\mathbb{H}$.

(ii)

$$\sum_j \langle \widetilde{e}_j|A\widetilde{e}_j\rangle = \sum_j \langle Ue_j|AUe_j\rangle = \sum_j \langle e_j|U^*AUe_j\rangle = \sum_j (U^*AU)_{jj} = \sum_{j,k,l} U^*_{jk}A_{kl}U_{lj}$$

$$= \sum_{k,l} A_{kl} \sum_j U_{lj}U^*_{jk} = \sum_{k,l} A_{kl} \underbrace{\sum_j (UU^*)_{lk}}_{=\delta_{lk}} = \sum_k A_{kk}$$

$$= \sum_k \langle e_k|Ae_k\rangle.$$

Solution 2.16

(i) For any $A, B \in L(\mathbb{H})$ we have

$$\text{tr}\,(AB) = \sum_j (AB)_{jj} = \sum_{j,k} A_{jk}B_{kj} = \sum_{j,k} B_{kj}A_{jk} = \sum_k (AB)_{kk} = \text{tr}\,(BA)\,.$$

(ii) The implication $\Leftarrow$ is obvious. To prove $\Rightarrow$ suppose for all $A \in L(\mathbb{H})$ we have $\text{tr}\,(AB) = 0$. Define A^{rs} as the operator that has as the j,k-th matrix element $(A^{rs})_{jk} = \delta_{rj}\delta_{sk}$. Then we find for any r, s

$$0 = \text{tr}\,(A^{rs}B) = \sum_{j,k} (A^{rs})_{jk}B_{kj} = \sum_{j,k} \delta_{rj}\delta_{sk}B_{kj} = B_{rs},$$

such that $B = 0$.

Solution 2.17 Let $t \mapsto U(t,t_0)$ be a solution of (2.69). To show the unitarity of $U(t,t_0)$ consider

$$\frac{d}{dt}||U(t,t_0)\psi||^2 \underset{(2.5)}{=} \frac{d}{dt}\langle U(t,t_0)\psi|U(t,t_0)\psi\rangle$$

$$= \langle \frac{d}{dt}U(t,t_0)\psi|U(t,t_0)\psi\rangle + \langle U(t,t_0)\psi|\frac{d}{dt}U(t,t_0)\psi\rangle$$

$$\underset{(2.69)}{=} \langle -iH(t)U(t,t_0)\psi|U(t,t_0)\psi\rangle + \langle U(t,t_0)\psi|-iH(t)U(t,t_0)\psi\rangle$$

$$\underset{(2.4),(2.6)}{=} i\big(\langle H(t)U(t,t_0)\psi|U(t,t_0)\psi\rangle - \langle U(t,t_0)\psi|H(t)U(t,t_0)\psi\rangle\big)$$

$$\underset{H(t)^*=H(t)}{=} i\big(\langle H(t)^*U(t,t_0)\psi|U(t,t_0)\psi\rangle - \langle U(t,t_0)\psi|H(t)U(t,t_0)\psi\rangle\big)$$

$$\underset{(2.30)}{=} 0.$$

Hence, for any $|\psi\rangle \in \mathbb{H}$ and $t \geq t_0$

$$||U(t,t_0)\psi|| = \text{const} = ||U(t_0,t_0)\psi|| \underset{(2.69)}{=} ||\psi||$$

and (2.37) implies that $U(t,t_0) \in \mathcal{U}(\mathbb{H})$ for all $t \geq t_0$.

Suppose $t \mapsto V(t,t_0)$ is another solution of (2.69). The proof that it has to be equal to $U(t,t_0)$ is a repetition of the above arguments with U replaced by $U - V$.

$$\frac{d}{dt}\big|\big|\big(U(t,t_0) - V(t,t_0)\big)\psi\big|\big|^2$$

$$\underset{(2.5)}{=} \frac{d}{dt}\langle \big(U(t,t_0) - V(t,t_0)\big)\psi|\big(U(t,t_0) - V(t,t_0)\big)\psi\rangle$$

$$\vdots$$

$$\underset{(2.69),(2.4),(2.6)}{=} i\big(\langle H(t)\big(U(t,t_0) - V(t,t_0)\big)\psi|\big(U(t,t_0) - V(t,t_0)\big)\psi\rangle$$
$$- \langle \big(U(t,t_0) - V(t,t_0)\big)\psi|H(t)\big(U(t,t_0) - V(t,t_0)\big)\psi\rangle\big)$$

$$\underset{H(t)^*=H(t),(2.30)}{=} 0.$$

Consequently, for any $|\psi\rangle \in \mathbb{H}$ and $t \geq t_0$

$$\big|\big|\big(U(t,t_0) - V(t,t_0)\big)\psi\big|\big| = \text{const} = \big|\big|\big(U(t_0,t_0) - V(t_0,t_0)\big)\psi\big|\big| \underset{(2.69)}{=} 0,$$

which implies that $U(t,t_0) = V(t,t_0)$ for all $t \geq t_0$.

Solution 2.18 Since for all $t \geq t_0$ we have

$$U(t,t_0)^*U(t,t_0) = \mathbf{1} = U(t,t_0)U(t,t_0)^*, \tag{G.65}$$

it follows that for $t = t_0$

$$\mathbf{1} = U(t_0,t_0)^*U(t_0,t_0) \underbrace{=}_{(2.71)} U(t_0,t_0)^*,$$

verifying the initial condition. Taking the derivative with respect to t on both sides in (G.65) we find

$$\frac{d}{dt}\left(U(t,t_0)^*\right)U(t,t_0) = -U(t,t_0)^*\frac{d}{dt}\left(U(t,t_0)\right).$$

Multiplying both sides with i and with $U(t,t_0)^*$ from the right and using again (G.65) we obtain

$$
\begin{aligned}
i\frac{d}{dt}U(t,t_0)^* &= -U(t,t_0)^*\left(i\frac{d}{dt}\left(U(t,t_0)\right)\right)U(t,t_0)^* \\
&\underbrace{=}_{(2.71)} -U(t,t_0)^*\left(\mathsf{H}(t)U(t,t_0)\right)U(t,t_0)^* \\
&\underbrace{=}_{(G.65)} -U(t,t_0)^*\mathsf{H}(t),
\end{aligned}
$$

proving (2.73).

Solution 2.19

(i) To begin with, one has

$$\sigma_1^2 = \begin{pmatrix} 0 & 1 \\ 1 & 0 \end{pmatrix}\begin{pmatrix} 0 & 1 \\ 1 & 0 \end{pmatrix} = \begin{pmatrix} 1 & 0 \\ 0 & 1 \end{pmatrix} = \mathbf{1}$$

and finds analogously $\sigma_2^2 = \mathbf{1} = \sigma_3^2$. Furthermore

$$
\begin{aligned}
\sigma_1\sigma_2 &= \begin{pmatrix} 0 & 1 \\ 1 & 0 \end{pmatrix}\begin{pmatrix} 0 & -i \\ i & 0 \end{pmatrix} = i\begin{pmatrix} 1 & 0 \\ 0 & -1 \end{pmatrix} = i\sigma_3 = i\varepsilon_{123}\sigma_3 \\
\sigma_2\sigma_1 &= \begin{pmatrix} 0 & -i \\ i & 0 \end{pmatrix}\begin{pmatrix} 0 & 1 \\ 1 & 0 \end{pmatrix} = -i\begin{pmatrix} 1 & 0 \\ 0 & -1 \end{pmatrix} = -i\sigma_3 = i\varepsilon_{213}\sigma_3 \\
\sigma_1\sigma_3 &= \begin{pmatrix} 0 & 1 \\ 1 & 0 \end{pmatrix}\begin{pmatrix} 1 & 0 \\ 0 & -1 \end{pmatrix} = -i\begin{pmatrix} 0 & -i \\ i & 0 \end{pmatrix} = -i\sigma_2 = i\varepsilon_{132}\sigma_2 \\
&= -\sigma_3\sigma_1
\end{aligned}
$$

$$\sigma_2\sigma_3 = \begin{pmatrix} 0 & -i \\ i & 0 \end{pmatrix} \begin{pmatrix} 1 & 0 \\ 0 & -1 \end{pmatrix} = i \begin{pmatrix} 0 & 1 \\ 1 & 0 \end{pmatrix} = i\sigma_1 = i\varepsilon_{231}\sigma_1$$

$$= -\sigma_2\sigma_3$$

From this and $\sigma_j^2 = \mathbf{1}$ follows $\sigma_j\sigma_k = \delta_{jk}\mathbf{1} + i\varepsilon_{jkl}\sigma_l$.

(ii) Hence

$$[\sigma_j, \sigma_k] = \sigma_j\sigma_k - \sigma_k\sigma_j = \delta_{jk}\mathbf{1} + i\varepsilon_{jkl}\sigma_l - \underbrace{\delta_{kj}}_{\delta_{jk}}\mathbf{1} - i\underbrace{\varepsilon_{kjl}}_{-\varepsilon_{jkl}}\sigma_l = 2i\varepsilon_{jkl}\sigma_l .$$

(iii) Again with $\sigma_j\sigma_k = \delta_{jk}\mathbf{1} + i\varepsilon_{jkl}\sigma_l$ it follows that

$$\{\sigma_j, \sigma_k\} = \sigma_j\sigma_k + \sigma_k\sigma_j = \delta_{jk}\mathbf{1} + i\varepsilon_{jkl}\sigma_l + \delta_{kj}\mathbf{1} + i\underbrace{\varepsilon_{kjl}}_{=-\varepsilon_{jkl}}\sigma_l = 2\delta_{jk}\mathbf{1} .$$

(iv) That $\sigma_j^* = \sigma_j$ holds is easily verified by using the defining matrices of (2.74) and (2.35). It follows that

$$\sigma_j^*\sigma_j = \sigma_j^2 \underbrace{=}_{(2.76)} \mathbf{1}$$

and from (2.37) that $\sigma_j \in U(2)$.

Solution 2.20 To determine the eigenvalues of σ_x we solve

$$\det(\sigma_x - \lambda\mathbf{1}) = \det\begin{pmatrix} -\lambda & 1 \\ 1 & -\lambda \end{pmatrix} = \lambda^2 - 1 = 0$$

to find $\lambda_\pm = \pm 1$. Let $|\uparrow_{\hat{x}}\rangle = \begin{pmatrix} v_1^+ \\ v_2^+ \end{pmatrix}$ be eigenvector for the eigenvalue $\lambda_+ = +1$ and $|\downarrow_{\hat{x}}\rangle = \begin{pmatrix} v_1^- \\ v_2^- \end{pmatrix}$ be eigenvector for the eigenvalue $\lambda_- = -1$, that is, one has

$$\sigma_x\begin{pmatrix} v_1^\pm \\ v_2^\pm \end{pmatrix} = \begin{pmatrix} 0 & 1 \\ 1 & 0 \end{pmatrix}\begin{pmatrix} v_1^\pm \\ v_2^\pm \end{pmatrix} = \begin{pmatrix} v_2^\pm \\ v_1^\pm \end{pmatrix} = \lambda_\pm\begin{pmatrix} v_1^\pm \\ v_2^\pm \end{pmatrix} .$$

Then $v_1^\pm = \pm v_2^\pm$ holds, and with the normalization condition $(v_1^\pm)^2 + (v_2^\pm)^2 = 1$ we find

$$|\uparrow_{\hat{x}}\rangle = \frac{1}{\sqrt{2}}\begin{pmatrix} 1 \\ 1 \end{pmatrix} = \frac{|0\rangle + |1\rangle}{\sqrt{2}} \quad \text{and} \quad |\downarrow_{\hat{x}}\rangle = \frac{1}{\sqrt{2}}\begin{pmatrix} 1 \\ -1 \end{pmatrix} = \frac{|0\rangle - |1\rangle}{\sqrt{2}} \qquad \text{(G.66)}$$

as eigenvectors, and thus

$$|\langle \uparrow_{\hat{\mathbf{x}}} |0\rangle|^2 = \frac{1}{2} = |\langle \downarrow_{\hat{\mathbf{x}}} |0\rangle|^2$$

for the probabilities.

Solution 2.21 For σ_x we have determined in the Solution 20 of Exercise 2.20 the eigenvalues $+1, -1$ and the eigenvectors $| \uparrow_{\hat{\mathbf{x}}} \rangle = \frac{1}{\sqrt{2}} \begin{pmatrix} 1 \\ 1 \end{pmatrix}, | \downarrow_{\hat{\mathbf{x}}} \rangle = \frac{1}{\sqrt{2}} \begin{pmatrix} 1 \\ -1 \end{pmatrix}$. With these we obtain

$$
\begin{aligned}
\sigma_x &= (+1)| \uparrow_{\hat{\mathbf{x}}} \rangle\langle | \uparrow_{\hat{\mathbf{x}}} \rangle| + (-1)| \downarrow_{\hat{\mathbf{x}}} \rangle\langle | \downarrow_{\hat{\mathbf{x}}} \rangle| \\
&= \frac{1}{\sqrt{2}} \begin{pmatrix} 1 \\ 1 \end{pmatrix} \frac{1}{\sqrt{2}} (1\ 1) - \frac{1}{\sqrt{2}} \begin{pmatrix} 1 \\ -1 \end{pmatrix} \frac{1}{\sqrt{2}} (1\ -1) \\
&\underset{(2.27)}{=} \frac{1}{2} \begin{pmatrix} 1 & 1 \\ 1 & 1 \end{pmatrix} - \frac{1}{2} \begin{pmatrix} 1 & -1 \\ -1 & 1 \end{pmatrix} = \begin{pmatrix} 0 & 1 \\ 1 & 0 \end{pmatrix} .
\end{aligned}
$$

Solution 2.22 Recall from (2.83) that $\rho_i \in D(\mathbb{H})$ for $i \in \{1,2\}$ implies

$$\rho_i^* = \rho_i \qquad\qquad\qquad\qquad\qquad\quad \text{(G.67)}$$
$$\langle \psi | \rho_i \psi \rangle \geq 0 \qquad \forall | \psi \rangle \in \mathbb{H} \qquad\quad \text{(G.68)}$$
$$\mathrm{tr}(\rho_i) = 1 \qquad\qquad\qquad\qquad\qquad \text{(G.69)}$$

such that for $u \in [0,1]$

$$
\left(u\rho_1 + (1-u)\rho_2 \right)^* = u\rho_1^* + (1-u)\rho_2^* \underset{(G.67)}{=} u\rho_1 + (1-u)\rho_2
$$

$$
\langle \psi | \left(u\rho_1 + (1-u)\rho_2 \right) \psi \rangle \underset{(2.4)}{=} u\langle \psi | \rho_1 \psi \rangle + (1-u)\langle \psi | \rho_2 \psi \rangle \underset{(G.68)}{\geq} 0 \quad \forall | \psi \rangle \in \mathbb{H}
$$

$$
\underset{\text{Def. 2.12}}{\Rightarrow} u\rho_1 + (1-u)\rho_2 \geq 0
$$

$$
\mathrm{tr}\left(u\rho_1 + (1-u)\rho_2 \right) = u\,\mathrm{tr}(\rho_1) + (1-u)\,\mathrm{tr}(\rho_2) \underset{(G.69)}{=} 1
$$

and thus $u\rho_1 + (1-u)\rho_2 \in D(\mathbb{H})$.

Solution 2.23 In order to show (2.84), we need to prove that $U\rho U^*$ satisfies all defining properties (2.80)–(2.82).

To begin with, we have

$$(U\rho U^*)^* \underbrace{=}_{(2.47)} (U^*)^*\rho^* U^* \underbrace{=}_{(2.31)} U\rho^* U^* \underbrace{=}_{(2.80)} U\rho U^*$$

proving that $U\rho U^*$ is self-adjoint.

As for positivity, we observe that for any $|\psi\rangle \in \mathbb{H}$

$$\langle\psi|U\rho U^*\psi\rangle \underbrace{=}_{(2.30)} \langle U^*\psi|\rho U^*\psi\rangle \underbrace{\geq}_{\text{Def. 2.12 and (2.81)}} 0$$

and thus, again according to Definition 2.12, it follows that $U\rho U^* \geq 0$.

Lastly,

$$\text{tr}(U\rho U^*) \underbrace{=}_{(2.58)} \text{tr}(U^* U\rho) \underbrace{=}_{(2.37)} \text{tr}(\rho) \underbrace{=}_{(2.82)} 1$$

verifying (2.82) for $U\rho U^*$ as well.

Solution 2.24 Let the density operator be given as

$$\rho_\psi = |\psi\rangle\langle\psi| \tag{G.70}$$

for a pure state $|\psi\rangle$ and let

$$A = \sum_k \lambda_k |e_k\rangle\langle e_k| \tag{G.71}$$

be an observable given in diagonal form with the eigenvalues $\{\lambda_k\}$ and $\{|e_k\rangle\}$ an ONB consisting of the respective eigenvectors. Moreover, let P_λ denote the projection onto the eigenspace of $\lambda \in \{\lambda_k\}$. We proceed to verify that the generalizations for the expectation value, measurement probability, projection onto the state after measurement and time evolution given in Postulate 6 for ρ_ψ coincide with the statements made for a pure state $|\psi\rangle$ in the Postulates 1–4.

Expectation Value

$$\langle A \rangle_{\rho_\psi} \underbrace{=}_{(2.85)} \mathrm{tr}\left(\rho_\psi A\right) \underbrace{=}_{(G.70)} \mathrm{tr}\left(|\psi\rangle\langle\psi|A\right) \underbrace{=}_{(G.71)} \mathrm{tr}\left(|\psi\rangle\langle\psi|\sum_k \lambda_k |e_k\rangle\langle e_k|\right)$$

$$= \sum_k \lambda_k \mathrm{tr}\left(|\psi\rangle\langle\psi|e_k\rangle\langle e_k|\right) \underbrace{=}_{(2.57)} \sum_{k,j} \lambda_k \langle e_j|\psi\rangle\langle\psi|e_k\rangle \underbrace{\langle e_k|e_j\rangle}_{=\delta_{jk}}$$

$$= \sum_k \lambda_k \langle\psi|e_k\rangle\langle e_k|\psi\rangle = \langle\psi|\sum_k \lambda_k |e_k\rangle\langle e_k|\psi\rangle \underbrace{=}_{(G.71)} \langle\psi|A\psi\rangle$$

$$\underbrace{=}_{(2.60)} \langle A \rangle_\psi$$

Measurement Probability

$$\mathbf{P}_{\rho_\psi}(\lambda) \underbrace{=}_{(2.86)} \mathrm{tr}\left(\rho P_\lambda\right) \underbrace{=}_{(G.70)} \mathrm{tr}\left(|\psi\rangle\langle\psi|P_\lambda\right) \underbrace{=}_{\text{Def. 2.11}} \mathrm{tr}\left(|\psi\rangle\langle\psi|P_\lambda^2\right)$$

$$\underbrace{=}_{(2.58)} \mathrm{tr}\left(P_\lambda|\psi\rangle\langle\psi|P_\lambda\right) \underbrace{=}_{(2.57)} \sum_k \langle e_k|P_\lambda\psi\rangle\langle\psi|P_\lambda e_k\rangle$$

$$\underbrace{=}_{\text{Def. 2.11}} \sum_k \langle e_k|P_\lambda\psi\rangle\langle P_\lambda\psi|e_k\rangle = \langle P_\lambda\psi|\underbrace{\sum_k |e_k\rangle\langle e_k|P_\lambda\psi\rangle)}_{=P_\lambda|\psi\rangle}$$

$$= \langle P_\lambda\psi|P_\lambda\psi\rangle = ||P_\lambda\psi||^2$$

$$\underbrace{=}_{(2.62)} \mathbf{P}_\psi(\lambda) \tag{G.72}$$

Projection   From (G.72) we see that

$$\mathrm{tr}\left(\rho_\psi P_\lambda\right) = ||P_\lambda\psi||^2 \tag{G.73}$$

and thus

$$\frac{P_\lambda\rho_\psi P_\lambda}{\mathrm{tr}\left(\rho_\psi P_\lambda\right)} \underbrace{=}_{(G.70),(G.73)} \frac{P_\lambda|\psi\rangle\langle\psi|P_\lambda}{||P_\lambda\psi||^2} \underbrace{=}_{\text{Def. 2.11}} \frac{P_\lambda|\psi\rangle\langle\psi|P_\lambda^*}{||P_\lambda\psi||^2} \underbrace{=}_{(2.33)} \frac{P_\lambda|\psi\rangle\langle P_\lambda\psi|}{||P_\lambda\psi||^2}$$

$$= \frac{P_\lambda|\psi\rangle}{||P_\lambda\psi||}\frac{\langle P_\lambda\psi|}{||P_\lambda\psi||} = \rho_{\frac{P_\lambda|\psi\rangle}{||P_\lambda\psi||}},$$

that is, $\frac{P_\lambda\rho P_\lambda}{\mathrm{tr}(\rho P_\lambda)}$ is the density operator of the pure state $\frac{P_\lambda|\psi\rangle}{||P_\lambda\psi||}$.

Time Evolution   Let $\rho(t_0) = \rho_{\psi(t_0)} = |\psi(t_0)\rangle\langle\psi(t_0)|$ be the initial state and $\rho(t)$ be the state at time t. Then we have

$$\rho(t) \underbrace{=}_{(2.88)} U(t,t_0)\rho(t_0)U(t,t_0)^* = U(t,t_0)|\psi(t_0)\rangle\langle\psi(t_0)|U(t,t_0)^*$$

$$\underbrace{=}_{(2.33)} |U(t,t_0)\psi(t_0)\rangle\langle U(t,t_0)\psi(t_0)| \underbrace{=}_{(2.70)} \rho_{U(t,t_0)|\psi(t_0)\rangle}$$

$$\underbrace{=}_{(2.71)} \rho_{\psi(t)},$$

that is, $\rho(t)$ is the density operator of the pure state $|\psi(t)\rangle = U(t,t_0)|\psi(t_0)\rangle$.

Solution 2.25 We have

$$\langle A\rangle_\rho \underbrace{=}_{(2.85)} \mathrm{tr}(\rho A) \underbrace{=}_{(2.100)} \mathrm{tr}\left(\sum_j p_j|\psi_j\rangle\langle\psi_j|A\right) = \sum_j p_j\,\mathrm{tr}(|\psi_j\rangle\langle\psi_j|A)$$

$$\underbrace{=}_{(2.57)} \sum_j p_j \sum_k \underbrace{\langle\psi_k|\psi_j\rangle}_{=\delta_{jk}}\langle\psi_j|A\psi_k\rangle = \sum_j p_j\langle\psi_j|A\psi_j\rangle \qquad (G.74)$$

and thus

$$\langle A\rangle_\psi \underbrace{=}_{(2.60)} \langle\psi|A\psi\rangle \underbrace{=}_{(2.100)} \langle\sum_j \sqrt{p_j}\psi_j|A\sum_k \sqrt{p_k}\psi_k\rangle \underbrace{=}_{(2.4),(2.6)} \sum_{j,k} \sqrt{p_jp_k}\langle\psi_j|A\psi_k\rangle$$

$$= \sum_j p_j\langle\psi_j|A\psi_j\rangle + \sum_{j\neq k} \sqrt{p_jp_k}\langle\psi_j|A\psi_k\rangle \underbrace{=}_{(G.74)} \langle A\rangle_\rho + \sum_{j\neq k} \sqrt{p_jp_k}\langle\psi_j|A\psi_k\rangle.$$

Solution 2.26 To begin with, we have

$$\rho = \frac{2}{5}|\uparrow_{\hat{x}}\rangle\langle\uparrow_{\hat{x}}| + \frac{3}{5}|0\rangle\langle 0| \underbrace{=}_{(G.66),(2.78)} \frac{2}{5}\frac{1}{\sqrt{2}}\binom{1}{1}\frac{1}{\sqrt{2}}(1\ 1) + \frac{3}{5}\binom{1}{0}(1\ 0)$$

$$\underbrace{=}_{(2.29)} \frac{1}{5}\begin{pmatrix}1 & 1\\ 1 & 1\end{pmatrix} + \frac{3}{5}\begin{pmatrix}1 & 0\\ 0 & 0\end{pmatrix} = \frac{1}{5}\begin{pmatrix}4 & 1\\ 1 & 1\end{pmatrix}.$$

Obviously $\mathrm{tr}(\rho) = 1$. To determine the eigenvalues $p_{1,2}$, we solve

$$\det(\rho - \lambda\mathbf{1}) = \det\left(\frac{1}{5}\begin{pmatrix}4-\lambda & 1\\ 1 & 1-\lambda\end{pmatrix}\right) = \lambda^2 - \lambda + \frac{3}{25} = 0$$

and find $p_\pm = \frac{1}{2} \pm \frac{\sqrt{13}}{10}$. Let $|\psi_\pm\rangle = \begin{pmatrix} u_\pm \\ v_\pm \end{pmatrix}$ be the eigenvectors for the eigenvalues p_i, that is,

$$\frac{1}{5}\begin{pmatrix} 4 & 1 \\ 1 & 1 \end{pmatrix}\begin{pmatrix} u_\pm \\ v_\pm \end{pmatrix} = \frac{1}{5}\begin{pmatrix} 4u_\pm + v_\pm \\ u_\pm + v_\pm \end{pmatrix} = p_\pm \begin{pmatrix} u_\pm \\ v_\pm \end{pmatrix}.$$

As a solution of that we find after normalization

$$|\psi_\pm\rangle = \frac{1}{\sqrt{26 \mp 6\sqrt{13}}}\begin{pmatrix} 2 \\ \pm\sqrt{13}-3 \end{pmatrix}.$$

With this we obtain

$$p_+|\psi_+\rangle\langle\psi_+| + p_-|\psi_-\rangle\langle\psi_-|$$

$$= \frac{\frac{1}{2}+\frac{\sqrt{13}}{10}}{26-6\sqrt{13}}\begin{pmatrix} 2 \\ \sqrt{13}-3 \end{pmatrix}\begin{pmatrix} 2\sqrt{13}-3 \end{pmatrix}$$

$$+ \frac{\frac{1}{2}-\frac{\sqrt{13}}{10}}{26+6\sqrt{13}}\begin{pmatrix} 2 \\ -\sqrt{13}-3 \end{pmatrix}\begin{pmatrix} 2 -\sqrt{13}-3 \end{pmatrix}$$

$$= \frac{\frac{1}{2}+\frac{\sqrt{13}}{10}}{26-6\sqrt{13}}\begin{pmatrix} 4 & 2\sqrt{13}-6 \\ 2\sqrt{13}-6 & 22-6\sqrt{13} \end{pmatrix}$$

$$+ \frac{\frac{1}{2}-\frac{\sqrt{13}}{10}}{26+6\sqrt{13}}\begin{pmatrix} 4 & -2\sqrt{13}-6 \\ -2\sqrt{13}-6 & 22+6\sqrt{13} \end{pmatrix}$$

$$= \dots$$

$$= \frac{1}{10}\begin{pmatrix} 8 & 2 \\ 2 & 2 \end{pmatrix} = \frac{1}{5}\begin{pmatrix} 4 & 1 \\ 1 & 1 \end{pmatrix} = \rho.$$

To show that $\rho > \rho^2$, we first observe that

$$\rho - \rho^2 = \frac{1}{5}\begin{pmatrix} 4 & 1 \\ 1 & 1 \end{pmatrix} - \left(\frac{1}{5}\begin{pmatrix} 4 & 1 \\ 1 & 1 \end{pmatrix}\right)^2 = \frac{1}{25}\begin{pmatrix} 20 & 5 \\ 5 & 5 \end{pmatrix} - \frac{1}{25}\begin{pmatrix} 17 & 5 \\ 5 & 2 \end{pmatrix}$$

$$= \frac{1}{25}\begin{pmatrix} 3 & 0 \\ 0 & 3 \end{pmatrix}.$$

Now let $|\phi\rangle = \begin{pmatrix} \phi_1 \\ \phi_2 \end{pmatrix} \in \mathbb{H} \smallsetminus \{0\}$ be arbitrary. Then

$$\langle\phi|(\rho-\rho^2)\phi\rangle = (\phi_1\ \phi_2)\frac{1}{25}\begin{pmatrix} 3 & 0 \\ 0 & 3 \end{pmatrix}\begin{pmatrix} \phi_1 \\ \phi_2 \end{pmatrix} = 3\frac{(\phi_1)^2+(\phi_2)^2}{25}$$

$$> 0 \qquad \forall|\phi\rangle \in \mathbb{H} \smallsetminus \{0\}$$

and thus $\rho > \rho^2$.

Solution 2.27 Let $|\varphi\rangle, |\psi\rangle \in \mathbb{H}$. On the one hand, we have

$$
\begin{aligned}
\rho_{\varphi+\psi} &= \left(|\varphi\rangle + |\psi\rangle\right)\left(\langle\varphi| + \langle\psi|\right) \\
&= |\varphi\rangle\langle\varphi| + |\varphi\rangle\langle\psi| + |\psi\rangle\langle\varphi| + |\psi\rangle\langle\psi|
\end{aligned}
$$

and on the other hand

$$
\begin{aligned}
\rho_{\varphi+e^{i\alpha}\psi} &= \left(|\varphi\rangle + |e^{i\alpha}\psi\rangle\right)\left(\langle\varphi| + \langle e^{i\alpha}\psi|\right) = \left(|\varphi\rangle + e^{i\alpha}|\psi\rangle\right)\left(\langle\varphi| + e^{-i\alpha}\langle\psi|\right) \\
&= |\varphi\rangle\langle\varphi| + e^{-i\alpha}|\varphi\rangle\langle\psi| + e^{i\alpha}|\psi\rangle\langle\varphi| + |\psi\rangle\langle\psi| ,
\end{aligned}
$$

such that

$$
\rho_{\varphi+e^{i\alpha}\psi} - \rho_{\varphi+\psi} = \left(e^{-i\alpha} - 1\right)|\varphi\rangle\langle\psi| + \left(e^{i\alpha} - 1\right)|\psi\rangle\langle\varphi| .
$$

Solution 2.28 Generally, we have for the probability to measure the eigenvalue λ_i of $A = \sum_j |e_j\rangle\lambda_j\langle e_j|$ in the state ρ

$$
\langle P_{e_i}\rangle_\rho \underbrace{=}_{(2.86)} \mathrm{tr}\left(P_{e_i}\rho\right) . \tag{G.75}
$$

For pure states $\rho_\psi = |\psi\rangle\langle\psi|$ this becomes

$$
\langle P_{e_i}\rangle_{\rho_\psi} \underbrace{=}_{(2.101)} |\langle e_i|\psi\rangle|^2 . \tag{G.76}
$$

For $A = \sigma_z$ one then has $\lambda_1 = +1, \lambda_2 = -1, |e_1\rangle = |0\rangle = \begin{pmatrix} 1 \\ 0 \end{pmatrix}, |e_2\rangle = |1\rangle = \begin{pmatrix} 0 \\ 1 \end{pmatrix}$.

(i) With $|\psi\rangle = |\uparrow_{\hat{\mathbf{x}}}\rangle = \frac{|0\rangle + |1\rangle}{\sqrt{2}}$, Eq. (G.76) becomes

$$
\langle P_{e_1}\rangle_{\rho_{|\uparrow_{\hat{\mathbf{x}}}\rangle}} = |\langle e_1|\uparrow_{\hat{\mathbf{x}}}\rangle|^2 = \left|\langle 0|\frac{|0\rangle + |1\rangle}{\sqrt{2}}\rangle\right|^2 = \frac{1}{2} .
$$

(ii) With $|\psi\rangle = |\downarrow_{\hat{\mathbf{x}}}\rangle = \frac{|0\rangle - |1\rangle}{\sqrt{2}}$, it follows similarly that

$$
\langle P_{e_1}\rangle_{\rho_{|\downarrow_{\hat{\mathbf{x}}}\rangle}} = |\langle e_1|\downarrow_{\hat{\mathbf{x}}}\rangle|^2 = \left|\langle 0|\frac{|0\rangle - |1\rangle}{\sqrt{2}}\rangle\right|^2 = \frac{1}{2} .
$$

(iii) Likewise, with $|\psi\rangle = \frac{1}{\sqrt{2}}(|\uparrow_{\hat{x}}\rangle + |\downarrow_{\hat{x}}\rangle) = |0\rangle$, it follows from (G.76) that

$$\langle P_{e_1}\rangle_{\rho_{|0\rangle}} = |\langle e_1|0\rangle|^2 = |\langle 0|0\rangle|^2 = 1.$$

(iv) Finally, one has with $\rho = \frac{1}{2}(|\uparrow_{\hat{x}}\rangle\langle\uparrow_{\hat{x}}| + |\downarrow_{\hat{x}}\rangle\langle\downarrow_{\hat{x}}|)$ and $|\uparrow_{\hat{x}}\rangle = \frac{1}{\sqrt{2}}(|0\rangle + |1\rangle) = \frac{1}{\sqrt{2}}\begin{pmatrix} 1 \\ 1 \end{pmatrix}$ as well as $|\downarrow_{\hat{x}}\rangle = \frac{1}{\sqrt{2}}(|0\rangle - |1\rangle) = \frac{1}{\sqrt{2}}\begin{pmatrix} 1 \\ -1 \end{pmatrix}$ that

$$\begin{aligned}
\rho &= \frac{1}{2}\left(\frac{1}{\sqrt{2}}\begin{pmatrix} 1 \\ 1 \end{pmatrix}\frac{1}{\sqrt{2}}(1\ 1) + \frac{1}{\sqrt{2}}\begin{pmatrix} 1 \\ -1 \end{pmatrix}\frac{1}{\sqrt{2}}(1\ -1)\right) \\
&= \frac{1}{4}\left(\begin{pmatrix} 1 & 1 \\ 1 & 1 \end{pmatrix} + \begin{pmatrix} 1 & -1 \\ -1 & 1 \end{pmatrix}\right) = \frac{1}{2}\begin{pmatrix} 1 & 0 \\ 0 & 1 \end{pmatrix} \\
&= \frac{1}{2}\mathbf{1}.
\end{aligned}$$

Then (G.75) implies

$$\langle P_{e_1}\rangle_\rho = \mathrm{tr}\left(P_{e_1}\frac{1}{2}\mathbf{1}\right) = \frac{1}{2}.$$

Solution 2.29 From (2.76) in Exercise 2.19 we know that $\sigma_j\sigma_k = \delta_{jk}\mathbf{1} + i\varepsilon_{jkl}\sigma_l$. Using this, we find

$$\begin{aligned}
(\mathbf{a}\cdot\sigma)(\mathbf{b}\cdot\sigma) &= \sum_{j,k}a_jb_k\sigma_j\sigma_k \underbrace{=}_{(2.76)} \sum_{j,k}a_jb_k\left(\delta_{jk}\mathbf{1} + i\varepsilon_{jkl}\sigma_l\right) \\
&= \left(\sum_{j,k}a_jb_k\delta_{jk}\right)\mathbf{1} + i\sum_{j,k}a_jb_k\varepsilon_{jkl}\sigma_l \\
&= (\mathbf{a}\cdot\mathbf{b})\mathbf{1} \\
&\quad + i(a_1b_2\varepsilon_{123}\sigma_3 + a_2b_1\varepsilon_{213}\sigma_3 \\
&\quad + a_1b_3\varepsilon_{132}\sigma_2 + a_3b_1\varepsilon_{312}\sigma_2 \\
&\quad + a_2b_3\varepsilon_{231}\sigma_1 + a_3b_2\varepsilon_{321}\sigma_1) \\
&= (\mathbf{a}\cdot\mathbf{b})\mathbf{1} \\
&\quad + i(a_1b_2 - a_2b_1)\sigma_3 \\
&\quad + i(a_3b_1 - a_1b_3)\sigma_2 \\
&\quad + i(a_2b_3 - a_3b_2)\sigma_1 \\
&= (\mathbf{a}\cdot\mathbf{b})\mathbf{1} + i(\mathbf{a}\times\mathbf{b})\cdot\sigma.
\end{aligned}$$

Solution 2.30 We have

$$
\begin{aligned}
\operatorname{tr}(\rho_{\mathbf{x}}\sigma_j) \underset{(2.127)}{=}& \operatorname{tr}\left(\frac{1}{2}(\mathbf{1}+\mathbf{x}\cdot\boldsymbol{\sigma})\sigma_j\right) = \frac{1}{2}\operatorname{tr}\left(\sigma_j + \sum_{k=1}^{3} x_k \sigma_k \sigma_j\right) \\
=& \ \frac{1}{2}\underbrace{\operatorname{tr}(\sigma_j)}_{=0} + \frac{1}{2}\sum_{k=1}^{3} x_k \operatorname{tr}(\sigma_k \sigma_j) \\
\underset{(2.76)}{=}& \ \frac{1}{2}\sum_{k=1}^{3} x_k \operatorname{tr}\left(\mathbf{1}\delta_{kj} + i\varepsilon_{kjl}\sigma_l\right) = \frac{1}{2}\sum_{k=1}^{3} x_k\left(\delta_{kj}\underbrace{\operatorname{tr}(\mathbf{1})}_{=2} + i\varepsilon_{kjl}\underbrace{\operatorname{tr}(\sigma_l)}_{=0}\right) \\
=& \ x_j.
\end{aligned}
$$

Solution 2.31 With $A^2 = \mathbf{1}$ we have

$$
\begin{aligned}
e^{i\alpha A} &= \sum_{n=0}^{\infty} \frac{(i\alpha)^n}{n!} A^n \\
&= \sum_{k=0}^{\infty} \frac{(i\alpha)^{2k}}{2k!}\underbrace{A^{2k}}_{=1} + \sum_{j=0}^{\infty} \frac{(i\alpha)^{2j+1}}{(2j+1)!}\underbrace{A^{2j+1}}_{=A} \\
&= \mathbf{1}\underbrace{\sum_{k=0}^{\infty} \frac{(i\alpha)^{2k}}{2k!}}_{=\cos\alpha} + A\underbrace{\sum_{j=0}^{\infty} \frac{(i\alpha)^{2j+1}}{(2j+1)!}}_{=i\sin\alpha} \\
&= \cos\alpha\,\mathbf{1} + i\sin\alpha A.
\end{aligned}
$$

Solution 2.32 Let $\hat{\mathbf{n}} \in S^1_{\mathbb{R}^3}$ and $\alpha, \beta \in \mathbb{R}$. Then we have

$$
\begin{aligned}
D_{\hat{\mathbf{n}}}(\alpha)D_{\hat{\mathbf{n}}}(\beta) \underset{(2.130)}{=}& \left(\cos\frac{\alpha}{2}\mathbf{1} - i\sin\frac{\alpha}{2}\hat{\mathbf{n}}\cdot\boldsymbol{\sigma}\right)\left(\cos\frac{\beta}{2}\mathbf{1} - i\sin\frac{\beta}{2}\hat{\mathbf{n}}\cdot\boldsymbol{\sigma}\right) \\
=& \ \cos\frac{\alpha}{2}\cos\frac{\beta}{2}\mathbf{1} - \sin\frac{\alpha}{2}\sin\frac{\beta}{2}\underbrace{(\hat{\mathbf{n}}\cdot\boldsymbol{\sigma})^2}_{=1} \\
& - i\underbrace{\left(\cos\frac{\alpha}{2}\sin\frac{\beta}{2} + \sin\frac{\alpha}{2}\cos\frac{\beta}{2}\right)}_{=\sin\frac{\alpha+\beta}{2}}\hat{\mathbf{n}}\cdot\boldsymbol{\sigma}
\end{aligned}
$$

$$= \left(\cos \frac{\alpha}{2} \cos \frac{\beta}{2} - \sin \frac{\alpha}{2} \sin \frac{\beta}{2} \right) \mathbf{1} - \mathrm{i} \sin \frac{\alpha+\beta}{2} \hat{\mathbf{n}} \cdot \sigma$$

$$\underbrace{\qquad\qquad\qquad\qquad\qquad\qquad}_{=\cos \frac{\alpha+\beta}{2}}$$

$$= \cos \frac{\alpha+\beta}{2} \mathbf{1} - \mathrm{i} \sin \frac{\alpha+\beta}{2} \hat{\mathbf{n}} \cdot \sigma$$

$$= D_{\hat{\mathbf{n}}}(\alpha+\beta).$$

Solution 2.33 From Lemma 2.32 we know already that there exist $\alpha, \beta, \gamma, \delta \in \mathbb{R}$ such that in the standard basis $\{|0\rangle, |1\rangle\}$ the matrix of U is given by

$$U \underset{(2.133)}{=} \mathrm{e}^{\mathrm{i}\alpha} \begin{pmatrix} \mathrm{e}^{-\mathrm{i}\frac{\beta+\delta}{2}} \cos \frac{\gamma}{2} & -\mathrm{e}^{\mathrm{i}\frac{\delta-\beta}{2}} \sin \frac{\gamma}{2} \\ \mathrm{e}^{\mathrm{i}\frac{\beta-\delta}{2}} \sin \frac{\gamma}{2} & \mathrm{e}^{\mathrm{i}\frac{\beta+\delta}{2}} \cos \frac{\gamma}{2} \end{pmatrix}. \tag{G.77}$$

On the other hand, we have

$$D_{\hat{\mathbf{z}}}(\delta) \underset{(2.31)}{=} \cos \frac{\delta}{2} \mathbf{1} - \mathrm{i} \sin \frac{\delta}{2} \hat{\mathbf{z}} \cdot \sigma = \cos \frac{\delta}{2} \mathbf{1} - \mathrm{i} \sin \frac{\delta}{2} \sigma_z$$

$$= \begin{pmatrix} \cos \frac{\delta}{2} - \mathrm{i} \sin \frac{\delta}{2} & 0 \\ 0 & \cos \frac{\delta}{2} + \mathrm{i} \sin \frac{\delta}{2} \end{pmatrix} = \begin{pmatrix} \mathrm{e}^{-\mathrm{i}\frac{\delta}{2}} & 0 \\ 0 & \mathrm{e}^{\mathrm{i}\frac{\delta}{2}} \end{pmatrix}$$

and

$$D_{\hat{\mathbf{y}}}(\gamma) \underset{(2.31)}{=} \cos \frac{\gamma}{2} \mathbf{1} - \mathrm{i} \sin \frac{\gamma}{2} \hat{\mathbf{y}} \cdot \sigma = \cos \frac{\gamma}{2} \mathbf{1} - \mathrm{i} \sin \frac{\gamma}{2} \sigma_y$$

$$= \begin{pmatrix} \cos \frac{\gamma}{2} & -\sin \frac{\gamma}{2} \\ \sin \frac{\gamma}{2} & \cos \frac{\gamma}{2} \end{pmatrix}$$

such that

$$D_{\hat{\mathbf{z}}}(\beta) D_{\hat{\mathbf{y}}}(\gamma) D_{\hat{\mathbf{z}}}(\delta) = \begin{pmatrix} \mathrm{e}^{-\mathrm{i}\frac{\beta}{2}} & 0 \\ 0 & \mathrm{e}^{\mathrm{i}\frac{\beta}{2}} \end{pmatrix} \begin{pmatrix} \cos \frac{\gamma}{2} & -\sin \frac{\gamma}{2} \\ \sin \frac{\gamma}{2} & \cos \frac{\gamma}{2} \end{pmatrix} \begin{pmatrix} \mathrm{e}^{-\mathrm{i}\frac{\delta}{2}} & 0 \\ 0 & \mathrm{e}^{\mathrm{i}\frac{\delta}{2}} \end{pmatrix}$$

$$= \begin{pmatrix} \mathrm{e}^{-\mathrm{i}\frac{\beta+\delta}{2}} \cos \frac{\gamma}{2} & -\mathrm{e}^{\mathrm{i}\frac{\delta-\beta}{2}} \sin \frac{\gamma}{2} \\ \mathrm{e}^{\mathrm{i}\frac{\beta-\delta}{2}} \sin \frac{\gamma}{2} & \mathrm{e}^{\mathrm{i}\frac{\beta+\delta}{2}} \cos \frac{\gamma}{2} \end{pmatrix}.$$

Together with (G.77) this results in $U = \mathrm{e}^{\mathrm{i}\alpha} D_{\hat{\mathbf{z}}}(\beta) D_{\hat{\mathbf{y}}}(\gamma) D_{\hat{\mathbf{z}}}(\delta)$.

Solution 2.34

$$\sigma_x D_{\hat{y}}(\eta)\sigma_x \underbrace{=}_{(2.130)} \sigma_x\Big(\cos\frac{\eta}{2}\mathbf{1} - i\sin\frac{\eta}{2}\hat{y}\cdot\sigma\Big)\sigma_x$$

$$= \cos\frac{\eta}{2}\underbrace{\sigma_x^2}_{=\mathbf{1}} - i\sin\frac{\eta}{2}\underbrace{\sigma_x\sigma_y\,\sigma_x}_{=i\sigma_z}$$

$$= \cos\frac{\eta}{2}\mathbf{1} + \sin\frac{\eta}{2}\underbrace{\sigma_z\sigma_x}_{=i\sigma_y} = \cos\frac{\eta}{2}\mathbf{1} + i\sin\frac{\eta}{2}\sigma_y$$

$$\underbrace{=}_{(2.131)} D_{\hat{y}}(-\eta).$$

Similarly, one shows the second equation in (2.151).

Solution 2.35 Let

$$A = \begin{pmatrix} a & b \\ c & d \end{pmatrix} \in L(\mathbb{H}).$$

Setting

$$z_0 = \frac{a+d}{2}, \quad z_1 = \frac{b+c}{2}, \quad z_2 = i\frac{b-c}{2}, \quad z_3 = \frac{a-d}{2},$$

yields

$$\sum_{\alpha=0}^{3} z_\alpha\sigma_\alpha = \begin{pmatrix} z_0+z_3 & z_1-iz_2 \\ z_1+iz_2 & z_0-z_3 \end{pmatrix} = \begin{pmatrix} a & b \\ c & d \end{pmatrix} = A.$$

From Lemma 2.35 we know that there exist $\alpha,\xi\in\mathbb{R}$ and $\hat{n}\in S^1_{\mathbb{R}^3}$ such that for any $A\in\mathcal{U}(\mathbb{H})$

$$A = e^{i\alpha}D_{\hat{n}}(\xi) \underbrace{=}_{(2.130)} e^{i\alpha}\Big(\cos\frac{\xi}{2}\mathbf{1} - i\sin\frac{\xi}{2}\hat{n}\cdot\sigma\Big) = z_0\mathbf{1} + \mathbf{z}\cdot\sigma,$$

where we now have set

$$z_0 = e^{i\alpha}\cos\frac{\xi}{2}, \qquad \mathbf{z} = -i\sin\frac{\xi}{2}\hat{n},$$

such that

$$|z_0|^2 + |\mathbf{z}|^2 = \cos^2\frac{\xi}{2} + \sin^2\frac{\xi}{2}|\hat{n}|^2 = \cos^2\frac{\xi}{2} + \sin^2\frac{\xi}{2} = 1.$$

Solutions to Exercises from Chapter 3

Solution 3.36

$$\Big((a|\varphi\rangle)\otimes|\psi\rangle\Big)(\xi,\eta)\underbrace{=}_{(3.1)}\langle\xi|a\varphi\rangle\langle\eta|\psi\rangle\underbrace{=}_{(2.4)}a\langle\xi|\varphi\rangle\langle\eta|\psi\rangle\underbrace{=}_{(3.1)}a\Big(|\varphi\rangle\otimes|\psi\rangle\Big)(\xi,\eta)$$

Similarly, one shows $|\varphi\rangle\otimes(a|\psi\rangle)=a(|\varphi\rangle\otimes|\psi\rangle)$. Next we have

$$\Big(a(|\varphi\rangle\otimes|\psi\rangle)+b(|\varphi\rangle\otimes|\psi\rangle))\Big)(\xi,\eta)\underbrace{=}_{(3.3)}a(|\varphi\rangle\otimes|\psi\rangle)(\xi,\eta)+b(|\varphi\rangle\otimes|\psi\rangle)(\xi,\eta)$$

$$\underbrace{=}_{(3.1)}a\langle\xi|\varphi\rangle\langle\eta|\psi\rangle+b\langle\xi|\varphi\rangle\langle\eta|\psi\rangle$$

$$=(a+b)\langle\xi|\varphi\rangle\langle\eta|\psi\rangle$$

$$\underbrace{=}_{(3.1)}(a+b)\Big(|\varphi\rangle\otimes|\psi\rangle\Big)(\xi,\eta)$$

and

$$\Big((|\varphi_1\rangle+|\varphi_2\rangle)\otimes|\psi\rangle\Big)(\xi,\eta)\underbrace{=}_{(3.1)}\langle\xi|(|\varphi_1\rangle+|\varphi_2\rangle)\langle\eta|\psi\rangle=(\langle\xi|\varphi_1\rangle+\langle\xi|\varphi_2\rangle)\langle\eta|\psi\rangle$$

$$=\langle\xi|\varphi_1\rangle\langle\eta|\psi\rangle+\langle\xi|\varphi_2\rangle\langle\eta|\psi\rangle$$

$$\underbrace{=}_{(3.1)}\Big(|\varphi_1\rangle\otimes|\psi\rangle+|\varphi_2\rangle\otimes|\psi\rangle\Big)(\xi,\eta).$$

Similarly, one shows

$$|\varphi\rangle\otimes(|\psi_1\rangle+|\psi_2\rangle)=|\varphi\rangle\otimes|\psi_1\rangle+|\varphi\rangle\otimes|\psi_2\rangle.$$

Solution 3.37 Suppose the $\Psi_{ab}\in\mathbb{C}$ are such that

$$\sum_{a,b}\Psi_{ab}|e_a\otimes f_b\rangle=0\in\mathbb{H}^A\otimes\mathbb{H}^B.$$

Hence, (3.1) implies that for any $(\xi,\eta)\in\mathbb{H}^A\times\mathbb{H}^B$

$$\left(\sum_{a,b}\Psi_{ab}|e_a\otimes f_b\rangle\right)(\xi,\eta)\underbrace{=}_{(3.3),(3.1)}\sum_{a,b}\Psi_{ab}\langle e_a|\xi\rangle\langle f_b|\eta\rangle=0$$

and in particular for every $(\xi,\eta)=(e_{a'},f_{b'})$

$$0=\sum_{a,b}\Psi_{ab}\underbrace{\langle e_a|e_{a'}\rangle}_{\delta_{a,a'}}\underbrace{\langle f_b|f_{b'}\rangle}_{\delta_{b,b'}}=\Psi_{a',b'}.$$

According to Definition 2.3 the vectors in the set $\{|e_a\otimes f_b\rangle\}$ are then linearly independent.

Solution 3.38 Let $\{|e_a\rangle\}$ be an ONB in $\mathbb{H}^A$ and $\{|f_b\rangle\}$ be an ONB in $\mathbb{H}^B$. The expression $\langle\Psi|\Phi\rangle$ as defined in (3.7) is positive-definite since for any $|\Psi\rangle=\sum_{a,b}\Psi_{ab}|e_a\otimes f_b\rangle$ we have

$$\langle\Psi|\Psi\rangle=\sum_{a,b}|\Psi_{ab}|^2\geq 0$$

and thus

$$\langle\Psi|\Psi\rangle=0\quad\Leftrightarrow\quad\Psi_{ab}=0\ \forall a,b\quad\Leftrightarrow\quad|\Psi\rangle=0.$$

Moreover, let

$$\{|\widetilde{e_a}\rangle:=U^A|e_a\rangle=\sum_{a_1}\langle e_{a_1}|U^Ae_a\rangle|e_{a_1}\rangle=\sum_{a_1}U^A_{a_1a}|e_{a_1}\rangle\}\subset\mathbb{H}^A$$
$$\{|\widetilde{f_b}\rangle:=U^B|f_b\rangle=\sum_{b_1}\langle f_{b_1}|U^Bf_b\rangle|f_{b_1}\rangle=\sum_{b_1}U^B_{b_1b}|f_{b_1}\rangle\}\subset\mathbb{H}^B$$

be other ONBs in $\mathbb{H}^A$, resp. $\mathbb{H}^B$. From Exercise 2.15 we know that then the maps $U^A:\mathbb{H}^A\to\mathbb{H}^A, U^B:\mathbb{H}^B\to\mathbb{H}^B$ are necessarily unitary. Thus, we have

$$|\Phi\rangle=\sum_{a_1,b_1}\Phi_{a_1b_1}|e_{a_1}\otimes f_{b_1}\rangle=\sum_{a,b}\widetilde{\Phi_{ab}}|\widetilde{e_a}\otimes\widetilde{f_b}\rangle$$
$$=\sum_{a,b}\widetilde{\Phi_{ab}}\sum_{a_1}U^A_{a_1a}|e_{a_1}\rangle\otimes\sum_{b_1}U^B_{b_1b}|f_{b_1}\rangle=\sum_{a_1,b_1}\sum_{a,b}U^A_{a_1a}U^B_{b_1b}\widetilde{\Phi_{ab}}|e_{a_1}\otimes f_{b_1}\rangle$$

from which it follows that

$$\Phi_{a_1b_1}=\sum_{a,b}U^A_{a_1a}U^B_{b_1b}\widetilde{\Phi_{ab}}.$$

Similarly, we obtain

$$\Psi_{a_1 b_1} = \sum_{a,b} U^A_{a_1 a} U^B_{b_1 b} \widetilde{\Psi}_{ab}$$

and thus finally

$$
\begin{aligned}
\sum_{a_1,b_1} \overline{\Psi_{a_1 b_1}} \Phi_{a_1 b_1} &= \sum_{a_1,b_1} \overline{\sum_{a,b} U^A_{a_1 a} U^B_{b_1 b} \widetilde{\Psi}_{ab}} \sum_{a_2,b_2} U^A_{a_1 a_2} U^B_{b_1 b_2} \widetilde{\Phi}_{a_2 b_2} \\
&= \sum_{a,b} \sum_{a_2,b_2} \sum_{a_1,b_1} \overline{U^A_{a_1 a}} U^A_{a_1 a_2} \overline{U^B_{b_1 b}} U^B_{b_1 b_2} \overline{\widetilde{\Psi}_{ab}} \widetilde{\Phi}_{a_2 b_2} \\
&= \sum_{a,b} \sum_{a_2,b_2} \underbrace{\left(\sum_{a_1} U^{A*}_{a a_1} U^A_{a_1 a_2} \right)}_{=\delta_{a a_2}} \underbrace{\left(\sum_{b_1} U^{B*}_{b b_1} U^B_{b_1 b_2} \right)}_{=\delta_{b b_2}} \overline{\widetilde{\Psi}_{ab}} \widetilde{\Phi}_{a_2 b_2} \\
&= \sum_{a,b} \overline{\widetilde{\Psi}_{ab}} \widetilde{\Phi}_{ab},
\end{aligned}
$$

that is, $\langle \Psi | \Phi \rangle$ as defined in (3.7) does not depend on the choice of the ONBs $\{e_a\} \subset \mathbb{H}^A$ and $\{f_b\} \subset \mathbb{H}^B$.

Solution 3.39

$$
\begin{aligned}
\langle \Phi^+ | \Phi^+ \rangle &= \frac{1}{2} \langle 00 + 11 | 00 + 11 \rangle = \frac{1}{2} \Big(\langle 00 | 00 \rangle + \langle 11 | 00 \rangle + \langle 00 | 11 \rangle + \langle 11 | 11 \rangle \Big) \\
&\underset{(3.4)}{=} \frac{1}{2} \Big(\underbrace{\langle 0 | 0 \rangle \langle 0 | 0 \rangle}_{=1} + \underbrace{\langle 1 | 0 \rangle \langle 1 | 0 \rangle}_{=0} + \underbrace{\langle 0 | 1 \rangle \langle 0 | 1 \rangle}_{=0} + \underbrace{\langle 1 | 1 \rangle \langle 1 | 1 \rangle}_{=1} \Big) \\
&= 1.
\end{aligned}
$$

$$
\begin{aligned}
\langle \Phi^+ | \Phi^- \rangle &= \frac{1}{2} \langle 00 + 11 | 00 - 11 \rangle = \frac{1}{2} \Big(\underbrace{\langle 00 | 00 \rangle}_{=1} - \underbrace{\langle 00 | 11 \rangle}_{=0} + \underbrace{\langle 11 | 00 \rangle}_{=0} - \underbrace{\langle 11 | 11 \rangle}_{=1} \Big) \\
&= 0.
\end{aligned}
$$

Analogously, one shows

$$
\begin{aligned}
\langle \Phi^- | \Phi^- \rangle &= 1 = \langle \Psi^\pm | \Psi^\pm \rangle \\
\langle \Psi^+ | \Psi^- \rangle &= 0 = \langle \Phi^\pm | \Psi^\pm \rangle = \langle \Phi^\mp | \Psi^\pm \rangle.
\end{aligned}
$$

Solution 3.40 For $i \in \{1,2\}$ let $|\varphi_i\rangle \in \mathbb{H}^A$ and $|\psi_i\rangle \in \mathbb{H}^B$. Then we have

$$\langle (M^A \otimes M^B)^* \varphi_1 \otimes \psi_1 | \varphi_2 \otimes \psi_2 \rangle \underbrace{=}_{(2.30)} \langle \varphi_1 \otimes \psi_1 | (M^A \otimes M^B) \varphi_2 \otimes \psi_2 \rangle$$

$$= \langle \varphi_1 \otimes \psi_1 | M^A \varphi_2 \otimes M^B \psi_2 \rangle$$

$$\underbrace{=}_{(3.4)} \langle \varphi_1 | M^A \varphi_2 \rangle \langle \psi_1 | M^B \psi_2 \rangle$$

$$\underbrace{=}_{(2.30)} \langle (M^A)^* \varphi_1 | \varphi_2 \rangle \langle (M^B)^* \psi_1 | \psi_2 \rangle$$

$$\underbrace{=}_{(3.4)} \langle (M^A)^* \varphi_1 \otimes (M^B)^* \psi_1 | \varphi_2 \otimes \psi_2 \rangle$$

$$= \langle ((M^A)^* \otimes (M^B)^*) \varphi_1 \otimes \psi_1 | \varphi_2 \otimes \psi_2 \rangle .$$

Solution 3.41 Let $\{|\widetilde{e}_a\rangle\} \subset \mathbb{H}^A$ and $\{|\widetilde{f}_b\rangle\} \subset \mathbb{H}^B$ be two other ONBs. From Exercise 2.15 we know that then there exist unitary operators $U^A \in \mathcal{U}(\mathbb{H}^A)$ and $U^B \in \mathcal{U}(\mathbb{H}^B)$ such that

$$\begin{aligned} |\widetilde{e}_a\rangle &= U^A |e_a\rangle = \sum_{a'} U^A_{a'a} |e_{a'}\rangle \\ |\widetilde{f}_b\rangle &= U^B |f_b\rangle = \sum_{b'} U^B_{b'b} |f_{b'}\rangle . \end{aligned} \tag{G.78}$$

Let $\widetilde{M}_{a_1 b_1, a_2 b_2}$ be the matrix of M in the ONB $\{|\widetilde{e}_a \otimes \widetilde{f}_b\rangle\}$. Then

$$\widetilde{M}_{a_1 b_1, a_2 b_2} \underbrace{=}_{(2.22)} \langle \widetilde{e}_{a_1} \otimes \widetilde{f}_{b_1} | M(\widetilde{e}_{a_2} \otimes \widetilde{f}_{b_2}) \rangle$$

$$\underbrace{=}_{(G.78)} \sum_{a_1' b_1' a_2' b_2'} \langle U^A_{a_1' a_1} |e_{a_1'}\rangle \otimes U^B_{b_1' b_1} |f_{b_1'}\rangle | M(U^A_{a_2' a_2} |e_{a_2'}\rangle \otimes U^B_{b_2' b_2} |f_{b_2'}\rangle)\rangle$$

$$\underbrace{=}_{(2.22)} \sum_{a_1' b_1' a_2' b_2'} \overline{U^A_{a_1' a_1} U^B_{b_1' b_1}} U^A_{a_2' a_2} U^B_{b_2' b_2} \langle e_{a_1'} \otimes f_{b_1'} | M(e_{a_2'} \otimes f_{b_2'}) \rangle$$

$$= \sum_{a_1' b_1' a_2' b_2'} \overline{U^A_{a_1' a_1} U^B_{b_1' b_1}} U^A_{a_2' a_2} U^B_{b_2' b_2} M_{a_1' b_1', a_2' b_2'} ,$$

where we can use that (2.35) implies

$$\overline{U^A_{a_1' a_1} U^B_{b_1' b_1}} = (U^A)^*_{a_1 a_1'} (U^B)^*_{b_1 b_1'}$$

and we obtain

$$\sum_{a_1 a_2 b} \tilde{M}_{a_1 b, a_2 b} |\tilde{e}_{a_1}\rangle\langle\tilde{e}_{a_2}|$$

$$= \sum_{a_1 a_2 b a_1' b_1' a_2' b_2'} \left(U^A\right)^*_{a_1 a_1'} \left(U^B\right)^*_{b b_1'} U^A_{a_2' a_2} U^B_{b_2' b} M_{a_1' b_1', a_2' b_2'} |\tilde{e}_{a_1}\rangle\langle\tilde{e}_{a_2}| \tag{G.79}$$

$$= \sum_{a_1' b_1' a_2' b_2'} \left(\sum_b U^B_{b_2' b} \left(U^B\right)^*_{b b_1'}\right) M_{a_1' b_1', a_2' b_2'} \left(\sum_{a_1} \left(U^A\right)^*_{a_1 a_1'} |\tilde{e}_{a_1}\rangle\right) \left(\sum_{a_2} U^A_{a_2' a_2} \langle\tilde{e}_{a_2}|\right).$$

Here we can use that U^A and U^B are unitary, hence $U^B(U^B)^* = \mathbf{1}^B$ and thus

$$\sum_b U^B_{b_2' b} \left(U^B\right)^*_{b b_1'} = \delta_{b_2 b_1} \tag{G.80}$$

as well as $U^A(U^A)^* = \mathbf{1}^A$ and thus

$$\sum_{a_1} \left(U^A\right)^*_{a_1 a_1'} |\tilde{e}_{a_1}\rangle \underbrace{=}_{(G.78)} \sum_{a_1 a'} \left(U^A\right)^*_{a_1 a_1'} U^A_{a' a_1} |e_{a'}\rangle = \sum_{a'} \underbrace{\left(\sum_{a_1} U^A_{a' a_1} \left(U^A\right)^*_{a_1 a_1'}\right)}_{=\delta_{a' a_1'}} |e_{a'}\rangle$$

$$= |e_{a_1'}\rangle.$$

Likewise, we have

$$\sum_{a_2} U^A_{a_2' a_2} \langle\tilde{e}_{a_2}| = \langle e_{a_2'}|. \tag{G.81}$$

Inserting (G.80)–(G.81) into (G.79) yields

$$\sum_{a_1 a_2 b} \tilde{M}_{a_1 b, a_2 b} |\tilde{e}_{a_1}\rangle\langle\tilde{e}_{a_2}| = \sum_{a_1' b_1' a_2' b_2'} M_{a_1' b_1' a_2' b_2'} \delta_{b_1' b_2'} |e_{a_1'}\rangle\langle e_{a_2'}| = \sum_{a_1 a_2} M_{a_1 b, a_2 b} |e_{a_1}\rangle\langle e_{a_2}|,$$

which shows that in (3.46) the right side in the equation for $\mathrm{tr}^B(M)$ does not depend on the choice of the the ONBs $\{|e_a\rangle\}$ and $\{|f_b\rangle\}$.

Solution 3.42 We have

$$\mathrm{tr}\left(\mathrm{tr}^B(M)\right) = \mathrm{tr}\left(\mathbf{1}^A \, \mathrm{tr}^B(M)\right) \underbrace{=}_{(3.48)} \mathrm{tr}\left((\mathbf{1}^A \otimes \mathbf{1}^B)M\right) = \mathrm{tr}(M),$$

and the proof of the second identity is similar.

Solution 3.43 Let $\{|e_a\rangle\}$ be an ONB in $\mathbb{H}^A$ and $\{|f_b\rangle\}$ be an ONB in $\mathbb{H}^B$. Then $\{|e_a \otimes f_b\rangle\}$ is an ONB in $\mathbb{H}^A \otimes \mathbb{H}^B$ and the matrix of $M^A \otimes M^B$ in this basis is given by the right side of (3.35). Consequently

$$\mathrm{tr}\left(M^A \otimes M^B\right) \underset{(2.57)}{=} \sum_{a,b}\left(M^A \otimes M^B\right)_{ab,ab} \underset{(3.33)}{=} \left(\sum_a M^A_{aa}\right)\left(\sum_b M^B_{bb}\right)$$
$$\underset{(2.57)}{=} \mathrm{tr}\left(M^A\right)\mathrm{tr}\left(M^B\right).$$

Now

$$\mathrm{tr}^B\left(M^A \otimes M^B\right) = \sum_{a_1,a_2}\left(\mathrm{tr}^B\left(M^A \otimes M^B\right)\right)_{a_1 a_2}|e_{a_1}\rangle\langle e_{a_2}|, \qquad (\mathrm{G}.82)$$

where

$$\left(\mathrm{tr}^B\left(M^A \otimes M^B\right)\right)_{a_1 a_2} \underset{(3.52)}{=} \sum_b\left(M^A \otimes M^B\right)_{a_1 b, a_2 b} \underset{(3.33)}{=} M^A_{a_1 a_2}\sum_b M^B_{bb}$$
$$\underset{(2.57)}{=} M^A_{a_1 a_2}\mathrm{tr}\left(M^B\right),$$

and thus (G.82) becomes

$$\mathrm{tr}^B\left(M^A \otimes M^B\right) = M^A\,\mathrm{tr}\left(M^B\right).$$

The proof for $\mathrm{tr}^A\left(M^A \otimes M^B\right) = M^B\,\mathrm{tr}\left(M^A\right)$ is, of course, similar.

Solution 3.44 According to (3.44) one has in general for a $|\Psi\rangle \in \mathbb{H}^A \otimes \mathbb{H}^B$ that

$$\rho^A(\Psi) = \sum_{a_1,a_2,b}\overline{\Psi_{a_2 b}}\Psi_{a_1 b}|e_{a_1}\rangle\langle e_{a_2}|,$$

where the $\{|e_{a_j}\rangle\}$ are an ONB in $\mathbb{H}^A$. With $\mathbb{H}^A = {}^{\P}\mathbb{H} = \mathbb{H}^B, |e_0\rangle = |0\rangle^A, |e_1\rangle = |1\rangle^A$ as ONB in $\mathbb{H}^A$ this becomes

$$\rho^A(\Psi) = \left(\overline{\Psi_{00}}\Psi_{00} + \overline{\Psi_{01}}\Psi_{01}\right)|0\rangle^A\langle 0|$$
$$+ \left(\overline{\Psi_{00}}\Psi_{10} + \overline{\Psi_{01}}\Psi_{11}\right)|1\rangle^A\langle 0|$$
$$+ \left(\overline{\Psi_{10}}\Psi_{00} + \overline{\Psi_{11}}\Psi_{01}\right)|0\rangle^A\langle 1| \qquad (\mathrm{G}.83)$$
$$+ \left(\overline{\Psi_{10}}\Psi_{10} + \overline{\Psi_{11}}\Psi_{11}\right)|1\rangle^A\langle 1|.$$

For the BELL states

$$|\Phi^\pm\rangle = \frac{1}{\sqrt{2}}\left(|00\rangle \pm |11\rangle\right)$$

$$|\Psi^\pm\rangle = \frac{1}{\sqrt{2}}\left(|01\rangle \pm |10\rangle\right)$$

we find

$$\Phi_{00}^\pm = \pm\Phi_{11}^\pm = \Psi_{01}^\pm = \pm\Psi_{10}^\pm = \frac{1}{\sqrt{2}} \tag{G.84}$$

$$\Phi_{01}^\pm = \Phi_{10}^\pm = \Psi_{00}^\pm = \Psi_{11}^\pm = 0. \tag{G.85}$$

Inserting (G.84) and (G.85) into (G.83) results in

$$\rho^A(\Phi^\pm) = \rho^A(\Psi^\pm) = \frac{1}{2}\left(|0\rangle^A\langle 0| + |1\rangle^A\langle 1|\right) = \frac{1}{2}\mathbf{1}^A.$$

Similarly, one finds using (3.45), that

$$\rho^B(\Phi^\pm) = \rho^B(\Psi^\pm) = \frac{1}{2}\left(|0\rangle^B\langle 0| + |1\rangle^B\langle 1|\right) = \frac{1}{2}\mathbf{1}^B.$$

Solution 3.45 Any basis $\{\mathbf{v}_j \mid j \in \{1,\dots \dim \mathbb{V}\}\}$ of $\mathbb{V}$ can be used to form a basis $\{\mathbf{v}_{j_1} \otimes \cdots \otimes \mathbf{v}_{j_n}\}$ of $\mathbb{V}^{\otimes n}$ such that any vector $\mathbf{w} \in \mathbb{V}^{\otimes n}$ can be written in the form

$$\mathbf{w} = \sum_{j_1\cdots j_n} w_{j_1\dots j_n}\mathbf{v}_{j_1} \otimes \cdots \otimes \mathbf{v}_{j_n}.$$

This allows us for any set of $A_1,\dots,A_n \in L(\mathbb{V})$ to define the action of $A_1 \otimes \cdots \otimes A_n$ on any $\mathbf{w} \in \mathbb{V}^{\otimes n}$ by

$$\left(A_1 \otimes \cdots \otimes A_n\right)\mathbf{w} = \sum_{j_1\cdots j_n} w_{j_1\dots j_n}\left(A_1\mathbf{v}_{j_1}\right) \otimes \cdots \otimes \left(A_n\mathbf{v}_{j_n}\right).$$

such that

$$A_1 \otimes \cdots \otimes A_n \in L\left(\mathbb{V}^{\otimes n}\right)$$

and it follows that

$$L(\mathbb{V})^{\otimes n} \subset L\left(\mathbb{V}^{\otimes n}\right). \tag{G.86}$$

Using that

$$\dim L(\mathbb{V}) = \left(\dim \mathbb{V}\right)^2 \tag{G.87}$$

and

$$\dim \mathbb{V}^{\otimes n} = \left(\dim \mathbb{V}\right)^n, \tag{G.88}$$

we find

$$\dim L(\mathbb{V})^{\otimes n} \underbrace{=}_{\text{(G.88)}} \left(\dim L(\mathbb{V})\right)^n \underbrace{=}_{\text{(G.87)}} \left(\dim \mathbb{V}\right)^{2n} \tag{G.89}$$

and

$$\dim L(\mathbb{V}^{\otimes n}) \underbrace{=}_{\text{(G.87)}} \left(\dim \mathbb{V}^{\otimes n}\right)^2 \underbrace{=}_{\text{(G.88)}} \left(\dim \mathbb{V}\right)^{2n} \tag{G.90}$$

and (G.86) together with (G.89) and (G.90) imply that $L(\mathbb{V})^{\otimes n} = L(\mathbb{V}^{\otimes n})$.

Solution 3.46 First note that for any $|\psi\rangle \in \mathbb{H}^A \smallsetminus \{0\}$ we have

$$(|\psi\rangle\langle\psi| \otimes \mathbf{1}^B)^2 = |\psi\rangle\langle\psi|\psi\rangle\langle\psi| \otimes \mathbf{1}^B \underbrace{=}_{\text{(2.5)}} ||\psi||^2 (|\psi\rangle\langle\psi| \otimes \mathbf{1}^B). \tag{G.91}$$

Using an ONB $\{|e_a\rangle\} \subset \mathbb{H}^A$, we find then

$$\langle\psi| \sum_{b_1,b_2} K^*_{(b_1,b_2)} K_{(b_1,b_2)} \psi\rangle$$

$$\underbrace{=}_{\text{(2.43)}} \sum_{a_1,a_2} \langle\psi|e_{a_1}\rangle\langle e_{a_1}| \sum_{b_1,b_2} K^*_{(b_1,b_2)} K_{(b_1,b_2)} e_{a_2}\rangle\langle e_{a_2}|\psi\rangle$$

$$= \sum_{a_1,a_2} \langle e_{a_2}|\psi\rangle\langle\psi|e_{a_1}\rangle\langle e_{a_1}| \sum_{b_1,b_2} K^*_{(b_1,b_2)} K_{(b_1,b_2)} e_{a_2}\rangle$$

$$\underbrace{=}_{\text{(2.57)}} \mathrm{tr}\left(|\psi\rangle\langle\psi| \sum_{b_1,b_2} K^*_{(b_1,b_2)} K_{(b_1,b_2)}\right)$$

$$\underbrace{=}_{\text{(3.84)}} \mathrm{tr}\left(|\psi\rangle\langle\psi| \mathrm{tr}^B\left((\mathbf{1}^A \otimes \sqrt{\rho^B})V^*V(\mathbf{1}^A \otimes \sqrt{\rho^B})\right)\right)$$

$$\underbrace{=}_{\text{(3.47)}} \mathrm{tr}\left((|\psi\rangle\langle\psi| \otimes \mathbf{1}^B)(\mathbf{1}^A \otimes \sqrt{\rho^B})V^*V(\mathbf{1}^A \otimes \sqrt{\rho^B})\right)$$

$$\underbrace{=}_{\text{(G.91)}} \frac{1}{||\psi||^2} \mathrm{tr}\left((|\psi\rangle\langle\psi| \otimes \mathbf{1}^B)^2(\mathbf{1}^A \otimes \sqrt{\rho^B})V^*V(\mathbf{1}^A \otimes \sqrt{\rho^B})\right)$$

$$\underbrace{=}_{\text{(2.58)}} \frac{1}{||\psi||^2} \mathrm{tr}\left((|\psi\rangle\langle\psi| \otimes \mathbf{1}^B)(\mathbf{1}^A \otimes \sqrt{\rho^B})V^*V(\mathbf{1}^A \otimes \sqrt{\rho^B})(|\psi\rangle\langle\psi| \otimes \mathbf{1}^B)\right)$$

$$= \frac{1}{||\psi||^2} \mathrm{tr}\left((|\psi\rangle\langle\psi| \otimes \sqrt{\rho^B})V^*V(|\psi\rangle\langle\psi| \otimes \sqrt{\rho^B})\right).$$

Solution 3.47 Using an ONB $\{|e_a \otimes f_b\rangle\} \subset \mathbb{H}^A \otimes \mathbb{H}^B$, we obtain

$$\mathrm{tr}\left((|\psi\rangle\langle\psi| \otimes \sqrt{\rho^B})V^*V(|\psi\rangle\langle\psi| \otimes \sqrt{\rho^B})\right)$$

$$\underset{(2.57)}{=} \sum_{a,b} \langle e_a \otimes f_b|(|\psi\rangle\langle\psi| \otimes \sqrt{\rho^B})V^*V(|\psi\rangle\langle\psi| \otimes \sqrt{\rho^B})e_a \otimes f_b\rangle$$

$$\underset{(2.30)}{=} \sum_{a,b} \langle(|\psi\rangle\langle\psi| \otimes \sqrt{\rho^B})^* e_a \otimes f_b|V^*V(|\psi\rangle\langle\psi| \otimes \sqrt{\rho^B})e_a \otimes f_b\rangle$$

$$\underset{(3.31)}{=} \sum_{a,b} \langle(|\psi\rangle\langle\psi|^* \otimes \sqrt{\rho^B}^*)e_a \otimes f_b|V^*V(|\psi\rangle\langle\psi| \otimes \sqrt{\rho^B})e_a \otimes f_b\rangle$$

$$\underset{(3.81),(2.36)}{=} \sum_{a,b} \langle(|\psi\rangle\langle\psi| \otimes \sqrt{\rho^B})e_a \otimes f_b|V^*V(|\psi\rangle\langle\psi| \otimes \sqrt{\rho^B})e_a \otimes f_b\rangle$$

$$= \sum_{a,b} \langle|\psi\rangle\langle\psi|e_a\rangle \otimes \sqrt{\rho^B}|f_b\rangle|V^*V(|\psi\rangle\langle\psi|e_a\rangle \otimes \sqrt{\rho^B}|f_b\rangle)\rangle$$

$$\underset{(2.4),(2.6)}{=} \sum_{a,b} |\langle e_a|\psi\rangle|^2 \langle|\psi\rangle \otimes \sqrt{\rho^B}|f_b\rangle|V^*V(|\psi\rangle \otimes \sqrt{\rho^B}|f_b\rangle)\rangle$$

$$\underset{(3.73)}{\leq} \kappa \sum_{a,b} |\langle e_a|\psi\rangle|^2 \langle|\psi\rangle \otimes \sqrt{\rho^B}|f_b\rangle||\psi\rangle \otimes \sqrt{\rho^B}|f_b\rangle\rangle$$

$$\underset{(2.12),(3.4)}{=} \kappa \|\psi\|^2 \sum_b \langle\psi|\psi\rangle\langle\sqrt{\rho^B}|f_b\rangle|\sqrt{\rho^B}|f_b\rangle\rangle$$

$$\underset{(2.5)}{=} \kappa \|\psi\|^4 \sum_b \left\|\sqrt{\rho^B}f_b\right\|^2 \underset{(3.79)}{=} \kappa \|\psi\|^4 \sum_b \|\sqrt{q_b}f_b\|^2$$

$$\underset{(2.7)}{=} \kappa \|\psi\|^4 \sum_b q_b \underbrace{\|f_b\|^2}_{=1} = \kappa \|\psi\|^4 \underbrace{\sum_b q_b}_{(3.77)} = \kappa \|\psi\|^4 .$$

Solution 3.48 Combining the ONB $\{|f_b\rangle\}$ of $\mathbb{H}^B$ used in (3.87) to define $\check{V}$ with an ONB $\{|e_a\rangle\}$ of $\mathbb{H}^A$ to form an ONB $\{|e_a \otimes f_b\rangle\}$ of $\mathbb{H}^A \otimes \mathbb{H}^B$ we find

$$\langle\check{V}^*(e_a \otimes f_b|\psi \otimes f_1\rangle \underset{(2.30)}{=} \langle e_a \otimes f_b|\check{V}(\psi \otimes f_1)\rangle \underset{(3.87)}{=} \langle e_a \otimes f_b|\sum_l K_l|\psi\rangle \otimes |f_l\rangle\rangle$$

$$\underset{(3.4)}{=} \sum_l \langle e_a|K_l\psi\rangle \underbrace{\langle f_b|f_l\rangle}_{=\delta_{bl}} = \langle e_a|K_b\psi\rangle . \tag{G.92}$$

With this we obtain for every $|\psi \otimes f_1\rangle \in \iota\{\mathbb{H}^A\}$

$$\langle \psi \otimes f_1 | \check{V}^*(e_a \otimes f_b) \rangle \underbrace{=}_{(2.1)} \overline{\langle \check{V}^*(e_a \otimes f_b) | \psi \otimes f_1 \rangle} \underbrace{=}_{(G.92)} \overline{\langle e_a | K_b \psi \rangle}$$

$$\underbrace{=}_{(2.1)} \langle K_b \psi | e_a \rangle \underbrace{=}_{(2.30),(2.31)} \langle \psi | K_b^* e_a \rangle$$

$$= \sum_l \langle \psi | K_l^* e_a \rangle \underbrace{\langle f_l | f_b \rangle}_{=\delta_{lb}}$$

$$\underbrace{=}_{(3.4)} \left(\sum_l \langle \psi | K_l^* \otimes \langle f_l | \right) \left(|e_a \otimes f_b \rangle \right)$$

proving the claim (3.88).

Solution 3.49 Let $\mathbf{a}_1, \ldots, \mathbf{a}_m, \mathbf{b}_1, \ldots, \mathbf{b}_{n-m} \in \mathbb{C}^n$ be such that

$$A = \begin{pmatrix} \mathbf{a}_1 & \cdots & \mathbf{a}_m \end{pmatrix}, \qquad B = \begin{pmatrix} \mathbf{b}_1 & \cdots & \mathbf{b}_{n-m} \end{pmatrix},$$

where the $\mathbf{a}_j$ are given and the $\mathbf{b}_j$ are yet to be determined. Moreover, set

$$V = \begin{pmatrix} A & B \end{pmatrix}.$$

Then we have

$$V^*V = \begin{pmatrix} \overline{\mathbf{a}}_1 \cdot \mathbf{a}_1 & \cdots & \overline{\mathbf{a}}_1 \cdot \mathbf{a}_m & \overline{\mathbf{a}}_1 \cdot \mathbf{b}_1 & \cdots & \overline{\mathbf{a}}_1 \cdot \mathbf{b}_{n-m} \\ \vdots & & \vdots & \vdots & & \vdots \\ \overline{\mathbf{a}}_m \cdot \mathbf{a}_1 & \cdots & \overline{\mathbf{a}}_m \cdot \mathbf{a}_m & \overline{\mathbf{a}}_m \cdot \mathbf{b}_1 & \cdots & \overline{\mathbf{a}}_m \cdot \mathbf{b}_{n-m} \\ \overline{\mathbf{b}}_1 \cdot \mathbf{a}_1 & \cdots & \overline{\mathbf{b}}_1 \cdot \mathbf{a}_m & \overline{\mathbf{b}}_1 \cdot \mathbf{b}_1 & \cdots & \overline{\mathbf{b}}_1 \cdot \mathbf{b}_{n-m} \\ \vdots & & \vdots & \vdots & & \vdots \\ \overline{\mathbf{b}}_{n-m} \cdot \mathbf{a}_1 & \cdots & \overline{\mathbf{b}}_{n-m} \cdot \mathbf{a}_m & \overline{\mathbf{b}}_{n-m} \cdot \mathbf{b}_1 & \cdots & \overline{\mathbf{b}}_{n-m} \cdot \mathbf{b}_{n-m} \end{pmatrix},$$

where $\overline{\mathbf{u}} \cdot \mathbf{v} = \sum_{j=1}^n \overline{u}_j v_j$ denotes the scalar product in $\mathbb{C}^n$. For V^*V to have form given in (3.90) the $\mathbf{b}_j$ have to satisfy

$$\overline{\mathbf{a}}_l \cdot \mathbf{b}_k = 0 \qquad \forall l \in \{1, \ldots, m\}; \; k \in \{1, \ldots, n-m\} \tag{G.93}$$

$$\overline{\mathbf{b}}_j \cdot \mathbf{b}_k = c \delta_{jk} \qquad \forall j, k \in \{1, \ldots, n-m\}. \tag{G.94}$$

Each of the $n - m$ vectors $\mathbf{b}_j$ has n components giving us altogether $n(n - m)$ unknowns. Equation (G.93) gives m equations whereas due to the symmetry (G.94) provides $\frac{(n-m)(n-m+1)}{2}$ equations. As long as

$$\frac{(n - m)(n - m + 1)}{2} + m \leq n(n - m) \tag{G.95}$$

we can find the $\mathbf{b}_j$ and thus the matrix $B \in \mathrm{Mat}(n \times (n - m), \mathbb{C})$ delivering the required form (3.90) for V^*V. Rearranging terms shows that (G.95) is equivalent to $m(m + 1) \leq n(n - 1)$. Since by assumption $n > m$ it follows that $n - 1 \geq m$ and $n \geq m + 1$ which guarantees $m(m + 1) \leq n(n - 1)$ and thus (G.95).

Solution 3.50 Since an orthogonal projection P^B satisfies

$$\mathbf{1}^A \otimes P^B \underbrace{=}_{\text{Def. 2.11}} \mathbf{1}^A \otimes (P^B)^2 = (\mathbf{1}^A \otimes P^B)^2 \tag{G.96}$$

and

$$\mathbf{1}^A \otimes P^B \underbrace{=}_{\text{Def. 2.11}} \mathbf{1}^A \otimes (P^B)^* \underbrace{=}_{(3.31)} (\mathbf{1}^A \otimes P^B)^* \tag{G.97}$$

we have

$$\mathrm{tr}\left((\mathbf{1}^A \otimes P^B)U(\rho^A \otimes \rho^B)U^*\right) \underbrace{=}_{(\text{G.96})} \mathrm{tr}\left((\mathbf{1}^A \otimes P^B)^2 U(\rho^A \otimes \rho^B)U^*\right)$$

$$\underbrace{=}_{(2.58)} \mathrm{tr}\left((\mathbf{1}^A \otimes P^B)U(\rho^A \otimes \rho^B)U^*(\mathbf{1}^A \otimes P^B)\right)$$

$$\underbrace{=}_{(\text{G.97})} \mathrm{tr}\left((\mathbf{1}^A \otimes P^B)U(\rho^A \otimes \rho^B)U^*(\mathbf{1}^A \otimes P^B)^*\right)$$

$$\underbrace{=}_{(2.47)} \mathrm{tr}\left((\mathbf{1}^A \otimes P^B)U(\rho^A \otimes \rho^B)\left((\mathbf{1}^A \otimes P^B)U\right)^*\right)$$

$$\underbrace{=}_{(3.101)} \mathrm{tr}\left(V(\rho^A \otimes \rho^B)V^*\right)$$

$$\underbrace{=}_{(3.49)} \mathrm{tr}\left(\mathrm{tr}^B\left(V(\rho^A \otimes \rho^B)V^*\right)\right)$$

$$\underbrace{=}_{(3.100)} \mathrm{tr}\left(K(\rho^A)\right).$$

Solution 3.51 For any $\mathbf{x}_1, \mathbf{x}_2 \in B^1_{\mathbb{R}^3}$ and $\mu \in [0, 1]$ we have

$$
\rho_{\mu\mathbf{x}_1+(1-\mu)\mathbf{x}_2} \underset{(3.107)}{=} \frac{1}{2}\Big(\mathbf{1}+\big(\mu\mathbf{x}_1+(1-\mu)\mathbf{x}_2\big)\cdot\sigma\Big)
$$

$$
= \mu\frac{1}{2}\Big(\mathbf{1}+\mathbf{x}_1\cdot\sigma\Big)+(1-\mu)\frac{1}{2}\Big(\mathbf{1}+\mathbf{x}_2\cdot\sigma\Big)
$$

$$
\underset{(3.107)}{=} \mu\rho_{\mathbf{x}_1}+(1-\mu)\rho_{\mathbf{x}_2} \tag{G.98}
$$

and thus

$$
\widehat{K}\big(\mu\mathbf{x}_1+(1-\mu)\mathbf{x}_2\big) \underset{(3.108)}{=} \mathrm{tr}\big(K(\rho_{\mu\mathbf{x}_1+(1-\mu)\mathbf{x}_2})\sigma\big) \underset{(G.98)}{=} \mathrm{tr}\big(K(\mu\rho_{\mathbf{x}_1}+(1-\mu)\rho_{\mathbf{x}_2})\sigma\big)
$$

$$
\underset{(3.106)}{=} \mathrm{tr}\big(\mu K(\rho_{\mathbf{x}_1})\sigma+(1-\mu)K(\rho_{\mathbf{x}_2})\sigma\big)
$$

$$
= \mu\,\mathrm{tr}\big(K(\rho_{\mathbf{x}_1})\sigma\big)+(1-\mu)\,\mathrm{tr}\big(K(\rho_{\mathbf{x}_2})\sigma\big)
$$

$$
\underset{(3.108)}{=} \mu\widehat{K}(\mathbf{x}_1)+(1-\mu)\widehat{K}(\mathbf{x}_2).
$$

Solutions to Exercises from Chapter 4

Solution 4.52 According to the starting assumption, ρ^A and ρ^B are given as in Definition 4.1, that is, each of them is self-adjoint, positive and has trace 1. From (3.32) it follows that then $\rho^A \otimes \rho^B$ is self-adjoint. In order to show the positivity of the $\rho^A \otimes \rho^B$ note at first that

$$
\rho^A \otimes \rho^B = \big(\rho^A \otimes \mathbf{1}\big)\big(\mathbf{1} \otimes \rho^B\big) = \big(\mathbf{1} \otimes \rho^B\big)\big(\rho^A \otimes \mathbf{1}\big), \tag{G.99}
$$

where also

$$
\big(\rho^A \otimes \mathbf{1}\big)^* = \rho^A \otimes \mathbf{1}
$$

$$
\big(\mathbf{1} \otimes \rho^B\big)^* = \mathbf{1} \otimes \rho^B.
$$

Both $\rho^A \otimes \mathbf{1}$ as well as $\mathbf{1} \otimes \rho^B$ are positive because for an arbitrary vector

$$|\Psi\rangle = \sum_{a,b} \Psi_{ab}|e_a\rangle \otimes |f_b\rangle \in \mathbb{H}^A \otimes \mathbb{H}^B$$

we find that

$$
\langle\Psi|\left(\rho^A \otimes \mathbf{1}\right)\Psi\rangle \underset{(3.8)}{=} \sum_{a_1 a_2, b_1 b_2} \overline{\Psi_{a_1 b_1}}\Psi_{a_2 b_2}\langle e_{a_1} \otimes f_{b_1}|\left(\rho^A \otimes \mathbf{1}\right)e_{a_2} \otimes f_{b_2}\rangle
$$

$$
\underset{(3.29)}{=} \sum_{a_1 a_2, b_1 b_2} \overline{\Psi_{a_1 b_1}}\Psi_{a_2 b_2}\langle e_{a_1}|\rho^A e_{a_2}\rangle \underbrace{\langle f_{b_1}|f_{b_2}\rangle}_{=\delta_{b_1 b_2}}
$$

$$
= \sum_{a_1 a_2, b} \overline{\Psi_{a_1 b}}\Psi_{a_2 b}\langle e_{a_1}|\rho^A e_{a_2}\rangle
$$

$$
= \sum_{b} \langle \underbrace{\sum_{a_1}\Psi_{a_1 b}e_{a_1}}_{=:\psi_b}|\rho^A \sum_{a_2}\Psi_{a_2 b}e_{a_2}\rangle
$$

$$
= \sum_{b} \underbrace{\langle\psi_b|\rho^A\psi_b\rangle}_{\geq 0}
$$

$$
\geq \ 0,
$$

where the positivity of ρ^A was used in the penultimate line. Similarly, one shows that $\mathbf{1}\otimes\rho^B$ is positive. Since the $\rho^A \otimes \mathbf{1}$ as well as the $\mathbf{1}\otimes\rho^B$ are self-adjoint and positive and according to (G.99) commute, it follows that for every pair $\rho^A \otimes \mathbf{1}, \mathbf{1}\otimes\rho^B$ there exists an ONB $|e_a \otimes f_b\rangle$ in which both are diagonal

$$
\rho^A \otimes \mathbf{1} = \sum_{a,b} \lambda^A_{a,b}|e_a \otimes f_b\rangle\langle e_a \otimes f_b|
$$

$$
\mathbf{1}\otimes\rho^B = \sum_{a,b} \lambda^B_{a,b}|e_a \otimes f_b\rangle\langle e_a \otimes f_b|,
$$

where due to the positivity of the $\rho^A \otimes \mathbf{1}, \mathbf{1}\otimes\rho^B$ we also have

$$
\lambda^X_{a,b} \geq 0 \quad \text{for } X \in \{A, B\}. \tag{G.100}
$$

With (G.99) one obtains thus

$$
\rho^A \otimes \rho^B = \sum_{a,b} \lambda^A_{a,b}\lambda^B_{a,b}|e_a \otimes f_b\rangle\langle e_a \otimes f_b|
$$

and because of (G.100) it follows that $\rho^A \otimes \rho^B$ is positive.

Finally, the trace property for $\rho^A \otimes \rho^B$ follows from

$$
\text{tr}\left(\rho^A \otimes \rho^B\right) \underset{(3.57)}{=} \underbrace{\text{tr}\left(\rho^A\right)}_{=1}\underbrace{\text{tr}\left(\rho^A\right)}_{=1} = 1.
$$

Solution 4.53 With the results of Exercise 2.20 as given in (G.66) we have

$$|\uparrow_{\hat{x}}\rangle = \frac{1}{\sqrt{2}}(|0\rangle + |1\rangle) \qquad \text{and} \qquad |\downarrow_{\hat{x}}\rangle = \frac{1}{\sqrt{2}}(|0\rangle - |1\rangle)$$

such that

$$
\begin{aligned}
|\uparrow_{\hat{x}}\rangle \otimes |\uparrow_{\hat{x}}\rangle + |\downarrow_{\hat{x}}\rangle \otimes |\downarrow_{\hat{x}}\rangle &= \frac{|0\rangle + |1\rangle}{\sqrt{2}} \otimes \frac{|0\rangle + |1\rangle}{\sqrt{2}} + \frac{|0\rangle - |1\rangle}{\sqrt{2}} \otimes \frac{|0\rangle - |1\rangle}{\sqrt{2}} \\
&= \frac{1}{2}(|00\rangle + |01\rangle + |10\rangle + |11\rangle) \\
&\quad + \frac{1}{2}(|00\rangle - |01\rangle - |10\rangle + |11\rangle) \\
&= |00\rangle + |11\rangle.
\end{aligned}
$$

Solution 4.54 Per definition (2.125) and (2.126) one has

$$
|\uparrow_{\hat{n}}\rangle = \begin{pmatrix} e^{-i\frac{\phi}{2}}\cos\frac{\theta}{2} \\ e^{i\frac{\phi}{2}}\sin\frac{\theta}{2} \end{pmatrix} = e^{-i\frac{\phi}{2}}\cos\frac{\theta}{2}|0\rangle + e^{i\frac{\phi}{2}}\sin\frac{\theta}{2}|1\rangle
$$

$$
|\downarrow_{\hat{n}}\rangle = \begin{pmatrix} -e^{-i\frac{\phi}{2}}\sin\frac{\theta}{2} \\ e^{i\frac{\phi}{2}}\cos\frac{\theta}{2} \end{pmatrix} = -e^{-i\frac{\phi}{2}}\sin\frac{\theta}{2}|0\rangle + e^{i\frac{\phi}{2}}\cos\frac{\theta}{2}|1\rangle
$$

and thus

$$
\begin{aligned}
&|\uparrow_{\hat{n}}\rangle \otimes |\downarrow_{\hat{n}}\rangle - |\downarrow_{\hat{n}}\rangle \otimes |\uparrow_{\hat{n}}\rangle \\
&= \left(e^{-i\frac{\phi}{2}}\cos\frac{\theta}{2}|0\rangle + e^{i\frac{\phi}{2}}\sin\frac{\theta}{2}|1\rangle \right) \otimes \left(-e^{-i\frac{\phi}{2}}\sin\frac{\theta}{2}|0\rangle + e^{i\frac{\phi}{2}}\cos\frac{\theta}{2}|1\rangle \right) \\
&\quad - \left(-e^{-i\frac{\phi}{2}}\sin\frac{\theta}{2}|0\rangle + e^{i\frac{\phi}{2}}\cos\frac{\theta}{2}|1\rangle \right) \otimes \left(e^{-i\frac{\phi}{2}}\cos\frac{\theta}{2}|0\rangle + e^{i\frac{\phi}{2}}\sin\frac{\theta}{2}|1\rangle \right) \\
&= -e^{-i\phi}\cos\frac{\theta}{2}\sin\frac{\theta}{2}|00\rangle + \cos^2\frac{\theta}{2}|01\rangle - \sin^2\frac{\theta}{2}|10\rangle + e^{i\phi}\cos\frac{\theta}{2}\sin\frac{\theta}{2}|11\rangle \\
&\quad - \left(-e^{-i\phi}\cos\frac{\theta}{2}\sin\frac{\theta}{2}|00\rangle - \sin^2\frac{\theta}{2}|01\rangle + \cos^2\frac{\theta}{2}|10\rangle + e^{i\phi}\cos\frac{\theta}{2}\sin\frac{\theta}{2}|11\rangle \right) \\
&= |01\rangle - |10\rangle \underbrace{=}_{(3.28)} \sqrt{2}|\Psi^-\rangle.
\end{aligned}
$$

Solution 4.55 Since $\hat{\mathbf{n}}$ in the result (4.25) of Exercise 4.54 is arbitrary, we can choose $\hat{\mathbf{n}} = \hat{\mathbf{n}}^A$ and write $|\Psi^-\rangle$ as

$$|\Psi^-\rangle = \frac{1}{\sqrt{2}}\left(|\uparrow_{\hat{\mathbf{n}}^A}\rangle \otimes |\downarrow_{\hat{\mathbf{n}}^A}\rangle - |\downarrow_{\hat{\mathbf{n}}^A}\rangle \otimes |\uparrow_{\hat{\mathbf{n}}^A}\rangle\right). \tag{G.101}$$

with $\Sigma^A_{\hat{\mathbf{n}}^A} = \hat{\mathbf{n}}^A \cdot \boldsymbol{\sigma}$ and $\Sigma^B_{\hat{\mathbf{n}}^B} = \hat{\mathbf{n}}^B \cdot \boldsymbol{\sigma}$ it then follows that

$$\left\langle \Sigma^A_{\hat{\mathbf{n}}^A} \otimes \Sigma^B_{\hat{\mathbf{n}}^B}\right\rangle_{\Psi^-} = \langle\Psi^-|(\hat{\mathbf{n}}^A\cdot\boldsymbol{\sigma}\otimes\hat{\mathbf{n}}^B\cdot\boldsymbol{\sigma})\Psi^-\rangle \tag{G.102}$$

$$= \frac{1}{\sqrt{2}}\langle\Psi^-|\underbrace{\hat{\mathbf{n}}^A\cdot\boldsymbol{\sigma}|\uparrow_{\hat{\mathbf{n}}^A}\rangle}_{=+|\uparrow_{\hat{\mathbf{n}}^A}\rangle}\otimes\hat{\mathbf{n}}^B\cdot\boldsymbol{\sigma}|\downarrow_{\hat{\mathbf{n}}^A}\rangle - \underbrace{\hat{\mathbf{n}}^A\cdot\boldsymbol{\sigma}|\downarrow_{\hat{\mathbf{n}}^A}\rangle}_{=-|\downarrow_{\hat{\mathbf{n}}^A}\rangle}\otimes\hat{\mathbf{n}}^B\cdot\boldsymbol{\sigma}|\uparrow_{\hat{\mathbf{n}}^A}\rangle\rangle$$

$$= \frac{1}{\sqrt{2}}\langle\Psi^-|(|\uparrow_{\hat{\mathbf{n}}^A}\rangle\otimes\hat{\mathbf{n}}^B\cdot\boldsymbol{\sigma}|\downarrow_{\hat{\mathbf{n}}^A}\rangle + |\downarrow_{\hat{\mathbf{n}}^A}\rangle\otimes\hat{\mathbf{n}}^B\cdot\boldsymbol{\sigma}|\uparrow_{\hat{\mathbf{n}}^A}\rangle)\rangle.$$

In the last term we can use the following identity

$$\begin{aligned}\hat{\mathbf{n}}^B\cdot\boldsymbol{\sigma}|\downarrow_{\hat{\mathbf{n}}^A}\rangle &= \hat{\mathbf{n}}^B\cdot\boldsymbol{\sigma}\left(-\hat{\mathbf{n}}^A\cdot\boldsymbol{\sigma}\right)|\downarrow_{\hat{\mathbf{n}}^A}\rangle\\ &= -\left(\hat{\mathbf{n}}^B\cdot\boldsymbol{\sigma}\right)\left(\hat{\mathbf{n}}^A\cdot\boldsymbol{\sigma}\right)|\downarrow_{\hat{\mathbf{n}}^A}\rangle\\ &\underset{(2.121)}{=} -\left((\hat{\mathbf{n}}^B\cdot\hat{\mathbf{n}}^A)\mathbf{1}+\mathrm{i}(\hat{\mathbf{n}}^B\times\hat{\mathbf{n}}^A)\cdot\boldsymbol{\sigma}\right)|\downarrow_{\hat{\mathbf{n}}^A}\rangle. \tag{G.103}\end{aligned}$$

Analogously, one shows

$$\hat{\mathbf{n}}^B\cdot\boldsymbol{\sigma}|\uparrow_{\hat{\mathbf{n}}^A}\rangle = \left((\hat{\mathbf{n}}^B\cdot\hat{\mathbf{n}}^A)\mathbf{1}+\mathrm{i}(\hat{\mathbf{n}}^B\times\hat{\mathbf{n}}^A)\cdot\boldsymbol{\sigma}\right)|\uparrow_{\hat{\mathbf{n}}^A}\rangle. \tag{G.104}$$

Inserting (G.103) and (G.104) in (G.102) yields

$$\begin{aligned}\left\langle \Sigma^A_{\hat{\mathbf{n}}^A} \otimes \Sigma^B_{\hat{\mathbf{n}}^B}\right\rangle_{\Psi^-} &= \frac{-\hat{\mathbf{n}}^B\cdot\hat{\mathbf{n}}^A}{\sqrt{2}}\langle\Psi^-|(|\uparrow_{\hat{\mathbf{n}}^A}\rangle\otimes|\downarrow_{\hat{\mathbf{n}}^A}\rangle - |\downarrow_{\hat{\mathbf{n}}^A}\rangle\otimes|\uparrow_{\hat{\mathbf{n}}^A}\rangle)\rangle\\ &\quad -\frac{\mathrm{i}}{\sqrt{2}}\langle\Psi^-|(|\uparrow_{\hat{\mathbf{n}}^A}\rangle\otimes(\hat{\mathbf{n}}^B\times\hat{\mathbf{n}}^A)\cdot\boldsymbol{\sigma}|\downarrow_{\hat{\mathbf{n}}^A}\rangle)\rangle\\ &\quad +\frac{\mathrm{i}}{\sqrt{2}}\langle\Psi^-|(|\downarrow_{\hat{\mathbf{n}}^A}\rangle\otimes(\hat{\mathbf{n}}^B\times\hat{\mathbf{n}}^A)\cdot\boldsymbol{\sigma}|\uparrow_{\hat{\mathbf{n}}^A}\rangle)\rangle\\ &\underset{(G.101)}{=} -\hat{\mathbf{n}}^B\cdot\hat{\mathbf{n}}^A\underbrace{\langle\Psi^-|\Psi^-\rangle}_{=1}\\ &\quad -\frac{\mathrm{i}}{2}\langle\uparrow_{\hat{\mathbf{n}}^A}\otimes\downarrow_{\hat{\mathbf{n}}^A} - \downarrow_{\hat{\mathbf{n}}^A}\otimes\uparrow_{\hat{\mathbf{n}}^A}|\uparrow_{\hat{\mathbf{n}}^A}\otimes(\hat{\mathbf{n}}^B\times\hat{\mathbf{n}}^A)\cdot\boldsymbol{\sigma}|\downarrow_{\hat{\mathbf{n}}^A}\rangle\rangle\\ &\quad +\frac{\mathrm{i}}{2}\langle\uparrow_{\hat{\mathbf{n}}^A}\otimes\downarrow_{\hat{\mathbf{n}}^A} - \downarrow_{\hat{\mathbf{n}}^A}\otimes\uparrow_{\hat{\mathbf{n}}^A}|\downarrow_{\hat{\mathbf{n}}^A}\otimes(\hat{\mathbf{n}}^B\times\hat{\mathbf{n}}^A)\cdot\boldsymbol{\sigma}|\uparrow_{\hat{\mathbf{n}}^A}\rangle\rangle\end{aligned}$$

$$\underbrace{=}_{(3.4)} -\hat{\mathbf{n}}^B \cdot \hat{\mathbf{n}}^A \tag{G.105}$$

$$-\frac{i}{2}\left(\langle \downarrow_{\hat{\mathbf{n}}^A} |(\hat{\mathbf{n}}^B \times \hat{\mathbf{n}}^A)\cdot\boldsymbol{\sigma}| \downarrow_{\hat{\mathbf{n}}^A}\rangle + \langle \uparrow_{\hat{\mathbf{n}}^A} |(\hat{\mathbf{n}}^B \times \hat{\mathbf{n}}^A)\cdot\boldsymbol{\sigma}| \uparrow_{\hat{\mathbf{n}}^A}\rangle\right).$$

In order to show

$$\left\langle \Sigma^A_{\hat{\mathbf{n}}^A} \otimes \Sigma^B_{\hat{\mathbf{n}}^B} \right\rangle_{\Psi^-} = -\hat{\mathbf{n}}^B \cdot \hat{\mathbf{n}}^A , \tag{G.106}$$

we prove that in general for $\hat{\mathbf{m}}, \hat{\mathbf{n}} \in S^1_{\mathbb{R}^3}$

$$\langle \downarrow_{\hat{\mathbf{n}}} |\hat{\mathbf{m}}\cdot\boldsymbol{\sigma}\, \downarrow_{\hat{\mathbf{n}}}\rangle + \langle \uparrow_{\hat{\mathbf{n}}} |\hat{\mathbf{m}}\cdot\boldsymbol{\sigma}\, \uparrow_{\hat{\mathbf{n}}}\rangle = 0 \tag{G.107}$$

holds. With $\hat{\mathbf{m}} = \hat{\mathbf{n}}^B \times \hat{\mathbf{n}}^A$ and $\hat{\mathbf{n}} = \hat{\mathbf{n}}^A$ it then follows that the second term in (G.105) vanishes. To show (G.107), we consider first $\hat{\mathbf{m}}\cdot\boldsymbol{\sigma}| \uparrow_{\hat{\mathbf{n}}}\rangle$ in the ONB $\{| \uparrow_{\hat{\mathbf{n}}}\rangle, | \downarrow_{\hat{\mathbf{n}}}\rangle\}$:

$$\hat{\mathbf{m}}\cdot\boldsymbol{\sigma}| \uparrow_{\hat{\mathbf{n}}}\rangle = a_{\hat{\mathbf{m}}}| \uparrow_{\hat{\mathbf{n}}}\rangle + b_{\hat{\mathbf{m}}}| \downarrow_{\hat{\mathbf{n}}}\rangle . \tag{G.108}$$

If $b_{\hat{\mathbf{m}}} = 0$, it follows that

$$\hat{\mathbf{m}}\cdot\boldsymbol{\sigma}| \uparrow_{\hat{\mathbf{n}}}\rangle = a_{\hat{\mathbf{m}}}| \uparrow_{\hat{\mathbf{n}}}\rangle , \tag{G.109}$$

and $a_{\hat{\mathbf{m}}}$ is an eigenvalue of $\hat{\mathbf{m}}\cdot\boldsymbol{\sigma}$ with eigenvector $| \uparrow_{\hat{\mathbf{n}}}\rangle$. From (2.29) it follows immediately, that $(\hat{\mathbf{m}}\cdot\boldsymbol{\sigma})^2 = \mathbf{1}$, and thus the eigenvalues of $\hat{\mathbf{m}}\cdot\boldsymbol{\sigma}$ are given by ± 1. The eigenspace for the eigenvalue $-a_{\hat{\mathbf{m}}}$ is one-dimensional and orthogonal to the eigenvector $| \downarrow_{\hat{\mathbf{n}}}\rangle$ for the eigenvalue $a_{\hat{\mathbf{m}}}$. Hence,

$$\hat{\mathbf{m}}\cdot\boldsymbol{\sigma}| \downarrow_{\hat{\mathbf{n}}}\rangle = -a_{\hat{\mathbf{m}}}| \downarrow_{\hat{\mathbf{n}}}\rangle , \tag{G.110}$$

and (G.107) follows from (G.109) and (G.110).

In case $b_{\hat{\mathbf{m}}} \neq 0$, we obtain from (G.108) because of $\langle \downarrow_{\hat{\mathbf{n}}} | \uparrow_{\hat{\mathbf{n}}}\rangle = 0$ first

$$\langle \downarrow_{\hat{\mathbf{n}}} |\hat{\mathbf{m}}\cdot\boldsymbol{\sigma}\, \downarrow_{\hat{\mathbf{n}}}\rangle = a_{\hat{\mathbf{m}}} . \tag{G.111}$$

On the other hand, because of $(\hat{\mathbf{m}}\cdot\boldsymbol{\sigma})^2 = \mathbf{1}$ it follows from (G.108) also that

$$\begin{aligned}| \uparrow_{\hat{\mathbf{n}}}\rangle &= a_{\hat{\mathbf{m}}}\hat{\mathbf{m}}\cdot\boldsymbol{\sigma}| \uparrow_{\hat{\mathbf{n}}}\rangle + b_{\hat{\mathbf{m}}}\hat{\mathbf{m}}\cdot\boldsymbol{\sigma}| \downarrow_{\hat{\mathbf{n}}}\rangle \\ &= a_{\hat{\mathbf{m}}}\left(a_{\hat{\mathbf{m}}}| \uparrow_{\hat{\mathbf{n}}}\rangle + b_{\hat{\mathbf{m}}}| \downarrow_{\hat{\mathbf{n}}}\rangle\right) + b_{\hat{\mathbf{m}}}\hat{\mathbf{m}}\cdot\boldsymbol{\sigma}| \downarrow_{\hat{\mathbf{n}}}\rangle .\end{aligned}$$

Taking on both sides the scalar product with $\langle \downarrow_{\hat{\mathbf{n}}} |$ yields, because of $b_{\hat{\mathbf{m}}} \neq 0$ and $\langle \downarrow_{\hat{\mathbf{n}}} | \uparrow_{\hat{\mathbf{n}}}\rangle = 0$, thus

$$\langle \downarrow_{\hat{\mathbf{n}}} |\hat{\mathbf{m}}\cdot\boldsymbol{\sigma}\, \downarrow_{\hat{\mathbf{n}}}\rangle = -a_{\hat{\mathbf{m}}} . \tag{G.112}$$

From (G.111) and (G.112) follows (G.107) also in the case $b_{\hat{\mathbf{m}}} \neq 0$ and thus finally (G.106).

Alternatively, one can verify (G.106) also by an explicit calculation, making use of the representations $\hat{\mathbf{n}}, |\uparrow_{\hat{\mathbf{n}}^A}\rangle| \downarrow_{\hat{\mathbf{n}}^A}\rangle, \hat{\mathbf{n}}^A \cdot \sigma$ in (2.122)–(2.123) and (2.125)–(2.126). But that is equally lengthy.

Solution 4.56 Since $\hat{\mathbf{n}}$ in the result (4.25) of Exercise 4.54 is arbitrary, we can choose $\hat{\mathbf{n}} = \hat{\mathbf{n}}^A$ and represent $|\Psi^-\rangle$ as

$$|\Psi^-\rangle = \frac{1}{\sqrt{2}} \left(|\uparrow_{\hat{\mathbf{n}}^A}\rangle \otimes |\downarrow_{\hat{\mathbf{n}}^A}\rangle - |\downarrow_{\hat{\mathbf{n}}^A}\rangle \otimes |\uparrow_{\hat{\mathbf{n}}^A}\rangle \right) . \qquad (G.113)$$

With $\Sigma^A_{\hat{\mathbf{n}}^A} = \hat{\mathbf{n}}^A \cdot \sigma$ we thus have

$$
\begin{aligned}
\left\langle \Sigma^A_{\hat{\mathbf{n}}^A} \otimes \mathbf{1} \right\rangle_{\Psi^-} &= \langle \Psi^- | (\hat{\mathbf{n}}^A \cdot \sigma \otimes \mathbf{1}) \Psi^- \rangle \\
&\underset{(G.113)}{=} \frac{1}{\sqrt{2}} \langle \Psi^- | \underbrace{\hat{\mathbf{n}}^A \cdot \sigma | \uparrow_{\hat{\mathbf{n}}^A}\rangle}_{=+|\uparrow_{\hat{\mathbf{n}}^A}\rangle} \otimes |\downarrow_{\hat{\mathbf{n}}^A}\rangle - \underbrace{\hat{\mathbf{n}}^A \cdot \sigma | \downarrow_{\hat{\mathbf{n}}^A}\rangle}_{=-|\downarrow_{\hat{\mathbf{n}}^A}\rangle} \otimes |\uparrow_{\hat{\mathbf{n}}^A}\rangle \rangle \\
&= \frac{1}{\sqrt{2}} \langle \Psi^- | \uparrow_{\hat{\mathbf{n}}^A} \otimes \downarrow_{\hat{\mathbf{n}}^A} + \downarrow_{\hat{\mathbf{n}}^A} \otimes \uparrow_{\hat{\mathbf{n}}^A} \rangle \\
&\underset{(G.113)}{=} \frac{1}{2} \langle \uparrow_{\hat{\mathbf{n}}^A} \otimes \downarrow_{\hat{\mathbf{n}}^A} - \downarrow_{\hat{\mathbf{n}}^A} \otimes \uparrow_{\hat{\mathbf{n}}^A} | \uparrow_{\hat{\mathbf{n}}^A} \otimes \downarrow_{\hat{\mathbf{n}}^A} + \downarrow_{\hat{\mathbf{n}}^A} \otimes \uparrow_{\hat{\mathbf{n}}^A} \rangle \\
&= 0,
\end{aligned}
$$

where we used $\langle \uparrow_{\hat{\mathbf{n}}^A} | \uparrow_{\hat{\mathbf{n}}^A}\rangle = 1$ and $\langle \uparrow_{\hat{\mathbf{n}}^A} | \downarrow_{\hat{\mathbf{n}}^A}\rangle = 0$ in the last step.

Analogously, one shows that $\left\langle \mathbf{1} \otimes \Sigma^B_{\hat{\mathbf{n}}^B} \right\rangle_{\Psi^-} = 0$.

Solution 4.57 With

$$
\begin{aligned}
\Sigma_{\hat{\mathbf{n}}} &= \hat{\mathbf{n}} \cdot \sigma \\
\hat{\mathbf{m}} \cdot \sigma | \uparrow_{\hat{\mathbf{m}}}\rangle &\underset{(2.124)}{=} |\uparrow_{\hat{\mathbf{m}}}\rangle \qquad\qquad (G.114) \\
(\hat{\mathbf{m}} \cdot \sigma)^* &= \hat{\mathbf{m}} \cdot \sigma \qquad\qquad (G.115)
\end{aligned}
$$

one has

$$
\begin{aligned}
\langle \Sigma_{\hat{\mathbf{n}}} \rangle_{|\uparrow_{\hat{\mathbf{m}}}\rangle} &= \langle \uparrow_{\hat{\mathbf{m}}} | (\hat{\mathbf{n}} \cdot \sigma) \uparrow_{\hat{\mathbf{m}}} \rangle \\
&\underset{(G.114)}{=} \frac{1}{2} \Big[\langle (\hat{\mathbf{m}} \cdot \sigma) \uparrow_{\hat{\mathbf{m}}} | (\hat{\mathbf{n}} \cdot \sigma) \uparrow_{\hat{\mathbf{m}}} \rangle + \langle \uparrow_{\hat{\mathbf{m}}} | (\hat{\mathbf{n}} \cdot \sigma)(\hat{\mathbf{m}} \cdot \sigma) \uparrow_{\hat{\mathbf{m}}} \rangle \Big] \\
&\underset{(G.115)}{=} \frac{1}{2} \langle \uparrow_{\hat{\mathbf{m}}} | \big[(\hat{\mathbf{m}} \cdot \sigma)(\hat{\mathbf{n}} \cdot \sigma) + (\hat{\mathbf{n}} \cdot \sigma)(\hat{\mathbf{m}} \cdot \sigma) \big] \uparrow_{\hat{\mathbf{m}}} \rangle \\
&\underset{(2.121)}{=} \frac{1}{2} \langle \uparrow_{\hat{\mathbf{m}}} | \big[(\hat{\mathbf{m}} \cdot \hat{\mathbf{n}}) \mathbf{1} + \mathrm{i}((\hat{\mathbf{m}} \times \hat{\mathbf{n}}) \cdot \sigma) + (\hat{\mathbf{n}} \cdot \hat{\mathbf{m}}) \mathbf{1} + \mathrm{i}((\hat{\mathbf{n}} \times \hat{\mathbf{m}}) \cdot \sigma) \big] \uparrow_{\hat{\mathbf{m}}} \rangle \\
&= \hat{\mathbf{n}} \cdot \hat{\mathbf{m}} + \frac{\mathrm{i}}{2} \langle \uparrow_{\hat{\mathbf{m}}} | \big(\underbrace{\big[\hat{\mathbf{m}} \times \hat{\mathbf{n}} + \hat{\mathbf{n}} \times \hat{\mathbf{m}} \big]}_{=0} \cdot \sigma \big) \uparrow_{\hat{\mathbf{m}}} \rangle \\
&= \hat{\mathbf{n}} \cdot \hat{\mathbf{m}}.
\end{aligned}
$$

Solution 4.58 To begin with, we have

$$
P_{\lambda_a} |\Psi\rangle \underset{(4.49)}{=} \big(|e_a\rangle\langle e_a| \otimes \mathbf{1}^B \big) \sum_{a_1,b} \Psi_{a_1 b} |e_{a_1}\rangle \otimes |f_b\rangle = \sum_{a_1,b} \Psi_{a_1 b} |e_a\rangle \underbrace{\langle e_a | e_{a_1} \rangle}_{=\delta_{a a_1}} \otimes |f_b\rangle
$$

$$
= \sum_b \Psi_{ab} |e_a\rangle \otimes |f_b\rangle ,
$$

which implies

$$
\begin{aligned}
P_{\lambda_a} |\Psi\rangle\langle\Psi| P_{\lambda_a} &\underset{(3.8)}{=} \sum_{b_1,b_2} \Psi_{ab_1} \overline{\Psi_{ab_2}} |e_a\rangle \otimes |f_{b_1}\rangle \langle e_a| \otimes \langle f_{b_2}| \\
&\underset{(3.36)}{=} \sum_{b_1,b_2} \Psi_{ab_1} \overline{\Psi_{ab_2}} |e_a\rangle\langle e_a| \otimes |f_{b_1}\rangle\langle f_{b_2}| .
\end{aligned} \tag{G.116}
$$

Inserting (G.116) into (4.51) yields

$$
\rho = \sum_a \sum_{b_1,b_2} \Psi_{ab_1} \overline{\Psi_{ab_2}} |e_a\rangle\langle e_a| \otimes |f_{b_1}\rangle\langle f_{b_2}| \tag{G.117}
$$

for the density operator of the composite system. From Corollary 3.20 we know that then the sub-system in $\mathbb{H}^B$ is described by

$$\rho^B(\rho) \underset{(3.56)}{=} \mathrm{tr}^A(\rho) \underset{(G.117)}{=} \mathrm{tr}^A\left(\sum_a \sum_{b_1,b_2} \Psi_{ab_1}\overline{\Psi_{ab_2}}|e_a\rangle\langle e_a| \otimes |f_{b_1}\rangle\langle f_{b_2}|\right)$$

$$= \sum_a \sum_{b_1,b_2} \Psi_{ab_1}\overline{\Psi_{ab_2}}\,\mathrm{tr}^A\left(|e_a\rangle\langle e_a| \otimes |f_{b_1}\rangle\langle f_{b_2}|\right)$$

$$\underset{(3.57)}{=} \sum_a \sum_{b_1,b_2} \Psi_{ab_1}\overline{\Psi_{ab_2}}\underbrace{\mathrm{tr}\left(|e_a\rangle\langle e_a|\right)}_{=1}|f_{b_1}\rangle\langle f_{b_2}|$$

$$= \sum_a \sum_{b_1,b_2} \Psi_{ab_1}\overline{\Psi_{ab_2}}|f_{b_1}\rangle\langle f_{b_2}|,$$

where we used in the last equation that $\mathrm{tr}\left(|e_a\rangle\langle e_a|\right) = \mathrm{tr}\left(\rho_{e_a}\right) = 1$ for any pure state $|e_a\rangle$.

Solutions to Exercises from Chapter 5

Solution 5.59 To begin with, we have for any $V \in \mathcal{U}(\mathbb{H})$

$$(V^* - \mathbf{1})(V - \mathbf{1}) = V^*V - V^* - V + \mathbf{1} \underset{(2.37)}{=} 2\mathbf{1} - V^* - V \qquad (G.118)$$

as well as

$$\left(\mathbf{1}^{\otimes n+1} + |a\rangle\langle a| \otimes (V - \mathbf{1}) \otimes |b\rangle\langle b|\right)^*$$
$$\underset{(3.31)}{=} \mathbf{1}^{\otimes n+1} + \left(|a\rangle\langle a|\right)^* \otimes (V - \mathbf{1})^* \otimes \left(|b\rangle\langle b|\right)^*$$
$$\underset{(2.35)}{=} \mathbf{1}^{\otimes n+1} + |a\rangle\langle a| \otimes (V^* - \mathbf{1}) \otimes |b\rangle\langle b|. \qquad (G.119)$$

With this we obtain

$$\left(\mathbf{1}^{\otimes n+1} + |a\rangle\langle a| \otimes (V-\mathbf{1}) \otimes |b\rangle\langle b|\right)^* \left(\mathbf{1}^{\otimes n+1} + |a\rangle\langle a| \otimes (V-\mathbf{1}) \otimes |b\rangle\langle b|\right)$$

$$\underset{\text{(G.119)}}{=} \left(\mathbf{1}^{\otimes n+1} + |a\rangle\langle a| \otimes (V^*-\mathbf{1}) \otimes |b\rangle\langle b|\right) \left(\mathbf{1}^{\otimes n+1} + |a\rangle\langle a| \otimes (V-\mathbf{1}) \otimes |b\rangle\langle b|\right)$$

$$= \mathbf{1}^{\otimes n+1} + |a\rangle\langle a| \otimes (V-\mathbf{1}) \otimes |b\rangle\langle b|$$
$$+ |a\rangle\langle a| \otimes (V^*-\mathbf{1}) \otimes |b\rangle\langle b|$$
$$+ |a\rangle\langle a| \otimes (V^*-\mathbf{1})(V-\mathbf{1}) \otimes |b\rangle\langle b|$$

$$= \mathbf{1}^{\otimes n+1} + |a\rangle\langle a| \otimes \left((V^*-\mathbf{1})(V-\mathbf{1}) + V + V^* - 2\mathbf{1}\right) \otimes |b\rangle\langle b|$$

$$\underset{\text{(G.118)}}{=} \mathbf{1}^{\otimes n+1}.$$

Solution 5.60 First we show (5.15). Per definition one has

$$\Lambda^1(V) = \mathbf{1}^{\otimes 2} + |1\rangle\langle 1| \otimes (V-\mathbf{1})$$
$$= \mathbf{1} \otimes \mathbf{1} + |1\rangle\langle 1| \otimes V - |1\rangle\langle 1| \otimes \mathbf{1}$$
$$= \left(|0\rangle\langle 0| + |1\rangle\langle 1|\right) \otimes \mathbf{1} + |1\rangle\langle 1| \otimes V - |1\rangle\langle 1| \otimes \mathbf{1}$$
$$= |0\rangle\langle 0| \otimes \mathbf{1} + |1\rangle\langle 1| \otimes V. \tag{G.120}$$

The proof of (5.16) with projections $|0\rangle\langle 0|, \ldots$ is a cumbersome writing down of many terms and lengthy. A more concise alternative proof can be given if we use the matrix representation in the computational basis. In this one has with (G.120) at first

$$\Lambda^1(X) = |0\rangle\langle 0| \otimes \mathbf{1} + |1\rangle\langle 1| \otimes X$$

$$= \begin{pmatrix} 1 \\ 0 \end{pmatrix} (1,0) \otimes \begin{pmatrix} 1 & 0 \\ 0 & 1 \end{pmatrix} + \begin{pmatrix} 0 \\ 1 \end{pmatrix} (0,1) \otimes \begin{pmatrix} 0 & 1 \\ 1 & 0 \end{pmatrix}$$

$$= \begin{pmatrix} 1 & 0 \\ 0 & 0 \end{pmatrix} \otimes \begin{pmatrix} 1 & 0 \\ 0 & 1 \end{pmatrix} + \begin{pmatrix} 0 & 0 \\ 0 & 1 \end{pmatrix} \otimes \begin{pmatrix} 0 & 1 \\ 1 & 0 \end{pmatrix}$$

$$= \begin{pmatrix} 1 & 0 & 0 & 0 \\ 0 & 1 & 0 & 0 \\ 0 & 0 & 0 & 0 \\ 0 & 0 & 0 & 0 \end{pmatrix} + \begin{pmatrix} 0 & 0 & 0 & 0 \\ 0 & 0 & 0 & 0 \\ 0 & 0 & 0 & 1 \\ 0 & 0 & 1 & 0 \end{pmatrix}$$

$$= \begin{pmatrix} 1 & 0 & 0 & 0 \\ 0 & 1 & 0 & 0 \\ 0 & 0 & 0 & 1 \\ 0 & 0 & 1 & 0 \end{pmatrix}. \tag{G.121}$$

Analogously, we have

$$\Lambda_1(X) = \mathbf{1} \otimes \mathbf{1} + (X - \mathbf{1}) \otimes |1\rangle\langle 1|$$

$$= \begin{pmatrix} 1 & 0 \\ 0 & 1 \end{pmatrix} \otimes \begin{pmatrix} 1 & 0 \\ 0 & 1 \end{pmatrix} + \begin{pmatrix} -1 & 1 \\ 1 & -1 \end{pmatrix} \otimes \begin{pmatrix} 0 & 0 \\ 0 & 1 \end{pmatrix}$$

$$= \begin{pmatrix} 1 & 0 & 0 & 0 \\ 0 & 1 & 0 & 0 \\ 0 & 0 & 1 & 0 \\ 0 & 0 & 0 & 1 \end{pmatrix} + \begin{pmatrix} 0 & 0 & 0 & 0 \\ 0 & -1 & 0 & 1 \\ 0 & 0 & 0 & 0 \\ 0 & 1 & 0 & -1 \end{pmatrix}$$

$$= \begin{pmatrix} 1 & 0 & 0 & 0 \\ 0 & 0 & 0 & 1 \\ 0 & 0 & 1 & 0 \\ 0 & 1 & 0 & 0 \end{pmatrix} \tag{G.122}$$

and

$$H^{\otimes 2} = \frac{1}{\sqrt{2}} \begin{pmatrix} 1 & 1 \\ 1 & -1 \end{pmatrix} \otimes \frac{1}{\sqrt{2}} \begin{pmatrix} 1 & 1 \\ 1 & -1 \end{pmatrix}$$

$$= \frac{1}{2} \begin{pmatrix} 1 & 1 & 1 & 1 \\ 1 & -1 & 1 & -1 \\ 1 & 1 & -1 & -1 \\ 1 & -1 & -1 & 1 \end{pmatrix}. \tag{G.123}$$

With (G.121) and (G.123) one then obtains

$$H^{\otimes 2}\Lambda^1(X)H^{\otimes 2} = \frac{1}{4} \begin{pmatrix} 1 & 1 & 1 & 1 \\ 1 & -1 & 1 & -1 \\ 1 & 1 & -1 & -1 \\ 1 & -1 & -1 & 1 \end{pmatrix} \begin{pmatrix} 1 & 0 & 0 & 0 \\ 0 & 1 & 0 & 0 \\ 0 & 0 & 0 & 1 \\ 0 & 0 & 1 & 0 \end{pmatrix} \begin{pmatrix} 1 & 1 & 1 & 1 \\ 1 & -1 & 1 & -1 \\ 1 & 1 & -1 & -1 \\ 1 & -1 & -1 & 1 \end{pmatrix}$$

$$= \underbrace{\begin{pmatrix} 1 & 0 & 0 & 0 \\ 0 & 0 & 0 & 1 \\ 0 & 0 & 1 & 0 \\ 0 & 1 & 0 & 0 \end{pmatrix}}_{\text{(G.122)}} = \Lambda_1(X).$$

The proof of (5.17) is simpler in the operator-representation. With (G.120) it follows that

$$\Lambda^1(M(\alpha)) = |0\rangle\langle 0| \otimes \mathbf{1} + |1\rangle\langle 1| \otimes M(\alpha)$$
$$= |0\rangle\langle 0| \otimes \mathbf{1} + |1\rangle\langle 1| \otimes e^{i\alpha}\mathbf{1}$$
$$= \left(|0\rangle\langle 0| + e^{i\alpha}|1\rangle\langle 1|\right) \otimes \mathbf{1} = P(\alpha) \otimes \mathbf{1}.$$

Solution 5.61 Since complex numbers can be multiplied to any factor in a tensor product, that is, since for any $c \in \mathbb{C}$

$$\cdots \otimes c |\psi\rangle \otimes \cdots \otimes |\varphi\rangle \otimes \cdots = \cdots \otimes |\psi\rangle \otimes \cdots \otimes c |\varphi\rangle \otimes \ldots$$

holds, one has

$$S_{jk}^{(n)} \bigotimes_{l=n-1}^{0} |\psi_l\rangle$$

$$= |\psi_{n-1} \ldots \psi_{j+1}\rangle \otimes |0\rangle \langle 0|\psi_j\rangle \otimes |\psi_{j-1} \ldots \psi_{k+1}\rangle \otimes |0\rangle \langle 0|\psi_k\rangle \otimes |\psi_{k-1} \ldots \psi_0\rangle$$
$$+ |\psi_{n-1} \ldots \psi_{j+1}\rangle \otimes |1\rangle \langle 1|\psi_j\rangle \otimes |\psi_{j-1} \ldots \psi_{k+1}\rangle \otimes |1\rangle \langle 1|\psi_k\rangle \otimes |\psi_{k-1} \ldots \psi_0\rangle$$
$$+ |\psi_{n-1} \ldots \psi_{j+1}\rangle \otimes |0\rangle \langle 1|\psi_j\rangle \otimes |\psi_{j-1} \ldots \psi_{k+1}\rangle \otimes |1\rangle \langle 0|\psi_k\rangle \otimes |\psi_{k-1} \ldots \psi_0\rangle$$
$$+ |\psi_{n-1} \ldots \psi_{j+1}\rangle \otimes |1\rangle \langle 0|\psi_j\rangle \otimes |\psi_{j-1} \ldots \psi_{k+1}\rangle \otimes |0\rangle \langle 1|\psi_k\rangle \otimes |\psi_{k-1} \ldots \psi_0\rangle$$
$$= |\psi_{n-1} \ldots \psi_{j+1}\rangle \otimes |0\rangle \langle 0|\psi_k\rangle \otimes |\psi_{j-1} \ldots \psi_{k+1}\rangle \otimes |0\rangle \langle 0|\psi_j\rangle \otimes |\psi_{k-1} \ldots \psi_0\rangle$$
$$+ |\psi_{n-1} \ldots \psi_{j+1}\rangle \otimes |1\rangle \langle 1|\psi_k\rangle \otimes |\psi_{j-1} \ldots \psi_{k+1}\rangle \otimes |1\rangle \langle 1|\psi_j\rangle \otimes |\psi_{k-1} \ldots \psi_0\rangle$$
$$+ |\psi_{n-1} \ldots \psi_{j+1}\rangle \otimes |0\rangle \langle 0|\psi_k\rangle \otimes |\psi_{j-1} \ldots \psi_{k+1}\rangle \otimes |1\rangle \langle 1|\psi_j\rangle \otimes |\psi_{k-1} \ldots \psi_0\rangle$$
$$+ |\psi_{n-1} \ldots \psi_{j+1}\rangle \otimes |1\rangle \langle 1|\psi_k\rangle \otimes |\psi_{j-1} \ldots \psi_{k+1}\rangle \otimes |0\rangle \langle 0|\psi_j\rangle \otimes |\psi_{k-1} \ldots \psi_0\rangle$$
$$= |\psi_{n-1} \ldots \psi_{j+1}\rangle \otimes |0\rangle \langle 0|\psi_k\rangle \otimes |\psi_{j-1} \ldots \psi_{k+1}\rangle \otimes |\psi_j\rangle \otimes |\psi_{k-1} \ldots \psi_0\rangle$$
$$+ |\psi_{n-1} \ldots \psi_{j+1}\rangle \otimes |1\rangle \langle 1|\psi_k\rangle \otimes |\psi_{j-1} \ldots \psi_{k+1}\rangle \otimes |\psi_j\rangle \otimes |\psi_{k-1} \ldots \psi_0\rangle$$
$$= |\psi_{n-1} \ldots \psi_{j+1}\rangle \otimes |\psi_k\rangle \otimes |\psi_{j-1} \ldots \psi_{k+1}\rangle \otimes |\psi_j\rangle \otimes |\psi_{k-1} \ldots \psi_0\rangle .$$

This proves (5.31). From that (5.32) follows since the second application of $S_{jk}^{(n)}$ reverses the exchange of the qubits $|\psi_j\rangle$ and $|\psi_k\rangle$.

As $S_{jk}^{(n)}$ acts only on the factor spaces $\mathbb{H}_j$ and $\mathbb{H}_k$, and $S_{lm}^{(n)}$ acts only on the factor spaces $\mathbb{H}_l$ and $\mathbb{H}_m$, it follows for $j,k \notin \{l,m\}$ that $S_{jk}^{(n)} S_{lm}^{(n)} = S_{lm}^{(n)} S_{jk}^{(n)}$, proving (5.33). For the same reason, (5.34) follows directly from the successive application of the $S_{n-1-j\,j}^{(n)}$ in $S^{(n)}$.

Solution 5.62 For the proof of (5.46) one obtains from Definition 5.18 that

$$T_{|x\rangle|y\rangle}(V) \, T_{|x\rangle|y\rangle}(W)$$

$$\underset{(5.44)}{=} \left(\sum_{\substack{z=0 \\ z \neq x,y}}^{2^n-1} |z\rangle \langle z| + v_{00}|x\rangle \langle x| + v_{01}|x\rangle \langle y| + v_{10}|y\rangle \langle x| + v_{11}|y\rangle \langle y| \right)$$

$$\left(\sum_{\substack{z=0 \\ z \neq x,y}}^{2^n-1} |z\rangle \langle z| + w_{00}|x\rangle \langle x| + w_{01}|x\rangle \langle y| + w_{10}|y\rangle \langle x| + w_{11}|y\rangle \langle y| \right) .$$

Taking into account that vectors of the computational basis $|x\rangle$ and $|y\rangle$ satisfy $\langle x|y\rangle = \delta_{xy}$, this then becomes

$$T_{|x\rangle|y\rangle}(V)\,T_{|x\rangle|y\rangle}(W) = \sum_{\substack{z=0\\z\neq x,y}}^{2^n-1}|z\rangle\langle z|$$
$$+(v_{00}w_{00}+v_{01}w_{10})|x\rangle\langle x|+(v_{00}w_{01}+v_{01}w_{11})|x\rangle\langle y|$$
$$+(v_{10}w_{00}+v_{11}w_{10})|y\rangle\langle x|+(v_{10}w_{01}+v_{11}w_{11})|y\rangle\langle y|$$
$$= \sum_{\substack{z=0\\z\neq x,y}}^{2^n-1}|z\rangle\langle z|$$
$$+(VW)_{00}|x\rangle\langle x|+(VW)_{01}|x\rangle\langle y|$$
$$+(VW)_{10}|y\rangle\langle x|+(VW)_{11}|y\rangle\langle y|$$
$$\underbrace{=}_{(5.44)} T_{|x\rangle|y\rangle}(VW).$$

To prove (5.47), one uses that the matrix representation of V^* is given in the computational basis by

$$V^* = \begin{pmatrix} \overline{v_{00}} & \overline{v_{10}} \\ \overline{v_{01}} & \overline{v_{11}} \end{pmatrix} \tag{G.124}$$

and that $|a\rangle\langle b|^* = |b\rangle\langle a|$ holds. With this we thus have

$$T_{|x\rangle|y\rangle}(V)^*$$
$$\underbrace{=}_{(5.44)} \sum_{\substack{z=0\\z\neq x,y}}^{2^n-1}(|z\rangle\langle z|)^*+(v_{00}|x\rangle\langle x|)^*+(v_{01}|x\rangle\langle y|)^*+(v_{10}|y\rangle\langle x|)^*+(v_{11}|y\rangle\langle y|)^*$$
$$\underbrace{=}_{(2.32),(2.36)} \sum_{\substack{z=0\\z\neq x,y}}^{2^n-1}|z\rangle\langle z|+\overline{v_{00}}|x\rangle\langle x|+\overline{v_{01}}|y\rangle\langle x|+\overline{v_{10}}|x\rangle\langle y|+\overline{v_{11}}|y\rangle\langle y|$$
$$\underbrace{=}_{(5.44),(G.124)} T_{|x\rangle|y\rangle}(V^*).$$

In order to prove (5.48) we exploit (5.46) and (5.47)

$$T_{|x\rangle|y\rangle}(V)\,T_{|x\rangle|y\rangle}(V)^* \underbrace{=}_{(5.47)} T_{|x\rangle|y\rangle}(V)\,T_{|x\rangle|y\rangle}(V^*) \underbrace{=}_{(5.46)} T_{|x\rangle|y\rangle}(VV^*) = T_{|x\rangle|y\rangle}(\mathbb{1})$$
$$\underbrace{=}_{(5.44)} \mathbb{1}^{\otimes n}.$$

Solution 5.63 Let

$$x = \sum_{j=0}^{2^n-1} x_j 2^j < \sum_{j=0}^{2^n-1} y_j 2^j = y$$

with $x_j, y_j \in \{0,1\}$ and

$$L_{01} := \left\{ j \in \{0,\dots,n-1\} \mid x_j = 0 \text{ and } y_j = 1 \right\} = \{h_1,\dots,h_{|L_{01}|}\}$$
$$L_{10} := \left\{ j \in \{0,\dots,n-1\} \mid x_j = 1 \text{ and } y_j = 0 \right\} = \{k_1,\dots,k_{|L_{10}|}\}.$$

The set L_{01} cannot be empty since otherwise $x < y$ would not hold. Set $g^0 = x$ and for $l \in \{1,\dots,|L_{10}|\}$

$$|g^l\rangle = \mathbf{1}^{\otimes n - k_l} \otimes X \otimes \mathbf{1}^{\otimes k_l - 1} |g^{l-1}\rangle.$$

Then set for $l \in \{1,\dots,|L_{01}|\}$

$$|g^{l+|L_{10}|}\rangle = \mathbf{1}^{\otimes n - h_l} \otimes X \otimes \mathbf{1}^{\otimes h_l - 1} |g^{l+|L_{10}|-1}\rangle.$$

The $|g^l\rangle$ thus constructed start with $|x\rangle$, and, by construction, two consecutive elements differ in only one qubit until all qubits in which $|x\rangle$ differs from $|y\rangle$ have been reset to equal the values for $|y\rangle$. The last element is thus $|y\rangle$. Consequently, the $|g^l\rangle$ constitute a GRAY-coded transition from $|x\rangle$ to $|y\rangle$.

Solution 5.64 Let

$$|\Psi\rangle = \sum_{x=0}^{2^n-1} \sum_{y=0}^{2^m-1} \Psi_{xy} |x\rangle \otimes |y\rangle \in \mathbb{H}^A \otimes \mathbb{H}^B$$

be arbitrary, where we have made use of the computational basis in $\mathbb{H}^A = \mathbb{\Pi}^{\otimes n}$ and $\mathbb{H}^B = \mathbb{\Pi}^{\otimes m}$. Then we have

$$\|U_f \Psi\|^2 \underset{(2.4)}{=} \langle U_f \Psi | U_f \Psi \rangle \underset{(5.82)}{=} \left\langle \sum_{x,y} \Psi_{xy} |x\rangle \otimes |y \boxplus f(x)\rangle \,\Big|\, \sum_{a,b} \Psi_{ab} |a\rangle \otimes |b \boxplus f(a)\rangle \right\rangle$$

$$\underset{(2.6)}{=} \sum_{x,y,a,b} \overline{\Psi_{xy}} \Psi_{ab} \langle |x\rangle \otimes |y \boxplus f(x)\rangle \| |a\rangle \otimes |b \boxplus f(a)\rangle \rangle$$

$$\underset{(3.4)}{=} \sum_{x,y,a,b} \overline{\Psi_{xy}} \Psi_{ab} \underbrace{\langle x|a\rangle}_{=\delta_{xa}} \langle y \boxplus f(x) | b \boxplus f(a)\rangle$$

$$\underset{(3.24)}{=} \sum_{x,y,b} \overline{\Psi_{xy}} \Psi_{xb} \langle y \boxplus f(x) | b \boxplus f(x)\rangle.$$

Now,

$$\langle y \boxplus f(x) | b \boxplus f(x) \rangle$$

$$\underbrace{=}_{(5.80)} \langle y_{n-1} \overset{2}{\oplus} f(x)_{n-1} \otimes \cdots \otimes y_0 \overset{2}{\oplus} f(x)_0 | b_{n-1} \overset{2}{\oplus} f(x)_{n-1} \otimes \cdots \otimes b_0 \overset{2}{\oplus} f(x)_0 \rangle$$

$$\underbrace{=}_{(3.4)} \prod_{j=0}^{n-1} \underbrace{\langle y_j \overset{2}{\oplus} f(x)_j | b_j \overset{2}{\oplus} f(x)_j \rangle}_{=\delta_{y_j b_j}}$$

$$\underbrace{=}_{(3.24)} \delta_{yb}$$

and thus

$$||U_f \Psi||^2 = \sum_{x,y} |\Psi_{xy}|^2 = ||\Psi||^2$$

for any $|\Psi\rangle \in \mathbb{H}^A \otimes \mathbb{H}^B$. It follows from (5.7) that U_f is unitary.

Solution 5.65 To prove the claim, it suffices to show that $U_c^* U_c$ maps any vector of the computational basis in $\mathbb{H}^{\otimes 4}$ onto itself. This can be seen with the help of (5.97) and (5.99) as follows:

$$U_c^* U_c \left(|x_3\rangle \otimes |x_2\rangle \otimes |x_1\rangle \otimes |x_0\rangle \right)$$

$$\underbrace{=}_{(5.97)} U_c^* \left(|\underbrace{x_0(x_1 \overset{2}{\oplus} x_2) \overset{2}{\oplus} x_1 x_2 \overset{2}{\oplus} x_3}_{=x_3'}\rangle \otimes |\underbrace{x_1 \overset{2}{\oplus} x_2}_{=x_2'}\rangle \otimes |x_1\rangle \otimes |x_0\rangle \right)$$

$$\underbrace{=}_{(5.99)} |(x_0 \overset{2}{\oplus} x_1) x_2' \overset{2}{\oplus} x_1 \overset{2}{\oplus} x_3'\rangle \otimes |x_1 \overset{2}{\oplus} x_2'\rangle \otimes |x_1\rangle \otimes |x_0\rangle$$

$$= |(x_0 \overset{2}{\oplus} x_1)\underbrace{(x_1 \overset{2}{\oplus} x_2)}_{=x_2'} \overset{2}{\oplus} x_1 \overset{2}{\oplus} \underbrace{x_0(x_1 \overset{2}{\oplus} x_2) \overset{2}{\oplus} x_1 x_2 \overset{2}{\oplus} x_3}_{=x_3'}\rangle$$

$$\otimes |\underbrace{x_1 \overset{2}{\oplus} x_1 \overset{2}{\oplus} x_2}_{=0}\rangle \otimes |x_1\rangle \otimes |x_0\rangle$$

$$= |(x_0 x_1 \overset{2}{\oplus} x_0 x_2 \overset{2}{\oplus} x_1 \overset{2}{\oplus} x_1 x_2 \overset{2}{\oplus} x_1 \overset{2}{\oplus} x_0 x_1 \overset{2}{\oplus} x_0 x_2 \overset{2}{\oplus} x_1 x_2 \overset{2}{\oplus} x_3\rangle$$

$$\otimes |x_2\rangle \otimes |x_1\rangle \otimes |x_0\rangle$$

$$= |x_3\rangle \otimes |x_2\rangle \otimes |x_1\rangle \otimes |x_0\rangle .$$

Solution 5.66 From Definition 5.48 it follows that one has for an arbitrary vector $|u\rangle$ of the computational basis

$$F|u\rangle = \frac{1}{2^{\frac{n}{2}}} \sum_{x,y=0}^{2^n-1} \exp\left(2\pi i \frac{xy}{2^n}\right) |x\rangle \underbrace{\langle y|u\rangle}_{=\delta_{yu}} = \frac{1}{2^{\frac{n}{2}}} \sum_{x=0}^{2^n-1} \exp\left(2\pi i \frac{xu}{2^n}\right) |x\rangle .$$

For arbitrary vectors $|u\rangle$ and $|v\rangle$ of the computational basis this implies

$$\langle Fu|Fv\rangle = \frac{1}{2^n} \langle \sum_{x=0}^{2^n-1} \exp\left(2\pi i \frac{xu}{2^n}\right) |x\rangle | \sum_{y=0}^{2^n-1} \exp\left(2\pi i \frac{yv}{2^n}\right) |y\rangle\rangle$$

$$\underbrace{=}_{(2.4),(2.6)} \frac{1}{2^n} \sum_{x,y=0}^{2^n-1} \exp\left(2\pi i \frac{yv-xu}{2^n}\right) \underbrace{\langle x|y\rangle}_{=\delta_{xy}}$$

$$= \frac{1}{2^n} \sum_{x=0}^{2^n-1} \exp\left(2\pi i x \frac{v-u}{2^n}\right)$$

$$= \frac{1}{2^n} \sum_{x=0}^{2^n-1} \left(\exp\left(2\pi i \frac{v-u}{2^n}\right)\right)^x$$

$$= \begin{cases} 1 & \text{if } u=v \\ \frac{1-\left(\exp\left(2\pi i \frac{v-u}{2^n}\right)\right)^{2^n}}{1-\exp\left(2\pi i \frac{u-v}{2^n}\right)} = 0 & \text{if } u \neq v \end{cases}$$

$$= \delta_{uv} . \tag{G.125}$$

For arbitrary

$$|\varphi\rangle = \sum_{u=0}^{2^n-1} \varphi_u |u\rangle \quad \text{and} \quad |\psi\rangle = \sum_{v=0}^{2^n-1} \psi_v |v\rangle \tag{G.126}$$

in $\mathbb{H}^{\otimes n}$ thus

$$\langle F\varphi|F\psi\rangle \underbrace{=}_{(G.126)} \sum_{u,v=0}^{2^n-1} \overline{\varphi_u}\psi_v \underbrace{\langle Fu|Fv\rangle}_{=\delta_{uv}} \underbrace{=}_{(G.125)} \sum_{u,v=0}^{2^n-1} \overline{\varphi_u}\psi_v \underbrace{=}_{(2.13)} \langle \varphi|\psi\rangle$$

holds and by Definition 2.9 F is unitary.

Solutions to Exercises from Chapter 6

Solution 6.67 We show this by induction in n. For $n=1$ we have

$$H|x\rangle = H|x_0\rangle \underbrace{=}_{(2.162)} \frac{1}{\sqrt{2}}\left(|0\rangle + (-1)^{x_0}|1\rangle\right)$$

$$= \frac{1}{\sqrt{2}} \sum_{y=0}^{1} (-1)^{x \overset{2}{\odot} y}|y\rangle,$$

which proves the claim for $n = 1$. Suppose then the claim is true for n, that is,

$$H^{\otimes n}|x\rangle = \frac{1}{2^{\frac{n}{2}}} \sum_{y=0}^{2^n-1} (-1)^{x \overset{2}{\odot} y}|y\rangle \qquad (G.127)$$

holds for a given n. For $|x\rangle \in \mathbb{H}^{\otimes n+1}$ we use the notation $|x\rangle = |x_n \ldots x_0\rangle = |x_n\rangle \otimes |\check{x}\rangle$, where $|\check{x}\rangle \in \mathbb{H}^{\otimes n}$. Then it follows that

$$
\begin{aligned}
H^{\otimes n+1}|x\rangle &= H|x_n\rangle \otimes H^{\otimes n}|\check{x}\rangle \\
&\underbrace{=}_{(G.127)} H|x_n\rangle \otimes \frac{1}{2^{\frac{n}{2}}} \sum_{y=0}^{2^n-1} (-1)^{\check{x} \overset{2}{\odot} y}|y\rangle \\
&\underbrace{=}_{(2.162)} \frac{1}{\sqrt{2}}\left(|0\rangle + (-1)^{x_n}|1\rangle\right) \otimes \frac{1}{2^{\frac{n}{2}}} \sum_{y=0}^{2^n-1} (-1)^{\check{x} \overset{2}{\odot} y}|y\rangle \\
&= \frac{1}{2^{\frac{n+1}{2}}} \sum_{y=0}^{2^n-1} \left((-1)^{\check{x} \overset{2}{\odot} y}|0 y_{n-1} \ldots y_0\rangle + (-1)^{x_n + \check{x} \overset{2}{\odot} y}|1 y_{n-1} \ldots y_0\rangle\right) \\
&= \frac{1}{2^{\frac{n+1}{2}}} \sum_{y=0}^{2^{n+1}-1} (-1)^{x_n y_n + \check{x} \overset{2}{\odot} y}|y_n y_{n-1} \ldots y_0\rangle \\
&= \frac{1}{2^{\frac{n+1}{2}}} \sum_{y=0}^{2^{n+1}-1} (-1)^{x \overset{2}{\odot} y}|y\rangle,
\end{aligned}
$$

where in the last step we used that $a, b \in \{0, 1\}$ implies $(-1)^{a+b} = (-1)^{a \overset{2}{\oplus} b}$. Hence, the claim holds for $n + 1$ as well and the induction is complete.

Solution 6.68 Because of (2.76), one has $\mathbf{1} = \mathbf{1}^2 = \sigma_x^2 = \sigma_z^2$. With the definition of the PAULI matrices (2.74) and (2.35) one can easily verify that $\mathbf{1}^* = \mathbf{1}, \sigma_x^* = \sigma_x, \sigma_z^* = \sigma_z$ and thus

$$
\begin{aligned}
\sigma_z \sigma_x \left(\sigma_z \sigma_x\right)^* &= \sigma_z \sigma_x \sigma_x^* \sigma_z^* = \sigma_z \sigma_x^2 \sigma_z \\
&= \sigma_z^2 = \mathbf{1}.
\end{aligned}
$$

Hence, we have that for every $U^A \in \{1, \sigma_x^A, \sigma_z^A, \sigma_z^A \sigma_x^A\}$ then $U^A U^{A*} = 1^A$ holds and consequently,

$$\left(U^A \otimes 1^B\right)\left(U^A \otimes 1^B\right)^* = \left(U^A \otimes 1^B\right)\left(U^{A*} \otimes 1^B\right)$$
$$= \left(U^A U^{A*} \otimes 1^B\right) = 1^A \otimes 1^B$$
$$= 1^{AB}.$$

Furthermore, one has

$$\left(\sigma_x^A \otimes 1^B\right)|\Phi^+\rangle = \left(\sigma_x^A \otimes 1^B\right)\frac{1}{\sqrt{2}}\left(|00\rangle + |11\rangle\right)$$
$$= \frac{1}{\sqrt{2}}\left(\sigma_x^A|0\rangle \otimes |0\rangle + \sigma_x^A|1\rangle \otimes |1\rangle\right)$$
$$= \frac{1}{\sqrt{2}}\left(|1\rangle \otimes |0\rangle + |0\rangle \otimes |1\rangle\right) = \frac{1}{\sqrt{2}}\left(|10\rangle + |01\rangle\right)$$
$$= |\Psi^+\rangle$$

and

$$\left(\sigma_z^A \sigma_x^A \otimes 1^B\right)|\Phi^+\rangle = \left(\sigma_z^A \sigma_x^A \otimes 1^B\right)\frac{1}{\sqrt{2}}\left(|00\rangle + |11\rangle\right)$$
$$= \frac{1}{\sqrt{2}}\left(\sigma_z^A \sigma_x^A|0\rangle \otimes |0\rangle + \sigma_z^A \sigma_x^A|1\rangle \otimes |1\rangle\right)$$
$$= \frac{1}{\sqrt{2}}\left(\sigma_z^A|1\rangle \otimes |0\rangle + \sigma_z^A|0\rangle \otimes |1\rangle\right)$$
$$= \frac{1}{\sqrt{2}}\left(-|1\rangle \otimes |0\rangle + |0\rangle \otimes |1\rangle\right) = \frac{1}{\sqrt{2}}\left(|01\rangle - |10\rangle\right)$$
$$= |\Psi^-\rangle.$$

Solution 6.69 Let r be the period of the function $f_{b,N}(n) = b^n \bmod N$. Definition 6.7 then implies that we have for all $n \in \mathbb{N}$ that $f_{b,N}(n+r) = f_{b,N}(n)$ holds. In particular, for $n = 0$ it follows that

$$b^r \bmod N = f_{b,N}(0+r) = f_{b,N}(0) = 1. \tag{G.128}$$

According to Definition D.20, the order $\mathrm{ord}_N(b)$ of b modulo N is the smallest number that satisfies (G.128). This implies

$$r \geq \mathrm{ord}_N(b). \tag{G.129}$$

On the other hand, we have for all $n \in \mathbb{N}_0$ that

$$
\begin{aligned}
f_{b,N}(n + \operatorname{ord}_N(b)) &= b^{n + \operatorname{ord}_N(b)} \bmod N \\
&\underset{\text{(D.20)}}{=} b^n \left(b^{\operatorname{ord}_N(b)} \bmod N \right) \bmod N \\
&\underset{\text{Def. D.20}}{=} b^n \bmod N \\
&= f_{b,N}(n) \,.
\end{aligned}
$$

Since the period r is the smallest number with the property $f_{b,N}(n + r) = f_{b,N}(n)$, it follows that

$$
r \le \operatorname{ord}_N(b) \,. \tag{G.130}
$$

The claim $r = \operatorname{ord}_N(b)$ then follows from (G.129) and (G.130).

Solution 6.70 Choosing an ONB $\{|\Phi_0\rangle = |\Psi_3\rangle, |\Phi_1\rangle, \dots\}$ in $\mathbb{H}^A \otimes \mathbb{H}^B$ which contains $|\Psi_3\rangle$ to calculate this trace, we obtain

$$
\begin{aligned}
\operatorname{tr}\left(|\Psi_3\rangle\langle\Psi_3|\left(|z\rangle\langle z| \otimes \mathbf{1}^B\right)\right) &\underset{\text{(2.57)}}{=} \sum_j \underbrace{\langle\Phi_j|\Psi_3\rangle}_{=\delta_{j0}} \langle\Psi_3|\left(|z\rangle\langle z| \otimes \mathbf{1}^B\right)\Phi_j\rangle \\
&= \langle\Psi_3|\left(|z\rangle\langle z| \otimes \mathbf{1}^B\right)\Psi_3\rangle
\end{aligned} \tag{G.131}
$$

Since the projection $\left(|z\rangle\langle z| \otimes \mathbf{1}^B\right)$ has the properties

$$
\left(|z\rangle\langle z| \otimes \mathbf{1}^B\right)^2 = |z\rangle \underbrace{\langle z|z\rangle}_{=1} \langle z| \otimes \mathbf{1}^B \tag{G.132}
$$

$$
\left(|z\rangle\langle z| \otimes \mathbf{1}^B\right)^* \underset{\text{(3.31)}}{=} \left(|z\rangle\langle z|\right)^* \otimes \mathbf{1}^B \underset{\text{(2.36)}}{=} |z\rangle\langle z| \otimes \mathbf{1}^B \,, \tag{G.133}
$$

we have

$$
\begin{aligned}
\langle\Psi_3|\left(|z\rangle\langle z| \otimes \mathbf{1}^B\right)\Psi_3\rangle &\underset{\text{(G.132)}}{=} \langle\Psi_3|\left(|z\rangle\langle z| \otimes \mathbf{1}^B\right)^2\Psi_3\rangle \\
&\underset{\text{(2.30)}}{=} \langle\left(|z\rangle\langle z| \otimes \mathbf{1}^B\right)^*\Psi_3|\left(|z\rangle\langle z| \otimes \mathbf{1}^B\right)\Psi_3\rangle \\
&\underset{\text{(G.133)}}{=} \langle\left(|z\rangle\langle z| \otimes \mathbf{1}^B\right)\Psi_3|\left(|z\rangle\langle z| \otimes \mathbf{1}^B\right)\Psi_3\rangle \\
&\underset{\text{(2.5)}}{=} \left\| \left(|z\rangle\langle z| \otimes \mathbf{1}^B\right)|\Psi_3\rangle \right\|^2
\end{aligned} \tag{G.134}
$$

and inserting (G.134) in (G.131) yields the claim (6.45).

Solution 6.71 If $\frac{2^L}{r} =: m \in \mathbb{N}$ holds, it follows from (6.36) that

$$J = \left\lfloor m - \frac{1}{r} \right\rfloor = m - 1$$

as well as from (6.37) that

$$R = 2^L - 1 \bmod r = 2^L - 1 - \left\lfloor \frac{2^L - 1}{r} \right\rfloor r = 2^L - 1 - (m-1)r = r - 1.$$

Furthermore, with (6.38) we have for all $k \in \mathbb{N}$ with $0 \le k \le r - 1 = R$ that

$$J_k = J = m - 1.$$

Inserting this into (6.46) this yields

$$W(z) = \begin{cases} \frac{1}{2^{2L}} \sum_{k=0}^{r-1} m^2 & \text{if } \frac{z}{m} \in \mathbb{N} \\ \frac{1}{2^{2L}} \sum_{k=0}^{r-1} \left| \frac{1 - e^{2\pi i \frac{z}{m} m}}{1 - e^{2\pi i \frac{z}{m}}} \right|^2 & \text{else} \end{cases}$$

$$= \begin{cases} \frac{r}{2^{2L}} \left(\frac{2^L}{r} \right)^2 & \text{if } \frac{z}{m} \in \mathbb{N} \\ 0 & \text{else} \end{cases}$$

$$= \begin{cases} \frac{1}{r} & \text{if } \frac{z}{m} \in \mathbb{N} \\ 0 & \text{else.} \end{cases}$$

Solution 6.72 For $n \in \mathbb{N}_0$ we obtain, by equating the imaginary parts of the left side to the right side of

$$\cos(n\alpha) + i\sin(n\alpha) = e^{in\alpha} = \left(e^{i\alpha} \right)^n = \left(\cos\alpha + i\sin\alpha \right)^n,$$

that

$$\sin(n\alpha) = \sum_{l=0}^{\lfloor \frac{n}{2} \rfloor} (-1)^l \binom{n}{2l+1} \cos^{n-2l-1}\alpha \sin^{2l+1}\alpha.$$

From this we have

$$\frac{\sin(n\alpha)}{\sin\alpha} = \sum_{l=0}^{\lfloor \frac{n}{2} \rfloor} (-1)^l \binom{n}{2l+1} \cos^{n-2l-1}\alpha \sin^{2l}\alpha$$

and also obtain

$$\left(\frac{\sin(n\alpha)}{\sin\alpha}\right)' = \sum_{l=0}^{\lfloor\frac{n}{2}\rfloor}(-1)^{l+1}\binom{n}{2l+1}(n-2l-1)\cos^{n-2l-2}\alpha\sin^{2l+1}\alpha$$
$$+\sum_{l=1}^{\lfloor\frac{n}{2}\rfloor}(-1)^{l}\binom{n}{2l+1}2l\cos^{n-2l}\alpha\sin^{2l-1}\alpha$$

as well as

$$\left(\frac{\sin(n\alpha)}{\sin\alpha}\right)'' = \sum_{l=0}^{\lfloor\frac{n}{2}\rfloor}(-1)^{l}\binom{n}{2l+1}(n-2l-1)(n-2l-2)\cos^{n-2l-3}\alpha\sin^{2l+2}\alpha$$
$$-\sum_{l=0}^{\lfloor\frac{n}{2}\rfloor}(-1)^{l}\binom{n}{2l+1}(n-2l-1)(2l+1)\cos^{n-2l-1}\alpha\sin^{2l}\alpha$$
$$+\sum_{l=1}^{\lfloor\frac{n}{2}\rfloor}(-1)^{l}\binom{n}{2l+1}2l(2l-1)\cos^{n-2l+1}\alpha\sin^{2l-2}\alpha$$
$$-\sum_{l=1}^{\lfloor\frac{n}{2}\rfloor}(-1)^{l}\binom{n}{2l+1}2l(n-2l)\cos^{n-2l-1}\alpha\sin^{2l}\alpha.$$

At $\alpha = 0$ one thus has

$$\frac{\sin(n\alpha)}{\sin\alpha}\Big|_{\alpha=0} = n$$
$$\left(\frac{\sin(n\alpha)}{\sin\alpha}\right)'\Big|_{\alpha=0} = 0 \tag{G.135}$$
$$\left(\frac{\sin(n\alpha)}{\sin\alpha}\right)''\Big|_{\alpha=0} = \frac{n}{3}(1-n^2).$$

In $s(\alpha) = \frac{\sin(\alpha\tilde{J}_k)}{\sin\alpha}$ we have $\tilde{J}_k \in \mathbb{N}$ and $\tilde{J}_k = J_k + 1 \geq \left\lfloor\frac{2^L-1}{r}\right\rfloor > \left\lfloor 2^{\frac{L}{2}} - \frac{1}{r}\right\rfloor > 1$ since we can assume $L > 2$. Hence, it follows from (G.135) that

$$s(0) = \tilde{J}_k$$
$$s'(0) = 0$$
$$s''(0) = \frac{\tilde{J}_k}{3}(1-\tilde{J}_k) < 0,$$

that is, s has a maximum at $\alpha = 0$. That s has no further extrema in $]0, \frac{\pi r}{2^{L+1}}[$ can be seen as follows. First, we obtain by explicit computation

$$s'(\alpha) = \frac{\tilde{J}_k \cos\left(\alpha \tilde{J}_k\right)\sin\alpha - \sin\left(\alpha \tilde{J}_k\right)\cos\alpha}{\sin^2\alpha}. \tag{G.136}$$

If $\alpha\tilde{J}_k = \frac{\pi}{2}$, it follows that $s'(\alpha) < 0$. This value is then not an extremum. Suppose now that $\alpha \in]0, \frac{\pi r}{2^{L+1}}[$ and $\alpha\tilde{J}_k \neq \frac{\pi}{2}$. Because of (6.51) and (6.52), one has then

$$\alpha < \frac{\pi}{2^{\frac{L}{2}+1}}$$
$$\alpha\tilde{J}_k < \frac{\pi}{2} + \frac{\pi}{2^{\frac{L}{2}+1}}.$$

This implies that in the case $\alpha\tilde{J}_k < \frac{\pi}{2}$ as well as $\alpha\tilde{J}_k > \frac{\pi}{2}$

$$\tan\alpha < \tan(\alpha\tilde{J}_k)$$

holds and thus also

$$(\tilde{J}_k \tan\alpha)' = \tilde{J}_k(1 + \tan^2\alpha) < \tilde{J}_k(1 + \tan^2(\alpha\tilde{J}_k)) = \tan(\alpha\tilde{J}_k)'.$$

This, together with $\tilde{J}_k \tan\alpha|_{\alpha=0} = \tan(\alpha\tilde{J}_k)|_{\alpha=0}$, yields

$$\tilde{J}_k \tan\alpha < \tan(\alpha\tilde{J}_k)$$

and thus

$$\tilde{J}_k \cos\left(\alpha\tilde{J}_k\right)\sin\alpha < \sin\left(\alpha\tilde{J}_k\right)\cos\alpha.$$

Because of (G.136) the latter is equivalent to $s'(\alpha) < 0$. This completes the proof of $s'(\alpha) < 0$ for $\alpha \in]0, \frac{\pi r}{2^{L+1}}]$. Since $s(-\alpha) = s(\alpha)$, it also follows that $s'(\alpha) > 0$ for $\alpha \in [-\frac{\pi r}{2^{L+1}}, 0[$. In the interval $[-\frac{\pi r}{2^{L+1}}, \frac{\pi r}{2^{L+1}}]$ the function $s(\alpha)$ thus takes a maximum at $\alpha = 0$ and decreases to the left and right of $\alpha = 0$. Hence, inside the interval it is greater than at the boundaries $\pm\frac{\pi r}{2^{L+1}}$. Due to $s(-\alpha) = s(\alpha)$ we can choose $\alpha_{min} = \frac{\pi r}{2^{L+1}}$. Finally, one has in the given interval that $s(\alpha) \geq 0$ so that there also $s(\alpha)^2 \geq s(\alpha_{min})^2$ holds.

Solution 6.73 Per Definition 6.15 we have

$$f \text{ hides } \mathcal{H} \quad \Leftrightarrow \quad \forall g_1, g_2 \in \mathcal{G} \quad f(g_1) = f(g_2) \Leftrightarrow g_1^{-1}g_2 \in \mathcal{H}$$

and since

$$
\begin{aligned}
g_1^{-1} g_2 \in \mathcal{H} \quad &\Leftrightarrow \quad \exists h \in \mathcal{H} : \; g_1^{-1} g_2 = h \\
&\Leftrightarrow \quad \exists h \in \mathcal{H} : \; g_2 = g_1 h \\
&\underbrace{\Leftrightarrow}_{(\text{F.25})} \quad g_2 \mathcal{H} = g_1 \mathcal{H}
\end{aligned}
$$

the claim (6.88) follows.

Solution 6.74 For any $g_1, g_2 \in \mathcal{G}$ we have

$$
\langle \Psi^A_{[g_1]_{\mathcal{H}}} | \Psi^A_{[g_2]_{\mathcal{H}}} \rangle \underset{(6.99)}{=} \frac{1}{|\mathcal{H}|} \sum_{k_1 \in [g_1]_{\mathcal{H}}} \sum_{k_2 \in [g_2]_{\mathcal{H}}} \langle k_1 | k_2 \rangle \underset{(6.89)}{=} \frac{1}{|\mathcal{H}|} \sum_{k_1 \in [g_1]_{\mathcal{H}}} \sum_{k_2 \in [g_2]_{\mathcal{H}}} \delta_{k_1, k_2}.
$$
$$(G.137)$$

From Lemma F.20 we know that the two cosets $[g_1]_{\mathcal{H}}$ and $[g_2]_{\mathcal{H}}$ are either identical or disjoint, such that

$$
\begin{aligned}
\sum_{k_1 \in [g_1]_{\mathcal{H}}} \sum_{k_2 \in [g_2]_{\mathcal{H}}} \delta_{k_1, k_2} &= \begin{cases} \sum_{k_1, k_2 \in [g_1]_{\mathcal{H}}} \delta_{k_1, k_2} & \text{if } [g_1]_{\mathcal{H}} = [g_2]_{\mathcal{H}} \\ 0 & \text{if } [g_1]_{\mathcal{H}} \neq [g_2]_{\mathcal{H}} \end{cases} \\
&= \begin{cases} \sum_{k \in [g_1]_{\mathcal{H}}} 1 & \text{if } [g_1]_{\mathcal{H}} = [g_2]_{\mathcal{H}} \\ 0 & \text{if } [g_1]_{\mathcal{H}} \neq [g_2]_{\mathcal{H}} \end{cases} \\
&= \begin{cases} |\mathcal{H}| & \text{if } [g_1]_{\mathcal{H}} = [g_2]_{\mathcal{H}} \\ 0 & \text{if } [g_1]_{\mathcal{H}} \neq [g_2]_{\mathcal{H}}, \end{cases}
\end{aligned}
\qquad (G.138)
$$

where in the last equation we used Theorem F.21, which tells us that the number of distinct cosets of $\mathcal{H}$ is equal to $|\mathcal{H}|$. Inserting (G.138) into (G.137) then yields the claim (6.101).

Solution 6.75 To prove the claim, we have to show that $\mathcal{H}$ as defined in (6.115) satisfies the requirements of Definition F.6. Clearly, this set $\mathcal{H}$ is a subset of $\mathcal{G} = \mathbb{Z}_N \times \mathbb{Z}_N$. With the choice $u = 0$ it also contains the neutral element $e_\mathcal{G} = ([0]_{N\mathbb{Z}}, [0]_{N\mathbb{Z}})$ of that group, verifying (F.14).

For any two elements $([u_i]_{N\mathbb{Z}}, [-du_i]_{N\mathbb{Z}}) \in \mathcal{H}$, where $i \in \{1,2\}$, we have

$$
\begin{aligned}
& \left([u_1]_{N\mathbb{Z}}, [-du_1]_{N\mathbb{Z}}\right) +_\mathcal{G} \left([u_2]_{N\mathbb{Z}}, [-du_2]_{N\mathbb{Z}}\right) \\
& \underbrace{=}_{(\text{F.35})} \left([u_1 +_{\mathbb{Z}_N} u_2]_{N\mathbb{Z}}, [-du_1 +_{\mathbb{Z}_N} (-du_2)]_{N\mathbb{Z}}\right) \\
& \underbrace{=}_{(\text{F.35})} \left([(u_1 + u_2) \bmod N]_{N\mathbb{Z}}, [-(du_1 + du_2) \bmod N]_{N\mathbb{Z}}\right) \\
& \underbrace{=}_{(\text{F.35})} \left([u_1 + u_2]_{N\mathbb{Z}}, [(-d(u_1 + u_2)) \bmod N]_{N\mathbb{Z}}\right) \\
& \underbrace{=}_{(\text{D.20})} \left([u_1 + u_2]_{N\mathbb{Z}}, [(-d(u_1 + u_2) \bmod N) \bmod N]_{N\mathbb{Z}}\right) \\
& \underbrace{=}_{(\text{F.35})} \left([u_1 + u_2]_{N\mathbb{Z}}, [(-d(u_1 +_{\mathbb{Z}_N} u_2)]_{N\mathbb{Z}}\right) \in \mathcal{H},
\end{aligned}
$$

proving (F.16).

Lastly, for any $[u]_{N\mathbb{Z}} \in \mathbb{Z}_N$ we have that $([u]_{N\mathbb{Z}}, [-du]_{N\mathbb{Z}}) \in \mathcal{H}$, and that its inverse $([-u]_{N\mathbb{Z}}, [du]_{N\mathbb{Z}})$ is also an element of $\mathcal{H}$, verifying (F.15).

Solution 6.76 From (6.119) we know that any $\chi \in \mathcal{H}^\perp$ is of the form $\chi_{dn \bmod N, n}$. For such characters we have for any $([x]_{N\mathbb{Z}}, [y]_{N\mathbb{Z}}) \in \mathcal{G}$

$$
\begin{aligned}
\chi_{dn \bmod N, n}([x]_{N\mathbb{Z}}, [y]_{N\mathbb{Z}}) \underbrace{=}_{(6.117)} & \; e^{2\pi i \frac{(dn \bmod N)x + ny}{N}} = e^{2\pi i \frac{dnx + ny}{N}} = \left(e^{2\pi i \frac{dx + y}{N}}\right)^n \\
\underbrace{=}_{(6.117)} & \; \left(\chi_{d,1}([x]_{N\mathbb{Z}}, [y]_{N\mathbb{Z}})\right)^n \\
\underbrace{=}_{(\text{F.61})} & \; \chi_{d,1}^n([x]_{N\mathbb{Z}}, [y]_{N\mathbb{Z}}),
\end{aligned}
$$

which shows that every element of $\mathcal{H}^\perp$ is some power of $\chi_{d,1}$, implying $\mathcal{H}^\perp = \langle \chi_{d,1} \rangle$.

Solution 6.77 We infer from (6.75) that

$$
\begin{aligned}
\mathcal{H} &= \left\{ ([m]_{6\mathbb{Z}}, [-3m]_{6\mathbb{Z}}) \in \mathbb{Z}_6 \times \mathbb{Z}_6 \mid [m]_{6\mathbb{Z}} \in \mathbb{Z}_6 \right\} \\
&= \left\{ ([0]_{6\mathbb{Z}}, [0]_{6\mathbb{Z}}), \, ([1]_{6\mathbb{Z}}, [3]_{6\mathbb{Z}}), \, ([2]_{6\mathbb{Z}}, [0]_{6\mathbb{Z}}), \right. \\
&\qquad \left. ([3]_{6\mathbb{Z}}, [3]_{6\mathbb{Z}}), \, ([4]_{6\mathbb{Z}}, [0]_{6\mathbb{Z}}), \, ([5]_{6\mathbb{Z}}, [3]_{6\mathbb{Z}}) \right\}
\end{aligned}
$$

and from (6.119) that

$$
\mathcal{H}^\perp = \left\{ \chi_{3n,n} \in \widehat{\mathbb{Z}_6} \times \widehat{\mathbb{Z}_6} \mid [n]_{6\mathbb{Z}} \in \mathbb{Z}_6 \right\} = \left\{ \chi_{0,0}, \chi_{3,1}, \chi_{0,2}, \chi_{3,3}, \chi_{0,4}, \chi_{3,5} \right\}.
$$

For these characters we find

$$
\begin{aligned}
\mathrm{Ker}(\chi_{0,0}) &= \mathbb{Z}_6 \times \mathbb{Z}_6 = \mathcal{G} \\
\mathrm{Ker}(\chi_{3,1}) &= \left\{ ([x]_{6\mathbb{Z}}, [y]_{6\mathbb{Z}}) \in \mathbb{Z}_6 \times \mathbb{Z}_6 \mid [3x+y]_{6\mathbb{Z}} = 0 \right\} \\
&= \left\{ ([0]_{6\mathbb{Z}}, [0]_{6\mathbb{Z}}), \, ([1]_{6\mathbb{Z}}, [3]_{6\mathbb{Z}}), \, ([2]_{6\mathbb{Z}}, [0]_{6\mathbb{Z}}), \, ([3]_{6\mathbb{Z}}, [3]_{6\mathbb{Z}}), \right. \\
&\qquad \left. ([4]_{6\mathbb{Z}}, [0]_{6\mathbb{Z}}), \, ([5]_{6\mathbb{Z}}, [3]_{6\mathbb{Z}}) \right\} = \mathcal{H} \\
\mathrm{Ker}(\chi_{0,2}) &= \left\{ ([x]_{6\mathbb{Z}}, [y]_{6\mathbb{Z}}) \in \mathbb{Z}_6 \times \mathbb{Z}_6 \mid [2y]_{6\mathbb{Z}} = 0 \right\} \\
&= \left\{ ([0]_{6\mathbb{Z}}, [0]_{6\mathbb{Z}}), \, ([1]_{6\mathbb{Z}}, [3]_{6\mathbb{Z}}), \, ([2]_{6\mathbb{Z}}, [0]_{6\mathbb{Z}}), \, ([3]_{6\mathbb{Z}}, [3]_{6\mathbb{Z}}), \right. \\
&\qquad ([4]_{6\mathbb{Z}}, [0]_{6\mathbb{Z}}), \, ([5]_{6\mathbb{Z}}, [3]_{6\mathbb{Z}}), \, ([0]_{6\mathbb{Z}}, [3]_{6\mathbb{Z}}), \, ([1]_{6\mathbb{Z}}, [0]_{6\mathbb{Z}}), \\
&\qquad \left. ([2]_{6\mathbb{Z}}, [3]_{6\mathbb{Z}}), \, ([3]_{6\mathbb{Z}}, [0]_{0\mathbb{Z}}), \, ([4]_{6\mathbb{Z}}, [3]_{6\mathbb{Z}}), \, ([5]_{5\mathbb{Z}}, [0]_{6\mathbb{Z}}) \right\} \\
\mathrm{Ker}(\chi_{3,3}) &= \left\{ ([x]_{6\mathbb{Z}}, [y]_{6\mathbb{Z}}) \in \mathbb{Z}_6 \times \mathbb{Z}_6 \mid [9x+3y]_{6\mathbb{Z}} = 0 \right\} \\
&= \left\{ ([0]_{6\mathbb{Z}}, [0]_{6\mathbb{Z}}), \, ([1]_{6\mathbb{Z}}, [3]_{6\mathbb{Z}}), \, ([2]_{6\mathbb{Z}}, [0]_{6\mathbb{Z}}), \, ([3]_{6\mathbb{Z}}, [3]_{6\mathbb{Z}}), \right. \\
&\qquad ([4]_{6\mathbb{Z}}, [0]_{6\mathbb{Z}}), \, ([5]_{6\mathbb{Z}}, [3]_{6\mathbb{Z}}), \, ([0]_{6\mathbb{Z}}, [2]_{6\mathbb{Z}}), \, ([1]_{6\mathbb{Z}}, [1]_{6\mathbb{Z}}), \\
&\qquad \left. ([2]_{6\mathbb{Z}}, [2]_{6\mathbb{Z}}), \, ([3]_{6\mathbb{Z}}, [1]_{0\mathbb{Z}}), \, ([4]_{6\mathbb{Z}}, [2]_{6\mathbb{Z}}), \, ([5]_{5\mathbb{Z}}, [1]_{6\mathbb{Z}}) \right\} \\
\mathrm{Ker}(\chi_{0,4}) &= \left\{ ([x]_{6\mathbb{Z}}, [y]_{6\mathbb{Z}}) \in \mathbb{Z}_6 \times \mathbb{Z}_6 \mid [4y]_{6\mathbb{Z}} = 0 \right\} \\
&= \left\{ ([0]_{6\mathbb{Z}}, [0]_{6\mathbb{Z}}), \, ([1]_{6\mathbb{Z}}, [3]_{6\mathbb{Z}}), \, ([2]_{6\mathbb{Z}}, [0]_{6\mathbb{Z}}), \, ([3]_{6\mathbb{Z}}, [3]_{6\mathbb{Z}}), \right. \\
&\qquad ([4]_{6\mathbb{Z}}, [0]_{6\mathbb{Z}}), \, ([5]_{6\mathbb{Z}}, [3]_{6\mathbb{Z}}), \, ([0]_{6\mathbb{Z}}, [3]_{6\mathbb{Z}}), \, ([1]_{6\mathbb{Z}}, [3]_{6\mathbb{Z}}), \\
&\qquad \left. ([2]_{6\mathbb{Z}}, [3]_{6\mathbb{Z}}), \, ([3]_{6\mathbb{Z}}, [0]_{0\mathbb{Z}}), \, ([4]_{6\mathbb{Z}}, [3]_{6\mathbb{Z}}), \, ([5]_{5\mathbb{Z}}, [0]_{6\mathbb{Z}}) \right\} \\
\mathrm{Ker}(\chi_{3,5}) &= \left\{ ([x]_{6\mathbb{Z}}, [y]_{6\mathbb{Z}}) \in \mathbb{Z}_6 \times \mathbb{Z}_6 \mid [9x+5y]_{6\mathbb{Z}} = 0 \right\} \\
&= \left\{ ([0]_{6\mathbb{Z}}, [0]_{6\mathbb{Z}}), \, ([1]_{6\mathbb{Z}}, [3]_{6\mathbb{Z}}), \, ([2]_{6\mathbb{Z}}, [0]_{6\mathbb{Z}}), \, ([3]_{6\mathbb{Z}}, [3]_{6\mathbb{Z}}), \right. \\
&\qquad \left. ([4]_{6\mathbb{Z}}, [0]_{6\mathbb{Z}}), \, ([5]_{6\mathbb{Z}}, [3]_{6\mathbb{Z}}) \right\} = \mathcal{H}.
\end{aligned}
$$

With

$$
\chi_{3,1}([x]_{6\mathbb{Z}}, [y]_{6\mathbb{Z}}) \underset{(6.117)}{=} e^{2\pi i \frac{3x+y}{6}} = e^{\pi i \left(x + \frac{y}{3} \right)}
$$

we also have

$$\chi_{0,0}([x]_{6\mathbb{Z}},[y]_{6\mathbb{Z}}) = 1 \qquad\qquad = \left(\chi_{3,1}([x]_{6\mathbb{Z}},[y]_{6\mathbb{Z}})\right)^{0}$$

$$\chi_{3,1}([x]_{6\mathbb{Z}},[y]_{6\mathbb{Z}}) = e^{\pi i\left(x+\frac{y}{3}\right)} \qquad = \left(\chi_{3,1}([x]_{6\mathbb{Z}},[y]_{6\mathbb{Z}})\right)^{1}$$

$$\chi_{0,2}([x]_{6\mathbb{Z}},[y]_{6\mathbb{Z}}) = e^{\pi i\frac{2y}{3}} \qquad\quad = \left(\chi_{3,1}([x]_{6\mathbb{Z}},[y]_{6\mathbb{Z}})\right)^{2}$$

$$\chi_{3,3}([x]_{6\mathbb{Z}},[y]_{6\mathbb{Z}}) = e^{\pi i(x+y)} \qquad\; = \left(\chi_{3,1}([x]_{6\mathbb{Z}},[y]_{6\mathbb{Z}})\right)^{3}$$

$$\chi_{0,4}([x]_{6\mathbb{Z}},[y]_{6\mathbb{Z}}) = e^{\pi i\frac{4y}{3}} \qquad\quad = \left(\chi_{3,1}([x]_{6\mathbb{Z}},[y]_{6\mathbb{Z}})\right)^{4}$$

$$\chi_{3,5}([x]_{6\mathbb{Z}},[y]_{6\mathbb{Z}}) = e^{\pi i\left(x+\frac{5y}{3}\right)} \quad = \left(\chi_{3,1}([x]_{6\mathbb{Z}},[y]_{6\mathbb{Z}})\right)^{5},$$

confirming (6.120).

Solution 6.78 To calculate the trace in $\mathbb{H}^{A}$ we use that, because of (6.113), we can utilize the basis

$$\left\{|r\rangle \otimes |s\rangle \mid r,s \in \{0,\ldots,N-1\}\right\} \subset \mathbb{H}^{A}$$

such that

$$\operatorname{tr}\left(P_{u,v}F_{\mathcal{G}}\rho^{A}F_{\mathcal{G}}^{*}\right) \underset{(2.57)}{=} \sum_{r,s\in\{0,\ldots,N-1\}} \langle r\otimes s|(P_{u,v}F_{\mathcal{G}}\rho^{A}F_{\mathcal{G}}^{*})r\otimes s\rangle$$

$$\underset{(6.123)}{=} \sum_{r,s\in\{0,\ldots,N-1\}} \underbrace{\langle r|u\rangle}_{=\delta_{ru}} \underbrace{\langle s|v\rangle}_{=\delta_{sv}} \langle u\otimes v|(F_{\mathcal{G}}\rho^{A}F_{\mathcal{G}}^{*})r\otimes s\rangle = \langle u\otimes v|(F_{\mathcal{G}}\rho^{A}F_{\mathcal{G}}^{*})u\otimes v\rangle$$

$$\underset{(6.122)}{=} \frac{1}{N^{2}} \sum_{\substack{[g]_{\mathcal{H}}\in\mathcal{G}/\mathcal{H}\\ [n]_{N\mathbb{Z}},[m]_{N\mathbb{Z}}\in\mathbb{Z}_{N}}} e^{2\pi i\frac{dx+y}{N}(n-m)} \langle u|dn\bmod N\rangle\langle dm\bmod N|u\rangle \underbrace{\langle v|n\rangle}_{=\delta_{vn}} \underbrace{\langle m|v\rangle}_{=\delta_{mv}}$$

$$= \frac{1}{N^{2}} \sum_{[g]_{\mathcal{H}}\in\mathcal{G}/\mathcal{H}} |\langle u|dv\bmod N\rangle|^{2} = \frac{|\langle u|dv\bmod N\rangle|^{2}}{N^{2}} \sum_{[g]_{\mathcal{H}}\in\mathcal{G}/\mathcal{H}} 1$$

$$= \frac{|\langle u|dv\bmod N\rangle|^{2}}{N^{2}} |\mathcal{G}/\mathcal{H}| \underset{(F.36)}{=} \frac{|\langle u|dv\bmod N\rangle|^{2}}{N^{2}} \frac{|\mathcal{G}|}{|\mathcal{H}|}$$

$$\underset{(6.111),(6.116)}{=} \frac{|\langle u|dv\bmod N\rangle|^{2}}{N}.$$

Solution 6.79 For the proof of (6.152) note that (6.149) implies

$$\sin\theta_{0} = \sqrt{\frac{m}{N}} \quad\text{and}\quad \cos\theta_{0} = \sqrt{1-\frac{m}{N}}. \tag{G.139}$$

It follows then that

$$|\Psi_0\rangle \underset{(6.148)}{=} \frac{1}{\sqrt{N}}\sum_{x=0}^{N-1}|x\rangle \underset{(6.138)}{=} \frac{1}{\sqrt{N}}\left(\sqrt{N-m}|\Psi_{S\perp}\rangle + \sqrt{m}|\Psi_S\rangle\right)$$
$$\underset{(G.139)}{=} \cos\theta_0|\Psi_{S\perp}\rangle + \sin\theta_0|\Psi_S\rangle,$$

proving (6.152).

Since $\|\Psi_0\| = 1$, the projector onto the subspace $\mathbb{H}_{sub} = \mathrm{Span}\{|\Psi_0\rangle\}$ is $P_{sub} = |\Psi_0\rangle\langle\Psi_0|$. Consequently R_{Ψ_0} as defined in (6.150) coincides with the definition of a reflection about $|\Psi_0\rangle$ given in (6.147).

Solution 6.80 With

$$\cos((2j+1)\alpha) = \frac{e^{i(2j+1)\alpha} + e^{-i(2j+1)\alpha}}{2} = \frac{e^{i\alpha}}{2}\left(e^{2i\alpha}\right)^j + \frac{e^{-i\alpha}}{2}\left(e^{-2i\alpha}\right)^j$$

we obtain

$$\sum_{j=0}^{J-1}\cos((2j+1)\alpha) = \frac{e^{i\alpha}}{2}\underbrace{\sum_{j=0}^{J-1}\left(e^{2i\alpha}\right)^j}_{=\frac{1-e^{i2J\alpha}}{1-e^{i2\alpha}}} + \frac{e^{-i\alpha}}{2}\underbrace{\sum_{j=0}^{J-1}\left(e^{-2i\alpha}\right)^j}_{=\frac{1-e^{-i2J\alpha}}{1-e^{-i2\alpha}}}$$
$$= \frac{e^{i\alpha}}{2}\frac{1-e^{i2J\alpha}}{1-e^{i2\alpha}} + \frac{e^{-i\alpha}}{2}\frac{1-e^{-i2J\alpha}}{1-e^{-i2\alpha}}$$
$$= \frac{e^{iJ\alpha}}{2}\frac{e^{-iJ\alpha}-e^{iJ\alpha}}{e^{-i\alpha}-e^{i\alpha}} + \frac{e^{-iJ\alpha}}{2}\frac{e^{iJ\alpha}-e^{-iJ\alpha}}{e^{i\alpha}-e^{-i\alpha}}$$
$$= \frac{e^{iJ\alpha}+e^{-iJ\alpha}}{2}\frac{e^{iJ\alpha}-e^{-iJ\alpha}}{e^{i\alpha}-e^{-i\alpha}}$$
$$= \cos(J\alpha)\frac{\sin(J\alpha)}{\sin\alpha} = \frac{2\cos(J\alpha)\sin(J\alpha)}{2\sin\alpha}$$
$$= \frac{\sin(2J\alpha)}{2\sin\alpha}.$$

Solutions to Exercises from Chapter 7

Solution 7.81 Recalling from (7.1) that for $a,b \in \mathbb{F}_2$

$$a +_{\mathbb{F}_2} b = a \overset{2}{\oplus} b = (a+b) \bmod 2 = \begin{cases} 0 & \text{if } a = b \\ 1 & \text{if } a \neq b, \end{cases}$$

the claims (i) - (iii) are obvious. To show (iv), consider

$$d_H(\mathbf{u}, \mathbf{v}) + d_H(\mathbf{v}, \mathbf{w}) - d_H(\mathbf{u}, \mathbf{w}) = \sum_{j=1}^{k} \underbrace{u_j \overset{2}{\oplus} v_j + v_j \overset{2}{\oplus} w_j - u_j \overset{2}{\oplus} w_j}_{=:a_j},$$

where $u_j \overset{2}{\oplus} v_j, v_j \overset{2}{\oplus} w_j, u_j \overset{2}{\oplus} w_j \in \{0,1\}$ and thus for any $j \in \{1, \ldots, k\}$

$$\begin{aligned} u_j = w_j &\Rightarrow & a_j \geq 0 \\ u_j \neq w_j \text{ and } u_j = v_j &\Rightarrow & v_j \neq w_j &\Rightarrow & a_j = 0 \\ u_j \neq w_j \text{ and } u_j \neq v_j &\Rightarrow & v_j = w_j &\Rightarrow & a_j = 0. \end{aligned}$$

Hence, we obtain $d_H(\mathbf{u}, \mathbf{v}) + d_H(\mathbf{v}, \mathbf{w}) - d_H(\mathbf{u}, \mathbf{w}) \geq 0$.

Solution 7.82

(i) Recall that the kernel of any linear map $F : \mathbb{V} \to \mathbb{W}$ between finite-dimensional vector spaces $\mathbb{V}$ and $\mathbb{W}$ is defined as

$$\mathrm{Ker}(F) = \{ \mathbf{w} \in \mathbb{V} \mid F\mathbf{w} = 0 \}. \tag{G.140}$$

For any $\mathbf{w} \in \mathbb{F}_2^k$ it follows from (7.8) that $\mathrm{G}\mathbf{w} \in \mathrm{Ker}(\mathrm{H})$ and thus

$$\mathrm{HG}\mathbf{w} \underset{(\mathrm{G}.140)}{=} 0,$$

verifying that $\mathrm{HG} = 0$.

(ii) We use the following result from basic linear algebra: for any linear map $F : \mathbb{V} \to \mathbb{W}$ between finite-dimensional vector spaces $\mathbb{V}$ and $\mathbb{W}$ we have

$$\dim F\{\mathbb{V}\} = \dim \mathbb{V} - \dim \mathrm{Ker}(F).$$

Applying this to $\mathrm{H} : \mathbb{F}_2^n \to \mathbb{F}_2^{n-k}$, we find

$$\dim \mathrm{H}\{\mathbb{F}_2^n\} = \dim \mathbb{F}_2^n - \dim \mathrm{Ker}(\mathrm{H}) \underset{(7.8)}{=} n - \dim \mathrm{G}\{\mathbb{F}_2^k\} = n - k,$$

where in the last equation we used that per definition G is of maximal rank k.

(iii) Let

$$H = \begin{pmatrix} \mathbf{h}_1^T \\ \vdots \\ \mathbf{h}_{n-k}^T \end{pmatrix},$$

where the $\mathbf{h}_j \in \mathbb{F}_2^n$ for $j \in \{1,\ldots,n-k\}$ are linearly independent. Define

$$\widetilde{\mathbf{h}}_j = \begin{cases} \mathbf{h}_1 + \mathbf{h}_2 & \text{if } j = 1 \\ \mathbf{h}_j & \text{if } j \neq 1. \end{cases} \tag{G.141}$$

Then the $\widetilde{\mathbf{h}}_j$ are linearly independent. To prove this, suppose $a_j \in \mathbb{F}_2$ for $j \in \{1,\ldots,n-k\}$ are such that

$$0 = \sum_{j=1}^{n-k} a_j \widetilde{\mathbf{h}}_j \underset{(G.141)}{=} a_1(\mathbf{h}_1 + \mathbf{h}_2) + \sum_{j=2}^{n-k} a_j \mathbf{h}_j$$

$$= a_1 \mathbf{h}_1 + (a_1 + a_2)\mathbf{h}_2 + a_3 \mathbf{h}_3 + \cdots + a_{n-k}\mathbf{h}_{n-k}.$$

Since the $\mathbf{h}_j$ are linearly independent, we must have $a_1 = a_1 + a_2 = a_3 = \cdots = a_{n-k} = 0$ from which it follows that $a_j = 0$ for all $j \in \{1,\ldots,n-k\}$. Thus, we have shown that

$$\sum_{j=1}^{n-k} a_j \widetilde{\mathbf{h}}_j = 0 \quad \Rightarrow \quad a_j = 0 \quad \forall j \in \{0,\ldots,n-k\},$$

which means that the $\widetilde{\mathbf{h}}_j$ are linearly independent. Therefore,

$$\widetilde{H} = \begin{pmatrix} \widetilde{\mathbf{h}}_1^T \\ \vdots \\ \widetilde{\mathbf{h}}_{n-k}^T \end{pmatrix}$$

has maximal rank $\dim \widetilde{H}\{\mathbb{F}_2^n\} = n-k$, and we have

$$\dim \mathrm{Ker}(\widetilde{H}) = n - \dim \widetilde{H}\{\mathbb{F}_2^n\} = k = \dim \mathrm{Ker}(H). \tag{G.142}$$

Moreover,

$$\mathbf{u} \in \mathrm{Ker}(H) \quad \Leftrightarrow \quad \sum_{l=1}^{n} (\mathbf{h}_j)_l u_l = 0 \quad \forall j \in \{1,\ldots,n-k\}$$

$$\Rightarrow \quad \sum_{l=1}^{n} (\widetilde{\mathbf{h}}_j)_l u_l = 0 \quad \forall j \in \{1,\ldots,n-k\}$$

$$\Rightarrow \quad \mathbf{u} \in \mathrm{Ker}(\widetilde{H}),$$

which implies $\mathrm{Ker(H)} \subset \mathrm{Ker(\widetilde{H})}$, and together with (G.142) this gives

$$\mathrm{Ker}(\widetilde{H}) \underset{(7.8)}{=} \mathrm{Ker}(H) = G\{\mathbb{F}_2^k\}.$$

Consequently, $\widetilde{H}$ is a parity check matrix as well. But $\widetilde{H}$ cannot be equal to H since this would require $\widetilde{\mathbf{h}}_1 = \mathbf{h}_1$ and (G.141) shows that then $\mathbf{h}_2 = 0$, which is impossible since the $\mathbf{h}_j$ are assumed linearly independent.

Solution 7.83 For any $\mathbf{a}, \mathbf{b} \in \mathbb{F}_2^n$ we have

$$\mathrm{syn}_c(\mathbf{a} \overset{2}{\oplus} \mathbf{b}) \underset{(7.11)}{=} H(\mathbf{a} \overset{2}{\oplus} \mathbf{b}) = H\mathbf{a} \overset{2}{\oplus} H\mathbf{b} \underset{(7.11)}{=} \mathrm{syn}_c(\mathbf{a}) \overset{2}{\oplus} \mathrm{syn}_c(\mathbf{b}).$$

Solution 7.84 Evaluating the encoding map for a $|\psi\rangle = a|0\rangle + b|1\rangle \in \mathbb{H}$ with $a, b \in \mathbb{C}$, we find first

$$A_1\iota|\psi\rangle \underset{(7.30)}{=} A_1\Big((a|0\rangle + b|1\rangle) \otimes |0\rangle^8 \Big)$$

$$\underset{(7.31)}{=} \Big(|1\rangle\langle 1| \otimes \mathbf{1}^{\otimes 2} \otimes X \otimes \mathbf{1}^{\otimes 2} \otimes X \otimes \mathbf{1}^{\otimes 2} \Big) \Big((a|0\rangle + b|1\rangle) \otimes |0\rangle^8 \Big)$$

$$+ \Big(|0\rangle\langle 0| \otimes \mathbf{1}^{\otimes 8} \Big) \Big((a|0\rangle + b|1\rangle) \otimes |0\rangle^8 \Big)$$

$$= a|0\rangle^9 + b|1\rangle \otimes |0\rangle^2 \otimes |1\rangle \otimes |0\rangle^2 \otimes |1\rangle \otimes |0\rangle^2$$

$$= a|0\rangle^9 + b(|100\rangle)^{\otimes 3}, \tag{G.143}$$

where we used $X|0\rangle = \sigma_x|0\rangle = |1\rangle$. Next, we have

$$A_2A_1\iota|\psi\rangle \underset{(7.31),(G.143)}{=} \big(H \otimes \mathbf{1}^{\otimes 2}\big)^{\otimes 3} \Big(a|0\rangle^9 + b(|100\rangle)^{\otimes 3} \Big) \tag{G.144}$$

$$\underset{(2.160),(2.161)}{=} a\left(\frac{|0\rangle + |1\rangle}{\sqrt{2}} \otimes |0\rangle^2 \right)^{\otimes 3} + b\left(\frac{|0\rangle - |1\rangle}{\sqrt{2}} \otimes |0\rangle^2 \right)^{\otimes 3}$$

and finally,

$$
\begin{aligned}
C_q|\psi\rangle &= A_3A_2A_1\iota|\psi\rangle \\
&\underbrace{=}_{(7.31),(G.144)} \left(|1\rangle\langle1|\otimes X\otimes X+|0\rangle\langle0|\otimes\mathbf{1}^{\otimes2}\right)^{\otimes3} \\
&\qquad \left[a\left(\frac{|0\rangle+|1\rangle}{\sqrt{2}}\otimes|0\rangle^2\right)^{\otimes3}+b\left(\frac{|0\rangle-|1\rangle}{\sqrt{2}}\otimes|0\rangle^2\right)^{\otimes3}\right] \\
&= a\left[\left(|1\rangle\langle1|\otimes X\otimes X+|0\rangle\langle0|\otimes\mathbf{1}^{\otimes2}\right)\left(\frac{|0\rangle+|1\rangle}{\sqrt{2}}\otimes|0\rangle^2\right)\right]^{\otimes3} \\
&\quad +b\left[\left(|1\rangle\langle1|\otimes X\otimes X+|0\rangle\langle0|\otimes\mathbf{1}^{\otimes2}\right)\left(\frac{|0\rangle-|1\rangle}{\sqrt{2}}\otimes|0\rangle^2\right)\right]^{\otimes3} \\
&= a\left[\left(|1\rangle\langle1|\otimes X\otimes X+|0\rangle\langle0|\otimes\mathbf{1}^{\otimes2}\right)\frac{|000\rangle+|100\rangle}{\sqrt{2}}\right]^{\otimes3} \\
&\quad +b\left[\left(|1\rangle\langle1|\otimes X\otimes X+|0\rangle\langle0|\otimes\mathbf{1}^{\otimes2}\right)\frac{|000\rangle-|100\rangle}{\sqrt{2}}\right]^{\otimes3} \\
&= a\left(\frac{|000\rangle+|111\rangle}{\sqrt{2}}\right)^{\otimes3}+b\left(\frac{|000\rangle-|111\rangle}{\sqrt{2}}\right)^{\otimes3}.
\end{aligned}
$$

Solution 7.85 In the following we use that

$$
\begin{aligned}
\langle0|1\rangle &= 0=\langle1|0\rangle & \text{(G.145)} \\
|0\rangle\langle0|+|1\rangle\langle1| &= \mathbf{1} & \text{(G.146)} \\
X^2 &= \sigma_x^2\underbrace{=}_{(2.76)}\mathbf{1} & \text{(G.147)} \\
H^2 &\underbrace{=}_{(2.163)}\mathbf{1}. & \text{(G.148)}
\end{aligned}
$$

Therefore, we have first

$$
\begin{aligned}
A_1^2 &\underbrace{=}_{(7.31)} \left(|1\rangle\langle1|\otimes\mathbf{1}^{\otimes2}\otimes X\otimes\mathbf{1}^{\otimes2}\otimes X\otimes\mathbf{1}^{\otimes2}+|0\rangle\langle0|\otimes\mathbf{1}^{\otimes8}\right)^2 \\
&\underbrace{=}_{(G.145)} \left(|1\rangle\langle1|\otimes\mathbf{1}^{\otimes2}\otimes X\otimes\mathbf{1}^{\otimes2}\otimes X\otimes\mathbf{1}^{\otimes2}\right)^2+\left(|0\rangle\langle0|\otimes\mathbf{1}^{\otimes8}\right)^2 \\
&\underbrace{=}_{(G.147)} |1\rangle\langle1|\otimes\mathbf{1}^{\otimes8}+|0\rangle\langle0|\otimes\mathbf{1}^{\otimes8}=(|1\rangle\langle1|+|0\rangle\langle0|)\otimes\mathbf{1}^{\otimes8} \\
&\underbrace{=}_{(G.146)} \mathbf{1}^{\otimes9}
\end{aligned}
$$

and then

$$A_2^2 \underset{(7.31)}{=} \left(H \otimes \mathbf{1}^{\otimes 2}\right)^{\otimes 3} \left(H \otimes \mathbf{1}^{\otimes 2}\right)^{\otimes 3} = \left(H^2 \otimes \mathbf{1}^{\otimes 2}\right)^{\otimes 3} \underset{(\text{G}.148)}{=} \left(\mathbf{1}^{\otimes 3}\right)^{\otimes 3}$$

$$= \mathbf{1}^{\otimes 9}.$$

Finally, we obtain

$$A_3^2 \underset{(7.31)}{=} \left(|1\rangle\langle 1| \otimes X \otimes X + |0\rangle\langle 0| \otimes \mathbf{1}^{\otimes 2}\right)^{\otimes 3} \left(|1\rangle\langle 1| \otimes X \otimes X + |0\rangle\langle 0| \otimes \mathbf{1}^{\otimes 2}\right)^{\otimes 3}$$

$$= \left(|1\rangle\langle 1| \otimes X^2 \otimes X^2 + |0\rangle\langle 0| \otimes \mathbf{1}^{\otimes 2}\right)^{\otimes 3} \underset{(\text{G}.147)}{=} \left(|1\rangle\langle 1| \otimes \mathbf{1}^{\otimes 2} + |0\rangle\langle 0| \otimes \mathbf{1}^{\otimes 2}\right)^{\otimes 3}$$

$$= \left((|1\rangle\langle 1| + |0\rangle\langle 0|) \otimes \mathbf{1}^{\otimes 2}\right)^{\otimes 3} \underset{(\text{G}.146)}{=} \left(\mathbf{1}^{\otimes 3}\right)^{\otimes 3}$$

$$= \mathbf{1}^{\otimes 9}.$$

Solution 7.86 Keeping in mind that the U_{ab} are just complex numbers, we have

$$\widetilde{\mathcal{E}}_a^* \underset{(7.41)}{=} \left(\sum_{b=1}^{m} U_{ab} \mathcal{E}_b\right)^* = \sum_{b=1}^{m} (U_{ab} \mathcal{E}_b)^* \underset{(2.32)}{=} \sum_{b=1}^{m} \overline{U_{ab}} \, \mathcal{E}_b^*$$

$$\underset{(2.34)}{=} \sum_{b=1}^{m} U_{ba}^* \mathcal{E}_b^*, \qquad\qquad (\text{G}.149)$$

which implies

$$\sum_a \widetilde{\mathcal{E}}_a \rho \widetilde{\mathcal{E}}_a^* \underset{(7.41),(\text{G}.149)}{=} \sum_a \left(\sum_b U_{ab} \mathcal{E}_b\right) \rho \left(\sum_c U_{ca}^* \mathcal{E}_c^*\right)$$

$$= \sum_{b,c} \left(\sum_a U_{ca}^* U_{ab}\right) \mathcal{E}_b \rho \mathcal{E}_c^* = \sum_{b,c} \underbrace{(U^* U)_{cb}}_{=\delta_{cb}} \mathcal{E}_b \rho \mathcal{E}_c^*$$

$$= \sum_b \mathcal{E}_b \rho \mathcal{E}_b^*,$$

and similarly,

$$\sum_a \widetilde{\mathcal{E}}_a^* \widetilde{\mathcal{E}}_a \underbrace{=}_{(7.41),(G.149)} \sum_a \left(\sum_b U_{ba}^* \mathcal{E}_b^* \right) \left(\sum_c U_{ac} \mathcal{E}_c \right)$$

$$= \sum_{b,c} \left(\sum_a U_{ba}^* U_{ac} \right) \mathcal{E}_b^* \mathcal{E}_c = \sum_{b,c} \underbrace{(U^* U)_{bc}}_{=\delta_{bc}} \mathcal{E}_b^* \mathcal{E}_c$$

$$= \sum_b \mathcal{E}_b^* \mathcal{E}_b .$$

Solution 7.87 As quantum operations, S and T are convex-linear, and we have for every $\rho_1, \rho_2 \in D(\mathbb{H})$ and $\mu \in [0,1]$

$$S(\mu\rho_1 + (1-\mu)\rho_2) = \mu S(\rho_1) + (1-\mu)S(\rho_2). \tag{G.150}$$

Let

$$T\left(\frac{S(\rho)}{\mathrm{tr}(S(\rho))} \right) = \rho \quad \forall \rho \in D(\mathbb{H}). \tag{G.151}$$

Using the operator sum representation for T with T_l denoting the operation elements, we have for any $\rho \in D(\mathbb{H})$

$$\frac{1}{\mathrm{tr}(S(\rho))} \sum_l T_l S(\rho) T_l^* = \sum_l T_l \frac{S(\rho)}{\mathrm{tr}(S(\rho))} T_l^* \underbrace{=}_{(3.97)} T\left(\frac{S(\rho)}{\mathrm{tr}(S(\rho))} \right) \underbrace{=}_{(G.151)} \rho,$$

which implies

$$\sum_l T_l S(\rho) T_l^* = \mathrm{tr}(S(\rho)) \rho. \tag{G.152}$$

Consequently, we have for any $\rho_1, \rho_2 \in D(\mathbb{H})$ and $\mu \in [01,]$

$$\mu\rho_1 + (1-\mu)\rho_2 \underbrace{=}_{(G.151)} T\left(\frac{S(\mu\rho_1 + (1-\mu)\rho_2)}{\mathrm{tr}(S(\mu\rho_1 + (1-\mu)\rho_2))} \right)$$

$$\underbrace{=}_{(G.150)} T\left(\frac{\mu S(\rho_1) + (1-\mu)S(\rho_2)}{\mathrm{tr}(\mu S(\rho_1) + (1-\mu)S(\rho_2))} \right)$$

$$\underbrace{=}_{(3.97)} \sum_l T_l \frac{\mu S(\rho_1) + (1-\mu)S(\rho_2)}{\mathrm{tr}(\mu S(\rho_1) + (1-\mu)S(\rho_2))} T_l^*$$

$$= \frac{\mu \sum_l T_l S(\rho_1) T_l^* + (1-\mu) \sum_l T_l S(\rho_2) T_l^*}{\mu \operatorname{tr}(S(\rho_1)) + (1-\mu) \operatorname{tr}(S(\rho_2))}$$

$$\underset{\text{(G.152)}}{=} \frac{\mu \operatorname{tr}(S(\rho_1)) \rho_1 + (1-\mu) \operatorname{tr}(S(\rho_2)) \rho_2}{\mu \operatorname{tr}(S(\rho_1)) + (1-\mu) \operatorname{tr}(S(\rho_2))},$$

which implies

$$\mu \left(\frac{\operatorname{tr}(S(\rho_1))}{\mu \operatorname{tr}(S(\rho_1)) + (1-\mu) \operatorname{tr}(S(\rho_2))} - 1 \right) \rho_1$$

$$= (1-\mu) \left(1 - \frac{\operatorname{tr}(S(\rho_2))}{\mu \operatorname{tr}(S(\rho_1)) + (1-\mu) \operatorname{tr}(S(\rho_2))} \right) \rho_2 .$$

Since ρ_1 and ρ_2 and $\mu \in [0,1]$ are arbitrary, it follows that the terms in the parenthesis have to vanish from which it follows that $\operatorname{tr}(S(\rho_1)) = \operatorname{tr}(S(\rho_2))$ and thus $\operatorname{tr}(S(\rho)) = \text{const}$ as claimed.

Solution 7.88

$$\left\| (A - \langle \psi|A\psi \rangle) |\psi \rangle \right\|^2 \underset{\text{(2.5)}}{=} \langle A\psi - \langle \psi|A\psi \rangle \psi | A\psi - \langle \psi|A\psi \rangle \psi \rangle$$

$$\underset{\text{(2.4),(2.6)}}{=} \langle A\psi|A\psi \rangle - \langle \psi|A\psi \rangle \langle A\psi|\psi \rangle - \langle A\psi|\psi \rangle \langle \psi|A\psi \rangle$$

$$+ \langle A\psi|\psi \rangle \langle \psi|A\psi \rangle \langle \psi|\psi \rangle$$

$$\underset{\text{(2.6)}}{=} \langle A\psi|A\psi \rangle - |\langle \psi|A\psi \rangle|^2 \, (2 - \underbrace{\|\psi\|^2}_{=1})$$

$$\underset{\text{(2.31),(2.30)}}{=} \langle \psi|A^*A\psi \rangle - |\langle \psi|A\psi \rangle|^2$$

Solution 7.89 The claim $\Leftarrow$ is trivially true. To prove $\Rightarrow$, let

$$\tilde{a} \colon S_{\mathbb{H}}^1 \longrightarrow \mathbb{C}$$
$$|\psi \rangle \longmapsto \langle \psi|A\psi \rangle$$

and for all $|\psi \rangle \in \mathbb{H}$ with $\|\psi\| = 1$ let

$$A|\psi \rangle = \tilde{a}(|\psi \rangle)|\psi \rangle . \tag{G.153}$$

Moreover, for $i \in \{1,2\}$ let $|\psi_i\rangle \in \mathbb{H}$ be two linearly independent vectors satisfying $\||\psi_i\|| = 1$, and let $z_1, z_2 \in \mathbb{C}$ with $z_1 z_2 \neq 0$ be such that $\||z_1|\psi_1\rangle + z_2|\psi_2\rangle\|| = 1$ as well. Then it follows that

$$A\big(z_1|\psi_1\rangle + z_2|\psi_2\rangle\big) \underbrace{=}_{(G.153)} \widetilde{a}(z_1|\psi_1\rangle + z_2|\psi_2\rangle)\big(z_1|\psi_1\rangle + z_2|\psi_2\rangle\big) \qquad (G.154)$$

Since A is a linear operator, we also have

$$A\big(z_1|\psi_1\rangle + z_2|\psi_2\rangle\big) = z_1 A|\psi_1\rangle + z_2 A|\psi_2\rangle \underbrace{=}_{(G.153)} \widetilde{a}(|\psi_1\rangle)z_1)|\psi_1\rangle + \widetilde{a}(|\psi_2\rangle)z_2)|\psi_2\rangle .$$

$$(G.155)$$

Equating the right sides of (G.154) and (G.155) and using that the $|\psi_i\rangle$ are linearly independent, it follows that their respective coefficients have to coincide and thus

$$\widetilde{a}(|\psi_1\rangle) = \widetilde{a}(z_1|\psi_1\rangle + z_2|\psi_2\rangle) = \widetilde{a}(|\psi_2\rangle) .$$

This implies that $\widetilde{a}$ is constant on the unit sphere $S_{\mathbb{H}}^1$, that is, $\widetilde{a}(|\psi\rangle) = \text{const} = a \in \mathbb{C}$ and thus

$$A|\psi\rangle = a|\psi\rangle \quad \forall |\psi\rangle \in S_{\mathbb{H}}^1 . \qquad (G.156)$$

But then for any $|\varphi\rangle \in \mathbb{H} \smallsetminus \{0\}$ we have that $\frac{|\varphi\rangle}{\||\varphi\||} \in S_{\mathbb{H}}^1$ which implies

$$A\frac{|\varphi\rangle}{\||\varphi\||} \underbrace{=}_{(G.156)} a\frac{|\varphi\rangle}{\||\varphi\||} ,$$

and it follows that $A|\varphi\rangle = a|\varphi\rangle$ for all $|\varphi\rangle \in \mathbb{H}$ since this is also trivially true for $|\varphi\rangle = 0$.

Solution 7.90 We first show that there is a $g \in \mathcal{P}_9$ with $w_{\mathcal{P}}(g) = 3$ such that $\langle \Psi_0 | g \Psi_1 \rangle \neq 0 = f(g)\delta_{01}$, where the basis codewords $|\Psi_0\rangle$ and $|\Psi_1\rangle$ are given by (7.33). For this consider

$$g = (1^{\otimes 2} \otimes Z)^{\otimes 3} = 1 \otimes 1 \otimes Z \otimes 1 \otimes 1 \otimes Z \otimes 1 \otimes 1 \otimes Z , \qquad (G.157)$$

which satisfies $w_{\mathcal{P}}(g) = 3$. Noting that

$$
\begin{aligned}
(1^{\otimes 2} \otimes Z)|000\rangle &= |0\rangle \otimes |0\rangle \otimes Z|0\rangle = |0\rangle \otimes |0\rangle \otimes |0\rangle = |000\rangle \\
(1^{\otimes 2} \otimes Z)|111\rangle &= |1\rangle \otimes |1\rangle \otimes Z|1\rangle = |1\rangle \otimes |1\rangle \otimes (-|1\rangle) = -|111\rangle ,
\end{aligned}
\qquad (G.158)
$$

we obtain

$$g|\Psi_1\rangle \underset{\text{(G.157),(7.33)}}{=} (1^{\otimes 2} \otimes Z)^{\otimes 3} \left(\frac{|000\rangle - |111\rangle}{\sqrt{2}} \right)^{\otimes 3}$$

$$= \left(\frac{(1^{\otimes 2} \otimes Z)|000\rangle - (1^{\otimes 2} \otimes Z)|111\rangle}{\sqrt{2}} \right)^{\otimes 3} \underset{\text{(G.158)}}{=} \left(\frac{|000\rangle + |111\rangle}{\sqrt{2}} \right)^{\otimes 3}$$

$$\underset{\text{(7.33)}}{=} |\Psi_0\rangle.$$

Therefore, we have

$$\exists g \in \mathcal{P}_9 : \; \mathrm{w}_{\mathcal{P}}(g) = 3 \text{ and } \langle \Psi_0 | g \Psi_1 \rangle = \langle \Psi_0 | \Psi_0 \rangle = 1 \neq 0 = f(g)\delta_{01}.$$

It remains to show that for any $h \in \mathcal{P}_9$ with $\mathrm{w}_{\mathcal{P}}(h) \leq 2$ and $x, y \in \{0, 1\}$ we have, instead,

$$\langle \Psi_x | h \Psi_y \rangle = f(h)\delta_{xy}.$$

We do this in two steps, first addressing the case $x \neq y$, and in a second step we show that it holds for the case $x = y$ as well. To begin with, we note that with Σ_α^j as defined in (7.43) any $h \in \mathcal{P}_9$ with $\mathrm{w}_{\mathcal{P}}(h) \leq 2$ is of the form

$$h = \mathrm{i}^c \Sigma_\alpha^j \Sigma_\beta^l \tag{G.159}$$

with $c, \alpha, \beta \in \{0, \ldots, 3\}$ and $j, l \in \{0, \ldots, 8\}$. We also introduce the following intuitive and helpful notations.

$$|\psi_\pm\rangle := \frac{|000\rangle \pm |111\rangle}{\sqrt{2}} \in \mathbb{H}^{\otimes 3}$$

$$|\Psi_+\rangle := |\Psi_0\rangle = \left(\frac{|000\rangle + |111\rangle}{\sqrt{2}} \right)^{\otimes 3} = |\psi_+\rangle^{\otimes 3}$$

$$|\Psi_-\rangle := |\Psi_1\rangle = \left(\frac{|000\rangle - |111\rangle}{\sqrt{2}} \right)^{\otimes 3} = |\psi_-\rangle^{\otimes 3}.$$

For $j \in \{0, \ldots, 8\}$ we set $\hat{j} := \lfloor \frac{j}{3} \rfloor$ and $\check{j} := j \bmod 3$. Therefore, we obtain

$$\Sigma_\alpha^j |\Psi_\pm\rangle = \delta_{\hat{j}2} \left(\Sigma_\alpha^{\check{j}} |\psi_\pm\rangle \right) \otimes |\psi_\pm\rangle \otimes |\psi_\pm\rangle$$

$$+ \delta_{\hat{j}1} |\psi_\pm\rangle \otimes \left(\Sigma_\alpha^{\check{j}} |\psi_\pm\rangle \right) \otimes |\psi_\pm\rangle \tag{G.160}$$

$$+ \delta_{\hat{j}0} |\psi_\pm\rangle \otimes |\psi_\pm\rangle \otimes \left(\Sigma_\alpha^{\check{j}} |\psi_\pm\rangle \right).$$

It follows that

$$\langle \Sigma_\alpha^j \Psi_+ | \Sigma_\beta^l \Psi_- \rangle = 0 \tag{G.161}$$

since, due to (3.4), each of the nine scalar products formed by using (G.160) in the left side of (G.161) will always involve scalar products of tensor factors in $\mathbb{H}^{\otimes 3}$ of the form $\langle \psi_+ | \psi_- \rangle = 0$. Hence, we have for any $h \in \mathcal{P}_9$ with $w_{\mathcal{P}}(h) \leq 2$ that

$$\langle \Psi_0 | h \Psi_1 \rangle \underbrace{=}_{\text{(G.159),(G.160)}} \langle \Psi_+ | i^c \Sigma_\alpha^j \Sigma_\beta^l \Psi_- \rangle \underbrace{=}_{\text{(2.4),(2.30)}} i^c \langle (\Sigma_\alpha^j)^* \Psi_+ | \Sigma_\beta^l \Psi_- \rangle$$

$$\underbrace{=}_{(\Sigma_\alpha^j)^* = \Sigma_\alpha^j} i^c \langle \Sigma_\alpha^j \Psi_+ | \Sigma_\beta^l \Psi_- \rangle \underbrace{=}_{\text{(G.161)}} 0$$

$$= f(h)\delta_{01}$$

Finally, we show that for any $h \in \mathcal{P}_9$ with $w_{\mathcal{P}}(h) \leq 2$ we have $\langle \Psi_0 | h \Psi_0 \rangle = f(h) = \langle \Psi_1 | h \Psi_1 \rangle$. For this note first that

$$\langle \Sigma_\alpha^j \Psi_\pm | \Sigma_\beta^l \Psi_\pm \rangle \underbrace{=}_{\text{(G.160)}} \delta_{\hat{j}\hat{l}} \langle \Sigma_\alpha^{\check{j}} \Psi_\pm | \Sigma_\beta^{\check{l}} \Psi_\pm \rangle \tag{G.162}$$

$$+ (1 - \delta_{\hat{j}\hat{l}}) \langle \Sigma_\alpha^{\check{j}} \Psi_\pm | \psi_\pm \rangle \langle \psi_\pm | \Sigma_\beta^{\check{l}} \Psi_\pm \rangle.$$

Here we have

$$\langle \psi_\pm | \Sigma_\beta^{\check{l}} \psi_\pm \rangle$$

$$\underbrace{=}_{\text{(G.160)}} \frac{1}{2} \left(\langle 000 | \Sigma_\beta^{\check{l}} 000 \rangle + \langle 111 | \Sigma_\beta^{\check{l}} 111 \rangle \pm \langle 000 | \Sigma_\beta^{\check{l}} 111 \rangle \pm \langle 111 | \Sigma_\beta^{\check{l}} 000 \rangle \right)$$

$$\underbrace{=}_{\text{(3.4)}} \frac{1}{2} \left(\underbrace{\langle 0 | \sigma_\beta 0 \rangle + \langle 1 | \sigma_\beta 1 \rangle}_{= 2\delta_{\beta 0}} \pm \langle 0 | \sigma_\beta 1 \rangle \underbrace{\langle 0 | 1 \rangle^2}_{= 0} \pm \langle 1 | \sigma_\beta 0 \rangle \underbrace{\langle 1 | 0 \rangle^2}_{= 0} \right)$$

$$= \delta_{\beta 0}. \tag{G.163}$$

Likewise, we obtain

$$\langle \Sigma_\alpha^{\check{j}} \psi_\pm | \Sigma_\beta^{\check{l}} \psi_\pm \rangle \underbrace{=}_{(\Sigma_\alpha^j)^* = \Sigma_\alpha^j} \langle (\Sigma_\alpha^{\check{j}})^* \psi_\pm | \Sigma_\beta^{\check{l}} \psi_\pm \rangle \underbrace{=}_{\text{(2.30)}} \langle \psi_\pm | \Sigma_\alpha^{\check{j}} \Sigma_\beta^{\check{l}} \psi_\pm \rangle$$

$$\underbrace{=}_{\text{(G.160)}} \frac{1}{2} \left(\langle 000 | \Sigma_\alpha^{\check{j}} \Sigma_\beta^{\check{l}} 000 \rangle + \langle 111 | \Sigma_\alpha^{\check{j}} \Sigma_\beta^{\check{l}} 111 \rangle \right.$$

$$\left. \pm \langle 000 | \Sigma_\alpha^{\check{j}} \Sigma_\beta^{\check{l}} 111 \rangle \pm \langle 111 | \Sigma_\alpha^{\check{j}} \Sigma_\beta^{\check{l}} 000 \rangle \right),$$

where

$$\langle 000|\Sigma_\alpha^{\check{j}}\Sigma_\beta^{\check{l}}111\rangle = \delta_{\check{j}\check{l}}\langle 0|\sigma_\alpha\sigma_\beta 1\rangle\underbrace{\langle 0|1\rangle^2}_{=0} + (1-\delta_{\check{j}\check{l}})\langle 0|\sigma_\alpha 1\rangle\langle 0|\sigma_\beta 1\rangle\underbrace{\langle 0|1\rangle}_{=0}$$

$$= 0$$

and, similarly, $\langle 111|\Sigma_\alpha^{\check{j}}\Sigma_\beta^{\check{l}}000\rangle = 0$, such that

$$\langle\Sigma_\alpha^{\check{j}}\psi_\pm|\Sigma_\beta^{\check{l}}\psi_\pm\rangle = \frac{1}{2}\left(\langle 000|\Sigma_\alpha^{\check{j}}\Sigma_\beta^{\check{l}}000\rangle + \langle 111|\Sigma_\alpha^{\check{j}}\Sigma_\beta^{\check{l}}111\rangle\right) =: \check{C}_{(\alpha,\check{j}),(\beta,\check{l})} \quad \text{(G.164)}$$

Inserting (G.163) and (G.164) into (G.162) gives

$$\langle\Sigma_\alpha^j\Psi_\pm|\Sigma_\beta^l\Psi_\pm\rangle = \delta_{\check{j}\check{l}}\check{C}_{(\alpha,\check{j}),(\beta,\check{l})} + (1-\delta_{\check{j}\check{l}})\delta_{\alpha 0}\delta_{\beta 0} =: C_{(\alpha,j),(\beta,l)} \quad \text{(G.165)}$$

such that, finally,

$$\langle\Psi_0|h\Psi_0\rangle \underbrace{=}_{\text{(G.159),(G.160)}} \langle\Psi_+|\mathrm{i}^c\Sigma_\alpha^j\Sigma_\beta^l\Psi_+\rangle \underbrace{=}_{\text{(2.4),(2.30)}} \mathrm{i}^c\langle(\Sigma_\alpha^j)^*\Psi_+|\Sigma_\beta^l\Psi_+\rangle$$

$$\underbrace{=}_{(\Sigma_\alpha^j)^*=\Sigma_\alpha^j} \mathrm{i}^c\langle\Sigma_\alpha^j\Psi_+|\Sigma_\beta^l\Psi_+\rangle \underbrace{=}_{\text{(G.165)}} \mathrm{i}^c\langle\Sigma_\alpha^j\Psi_-|\Sigma_\beta^l\Psi_-\rangle$$

$$\underbrace{=}_{(\Sigma_\alpha^j)^*=\Sigma_\alpha^j} \mathrm{i}^c\langle(\Sigma_\alpha^j)^*\Psi_-|\Sigma_\beta^l\Psi_-\rangle \underbrace{=}_{\text{(2.4),(2.30)}} \langle\Psi_-|\mathrm{i}^c\Sigma_\alpha^j\Sigma_\beta^l\Psi_-\rangle$$

$$\underbrace{=}_{\text{(G.159),(G.160)}} \langle\Psi_1|h\Psi_1\rangle.$$

Solution 7.91 From Fig. 7.6 we have

$$A_1 = |1\rangle\langle 1|\otimes\mathbf{1}\otimes X\otimes\mathbf{1} + |0\rangle\langle 0|\otimes\mathbf{1}\otimes\mathbf{1}\otimes\mathbf{1}$$
$$A_2 = \mathbf{1}\otimes|1\rangle\langle 1|\otimes\mathbf{1}\otimes X\otimes\mathbf{1} + \mathbf{1}\otimes|0\rangle\langle 0|\otimes\mathbf{1}\otimes\mathbf{1}\otimes\mathbf{1}$$
$$B_1 = |1\rangle\langle 1|\otimes\mathbf{1}\otimes\mathbf{1}\otimes\mathbf{1}\otimes X + |0\rangle\langle 0|\otimes\mathbf{1}\otimes\mathbf{1}\otimes\mathbf{1}\otimes\mathbf{1}$$
$$B_2 = \mathbf{1}\otimes\mathbf{1}\otimes|1\rangle\langle 1|\otimes\mathbf{1}\otimes X + \mathbf{1}\otimes\mathbf{1}\otimes|0\rangle\langle 0|\otimes\mathbf{1}\otimes\mathbf{1}$$

such that

$$A_2A_1 = |1\rangle\langle 1|\otimes|1\rangle\langle 1|\otimes\mathbf{1}\otimes\mathbf{1}\otimes\mathbf{1} + |1\rangle\langle 1|\otimes|0\rangle\langle 0|\otimes\mathbf{1}\otimes X\otimes\mathbf{1}$$
$$+ |0\rangle\langle 0|\otimes|1\rangle\langle 1|\otimes\mathbf{1}\otimes X\otimes\mathbf{1} + |0\rangle\langle 0|\otimes|0\rangle\langle 0|\otimes\mathbf{1}\otimes\mathbf{1}\otimes\mathbf{1}$$
$$B_2B_1 = |1\rangle\langle 1|\otimes\mathbf{1}\otimes|1\rangle\langle 1|\otimes\mathbf{1}\otimes\mathbf{1} + |1\rangle\langle 1|\otimes\mathbf{1}\otimes|0\rangle\langle 0|\otimes\mathbf{1}\otimes X$$
$$+ |0\rangle\langle 0|\otimes\mathbf{1}\otimes|1\rangle\langle 1|\otimes\mathbf{1}\otimes X + |0\rangle\langle 0|\otimes\mathbf{1}\otimes|0\rangle\langle 0|\otimes\mathbf{1}\otimes\mathbf{1}.$$

After multiplying out term by term in $B_2 B_1 A_2 A_1$ and using $\langle 0|1 \rangle = 0, \langle 1|1 \rangle = 1 = \langle 0|0 \rangle$ as well as $X^2 = \mathbf{1}$ we find

$$
\begin{aligned}
S = \ & |1\rangle\langle 1| \otimes |1\rangle\langle 1| \otimes |1\rangle\langle 1| \otimes \mathbf{1} \otimes \mathbf{1} + |1\rangle\langle 1| \otimes |1\rangle\langle 1| \otimes |0\rangle\langle 0| \otimes \mathbf{1} \otimes X \\
& + |1\rangle\langle 1| \otimes |0\rangle\langle 0| \otimes |1\rangle\langle 1| \otimes X \otimes \mathbf{1} + |1\rangle\langle 1| \otimes |0\rangle\langle 0| \otimes |0\rangle\langle 0| \otimes X \otimes X \\
& + |0\rangle\langle 0| \otimes |1\rangle\langle 1| \otimes |1\rangle\langle 1| \otimes X \otimes X + |0\rangle\langle 0| \otimes |1\rangle\langle 1| \otimes |0\rangle\langle 0| \otimes X \otimes \mathbf{1} \\
& + |0\rangle\langle 0| \otimes |0\rangle\langle 0| \otimes |1\rangle\langle 1| \otimes \mathbf{1} \otimes X + |0\rangle\langle 0| \otimes |0\rangle\langle 0| \otimes |0\rangle\langle 0| \otimes \mathbf{1} \otimes \mathbf{1}
\end{aligned}
$$

$$
\overset{(3.36)}{=} \quad \begin{aligned}
& \left(|111\rangle\langle 111| + |000\rangle\langle 000| \right) \otimes \mathbf{1} \otimes \mathbf{1} \\
& + \left(|110\rangle\langle 110| + |001\rangle\langle 001| \right) \otimes \mathbf{1} \otimes X \\
& + \left(|101\rangle\langle 101| + |010\rangle\langle 010| \right) \otimes X \otimes \mathbf{1} \\
& + \left(|100\rangle\langle 100| + |011\rangle\langle 011| \right) \otimes X \otimes X ,
\end{aligned}
$$

which is (7.109).

Solution 7.92 Let $h_1, h_2 \in \mathcal{P}_n$, and let g_j be one of the $n-k$ generators of $\mathcal{S}$. Then we have

$$
(-1)^{l_j(h_1 h_2)} g_j h_1 h_2 \underset{(7.122)}{=} h_1 h_2 g_j \underset{(7.122)}{=} h_1 (-1)^{l_j(h_2)} g_j h_2 \underset{(7.122)}{=} (-1)^{l_j(h_1)+l_j(h_2)} g_j h_1 h_2
$$

such that

$$
l_j(h_1 h_2) = \left(l_j(h_1) + l_j(h_2) \right) \bmod 2 \underset{(5.2)}{=} l_j(h_1) \overset{2}{\oplus} l_j(h_2) \tag{G.166}
$$

and thus

$$
\begin{aligned}
\mathrm{syn}_q(h_1 h_2) \ & \underset{(7.121)}{=} \ \left(l_1(h_1 h_2), \ldots, l_{n-k}(h_1 h_2) \right) \\
& \underset{(G.166)}{=} \ \left(l_1(h_1) \overset{2}{\oplus} l_1(h_2), \ldots, l_{n-k}(h_1) \overset{2}{\oplus} l_{n-k}(h_2) \right) \\
& \underset{(7.121)}{=} \ \mathrm{syn}_q(h_1) \overset{2}{\oplus} \mathrm{syn}_q(h_2) .
\end{aligned}
$$

Solution 7.93 Per Definition F.16 we have that $h \in N(\mathcal{S})$ means that $h\mathcal{S} = \mathcal{S}h$ and thus for any $g \in \mathcal{S}$ there exists a $\widetilde{g} \in \mathcal{S}$ such that

$$h\widetilde{g} = gh. \tag{G.167}$$

Consequently, for any $|\Psi\rangle \in \mathbb{H}^{C_q}$

$$gh|\Psi\rangle \underbrace{=}_{\text{(G.167)}} h\widetilde{g}|\Psi\rangle \underbrace{=}_{\widetilde{g} \in \mathcal{S}} h|\Psi\rangle,$$

which means that $h|\Psi\rangle$ is left unchanged by the action of any $g \in \mathcal{S}$. But $\mathbb{H}^{C_q}$ is the subspace of all vectors left unchanged by every element of $\mathcal{S}$, which implies that $h|\Psi\rangle \in \mathbb{H}^{C_q}$.

Solutions to Exercises from Chapter 8

Solution 8.94 As per Definition 2.8, we have for any $|\psi\rangle, |\varphi\rangle \in \mathbb{H}$

$$\langle \psi | A(s)\varphi \rangle = \langle A(s)^* \psi | \varphi \rangle.$$

Consequently

$$\frac{d}{ds}\langle \psi | A(s)\varphi \rangle = \frac{d}{ds}\langle A(s)^* \psi | \varphi \rangle. \tag{G.168}$$

The linearity and continuity properties (see Definition 2.1 and Exercise 2.5) of the scalar product allow us to pull the derivatives inside so that (G.168) implies

$$\langle \psi | \dot{A}(s)\varphi \rangle = \langle \frac{d}{ds}\left(A(s)^*\right)\psi | \varphi \rangle \underbrace{=}_{\text{(2.30)}} \langle \psi | \left(\frac{d}{ds}\left(A(s)^*\right)\right)^* \varphi \rangle$$

for any $|\psi\rangle, |\varphi\rangle \in \mathbb{H}$ and thus

$$\dot{A}(s) = \left(\frac{d}{ds}\left(A(s)^*\right)\right)^*$$

$$\Rightarrow \quad \left(\dot{A}(s)\right)^* = \left[\left(\frac{d}{ds}\left(A(s)^*\right)\right)^*\right]^* \underbrace{=}_{\text{(2.31)}} \frac{d}{ds}\left(A(s)^*\right).$$

Solution 8.95 From (8.8) we have

$$
\varepsilon^2 \;\geq\; \underbrace{\big|\big|\,|\Phi\rangle - |\Psi\rangle\,\big|\big|^2}_{(2.5)} = \langle \Phi - \Psi | \Phi - \Psi \rangle
$$

$$
\underbrace{=}_{(2.5)} ||\Phi||^2 + ||\Psi||^2 - \langle\Phi|\Psi\rangle - \underbrace{\langle\Psi|\Phi\rangle}_{(2.1)} = 2 - 2\,\mathrm{Re}\left(\langle\Phi|\Psi\rangle\right),
$$

where we also used the assumption $||\Phi|| = 1 = ||\Psi||$ in the last equation. Thus,

$$
\mathrm{Re}\left(\langle\Phi|\Psi\rangle\right) \geq 1 - \frac{\varepsilon^2}{2}
$$

with which we obtain

$$
\begin{aligned}
|\langle\Phi|\Psi\rangle|^2 &= \mathrm{Re}\left(\langle\Phi|\Psi\rangle\right)^2 + \mathrm{Im}\left(\langle\Phi|\Psi\rangle\right)^2 \\
&\geq \mathrm{Re}\left(\langle\Phi|\Psi\rangle\right)^2 \geq \left(1 - \frac{\varepsilon^2}{2}\right)^2 = 1 + \frac{\varepsilon^4}{4} - \varepsilon^2 \\
&\geq 1 - \varepsilon^2.
\end{aligned}
$$

Solution 8.96 For any computational basis vector $|x\rangle = |x_{n-1}\ldots x_0\rangle \in \mathbb{H}^{\otimes n}$ we have

$$
\mathsf{H}_{\mathrm{ini}}|x\rangle \underbrace{=}_{(8.24)} \sum_{j=0}^{n-1} \Sigma_z^j |x\rangle \underbrace{=}_{(8.26)} \left(n - 2\sum_{j=0}^{n-1} x_j\right) |x\rangle .
$$

Hence, each $|x\rangle$ is an eigenvector of $\mathsf{H}_{\mathrm{ini}}$ with eigenvalue

$$
E_{\mathrm{ini},x} = n - 2\sum_{j=0}^{n-1} x_j , \tag{G.169}
$$

where $x_j \in \{0,1\}$, and we can determine the eigenvalue $E_{\mathrm{ini},x}$ by the number

$$
l_x = n - \sum_{j=0}^{n-1} x_j \tag{G.170}
$$

of the x_j in $|x\rangle = |x_{n-1}\ldots x_0\rangle$ which satisfy $x_j = 0$. The lowest such number is $l_{2^n-1} = 0$ with eigenvalue $E_{\mathrm{ini},l_{2^n-1}} = -n$ and eigenvector $|2^n - 1\rangle = |1\ldots 1\rangle$. The highest l_x is $l_0 = n$ with eigenvalue $E_{\text{ini},l_0} = n$ and eigenvector $|0\ldots0\rangle$. Consequently, the eigenvalues are of the form

$$E_{\text{ini},l} \underbrace{=}_{\text{(G.169),(G.170)}} 2l - n \qquad \text{for } l \in \{0,\ldots,n\},$$

and for a given l there are $\binom{n}{l}$ distinct $|x_{n-1}\ldots x_0\rangle$ such that $l = n - \sum_{j=0}^{n-1} x_j$.

Solution 8.97 For any computational basis vector $|x\rangle = |x_{n-1}\ldots x_0\rangle \in \mathbb{H}^{\otimes n}$ we have

$$
\begin{aligned}
\mathsf{H}_{\text{fin}}|x\rangle \underbrace{=}_{\text{(8.24)}} & \left[\sum_{\substack{i,j=0 \\ i \neq j}}^{n-1} J_{ij} \Sigma_z^i \Sigma_z^j + \sum_{j=0}^{n-1} K_j \Sigma_z^j + c\mathbf{1}^{\otimes n} \right] |x\rangle \\
\underbrace{=}_{\text{(8.26)}} & \left[\sum_{\substack{i,j=0 \\ i \neq j}}^{n-1} J_{ij}(1 - 2x_i)(1 - 2x_j) + \sum_{j=0}^{n-1} K_j(1 - 2x_j) + c\mathbf{1}^{\otimes n} \right] |x\rangle \\
= & \left[4 \sum_{\substack{i,j=0 \\ i \neq j}}^{n-1} x_i J_{ij} x_j - 2 \sum_{\substack{i,j=0 \\ i \neq j}}^{n-1} (x_i J_{ij} + J_{ij} x_j) - 2 \sum_{j=0}^{n-1} K_j x_j + \sum_{\substack{i,j=0 \\ i \neq j}}^{n-1} J_{ij} + \sum_{j=0}^{n-1} K_j + c \right] |x\rangle .
\end{aligned}
$$

Using (8.25), we obtain

$$
\begin{aligned}
& 4 \sum_{\substack{i,j=0 \\ i \neq j}}^{n-1} x_i J_{ij} x_j - 2 \sum_{\substack{i,j=0 \\ i \neq j}}^{n-1} (x_i J_{ij} + J_{ij} x_j) - 2 \sum_{j=0}^{n-1} K_j x_j + \sum_{\substack{i,j=0 \\ i \neq j}}^{n-1} J_{ij} + \sum_{j=0}^{n-1} K_j + c \\
= \ & 4 \sum_{\substack{i,j=0 \\ i \neq j}}^{n-1} x_i \frac{Q_{ij}}{4} x_j - 2 \sum_{\substack{i,j=0 \\ i \neq j}}^{n-1} \left(\frac{x_i Q_{ij} + Q_{ij} x_j}{4} \right) \\
& - 2 \sum_{j=0}^{n-1} \left(-\frac{1}{4} \sum_{\substack{i=0 \\ i \neq j}}^{n-1} (Q_{ij} + Q_{ji}) - \frac{1}{2} Q_{jj} \right) x_j \\
& + \sum_{\substack{i,j=0 \\ i \neq j}}^{n-1} \frac{Q_{ij}}{4} + \sum_{j=0}^{n-1} \left(-\frac{1}{4} \sum_{\substack{i=0 \\ i \neq j}}^{n-1} (Q_{ij} + Q_{ji}) - \frac{1}{2} Q_{jj} \right) \\
& + \frac{1}{4} \sum_{\substack{i,j=0 \\ i \neq j}}^{n-1} Q_{ji} + \frac{1}{2} \sum_{j=0}^{n-1} Q_{jj}
\end{aligned}
$$

$$= \sum_{\substack{i,j=0 \\ i \neq j}}^{n-1} x_i Q_{ij} x_j + \sum_{j=0}^{n-1} Q_{jj} x_j \underbrace{=}_{x_j \in \{0,1\}} \sum_{\substack{i,j=0 \\ i \neq j}}^{n-1} x_i Q_{ij} x_j + \sum_{j=0}^{n-1} Q_{jj} x_j^2 = \sum_{i,j=0}^{n-1} x_i Q_{ij} x_j$$

$$\underbrace{=}_{(8.23)} B(x).$$

Solution 8.98 From (2.36) we know that $|\Psi\rangle\langle\Psi|^* = |\Psi\rangle\langle\Psi|$ for all $|\Psi\rangle \in \mathbb{H}$ and since $f(s) \in \mathbb{R}$ it follows that $\mathsf{H}_{\text{ini}}, \mathsf{H}_{\text{fin}}$ and $\mathsf{H}_T(s)$ are all self-adjoint.

By definition $||\Psi_0||^2 = 1$. Hence, we have for any $|\Phi\rangle \in \mathbb{H}$

$$|\langle\Phi|\Psi_0\rangle|^2 \underbrace{\leq}_{(2.16)} ||\Phi||^2 \underbrace{||\Psi_0||^2}_{=1} = ||\Phi||^2 \underbrace{=}_{(2.5)} \langle\Phi|\Phi\rangle$$

and thus

$$\begin{aligned}
0 &\leq \langle\Phi|\Phi\rangle - |\langle\Phi|\Psi_0\rangle|^2 \\
&\underbrace{=}_{(2.1)} \langle\Phi|\Phi\rangle - \langle\Phi|\Psi_0\rangle\langle\Psi_0|\Phi\rangle = \langle\Phi|(1 - |\Psi_0\rangle\langle\Psi_0|)\Phi\rangle \\
&\underbrace{=}_{(8.28)} \langle\Phi|\mathsf{H}_{\text{ini}}\Phi\rangle
\end{aligned}$$

proving the positivity of H_{ini}.

To show this for H_{fin} we recall that any orthogonal projection P satisfies per definition $P^2 = P$ and $P^* = P$ as well as from (2.55) that $||P|| = 1$. Consequently, for any $|\Phi\rangle \in \mathbb{H}$

$$\begin{aligned}
\langle\Phi|P_{\text{S}}\Phi\rangle &= \langle\Phi|P_{\text{S}}^2\Phi\rangle = \langle\Phi|P_{\text{S}}^* P_{\text{S}}\Phi\rangle = \langle P_{\text{S}}\Phi|P_{\text{S}}\Phi\rangle \underbrace{=}_{(2.5)} ||P_{\text{S}}\Phi||^2 \\
&\underbrace{\leq}_{(2.51)} ||P_{\text{S}}||^2 ||\Phi||^2 \underbrace{=}_{(2.55)} ||\Phi||^2 \underbrace{=}_{(2.5)} \langle\Phi|\Phi\rangle
\end{aligned}$$

and thus

$$\begin{aligned}
0 &\leq \langle\Phi|\Phi\rangle - \langle\Phi|P_{\text{S}}\Phi\rangle = \langle\Phi|(1 - P_{\text{S}})\Phi\rangle \\
&\underbrace{=}_{(8.30)} \langle\Phi|\mathsf{H}_{\text{fin}}\Phi\rangle,
\end{aligned}$$

proving the positivity of H_{fin}.

As a consequence of the positivity of H_{ini} and H_{fin} and the properties of the schedule $f : [0,1] \to [0,1]$, we have thus for any $|\Phi\rangle \in \mathbb{H}$ that

$$\langle\Phi|H_T(s)\Phi\rangle \underset{(8.32)}{=} \underbrace{(1-f(s))}_{\geq 0}\underbrace{\langle\Phi|H_{ini}\Phi\rangle}_{\geq 0} + \underbrace{f(s)}_{\geq 0}\underbrace{\langle\Phi|H_{fin}\Phi\rangle}_{\geq 0}$$
$$\geq \; 0.$$

Solution 8.99 Note that for $s \in]0,1[$ we have

$$f(s) < 1$$
$$\Rightarrow \qquad f(s) - \frac{1}{2} < \frac{1}{2}$$
$$\Rightarrow \qquad \tilde{m}\left(f(s)-\frac{1}{2}\right)^2 < \frac{\tilde{m}}{4}$$
$$\Rightarrow \qquad \left(f(s)-\frac{1}{2}\right)^2 < \frac{1}{4}\left[\tilde{m}+4(1-\tilde{m})\left(f(s)-\frac{1}{2}\right)^2\right]$$

and thus

$$-\frac{1}{2}\sqrt{\tilde{m}+4(1-\tilde{m})\left(f(s)-\frac{1}{2}\right)^2} < f(s)-\frac{1}{2} < \frac{1}{2}\sqrt{\tilde{m}+4(1-\tilde{m})\left(f(s)-\frac{1}{2}\right)^2},$$

which implies

$$\underbrace{\frac{1}{2}-\frac{1}{2}\sqrt{\tilde{m}+4(1-\tilde{m})\left(f(s)-\frac{1}{2}\right)^2}}_{=E_-(s)} < \underbrace{1-f(s)}_{=E_1(s)}$$

$$\bullet \qquad\qquad < \underbrace{\frac{1}{2}+\frac{1}{2}\sqrt{\tilde{m}+4(1-\tilde{m})\left(f(s)-\frac{1}{2}\right)^2}}_{=E_+(s)}.$$

Finally, also using $0 < \tilde{m} < 1$, we have

$$f(s) < 1$$

$$\Rightarrow \qquad (1 - \tilde{m})\left(f(s)^2 - f(s)\right) < 0$$

$$\Rightarrow \qquad \frac{1}{4}\left[\tilde{m} + 4(1 - \tilde{m}))\left(f(s)^2 - f(s) + \frac{1}{4}\right)\right] < \frac{1}{4}$$

$$\Rightarrow \qquad \underbrace{\frac{1}{2} + \frac{1}{2}\sqrt{\tilde{m} + 4(1 - \tilde{m}))\left(f(s) - \frac{1}{2}\right)^2}}_{=E_+(s)} < 1 = E_2(s).$$

Solution 8.100 This is shown by a series of straightforward implications:

$$0 \le f(s) \le 1$$

$$\Rightarrow \qquad 0 \le f(s)\left(1 - f(s)\right)$$

$$\Rightarrow \qquad 0 \le -\tilde{m}\left(f(s)^2 - f(s)\right)$$

$$\Rightarrow \qquad 0 \le \frac{\tilde{m}}{4} - \tilde{m}\left(f(s) - \frac{1}{2}\right)^2$$

$$\Rightarrow \qquad \left(f(s) - \frac{1}{2}\right)^2 \le \frac{1}{4}\left[\tilde{m} + 4(1 - \tilde{m})\left(f(s) - \frac{1}{2}\right)^2\right]$$

$$\Rightarrow \qquad f(s) - \frac{1}{2} \le \underbrace{\frac{1}{2}\sqrt{\tilde{m} + 4(1 - \tilde{m})\left(f(s) - \frac{1}{2}\right)^2}}_{=\frac{1}{2} - E_-(s)}$$

$$\Rightarrow \qquad 0 \le 1 - f(s) - E_-(s).$$

Solution 8.101 From the definition of P_{sub} in (8.60) and the definition of the $|\widehat{\Phi}_\pm(s)\rangle$ in (8.49) it follows that

$$P_{\mathrm{sub}}|\widehat{\Phi}_\pm(s)\rangle = |\widehat{\Phi}_\pm(s)\rangle. \qquad (\mathrm{G}.171)$$

Hence, we have

$$\mathsf{H}_T(s)\big|_{\mathbb{H}_{\mathrm{sub}}}|\widehat{\Phi}_\pm(s)\rangle \underbrace{=}_{(8.64)} P_{\mathrm{sub}}\mathsf{H}_T(s)P_{\mathrm{sub}}|\widehat{\Phi}_\pm(s)\rangle \underbrace{=}_{(\mathrm{G}.171)} P_{\mathrm{sub}}\mathsf{H}_T(s)|\widehat{\Phi}_\pm(s)\rangle$$

$$\underbrace{=}_{\text{Thm. 8.11}} P_{\mathrm{sub}}E_\pm(s)|\widehat{\Phi}_\pm(s)\rangle \underbrace{=}_{(\mathrm{G}.171)} E_\pm(s)|\widehat{\Phi}_\pm(s)\rangle,$$

such that $\{E_\pm(s)\} \subset \sigma\left(\mathsf{H}_T(s)\big|_{\mathbb{H}_{\mathrm{sub}}}\right)$ and $|\widehat{\Phi}_\pm(s)\rangle$ are two orthonormal eigenvectors of $\mathsf{H}_T(s)\big|_{\mathbb{H}_{\mathrm{sub}}}$. But $\mathbb{H}_{\mathrm{sub}}$ is a two-dimensional HILBERT space, and it follows that $\mathsf{H}_T(s)\big|_{\mathbb{H}_{\mathrm{sub}}}$ cannot have any other eigenvalues than $E_\pm(s)$, which yields the claim $\{E_\pm(s)\} = \sigma\left(\mathsf{H}_T(s)\big|_{\mathbb{H}_{\mathrm{sub}}}\right)$.

Solution 8.102 To begin with, we have

$$
\begin{aligned}
\left(\mathsf{H}_{\mathrm{fin}} - \mathsf{H}_{\mathrm{ini}}\right)\big|_{\mathbb{H}_{\mathrm{sub}}} \underbrace{=}_{(8.64)} & \; P_{\mathrm{sub}}\left(\mathsf{H}_{\mathrm{fin}} - \mathsf{H}_{\mathrm{ini}}\right)P_{\mathrm{sub}} \\
\underbrace{=}_{(8.28),(8.30)} & \; P_{\mathrm{sub}}\left(|\Psi_0\rangle\langle\Psi_0| - P_{\mathrm{S}}\right)P_{\mathrm{sub}} \\
\underbrace{=}_{(8.61),(8.63)} & \; |\Psi_0\rangle\langle\Psi_0| - |\Psi_{\mathrm{S}}\rangle\langle\Psi_{\mathrm{S}}| \\
\underbrace{=}_{(8.59)} & \; (1-\widetilde{m})|\Psi_{\mathrm{S}\perp}\rangle\langle\Psi_{\mathrm{S}\perp}| + (\widetilde{m}-1)|\Psi_{\mathrm{S}}\rangle\langle\Psi_{\mathrm{S}}| \\
& + \sqrt{\widetilde{m}(1-\widetilde{m})}\left(|\Psi_{\mathrm{S}\perp}\rangle\langle\Psi_{\mathrm{S}}| + |\Psi_{\mathrm{S}}\rangle\langle\Psi_{\mathrm{S}\perp}|\right).
\end{aligned}
$$

Now let $|\Psi\rangle \in \mathbb{H}_{\mathrm{sub}}$ with $||\Psi||^2 = 1$, which means

$$
|\Psi\rangle = a|\Psi_{\mathrm{S}\perp}\rangle + b|\Psi_{\mathrm{S}}\rangle
$$

with

$$
|a|^2 + |b|^2 = 1. \tag{G.172}
$$

Consequently,

$$
\begin{aligned}
\left(\mathsf{H}_{\mathrm{fin}} - \mathsf{H}_{\mathrm{ini}}\right)\big|_{\mathbb{H}_{\mathrm{sub}}}|\Psi\rangle = & \left[(1-\widetilde{m})a + \sqrt{\widetilde{m}(1-\widetilde{m})}b\right]|\Psi_{\mathrm{S}\perp}\rangle \\
& + \left[(\widetilde{m}-1)b + \sqrt{\widetilde{m}(1-\widetilde{m})}a\right]|\Psi_{\mathrm{S}}\rangle
\end{aligned}
$$

such that

$$
\begin{aligned}
\left|\left|\left(\mathsf{H}_{\mathrm{fin}} - \mathsf{H}_{\mathrm{ini}}\big|_{\mathbb{H}_{\mathrm{sub}}}|\Psi\rangle\right)\right|\right|^2 \underbrace{=}_{(2.14)} & \; \left|(1-\widetilde{m})a + \sqrt{\widetilde{m}(1-\widetilde{m})}b\right|^2 \\
& + \left|(\widetilde{m}-1)b + \sqrt{\widetilde{m}(1-\widetilde{m})}a\right|^2 \\
= & \; \left((1-\widetilde{m})^2 + \widetilde{m}(1-\widetilde{m})\right)\left(|a|^2 + |b|^2\right) \\
\underbrace{=}_{(G.172)} & \; 1 - \widetilde{m}.
\end{aligned}
$$

Together with Definition 2.12 of the operator norm this then implies the claim (8.66).

Solution 8.103 The claim is shown by straightforward computation.

$$
A\mathbf{e}_0 =
\begin{pmatrix}
1 & -(a_1)^{-1} & 0 & & & \cdots & & & 0 \\
-a_1 & 2 & -(a_2)^{-1} & 0 & & & & & \\
0 & -a_2 & 2 & -(a_3)^{-1} & & & & & \\
\vdots & & \ddots & \ddots & \ddots & & & & \vdots \\
& & & -a_{j-1} & 2 & & -(a_j)^{-1} & & \\
\vdots & & & & \ddots & \ddots & & \ddots & \vdots \\
& & & & & -a_{L-2} & 2 & -(a_{L-1})^{-1} & 0 \\
& & & & & & -a_{L-1} & 2 & -(a_L)^{-1} \\
0 & & & \cdots & & & 0 & -a_L & 1
\end{pmatrix}
$$

$$
\times
\begin{pmatrix}
1 \\
a_1 \\
a_2 a_1 \\
\vdots \\
a_{j-1} a_{j-2} \cdots a_2 a_1 \\
\vdots \\
a_{L-2} a_{L-3} \cdots a_2 a_1 \\
a_{L-1} a_{L-2} \cdots a_2 a_1 \\
a_L a_{L-1} \cdots a_2 a_1
\end{pmatrix}
$$

$$
=
\begin{pmatrix}
1 - (a_1)^{-1} a_1 \\
-a_1 + 2a_1 - (a_2)^{-1} a_2 a_1 \\
\\
\vdots \\
-a_{j-1} a_{j-2} \cdots a_2 a_1 + 2 a_{j-1} a_{j-2} \cdots a_2 a_1 - (a_j)^{-1} a_j a_{j-1} \cdots a_1 \\
\vdots \\
\\
-a_L a_{L-1} \cdots a_2 a_1 + a_L a_{L-1} \cdots a_2 a_1
\end{pmatrix}
=
\begin{pmatrix}
0 \\
0 \\
\vdots \\
0 \\
\vdots \\
0
\end{pmatrix}.
$$

Solution 8.104 To prove (8.96), we note that

$$||\Xi(l)||^2 \underbrace{=}_{(2.5)} \langle \Xi(l)|\Xi(l)\rangle$$

$$\underbrace{=}_{(8.93)} \langle U_l \cdots U_1 \Psi_{\text{ini}}|U_l \cdots U_1 \Psi_{\text{ini}}\rangle \underbrace{=}_{(2.30)} \langle \Psi_{\text{ini}}|U_1^* \cdots U_l^* U_l \cdots U_1 \Psi_{\text{ini}}\rangle$$

$$\underbrace{=}_{(2.37)} \langle \Psi_{\text{ini}}|\Psi_{\text{ini}}\rangle \underbrace{=}_{(2.5)} ||\Psi_{\text{ini}}||^2 = 1.$$

To show (8.97), we have

$$\langle \Gamma(l)|\Gamma(m)\rangle \underbrace{=}_{(8.94)} \langle \Xi(l) \otimes x(l)|\Xi(m) \otimes x(m)\rangle$$

$$\underbrace{=}_{(3.4)} \langle \Xi(l)|\Xi(m)\rangle\langle x(l)|x(m)\rangle \underbrace{=}_{(8.91)} \langle \Xi(l)|\Xi(m)\rangle \delta_{lm} = \langle \Xi(l)|\Xi(l)\rangle \delta_{lm}$$

$$\underbrace{=}_{(2.5)} ||\Xi(l)||^2 \delta_{lm} \underbrace{=}_{(8.96)} \delta_{lm}.$$

Solution 8.105 Starting with (8.115), we have

$$\left(|a0\rangle\langle b0| \otimes \mathbf{1}^{\otimes L-2}\right)|x(m)\rangle = |a\rangle\langle b|x(m)_{L-1}\rangle \otimes |0\rangle\langle 0|x(m)_{L-2}\rangle \otimes |x(m)_{L-3}\ldots x(m)_0\rangle$$

$$= \delta_{b,x(m)_{L-1}} \delta_{0,x(m)_{L-2}}|a0x(m)_{L-3}\ldots x(m)_0\rangle. \qquad \text{(G.173)}$$

Recalling (8.90), we obtain

$$x(m)_{L-2} = 0 \quad \Leftrightarrow \quad m < 2.$$

So only $m = 0$ or $m = 1$ give non-zero values for the right side of (G.173). From (8.92) we also know that

$$|x(0)\rangle = |0\ldots 0\rangle \quad \text{and} \quad |x(1)\rangle = |10\ldots 0\rangle.$$

We can thus re-write (G.173) as

$$
\begin{aligned}
\left(|a0\rangle\langle b0| \otimes \mathbf{1}^{\otimes L-2} \right)|x(m)\rangle &= \delta_{b,x(m)_{L-1}}\left(\delta_{m,0}+\delta_{m,1}\right)|a0\ldots 0\rangle \\
&= \left(\delta_{b,x(0)_{L-1}}\delta_{m,0}+\delta_{b,x(1)_{L-1}}\delta_{m,1}\right)|a0\ldots 0\rangle \\
&= \left(\delta_{b,0}\delta_{m,0}+\delta_{b,1}\delta_{m,1}\right)|a0\ldots 0\rangle \\
&= \left(\delta_{b,0}\delta_{m,0}+\delta_{b,1}\delta_{m,1}\right)\left(\delta_{a,0}|x(0)\rangle+\delta_{a,1}|x(1)\rangle\right) \\
&= \delta_{b,0}\delta_{m,0}\delta_{a,0}|x(0)\rangle + \delta_{b,1}\delta_{m,1}\delta_{a,0}|x(0)\rangle \\
&\quad + \delta_{b,0}\delta_{m,0}\delta_{a,1}|x(1)\rangle + \delta_{b,1}\delta_{m,1}\delta_{a,1}|x(1)\rangle \\
&= \delta_{b,0}\delta_{m,0}\delta_{a,0}|x(m)\rangle + \delta_{b,1}\delta_{m,1}\delta_{a,0}|x(m-1)\rangle \\
&\quad + \delta_{b,0}\delta_{m,0}\delta_{a,1}|x(m+1)\rangle + \delta_{b,1}\delta_{m,1}\delta_{a,1}|x(m)\rangle ,
\end{aligned}
$$

which is (8.115). For (8.116) we have

$$
\left(\mathbf{1}^{\otimes l-2} \otimes |1a0\rangle\langle 1b0|\mathbf{1}^{\otimes L-l-1}\right)|x(m)\rangle \tag{G.174}
$$
$$
= \delta_{1,x(m)_{L-l+1}}\,\delta_{b,x(m)_{L-l}}\,\delta_{0,x(m)_{L-l-1}}|x(m)_{L-1}\ldots x(m)_{L-l+2}1a0x(m)_{L-l-2}\ldots x(m)_0\rangle .
$$

Using again (8.90), we find

$$
x(m)_{L-l+1} = 1 \quad\Leftrightarrow\quad m \geq l-1 \quad\text{and}\quad x(m)_{L-l-1} = 0 \quad\Leftrightarrow\quad m < l+1 .
$$

Hence, for the right side of (G.174) not to vanish, we must have $l-1 \leq m < l = 1$, that is, $m = l-1$ or $m = l$ only. Recalling again (8.92), we know that

$$
|x(l-1)\rangle = |\underbrace{1\ldots 1}_{l-1\ \text{times}} 0\ldots 0\rangle \quad\text{and}\quad |x(l)\rangle = |\underbrace{1\ldots 1}_{l\ \text{times}}0\ldots 0\rangle .
$$

We can thus re-write (G.174) as

$$
\begin{aligned}
&\left(\mathbf{1}^{\otimes l-2} \otimes |1a0\rangle\langle 1b0|\mathbf{1}^{\otimes L-l-1}\right)|x(m)\rangle \\
&= \delta_{b,x(m)_{L-l}}\left(\delta_{m,l-1}+\delta_{m,l}\right)|\underbrace{1\ldots 1}_{l-1\ \text{times}} a0\ldots 0\rangle \\
&\underset{(8.90)}{=} \left(\delta_{b,0}\delta_{m,l-1}+\delta_{b,1}\delta_{m,l}\right)|\underbrace{1\ldots 1}_{l-1\ \text{times}} a0\ldots 0\rangle \\
&= \left(\delta_{b,0}\delta_{m,l-1}+\delta_{b,1}\delta_{m,l}\right)\left(\delta_{a,0}|x(l-1)\rangle+\delta_{a,1}|x(l)\rangle\right) \\
&= \delta_{a,0}\delta_{b,0}\delta_{m,l-1}|x(m)\rangle + \delta_{a,1}\delta_{b,1}\delta_{m,l}|x(m)\rangle \\
&\quad + \delta_{a,1}\delta_{b,0}\delta_{m,l-1}|x(m+1)\rangle + \delta_{a,0}\delta_{b,1}\delta_{m,l}|x(m-1)\rangle ,
\end{aligned}
$$

which proves (8.116). For (8.117) consider

$$
\left(\mathbf{1}^{\otimes L-2} \otimes |1a\rangle\langle 1b|\right)|x(m)\rangle = \delta_{1,x(m)_1}\delta_{b,x(m)_0}|x(m)_{L-1}\ldots x(m)_2 1a\rangle .
$$

Once more, (8.90) implies

$$x(m)_1 = 1 \quad \Leftrightarrow \quad m = L-1 \quad \text{or} \quad m = L.$$

Proceeding as before, this yields

$$
\begin{aligned}
\left(1^{\otimes L-2} \otimes |1a\rangle\langle 1b|\right)|x(m)\rangle &= \delta_{b,x(m)_0}\left(\delta_{m,L-1} + \delta_{m,L}\right)|x(m)_{L-1}\ldots x(m)_2 1a\rangle \\
&\underset{(8.90)}{=} \left(\delta_{b,0}\delta_{m,L-1} + \delta_{b,1}\delta_{m,L}\right)|1\ldots 11a\rangle \\
&= \left(\delta_{m,L-1}\delta_{b,0} + \delta_{m,L}\delta_{b,1}\right)\left(\delta_{a,0}|x(L-1)\rangle + \delta_{a,1}|x(L)\rangle\right) \\
&= \left(\delta_{a,0}\delta_{b,0}\delta_{m,L-1} + \delta_{a,1}\delta_{b,1}\delta_{m,L}\right)|x(m)\rangle \\
&\quad + \delta_{a,0}\delta_{b,1}\delta_{m,L}|x(m-1)\rangle + \delta_{a,1}\delta_{b,0}\delta_{m,L-1}|x(m+1)\rangle,
\end{aligned}
$$

which completes the proof of (8.117).

Solution 8.106 Making use of the result (2.48) and the fact that both $H_T(s)$ and P_{sub} are self-adjoint, we see that the claim (8.129) is proven, if we can show that their product $H_T(s)P_{\text{sub}}$ is self-adjoint. From (8.122) it is evident that this is the case if $H_{\text{ini}}P_{\text{sub}}$ and $H_{\text{fin}}P_{\text{sub}}$ are self-adjoint. From (8.99) and (8.100) it follows that the latter is true if in turn each $H_X P_{\text{sub}}$ is self-adjoint for the various $X \in \{\text{c-ini, input, clock}, l\}$. We now proceed to show this. First, we consider

$$
\begin{aligned}
H_{\text{c-ini}}P_{\text{sub}} &\underset{(8.129)}{=} H_{\text{c-ini}} \sum_{m=0}^{L} |\Gamma(m)\rangle\langle\Gamma(m)| = \sum_{m=0}^{L} H_{\text{c-ini}}|\Gamma(m)\rangle\langle\Gamma(m)| \\
&\underset{(8.118)}{=} \sum_{m=0}^{L} (1-\delta_{m,0})|\Gamma(m)\rangle\langle\Gamma(m)| = \sum_{m=1}^{L} |\Gamma(m)\rangle\langle\Gamma(m)| \\
&= P_{\text{sub}} - |\Gamma(0)\rangle\langle\Gamma(0)|,
\end{aligned}
$$

where (2.36) assures us that the last expression is self-adjoint, implying that this is also the case for $H_{\text{c-ini}}P_{\text{sub}}$. Similarly, we have

$$H_{\text{input}}P_{\text{sub}} \underset{(8.129)}{=} \sum_{m=0}^{L} H_{\text{input}}|\Gamma(m)\rangle\langle\Gamma(m)| \underset{(8.119)}{=} 0$$

$$H_{\text{clock}}P_{\text{sub}} \underset{(8.129)}{=} \sum_{m=0}^{L} H_{\text{clock}}|\Gamma(m)\rangle\langle\Gamma(m)| \underset{(8.120)}{=} 0,$$

and since the zero operator is obviously self-adjoint, so are $H_{\text{input}}P_{\text{sub}}$ and $H_{\text{clock}}P_{\text{sub}}$.

Lastly, for $l \in \{1,\ldots,L\}$ consider

$$\underset{(8.129)}{\mathsf{H}_l P_{\text{sub}}} = \sum_{m=0}^{L} \mathsf{H}_l |\Gamma(m)\rangle\langle\Gamma(m)|$$

$$\underset{(8.121)}{=} \frac{1}{2} \sum_{m=0}^{L} (\delta_{m,l-1} + \delta_{m,l}) |\Gamma(m)\rangle\langle\Gamma(m)|$$

$$- \frac{1}{2} \sum_{m=0}^{L} \left(\delta_{m,l-1} |\Gamma(m+1)\rangle\langle\Gamma(m)| + \delta_{m,l} |\Gamma(m-1)\rangle\langle\Gamma(m)|\right)$$

$$= \frac{1}{2} \left(|\Gamma(l-1)\rangle\langle\Gamma(l-1)| + |\Gamma(l)\rangle\langle\Gamma(l)| \right.$$
$$\left. - |\Gamma(l)\rangle\langle\Gamma(l-1)| - |\Gamma(l-1)\rangle\langle\Gamma(l)|\right),$$

which is self-adjoint due to (2.36). Consequently, $\mathsf{H}_{\text{prop}} P_{\text{sub}} = \frac{1}{2}\sum_{l=1}^{L} \mathsf{H}_l P_{\text{sub}}$ is self-adjoint and thus ultimately $\mathsf{H}_T(s) P_{\text{sub}}$ as well. As already mentioned above, the claim (8.130) then follows from (2.48).

Solution 8.107 Let

$$|\Psi\rangle = \sum_{m=0}^{L} \Psi_m |\Gamma(m)\rangle$$

be an arbitrary vector of $\mathbb{H}_{\text{sub}}$. Then we have

$$\underset{(8.104)}{\langle\Psi|\mathsf{H}_{\text{prop}}\Psi\rangle} = \sum_{k,m=0}^{L} \overline{\Psi_k}\Psi_m \frac{1}{2} \sum_{l=1}^{L} \langle\Gamma(k)|\mathsf{H}_l\Gamma(m)\rangle$$

$$\underset{(8.121)}{=} \frac{1}{2} \sum_{k,m=0}^{L} \sum_{l=1}^{L} \overline{\Psi_k}\Psi_m \left((\delta_{m,l-1} + \delta_{m,l})\langle\Gamma(k)|\Gamma(m)\rangle\right.$$

$$\left. - \delta_{m,l-1}\langle\Gamma(k)|\Gamma(m+1)\rangle - \delta_{m,l}\langle\Gamma(k)|\Gamma(m-1)\rangle\right)$$

$$\underset{(8.97)}{=} \frac{1}{2} \sum_{k,m=0}^{L} \sum_{l=1}^{L} \overline{\Psi_k}\Psi_m \left((\delta_{m,l-1} + \delta_{m,l})\delta_{k,m} - \delta_{m,l-1}\delta_{k,m+1} - \delta_{m,l}\delta_{k,m-1}\right)$$

$$= \frac{1}{2} \sum_{k=0}^{L} \sum_{l=1}^{L} \left(\overline{\Psi_k}\Psi_k(\delta_{k,l-1} + \delta_{k,l}) - \overline{\Psi_k}\Psi_{l-1}\delta_{k,l} - \overline{\Psi_k}\Psi_l\delta_{k,l-1}\right)$$

$$= \frac{1}{2} \sum_{l=1}^{L} \left(\overline{\Psi_{l-1}}\Psi_{l-1} + \overline{\Psi_l}\Psi_l - \overline{\Psi_l}\Psi_{l-1} - \overline{\Psi_{l-1}}\Psi_l\right)$$

$$= \frac{1}{2} \sum_{l=1}^{L} |\Psi_{l-1} - \Psi_l|^2 \geq 0. \tag{G.175}$$

For an arbitrary $|\Phi\rangle \in \mathbb{H}^U \otimes \mathbb{H}^C$ we have $P_{\text{sub}}|\Phi\rangle \in \mathbb{H}_{\text{sub}}$ and thus

$$\langle\Phi|H_{prop}|_{H_{sub}}\Phi\rangle \underbrace{=}_{(8.131)} \langle\Phi|P_{sub}H_{prop}P_{sub}\Phi\rangle \underbrace{=}_{(2.30),P_{sub}^*=P_{sub}} \langle P_{sub}\Phi|H_{prop}P_{sub}\Phi\rangle$$

$$\underbrace{\geq}_{(G.175)} 0,$$

and Definition 2.12 implies (8.132).

Solution 8.108 To prove the claim, it is helpful to use the abbreviating notations $b = b(s,E(s))$ and $\theta_j = \theta_j(s,E(s))$ and to introduce the following variables for $j \in \{1,2\}$

$$\begin{aligned} u_j &= m\theta_j \\ v_j &= (m-2)\theta_j, \end{aligned} \tag{G.176}$$

which satisfy

$$\begin{aligned} \frac{u_j + v_j}{2} &= (m-1)\theta_j \\ \frac{u_j - v_j}{2} &= \theta_j. \end{aligned} \tag{G.177}$$

We then have

$$b\Phi(s)_{m-1} - \Phi(s)_{m-2} \underbrace{=}_{(8.148)} A_j\big[b\,\mathrm{Co}_j\left((m-1)\theta_j\right) - \mathrm{Co}_j\left((m-2)\theta_j\right)\big]$$

$$+ B_j\big[b\,\mathrm{Si}_j\left((m-1)\theta_j\right) - \mathrm{Si}_j\left((m-2)\theta_j\right)\big]$$

$$\underbrace{=}_{(8.147)} A_j\big[2\,\mathrm{Co}_j(\theta_j)\,\mathrm{Co}_j\left((m-1)\theta_j\right) - \mathrm{Co}_j\left((m-2)\theta_j\right)\big]$$

$$+ B_j\big[2\,\mathrm{Co}_j(\theta_j)\,\mathrm{Si}_j\left((m-1)\theta_j\right) - \mathrm{Si}_j\left((m-2)\theta_j\right)\big]$$

$$\underbrace{=}_{(G.177)} A_j\Big[2\,\mathrm{Co}_j\left(\frac{u_j - v_j}{2}\right)\mathrm{Co}_j\left(\frac{u_j + v_j}{2}\right) - \mathrm{Co}_j\left(v_j\right)\Big]$$

$$+ B_j\Big[2\,\mathrm{Co}_j\left(\frac{u_j - v_j}{2}\right)\mathrm{Si}_j\left(\frac{u_j + v_j}{2}\right) - \mathrm{Si}_j\left(v_j\right)\Big].$$

Here we can use the trigonometric identities

$$\begin{aligned} 2\,\mathrm{Co}_j\left(\frac{u+v}{2}\right)\mathrm{Co}_j\left(\frac{u-v}{2}\right) &= \mathrm{Co}_j(u) + \mathrm{Co}_j(-v) = \mathrm{Co}_j(u) + \mathrm{Co}_j(v) \\ 2\,\mathrm{Co}_j\left(\frac{u+v}{2}\right)\mathrm{Si}_j\left(\frac{u-v}{2}\right) &= \mathrm{Si}_j(u) - \mathrm{Si}_j(-v) = \mathrm{Si}_j(u) + \mathrm{Si}_j(v), \end{aligned} \tag{G.178}$$

which hold for $j = 1$ and $j = 2$. Hence, we obtain

$$b\Phi(s)_{m-1} - \Phi(s)_{m-2} = A_j\operatorname{Co}_j(u_j) + B_j\operatorname{Si}_j(u_j)$$
$$\underbrace{=}_{(G.176)} A_j\operatorname{Co}_j(m\theta_j) + B_j\operatorname{Si}_j(m\theta_j)$$
$$\underbrace{=}_{(8.148)} \Phi(s)_m,$$

which proves the claim.

Solution 8.109 We will first show that the solution (8.148) of the recursion together with the boundary conditions (8.142) and (8.144) leads to (8.150), where Ta$_j$ is as defined in (8.149).

We use again the abbreviating notation $\theta_j = \theta_j\big(s, E(s)\big)$. The boundary condition (8.142) requires

$$\Phi(s)_1 = a\Phi(s)_0,$$

which implies

$$\frac{A_j}{B_j} = \frac{\operatorname{Si}_j(\theta_j)}{a - \operatorname{Co}_j(\theta_j)}. \tag{G.179}$$

Whereas (8.144) requires

$$\Phi(s)_L = c\Phi(s)_{L-1},$$

which implies

$$\frac{A_j}{B_j} = \frac{c\operatorname{Si}_j\big((L-1)\theta_j\big) - \operatorname{Si}_j(L\theta_j)}{\operatorname{Co}_j(L\theta_j) - \operatorname{Co}_j\big((L-1)\theta_j\big)}. \tag{G.180}$$

Hence, we must have

$$\frac{\operatorname{Si}_j(\theta_j)}{a - \operatorname{Co}_j(\theta_j)} \underbrace{=}_{(G.179),(G.180)} \frac{c\operatorname{Si}_j\big((L-1)\theta_j\big) - \operatorname{Si}_j(L\theta_j)}{\operatorname{Co}_j(L\theta_j) - \operatorname{Co}_j\big((L-1)\theta_j\big)},$$

which leads to

$$\operatorname{Si}_j(\theta_j)\operatorname{Co}_j(L\theta_j) - \operatorname{Si}_j(L\theta_j)\operatorname{Co}_j(\theta_j) = ac\operatorname{Si}_j\big((L-1)\theta_j\big) - a\operatorname{Si}_j(L\theta_j)$$
$$+ c\big[\operatorname{Si}_j(\theta_j)\operatorname{Co}_j\big((L-1)\theta_j\big) - \operatorname{Si}_j\big((L-1)\theta_j\big)\operatorname{Co}_j(\theta_j)\big].$$

On the left side and in the last term on the right side we use the identity

$$\operatorname{Si}_j(u)\operatorname{Co}_j(v) - \operatorname{Co}_j(u)\operatorname{Si}_j(v) = \operatorname{Si}_j(u - v), \tag{G.181}$$

which holds for $j = 1$ and $j = 2$, to obtain

$$(1 + ac)\,\mathrm{Si}_j\left((L-1)\theta_j\right) = a\,\mathrm{Si}_j(L\theta_j) - c\,\mathrm{Si}_j\left((L-2)\theta_j\right). \tag{G.182}$$

Here we can use the second of the identities (G.178) with $u = L\theta_j$ and $v = -(L - 2)\theta_j$, which yields

$$\mathrm{Si}_j(L\theta_j) \underset{(\mathrm{G.178})}{=} 2\mathrm{Co}_j(\theta_j)\,\mathrm{Si}_j\left((L-1)\theta_j\right) - \mathrm{Si}_j\left((L-2)\theta_j\right)$$

$$\underset{(8.147)}{=} b\,\mathrm{Si}_j\left((L-1)\theta_j\right) - \mathrm{Si}_j\left((L-2)\theta_j\right).$$

Inserting this into (G.182) gives

$$(1 + ac - ab)\,\mathrm{Si}_j\left((L-1)\theta_j\right) = (c - a)\,\mathrm{Si}_j\left((L-2)\theta_j\right).$$

Using the identity (G.181) with $u = (L-1)\theta_j$ and $v = \theta_j$, we find

$$\mathrm{Si}_j\left((L-2)\theta_j\right) \underset{(\mathrm{G.181})}{=} \mathrm{Si}_j\left((L-1)\theta_j\right)\mathrm{Co}_j(\theta_j) - \mathrm{Co}_j\left((L-1)\theta_j\right)\mathrm{Si}_j(\theta_j)$$

$$\underset{(8.147)}{=} \mathrm{Si}_j\left((L-1)\theta_j\right)\frac{b}{2} - \mathrm{Co}_j\left((L-1)\theta_j\right)\mathrm{Si}_j(\theta_j).$$

Inserting this into (G.182) yields, after some re-arrangements,

$$\mathrm{Ta}_j\left((L-1)\theta_j\right) = \frac{2(c-a)\,\mathrm{Si}_j(\theta_j)}{b(c+a) - 2ac - 2}. \tag{G.183}$$

In deriving (G.183) we have ignored the fact that $\mathrm{Co}_j\left((L-1)\theta_j\right)$ as well as $b(c + a) - 2ac - 2$ can be zero. We will continue to do so, but shall discuss the points (s, E) where this happens in more detail in the proof of Theorem 8.26.

Now, (8.147) also implies

$$\mathrm{Si}_j(\theta_j) \underset{(8.147)}{=} \begin{cases} \sinh\theta_1 & \text{for } j = 1 \\ \sin\theta_2 & \text{for } j = 2 \end{cases} = \begin{cases} \sqrt{\cosh^2\theta_1 - 1} & \text{for } j = 1 \\ \sqrt{1 - \cos^2\theta_2} & \text{for } j = 2 \end{cases}$$

$$\underset{(8.147)}{=} \begin{cases} \frac{1}{2}\sqrt{b^2 - 4} & \text{for } j = 1 \\ \frac{1}{2}\sqrt{4 - b^2} & \text{for } j = 2 \end{cases}$$

$$= \frac{1}{2}\sqrt{(-1)^j(4 - b^2)},$$

which, inserted into (G.183), yields (8.149). We have thus shown that a solution of (8.142)–(8.144) implies (8.150). From Lemma 8.25 we know that a solution of (8.142)–(8.144) is equivalent to $|\Phi(s)\rangle$ being an eigenvector with eigenvalue $E(s)$. Consequently, $E(s)$ has to be a solution of (8.150).

Solution 8.110 The proof of the claim (8.152) is accomplished by making use of (8.145) and (8.151) and straightforward, albeit lengthy, calculations, which we give here without many of the intermediate steps. Again, we ignore here the fact that $b(c+a)-2ac-2$ can be zero. We will continue to do so when proving (8.152), but shall discuss the points (s,E) where this happens in more detail in the proof of Theorem 8.26. To begin with, we have from (8.145)

$$a+c = \frac{4E(E-1)+2s}{s(2-2E-s)}.$$

Here and in what follows we use the abbreviation $E = E(s)$. We then have

$$b(a+c) = \frac{4s(1-E)-8E(E-1)^2}{s^2(2-2E-s)}$$

as well as

$$2ac = \frac{2s^3-4Es^2}{s^2(2-2E-s)}$$

and

$$c-a = \frac{2s(s-1)+4E(1-E)}{s(2-2E-s)}. \tag{G.184}$$

Hence, we find

$$b(a+c)-2ac-2 = 4\frac{s^2(2E-1)+s(1-E)-2E(E-1)^2}{s^2(2-2E-s)}. \tag{G.185}$$

On the other hand, (8.151) gives

$$2(s_- - E)(E-p_+)(E-p_-) = s^2(2E-1)+s(1-E)-2E(E-1)^2$$

as well as

$$(E-z_+)(E-z_-) = -\frac{1}{4}\left(2s(s-1)+4E(1-E)\right)$$

such that (G.185) becomes

$$b(a+c)-2ac-2 = \frac{8(s_- - E)(E-p_+)(E-p_-)}{s^2(2-2E-s)}, \tag{G.186}$$

whereas (G.184) becomes

$$c - a = \frac{-4(E - z_+)(E - z_-)}{s(2 - 2E - s)}. \tag{G.187}$$

Using (G.186) and (G.187) yields

$$\frac{(c-a)\sqrt{(-1)^j(4-b^2)}}{b(a+c) - 2ac - 2} = \frac{(E-z_+)(E-z_-)}{(E-p_+)(E-p_-)} \frac{s\sqrt{(-1)^j(4-b^2)}}{2(E-s_-)}. \tag{G.188}$$

Here we can use that

$$(-1)^j(4-b^2) \underset{(8.145)}{=} (-1)^j \left(4 - 4\frac{(1-E)^2}{s^2} \right) = \frac{4(-1)^j}{s^2}\left(s^2 - (E-1)^2 \right)$$

$$= \frac{4(-1)^j}{s^2}\left(s - (E-1) \right)\left(s + (E-1) \right)$$

$$\underset{(8.151)}{=} \frac{4(-1)^j}{s^2}(s_+ - E)(E - s_-),$$

where $s_+ - E > 0$ in both cases and $s_- - E > 0$ in Case $j = 1$ and $s_- - E \leq 0$ in Case $j = 2$. Consequently, we have

$$\frac{s\sqrt{(-1)^j(4-b^2)}}{2(E - s_-)} = \sqrt{s_+ - E}\, \frac{\sqrt{(-1)^j(E - s_-)}}{(E - s_-)},$$

where

$$\frac{\sqrt{(-1)^j(E - s_-)}}{(E - s_-)} = \begin{cases} \frac{\sqrt{s_- - E}}{(E - s_-)} & \text{if } j = 1 \\ \frac{\sqrt{E - s_-}}{(E - s_-)} & \text{if } j = 2 \end{cases} = \begin{cases} -\frac{1}{\sqrt{s_- - E}} & \text{if } j = 1 \\ \frac{1}{\sqrt{E - s_-}} & \text{if } j = 2 \end{cases}$$

$$= \frac{(-1)^j}{\sqrt{(-1)^j(E - s_-)}},$$

such that

$$\frac{s\sqrt{(-1)^j(4-b^2)}}{2(E - s_-)} = (-1)^j \sqrt{\frac{s_+ - E}{(-1)^j(E - s_-)}}.$$

Inserting this into (G.188), we obtain

$$\frac{(c-a)\sqrt{(-1)^j(4-b^2)}}{b(a+c) - 2ac - 2} = (-1)^j \frac{(E-z_+)(E-z_-)}{(E-p_+)(E-p_-)} \sqrt{\frac{s_+ - E}{(-1)^j(E - s_-)}} \underset{(8.151)}{=} h_j(s, E)$$

proving the claim (8.152).

Solution 8.111 The statements for $s = 0$ and $s = 1$ are easily obtained by inserting these values in (8.151).

For $s \in]0, 1[$ we first obtain

$$
\begin{aligned}
&(1+s)^2 > 1+s^2 &&> (1-s)^2 + s^2 = 1 - 2s + 2s^2 \\
\Rightarrow\quad &1+s > \sqrt{1+s^2} &&> \sqrt{1-2s+2s^2} \\
\Rightarrow\quad &2+2s > 1+s+\sqrt{1+s^2} &&> 1+\sqrt{1-2s+2s^2} &&\text{(G.189)} \\
\Rightarrow\quad &1+s > \tfrac{1}{2}\left(1+s+\sqrt{1+s^2}\right) &&> \tfrac{1}{2}\left(1+\sqrt{1-2s+2s^2}\right) \\
\Leftrightarrow\quad &s_+ > p_+ &&> z_+ .
\end{aligned}
$$

Similarly, we have

$$
\begin{aligned}
&&1 > s& \\
\Rightarrow\quad&& 2s > 2s^2& \\
\Rightarrow\quad&& 1-2s+2s^2 > 1-4s+4s^2 = (1-2s)^2& \\
\Rightarrow\quad&& \sqrt{1-2s+2s^2} > 1-2s \quad &> -\sqrt{1-2s+2s^2} \\
\Rightarrow\quad&& 1+\sqrt{1-2s+2s^2} > 2(1-s) \quad &> 1-\sqrt{1-2s+2s^2} \\
\Rightarrow\quad&& \tfrac{1}{2}\left(1+\sqrt{1-2s+2s^2}\right) > 1-s \quad &> \tfrac{1}{2}\left(1-\sqrt{1-2s+2s^2}\right) \\
\Leftrightarrow\quad&& z_+ > s_- \quad &> z_-
\end{aligned}
$$

and

$$
\begin{aligned}
&&\sqrt{1+s^2} > 1& \\
\Rightarrow\quad&& -2s > -2s\sqrt{1+s^2}& \\
\Rightarrow\quad&& 1-2s+2s^2 > 1+2s^2 - 2s\sqrt{1+s^2} = (s-\sqrt{1+s^2})^2& \\
\Rightarrow\quad&& \sqrt{1-2s+2s^2} > s-\sqrt{1+s^2} \quad &> -\sqrt{1-2s+2s^2} \\
\Rightarrow\quad&& 1+\sqrt{1-2s+2s^2} > 2(1-s) \quad &> 1-\sqrt{1-2s+2s^2} \\
\Rightarrow\quad&& \tfrac{1}{2}\left(1+\sqrt{1-2s+2s^2}\right) > \tfrac{1}{2}\left(1+s-\sqrt{1+s^2}\right) \quad &> \tfrac{1}{2}\left(1-\sqrt{1-2s+2s^2}\right) \\
\Leftrightarrow\quad&& z_+ > p_- \quad &> z_- .
\end{aligned}
$$

$$\text{(G.190)}$$

Together (G.189)–(G.190) imply for $0 < s < 1$ that $z_- < p_-, s_- < z_+ < p_+ < s_+$. For $0 < s < \tfrac{3}{4}$ we find

$$
\begin{aligned}
&&\tfrac{3}{4} > s& \\
\Rightarrow\quad&& 6s > 8s^2& \\
\Rightarrow\quad&& 1+s^2 > 9s^2 - 6s + 1 = (1-3s)^2& \\
\Rightarrow\quad&& 1-3s > -\sqrt{1+s^2}& \\
\Rightarrow\quad&& 2-2s > 1+s-\sqrt{1+s^2}& \\
\Rightarrow\quad&& 1-s > \tfrac{1}{2}\left(1+s-\sqrt{1+s^2}\right)& \\
\Leftrightarrow\quad&& s_- > p_-&
\end{aligned}
$$

and conversely for $\frac{3}{4} \leq s < 1$ that $p_- \leq s_-$. Altogether thus for $0 < s < 1$

$$z_- < p_- < s_- < z_+ < p_+ < s_+ \quad \text{if } s < \frac{3}{4}$$

$$z_- < s_- \leq p_- < z_+ < p_+ < s_+ \quad \text{if } s \geq \frac{3}{4}$$

as claimed in (8.154).

Solution 8.112 Let $s \in]0, 1]$. Then we have

$$
\begin{aligned}
& s \lessgtr \tfrac{3}{4} \\
\Leftrightarrow\quad & 0 \lessgtr 3 - 4s \\
\Leftrightarrow\quad & 1 + s^2 \lessgtr s^2 - 4s + 4 = (2-s)^2 \\
\Leftrightarrow\quad & \sqrt{1+s^2} \lessgtr 2 - s \\
\Leftrightarrow\quad & 1 + s^2 \lessgtr (2-s)\sqrt{1+s^2} \\
\Leftrightarrow\quad & 1 - (2-s)\sqrt{1+s^2} \lessgtr -s^2 \\
\Leftrightarrow\quad & (2-s)(1+s - \sqrt{1+s^2}) \lessgtr 1 + s - 2s^2 \\
\Leftrightarrow\quad & p_- \underbrace{= \tfrac{1}{2}\left(1+s-\sqrt{1+s^2}\right)}_{(8.151)} \lessgtr \underbrace{\tfrac{1+s-2s^2}{2(2-s)} = \hat{E}}_{(8.157)}
\end{aligned}
$$

and similarly

$$
\begin{aligned}
& s \lessgtr \tfrac{3}{4} = \tfrac{6}{8} \\
\Leftrightarrow\quad & \tfrac{1}{8} \lessgtr \tfrac{7}{8} - s \\
\Leftrightarrow\quad & \tfrac{1}{64} \lessgtr \left(\tfrac{7}{8}-s\right)^2 = \tfrac{49}{64} - \tfrac{7}{4}s + s^2 \\
\Leftrightarrow\quad & 0 \lessgtr 3 - 7s + 4s^2 \\
\Leftrightarrow\quad & 1 + s - 2s^2 \lessgtr 4 - 6s + 2s^2 = 2(2-s)(1-s) \\
\Leftrightarrow\quad & \hat{E} \underbrace{= \tfrac{1+s-2s^2}{2(2-s)}}_{(8.157)} \lessgtr \underbrace{1 - s = s_-}_{(8.151)}.
\end{aligned}
$$

Lastly,

$$s_-\left(\frac{3}{4}\right) = \frac{1}{4} = \hat{E}\left(\frac{3}{4}\right) = p_-\left(\frac{3}{4}\right)$$

is easily verified from the definitions (8.151) and (8.157) of $s_-, \hat{E}$ and p_-.

Solution 8.113 Let $p_-(s) \in I_{q_-}(s)$ and $p_-(s) \neq \overline{E}_{s,q_- -2}$. Then we have $\overline{E}_{s,q_- -2} < p_-(s)$. It follows from (8.161) and (8.151) that then

$$1 - s\cos\alpha_{q_--2} < \frac{1}{2}\left(1 + s - \sqrt{1+s^2}\right) < 1,$$

which implies $\cos\alpha_{q_--2} > 0$. Thus

$$\alpha_{q_--2} \underbrace{=}_{(8.160)} \frac{2(q_--2)+1}{2(L-1)}\pi < \frac{\pi}{2},$$

from which we obtain $q_- < \frac{L}{2} + 1$.

Next, let $p_+(s) \in I_{q_+}(s)$. Then we must have $p_+(s) < \overline{E}_{s,q_+-1}$. It follows again from (8.161) and (8.151) that then

$$1 < \frac{1}{2}\left(1 + s + \sqrt{1+s^2}\right) < 1 - s\cos\alpha_{q_+-1},$$

which implies $\cos\alpha_{q_+-1} < 0$. Thus

$$\frac{\pi}{2} < \alpha_{q_+-1} \underbrace{=}_{(8.160)} \frac{2(q_+-1)+1}{2(L-1)}\pi < \pi,$$

Solution 8.114 For $p_\pm(s) \in I_q(s)$ to be the case, (8.163) requires $\frac{L}{2} < q < \frac{L}{2}+1$. If $L = 2k$, this would imply $k < q < k+1$, which has no solution for $q \in \mathbb{N}_0$. Hence, $p_\pm(s) \in I_q(s)$ is only possible for $L = 2k+1$ in which case $q = k+1$. Moreover, $p_\pm(s) \in I_q(s)$ also implies that p_+ and p_- cannot be further apart than the width of the interval, that is, we must have $p_+(s) - p_-(s) < \overline{E}_{s,q-1} - \overline{E}_{s,q-2}$. It then follows from (8.161) and (8.151) that

$$\sqrt{1+s^2} < s(\cos\alpha_{q-2} - \cos\alpha_{q-1}). \tag{G.191}$$

Using that $L = 2k+1, q = k+1$ and (8.160), we find

$$\alpha_{q-2} = \frac{2(q-2)+1}{2(L-1)}\pi = \left(\frac{1}{2} - \frac{1}{4k}\right)\pi$$

$$\alpha_{q-1} = \frac{2(q-1)+1}{2(L-1)}\pi = \left(\frac{1}{2} + \frac{1}{4k}\right)\pi$$

such that $\alpha_{q-1} + \alpha_{q-2} = \pi$ and thus $\cos\alpha_{q-1} = -\cos\alpha_{q-2}$. This together with (G.191) and $s \in]0,1]$ implies

$$1 < \sqrt{1+s^2} < 2s\cos\alpha_{q-2} \le 2\cos\alpha_{q-2}.$$

Hence, for $p_\pm(s) \in I_q(s)$ to be true, we must have $\cos\alpha_{q-2} > \frac{1}{2}$ and thus

$$\alpha_{q-2} = \left(\frac{1}{2} - \frac{1}{4k} \right) \pi < \frac{\pi}{3},$$

which requires $k < \frac{3}{2}$, that is, $L = 2k + 1 < 4$. Since we know already that L has to be odd, and we assume $L > 1$, the only remaining possibility for $p_\pm(s) \in I_q(s)$ to hold would be the case $k = 1$ and thus $L = 3$ and $q = k + 1 = 2$. Consequently, $\alpha_{q-2} = \alpha_0 = \frac{\pi}{4}$ and $\cos \alpha_{q-2} = \frac{1}{\sqrt{2}}$, and the only remaining possibility is

$$p_\pm \in I_2(s) \underbrace{=}_{(8.162)} \left] 1 - \frac{s}{\sqrt{2}}, 1 + \frac{s}{\sqrt{2}} \right].$$

It follows from the definition (8.151) of the $p_\pm$ that in this case

$$1 - \frac{s}{\sqrt{2}} < \frac{1}{2} \left(1 + s \pm \sqrt{1 + s^2} \right) \leq 1 + \frac{s}{\sqrt{2}}.$$

The first of these inequalities implies

$$
\begin{aligned}
&\sqrt{1 + s^2} & &< (\sqrt{2} + 1)s - 1 \\
\Rightarrow \quad &1 + s^2 & &< (2 + 1 + 2\sqrt{2})s^2 + 1 - 2s(\sqrt{2} + 1) \\
\Rightarrow \quad &1 & &< s,
\end{aligned}
$$

which is outside the range to which we have restricted s. Consequently, it is impossible to have $p_\pm(s) \in I_q(s)$ for any $L > 1$.

Solution 8.115 To begin with, we have

$$\left\| \hat{H}(t) - H(t) \right\| \underbrace{=}_{(8.210)} \left\| \hat{H}_{\kappa(t)} - H(t) \right\|$$

$$\underbrace{=}_{(8.206),(8.208)} \left\| \left(1 - \frac{\kappa(t)}{J} \right) H_{\text{ini}} + \frac{\kappa(t)}{J} H_{\text{fin}} - \left(1 - \frac{t - t_{\text{ini}}}{T} \right) H_{\text{ini}} - \frac{t - t_{\text{ini}}}{T} H_{\text{fin}} \right\|$$

$$= \left\| \left(\frac{\kappa(t)}{J} - \frac{t - t_{\text{ini}}}{T} \right) H_{\text{fin}} - \left(\frac{\kappa(t)}{J} - \frac{t - t_{\text{ini}}}{T} \right) H_{\text{ini}} \right\|$$

$$\underbrace{=}_{(2.7)} \left| \frac{\kappa(t)}{J} - \frac{t - t_{\text{ini}}}{T} \right| \left\| H_{\text{fin}} - H_{\text{ini}} \right\|. \tag{G.192}$$

Using that $T = J\Delta t$, we obtain

$$
\frac{\kappa(t)}{J} - \frac{t - t_{\text{ini}}}{T} = \frac{\kappa(t)}{J} - \frac{t - t_{\text{ini}}}{J\Delta t} = \frac{1}{J}\left(\kappa(t) - \frac{t - t_{\text{ini}}}{\Delta t}\right)
$$

$$
\underset{(8.209)}{=} \frac{1}{J}\underbrace{\left(\left[\frac{t - t_{\text{ini}}}{\Delta t}\right] - \frac{t - t_{\text{ini}}}{\Delta t}\right)}_{\in [0,1[}
$$

$$
\leq \frac{1}{J} \tag{G.193}
$$

so that

$$
||\hat{H}(t) - H(t)|| \underset{(G.192)}{=} \left|\frac{\kappa(t)}{J} - \frac{t - t_{\text{ini}}}{T}\right| \, ||H_{\text{fin}} - H_{\text{ini}}||
$$

$$
\underset{(G.193)}{\leq} \frac{1}{J}||H_{\text{fin}} - H_{\text{ini}}|| \, .
$$

Solutions to Exercises from Appendix B

Solution B.116

(i) We show $\hat{c}_j^- \in \{0, -1\}$ by induction in j. The induction-start at $j = 0$ is given
by the starting assumption $\hat{c}_0^- = 0$.
For the inductive step from j to $j+1$ we suppose that $\hat{c}_j^- \in \{0, -1\}$ holds true
for j. Then the possible values for $\hat{c}_{j+1}^-$ as a function of the possible values of
the a_j, b_j and $\hat{c}_j^-$ are as shown in Table G.1 and $\hat{c}_{j+1}^- \in \{0, -1\}$ is satisfied.

(ii) To prove (B.8), note that the assumptions $0 \leq a, b < 2^n$ and $\hat{d}_j \in \{0, 1\}$ as well
as (B.5) imply

$$
-2^n < b - a = \underbrace{\sum_{j=0}^{n-1} \hat{d}_j 2^j}_{\geq 0} + \hat{c}_n^- 2^n < 2^n . \tag{G.194}
$$

Hence, in the case $b \geq a$ it follows that $\hat{c}_n^-$ has to take the value 0 and in the
case $b < a$ it has to take the value -1. Conversely, it follows from (G.194) that
$\hat{c}_n^- = 0$ implies $b \geq a$ and that $\hat{c}_n^- = -1$ implies $b < a$.

Table G.1 Table of values for $\hat{c}_{j+1}^-$ as function of a_j, b_j and $\hat{c}_j^-$

a_j	b_j	$\hat{c}_j^-$	$\hat{c}_{j+1}^- = \left\lfloor \frac{b_j - a_j + \hat{c}_j^-}{2} \right\rfloor$
0	0	0	0
0	0	−1	−1
0	1	0	0
0	1	−1	0
1	0	0	−1
1	0	−1	−1
1	1	0	0
1	1	−1	−1

Solutions to Exercises from Appendix C

Solution C.117 The assumption $f_i(N) \in O(g_i(N))$ for $i \in \{1,2\}$ means that there exist $C_i \in \mathbb{R}$ and $M_i \in \mathbb{N}$ such that

$$\forall N > M_i \quad |f_i(N)| \leq C_i |g_i(N)| \qquad N \to \infty.$$

For $\hat{M} := \max\{M_1, M_2\}$ we then have for all $N > \hat{M}$:

(i)

$$\begin{aligned} |f_1(N) + f_2(N)| &\leq |f_1(N)| + |f_2(N)| \\ &\leq C_1 |g_1(N)| + C_2 |g_2(N)| \\ &\leq \max\{C_1, C_2\} \left(|g_1(N)| + |g_2(N)| \right) \end{aligned}$$

and thus $f_1(N) + f_2(N) \in O(|g_1(N)| + |g_2(N)|)$.

(ii)

$$\begin{aligned} |f_1(N) f_2(N)| &\leq |f_1(N)| |f_2(N)| \\ &\leq C_1 |g_1(N)| C_2 |g_2(N)| \\ &= C_1 C_2 |g_1(N) g_2(N)| \end{aligned}$$

and thus $f_1(N) f_2(N) \in O(g_1(N) g_2(N))$.

(iii) For $N > M$ we have, by assumption, $|g_1(N)| < |g_2(N)|$, which implies

$$\begin{aligned} |f_1(N) + f_2(N)| &\leq |f_1(N)| + |f_2(N)| \\ &\leq C_1 |g_1(N)| + C_2 |g_2(N)| \\ &\leq (C_1 + C_2) |g_2(N)|, \end{aligned}$$

such that $f_1(N) + f_2(N) \in O(g_2(N))$.

Solutions to Exercises from Appendix D

Solution D.118 To show $\Rightarrow$, note that

$$a \bmod N = b \bmod N \underbrace{\Rightarrow}_{(D.1)} a - \left\lfloor \frac{a}{N} \right\rfloor N = b - \left\lfloor \frac{b}{N} \right\rfloor N,$$

and it follows that N divides $a - b$, which implies $(a - b) \bmod N = 0$.

To show $\Leftarrow$, let $(a - b) \bmod N = 0$. Then it follows from (D.1) that there exists a $z \in \mathbb{Z}$ such that $a - b = zN$. Consequently,

$$b \bmod N \underbrace{=}_{(D.1)} b - \left\lfloor \frac{b}{N} \right\rfloor N = a - zN - \left\lfloor \frac{a - zN}{N} \right\rfloor N = a - \left\lfloor \frac{a}{N} \right\rfloor \underbrace{=}_{(D.1)} a \bmod N.$$

Solution D.119 Let $a, N \in \mathbb{N}$ with $a > N$. For any $x \in \mathbb{R}$ we have

$$0 \leq x - \lfloor x \rfloor < 1$$

such that $\frac{a}{N} - \left\lfloor \frac{a}{N} \right\rfloor < 1$ and thus

$$a \bmod N \underbrace{=}_{(D.1)} a - \left\lfloor \frac{a}{N} \right\rfloor N < N.$$

On the other hand, one can easily convince oneself with a graph of the functions that $x \geq 1$ implies $\frac{1}{2}x < \lfloor x \rfloor$. Because $a > N$ this in turn implies $\frac{1}{2}\frac{a}{N} < \left\lfloor \frac{a}{N} \right\rfloor$ from which it follows that $\frac{a}{2} < \left\lfloor \frac{a}{N} \right\rfloor N$ and thus

$$a \bmod N \underbrace{=}_{(D.1)} a - \left\lfloor \frac{a}{N} \right\rfloor N < \frac{a}{2}.$$

Solution D.120 Let $u, v, u_j \in \mathbb{Z}$ and $k, a, N \in \mathbb{N}$. We first show (D.20).

$$u(v \bmod N) \bmod N \underbrace{=}_{(D.1)} u(v \bmod N) - \left\lfloor \frac{u(v \bmod N)}{N} \right\rfloor N$$

(D.1)

$$\underbrace{=}_{(D.1)} u\left(v - \left\lfloor \frac{v}{N} \right\rfloor N\right) - \left\lfloor \frac{u\left(v - \left\lfloor \frac{v}{N} \right\rfloor N\right)}{N} \right\rfloor N$$

(D.1)

$$= uv - u\left\lfloor \frac{v}{N} \right\rfloor N - \left\lfloor \frac{uv}{N} - u\left\lfloor \frac{v}{N} \right\rfloor \right\rfloor N$$

$$= uv - u\left\lfloor \frac{v}{N} \right\rfloor N - \left\lfloor \frac{uv}{N} \right\rfloor N + u\left\lfloor \frac{v}{N} \right\rfloor N$$

$$= uv - \left\lfloor \frac{uv}{N} \right\rfloor N$$

$$\underbrace{=}_{(D.1)} uv \bmod N.$$

(D.1)

Repeated application of (D.20) then yields (D.21):

$$\left(\prod_{j=1}^{k}(u_j \bmod N) \right) \bmod N \underbrace{=}_{(D.20)} \left(\prod_{j=1}^{k-1}(u_j \bmod N)u_k \right) \bmod N$$

$$= \quad \ldots$$

$$= \left(\prod_{j=1}^{k} u_j \right) \bmod N.$$

With $u_j = u$, then (D.22) follows as a special case of (D.21). To prove (D.23), it suffices to show this for u_1 and u_2. The claim then follows from repeated application of the statement for u_1 and u_2. For the left side one has for u_1 and u_2 by definition

$$\left(u_1 \bmod N + u_2 \bmod N \right) \bmod N = \left(u_1 - \left\lfloor \frac{u_1}{N} \right\rfloor N + u_2 - \left\lfloor \frac{u_2}{N} \right\rfloor N \right) \bmod N$$

$$= u_1 - \left\lfloor \frac{u_1}{N} \right\rfloor N + u_2 - \left\lfloor \frac{u_2}{N} \right\rfloor N$$

$$- \left\lfloor \frac{u_1 - \left\lfloor \frac{u_1}{N} \right\rfloor N + u_2 - \left\lfloor \frac{u_2}{N} \right\rfloor N}{N} \right\rfloor N$$

$$= u_1 + u_2 - \left\lfloor \frac{u_1}{N} \right\rfloor N - \left\lfloor \frac{u_2}{N} \right\rfloor N$$

$$- \left\lfloor \frac{u_1 + u_2}{N} - \left\lfloor \frac{u_1}{N} \right\rfloor - \left\lfloor \frac{u_2}{N} \right\rfloor \right\rfloor N$$

$$= u_1 + u_2 - \left\lfloor \frac{u_1}{N} \right\rfloor N - \left\lfloor \frac{u_2}{N} \right\rfloor N$$

$$- \left\{ \left\lfloor \frac{u_1 + u_2}{N} \right\rfloor - \left\lfloor \frac{u_1}{N} \right\rfloor - \left\lfloor \frac{u_2}{N} \right\rfloor \right\} N$$

$$= u_1 + u_2 - \left\lfloor \frac{u_1 + u_2}{N} \right\rfloor$$

$$= (u_1 + u_2) \bmod N.$$

Solutions to Exercises from Appendix F

Solution F.121 Suppose first that besides e there is an $\tilde{e} \in \mathcal{G}$ such that for every $g \in \mathcal{G}$ we also have

$$g\tilde{e} = g \tag{G.195}$$

$$gg^{-1} = \tilde{e}. \tag{G.196}$$

This implies

$$g^{-1}\underbrace{\tilde{e} = }_{(G.195)} g^{-1} \quad \Rightarrow \quad gg^{-1}\tilde{e} = gg^{-1} \underbrace{=}_{(F.3)} e$$

$$\Rightarrow \quad e \underbrace{=}_{(G.196)} \tilde{e}\tilde{e} \underbrace{=}_{(G.195)} \tilde{e},$$

showing that e is unique.

Now, suppose that h_1 and h_2 are two inverses of $g \in \mathcal{G}$, that is,

$$gh_1 = e = gh_2. \tag{G.197}$$

Then it follows for $i \in \{1, 2\}$ that

$$g \underbrace{=}_{(F.2)} ge \underbrace{=}_{(F.3)} gh_ih_i^{-1} \underbrace{=}_{(G.197)} eh_i^{-1} \tag{G.198}$$

and thus

$$h_ig \underbrace{=}_{(G.198)} h_ieh_i^{-1} \underbrace{=}_{(F.2)} h_ih_i^{-1} \underbrace{=}_{(F.3)} e. $$

Consequently, we also have

$$h_1g = e = h_2g, \tag{G.199}$$

which finally implies

$$h_1 \underset{(F.2)}{=} h_1 e \underset{(F.3)}{=} h_1 g g^{-1} \underset{(G.199)}{=} h_2 g g^{-1} \underset{(F.3)}{=} h_2 e \underset{(F.2)}{=} h_2.$$

Solution F.122 From (F.2) we have for $g = e$ that $ee = e$ and from (F.3) we see that then $e = e^{-1}$, proving (F.4).

Next, consider

$$gh(h^{-1}g^{-1}) = ghh^{-1}g^{-1} \underset{(F.3)}{=} geg^{-1} \underset{(F.2)}{=} gg^{-1} \underset{(F.3)}{=} e,$$

which proves (F.5). To show (F.6) note that (F.3) implies $e = g^{-1}(g^{-1})^{-1}$. Multiplying both sides with g and using that, because of (F.2), we have on the left side $ge = g$, we obtain

$$g = gg^{-1}(g^{-1})^{-1} \underset{(F.3)}{=} e(g^{-1})^{-1} \underset{(F.4)}{=} e^{-1}(g^{-1})^{-1} \underset{(F.5)}{=} (g^{-1}e)^{-1} \underset{(F.2)}{=} (g^{-1})^{-1},$$

which proves (F.6). Next, we have

$$h = g \quad \Rightarrow \quad hg^{-1} = gg^{-1} \underset{(F.3)}{=} e$$

$$\underset{(F.3)}{\Rightarrow} \quad h^{-1} = g^{-1} \tag{G.200}$$

$$\underset{(G.200)}{\Rightarrow} \quad (h^{-1})^{-1} = (g^{-1})^{-1} \underset{(F.6)}{\Rightarrow} \quad h = g$$

proving (F.7).

Solution F.123 To prove (F.8), we note that

$$(g^{-1}g)^{-1} \underset{(F.5)}{=} g^{-1}(g^{-1})^{-1} \underset{(F.6)}{=} g^{-1}g, \tag{G.201}$$

which implies

$$e \underset{(F.3)}{=} g^{-1}g(g^{-1}g)^{-1} \underset{(G.201)}{=} g^{-1}gg^{-1}g \underset{(F.3)}{=} g^{-1}eg \underset{(F.2)}{=} g^{-1}g,$$

proving (F.8). To show (F.9), we note that

$$eg \underset{(F.7)}{=} \left((eg)^{-1}\right)^{-1} \underset{(F.5)}{=} \left(g^{-1}e^{-1}\right)^{-1} \underset{(F.4)}{=} \left(g^{-1}e\right)^{-1} \underset{(F.2)}{=} \left(g^{-1}\right)^{-1} \underset{(F.7)}{=} g.$$

To show (F.10), let $gh = gk$ for some $g, h, k \in \mathcal{G}$. Multiplying both sides from the left by g^{-1} and using (F.8) implies $h = k$. Multiplying this in turn by g from the left yields $gh = gk$. Hence, $gh = gk$ if and only if $h = k$. The proof of the equivalence of $hg = kg$ with $h = k$ is similar.

Solution F.124 Any subgroup contains the neutral element $e_\mathcal{G}$, which is then also contained in $\mathcal{H}_\cap$.

Let $g_1, g_2 \in \mathcal{H}_\cap$. Then $g_1, g_2 \in \mathcal{H}_j$ for every $j \in I$. Since each $\mathcal{H}_j$ is a subgroup, we thus have that also $g_1 g_2 \in \mathcal{H}_j$ for every $j \in I$, and it follows that $g_1 g_2 \in \mathcal{H}_\cap$.

Likewise, if $g \in \mathcal{H}_\cap$, then $g \in \mathcal{H}_j$ and thus $g^{-1} \in \mathcal{H}_j$ for each $j \in I$, which implies that $g^{-1} \in \mathcal{H}_\cap$ and completes the proof that $\mathcal{H}_\cap$ is a subgroup.

Solution F.125 Since $e_\mathcal{G} g = g e_\mathcal{G}$ for all $g \in \mathcal{G}$, it follows that $e_\mathcal{G} \in \text{Clz}_\mathcal{G}(S)$.

Let $h_1, h_2 \in \text{Clz}_\mathcal{G}(S)$. Then we have for every $g \in S$

$$(h_1 h_2)g = h_1(h_2 g) \underset{(F.17)}{=} h_1 g h_2 \underset{(F.17)}{=} g h_1 h_2$$

and thus $h_1 h_2 \in \text{Clz}_\mathcal{G}(S)$.

Finally, for any $h \in \text{Clz}_\mathcal{G}(S)$ and $g \in S$

$$hg \underset{(F.17)}{=} gh \quad \Rightarrow \quad gh^{-1} = h^{-1}g \quad \underset{(F.17)}{\Rightarrow} \quad h^{-1} \in \text{Clz}_\mathcal{G}(S).$$

Solution F.126 Let g be any element of the group $\mathcal{G}$. Since $e \in \mathcal{H}$, it follows from (F.19) that $e \in \mathcal{H}^g$ as well and (F.14) is satisfied for $\mathcal{H}^g$.

Now let $h' \in \mathcal{H}^g$, that is, there exists an $h \in \mathcal{H}$ such that $h' = ghg^{-1}$. But we also have from (F.19) that $gh^{-1}g^{-1} \in \mathcal{H}^g$. Hence,

$$h'\left(gh^{-1}g^{-1}\right) = ghg^{-1}gh^{-1}g^{-1} = e$$

and thus $(h')^{-1} = gh^{-1}g^{-1} \in \mathcal{H}^g$, which shows that (F.15) is satisfied for $\mathcal{H}^g$.

Finally, let $h'_1, h'_2 \in \mathcal{H}^g$, which, due to (F.19), means that there exist $h_1, h_2 \in \mathcal{H}$ such that $h'_j = gh_j g^{-1}$ for $j \in \{1, 2\}$. Since $h_1 h_2 \in \mathcal{H}$, it follows that

$$h'_1 h'_2 = \left(gh_1 g^{-1}\right)\left(gh_2 g^{-1}\right) = gh_1 h_2 g^{-1} \in \mathcal{H}^g,$$

which shows that (F.16) is also satisfied for $\mathcal{H}^g$.

Solution F.127 The neutral element e of $\mathcal{G}$ clearly satisfies $S^e = eSe^{-1} = S$. Consequently, $e \in \text{Nor}_{\mathcal{G}}(S)$, and (F.14) is satisfied.

For any $h_1, h_2 \in \text{Nor}_{\mathcal{G}}(S)$ we have

$$S^{h_1 h_2} \underbrace{=}_{\text{(F.18)}} h_1 h_2 S (h_1 h_2)^{-1} = h_1 h_2 S h_2^{-1} h_1^{-1} \underbrace{=}_{\text{(F.18)}} h_1 S^{h_2} h_1^{-1} \underbrace{=}_{\text{(F.20)}} h_1 S h_1^{-1} \underbrace{=}_{\text{(F.18)}} S^{h_1}$$

$$\underbrace{=}_{\text{(F.20)}} S$$

such that $h_1 h_2 \in \text{Nor}_{\mathcal{G}}(S)$, verifying (F.16).

Finally, for any $h \in \text{Nor}_{\mathcal{G}}(S)$ we have

$$h S h^{-1} \underbrace{=}_{\text{(F.18)}} S^h \underbrace{=}_{\text{(F.20)}} S$$

$$\Rightarrow \quad S = h^{-1} S (h^{-1})^{-1} \underbrace{=}_{\text{(F.18)}} S^{h^{-1}}$$

such that $h^{-1} \in \text{Nor}_{\mathcal{G}}(S)$, and (F.15) is satisfied as well.

Solution F.128 The neutral element e of $\mathcal{G}$ clearly satisfies $eg = ge$ and so $e \in \text{Ctr}(\mathcal{G})$, verifying (F.14).

For any $h_1, h_2 \in \text{Ctr}(\mathcal{G})$ we have

$$h_1 h_2 g \underbrace{=}_{\text{(F.21)}} h_1 g h_2 \underbrace{=}_{\text{(F.21)}} g h_1 h_2$$

and thus $h_1 h_2 \in \text{Ctr}(\mathcal{G})$, showing that (F.16) holds.

Moreover, for any $h \in \text{Ctr}(\mathcal{G})$ and $g \in \mathcal{G}$ we have $hg = gh$ which implies $gh^{-1} = h^{-1}g$ and thus $h^{-1} \in \text{Ctr}(\mathcal{G})$ as well, verifying (F.15).

Finally, we find for any $g \in \mathcal{G}$ that

$$\text{Ctr}(\mathcal{G})^g \underbrace{=}_{\text{(F.18)}} \left\{ ghg^{-1} \mid h \in \text{Ctr}(\mathcal{G}) \right\} \underbrace{=}_{\text{(F.21)}} \left\{ hgg^{-1} \mid h \in \text{Ctr}(\mathcal{G}) \right\} = \text{Ctr}(\mathcal{G}),$$

proving that $\text{Ctr}(\mathcal{G})$ is normal.

Solution F.129 Definition F.15 of a normal subgroup $\mathcal{H}$ implies

$$\mathcal{H} \text{ is normal}$$

$$\Leftrightarrow \quad \forall g \in \mathcal{G} \quad \forall h \in \mathcal{H} \quad \exists \tilde{h} \in \mathcal{H} \text{ and } \forall \tilde{h} \in \mathcal{H} \quad \exists h \in \mathcal{H}: \quad ghg^{-1} = \tilde{h}$$

$$\Leftrightarrow \quad \forall g \in \mathcal{G} \quad \forall h \in \mathcal{H} \quad \exists \tilde{h} \in \mathcal{H} \text{ and } \forall \tilde{h} \in \mathcal{H} \quad \exists h \in \mathcal{H}: \quad gh = \tilde{h}g$$

$$\underbrace{\Leftrightarrow}_{\text{(F.22),(F.23)}} \quad \forall g \in \mathcal{G} \quad g\mathcal{H} = \mathcal{H}g.$$

Solution F.130 Associativity of the product follows from the associativity of of the products in each group. The neutral element is given by

$$e_{\mathcal{G}_1 \times \mathcal{G}_2} := (e_{\mathcal{G}_1}, e_{\mathcal{G}_2}) \tag{G.202}$$

since for any $(g_1, g_2) \in \mathcal{G}_1 \times \mathcal{G}_2$

$$(g_1, g_2) \cdot_\times (e_{\mathcal{G}_1}, e_{\mathcal{G}_2}) \underbrace{=}_{\text{(F.37)}} (g_1 e_{\mathcal{G}_1}, g_2 e_{\mathcal{G}_2}) = (g_1, g_2).$$

Similarly, any $(g_1, g_2) \in \mathcal{G}_1 \times \mathcal{G}_2$ has the inverse

$$(g_1, g_2)^{-1} = (g_1^{-1}, g_2^{-1})$$

since

$$(g_1, g_2) \cdot_\times (g_1, g_2)^{-1} \underbrace{=}_{\text{(F.37)}} (g_1 g_1^{-1}, g_2 g_2^{-1}) = (e_{\mathcal{G}_1}, e_{\mathcal{G}_2}) \underbrace{=}_{\text{(G.202)}} e_{\mathcal{G}_1 \times \mathcal{G}_2}.$$

If $\mathcal{G}_1$ and $\mathcal{G}_2$ are both finite with their number of elements $|\mathcal{G}_1|$ and $|\mathcal{G}_2|$, then the set $\mathcal{G}_1 \times \mathcal{G}_2$ is finite as well and has the number of elements $|\mathcal{G}_1||\mathcal{G}_2|$.

Solution F.131 By the defining property (F.40) of a stabilizer, we have for every $m \in M$ that $e.m = m$, which implies that $e \in \text{Sta}_\mathcal{G}(Q)$ and thus verifies (F.14).

Let $h, g \in \text{Sta}_\mathcal{G}(Q)$ and $m \in M$ be arbitrary. Then we have

$$(hg).m \underbrace{=}_{\text{(F.41)}} h.(g.m) \underbrace{=}_{\text{(F.42)}} h.m \underbrace{=}_{\text{(F.42)}} m$$

such that $hg \in \text{Sta}_\mathcal{G}(Q)$, which proves (F.16). Finally, we find

$$m \underbrace{=}_{(F.40)} e.m \underbrace{=}_{(F.8)} (g^{-1}g).m \underbrace{=}_{(F.41)} g^{-1}.(g.m) \underbrace{=}_{(F.42)} g^{-1}.m,$$

and thus $g^{-1} \in \mathrm{Sta}_{\mathcal{G}}(Q)$, which shows (F.15) and completes the proof that $\mathrm{Sta}_{\mathcal{G}}(G)$ is a subgroup of $\mathcal{G}$.

Solution F.132 For any $g \in \mathcal{G}_1$ we have

$$\varphi(g)\varphi(g^{-1}) \underbrace{=}_{(F.44)} \varphi(gg^{-1}) = \varphi(e_{\mathcal{G}_1}) \underbrace{=}_{(F.47)} e_{\mathcal{G}_2}$$

and thus $\varphi(g^{-1}) = \varphi(g)^{-1}$.

Solution F.133 For $i \in \{1,2\}$ let $\chi_i \in \widehat{\mathcal{G}_i}$ and define

$$(\chi_1,\chi_2) : \mathcal{G}_1 \times \mathcal{G}_2 \longrightarrow \mathrm{U}(1) \\ (g_1,g_2) \longmapsto \chi_1(g_1)\chi_2(g_2), \tag{G.203}$$

which is a homomorphism, since for any $g_i, g_i' \in \mathcal{G}_i$

$$(\chi_1,\chi_2)\big((g_1,g_2) \cdot_\times (g_1',g_2')\big) \underbrace{=}_{(F.39)} (\chi_1,\chi_2)(g_1g_1', g_2g_2') \underbrace{=}_{(G.203)} \chi_1(g_1g_1')\chi_2(g_2g_2')$$
$$\underbrace{=}_{(F.44)} \chi_1(g_1)\chi_1(g_1')\chi_2(g_2)\chi_2(g_2')$$
$$\underbrace{=}_{(G.203)} (\chi_1\chi_2)(g_1,g_2)(\chi_1\chi_2)(g_1',g_2').$$

Hence, $(\chi_1,\chi_2) \in \widehat{\mathcal{G}_1 \times \mathcal{G}_2}$, and since we know from Theorem F.34 that $\widehat{\mathcal{G}_1}$, $\widehat{\mathcal{G}_2}$ and $\widehat{\mathcal{G}_1 \times \mathcal{G}_2}$ are groups and from Exercise F.130 that $\widehat{\mathcal{G}_1} \times \widehat{\mathcal{G}_2}$ is a group, thus

$$\widehat{\mathcal{G}_1} \times \widehat{\mathcal{G}_2} \le \widehat{\mathcal{G}_1 \times \mathcal{G}_2}. \tag{G.204}$$

Moreover, we have

$$|\widehat{\mathcal{G}_1} \times \widehat{\mathcal{G}_2}| \underbrace{=}_{(F.38)} |\widehat{\mathcal{G}_1}||\widehat{\mathcal{G}_2}| \underbrace{=}_{(F.70)} |\mathcal{G}_1||\mathcal{G}_2| \underbrace{=}_{(F.38)} |\mathcal{G}_1 \times \mathcal{G}_2| \underbrace{=}_{(F.70)} |\widehat{\mathcal{G}_1 \times \mathcal{G}_2}|,$$

which, together with (G.204), implies

$$\widehat{\mathcal{G}_1} \times \widehat{\mathcal{G}_2} = \widehat{\mathcal{G}_1 \times \mathcal{G}_2}.$$

Solution F.134 Let $\chi \in \widehat{\mathcal{G}}$, and set

$$S_\chi = \sum_{h \in \mathcal{H}} \chi(h).$$

Then for any $\check{h} \in \mathcal{H}$

$$S_\chi \chi(\check{h}) = \sum_{h \in \mathcal{H}} \chi(h)\chi(\check{h}) = \sum_{h \in \mathcal{H}} \chi(h\check{h}) = S_\chi, \qquad (G.205)$$

where in the penultimate equation we have used that $\chi \in \mathrm{Hom}(\mathcal{G}, \mathrm{U}(1))$ and thus satisfies (F.43). It follows from (G.205) that for every $\chi \in \widehat{\mathcal{G}}$ and $\check{h} \in \mathcal{H}$

$$S_\chi\big(\chi(\check{h}) - 1\big) = 0. \qquad (G.206)$$

Now, let $\chi \in \mathcal{H}^\perp$. Then (F.72) implies that $\mathcal{H} \subset \mathrm{Ker}(\chi)$, and it follows that $\chi(h) = 1$ for all $h \in \mathcal{H}$. This implies $S_\chi = |\mathcal{H}|$.

On the other hand, if there exists an $\check{h} \in \mathcal{H}$ such that $\chi(\check{h}) \neq 1$, then this is equivalent to $\mathcal{H} \not\subset \mathrm{Ker}(\chi)$ and this in turn to $\chi \notin \mathcal{H}^\perp$. Moreover, (G.206) implies that in this case we must have $S_\chi = 0$.

Solution F.135 By assumption, $\mathcal{H}$ is a subgroup of $\mathcal{G}$ and from Lemma F.30 we know that the kernel of a homomorphism is a subgroup. Since each character is by definition a homomorphism, its kernel is also a subgroup and from Exercise F.124 we know that any intersection of a set of subgroups is again a subgroup. Consequently, $\bigcap_{\chi \in \mathcal{H}^\perp} \mathrm{Ker}(\chi)$ is a subgroup of $\mathcal{G}$.

Now, let $h \in \mathcal{H}$. Then by definition (F.72) of $\mathcal{H}^\perp$ we have for any $\chi \in \mathcal{H}^\perp$ that $\mathcal{H} \subset \mathrm{Ker}(\chi)$. Hence,

$$\mathcal{H} \subset \bigcap_{\chi \in \mathcal{H}^\perp} \mathrm{Ker}(\chi)$$

and each of them being a subgroup, the claim (F.77) is proven.

Solution F.136 We have

$$
F_{\mathcal{G}}^{*} \underset{(\text{F.107})}{=} \left(\frac{1}{\sqrt{|\mathcal{G}|}} \sum_{\substack{g \in \mathcal{G} \\ \chi \in \widehat{\mathcal{G}}}} \chi(g)(|\chi\rangle\langle g|) \right)^{*} \underset{(2.32)}{=} \frac{1}{\sqrt{|\mathcal{G}|}} \sum_{\substack{g \in \mathcal{G} \\ \chi \in \widehat{\mathcal{G}}}} \overline{\chi(g)}(|\chi\rangle\langle g|)^{*}
$$

$$
\underset{(2.36)}{=} \frac{1}{\sqrt{|\mathcal{G}|}} \sum_{\substack{g \in \mathcal{G} \\ \chi \in \widehat{\mathcal{G}}}} \overline{\chi(g)}|g\rangle\langle \chi| \underset{(\text{F.62})}{=} \frac{1}{\sqrt{|\mathcal{G}|}} \sum_{\substack{g \in \mathcal{G} \\ \chi \in \widehat{\mathcal{G}}}} \chi^{-1}(g)|g\rangle\langle \chi| \qquad (\text{G.207})
$$

such that

$$
F_{\mathcal{G}} F_{\mathcal{G}}^{*} \underset{(\text{F.107}),(\text{G.207})}{=} \frac{1}{|\mathcal{G}|} \sum_{\substack{g_1,g_2 \in \mathcal{G} \\ \chi_1,\chi_2 \in \widehat{\mathcal{G}}}} \chi_1(g_1)\chi_2^{-1}(g_2)|\chi_1\rangle \underbrace{\langle g_1|g_2\rangle}_{= \delta_{g_1 g_2} \atop (\text{F.106})} \langle \chi_2|
$$

$$
= \frac{1}{|\mathcal{G}|} \sum_{\chi_1,\chi_2 \in \widehat{\mathcal{G}}} \left(\sum_{g \in \mathcal{G}} \chi_1(g)\chi_2^{-1}(g) \right) |\chi_1\rangle\langle \chi_2|
$$

$$
\underset{(\text{F.76})}{=} \sum_{\chi \in \widehat{\mathcal{G}}} |\chi\rangle\langle \chi|
$$

$$
\underset{(\text{F.106})}{=} \mathbf{1}_{\mathbb{H}},
$$

and it follows from (2.37) that $F_{\mathcal{G}}$ is unitary.

Solution F.137 With the change of variables

$$
\begin{aligned}
u &= r_1 + r_2 \\
v &= r_1 r_2,
\end{aligned} \qquad (\text{G.208})
$$

it follows first from (F.109) that $r_3 = -u$ and then

$$
(r_1 - r_2)(r_1 - r_3)(r_2 - r_3) = (r_1 - r_2)(r_1 + u)(r_2 + u) = (r_1 - r_2)(2u^2 + v) \quad (\text{G.209})
$$

as well as

$$
A \underset{(\text{F.110})}{=} -(u^2 + v) \qquad (\text{G.210})
$$

$$
B \underset{(\text{F.111})}{=} uv. \qquad (\text{G.211})
$$

Consequently, we have

$$\left((r_1 - r_2)(r_1 - r_3)(r_2 - r_3)\right)^2 \underset{\text{(G.209)}}{=} (r_1 - r_2)^2(2u^2 + v)^2$$

$$\underset{\text{(G.208)}}{=} (u^2 - 4v)(2u^2 + v)^2$$

$$= 4u^6 - 15u^2v^2 - 12vu^4 - 4v^3 \quad \text{(G.212)}$$

as well as

$$-4A^3 \underset{\text{(G.210)}}{=} -4(u^2 + v)^3 = 4u^6 + 12v^2u^2 - 12vu^4 - 4v^3$$

$$\hspace{6cm} \text{(G.213)}$$

$$-27B^2 \underset{\text{(G.211)}}{=} -27u^2v^2$$

such that

$$-\Delta_E \underset{\text{(F.108)}}{=} -(4A^3 + 27B^2)$$

$$\underset{\text{(G.213)}}{=} 4u^6 - 15u^2v^2 - 12vu^4 - 4v^3$$

$$\underset{\text{(G.212)}}{=} \left((r_1 - r_2)(r_1 - r_3)(r_2 - r_3)\right)^2.$$

Solution F.138 Clearly, the matrix product is associative and $\sigma_0 = \mathbf{1} = e_{\mathcal{P}} \in \mathcal{P}$ is a neutral element under matrix multiplication.

For any two elements $i^a\sigma_\alpha, i^b\sigma_\beta \in \mathcal{P}$, we have for their product

$$i^a\sigma_\alpha i^b\sigma_\beta = i^{a+b}\sigma_\alpha\sigma_\beta \underset{\text{(2.76)}}{=} \begin{cases} i^{a+b}\sigma_\beta & \text{if } \alpha = 0 \\ i^{a+b}\sigma_\alpha & \text{if } \beta = 0 \\ i^{a+b}(\delta_{\alpha\beta}\sigma_0 + i\varepsilon_{\alpha\beta\gamma}\sigma_\gamma) & \text{if } \alpha \neq 0 \neq \beta \end{cases}$$

$$= \begin{cases} i^{a+b}\sigma_\beta & \text{if } \alpha = 0 \\ i^{a+b}\sigma_\alpha & \text{if } \beta = 0 \\ i^{a+b}\sigma_0 & \text{if } \alpha = \beta \neq 0 \\ i^{a+b+c}\sigma_\gamma \text{ with } c = 0 \text{ or } 1 & \text{if } \alpha \neq 0 \neq \beta \neq \alpha, \end{cases} \quad \text{(G.214)}$$

and, noting that $i^d = i^{d \bmod 4}$, we see that

$$i^a\sigma_\alpha i^b\sigma_\beta \in \mathcal{P} = \left\{ i^c\sigma_\gamma \,\middle|\, c, \gamma \in \{0, \ldots, 3\} \right\}.$$

It follows from (G.214) that for any $i^a \sigma_\alpha \in \mathcal{P}$

$$\left(i^a \sigma_\alpha\right)^* i^a \sigma_\alpha = \overline{i^a} i^a \sigma_\alpha^* \sigma_\alpha = \sigma_0 = \mathbf{1},$$

proving not only that any element in $\mathcal{P}$ has an inverse in $\mathcal{P}$, but that also the inverse is given by its adjoint and thus $\mathcal{P} < U(2)$. By selecting a basis in $\mathbb{H}$, we can identify every element in the matrix group $U(2)$ bijectively with an element in $\mathcal{U}(\mathbb{H})$.

The elements $i^a \sigma_\alpha \in \mathcal{P}$ are given by specifying one of the four exponents $a \in \{0,\ldots,3\}$ together with one of the four indices $\alpha \in \{0,\ldots,3\}$, giving altogether 16 distinct elements and proving that $|\mathcal{P}| = 16$.

Solution F.139 To prove (F.127), we first need to show that any element $i^a \sigma_\alpha \in \mathcal{P}$ can be written in the form

$$i^a \sigma_\alpha = \prod_{p \geq 0} \sigma_x^{u_p} \sigma_y^{v_p} \sigma_z^{w_p}$$

with $u_p, v_p, w_p \in \mathbb{N}_0$. First, we show how $i^a \sigma_0$ for $a \in \{0,\ldots,3\}$ can be written in this form:

$$\sigma_0 = \sigma_x^2 = \sigma_x^0 \qquad i\sigma_0 = \sigma_z \underbrace{\sigma_x \sigma_y}_{=i\sigma_z}$$

$$-\sigma_0 = (\sigma_z \sigma_x \sigma_y)^2 \qquad -i\sigma_0 = (\sigma_z \sigma_x \sigma_y)^3.$$

With this, we can generate $i^a \sigma_j$ for $j \in \{1,2,3\}$ by multiplying σ_j with the powers of $\sigma_z \sigma_x \sigma_y$ given above. For example,

$$i\sigma_y = \sigma_z \sigma_x, \quad -\sigma_y = \sigma_z \sigma_x (\sigma_z \sigma_x \sigma_y), \quad -i\sigma_y = \sigma_z \sigma_x (\sigma_z \sigma_x \sigma_y)^2,$$

which shows that we can generate $i^a \sigma_y$ for $a \in \{1,2,3\}$ and in the same way this can be done for σ_x and σ_z.

To show (F.128), we first note that as a set

$$\langle i\sigma_0 \rangle = \{\pm\sigma_0, \pm i\sigma_0\}. \tag{G.215}$$

Now, let $g = i^a \sigma_\alpha \in \mathrm{Ctr}(\mathcal{P})$. This implies $g i^b \sigma_\beta = i^b \sigma_\beta g$ for all $b, \beta \in \{0,\ldots,3\}$. Hence, σ_α has to be such that

$$\sigma_\alpha \sigma_\beta = \sigma_\beta \sigma_\alpha \quad \forall \beta \in \{0,\ldots,3\},$$

which can only be satisfied by $\alpha = 0$. It follows that $g = i^a \sigma_0$ with $a \in \{0,\ldots,3\}$, which implies $g \in \{\pm\sigma_0, \pm i\sigma_0\}$. With (G.215), we thus have

$$\mathrm{Ctr}(\mathcal{P}) \subset \langle i\sigma_0 \rangle.$$

Since $\sigma_0 = \mathbf{1}$, the converse inclusion is obvious, and (F.128) is proven.

Solution F.140 With

$$g \underbrace{=}_{\text{(F.141)}} i^{c(g)} \Sigma_x\big(\mathbf{x}(g)\big) \Sigma_z\big(\mathbf{z}(g)\big) \tag{G.216}$$

$$h \underbrace{=}_{\text{(F.141)}} i^{c(h)} \Sigma_x\big(\mathbf{x}(h)\big) \Sigma_z\big(\mathbf{z}(h)\big) \tag{G.217}$$

$$gh \underbrace{=}_{\text{(F.141)}} i^{c(gh)} \Sigma_x\big(\mathbf{x}(gh)\big) \Sigma_z\big(\mathbf{z}(gh)\big) \tag{G.218}$$

we have

$$gh \underbrace{=}_{\text{(G.216),(G.217)}} i^{c(g)+c(h)} \Sigma_x\big(\mathbf{x}(g)\big) \Sigma_z\big(\mathbf{z}(g)\big) \Sigma_x\big(\mathbf{x}(h)\big) \Sigma_z\big(\mathbf{z}(h)\big)$$

$$\underbrace{=}_{\text{(F.140)}} i^{c(g)+c(h)} (-1)^{\mathbf{z}(g) \overset{2}{\odot} \mathbf{x}(h)} \Sigma_x\big(\mathbf{x}(g)\big) \Sigma_x\big(\mathbf{x}(h)\big) \Sigma_z\big(\mathbf{z}(g)\big) \Sigma_z\big(\mathbf{z}(h)\big)$$

$$\underbrace{=}_{\text{(F.139)}} i^{c(g)+c(h)+2\mathbf{z}(g) \overset{2}{\odot} \mathbf{x}(h)} \Sigma_x\big(\mathbf{x}(g) \overset{2}{\oplus} \mathbf{x}(h)\big) \Sigma_z\big(\mathbf{z}(g) \overset{2}{\oplus} \mathbf{z}(h)\big)$$

$$\underbrace{=}_{\text{(G.218)}} i^{c(gh)} \Sigma_x\big(\mathbf{x}(gh)\big) \Sigma_z\big(\mathbf{z}(gh)\big) \, ,$$

and since Lemma F.68 tells us that the representation of gh with the help of $c(gh), \mathbf{x}(gh)$ and $\mathbf{z}(gh)$ is unique, the claims (F.143)–(F.145) follow.

Solution F.141 First, we show that $\mathrm{Nor}_{\mathcal{P}_n}(\mathcal{S}) \subset \mathrm{Clz}_{\mathcal{P}_n}(\mathcal{S})$. For this let $g \in \mathrm{Nor}_{\mathcal{P}_n}(\mathcal{S})$. By Definition F.16 this implies $g\mathcal{S} = \mathcal{S}g$, which means that

$$\forall h \in \mathcal{S} \; \exists \widetilde{h} \in \mathcal{S}: \; gh = \widetilde{h}g. \tag{G.219}$$

From Proposition F.70 we know that for any $h, g \in \mathcal{P}_n$ we have either $hg = gh$ or $hg = -gh$. Hence, (G.219) becomes

$$\forall h \in \mathcal{S} \; \exists \widetilde{h} \in \mathcal{S}: \; \pm hg = \widetilde{h}g,$$

and we must have either $\widetilde{h} = h$ or $\widetilde{h} = -h$. Suppose $\widetilde{h} = -h$. Since $\mathcal{S}$ is a subgroup and $\widetilde{h}, h \in \mathcal{S}$, it follows that then $\widetilde{h}h^{-1} = -\mathbf{1}^{\otimes n} \in \mathcal{S}$, which we have excluded. Therefore, we must have $\widetilde{h} = h$, and (G.219) implies

$$gh = hg \quad \forall g \in \mathrm{Nor}_{\mathcal{P}_n}(\mathcal{S}) \text{ and } h \in \mathcal{S}$$

and thus

$$\mathrm{Nor}_{\mathcal{P}_n}(\mathcal{S}) \subset \mathrm{Clz}_{\mathcal{P}_n}(\mathcal{S}). \tag{G.220}$$

To prove the converse inclusion, let $g \in \mathrm{Clz}_{\mathcal{P}_n}(\mathcal{S})$. Then it follows from the definition of Clz in (F.17) that

$$gh = hg \quad \forall h \in \mathcal{S}$$

and thus $g\mathcal{S} = \mathcal{S}g$, which implies $g \in \mathrm{Nor}_{\mathcal{P}_n}(\mathcal{S})$. Hence,

$$\mathrm{Nor}_{\mathcal{P}_n}(\mathcal{S}) \supset \mathrm{Clz}_{\mathcal{P}_n}(\mathcal{S}),$$

and this together with (G.220) completes the proof of the claim.

Solutions to Exercises from Appendix G

Solution G.142 We have for every $j \in I$ and $z \in \mathbb{C} \setminus \sigma(A)$

$$(A - z\mathbf{1}) \sum_{j \in I} \frac{P_j}{\lambda_j - z} \underset{(2.42)}{=} \left(\sum_{k \in I} \lambda_k P_k - z\mathbf{1} \right) \sum_{j \in I} \frac{P_j}{\lambda_j - z}$$

$$= \sum_{k,j \in I} \frac{\lambda_k P_k P_j}{\lambda_j - z} - \sum_{j \in I} \frac{z P_j}{\lambda_j - z}$$

$$\underset{(2.44)}{=} \sum_{k,j \in I} \frac{\lambda_k \delta_{kj} P_j}{\lambda_j - z} - \sum_{j \in I} \frac{z P_j}{\lambda_j - z} = \sum_{j \in I} \frac{\lambda_j P_j}{\lambda_j - z} - \sum_{j \in I} \frac{z P_j}{\lambda_j - z}$$

$$= \sum_{j \in I} P_j \underset{(2.43)}{=} \mathbf{1}.$$

In the same way one verifies that $\sum_{j \in I} \frac{P_j}{\lambda_j - z}(A - z\mathbf{1}) = \mathbf{1}$. Consequently, we have

$$\sum_{j \in I} \frac{P_j}{\lambda_j - z} = (A - z\mathbf{1})^{-1} \underset{(G.1)}{=} R_z(A)$$

Solution G.143 From (G.1) in Definition G.1 of the resolvent we know that

$$R_z\big(\mathsf{H}_T(s)\big)\Big(\mathsf{H}_T(s)-z\mathbf{1}\Big)=\mathbf{1}. \tag{G.221}$$

Since we require H_T to satisfy the Adiabatic Assumption (AA), the function $s \mapsto \mathsf{H}_T(s)$ is twice differentiable. Taking the derivative with respect to s on both sides of (G.221) implies

$$\left(\frac{d}{ds}R_z\big(\mathsf{H}_T(s)\big)\right)\underbrace{\Big(\mathsf{H}_T(s)-z\mathbf{1}\Big)}_{\underset{(\text{G.1})}{=}\big(R_z(\mathsf{H}_T(s))\big)^{-1}}=-R_z\big(\mathsf{H}_T(s)\big)\dot{\mathsf{H}}_T(s)$$

from which the claim (G.7) follows.

Solution G.144 The claim actually follows from the results of Exercise 2.18, but we give the proof here once more for $U_{A,j}(s)$.

Since for all $j \in I$ and $s \in [0,1]$ we have

$$U_{A,j}(s)^* U_{A,j}(s) = \mathbf{1} = U_{A,j}(s) U_{A,j}(s)^*, \tag{G.222}$$

it follows that

$$\mathbf{1} = U_{A,j}(0)^* U_{A,j}(0) \underset{(\text{G.21})}{=} U_{A,j}(0)^*,$$

proving that $U_{A,j}(0)^*$ satisfies the initial condition in (G.22). Taking the derivative with respect to s on both sides in (G.222), we find

$$\dot{U}_{A,j}(s)^* U_{A,j}(s) + U_{A,j}(s)^* \dot{U}_{A,j}(s) = 0.$$

Multiplying both sides with i, with $U_{A,j}(s)^*$ from the right, and using again (G.222), we obtain

$$\begin{aligned}
\mathrm{i}\dot{U}_{A,j}(s)^* &= -U_{A,j}(s)^*\Big(\mathrm{i}\dot{U}_{A,j}(s)\Big)U_{A,j}(s)^* \\
&\underset{(\text{G.21})}{=} -U_{A,j}(s)^*\Big(\mathsf{H}_{A,j}(s)U_{A,j}(s)\Big)U_{A,j}(s)^* \\
&\underset{(\text{G.222})}{=} -U_{A,j}(s)^*\mathsf{H}_{A,j}(s),
\end{aligned}$$

proving (G.22).

Solution G.145 From (G.24) we see that

$$(A - \lambda_j 1)\big|_{P_j\{\mathbb{H}\}^\perp} \sum_{k\in I\smallsetminus\{j\}} \frac{1}{\lambda_k - \lambda_j} P_k\big|_{P_j\{\mathbb{H}\}^\perp} \underset{(G.24)}{=} \sum_{k,l\in I\smallsetminus\{j\}} \frac{\lambda_l - \lambda_j}{\lambda_k - \lambda_j}\left(P_l P_k\right)\big|_{P_j\{\mathbb{H}\}^\perp}$$

$$\underset{(2.44)}{=} \sum_{k\in I\smallsetminus\{j\}} P_k\big|_{P_j\{\mathbb{H}\}^\perp}$$

$$= 1\big|_{P_j\{\mathbb{H}\}^\perp}.$$

Consequently, we have

$$\left(\left(A - \lambda_j 1\right)\big|_{P_j\{\mathbb{H}\}^\perp}\right)^{-1} = \sum_{k\in I\smallsetminus\{j\}} \frac{1}{\lambda_k - \lambda_j} P_k\big|_{P_j\{\mathbb{H}\}^\perp}. \tag{G.223}$$

Now, (2.44) and $P_j^* = P_j$ imply that for all $|\Phi\rangle, |\Psi\rangle \in \mathbb{H}$ one has $\langle(1 - P_j)\Phi|P_j\Psi\rangle = 0$ and thus

$$(1 - P_j) : \mathbb{H} \to P_j\{\mathbb{H}\}^\perp \subset \mathbb{H}.$$

Using this in (G.223), gives

$$(1 - P_j)\left(\left(A - \lambda_j 1\right)\big|_{P_j\{\mathbb{H}\}^\perp}\right)^{-1}(1 - P_j)$$

$$= \sum_{k\in I\smallsetminus\{j\}} \frac{1}{\lambda_k - \lambda_j}(1 - P_j)P_k\big|_{P_j\{\mathbb{H}\}^\perp}(1 - P_j)$$

$$= \sum_{k\in I\smallsetminus\{j\}} \frac{1}{\lambda_k - \lambda_j}(1 - P_j)P_k(1 - P_j)$$

$$= \sum_{k\in I\smallsetminus\{j\}} \frac{1}{\lambda_k - \lambda_j}(P_k - P_j P_k - P_k P_j + P_j P_k P_j)$$

$$\underset{(2.44)}{=} \sum_{k\in I\smallsetminus\{j\}} \frac{1}{\lambda_k - \lambda_j}P_k,$$

which proves (G.25).

The claim (G.26) then follows from (G.25) and (2.44), whereas (G.27) follows immediately from (G.26).

Solution G.146 We have

$$\left|\left|\check{R}_j\Big(\big(\dot{H}_T P_j A + A P_j \dot{H}_T\big)\check{R}_j + \big(\dot{H}_T \check{R}_j A + A \check{R}_j \dot{H}_T\big)P_j\Big)\Psi\right|\right|^2$$

$$\underset{(2.51)}{\leq} \quad ||\check{R}_j||^2 \left|\left|\Big(\big(\dot{H}_T P_j A + A P_j \dot{H}_T\big)\check{R}_j + \big(\dot{H}_T \check{R}_j A + A \check{R}_j \dot{H}_T\big)P_j\Big)\Psi\right|\right|^2$$

$$\underset{(2.18)}{\leq} \quad ||\check{R}_j||^2 \left(\left|\left|\big(\dot{H}_T P_j A + A P_j \dot{H}_T\big)\check{R}_j \Psi\right|\right| + \left|\left|\big(\dot{H}_T \check{R}_j A + A \check{R}_j \dot{H}_T\big)P_j\Psi\right|\right|\right)^2$$

$$\underset{(2.51)}{\leq} \quad ||\check{R}_j||^2 \left(\left|\left|\big(\dot{H}_T P_j A + A P_j \dot{H}_T\big)\right|\right| \, ||\check{R}_j \Psi|| + \left|\left|\big(\dot{H}_T \check{R}_j A + A \check{R}_j \dot{H}_T\big)\right|\right| \, ||P_j\Psi||\right)^2$$

$$\underset{(2.51),(2.53)}{\leq} \quad \left(2\left|\left|\dot{H}_T\right|\right| \, ||A|| \, ||\check{R}_j||\right)^2 \left(||P_j|| \, ||\check{R}_j \Psi|| + ||\check{R}_j|| \, ||P_j\Psi||\right)^2$$

$$\underset{||P_j||=1,(G.27)}{\leq} \quad \left(2\left|\left|\dot{H}_T\right|\right| \, ||A|| \, ||\check{R}_j||\right)^2 \left(||\check{R}_j(1-P_j)\Psi|| + ||\check{R}_j|| \, ||P_j\Psi||\right)^2$$

$$\underset{(2.53)}{\leq} \quad \left(2\left|\left|\dot{H}_T\right|\right| \, ||A|| \, ||\check{R}_j||^2\right)^2 \left(||(1-P_j)\Psi|| + ||P_j\Psi||\right)^2 . \qquad (G.224)$$

Here we can further use the estimate

$$
\begin{aligned}
\big(||(1-P_j)\Psi|| + ||P_j\Psi||\big)^2 &= ||(1-P_j)\Psi||^2 + ||P_j\Psi||^2 + 2||(1-P_j)\Psi|| \, ||P_j\Psi|| \\
&\underset{(2.15)}{\leq} ||\Psi||^2 + 2\underbrace{||(1-P_j)||}_{=1} \underbrace{||P_j||}_{=1} ||\Psi||^2 \\
&= 3||\Psi||^2 ,
\end{aligned}
$$

which, inserted into (G.224), yields the claim (G.41).

References

1. C.H. Bennett, G. Brassard, in *Proceedings of IEEE International Conference on Computers, Systems, and Signal Processing*, Bangalore, India (New York, 1984)
2. J.F. Clauser, M.A. Horne, A. Shimony, R.A. Holt, Phys. Rev. Lett. **25**, 15 (1969)
3. A. Ekert, Phys. Rev. Lett. **67**, 661 (1991)
4. A. Einstein, B. Podolsky, N. Rosen, Phys. Rev. **47**, 777 (1935)
5. M. Planck, Annalen der Physik **4**(3), 553 (1901)
6. A. Einstein, Annalen der Physik **17**, 132 (1905)
7. E. Schrödinger, Naturwissenschaften Heft **48**, 807 (1935)
8. J.S. Bell, Rev. Mod. Phys. **38**(3), 447 (1966)
9. A. Aspect, J. Dallibard, G. Roger, Phys. Rev. Lett. **49**, 1804 (1982)
10. N. Wiener, *Cybernetics: or Control and Communication in the Animal and the Machine* (The MIT Press, 1948)
11. C.E. Shannon, Bell Sys. Tech. J. **27**, 379 (1948)
12. R. Feynman, Int. J. Theor. Phys. **21**(6/7), 467 (1982)
13. A. Turing, Proc. Lond. Math. Soc. **42**, 230 (1936)
14. P. Benioff, Phys. Rev. Lett. **48**(23), 1581 (1982)
15. W.K. Wootters, W.H. Zurek, Nature **299**, 802 (1982)
16. D. Deutsch, Proc. R. Soc. Lond. Ser. A **400**, 97 (1985)
17. D. Deutsch, Proc. R. Soc. Lond. Ser. A **425**(1868), 73 (1989)
18. C.H. Bennet, G. Brassard, C. Crépeau, R. Jozsa, A. Perez, W.K. Wootters, Phys. Rev. Lett. **70**(13), 1895 (1993)
19. P. Shor, in *Proceedings of the 35th Annual Symposium on Foundations of Computer Science* (IEEE Computer Society Press, 1994), pp. 124–134
20. P. Shor, SIAM J. Comput. **26**(5), 1484 (1997)
21. L.K. Grover, in *Proceedings of of the 28th Annual ACM Symposiium on Theory of Computing* (1996), pp. 212–219
22. L. Grover, Phys. Rev. Lett. **79**(2), 325 (1997)
23. D. Bouwmeester, J.W. Pan, K. Mattle, M. Eibl, H. Weinfurter, A. Zeilinger, Nature **390**, 575 (1997)
24. I.L. Chuang, N. Gershenfeld, M. Kubinec, Phys. Rev. Lett. **80**(15), 3408 (1998)
25. L.M.K. Vandersypen, G. Breyta, M. Steffen, C.S. Yannoni, M.H. Sherwood, I.L. Chuang, Nature **414**(6866), 883 (2001)
26. X.S. Ma, T. Herbst, T. Scheidl, D. Wang, S. Kropatschek, W. Naylor, B. Wittmann, A. Mech, J. Kofler, E. Anisimova, V. Makarov, T. Jennewein, R. Ursin, A. Zeilinger, Nature **489**, 269 (2012)
27. J.G. Ren, P. Xu, H.L. Yong, L. Zhang, S.K. Liao, Nature **549**, 70 (2017). https://doi.org/10.1038/nature23675

© Springer Nature Switzerland AG 2019
W. Scherer, *Mathematics of Quantum Computing*,
https://doi.org/10.1007/978-3-030-12358-1

28. A. Calderbank, P. Shor, Phys. Rev. A **54**(2), 10 (1996)
29. A.M. Steane, Phys. Rev. Lett. **77**(5), 793 (1997)
30. A. Barenco, C.H. Bennett, R. Cleve, D.P. DiVincenzo, N. Margolus, P. Shor, T. Sleator, J.A. Smolin, H.H. Weinfurter, Phys. Rev. A **52**(5), 4083 (1995)
31. D.P. DiVincenzo, Phys. Rev. A **51**(2), 1015 (1995)
32. B. Apolloni, C. Carvalho, D. de Falco, Stoch. Process. Their Appl. **33**(2), 233 (1989)
33. W. van Dam, M. Mosca, U. Vazirani, in *Proceedings of the 42nd Annual Symposium on Foundations of Computer Science* (2001), pp. 279–287. https://doi.org/10.1109/SFCS.2001.959902
34. D. Aharonov, W. van Dam, J. Kempe, Z. Landau, S. LLoyd, O. Regev, SIAM Rev. **50**(4), 755 (2008). https://doi.org/10.1137/080734479
35. M. Freedman, A. Kitaev, M. Larsen, Z. Wang, Bull. Am. Math. Soc. **40**, 31 (2003)
36. A. Kitaev, Ann. Phys. **303**, 2 (2003)
37. A. Galindo, P. Pascal, *Quantum Mechanics*, vol. I & II (Springer, 2012)
38. A. Messiah, *Quantum Mechanics*, vol. I & II (Dover, 2017)
39. C. Cohen-Tannoudji, B. Diu, F. Laloe, *Quantum Mechanics*, vol. I & II (Wiley, 1991)
40. J.J. Sakurai, J.J. Napolitano, *Modern Quantum Mechanics*, 2nd edn. (Pearson, 2013)
41. J.T. Cushing, *Quantum Mechanics: Historical Contingency and the Copenhagen Hegemony* (The Univeristy of Chicago Press, 1994)
42. R. Omnès, *Understanding Quantum Mechanics* (Princeton University Press, 1999)
43. B. d'Espagnat, *Veiled Reality: An Analysis of Quantum Mechanical Concepts* (Westview Press, 2003)
44. J.S. Bell, *Speakable and Unspeakable in Quantum Mechanics: Collected Papers on Quantum Philosophy*, revised edn. (Cambridge University Press, 2004)
45. F. Laloë, *Do We Really Understand Quantum Mechanics?* (Cambridge University Press, 2012)
46. S. Gao, *The Meaning of the Wave Function* (Cambrideg University Press, 2017)
47. F.T. Boge, *Quantum Mechanics Between Ontology and Epistemology* (Springer, 2018)
48. R.A. Meyers (ed.), *Encyclopedia of Complexity and Systems Science* (Springer, 2009)
49. C. Nayak, S.H. Simon, A. Stern, M. Freedman, S.D. Sarma, Rev. Mod. Phys. **80**, 1083 (2008)
50. M. Reed, B. Simon, *Methods of Modern Mathematical Physics I: Functional Analysis* (Academic Press, New York City, 1978)
51. D. Werner, *Funktionalanalysis* (Springer, 2011)
52. A.M. Gleason, Indiana Univ. Math. J. **6**, 88517893 (1957)
53. B.P. Rynne, M.A. Youngson, *Linear Functional Analysis* (Springer, 2008)
54. E. Kreyszig, *Introductory Functional Analysis with Applications* (Wiley, 1989)
55. Y. Choquet-Bruhat, C. DeWitt-Morette, *Analysis, Manifolds and Physics* (North-Holland, 1982)
56. A. Lucas, Front. Phys. **2**, 5 (2014). https://doi.org/10.3389/fphy.2014.00005
57. K. Kraus, *Effects and Operations: Fundamental Notions of Quantum Theory* (Springer, 1983)
58. K. Kraus, Ann. Phys. **64**, 311 (1971)
59. I. Bengtsson, K. Zyczkowski, *Geometry of Quantum States* (Cambridge University Press, 2006)
60. W.F. Stinespring, in *Proceedings of the American Mathematical Society* (1955), pp. 211–216
61. M.A. Nielsen, I. Chuang, *Quantum Computation and Quantum Information* (Cambridge University Press, 2010)
62. K. Parthasarathy, *An Introduction to Quantum Stochastic Calculus* (Birkhäuser Basel, 1991)
63. J.S. Bell, Physics **1**, 195 (1964)
64. R. Werner, Phys. Rev. A **40**, 4277 (1989)
65. M. Zukowski, A. Zeilinger, M. Horne, A. Ekert, Phys. Rev. Lett. **71**(26), 4287 (1993). https://doi.org/10.1103/PhysRevLett.71.4287
66. S. Bose, V. Vedral, P. Knight, Phys. Rev. A **57**(2), 822 (1998). https://doi.org/10.1103/PhysRevA.57.822
67. J.W. Pan, D. Bouwmeester, H. Weinfurter, A. Zeilinger, Phys. Rev. Lett. **80**(18), 3891 (1998)

68. D. Dieks, Phys. Lett. A **92**, 271 (1982)
69. J. Audretsch, *Entangled Systems* (Wiley-VCH, 2007)
70. M. Fernández, *Models of Computation: An Introduction to Computability Theory* (Springer, 2009)
71. M. Reck, A. Zeilinger, H.J. Bernstein, P. Bertani, Phys. Rev. Lett. **73**(1), 58 (1994)
72. A. Vedral, V. Barenco, A. Ekert, Phys. Rev. A **54**, 147 (1996)
73. R. Jozsa, Proc. R. Soc. Lond. A **454**, 323 (1998)
74. N.D. Mermin, *Quantum Computer Science: An Introduction* (Cambridge University Press, 20017)
75. D. Deutsch, R. Jozsa, Proc. R. Soc. Lond. Ser. A **439**(1907), 553 (1992)
76. C. Bennett, S. Wiesner, Phys. Rev. Lett. **69**, 2881 (1992)
77. C.H. Bennet, G. Brassard, A.K. Ekert, Sci. Am. **70**, 26 (1992)
78. R.L. Rivest, A. Shamir, L.M. Adleman, Commun. ACM **51**(2), 2738 (1978)
79. J. Hoffstein, J. Pipher, J.H. Silverman, *An Introduction to Mathematical Cryptography* (Springer, 2014)
80. Wikipedia. http://en.wikipedia.org/wiki/RSA_numbers
81. T. Kleinjung, K. Aoki, J. Franke, A.K. Lenstra, E. Thomé, J.W. Bos, P. Gaudry, A. Kruppa, P.L. Montgomery, D. Osvik, H. te Riele, A. Timofeev, P. Zimmermann, in *Advances in Cryptology '96 CRYPTO 2010*, ed. by T. Rabin, Lecture Notes in Computer Science, vol. 6223 (Springer, 2010), pp. 333–350. http://dx.doi.org/10.1007/978-3-642-14623-7_18
82. E. Corporation. RSA-768 is Factored! http://www.emc.com/emc-plus/rsa-labs/historical/rsa-768-factored.htm
83. A.K. Lenstra, W. Hendrik Jr. (eds.), *The Development of the Number Field Sieve*. No. 1554 in Lecture Notes in Mathematics (Springer, 1993)
84. C. Pomerance, Not. AMS **43**(12), 1476 (1996)
85. R. Crandall, C. Pomerance, *Prime Numbers: A Computational Perspective*. Lecture Notes in Statistics (Springer, 2006). https://books.google.co.uk/books?id=ZXjHKPS1LEAC
86. J.B. Rosser, L. Schoenfeld, Ill. J. Math. **6**, 64 (1962)
87. A. Childs, W. van Dam, Rev. Mod. Phys. **82**, 1 (2010)
88. M. Mosca, in *Encyclopedia of Complexity and Systems Science* (Springer, 2008). arXiv:0808.0369v1
89. W. Diffie, M.E. Hellman, IEEE Trans. Inf. Theory **22**(6), 644 (1976)
90. G.H. Hardy, E.M. Wright, *An Introduction to the Theory of Numbers* (Oxford University Press, 2008)
91. A.M. Antonopoulos, *Mastering Bitcoin*, 2nd edn. (O'Reilly, 2017)
92. A. Khalique, K. Singh, S. Sood, Int. J. Comput. Appl. **2**(2), 21 (2010). https://doi.org/10.5120/631-876
93. D.L.R. Brown. *Sec 2: Recommended Elliptic Curve Domain Parameters*. http://www.secg.org/sec2-v2.pdf
94. M. Boyer, G. Brassard, P. Høyer, A. Tapp, Fortschritte der Physik **49**, 493 (1998)
95. S. Jordan. *Quantum Algorithm Zoo*. http://math.nist.gov/quantum/zoo/
96. W. van Dam, Y. Sasaki, in *Diversities in Quantum Computation and Quantum Information* (World Scientific, 2012), pp. 79–105. https://doi.org/10.1142/9789814425988_0003
97. L.C. Washington, *Elliptic Curves*, 2nd edn. (Chapman & Hall/CRC, 2008)
98. S.J. Devitt, W.J. Munro, K. Nemoto, Rep. Prog. Phys. **76**(7), 076001 (2013)
99. P. Shor, Phys. Rev. A **52**(5), R2493 (1995)
100. D. Gottessmann, in *Quantum Information Science and Its Contributions to Mathematics*, vol. 68 (American Mathematical Society, 2010), pp. 13 – 58. arXiv:0904.2557v1
101. D.A. Lidar, T.A. Brun (eds.), *Quantum Error Correction* (Cambridge University Press, 2013)
102. D. Gottessmann, Phys. Rev. A **54**, 1862 (1996)
103. A.R. Calderbank, E.M. Rains, P.W. Shor, N.J.A. Sloane, Phys. Rev. Lett. **78**, 405 (1997)
104. V. Pless, *Introduction to the Theory of Error-Correcting Codes*, 3rd edn. (Wiley-Interscience, 1998)

105. A.M. Steane, in *Quantum Computers, Algorithms and Chaos*, ed. by P.Z.G. Casati, D.L. Shepelyansky (2006)
106. F. Gaitan, *Quantum Error Correction and Fault Tolerant Quantum Computing* (CRC Press, 2008)
107. A. Kitaev, in *Proceedings of the 3rd International Conference of Quantum Communication and Measurement*, ed. by O. Hirota, A.S. Holevo, C.M. Caves (Plenum Press, 1997), pp. 181–188
108. A. Kitaev, *Topological Quantum Codes and Anyons*, vol. 58 (American Mathematical Society, 2002), pp. 267–272
109. M. Born, V. Fock, Zeitschrift für Physik **51**(3–4), 165 (1928). https://doi.org/10.1007/BF01343193
110. T. Kato, J. Phys. Soc. Jpn. **5**, 435 (1950). https://doi.org/10.1143/JPSJ.5.435
111. S. Jansen, M.B. Ruska, R. Seiler, J. Math. Phys. **48**, 102111 (2007). https://doi.org/10.1063/1.2798382
112. E. Farhi, J. Goldstone, S. Gutmann, M. Sipser, Quantum computation by adiabatic evolution (2000). arXiv:0001106v1
113. J. Roland, N.J. Cerf, Phys. Rev. A **65** (2002)
114. T. Albash, D.A. Lidar, Adiabatic quantum computing. ArXiv:1611.04471v1
115. M.W. Hirsch, S. Smale, *Differential Equations, Dynamical Systems and Linear Algebra* (Academic Press, 1974)
116. P. Deift, M.B. Ruskai, W. Spitzer, Quantum Inf. Process. **6**, 121 (2007)
117. C. Moler, C.V. Loan, SIAM Rev. **45**(1), 1 (2003)
118. K.L. Pudenz, D.A. Lidar, Quantum Inf. Process. **12**(5) (2011). arXiv:1109.0325v1
119. A. Das, S. Suzuki, Eur. Phys. J. Spec. Top. **224**, 5 (2015)

Index

A

Abelian group, 560
Abelian Hidden Subgroup Problem (AHSP), 302
Adiabatic
   assumption, 409
   generator, 627
   generic algorithm, 413
   intertwiner, 628
   schedule, 414
   search algorithm, 420
   theorem, 636
Algorithm
   adiabatic generic, 413
   DEUTSCH–JOZSA, 250
   EUCLID, 517
   GROVER, 335
   SHOR, 274
Alice, 91
Amplitude amplification, 324
Ancilla, 204
   register, 204
Anti-linear, 13
Asymmetric cipher, 256
Auxiliary register, 204

B

Basis
   BELL, 89
   codeword, 356
   computational, 87
   expansion of vector, 16
   HILBERT space, 15
   orthonormal, 15
BB84, 258
BELL
   basis, 89
   inequality, 144
      CHSH-generalization, 148
   telephone, 154
Binary
   addition, 162
      factor-wise, 206
   exponentiation, 233
   fractions, 239
   quantum gate, 169
   representation, 86
Bit-flip error, 367
Bit-phase-flip error, 367
BLOCH representation, 62
Bob, 91
Bra-vector, 18

C

Center, 569
Centralizer, 567
Character
   conjugate, 580
   of abelian group, 580
   trivial, 581
Character group, 583
Cipher, 256
   asymmetric, 256
   symmetric, 256
Circuit
   classical, 165
   quantum
      length of, 201
      plain, 201
      with ancilla, 204
Classical
   AND-gate, 162

© Springer Nature Switzerland AG 2019
W. Scherer, *Mathematics of Quantum Computing*,
https://doi.org/10.1007/978-3-030-12358-1

circuit, 165
code, 349
    generator of, 349
codeword, 349
computational process, 161
error detection and correction protocol,
    354
NOT-gate, 162
OR-gate, 163
TOFFOLI-gate, 163
XOR-gate, 163
Clock
    space, 443
    states, 443
Code
    classical, 349
        distance of, 349
    quantum, 356
        distance of, 381
Codeword
    classical, 349
    quantum, 356
Coherent, 55
Commutator, 26
Compatible, 35
Computational
    basis, 87
    process
        classical, 161
        quantum mechanical, 169
Conjugate, 567
    character, 580
    subgroup, 568
Continued fraction
    convergent, 545
    finite, 545
    sequence of convergents, 545
Controlled quantum gate, 176
Convergent of continued fraction, 545
Coprime, 517
Correctable quantum error, 369
Correlation, 504
    EPR, 146
Coset
    left, 569
    right, 569
Counter states, 443
Covariance, 504
Cyclic group, 565

D
Decoding

classical
    maximum-likelihood, 354
    quantum, 356
Decoherence, 56
Decryption, 256
Degenerate, 23
Dense quantum coding, 251
Density operator, 44
    reduced, 103
DEUTSCH's problem, 247
Diagonal form, 24
Diagonalizable, 24
Direct product group, 575
Discarding the ancilla, 205
Discrete
    group, 560
    logarithm, 310
Discrete Logarithm Problem (DLP), 310
Discriminant of elliptic curve, 601
Distance, 347
    HAMMING, 348
    of a quantum code, 381
    of classical code, 349
Divisible, 516
Dual
    group, 583
    space, 18

E
Eigenspace, 23
Eigenstate, 34
Eigenvalue, 23
    gap, 623
Eigenvector, 23
EINSTEIN-PODOLSKY-ROSEN   paradox,
    137
EK91, 262
Elliptic curve, 601
    discriminant, 601
    WEIERSTRASS equation, 601
Encoding
    classical, 349
    quantum, 356
Encryption, 256
Entangled, 129
    maximally, 133
Entanglement swapping, 133
Environmental representation, 120
EPR
    correlations, 146
    paradox, 137
Equation

SCHRÖDINGER, 38
WEIERSTRASS, 601
Error
  bit-flip, 367
  bit-phase-flip, 367
  correcting code
    classical, 349
    quantum, 356
  m-qubit, 365
  operation in quantum code, 365
  operators, 365
  phase-flip, 367
EUCLID algorithm, 517
EULER
  function, 525
  theorem of, 526
Exchange operator, 181
Expectation value, 12
  of a random variable, 503
  quantum mechanical
    in a pure state, 30
    in mixed state, 45

F
Field, 600
Finite-dimensional, 14
Finite group, 560
First Group Isomorphism Theorem, 579
FOURIER transform
  discrete, 238
  on group, 599
  quantum, 237
Function
  EULER-, 525
  hash, 318
  measurable, 501

G
Gap of eigenvalues, 623
Gate
  classical, 161
    AND, 162
    NOT, 162
    OR, 163
    reversible, 161
    TOFFOLI, 163
    universal, 165
    XOR, 163
  quantum, 169
    binary, 169
    HADAMARD, 171

NOT, 171
    unary, 169
    universal, 175
General linear group, 562
Generator
  adiabatic, 627
  of cyclic group, 565
GRAY-coded, 193
Greatest common divisor, 517
Group, 559
  abelian, 560
    character, 580
  cyclic, 565
  direct product, 575
  discrete, 560
  dual, 583
  finite, 560
  finitely generated, 565
  FOURIER transform on, 599
  general linear, 562
  generated, 565
  generator, 565
  homomorphism, 576
  isomorphism, 576
  left action, 575
  of characters, 583
  order of, 560
  orthogonal, 565
  PAULI, 609
  quotient, 573
  special orthogonal, 565
  special unitary, 564
  unitary, 564
GROVER iteration, 330

H
HADAMARD-
  gate, 171
  transformation, 74
Half-prime, 525
Hamiltonian, 38
HAMILTON operator, 38
HAMMING
  distance, 348
  weight, 348
Hash function, 318
Hermitian, 22
Hidden Subgroup Problem (HSP), 302
Hiding a subgroup, 301
HILBERT space, 12
Homomorphism, 576

I

Incoherent, 55
Incompatible, 35
Independent group elements, 566
Inequality
   BELL, 144
      CHSH-generalization, 148
   SCHWARZ-, 17
Integer part, 515
Interference, 32
Invariant subgroup, 568
Isomorphism, 576

K

Kernel of a homomorphism, 576
Ket-vector, 18
k-local operator, 105
KRAUS operator, 120

L

LANDAU symbol
   big, 513
   little, 513
Left
   action, 575
   coset, 569
Length of quantum circuit, 201
Linearly independent, 14
Logical qubit, 356

M

Matrix
   element, 19
   of an operator, 19
   PAULI, 40
   representation, 19
Maximal subgroup, 563
Maximum-likelihood decoding, 354
Max-min principle, 474
Mean value, 11
Measurable
   function, 501
   space, 501
Measure, 501
   probability, 502
   space, 501
Measurement
   of a quantum register, 213
   probability, 30
   sharp, 34
Minimal subgroup, 563

Mixed state, 44
m-qubit error, 365
Multiplicative inverse modulo N, 523

N

n-fold PAULI group, 611
No-Cloning Theorem, 159
Non-degenerate, 23
Norm
   of a vector, 13
   of an operator, 26
Normalizer, 568
Normal subgroup, 568
Normed, 14

O

Observable, 12, 29
Operation element, 120
Operator, 22
   adjoint, 22
   bounded, 26
   density, 44
   diagonalizable, 24
   HAMILTON, 38
   k-local, 105
   KRAUS, 120
   norm, 26
   positive, 26
   recovery, 369
   resolvent, 621
   self-adjoint, 22
   strictly positive, 26
   super-, 22
   unitary, 23
Operator-sum representation, 120
Oracle, 327
Order
   modulo N, 528
   of a group, 560
   of group element, 560
   polynomial, 513
Orthogonal, 14
   group, 565
Orthonormal basis (ONB), 15

P

Paradox
   EINSTEIN-PODOLSKY-ROSEN, 137
   EPR, 137
Parity check matrix, 349
Partial trace, 100

PAULI
    group, 609
    matrices, 40
    n-fold group, 611
Phase shift
    conditional, 242
Phase-flip error, 367
Physical qubit, 356
Polynomial order, 513
Positive operator, 26
Primitive root modulo N, 528
Probability
    distribution, 502
        discrete, 502
    measure, 502
    space, 502
Problem
    Abelian Hidden Subgroup (AHSP), 302
    DEUTSCH's, 247
    Discrete Logarithm (DLP), 310
Product-state, 128
Projection
    operator, 25
    orthogonal, 25
    postulate, 37
Projector, 25
Proper subgroup, 563

Q
QECC, 356
q-register, 169
Quadratic Unconstrained Binary Optimization (QUBO), 417
Quantum
    adder modulo N, 226
    adiabatic theorem, 636
    channel, 121
    code, 356
    codeword, 356
    computational process, 169
    copier, 158
    decoding, 356
    encoding, 356
    encoding space, 356
    error
        correctable, 369
        correcting code, 356
        detection and correction protocol, 385
        detection and correction protocol for stabilizer QECC, 401
        operation in code, 365

    operators, 365
FOURIER transform, 237
    gate, 169
        binary, 169
        controlled, 176
        unary, 169
        universal, 175
    multiplier modulo N, 229
    No-Cloning Theorem, 159
    -NOT-gate, 171
    operation, 120
    parallelism, 212
    plain circuit, 201
    register, 169
        measurement, 213
        observation, 213
        read-out, 213
Qubit, 58
    logical, 356
    physical, 356
    space, 58
    state, 58
Quotient group, 573

R
Random variable, 502
    discrete, 502
Ray, 32
Read-out of a quantum register, 213
Recovery operation from quantum error, 369
Reduced resolvent, 629
Reflection about a subspace, 328
Register
    ancilla, 204
    auxiliary, 204
Relative frequency, 11
Remainder after division, 515
Resolvent
    of an operator, 621
    reduced, 629
Reversible classical gate, 161
Right coset, 569

S
Scalar product, 12
Schedule of adiabatic algorithm, 414
SCHMIDT decomposition, 107
SCHRÖDINGER equation, 38
SCHWARZ-inequality, 17
Search
    adiabatic algorithm, 420

GROVER algorithm, 335, 338
Separable, 129
    state, 128
σ-algebra, 501
Smallest common multiple, 517
Space
    HILBERT-, 12
    measurable, 501
    probability, 502
    qubit, 58
Span, 14
Special
    orthogonal group, 565
    unitary group, 564
Spectrum, 23
Spin, 39
    rotation, 64
Stabilizer, 575
    code, 393
State, 12
    entangled, 129
    mixed, 44
    product-, 128
    pure, 30
    qubit, 58
    separable, 128
    vector, 30
Subgroup, 563
    conjugate, 568
    invariant, 568
    maximal, 563
    minimal, 563
    normal, 568
    proper, 563
Superoperator, 22
Superposition principle, 32
Swap operator
    global, 181
Symmetric cipher, 256
Syndrome
    classical, 352
    detection operator, 384
    extraction, 385
    of quantum error, 385
    of stabilizer QECC, 396

T
Target-qubit, 176
Teleportation, 253
Tensor product, 81
Theorem
    Adiabatic, 636
    EULER, 526
    First Group Isomorphism, 579
    GLEASON, 46
    No-Cloning, 159
    PYTHAGORAS, 17
    RIESZ, 18
Time evolution, 38
Trace, 28
    partial, 100
    -preserving quantum operation, 121
Trivial character, 581

U
Unary quantum gate, 169
Uncertainty, 33
    relation, 35
        HEISENBERG, 36
Unitary
    group, 564
    operator, 23
Unit vector, 14
Universal quantum gate, 175

V
Variance, 504

W
Wave function, 37
    collapse of, 37
WEIERSTRASS equation, 601
Weight
    HAMMING, 348
    of an element of the PAULI group, 611

X
X, 171

Printed in the United States
By Bookmasters